Detlev Möller
Atmospheric Chemistry

Also of Interest

Chemistry of the Climate System. Volume 1: Fundamentals and Processes
Detlev Möller, 2019
ISBN 978-3-11-055975-0, e-ISBN (PDF) 978-3-11-056126-5,
e-ISBN (EPUB) 978-3-11-055992-7

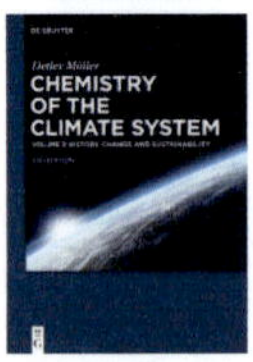

Chemistry of the Climate System. Volume 2: History, Change and Sustainability
Detlev Möller, 2020
ISBN 978-3-11-055985-9, e-ISBN (PDF) 978-3-11-056134-0,
e-ISBN (EPUB) 978-3-11-055996-5

Chemistry for Environmental Scientists
Detlev Möller, 2015
ISBN 978-3-11-040999-4, e-ISBN (PDF) 978-3-11-041001-3,
e-ISBN (EPUB) 978-3-11-041933-7
2nd edition to be published 2022

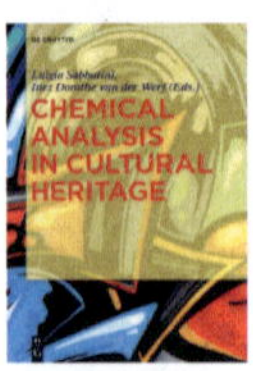

Chemical Analysis in Cultural Heritage
Luigia, Sabbatini, Inez Dorothé van der Werf, 2020
ISBN 978-3-11-045641-7, e-ISBN (PDF) 978-3-11-045753-7,
e-ISBN (EPUB) 978-3-11-045648-6

Green Chemistry. Principles and Designing of Green Synthesis
Syed Kazim Moosvi, Waseem Gulzar Naqash, Mohd. Hanief Najar, 2021
ISBN 978-3-11-075188-8, e-ISBN (PDF) 978-3-11-075189-5,
e-ISBN (EPUB) 978-3-11-075203-8

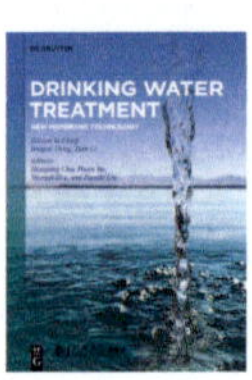

Drinking Water Treatment. New Membrane Technology
Bingzhi Dong, Tian Li, Huaqiang Chu, Huan He, Shumin Zhu, Junxia Liu (Eds.), 2021
ISBN 978-3-11-059559-8, e-ISBN (PDF) 978-3-11-059684-7,
e-ISBN (EPUB) 978-3-11-059315-0

Detlev Möller

Atmospheric Chemistry

A Critical Voyage Through the History

DE GRUYTER

Author
Prof. Dr. Detlev Möller
Brandenburgische Technische Universität
Cottbus und Senftenberg
Platz der Deutschen Einheit 1
03046 Cottbus, Germany
de-moe@t-online.de

ISBN 978-3-11-073739-4
e-ISBN (PDF) 978-3-11-073246-7
e-ISBN (EPUB) 978-3-11-073251-1

Library of Congress Control Number: 2021949664

Bibliographic information published by the Deutsche Nationalbibliothek
The Deutsche Nationalbibliothek lists this publication in the Deutsche Nationalbibliografie; detailed bibliographic data are available on the Internet at http://dnb.dnb.de.

Cover image: Flammarion, Camille (1888) L'atmosphére. Hachette, Paris, p. 146.
Back cover image: The author in front of the Keeling Building at Mauna Loa Observatory, Hawaii in 2002
Typesetting: VTeX UAB, Lithuania
Printing and binding: CPI books GmbH, Leck

www.degruyter.com

To the forgotten scientists and researchers

Geschichte schreiben ist eine Art, sich das Vergangene vom Halse zu schaffen.
[Writing history is a way of getting rid of the past]

Johann Wolfgang von Goethe (1749–1832)
Goethe's Werke. Vollständige Ausgabe. Vol. 49, Cotta, Stuttgart und Tübingen, 1833, p. 46

Foreword

In the year of editing this volume, 2022, we celebrate the 250th anniversary of the discovery of oxygen in atmospheric air by *Carl Wilhelm Scheele*. This is a good reason to go on a journey through the history of atmospheric chemistry. But you cannot see everything on a trip. This voyage starts with the Greek philosophers and ends in the late twentieth century. Most of the book deals with the history between 1750 and 1950 and is likely complete concerns issues in atmospheric chemistry. The history between 1950 and 1970 is still manageable but after that exponentially expanding. Thus, from the 1970s until about the year 2000 this journey stops only at highlights – seen subjectively.

> Because the history of science is a history of ideas and the people who had these ideas, it is a study of how these ideas were generated, what the scientists did and why they did it. When the history of science is considered in these terms, it is obvious that we are also talking about the philosophy of science. I think that in addition to the study of history of science having an intellectual fascination of its own, it provides a way for us to develop our own personal philosophy of science

writes the American geochemist and hydrologist *William Back* (1925–2008) in the foreword to "The History of Hydrology" edited by Landa and Ince (1987), and can also stand for this "History of Atmospheric Chemistry".

It is my aim to pay respect to all persons who studied the substances in the air, to those who made small, and to those who made giant contributions for the progress in atmospheric science. *Robert Angus Smith* (1817–1884), who wrote the first monograph on air with the title "Air and Rain. The Beginnings of a Chemical Climatology" (London 1872), writes (Smith 1872, p. 5):

> It is not my intention to give much historical matter. The history of our knowledge of air would make of itself a volume, and it might by every interesting one, and almost amusing. It must stand by itself.

Now, 150 years later, this volume, the *Atmospheric Chemistry* and its *History*, in your hand was also written in respect to *Smith*, the first *Air Chemist*, and to those (*Scheele*, *Priestley*, *Lavoisier*) who another 100 years earlier discovered the main chemical composition of our atmosphere. However, the story of understanding our atmosphere goes back to very ancient time. Sir *William Ramsay* wrote an excellent history of the investigation of our atmosphere (Ramsay 1896, 1907). *Ramsay* wrote that to present a historic overview on the theories, which did try to explain the nature of atmospheric air, would almost mean to write a history on chemistry and physics. Moreover, "[...] to separate physics and chemistry fundamentally is impossible because they deal with the very same task, the insight into matter," the German chemist *Jean D'Ans* (1881–1969) wrote in the preface to his "*Einführung in die allgemeine und anorganische Chemie*" (Berlin 1948). This is specifically true for understanding the atmosphere and *Atmospheric*

https://doi.org/10.1515/9783110732467-201

Chemistry. The German photochemist *Alfred Hermann Benrath* (1871–1969) wrote in his preface of "*Lehrbuch der Photochemie*" [Textbook of Photochemistry]; Benrath (1912, p. 2):[1]

> In no textbook the history of the sciences treated may be missing, because only if one knows the knowledge and the school of thought of this time, in which the discovery succeeded, one may assess their true value. Otherwise, historic studies motivate often to new researches. You can find some lost field of work which can be successful developed with the help of new methods, and some broken reasoning which you can implement having greater experience.

Sixty years ago, *Christian Junge* (1912–1996), one of the founding "fathers" of modern atmospheric chemistry, summarized the existing knowledge in his field of research. Junge (1963a) counted 388 research articles dealing with gases, aerosols, precipitation and pollution chemistry between 1915 and 1962 where 146 citations fall in the period 1958–1962. Nowadays the publication rate and number of researchers in atmospheric chemistry are orders of magnitude greater.

The history of science is the study of the historical development of science and scientific knowledge. Great scientific progress was made in the nineteenth century. The twentith century presumed explosive scientific growth – what do we expect in the present twenty-first century? Growth of science or "growth of knowledge," a term coined by *Karl Popper* (1902–1994),[2] were shown as "an exponential curve" by the American sociologist *William Fielding Ogburn* (1886–1959), based on data listed by Darmstaedter (1908) of discoveries and inventions (Ogburn (1922)). The British sociologist *Basil Bernstein* (1924–2000) said (Bernstein 2000, p. 93);

> The theory, however primitive, has always come before the research. Thus, by the time a piece of research was initiated the theory has already been subject to conceptual clarification as it engages with the empirical problem. And by the time it has finished there were further conceptual developments.

Imre Lakatos (1922–1974)[3] wrote (Lakatos 1981): "Philosophy of science without history is empty; history of science without philosophy is blind." *Colin Archibald Rus-*

---

1 Original German text: In keinem Lehrbuche darf die Geschichte der darin behandelten Wissenschaft fehlen, denn nur dann, wenn man das Wissen und die Geistesrichtung der Zeit, in welcher die Entdeckung gelungen ist, kennt, kann man diese ihrem wahren Wert nach beurteilen. Andererseits regt das geschichtliche Studium oft zu neuen Forschungen an. Man findet manches verlassene Arbeitsgebiet, das mit Hilfe neuer Methoden erfolgreich erschlossen werden kann, und manchen abgebrochenen Gedankengang, den man im Besitze größerer Erfahrung durchzuführen vermag.

2 *Sir Karl Raimund Popper*, Austrian–British philosopher who developed the *critical rationalism*. *Popper* pointed out is that you can never justify any scientific theory, but you can falsify it (Popper (1962)).

3 Birth name: *Lipschitz*, Hungarian mathematician, physicist and philosopher at the University of Debrecen, Hungary. In 1953, he fled to Vienna and finally to London to study at the University of Cambridge for a doctorate in philosophy. In 1960, *Lakatos* was appointed to the London School of Economics. Many works appeared after his death in editions.

sell (1928–2013), English chemist, writes (Russell 1984): "Science without its history is like a man without a memory. The results of such collective amnesia are dire."

This *voyage through the history* is written by a chemist, who spends 40 years in studying atmospheric chemistry and air pollution. When I began to learn what atmospheric chemistry means, it was the great time of sulfur dioxide pollution, the 1970s (Möller (1977)). It was soon clear to me – and first shown by Junge and Ryan (1958) – that aqueous-phase chemistry is dominant in overall conversion of SO_2 to sulfuric acid, i. e., the chemistry in clouds, fog and rain drops (Möller 1980). Being fascinated from the first publication on complex chemistry in raindrops (Graedel and Weschler (1981)), we began to study "acid rain" and deposition (Marquardt et al. (1984)) and to develop the first European model of aqueous-phase chemistry for raindrops and rain events (Mauersberger and Möller (1990), Möller and Mauersberger (1990)). Soon later, we learned that chemical conversion in rain is negligible comparing to chemistry in clouds while developing a cloud chemistry model (Möller and Mauersberger (1992)). Following the motto of *Gottfried Wilhelm Leibniz* (1646–1716) "*theoria cum praxis*," we build-up in 1991 a mountain cloud chemistry station at Mt. Brocken/Harz (Möller et al. (1993)), in continuous operation until 2010.

I'm not a historian who is able to present the past from a true perspective of their time – this also would not be my aim. Furthermore, I also will not "study the past with reference to the present," what the British historian *Herbert Butterfield* (1900–1979) termed the "Whig history," and criticized the "study of the past for the sake of the present" (Butterfield 1931). History, he stated, cannot be used to justify certain contents in the present. If possible, however, I try to interpret the past – almost limited to experimental findings in the nineteenth century – through current values, without dismissal of the problems and ideas of earlier scientists. In this way, it is possible to draw some ideas on the historical chemical state of the air. Hence, I name this voyage *critical*. However, nowhere in this book it is my attention to express my criticism to colleagues and scientific ancestors. Great scientists too were subject to errors; doing science consists from the permanent loop observation, interpretation, conclusion, and again testing against new observation. The Swiss physicist *Albrecht Neftel* (born 1951), great in data analysis and working in atmospheric chemistry, said in 1995 to me that he feels like a truffle dog – you have collected many different data from environmental measurements and will see nothing at the first glance, but suddenly you smell what you expect to observe. This ability nobody can learn at a university; it can be acquired only in the course of scientific life, and what we then name empirical knowledge. If this volume can contribute more than to be "a nice story" on atmospheric chemistry, than hopefully it inspires the reader to more *critical* reading of scientific publications, and not to forget the older one. Already in the nineteenth century – but until present – wrong statements on causes, relationships, etc., "variously modified, have been so often repeated, that at last they are universally received as a matter of fact" (Kingzett 1874, p. 511).

There is no book on atmospheric chemistry on the market dealing with its history but several books on the history of air pollution: Thorsheim (2006), Jacobson (2002), Brüggemeier (1996), Spelsberg (1988), Brimblecombe (1987), Spiegelberg (1984). There are many books on history of meteorology: Shaw (1942),[4] Khrgian (1970), Frisinger (1977), Brush and Landsberg (1985), Fleming (1990, 2012, 2016),[5] Harper (2008), Henson (2010), and in German: Hellmann (1883, 1904, 1908, 1914, 1917, 1920, 1921, 1922a,b, 1924, 1927), Schneider-Carius (1955),[6] Körber (1987). The German meteorologist *Johann Georg Gustav Hellmann* (1854–1939)[7] was the first who wrote many essays on history of meteorology and edited (1892–1907) together with *Julius von Hann* the "*Meteorologische Zeitschrift*."[8]

Interestingly that books on history of hydrology[9] – the science of the water cycle – cover rivers, aquifers, soil water, floods but not atmospheric water (clouds, fog and precipitation); only evaporation and precipitation (in terms of fluxes in time and space) at the earth-atmosphere interface is of interest to hydrologists. However, hydrology[10] has a long tradition: the ancient philosophers tried to resolve the mystery

4 *Sir William Napier Shaw* (1884–1945) English meteorologist; *Shaw* also studied air pollution, publishing his book *The Smoke Problem of Great Cities* in 1925.

5 *James Rodger Fleming* is the editor-in-chief of *History of Meteorology*, a peer-reviewed annual journal produced by the International Commission on History of Meteorology (ICHM); Vol. 1 in 2004. *Fleming* provides a "Guide to Historical Resources in Atmospheric Sciences. Archives, Manuscripts, and Special Collections in the Washington, D. C. area (1997): http://www.colby.edu/sts/97guide/. There is another journal on history: *Earth Sciences History: Journal of the History of the Earth Sciences Society*. USA, Vol. 1 (1982).

6 *Karl Schneider-Carius* (1896–1959) German meteorologist, director of the Geophysical Institute (University Leipzig) 1956–1959.

7 German meteorologist and climatologist; 1882–1885 director pro tem of the "*Preusische Meteorologische Institut*" [Prussian Meteorological Institute], 1886 part-time professor at the Berlin University. 1907–1922 director of the department for climatology of the Prussian Meteorological Institute. Known for invention of many meteorological instruments and works on history of meteorology.

8 "*Meteorologische Zeitschrift*" has a long and rich history; the predecessor journal, "*Zeitschrift der österreichischen Gesellschaft für Meteorologie*," first appeared in 1866. In 1884, one year after the foundation of the German Meteorological Society, the first volume of the "*Meteorologische Zeitschrift*" was published. Merged with the respective Austrian journal 2 years later, it existed for a first period until 1944. Following several separate publications in the post-war period after 1945 "*Meteorologische Zeitschrift*" was refounded in 1992 as a joint publication of the Austrian, the Swiss and the German Meteorological Societies. German was a well-accepted scientific language in meteorology until about the 1920s, when English started to become the dominant language (Emeis (2008)).

9 For example, Landa and Ince (1987), Biswas (1970).

10 The term hydrology as scientific discipline was used only after 1950.

of the hydrological cycle. The invention of the barometer (1643)[11] and thermometer (1714)[12] marks the dawn of the real study of the physics of the atmosphere, the quantitative study by which alone we are enabled to form any true conceptions of its true (Shaw 1942, p. 115). *Aristoteles* (384–322 BC)[13] "Meteorologica" is described in all books on history of meteorology but traditional the history of meteorological elements (wind, pressure, temperature, humidity, rainfall, evaporation) and its observation including weather forecast is in the center of interest (Aristoteles (1923)). Only Schneider-Carius (1955) goes on form of clouds (ibid., pp. 130–143), dew (ibid., pp. 206–209), and physics of clouds and precipitation (ibid., pp. 274–285). Excellent is the book "History of the Theories of Rain" (1965) by *William Edgar Knowles Middleton* (1903–1998),[14] which deals with theories of the hydrometeors (clouds, rain, snow, hail and dew) up to about 1914. But no book on history of meteorology deals with air chemistry. And no book on history of chemistry deals with air chemistry; but with substances found in the air.[15] In discovering the properties of the atmosphere, we may distinguish between the following epochs:

- Antiquity (about 600–300 BC): air phenomenology and earth's philosophy;
- Middles Ages and Renaissance (fifteenth–seventeenth century): understanding the air being a body, first meteorological measurements;
- Reconnaissance (eighteenth/nineteenth century): identification of air composition (including foreign bodies), basic relationships between plant, animal and air, establishing scientific disciplines like chemistry, meteorology, geography, agriculture economy;
- Industrial era (since about 1850): measurement of trace species in air, understanding of air pollution, cycling of matter;
- Modern era (after 1950): air pollution control (monitoring), understanding of atmospheric chemistry, biogeochemistry, global change.

---

11 *Evangelista Torricelli* (1608–1647) invented in 1643 the first mercury barometer which is used to measure atmospheric pressure.

12 The first exact mercury thermometer by *Daniel Gabriel Fahrenheit* (1686–1736), who introduced in 1709 the alcohol thermometer. He was a German glassblower, physicist and engineer, born in Danzig (modern Gdansk in Poland) but lived most of his life in Holland. In 1717, he settled in The Hague with the trade of glassblowing, making barometers, altimeters and thermometers. From 1718 onward, he lectured in chemistry in Amsterdam. There were some predecessors: in 1593 *Galileo Galilei* (1564–1641) invented a rudimentary water thermoscope and in 1612 *Santorio Santorio* (1561–1632) put a numerical scale on his thermoscope. Also termed *Sanctorius of Padu*, Italian physiologist, physician in Padua where he performed experiments on temperature, respiration and weight.

13 In Greek Ἀριστοτέλης; in English *Aristotle*, but I use the Greek *Aristoteles* in this volume.

14 *Middleton*, a Canadian meteorologist, wrote 15 books and many scientific papers related to the science of weather instruments and meteorological optics as well as their history. His major contribution was the book *Meteorological Instruments*, first published in 1941.

15 For example, Kopp (1843; see reprint in 1931) who wrote a section on "Gases; atmospheric air; oxygen; nitrogen" in Vol. 3 (pp. 175–218).

In studying old subject books, a problem occurs for a "modern man," the language, namely the old terms and formulas for chemical substances, which are not consistent with that what we today understand behind it. In some cases, it is impossible to identify a defined chemical compound and I leave it with the *historical term*. In the time before about 1850, obviously the number of scientists in the field were so limited that each of them was familiar with the others either personally or had access to all the books and journals; hence they did not cite any bibliography and only referred the surname; for those persons who are not so familiar I did try to find the given name and the living dates (not being successful in all cases) to express my respect to the ancestors. A full citation in books and journal papers began only at the end of the nineteenth century. However, I found many wrong and missing citations in works over the last 150 years; therefore, I never cite here a work without having seen and read it either printed or online in its *original* form. Only in a few cases, where it was not possible to get access to the original work – and I find it worth and credible to quote it here – I refer to the author who cited it. I also like to mention that the reader has to take care in the use of historic values, especially concentrations of chemical substances in waters and air. In publications before about 1880, no or only rudimentary details on the sampling and analytical methods were given; obviously that the methods (cited by its name) were familiar to all authors and readers at that time. It would be a separate scientific work to describe the old analytical methods and to evaluate historic values on its measurement trueness; some attempts have been made, e. g., Brimblecombe and Pitman (1980), Bojkov (1986), Volz and Kley (1988).

Today, an innumerable number of field sites is devoted to study the atmospheric chemical composition of the atmosphere, often only active for short periods, with sometimes barely more than a dozen samples were collected and analyzed for whatever purpose. The British economist *Josiah Charles Stamp* (1880–1941) wrote: "We are so obsessed with the delight and advantage of discovery of new things that we have no proportionate regard for the problems of arrangement and absorption of the things discovered" (Stamp 1937, p. 60). I hope that this *historical* atmospheric chemistry, at least since the systematic monitoring in the second half of the nineteenth century – which is certainly unknown to most modern atmospheric chemists – not only encourages respect for our scientific ancestors but may definitely help to avoid many scientifically meaningless studies of the kind that have appeared over the last few decades. The endeavor remains to learn from previous studies to ask the appropriate open questions and draw the right conclusions for further studies. We can learn from the past that all kinds of persons were interested in the subject from a philosophical perspective and/or with respect to the application of techniques (engineering) but always motivated by the specific problems (e. g., pollution) of their era. We also hold deep respect for our scientific ancestors for their brilliant conclusions, based on scientific experiments with very simple techniques and limited quantitative measurements. The great interest in historical data from the era before fossil fuel combustion lies also in deduction of background concentrations, in other words, the natural reference concentrations for assessing the human-influenced changes in the chemical composition of our atmosphere (*climate change*). On the other hand, recognition of the tremendous urban air pollution in past would also provide a scale for a more realistic assessment of the present air pollution levels, often still referred to as serious and life shortening by politicians and administrators.

The work in your hand contains three main chapters, covering the chemistry of the condensed phase in the atmosphere, first, the different forms of atmospheric waters (precipitation, fog and clouds, dew), and secondly dust, now mostly termed particulate matter, and more scientifically, atmospheric aerosol. A third section treats the gases in the atmosphere. An introductory chapter covers the roots of the term atmospheric chemistry in its relations to chemistry in general and biogeochemistry as the chemistry of the climate system. Furthermore, a brief overview of understanding chemical reactions in aqueous and gaseous phase is given.

Notwithstanding, this volume must limit and will thus not include topics (with a few exceptions in close relation to some substances) that are absolutely necessary for understanding atmospheric chemistry such as dry deposition, particle microphysics and dynamics, instruments and methods for sampling and determination of atmospheric substances and investigation of chemical processes in reactions chambers, as well as modeling of chemical reactions. Last but not least, I like to express my thanks for always promptly sending copies and bibliographic information to *Kerstin Rysanek* (library of the "*Umweltbundesamt*" Bad Elster, UBA), *Annette Dietrich*, *Andrea Lehnhardt* and *Yvonne Kurz* (library of "*Deutscher Wetterdienst*" Offenbach, DWD), *Sabine Mendelski* and *Marcus Bahler* (library of the Meteorological Institute, Free University Berlin). For biographic information, I thank *Karl-Heinz Bernhardt* (Berlin), *Uwe Feister* (Potsdam), *Eberhard Freydank* (Dresden), *Henning Rodhe* (Stockholm) and *Otto Wienhaus* (Tharandt). My very special thanks go to *Grete Cauer* (born 1934, daughter of *Hans Cauer*, Bielefeld), who authorized me to examine the personal and scientific estate of *Hans Cauer*. Further special thanks go to *Peter Filippovich Svistov* [Петр Филиппович Свистов] from St. Petersburg and my friends *Horváth László* (Budapest), *Alexey Grigorjevich Ryaboshapko* [Алексей Григорьевич Рябошапко] (1942–2021) from Moscow and *Leo Klasinc* (1937–2021) from Zagreb, all famous (atmospheric) chemists, who gave me many biographic and bibliographic information. *Kazuhiko Sekiguchi* (Saitama, Japan) was a great help to me in understanding some texts published in Japanese. Finally, I like to mention that in all citations I use the original spelling in all languages that changed over the time; for example, *rain water* (still used by Junge and Werby (1958)) instead of *rainwater* (already used by Junge (1963a)).

This book has 105 figures and 57 tables.

Berlin, March 2022 Detlev Möller

Contents

Foreword —— IX

1 Introduction —— 1
1.1 Atmospheric chemistry: a terminological and historical approach —— **2**
1.1.1 First uses of the term "atmospheric chemistry" —— **2**
1.1.2 From chemical meteorology and climatology to atmospheric chemistry —— **5**
1.1.3 *Hans Cauer* and the German pioneers of modern atmospheric chemistry —— **17**
1.2 Biogeochemistry: a terminological and historical approach —— **43**
1.2.1 From geochemistry to biogeochemistry —— **44**
1.2.2 From biosphere to noosphere —— **48**
1.2.3 Cycling of matter —— **52**
1.3 Chemistry: a terminological and historical approach —— **60**
1.3.1 On the origin of the word "chemistry" —— **62**
1.3.2 On the chemical nomenclature and symbols —— **64**
1.3.3 Alchemical views on airs and gases —— **72**
1.3.4 On the words air and atmosphere —— **80**
1.3.5 A brief history of chemical reactions —— **82**
1.3.5.1 From alchemy to quantum chemistry —— **83**
1.3.5.2 The roots of atmospheric chemistry —— **94**
1.3.5.3 Early history of light-induced chemical reactions: photochemistry —— **97**

2 Investigation of waters in the air —— 107
2.1 Introduction —— **111**
2.2 On condensation and evaporation of atmospheric water —— **118**
2.2.1 Ancient views —— **118**
2.2.2 Theories on condensation and evaporation 1600–1850 —— **121**
2.2.3 Theories on condensation after 1850 —— **129**
2.2.4 Vesicles versus droplets —— **132**
2.2.5 On fog and clouds —— **135**
2.3 Rainwater studies —— **142**
2.3.1 Rainwater studies for understanding the chemical composition of air —— **144**
2.3.1.1 Rainwater studies before 1800 —— **144**
2.3.1.2 Rainwater studies 1813–1863: an analytical approach —— **157**
2.3.2 Rainwater analysis for understanding agricultural chemistry (1851–1919) —— **173**
2.3.2.1 The role of agriculture experimental stations —— **173**
2.3.2.2 The "nitrogen period" —— **177**

2.3.3 Rainwater analysis for understanding cycling salts (1877–1956) — **201**
2.3.4 Rainwater analysis for understanding air pollution — **207**
2.3.4.1 *Robert Angus Smith*: air and rain (1845–1879) — **208**
2.3.4.2 The "sulfur and nitrogen period" (1869–1947) — **214**
2.3.4.3 Rainwater studies in Japan — **222**
2.3.4.4 Rainwater studies in Russia and the Soviet Union — **224**
2.3.4.5 Rainwater studies in India in the 1930s — **226**
2.3.4.6 On the pH of rainwater (1936–1960) — **227**
2.3.5 Precipitation chemistry after 1945 — **232**
2.3.5.1 The "global cycle period" (1947–1980) — **238**
2.3.5.2 Period of "acid rain" (1970–1990) — **243**
2.3.5.3 Period of "air pollution control": the East German case (1980–2000)[16] — **248**
2.3.5.4 An outlook — **251**
2.4 Fog and cloud water studies — **252**
2.4.1 Fog water studies before 1900 — **253**
2.4.2 Fog and cloud water studies: understanding condensation nuclei (1925–1955) — **255**
2.4.3 Fog and cloud water studies: understanding air pollution (1955–1970) — **259**
2.4.4 Fog and cloud water studies: understanding chemistry (1976–1990) — **263**
2.4.5 Fog and cloud water studies after 1990: an outlook — **268**
2.5 Dew water studies — **275**

3 Investigation of particulate matter in the air — 283
3.1 Dust, soot, and air pollution in a historical view — **285**
3.1.1 Air pollution — **285**
3.1.2 Smoke and fog: smog and the soot plague — **290**
3.2 On dust — **301**
3.2.1 *Ehrenberg*'s red dust and other dust fall — **301**
3.2.2 On falling stones and colored precipitation — **307**
3.2.3 On organic particulate matter — **312**
3.2.4 On dry fog — **314**
3.3 Condensation nuclei, ions and dust particles — **320**
3.3.1 Dust particles and haze formation: investigations before 1900 — **320**
3.3.2 Understanding the nature of condensation nuclei: research 1900–1960 — **326**

16 This chapter is dedicated to my colleagues who died too soon: *Wolfgang Rolle* (1934–1996), *Wolfgang Marquardt* (1933–2002), and *Eberhard Schaller* (1950–2015).

3.4 On sea salt — 331
3.4.1 Sea salt and condensation nuclei — 333
3.4.2 Sea salt and chlorine degassing — 335
3.4.3 Sea salt and nitrate deposits in Northern Chile — 342

4 Investigation of gases in the air — 345
4.1 A brief history of the discovery of gases in air — 346
4.2 The main gases of the air — 355
4.2.1 Nitrogen — 355
4.2.2 Oxygen — 357
4.2.3 Water — 370
4.2.4 Argon and the other novel gases — 374
4.2.5 Air analysis — 376
4.2.5.1 Determination the goodness of air: eudiometry — 379
4.2.5.2 Analysis of oxygen — 386
4.3 Carbon compounds in air — 394
4.3.1 Carbon dioxide — 394
4.3.1.1 Discovery of carbon dioxide — 394
4.3.1.2 Methods for determination of carbon dioxide in the air — 397
4.3.1.3 Carbon dioxide measurements in the nineteenth century — 406
4.3.1.4 The twentieth century rise of carbon dioxide — 424
4.3.1.5 Carbon dioxide and climate — 429
4.3.2 Carbon monoxide (CO) — 432
4.3.3 Organic matter and compounds — 434
4.3.3.1 Methane — 437
4.3.3.2 Other hydrocarbons and organic compounds — 440
4.4 Ozone (O_3) — 450
4.4.1 Early history and discovery of ozone — 455
4.4.2 On the nature of ozone — 462
4.4.2.1 Constitution — 462
4.4.2.2 Properties — 467
4.4.2.3 Sources — 471
4.4.3 Ozone and antozone — 481
4.4.4 Understanding atmospheric chemistry of oxygen compounds — 485
4.4.4.1 Gas-phase chemistry — 486
4.4.4.2 Aqueous-phase chemistry — 492
4.4.5 Studies of atmospheric ozone in the nineteenth century — 494
4.4.5.1 Introduction — 494
4.4.5.2 On the criticisms of the ozonometry — 507
4.4.5.3 Observations of atmospheric ozone in Germany — 518
4.4.5.4 Observations of atmospheric ozone in Austria — 533
4.4.5.5 Observations of atmospheric ozone in Switzerland and Italy — 541

4.4.5.6 Observations of atmospheric ozone in the United Kingdom — 542
4.4.5.7 Observations of atmospheric ozone in France — 551
4.4.5.8 Observations of atmospheric ozone in America — 560
4.4.5.9 Observations of atmospheric ozone in other countries — 567
4.4.6 Do ozonometric observations have value? — 570
4.4.7 Photophysics, photochemistry and formation of ozone 1878–1932 — 572
4.4.8 Atmospheric ozone research in the twentieth century — 578
4.4.8.1 Total and stratospheric ozone — 580
4.4.8.2 Troposphere: observations 1900–1944 — 593
4.4.8.3 Troposphere: after-war observations until about 1960 — 606
4.4.8.4 The beginning of systematic surface-near monitoring — 614
4.4.8.5 Tropospheric ozone research after 1970: an outlook — 621
4.4.8.6 Understanding tropospheric ozone formation and budget — 624
4.5 Hydrogen peroxide (H_2O_2) — 632
4.5.1 Early history and discovery — 632
4.5.2 Studies of hydrogen peroxide in the ninetheenth century — 634
4.5.2.1 On the nature of hydrogen peroxide — 634
4.5.2.2 Detection of hydrogen peroxide in the atmosphere — 638
4.5.3 Understanding hydrogen peroxide chemistry in the atmosphere — 647
4.5.3.1 Gas-phase chemistry — 649
4.5.3.2 Aqueous-phase and interfacial chemistry — 651
4.5.4 Studies of hydrogen peroxide after 1945 — 658
4.5.5 Understanding tropospheric H_2O_2 chemistry and budget — 668
4.6 Nitrogen compounds — 670
4.6.1 Ammonia (NH_3) — 670
4.6.1.1 Early history and discovery — 670
4.6.1.2 Early measurements of ammonia in the air — 672
4.6.1.3 Studies of atmospheric ammonia in the twentieth century — 676
4.6.2 Oxides and oxoacids of nitrogen — 679
4.6.2.1 Early history and discovery — 679
4.6.2.2 The compounds of oxidized nitrogen — 683
4.6.2.3 Dr. *Mitchill*'s septon theory — 688
4.6.2.4 Transformation of nitrogen in water — 689
4.6.2.5 Research in the nineeenth century: oxidation of nitrogen in air — 693
4.6.2.6 Research in the twentieth century — 700
4.7 Sulfur compounds — 706
4.7.1 Early history and discovery — 706
4.7.2 Sulfur dioxide (SO_2) — 710
4.7.2.1 Early research: town pollution and smelter smoke — 712
4.7.2.2 Early research: forest damage and SO_2 — 720
4.7.2.3 Atmospheric chemistry: oxidation of SO_2 — 722

4.7.3 A very short sketch on atmospheric sulfur research 1975–1995 — **727**

5 An outlook: from the past into future — 729

Abbreviations used in this volume — 731

Name index — 733

Bibliography — 755

Subject index — 905

1 Introduction

"Of all things on the face of the earth definitions are the most accursed," said *John Hunter* (1728–1793), a British physician (cited after Stedman 1920, p. V). The typical dictionary definition of an atmosphere is "the mixture of gases surrounding the Earth and other planets" or "the whole mass of an aeriform fluid surrounding the Earth". This definition is very close to that given by *Robert Boyle* (1863, p. 1):

> BY the Air I commonly understand that thin, fluid, diaphanous, compressible and dilatable Body in which we breathe, and wherein we move, which envelops the Earth on all sides to a great height above the highest Mountains.

In his textbook (Stöckhardt 1851, p. 4) *The Principles of Chemistry, Julius Adolph Stöckhardt* (1809–1886)[16] wrote: "Wherever we look upon our Earth, chemical action is seen taking place, on the land, in the air, or in the depths of the sea".[17] What is atmospheric chemistry and who is an "atmospheric chemist" [in German "*Luftchemiker*"]? The latter needs a definition. An atmospheric chemist is a person who most of his or her professional life deals with the chemical composition of atmosphere to understand the origin and further transformation of compounds and substances in gaseous, solid (in the form of particulates) and aqueous state, which are either naturally or artificially released into the air.[18] Thus, depending on the motivation for studying the atmosphere, these researchers can be naturalists, chemists, physicians, physicists, meteorologists, biologists, and others. I found probably the first phrase in literature about "chemical process in air" written by an anonymous writer:

> [...] die Stoffe [...] in unserem Luftkreise, als der großen Werkstatt der Natur, – entstehen, und eines chemischen Prozesses der Luft im Gewitter, oder im Regen oder Schnee auf die Erde herabkommen lassen. [... substances ... in our atmosphere, the great workshop of nature, – origi-

16 *Julius Adolph Stöckhardt* became a pharmacist in Berlin (1833), wrote a dissertation (in Latin) on methods of teaching in natural sciences (Leipzig 1837), worked as a teacher of chemistry in Chemnitz and became the first professor of agricultural chemistry in Tharandt (1847–1883). *Stöckhardt* was the pioneer in creation the German agricultural experiment stations and, in 1849, identified sulfur dioxide as the cause of smoke damages to plants.

17 First German Edition: Schule der Chemie oder erster Unterricht in der Chemie, versinnlicht durch einfache Versuche. Zum Schulgebrauch und zur Selbstbelehrung, insbesondere für angehende Apotheker, Landwirte, Gewerbetreibende, etc., Vieweg, Braunschweig 1846. German original: Wohin wir nur blicken auf unserer Erde, überall gewahren wir chemische Prozesse, auf dem Festlande, in der Luft, wie in den Tiefen des Meeres (6th ed. 1852, p. 2 and 19th ed. 1881, p. 4).

18 For long time, this research was carried out solely by measurements in external air; studying of air chemistry in laboratory measurements in suitable reaction chambers begun in the 1970s (with a few exceptions of simple experiments to understand photochemistry carried out back to the 1930s), followed by modeling (simulation of atmospheric chemistry) of progressively enlarged chemical reaction systems, which, finally, were linked with dispersion and impact models.

https://doi.org/10.1515/9783110732467-001

> nate, and by means of a chemical process in air or thunderstorm, or in rain and snow precipitate to earth] (Anonymous 1821, p. 3376; see also quotation on p. 311)

1.1 Atmospheric chemistry: a terminological and historical approach

1.1.1 First uses of the term "atmospheric chemistry"

Johann Friedrich Gmelin (1748–1804)[19] was the first who used the term "*Luftchemie*" [air chemistry] (Gmelin 1798, p. 6–7) in sense of a label for a new science:

> Dass übrigens Helmont in dieser Kindheit der Luftchemie Stoffe, [...] kaum einen anderen Unterschied zwischen den Gasarten gemacht, als dass einige bei Berührung mit brennenden Körpern sich entzünden, andere die Flamme sogleich auslöschen [Helmont, by the way, in that childhood of air chemistry, ... did not made another difference between the kinds of gases as that some ignite when contacting burning bodies and others the flame extinguishes immediately].

It is remarkable that in an English version of this paper (Gmelin 1801, p. 195), the phrase "*Luftchemie*" was not translated. *Augustus Allen Hayes* (1806–1882), an assayer to the Commonwealth of Massachusetts and a prominent but nowadays forgotten American chemist, in a Paper (Hayes 1851) writes as first the phrase about *chemistry of the atmosphere*:

> The chemistry of our atmosphere has, from the earliest tine, been deemed of high interest, and its connection with the phenomena of animal and vegetable life has led to its study the minds of the most eminent chemical investigators.

James Glaisher (1809–1903),[20] secretary of the British Meteorological Society, made another comment about the need of chemical investigations for understanding meteorological processes on the Sixth Annual General Meeting (on 27 May 1856), see (Barker 1856, p. 46):

> I rejoice that the persevering spirit of inquiry which distinguishes the present age showed have added another meteorological element of investigation to the preceding; one too, which, if somewhat verging upon the field of chemical inquiry, promises to be a subtle and important agent in and of this research into the nature and extend of meteorological influence [...]

---

19 *Johann Friedrich Gmelin* was a German physician, chemist, and naturalist (father of *Leopold Gmelin*) in Tübingen and Göttingen. *Wilhelm August Lampadius* was one of his students.

20 *James Glaisher* was a pioneer in weather forecasting and photography. In 1829–1830, an assistant on the Trigonometrical Survey of Ireland In 1833, he was taken on as an assistant by the astronomer *George Biddell Airy* (1801–1899) at the Cambridge University Observatory,since 1835 director of the Greenwich Royal Observatory), and since 1835 in Greenwich at the Observatory. He edited the *Proceeding of the Meteorological Society*.

In French, "*chimique de l'atmosphère*" was first used by Boussingault (1834, p. 148), and by Houzeau (1872a, p. 26) as "*chimique de l'air*". In other languages, the phrase "chemistry of the atmosphere" in headings appeared for the first time in Spain "*quimicos sobre el Aire atmosférico*" by *Ramon Torres Muñoz de Luna* (1822–1890) and the French translation by *Henri-François Gaultier de Claubry* (1792–1878) "*chimique sur l'air atmosphérique*" in 1861 (Luna 1860). The phrase "*Chemie der atmosphärischen Niederschläge*" [chemistry of atmospheric precipitation] in heading of an article was first used by *Christoph Friedrich Goppelsroeder* (1837–1919), a Swiss chemist who worked and lived in Basel (Goppelsroeder 1871, p. 139). *Edmund Monroe Pendleton* (1815–1884), professor for agriculture at University of Georgia (USA), headed a chapter in his textbook of scientific agriculture (Pendleton 1874, pp. 141–159) "chemistry of the atmosphere".

In the review "*Jahresbericht über die Fortschritte auf dem Gesammtgebiete der Agrikultur-Chemie*" [Annual Reviews on Progresses in the Whole Field of Agriculture Chemistry],[21] containing also annual reviews on air and atmospheric research (meteorology, chemistry, and physics), without a doubt the best in the world, the term "*Die Chemie der Luft*" [chemistry of air] for the first time was applied as a heading by *Theodor Dietrich* (1833–1917).[22] In 1889, *Karl Richard Hornberger* (1849–1918)[23] first used the term "*Chemie der Atmosphäre und atmosphärischen Niederschläge*" [chemistry of the atmosphere and atmospheric precipitation]. The term "*Chemie der Atmosphäre*" [chemistry of the atmosphere][24] was applied by *Anton Baumann* (1856–1912)[25] in 1892.

---

21 This review journal appeared in 1860–1865 as Jahresbericht über die Fortschritte der Agriculturchemie (Vol. 1–7) and in 1866–1919 as Jahresbericht über die Fortschritte auf dem Gesammtgebiete der Agrikultur-Chemie (Vol. 8–60) with a subheading "Luft" [air] 1860–1874 and "Atmosphäre" (1876–1919) in the table of contents; the section heading "Die Chemie der Luft" appeared only in Vol. 13–15 (for the years 1870/72) 1874 and 16–17 (for the years 1873/74) 1876, the section heading "Chemie der Atmosphäre und atmosphärischen Niederschläge" only in Vol. 32 and 33 for the years 1889 and 1890 (rapporteur *Richard Hornberger*), and the heading "Chemie der Atmosphäre" only in Vol. 35 for the year 1892 (rapporteur *Anton Baumann*) and Vol. 36 for 1893 by rapporteur *Fritz Erk* (1857–1919), a German meteorologist in Munich.

22 *Theodor Dietrich* was a German chemist, assistant at *Stöckhardt* in Tharandt (1852–1857) and head of the Experimental Station Kurhessen [Electoral Hesse] in Haidau and since 1865 in Altmorschen and 1880 in Marburg.

23 *Karl Richard Hornberger* was a German chemist, professor of mineralogy and soil sciences at Forest Academy in Münden; he was director of the agricultural research station Kuschen (1875–1877), where *Ewald Wollny* (1864–1865) analyzed rainwater (see Table 2.7).

24 A second subheading was "Physics of the Atmosphere" (only 1892 and 1893).

25 *Anton Baumann* was a German chemist at "*Forstliche Versuchsanstalt*" [forest experimental station] of the University Munich; from 1898 director of the first "*Königliche Bayerische Moorkulturanstalt*" [Royal Bavarian Peatland Management Academy].

In his book (Ramsay 1896, p. 18), Sir *William Ramsay* used the term "the gases of the atmosphere" in the following phrase: "Mayow's contributions to the *chemistry of the atmosphere* [...]". In 1897, *Henriet*'s book[26] "*Les gaz de l'atmosphére*" was published in Paris in a style of a modern monograph on atmospheric chemistry (Henriet 1897). He also used the term "*chimique de l'atmosphére*". His introduction begins to emphasize air chemistry monitoring:

> [...] nécessite des recherches qu'il est indispensable de poursuivre pendant un temps très long, afin de ne pas donner l'importance d'une loi à des phénomènes qui peuvent n'être qu'accidentels.

This was likely fully forgotten by "modern" atmospheric chemists and the term "atmospheric chemistry" was used again 66 years later (in German as "*Chemie der Atmosphäre*") by *Hans Cauer* in 1948 (Cauer 1948b) and as "*chemisch-meteorologische Forschung*" [chemical meteorological research] in 1949 (Cauer 1949a); the latter was translated into English using the term *atmospheric chemistry* (Cauer 1951a). However, *Hans Drischel* (1915–1980),[27] who was in close contact with *Cauer*, wrote (Drischel 1940, p. 331–332):

> Gewiß sind unsere Kenntnisse von der Chemie der Atmosphäre noch sehr lückenhaft. Aber planmäßige Untersuchungen werden ein Beobachtungsmaterial fördern, das geeignet ist, die noch offenen Fragen zu klären [Indeed, our knowledge of the chemistry of the atmosphere is still very sketchy, but scheduled examinations will produce observation material that is suitable to clarify the outstanding question].

Similar to *Hans Cauer* in Germany, in the Soviet Union, in Odessa, *Evgeny Samoilovich Burkser* [Евгений Самойлович Бурксер] (1887–1965), first Russian-Ukrainian radiochemist and geochemist in *Vernadsky*'s tradition, introduced the term "*аэрохимия*" [air chemistry][28] as a section of geochemistry of the atmosphere; since the 1920s *Burkser*

---

26 *Henri Henriet* (no life data known) published many papers on ozone and other air constituents. In *Nature* 55 (1897) 579, a fair review appeared: "*Les gaz de l'atmosphére* by H. Henriet, is an excellent little volume on the chemistry of our atmosphere. The author is chemist at the Montsouris Observatory, and the methods of analysis described by him, as well as the results of investigation into the composition of the air at different places and at different times, make this little book very valuable for meteorologists as well chemists".

27 *Hans Drischel* was a German physician, physiologist and internationally known biocyberneticist. Studied medicine (1934–1941) in Breslau and studied in his dissertation rainwater from balneological point of view (Breslau 1941). In 1941–1949 war physician and Soviet prisoner, in 1951–1958 at university Greifswald, since 1959 professor for physiology at Leipzig University.

28 After 1960 in the USSR, only the term "атмохимический" [atmochimitscheskij – air chemical] parallel to geochemical, hydrochemical, etc., was used but never "атмохимия" [atmochimija – air chemistry], the only equivalent in Russian to "atmospheric chemistry" or "chemistry of the atmosphere". At the beginning of the 1920s, in Schichany (Шиханы), Saratov area (USSR), an air chemical station [аэрохимическая станция] was established (classified object "Томка" [Tomka]) for the investigation of chemical weapon (with the participation of Germans) from 1928 was a part of the 33th

studied the atmospheric chemical composition with respect to balneological aptitude (Burkser 1924, 1937, 1940, Burkser and Gernet 1939, Burkser and Burkser 1951).

It was soon used as the label for a new discipline. The first modern monograph *Air Chemistry* in the field of this new discipline was written, in 1963, by *Christian Junge*, soon after he had published chapters entitled "Atmospheric Chemistry" (Junge 1956, 1958b). The first monograph "Photochemistry of Air Pollution" in the field of this new discipline was written by *Philipp Albert Leighton* (1897–1983) in 1961, based on the results gained from the Los Angeles smog studies. Books (monographs and textbooks) on the chemical composition of the atmosphere and air chemistry before *Leigthon* and *Junge*: Priestley (1774),[29] Scheele (1777), Smith (1872), Fox (1873), Henriet (1897), and Ramsay (1896); and after *Junge*'s monograph: Butcher and Charlson (1972), Rasool (1973), Heicklen (1976), Graedel (1978), Butler (1979), Mészáros (1981), Isidorov (2001, 1990), Seinfeld (1986), Finlayson-Pitts and Pitts (1986, 2000), Warneck (1988), Graedel and Crutzen (1993, 1995), Hobbs (1995, 2000),[30] Brimblecombe (1996), Seinfeld and Pandis (1998), Möller (2003, 2010, 2019).

1.1.2 From chemical meteorology and climatology to atmospheric chemistry

Before the 1950s, this "discipline" was almost termed "chemical meteorology". This term, "*Chimie Météorologique*", was first used as a classification of presentations in categories in the journal "*Comptes rendus hebdomadaires des séances de l'Académie des sciences*" of the French *Académie des Sciences* in 1856. The expression "*Chemische Meteorologie*" in German, was probably first used by the Swiss chemist *Armand Müller* (unknown living data) as a heading in a booklet, dealing with chemical reactions in the atmosphere under the influence of light (Müller 1873, p. 44). The French physicist *Alfred Cornu* (1841–1902) used the phrase "*Physique et de la Chimie météorologiques*" (Cornu 1879b, p. 1290).

However, since the late 1920s, chemical meteorology was mainly looking for the relationship between condensation nuclei, its chemical composition, and the formation of clouds and rain. That term is still in use as a subtitle for the journal "*Tellus*" (Series B: Chemical and Physical Meteorology, started in 1982 when "*Tellus*", created in 1949, was split into Series A and B) and for a professorship at the University of Stock-

---

Central Research Institute of the Ministry of Defense (testing institute). The Central Air Chemical Museum opened in Moscow in 1927 [Центральный аэрохимический музей], later (1941) renamed in Museum of Aviation.

29 It is a six-volume work (1774–1786) in different editions. First printing: Observations on different kinds of air. In: Philosophical Transactions, Vol. 62, 1772, pp. 147–264.

30 *Peter Victor Hobbs* (1936–2005) was a British-born professor of atmospheric sciences and director of the Cloud and Aerosol Research Group at the University of Washington.

holm[31] (*Henning Rodhe* held the chair from 1980 to 2008, continued by *Caroline Leck* since 2002), created in 1979 by *Bert Bolin* (1925–2007).[32] *Bolin* defined the task of atmospheric chemistry as "interplay between the atmosphere, the surface of the continents and oceans, biological activities, man and last but not least the ever moving atmosphere itself" (Bolin 1959).

Another term, "chemical climatology" came into use with the famous book *Air and Rain – The Beginning of a Chemical Climatology* (London 1872) by *Robert Angus Smith* (1817–1884).[33] *Smith*, in the paper entitled "Chemical Climatology" (Smith 1873), who had also defined climate ten years before Hann (1883) did as "the average condition is the climate of a place" and had already attributed to the *antropoghenic climate change*: "We are exposed to great changes of climates arising from the conditions of our civilization, [. . .]", explains his aim "[. . .] to examine the subject rather from the point of view which may be termed chemical, as distinguished from the physical and medical" (Smith 1873, p. 2), and finally proposing "[. . .] perhaps to the establishment of a department at some Observatories for Chemical Climatology and Meteorology" (Smith 1873, p. 11). At that time nothing was known on chemical processes in the air,[34] hence the book's focus was on the description of concentrations of air constituents in time and space similary to physical parameters ("meteorological elements" like temperature, pressure, wind, etc.). It cannot be stressed enough how far-sighted *Robert Angus Smith* the idea of a *Chemical Climatology* emphasized; he wrote (already in 1869) (Smith 1877):

> For a satisfactory investigation of the subject, one must look to the multiplication of these experiments, and perhaps to the establishment of a department at some Observatories for Chemical Climatology and Meteorology [. . .] The importance of having a chemical department to out observatories cannot be long overlooked.

Smith, obviously disappointed, writes further:

31 *Carl-Gustaf Rossby* (1898–1957) established the Department of Meteorology at Stockholm University (MISU) in 1947. After his death, *Bert Bolin* took over the leadership in 1957, but *Rossby*'s professorship was a personal one. The position of a permanent chair in meteorology was acquired in 1961, and *Bolin* hold the new position then for almost 30 years. Among *Bolin*'s supervised PhD students were *Henning Rodhe*, *Paul Crutzen* and *Lennart Granat*, who became outstanding atmospheric chemists.

32 *Bert Rickard Johannes Bolin* was a Swedish meteorologist and professor at Stockholm University and involved in international climate research cooperation from the 1960s; the first chairman of the The Intergovernmental Panel on Climate Change (IPCC) from 1988 to 1998 (see Rodhe 2013).

33 *Robert Angus Smith* was a Scottish chemist, who investigated numerous environmental issues. He was a scholar of *Liebig* and studied the rain chemistry in 1848 in Manchester, for the first time from air pollution point of view. He was appointed Queen Victoria's first inspector under the *Alkali Acts Administration* of 1863. Further reading on *Smith* see Gibson and Farrar (1974).

34 However, see p. 311 as a first phrase upon "chemical process in air" by an unknown author in 1821.

> I should were glad had may work caused in this country a beginning like was lately at the observatory Montsouris, at Paris, or at least resembling it: and I blame myself for not pushing forward the idea, although my numerous engagements may well form kind of apology.

63 years after *Smith* (in 1935), *Hans Cauer* re-introduced the term "chemical climatology" (he became known with *Smith*'s work only in 1938). Notwithstanding, another term arose more than 100 years ago, *bioclimatology*, and later *biometeorology*.[35] In Sweden, *Carl-Gustaf Arvid Rossby* (1898–1957) and *Hans Gabriel Egnér* (1896–1989) first used the term *chemical climate* for air masses having a characteristic chemical feature (Rossby and Egnér 1955).[36] Independently from *Cauer*, *Helmut Mrose* published "*Chemische Aufgaben der Meteorologie*" [chemical tasks of meteorology] and emphasized that for an understanding of atmospheric processes not only physical but also chemical methods are essential (Mrose 1949). The British physicist *Sydney Chapman*, known for the first theory of ozone formation in the upper atmosphere (1930), regarded atmospheric chemistry in 1951 only to be important for the upper atmosphere, and wrote (Chapman 1951, p. 262):

> The meteorology of the lower atmosphere, the region of weather, is essentially a *physical* science, in which chemistry plays practically no part; chemists may become good meteorologists, but meteorologists need not know much chemistry.
> Chemists are indeed interested in the composition of air, but their work has generally been supposed to end with the chemical analysis.
> [...] the main fact of tropospheric chemistry is the uniformity of the composition of dry air all over the globe, near the ground, [...]

Junge (1956, 1963b) regarded air chemistry as a branch of meteorology. Scharnow et al. (1965)[37] defined meteorology as the physics and chemistry of the atmosphere, but meteorology was also reduced to just the physics of the atmosphere (Liljequist and Cehak 1984).[38] *Wolfgang Warmbt* (1958, p. 30) writes (translation from German):

> Air chemistry is still a very young research area, on their results the natural sciences and medicine are very interested.

---

35 In the Soviet Meteorological Dictionary (Chromov and Mamontova 1974) I found the best definition of biometeorology: science of the interactions between physical and chemical processes in the atmosphere and living organisms, especially on the impact of weather and climate on humans, animals and plants. From biometeorology the bioclimatology is derived, dealing with the impact of the climate in sense of the statistical regime of the atmosphere, on living organisms.

36 Understanding of a chemical climate and hence a chemical climatology is thus in line with *Robert Angus Smith* (Smith 1872), *Hans Cauer* (Cauer 1934), *Hans Egnér* and further developed by Möller (2006).

37 *Ulrich Scharnow* (1926–1999) was a German captain and professor for nautical science.

38 *Gösta Hjalmar Liljequist* (1914–1995) was a Swedish meteorologist.

A more general definition, but one that is appealing as a wonderful phrase, is given by *Christian Junge*: "Air chemistry is defined [...] as the branch of atmospheric science concerned with the constituents and chemical processes of the atmosphere [...]" (Junge 1963a). Atmospheric chemistry is now widely defined as the discipline dealing with the origin, distribution, transformation, and deposition of gaseous, dissolved and solid substances in air (Möller 2011). However, this is more than chemistry, as it includes apart from chemical reactions the investigation of sources of atmospheric constituents (biology, geochemistry, technology, etc.), transport, and transfer processes (physics, meteorology). Biogeochemistry, already introduced in the 1930s, was gradually institutionalized in the 1970s as an interdisciplinary subject to atmospheric chemistry, dealing with

- surface (vegetation, soil, sea) – atmosphere exchange,
- budgets and cycles of trace species.

Another important discipline, which is essential for an understanding of atmospheric chemistry, is atmospheric physics, dealing with

- radiation transfer,
- precipitation and cloud physics,
- aerosol particle physics.

Otherwise, atmospheric chemistry is essential for these disciplines; hence the need for interdisciplinary cooperation led not only to the International Geophysical Year 1958 but also, in the 1980s, to the International Geosphere-Biosphere Programme (IGBP), launched in 1987 to coordinate international research on global-scale and regional-scale interactions between earth's biological, chemical, and physical processes and their interactions with human systems, closed in 2015. Thought the impact of atmospheric constituents on climate, humans, vegetation, etc., is not within the task of atmospheric chemistry (it is bioclimatology and medical meteorology), but it has a strong motivation to deal with it.

Description of air pollution and thus air quality was already the focus of *Smith*'s chemical climatology. The beginning, however, goes back to *Humboldt*'s definition of the climate (see Möller 2020, p. 285), which was the basis for *Carl Dorno* (1865–1942), German chemist and climatologist, who established the modern (in contrast to the ancient idea of *Hippocrates*) *bioclimatology* in 1906. It was further developed by *Adolf Loewy* (1862–1937), German physiologist, who defines climate (in a physiological sense), Loewy (1924):

> [...] im physiologischen Sinne unter Klima die Summe aller für einen Ort typischen atmosphärischen und terrestrischen Zustände, durch die unser Befinden unmittelbar beeinflußt wird [... the sum of all atmospheric and terrestrial states, typical for a given location and influencing human feeling].

However, it is worth noting that Loewy (1924) includes not only the physical climate factors (which he considered to be the most important) but also *chemical* climate factors (in terms of constituents in air). *Loewy* used the term "*Klimaphysiologie*" [climate physiology] and noticed (Loewy 1924, p. 75) that exclusively the impurities have a negative impact on human health. Already Renk (1886, pp. 170–172)[39] concluded that atmospheric ozone has no hygienic importance but irrespirable gases and dust.

The aim of climatology is to establish the mean states of the atmosphere over the different parts of the earth's surface, including the describtion of its variations (anomalies) within longer periods for the same location (Hann 1883).[40] *Hann*'s definition was principally accepted by all great succeeding climatologists like Köppen (1899, 1923, 1931)[41] and Schneider-Carius (1961). *Hann* also introduced the term "*Faktoren des Klimas*" [*climatic elements*] emphasizing that measurements must characterize these through numerical values. The climatic elements *Hann* listed in the following order: temperature (air and soil, radiation), atmospheric humidity (including water vapor and precipitation), cloudiness, winds, air pressure (being less important in contrast to weather), evaporation, and air composition. Measurements of minor constituents in air were very rare and uncertain at that time, but *Hann* noted explicitly in the third edition of his "*Handbuch der Klimatologie*" (1908) dust, organic compounds, ozone, hydrogen peroxide, in reference to the hygienic state of the air. Thus Köppen (1906, p. 9) writes:[42]

> Mit dem Fortschritt des Wissens werden neue Gegenstände in die Zahl der klimatischen Elemente aufgenommen, wenn deren geographische Züge entschleiert werden [With the progress of knowledge, new items will be included in the number of climatological elements, when their geographic feature were unveiled].

The German hygienist *Max Rubner* (1854–1931) defined climate as "all through the special location caused influences on human health" (Rubner 1907). However, the term *bioclimatology* was probably first used, in 1912, by *Wilhelm Richard Eckardt* (1879–1930), German paleoclimatologist and scholar of *Dorno* in Jena (Eckardt 1912).

The term *Medical Meteorology* was introduced by Moffat (1853) in England, see p. 542. Today, *Adolf Albrecht Mühry* (1810–1888), German physician at Göttingen, is termed as the first German *bioclimatologist* (however, this term was not used at that time). In the 1930s, *bioclimatology* was institutionalized in Austria and Germany with

---

39 *Friedrich Georg Renk* (1850–1928), scholar of *Pettenkofer* in Munich; hygienist in Berlin since 1887.

40 *Julius von Hann* (1839–1921) was Professor of Physical Geography (1874–1897) at University of Vienna, in 1897–1900 Professor of Meteorology in Graz, and afterwards until 1910 again in Vienna.

41 *Wladimir Peter Köppen* (1846–1940), German-Russian climatologist and head of the "*Seewetterdienst*" [Division of Marine Meteorology] (1875–1919) at the "*Seewetterwarte Hamburg*" (whose first director was *Georg von Neumayer*).

42 It is worth to note that in following editions (1923 and 1930) the same phrase unchanged is printed, suggesting that *no* progress in adopting new (more or other) climatological elements happened.

scientific journals,[43] institutes[44] and just after World War II.[45] Whereas between 1930 and about 1953 an important topic of bioclimatology was studying the condensation nuclei (Findeisen 1938, Burckhardt and Flohn 1939;[46] almost its number but also its composition and origin; Chapter 3.3.2), that gives birth to modern air chemistry as seen in papers by *Christian Junge* (Junge 1936, 1951, 1956); and, in the 1950s, it was developed to plant and medical meteorology (Glaser 1957).[47]

---

43 In Austria, "*Bioklimatische Beiblätter der Meteorologischen Zeitschrift*" was founded in 1934 by the "*Österreichische Gesellschaft für Meteorologie*" (until 1937) and published until 1943. *Ferdinand Steinhauser* (1905–1991), who was already one of the editors of the latter journal, founded in 1949 "*Archiv für Meteorologie, Geophysik und Bioklimatologie*" (until 1980, subdivided in Serie A – meteorology and geophysics, and Serie B – climatology, environmental meteorology, radiation research). In 1948 the Austrian Society for Meteorology established "*Wetter und Leben – Zeitschrift für Bioklimatologie*" (1964 renamed in "*Zeitschrift für angewandte Meteorologie*").

44 The first "*bioklimatische Forschungsanstalt*" [bioclimatic research institute] was established by the German physicians *Karl Gmelin* (1863–1941) and *Carl Häberlin* (1870–1954) in 1928 in Wyk (on Island Föhr, Baltic Sea) which was related to the University Kiel in 1931 with the physician *Alfred Otto Heinrich Pfleiderer* (1900–1973) as head and the geophysicist *Konrad Johannes Karl Büttner* (1903–1970). Both scientists were Nazis and did knew from the hypothermic experiments in the Dachau concentration camp since 1942; *Büttner* went in 1950 to USA as meteorologist at Air Force and since 1953 at University Seattle. *Pfleiderer* became the director of the "*Institut für Bioklimatologie und Meeresheilkunde*" [institute of bioclimatology and marine medicine] of the University Kiel in Westerland (Sylt) and held this position after war. Another "*bioklimatische Forschungsstelle*" was established in 1939 at Königstein (Taunes) together with the Meteorological Institute of the University Frankfurt/Main; director was the German physician and balneologist *Walter Amelung* (1894–1988). Historical interesting is that *Cauer* worked together with *Amelung* but had a bad opinion of *Pfleiderer*.

45 *Reinhold Reiter* (1920–1998), German meteorologist, founded in 1949 in Munich the "*physikalisch-bioklimatische Forschungsstelle*" [physical bioclimatic institute] and in 1953 established in Garmisch-Partenkirchen (mainly research on atmospheric electricity). At Freiburg, the "*Medizinmeteorologische Forschungsstelle*" [medical meteorological institute] belonged to the "*Bioklimainstitut*" Freiburg, was founded by *Heinz Lossnitzer* (1904–1964), in 1952 integrated as Meteorological Observatory of the German Weather Service, in 1964 split in a Department of Biometeorology (headed by *Robert Neuwirth*) and from 1970 until today continues as "*Medizinmeteorologische Forschungsstelle*" of the German Weather Service; this institute was focused on air quality (pollution, radiation, and radioactivity) in relation to human health).

46 *Hans Burckhardt* (1910–1972) was a German meteorologist at the "*Reichsamt für Wetterdienst*", after the war "*Deutscher Wetterdienst*" (Bad Kissingen) and "*Agrarmeteorologische Versuchs- und Beratungsstelle*" in Neustadt (Rhineland-Palatinate). *Flohn* see p. 606.

47 This change is clearly seen in the publications from different bioclimatic research institutions in East Germany between 1946 and 1963, listed by *Heinz Würfel* (1920–1992), "*Forschungsinstitut für Bioklimatologie*" [research institute for bioclimatology] at Berlin-Buch (Würfel 1965), subdivided into the following categories (in parenthesis number of publications): medical biometeorology (66), medical bioclimatology (87), air chemistry within biometeorology (42), experimental biometeorology (7). All meteorologists who contributed to air chemistry, *Emil Friedrich Flach* (1905–1995), *Gerhard Hentschel* (1920–1995), who was the director of the institute in Berlin-Buch until 1987, *Helmut Mrose* (1910–1999), *Rolf Noack* (1938–1967), *Wolfgang Warmbt* (1916–2008), and *Helmut Zenker* (1921–1978), also contributed to other fields of "biometeorology and bioclimatology". It is remarkable that the very impor-

Cauer used also the terms "*chemisch-bioklimatisch*" [chemical bioclimatic] and "*chemisch-lufthygienisch*" [chemical-air hygienic] for his studies of atmospheric chemistry, according to the "application" (health or air pollution). However, accepting the general definition of climate (and thus climatology), all these terms (medical meteorology, chemical meteorology, biometeorology) describe nothing more than air chemical studies concerning the impacts on living organisms. Any air chemical study, however, is not restricted to any application in impact research, either living organisms or nonliving things. In the late 1960s, in German-speaking countries (Germany, Austria, Switzerland) the term "*Immissionsklimatologie*" arose, used from meteorologists and later geographers to describe the concentrations fields (by modeling and measurements) due to pollutant dispersion in cities, industrial, and orographic areas.

With the discovery of the *main* chemical composition of air, different disciplines arose dealing with the *minor* chemical composition of air:
- analytical chemistry,
- agricultural chemistry,
- sanitary chemistry,
- pollution chemistry,
- atmospheric chemistry.

The general view on air pollution and atmospheric chemistry at the end of the nineteenth century is nicely summarized by Renk (1886, p. 55); translated from the German original:

> All gases to be released from the earth's surface, escape into the atmosphere where they either be converted into other chemical compounds – as for example gaseous ammonia, that forms salts with carbonic acid, nitric acid or nitrous acid – or they remain unchanged; in neither of the two cases, however, their residence in the atmosphere is of long duration because certain atmospheric processes permanently remove them and thus keep the atmosphere clean. These are particularly atmospheric precipitation and life processes of the vegetation.

Renk refers the following pollutants:[48] carbonic acid, ammonia, nitric acid, and nitrous acid of natural origin, sulfur dioxide, hydrochloric acid, carbon disulfide, phosphoric acid, arsenic, and chlorine from different commercial enterprises. Hydrogen sulfide was known from mineral springs and rotting organic material. In marsh gases,

---

tant publications between 1952 and 1957 on atmospheric ozone, dust, and precipitation chemistry by *Friedrich Teichert* (1905–1986) are not listed in Würfel (1965) due to political reasons (*Teichert* left East to West Germany in 1956); see also footnote 83 on p. 33. The publications in "air chemistry" are concerned condensation nuclei (*Flach, Hentschel, Mrose, Würfel, Zenker*) and air pollution (*Hentschel, Mrose, Noack*). *Cauer* was in personal contact and correspondence with *Hentschel, Mrose, Warmbt,* and *Teichert* after his escape.

48 Lehmann and Heller (1940) refer as air pollutants (and describe their impacts and measurement): CO_2, CO, SO_2, HCl, Cl_2, HNO_3, HNO_2, NO, NO_2, NH_3, H_2SO_4, HF, H_2S, CS_2, HCN, AsH_3, C_2H_4O (ethylene oxide), smelling substances, benzene, benzine, dust, soot, and microorganisms.

methane and phosphine where found. From places where cadaver and organic waste is rotting, putrefaction gases were known which include carbonic acid, hydrogen sulfide, ammonia, amines, methane, volatile fatty acids, and other smelling gases. A further important source of pollutants in the late nineteenth century was the use of town gas containing hydrogen, methane, carbon monoxide, heavy hydrocarbons (ethylene, propylene, butylene, acetylene, allylene, and benzene). Other kinds of illumination such es candles and petrol lamps were identified as sources of hydrocarbons, volatile fatty acids and acrolein, etc.; *Renk* further refers (Renk 1886, p. 61) that perspiration contains formic acid, acetic acid, butyric acid, and propionic acid, and, furthermore, ammonia, valeric acid, caproic acid, caprylic acid, and other specifically smelling substances. Finally, *Renk* refers industrial gases having an eminent importance for human health as irrespirable gases like sulfurous acid, sulfuric acid, nitrous acid, hydrochloric acid, ammonia, chlorine, and as noxious gases like carbon dioxide, carbon monoxide, hydrogen sulfide, carbon disulfide, arsine, phosphine and, furthermore, iodine and bromine, turpentine vapor, oil vapor, tar, petrol vapors, and others. Soon after the discovery of trace gases in the air (and exhaust gases from industry and heating), it was recognized the harmfulness of gases (Eulenberg 1865, Fischer 1878, Hirt 1871, Henderson and Haggard 1927, Flury and Zernik 1931).[49] Fox (1886, p. 257) writes:

> Such as hydrochloric acid gas from alkali works, arsenical vapours from copper-smelting works, hydrofluoric acid from superphosphate manufactories, etc., injure animal and vegetable life, sometimes destroying all trace of the latter for miles around. Then the air is vitiated by bisulphide of carbon from indiarubber works; chlorine, sulphurous and sulphuric acids from bleaching works, hydrogen sulphide from chemical works where ammonia is manufactured. It is poisoned also by carbonic acid, carbonic oxide, and hydrogen sulphide, from brickfields and cement works; by organic vapours from the refiners, bone burners, slaughter-houses, etc., and by the fumes of oxide of zinc, producing "brassfounder" argue.

A paper entitled "*Composition de l'atmospère*" by *Jean-Baptiste-Alphonse Chevallier* (1793–1879), a French pharmacist and student of *Vauquelin*, the finest analytical chemist of his days in France, is likely the first written document (without any details on sampling and analysis) concerning trace gases in the air (Chevallier 1834):

- in general, the air in Paris and many other sites contains ammonia and organic matter,

49 *Hermann Eulenberg* (1814–1902) was a German physician; *Ferdinand Fischer* (1842–1916) was a German chemist specialized in chemical technology; *Ludwig Hirt* (1844–1907) was a German physician, founder of occupational medicine; *Yandell Henderson* (1873–1944) was an American physiologist, termed as "expert on gases"; *Howard Howard Wilcox Haggard* (1891–1959) was an American physician; *Ferdinand Flury* (1877–1942) was a German pharmacologist, international known for research on chemical warfare agents; *Franz Zernik* (1876–1941) was a German chemist and pharmacist, chosen suicide to avoid deportation; he was consultor of the "*Reichswehr*" [Armed Forces of the Weimar Republic] on chemical weapons.

- dew contains ammonia and organic matter,
- the composition of the air varies not much because of the overall combustion and decomposition of vegetable and animal matter, etc.,
- the air of London contains sulfuric acid,
- the air of sewage in Paris contains ammonium acetate and hydrogen sulfate,
- the air in the valley of Montsaucon (in the North of Paris) contains ammonium hydrogen sulfate.

Between 1920 and 1940, atmospheric physics, like ozone research and nucleation theory, was developed mostly only by physicists and meteorologists but to a lesser extent by chemists. Despite some ideas on chemical processes in air (mostly by electrical discharges as shown by *Cavendish*) at the very beginning of the nineteenth century, only the formation of ozone under strong ultra-violet (UV) irradiation had been roughly understood by the 1930s. Before World War II, ozone was believed to be produced only naturally in the stratosphere. In 1924, *Gordon Dobson* (1889–1975)[50] designed a spectrometer, which became the standard instrument for measuring both total column ozone and the profiles of ozone. With *Dobson* spectrophotometers the British geophysicist *Joseph Charles* [*Joe*] *Farman* (1930–2013) and his team on the British Antarctic survey discovered the ozone hole in 1984 (Farman et al. 1985).

In this period (1920–1940), research on photochemistry and reaction kinetics (most radicals and reactions to be found later relevant in atmospheric air were identified) also developed, but it was exclusively motivated by industrial and military application, namely, combustion research. All these investigations were the roots of atmospheric chemistry research in the 1950s due to the Los Angeles Pollution in the late 1940s and the London Smog event in 1952. In 1944, plant injuries had been observed in the Los Angeles area, which for the first time were not related to "classical" pollutants (like SO_2 or fluorine compounds), already described in the late nineteenth century. In the early 1950s, *Arie Jan Haagen-Smit* (1900–1977) and co-workers made automobile exhaust gases responsible for surface-near formation of ozone, which then were considered as the impact species. Whereas many radicals were proposed during the study of atomic reactions in combustion and explosive processes since about 1925; fast kinetic observations and intermediate detection became possible only in the 1940s with spectroscopic techniques. Scientific literature in the 1950s was full of studies on this topic. A complete description of radical reactions with citation of original sources can be found in Leighton (1961). Summer and winter smog strongly stimulated the progress in ozone and sulfur chemistry. Junge and Ryan (1958) first pointed out the importance of the atmospheric aqueous phase for SO_2 oxidation. Al-

50 *Gordon Miller Bourne Dobson*, a British physicist and meteorologist. He built the first ozone spectrophotometer and studied the results over many years (now termed Dobson spectrometer or Dobsonmeter giving the Dobson unit). See Walshaw (1989) for biography.

though acidic rain (Smith 1872),[51] the bleaching properties of dew (Prout 1834, 1836) and hydrogen peroxide in the rain (Meissner 1863b, Schöne 1874) were known as phenomena many years ago, atmospheric aqueous phase chemistry has long been ignored compared with gas-phase chemistry. However, it must be mentioned another "root" of atmospheric chemistry, i. e., the atomic bomb tests since the early 1950s and the related issues of the distribution, transportation, and removal of radioactive species and atmospheric aerosols.

Notwithstanding that between 1950 and 1990 (namely with the development of sophisticated laboratory techniques) reaction and photolysis rates of chemical reactions were studied repeatedly. This, the development of computer techniques, and the political task for air pollution control (clean air strategy) led in the late 1970s to the development of progressively complex atmospheric chemistry models, stepwise including transportation and distribution from local to global scale, and finally linking with climate models.

Except for the discovery the main chemical composition of air in the second half of the eighteenth century, the history of atmospheric chemistry cannot be simply divided into periods, because all other research topics can actually be named from at the beginning, but there are ongoing investigations:

- research on chemical composition of atmospheric waters (since the first half of the eighteenth century).
- discovery and studying of trace compounds in the air (since the late eighteenth century),
- research on air pollution (since about 1850),
- research on dust in air (since the 1870s),
- research on atmospheric aerosol (since the 1920s),
- research on stratospheric ozone (since the 1920s).

Between 1950 and 2000, we can subdivide the atmospheric chemistry research into the following issues (some of them are overlapping):

- fundamental research (chemical kinetics and mechanisms; laboratory studies),
- research on global cycling (sulfur, chlorine, nitrogen, carbon),
- precipitation chemistry ("acid rain"),
- plume chemistry (pollutant dispersion and conversion),
- cloud and fog chemistry (aqueous-phase and interfacial chemistry),
- tropospheric ozone and ROS research (photochemistry, "summer smog"),
- stratospheric ozone research (ozone depletion, "ozone whole"),

51 Smith (1872) used the term "acid rain" only twice in his book (on page 444) in connection with effects of the atmosphere on stones and iron: "I was led to attribute this effect to the slow, but constant, action of the acid rain" and "[...] iron oxidizes readily, [...] where the acid rain [...]". However, the term acid rain was first used in French "*pluie acide*" by Ducros (1845), see p. 169.

– atmospheric aerosol research (heterogeneous chemistry, SOA and CCN formation).

Besides laboratory reactor studies under defined conditions, monitoring and field experiments were the essential tools. Unfortunately, long-term environmental measurements were underappreciated and underfunded, because they are seen neither as basic measurements to test scientific hypothesis nor as challenging high-tech opportunities for profit by commercial interests. Other air pollution studies (like heavy metals, POP's, greenhouse gases) included less chemistry but mostly physics, and toxicology. It is obvious that, with the exception of laboratory fundamental research, physical and especially meteorological research is related to atmospheric chemistry: transport, mixing, and conversion occur simultaneously. The aim of atmospheric chemistry was an understanding of time-spatial dependent and changing toxicity, acidity, oxidation/reduction capacity, composition, and size distribution of particles to explain impacts like acidification of soils and lakes, plant injures, forest decline, human diseases, climate forcing and change.

Studying the chemistry of a natural system (like the atmosphere) is done by chemical quantitative analysis of the different matters. An uncountable number of analyses are needed to gain a three-dimensional distribution of the concentration of specific substances or, in other words, the chemical composition of the natural system. This is the static approach to obtain an averaged chemical composition over a given time period, depending on the lifetime of the substances. However, concentration changes occur with time because of transportation (motion), transformation (chemical reactions), and emission (seasonal and diurnal variation). Hence, a dynamic approach is also needed (chemical monitoring and modeling). Chemical reactions can be studied – which means establishing the kinetics and mechanisms – only in the laboratory under controlled (and hence replicable) conditions in a large variety of reaction chambers. Finally, it is not important whether atmospheric chemistry is seen as a part of geochemistry, biogeochemistry or meteorology.

Until the 1960s, measurement of atmospheric trace gases (with the exception of ozone) was not seen as an academic mission but one as air pollution monitoring ("atmospheric mud") by hygienic institutions and later by environmental agencies. Nevertheless, trace gas monitoring (the best-known icon remains the CO_2 Mauna Loa record) has been and remains an important part of atmospheric chemistry ("Chemical Climatology"). The origins of modern atmospheric chemistry, however, lie in the precipitation chemistry (within the chain rain–cloud–condensation nucleus) and the aerosol chemistry (namely, in condensation of nuclei), linking physical and chemical processes in the atmosphere.

Since then, many significant advances were made. For example, in the early 1970s, the role of hydroxyl radicals (OH) in oxidizing gaseous pollutant was discovered, which, finally, led to their removal from the atmosphere. The essential ingredients for the production of OH radicals are ozone, water vapor, and UV solar radiation.

The catalytic role of nitrogen oxides in ozone production was also determined at the beginning of the 1970s. Until then, it was generally believed that tropospheric ozone was produced in the stratosphere and transported downwards into the troposphere. The feedstocks for the creation of ozone are carbon monoxide, methane, and many volatile organic compounds. Both natural and anthropogenic processes are responsible for their emissions. Although the main photochemical chain reactions are reasonably well known, their quantification needs further research. They all are parts of the biogeochemical cycles of carbon, nitrogen, and sulfur. They can also play a role in climate, as does particulate matter, which, unlike the greenhouse gases (CO_2, CH_4), tends to cool the Earth and atmosphere.

The proceedings of the Dahlem Workshop on Atmospheric Chemistry, to be held in Berlin, May 1982 (Goldberg 1982) not only reads like who is who in atmospheric chemistry[52] of that time but also reflects the achievements of research in atmospheric chemistry after 1960. It included four group reports (aqueous chemistry in the atmosphere, changes in atmospheric composition, biogenic contribution to atmospheric chemistry, tropospheric gases, aerosols, and photochemical reactions) supported by lecture.

Like every science, atmospheric chemistry stands on two columns, praxis and theory, and both parts have developed historically independently, but at the same time in interconnection with each other. Before having scientific instruments to measure values and obtain data, theories emerged from simple visual (but also from the sense of smell and taste) observations. A theory (or model) was the first in the history (already in ancient Greek) verbal description of a process. At the beginning of the nineteenth century, chemical equations, their sets, and then a mathematical equation or their

52 Atmospheric chemists (in alphabetical order): *Meinrath O. Andreae* (born 1949), *Greg P. Ayers* (born 1950), *Heinz G. Bingemer* (born 1951), *Ralph J. Cicerone* (1943–2016), *Paul J. Crutzen* (1933–2021), *Robert A. Duce* (born 1935), *Dieter H. Ehhalt* (born 1935), *Hans-Walter Georgii* (1924–2018), *Thomas Eldon Graedel* (born 1938), *Jürgen Hahn* (born 1936), *Ruprecht Jaenicke* (born 1940), *Dieter Klockow* (1934–2018), *Peter S. Liss* (born 1942), *Franz X. Meixner* (born 1951), *Mario J. Molina* (born 1943), *James J. Morgan* (born 1932), *Hiromi Niki* (1937–1995), *Stuart Arthur Penkett* (born 1939), *Henning Rodhe* (born 1941), *Frank Sherwood Rowland* (1927–2012), *Jochen Rudolph* (born 1947), *Ulrich Schmidt* (born 1943), *Wolfgang Seiler* (born 1940), *F. Georg N. Slinn* (born 1938), *Reinhard E. Zellner* (born 1944). The author (DM), living in Berlin 20 km to the east from Dahlem at that time, was not invited, and, even if so, would not be able to attend the workshop due to the iron curtain, the "*Berliner Mauer*" [Berlin Wall]; but with the exception of *Cicerone, Molina, Morgan* and *Rowland*, he met all other scientists later. DM was elected in 1983 as a member of the Commission on Atmospheric Chemistry and Global Pollution (CACGP), founded as Commission on Atmospheric Chemistry and Radioactivity in 1957, changed its name to Commission on Atmospheric Chemistry and Global Pollution in 1971 (president: *C. Junge*). Members in 1983: *Charlson, Cicerone, Cox, Crutzen, Danielsen, Ehhalt, Goldsmith, Keeling, Merlivat, Miller, Pruppacher, Rowland, Ryaboshapko, Siegenthaler, Spedding, Varhelyi, Whelpdale, Brasseur, Madelaine, Ono, Pearman, Petrenchuk*, and the honorary members *Junge, Bolin*, and *Martell*. However, DM could first participate at the Commission Meeting and Symposium on Global Atmospheric Chemistry, hold in Chamrouse, France 1990.

set gradually arose. Thus, the kinetics of a system of chemical reactions (the mechanism, including not only a few but up to thousands of reactions) can be computed in simple cases analytically and in complex cases numerically. Modeling (or simulation) means simply the prediction different scenarios by varying the initial values, such as reactants, their ratio and concentration, and conditions, such as radiation, temperature, pressure, etc., in the system. In chemistry, the main aim is to determine the changes in concentrations with time. In atmospheric modeling, the distribution (or dispersion) is an additional important issue, i. e., representation of concentration in time and space, in the past and also in the future, in short-time and long-time scale with different spatial resolutions. *Modeling* is one tool of the practical investigation in atmospheric chemistry, but not part of this "History" because of limitations. The other two columns, on which atmospheric chemistry stands, are investigations in the laboratory and the atmosphere, i. e., chemical observations under *controlled* (and thus repeatable conditions to obtain fundamental data) and *natural* conditions. Historically, observations of properties of the atmospheric air were first, and cover the major part of this "History".

1.1.3 *Hans Cauer* and the German pioneers of modern atmospheric chemistry

The motivation to carry out chemical investigations of the atmosphere is very different, ranging from pure academic interest to applied research. Hence, scientists and engineers who are doing such investigations, are not *per se* atmospheric chemists, not chemists necessarily, but often meteorologists, physicists, physicians, biologists, geographers, etc. The beginning of research has been and remains scientific curiosity and a strong desire to find answers to the question about how the natural world works. In the middle of the nineteenth century, the air pollution phenomenon in London (and other British towns) gave the birth *Air Pollution Chemistry* (later often called *Smog Chemistry*), it was started by *Robert Angus Smith*, the first atmospheric and air pollution chemist worldwide and then continued by the French chemist *Albert Lévy* (1844–1907), chief-chemist at the observatory Montsouris in Paris. The German chemist *Herrmann Emil Schöne* (1838–1896)[53] in Moscow, who was in his official job as an agricultural chemist, must be named among the early atmospheric chemists because of his fundamental research devoted to the atmospheric hydrogen peroxide and ozone. Studying the chemistry of London smoke, the British chemist *Julius Berend Cohen* (1859–1935) did a pioneering work.

After 1930, but before World War II, further pioneering work in atmospheric chemistry was done by German chemists:[54] *Wilhelm Liesegang* (1894–1953) and *Hans*

---

53 See footnote 1010 on p. 515 on his biography.

54 Soon after World War II, the US American chemist of Dutch origin *Arie Jan Haagen-Smit* (1900–1977) was the first in a long line of American atmospheric chemists of the first generation (born between 1897

Cauer (1899–1962), the natural scientist *Helmut Mrose* (1910–1999), and meteorologist *Christian Junge* (1912–1996). *Smith* and *Cauer* coined independently the term *chemical climatology*, expressing that the chemical composition (or the concentration of a substance in air) of the atmosphere is the result of a complex process, which involves large-scale distribution and meteorological processes. However, *Smith*, *Lévy*, *Liesegang*, *Cauer*, and *Mrose* spend much of their working time on developing the methods for air sampling and chemical analysis; their options in the understanding of the atmospheric budgets of trace substances was still very limited. *Junge*, a meteorologist, who began his scientific career with investigation of condensation nuclei in the mid-1930s, was the first who shortly after the war recognized the interaction between aerosols and gaseous trace substances in the understanding of budgets; he coined the German term "*Spurenstoffhaushalt*" (Junge 1953).[55] It is remarkable that from this time (with some exceptions) most "atmospheric chemists" who conducted research in the field of cycling and budgeting of trace substances in the atmosphere were meteorologists, whereas those who studied chemical reactions related to the atmosphere in the laboratory were chemists and physicists. Junge (1953, p. 25) still writes that the "sensitivity of spectroscopic methods is not enough comparing with air chemical methods so that from the first hardly any help is to be expected". Only in the 1970s, physical detection principles were successfully developed or adopted for atmospheric measurements, and, thus, a majority of "atmospheric chemists" became physicists and chemists.

Cauer (1948b, 1951a) emphasized that analytical chemists must be trained to conduct chemical analysis of atmospheric trace gases and condensation nuclei together with physical measurements like level of UV radiations and number of concentrations. *Cauer* introduced microchemical analytical methods in the 1930s and hold lifelong suspicion to measurements taken by physicians. On contrary, from *Cauer*'s discussion on chemical processes in the atmosphere (Cauer 1938, p. 415) we can learn that only curious ideas existed in that time. Correctly he notified that colloid-chemical processes (citing Schmauß and Wigand 1929) are not sufficient to explain the "*Chemismus der Luft*" [chemistry of the air]. He writes that the chemical processes of the air proceed exclusively according to the natural laws, and the compound "*Dreckoxyd*" [dreck oxide] does not exist, he writes ironically. *Cauer* further writes that ozone is produced through photodissociation of oxygen, but without precise analytical work concerning

and 1923 and died between 1977 and 2016), studying the Los Angeles Air Pollution, of which *Leighton*, *Cadle*, *Pitts*, and *Calvert* became the leading US American atmospheric chemists for the following decades, establishing eminent scientific schools, creating another generation of famous atmospheric chemists. See also p. 490 for American pioneers in photochemistry of the 1940s and 1950s.

55 However, his knowledge of the sources was still limited; from the hitherto known atmospheric values of sulfate content, he concluded that, besides anthropogenic (SO_2) sources, there are also very effective natural ones, such as volcanoes and "decomposition of albumen of organic substances" (Junge 1953, p. 24).

the chemical composition of the air and specific conclusions about the occurred processes. Remarkably, he refers Vernadsky (1930), who emphasized the importance of radioactivity for chemical reactions in the atmosphere, but more important *Cauer* noted the idea of Audubert (1938)[56] that all chemical reactions are accompanied by the emission of ultraviolet radiation. Without given further comments by *Cauer* to this idea, we must assume that he had no idea about chemical reactions in the atmosphere; for example, his ideas about chlorine formation by ozone from the sea surface (Chapter 3.4.2 and namely p. 338) were curious.

Nevertheless, there were meteorologists after the war (*Junge* and *Georgii* in Western Germany and *Teichert*, *Mrose* and *Warmt* in Eastern Germany) who made significant contributions in atmospheric chemistry. On contrary, laboratory investigations (in the 1970s) of chemical reactions related to the atmosphere has become the prerogative of chemists and physicists.

Despite a knowledge of chemical reactions in the air remained rather limited before the 1950s, *Hans Cauer* is to be regarded as the German founder of a "chemical climatology" (Cauer 1934); he introduced the term "*chemische Klimafaktoren*" [chemical climate factors] and devided local, nonlocal, and anthropogenic factors according to their origin (salts and gases from different sources). Three years later (Cauer 1937a) he refers the book by *Robert Angus Smith* and noted that since *Smith* no progress was made in chemical climatology and this term was replaced by "*Lufthygiene*" [air quality]; *Cauer* also used the terms chemical meteorology and air chemistry synonymously. Cauer (1949a) wrote that the chemical meteorology has arisen from the field of chemical climatology, principally concerned with the problem of the formation of clouds, and thus the origin and the nature of the condensation nuclei. In 1956, *Cauer* denoted[57] chemical climatology as "applied air chemistry". In a letter from 13 September 1960 to the president of the Max-Planck-Society, *Adolf Butenandt*, he wrote that the fundamental research in the field of air chemistry must be essentially developed before making legal conclusions and statutory orders in air pollution control, and that, unfortunately, the chemical climatology has not yet found a permanent scientific abode (he did not obtain an answer).

Hans Cauer (Fig. 1.1) was born in Bad Kreuznach,[58] Rhineland-Palatinate, on 7 July 1899, the son of the sculptor *Hugo Cauer* (1864–1918) in a family of six generations of well-known German sculptors and painters at Bad-Kreuznach since the early nineteenth century founded by *Emil Cauer* (1800–1867). *Hans Cauer* died after a serious

---

56 *René Audubert* (1892–1957) was a French chemist and director of the laboratory of physical chemist and electrochemist at the Paris University.

57 In a letter to Dr. *Glupe* from Dept. of Research, Ministry of Defense in Bonn, 17 April 1956.

58 His full given names: *Johannes Robert Paul Stanislaus Hans* (names of his ancestors), he never used it in any document. Source: Biographisch-literarisches Handwörterbuch der exakten Naturwissenschaften Bd. 7a (Berichtsjahre 1932 bis 1953) Teil 1 A-E, Poggendorff-Redaktion der Sächsischen Akademie der Wissenschaften, Leipzig (1956), pp. 588–589.

Figure 1.1: *Hans Cauer* at work, about late 1940s (estate of *Grete Cauer*).

illness on 18 January 1962 in Berlin, but is buried in the family tomb in Bad Kreuznach.

Just after finishing the Gymnasium in 1917, he became a war volunteer, was a French war prisoner until January 1920. Before entering the Agricultural College Weihenstephan in 1921, he spent an agricultural internship. After receiving the agricultural Diploma, he studied chemistry and geology at the University of Giessen and defended his dissertation on the occurrence of iodine in nature and its relation to goiter (Cauer 1929) under *Wilhelm Meigen* (1873–1934) supervision. In the same year, he married *Grete von Mittelstaedt* (1900–2005), who studied together with *Cauer* in Weihenstephan. In the following years until 1934, *Cauer* continued his studies of iodine (partly as an assistant in Giessen until 1931 and by getting different grants, namely from the *Kerckhoff* Institute in Bad Nauheim) (Cauer 1932) at several sites (observatory Sonnblick, Jungfraujoch, High Tatra, at Helgoland and other locations like thorn houses at Bad Kreuznach, Bad Salzungen and Bad Nauheim) with the first investigations of fog nuclei [*Nebelkerne*]. This aerosol research accompanied throughout his life, namely in the use of inhaling facilities against silicosis in the 1950s. However, *Cauer*'s most important result from the iodine and aerosol studies was the idea of a *chemical climatology*, i. e., an understanding that chemical investigations (he also called it chemical-bioclimatic) in atmospheric air must be considered in terms of climate (Cauer 1934). In 1934, he moves to Berlin as co-worker at the "*Reichsgesundheitsamt*" [Reich Health Office]. Already in Bad Nauheim, *Cauer* began to develop a method for determination of ozone in air, extended to nitrite (i. e., the total amount of nitrous gases and nitrous acid in air) and the so-called "*Gesamtoxydationswert*" [total oxidation level] (Cauer 1935b). Together with *Eugen Quitmann* (1884–1945)[59] he developed sampling devices for atmospheric trace species in combination with microanalytical techniques (Quitmann 1935, 1938, 1940). In the 1940s, *Cauer* developed together with *Quitmann* a method for the determinination of "*Gesamtreduktionswert*" [total reduction value][60] (Quitmann 1938), namely to an assessment of the CO content in air of closed room, which was of strategic interest of the "*Wehrmacht*" [armed forces].

In 1936, *Cauer* became a public official and in 1943 "*Chemierat*" [Chemical Council]. The studies at the different Mountain sites (and also at Pic du Midi; Fig. 1.2) *Cauer* conducted between 1931 and 1942 and after the war often together with his wife *Grete Cauer* (Cauer 1936, 1937a,b, 1938, Cauer and Cauer 1941, 1942, 1943). Since 1935 he also worked at Bad Reinerz and later Oberschreiberhau (also written Ober-Schreiberhau). In 1942, *Cauer* became the head of the "*Institut für chemische Klimatologie*" [Institute for Chemical Climatology] in Oberschreiberhau (now Szklarska Poręba Gorna in Poland) associated with the "*Reichsanstalt für das deutsche Bäderwesen an der Universität Breslau*" [Reich Institute for the German Balneology at University Breslau]. Director of the Reich Institute was *Heinrich Vogt* (1875–1957), physician and balneologist and the founder of the Journal "Der Balneologe" that appeared 1934–1944. In 1936, *Cauer* not only measured at "*Glatzer Bergland*" [Glatz Hills] (Today *Kraj górski Glatzer* in Poland) ozone but also continued his iodine research that he extended to the Bretagne in the 1936–1941 years; he found that burning of seaweed was the largest source of atmospheric iodine. However, more and more *Cauer* dealt with sampling and analyzing of condensation nuclei. Among his tasks was also the measurements of indoor pollution of public houses and factories. Increasingly, he was engaged in military research, studying supercooled droplets with respect to airplane icing, air degradation (and restoration) in closed rooms (bunkers, submarine).[61] In 1943, he was sent to *Manfred Curry*[62] in Riederau (see also p. 603) to conduct climatological air

59 *Eugen Quitmann* was a German chemist in Berlin at the "*Hygienisches Institut* (*Gesundheitsamt*)"; he died in Russian war imprisonment.

60 A few years before the Hungarian chemist *Lajos (Ludwig Wilhelm) Winkler* (1863–1939) proposed a method based on the consumption of potassium permanganate given as relative values for the estimation of "*Reduktionsvermögen*" [reduction capability] of polluted air, sampling condensed water in a porcelain bowl in which there was a bottle filled with ice water (Winkler 1935); he found in the summer air in Budapest 2° and in rooms 15–18°. *Winkler* became in 1909 the successor of *Carl von Than* as head of the Chemical Institute at the Budapest University.

61 It is notable that after the war *Cauer* transferred his knowledge to the U. S. Fleet who acknowledged his contributions; in the early 1950s, the Swedish Navy also contacted *Cauer* with this concern.

62 *Manfred Curry* (1899–1953) was a German physician of American citizenship, but most famous as sailor and constructer, founded in 1942 the American Bioclimatic Research Institute in Riederau (Ammersee) near Munich and published the two-volume "*Bioklimatik*" (Curry 1946), proposing the "aran hypothesis". *Hans-Adolf Hänschen* (unknown living data), former coworker of *Curry* and physician

Figure 1.2: View of the Observatory Pic-du-Midi about late 1940s (estate of *Grete Cauer*).

examination.[63] His investigations of "stinking air" (in colloquial German "*Mief*") in bunkers, submarines, and caves (*Cauer* named it "*Vermiefung*") led to his nickname "*Miefdoktor*" after the war (Journal "*Stern*" 1956, No. 9, pp. 6–9).

The 1930s were *Cauer*'s most creative years; he (partly together with *Quitmann*) developed sampling methods for gases (different aspirators called "*Waschrohr*";

from Norderney, who succeeded *Curry* as institutes head, wrote a letter to *Cauer* (23 February 1954) to inform him that the "*Aranmessgerät*", patented in Germany and the USA is sold for DM 810.

63 From a letter of *Cauer* to "*Generaloberstabsarzt*" Professor Dr. *Hippke* from January 6, 1943; *Erich Hippke* (1888–1969) was Chief Medical Officer of the Luftwaffe. In this letter *Cauer* also noted that "*Staatssekretär*" [state secretary] *Esser* gave the instruction to establish the Institute for Chemical Climatology on 1 April 1943; *Hermann Esser* (1900–1981) was a follower of *Hitler*. In a letter from *Esser* to *Cauer* dated 15 June 1943, *Cauer* was asked to terminate all investigations "*bezüglich des Ihnen bekannten bioklimatischen Stoffes*" [regarding the bioclimatic substance known to you] that was declared to secret commando operation. This substance was obviously so secret that it was not named in this letter and not found belong any estate of *Cauer* (*Cauer* was a decent person who retained alphabetically all his correspondence after the war in five ring binders, but from the time before 1945, only a few letters remained; all documents I sighted with thanks to his daughter *Grete Cauer*). We can only speculate that this substance was a poison gas. New gas chambers in crematorium I in the Auschwitz main camp went into operation in summer 1943, and it is known that gas detectors for hydrogen cyanide were ordered (the biocide "Zyklon B" contained no longer smelling warning substances). Thus, it is not unlikely that HCN was behind "the bioclimatic substance". No report on the works of *Cauer* with *Curry* is available and there is no notice of *Cauer* on it. It is known, the Nazi's did try to cover up their crimes, the gassings with Zyklon B were stopped in autumn 1944, and then the dismantling of the gas chambers began.

Quitmann (1935)) and nuclei (dew condensor, see p. 256), and microanalytical methods. His "*Waschvorrichtung für chemische Analysen*" [scrubber for chemical analysis] was patented in 1932 (no. 641844); Fig. 1.4 and 1.5. Most of his results were published after the war (Cauer 1948a,b, 1949a,b, 1950, 1951a,b,c, 1956). In his first after-war publication, *Cauer* summarized the research on pH of nuclei and ozone carried out in the 1930s, but he still very focused on bioclimatology, i. e., the impact on humans, and concluded from his experience for further research on "chemical climatology" (Cauer 1948a, p. 67):

> Ohne die exakteste Durchführung der chemischen und physikochemischen Arbeiten ist ein Fortschreiten auf diesem Gebiet völlig ausgeschlossen. Die chemischen Untersuchungen sind bisher in der Hand von Ungeübten, auch sogenannten Schmalspurchemikern, stets rasch zum Stillstand gekommen oder haben unbrauchbare Ergebnisse erbracht. Um unnütze Geldausgaben, Geldverschwendung zu vermeiden, wird es daher nicht zu umgehen sein, daß diese subtilen Arbeiten unbedingt in der Hand des Mikrochemikers bleiben bezw. unter seiner genauesten Aufsicht [Without the most exact execution of chemical and physicochemical works, progress in this field is completely excluded. Chemical investigations so far in hand of untrained, also so-called small-time chemists, always rapidly ended or produced unusable results. To avoid useless expenditures, waste of money, it will therefore be inevitable that these subtle works remain absolutely in hand of microchemists, or under his meticulous supervision].

In front of the advancing Soviet troops, in early 1945 *Cauer* moved from Oberschreiberhau via Prague to Franzensbad (today Františkovy Lázně in the Czech Republic) where he further worked in the "*Moorforschungsinstitut*" [institute of peatland science],[64] and from there, American troops "arrest" him because of his air conditioning research for submarines to Bavaria (Hohenheim); his results were included in a "monograph on submarine medicine"[65] (for some 5 month 1947/48 he was salaried in Heidelberg at the U. S. Fleet).[66] In Hohenheim, *Cauer* re-established privately the "*Institut für chemische Klimatologie*" in 1946 (Fig. 1.3 and 1.6). Despite *Cauer* was included in "*Kürschners Deutscher Gelehrtenkalender*" [Kürschner's Encyclopedia of German Scholars] in 1950 (p. 285) and in 1961 (pp. 279–270), he fought unsuccessfully until his death for inclu-

---

64 Founded in 1941 as part of a working group ("*Arbeitsgemeinschaft*") belong to the "*Reichsanstalt für das deutsche Bäderwesen Breslau*" (many other institutes were included, from which the following are referred in the volume (in parenthesis name of heads in 1943): "*Rheumainstitut* Bad Elster" (*Sorgenfrei*), "Kerckhoff-Institut Bad Nauheim" (*Koch*), "*Rheumaforschungsstelle* Bad Warmbrunn" (*Tichy*), "*Bioklimatische Forschungsstelle* Wyk" (*Pfleiderer*).

65 Cauer, H. (1948) Chemical climatology of the trace substances in the air of submarines. In: monograph on submarine medicine. Folio II, Germany, U. S. Zone, Office of naval advisors.pp. B: II-1-B; II-108. *Cauer* developed 1943 a method for removal of CO without using filters in submarines with long dive time.

66 The Swedish Navy ("*Kungl Marinförvaltningen*") in a letter to *Cauer* from 30 April 1954) studied *Cauer*'s report and told him that only the investigations on the occurrence of formaldehyde and CO in submarine could be of interest.

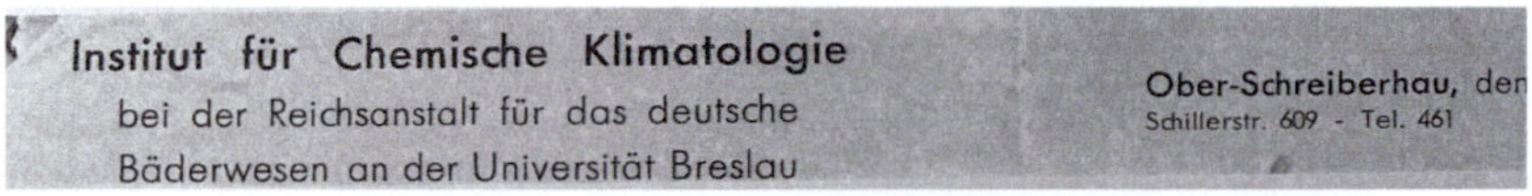

Institut für Chemische Klimatologie
bei der Reichsanstalt für das deutsche
Bäderwesen an der Universität Breslau

Ober-Schreiberhau, den
Schillerstr. 609 - Tel. 461

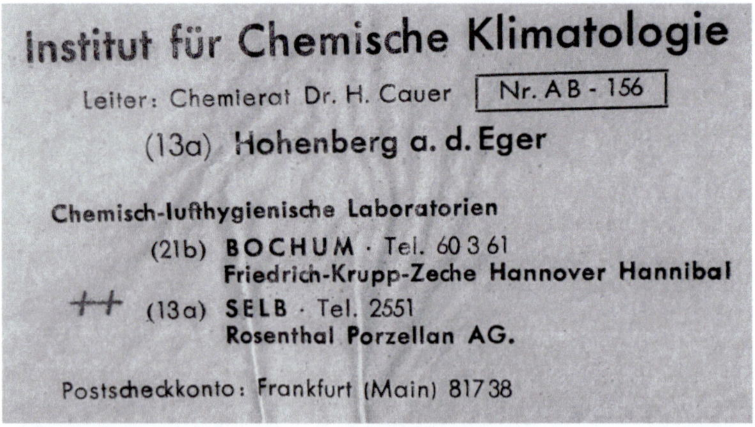

Institut für Chemische Klimatologie
Leiter: Chemierat Dr. H. Cauer | Nr. AB - 156 |
(13a) Hohenberg a. d. Eger

Chemisch-lufthygienische Laboratorien
(21b) BOCHUM · Tel. 60 3 61
Friedrich-Krupp-Zeche Hannover Hannibal
(13a) SELB · Tel. 2551
Rosenthal Porzellan AG.

Postscheckkonto: Frankfurt (Main) 81738

Figure 1.3: Letterheaded paper from *Cauer*'s institute in Oberschreiberhau (1942–1945) and the building (foto dated 2021), now in ul. Słowackiego 1, Szklarska Poręba Górna, very close to the railway station (the building was constructed around 1850 and used in the late war years as military hospital and after the war as kindergarten (now private)). Re-founded after war at Hohenberg (1946–1960); see Fig. 1.6. Note that he had "branches" in Bochum and Selb. In Hohenheim, the "institute" was letter-box company in the 1950s.

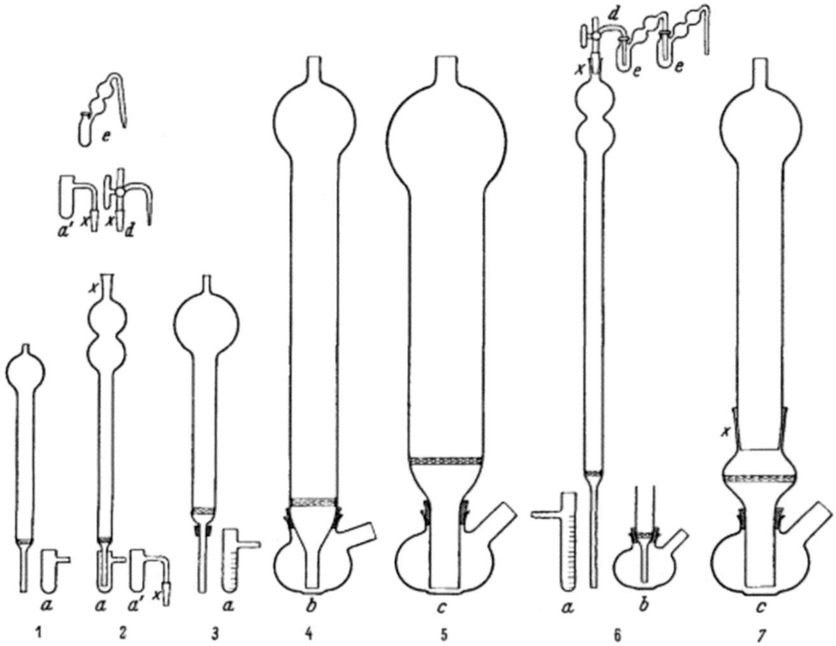

Abb. 14. Waschrohre Typ 1—7.

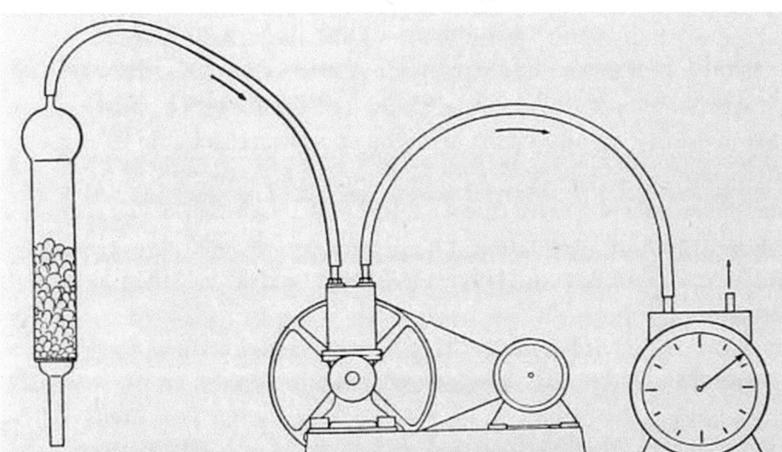

FIG. 1.—Schematic representation of apparatus for detecting traces of gaseous materials; (*a*) Cauer absorption tubes without separate charging vessel; (*b*) motor and pump; (*c*) dry-gas meter.

Figure 1.4: Above: Schematic representation of different "*Waschrohre*" [absorption tubes] for sampling traces of gaseous materials (Cauer 1935a, p. 169). Down: The whole apparatus (Cauer 1951a, p. 1126).

Figure 1.5: Air sampling with aspirators (*Cauers*'s *Waschrohr*) at Sonnblick Observatory in the early 1940s (estate of *Grete Cauer*).

sion of his institute into the governmental sector. In a letter dated 25 May 1954 to *Karl Knoch*, he proposed to establish an "Institute for Air and Aerosol Chemistry".

To ensure a living for his family, he did many small jobs for the industry, workplace examinations, namely dust control and later inhaling techniques for miners. Fortunately for the air chemistry, he also continued atmospheric aerosol studies at mountain locations (Nebelhorn, Zugspitze, Pic du Midi) and measurements of ozone and other substances at several spa towns. However, his main task after the war remained silicosis research at Rosenthal Porcelain factory in Selb, "*Friedrich Krupp Zeche*" [coal mine] Hannover/Hannibal in Bochum (1947–1954), Max-Planck Institute for Silicosis Research in Berlin.

In 1955 finally, his status as "*Beamter*" was re-confirmed, and he got a job at the Institute for Air and Water Hygiene belonging to the Senate Administration of Health Security in Berlin, but in 1956 he was on leave to Bochum (at the State Medical Doctor) with the task to include chemical climatology within commercial medicine, but to no avail. Unfortunately, his task was not the science but control and reviewing. Thus, he sent letters to the Max Planck Society for funding a project on "physical-chemical principles of natural aerosols" (May 1957), or, to the president[67] of "*Bundesgesundheitsamt*" [Federal Health Office] for admittance of "himself and his small chemical-air hygienic institute", the ministry of defense, politicians and other – without any positive response. In a private letter from June 1957, *Cauer* writes that his administrative

67 *Wilhelm Hagen* (1893–1982), a letter from 18 May, 1957. It was the successor organization (founded in 1952 and terminated in 1994) of the "*Reichsgesundheitsamt*" were *Cauer* was employed before 1945.

Figure 1.6: Above: The laboratory of *Cauer* about 1949 [chemical and air hygienic laboratory to fight silicosis at the Rosenthal Porcelain corporation]. Bottom: *Hans* and *Grete Cauer* at air sampling at an industrial furnace, around 1950 (estate of *Grete Cauer*).

job in Bochum made him unsatisfied and he hopes to obtain in a few years enough pension to work only for the Max-Planck Society.[68] Further he lamented that he needs the time for his book (see below) which is much more important in his affects than his official duty.

At the end of 1950s, *Cauer* clearly recognized that it is not enough to reduce "visible" air pollution such as smoke and dust; he put attention to small-sized particles being of biological importance and to an understanding of complex atmospheric reactions pathways, including oxidation, reduction, acidification, alkalinization, etc. However, first of all, *Cauer* writes, the detection methods must be improved (because there was no progress in the last 30 years), and funding of fundamental sciences by independent researchers is needed (in a letter to the journal "Hobby" from 7 August 1959).

The missing funding for fundamental atmospheric research in West Germany in the 1950s was also the reason for the leave of *Christian Junge* to the USA in 1953, although *Junge* continued his academic career in 1950–1953 in Frankfurt/Main in contrast to *Cauer* and *Mrose*. Whereas *Helmut Mrose* in East Germany after the war had a secured position, *Cauer* remained until 1955 in unsecured employment. The colloid physicist *Johann-Ludwig von Eichborn* (1907–?) emphasized in a report "on the state of aerosol research in Germany (March 1956, 18 pp.) the unique activities of *Cauer* in investigation of the interactions between gaseous trace substances and the particulate aerosol in the atmosphere;[69] he notified that the Fraunhofer Society (which *Cauer* funded a time long) considered the re-founding of his institute for chemical climatology.

Nevertheless, *Hans Cauer* was very active in scientific cooperation.[70] The list of scientists with whom *Cauer* contacted reads like who's who in the atmospheric and environmental research in the 1950s:[71] *Walther Amelung* (1894–1988), *Max Bider*

---

68 *Cauer* was external employee at The Institute for Silicate Research in Berlin-Dahlem and was in close collaboration with its head, chemist *Luise Holzapfel* (1900–1963). *Holzapfel* was an excellent scientist (supported by *Max von Laue*, *Peter Adolf Thiessen*, *Wilhelm Eitel* and others), never in the Nazi party, worked since 1939 in the silicate institute (Kaiser-Wilhelm Society), made his habilitation in 1943 and could continue his research as department head in 1945. The Max-Planck administration, however, did not show any respect to it and closed the Berlin branch (the institute was meanwhile in Würzburg) in 1960. Since 1952 *Holzapfel*'s research was focused on silicosis. *Cauer* headed an aerosol research project between 1956 and 1959 together with *Holzapfel*, *Reiter* (from Farchant) and *Fritz Hesse* (balneologist at Norderney; unknown living data)).

69 Finally, it was *Junge*'s merit to develop the atmospheric aerosol research. Whereas Junge (1952) introduced the terms "active" and inactive" condensation nuclei, Cauer (1954) used the term "potential" aerosol.

70 In the 1950s, *Cauer* had also lost contact with *Hans-Walter Georgii*, *Helmut Mrose*, *Wolfgang Warmt*, and *Friedrich Teichert*. In the early 1950s, *Cauer* also made experiments together with *Junge* (Junge 1953, p. 20, see also p. 256).

71 It is interesting to note that *Cauer* was unable to read and speak English (he let translate his correspondence), but he was rather fluent in French.

(1905–1979), *Alan West Brewer*, *Alexander Goetz* (1897–1970), *Eville Gorham*, *Johannes Grunow*, *Robert Guderian* (1928–2005), *Herrmann Harrassowitz*, *Kurt Albert Edmund Theodor Kalle* (1898–1975), *Karl Heinrich Knoch*, *Rudolf Lessing* (1878–1963), *Walter Mörikofer* (1892–1976), *Leo Wenzel Pollak* (1888–1964), *Reinhold Reiter*, *Carl-Gustaf Rossby*, *Květoslav Rudolph Spurný* (1923–1999), *Hans-Werner Schlipköter* (1924–2010), *Harry Wexler*, and other, namely physicians.[72]

Cauer wrote (about 1955) in a prologue of his planned book[73] "*Chemische Klimatologie*" that during his studies since 1926 on the occurrence of iodine in minerals, waters, and air the thought did come about the need of a comprehensive physicochemical treatment of the atmosphere, focusing on bioclimatic relevant substances, its origin, physical and chemical behavior in the atmosphere and its cycling. In the preface of this manuscript *Cauer* writes (translation from German) that "the chemical climatology is not contrary to physical climatology. But the first was ignored in the past 50 years so that the chemical part must be consciously emphasized. In future, however, there can only be one climatology, the physicochemical climatology, which together with physicochemical meteorology inextricable represents the subject

72 *Amelung* was a German physician, bioclimatologist, and balneologist in Königsstein (Taunus). *Bider* was a Swiss meteorologist in Basel. *Goetz* was a German physicist, who immigrated to the USA in 1930 and became professor at the California Institute of Technology, most known for his study of aerosols and atmospheric pollution; the Austrian internationally known aerosol researcher *Othmar Preining* (1927–2007) was his scholar. *Guderian* was a German agricultural scientist and pioneering in air pollution research in the 1950s. *Kalle* was a German oceanographer at the Hamburg University. *Lessing* was a German-British chemist who studied in Munich and was assistant of *Willstätter*; in 1903 he came to London, first to the Owens College and soon worked as industrial chemist who was since 1921 concerned with air pollution (member of the Coal Research Club, president of the National Society for Clean Air from 1956 until 1958); *Lessing* wrote letters to *Cauer* in English! *Mörikofer* was a Swiss physicist and head of the physical-meteorological observatory at Davos 1929–1966. *Pollak* was a German-Austrian geophysicist and meteorologist at the geophysical institute of the German University in Prague; in 1939 he immigrated to Dublin; *Pollak* wrote letters to *Cauer* in German. *Spurný* was a Czech aerosol researcher who immigrated in 1972 to West Germany. *Schlipköter* was German physician, called "father of environmental medicine".

73 Unfortunately, due to the early death of *Cauer*, this book was not finished. About 80 pages of his manuscript are in my hands, several copies of letters, and the contract from 1955 (based on a first contract from 1939) with the publishing house "*Akademie-Verlag*" Berlin; it was planned with 480 pages and a circulation of 1500 copies. Because this publishing house was after the war in East Berlin, *Cauer* asked different publishers in West Germany and received rejections. It is notable that DeGruyter in a letter to *Cauer* from 29 November 1955, expresses the willingness to publish his book (today, in the archive of "my" publishing house, no documents have been found). However, in a letter from 1 December 1955, the "*Akademie-Verlag*", referring the positive opinion of geophysicist *Hans Ertel* (1904–1971), was willing to edit *Cauer*'s book, but including some explanation of impacts on plants by air pollution, likely involving another specialist. In a letter from 13 February 1956 to the senator of health, *Hans Schmirjan* (1901–1961), *Cauer* expresses his concern that his book cannot be published in Germany if it is not allowed to publish in East Berlin (by courtesy of his daughter *Grete Cauer*, Bielefeld, 2020).

of physical chemistry of the atmosphere". Consequently, he included into such climatology[74]

- temperature conditions,
- air movement,
- chemical composition of air and its chemical metabolism,
- suspended matter (dust, aerosols),
- electrical conditions (including radioactivity),
- pressure variations.

In summary, *Cauer*'s contributions to atmospheric chemistry are briefly

- developing of sampling and analytical methods for different trace substances,
- systematic ozone measurements,
- first hypothesis of release of HCl from sea salt particles under the influence of O_3,
- first chemical analysis of condensation nuclei,
- introduction of atmospheric oxidation and reduction capacity as "climate elements".

Between 1955 and 1961, *Cauer* had a frequent contact with East German scientists in his research field (*Mrose*, *Warmbt*, *Renger*, *Hentschel*).[75] It is notable that *Cauer* writes in a letter to *Friedrich Renger* dated 22 November 1957 that "the most important investigations on ozone were made in your part of Germany" [GDR]. In *Cauer*'s letter to *Warmt* (6 February 1961) he states that "a new method was developed, not only improved but essentially simplified" (in relation to the methods by *Cauer* and *Ehmert*).

Another German pioneer in air chemistry with contributions before 1945 was *Friedrich Helmut Mrose* (Fig. 1.7), born on 19 May 1910 in Schneeberg, a small city in Erzgebirge; he died on 19 June 1999 in Oschatz (Saxony).[76] His father Prof. Dr. *Hermann Mrose* (1876–1960) was director of studies of the State "*Realgymnasium*" in Schneeberg, retired in 1935 and moved to Oschatz where he again worked as a teacher in 1946–1950 (it is notable that obviously he stayed together with his son in 1950–1958 in Friedrichroda and afterwards in Radebeul). In 1930–1935, *Helmut Mrose*

74 Written in a letter to *Klumb* from 11 January 1955; *Hans Klumb* (1902–1980) was a professor for experimental physics at the University of Mainz. Note that *Cauer* studied since about 1940 until his death in 1962 mostly air in closed rooms (submarines, bunkers, caves, mines, shop floors, and salt-breath therapy rooms), today named indoor climate.

75 *Hentschel* organized a Bioclimatic Conference (April 3–5, 1959) in Berlin ("democratic sector"); *Warmbt* invited *Cauer* (a letter from 7.2.59) and communicated that *Mrose* will held a talk on air chemistry. However, *Cauer* not participated.

76 Little is known on *Mrose* and almost nothing to find in the internet. Sources of my information were: archives of the universities Dresden, Jena and Rostock, the "*Herder-Gymnasium*" Schneeberg, local resident's registration offices in Radebeul and Oschatz, address book Oschatz 1937, study group ornithology and nature protection Radebeul (he was its head 1970–1980). Furthermore, the correspondence between *Mrose* and *Cauer* (courtesy by *Grete Cauer*).

Figure 1.7: *Helmut Mrose* about 1965 (Source: Herrmann et al. 1991, p. 7).

as prospective teacher taught the natural sciences in Jena, Rostock, Munich, and Dresden. His dissertation on "Evaporation measurements on open water surfaces. With an appendix on dew measurement"[77] was defended at the Technical University of Dresden in 1937. Afterwards, he was employed as a meteorological observer at the weather station Collmberg near Oschatz, and in 1938–1945 at the Hygienic Institute of the University Jena. After the war he was again employed at the Collberg station (this station was belonging to the Leipzig University), and in 1950–1958 at the "*Bioklimatische Forschungsstation*" [bioclimatic research station] Friedrichroda (Thuringia)[78] before he moved to the Observatory Wahnsdorf (Dresden-Radebeul). Little is known on the life and work of *Mrose*, and we can compile only from his publications. In 1939, like *Cauer* and *Junge* he began the atmospheric research with rime sampling and its chemical analysis (Mrose 1940) to investigate the nature of condensation nuclei. Whereas *Cauer* was still focused on chloride from sea salt as dominant part of nuclei and *Junge* made so far only particle number measurements, *Mrose* as the first who notified that sulfate from combustion is the main part of nuclei.[79] The negative influence of atmospheric SO_2 on lichen Mrose (1941) evaluated not only because of

---

77 Referees were the meteorologist *Willi König* (1884–1955), scholar of *Hellmann*, and the experimental physicist *Rudolf Tomaschek* (1895–1966), suspended from the Dresden University in 1945 as follower of "*Deutsche Physik*".

78 Belong the "*Reichsamt für Wetterdienst*" [Reich Office for Weather Service] were the bioclimatic research stations in Bad Elster, Braunlage, Friedrichroda and Wyk/Föhr. The year of founding of the station Friedrichroda is unknown; head was *Johann Nippoldt* (1874–1936) and afterwards by *Wilhelm Josef Müller* (1907–1969) see also Burckhardt and Flohn (1939, p. 79).

79 *Mrose* cites Houghton and Bemis (1939) who collected fog water 1938 and 1939 at Mt. Washington, finding "almost sulfur compounds beside traces of chlorine compounds"; in the original paper (what

the almost extinction of large tree lichen but also due to its sulfate content. In his first publication after the war (Mrose 1949) he reported that he found sulfate-free rime only at Hoher Sonnblick, and that besides sulfuric acid, compounds of nitrogen can also act as nucleates due to their large hygroscopicity; he further wrote that ammonia and compounds of nitric acid have been regularly found. On Crete, however, he found no nitrate, concluding that the source of nitrates are originates from industrial sources. Furthermore, in snow at Hoher Sonnblick he found hydrogen peroxide only during daytime but never at night, concluding that H_2O_2 may be gained via reactions occurred under UV irradiation. Thus, we can assume that *Mrose* was able (likely as meteorological observer) to continue limited research during the war. In Friedrichroda, *Mrose* headed his chemical laboratory and continued field and (limited) laboratory studies on nuclei (Mrose 1952, 1954) and began with fog sampling[80] at the beginning of 1954 in Mt. Inselsberg (see pp. 259). In a letter to *Cauer* dated 7 July 1955, *Mrose* reports that fog sampling and its chemical analysis is a useful tool for estimating the content of pollutants in the air, when simultaneous the liquid water content is determined; however, he was not very successful with measurement of vapor pressure during fog and after its dissipation. *Mrose* was known with the article of Houghton (1955) who proved only the presence of sulfates and chlorides and determined the pH in fog water samples (see p. 257). Between 1955 and 1961, *Mrose* and *Cauer* exchanged many letters (10 from *Mrose* and 5 from *Cauer* are in my hand) not only with technical information but also with private one (thus I learned that *Mrose* was an enthusiastic motorcyclist, also in winter time). They also had met in West Berlin and at a conference in Mainz in 1956. *Mrose* repeated *Cauer*'s experiment with the dew-cup and estimated the reduction value, and adopted *Cauer*'s microanalytical methods. In summer 1955, *Mrose* determinated similary to *Cauer*'s dew-cup (using simply a flask filled with ice) 0.01 mg m^{-3} NH_3, with an aspirator ("*Waschflasche*") of the same value and in fog water 10 mg L^{-1}, respectively, if assuming LWC of 1 g m^{-3}. From the correspondence the technical laboratory problems became clear: missing glass electrode for pH measurements, rotameter, etc.[81] Mrose (1958) gives a survey on his experiments made in Friedrichroda.

---

was in the hand of *Mrose* because he used a review (Keil, K. (1940) *Z. angew. Met.* **57**, 65) it is written: "[...] the principal solutes are sulfates, although traces of chlorides were also observed".

80 Mrose, H. (1955) Atmosphärische Verunreinigung und ihre Feststellung. Aus der Arbeit der Bioklimatischen Forschungsstation Friedrichroda. "Neues Deutschland", September 9, 1955.

81 *Mrose* carried the though moving to West Germany in early 1958, and *Cauer* asked the "*Steinkohlenbergbauverein*" [Coal Mining Association] in Essen for a laboratory job, but because of *Mrose*'s age there was no interest. Fortunately, because *Mrose* had to work in West Germany on laboratory technician level. In 1980 he was honoured with the "*Reinhard-Süring-Plakette*" in silver; this honour was introduced 1978 by the Meteorological Service (MD) of the GDR and after 1990 continued by the German Weather Service (DWD).

The Wahnsdorf's departure from *Teichert* in 1956 from the Observatory caused obviously then the problem of a missing analytical specialist, and *Mrose* was transferred from Friedrichroda to Radebeul in May 1958. *Mrose* first developed[82] a manual atmospheric SO_2 analyzing method (based on *Cauer*) and performed a two-year measurement (1958 and 1959); Mrose (1962), Mrose and Warmbt (1964); moreover, he measured wet-chemically using an aspirator CO_2, NH_3 (together with ammonium ions) and HNO_3 (together with nitrate ions). And finally, he reports to *Cauer* that the working conditions are much better in Radebeul. In summer 1960, at the observatory *Mrose* and *Warmbt* conducted the intercomparison measurements by the condensation and the filter method; he found that the dew is soot free even in very polluted air and that the dew condensation process collects only gases. *Mrose* obviously had large respect to *Cauer*'s scientific contributions; he pushed him in a letter to write his book with all his experimental experiences. In his own small booklet (Mrose 1955), he devoted some pages (pp. 39–41) to the present "*Cauer's Waschflasche*" with a figure, and his dew condensation method, but concluded that sampling of air using glass wool filters is effective in sampling of particulate matter, done at his working place in Friedrichroda. The 1950s were still relatively "relaxed" for scientists from the GDR, concerning the contacts with western researchers and visits. *Mrose* and *Hentschel* planned to attend the 4th German Aerosol Conference hosted by the Fraunhofer Society in April 1961 in Bad Lippspringe, but the travel was not allowed by East German officials because in the invitation it was written "*Ostzone*" [east zone] instead of "DDR" [GDR]. In a letter dated 11 February 1961, *Mrose* communicates that he visited *Klumb*'s meeting ("*Arbeitstagung Schwebstaubtechnik*") in October 1960 in Mainz (*Cauer* was not present and writes later that his interest in atmospheric aerosols no longer concerns this community), and visited *Georgii* in Frankfurt am Main and the Taunus Observatory. *Cauer* met *Warmbt* in 1959 at the Clean Air Conference in London.

Thus, *Mrose*'s research work was a significant contribution to the atmospheric chemistry in the 1950s and 1960s in the eastern part of Germany (in the western part *Georgii* and *Junge* contributed since the end of 1950s; in January 1953, *Junge* left Germany for the USA and returned only in 1961). The Meteorological Observatory Wahnsdorf near the city of Dresden in East Germany (German Democratic Republic) was a pioneering one in air chemical research after the war, it was started by *Johannes Goldschmidt* (1894–1952), physician and head of the observatory (1936–1952) who worked there since 1928. *Goldschmidt* initiated the surface ozone measurements at the Meteorological Observatory Wahnsdorf[83] started in 1952 by *Friedrich Teichert*

82 Just after arriving in Wahnsdorf, he also developed a method for determination of soot in air (Mrose 1959a).

83 The weather station Wahnsdorf was founded in 1916 and is located suburb of Dresden (Radebeul) on a hillside 246 m a. s. l. (Dresden lies in a valley) hence influenced by city pollution. It was one of three historical meteorological observatories in the GDR (beside Potsdam and Lindenberg). In 1974, the

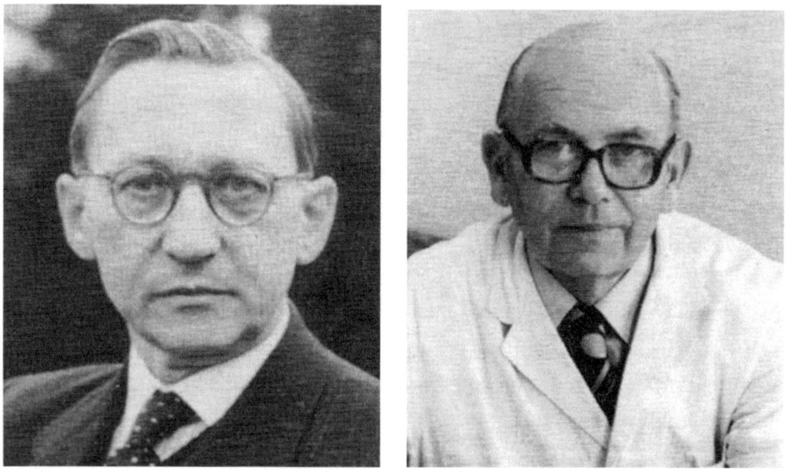

Figure 1.8: *Friedrich Teichert* about 1955 (left) and *Wolfgang Warmbt* about 1975 (right); source: Herrmann et al. (1991, pp. 7 and 17).

(1905–1986)[84] (Fig. 1.8), who became the head of the observatory (with 18 workers) after *Goldschmidt*'s early death. After *Teichert*'s left to West Germany, *Wolfgang Warmbt* (1916–2008)[85] (Fig. 1.8) continued this work. *Warmbt* expanded the surface ozone

Meteorological Service appointed the observatory as "*Forschungsdienststelle für meteorologische Aspekte der Luftreinhaltung*" [research department for meteorological aspects of air pollution control]. With the ending GDR it was closed end of 1991 (the same fate suffered the well-known observatory Potsdam; only Lindenberg further operates). In West Germany only Hohenpeißenberg operates today as another observatory with long-time competence in air chemistry (1993 the Hamburg observatory was also closed). Since 1994 the former buildings of the Wahnsdorf Observatory are used by a private company ("*Staatliche Umweltbetriebsgesellschaft*") that among others operates the air pollution monitoring network in Saxony.

84 *Teichert* also started monitoring of air constituents including dust, chemical compounds, and radioactivity of precipitable water (Teichert 1952, 1953, 1955, Teichert and Warmbt 1956, Teichert and Greifenhagen 1954, Teichert 1956a,b, 1957a,b). Nothing is known on the past of *Teichert* before his work as meteorologist in Wahnsdorf since 1946; he left the GDR in 1956 (not 1957 as published) for West Germany and asked first *Hans Cauer* to get a job. Only in 1960 he became a lecturer for physics and mathematics at the "*Oskar-von-Miller-Polytechnikum*" (a higher technical college) in Munich, and became in 1961 an established official (letter of *Teichert* to *Cauer* from 13 April 1961).

85 *Wolfgang Warmbt* was a German meteorologist and bioclimatologist; studied in Leipzig and Hamburg, dissertation (1942) in Leipzig "*Beiträge zur Häufigkeitsklimatologie des Ostseeraumes*" [contributions to the frequency climatology of the Baltic area]. In 1942–1945, he worked at "*Ionosphären-Beobachtungsstation bei der Erprobungsstelle der Luftwaffe*" in Rechlin (Mecklenburg) [Ionosphere Observational Station belong the Testing Center of the Air Force] (founded 1934 by *Walter Dieminger*) of the "*Zentralstelle für Funkberatung*" [Central Department of Radio Advice]. Obviously, he met *Hans Tichy* in Bad Warmbrunn and *Hans Cauer* (with whom he corresponded after the war) in Oberschreiberhau. Habilitation 1963 "*Luftchemische Untersuchung des bodennahen Ozone 1952–1961.*

measurements by initiating a first worldwide network of sites, and after *Warmbt*'s retirement, it was continued by *Uwe Feister* (born 1950), who was mainly responsible for total ozone and UV measurements at the Potsdam observatory.[86] *Mrose*'s historical valuable contributions to atmospheric chemistry are

- first evidence in Europe[87] that in rime samples (frozen fog or cloud droplets) sulfate is dominant and must be seen (as the result of coal combustion) as artificially condensation nucleus (Mrose 1940),
- first proposal and evidence (Mrose 1952, 1954) that solar light causes condensation nuclei,[88]
- beginning of fog and cloud chemistry monitoring in 1950 (Mrose 1955, 1966),
- beginning of first rain chemistry monitoring in Germany after the war (1955).

The pioneering research in dust, fog, precipitation, ozone, and sulfur dioxide by *Teichert*, *Mrose* (1975 retired) and *Warmbt* (1981 retired) was continued by new workers of the observatory: The meteorologist *Wolfgang Marquardt* (1933–2002)[89] worked at Wahnsdorf in 1959–1978 and was responsible for the radioactivity measurement network in air and precipitation. The chemist *Günter Herrmann* (born 1933) continued *Mrose*'s measurements of SO_2 and developed an automatic analyzer (Herrmann

---

Methoden und Ergebnisse" [Air chemical studies of surface near ozone 1952–1961. Methods and results] at University Dresden. 1946–1984 he worked at the Observatory Wahnsdorf near Dresden. *Warmbt* worked since 1946 in Wahnsdorf and 1950–1964 in the associated "*bioklimatische Forschungseinrichtung*" [bioclimatic research institute], but he was soon involved in the ozone measurements together with *Teichert*. After *Teichert*'s leaving he was the provisional head until *Hinzpeter* took over the heading in 1958 (he left in 1961 the GDR); succeeded by *Gerhard Dietze* (1921–2016) as observatory head (1961–1986). *Dietze*, who was a specialist in optics of the atmosphere, supported the further air chemistry research: *Warmbt* became in 1964 the head of the working group "*Luftchemie*".

86 Detailed descriptions of the methods of measurement, the instruments applied in the network, and results of measurements were published by Teichert and Warmbt (1956), Warmbt (1962, 1964, 1966, 1979), Grasnick and Warmbt (1973), Feister and Warmbt (1987), Feister et al. (1989). Surface ozone was also measured by foreign ships (Warmbt 1965) on the Antarctic continent (Warmbt and Kolbig 1978) and in mountainous regions (Warmbt 1980).

87 Before this investigation, sea salt compounds (especially chlorides) has been seen as nuclei (see p. 255–256). However, Mrose (1940, p. 243) refers Houghton and Bemis (1939) who also found mainly sulfate in fog at Mt. Washington.

88 *Mrose* cited Schulz (1939) who found an increase in the number of nuclei in the warm season but rejected the idea of formation by solar radiation (assuming soil substances); Schulz (1939, p. 115) refers *Georg Jenrich* (unknown living data), student of *Albert Wigand* in Halle, who concluded from the measurements on a direct formation of nuclei by solar light (Jenrich 1914). *Leo Schulz* (1908–1970) was head of the bioclimatic research station and weather station, respectively, in Braunlage since 1934.

89 In 1979, *Marquardt* moved to the "*Institute für Energetik*" in Leipzig where he began to design (with *Peter Ihle*) a wet-only sampler and a rainwater sampling background network (Seehausen 1982–2002, Greifswald 1984–1989 and Wiesenburg 1988–1989). *Peter Ihle* (born 1942), a German physicist in Leipzig (Marquardt and Ihle 1988).

1962, 1963, 1972). The physicist *Manfred Zier* (born 1936), since 1962 in Wahnsdorf, was responsible for dust monitoring, developed around 1985 a bulk-wet sampler for precipitations and organized a precipitation chemistry network belonged to the Meteorological service (32 stations since 1988); in the mid-1980s *Zier* constructed a passive cloud water sampler and conducted measurements in Zinnwald in 1987 (Zier 1992).

Monitoring of air pollution in East Germany was carried out by many persons and institutions (see Möller 2020, pp. 424–446), but atmospheric chemistry, as a new scientific discipline, was only further developed in West Germany after the pioneering work of *Hans Cauer* in the 1960s by *Hans-Walter Georgii* and *Christian Junge* and his scholars (see below). In the GDR, after the mid-1970s, two "schools" in air chemistry appeared, one in Leipzig organized by chemist *Wolfgang Rolle* (1934–1996) at the "*Forschungstelle für Chemische Toxikologie*" (FcT) [research center for chemical toxicology][90] and the other in Berlin organized by chemist *Detlev Möller*.[91]

---

90 Director was *Karlheinz Lohs* (1929–1996), chemist and international known toxicologist. *Rolle* studied chemistry in Leipzig and Merseburg (1953–1959) and after an employment at the petrochemical factory Leuna he worked (1961–1971) at the Institute of Stable Isotopes (ISI) at the Leipzig research campus of the Academy of sciences (AdW), and moved than to the "*Forschungsstelle für chemische Toxikologie*" FCT [research institute of chemical toxicology] at the same location. In 1992, he became the head of the Department Atmospheric Chemistry of the Institute for Tropospheric Research (IfT) in Leipzig and professor for atmospheric chemistry at the institute for meteorology (University Leipzig). *Rolle* habilitated in 1980 with a work on nitroso compounds in the atmosphere. Since then, he worked together with the physicist *Eberhard Renner* (born 1945) on modeling gas-phase ozone chemistry. Other co-workers (*Erika Brüggemann*, born 1948 and *Thomas Gnauck*, born 1949) developed analytical methods and conducted the chemical analysis of the rainwater collected by *Marquadt* and *Ihle* (their institute was only a few hundred meters away located); *Gerald Spindler* (born 1956) who joint *Rolle*'s group in 1987 became after 1992 the leading scientist of the Melpitz research station (near Leipzig) of the IfT.

91 *Detlev Möller* (born 1947) began in 1965 to study chemistry at the Humboldt University in Berlin and made his dissertation (1972) in the field of electrochemistry. Since 1974 belonged to the Academy of Sciences in Berlin, for many years he first worked theoretically, modeling atmospheric chemistry of sulfur. His habilitation in 1982 is the only one in German speaking countries that is denoted to Atmospheric Chemistry (on the sulfur cycle). *Möller* coordinated East German research activities since 1980 in atmospheric chemistry with a working group belonging to the Engineering Commission Air Pollution Control ["*Zentrale Arbeitsgruppe Reinhaltung der Luft bei der Kammer der Technik*" headed by *Herbert Mohry* (1930–2021)]. With *Rolle* and *Marquardt*, a close cooperation on "acid rain" began (Marquardt et al. 1984). From the beginning of 1980s he extended modeling on the field of aqueous-phase chemistry with the mathematician *Günther Mauersberger* (born 1946), built-up new chemical laboratories for analysis of atmospheric trace compounds, introduced denuder for sampling and developed an automatic rainwater sampler with his coworkers *Renate Auel* (born 1943), *Wolfgang Wieprecht* (born 1949), *Karin Acker* (born 1958), *Jürgen Hofmeister* (born 1952) and since 1992 *Dieter Kalaß* (born 1952). In 1989, *Möller* became a head of the department "Atmospheric Chemistry" at the Heinrich-Hertz Institute for Atmospheric Research and Geomagnetism (HHI). He rejected the recommendation of the West German evaluation commission in 1991 to go to the new founded tropospheric research institute in Leipzig and instead recommended the group of *Rolle* for the new Leipzig institute. In 1992, he became

Figure 1.9: *Cristian Junge* 1969 at Tenerife during a measurement campaign; photo by *Ruprecht Jaenicke*.

The third and most eminent person, belonged to the German pioneers[92] (and first generation in line with *Cauer* and *Mrose*) in atmospheric chemistry, was the meteorologist *Christian Eduard Friedrich Junge*, born on 2 July 1912 in Elmshorn (a small city in the north of Hamburg); he died on 18 June 1996 in Überlingen (Lake Constance); Fig. 1.9. The outstanding importance of *Junge* in the development of modern atmospheric chemistry is due to its connection with physical (small scale processes), meteorological (up to global circulation scale), and chemical processes; he also was a quite good experimentalist to apply *Cauer*'s microanalytical methods (such as *Mrose*), but he was much more experienced as observational meteorologist (which was fully missing at *Cauer*) than *Mrose*. Furthermore, he had the good fortune after the war (1950–1953) to continue academic research (habilitation at University Frankfurt[93]) and afterwards to work (1953–1961) in the USA at the Air Force Cambridge Research Laboratory with excellent technical facilities. Finally, as professor for meteorology (1961–1968) at the University of Mainz, and namely as a head of the newly established department for chemistry of the atmosphere and physical chemistry of isotopes (1968–1978) at the Max Planck Institute for Chemistry in Mainz, he was able to create a new generation of atmospheric chemists, who became internationally known and obtained leading positions (see below). Since *Crutzen* became the head, the department was simply named "*Abteilung Luftchemie*" [department for air chemistry].

---

the head of the Berlin branch for air chemistry of the "*Institut für atmosphärische Umweltforschung*" (head *Wolfgang Seiler*, born 1940) of the Fraunhofer Society, and, in 1994, he was appointed to a full professor for "*Luftreinhaltung und Luftchemie*" [air pollution control and air chemistry] at the Technical University Cottbus but remained with his new laboratories mostly in Berlin.

92 *Robert Duce* writes in his necrology (Duce 1997): "Junge is generally credited with being the father of the modern discipline of atmospheric chemistry". Without wanting to diminish *Junge*'s merits, this statement arose because of missing knowledge of *Cauer* abroad (who was known only in Germany) and *Mrose* (who was almost unknown before this "History").

93 He was assistant of *Ratje Mügge* (1896–1975), a German meteorologist, geophysicist and excellent teacher.

Junge studied meteorology and geophysics (1931–1935) in Graz, Hamburg, and Frankfurt am Main; he made his dissertation entitled "*Übersättigungsmessungen an atmosphärischen Kondensationskernen*" [measurements of supersaturation on atmospheric condensation nuclei] in 1935, and then he was an assistant (1935–1937) of *Franz Linke* (1878–1944).[94] The 1930s were a fruitful research period in Frankfurt am Main (including the Taunus observatory) concerning atmospheric nuclei and its bioclimatic relevance: *Junge*'s doctor father *Linke* also contributed to the atmospheric aerosols (Linke 1942), *Walther Amelung* and *Helmut Erich Landsberg* (1906–1985)[95] conducted measurements of nuclei (Amelung and Landsberg 1934, Landsberg 1934, 1938), and other scientists at different location (e. g., Israel and Schulz 1932).[96] Thus, Junge (1935, 1936) already created his significance for atmospheric aerosol research,[97] which rests on three subjects: aerosol size distribution (Junge 1951, 1952, 1953, 1954, 1955), aerosol and the link with trace gases and precipitation chemistry (Junge 1952, 1956, 1957, 1958a,b, Junge and Gustafson 1956), and the stratospheric (sulfate) aerosol layer, later named *Junge* layer (Junge et al. 1961).[98]

Since 1937 *Junge* was employed by the German Weather Service ["*Reichswetterdienst*"] at the Office for Instruments in Hamburg. *Junge* spent the entire war as weather forecaster on duty in many parts of the theatres of war: France, North Africa, Crete, and Italy. For two years he had been an American military prisoner; afterwards, he again joined the "*Deutscher Wetterdienst*" in Hamburg until 1950. During this period (1950–1952) at the meteorological institute in Frankfurt, *Junge* met *Cauer* for the first time and they conducted experiments together, but most important, *Junge* (after some studies of the chemical nature of nuclei by electron diffraction) concluded that only microanalytical methods are helpful for particle analysis. Conducting particle sam-

94 German meteorologist and geophysicist who organized worldwide the first aviation weather service 1909 in Frankfurt; he was also interested in bioclimatology.

95 *Helmut Erich Landsberg* a German-American meteorologist, born in Frankfurt am Main who moved (1934) to the USA where he became an influential climatologist. From 1930 to 1934 he was the supervisor of the *Taunus* Observatory at Frankfurt's Institute of Meteorology and Geophysics.

96 *Hans Israel* (1902–1970) was a German meteorologist who worked at *Linke* in Frankfurt (1928–1932) and was mainly interested in ions and later atmospheric electricity: Israel, H. (1931) Zur Theorie und Methodik der Grössenbestimmung von Luftionen. *Gerlands Beiträge zur Geophysik* **31**, 171–216. See also: Schwere Ionen der Atmosphäre. *Zeitschrift für Geophysik* **7** (1931) 127–133. Ionen und Kerne; eine kritische Studie. *Gerlands Beiträge zur Geophysik* **57** (1941) 261–282. Israel (1970).

97 In the introduction to the book "The atmospheric condensations nuclei" (Burckhardt and Flohn 1939), without any doubt the best presentation in this field before World War II, the authors acknowledged *Junge* (Berlin) and *Findeisen* (Friedrichshafen) for valuable comments.

98 *Ruprecht Jaenicke* (born 1940) wrote biographies of Junge (Jaenicke 2012). *Jaenicke* worked in *Junge*'s department in Mainz 1969—1979 and became then professor for meteorology at the University of Mainz. He is most known for the mathematical description of the aerosol size distribution (Jaenicke and Davis 1976). See also Lax (2018) on the history of the Max Planck Institute for Chemistry in Mainz.

pling in Frankfurt (city and Taunus observatory), Zugspitze and Wyk with a konimeter[99] in two fractions (below and above 0.9 µm) were able to show that the small particle fraction consisted mainly of ammonium sulfate and the larger one of chloride, but that chloride is not only originated from sea salt but also from artificial pollution sources. His measurements (including those after 1953 in the USA at Round Hill Field Station and Hawaii) result in the first continues size distributions of atmospheric aerosol and its first graphical and numerical presentation in 1952 (later to be known as *Junge* distribution). After his habilitation in 1953, *Junge*'s scientific future appeared to be dim in Germany during the 1950s. Therefore, he accepted an invitation by *Helmut Landsberg* and moved to the US.

Junge, who published in 1963 his pioneering monograph[100] "Air Chemistry and Radioactivity", "was instrumental in bridging the gap between classical chemistry and meteorology, tow disciplines that had previously barely recognized each other" writes Duce (1997). *Junge*'s merit is that he "could conceptualize the earth's atmosphere as a complex but integrated chemical and physical system, closely connected to human activities and ither biological processes on the land and in the sea" (Duce 1997). It is often said that successful science is a mixture of luck, competence, and hard work. Surely, *Christian Junge* lost 10 years of scientific work through the war, but he was luck in contrast to *Cauer* and *Mrose*. *Hans Cauer* remains unforgotten because of his view of a "chemical climatology", and *Helmut Mrose* must be placed on the podium of pioneers, showing soon after the war the importance of "*Chemische Aufgaben der Meteorologie*" [chemical tasks of meteorology] (Mrose 1949).

Junge was not a pioneer in recognizing the cycling of atmospheric gases (Junge 1972); it was *Vernadsky*. After the war (see also the history of atmospheric chemistry and research in West Germany: Lax 2016, 2018 and Warneck 2009) *Junge*'s first scholar (although officially *Mügge* was the supervisor) *Hans-Walter Georgii* (1924–2018)[101]

99 A measured volume of air is drawn through a jet, to impinge on a surface coated with a glycerin jelly. The dust adheres to the surface and the particles are counted under a microscope. Thus, it is one-stage impactor, first constructed by Sir *Robert Nelson Kotze* (1879–1953) in 1919, appointed Government Mining Engineer in the Union of South Africa. This konimeter called apparatus was first commercial produced in the Carl Zeiss Company in Jena in 1929, constructed by *Fritz Löwe* (1874–1955) further produced in the 1950s and use until the 1970s.

100 A "better" title would be "air chemistry and meteorology" (used in his German publication Junge (1963b)), but "atmospheric radioactivity" as a result of atom bombing test in the 1950s was in the USA a reason for funding atmospheric research; however, in the late 1950s, many U. S. laboratories turned toward space research, and *Junge*'s research upon "terrestrial" caused atmospheric research was terminated, a reason to went back to Germany.

101 His father *Walter Georgii* (1888–1968) was meteorologist and head of the aviation meteorological service in Frankfurt. Originally, *Hans-Walter Georgii* wanted to study chemistry (Lax 2016), but because of the wrong conditions in Frankfurt he begun 1948 in Munich, and later in Frankfurt, to study meteorology, physics and chemistry, made his diploma in 1953 under the supervision of *Junge*, became assistant in Frankfurt and made his habilitation 1959 in meteorology in Frankfurt. 1964 he became was the first who was interested in cycles of atmospheric trace gases. He installed in the garden of the Institute in Frankfurt/Main (Feldbergstr. 47) the "first German air measuring station",[102] and established in 1968 together with *Junge* and *Kurt Bullrich* (1920–2010), professor for meteorology at the University of Mainz, a "*Sonderforschungsbereich*" 73 [Special Research Field] "Atmospheric Trace Substances", continued until 1986 and just followed until 1997 by another "Chemistry and Dynamics of Hydrometeors" ["*Chemie und Dynamik der Hydrometeore*"].

Thus, *Junge* and *Georgii* created without a doubt the largest scientific school and thought of atmospheric chemistry in Europe with another new generation (born between 1940 and 1955) of atmospheric scientists. *Junge* created in 1970 four working groups in his institute: chemical mechanisms and photochemistry with *Peter Warneck* (1928–2019),[103] aerosol research with *Ruprecht Jaenicke*, trace gases with *Wolfgang Seiler*, and biogeochemistry with *Manfred Schidlowski* (1933–2012). The successor of *Junge* at the MPI in Mainz became *Paul Crutzen* (1933–2021)[104] (Fig. 1.10) in 1980 who introduced modeling as research tool in atmospheric chemistry; *Ruprecht Jaenicke* already left 1979 *Junge*'s groups as professor of meteorology at the university Mainz and

---

professor of meteorology in Münster, 1965 in Cologne and two months later he took over the chair position for meteorology in Frankfurt until his retirement in 1993. *Georgii* worked 1958–1960 as scientific assistant of the City Frankfurt to investigate the urban climate and air pollution; in that time, he was in contact with *Cauer*.

102 Necrology by *Klaus Vogel* and *Heinz Bingemer* (2018); however, the *first* German air pollution measuring station *after* the war was established in the 1950s at the Observatory Wahnsdorf near Dresden. See also Däßler and Stein (1968) concerns the early monitoring network in Saxony.

103 *Warneck* received the diploma in 1954 and the doctorate in 1956 in physical chemistry at the university in Bonn. 1959–1970 he studied chemical reaction in planet's atmospheres in the laboratory at Geophysics Corporation of America (GCA) in Bedford (Massachusetts). In 1970, he became the head of the chemical kinetics group, Max Planck Institute for Chemistry, with *Junge* as director. In 1974 he also became professor of physical chemistry at the university in Mainz. 1992 he was appointed the founding director of the new Institute for Tropospheric Research in Leipzig (IfT, later abbreviated by TROPOS). *Warneck*'s research included laboratory studies of chemical mechanisms and photochemistry as well as the development of analytical techniques for field measurements. Since 1990, his interests are focused on chemical reactions in clouds.

104 *Crutzen* was if Dutch origin, born in Amsterdam, who studied first constructional engineering. In 1959, he moved to Stockholm as computer programmer at the Institute of Meteorology, where he also studied meteorology and obtained a PhD in 1968 and habilitated in 1973 (On the photochemistry of ozone in the stratosphere and troposphere and pollution of the stratosphere by high-flying aircraft). In 1974–1980, he worked at the National Oceanic and Atmospheric Administration (NCAR) in Boulder, 1980–2000 as director of the Department of Atmospheric Chemistry (MPI) in Mainz (successor of *Christian Junge*). He was professor at the following Universities: Colorado State University, of California, of Chicago, of Utrecht, and of Stockholm. In 1995 Nobel prize winner (together with *Rowland* and *Molina*) for his research on stratospheric ozone. I (DM) met him first time in 1990 and owes him support and recognition after German unification. I got to know him warmhearted and collegially: after receipt of the Nobel award in 1995 he was so busy and frequently invited worldwide, but he followed my invitation to a summer colloquium hold in 1996 in Cottbus, the former East Germany.

Wolfgang Seiler in 1986 as director of the Fraunhofer Institute for Atmospheric Environmental Research in Garmisch-Partenkirchen; *Peter Warneck* stayed with an independent working group at the MPI Mainz.

Figure 1.10: *Paul Crutzen* at the International *Quadrennial Ozone Symposium* in the island of Kos, Greece 2004; photo by *Detlev Möller*.

In *Georgii*'s group (all meteorologists) *Gode Gravenhorst*, *Heinz Bingemer*, *Franz Meixner*, *Harald Berresheim* must be named having important contributions to atmospheric chemistry, and others after the 1980s. Likely in remembering *Cauer*'s statement that chemical analysis and namely the development of new methods should be in the hand of a chemist, the chemist *Wolfgang Jaeschke* (born 1942)[105] joint *Georgii*'s group in 1974.

105 *Jaeschke* studied chemistry at University Frankfurt/Main and earned his PhD on application of chemiluminescence for atmospheric measurements in the institute of physical biochemistry at *Joachim Stauff* (1911–?); Jaeschke and Stauff (1978). He left *Georgii*'s group in 1979 and worked under the president *Hartwig Kelm* (1933–2012), organizing for the NATO Advanced Study Institute the conference "Chemistry of the Unpolluted and Polluted Troposphere", held on the Island of Corfu, Greece (1981) and on "Chemistry of Multiphase Atmospheric Systems" two years later (Corfu 1983). In 1980 the "*Zentrum für Umweltforschung*" [Center for Environmental Research] was founded at the Goethe University Frankfurt/Main with *Jaeschke* as Technical Director who habilitated in 1986 and became later professor.

Like in the USA,[106] studying chemical reactions in the atmosphere evolved from another discipline, the classical kinetics of gas reactions. A German school of atmospheric physical chemists was the physical chemistry institute at University Bonn, headed since 1954 by *Wilhelm Groth* (1904–1977)[107] who was famous for his research on gas-phase photochemistry in his especially built “Bonner Kugel”: *Peter Warneck* (1928–2019), *Karl-Heinz Becker* (born 1935), *Franz-Josef Comes* (1928–2015), *Ulrich Schurath* (born 1939), *Reinhard Zellner* (born 1944), *Karsten Levsen* (born 1939) and *Dieter Kley* (born 1937). Last but not least, *Dieter Ehhalt* (born 1935), graduated as physician in Heidelberg, became in 1974 the director of the Institute for Atmospheric Chemistry in Jülich. *Becker* became professor for physical chemistry in 1974 at the “*Gesamthochschule*” in Wuppertal and studied atmospheric reaction systems with his coworkers (to name *Ian Barnes* and *Peter Wiesen*) since the end of the 1970s. *Comes* became professor at the University Frankfurt/Main in 1970 and studied traces (e. g., OH) by laser methods. *Schurath* became in 1978 professor for physical chemistry in Bonn and 1995 in Heidelberg, together with a working group on aerosol chemistry (AIDA chamber) at the Institute for Meteorology and Climate Research of the Research Center Karlsruhe.[108] *Zellner* studied in Göttingen where he became professor for physical chemistry in 1983 and 1991 at the University Duisburg-Essen. *Zellner*'s most famous student in atmospheric chemistry was *Hartmut Herrmann* (born 1961) who succeeded *Wolfgang Rolle* (1934–1996) as professor for atmospheric chemistry in Leipzig in 1998 (TROPOS – institute for tropospheric research). *Levsen* became a

106 In the 1950s, pioneering work were conducted on photochemical smog chemistry by many US American scientists; apart from *Pitts*, *Calvert*, *Cadle* and *Leighton* (see also pp. 490–491) I like to refer here (no biographic data available) the chemists *G. J. Doyle*, *Philipp Lincoln Hanst* (1924–2007), *J. T. Middleton*, *E. A. Schuck*, and *E. R. Stephans*, developing the conception of hydrocarbon reactivity (for references see Leighton 1961), further developed in the 1960s including the chemistry of nitrogen oxides by *Aubrey P. Altshuller* (born 1927) and *Joseph J. Bufalini* (born 1931), and *Julian P. Heicklen* (born 1932); Altshuller and Bufalini (1965, 1971). Altshuller (1983) pointed out the role of natural emitted volatile organic compounds (VOC) in smog formation. *Roger Atkinson* (born 1945) and his coworkers carried out a large number of kinetic and mechanistic studies of the reaction of the OH radical with organic compounds in the gas phase; they also introduced the conception of the “maximum incremental reactivity” (MIT); Atkinson and Carter (1984), Atkinson (1986), and further publications until 2008. *John Hersh Seinfeld* (born 1942) and his group formulated a chemical mechanism for ozone formation and in 1973 developed the first urban-scale atmospheric chemical-transport model, which was applied to the Los Angeles basin (Seinfeld 1986). However, the research on VOC degradation is a “never-ending story” (Calvert et al. 2008, Wallington et al. 2018).

107 *Groth* (together with *Suess* and *Harteck*) worked (1939–1945) on the German nuclear energy project, also known as the “*Uranverein*” [Uranium Club]. Studied physics in Munich and Tübingen, since 1932 assistant of *Harteck* in Hamburg, habilitation 1938 and 1950 professor for physical chemistry; since 1952 at Bonn.

108 Since 1971 *Schurath* was in close cooperation with *Hans Güsten* (1931–?) who studied chemistry in Mainz and worked since 1964 in Karlsruhe and headed different working groups in the field of photochemistry and environmental chemistry.

professor of physical chemistry at the University of Bonn in 1979 and 1985 head of the Department of Analytical Chemistry, Fraunhofer Institute of Toxicology and Experimental Medicine, Hannover. *Kley*, after a long stay at the Colorado State University, became an institute director at the Nuclear Research Center Jülich, such as *Ehhalt*. Beside *Groth* at Bonn, another "school" was at the Ruhr-Universät Bochum under the heading of *Friedrich Stuhl* (1937–2005), were *Cornelius Zetzsch* (born 1945) and *Andreas Wahner* (born 1958) began to study specific reactions between the OH radical and organic compounds; *Wahner* succeeded *Ehhalt* in 2000, and *Zetzsch* who studied in Göttingen and habilitated (1978) in Bochum, became afterwards professor and head of a working group atmospheric chemistry at the similar institute like *Levsen* in Hannover, but moved in 1999 to the University of Bayreuth with his equipment. At the Max-Planck-Institute for chemistry in Mainz, the Belgian *Geert Moortgat* (born 1941) begun to study radical reactions with organic compounds. In the 1990s, the European Photoreactor (EUPHORE) in Valencia (Spain), the largest "smog chamber" in Europe, was opened for international research work.[109]

1.2 Biogeochemistry: a terminological and historical approach

Chemistry was a priori the science of mineral, animal, and vegetable matter. The chemistry of the earth system – when not considering life – is *geochemistry*. The term *geochemistry*, like many other scientific terms, has variable connotations. If geochemistry means simply the chemical study of the Earth or parts of the Earth, then geochemistry must be as old as chemistry itself, and dates from the attempts of Babylonian and Egyptian metal-workers and potters to understand the nature and properties of their materials (Tomkeiev 1944). Schönbein (1838, p. 280) used the term "*Geochemie*" for the first time, writing that an understanding of chemical evolution (he writes in German "*chemische Bildungsepochen*"), similar to the biological development, needs a chemical-geological consideration as a new branch of chemistry

109 In Europe, due to the focus on "classical" air pollution problems, like winter smog, acidification and *Waldsterben*, research on ozone formation and thus the role of organic compounds, namely VOC degradation using "smog chambers" begun much later with the research of the British chemist *Richard Anthony* ("Tony") *Cox* (born 1941), who had a close collaboration with *Roger Atkinson*, in the late 1970s. *Cox* closely worked together with *Richard G.* ("Dick") *Derwent* (born 1945), past scientist with Britain Meteorological Office, who was the first in Europe in the middle of the 1970s to begin developing photochemical models. Together with *Michael E. Jenkin* and others (namely *Donald L. Baulch* and *Sandra Saunders*) he developed in the 1990s the photochemical ozone creation potential (POCP) and related everything to ozone-producing (Derwent et al. 2017). Together with *Michael J.* ("Mike") *Pilling* (born 1942) and others *Derwent* created in the 1990s the Master Chemical Mechanism (MCM), that is a near-explicit chemical mechanism describing the degradation of CH_4 and 142 VOCs in the troposphere (the organic component of the version MCMv3 contains in the region of 12600 reactions and 4500 chemical species).

(Schönbein 1838, p. 275). It is worth to cited a note that did appear in *Nature* (Anonymous 1931):[110]

> Clarke's invaluable "Data of Geochemistry" is justly so highly esteemed, especially in English-speaking countries, that there was a natural tendency to regard it not only as an indispensable compilation of data but also as a standard exposition of geochemistry itself. Only recently has this tendency been checked by growing acquaintance with the brilliant work of Fersman, Vernadsky, and Goldschmidt. A most useful and illuminating little book was published in French by Vernadsky in 1924 under the title "La Géochimie". In 1927 a revised translation appeared in Russian, and now a third and greatly enlarged edition has become available, this time in German.

It is obvious that *biochemistry* deals with chemical processes in organisms and thus the chemical interaction between organisms occurs via geochemical processes; consequently, biogeochemistry is *the* chemistry of the climate system which is defined (Möller 2020) as that part of the earth system affecting life. Subdividing the geogenic part of the climate system into "other" systems, we have the atmosphere, hydrosphere, cryosphere, and lithosphere; thus atmospheric, aquatic and soil chemistry are the sub-disciplines of geochemistry.

1.2.1 From geochemistry to biogeochemistry

The Russian geochemist *Vladimir Ivanovich Vernadsky* [Владимир Иванович Вернадский] (1863–1945)[111] is the founder of the new science *biogeochemistry*. Between 1909 and 1910 he established geochemistry as an independent branch of

110 *Frank Wigglesworth Clarke* (1847–1931) was an American chemist who is credited with having determined composition of the earth crust. *Alexander Jevgenjevich Fersman* [Александр Евгеньевич Ферсман] (1883–1945) was a prominent Russian-Soviet geochemist and mineralogist; since 1924 scholar of *Vernadsky* (Fersman 1923). *Victor Moritz Goldschmidt* (1888–1947), born in Switzerland, moved with his father (who was a famous professor of chemistry) in 1901 to Oslo (termed Kristiania until 1924), was a geochemist who published in German 1911–1937 in the Norwegian Journal "*Skrifter utgit av Videnskapsselskapet i Kristiania. I, Matematisk-naturvidenskabelig klasse*".

111 Also written as *W.* (*Wladimir*) *Vernadskij* and *Wernadskij*; Russian/Ukrainian geologist. His research ranged from meteorites and cosmic dust to microbiology and migration of microelements via living organisms in ecosystems. In line with *Humboldt*, *Vernadsky* was a polymath and genius. There exist a large number of publications on geochemical studies in Russia and abroad. Complete biography of *Vernadsky*: http://www.sgm.ru/DOWNLOAD/ABOUT/Nauchnye_trudy_Vernadskogo.pdf. Many publications are hardly to obtain, namely the special prints before 1930 from Petrograd or Leningrad, resp., despite many collected editions in Russian were published after *Vernadsky*'s death. In the former Soviet Union before "glasnost" (e. g., before 1988), *Vernadsky* has placed on a political pedestal, and only after that time a few publications did appear (e. g., Krüger et al. 1990, Ghilarov 1995, Levit 2001) that have shown his complex thinking and that he was way ahead of his time. However, there is still a need of annotated editions in English of *Vernadsky*'s works that show his way of thinking in combining geo- with biochemistry, the transfer from bio- to noosphere, and the role of biogeochemistry in global environmental research. See also: Vernadsky (2000–2001, 2005).

science, independently from *Victor Moritz Goldschmidt*, who pointed out on 1937 the importance of the *Vernadsky*'s work (Vinogradov 1943, p. 660). Whereas *Frank Wigglesworth Clarke* collected an extensive geochemical data collection, *Vernadsky* treated the ways of a dynamic geochemistry, already seeing biogeochemical cycling (yet without using this term), writing that understanding the chemical composition of the earth's crust needs the investigation of the natural processes and the role of chemical elements in the earth's history (Vernadsky 1910, pp. 74–75).[112] In his studies he recognized that almost all geologic processes that occur on the earth's surface are affected by biological activities (Vernadsky 1921, 1928). From his lectures at the Sorbonne during 1922–1923, with some amplification, a book of four chapters in French results (Vernadsky 1924), later in several editions in Russian (Vernadsky 1927a,c), and in German (Vernadsky 1930). His idea to establish a biogeochemical laboratory was first published in English (Vernadsky 1923) and later in Russian (Vernadsky 1927b). His ideas on the biosphere he published in 1926, and in 1929 in French (but not in German),[113] and in English first in a reduced version in 1986 (Vernadsky 1926a). Several publications deal with biogeochemistry (Vernadsky 1926b, 1927a,b, 1931, 1932, 1934a, 1945a).

Long before institutionalizing biogeochemistry in the Western World, *Vernadsky* founded in 1927 at the KEPS[114] a "department of living matter" [*Отдел живого вещества*] which was, later (October 1928) established as an independent "Biogeochemical Laboratory" belonging to the Soviet Academy of Sciences [*Биогеохимическая лаборатория АН СССР (БИОГЕЛ)*]; *Vernadsky* remained director until he died in 1945. This laboratory was then integrated in the "Institute of Geochemistry and Analytical Chemistry V. I. Vernadsky", founded in 1947 [*Институт геохимии и аналитической химии им. В. И. Вернадского (ГЕОХИ АН СССР)*]. Its first director was *Alexander Pavlovitch Vinogradov* [Алекса́ндр Па́влович Виногра́дов] (1895–1975), who became in 1930 the deputy of *Vernadsky*'s laboratory (see also Vinogradov (1943) on biogeochemical research in the USSR). Despite biogeochemistry was at that time primarily focused on the investigation of distribution of elements in soils, organ-

112 In his later works on geochemistry, he defined (cited after Krüger et al. 1990) "geochemistry studies scientifically the chemical elements, i. e., the atoms of the earth's crust, and, so far as possible, of the whole planet. It studies its history, distribution and movement in space and time as well as its genetic interrelations on our planet".

113 The German chemist *Ernst Karl Ferdinand Kordes* (1900–1976), born in St. Petersburg, professor at Leipzig, Posen, Jena and Bonn, translated (from the Russian version) and edited *Vernadsky*'s "geochemistry" (Vernadsky 1927c), in which obviously already his ideas on the biosphere and biogeochemistry had been included (Vernadsky 1930).

114 (КЕПС) Комиссия по изучению естественных производительных сил России [Commission for studying the natural productive forces of Russia], established in 1915 at the St. Petersburg Academy of Sciences on recommendation by *Vernadsky*.

isms and waters, Vernadsky (1924, 1930, 1935) included the atmosphere (and thus its chemistry) as part of geochemistry; thus Vinogradov (1943) refers the geochemists *Burkser* in Odessa (see pp. 4 and 224) and *Bertrand* in Paris (see p. 221), who carried out chemical rainwater studies, in *Vernadsky*'s tradition.

It was not until the 1980s that western scientists were aware of *Vernadsky*'s work. Still *Bert Bolin*, who likely was the first who used in the West the term "biogeochemistry" in sense of a science (Bolin 1983) did not cite *Vernadsky*. The first institution with "biogeochemistry" in its name was the "*Abteilung Biogeochemie*", established at the Max-Planck Institute for Chemistry (MPI) in Mainz by the proposal of *Paul Crutzen* and headed (1987–2017) by *Meinrath O.* ("*Andy*") *Andreae* (born 1949); see Andreae (2012).

The first "biogeochemist" after Vernadsky was the Soviet biologist *Viktor Vladislavovich Kovalskij* [Ви́ктор Владисла́вович Кова́льский] (1899–1984), who was the head (1954–1984) of the Biogeochemical Laboratory of ГЕОХИ (see above) which was renamed in 1988 in Laboratory of Biogeochemistry of the Environment [*лаборатория биогеохимии окружающей среды*]. Another early Soviet institution of biogeochemistry was the "Laboratory of Microbial Biogeochemistry in the Institute of Biochemistry and Physiology of Microorganism" belonged to the Academy of Sciences [*лаборатории микробной биогеохимии Институт биохимии и физиологии микроорганизмов АН СССР*] at Puschtschino [Пущино] near Moscow, headed (1969–1984) by the Soviet microbiologist *Mikhail Vladimirovich Ivanov* [Михайл Влади́мирович Ивано́в] (1930–2018); the director of this institute was *Georgij Konstantinovich Skryabin* [Георгий Константинович Скрябин] (1917–1989).

The idea to establish a special project on global biogeochemical cycles within SCOPE (Scientific Committee of Problems of the Environment, founded 1969) arose in 1973 on a session in Paris, where *Bert Bolin* played a leading role in its project on biogeochemical cycles. This was soon adopted by the National Soviet SCOPE Committee in 1974, and the first publication entitled "Biogeochemical Cycles" (in Russian) appeared (Kovda 1976), containing contributions on the cycles of carbon, nitrogen, sulfur, oxygen, heavy metals, phosphorus, and calcium. In December 1975, the Swedish SCOPE Committee organized the first international conference on biogeochemical cycles hold in Orsundsbro, Sweden, published as SCOPE report No. 7 "Nitrogen, phosphorus and sulphur – global cycles" (Svensson and Søderlund 1976). In the SCOPE report No. 13, Bolin et al. (1979) contributed with "The global biogeochemical carbon cycle". In October 1976, the World Meteorological Organization (WMO) meeting of experts to review criteria for siting background air-pollutions stations was convened at the MPI in Mainz, argued to understand the atmospheric part of biogeochemical cycles (WMO Bulletin Vol. 26. 1977, p. 28). Between 1980 and 1992, 6 volumes of a series entitled "The Natural Environment and the Biogeochemical Cycles" were published as part of the "The Handbook of Environmental Chemistry", edited by *Otto Hutzinger*

(1933–2012).[115] *Skryabin, Ivanov*, and *Freney*[116] edited in 1983 (in Russian) "The Global Biogeochemical Sulphur Cycle" (Skryabin et al. 1983), soon (1985) followed by "The Biogeochemical cycling of Sulfur and Nitrogen in the Remote Atmosphere", edited by *James N. Galloway* (born 1944), *Robert Jay Charlson* (born 1936), *Meinrath O. Andreae* (born 1949), and *Henning Rodhe* (born 1941), based on a workshop held in 1984.[117]

There are several definitions of biogeochemistry; I like this one: "Biogeochemistry is the study of how chemical elements flow through living systems and their physical environments. It investigates the factors that influence cycles of key elements like carbon, nitrogen and phosphorus" (https://www.nature.com/subjects/biogeochemistry). *Eville Gorham* (1925–2020), a Canadian-American biologist, also termed the "grandfather of acid rain" (see p. 231), defined biogeochemistry very similar in his extended history on this subject (Gorham 1981, 1991). The journal *Biogeochemistry*, founded in 1984, describes the subject as follows "[...] dealing with biotic controls on the chemistry of the environment, or with the geochemical control of the structure and function of ecosystems. Cycles are considered, either of individual elements or of specific classes of natural or anthropogenic compounds in ecosystems. Particular emphasis is given to coupled interactions of element cycles [...] from the molecular to global scales to elucidate the mechanisms driving patterns in biogeochemical cycles through space and time". The journal *Global Biogeochemical Cycles* was founded later in 1987.

As the term "biogeochemistry" says, it is almost biochemistry and geochemistry of the lithosphere, pedosphere, and hydrosphere in interaction with living organism – mostly microorganisms, to a lesser importance plant and nearly negligible on global scale, animals. However, humans became able to disturb and interrupt *natural* biogeochemical cycles, and thus artificial emissions also became a part of biogeochemical cycling. The atmosphere is a part of biogeochemical cycles; in short, *biogeochemistry* meets *atmospheric chemistry* at the borderline between the atmosphere and the earth surface, studying the mass balance (or budget) including fluxes, turnovers and metabolism. The common interest concerns the sea-atmosphere, soil-atmosphere, and vegetation-atmosphere exchange and the impacts like acidification, "*Waldsterben*", eutrophication, global and climate change.

---

115 An Austrian chemist who worked (1958–1974) in Canada, in 1973–1983 in Amsterdam and since 1983 in Bayreuth (Chair for Ecological Chemistry and Geochemistry at the University). Hutzinger (1980).

116 *John Raymond Freney* (1928–2015) was an Australian agriculture scientist who edited several books on biogeochemical cycling in soils and aquatic environment.

117 The author used the term "biogeochemical cycling" probably the first in Germany in 1982 (Möller and Schieferdecker 1982) and later (Möller 1985, 1988, 1996); he also coined the term "*geoökologischer Stoffkreislauf*" [geoecological cycling] in 1982.

1.2.2 From biosphere to noosphere

Humans – by decoupling their life cycle from natural conditions – have altered "natural" biogeochemical cycles. In recent decades, humans have become a very important force in the earth system, demonstrating that emissions and land use change are the causes of many of our environmental issues. These emissions are responsible for the major global reorganization of biogeochemical cycles. With humans as part of nature and the evolution of a man-made changed earth system, we also have to accept that we are unable to remove the present system into a preindustrial or even prehuman state because this means disestablishing humans. The key question is, which parameters of the climate system allow the existence of humans under which specific conditions.

During his works abroad (1935 in Paris and London) *Vladimir Ivanovich Vernadsky* wrote "*Научная мысль как планетное явление*" [Scientific though as global phenomenon], published only in 1977[118] and held in a lecture in January 1942 during his evacuation from Moscow to Kazakhstan, entitled "*О геологических оболочках Земли как планеты*" [On geological envelopes of the planet Earth]; he created the basic ideas that decades later were published in the Western World as "global thinking". *Vernadsky* understood by *noosphere* (termed the *anthroposphere* by *Paul Crutzen*) a new dimension of the biosphere, developing under the evolutionary influence of humans on natural processes (Vernadsky 1926a,b); consequently, Crutzen and Stoermer (2000) proposed name to the present epoch *Anthropocene*. "The Great Acceleration is reaching criticality. Whatever unfolds, the next few decades will surely be a tipping point in the evolution of the Anthropocene" wrote Steffen et al. (2007). Now it seems that humankind enters a "*nouveau regime climatique*" [new climatic regime], according to French sociologist and philosopher *Bruno Latour* (born 1947); Latour (2015).

The British scientist *James Lovelock* (born 1919) together with *Lynn Margulis* (1938–2011) developed the hypothesis that the Earth is a self-controlling system (*Gaia*: the earth goddess in Greek) and proposed that our present atmosphere is far from the chemical equilibrium that is assumed for other planets (Lovelock and Margulis 1974, Crutzen 2002). One expression of this is the difference in the redox potentials between biosphere (reducing medium) and atmosphere (oxidizing medium). It is believed that living organisms are responsible (to a large extent) for the chemical composition of the present atmosphere and, from the opposite point of view, the chemical composition of the atmosphere determines the biota. Remarkably, the composition of the biosphere is similar to that of the present atmosphere. The *Gaia hypothesis* is an ecological hypothesis proposing that the biosphere and the physical components of the

118 Размышления натуралиста. Научная мысль как планетное явление. В 2-х кн. Кн.2., М.: Наука, 192 с. [Reflexions of a naturalist. scientific though as global phenomenon. In two books, book two. Publ. Nauka, Moscow, 192 pp.].

Earth (atmosphere, cryosphere, hydrosphere and lithosphere) are closely integrated to form a complex interacting system that maintains the climate and biogeochemical conditions on the Earth in a preferred *homeostasis*, originally proposed by *James Lovelock* as the earth-feedback hypothesis (Lovelock 1979). In short, the biosphere itself becomes an organism; it acts more like an organism and appears to be internally controlled. Gaia is the "invisible hand" of biogeochemistry. *Lovelock* suggests that life processes regulate the radiation balance of the Earth to keep it habitable. As a well-known example, a self-controlling cycle between marine sulfur emissions and air temperature (CLAW hypothesis) was established by Charlson et al. (1987), initiating more useful air chemical and climate research. Behind this relationship is the so-termed *Twomey* effect (Twomey 1974), which describes how cloud condensation nuclei (CCN) from anthropogenic pollution may increase the amount of solar radiation reflected by clouds.

Natura non facit saltus (Latin for "nature does not make jumps/leaps") was a principle of natural philosophy since at least *Aristoteles*'s time and then used as an axiom by *Gottfried Wilhelm Leibniz* (1646–1716),[119] *Isaac Newton* (1642–1726),[120] *Charles Darwin* (1809–1882)[121] and others. A modern understanding of evolution includes not only continuing development but also leaps (catastrophes). This is referred to as "transformation of quantity into quality" (*dialectic leap*) and may characterize the current discussion on the impacts of climate change.

The term "biosphere" was coined in 1875 by the famous Austrian geologist *Eduard Suess* (1831–1914), (Suess 1875; quoted after Smil 2002):

> [...] one thing seems to be foreign on this large celestial body consisting of spheres, namely, organic life. But this life is limited to a determined zone at the surface of the lithosphere. The plant, whose deep roots plunge into the soil to feed, and which at the same time rises into the air to breathe, is a good illustration of organic life in the region of interaction between the upper sphere and the lithosphere, and on the surface of continents it is possible to single out an independent biosphere.

In his interpretation, the "biosphere" is an envelope of life, which "is limited to a determined zone at the surface of the lithosphere". The term was never given a definition or elaborated upon until *Vladimir Ivanovich Vernadsky*. *Vernadsky*, who had met *Suess* in 1911, popularized the term biosphere in his book *The Biosphere* (first published in Russian in 1926 and in French in 1929, not translated into a full English version until

---

119 *Gottfried Wilhelm Leibniz* was a German philosopher, polymath and mathematician who wrote primarily in Latin and French; belong *Humboldt* the greatest German polymath.

120 *Isaac Newton* was an English physicist, mathematician, astronomer, natural philosopher, alchemist, and theologian who is perceived and considered by a substantial number of scholars and the general public as one of the most influential men in history.

121 *Charles Darwin* was an English naturalist, geologist, and biologist, best known for his contributions to the science of evolution.

more than 60 years later), hypothesizing that life is the geological force that shapes the Earth (Vernadsky 1926a, 1998). In his words (Vernadsky 1944):

> [...] a definite geological envelope markedly distinguished from all other geological envelopes of our planet. This is only because it is inhabited by living matter, which reveals itself as a geological force of immense proportions, completely remaking the biosphere and changing its physical, chemical, and mechanical properties, but also because the biosphere is the only envelope of the planet into which energy permeates in a notable way, changing it even more than does living matter.

Hutchinson (1970)[122] wrote that it is essentially *Vernadsky*'s concept of the biosphere that we accept today. *Vernadsky* was the first to propose the idea of biogeochemical cycling (having asked: what is the impact of life on geology and chemistry of the Earth?). He wrote (Vernadsky 1998):

> The biogeochemical energy of living matter is determined primarily by the reproduction of organisms, by their unremitting endeavor (determined by the energetics of the planet) to achieve a minimum of free energy – determined by the fundamental laws of thermodynamics corresponding to the existence and stability of the planet.

The term "ecosystem" was coined in 1930 by *Arthur Roy Clapham* (1904–1990)[123] to denote the combined physical and biological components of an environment (Willis 1997). The British ecologist *Sir Arthur George Tansley* (1871–1955) later refined the term, describing it as: "The whole system, [...] including not only the organism-complex, but also the whole complex of physical factors forming what we call the environment" (Tansley 1935). He championed the term *ecosystem* in 1935 and *ecotope* in 1939 (Cooper 1957). *Vernadsky* defined "ecology" (originally intended as the "economy of nature") as the science of the biosphere.

In *Vernadsky*'s theory of how the Earth develops, the *noosphere* is the third stage in a succession of phases of development of the planet, after the geosphere (inanimate matter) and the biosphere (biological life). He was one of the first scientists to recognize that oxygen, nitrogen, and carbon dioxide in the earth's atmosphere results from biological processes. In the 1920s, he published works arguing that living organisms could reshape the planets as surely as any physical force. The term "noosphere" was first coined by the French mathematician and philosopher, *Edouard Louis Emmanuel Julien Le Roy* (1870–1954), Le Roy (1928). *Le Roy*, building on *Vernadsky*'s ideas and on discussions with *Pierre Teilhard de Chardin* (1881–1955) – both attended *Vernadsky*'s lectures on biogeochemistry at the Sorbonne in 1922–1923 – came up with the term "noosphere", which he introduced in his lectures at the Collège de France in 1927

122 *George Evelyn Hutchinson* (1903–1991) was a British limnologist and ecologist.

123 *Arthur Roy Clapham* was a British botanist and worked at Rothamsted Experimental Station as a crop physiologist (1928–1930), professor of botany at Sheffield University 1944–1969.

(Le Roy 1927). *Vernadsky* saw the concept as a natural extension of his own ideas, predating *Le Roy*'s choice of the term (Smil 2002, p. 13).

Teilhard de Chardin was a Jesuit who introduced in 1925 the concept of noosphere, the "sphere of mind" superimposed on the biosphere or the sphere of life. Not until after the appearance of humans on the Earth did the *noogenesis* begin. The term *noogenesis* comes from the Greek word *νοῦς* (noos) = mind. *Teilhard de Chardin* wrote in an essay entitled "Hominisation" (1925): "And this amounts to imagining, in one way or another, above the animal biosphere a human sphere, a sphere of reflection, of conscious invention, of conscious souls (the noosphere, if you will)". He later wrote: "Man discovers that he is nothing else than evolution become conscious of itself. The consciousness of each of us is evolution looking at itself and reflecting upon itself." (Teilhard de Chardin 1961, p. 221). This conception is based on philosophical writings, however, and completely neglects *Vernadsky*'s biogeochemical approach. The divergence is perhaps best expressed as an opposition between the anthropocentric view of life (*Teilhardian* biosphere) and the biocentric view of the natural economy (*Vernadskian* biosphere).

Vernadsky first took up the term "noosphere" in 1931, as a new dimension of the biosphere under the evolutionary influence of humankind (Vernadsky 1944). He wrote: "The Noosphere is the last of many stages in the evolution of the biosphere in geological history" (Vernadsky 1945b). The biosphere became a real geological force that is changing the face of the Earth, and the biosphere is changing into the noosphere. In *Vernadsky*'s interpretation (1945), the noosphere is a new evolutionary stage of the biosphere, when human reason will provide further sustainable development both of humanity and of the global environment (Vernadsky 1945b, p. 10):

> In our century the biosphere has acquired an entirely new meaning; it is being revealed as a planetary phenomenon of cosmic character [...] In the twentieth century, man, for the first time in the history of Earth, knew and embraced the whole biosphere, completed the geographic map of the planet Earth, and colonized its whole surface. Mankind became a single totality in the life on Earth [...] The noosphere is the last of many stages in the evolution of the biosphere in geological history.

At present, under the aspect of climate change, the following phrase of *Vernadsky* is topical as never before (Vernadsky 1954, p. 10, Vernadsky 1991, p. 18).[124]

> Мы как раз переживаем ее яркое вхождение в геологическую историю планеты. В последние тысячелетия наблюдается интенсивный рост влияния одного видового живого вещества – цивилизованного человечества – на изменение биосферы. Под влиянием научной мысли и человеческого труда биосфера переходит в новое состояние – в *ноосферу*.

124 From Chapter 1, § 19: Научная мысль и научная работа как геологическая сила в биосфер [Scientific thought and scientific work as geological force in the biosphere]. Note that there are several editions of *Vernadsky*'s works before 1945. See also electronic archive: http://vernadsky.lib.ru

> [We are living in a brand new, bright geological epoch. Man, through his labor – and his conscious relationship to life – is transforming the envelope of the Earth – the geological region of life, the biosphere. Man is shifting it into a new geological state: through his labor and his consciousness, the biosphere is in a process of transition to the *noosphere*].
> Note that this is a rather free translation from Russian (Vernadsky 2000–2001, p. 22). It follows my translation which is much closer to the Russian version: We just undergo within a bright geological epoch. In the past thousands of years, one observed an intensive growth of the influence of one specific living organisms – the civilized humanity – on a changing biosphere. Under the action of scientific thought and human work the biosphere transfer into a new state – the *noosphere*.

Vernadsky's ideas on atmosphere-biosphere interactions have been "rediscovered" several decades later (Gorham 1981) – without knowing him. Today the term "Anthroposphere" is used. The idea of a close interrelation between humans and the biosphere is topical in understanding the "earth system", i. e., climate change, and is used by Schellnhuber (1999) with the terminology "global mind" and by Crutzen and Stoermer (2000) with Anthropocene to characterize the present epoch. Zalasiewicz et al. (2008) published the first proposal for the formal adoption of the Anthropocene epoch by geologists.[125]

1.2.3 Cycling of matter

> Nach den neuesten Forschungsergebnissen auf dem Gebiete der Kosmogonie besteht der ganze Weltbetrieb in einem ewigen sich wiederholenden Kreislauf dessen, was wir als Energie und Materie bezeichnen [According to the newest results of research in the field of cosmogony, the whole global operation consists in an eternal repetitive cycle, what we call energy and matter] (Engler 1911, p. 3).

The Russian biochemist *Alexander Ivanovich Oparin* [Александр Иванович Опарин] (1894–1980) who established a theory of the origin of life on the Earth (Oparin 1924, 1936), independently from *John Burdon Sanderson Haldane* (1892–1960 (see Möller 2020, p. 243), wrote (Oparin 1924):

> There is no fundamental difference between a living organism and lifeless matter. The complex combination of manifestations and properties so characteristic of life must have arisen in the process of the evolution of matter.

Vernadsky, together with *Oparin*, first formulated biogeochemical cycling (Vernadsky and Oparin 1930). However, the fundaments of cycling of matter were created by *Boussingault* (see p. 57) in studying assimilation of gases and the uptake of nitrogen

125 The Anthropocene Working Group of the International Union of Geologic Sciences (IUGS) voted in May 2019 to treat the Anthropocene as a formal chrono-stratigraphic unit.

compounds dissolved in water by plants and the decomposition of dead biomass as source of gases, which led to a first understanding of matter cycles by early agricultural chemists (Knop 1868).

Today, we distinguish between four types of global cycles of material, which are interlinked together: the water (or hydrological) cycle, the sea salt cycle, the biogeochemical cycles (of carbon, nitrogen and sulfur),[126] and the chemical cycles of volcanic emanations (including rock weathering) and anthropogenic pollutants. The water cycle was well described already in ancient Greek by *Aristoteles* (Chapter 2.2.1). We can assume that cycling water was also recognized in ancient times before *Aristoteles* as the link between earth, heaven and water, controlled by gods. However, as described in Chapter 2.2.3, a physical understanding of evaporation, condensation (cloud formation) and formation of precipitation were not reached until the end of the nineteenth century, even today the aerosol-cloud interaction is not fully understood.

Understanding of cycling of other substances could not begin before the discovery of gases and particulate matter (dust) in the atmosphere. Any cycle starts (more or less by convention) with the emissions of a substance into the atmosphere. The emission of "foreign bodies" like fumes, smoke and dust was surely observed in ancient times and even their distribution and dissipation and removal by deposition, giving the simplest form of a "cycle" or "budget" (the link between deposition and emission remains open).

The emission of "bodies" like sands from North Africa, ashes from volcanoes, vapors, smoke and soot into the air was for the first time clearly expressed by *Boerhaave* in his "Elements of Chemistry", but also its deposition and circulating (Boerhaave 1735):

> For as the Earth, considered in its whole extent, receives every thing that falls out of the Air; so on the other hand, the Air receives every thing again from the Earth; and thus between these two elements, there happens, as it were, a perpetual revolution and distillation of all things. (p. 282) [. . .] odoriferous Spirits are at any time by Nature produced in Plants, all these are, certainly, at length contained in the Air alone. [. . .] these Spirits should afterwards return with the Water of the Air [. . .] (p. 282)
> All these Spirits of vegetation, pinguious Corpuscles of the Earth... dispersed into the aerial Chaos... return in their time, [. . .] and by thus circulating backwards and forwards, [. . .] being deposited for a short time, return into the Air again. (p. 283)

However, many years before, in 1676, *Edme Mariotte* (1620–1684) perceived the role of the atmosphere, though again without an experimental basis (cited after Aulie 1970):

126 These are the main cycles; additional there are some minor cycles like those of phosphorus and some trace metals. Cycling of hydrogen and oxygen is combined with other elements; all cycles are more or less interlinked.

> [. . .] these volatile salts, etc., are mixed in the air with aqueous vapors, etc., and fall again with the rain formed with these vapors onto the surface of the ground. There they penetrate together as far as then roots of plants, where they enter with some particles of soil [. . .]

In the nineteenth century, the atmosphere was still seen as the medium for dissipation and dilution of air pollutants. The German chemist *Clemens Alexander Winkler* (1838–1904), who developed the first desulfurization in factory in Saxony (Winkler 1880), wrote poetically on the self-cleaning capacity of the atmosphere and the final oceanic sink of all pollutants (Winkler 1896):

> Die chemische Fabrik da draußen, welche wir Natur nennen, arbeitet seit undenklicher Zeit rastlos Tag und Nacht und bleibt trotzdem blank und blitzsauber. Sie zeigt keinen häßlichen Kehrichtwinkel, keinen schmutzigen Schutthaufen und ihre Wasserläufe hält sie klar, ihren Dunstkreis neutral, hell, rauchfrei. Das hat aber seinen guten Grund und ist, so vermessen das klingen mag, eigentlich gar kein Kunststück, denn hinter der Erde steht als mächtige, unversiegbare Kraftquelle die Sonne. Vor dieser dreht sie sich im rastlosen Wirbel, ihr in jedem Augenblick eine neue Flanke darbietend. Hier wallt der Dampf auf, dort schlägt er sich wieder nieder, alles Fremdartige mit sich nehmend; da Riesenverdunstung, dort Riesenkondensation – ein einziger, großartiger, rings um die Erde laufender Destillationsprozeß! Und so badet sich denn diese unsere Erde fortwährend im eigenen Destillate und was dieses herunterwäscht aus ihrer Atmosphäre und von ihrem Felsenleibe, das fließt ohn Unterlaß dem mächtigen Sammelbecken des Meeres zu, um in seinen Tiefen unmerklich zu verschwinden. [The chemical factory out there, that we call Nature, works since time immemorial restless day and night, and stays anyway blank and sparkling clean. She shows no ugly garbage angle, no dirty heap of rubble, and her watercourses keep it clear, the orbit neutral, light and smokeless. But that has a good reason, as presumptuous as it may sound, actually not feat, because behind the Earth stands the Sun as powerful, inexhaustible source of power. Before this she turns in a restless vortex, presenting her in every moment a new flank. Here the steam flows, there he beats down again, taking everything foreign; here the giant evaporation and there the giant condensation – a unique, monumental distillation process, running around the Earth! And so, bathes our Earth constantly in their own distillates, and what these washes down from the atmosphere and from her rock body flows without cease to the mightful melting pot of the ocean to disappear in his depths].

Likely the first artificial emission estimate and the conclusion on its controlling was made by *Eberhard August Wilhelm von Zimmermann* (1743–1815), German geographer, who visited the Oker steel works near Goslar. He presents a report of an officer[127] of the factory who carried out experiments to learn how much of the metals escape into the atmosphere. The balance he determined by weighting the ores and coal giving into the furnace and the gaining metals (lead and zinc) and remaining ashes and clinkers in a work shift: he found that from (recalculated from old pounds) 8513 kg ores and coal, 5964 kg metals and ashes remained; thus 2544 kg of "minerals" escape into air, and extrapolated to all metal works around Goslar, some 40.000 centner yearly

127 "*Hr. Zehntgegenschreiber*" *Volkmar* [tenth counter clerk Mr. Volkmar].

(Zimmermann 1775, p. 50–53). *Zimmermann* emphasized that only an improvement of the furnaces can reduce this "loss" and concludes:

> Wie erstaunlich muß die Atmosphäre mit solchen mineralischen Theilen ausgefüllet seyn. Alles dies aber kehrt durch den Niederschlag zur Erde zurück, und allein daraus können sich hinreichend neue Mineralien erzeugen, und erzeugen sich gewiß [How amazing the atmosphere must be filled with such mineral parts, and alone out of it could be generated adequately new minerals, and certainly produce].

However, it was unknown at that time that apart from dust emission, gases such as CO_2 and SO_2 were the mainly emitted compounds. With the step-wise understanding of plant growth, the carbon (CO_2) and nitrogen uptake and release was recognized and the science of agriculture chemistry arose. It is worth to mention that the nineteenth century is widely characterized through an enthusiastic accentuation of the philosophy of nature: everything makes sense in nature. With the increasing scientific discoveries and insights in natural processes, a materialistic view arose. This is wonderful expressed by *Ludwig Büchner* (1824–1899) in his book "Kraft and Licht" (English edition "Force and Matter": Büchner 1884, p. 30):

> [. . .] no motion in Nature arises from nothing nor passes into nothing, and as it the material universe each individual form can only realize its existence by drawing upon a vast and never changing store of matter, so each motion forms the basis of its existence out of an incommensurable, never changing store of force of energy sooner or later in some fashion or other to the Whole.

In the first German edition (Büchner 1855, p. 18), the word "*Kreislauf*" was used in connection with the motion in nature. *Jacob Moleschott* (1822–1893)[128] wrote: "*Kreislauf des Lebens*" (Moleschott 1852) as a criticism of *Liebig*'s mineral theory, where the sixth chapter is entitled "*Kreislauf des Stoffs*" [cycle of matter]. *Moleschott* writes (1852, p. 143):

> Stickstoff, Kohlenstoff, Wasserstoff und Sauerstoff sind zum Theil als freier Stickstoff und freier Sauerstoff, zum Theil als Kohlensäure, Wasser und Ammoniak in dem Dampfkreis unserer Erde vorhanden [Nitrogen, carbon, hydrogen and oxygen are partly present as free nitrogen and as free oxygen, and partly present as carbonic acid, water and ammonia in the atmosphere of our Earth].

And (Moleschott 1852, p. 82):

> Aber die Kohlensäure stammt vom athmenden Menschen und Tieren, von dem Holz und den Steinkohlen, die wir verbrennen. Die Pflanze führt den Kohlenstoff in den Kreis des Lebens zurück

---

128 *Jacob Moleschott* was a Dutch physicist and physiologist in Utrecht, Heidelberg and Rome. To my knowledge, there is no English edition (the title is found in literature where "*Kreislauf*" is translated into English by several terms: cycle, loop, and circuit), a French (1866) and Italian (1870) edition appeared.

[But the carbonic acid stems from breathing humans and animals, from the wood and the hard coals we burn. The plant transfers the carbon in the cycle of life back].

The Scottish agricultural chemist and mineralogist *James Finlay Weir Johnston* (1796–1855) called the circulation of matter, the transformation of dust and ashes, and particularly bodily dust and ashes, into new forms (Johnston 1853, p. 558):

In the face of this clear knowledge, how untrue to nature, how irrational, how misleading are the views which many conscientious divines have promulgated with regard to the resurrection of man. As if the same matter which forms our body, when we are laid in the grave, and which, after a brief residence there, makes its way, through some nutritive plant, into the body of another man, and forms part of his body still when he is buried – as if this matter, which is neither his nor mine, has already "been slave to thousands," and may be buried with ten thousand bodies more, before the resurrection comes – as if this matter were meant to form the clothing of the disembodied spirit, when, in visible form and sensible identity, it shall be raised on the day when "small and great" shall appear before the dread tribunal.

However, *Wilhelm Knop* (1817–1891)[129] is likely the first who defines a "cycle of matter" in his book "*Der Kreislauf des Stoff. Lehrbuch der Agricultur-Chemie*", very literary but correct (Knop 1868, p. 3):

Wir beobachten, wie ein Quantum des wägbaren Stoffs unseres Planeten, getrieben von Kräften, welche die Sonne in Licht und Wärme äussert, in stetem Kreislauf vom Mineralreich zum Planzenreich und durch dieses zum Thierreich aufsteigt, um mit dem Untergang der Körper beider Reiche der organisirten Welt desorganisiert und dem Mineralreich, den Gewässern und der beweglichen Atmospäre überliefert zu werden, und von da aus den Kreislauf von Neuem zu beginnen [We observe how a part of weighable matter of our planet, forced by powers, which are expressed by the Sun in light and heat, in a permanent cycle ascent from the mineral kingdom to the vegetable kingdom and through them to the animal kingdom for the purpose of being disorganized after decay of both kingdoms of the organized world, and transported into the mineral kingdom, the waters and the versatile atmosphere, and to start from there the cycle anew].

František (*Franz*) *Pošepný* (1836–1895), Czech geologist, clearly expressed the "salt cycle" and the formation of sea salt from sea water (Pošepný 1878); see also p. 202 with another citation:

Wir kommen also auch auf diesem Wege zu der Annahme einer anderen, unerschöpflichen, sich stets erneuernden Quelle der Chlorverbindungen. Nun kennen wir aber keine andere solche Quelle, welche so ansehnliche Quantitäten liefern könnte, als das Meer [. . .] Wird das Meerwasser in die feinsten Theilchen zerschlagen, und diese können ebenso gut fortgerissen werden, wie feinvertheilte feste Körper, von denen es bekannt ist, dass sie auf ansehnliche Distanzen von der

129 *Johann August Ludwig Wilhelm Knop*, German professor for agricultural chemistry in Leipzig (1861–1882). Studied in Göttingen and was assistant of *Wöhler* (1841–1843), of *Leopold Gmelin* in Heidelberg (1844) and of *Erdmann* in Leipzig (1845–1847). Teacher at "*Handelslehranstalt*" [Commercial College] in Leipzig (1847–1856) and successor of *Wolff* as director of the agriculture experimental station (1856–1866). He wrote several books on fertilizing and agricultural chemistry.

> Atmosphäre transportirt werden [. . .] Durch die Annahme eines salzhaltigen atmosphärischen Niederschlages erklären sich manche Erscheinungen [So we come in this way to the assumption of another, always renewing source of compounds of chlorine. But we do not know another one source, which can deliver such respectable quantities then the sea [. . .] will that sea water smashed into fine dispersed particles, from which it is known that they will be transported by the atmosphere over large distances, so it will explain some phenomena] (Pošepný 1878, pp. 194–195).
> In offenen Becken werden die atmosphärischen Niederschläge sammt den in ihnen ursprünglich vorhanden gewesenen, sowie den auf ihrer Circulation an der Oberfläche und durch das Gestein nachträglich aufgenommenen und vielfach veränderten Substanzen wieder ins Meer zurück [in open basins the atmospheric precipitations, together with those substances originally being present in them, and those subsequent incorporated through the circulation from the surface and the rocks, and multiple changed, move back to the sea] (Pošepný 1878, p. 199).

Furthermore, Müntz (1891) has pointed out that without this circulation of salt, and its replenishment of the land, the latter would soon be drained of its chlorides, and its replenishment of the land, the latter would soon be drained of its chlorides.

French agriculture chemist *Jean-Baptiste Boussingault* (1802–1887)[130] identified nitrogen as an important substance for plants. In connection with the growth of plants, it was long time disputed where the nitrogen comes from. *Liebig* believed that plants only need a limited number of nutrients to grow. These nutrients are present in the soil and, having been used up following intensive cultivation, need to be added to the soil again in order to guarantee the next crop. He was, therefore, the founder of the artificial fertilizer industry. However, confusion about where the necessary nitrogen came from remained. French chemist *Jules Reiset* (1818–1896) recognized in 1856 that decaying organic matter releases nitrogen. Two decades later, the processes of nitrification, biological nitrogen fixation and denitrification were discovered. The process of nitrification was discovered in 1877 by *Theophile Schloesing* (1824–1919) and *Achille Müntz* (1846–1917). So, by the end of the nineteenth century, the basic building blocks of the nitrogen cycle had been discovered (Table 1.1). *Joseph König* (1834–1930), a German agriculture chemist and head of the experimental station in Münster, likely wrote the first short monograph on the nitrogen cycle: "*Der Kreislauf des Stickstoffs und seine Bedeutung für die Landwirthschaft*" [The cycle of nitrogen and its importance for agriculture] (König 1878). *Paul Ehrenberg* (1907),[131] another German agricultural chemist in Breslau wrote (habilitation dissertation) "*Die Bewegung des Ammoniakstickstoffs in der Natur*" [the movement of nitrogen in nature] (Ehrenberg 1907). With the discovery of main microbial redox processes at the end of the nineteenth and the beginning of twentieth century, the carbon, nitrogen and sulfur cycle were roughly understood,

---

130 *Jean Baptiste Joseph Dieudonne Boussingault*, French agricultural chemist, was the author of "*Traite d'economie rurale*" (1844), which was remodeled as "*Agronomie, chimie agricole, et physiologie*" (5 vols., 1860–1874; 2nd ed., 1884), translated into many languages, e. g., as "rural economy".
131 Not familiar with *Christian Gottlieb Ehrenberg* (1795–1867).

Table 1.1: Milestones in describing the nitrogen cycle (after Galloway et al. 2013).

year	discoverer	work
1836	Jean-Baptiste Boussingault	nitrogen as a nutrient for plants
1838	Jean-Baptiste Boussingault	legumes could fix their nitrogen (but he did not know how)
1840	Justus von Liebig	advocated the addition of certain nutrients to the soil for plant growth, making him the founder of the artificial fertilizer industry
1843	John Bennet Lawes and Joseph Henry Gilbert	confirmed that nitrogen helps plants grow and that nitrogen comes from sources other than precipitation
1856	Jules Reiset	recognized that decaying matter releases nitrogen, providing the basis for the nitrogen cycle
1877	Theophile Schloesing and Achille Müntz	discover of the process of nitrification
1880	Hermann Hellriegel[a] and Hermann Wilfarth[b]	discovered the process of biological nitrogen fixation

[a](1831–1895), German agriculture chemist; assistant of *Stöckhardt* at Tharandt (1851–1856), head of the agricultural research station Dahme (1857–1873). In 1873, he moved to Beerenburg and founded there in 1882 an agricultural research station.
[b](1853–1904), German agriculture chemist; 1889–1882 assistant of *Fittbogen* at the Dahme station and then he followed *Hellriegel* to Bernburg where he became his successor in 1895.

as described for example by the German physiologist *Amandus Hahn* (1889–1952), Hahn (1941). Quantification of the cycles (fluxes, concentrations, residence times), however, began only after 1950 with extended measurements. An understanding of the cycling of volcanic and anthropogenic emissions, however, was only possible after understanding biogeochemical cycling and tectonic processes in the late twentieth century.

The first notion of a sulfur cycle is apparently in the work of Kossovich (1913);[132] who describes the sulfur cycle on a regional scale. Soon later described as the "story of sulfur" by *Waldemar Lindgren* (1860–1939), Swedish-American geologist (Lindgren 1923), and elaborated in books on geochemistry like Rankama and Sahama (1950). *Lindgren* has well emphasized the fact that in nature sulfur is continuously in movement, changing from sulfide to sulfate, and back from sulfate to sulfide. He writes that all active volcanoes give up enormous quantities of hydrogen sulfide (it was not yet known that volcanoes emit almost SO_2):

> [...] some of which is oxidized to form sulphur dioxide and then the trioxide. The resulting sulphuric acid descends to the earth with rain to form sulphates by reacting with basic rocks. A part of hydrogen sulphide is reduced to native sulphur [...] In the average river waters, sulphates are present in relatively large amounts, and enormous quantities are discharged into the ocean, so

132 *Peter Samjonovich Kossovich* [in German transcription also written *Kossowitsch*] (1862–1915); see p. 205.

that the sulphates would predominate over the chlorides in the waters of the sea were it not for the continuous reduction of sulphates to sulphides [. . .] (cited after Mellor, X, p. 9).

The first sulfur bacteria were discovered at the ending nineteenth century; the Swiss bacteriologist *Max Düggeli* (1878–1941) published the first review and gives a large reference list (Düggeli 1919).[133]

The motivation, to establish networks after 1950 for air constituents was no longer exclusively from agriculture chemistry but now from air pollution by sulfur and nitrogen oxides. Eriksson (1959, 1960), Kellog et al. (1972), Søderlund and Svensson (1976), and other published the first quantified cycles of sulfur and nitrogen, which were improved by gaining more and more data in the following 20 years. But these western authors did not know about the famous paper by Kossovich (1913) on the cycle of sulfur and chlorine. In the late 1980s, chemistry-transport modeling (CTM) became a tool for describing global cycles, and with providing emissions, highly resolved in time and space, back to 1750, and including future scenarios, modeled global cycles are anew *en vogue*. Here are referred further publications (before the year 2003) on cycling matter: on the nitrogen cycle (Delwiche 1970, Jaffe 1992, Möller 1996), the sulfur cycle (Eriksson 1963, Georgii 1970, Friend 1973, Granat et al. 1976, Rodhe 1978, Möller 1982, 1985, 1996, Ivanov and Freney 1983, Ryaboshapko 1983, Pham et al. 1995, Brimblecombe 2003), the carbon cycle (Bolin 1970, Scurlock and Hall 1991, Holmén 1992), the phosphorus cycle (Pierrou 1976), the oxygen cycle (Cloud and Gibor 1970), and the chlorine cycle (Graedel and Keene 1990, Möller 1990), and in general (Boresch 1931, Kovda 1976).[134]

---

133 Pioneers in discovering sulfur bacteria were (among others) the Dutch microbiologist and botanist *Martinus Willem Beijerinck* (1851–1937) and the Russian scientists *Sergej Nikolaewitsch Winogradsky* [Сергей Николаевич Виноградский] (1856–1953) who was a microbiologist who immigrated 1917 and worked since 1922 in Paris at the Institute Pasteur, *Efrem Moisseevich Brusilovsky* [Ефрем Моисеевич Брусиловский] (1854–1933) who was a physician and balneologist in Odessa, *Nikolai Dmitrievich Selinsky* (Николай Дмитриевич Зелинский] (1861–1953) who was a chemist in Moscow, and *Vasili Vasilievich Zavyalov* [Василий Васильевич Завьялов] (1873–1930) who was physiologist at Odessa. Additional I like to refer *Alexander Nathanson* (1878–1940) who was a German biologist and professor at the University Leipzig and who made his habilitation 1902 upon a new group of sulfur bacteria in the age of 24 years.

134 Biographic information on authors who are not further cited in this volume: *Karl Boresch* (1886–1947) was a German botanist. *Constantin Collin Delwiche* (1919–2001) was a US soil biogeochemist. *Aharon Gibor* (1925–2015) was US biologist. *Preston Ercelle Cloud* (1912–1991) was US paleontologist. *Allen L. Lazrus* (born 1933) was the director of the NCAR sampling program. *Edward Ambrose Martell* (1918–1995) was a radiochemist at NCAR. *Alexey Grigorjevich Ryaboshapko* (1942–2021) Russian chemist in air pollution control in Moscow at the institute for global climate and ecology [*Институт глобального климата и экологии имени академика Ю. А. Израэля*]. *James P. Friend* (born likely 1933), US chemist. *William Welch Kellog* (1917–2007) was an American meteorologist and climatologist. *David Hall* (1935–1999) was Professor of Biology, King's College, London. *Kim Holmén* is meteorologist at MISU, Stockholm. *William C. Keene* is US environmental chemist. *Vik-*

1.3 Chemistry: a terminological and historical approach

In the *Encyclopaedia Britannica* published in Edinburgh in 1771 (shortly before the discovery of the chemical composition of air), chemistry is defined as (see also *Scheele*'s definition in footnote 697 on p. 350) "to separate the different substances that enter into the composition of bodies [analytical chemistry in modern terms]; to examine each of them apart; to discover their properties and relations [physical chemistry in modern terms]; to decompose those very substances, if possible; to compare them together, and combine them with others; to reunite them again into one body, so as to reproduce the original compound with all its properties; or even to produce new compounds that never existed among the works of nature, from mixtures of other matters differently combined [synthesis chemistry in modern terms]". This definition further evolved until, in 1947, it came to mean the science of substances: their structure, their properties, and the reactions that change them into other substances. A characterization accepted by *Linus Pauling* (1901–1994) in his book "*General Chemistry*" (Dover Publications 1947), revolutionizing the teaching of chemistry by presenting it in terms of unifying principles instead of as a body of unrelated facts. However, *Wilhelm Ostwald* (1853–1932) already used such principles of generalizing in his book "*Prinzipien der Chemie*" (Leipzig 1907), subdividing the chemistry into chapters of state of matter and properties of bodies, phase equilibrium, solution and ions, chemical processes and reaction rate.

As a short definition, chemistry is the scientific study of matter, its properties and interactions with other matter and with energy. Consequently, inorganic and organic chemistry is the science of matter, physical and theoretical chemistry is the science of properties and interactions, and analytical chemistry is the science studying the composition and structure of bodies. As we can see, all disciplines overlap. The original focus of *geology* as the science of the (solid) earth, *hydrology* as the science of liquid water, and *meteorology* as the science of the atmosphere, is still valid and should not be diffused. However, the modern study of earth sciences looks at the planet as a large, complex network of physical, chemical, and biological interactions. This is known as a systems approach to the study of the global environment. The chemistry of the earth system (global environment) – when not considering life – is *geochemistry*.

Much of early chemistry was analytical chemistry, since the questions of which elements and chemicals are present in the world around us and what their fundamental nature is, are very much in the realm of analytical chemistry (Szabadváry 1966).

---

tor Abramovich Kovda (1917–1991) [Ви́ктор Абра́мович Ко́вда] was a Russian soil scientist. On *Pierrou* is only known that he worked at that time in the Department of Limnology University of Uppsala.

Before 1800, the German term for analytical chemistry[135] was "*Scheidekunst*" [separation craft]; in Dutch, chemistry is still generally termed "*scheikunde*". Before developing reagents to identify substances by specific reactions, simple knowledge of the features of the chemicals (smell, color, flavor, crystalline structure, etc.) was used to "identify" substances. With *Lavoisier*'s modern terminology of substances (1789) and his law of the conservation of mass, chemists acquired the basis for chemical analysis (and synthesis).

The German analytical chemist *Carl Remigius Fresenius* (1818–1897) wrote the first textbook[136] on analytical chemistry (1841 on qualitative, and 1847 on quantitative analysis) which is still generally valid, rewritten by *Gerhard Jander* (1892–1961)

---

135 In Vol. 1 (p. 142) of the *Encyclopædia Britannica* (1871) the following definition is given: Analysis, in chemistry, the reducing of a heterogeneous or mixt body, into its original principles or component parts.

136 The first textbook on this issue was written by *Johann Friedrich Gmelin* (1848–1804), father of *Leopold Gmelin*, "*Chemische Grundsätze der Probier- und Schmelzkunst*" [Chemical principles of assaying and the art of melting] (Halle 1786) and soon later by *Johann Friedrich August Göttling* (1753–1809) entitled "*Vollständiges chemisches Probekabinett*" [Complete chemical sample cabinet] (Jena 1790), published in a second edition already with a more scientific title "*Praktische Anleitung zur prüfenden und zerlegenden Chemie*" [Practical instruction for testing and separating chemistry] (Jena 1802). Furthermore: *Nicolas Louis Vauquelin* (1763–1826) "*Manuel de l'essayeur*" (Paris 1799), translated into German "*Handbuch der Probierkunst*" [handbook of assaying] (Königsberg 1800). *Wilhelm August Lampadius* (1772–1842) declared in his "*Handbuch zur chemischen Analyse der Mineralkörper*" [Handbook for chemical analysis of mineral bodies] (Freyberg 1801) the analytical chemistry as an independent discipline; he writes in the preface that the "*Analysis der genannten Körper*" [analysis of the mentioned bodies] should be named "*vorzugsweise analytische Chemie*" [preferably analytical chemistry]. *Johann Heinrich Kopp* (1777–1858), father of *Hermann Kopp*, wrote "*Grundriß der chemischen Analyse mineralogischer Körper*" [Fundamentals of chemical analysis of mineral bodies] (Frankfurt/M. 1805) and *Johann Friedrich John* (1772–1847) "*Chemisches Laboratorium, oder Anwendung der chemischen Analyse der Naturalien. Mit einer Vorrede von M. H. Klaproth*" (Berlin 1808). *Thomas Balthasar Fabricius* (1774–1851) "*Anleitung zur chemischen Analyse unorganischer Naturkörper*" [Treatise to analyse inorganic natural bodies] (Kiel 1810). From the famous "*Traité de Chimie élémentaire théorique et pratique*" (in 4 Volumes) by *Louis Jacques Thénard* (1777–1857), Vol. 4 ("*Suivi d'un Essai sur la philosophie chimique et d'un Précis sur l'analyse*", Paris 1816) appear as "*Anleitung zur chemischen Analyse dem gegenwärtigen Zustand der Wissenschaft gemäss; nach L. J. Thénard's Handbuch der theoretischen und praktischen Chemie; aus dem Franz. übersetzt und mit Anmerkungen begleitet von Johann Bartholmäus Trommsdorff*" (Erfurt 1817). All these books were focused on analysis of mineral matter. The first more general textbook on analytical chemistry was written by *Christian Heinrich Hermbstädt* (1753–1852) "*Handbuch der analytischen Chemie*" [Handbook of analytical chemistry] in two volumes (Altona 1821 and 1822) and soon later "*Handbuch der analytischen Chemie*" in two volumes (Berlin 1829; in 6 edns. until 1867) by *Heinrich Rose* (1795–1864) who first describes the separation process ("*Trennungsgang*" in German), but very circumstantial; hence this book by *Rose* was the reason that *Fresenius* wrote a more systematic approach of the separation process. *Johann Andreas Buchner* (1783–1852), German pharmacologist, best comments the fast development in chemistry at that time (Buchner 1826, p. 27) that "the chemical textbooks until beginning of the nineteenth century have mostly only a historic value". The books of *Rose* and *Fresenius* were translated into English. In Britain, the first book on analytical chemistry was written by *James Sheridan Muspratt* (1821–1871), an Irish-born chemist and

and *Ewald Blasius* (1921–1987), normally only termed "Jander-Blasius", is still used at German universities.[137]

Notwithstanding, the history of chemistry is to a large extent the history of atmospheric chemistry. The history of chemistry may be divided into periods, listed following (Bauer 1914, 1921);[138] the era since *Lavoisier* is also named "era of quantitative research" (Kopp 1931):

- chemistry of the ancients (until fours century AD),
- era of alchemy (from fours to sixteenth century),
- era of iatrochemistry (sixteenth and seventeenth century),
- era of phlogiston theory (1700–1774),
- era of *Lavoisier* (1774–1828),
- modern era of chemistry (from 1828 until present).

1.3.1 On the origin of the word "chemistry"

There are several ideas published on the origin of the term "chemistry" since *Hermann Kopp* (1817–1892) wrote his "history of chemistry" between 1842 and 1847. Until now, the etymology of the word "chemistry" is still unsolved (Pommerening 2018),[139] but an etymological origin from Egypt is likely. The word "Chemia" is first used by *Plutarch* (45–127) in his "*Isis et Osisis*" in connection with Egypt:

> Præterea Ægyptum,quæ vel maxime nigram habet terram, tanquamnigram oculi partem, *Chemia* vocant [Furthermore Egypt, where even the earth is so dark like black olives, is called *Chemia*], cited from Gren (1800, p. 4).

teacher, "Chemistry, theoretical, practical and analytical as applied and relating to the arts and manufactures" (in 2 vol. 1858–1860). The German-born chemist *Wilhelm* (*William*) *Dittmar* (1833–1899) published "A manual of qualitative and quantitative chemical analysis" (Manchester 1874) and "Analytical chemistry. A series of laboratory exercise" (London and Edinburgh 1879). *William E. Pink* and *George E. Webster* wrote "A course of analytical chemistry, qualitative and quantitative" (London 1874). The first American (very small) textbook is by *Henry Trimble* (1858–1898) "Practical and analytical chemistry: a complete course in chemical analysis" (Philadelphia 1885).

137 Jander, G. and E. Blasius (1949–1990) Einführung in das anorganisch-chemische Praktikum (Einschließlich der quantitativen Analyse) [Introduction into inorganic chemical practicum (Including quantitative analysis)]. Hirzel, Leipzig and Stuttgart (1st–13st edns). Jander, G. and E. Blasius (1951–2006) Lehrbuch der analytischen und präparativen anorganischen Chemie (Mit Ausnahme der quantitativen Analyse) [Textbook of analytical and preparative inorganic chemistry (With the exception of quantitative analysis)]. Hirzel, Leipzig and Stuttgardt (1st–16th eds).

138 *Karl Hugo Friedrich Bauer* (1874–1944) was a German chemist in Leipzig.

139 *Tanja Pommerening* (2018) Institut für Altertumswissenschaften (Institute for Classical Studies), University Mainz, pers. comm. Present Affiliation: Philipps University Marburg, Institut für Geschichte der Pharmazie und Medizin.

The term "*Khem*" appears in different relations with ancient Egypt. Letopolis (Greek: *Λητοῦς Πόλις*) was an ancient Egyptian city, the capital of the second Nome of Lower Egypt; its Egyptian name was *Khem* (*Ḫm*); Lichtheim (1980, p. 84). Akhmim (or Achmim) is a city in Upper Egypt located on the east bank of the Nile, referred to by the ancient Greeks as Khemmis, Chemmis and by the Coptics Chemin. In ancient Egypt, this city was the capital of the province Chemmite. The name of this city is derived from Egypt "*ḫnt-mnw*" (*Chenet-Min*); the Egyptian god *Min*, however, according to newer interpretations (Pommerening 2018), is not pronounced as "chem" as proposed by Egyptologist such *Francis Llewellyn Griffith* (1862–1934) in the nineteenth century.

In the Psalms, Egypt is termed "Terra Chami" (the land of Ham); Ham (Greek *Kham*) is the youngest son of Noah. *Ham* was a sun of god; Egypt "*ḫm*" means "to be hot" (Pommerening 2018). It is without any doubt that the Sun was associated with light and heat. Since the seventeenth century, several suggestions were made that relate to the word *Ham* is believed to come from the word "*Khawm*" (and *chamam*) which means "black, hot, and burnt" in Hebrew. Smith (1882) means that the Hebrew word חמם (pronunciation *khema*, or *chom* according to Modern Hebrew) means "heat" (in modern dictionaries it also means "brown", pronounced *chum*).

In all probability, the first use of chemistry was for obtaining metals from their ores. Hence, it is likely that the Egyptians used one word to designate the strange phenomena produced in substances using heat (e. g., burning, extraction, distillation, etc.). From that, it can be derived that *χημία* was the "Egyptian art", for more than 2000 years celebrated by the temple priest. Another explanation given by Lockemann (1950) that the root of word "black" led to a "black art" (alchemy) is not likely because behind "black art" always religious ceremonies were understood and not chemical operations. Nevertheless, there is consensus that *χημία* was the ancient name for Egypt, meaning "black earth".

Before the New-Platonic, the word *Chemia* is never mentioned in connection with chemical operations, neither by the Greek nor by the Romans. The word "chemistry" is only known since the fourth century in sense of a "discipline" dealing with substances and their transmutation namely the generation of gold and silver from ignoble metals. *Julius Maternicus Firmicus* (lived under Constantine the Great about 340 AD) for the first time used the word "Alchemiæ" appeared (cited from Gren 1800, p. 4):[140]

> Si fuerit hæc domus Mercurii, dabit Astronomiam; si Veneris, cantilenas et lætitiam; si Martis, opus armorum et instrumentorum; si Jovis, divinum cultum, scientiamque in lege; si Saturni, Scientam Alchemiæ.

140 *Friedrich Albrecht Carl Gren* (1760–1798) was a German chemist and author of "*Systematische Handbuch der Gesammten Chemie*" (1787–1790) in 3 volumes. According to Schorlemmer (1882) *Gren* stated that other editions of this work have also "scientia alchimiae". The manuscript in the library of the Vatican has "chymiæ" and not "alchymiæ" (Schorlemmer 1882).

In fragments of the alchemist and Hermetic philosopher, *Zosimos of Panopolis* (460–320), referring back to the Book of Enoch the word *χημεõ* "chemeu" (Zosimus Book of Chemeu) is frequently used for the work of metalworker *χυμδύτης* (metal casters); Hermann (1954), Fraser (2004).

For long time, for alchemy the meaning *χρυσοποïα* (the art of making gold) were used (Thomson 1830, Meyer 1914). Without doubt, the Arabians prefixed their article to the Greek *χημία* into the Arabic *al-kīmīā*. Thus, the terms Alchemy, Alchimy or Alchymy were introduced.[141] *Olaus Borrichius* (Latin) or *Ole Borch* (1626–1690), a Danish naturalist, likely was the first who used the term *Chemiæ* in a modern sense in his dissertation because he quoted *Zosimus* (460–520), who said that the term *Chemia* was never used before to signify the science of nature.

1.3.2 On the chemical nomenclature and symbols

Different names of gases and substances being relevant to the atmosphere will be mentioned in this volume, given already by alchemists, than by the discoverers and later by following researchers (Tables 1.2 to 1.4). Old literature (before the middle of the nineteenth century) has used terms, which are today accurately defined in atmospheric sciences, synonymously but also with a different sense. Moreover, it is difficult when reading old literature because the same phenomenon might be described by many different terms. Before the middle of the seventeenth century, almost all scientific books were written in Latin, and Latin terms for phenomena mostly originate from Old Greek. Only since the seventeenth century with an increasing number of scientific books in modern languages (German, French, and English), Latin terms were rendered differently, through existing words from the colloquial language, by different authors explaining the variety of terms. However, the decrease in the number of terms (i. e., exactly one term for one phenomenon) together with its accurate scientific definition did not begin stepwise before the middle of the nineteenth century.

The knowledge of the chemical constitution of substances was very limited, some compounds were confused and some terms were used for different substances. *Antoine Lavoisier*, who revolutionized the science of chemistry in the eighteenth century and replaced the mythical "phlogiston" with the term (and concept) of oxygen, clearly understood the importance of accurate definitions. In his words: "We cannot improve the language of any science without at the same time improving the science itself; nor can we, on the other hand, improve a science without improving the language or

141 There is no consensus in the literature whether different spelling is due to various translators (among Greek, Arabian and Latin) or through different original Greek root term like χημία, χημεία or χυμία (Kopp 1931, 1869, Schorlemmer 1882); in Latin: Chimia, Chymia.

Table 1.2: New names according to Lavoisier's New Nomenclature of "simple substances or like have not hitherto been decompounded" (after Mitchill's nomenclature of the new chemistry), cited from Duveen and Klickstein 1954, p. 292)[a]. See also Tables 1.3 and 1.4. Note that in the 1790s often the element (as considered in compounds) was differentiated from that in pure gaseous form, for example gaz hydrogène – Wasserstoffgas, gaz oxygéne – Sauerstoffgas.

French[c]	**German**	**English**
Lumière	Lichtstoff	Light[d]
Calorique	Wärmestoff	Caloric[d, e]
Oxygéne	Sauerstoff	Oxygen
Azote	Stickstoff[b]	Nitrogen[f]
Hydrogène	Wasserstoff	Hydrogen
Carbone	Kohlenstoff	Carbon
Soufre	Schwefel	Sulphur
Phosphorous	Phosphor	Phosphorus

[a]The first attempt to translate the French nomenclature into German was presented in the "*Taschenbuch für Scheidekünstler und Apotheker*" (11. Jahrgang, 1790, pp. 147–155) and is entitled "*Alphabetisches Verzeichnis der neuen französischen Nomenklatur*". The second effort to adopt the new terms to the German language was by *Christoph Girtanner* (1760–1800) in "*Neue chemische Nomenklatur für die deutsche Sprache*" (Berlin 1791). *Johann Andreas Scherer* published "*Versuch einer neuen Nomenclatur für Deutsche Chymisten*" (Wien 1792). *Samuel Latham Mitchill* (1764–1831), American naturalist and chemist, notes "the German words, derived from French, but they are in no way more appropriate for general use than the older ones, and some terms even longer and more cumbersome" (cited from Duveen and Klickstein 1954).
[b]Or "*Salpeterstoffgas*" (*Girtanner*) from "*gaz azote*" and "*Salpeterluft*" (in *Mitchill*). Scherer (1792, p. 17–18) proposed for the first "*Stickstoff*" and "*Stickstoffgas*".
[c]In the French edition *Radical muriatique* (chlorine Cl) have not been taken into other languages. Berzelius (1813b) named it muriatic radical.
[d]Light (*Lichtstoff, lumière*) and Caloric (*Wärmestoff, calorique*) was still assumed to be matter.
[e] Matter of heat (old terms: igneous fluid, matter of fire, element of heat).
[f]Berzelius (1813a) termed it nitric radical.

nomenclature" (Lavoisier 1789). From 1782 onwards *Lavoisier*, *Claude-Louis Berthollet* (1748–1821)[142] and *Antoine François Comte de Fourcroy* (1755–1809) collaborated with *Louis Bernard Guyton de Morveau* (1737–1816) who first proposed a table of chemical Nomenclature, in devising a new system (*Méthode de Nomenclatura chemique*),[143] read to the French Academy on 18 April 1787.[144] *Jean Henri Hassenfratz* (1755–1827) and *Pierre August Adet* (1763–1832) appended to this publication a system of symbolism to

142 Savoyard-French chemist; he was one of the first chemists to recognize the characteristics of a reverse reaction, and hence, chemical equilibrium; along with *Antoine Lavoisier* and others, he devised a chemical nomenclature.

143 J. Phys. (1782) 19, 310 and Ann. Chim. Phys 1798, 1, 24.

144 Methodé de Nomenclature Chemique. Chez Cuchet, Libraire, Paris, 1787, 314 pp. At least seven French issues and editions, one English, two Germans, one Spain and one Italian appeared, all of the full work (Duveen and Klickstein 1954). German translation by Karl von Meidinger (1793) *Methode*

Table 1.3: Historic terms of gases (airs, kinds of airs, vapors) in English, German, French and Latin; note instead of air (Latin aer or āër and aër) also gas (French gas) and instead of Luft also Gas is used (in early alchemistic times "spirit" [in German Geist] were used as general term). After Macquer (1788), Gehler (1789), Stromeyer (1808)[a] and Coxe (1808)[b]. Note that the formulas (chemical composition of molecule) where almost unknown or uncertain before 1800.

formula	term
O_2	oxygen gas, good air, pure air, vital air, fire air, dephlogisticated air, empyreal air, factius air
	Oxygengas, dephlogistisirte[c] Luft, dephlogistisirtes Gas, Lebensluft, gute Luft, reine Luft, einathembare Luft, entbrennbare Luft, brennstoffleere Luft, Feuerluft, Lebensluft, künstliche Luft, reine künstliche Luft, wahre künstliche Luft, Empyrealluft, säurezeugendes Gas, säurendes Gas, Sauerluft, Sauerstoffgas, respirable Luft, Mayows luftiger Salpetergeist
	gaz oxygène, air déphlogistiqué, air defeu de Scheele, air vital, air pur
	gas oxygenium, aër dephlogisticatus, gas dephlogisticatum, aër igneous, aër vitalis, aër purissimus, aër verus factitius, spiritus nitro- aëreous Mayowill
H_2S	sulphuretted hydrogen gas, hepatic air, heptic air, inflammable sulphureous air, sulphuretted inflammable air,
	schwefelhaltiges Wasserstoffgas, schwefliges Wasserstoffgas, Schwefelgas, hepatische Luft, Leberluft, stinkende Schwefelluft, Schwefelleberluft, geschwefeltes Wasserstoffgas, gasförmiger sulphurisirter Wasserstoff, Hydrothiongas, hydrothionsaures Gas, Hydrothion, Schwefelwasserstoff, Schwefelsulfid, Schwefelwasserstoffsäure
	gaz hydrogène sulfuré, gaz hepatique, air hepatique, air puant du soufre
	gas hydrogenium sulphuratum, gas hepaticum, aër hepaticus, mephitis hepatica, acidum hydrothionicum
NH_3	ammonia gas, ammoniacal air, alkaline air, volatile air, spirit of hartshorn
	Ammoniacgas, gasförmiges Ammoniac, ammoniacalisches Gas, flüchtige alkalische Luft, laugenartige Luft, alkalische Luft, urinöse Luft, laugensalzige Luft, flüchtig-alkalische Luft
	gaz ammoniaque, gaz ammoniacal, gaz alcali-volatil, gas alcalin, esprit alcalinum volatile
	gas ammonium, gas ammoniacale, aër alcalinus, mephitis urinosa, gas alkalin volatil
PH_3	phosphuretted hydrogen gas, phosphorated hydrogen gas, phosphorized hydrogen gas, phosphoric inflammable air, phosphoric gas
	Phosphorwasserstoffgas, Phosphorluft, phosphorische Luft, gephosphortes Wasserstoffgas, phosphorisirtes Wasserstoffgas, phosphorhaltiges Wasserstoffgas, gasförmiges phosphorisirter Wasserstoff, entzündliches Phosphorgas, phosphorisch-hepatische Luft, Phosphorleberluft
	air phosphorique, gaz hydrogène phosphorisé
	gas phosphoricum, mephitis phosphorica, gas hydrogenium phosphorisatum

Table 1.3: (continued).

formula	term
HF[d]	fluoric acid gas, gaseous fluoric acid, fluoracid gas, acid of spar, acid of sparry, sparry acid gas
	fluʃssaures Gas, gasförmige Fluʃssäure, spathsaure Luft, spathsaures Gas, spathgesäuertes Gas, fluʃsspathsaure Luft, Fluʃsspathgas, luftige Fluʃsspathsäure
	gaz acide fluorique, air acide spathique
	gas acidum fluoricum, acidum fluoricum gas fluoris mineralis, gas acidum spathosum, aer acidus spathosus, mephitis fluoris mineralis
H_2[r]	hydrogen gas, air inflammable, inflammable air, phlogiston of Kirwan[e]
	Wasserstoffgas, wassererzeugendes Gas, wasserbildendes Gas, inflammable Luft, brennbare Luft, entzündbare Luft, entzündliche Luft, brennende Luft, Brennluft, gemeine brennende Luft, brennbares mephitisches Gas, Kirwans Phlogiston
	gaz hydrogène, gaz inflammabile, phlogistique de Kirwan
	aer inflammabilis, mephitis inflammabilis, cas carbonum, gas pingue (Helmont), gas inflammable, aer inflammabilis
CH_4	marsh gas, swamp gas, carburetted hydrogen
	Sumpfluft, schlechte Luft
	gaz palustre, gaz inflammable des marais
	gas pastu paludism virecta
VOC[f]	carburetted carbon gas, carbonaceous inflammable gas, hydro-carbonate gas
	Kohlenstoffwasserstoffgas, kohliges Wasserstoffgas, kohlehaltige endzündbare Luft, schwere brennbare Luft, ölbildendes Gas
	gaz hydrogèn carburé, gaz inflammable charbonneux, gaz olefiant, gaz inflammable mephitisé
	gas hydrogenium carbonium
CO_2	carbonic acid gas, fixed air, mephitic air, acid air, acid of aerial, calcareous gas, acid of chalk, acid of charcoal, gas sylvestre, air fixe, factious air, air of Hales
	kohlenstoffsaures Gas, gasförmige Kohlenstoffsäure, kohlengesäuertes Gas, Kohlensäure, fixe Luft, Luftsäure, luftsaures Gas, Sauerluft, Kalkgas[g], Kreidensäure, kreidensaures Gas, wildes Gas, wilder Geist, Gährungsluft, Weingas, weinigtes Gas, weinigter Schwaden, Mostgas, mineralischer Brunnengeist, elastisches Mineralgas, künstliche Luft, feste Luft, Kalksäure, Kalkspathsäure, mephitisches Gas
	gaz acide carbonique, acide méphitique, gaz méphitique, gaz acide crayeux, air factice, air solide de Hales
	gas acidum carbonicum, aër fixus, aër fixatus, aer factitius, gas äereum, aër mephiticum, gas calcareum, gas silvestre, spiritus sylvestris, gas vinosum, mephitis vinosa, gas calcareum, gas musti, aër fermentationis, acidula, acidum mephiticum, acidum aëreum, acidum atmosphaericum, acidum cretae

Table 1.3: (continued).

formula	term
CO	reduced fixed air reduzierte fixe Luft [h] [h]
HCl[k]	marine acid gas, gaseous muriatic air, marine acid air, spirit of salt salzsaures Gas, gasförmige Salzsäure, seesaure Luft, kochsalzsaure Luft, luftige Salzsäure, gemeines salzsaures Gas, Salzgas gaz acide muriatique, gaz acide marin, air acid marin gas acidum muriaticum, mephitis muriatica, gas muriaticum, aër muriaticus, aër acidus salinus, gas muriatosum Grenii[g]
Cl_2[k]	dephlogisticated marine acid (air), oxymuriatic air dephlogistisirte Salzsäure [h] [h]
CH_3COOH	acetic acid gas, acetous air, vegetable acid air essigsaure Luft, gasförmige Essigsäure, Essigluft, vegetabilisch-saure Luft gaz acide acétique, gaz acide acéteux gas acidum aceticum, acetosum, aër acidus vegetabilis, mephitis acetosa
SO_2[m]	sulphureous acid gas, gaseous sulphureous acid, (volatile) vitriolic acid air, phlogisticated acid of vitriol schweflichtsaures Gas, schwefelsäurichtes Gas, flüchtiges schwefelsaures Gas, schwefelsaures Gas, Schwefelluft, vitriolische Luft, vitriolsaures Gas, luftförmige phlogistisirte Vitriolsäure, unvollkommene Schwefelsäure in Dampfgestalt, luftförmige Schwefelsäure gaz acide sulfureux, gaz acide vitriolique, acide de soufre aëriforme gas acidum sulphurosum, gas acidum vitriolicum, gas acidum sulphureum volatile, aer acidus vitriolicus, acidum vitrioli phlogisticatum aëriforme, mephitis acida sulphuris, gas acido sulfureo
SO_3[m]	vitriolic acid air, acid of vitriol, spirit of vitriol (*vitriol* = sulfate), vitriolsaure Luft, vitriolsaures Gas, flüchtiges schwefelsaures Gas, luftförmige Schwefelsäure, luftförmige phlogistisirte Vitriolsäure, Schwefelluft air acide vitriolique spiritus vitrioli

Table 1.3: (continued).

formula	term
HNO_3 [n]	spirit of nitre, acid of nitre, dephlogisticated nitrous air, dephlogisticated nitre, nitric acid air, acid of saltpetre, septic (nitric) acid[s] Salpetergeist, Salpeterluft, Salpetersäure, Luftsäure, gemeine Salpeterluft, salpetersaure Luft, phlogistisirte Salpetersäure, Salpeterdämpfe acide nitrique, gaz acide-nitreux gas acidum nitrosum, acidum nitri phlogisticum, mephitis acida nitri, spiritus nitri (*nitre* = KNO_3 and/or $NaNO_3$, and *nitres* = nitrates).
HNO_2 [p]	acid of nitre, phlogisticated nitrous air, phlogisticated nitre, septous (nitrous) acid[s], septic acid gas[s] salpetersaures Gas, unvollkommene Salpetersäure in Dampfgestalt, dephlogistisirte Salpeterluft, dephlogistisirte Salpetersäure gaz nitreux oxygèné, gaz nitrique, oxide gaseux d'azote h
NO [n]	nitrous oxide, gaseous oxide of nitrogen, gaseous oxide of azote, gaseous oxide of septon, septic gas, nitrous gas, dephlogisticated nitrous gas salpeterartige Luft, salpeterhalbsaures Gas, nitröse Luft, oxydirter Salpeterstoff, salpetrige Luft gaz nitreux oxygèné, gaz nitrique, oxide gaseux d' azote gas acidum nitrosum, gas acido nitroso, mephitis acida nitri, acidum nitrosum, nitri phlogisticatum
NO_2 [n]	gaseous nitric oxide, nitrous gas, nitric air, septous (nitric) gas oxydiertes Salpeterstoffgas, gasförmiger oxydulierter Salpeterstoff, gasförmiges Salpeterstoffoxydule, oxydirtes Stickstoffgas, oxydirtes Stickgas, dephlogistisirte Salpeterluft, gasförmige azotische Halbsäure, sauerstoffhaltiges Stickgas, salpetersaure Luft gaz nitreux, air acide-nitreux gas oxydum nitrogenii, gas nitrosum, aër nitrous, mephitis nitri phlogistica
N_2O	phlogisticated nitrous air phlogistirte salpetrige Luft air nitreux complex h

Table 1.3: (continued).

formula	term
N_2	mephitic air, impure air, vitiated air, phlogisticated air, inflammable air, septon, azotic gas
	phlogistisirte (oder phlogistische) Luft, verdorbene Luft, unreine Luft, Stickluft, Stickgas, Salpeterstoff, Salpeterstoffgas, azotisches Gas, Stickstoff, tödtliches Gas
	gaz ou air phlogistiqué, gaz azotique, azote
	aër phlogislicatus, aër vitiatus, mephitis aëris phlogistica, gas phlogislicatum, gas azoticum, azoticum

[a]*Friedrich Stromeyer* (1776–1835) was a German chemist in Göttingen.
[b]*John Redman Coxe* (1773–1864), Scholar, collector, writer and teacher of "materia medica" in Philadelphia, USA.
[c]Both spellings were used: (*de*)*phlogistirt* and (*de*)*phlogistisirt*. In modern German it is (false) written as (*de*)*phlogistisiert*.
[d]Discovered by *Scheele* in 1773; *Priestly* observed that fluoric acid gas (HF) corrodes and penetrates common glass.
[e]*Richard Kirwan* (1733–1812) was an Irish chemist, adherent of phlogiston theory, as described in his 1787 *Essay on Phlogiston and the Constitution of Acids*, wherein he identified phlogiston with hydrogen; by 1791, however, he seems to have abandoned phlogiston theory in favor of *Antoine Lavoisier*'s caloric theory.
[f]Under this substance hydrocarbons (generally organic substances) in modern terms are meant.
[g]German spelling of "*Kalk*" before 1800: "*Kalch*", e. g., "*Kalcherde*" = quicklime (CaO).
[h]Unknown in time when Latin was still used for scientific publications.
[k]Hydrochloric acid (HCl) was discovered by the alchemist *Jabir Ibn Hayyan* around the year 800 AD. *Scheele* discovered in 1774 dephlogisticated muriatic acid (chlorine, later renamed into oxygenated muriatic acid and simply oxymuriatic acid by *Kirwan*). *Humphry Davy* (1778–1829) recognized oxymuriatic acid to be an element and gave the name chlorine, derived from the Greek χλωρος (chlōros), meaning green-yellow. HOCl was discovered by the French chemist *Antoine Jérôme Balard* (1802–1876) in 1834 who gave it the name hypochlorous acid; however, the bleaching properties of solutions of oxymuriatic acid in water were recognized already in the late eighteenth century by *Berthollet* and others.
[m]SO_2 (H_2SO_3) and SO_3 (H_2SO_4) have not been separated before 1800 (the gas obtained by burning of sulfur and that from concentrated sulfuric acid – oleum). Fourcroy (1790, p. 380) writes "Sulphuric acid, when heated in a retort, soon loses part of its water, a very odorous and penetrating gas is disengaged [...] the sulphureous acid gas". From oleum SO_3 escapes (forming again H_2SO_4 in air) but sulfurous acid gas is that gained from burning sulfur (SO_2).
[n]HNO_3, NO and NO_2 often confused (the constitution of NO_2 was known already around 1790). *Bergman* distinguished between two states of the acid of nitre: dephlogisticated (nitric acid) and phlogisticated (nitrous acid).
[p]It remains unclear whether HNO_2 was separated from HNO_3 before 1800.
[r]Often confused with CH_4 as inflammable air.
[s]Nomenclature introduced by *Samuel Latham Mitchill*: septon = nitrogen, septic acid = nitric acid, septous acid = nitrous acid, septate = nitrate, septite = nitrite, septic gas = nitrous gas, not defined to be NO and/or NO_x, septous gas = azotic gas, i. e., nitrogen, gaseous oxide of septon = dephlogisticated nitrous air, i. e., HNO_3). *Mitchill* advocated the new French nomenclature in chemistry (Chapter 1.3.2) in America in 1792 (Nomenclature of the new chemistry, New York, 1794); most remarkable is his emphasis the teaching of an *anti-phlogistic* chemistry. Note, salpetre (British) and saltpeter (American).

Table 1.4: Historic general terms of gases in Latin.

term	formula or comment
gas acidum acetosum	CH_3COOH
gas acidum fluoricum	HF
gas acidum muriaticum	HCl
gas acidum nitrosum	NO_y (NO, NO_2, HNO_2, HNO_3)
gas acidum regale	HNO_3 + HCl (volatilized aqua regia)
gas acidum spathosum	HF
gas acidum sulfureum	SO_2
gas acidum vitriolicum	SO_3
gas ammoniacale	NH_3
gas hydrogenium sulphuratum	H_2S
gas inflammable	all inflammable gases
gas mephiticum	CO_2
gas nitrosum	NO_x
gas oxygenium	O_2
gas palustre	CH_4
gas phlogisticum	N_2
gas phosphoricum	PH_3
gas pingue	H_2
gas siccum	inflammable gas
gas silvestre	CO_2
gas ventosum	common air
gas vinosum	CO_2

be employed with the new term (but still close to the alchemistic symbols). An excellent book on "historical studies in the language of chemistry" is written by Crosland (1962).

John Dalton (1766–1844) was an English meteorologist who switched to chemistry when he saw the applications for chemistry of his ideas about the atmosphere. He proposed the Atomic Theory in 1803 which stated that

1. all matter was composed of small indivisible particles termed atoms,
2. atoms of a given element possess unique characteristics and weight, and
3. three types of atoms exist: simple (elements), compound (simple molecules), and complex (complex molecules).

Dalton's theory was presented in "*New System of Chemical Philosophy*"[145] since the old alchemical symbols were not fit to use in his theory, he proposed a new set of standard symbols for the chemical elements in the first volume of his *New System*.

der chemischen Nomenklatur für das antiphlogistische System, Wien 1793 and in shortened form by Christoph Girtanner (1792) *Anfangsgründe der antiphlogistischen Chemie*, Unger, Berlin, 494 pp.

145 Manchester, S. Russell for R. Bickerstaff, in three Volumes (1808–1827), 560 pp.

The Swedish chemist *Jöns Jacob Berzelius* (1779–1848) was one of the first European scientists to accept *John Dalton*'s atomic theory and to recognize the need for a new system of chemical symbols. He suggested just using letters, arguing those are easier to write and print (Berzelius 1814); he already named 47 elements:

> The chemical signs ought to be letters, for the greater facility of writing, and not to disfigure a printed book. Though this last circumstance may not appear of any great importance, it ought to be avoided whenever it can be done. I shall take, therefore, for the chemical sign, the *initial letter of the Latin name of each elementary substance*: but as several have the same initial letter, I shall distinguish them in the following manner:
> 1. In the class which I call metalloids, I shall employ the initial letter only, even when this letter is common to the metalloid and some metal.
> 2. In the class of metals, I shall distinguish those that have the same initials with another metal, or a metalloid, by writing the first two letters of the word.
> 3. If the first two letters be common to two metals, I shall, in that case, add to the initial letter the first consonant which they have not in common.

Writing the formula of a compound, *Berzelius* used small numbers for the number of atoms, but he placed them above the symbol, e. g., sulfur dioxide SO^2. Apart from errors due to faulty analysis, there were several weaknesses in *Berzelius*'s system. *Edward Turner* (1798–1837), who used the symbols of *Berzelius* still in his 4th edition of "Elements of Chemistry (1834) but most notably *James Finlay Weir Johnston* and *Liebig* brings us closer to the symbolism of to-day by writing the number of atoms as a subscript instead of as an index (Chemical Tables, Part 1, Edinburgh, 1836, donation to the Proceedings of the Royal Society of Edinburgh 10, 147).

1.3.3 Alchemical views on airs and gases

> An accurate knowledge, therefore, of the Air, by which its actuating properties may be understood, is absolutely necessary for the Chemist, Physician, and natural Philosopher (Boerhaave 1735, p. 349).

Air and water had been regarded as "elements" convertible into each other since *Aristoteles* (Chapter 2.2.1). The statement by *René Descartes*, the French philosopher, mathematician, scientist and writer, that water vapor is not (atmospheric) air, is remarkable as this was 15 years before the introduction of the term "gas" by *van Helmont*. The gaseous substances that were observed in alchemical experiments were named fumes, vapors and airs. Atmospheric air (termed common air) was still regarded as a uniform chemical substance.

Air, water and soil[146] (or in modern terms the atmosphere, hydrosphere and pedosphere) were not only the ancient "elements" forming the Earth with all its forms and phenomena but also characterize the states of matter (gas, liquid and solid). Humboldt (1850–52, p. 311)[147] writes

> The relative quantities of the substances composing the strata of air accessible to us have, since the beginning of the nineteenth century, become the object of investigations, in which Gay-Lussac and myself have taken an active part [. . .]

Whereas the (qualitative) description of air quality using terms like: clean ["*reine*"], foul ["*unreine*"], good ["*gute*"], bad ["*böse*"], stuffy ["*stickige*"], corrupted [*verderbte*], cold ["*kalte*"], cool ["*kühle*"] and warm ["*warme*"] were used verbally since Biblical times, different kinds of air ["*Luftarten*" or "*Luftgattungen*"] were discovered first by alchemists. This was not yet a complete analysis of atmospheric air but the identification of the formation of different gases (termed *airs* at that time) as a result of alchemistic experiments. The term *gas* was still unknown. Obviously, there was a requirement to distinguish the vapors[148] and airs found in chemical experiments from (atmospheric or common) air through a new term.

It is notable that at that time, atmospheric air was still regarded as a consistent chemical body. The Swiss-German Renaissance physician, botanist, alchemist, astrologer and general occultis*Philippus Aureolus Theophrastus Bombastus von Hohenheim*,[149] who was known under the name *Paracelsus* (1493–1541), in that sense, termed the "airspace", chaos. Air and chaos were a synonym to him (Loewe 1936). The primordial Greek term *χάος* denotes[150] an empty space and the beginning. However, emptiness cannot be identified with nothing. According to ancient cosmogony, after which

---

146 Today, "soil" means the upper part (pedosphere) of the solid earth whereas the rocky part (lithosphere) is situated below. In ancient times, "soil" was the synonym for the solid earth.

147 This work "Cosmos" by *Humboldt* was first published in German (1845): Humboldt, A. von (1845) Kosmos. Entwurf einer physischen Weltbeschreibung. J. G. Cotta'scher Verlag, Stuttgart und Tübingen, in 5 Vols., over 3600 pp. [First English Edition: Cosmos – a General Survey of the Physical Phenomena of the Universe. Hippolyte Bailliere Publisher, London 1845].

148 See also Chapter 1.3.4: *Dämpfe*, *Dünste* and *Lüfte*, also Fumus, Vapour, Exhalatio; Old High German: *dampf*; Middle High German: *tampf*; Danish, English, Dutch and Low German: *damp*; Old North German: *dampi*; Polish: *dim*. It belongs to the strong verb *dimpfen* (reek, smoke). It is related with Old High German *daum*, Middle High German *toum* and Austrian *Dam* (Ausdünstung) and Swedish *Dam* (flushy cloudy dust).

149 *Paracelsus*, practically *Philippus Aureolus Theophrastus Bombastus von Hohenheim* (1493–1541). The 16th and seventeenth century are therefore considered as the iatrochemical era. He became famous for the phrase "*Alle Ding' sind Gift und nichts ohn' Gift; allein die Dosis macht, das ein Ding' kein Gift ist*" [All things are poison and nothing is without poison; only the dose permits something not to be poisonous] but mostly in a shortened version sola dosis facit venenum. See: Paracelsus (1616). He is also termed "father of toxicology".

150 In ancient poetry also used for "airspace".

the world was born from the chaos and hence, chaos was creativity; having all opportunities (Genz 1994). From the primordial chaos (or mysterium magnum) arose through "separatio" the four elements: water, fire, soil and air. These elements caused further emergence and subsequent decomposition:[151]

> Dis mysterium magnum ist ein muter gewesen aller elementen und gleich in solchen auch ein grossmuter aller stern, beumen und der creaturen des fleischs [...] und ein element ist ein muter, deren seind vier, luft, feur, wasser, erden; aus den vier mutern werden alle ding geboren der ganzen welt [...] [This mysterium magnum was a mother of all elements and contemporary a grant mother of all stars, trees and creatures of bodies [...] and an element is a mother, of that are four, air, fire, water, soil; from these four mothers all things of the whole word were born [...].[152]

The great achievement of *Paracelcus* was to use alchemy to search for medical substances. Perhaps he was the first who considered substance-forming properties to be more chemical then philosophical, creating a preconception of chemical elements (however, only introduced scientifically by *Boyle*). Vital processes, telluric and cosmic physics (for *Paracelsus*, the atmosphere acts as the special matter "chaos") had already been connected through substantial relationships[153] (Bugge 1929, p. 96).

Van Helmont was the first who distinguished different airs or gases, being relates to atmospheric air (gas ventosum – common air) but recognized not identical due to different properties. He deals with gas in his treatise "De flatibus" in which he speaks of gas ventosum, gas pingue, gas siccum, gas fuliginosum sire endemicum, gas sylvester (sive incoercible, quod in corpus cogi non potest visibile),[154] gas sulphureum, gas uva, gas vini, gas musti, gas flammeum, etc.; some of which are really the same. He was the first clearly to realize the production of gas in various chemical processes (Helmont 1662, p. 106; Leicester and Klickstein 1952, p. 25):

---

151 In the lifetime of *Paracelsus* only very few books were published. Only since 1560 have (partly in alchemistic compilations) *Paracelsus*'s essays (several Hundreds) been published as "Liber Natura, sive Chaos veterum; generalem metallorum generationem, etc. demonstrans". In: Liber vexacionen. John Stacy (1656) pp. 83–89 (Glasgow University, bibliography MS Ferguson 237).

152 From page 1 in: Achter Theil der Bücher und Schriften / des Edlen / Hochgelehrten unnd Beiwehrten PHILOSOPHI unnd MEDICI, PHILIPPI THEOPHRASTI Bombast von Hohenheim / Paracelsi genannt: Jetzt auffs new auß den Originalien / und Theophrasti eygener Handschrift / so viel derselbigen zubekommen gewesen / auffs trewlichst und fleissigst an Tag geben: Durch Ionnem Huserum Brisgoium, Churfürstlichen Cöllnischen Rath und Medicum. In diesem Tomo (welcher als Erstes unter den Philosophischen) werden solche Bücher begriffen / darinnen fürnemlich die Philosophie de Generationibus & Fructibus quatuor Elementorum beschrieben wirde. Joh. Wechels Erben, Franckfort am Meyn (1603), 240 pp.

153 However, he also retains the three basic substances in alchemy: sulfur, mercurius and sal, which corresponds to the physical phenomena of combustibility (oiliness), liquefaction (evaporation) and solidification (solidity).

154 Hoefer, J. C. F. (1843) Historie de la chemie. Tome 2, Chez L. Hachette, Paris, p. 145.

> Suppose thou, of 62 pounds of Oaken coal, one pound of ashes is composed: Therefore, the remaining 61 pounds, are the wild spirit, which also being fired, cannot depart, the Vessel being shut.
> I call this Spirit, unknown hitherto, by the new name of Gas, which can neither be constrained by Vessels, nor reduced into a visible body, unless the seed being first extinguished.

Helmont observed also the *Gas of Wines* when grapes and other fruits are fermented and a gas from burning gun-powder (saltpeter, sulfur and coal). *Van Helmont* says more than once that he was the "inventor" of gas (halitum illum Gas vocavi), which *Paracelsus* was ignorant of (ignoravit [...] quidditatem Gas, meum scil. Inventum), and there is no doubt that *Paracelsus* had no such ideas on gases as he. *Van Helmont* definitely distinguishes gases from condensable vapors and from air, and from one another. He says that gas is composed of invisible atoms which can come together by intense cold and condense to minute liquid drops (atomi Gas, ob nimiam exiguitatem invisibiles [...] frigori excessum, in minimas rursus guttulas concidant), Partington (1936). As before mentioned, contemporaries like *Otto von Guericke* and *Robert Boyle* did not believe that (common) air is an element. From observations on reducing the air volume while combustion, calcinations and animal breathing it was right concluded that some part of air must be responsible for such processes. Moreover, from the weight increase during metal calcinations it was concluded that either from the fire or the air some material must be added. This was also observed by other scientists like *Jean Rey* (ca. 1590–1645), *Samuel Cottereau Du Clos* (1598–1685)[155] and *Henri-Louis Duhamel du Monceau* (1700–1782), all French chemists.

The meaning of different terms in different languages (e. g., French, English and German) was changing over time; the words were used in slightly different senses by various scientists (Table 1.3 and 1.4). There was obviously a need for a new word to name and distinguish the laboratory airs (i. e., gaseous substances) from atmospheric (common) air. The new word was proposed by *van Helmont* in his posthumously published book, "Ortus medicinae i. e. initia physicae inaudita" (Amsterdam 1652, p. 86): "hunk spiritum, incognitum hactenus, novo nomine Gas voco" [I call this entity, unknown hitherto, by the new name of Gas], and "ideo paradoxi licentia, in nominis egestate, halitum illum Gas vocavi, non longer a Chao veteran secretum"; in following the English text from Helmont (1662, p. 69):

> But because the water which is brought into vapour by cold, is another condition, than a vapour raised by heat. Therefore by the Licence of a Paradox, for want of a name, I have termed that vapour, Gas, being not far severed from the Chaos of the Auntients. In the mean time, it is sufficient for me to know. That Gas, is a far more subtile or fine thing than a vapour, mist, or distilled Oylinesses, although as yet, it by many times thicker than Air.[156]

155 For *Du Clos* was chemistry the science of substances, the physics of qualities (Franckowiak 2011).
156 The following phrase is very alchemistic and fully unclear: But Gas itself, materially taken, is water as yet masked with the ferment of Composed Bodies.

Johann Christoph Adelung (1732–1806) explains why this new term is necessary (Adelung 1796, p. 425):

> [. . .] dass unsere Naturkundige ein schicklicheres Wort, welches nicht so sehr das Gepräge der Alchymie an sich hätte, ausfündig machten [. . . that our natural scientists might find a seemlier word, not so much having the imprint of alchemy].

Adelung believed that *Helmont* had derived the word "gas" from the Dutch *Geest* (ghost). Carbon dioxide, hitherto termed spiritus sylvestris (wild spirit), was renamed by *Helmont* as "gas sylvestre". There are other ideas on the origin of the word gas, however, – air and chaos were synonymous for him. The Greek word χάος denotes both an empty sphere and the initiation. Emptiness is not synonymous with nothing because the Greek philosophers stated that the world is born from chaos or, in other words, the chaos is creative of life (Genz 1994). The Dutch pronunciation of "chaos" is close to "gas" when the letter "o" is omitted (Egli 1947).

Today, there is also the belief that *Helmont* introduced the term *gas* from the term "chaos" (the essays by *Paracelsus* were well known to him), according to Dutch pronunciation; omitting the "*o*" and pronouncing "*ch*" like "*g*" (Egli 1947). The Flemish text has: "*Gas-maeckinge: uyt bet water eenen gas* (*dat is eenen griexschen water-chaos*)", Partington (1936). However, it is also possible that the term is derived from "*Geist*"[157] (ghost or spirit, Lat. spiritus), which at that time was a common name for gases and airs in alchemy. Kopp (1931, Vol. 3. P. 178) writes:

> Woher das Wort zunächst gekommen ist, weisz man nicht; nach Juncker, den bekannten Schüler Stahl's, soll es aus *Gäscht*, dem bei der Gährung entstehenden, Schaume, abgeleitet sein [Where the word originally comes from, we do not know; according to Juncker, the well-known scholar of Stahl, it should stem from *Gäscht*, the foam forming during fermentation].

Antoine-Laurent Lavoisier writes in his work "*Opuscules physiques et chimiques*" (2e éd., Paris, 1802, p. 5):

> *Gas* vient du mot hollandais *Ghoast*, qui signifie *Esprit*. Les Anglais expriment la même idée par le mot *Ghost*, et les Allemands par le mot *Geist* qui se prononce *Gaistre*. Ces mots ont trop de rapport avec celui de *Gas*, pour qu'on puisse douter qu'il ne leur doive son origine.

In "*Deutsches Wörterbuch*" Vol. 4 (1897), the term *gas* is defined as follows:

> Gattungsname für Luftarten, oder luftförmige Flüssigkeiten wie die Wissenschaft den Begriff bestimmt, die sich von den Dämpfen unterscheiden durch die Unmöglichkeit oder Schwierigkeit sie

157 Dutch and Low North German: *geest*; Anglo-Saxon: *gâst* (also Old Friesian) *gæst*. The origin is seen in whiff [*Hauch*] and breath [*Atem*]. *Luther* writes (Hiob 4, 9): der himel ist durchs wort des herrn gemacht und all sein heer durch den geist seines munds [By the word of the LORD were the heavens made; and all the host of them by the breath of his mouth]. Insofar as the synonymy between breath, spirit, vapour, wind and kinds of air is given.

in tropfbare Gestalt zu bringen; auch von der gewöhnlichen Luft sind sie verschieden und wurden im Gegensatz zu ihr zuerst erkannt, während dieselbe jetzt selber von der Wissenschaft als gasförmig, als ein Gasgemenge bezeichnet wird [Common noun for kinds of air or aerial liquids as science defines that term, distinguished from vapours through the impossibility or difficulty to bring them into a drop-able form; but also that they differ from common air and in contrast to it they were at first perceived, while the same now is described by science gaseous, as gaseous mixture].

Helmont is said to have termed (after Adelung 1796) common air ["*gemeine Luft*"] a gas, or more specifically, **gas ventosum**. After *Gehler*, the terms **gas atmosphericum** and "*Dunstkreisluft*" [spherical air] were used synonymously.[158] In Krünitz (1779, Vol. 16, p. 404), it is written:

Das Gas nennen Helmont und andere Chemiker die unsichtbaren flüchtigen Theile, welche von selbst aus gewissen Körpern ausdampfen [...] z. e. die Dämpfe der in eine spirituöse oder in eine faulige Gährung gerathenen Materien, tödtliche Dämpfe aus brennenden Kohlen, die Schwaden in Bergwerken, u. s. w., und selbst den spiritus rector gewisser Substanzen, z. e. des Bisams, denn es wurde zuerst unter die spiritus oder Geister der Dinge gezählt [As gas, *Helmont* and other chemists call the invisible volatile parts that escape spontaneously from certain bodies [...], e. g., vapors of a spirituous or in fermentation processing matter, deadly vapors from burning coals, damp in mining, etc., and even spiritus rector of certain substances, e. g., that of musk because it was termed first among the spirits and ghosts of things].

However, for the next hundred years after *Helmont*, the term "gas" was not used by others. *Robert Boyle* describes in his book "An Experimental Discourse Of some UNHEEDED CAUSES OF THE *Insalubrity* and *Salubrity* OF THE AIR (1690) several properties of atmospheric substances (termed effluvia, exhalations, damps, corpuscles, smoak) concerning air quality, diseases and chemical operations. He writes (Boyle 1690, pp. 1–3):

The insalubrity and salibrity of the Air depends, [...] from *Effluvia* [...] some are almost constantly or daily sent up into the Air, and those I call *Ordinary Emissions*; and others [...] *Extraordinary Emissions*; [...] at stated times, and so deserve the title of *Periodicals*, or else uncertainly, [...] *irregular*.

In a certain sense *Boyle* writes on air chemical actions (Boyle 1690, pp. 64–65) when different "subterraneal exhalations" in the air "fit to associate with them" making them more hurtful:[159]

158 *Humboldt* writes on kinds of air ["*Gasarten*"] in "*Versuche über die chemische Zerlegung des Luftkreises*" [Experiments on chemical decomposition of air] (Braunschweig, 1799): "*Doch ist im Buche selber noch immer mehr von Luft als Gas die Rede*" (Deutsches Wörterbuch) [However, in this book itself is more talk of the term of air than of gas].

159 The term *Particles* and *Corpuscle* should not be regarded as "particle" in today's language (of dust, e. g.) but in sense of a substance and molecule. Aqua regia = $HCl + HNO_3$. It remains speculative whether *Boyle* considers the reaction of NO_y with sea salt (NaCl).

> This may be somewhat illustrated by considering, that the spirituous steams of Salt-peter are not wont sensibly to work on Gold, nor yet the spirituous Parts that the Fire raises from *Sal-armoniac*; and yet when these two sorts of Particles convene, there results from their Coalitions certain Corpuscles of a new nature, that compose the liquor Chymists call *Aqua Regis*.

This is not only a statement of "gas-to-particle conversion" in modern terms but also an expression of a synergistic effect (Boyle 1690, p. 65):

> By Analogy to this we may conceive, that sometimes the Subterraneal *Effluvia* may find the Air already impregnated with such Corpuscles, that by associating themselves therewith they may compose Corpuscles far more capable, than themselves were whilst apart, of having ill Effects [...]

A few pages later *Boyle* likely gives the first mention of vegetation damages in print (Boyle 1690, p. 67):

> On which occasion, I remember, that a great many Trees in some Land that belongs to me, having been suddenly much endamag'd [...] Leaves were some moiré, some less blasted: [...] that some Arsenical or other corrosive or poisonous Exhalations, being suddenly emitted from the Subterranean parts into the Air [...]

Furthermore, *Boyle* writes on "self-cleaning" the air (Boyle 1690, p. 71):

> [...] Subterranean Bodies, be sent up into the Air, store of Expirations of another kind, which meeting with those that formerly impregnated it, may either precipitate them, and so free the Air from them; or by other operations on them, and sometimes even by Coalitions with them, so alter their nature as to disable them from doing any farther mischief.

Boyle, however, had the wrong belief (the theory of miasma that lasted still further 150 years) that diseases like pestilence and fever are caused by "subterranean effluvia". In his book "*General History of the Air*" (1692)[160] *Boyle* describes the formation and study of different kinds of air but he writes nothing on its realization; he separated between artificial and performed air (aër artificial vel factitious). Here are some phrases (Boyle 1692) from the chapter "TITLE XI. Of Salts in the Air":[161]

---

160 This posthumously published work "The General History of the Air, Designed and Begun by the Honble. Robert Boyle Esq." (Awnsham and John Churchill, London, 1692, 259 pp.) on the nature of gases, seen through the press by *Boyle*'s friend *John Locke* (1632–1704) and containing some of *Locke*'s own early meteorological observations.

161 The different phrases in Boyle (1692, pp. 40–52) reads – transformed into today's English – like an introduction to air pollution chemistry. Despite nothing is known on experiments with external air, and *Boyle*'s knowledge is based on alchemistic laboratory experiments, he clearly recognized sources of emission (such as soils, vegetation, animals, volcanism, and fabrics), substances in atmospheric air (such sea salt, soil dust, nitrous oxides and acids, sulfuric acid), and its injurious properties. Furthermore, he recognized formation of new substances (salts not being acid) from primary emitted compounds.

> Amongst the effluviating Substances of the Terraqueous Globe, there are, as I have declared in another Paper, huge quantities of common or Marine Salt, besides Nitrous, Aluminous, Vitriolate, and perhaps other kinds of Salts (Boyle 1692, p. 40).
> [...] by those Vulcans, that have open Vents to discharge their Fumes into the Air; by those numerous Fires which burning in our Chimnies, produce much saline Smoak [...] I am prone to think, that the saline Particles of the Atmosphere are not all of one sort, but that there may be three or four differing kinds of Aerial Salts. (Boyle 1692, p. 41).
> [...] to the volatile Nitre of the Air; these Spirits being so far from being refreshing to the Nature of Animals, that they are exceeding corrosive (Boyle 1692, p. 42).
> Besides the hitherto-mention'd kinds of Salts, it seems not improbable to me, that the Air (especially about great Towns, and some other particular Places) may be impregnated with volatile Salts, that are of a Nature contrary to Acids (Boyle 1692, p. 44).
> [...] not only in the parts of Animals, but also in those of many Vegetables, that Putrefaction may either extricate or produce volatile Salts (Boyle 1692., p. 45) [*Boyle* means "Spirit of Hartshorn", i. e., ammonia].
> [...] differing kinds of Salts, as Nitrous, Salino-Sulphureous, &c. that we have shewn may be met with in the Air, some guesses may be made in a short time of this or that Salt [...] (Boyle 1692, p. 52).

Ramsay (1907) believed that having carried out such experiments, the composition of air had been discovered hundreds of years before. However, already 100 years earlier, Fischer (1802)[162] reviewed that *Boyle* (and other contemporaries) were still too much oriented on finding philosopher's stone while carrying out skillful experiments and gaining valuable observations but did overlook the real truth [... *die eigentliche Wahrheit, die oftmals so schön hervorleuchtete, übersahen*] (Fischer 1802, p. 187).

As already mentioned, the terms "air" and "gas" were used in parallel, but later on it was spoken also in the form of "kinds of air" and "aerial substances". Until the end of the eighteenth century, the known types of airs (gases) were subdivided into two main classes, these which support combustions and breathing (inhalable airs) and those which fade light and kill animals (mephitic airs). Among the former are included common air and dephlogisticated air (oxygen), which does not allow further subdivision. The second class (mephitic) is subdivided into inflammable and non-inflammable categories and further subdivided into those mixable with water (in other terms, water soluble) and those that cannot be mixed with water (Table 1.3). Priestley (1790, p. 8–9) writes in his section "Of the use of terms":[163]

> The only terms in common use were, *fixed air*, *mephistic*, and *inflammable*. The last [...] characterizes [...] that kind of air which takes fire, and explodes in the approach of flame, [...] The term *mephitic* is equally applicable to what is termed *fixed air*, and to many other kinds, since they are equally noxious, when breathed by animals [...]

162 *Johann Carl Fischer* (1760–1833), mathematician and physicist in Jena and Greifswald, wrote the first "history of physics" in several volumes (1801–1809).

163 Considerable confusion and error resulted from *Priestley*'s neglect to distinguish clearly between inflammable gases like hydrogen, carbon monoxide and hydrocarbons (Partington 1933, p. 350).

> I have [...] giving names to those kinds of air, to whom no names were given by others, as [...] *nitrous* air, *acid* air, *alkaline* air, *phlogisticated* air and *dephlogisticated* air, etc., using the term *air* as expressive [...]
> I therefore thing that term *gas*, which many uses, in this sense, to be unnecessary; the term *air*, as it had long been used by philosophers, being sufficient for the purpose [...]

The writer *Ferdinand Ludwig Strumpf*[164] included in his compendium of pharmacology a 40-pages chapter on atmospheric gases including an extended biography (Strumpf 1855):

> gas oxygenium, gas azoticum, gas azoticum oxydulatum, gas nitrosum (gas azoticum oxydatum), gas hydrogenium, Ozon (gas oxygenium allotropon), aer atmosphericum, aer atmosphericum mutatus, gas latrinum, gas canaliense.

1.3.4 On the words air and atmosphere

Alexander von Humboldt, the founder of modern geography and, in particular, physical geography, writes (Humboldt 1850–52, p. 302):

> The two envelopes of the solid surface of our planet – the liquid and the aëriform – exhibit, owing to the mobility of their particles, their currents, and their atmospheric relations, many analogies combined with the contrasts which arise from the great difference in the condition of their aggregation and elasticity.
> The Aërial Ocean rests partly on the solid earth [...], we find that the strata of air and water are subject to determinate laws of decrease of temperature.

Humboldt cites accurately the concentration of oxygen (20.8 %) and nitrogen (79.2)[165] and named as minor species "carbonic acid gas",[166] "carbureted hydrogen gas" (CH_4), "sulfuretted hydrogen gas" (H_2S) and "traces of ammoniacial vapors" (NH_3). *Humboldt* also clearly states (Humboldt 1850–52, pp. 312–313):

> Besides these substances, which we have considered as appertaining to the atmosphere [...] there are others accidentally mixed with them [...] Fogs, which have a peculiar smell at some seasons of the year [...] dust which darkens the air for an extended area [...]

164 Unknown living data; descendent of *Karl Christoph Strumpf* (1712–1754) who was professor for botany and chemistry at Halle.

165 Only at the end of the nineteenth century was it known that about 1 % of this value is due to novel gases.

166 Citing *Boussingault* and *Lewy*, *Humboldt* (1850–52, footnote on p. 311) mentioned that "the proportion of carbonic acid in the atmosphere [...] varied only between 0.00028 and 0.00031 in volume" – a very modern view.

Within these few phrases, *Humboldt* used six different terms (aëriform, aërial, atmosphere, air, vapours, and gas) for a phenomenon that we describe today simply by *air*. The British chemist *Thomas Thomson* (1773–1852)[167] writes (Thomson 1820, p. 162):

> The word AIR seems to were used at first to denote the atmosphere in general; but philosophers afterwards restricted it to the elastic fluid, which constitutes the greatest and the important part of the atmosphere, excluding the water and the foreign bodies which are occasionally found mixed with it.

The English and French word for "air" is derived directly from the Latin aer, which comes from the Greek ἀήρ.[168] Incidentally, the words "air" and "airs" were used for all gaseous chemical compounds ("kinds of airs") in addition to atmospheric air,[169] for a long time after *van Helmont* introduced the term *Gas* (gas, *gaz*). In the nineteenth century, "*Gas*" became the synonym for town gas.

The term "atmosphere", derived from Greek (ἀτμόσ = vapor, ἀτμις = vapor, damp, mist, *σφαίρα* = sphere, ball),[170] was unlikely to were used before the middle of the eighteenth century. The Dutch astronomer and mathematician *Willebrord van Roijen Snell* [or *Willebrord Snelius*] (1580–1626) translated the Old Dutch word "*damphooghde*" (in Latin altitudine vaporum, in German "*Dunsthöhe*")[171] into a new Latin term "atmosphæra" in 1608,[172] which at first merely described the altitude between the surface of the earth and the bottom of the clouds. The term "*atmosphaera*" was first named in England by astronomers in the context of moon observations[173] (Weekley 1967). *Gottfried Wilhelm Leibniz* used the term "*atmosphaera*" manifold in his letters.[174] *Robert Boyle* (1627–1691) often used the term "*Atmosphere*" (Boyle 1662,

---

167 Scottish chemist whose writings contributed to the early spread of *Dalton*'s atomic theory.

168 The German word "*Luft*" is a common Proto-Germanic word; in Old English *lyft*. The word "*Luft*" (also Danish, Swedish and Norwegian) is associated with brightness; the German "*Licht*" [light], an *air* (in atmospheric sense) without fog or clouds. For more information see Möller (2014a).

169 Atmospheric air (as gas mixture) has also been characterized with the following terms (Krünitz 1779): *atmosphärisches Gas, gemeine Luft*; Latin: Gas atmosphaericum, Aër atmosphaericus vulgaris, communis, Gas ventosum; French: *Gas atmosphérique, Air commun, Air de l'atmosphère.*

170 In old German books, instead of "*Atmosphäre*" [atmosphere], the term "*Dunstkreis*" [aërial sphere – air surrounding the earth or a part of it] and the words "*Luftmeer*" und "*Luftozean*" [ocean of air] are often used until end of the nineteenth century (Reimann 1857, Umlauft 1891).

171 "*Dunstglocke*", in Dutch "*mistigheid*" (see the discussion with the term mist below on p. 109).

172 *Snell* translated "*Wisconstighe Ghedachtenissen*" (printers Ian Bouwensz in Leyden) by *Simon Steven van Brugghe* (1548–1620) into Latin Hypomnemata mathematica (Leyden 1605–1608).

173 E. g., Boskovic, R. J. (1753) De luane atmosphaera dissertatio. Romae, Publ. G. Salomoni.

174 Sämtliche Schriften und Briefe (1662–1676): Reihe III. 2, S. 219, 220, 794, Ed. by Berlin-Brandenburgische Akademie der Wissenschaften und der Akademie der Wissenschaften in Göttingen. Mathematischer, naturwissenschaftlicher und technischer Briefwechsel: Reihe VI, 1, S. 56–58, VI, 2, S. 190, 233, 249f, 255, 271; VI, 3, S. 58, 229, 525, Ed. by Leibniz-Archiv der Niedersächsischen Landesbibliothek Hannover.

p. 18), partly synonymously with "Air" ([. . .] *the Air or Atmosphere* [. . .]; Boyle 1662, p. 64).[175] In most cases, *Boyle* used the world *Air* in terms of a substance:

> Since in that part of the Atmoſphere we live in, that which we call the free Air (and presume to be so uncompreſſed) is crouded into so very ſmall a part of that ſpace, which if it were not hindred it would poſſeſs. (Boyle 1662, p. 33–34)
> [. . .] the Air may conſiſt of any terrene or aqueous Corpuſcle, provided they be kept ſwimming in the interfluent Celeſtial Matter; it is obvious that Air may be as often generated, as Terreſtrial Particles; minute enough to be carried up and down, by the Celeſtial Matter, aſcend into the Atmoſphere. (Boyle 1662, p. 91–92)

Boyle also used the terms *Corpuscles of Air* (Boyle 1662, p. 19), *Ambient Air* (Boyle 1662, p. 24), *Aërial Particles* (Boyle 1662, p. 28) and *Aërial Corpuscles* (Boyle 1662, p. 70); the last one provided with the characteristics of "[. . .] *gravitating themselves* [. . .]", apparently aerosol particles in today's understanding as hydrometeors and dust particles.

1.3.5 A brief history of chemical reactions

Today, according to the IUPAC, we define a *chemical reaction* as a process that leads to the interconversion of chemical substances. Chemical reactions occur when chemical bonds between atoms are formed or broken. A chemical reaction is symbolically represented by a *chemical equation*, wherein one set of substances, termed the *reactants*, is converted into another set of substances, termed the *products*. The reactants and products are shown using their chemical formulas, and an arrow is used to indicate the direction of the reaction. The *rate* of a chemical reaction is a measure of how the concentration or pressure of the involved substances changes with time. The *mechanism* of a chemical reaction describes the sequence of elementary reactions that must occur to go from reactants to product.

The oxidation (or combustion) of carbon monoxide in air gaining carbon dioxide was known in the early nineteenth century, based on the balanced reaction $2\,CO + O_2 = 2\,CO_2$, but nothing was known on the elementary steps before the middle 1920s ($CO + OH \rightarrow CO_2 + H$ that includes downstream reactions to form OH from O_2 and subsequent reactions of H). To establish a balanced reaction (or gross reaction), the knowledge of the elementary composition of reactants and products is essential. To identify elementary reactions, the detection of (often short-living) intermediates is required. Thus, the history of analytical chemistry (and techniques) is in line with the history of understanding chemical reactions. Before the development of physical chemistry (the theory of molecule properties and interactions), only the phenomenology of chemical reactions was known. The history of chemical equations refers also to the historical

175 In contrast to Terrestrial Globe (Boyle 1692, p. 40) as the total solid earth (see p. 79) and Terraqueous Globe (Boyle 1692, p. 58) as the total aqueous earth (see p. 278).

development of the use of "symbols" and "characters" to represent chemicals (Chapter 1.3.2), Crosland (1962).

1.3.5.1 From alchemy to quantum chemistry

Alchemists knew that heat was the power to proceed conversions of substances; using alchemistic symbols and names, chemical reaction diagrams, are first known from the French iatrochemist *Jean Beguin* (about 1520–1620), in his "Tyrocinium Chymicum", a set of chemistry lecture notes started in 1610 in Paris and later published by several editors (first 1634 in Wittenbergæ[176] by C. Bergeri, 620 pp.). *Boyle* first combined in his studies on gases (relation between amount, pressure, and temperature) alchemy with the exact discipline physics. He introduced the scientific terms *element*, *chemical compound* and *chemical reaction*. In 1757, Scottish physician and chemist *William Cullen* (1710–1790), building on the new science of affinity chemistry, launched in 1718 by French chemist *Etienne Geoffroy* (1672–1731), pioneered the use of reaction arrow (→) to express or characterize the "elective" affinity preference (affinity force) of the reacting species. The next development of chemical equation theory, the use of reaction diagrams and affinity tables comes from *Cullen*'s student *Joseph Black*, in 1775 from the Swedish chemist *Tobern Bergman* (1735–1784), and in 1782, *Antoine Lavoisier*, who was employed, it seems for the first time, a "horizontal" (modern) mathematical-stylized (+, −) representation of chemical reactions, using ratios and products; moreover, in 1787 he was the first to make the first chemical equation with the sign of equality (=); Hartley (1971, p. 82). In about 1884, the Dutch physical chemist *Jacobus van't Hoff* (1852–1911) upgraded from the equal sign symbol (=) to the reversible reaction arrow symbol (⇆ or ⇌), signifying a reversible reaction.

Chemistry, first established as a scientific discipline around 1650 by *Robert Boyle* (1627–1691),[177] had been a non-scientific discipline (alchemy) until then (Boyle 1680); however, *Boyle* remained believing on alchemistic transmutation (More 1941). Even until the end of the eighteenth century, some researchers remained in its belief of transmutation, e. g., water into salt. Metallurgical chemistry is the oldest branch of chemistry. *Gellert* (1776) mentioned the seven ancient metals: gold, silver, copper, iron, tin, lead, and mercury, the two new metals, zinc and cobalt, as well as the three metalloids: arsenic, antimony, and bismuth. But he did not mention platinum although this metal should have already become known in Europe. Sulfur and carbon were known among the seven metals as the only other elements. Salts were known since ancient times, like salmiac (Sutton et al. 2008) and saltpeter.

176 Lutherstadt Wittenberg.

177 Irish-British natural philosopher, chemist, physicist, and inventor in Oxford and London; although he was an alchemist and believing the transmutation, for him chemistry was the science of the composition of substances, not merely an adjunct to the arts of the alchemist.

The earliest applications of chemical processes were concerned with the extraction and working of metals and the manufacture of pottery, which were forms of crafts practiced many centuries before the Bronze Age cultures of Egypt and Mesopotamia. It was the practical working life of ancient humans who accumulated knowledge not systematically but by accident. In ancient Egypt, apart from metal-working and pottery, dye works, glassmaking, beer brewing and preparation of drugs were cultivated on high levels where the knowledge was put into hands of the temple priest.

The ancient idea of a "materia prima" led later to the assumption of the existence of philosopher's stone, a "tincture universalis" or "magno sudarunt elixyre", which would make transmutation processes possible, e. g., formation of gold from ignoble metals and create a drug for live extension and all diseases. In Middle Age, however only three "elements" (base material) were regarded, namely Sulfur (characterizing the principle of combustibility: flammability), Mercurius (characterizing the principle of metallic properties: volatility and stability) and Sal (salty properties: solidity). Different mixtures of these "elements" would form the various substances found in nature. The task of alchemy was the search for philosopher's stone (Figs. 1.11 to 1.13); however, in the alchemistic period many observations and findings made too for a beginning understanding of transformations.

It is agreed that "chemical knowledge" was focused in Alexandria until the third century. In Baghdad under the caliphate from the eightth century, there is enthusiastic translation and study of Greek scientific texts. Arab alchemists,[178] in their pursuit of synthesized gold, make practical advances in techniques of distillation. And they identify several chemical substances. The experimental framework established by the polymath *Jābir* (*Dschabir*) *ibn Hayyān* [جابر بن حيان] (about 721–815, born in Persia), also known as *Geber*,[179] influenced alchemists as the discipline migrated through the Islamic world, and then to Europe in the twelfth century.

In the first half of the sixteenth century, at the same time of Reformation, chemistry (or alchemy) was now used as the exclusive basis for medicine and explanation of processes leading to health and diseases; now termed the iatrochemical era. *Paracelsus* was the main representative of this new direction. The chemical knowledge, originating from the Arabic world, now together with European findings, was

178 Another great centre of alchemical experiments in medieval Asia is China, where alchemical experiments have a slightly different purpose. The quarry is still gold, but as an elixir of eternal life. Daoist's make the most startling chemical discovery of the period, gunpowder.

179 *Jabir ibn Hayyan* is the supposed author of an enormous number and variety of works in Arabic often termed the Jabirian corpus. As early as the 10th century, the identity and exact corpus of works of *Jabir* was in dispute in Islamic circles. The authorship of all these works by a single figure, and even the existence of a historical *Jabir*, are also doubted by modern scholars. Instead, *Jabir ibn Hayyan* is seen more like a pseudonym to whom "underground writings" by various authors became ascribed. His name was Latinized as "Geber" in the Christian West and in 13th-century Europe an anonymous writer, usually referred to as Pseudo-Geber, produced alchemical and metallurgical writings under the pen-name *Geber*.

Figure 1.11: The alchemist; contemporary illustration (from Ellis 1850, p. 1); *Robert Ellis* (1823–1885), British physician and writer.

summarized by the German polymath *Andreas Libavius* (~1555–1615)[180] in his work "Alchemia" (Frankfurt 1597), the first systematic textbook on chemistry.

Although the iatrochemistry is the parent of pharmaceutical chemistry, the close relationship between medicine and chemistry, namely the search for the universal drug inhibited any systematic research in chemistry as natural science. Fourcroy (1790, p. 26) writes:

> All these authors have written on chemistry in a very obscure, confused manner. Though acquainted with some processes of solution, extraction, and purification, etc., their pretensions rose much higher than their knowledge; and scarce any advantages can be derived from a perusal of their work.

The medical chemistry remains dominant until the middle of the seventeenth century with the beginning of *Boyle's* new understanding of chemistry as a science. However, individual chemists developed important sub-disciplines of chemistry like *Georg Agri-*

180 Also known as *Libau* and *Basilius de Varna*; born in Halle who was a teacher, physician and chemist in Rothenburg and Coburg.

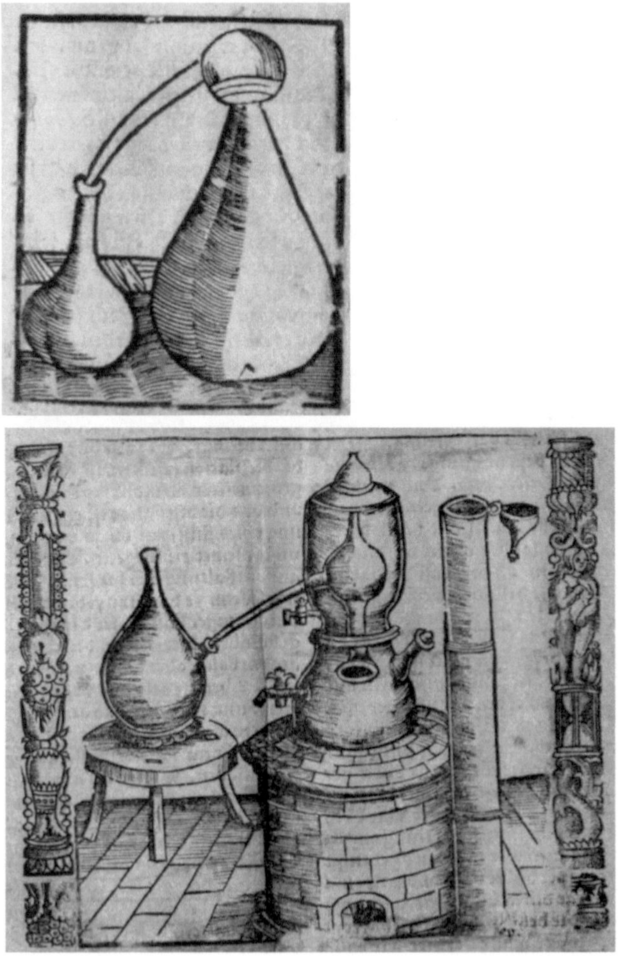

Figure 1.12: The distillation: alembic and receiver. Source: Geber (1530; above p. 59 and down p. 51). See also Fairley (1907) and Kockmann (2014) concerning early history of distillation.

cola[181] (1494–1555) the technical chemistry. Apart from *Agricola*, who contributed pioneering to metallurgy, *Bernard Palissy* (about 1510–1590) contributed to advances in ceramic and pottery. Furthermore, major progress in dye works, glassmaking, brandy distillery and beginning of agriculture chemistry were other characteristics of applied chemistry in the sixteenth century. One of the most important chemists of that time was *Johann Baptist (Jan) van Helmont* (1577–1644),[182] Fig. 1.14. He rejected *Paracelsus*

181 His German name was *Bauer*; he wrote the manual "De re metallica libri".

182 Flemish chemist, physiologist, and physician; he worked during the years just after *Paracelsus* and iatrochemistry, and is sometimes considered to be "the founder of pneumatic chemistry".

Figure 1.13: The mystery of alchemy: the last figure in Geber (1530, p. 136). The rack represents without doubt a dew sampler (see also Fig. 2.23 on p. 279), the function of the vessel remains secret.

Figure 1.14: *Jan Baptista van Helmont* (1577–1644); Line engraving (unknown author, 1666), source: Wikipedia Commons.

idea of the three "elements" (sulfur, mercurius and sal) and believed that all substances could be reduced to air and water. He was known for many careful observations and experiments and expressed for the first time the ideas of the *preservation of substance*. Apart from the iatrochemical research, *Johann Rudolf Glauber* (1604–1668) was the grand technical (or applied) chemist in that period. Whereas many salts, acids, and metals were produced und used in the sixteenth and seventeenth centuries, nothing was known on its chemical composition and systematic relations.

The last important exponent of medical chemistry was *Herman Boerhaave* (1668–1738)[183] who summarized the knowledge in his book "Elementa Chemiae" which is recognized as the first text on chemistry.[184] This work is organized into two Volumes, Vol. 1 contains Part I (containing the history of chemistry), Part II (con-

183 *Boerhaave* was variously professor of medicine, botany and chemistry at the University of Leiden from 1701 until his death in 1738. Prior to the invention of printed textbooks, the professor in a typical medieval university would read his lectures out loud to the students, who would, in turn, carefully transcribe them for their own personal use, a practice reflected in the modern German word for a lecture – "*Vorlesung*" – which literally means "to read in front of." In 1724 an unauthorized edition of *Boerhaave*'s popular chemical lectures, based on student lecture notes, was published anonymously in Paris under the title of "Institutiones et experimenta chemiae". The only way for him to preserve his reputation and prevent further sales of the spurious textbook was to publish an official version of his chemical lectures, which he finally did in 1732 under the title of "Elementa chemiae" (Jensen 2011). As it is known, many works attributed to *Boerhaave* were actually lectured notes published by former students.

184 Elementa chemiae, que anniversario labore docuit, in publicis, privatisque, scholius, Hermannus Boerhaave, qui continuet operationes chemiacae. In 2 Vol. (896+60 and 538+90 pp.) Leiden 1732; Bâle 1732, 1745, 1747; Tübingen 1732; Venice 1732, 1745, 1749, 1752, 1759, 1777 (J. R. Imhoff); Londini 1732 (Sumptubus S. K. et J. K.); Lipsiae 1732 (C. Fritsch, 744 + 470 pp.; containing only the theory of chemistry); Paris 1733, 1753 (G. Cavetier, 476+61 and 340+48 pp.). The Latin edition consists from 3 Parts (In Vegetantia, In Animalia, In Fossilia). Other Latin editions: Basileæ 1745, Venitiis 1749, Parisiis 1753. Translated into German (*Anfangsgründe der Chimie*, 1732–1734, 1738, 1762, 1782, 1791), French (*Eléments de Chymie*, 1748, 1752, 1754, 1755) and English (Elements of Chemistry 1727, 1735, 1741, 1753). Additional there were abridgements in English and German. The first English edition 1727, translated by P. Shaw and E. Chambers, based on the unauthorized edition 1724 from Paris, is entitled "A New Method of Chemistry including the Theory and Practices of that Art: Laid down on Mechanistic Principles and accommodated to the Uses and Life. The whole making a clear and rational System of Chemical Philosophy. A Critical History of Chemistry and Chemists; from the Origin of the Art to the present Time. Written by the Learned H. Boerhaave, Translated from the Printed Edition by P. Shaw and E. Chamgres, London, Printed for J. Osborn and T. Longman". A second edition (more careful and quite close to the Latin origin from 1732) in 1741 by *Peter Shaw* (1694–1736) was entitled "A new Method of Chemistry; including the History, Theory, and Practice of the Art. Translated from the Original Latin of Dr. Boerhaave's Elementa Chemiæ as published by himself. To which are added, Notes, and an Appendix, shewing The Necessary and Utility of Enlarging three Bounds of Chemistry" (Longman, London). With the same title appeared the third corrected edition in 1753. The official translation of *Boerhaave*'s textbook from 1732 is by *Timothy Dallowe* (Elements of Chemistry: being the Annual Lectures of Herman Boerhaave, Leiden 1735. Both can be claimed equally to reflect whatever it was *Boerhaave*'s Latin express (Christie 1994, Knoeff 2002). Volume I (528 pp.) is divided in two parts: Part I comprises a historical survey of chemistry; Part II, description of chemical notions and instruments, including 17 plates with illustrations of the latter. Volume II (376 pp.) is devoted to the description and explanation of 227 experiments (termed operations). In the following I'm citing only from the 1735 edition, Vol. 1. The German edition 1753: *Elementa Chemiae, Oder: Anfangs-Gründe der Chymie, Worinnen der Herr Auctor durch 227 Processe gründliche Anweisung gegeben, Auf was Art die natürlichen Cörper können Kunstmäßig analysirt, oder Chymisch aufgeschlossen und daraus heilsame Artzeneyen bereitet werden.* Leipzig, M. Blochberger, 1753 in 2 Vol., 1032 pp. Another German abridged edition: *Eines Engländers Auszug aus Hermann Boerhaave Anfangsgründen der Chemie übersetzt von *** Medic. Doctor.* Hannover, J. C. Richter, 1755, 588 pp. (based on the English edition London 1732).

taining the theory of chemistry), and Vol. 2 with Part III (containing the processes or operations of the art, including chemical operation upon vegetables, animals and minerals). Part II contains a large section "Of Air", containing subsections of the contents of air, dew in air and clouds whence in air, and a section "Of Water", containing a subsection "of rain water".

The time between *Boyle* and *Lavoisier* is often named the era of phlogiston theory. However, one man, the first chemist in Russia, who several times criticized wrong ideas of *Boyle*, must be noted here: *Michail Vasilievich Lomonossov* [Михаил Васильевич Ломоносов] (1711–1765), a genius, who had many good ideas, and if he had worked them out in detail, might have notably advanced science, but his duties as an academician, his variety of interests, and his irregular mode of life, hindered his progress in scientific work (Partington 1961, p. 204). *Lomonossov* can be regarded as the founder of the science of physical chemistry (Lomonossov 1752).

Especially behind the significant development of physics, the growing influence of the inductive method in science, the foundation of academic societies[185] in the second half of the seventeenth century with its publications contributed to a propagation of scientific results, benefit chemistry as science. However, still in this era the main focus remains on qualitative description of phenomena. Neglecting the weight ratios in chemical reactions (quantitative description) was the main cause for wide acceptance of the wrong phlogiston doctrine. Nevertheless, this period finally contributed to the elimination of the delusional idea of alchemy. Although a physician, *John Mayow* (1645–1679) is another British genius apart from *Boyle* whose earlier death delayed the understanding of combustion processes; doubtless he is termed the predecessor of *Lavoisier*. *Mayow* assumed that atmospheric air contains a substance (which he termed "spiritus igno-aereus" or "nitro-aereus"), which combines with metals during calcinations,[186] also is present in saltpeter,[187] supports breezing, and transforms venous into arterial blood.[188] In France, two scholars, but much less important in its scientific contribution, *Wilhelm Homberg* (1652–1715) and *Nicolas Lémery* (1645–1715) must be named. *Lémery* regarded chemistry as a demonstrative science and is one of the fathers of acid-base theory. In Germany, *Johann Kunckel* (1630–1703) and *Johann Joachim Becher* (1635–1685)[189] were protagonists between alchemy and modern chemistry. *Becher* "developed" the idea of *Paracelsus* further that instead of Mercury, Sal and Sulfur now all inorganic material consists from three "elements", mercurial,

185 Academia naturae curiosum in Wien (1652), Accademia del cimento in Florence (1657), Royal Society in London (1660), Académie royale de sciences in Paris (1666), Berliner Akademie (1700); later academic societies were founded in St. Petersburg (1725), Stockholm (1739) and Copenhagen (1743).
186 It is CO_2.
187 It is NO and/or NO_2.
188 It is oxygen.
189 German physician, alchemist, precursor of chemistry, scholar and adventurer, mainly in Mainz and Vienna; best known for his development of the phlogiston theory.

glass making and combustible (“terra pinguis”). When burning substances or calcinations[190] of metals, “terra pinguis” escapes.

This idea was accepted by *Georg Ernst Stahl* (1660–1734)[191] who explained combustion of any matter by release of a hypothetical fierily substance, which he named *Phlogiston*. In other words, he stated that every combustible *substance* contained a universal component of *fire* (it is still the old element idea of *Aristoteles*). This doctrine was accepted by almost all scientists for the next 120 years until *Lavoisier* finally rejected this idea in 1783. These days it is difficult to understand what “phlogiston” meant to eighteenth-century scientists. The phlogiston theory was to some extent derived from the old belief that there was a fire element and that all combustible bodies contained a common principle (element), *phlogiston* (which in Greek means “flammable” or “inflammable”), which is released in the process of combustion.[192] Substances, rich in phlogiston, like wood, burn almost completely; metals, which are low in phlogiston, burn less well. The phlogiston theory created great confusion and essentially embedded the understanding of the chemistry of phase-transfer processes and solid-gas reactions.

In *Stahl's* publications, phlogiston is never identified as a substance which can be prepared; it was a substantial support of a property, the capacity to combust. Conflicts arose from later ideas to regard phlogiston as a preparation like carbon or hydrogen. Kopp (1869, part III, pp. 222 ff.) summarizes *Stahl*'s ideas on phlogiston that combustion requires the presence of air or something similar, in which phlogiston dispenses (which also comes from rotting) and reached again into plants and from them into animals. Common metals contain phlogiston apart from an earthy component, which can be represented as metallic calces[193] or metallic ash. When heating with coal the phlogiston therein combines with them (carbon) and the metal appears again, hence, the calx[194] is a component of the metal. Another wrong conclusion by *Stahl* was the

---

190 The IUPAC defines calcination as heating to high temperatures in air or oxygen. The process of calcination derives its name from the Latin “calcinare” (to burn lime) due to its most common application, the decomposition of calcium carbonate (limestone) to calcium oxide (lime) and carbon dioxide. In alchemy, calcination was believed to be one of the 12 vital processes required for the transformation of a substance.

191 German chemist and physician in Weimar, Halle, and Berlin; known for his obsolete phlogiston theory.

192 The word Phlogiston is not by *Stahl* introduced in the chemical literature (as stated e. g., in Wikipedia and other sources). Kopp (1869, part III, p. 217, Footnote 462) cites various sources where Phlogiston were used before by *van Helmont*, *Becher* and *Sennert*. Derived from Greek φλογιστός = burnt. Phlogiston was equalized with caloricum, the matter of caloric or “*Wärmestoff*” in German (fire in old terms), i. e., the hypothetical principle of fire regarded formerly as a material substance.

193 Metallic calces (in German *Metallkalk*) are oxides, but at that time not separated from its carbonates or mixtures.

194 The product that resulted from heating a metal until it lost its metallic properties was termed a “calx of the metal”. More generally, a calx [plural: calces; in German: *Kalch* before 1800 and later *Kalk*] was any powdery substance yielded by heating a substance strongly in air.

idea that sulfur is composed from sulfurous acid and phlogiston. The conclusion that the individual combustion products must be weight less than the burned body as a composite was neglected. Even the (well observed) fact that some products are heavier like metal calces was also neglected. Finally, these observations were the reason for the fall of the phlogiston theory. *Stahl*'s merit, however, was the creation of a theory what we at present call oxidation-reduction; uptake of phlogiston means reduction and release of phlogiston is identical with the term oxidation.[195]

Nonetheless, this wrong theory did not hinder important progress by other chemists like *Joseph Black*, *Henry Cavendish*, *Andreas Sigismund Marggraf*, *Carl Wilhelm Scheele*, *Torbern Olof Bergman* and *Joseph Priestley* which all were phlogistians but contributed pioneering to our understanding of air chemical composition or the *chemistry of gases*. These chemists and their contributions will appear in later chapters.

The late era of phlogiston (1770–1790) shows significant progress in (qualitative) analytical chemistry and the development of apparatus for gas separations and volume measurements (pneumatic chemistry). With *Lavoisier*'s work (after about 1775) the *anti-phlogistic* system in chemistry developed, the modern understanding of combustion processes, namely uptake of oxygen (former escape of phlogiston) characterize combustion (oxidation) and release of oxygen (former uptake of phlogiston) is the inverse process (reduction). The contributions of *Lavoisier* to the development of a modern chemistry are immense; the interested reader should read special books. Understanding the role of oxygen can be seen to be the imprint benefit of *Lavoisier*.

From the ending of the eighteenth century, so many important chemists worked and their number increased continually (to name here the most important: *Martin Heinrich Klaproth*,[196] *John Dalton*, *Claude Louis Berthollet*, *Antoine François Comte de Fourcroy*, *Joseph Louis Proust*[197]). Berthollet (1805, p. 97–98) writes:

> Es läßt sich nicht oft genug wiederholen, daß sich die Wissenschaft *nur dadurch* weiter bringen läßt, daß man mit großer Schärfe die Thatsachen ausmittelt, und die Methoden immer mehr vervollkommnet, vermittelst deren man die die Erfahrungen einsammelt. Mangelt es an dieser Genauigkeit, so ist die Wissenschaft weiter nichts, als eine Sammlung unzusammenhängender Facta, auf welche sich ein System nach dem anderen bauen, und Meinungen, die sich mit einander im Widerspruche stehn, gründen lassen [It cannot repeated often enough that progress in science is only possible by detecting the facts with high precision, and by improving the methods with them one collect the experiences. Is there a lack of this accuracy, science is nothing more than a collection of incoherent facts, on which one system after another is built, and views, conflict which each other].

195 In Chapter 4.2.2 on the discovery of oxygen, we go into more details of phlogiston.

196 Klaproth (1742–1817) was one of the most eminent chemists (and apothecary) of his time in Berlin; discovered several elements.

197 Proust (1754–1826), French chemist, most important in chemical analysis if his time.

Leopold Gmelin (1788–1853),[198] a German chemist in Heidelberg, who edited the first modern compilation of chemistry, wrote in his "*Handbuch der theoretischen Chemie*" (Gmelin 1827, Vol. 1, p. 134):

> Die Zahl der bis jetzt bekannten einfachen wägbaren Stoffe beträgt 48; hiervon sind 38 metallischer Natur; die übrigen 10 erscheinen ohne metallisches Ansehen, theils gasförmig, theils in fester Gestalt. Sie sind die wichtigsten, welche durch die Mannigfaltigkeit ihrer Verbindungen die Einförmigkeit der Metalle aufheben [The number of hitherto known simple ponderable substances amounts 48; thereof are 38 of metallic nature; the remaining 10 appear without metallic view, partly gaseous, partly in solid form. They are to most important, which suspend the uniformity of the metals through the great diversity of its compounds].

In 1817, these 10 non-metallic elements included H, B, C, N, O, F, P, S, Cl, and I (only Br had been not yet known); with the exception of the rare boron.[199] In the nineteenth century, the terms ammonia, nitric acid, nitrous acid, sulfurous acid, sulfuric acid, etc., were used in the same sense for dissolved species (ammonium, nitrate, nitrite, sulfite, and sulfate) as well as anhydrites (e. g., SO_2). Ammonia (NH_3) was known as a result of putrefaction of vegetable and animal matter. Nitric acid (HNO_3) was "known" as a result of thunderstorms and was believed from oxidation of atmospheric ammonia. The various oxides of nitrogen had created some confusion in the early nineteenth century. In 1816, *Joseph Louis Gay-Lussac* (1878–1850)[200] had succeeded in distinguishing five oxides of nitrogen and had

Notwithstanding, the history of chemical reactions also corresponds with the progress in physical chemistry after 1850 in understanding thermodynamics and kinetics of chemical reactions and in the spectral analysis, the most important technique for investigation of chemical reactions (Table 1.5). However, the understanding of molecular structures, reactivities, and chemical bonding began slowly in the 1930s in the application of the theoretical basis, quantum chemistry. The phrase "*Quantenmechanik*" [quantum mechanics] was coined by a group of physicists at the University of Göttingen in the early 1920s including *Max Born*, *Werner Heisenberg*, and *Wolfgang Pauli*, and was first used in *Born*'s 1924 paper "*Zur Quantenmechanik*". The roots of quantum mechanics go back to *Michael Faraday* (1791–1867), who discovered in

---

198 German chemist (sun of *Johann Friedrich Gmelin*), born in Göttingen, studied in Göttingen and Tübingen; 1815 professor in Heidelberg. He wrote the "*Handbuch der Chemie*" (first edition 1817–1819).

199 Boron is found to around 90 % gaseous as H_3BO_3 and 10 % particulate in the atmosphere; volcanic exhalation is likely in the form of BF_3, which fast hydrolyse into borate. First Gast and Thompson (1959) proposed the sea as source, and after a long debate (Nishimura and Tanaka 1972), Fogg and Duce (1985) confirmed the sea as a main B source (about 80 %), other sources are anthropogenic (coal combustion) and volcanic. In rainwater, B is within the range 0.3–15 $\mu g\ L^{-1}$ (Fogg and Duce 1985, Demuth and Heumann 1999).

200 French chemist and physicist in Paris (from 1808 to 1832 at Sorbonne) is known mostly for two laws related to gases.

Table 1.5: Milestones and most important discoveries as fundamentals for understanding atmospheric chemistry (and chemistry and physics in general). Note that only the most prominent scientists are listed; several scientists contributed to different fields.

Gas laws	Jacques Charles (1746–1823), John Dalton (1766–1844), Joseph Gay-Lussac (1778–1852), Amedeo Avogadro (1776–1856), Josef Loschmidt (1821–1895)
absorption of gases in water	John Dalton (1766–1844), William Henry (1772–1836)
absorption of solar radiation by molecules	Johann Heinrich Lambert (1718–1777), August Beer (1825–1863)
properties of solutions	François Marie Raoult (1830–1901)
laws of thermodynamic	Hermann Heinrich Hess (1802–1850), Rudolf Clausius (1822–1888), Hermann von Helmholtz (1821–1894), Josiah Willard Gibbs (1839–1903), Walther Nernst (1864–1941)
law of mass action	Cato Maximilian Guldberg (1836–1902), Peter Waage (1833–1900), Jacobus van't Hoff (1852–1911)
spectral analysis	Robert Wilhelm Bunsen (1811–1899), Gustav Robert Kirchhoff (1824–1887)
laws of reaction kinetics	Ludwig Boltzmann (1844–1906), Svante Arrhenius (1859–1927)
electrolytic dissociation	Friedrich Wilhelm Ostwald (1856–1931), Svante Arrhenius (1859–1927)
theory of reaction rate	Max Ernst August Bodenstein (1871–1942), Cyril Norman Hinshelwood (1897–1967), Michael Polanyi (1891–1976), Henry Eyring (1901–1981)
quantum chemistry	Niels Bohr (1885–1962), Max Born (1882–1970), Erich Hückel (1896–1980), Robert Mulliken (1896–1986), Erwin Schrödinger (1887–1961), Enrico Fermi (1901–1952), Werner Heisenberg (1901–1976), Walter Heitler (1904–1981), Julius Robert Oppenheimer (1904–1967), Lew Davidovich Landau (1908–1968)
theory of chemical bonding	Fritz Wolfgang London (1900–1954), Linus Pauling (1901–1994)

1838 the cathode ray, *Gustav Robert Kirchhoff* (1824–1887), who described in 1859 the black-body radiation, *Ludwig Boltzmann* (1844–1906), who states in 1877 that all energy states in a physical system are discrete, *Heinrich Hertz* (1857–1894), who found the photoelectric effect, the quantum hypothesis presented in 1900 by *Max Planck* (1858–1947), and the postulation of *Albert Einstein* (1879–1955) in 1905 that light itself is made of individual quantum particles, termed photons in 1926 by *Gilbert Newton Lewis* (1875–1946). *Linus Pauling* made probably the greatest contribution to the field of chemical reactivity.

An understanding of spectroscopy, the light absorption by molecules (and atoms), leading to different photoactivated states (and finally photodecomposition or photodissociation) is closely linked with the German physicist *Marianus Czerny* (1896–1985) who recorded in 1925 the first rotation spectrum, the American physicist *Philip McCord Morse* (1903–1985) who proposed in 1929 a model for the potential energy curve for a molecular vibration which explicitly includes the effects of bond breaking, the German physicist *James Franck* (1882–1964)[201] and the American physicist *Edward Uhler Condon* (1902–1974), both contributing to a rule in spectroscopy that explains changes in electronic and vibrational energy levels of a molecule due to the absorption or emission of a photon (termed *Franck-Condon* principle), and others.

1.3.5.2 The roots of atmospheric chemistry

Chemical reactions in atmospheric air (in gas phase, in atmospheric water droplets, and on solid particles) are not different from those occurring in chemical reactors in laboratory or industrial plants. However, in the latter the energy to activate the reactants was provided since centuries by heat (thermochemistry), later in the nineteenth century by radiation (photochemistry), namely UV, and with the beginning of the twentieth century also by radioactive radiation (radiochemistry).

Due to the nature of the atmosphere, the only source of energy to activate molecules for chemical reactions is solar radiation. Thus, atmospheric chemistry is almost photochemistry, either through photolysis or photosensitization, and subsequent radical chemistry. There are only a few non-photochemical reaction types, but all-in link with a heterogeneous phase, like water or solid particles (p – particulate, g – gaseous):

- hydrolysis ($A + H_2O \rightarrow$ products, e. g. $A = CO_2, SO_2, NO_2, NH_3$),
- electrolytic dissociation (e. g., $HCl \rightarrow H^+ + Cl^-$),
- substitution (e. g., $NaCl(p) + HNO_3 \rightarrow NaNO_3 + HCl$), and
- addition (e. g., homogeneous nucleation $NH_3(g) + HCl(g) \rightarrow NH_4Cl(p)$).

Sunlight induced changes in materials (photochemistry) were observed thousands of years ago, like preparation of dye, and hundreds of years ago like bleaching, deleterious effects on beer, food and organic materials (Roth 1989). The history goes back to the early 18th century with first observations on light influenced chemical processes but termed *photochemistry* only at the end of the 19th century. Another root of this chemistry lies in the *combustion chemistry*, investigated since the 1920s, and exploring radical chain reactions.

250 years ago, the discovery of oxygen between 1772 and 1774 by *Scheele* and *Priestley* was the beginning of modern chemistry, established by *Lavoisier* a few years later. Within the same period, it was found that plants consume carbon dioxide and produce oxygen. The fact that plants do not grow without water was known long before. However, the principal relationships between water, oxygen, and carbon dioxide, and thus between the atmosphere and biosphere, now termed *photosynthesis*, became clear about 100 years later – still without knowing detailed chemical mechanisms (which were explored in the 1940s). Thus, the discovery of carbon dioxide and oxygen and the chemical composition of water (one part oxygen and two parts hydrogen) led to

201 1925 Nobel prizes together with *Gustav Hertz*; 1933 he immigrated with his family to the USA.

the global biogeochemical reaction ($[CH_2]$ stands for organic biomass and volatile organic compound, respectively)

$$CO_2 + H_2O \underset{\text{biosphere and atmosphere}}{\overset{\text{biosphere}}{\rightleftarrows}} [CH_2] + \frac{3}{2}O_2.$$

The oxidative capacity of the atmosphere was recognized already in the early 19th century, namely after the discovery of ozone and hydrogen peroxide. Hence the oxidation of ammonia to nitrous and further to nitric acid, hydrogen sulfide and sulfur dioxide to sulfuric acid, organic matter to carbon dioxide in the atmosphere was accepted without knowing any mechanism. After about 1850, it was believed that one atom of oxygen in ozone (O_2–O) and hydrogen peroxide (H_2O–O) is easily released for oxidation of other compounds. Furthermore, it was found that H_2O_2 can act as oxidizing as well as reducing agent. Fox (1873, p. 29) writes that ozone "[...] acts on most substances as an *oxidizing* agent of great power [...]". But nothing was known about the formation of ozone; it was produced by electric sparks (in ozone generators)[202] and thus also named electrified air and ozonized or electrified oxygen (*Schönbein*) and binoxide of oxygen (*Soret*). It was known that 16 cubic inches of pure ozone (O_3) yield 24 cubic inches of oxygen (O_2), giving the equation $2\,O_3 \rightarrow 3\,O_2$. Fox (1873, p. 14) presents for the formation of ozone the (incomplete) equation $4\,O + 8\,O_2 \rightarrow 8\,O_3$. *Fox* further proposed the hypothetical reaction (not existent but only as $O_3 + HO_2$ in gaseous and $O_3 + O_2^-$ in aqueous phase) $O_3 + H_2O_2 = 2\,O_2 + H_2O$. However, it was known (in aqueous phase) that ozone reduces hydrogen peroxide which become water, and itself converts into oxygen. Nothing was known on the formation of hydrogen peroxide; *Schönbein* believed that O_3 and H_2O_2 are simultaneous gained while electrification of air (he assumed hydrogen peroxide – termed antozone – to be only as another modification of oxygen), but *Babo* and *Nasse* and *Engler* proved that "antozone" is only produced when ozone suffers decomposition from the presence of water. In the middle of the 19th century most scientists believed the presence of ozone and hydrogen peroxide in the atmosphere. Whereas the presence of hydrogen peroxide was proved in 1868 by *Struve*, the French Academy of Science (because of decades of controversy concerning *Schönbein*'s ozonometer) in 1865 demolish the belief in atmospheric ozone. However, researchers like *Houzeau* and *Andrew* found enough evidence to place the matter beyond a doubt in the 1870s. Variations in the electrical condition of the atmosphere were concerned with furnishing the air with a principle that gives it chemical activity (Fox 1873, p. 50).

Nothing was known in the nineteenth century on the origin of ozone and hydrogen peroxide, but they were considered as natural constituents of the atmosphere. However, many speculations and curious reactions were proposed. Atmospheric chemistry

202 Other methods for ozone formation were the electrolysis of acidulated water, by air passing over phosphorus and by the action of strong sulfuric acid upon potassium permanganate.

was still reduced on "electric" activities in the atmosphere, mainly lightning. The evidence was given by *Cavendish*'s experiments on the formation of nitric acid in air through electric sparks and the finding of nitric acid in rain from thunderstorm by *Liebig*. *Schönbein*'s "finding" of formation of ammonium nitrite while water evaporation ($N_2 + 2\,H_2O = NH_4NO_2$) confused chemists for some decades and led to the speculation of ozone formation under conditions of waterfalls and graduation works. But *Schönbein* correctly proposed the formation of hydrogen peroxide by $3\,H_2O + O_3 = 3\,H_2O_2$ but speculated the reaction $2\,NH_3 + 3\,H_2O_2 = NH_4NO_2 + 4\,H_2O$. Thus, the presence of ammonium nitrite was assumed in all atmospheric water (dew, fog and rain) – and also detected. The formation of H_2O_2 due to electric discharges in air containing water vapor was first proposed by Boeke (1873)[203] and confirmed by Fischer and Ringe (1908).

The interest in understanding chemical reactions in the air began with air pollution, first with the London smog (Chapter 3.1.2) end of the 19th century. In that time, only ozone and hydrogen peroxide were known as the "reactive" species (oxidants) in air. Before 1900, its formation was believed to be linked with the electric discharges occurring in thunderstorm and, earlier in the nineteenth century, with the general atmospheric electricity. This view was based on the finding by *Cavendish* (formation of nitric acid; see p. 93) and *Schönbein* (formation of ozone; see p. 459) in air under the influence of electric sparks in laboratory devices. Hence it becomes the view that the atmosphere produces permanent nitrous and nitric acid, ozone and hydrogen peroxide from atmospheric nitrogen and oxygen. Already early in the nineteenth century, it was believed that ammonia, derived from putrefaction processes on earth, is oxidized in air to nitrous and nitric acid, and also combines with these acids to form salts (nitrite and nitrate, found in rainwater). Other oxidations of trace gases detected in atmospheric air like carbon monoxide, sulfur dioxide, hydrogen sulfide, methane and other hydrocarbons have not been considered in the nineteenth century. Without knowing the mechanism, the oxidation of nitrite to nitrate and sulfite to sulfate in rainwater and fog (smoky atmosphere) was recognized in the late nineteenth century.

Photochemistry was considered at the beginning of the 1920s, but only the O_3 formation from O_2 photolysis was described (Chapter 4.4.4.1). Radicals were detected in combustion gases in the 1930s, assumed to be in the atmosphere in the 1950s but detected only after 1970 (Table 1.6). The German physical chemist *Alfred Hermann Benrath* (1871–1969) presents in his book "*Lehrbuch der Photochemie*" (Benrath 1912) the best and almost complete history of photochemistry from the beginning until about 1910.

203 The first evidence of H_2O_2 in rain during a thunderstorm was provided by *Georg Meissner* in 1862.

Table 1.6: Discovery of chemical species in air (metals and metalloids are not listed); note that discovery of some elements and compounds in mineral, vegetable, and the animal matter was earlier.

	18th century	19th century	20th century
O	O_2	O_3, H_2O_2	O, OH, HO_2
N	N_2	HNO_3, NH_3, NO/NO_2 (NO_x)[a], ON[b]	N_2O, HNO_2, NO_3, N_2O_5, org. N
C	CO_2	carbonate, CH_4, OM[c], CO	NMVOC, POP, organic radicals
S		SO_2 (sulfite), SO_3 (sulfate), H_2S	CS_2, COS, org. S
Cl		HCl (chloride)	Cl_2, Cl, HOCl, ClO[d], org. Cl
Br		bromide	HBr, BrO[d], org. Br
I		iodide	HI, IO[d], org. I
H	H_2O	H_2	H, H^+, H_2O^-
P		phosphate, PH_3	other phosphanes, org. P

[a] Including nitrate, ammonium and nitrite
[b] Organic nitrogen ("albuminoid")
[c] Organic matter (unspecified)
[d] And many more oxides and oxo acids

1.3.5.3 Early history of light-induced chemical reactions: photochemistry

Count *Alexei Petrowitsch Graf Bestuschew-Rjumin* [Алексе́й Петро́вич Бесту́жев-Рю́мин] (1693–1766),[204] Chancellor of the Russian Empire, still under alchemistic influence, developed a "*tonico nervina*"[205] by dissolution of ferric chloride ($FeCl_3$) in alcohol and exposing the brown solution to solar radiation until it became colorless; however, in the dark with air access, the solution became yellow. *Bestuschew* did not know that he discovered the reduction of ferric (Fe^{3+}) to ferrous (Fe^{2+}). This solution was widely used in Europa until the beginning of the twentieth century (Biechle 1903, p. 288). Alchemistic speculations also led the German physician *Johann Heinrich Schulze* (1684–1744) to the discovery of the most important reaction in photography, the blackening of silver salts in 1727. *Carl Wilhelm Scheele* (1742–1786) describes in his "*Abhandlungen von der Luft und dem Feuer*" (1777) that the blackness which silver chloride gained in light, is reduced silver; he also distinguished between radiant heat and light.

Joseph-Louis Gay-Lussac and *Louis Jacques Thénard* (1777–1857) studied the influence of light on the reaction between equal volumes of chlorine and hydrogen. Their results indicated that in bright sunlight the gases reacted violently, shattering the flasks used; with a less intense light, the combination proceeded more slowly. The

204 In historic German literature *Bestusheff* and in English *Alexey Petrovich Bestuzhev-Ryumin*; in French *Alexy Comes de Bestuscheff Riumin.*

205 In German *Bestuscheff's Nerventinctur*, in French *teinture tonique et nervine de Bestucheff*, in Russian Бестужевские *капли* [Bestusheff's droplets], in pharmaceutical Latin Tinctura Ferri chlorati aethereal and Spiritus aetheris ferratus.

mixture lost its color and was converted into HCl, readily soluble in water (Gay-Lussac and Thénard 1809).

The first scientific description "*Ueber die chemische Wirksamkeit des Lichtes*" [On the chemical action of light] is given by the Baltic German *Theodor von Grotthuß* (1785–1822) in November 1818 (read to the Courland Society meeting); Grotthuß (1820, pp. 1–67). The English-born American scientist *John William Draper* (1811–1882) began in 1837 experimenting with photography and investigate the photochemically active rays with several publication between 1841 and 1844. *Draper* also studied the chlorine-hydrogen reaction and became convinced that light affected the chlorine alone, converting it from a passive to an active state (Wisniak 2013, and references therein). Independently from *Grotthuß*, *Draper* enunciated in 1841 the principle that only absorbed rays produce chemical change, now named as *Grotthuss-Draper* law. The German physicist *Wilhelm Constantin Wittwer* (1822–1908) who tried to develop an actinometer (not very successful and criticized concerning his experimental approach soon later by *Bunsen*), based on the influence of light on chlorine water, however, was the first who wrote the kinetic equation of a photochemical process in our today's understanding (Wittwer 1855, p. 600); s amount of chlorine, t time, and J light intensity (the symbol J is still used for the radiant and actinic flux, but in the equation replaced by the photolysis rate j which depends from the radiant flux, absorption cross section and quantum yield in dependence from the wavelength): $\mathrm{d}s = -sJ\mathrm{d}t$. *Robert Bunsen* (1811–1899) and *Henry Enfield Roscoe* (1833–1915) first studied more systematical the influence of light on chemical reactions (at the example of the chlorine hydrogen reaction)[206] and proposed the principle of "chemical extinction"; Bunsen and Roscoe (1855, 1857). *Bunsen* and *Roscoe* proved in extensive and careful experiment that the amount of converted substance m is proportional to the light intensity J and time t: $m = k \cdot J \cdot t$.

The German physicist *Ernst Pringsheim* (1859–1917) further investigated this photochemical reaction (Pringsheim 1887) and found evidence for the dissociation of Cl_2 into Cl (due to volume increase), that only certain wavelengths are absorbed and are responsible for chemical activity, and that the presence of water vapor accelerate or even makes the reaction possible. *Pringsheim*'s conclusion that intermediates like H, Cl and O (Pringsheim 1887 p. 426) are likely, and that they are probably gained through interaction with water molecules, is excellent for that time. The idea that light induced the dissociation of chlorine, was first expressed by *Emil Arnold Budde* (1841–1921), German physicist (Budde 1871), and soon later he wrote (Budde 1874, p. 477)

> Das Licht lockert und zersetzt die Chlormolecüle in freie Atome Cl [The light loses and disintegrates the chlorine molecules in free Cl].

206 In German Chlorknallgasreaktion.

The German chemist *Fritz Weigert* (1876–1947) made many pioneering contributions to photochemistry; in an early work he proposed the existence of "*Reaktionskerne*" [reaction nuclei] in photochemical processes, but wrote that the "catalytic effect of the produced substance by the light is of physical nature" (Weigert 1907, 1908); furthermore, he notes that all until hitherto known photochemical reactions are monomolecular (i. e., of first order),[207] and little temperature depending. He further studied the role of chlorine in photochemical reactions as "catalyst"; e. g., the acceleration of ozone decomposition in the presence of chlorine. Yet without knowing it, all these photochemical reactions studied with chlorine first gained Cl atoms as reactant with subsequent radical chains and that presence of humidity is essential for gaining reactive OH radicals.[208]

Hermann Wilhelm Vogel (1834–1898), German photochemist, found that the reduction of Fe(III) in potassium ferricyanide ($K_3[Fe(CH)_6]$) to Fe(II) in potassium ferrocyanide ($K_4[Fe(CH)_6]$), already known since a few decades,[209] is caused by light, gaining Prussian blue ($K[Fe^{III}Fe^{II}(CN)_6]$) and $Fe(OH)_3$ precipitate (Vogel 1871). The Russian-Soviet physicochemist *Vladimir Alexandrovich Kistiakowsky* (1865–1952)[210] in Petersburg[211] who also worked in Leipzig, discovered the first photochemical after-effect phenomena (in German "*Nachwirkungserscheinung*") during decomposition of hydrogen peroxide in the presence of prussiate of potash[212] (Kistiakowsky 1900). Two scientists must be named who contributed significant results in photochemistry and wrote books in this field of research: the Austrian photochemist *Josef Maria Eder* (1844–1944) and the Russian chemist *Ivan Stepanovich Plotnikov* (1878–1955);[213] Eder (1906), Plotnikov (1910, 1920).

---

207 First Bodenstein (1897) found the change of the reaction order under the influence of light in the decomposition reaction of HJ: in dark $2\,HJ = H_2 + J_2$ and in light $HJ = H + J$.

208 Today it is known that Cl_2 photolysis is the starting process, initiating a radical chain ($Cl + H_2$, $Cl_2 + H$, etc.); in the presence of humidity $H_2O + Cl = HCl + OH$ with subsequent OH induced reactions occur and in the presence of air (O_2), dependent from the radiation wavelength, O_3 is produce, gaining many intermediates and product (e. g., ClO, Cl_2O, HOCl, etc.).

209 The salts are known as "*Blutlaugensalz*" [blood-lye salt, prussiate of potash]; whereas Fe(III) is red and Fe(II) yellow. The blue color of Prussian blue, also termed iron blue [*Berliner Blau*] is due to charge-transfer Fe(III)-Fe(II).

210 Влади́мир Алекса́ндрович Кистяко́вский, cited as "Kistiakowsky, Wl." in literature.

211 Full name: Sankt Petersburg [Санкт-Петербург], Petrograd (1914–1924) and Leningrad (1924–1991); colloquial only Petersburg [Петербург] and Пи́тер [Piter].

212 A pure solution of H_2O_2 decomposes under light not noticeable. After addition of few drops of a mixture of red and yellow prussiate of potash, H_2O_2 just strongly decomposes (first-order low) with formation of gaseous oxygen and the reaction proceeds by the same rate law when darkening. *Kistiakowsky* proposed the formation of a catalyst without knowing the mechanism; see also Lal (1952). The charge-transfer complex Fe(III)-Fe(II) in Prussian blue initiates H_2O_2 reduction (to OH) as well oxidation (to HO_2) with subsequent radical chains, finally resulting in $2\,H_2O_2 = 2\,H_2O + O_2$.

213 Иван Степанович Плотников (in Germany also termed *Johannes Plotnikov*), studied at the Lomonosov University in Moscow, dissertation at *Ostwald* in 1905 and assistant in Leipzig until 1911,

At the turn of the twentieth century, photochemistry was barely developed as a science (Roth 2001). Although many photoreactions, namely in organic chemistry, were known, the underlying principles were poorly understood. The photochemical equivalence law, independently formulated the law between 1908 and 1913 by *Johannes Nikolaus Stark* (1874–1957) and *Albert Einstein*, was the beginning of a theoretical understanding of the photochemical process. *Emil Gabriel Warburg* (1846–1931)[214] first experimentally approved this law (Warburg 1911, 1914); he introduced the term "*photochemische Ausbeute*" [photochemical yield], i. e., that fraction of the radiation which is transformed into chemical energy, being depending on the wavelength.

Giacomo Luigi Ciamician (1857–1922) was an Italian photochemist (Nebbia and Kauffman 2007, Albini and Dichiarank 2009), who presented a comprehensive account on his ongoing work in molecular photochemistry and reported several reactions caused by light and ascertained that these were caused by light alone, not heat (Ciamician and Silver 1908). He also emphasized the utilization of solar energy which he deemed second only to nuclear energy (Ciamician 1912); he writes:

> Is fossil solar energy the only one that may be used in modern life and civilization? That is the question (Ciamician 1912, p. 385).
> [...] if our black and nervous civilization, based on coal, shall be followed by a quieter civilization based on the utilization of solar energy, that will not be harmful to progress and to human happiness (Ciamician 1912, p. 394).

The steady-state assumption, introduced by *Max Ernst August Bodenstein* (1871–1942)[215] in 1913 (*Z. phys. Chem.* **85**, 329), allowed the derivation of rate laws and their interpretation in terms of mechanism. Bodenstein (1913) developed on the example of the chlorine-hydrogen reaction a new theory of "*Lichtreaktionen*" [light or photochemical reactions] with the first idea (without using this term) of a chain mechanism (he used the German phrase "*Folge der angenommenen Vorgänge*"; Bodenstein 1913, p. 841), and showing that one photon can produce 10^6 molecules HCl (termed "*Lichtausbeute*", not to confuse with the modern term quantum yield). Despite *Budde* in 1874 already proposed the light-induced decomposition of the chlorine molecule, Bodenstein (1913) proposed that each photon produces two fractional parts of the molecule, a positive residual as "primary" and another, onto that the separated electron goes, as "secondary". He proposed the mechanisms in terms of a primary "light reaction", forming the activated molecule, the process of continuation, and a

established the first photochemical laboratory 1913 in Moscow, moved in 1917 to Berlin (at Agfa) and 1923 to Zagreb, where he established the institute of physics and technology at the university.

214 German physicist who studied in Heidelberg and Berlin; professor in Strasburg (1872), Freiburg (1876), and Berlin (1894); 1905–1922 head of "*Physikalisch-Chemische Reichsanstalt*", moved afterwards to Bayreuth.

215 German physical chemist in Göttingen, Leipzig, Hannover and Berlin; he was first to postulate a chain reaction mechanism.

deactivation reaction (the problem of non-existence of free electrons in gas-phase he "solves" by assuming a fast uptake of the electron by other molecule, e. g., $\ominus + O_2 = O_2^-$; the latter can form ozone, he writes):

$$Cl_2 + \text{light energy} = Cl_2^+ + \ominus$$
$$\ominus + Cl_2 = Cl_2^-$$
$$Cl_2^- + Cl_2^+ = 2Cl_2$$
$$Cl_2^- + H_2 = 2HCl + \ominus$$

Alternatively, *Bodenstein* (Bodenstein 1913, p. 846) formulated $Cl_2 = 2\,Cl$ as primary reaction; in the photochemical ozone decomposition he describes chlorine as the catalyst not being consumed and proposed (or other similar reactions): $Cl + O_3 = ClO_3$ and $2\,ClO_3 = Cl_2 + 3\,O_2$. Soon later, *Walther Nernst* states that the subsequent "dark" reactions must be separated from the first photochemical process (the *photolysis* he scribes) and formulates the reaction chain (Nernst 1919): $Cl + H_2 = HCl + H$ and $H + Cl_2 = HCl + Cl$, where small pollutions can also react with free hydrogen and chlorine atoms and thus reduce the "chain". *Nernst* shows further that bromine atoms cannot react with hydrogen molecules, and thus the photochemical equivalence law is valid for bromine photolysis.[216] He writes (Bodenstein 1913, p. 336):

> Es scheint, daß das Einsteinsche Gesetz berufen ist, in der Photolyse eine ahnliche Rollc zu spielen, wie das Faradaysche Gesetz in der Elektrolyse [It seems the Einstein's law is appointed to play a similar role as Faradays's law in electrolysis].

The Swiss chemist *Armand Müller* (unknown living data) in Zürich is likely the first who speculated on the formation of "*nasciertes Wasser*" [nascent water], denominated $\beta\,H_2O$ in contrast to "normal" $\alpha\,H_2O$ (Müller 1873, p. 45):

> [...] dass beim Auffall des Lichts durch das Nasciertwerden des Wassers die Verbindungen, je nach ihrer chemischen Natur zerlegt werden, indem sauerstoff- und wasserstoffhaltige Producte auftreten [... that by striking light through nascence of water the compounds decompose according to its chemical nature whilst gaining oxygen and hydrogen containing products].

Thus, *Müller* also explains the bleaching of linen under the influence of light and humidity as indirect water decomposition, proposing $3\beta\,H_2O + N_2 = O_3 + 2\,NH_3$ and $2\beta\,H_2O + 2\beta\,H_2O + O_2 = 2\,H_2O_2 + 2\alpha\,H_2O$. His termination "*chemische Intensität des Himmelslichtes*" [chemical intensity of sky light] can be regarded as the first idea of atmospheric photochemistry (in this connection he used the term *chemical meteorology*). Some 50 years later two other Swiss chemists (Baur and Neuweiler 1927, p. 906) write:

---

216 Bodenstein and Lütkemeier (1924) applied the stationary principle for the first in the photochemical reaction between bromine and hydrogen.

> Man nimmt allgemein an, dass bei der Rasenbleiche durch Lichtwirkung Sauerstoff aktiviert werde und dass der aktivierte Sauerstoff das Bleichmittel abgebe [. . .] Die Cellulose selbst dürfte ein Sensibilisator sein, [. . .] d. h. es bildet sich Hydroperoxyd zugleich mit irgendeiner Oxydation [It is generally assumed that at grass bleaching oxygen is activated through the action of light, and that the activated oxygen provides the bleaching agent, [. . .] The cellulose itself should be the sensitizer [. . .] i. e., hydroperoxide is formed together with someone oxidation].

The authors state in a footnote that obviously nobody did try to detect hydroperoxide[217] in the bleaching solution. *Emil Baur* (1873–1944), Swiss electrochemist in Zürich, understood (Baur 1918) *photolysis* – the photochemical process – as the formation of an electrical polarization within the molecule, hence as a "molecular electrolysis".[218] He explains the formation of ozone in the following steps:

$$\text{photopolarisation:}\quad O_2 + h\nu = O_2 \begin{cases} \oplus \\ \ominus \end{cases}$$

$$\text{dissociation:}\quad O_2 + h\nu = 2\ominus + O_2^{\bullet\bullet}$$

$$\text{ozone formation:}\quad O_2 \begin{Bmatrix} \oplus + \ominus \\ \ominus + \frac{1}{2} O_2^{\bullet\bullet} \end{Bmatrix} = O_3$$

This conception of "molecular electrolysis" *Baur* used for more than 10 years (Baur and Neuweiler 1927), namely for the process we call today *photosensitization*. *Emil Baur* termed a substance like ZnO and dyes "*Lichtempfänger*" [photoreceiver] and activated oxygen as cathodic gained hydroperoxide (OH' stands for OH^- and $H^\bullet$ for H^+, i. e., the process occurs in aqueous or interfacial wetted phase):

$$E \begin{cases} \oplus + OH' = \frac{1}{4} O_2 + \frac{1}{2} H_2O \\ \ominus + H^\bullet + \frac{1}{2} O_2 = \frac{1}{2} H_2O_2 \end{cases}$$

In modern terms, this equation corresponds to $E^\ominus + O_2 \rightarrow O_2^- \ (+H^+ \rightarrow HO_2)$. However, the role of the superoxide anion (O_2^-) in autoxidation processes became clear only after the 1950's, interestingly because of studying bleaching processes. Baur and Neuweiler (1927, p. 906) write:

---

217 It is hydrogen peroxide (H_2O_2) meant. Baeyer and Villiger (1900) proposed to rename "*Wasserstoffsuperoxyd*" [hydrogen peroxide] in "*Hydroperoxyd*" [hydroperoxide]. Note also that in literature before 1900, the formula of H_2O_2 was written as HO^2, not to confuse with the hydroperoxo radical HO_2. Obviously in the 1930s, the original name "*Wasserstoffsuperoxyd*" was introduced again. Other German names where "*Waserstoffperoxyd*" and (rarely) "*Wasserstoffhydroperoxid*". Since about 1950, only the term "*Wasserstoffperoxyd*" is used.

218 Today, photolysis (or photodissociation) is a chemical process by which chemical bonds are broken as the result of transfer of light energy (direct photolysis). Due to photosensitization processes (indirect photolysis) also compounds which do not absorb solar energy themselves are subject to photolytic degradation.

> Rückblickend können wir sagen, der Träger der bei aëroben Photolysen zu beobachtenden Sauerstoffaktivierung ist Hydroperoxyd. Seine Bildung beruht auf der Entstehung von naszentem Wasserstoff im photolytischen Prozess und verrät den kathodischen Zweig der von der Theorie geforderten Wasserphotolyse [Retrospectively we can state: the carrier of oxygen activation observed while aerobical photolysis is hydroperoxide. Its formation is based on the production of nascent hydrogen within the photolytic process and gives the cathodic branch of water photolysis according to the theory].

This phrase can be transferred into the reaction pathway $W + h\nu \rightarrow W^+ + H_2O^-$ ($\rightarrow OH^- + H$), following by $H + O_2 \rightarrow HO_2$. Baur and Neuweiler (1927, p. 907) interpret experiments carried out by Chakravarti and Dhar (1925) who found that dyes are fast bleached in the presence of sunlight, atmospheric oxygen and an oxidable substance, and similar experiments by Gaffron (1927) with chlorophyll, who defined the process as "*photochemische Sauerstoff-Übertragung*" [photochemical oxygen transfer] that the formation of hydroperoxide is favored under such conditions.[219] Benrath (1912, p. 137) clearly expresses the findings hitherto that during the photolysis of liquid water under ultra-violet light the reaction is not simply $2H_2O \rightarrow 2H_2 + O_2$ but that first hydrogen is gained and hydroperoxide can be detected, thus $2H_2O \rightarrow H_2O_2 + H_2$ (first proposed by Tian 1911), which also easily decomposes so that at a certain concentration its formation and decomposition is balanced and the escaping gas contains equivalent amounts of hydrogen and oxygen. This is also a clear evidence of intermediary formation of hydroxyl radicals: $H_2O \rightarrow H$ ($+ H \rightarrow H_2$) $+ OH$ ($+ OH \rightarrow H_2O_2$); hydrogen peroxide further decomposes to H_2O and O_2. However, under lower atmospheric conditions, direct water photolysis plays no role because of missing strong UV rays.

The term *photosensitization*, likely first used by the British chemist *Hugh Stott Taylor* (1890–1974) who mainly worked in catalysis research, was used to describe an initial photolysis (primary process) with subsequent radical chain reactions; he writes that *Weigert* sees in the photosensitization of ozone decomposition by chlorine, the simplest possible photo-process (Taylor 1926, p. 560).

The first idea (but only through today's interpretation) on formation of radicals comes from *Arthur Henry Downes* (1851–1938) and *Thomas Porter Blunt* (1842–1929), both British physicians who spend most of his life in public health service and are known for the first observation that sunlight destruct bacteria in water (Downes and Blunt 1877),[220] soon follows by another paper that this effect is due to the presence of oxygen in solution, and not depends from light *per se*, but that "destruction of an organic body in light is due to the oxidation of its hydrogen" (Downes and Blunt 1878, p. 210). Afterwards, Downes and Blunt (1879a) have shown that this decontamination

219 *Hans Gaffron* (1902–1979) was a German biochemist, who studied at *Wilhelm Traube* and worked at *Otto Heinrich Warburg* (1883–1970), sun of *Emil Warburg*, in Berlin. The Indian chemist *Dhar* (see p. 276) is the first who studied photochemical processes in atmospheric waters; see more details about aqueous-phase chemistry in Chapters 4.4.4.2 and 4.5.3.2.

220 An excellent contemporary review on the influence of light on organisms is given by Raum (1889).

occurs more efficient in presence of oxalic acid which is destroyed by sunlight in the presence of *external* oxygen; they proposed the formation of C_2O_4 "radicals" which decay into CO_2 and H_2O. The photoinduced decomposition of dissolved oxalates and oxalic acid were often investigated, for the first in detail by Berthelot (1914)[221] who found no effect under sunlight ($\lambda > 300$ nm) but with UV and detected CO_2 (>60 %), H_2 (around 30 %), CO (10 %) and traces of formic acid, corresponding to the equation COOH–COOH = $2\,CO_2 + H_2$. Baly et al. (1921)[222] studied the photosynthesis of hydrocarbons from carbon dioxide photolysis, and made two essential comments, first that in the photolysis of oxalic acid first an "activated" CO_2 molecule is gained and afterwards formaldehyde, and second that formation of formaldehyde is not only possible under strong UV light but also in the visible light region, but then in the presence of a *photocatalyst*:

> Under the influence of the light, the oxalic acid is decomposed to carbon dioxide, and this gives, in turn, formaldehyde [. . .] It is probable that the carbon dioxide at the moment of its formation is more active (Baly et al. 1921, p. 1030).
> Our results show conclusively, not only that it is possible to produce formaldehyde from carbon dioxide and water by the direct action of light of very short wave-length ($\lambda = 200$ nm), but that it is also possible to produce the formaldehyde from carbon dioxide and water in visible light in the presence of a photocatalyst (Baly et al. 1921, pp. 1032–1033).

Allmand and Reeve (1926) studied the decomposition of oxalic acid by UV light again, confirming the gross equation by *Berthelot*, and propose as primary product formic acid having high energy content (HCOOH*) that decays into $H_2 + CO_2$ or $H_2O + CO$. They adopt the idea of *Baly* et al. that first an "activated" CO_2 molecule is gained that reacts with water according to $CO_2^* + H_2O \rightarrow HCHO + O_2$ (Allmand and Reeve 1926, p. 2884).

Only in the early twentieth century, photochemistry (the photolytic dissociation of oxygen) was recognized to be the key for ozone formation in the upper atmosphere. The importance of oxygen radicals (like OH and HO_2, later termed reactive oxygen species ROS) in combustion processes were recognized in the late 1920s and that of photochemistry in the formation of the hydroxyl radical (OH) as a key oxidant in the atmosphere about three decades later:

$$H_2O + O_2(\leftrightarrow O_3) \leftrightarrow OH \leftrightarrow HO_2 \leftrightarrow H_2O_2 \rightarrow H_2O + O_2.$$

221 *Daniel Berthelot* (1865–1927), sun of *Marcellin Berthelot*. He did fundamental research on the conductivity of electrolyte solutions, thermodynamics, pyrometry, and photochemistry. *Daniel Berthelot* used ultraviolet radiation from the mercury vapor lamp to carry out a series of oxidation and polymerization reactions, and proved that many reactions carried on by electrolysis could also be performed by photolysis. In doing so he discovered a myriad of unknown phenomena (Wisniak 2010).

222 *Edward Charles Cyril Baly* (1871–1948) was a British chemist who worked in the field of photosynthesis.

The ubiquitous role of photocatalytic process in water chemistry, namely of interfacial electron transfer onto oxygen ($O_2 \rightarrow HO_2 \rightarrow H_2O_2$) and ozone ($O_3 \rightarrow OH$), becomes clear only at the end of the twentieth century:

$$H_2O + H_2O = H_2O^+[H^+ + OH] + H_2O^-[H + OH^-] = H_2O + H(+O_2 = HO_2) + OH(= 1/2\,H_2O_2).$$

Nonetheless, correct interpretation of early observations is only possible with today's knowledge of photochemistry, photocatalysis, radical reactions, and electron transfer processes.

2 Investigation of waters in the air

> Air and Water are primigenial or first-born Elements, and ever unchangeable, by cold, or heat into each other (Helmont 1662, p. 57)
> [. . .] we plainly perceive [. . .] that though Water may be rarefied into invisible Vapors, yet it is not really chang'd into Air, but onely devided by heat and scatter'd into very minute parts, which meeting together in the Alembick or Receiver, do presently return into such Water as they constituted before. And we also see, that even Spirit of Wine, and other subtle and fugitive Spirits, though they easily fly into the Air, and migle with it, do yet in the Glasses of Chymists easily lay aside the disguise of Air, and resume the divested form of Liquor. And so volatile Salts, as of Urine, Harts-horn etc [. . .] disperse themselves through the Air [. . .] (Boyle 1662, p. 83).
> The naturalist [. . .] recognizes that water is mass, in movement and as solvent, is one of the greatest agents of Nature and influences in a thousand ways its phenomena (Fourcroy 1800, p. 8).

Air and water were originally "elements" in ancient Greek and were transmutable; they represented two kinds of the "layer of mist" (atmosphere). Dark or thick air was mist or cloud, hiding the Gods (who lived in the upper air or sky; the Aether). Different terms are presented that describe fog and clouds in connection with the history of the process of understanding. It follows a list of historic terms for atmospheric waters (fog, vapors, drops, etc.), fume and exhalation in German, British English, French, and Latin (in the table on next page). In brackets American English translation of German historic terms (not used in historic British literature).

The ancient idea of transmutation (air ↔ water) survived until the end of the eighteenth century, when it was finally rejected by *Lavoisier*; however, Boyle (1662) already doubted it from his experiences of "New Experiments Physico-Mechanical Touching the Air" (Boyle 1662, p. 91): "[. . .] thought not that Air may be generated out of the water, yet that in general Air may be generated anew". For the first time, *Boyle* used the term *Æoliphile* (Boyle 1662, p. 85) as an "air-loving" substance,[223] in particular for water; "[. . .] that water may be rarefied into true Air".

While fog is always equated with "thick air" and darkness, the terms ὀμίχλη [mist], ἀχλυς[224] [gloom], αὔρα [breeze of air – *aura*], νεφέλη or νέφος (*Nephele* – cloud Nymph in Greek mythology) were in use. From *nephos*, the Latin terms: nūbēs, nūbilus and nebula are derived (expressing its gloom and darkness).[225] The German word "*Dunst*" is defined by *Johann Georg Krünitz* (1728–1796) in the "*Oeconomische Encyclopädie*" (Vol. 9, 1785) in the sense of fog, fume and vapors (in modern terms, "trace gases").[226]

---

223 Lampadius (1806) used in German for the first time the term "*Atmosphärilien*" [aerial matter] in the sense of all constituents of air with the exception of gases.

224 Acherōn is a mythological river in the underworld, identified by poets with the underworld (= darkness).

225 Old High German: *nebul*, Old Slavic: *nebo* (небо = sky in modern Russian), Sanskrit; *nabhas* [fog, vapour, clouds, air, heaven].

https://doi.org/10.1515/9783110732467-002

German	English	French	Latin
Wolke	cloud	nues	nubes
Nebel (light)[e]	mist, myst	bruillas, bruillasse	nebula
Nebel (dense)[e]	fog, brume	brouillard, brume	nebula
Dunst	haze, brume, damp	fumeés	nebula, vapor
Dampf	damp [vapor]	vapeur	vapor
Wasserdampf	steam	la vapeur d'eau	vapor aquae
Schwaden	damp, vapour	nuée	vapor
Brodem	vapour	buée	vapor
Dampfluft	vapoury air[a]	vapeur d'eau invisible	vapor caeli
Feuchtigkeit	moisture	humidité	humidus
tropfbares Wasser	drop-able water	gouttes d'eau	guttulis aquae
Dunstkörperchen	[haze particles]	petit corps opaques	–
Wassertröpfchen	[water droplets]	spères (d'eau)	stillæ
Bläschen, Bläsgen	vesicles, spherules	vésicles, spérules	vesicula, bullulae
Tropfen	globules, drops	gouttes (d'eau)	guttae
Tröpfchen	droplet	gouttelettes	guttula
Rauch[d]	smoke, smoak, fume	fumeé	fumus
Ausdünstung	exhalation, effluvia	exhalaison, effluence[b]	exhalatio, effluvium
Verdampfen	evaporation	evaporation	exhalo

[a]also: invisible vapour
[b]also: *émanation*
[c]also: vaporization; volatilization
[d]also: *Qualm*
[e]no separation in German

"*Grimm*'s *Wörterbuch*" defines "*Dunst*":[227]

> Im eigentlichen Sinne: eine Menge von Wasserdämpfen, die in tropfbar-flüssigem Zustande unmittelbar an oder über der Erdoberfläche schweben und die Luft mehr oder weniger undurchsichtig machen" [In a proper sense: a lot of water vapors that float in droplet-liquid state near to or at the earth's surface and making the air more or less non-transparent].

226 *Grimm* sets a linguistic relation between *Dunst* and the English (Old North German and Anglo-Saxon) dust = "*Staub*".

227 In English, there exist many terms for "*Dunst*": mist, haze, damp, vapour, brume, fume, aura. According to "Deutsches Wörterbuch" (Vol. 2, Col. 1559, shortened) also: "für dünne, nasse oder trockene Flüssigkeit die in die Luft steigt, meist sichtbar ist, doch auch nur durch den Geruch empfunden wird; vergl. *dampf, duft, brodem, qualm, schwaden*" [for thin, wet or dry liquid, rising into air, usually visible but also only conceived by smell; compare [. . .]; *dampf* = vapor, *duft* = flavor, *brodem* (no longer used in German) = vapor or steam, qualm troday's German: *Rauch*) = smoke, *schwaden* (in today's German only used for plumes of cooling towers) = plumes. Old High German *tunst*; Middle High German, Swedish and Danish *dunst*. In Gothic, Old-Saxon, Old Friesian, Low German and Dutch *dunst* does not exists; it is related to the Gothic *þinsan* and the lost *þinan* (to strech); Old North German, Anglo-Saxon and English *dust* (*Staub*). Nowadays, we translate atmospheric *Dunst* with haze.

The English "fog" is defined as (OED)[228] "thick, obscuring mist". The origin of this term is dated to around 1540 and it is linguistically similar to the Old Norwegian "*fok*", the Dutch "*vocht*" and the German "*Feucht*" [moist]. The Old English terms *nebule* and *nifol* [dark, gloomy] were taken directly from Latin. The differences between *haze*, *mist*, and *fog* are not reflected in German[229] where one term ["*Nebel*"] characterizes these words, which is likely to be an effect of the moist "English climate" on language (OED).

Obviously, the English term *mist* derives from ὀμίχλη;[230] as does the Swedish and Norwegian: "*mist*". In German, "*mistig*" means (in colloquial language) dirty.[231] Today, the *mist* is separated from *fog* mostly by droplet size: mist consists of larger drops that have a larger tendency to precipitate (more exactly: to sediment).[232] In Old English, instead of *mist*, the term *brume*, derived from the Latin bruma (winter) was used. In French, "*brume*" is the meteorological term for fog with visibility more than 1 km (which in German is termed "*Dunst*")[233] and "*brouillard*" for fog with visibility less than 1 km (in German "*Nebel*"). From that stem "*brumaille*" and "*brouillasie*" (fine thin fog and fog shower). German "*Seerauch*" ("*Meerrauch*", "*Flussrauch*") is evaporating fog above water (sea smoke, sea mist, water smoke, steam mist), which in French is "*fumée de mer*".

The Proto-German word "*Rauch*" [smoke, fume; Latin: fumus], which was always linked with a house fire and some burning material – in a chemical sense, a mixture from soot, gaseous (including water vapor), and other solid combustion products –

228 OED – Online Etymology Dictionary 2001–2012 Douglas Harper: http://www.etymonline.com

229 The chapter "Of Mists and Fogs" in Prout (1834 p. 312) is reduced in the German edition (Prout 1836, p. 214) to "*Vom Nebel*" [Of fog]. The opening sentence of this chapter in Prout (1834) "When mists, from other causes, are general and extend to considerable heights above the earth surface, they acquire the name of fogs" was completely declined in the German edition, because mist = fog = *Nebel*. Scotch mist means very thin rain (Lloyd and Noehden 1836). In many British publications of the nineteenth century the term mist was used instead of fog. Giberne (1890) wrote: "A mist is commonly distinguished from a fog as being made of rather larger drops, therefore feeling more wet".

230 In Lat. mingō and mejo (from that also derived the English "mist"; in its primordial meaning also urinate but in Sanskrit *mih*, *megha* (cloud, mist). The English "misty" (foggy) was in Old English "mistig" (see also next footnotes 231 and 232).

231 On the other hand, the German word *Mist* means dung but in a colloquial sense also "brass farthing" – again, evidence of "gloom" and "evil" in mist (fog).

232 Such weather in colloquial German is termed "*Mistwetter*": dark, cold and wet. Ehrenberg (1849, p. 122) cites the weather record of the British vessel Roxburgh on 4.2.1839 at Cape Verdean: "*Der Himmel war überzogen, das Wetter mistig* [obviously the English "misty" was used in the record] [...]" [Sky was overcast, weather misty]. The term "*mistig* = *neblig*" [foggy] is unusual in German; the adjective means: full of "*Mist*", dirty.

233 After Sachs (1911) *brume* means "thick fog", i. e., fog with visibility less than 1 km – also exactly reverse in modern meaning. "*Bruine*" is translated after Sachs (1911) as "*Staubregen*" [dust rain]; an unknown word in modern German. However, "*Staubregen*" (in French *pluies de poussières*) denotes dust fall events. Nowadays, *bruine* = *Sprühregen* = drizzle (identical with crachin); *brume* = *sèche* = haze = *Dunst* after METAR (meteorological aerodrome report).

was and still is in synonymous use with "*Dampf*" [vapor]. *Grimm* defines "*Dampf*" as (Deutsches Wörterbuch, Vol. 2, Col. 714, shortened):[234]

> [. . .] ein dichter, sichtbarer, feuchter Rauch oder Dunst, schwerer als Duft, leichter als Qualm und Schwaden, Fumus, Vapor, Exhalatio [. . . dense, visible, moist smoke or haze, heavier than flavor, lighter than fume and billow].

In English, *steam* is used only for water vapor, whereas *damp* (the similarity with German "*Dampf*" is significant: Old High German: *damph*), *vapor*(*s*), *smoke*, *vapeur*(*s*), *fumée*(*s*) are also used for other (evaporating, escaping) substances. The German term "*Ausdünstung*" (in modern usage, evaporation of water, exhalation and more generally, emission for all other substances) denotes the process (evaporation) as well as the product (vapor, gas); French: *émanation*, *exhalaision*, *effluence*, *evaporation*, *vaporisation*, *volatilisation*; English: *evaporation*, *effluvium*.

Atmospheric water physically includes water in the atmosphere in all aggregate states, i. e., as vapor, liquid (in droplet form), and solid (ice particles). In his historical valuable "Report on the Present State of Knowledge concerning Mineral and Thermal waters" *Charles Daubeny* includes water "[. . .] circulating through the atmosphere [. . .], and writes further "[. . .] atmospheric water, as being the purest form of any which nature presents, will supply us with the fewest materials for comment" (Daubeny 1836, p. 1). The historic term "atmospheric waters" means nowadays *hydrometeors*, i. e., meteoric water.[235] From historic reason, *dew* was termed among that "waters" before Wells (1814) stated that dew is not from water drops falling from the heaven. Gaseous water in the air (water vapor) is part of the main constituents (Chapters 2.2.2 and 4.2.3) and plays a crucial role in atmospheric chemistry: it is the source of hydroxyl radicals via the reaction $H_2O + O(^1D)$. Gaseous water is further essential in the process of nucleation (Chapter 3.3.2). Liquid water in the air, from the point of chemistry, is first a solvent for gaseous and solid substances, and secondly, it is a reagent in aqueous-phase chemistry, providing the hydrogen (H^+) and hydroxyl ions (OH^-). Liquid water in air exists only as droplets, and is formed through condensation of gaseous water

234 Interestingly, the German adjectives "*feucht*" [moist] (or "*wässrig*", "*nass*" – aqueous, wet) and "*trocken*" [dry] were used for the same term in the past to differentiate the state of matter of a general atmospheric phenomenon, that is, the "visible air" ["*sichtbare Luft*"]; the non-gaseous components for which all terms, like "*Nebel*", "*Dunst*", "*Dampf*" and "*Rauch*" were used. Today the adjectives "dry" [*trocken*, *sec*] and "wet" [*nass*] are used in atmospheric science only for the process of deposition to distinguish between dry and wet deposition. In German, a distinction is made between "*feuchter Dunst*" (moist haze, i. e., fog with visibility between 1 and 5 km) and "*trockener Dunst*" (dry haze, greater visibilities but still with a discernible opacity of the atmosphere). The latter represents cloud condensation nuclei in larger numbers, activated and exceeding the deliquescence point (almost 60–70 % relative humidity). However, it should be noted that "*Nebel*" ["*broullaird*"], "*Dunst*" ["*brume*"], "*Schwaden*" ["*nuée*"] and "*Brodem*" [""*buée*"] were used in French exclusively for water.

235 Today, meteoric water is a geological term that denotes ground water that has recently originated from the atmosphere.

onto particles, termed heterogeneous nucleation (dew is droplet-like water through condensation on surfaces of the earth). Liquid water as droplets but also frozen in different solid form removes from the atmosphere via precipitation, and evaporates back to gaseous water. Without understanding condensation and evaporation, the chemistry of rain, fog, cloud, and dew water cannot be understood.

2.1 Introduction

The *phenomena*, fog and clouds, precipitation (rain, snow, hail) as well as dew are well described since Antiquity (Möller 2008, 2014b,c, 2020). The word "dew" appears 34 times in the Bible, and it is primarily seen as a blessing. In Genesis 27:28, *Isaac* blesses *Jacob* with these words: "May God give you of the dew of heaven and of the fatness of the earth and plenty of grain and wine". The Roman physicians considered rainwater drinkable; *Marcus Vitruvius* (c 80–70 to 15 BC), Roman architect and civil engineer, wrote (De Architectura, Liber XIII, 2.1); cited after Erndtel (1730):[236]

> Itaque, quæ ex imbribus aqua colligitur, salubriores habet virtutes, quod eligitur ex omnibus fontibus levissimis subtilibusque tenuitatibus, deinde per æris exercitationem percolate tempestatibus liquescendo pervenit ad terram [Therefore rain-water has more wholesome qualities because it comes from the lightest and most finely reneous of all sources; then filtering through moving the air, it liquifies in storms and so reaches the earth].

Erndtel (1730, p. 122) further cites *Hippocrates*[237] who mentioned contaminated rainwater by thunder:

> Atque ex hac ipsa pluvial aqua Hippocrates 6, Epidem, S. 4. Eligit illam, quæ media aestate decidit, & ab æthere com tonitru excreata est [Thus rain-water, fallen in the summer and from the ether with exhalations from thunder, Hippocrates termed].

On the contrary, snow water they accounted to be fruitful for crops as the Roman writer *Aulus Gellicus* (125–180) writes:[238]

> Qui aquam nivale, frugibus et arboribus foecundam dicerent, sed hominibus potu nimis insalubrem ess, tabemque et morbus sensim, atque in diem longam visceribus inseminare [The snow water is fertile for trees but for men insolubrious when drinking it too much, and they became gradually ill, namely the intestines].

236 *Christian Heinrich Erndtel* (1676–1734) was a German physician, botanist and meteorologist in Dresden and Warshaw.

237 *Hippocrates* of Kos (about 460–360 BC) was a Greek physician.

238 Cited after "Des Flavius Vegetius Renatus fünf Bücher über Kriegswissenschaft und Kriegskunst der Römer. Aus dem Lateinischen übersetzt von Felix Joseph Lipowsky. J. E. v. Seidel, Sulzbach (1827), 320 pp." (p. 269). *Flavius Vegetius Renatus* was a writer of the Later Roman Empire (fourth century).

In line with naturalists and philosophers since *Aristoteles*, *John Baptiste van Helmont* "defined" terms like fume,[239] smoke, fog, mist, and cloud (Helmont 1662, p. 68):

> [...] now burns, and sends forth a fume or smoak.
> Not indeed, that fume is air, but is either a vapour, or drie exhalation, [...] Whatsoever exhalations therefore do from the Earth climbe upward, and are jouned in Clouds; for this cause also, those Clouds do stink, [...], because as many as have passed over the *Alps* with me [...]; but the rain water collected thences, how sweet and without favour it is, and almost incorruptible.

The English natural philosopher and courtier Sir *Kenelm Digby* (1603–1665) considered dew to be a source of nutriment (cited after Klein et al. 1988, p. 928):

> But above all, Dew makes all plants luxuriate and prosper most. Now, what may it be that endues these liquors with such profilic nitre? The meer water, which is common to all of them, cannot be it; there must be something else enclosed within it, to which the water servs but for a vehicle. Examine it by the Spagyrick [a Paracelcean term for alchemy] and you will find it is nothing else but a nitrous Salt which gives fecundity to all things; and from this salt, not only vegetables, but also all minerals draw their Origin.

We must assume that fragments of phrases like[240]
- halations subterranean [...] ascend yet higher, they dissolve either into rain, or congeal into snow, dew, and hoar frost [...],
- subterranean vapours, fruitful rain, distilled rain water [...],
- the vapours of the earth which rise, the same make or cause the rain, snow, and hoar-frost [...]

were transferred since the ancient time from one to another (alchemistic) writer. The "knowledge" that dew, rain, and snow are useful for plant growth was later (in the

239 The word "plume" was first used in 1878 in the meaning of "a longer stream of smoke" (onlineetymology). It is derived from Latin pluma, adopted in French *plume* for "soft feather". *Helmont* used the word nubes (Helmont 1648, p. 79), translated into "clouds" (Helmont 1662, p. 68) like also used by Evelyn (1661, p. 8) as "Clouds of Smoake" (in German "*Rauchwolke*" still named today) in the sense of todays "plume". Hence, the term "cloud" in this connection should not be assimilated with the meteorological cloud; however, Amsterdam was also known (like London) for frequent fog events and we can assume "town fog" (Chapter 3.1.2).

240 Found in: The last will and testament of Basil Valentine, monk of the Order of St. Bennet: which being alone, he hid under a table of marble, behind the high-altar of the cathedral church, in the imperial city of Erford, leaving it there to be found by him, whom God's providence should make worthy of it: to which is added two treatises, the first declaring his manual operations, the second shewing things natural and supernatural: never before published in English. Printed by S. G. and B. G. for Edward Brewster, London (1671), 530 pp. (translation from the German original). *Basil Valentine* is the Anglicized version of the name *Basilius Valentinus*, ostensibly a 15th-century alchemist, possibly Canon of the Benedictine Priory of Saint Peter in Erfurt, Germany but more likely a pseudonym used by one or several 16th-century German authors.

early eighteenth century) consolidated in the first textbooks (essays) on agriculture by *Christian Wolff*, *John Randall*, and *Francis Home*.[241]

Due to relatively simple ground-based sampling techniques (gauges, linen), rain, dew, and fog were sampled and investigated already by alchemists (Möller 2008, 2020). *Benedetto Castelli* (1578–1643),[242] a friend of *Galilei*, used the first rain gauge in 1639 to measure rainfall. *John Evelyn* (1620–1706), English educated country gentleman and author of several books, one of the founders of the Royal Society in London (1660) wrote (Evelyn 1661, p. 32):

> [. . .] this continual *Smoake*, which ascending in the day-time, is, by the descending Dew, and Cold, precipitated again at night: And this is manifest, if a piece of clean Linnen be spread all Night in any Court or Garden, the least infested as to appearance: but especially if it happens to rain, which carries it down in greater proportion, not only upon the earth but upon the *Water* also, [. . .]

This phrase can be interpreted as the first description of a "deposition gauge" (total precipitation sampler); Figs. 2.1 to 2.3. The "dew sampler" shown in Fig. 2.1 also may be used for rainwater collection when putting below a vessel or similar device for water sampling as described 100 years later by *Marggraf* (next chapter).

Evelyn expresses deposition ("[. . .] condens'd [. . .] Air [. . .] descending") of nitre from the air and emission ("[. . .] the Earth sends up [. . .]") from the earth (Evelyn 1676, pp. 173–174):

> In the mean time, of all *Waters*, that which descend from heaven, we find to be the richest, and properest in our work, as having been already *meteoriz'd*, and circulated in that great digestory, inrich'd and impregnated with astral influences from above at these propitious Seasons; whence that saying, *Annus fructificat, non Tellus*, has just Titles we every year Revolution behold and admire, when the sweet Dews of *Spring* and *Autumn* (hitherto constipated by cold, or consumed with too much heat) begin to be loosened, or moderately condens'd, by the more benign temper of the Air, impregnated the prepared Earth to receive the *Nitrous Spirits*, descending with their balmy pearls, yet with such difference of more or less benign, (as vapours haply, which the Earth sends up, may be sometimes qualified) that nothing is more uncertain.

Together with the phrase that follows, it is evidence for an early detecting of atmospheric nitrous nitrogen (ibid., p. 98):

---

241 Wolff, C. (1723) Vernünfftige Gedancken von den Würckungen der Natur. Renger, Halle, 523 pp. Randall, J. (1764) The semi-Virgilian husbandry, deduced from various experiments: or, an essay towards a new course of National Farming, with the Philosophy of Agriculture. Printed for B. Law, Thomas Field, and John Wilkie, London, 356 pp. Home, F. (1759) The principles of agriculture and vegetation. Printed for A. Millar, and A. Kincaid and J. Bell at Edinburgh, London, 207 pp. *Christian Wolff* (1679–1754), German enlightener and philosopher. *John Randall* (c. 1714–?), English agriculturalist. *Francis Home* (1719–1813) Scottish chemist and physician.

242 Born as *Antonio Castelli*, Italian mathematician in Pisa (replacing *Galileo*), and later in Rome; one of his students was *Evangelista Torricelli*.

Figure 2.1: Plate 4 from "*mutus liber*" (1677); Canseliet (1991). Compare this "sampler" with that shown in Fig. 2.23 on p. 279 used by *Wells*.

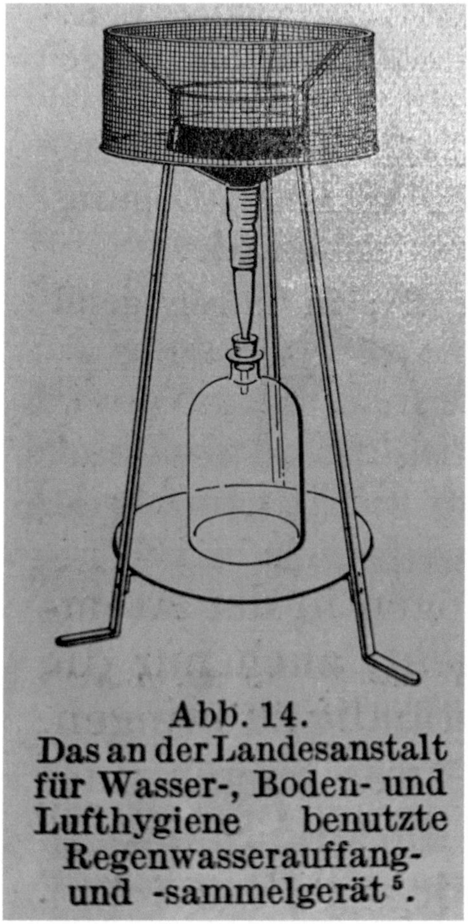

Figure 2.2: Rain gauge used by "*Landesanstalt für Wasser-, Boden- und Lufthygiene*" [State Institute of water, Soil and Air Hygiene] after *Liesegang* (Lehmann and Heller 1940, p. 588); the gauge was produced by Bergmann-Altman K.-G. [Ltd.] Berlin NW7.

> To which let me add, impregnating *Rains* and *Dews*, *cold* and *dry Winters*, with store of *Snow*, which I reckon equal to the richest manures, impregnated as they are with *Celestial Nitre*.

Evelyn was never an alchemist, but he must become familiar with (al)chemists or its tractates and draw his conclusions on "chemistry" of rain and dew, which he formulated already relatively clear and unconfused.[243] A more detailed history of the alchemistic treatment of atmospheric waters can be found in Möller (2020, pp. 154–162).

Collecting of rainwater, fog, and dew also goes back many hundreds of years to search after the philosopher's stone by alchemists; some remarkable observa-

243 Miller (1905a) cites this phrase, but notes that it must not be an indication of detection of nitrate, because at this time the term "nitrous air &c. were applied to what is now known as oxygen". I do not agree with that opinion, saltpeter and nitre was long known before but not oxygen in any kind of appearance.

Figure 2.3: Contemporary illustration of "*Regenmesser*" [ombrometer]; after Zimmermann (1865, p. 227).

tions were made that we can explain (or interpret) with our today's knowledge of atmospheric chemistry. In the eighteenth century, the chemical analysis ("*Scheidekunst* – the art of separation) developed with the increasing interest to study the composition of bodies; first quantitative analysis was carried out by gravimetry; however, stoichiometries as a precondition to describe chemical compounds developed only by the end of the eighteenth century.[244] In the eighteenth century, very few rainwater studies were made from the analytical chemistry point of view (Chapter 2.3.1.1), continued in the first half of the nineteenth century (Chapter 2.3.1.2). With that, applied interests appear, like water quality, agricultural chemistry (Chapter 2.3.2), air and water pollution, sanitary chemistry, public sanitation,[245] air pollu-

244 As a founder of stoichiometry applies *Jeremias Benjamin Richter* (1762–1807) who wrote "*Anfangsgründe der Stöchyometrie oder Meßkunst chymischer Elemente*" (Breslau and Hirschberg 1792).

245 In German, *sanitary* and *sanitation* means *Hygiene* (although the English term *hygiene* is also in use), see also next footnote. In the nineteenth century, the sanitary chemistry, and generally sanitary science and consequently sanitary institutes were established.

tion control,[246] and finally geochemistry, air chemistry, and biogeochemistry (Chapter 1.2.1).

The methods of classical qualitative analysis were completed with the first edition of *Fresenius* book (p. 61) "*Anleitung zur qualitativen chemischen Analyse*" (Fresenius 1841). Although all principles of gravimetry were known since *Berzelius*, *Fresenius* completed the quantitative analysis of mineral waters with his book "*Anleitung zur quantitativen chemischen Analyse*" (Fresenius 1847). After 1850, colorimetric methods were stepwise developed, and together with gravimetry, they remained the essential methods for analyzing rainwater until the 1980s. Electrochemical methods were also applied after 1960, but they remained too insensitive. The era of spectroscopic methods began after 1955 with atom absorption spectroscopy (AAS) and after 1970 inductively coupled plasma spectrometry (ICP) which is capable of detecting metals and several non-metals at very low concentrations. However, the ultimate method in analyzing rainwater is ion chromatography (IC), introduced after 1975 for anions like fluoride, chloride, nitrate, phosphate, and sulfate (and many more including organic anions, replacing the tedious, expensive, and often inaccurate wet chemical assay.

The first analysis of rainwater, described in detail, was carried out 1749–1751 by *Marggraf* of Berlin, motivated thought studying the quality of different waters (river and well). Then, 1813 began the era of analyzing rainwater at different locations, and soon later first reviews on the chemical composition of rainwater appeared (Liebig and Kopp 1853, Moleschott 1859, Ludwig 1862, Smith 1872).

In the nineteenth century, the systematic studies of deposition (precipitation chemistry) only began with *Justus von Liebig* (1803–1873),[247] known as the "father of the fertilizer industry", who discovered that plants assimilate (chemically fixed) nitrogen dissolved in rain and that plants take up nitrogen as ammonia from air for their nutrition (a wrong proposition). Since then, agricultural interests (deposition and drainage) have formed an important base for rainwater chemistry monitoring; seen from the soil and plant as in atmospheric input, first nitrogen as ammonia and nitric acid, and later sulfate (today, the number of stations is uncountable): Liebig (1840, 1843). From that time, the number of rain sampling and chemical analysis boomed.

The aim is to understand matter cycles, first on the local scale, initiated also precipitation chemistry in the 1840s, and on regional scales based on systematic research by monitoring from the 1950s.

---

246 In German, two terms were created, first "*Lufthygiene*" in the nineteenth century and "*Luftreinhaltung*" in the twentieth century as synonym that cannot translated directly [air pollution control].

247 German chemist; known as the "father of the fertilizer industry". He is regarded as one of the greatest chemistry teachers of all time. In 1824 at the age of 21 and with *Humboldt*'s recommendation, *Liebig* became a professor at the University of Giessen. He established the world's first major school of chemistry there.

Not only air pollution in urban areas but also damages to plants and forests (in England and Germany) stimulated several studies after the middle of the nineteenth century. In 1869–1870 in England and Scotland, *Robert Angus Smith* carried out numerous rainwater analyses gaining a systematic view on the geographic relations between chemical composition and sources (a first *chemical climatology*). From the second half of the nineteenth century, five monitoring series longer than two years (and partly over more than 10 years) are known. The first long-term monitoring started in 1853 in Rothamsted, an agricultural site in Britain, and in 1876 at the observatory Montsouris (Paris). Deposition studies (bulk sampling) due to the smoke problem started after 1910. In the middle of the 1970s, *Waldsterben* (new-type forest decline) were recognized and another booming period started including rain, fog, and cloud water research.

After World War II, precipitation chemistry monitoring began in Scandinavia first with the studying of chemical cycling (nitrogen, chlorine, and sulfur) and later with air pollution. The targets were investigation of a) trends (pH and compounds), b) regional differences, and c) pollution abatement.

2.2 On condensation and evaporation of atmospheric water

2.2.1 Ancient views

Our ancient views on air and water are based on the ideas of Greek philosophers, which began to develop since about 600 BC. One of the most fascinating ideas was the cycling among the "four elements", earth, water, air, and fire.

The close connection between air and water is founded on the ancient Greek. *Aristoteles* recognized that water evaporates from waters and the earth soil (ἀτμίς) and condenses in air ("solidifies" πύχνωσίς but also named "dense" δασύς by the ancients, from which the Latin dēnso and dēnsus are derived); in Modern Greek "to condense" means *συμπύκνωσν* (verbatim "contraction"). The Greek prefix συμ corresponds to the Latin con; hence, we see the origin of the modern "*condensare*". In ancient Greek, ἀτμις (atmis: water vapors) denotes the transfer of water (by evaporation) from the telluric form (hydrosphere) into ἀηρ (aer), the water vapor of the atmosphere and its return as precipitation to the earth, (with water) one of the two lower elements. We know from *Herodotus of Halicarnassus* (ca. 484–ca. 425 BC) that in the fifth century BC this theory was known and accepted and described by *Hippocrates*.

While the lower layer of the atmosphere (the celestial hemisphere from the ancient view) was characterized as ἀήρ, the upper layer was named αἰθήρ; both were regarded not only as different areas but also as different matters (since *Homer*). However, only air (ἀήρ) was seen as transmutable. Hence, ἀήρ exists as βαθύς (thick air) and appears to the eye as fog or cloud (Gilbert 1907, p. 18). Whereas αἰθήρ denotes the clean upper air or sky (the igneous sphere) and ἀήρ constitutes the lower layer of air, the "atmosphere" (in the German sense *Dunstkreis*, i. e., the misty sphere), filled with fog and clouds and being in darkness (... to mask the Gods). Initially, water (as precipitation) is celestial and feeds the terrestrial waters. Conversely, terrestrial water rising to heaven and there transforming into fog and clouds stands as a continuous process of becoming and metamorphosis of ἀήρ (Gilbert 1907, p. 25).

Before the sixth century BC, air was identified as emptiness. Greek natural philosophers assigned air and water apart from earth and fire to the four elements (in Latin, materia prima, primary matter). *Thales of Miletus* (624–546 BC) was the first person who is known to have tried to answer the question of how the universe could possibly be conceived as made not simply "by gods and daemons". He defined water (the liquid fluid) as a primary matter and regarded the Earth as a disc within the endless sea.

While *Homer* understood air as a gloomy substance (atmosphere) characterized by fog and clouds, *Anaximenes* (from Miletus, about 585–528 BC) recognized air as being invisible, only recognizable through heat and cold, wetness and motion (Gilbert 1907, p. 474). Insofar as the primordial term ἀήρ must be understood as fog (dimming). *Anaximenes* already understood πνεῦμα as compression of ἀήρ, whereas *Anaximander* from Miletus (about 611–546 BC), a student of *Thales*, identified πνεῦμα with the thinly dispersed substances of ἀήρ (what these might be remains beyond our imagination). With *Anaximander* and *Anaximenes*, the cycle of pre-Socratic philosophers is closed. *Anaximenes* assumed – in contrast to *Thales* – that air is a primary element (root or primordial matter) that can change its form according to density: diluted into fire, it may condense to wind and, by further condensation, into water and finally into soil and rocks. This was very likely the first "poetic" description of the idea that all material on Earth is subject to cycling, where "dilution" and "condensation" are the driving processes. *Empedocles of Acragas* (495–435 BC) introduced the four elements; earth, water, air and fire. The list was then extended by *Aristoteles* (384–322 BC) by a fifth one, the *æther* (explaining the heavenly, in Greek αιθέρας). Thus, the first to describe a number of weather phenomena and the water cycle was *Aristoteles* in his "*Meteorologica*" (Aristoteles 1923). He contributed many accurate explanations of atmospheric phenomena. The description of the water cycle (reasons for rain), as presented above, coulds were taken from a modern textbook.

In his book μετεωρολογιχά, *Aristoteles* placed the transformation of four elements (soil, water, air, and fire) in focus. Each of these elements occupies its own region, but one should understand that the matter of ἀήρ (air – aër) and ὕδωρ (water – idor) cannot be treated separately. The changing in states of elements is caused, according to the ancient philosophy, by two forces: heat and cold. Whereas older Greek philosophers treated water (*Thales*) or air (*Anaximenes*, *Heraclites*) as elementary bodies, *Aristoteles* (and his scholar *Platon*) did not consider the four elements as different basic materials but as carriers of different properties, belonging to a single primary matter (Meyer 1914). *Aristoteles* attributed to each element paired properties (warm, cold, dry, and wet): water is wet and cold and air wet and warm. However, he deemed the four elements insufficient to explain nature and therefore introduced οὐδία or *αἰθήρ* (Aether) as a fifth element, having an ethereal and more spiritual (the quinta essential in the Middle Ages) property. *Aristoteles* asked in his Meteorologica: “Since water is generated from air, and air from water, why are clouds not formed in the upper air?” He explained this as follows (Aristoteles 1923):

> But when the heat which was raising it leaves it, in part dispersing to the higher region, in part quenched through rising so far into the upper air, then the vapor cools because its heat is gone and because the place is cold, and condenses again and turns from air into water. And after the water has formed it falls down again to the earth. The exhalation of water is vapor: air condensing into water is cloud. Mist is what is left over when a cloud condenses into water, and is therefore rather a sign of fine weather than of rain; for mist might be termed a barren cloud. So we get a circular process that follows the course of the sun [. . .] From the latter [clouds] there fall three bodies condensed by cold, namely rain, snow, hail [. . .] When the water falls in small drops it is termed a drizzle; when the drops are larger it is rain [. . .] When this [vapor] cools and descends at night it is termed dew and hoar-frost.

Aristoteles further subdivided the lower layer of air. For an understanding of fog (and clouds), the area that immediately adjoins earth (nowadays termed the boundary layer) is of great importance, through “reflected radiance of solar heat” and is characterized through “rising water vapor” (ἀτμίς). *Aristoteles* denotes a cloud as πύχνωσις ἀέρος (thick air).

From his “Meteorologica” we know that *Aristoteles* believed that weather phenomena were caused by mutual interaction of the four elements (fire, air, water, earth), and the four prime contraries: hot, cold, dry, and moist. *Aristoteles* frequently argued against ideas that were actually closer to the truth than his own (Anthes et al. 1975). For example, he presented the views of *Anaxagoras* considering the cause of hail (Aristoteles 1923) as follows: some think that the cause and origin of hail are this: the cloud is thrust up into the colder upper atmosphere because the reflection of the sun’s rays from the earth ceases there, and upon its arrival there the water freezes. They [*Anaxagoras*] think this explains why hailstorms are more common in summer and warm countries.

The Greek philosopher *Anaxagoras* of Klazomenai (500–428 BC) came to Athens as a young man, more than 100 years before *Aristoteles*. Questioned on what he was born for, he answered: “To observe sun, moon, and heaven” (Diogenes 1921). His philosophy is based on the Eleats and *Empedokles*. With his doctrine that meteorological phenomena were caused by sun activities, he was in contradiction to the generally prevailing opinion. *Anaxagoras*’ theory is amazingly correct but *Aristoteles* wrote (Aristoteles 1923):

> [. . .] this is just opposite to what Anaxagoras says it is. He says that this happens when the cloud has risen into the cold air, whereas we say that this happens when the cloud has descended into the warm air [. . .].

Aristoteles, in contrast to this error, however, contributed many accurate explanations of atmospheric phenomena. The description of the water cycle (reasons for rain), as presented above, could be taken from a modern textbook.

2.2.2 Theories on condensation and evaporation 1600–1850

In ancient times, no further ideas on the form and constitution of vapors, fog, and clouds originated after *Aristoteles*. The idea of transmutation (air ↔ water) survived until the end of the eighteenth century when it was finally rejected by *Lavoisier* at the end of the eighteenth century. The milestones in the stepwise approach to explaining the formation of clouds and rain from the beginning seventeenth until ending of the nineteenth century are listed here:

- statement that atmospheric water is not air by Descartes (1637),[248]
- first artificial cloud/fog formation, "bubble theory" (vesicles) by Guericke (1672a,b,c),[249]
- cloud being a water suspension by Le Roy (1751),[250]
- first direct observation (walking in) of clouds by Saussure (1783),[251]
- first physical theory (no water dissolution in air) by Deluc (1787),[252]
- first cloud classification by Lamarck (1802)[253] and Howard (1803),[254]
- first microscopic examination of fog droplet and conclusion that they are not vesicles by Waller (1847a),[255]
- first collection of fog water for chemical analysis by Boussingault (1854),[256]
- The brothers Schlagintweit (1854)[257] estimated first the liquid water content in fog/cloud,

248 *René Descartes* (1596–1650), also known as *Renatus Cartesius*, French philosopher, mathematician, physicist, and writer who spent most of his adult life in the Netherlands. He was dubbed the "Father of Modern Philosophy".

249 *Otto von Guericke* (1602–1686) was a German scientist, inventor, and politician; his major scientific achievement was the establishment of the physics of vacuums; he served as the mayor of Magdeburg from 1646 to 1676.

250 *Charles Le Roy* (1726–1779) was French physician in Montpellier.

251 *Horace Bénédict de Saussure* (1740–1799) was a Swiss aristocrat, physicist and Alpine traveler; he directed his attention to the geology and physics of that region; he made experiments with various forms of hygrometer in all climates and at all temperatures. The father of *Théodore de Saussure*.

252 *Jean-André Deluc* [also *de Luc*] (1727–1817) was a Swiss geologist and meteorologist.

253 *Jean-Baptiste Pierre Antoine de Monet*, Chevalier de Lamarck (1744–1829), French biologist.

254 *Luke Howard* (1772–1864) was a British manufacturing chemist.

255 *Augustus Volney Waller* (1816–1870) was a British physiologist.

256 *Jean-Baptiste Boussingault*; see p. 57.

257 *Hermann* (von) *Schlagintweit* (1826–1882) and *Adolph Schlagintweit* (1829–1857), German Naturalists and Travellers.

- water condenses only on particles by Aitken (1881),
- final evidence that clouds consist from droplets and not vesicles by Aßmann (1885),[258]
- the London fog was first chemical examined by Cohen (1896).[259]

It should be noted that a *scientific* understanding of phase transfer processes (smelting, boiling, condensing, and freezing) did not exist before the middle of the nineteenth century (with the development of thermodynamics).

The question, of what fog and clouds consist of, was first asked by the great French philosopher and scientist *René Descartes* in his work "Les METEORES. Diſcours Premier" (Descartes 1637). He describes atmospheric phenomena empirically but based on careful observations (of course, at that time without measurements). *Descartes* distinguishes "*exhalaisons*" (in terms of vapors) and "*vapeurs*" (in terms of haze and steam and the German *Dunst*), whereas solely "*vapeurs*" represent water particles. "*Vapeurs*" (he did use it only in plural; the best German equivalent is "*Dunst*") he considers as being transparent. Only after "*condenſant & reſerrant*" (condensation and compression) are clouds ("*nuës*") and fog ("*brouillas*") formed. He writes (ibid., p. 122):

> [...] ſi elles s'eſtendent iuſques a la ſuperficie de la terre, on les nomme des brouillas; mais ſi elles demeurent ſuſpenduës plus haut, on les nomme des nuës [...] one named the vapors, dispersing at the earth's surface, fog; when vapors, however, hang on high, they are termed clouds].

Although the term "*vapeurs*" [German: *Dünste*] is used for water vapor as well as for water droplets, *Descartes* writes (Descartes 1637, p. 122):

> Et I leſt à remarquer que ce qui les fait ainſi deuenir moins tranſparentes, que l'air pur, c'eſt que lorſque leur mouuement's alentiſt, & que leurs parties ſont aſſés proches pour s'entretoucher, elles se ioignent & s'aſſemblent en diuers pétits tas, qui ſont autant de gouttes d'eau, oubien de parcelles de glace [It is taken into account that what vapors make more non-transparent than clean air is only based on the fact that its motion slows and its particles come so close to each other that they contact each other and combine into small heaps, being either water drops or ice particles].

This corresponds to our present molecular-mechanistic view of condensation. His "*petit parties des vapeurs*" are nothing more than water molecules in the air. *Descartes*

258 *Adolph Richard Aßmann* (1845–1918) was a German meteorologist (known for his aspiration psychrometer) and founder of aerology, first director (1905–1914) of the Royal Prussian Aeronautic Observatory Lindenberg (in State Brandenburg near Berlin).

259 *Julius Berend Cohen* (1859–1935) was a British chemist and a former *Owens* student (1878–1880) who upon *Roscoe*'s advice travelled to Munich (1882–1884) and took his PhD with *Adolf von Baeyer*; returned 1885 to Owens College in Manchester as demonstrator and lecturer like *G. H. Bailey*, and moved in 1890 to Yorkshire College (since 1904 University of Leeds).

describes the water particles as "[. . .] *longue, vnies, & gliſſantes, ainſi que de petites anguilles*" [long, interlinked and slippery as small eels]. *Descartes* also writes only on drops ("*gouttes*", from Latin guttae = drop, guttula = droplets, German "*Tropfen*" and as droplet "*Tröpfchen*") but never from "*bulles*" [vesica, German "*Blasé*"] or "*vésicules*" [small vesica, German "*Bläschen*"]; the water drops (unlike water particles, i. e., molecules) are "*exactement rondes*" [perfectly circular].

Only 20 years later, between 1650 and 1662, *Otto von Guericke*, mayor of Magdeburg, invented the air pump and worked on his famous experiments concerning the physics of the air, linking air-filled and air-less flasks. In addition, he studied the formation of clouds. In his famous book "Experimenta nova (ut vocantur) Magdeburgica de vacuo spatio" (1672), published more than 10 years after the experiments, he stated in the first chapter of the third book "De aere ejusque origine, natura & qualitatibus" [on the air, its origin, nature, and properties]; Guericke (1672a, p. 71), translated from the German edition (Guericke 1672b,c; Kraft 1996):

> The air, according to our idea, can be divided into steps or regions. Each kind of cloud, heavier or lighter, keeps to its own particular one of these regions, in which its weight matches that of the air. But if the air was compressed everywhere, it would be equally heavy above and below, so that clouds could not be formed in different ways in different regions; but as in water things either sink to the bottom or float, so the clouds would either descend to the earth, or goup to the highest part of the air [. . .]
> Air is not an elemental substance (*non est elementum*) [. . .] *Airis* nothing else then damping (*exspiratio*) or smell (odor) or effluence of waters, earth and othersubstances [. . .] Air and smell once generated from water or other things and will never be transformed back to water but remains air.

Due to the expansion (without recognizing that the saturation arose due to cooling) he observed the formation of fog (nebula) or a cloud (nubes). In section 11 ("*Versuch, mittels dessen Wolken, Wind und Regenbogenfarben in Glasgefäßen erzeugt werden können*")[260] he writes (Guericke 1672a):

> Quod tanto magis apparet, quantò magis vitrum interne humiditatibus refertum est; tunc enim plures ac copiosiores exurgunt bullulæ, ita ut [. . .] nebulam constituant; quæ per intromissionem aliquid aëris [. . .] tunc nebula illa in nubes dispergitur [This phenomenon becomes clearer the larger the humidity in the flask is; after that, more numerous and larger vesicles evolve so that a proper fog forms; but if there is free access to air, the clouds or fogs disappear. . .].

Shortly before that phrase, *Guericke* writes on guttulis minimis (small droplets), but later he uses the term bulla (in German "*Blase*") definitely for a bubble in water. It is unclear whether later scientists stem the term "*Bläschen*" (vesicle, in Latin vesicula) from bullulæ (bulla = water blister). Generalized, a "vesicular" is a more or less globular envelope filled with water. However, the fog vesicle is a "reverse" bulla, an aque-

260 Experiment to produce clouds, wind and rainbow-colors in glass vessels.

ous envelope filled with air. *Guericke* concludes from his experiments on cloud formation in the atmosphere,[261] where his explanation of "compression" is nothing else than the "thick air" in antiquity. It seems that he adopted the knowledge of *Descartes* without changes. From his several experiments, *Guericke* did not believe that air is an element – on the contrary, he found that due to fire, air losses some of its mass.

Christian Gottlieb Kratzenstein (1723–1795) writes in his "*Abhandlung von dem Aufsteigen der Dünste und Dämpfe*" (Kratzenstein 1744):[262]

> Dünste sind die kleinsten in der Luft schwimmenden Theilchen (wässerichten Materien) unterschieden den Dämpfen [. . .] Die Dünste bestehen aus kleinen Bläßgens [. . .] Die Dunstbläsgens, welche in der Luft schweben, sind inwendig mit Luft gefüllt [. . .]
> [*Dünste* are the smallest particles floating in air (aqueous matter), which differ from vapours [. . .] *Dünste* consist of small vesicles [. . .] The vesicles, floating in air, are internally filled with air. . .].

Charles Le Roy uses the term "*suspension de l'eau dans l'air*" to describe fog and clouds (Le Roy 1751). The paraphrase "suspension of water in air" for fog (and thus *Le Roy's* understanding of naturally "drop-able water") is already a modern scientific description of an aerosol;[263] suspendō (Lat.) = making or keeping floating.

Horace Bénédict de Saussure separates four types of "vapors": "*vapeur élastique pure, vapeur élastique dissolute*", both in the modern sense water vapor and then two "condensed types": "*vapeur vésiculaire, vapeur concrete*". He published his findings in "*Essais sur l'hygrométrie*" (1783).

In a chapter of "Magiae naturalis sive de miraculis rerum naturaliums" (Naples, 1589), contained by *Johann Baptista Porta* (1535–1615) on the extraction of water from the air, it is shown that if a large glass flask is filled with a mixture of ice and nitre, water condenses from the air to the outer walls of the vessel, and trickles down into a basin below as receiver (cited after Mellor, I, p. 81). Newton (1704) said that potassium carbonate deliquesces in air because of an attraction between the salt and the particles of moisture in the atmosphere, and asked: Why does not common salt or nitre deliquesce in the same way except for want of such an attraction? In *Saussure*'s "*Essais sur l'hygrométrie*" there is an excellent study of the moisture which is normally present in atmospheric air. He exposed "equal quantities of salt of tartar, quicklime, wood, lime, etc., all dried as perfectly as possible," to the same air, and found that

---

261 Although the Latinized term *atmosphaera* had already been introduced in 1608, *Guericke* used the term aerea sphaera (aerial sphere, *Lufthülle* in German).

262 "Treatise on ascending vapors" (it is impossible to find a translation for *Dünste und Dämpfe* – both are "vapors" in English).

263 The term aerosol (p. 284) was first introduced by Schmauß (1920). The terms sol, colloid and colloidal state were introduced by *Thomas Graham* in 1861. *Wolfgang Ostwald* – son of *Wilhelm Ostwald* – clearly saw that the system, which he characterized as heterogeneous or multiphase, must be studied and not only the colloid, i. e., the dispersed phase. Ostwald (1909) states fog as an example for the combination of gas-liquid and atmospheric dust for the combination of gas-solid.

they "imbibed water and increased in weight in unequal quantities." The salt of tartar took more than the lime, and the lime more than the wood. *Saussure* said that "these differences can only proceed from the different degrees of the affinity of these bodies for water," and he termed this affinity, the hygroscopic affinity of the bodies for the vapor so that the amount of vapor imbibed by different substances from the air "is proportional to their affinity for water vapor." *Saussure* also showed that the thirst or the attractive force of the body for aqueous vapor diminishes from moment to moment "in proportion as it drinks the vapor," otherwise expressed, the hygroscopic activity of the body diminishes in proportion as it approaches the point of saturation (cited after Mellor, I, p. 81). The Encyclopædiaa Britannica (1771) defines fog as follows:

> FOG, or Mist, a meteor, confiſting of groſs vapours, floating near the ſurface or any part therof.

Gehler (1833) notes that fog comprises

> [...] aus wässerigen Dunstbläschen, oder aus Wasserdunste, [...] [...from aqueous vapor vesicles or water vapor, ...].

Dunst has been already defined, writes Gehler (1826), under the keyword "*Dampf*" [vapor]. Diffusely, he tried to distinguish between gas and vapor, as well as to construct identities (by non-compliance with *Mariotte*'s law through vapors). However, today in physics, the vapor is identical to gas, but shows the feature that it is in contact with the liquid or solid phase (e. g., water); in colloquial language, water vapor is still equated with fog (correctly *steam*). Hence, *Johann Samuel Traugott Gehler* (1751–1795)[264] equates the terms "*Dampf*" and "*Dunst*", but notes that in English one differentiates between *vapor* and *steam*. Dalton (1802, p. 556) writes:

> The term *steam* or *vapour* is equally to those elastic fluids which, by cold and pressure of certain known degrees, are reduced wholly or in part into a liquid state. Such are the elastic fluids arising from water, alcohol, ether, ammonia, mercury &s. Other elastic fluids that cannot be reduced, or rather that have not yet been reduced, into a liquid state by the mutual agency of these two powers, are commonly denominated *gases*.

As discussed above, today we no longer define "*Dunst*" as "*Dampf*" (in the sense of gas) but as small dry, as well as wet, particles but also as very small aqueous particles suspended in air (haze). *Wilhelm August Lampadius* (1772–1842)[265] adapted *Deluc*'s explanation for evaporation of water, and writes (Lampadius 1806, p. 118):

---

264 German physicist and lawyer in Leipzig, familiar with ancient languages, French, English and Russian; but most known as editor of the "*Physikalisches Wörterbuch*" (1787–1795), continued 1825–1845 by *Brandes*, *Gmelin* and other.

265 *Wilhelm August Eberhard Lampadius* (1772–1842), German chemist and professor in Freiberg (Saxony), founder of modern metallurgy; well-known with *Humboldt* and *Goethe*.

> Das atmosphärische freye Feuer verbindet sich mit dem Wasser zu einem eigenen elastischen Fluidum, dem Wasserdampf; das Feuer ist fortleitendes Fluidum; das Wasser wägbare Substanz [Atmospheric free fire combines with water into its own elastic fluid, the water vapor; fire is the conducting fluid; water is the ponderable matter].

The ancient element "fire" was regarded in the sense of heat until the end of the eighteenth century (Deluc 1787). The ancient three elements Water – Fire – Air were connected for an understanding of evaporation and condensation until establishing thermodynamic laws in the late nineteenth century.

Since *Aristoteles* it has been known that dew only appears on calm and serene nights. Dew, fallen from the clear sky, was considered as matter from the Sun and even stars. Thus, alchemists treasured dew because they believed it to be *sideric* and were looking for the philosopher's stone, see also Fig. 2.1. *Christian Ludwig Gersten* (1701–1762), a German professor for mathematics in Giessen, was the first who concluded (based on observations) that dew is not fallen from heaven but is ascending from earth especially from plants (Gersten 1733). *Charles François de Cisternay du Fay* (1698–1739) a French chemist (known for the finding of two kinds of electricity), published in 1736 a paper[266] and wrote: "glass and porcelain collected much dew, while polished metal surfaces collected almost none". Also in 1736, *Pieter van Musschenbroek* (1692–1761) reported[267] on dew observations at Utrecht, and confessed that "he did not understand why dew collects on some surfaces far more than on others". He carried out many dew collections (and did store a sample 24 years in a flask without changes). *Le Roy* carried out dew experiments in 1751 in Paris (Le Roy 1751), and the Polish naturalist *Jan Michał Hube* (1737–1807) considered dew in sense of the "dissolution theory" (Hube 1790). Deluc (1787) and later *Lampadius*[268] were fighters against this "theory". *Lampadius* was likely the first who said that the temperature difference between earth and above air layer is important for dew formation (cited after *Gehler*'s).

William Charles Wells (1757–1817), born in South Carolina (USA) as a son of Scottish immigrants became a physician, philosopher, and printer, was the first to explain satisfactorily the phenomenon of dew (Wells 1814, Strachan 1866). After decisive experiments on dew, he published his book, "An Essay on Dew and several appearances connected with it", in London in 1814. This was the first and now still accepted scientific description of dew formation, coming after a long debate. By using the most decisive experiments *Wells* showed, that, apparently, all these phenomena (including

266 Mém. de Paris. (1736) p. 352 (cited after Gehler's *Physikalisches Wörterbuch*, Leipzig 1839, p. 667).

267 *Peter van Musschenbroek*; first published in Dutch: *Beginselen der Natuurkunde*, Leiden (1736) and translated into other languages: *Essai de physique*, 2 vols., Leyden, Chez S. Luchtmans (1739), Elements of natural philosophy. Translated from the Dutch by John Colson. 2 vols. London. (1744), *Grundlehren der Naturwissenschaft*, Ed. J. Chr. Gottsched, Leipzig (1747) 1242 pp.

268 Versuche und Beobachtungen über Elektrizität und Wärme der Atmosphäre angestellt im Jahre 1792 nach den Versuchen des Herrn de Lüc und einer Abhandlung über das Wasser. J. E. Hinrichs, Berlin and Stettin (1793) 200 pp.

hoar frost and mist too) were owing to the effects of radiation of heat from the earth's surface during the absence of sun. Finally, Aitken (1887) concluded from careful observations that dew never "falls" on the earth and that the great part of dew condensed on bodies is from vapor rising at night from the earth; further, that "dew-drops" formed on grass and other plants is not dew at all, but is formed of the exceeded sap of the plant (now termed guttation).

The British scientist and author *Charles Tomlinson* (1808–1897) reviewed in depth the phenomena of dew formation. Although it is nowadays accepted that *Wells* first explained dew formation correctly, *Tomlinson* concluded that *Well*'s crucial experiments had been anticipated by several investigators before (*Le Roy*, *Wilson*, *Six*, *Young*, *Pictet*, *Prévost*);[269] Tomlinson (1847, 1860). Aitken (1886), based on experiments, discusses the formation of dew results not (only) from the condensation of water vapor existing in the air but also from vapor evaporated from the warmer soil at night.

John Tyndall (1823–1893), an Irish physicist, known belong many other topics for his first explanation of atmospheric heat in terms of the capacities of various gases to absorb or transmit radiant heat, wrote the following very clear phrase (see Strachan (1866, p. 123)):

> Aqueous water is always diffused through the atmosphere. The clearest day is not exempt from it; indeed, in the Alps, the purest skies are often the most treacherous, the blue deepening with the amount of aqueous vapour in the air. Aqueous vapour is not visible; it is not fog; it is not cloud, it is not mist of any kind. These are formed of vapour which was condensed to water; but the true vapour is an impalpable transparent gas. It is diffused everywhere throughout the atmosphere, though in very different proportion.

Although *Lampadius* correctly observes that dew only occurs at (Lampadius 1806, p. 122) "*ruhiger und heiterer Luft*" [calme and serene air], he explains its formation wrong by a distillation process (evaporation – ascent – condensation – descent); hence its close relation to fog. The formation of fog *Lampadius* explains phenomenological correctly, but identifies fog partly with deep clouds because he means that (Lampadius 1806, p. 126) "*größern allgemeinen Nebel nehmen eine Höhe von 100 bis zu mehrern tausend Toisen von der Erdoberfläche an, ein*" [the larger general fogs occupy a height from 100 up to several thousand Toise above earth's surface]; 1 Toise equals to about 1.95 m. Consequently he defines (Lampadius 1806, p. 129) "*Wolken sind gar nichts anders als hohe Nebel* [...]" [clouds are nothing else then high fog]. *Lampadius*'s explanation on the formation of clouds and rain is cited here without further comment (Lampadius 1806, p. 129):

---

269 *Patrick Wilson* (1743–1811), Scottish astronomer in Glasgow and sun of *Alexander Wilson* (1714–1786) who was the first to use kites for meteorological observation; *James Six* (1730–1793) natural philosopher of Canterbury; *Thomas Young* (1773–1829) was a British polymath; *Pierre Prévost* (1751–1839) was a Genevan physicist; *Marc-Auguste Pictet* (1752–1825) of Geneva who was assisted *H.-B. Saussure*.

> Sie entstehen entweder bey dem Regen gewissermaßen als Abfall, oder sie geben Regen, oder es ist beydes zugleich. Beym Gewitter z. B. wird das Wasser aus der Luft erzeugt. Es fallen viele Wolken ab, da es electrische Materie in Menge giebt. Diese regnen selbst, wenn sich positive und negative vermengen. Viele dieser Wolken ziehen ab und verdunkeln eine andere Gegend. Zuweilen geht aber auch ein bloßer Nebel in Regen über [They either arise quasi as waste when it rains or they produce rain or it is both at the same time. During the thunderstorm, e. g., water is produced from air.[270] It will fall many clouds because there is a lot of electric matter. These are raining themselves when positive and negative mix. Many clouds pass away and darken other areas. But occasionally mere fog turns into rain].

The fog studies by *Kratzenstein* and *Saussure* have worked well into the nineteenth century as "textbook knowledge". *Ludwig Friedrich Kämtz* (1801–1867), who is regarded as the founder of modern meteorology, wrote still in his book "*Vorlesungen über Meteorologie*" (Kämtz 1840; the first edition is from 1831) two pages on the vesicle form of fog. However, *Kämtz* was the first who carried out systematic weather records to find regularities. He gives, for example, the annual cycle of the diameter of fog bubbles, which is larger (original in Parisian inch) in winter (about 27 μm) than in summer (about 16 μm). Hellmann (1917) referred *Kämtz*' textbook (Kämtz 1840) as to "the first great scientific textbook in meteorology"; it was translated into French (1843), English (1845), Italian (1846), Polish, and Russian. Even the well-known climatologist *Wladimir Peter Köppen* (1846–1940) cites in his "*Grundriss der Klimakunde*" (1931 p. 89) *Kämtz*' textbook, after which the vapor pressure does not follow *Dalton's* law because

> [. . .] in den oberen Regionen der Atmosphäre vermöge der Temperaturabnahme nicht so viel Dampf in elastischer Gestalt vorhanden sein kann, als dieser Ausdruck angibt [. . . in the upper regions of the atmosphere cannot exist so much vapor as this law expresses because of the temperature decrease].

Similar to *Saussure*, *Kämtz* states that fog forms only in saturated air; *Deluc* meant that fog also forms in dry air (likely because of his imperfect hygrometer). Kämtz (1840, p. 138) writes that the circumstances under which fog forms are "very different from those of dew formation" [. . . *entschiedenen Gegensatz zu denen bei der Entstehung des Thaues*]. *Kämtz* description of fog formation is completely right despite the assumption that fog particles are vesicles and not droplets (inside filled with air; before it was assumed that they are filled with "fire") but heavier than air; he also explains the floating and climbing in the air because of airflows and permanent cycling between evaporation and condensation. Even Prout (1834, p. 313) still means that "[. . .] mists and fogs [. . .] are [. . .] of minute hollow vesicles, having the quality of mutual repulsion [. . .]".

270 This (wrong) speculation is likely based on the experiments by Cavendish (1784) and Priestley (1785), where nitric acid is produced in moist air under the influence of electric sparks, being well-known at that time.

2.2.3 Theories on condensation after 1850

Since the experiments of *Guericke*, it was assumed that water vapor condenses when the dew point is fallen below. On the fundaments of the first kinetic gas theory by *Daniel Bernoulli* (1700–1782), *Rudolf Julius Clausius* (1822–1888)[271] developed a kinetic model of evaporation and condensation as well as the condition of a vapor-liquid equilibrium (Clausius 1864b). The French pharmacist *Paul-Jean Coulier* (1824–1890) and the Scottish physicist *John Aitken* (1839–1919)[272] carried out independently of each other first experiments with expansions chambers to study the process of water vapor condensation and found that "fine dust" must be present (Coulier 1875, Aitken 1881, 1883). *Coulier*, however, had some difficulties explaining the observations so that it is appropriate to name *Aitken* as discoverer of the condensation nuclei theory *Aitken* cited *Coulier* only in his second paper (1881) and derived the theory from *Coulier*'s and his own experimental results. Aitken (1881) draw the conclusions:

> (1) that whenever water vapour condenses in the atmosphere it always does so on some solid nucleus; (2) that dust-particles in the air form the nuclei on which vapour condenses; (3) that is there was no dust there would be no fogs, no clouds, no mists, and probably no rain [. . .]; (4) every puff of steam as it escapes into the air, shows the impure and dusty condition of our atmosphere.

However, a certain Dr. *Berger*[273] from Frankfurt/Main already doubt the "droplet theory" before *Aitken* and *Coulier* and proposed condensation onto dust particles to explain fog formation; Berger (1863, p. 459) writes:

> Zur Bildung eines [. . .] Nebels ist nothwendig, dass eine hinreichend gesättigte Luftmasse eine niedrigere Temperatur habe, als der Boden unter ihr [For the formation [. . .] of a fog it is essential that a sufficient saturated air mass has a lower temperature than the earth below].

Thus, he confirms the observations made by *Kämtz*. Whereas the celebrated *Clausius* derived from theoretical thoughts on light reflection[274] (Clausius 1849a,b) that "*diese*

271 German physicist and mathematician, graduated from the university of Berlin, professor in Zürich, Würzburg and Bonn.

272 He studied at the University in Glasgow and worked financial independent in a private research laboratory in Falkirk, Stirlingshire (where he was born and died). He began his research on condensation of water on airborne particles in 1875 and published his results from 1880 onwards.

273 I only found that he was a teacher at the "*Selectenschule*" (Catholic senior citizen school and progymnasium). Dr. *Berger* published some papers in "*Annalen der Physik*" on Leidenfrost effect (1863 and 1872), freezing of water and hail (1865) and "forest and weather" (1863).

274 In that time nothing was known on scattering; today we know that UV radiation is scattered on air molecules (*Rayleigh* scattering) and giving the "blue sky" (the cleaner and dry the bluer, as also recognized by *Clausius*), whereas the red light is scattered on dust particles (*Mie* scattering). Clausius (1849a, p. 188) assumed that reflection does not take place at non-transparent, in air floating corpuscles ["*undurchsichtigen, in der Atmosphäre schwebenden Körperchen*"] but these color appearances

Körperchen nicht massive Kügelchen seyn können" [these corpuscles cannot be massive globules], *Berger* agrees with the opinion of Brücke (1853)[275] who shows that

> [...] die Lichtstreuung mit der Teilchengröße rasch abnimmt, und dass es sich hiernach nur auf die Kleinheit und die gleichförmige Vertheilung der Dunstkörperchen, nicht aber auf ihre Beschaffenheit schließen lasse [... the light reflection decreases rapidly with the particles size, and that's hereafter one can conclude only on the smallness and uniform distribution of the fog corpuscles but not on their constitution].

Berger (1863, p. 466) writes further (it reads like a modern theory on nucleation):

> Der Wasserdampf schlägt sich an jedem Körper von genügend niedriger Temperatur nieder, mag dieser eine Fensterscheibe oder ein Stäubchen seyn. Warum sollte er sich nicht ebensogut und auf dieselbe Weise an erkalteten Lufttheilchen niederschlagen? Fallen erkaltete Lufttheilchen in einen gesättigten Raum herab, so wird dadurch also nicht eine Ausscheidung des Wasserdampfes bewirkt werden, sondern er wird sich an denselben condensieren [The water vapor condenses on each body of sufficient low temperature, either it is windowpane or a dust particle. Why it should not deposit in the same way on cooled down air particles? Fall cooled down air particles[276] into a saturated space, so it does not become a separation of the water vapor but it will condense onto them].

Robert von Helmholtz (1862–1889), chemist and son of *Herrmann von Helmholtz*, studied the formation of fog in a dust-free air and found that even for 10-fold supersatu-

can only be explained through "*die reflectierenden Körper dünne Platten mit parallelen Grenzflächen sind. Dadurch werden wir fast mit Nothwendigkeit zu der Annahme von feinen Dampfbläschen geführt, die selbst bei klarem Wetter noch in der Luft schweben und die Reflexion verursachen*" [the relecting body are thin plates with parallel interfaces. This will be led to the assumption of fine vapor vesicles that float even at calme weather in air and causes reflexion]. Deflection, reflection and scattering on small droplets is physically very difficult to describe. *Johann Silberschlag* (1721–1791) first described the "*Brockengespenst*" in 1780. This German word was accepted as technical term in meteorology [English: Brocken spectre, Brocken bow, mountain spectre or glockenspectre; French *Spectre de Brocken*]. It is a matter of an optical phenomenon where you see the very enlarged shadow of an observer against a fog- or cloudbank due to back scattering of sun light. Frequently the head of the shadow is surrounded by colored circles (that is a glory, a system of fine concentric interference fringes). First reports on glory observations are given during a French geodesic expedition to Perud 1737–1739 by the French scientist *Pierre Bouguer* (1698–1758) and the Spain captain *Antonio de Ulloa* (1716–1795); Bouguer (1749). Hellmann (1904) points out that this remarkable optic phenomenon was described already 600 years before by El-Kazwini or al Quazwini (1203–1283), quadi of Wasit (today in Iraq) in his book "Cosmography", cited by Hermann (1868).

275 In response, Clausius (1853) again repeats his argument from 1849 and emphasized the existence of haze spherules. However, it must be considered that these "haze spherules" would be very small ($\lambda/4$) to cause the blue sky according to *Clausius*, i. e. they cannot be cloud nor fog droplets. Such size (around 0.1) corresponds to condensation nuclei, which are not vesicles.

276 It remains unclear what "air particle" ["*Luftteilchen*"] means; we can speculate that simple air (an air parcel) is meant. Kober (1872) also writes that nobody understand how water condense onto (gaseous) air particles.

ration, no condensation occurs. Thus, the occurrence of clouds shows the presence of suspended dust in the atmosphere (Helmholtz 1886). The role of dust particles in fog formation describes the Belgian naturalist *Gustave Léonard van der Mensbrugghe* (1835–1911): "aqueous vapor condenses in the air only in the presence of solid particles around which the invisible vapor becomes a liquid" (Mensbrugghe 1892).

To the pioneers of nucleation theory in the nineteenth century, Vehkamäki (2006) termed *Pierre-Simon Laplace* (1749–1827), who derived in 1806 the conditions for the mechanical equilibrium of a surface separating in two phases and in 1870 by *William Thomson* [later Lord *Kelvin*] (1824–1907) together with *Josiah Willard Gibbs* (1839–1903) as a further step toward understanding the kinetics of phase change (*Gibbs*' droplet model), and later *Aitken, Helmholtz, Ostwald* and *Joseph John Thomson* (1865–1940). However, in the history of the classical nucleation theory, another scientist is fully forgotten, the German mathematician and physicist *Johann Friedrich Dellmann* (1805–1870) known for his studies on atmospheric electricity (Chapter 3.2.4 on dry fog), who wrote before *Aitken* found water vapor condensation onto dust particles, explaining why the air in a room full of tobacco smoke becomes drier (Dellmann 1869, p. 516):

> Entsprechend der Gastheorie bewegen sich die Dampfmoleküle bis sie an eine Wand oder gegen Rauchtheilchen stossen und an dieser hängen bleiben. Da letztere in der Luft ruhig schweben und nur der Luftströmung folgen, so werden immer mehr Damptheilchen der Luft entzogen. Die Rauchtheilchen werden durch das Aufnehmen der Dampfmoleküle immer schwerer, so dass sie endlich niederfallen, wodurch sich die Luft wieder allmählich reinigt [According to the gas theory molecules of vapor move until they collide onto a wall or smoke particles and stick there. Because the latter smoothly suspend and mover only with the air current, more and more vapor particles are removed from the air. Through the uptake of vapor molecules, the smoke particles become heavier, and finally precipitate whereas the air gradually cleanses].

The classical nucleation theory (CNT)[277] (in German "*Keimbildungstheorie*") was developed after the 1930s. First by Volmer and Weber (1926), who recognized that the metastability and defined the physical state of a nucleus (in German "*Keim*")[278] in pure vapors, soon later by Farkas (1927)[279] who first derived a kinetic model for the binary nucleating system, and finally by Becker and Döring (1935)[280] who derived the kinetics (*Becker-Döring* equations) for the formation of nuclei in supersaturated va-

---

277 This theory includes homogeneous nucleation (now termed gas-to-particle conversion) and heterogeneous nucleation (droplet formation); the latter is here considered.

278 This term is no longer used (*Keim* also means germ) and replaced by "*Kern*", namely "*Kondensationskern*".

279 *Ladislaus (László) Farkas* (1904–1945) was a Hungarian who studied chemistry in Vienna and became a co-worker of *Fritz Haber* in 1926; immigrated in 1933 to Cambridge (Britain) and 1935 to Israel. Died in an airplane crash.

280 *Richard Becker* (1887–1955) and *Werner Döring* (1811–2006) were German physicists.

pors. The fundamental theory was developed finally by the Soviet physicists Frenkel (1939) and Zeldovich (1942).[281]

The first treatment of binary nucleation dates back to *Håkon Flood* (1905–2001), a Norwegian chemist who worked together with *Max Volmer* (1885–1965)[282] in Berlin (Flood 1934); Volmer and Flood (1934) studied nucleation using similar chambers as *Wilson*. The American physicist-chemist, *Howard Reiss* (1922–2015), published a complete treatment of binary nucleation (Reiss 1950). The most important binary nucleation system in the atmosphere is that of sulfuric acid and water; Doyle's (1961) first publication predicted nucleation rates for this system. In that time, the process was almost termed and *vapor-to-liquid nucleation*; Stauffer and Kiang (1974) write:[283]

> To avoid misunderstandings, we list here various notations: "Gas-to-particle conversion" is a name for all phase transition processes by which gases are transformed into the liquid or solid state. One step in this gas-to-particle conversion process is "nucleation" which can occur as "homomolecular" or "heteromolecular" nucleation depending on whether only one gas or at least two different gases simultaneously come together to form a liquid or solid phase.

2.2.4 Vesicles versus droplets

After *Saussure*, the British physiologist *Augustus Volney Waller* was the first who observed fog droplets with a microscope and who concluded based on optical reasons that they are drops and not vesicles (Waller 1847a); he writes (Waller 1847a, p. 23):

> My experiments led me at that time to the conclusion, that the vapours, or rather fumes of condensed steam, consist of minute globules or spherules of water, and not of small vesicles according to the theory universally received at present, and expressed by the term *vesicular vapour*.

In England using microscope, Diness (1879) estimated the diameter of droplets in dense fog in the range of 16–127 μm. In 1884–85, *Adolph Richard Aßmann* carried out on Mt. Brocken several studies on the microphysics of clouds and estimated the droplet diameter in the range of 6–35 μm. *Aßmann* finalized the dispute over the nineteenth century about the "vesicle theory" by careful microscopic examinations of single droplets (Aßmann 1885). However, the findings just before that vapor condenses in the air only onto solid (hygroscopic) particles counteracted the belief that

281 *Jakow Borissowitsch Zeldovich* (or *Seldowitsch*) [Яков Борисович Зельдович] (1914–1987) studied gas kinetic reactions (see *Zeldovich* mechanism of NO formation) and was together with *Frenkel* (see p. 329) involved in the Soviet atomic bomb program.

282 German physical chemist, who made important contributions in electrochemistry. Although the *Volmer-Weber* equation or mechanisms is well-known, nothing is known on *A. Weber*.

283 The phrase "gas-to-particle conversion" was coined by *Volker Arnim Mohnen* (born 1937) in 1969 (Mohnen and Lodge 1969).

droplets are cavernous inside. *Aßmann* also did try to estimate the size of the residue (i. e., the condensation nucleus) after evaporation of a cloud droplet under the microscope – because of the limited resolution of the microscope, he did not see any residual and therefore concluded that it must be smaller than 1 µm (its size is really in the range of 0.1–0.5 µm). Kober (1872) wrote an enjoyable dissertation on the history of the "vesicle theory" and concluded "*Dunstkörperchen sind solide Tropfen*" [fog corpuscles are substantial droplets].

Because of the general urge for scientific education and the fascination with weather phenomena (Figs. 2.3 to 2.6) and exploring the atmosphere exerted to many people in the nineteenth century, many books were published to that issue (but nothing beyond the scientific knowledge already described by *Kämtz*): Anonymous (1847), Houzeau (1851),[284] Reimann (1857),[285] Zimmermann (1865),[286] Mangin (1866),[287] Reclus (1874),[288] Flammarion (1874, 1888), Giberne (1890),[289] Umlauft (1891), Marcuse (1896).[290] The French astronomer and author of many popular scientific books, *Camille Flammarion* (1842–1925) not yet adopted in his book "*l'atmosphère*" (1888)[291] the new insight that fog and clouds form only in the presence of condensation nuclei and that they do not consist of vesicles but from droplets (likely because the text from the first edition in 1874 was not reviewed). The Austrian geographer and teacher *Friedrich Umlauft* (1844–1923) probably has first published the new views in his textbook (Umlauft 1891, p. 244):

> Die Wasserdämpfe scheiden sich in Form von ganz kleinen Wasserkügelchen aus, die in der Luft frei schweben [...] Zur Nebelbildung scheint die Anwesenheit von Staubtheilchen in der Luft notwendige Voraussetzung zu sein [Water vapors deposit in form of small water globules, which freely float in air [...] For fog formation the presence of dust particles in air seems to be essential].

The German chemist *Hans Blücher* (1867–1927)[292] presents a modern definition in his book "*Luft*" (Blücher 1900, p. 62):

---

284 *Jean-Charles Houzeau* (1820–1888) was a Belgium astronomer – not to confuse with the French chemist *Jean Auguste Houzeau* (1829–1911), referred in literature with *Houzeau, A.* only.
285 *Ernst Julius Reimann* (unknown biographic data).
286 *W. F. A. Zimmermann* – Pseudonym of *Carl Gottfried Wilhelm Vollmer* (1797–1864), German author.
287 *Arthur Mangin* (1824–1887) was French author of popular scientific monographs.
288 *Jacques Élisée Reclus* (1830–1905) was a French geographer and anarchist.
289 *Agnes Giberne* (1845–1939) was a British author.
290 *Adolf Marcuse* (1860–1930) was a German astronomer.
291 First edition: Camille Flammarion: *L'atmosphère. Description des grands phénomènes de la nature* (Paris 1872); 1874 published in English translation and edited by the famous baloonist *James Glaisier*.
292 He worked in Gießen at the "*Chemisches Untersuchungsamt für die Provinz Oberhessen*" and in "*Hygienisches Institut*" at the university.

Fig. 69.—Above and below the rain cloud.

Figure 2.4: Contemporary illustration of rain: Above and below the rain cloud (after Flammarion 1874, p. 380).

> [...] die Nebelbildung ist Kondensation von Wasserdampf aus der Atmosphäre an festen Körpern (Staubteilchen), die in der Atmosphäre selbst schweben [... formation of fog is condensation of water vapor from the atmosphere onto solid bodies (dust particles), which float in the atmosphere].

2.2.5 On fog and clouds

Clouds – together with fog and precipitation – are surely the most fascinating weather phenomena of our atmosphere being observed by humans since antique eras (Figs. 2.4 to 2.6). In contrast to temperature, air pressure, humidity, wind, and rain, which can only be estimated by using of measurement instruments, the fog was recorded (until the end of the twentieth century) only by personal observation of the visibility at daily fixed-term values. Therefore, one can assume that such "atmospheric state" had been already recorded by first weather observations (Hellmann 1883). With the establishment of meteorological networks in many countries[293] in the second half of the nineteenth century, days on which fog appeared were noted. However, only a few scientific articles were published before 1900, namely on fogs in England and London, for example, by Marcet (1889), Clayden (1891), Brodie (1892), Scott (1893, 1896),[294] and Russell (1888, 1889, 1897). Scott (1893) writes: "It is impossible to obtain any information from our records as to the duration of fogs [...] for the three phenomena of "fog", "mist" and "haze" [...] there is, as yet, no clear distinction between them". Scott (1896) notes that fog on the British islands occurs frequently together with strong wind and rain, lasting sometimes up to a month.

Already our ancients separated fog from clouds by the criterion of the altitude above earth's surface. Microphysical, there is little difference between cloud and fog droplets; however, meteorologically the most important different parameter is the wind speed. Even today, there is no consensus in separation between mountain fog and clouds (Eugster 2008).[295] *William Marcet* (1828–1900), president of the British meteorological society, delivered an address on "Fogs", saying (Anonymous 1889, p. 311)

> Fogs and clouds are one and the same thing. A cloud is a fog when entered into, and a fog seen from a distance, suspended in the air, becomes a cloud.

Boussingault (1845, p. 492) writes

293 The first modern station of so-termed 1th order were established in Munich, Germany in 1846. The first series of observation (over 24 years) begun 1623 in Kassel by the Landgrave *Hermann von Hessen-Rotenburg* (1607–1658); Hellmann (1883). In 1664 begun the meteorological series of observation at the Parisian Observatory, which is the oldest and longest in operation in the world; unfortunately, the observation journals preserved only since 1785 (Hellmann 1927). A real meteorological network (65 stations) begun 1776 in France through the *Société Royal de Médicine*, but closed already in 1786.

294 *Arthur William Clayden* (1855–1944) was a British geologist; *Frederick J. Brodie* (unknown living dates) was British meteorologist in London; *Robert Henry Scott* (1833–1916) was an Irish meteorologist, who was president of the Royal Meteorological Society. On *Francis Albert Rollo Russell* see footnote 578 on p. 291.

295 My view is that fog is a "fixed" (not moving) cloud like an orographic cloud (cap cloud). Some scientist say that fog is a cloud with contact to the earth's surface. Of course, we know today from physical point of view that actual fog = cloud, and insofar fog could be regarded as a special type of a cloud.

> [. . .] a fog, as a celebrated naturalist said, is a cloud in which one is, and a cloud is a fog in which is not.

Flammarion (1888, p. 615) writes in this sense:

> Quoiqu'il n'y ait pas de différence *essentielle* entre les brouillards et les nuages, il y en a cependant une de fait: c'est qu'un brouillard est un lieu dans lequel la vapeur d'eau passe de l'état invisible á l'état visible, tandis qu'un nuage est un objet individuel, un groupement de vapeurs visible suivant une forme détermine [It exists no significant difference between fog and cloud, but it is fact that fog represents the transfer of water vapor from invisible to visible state whereas a cloud represents an aggregation of visible steam of a certain structure].

First attempts to estimate the liquid water content (LWC) of fog were made by using an aspirator, i. e., to suck air through sulfuric acid or calcium chloride where naturally also the humidity to a large extent were collected.[296] Such first examinations were carried out by the brothers *Hermann* and *Adolph Schlagintweit* (1854) on Monte Rosa (3152 m) next to the actual goal, the estimation of atmospheric carbon dioxide.[297] They found 3 g m^{-3} in dense clouds; but *Viktor Conrad* (1876–1962), Austrian geophysicist, comments that they used a wrong volumetric correction (on which Hann (1889) draw attention) so that the "true" value would be 0.4–2.0 g m^{-3} Conrad (1901). Diness[298] (1879) found in London fog at low temperature with 0.7 g m^{-3}. *William Henry Dines* (1855–1927) argued that one should not be quick to attribute increased mortality during the winter quarter to reduced temperatures, but rather to increased crowding and a heavily polluted atmosphere (Dines 1894). While the cold was a general event (affecting all cities and towns around London), the fogs and the spikes in mortality were not; they occurred only in London. *Eberhard Fugger* (1842–1919) found in Salzburg in fog 1.2–3.4 g m^{-3} (unpublished and cited by Conrad 1901). Conrad (1901), who first applied a method for exact LWC estimation (well described in Köhler 1926) on different mountains, like Schneeberg (1884 m), Schafberg (1798 m) and Hohen Sonnblick (3106 m), found LWC within the range 0.3–4.6 g m^{-3}; Köhler (1926) corrected these values down.

The nineteenth century ends with the knowledge of the essential presence of cloud condensation nuclei for water vapor condensation and the fact that fog and clouds consist from droplets. The role of dust particles in fog formation describes Mensbrugghe (1892):

> Aqueous vapor condenses in the air only in the presence of solid particles around which the invisible vapor becomes a liquid.

296 The absolute humidity at 100 % saturation is between 5 and 10 g m^{-3} for 0–10° C whereas the liquid water content does not depend from temperature and amounts around 0.3 g m^{-3} (Möller 2003).
297 By using the same method, they estimated CO_2 in the range of 400 to 1000 ppm – much too high.
298 *George Diness* (living dates unknown) was a British meteorologist – not to confuse with *W. H. Dines*.

Figure 2.5: Contemporary illustration of different clouds (after Zimmermann 1865, p. 192).

At this point, it is worth mentioning the first attempts in weather modification,[299] namely prevention of hail storms, which cause severe damages in agriculture. Popular superstition led already in the fiveteenth century to the idea of "*Wetterläuten*", that is, during a severe thunderstorm, the bells of churches are rung in the villages in order to rip open the clouds through their sound (Reith 2011). Since about 1600, in

299 Weather modification is the effort of man to change naturally occurring weather, for the benefit of someone. The best-known kind of weather modification is cloud seeding, with the goal of producing rain or snow, suppressing hail, or weakening hurricanes. Another important issue is fog dissipation. Since some years, the term geoengineering is used to compensate climate change; best known example is *Crutzen*'s proposal to inject sulfur dioxide in the stratosphere to create sulfate aerosol particles to cool the atmosphere through solar light reflexing. However, because of unpredictable response of the climate system, even with adverse impacts, its application is controversial discussed.

Styria (Austria), gun firing on clouds ["*Wetterschießen*"] was used to dissipate clouds (Fig. 2.6), but unsuccessful as already described by *Joseph Graf von Attems* (1754–1817); Attems (1787, p. 49):

> Sowohl Erfahrungen verschiedener Lande, als auch die einstimmigen Zeugnisse der berühmtesten Naturforscher haben seit mehreren Jahren erwiesen, daß das sognannte Wetterläuten, und Wetterschießen nicht allein gegen Ungewitter keine Hilfe verschaffe, sondern noch ehern deren stärkeren Ausbruch, und die gemeiniglich damit verbundenen unglücklichen Folgen befördere [Both experiences of different countries and unanimous certificates of most famous naturalists have shown since some years that so-termed Wetterläuten, and Wetterschießen not only help against thunderstorms but rather promotes ist stronger burst and thus fortify the impacts].[300]

Having no adequate translation of "*Wetterschießen*", this term was termed as "hailstorm artillery" and "hail shooting" (Shaw 1901), and soon later adopted in English (Anonymous 1902a); some papers and pamphlets appeared around 1900, which are curious from today's understanding (Stiger 1898, Strele 1898, Mack 1900, Szutsek 1900, Suschnig 1900a,b, 1901, Pernter 1900, Pernter and Trabert 1901, Shaw 1901, Girsberger and Stahel 1901). *Josef Maria Pernter* (1848–1808), Austrian physicist and director of the Central Meteorological Institute in Vienna since 1897, was one of the organizers of the "*Internationale Expertenkonferenz für Wetterschießen*", hold in Graz 1902 (Anonymous 1902b, 1903, Baillaud 1903), and writes (Pernter 1907):[301]

> Several times I got in the past two years inquiries on the progress and results of weather shooting. Because recently these were repeated, I will answer them below. I will begin from the state of the art at the time of the international conference of experts on weather shooting at Graz in 1902. On this I have reported at length in the Appendix of the report on this conference (see Annual Report over 1902 of the Royal Central Meteorological Institute in Vienna) and have stated the main result of this conference as follows: "The first result of the expert conference in Graz is that in the opinion of the experts the effectivity of weather shooting is not only dubious – as voiced by a serious majority of expert opinions – but, when all conditions and reservations of the opinions are considered, looks very dubious, even improbable". After this decisive important conference of experts in July 1902, further proof was looked for by execution of orderly correct shooting, as in particular is done at official shooting locations.

Note that artillery shells and anti-rail rockets became in use in the 1950s in many countries of the world, almost based on silver iodide nucleation. However, one suggestion is that the sound waves from the explosion induce cavitation in the hailstones, causing them to shatter and become "mushy" (Vittori 1960). The retired Civil War gen-

300 Nearly 200 years later, after several scientific operational programs of weather modification, the Word Meteorological Organization concluded (Neiburger 1969, p. 28): In the case of precipitation-augmentation and hail-suppression, it is not known under which circumstances and by what techniques operations will lead to success, and when opposite effects will result.

301 Translation by Jon Wieringa, De Bilt, February 2006; http://www.agrometeorology.org/filesfolder/repository/hailhistorywieringa.pdf

Figure 2.6: Gun firing on weather ["*Wetterschießen*"]. Contemporary illustration by F. Schlegel "*Das Wetterschießen in Steiermark*" from "*Das Buch für Alle*" (1900) p. 96 (after a copperplate print in the possession of the author).

eral and Chicago engineer *Edward Powers* (unknown living dates) published in 1871 the book "*War and the Weather*", describing his observations since 1841 that heavy gunfire caused rainfall. General *Robert Saint George Dyrenforth* (1844–1910), who represented the United States Department of Agriculture, based on his knowledge as a patent engineer, developed the idea of detonating balloons, filled with oxygen and hydrogen. He ordered to the American inventor *Carl Myers* (1847–1925) to construct a balloon for making rain for a first experiment in Texas in May 1891, and after successful rainmaking, the US Congress allocated $9000 for further experiments. The scientific world, however, tended to reject these theories and argued that in many cases rainfall did follow battles by the law of average (Spence 1961, p. 212).[302] The idea of weather modification continued until nowadays (Möller et al. 2001). The knowledge that in supercooled clouds ice particles form at the expense of liquid droplets (*Wegener-Bergeron-Findeisen* process, or "cold-rain process"), we acknowledge three researchers: *Alfred Wegener* (1880–1930) who, in 1904, attempted to rainout clouds by seeding with liquid air (Wegener 1911), *Tor Bergeron* (1891–1977) who concluded 1933 that for the rain formation, the simultaneous presence of ice crystals and supercooled droplets is essential (Bergeron 1935), and *Walter Findeisen* (1909–1945)[303] who continued *Wegener*'s experiments by artificial injection of ice into supercooled clouds (Findeisen 1938). In 1930, the Dutch, *August Willem Veraart* (1881–1947), had already tried to rain out clouds by injection of dry ice from aircraft into clouds, but without explaining (Veraart 1931). Houghton and Radford (1938a), American researchers, we meet later in fog and cloud chemistry research, conducted certainly the most advanced study on fog dissipation, proposing chemical seeding: "[. . .] the condensation of water vapor using calcium chloride is chosen as being probably the most practical of the fog dissipation methods considered". However, only in 1946 *Vincent Joseph Schaefer* (1906–1993)[304] find that dry ice injected in supercooled clouds produces a large number of new ice particles (Schaefer 1946). Soon later *Bernard Vonnegut* (1914–1997)[305] found that iodides of silver and lead gain the same effect (Vonnegut 1947).

Between 1921 and 1924, the Russian scientist *Vitold Ignatevich Vitkevich* [Витольд Игнатьевич Виткевич] (1888–1970) conducted a series of laboratory experiments in the Moscow Institute for Scientific Melioration [Московский научно-мелиоративный институт] with charged sand particles to accelerate the collision of water

---

302 K. Nodjimbaden (Smithsonian Magazine, Sept. 4, 2018) brings the unsuccessful rainmaking on the point: *Edward Powers* was not wrong in his observation that rain followed battle. But the likely explanation for this phenomenon is simply that generals tended to avoid fighting on rainy days. So, while *Dyrenforth* and the rainmakers of the 1890s may have conducted experiments on faulty assumptions, they are just one chapter in the long history of human interference in weather and climate.

303 German meteorologist, studied in Hamburg, PhD 1931; since 1940 head of weather office in Prague.

304 American meteorologist and chemist.

305 US atmospheric scientist, 1967 professor at University Albany.

droplets and tried to develop a seeding technique. A weather modification program to enhance precipitation was undertaken in a specially founded “Moscow Institute of Artificial Rain” [Институт искусственного дождевания]. In 1931, a department of that Institute was later transformed into the Leningrad “Institute of Experimental Meteorology”, with the main task to develop drought mitigation measures. At the institute, studies in the field of weather modification were headed by *Vladimir Nikolaevich Obolensky* [Владимир Николаевич Оболенский] (1877–1942). In 1939, field experiments to seed clouds using charged sand particles were carried out; furthermore, high-frequency electric discharges, radioactive minerals, calcium chloride, and crushed ice were used for seeding of clouds and fogs. After *Obolensky* died in 1942, the institute was integrated into the MGO (Voeikov Main Geophysical Observatory); Chernikov (2009). After the war, the “Central Aerological Observatory” (CAO) in Dolgoprudny [Долгопрудный] near Moscow took over the leading role in fog dissipation and cloud seeding with its director *Albert Alekseevich Chernikov* [Альберт Алексеевич Черников] (1936–2007). In the 20th century the following milestones can be listed in physical fog and cloud research

- Conrad (1901) developed the first exact method for LWC estimation, and Köhler (1926) carried out the first “exact” LWC measurements,
- the first patent for fog dissipation by Archenhold (1913),[306]
- Köhler (1923) first describes the nucleation process,
- first fog classification by Willett (1928)[307]
- Veraart (1931) first carried out experiment in cloud seeding by using of dry ice; Schaefer (1946) provides the theoretical explanation,
- Vonnegut (1947) studied nucleation of ice formation by silver iodide,
- Findeisen (1932) measured size and number of fog droplets,
- Köhler (1936) described the growth of condensation nuclei,
- Findeisen (1938) presents a theory of precipitation formation,
- Houghton and Radford (1938a) first carried out extended chemical analysis of fog and estimated droplet sizes distribution,
- A first sophisticated instrument (FSSP-100) for estimation of droplet size distribution and one (PVM-100) for LWC was introduced by Knollenberg (1981) and Gerber (1984), respectively.

306 *Friedrich Simon Archenhold* (1861–1946), German astronomer in Berlin. Likely one of the last recent patents in fog dissipation using dry ice (cold fog) and water ice (warm fog) blasting are by Möller et al. (2001, 2003).

307 *Hurd Curtis Willett* (1903–1992) was an American meteorologist, known for developing techniques in weather forecasting.

2.3 Rainwater studies

The harvesting and storing of rainwater go back thousands of years. The experience that rainwater is the purest natural water is probably as old as a sampling of it. We can only speculate that rainwater was widely used not only for drinking and washing but also in pharmaceutical and chemical operations. The theories concerned on how rain has formed the book by Middleton (1966) is highly recommended. Bibliographies on precipitation chemistry are given from Rigby and Sinha (1961), Asman and Conrads (1975), and for the Netherlands by Buijsman (2010, 2011).

Already in 1666, *Boyle* found while distilling and re-distilling pure rainwater, time and again, in glass vessels, a white powdery substance that obtained each time the water was evaporated; and the more the water distilled from a given glass vessel, the larger the amount of white powder (More 1941). He added that a friend, of unsuspected credit, had distilled water two hundred times "without finding the liquor grow weary of affording the white earth". It seemed to him as if water "might be very nigh totally brought into the earth, since out of an ounce of distilled rainwater he had already obtained nearly three-quarters of an ounce, if not more, of the often-mentioned earth" (cited after Mellor, I, p. 81).

Boerhaave (Fig. 2.7), who also sampled rain and snow (see Chapter 2.3.1.1), concluded that the earth, remaining after distillation of rainwater, is originated from dust which is permanent in the atmosphere and deposits into the vessel; he denied the idea of transmutation of water into the earth with the argument that the residue is very lit-

Figure 2.7: Portrait of *Herman Boerhaave* (Source: http://www.gutenberg.org/etext/15690).

tle comparing to the amount of distilled water. It is very interesting what he writes (Boerhaave 1735, p. 363):

> If you distill the purest water with gentle Fire, and in a very clean Vessel, to a perfect dryness, it will leave a slight spot at the bottom of the Glass. And this will always happen, repeat the Operation ever so often, with the same Water, but in fresh Vessels [...] white, light insipid Earth, that was fixed, heavy, and indissoluble in Water [...]
> Upon the authority of this account, the greatest Philosophers have laid down as certain, that water, by simple repeated distillation, may be absolutely transmuted into true Earth.
> A great quantity of this [...] Rain-water that was catches upon our observatory [...] I [...] distilled [...] into a very clean Receiver to perfect dryness. There remained then a white spot at the bottom of the Glass, but incredibly small in proportion to the quantity of Water.
> [...] the empty part of the Cucurbit, the Alembic, and the Receiver are full of the common Air of the Elaboratory where this Experiment is made: But this Air is always full of Dust [...]
> [...] I cannot certainly see that any Earth was really generated from the simple Body of the elementary Water in these Operations.

Marggraf (see below) also found alkaline earths after distillation of rainwater and snow – apart from nitrate and chloride in the residue. Animated by *Borrichius* experiments (see p. 147), *Marggraf* also distilled several times already distilled rainwater and found every time a turbid solution in the retort and finally weight earth, seemed to be quicklime. A small part of the residue, however, was insoluble in nitric acid [silica liberated from the glass]. *Boerhave*'s objection that the earth found in water is from atmospheric dust, *Marggraf* "disproved" by distillation in the hermetic closed alembic.

Only *Lavoisier* first proved (Lavoisier 1770) by exact weighting that the "earth" after long-time boiling of water in glass vessels is due to the dissolution of glass. *Lavoisier* traced the true source of this earth with the object of (cited after Mellor, I, p. 81) "settling by decisive experiments whether water can be changed into the earth as was thought by the old philosophers, and still is thought by some chemists of the day". By heating water in hermetically sealed glass vessels, after some days, the water became turbid, and little white specks separated from the water and floated about. The hermetically sealed glass vessels were weighed before and after the experiment. He proved that the earth does not come from outside the vessel, because the weight of the vessel and its contents do not alter. This is against *Boyle*'s hypothesis that fine igneous particles can pass through the glass, and are precipitated in the form of a white powder when they come in contact with water. Consequently, *Lavoisier* concludes that nothing can pass through the pores of the glass, and these little white particles, be they caused by what they may, are not caused by igneous particles passing through the glass. Still further, it was shown that the earth does not come from the water, because the weight of the water remains the same before and after the experiment. The earth comes from the vessel because the vessel loses in weight; and finally, the earth comes wholly from the vessel, because the loss in weight of the vessel is virtually equal to the weight of the earth formed. Hence, adds *Lavoisier*, "it follows from these experiments that the greater part, possibly the whole of the earth separated from rainwater by evaporation, is due to the solution of the vessels in which the water was collected and evaporated". Scheele (1777, p. 5), with certainty not knowing *Lavoisier's* experiments and publications, deduced a similar conclusion from other experiments. He analyzed the earth produced during the evaporation of water in glass vessels and showed that it has a similar composition to the stuff of which the vessel was made.[308] Table 2.1 lists the "highlights" in the rainwater investigation in chronological order.

2.3.1 Rainwater studies for understanding the chemical composition of air

2.3.1.1 Rainwater studies before 1800

> This, then O ye Chemists, is the true Nature of Rain-water, so far distant from a pure simplicity, and so tainted with various impurities (Boerhaave 1735, p. 349).

The purity of rainwater was long known and used to estimate the mineral content of other natural waters, first described by *Leonhard Thurneisser* (1530–1596), a successor of *Paracelsus*; Thurneisser (1572, p. IX) writes:

> [. . .] das Regenwasser hat auch ein vorsalzne eygenschafft in sich / welche es empfeht von den auffsteigenden dämpffen / die sich mit sampt dem Regen in der obersten Region versamlen [. . . rain water has a salty property obtaining from the ascending vapors which combine with the rain in upper layers].

According to Szabadváry (1966), he estimated the amount of dissolved substances in mineral water by weighting it with and comparing the value with the weight of the same volume of rainwater; the mineral water he distilled and evaporated and the residue he weighted and annealed, but his conclusions are rather confused.

Since naturalists had detected saltpeter in icicles and dew, they thought it came from "nitre in the air", which indeed became a prominent theme in seventeenth-century English literature. Sir *Kenelm Digby* claimed in 1661 he had augmented plant growth by adding saltpeter, maintained that plants received their nutrition from nitrous salts, and that "there is in the Aire a hidden food of life" and Sir *Thomas Browne* (1605–1682), for example, referred to a friend's trip to another country, where he might benefit "by change of Air, and imbibing the pure Aerial Nitre of these Parts [. . .]" (cited after Aulie 1970). *Francis Sylvius* (1614–1672)[309] proved nothing by experiments but he

308 *Reinhold von Walther* (1866–1945) had an interesting experiment, in 1915, to demonstrate the solubility of glass in water: 0.5 L of water are placed in a common litre flask with sufficient alizarine to produce a pale-yellow colour, the colour changes to a reddish-violet owing to the dissolution of alkali from the glass. By adding dilute sulfuric acid from a burette, the colour changes back to pale yellow when the alkali is neutralized. He found that after an hour's boiling, alkali equivalent to 0.018 L of centinormal sulfuric acid had been dissolved from a glass vessel (cited after Mellor, I, p. 82).

309 Latinized from *Franz de le Boë*, German-Dutch physician and naturalist who wrote "Praxeos medicae idea nova" (Leiden 1667).

Table 2.1: Milestones in investigating precipitation water. Note before 1850 all detections were qualitatively, hence also first quantitative determinations are listed (after 1850). First semiquantitative determinations were carried out Marggraf (1753) in Berlin for lime ($CaCO_3$), nitric acid (HNO_3) and hydrochloric acid (HCl), and Bertels (1842) for $CaCO_3$, NaCl, $CaSO_4$, FeO, Al_2O_3, SiO_2, K_2SO_4, organic nitrogen, humic acids); the first quantitative rainwater analysis carried out Pierre (1851) in Caen for Na, K, Mg, Ca, chloride and sulfate. In the table are further termed scientists who more systematically determined rainwater.

year	first findings ("highlights")	author
ca. 900	salty residue after distillation	unknown alchemists
ca. 1120	density to be 1 g cm^{-3}	Abd al-Rahman al-Khazini[g]
1665	saltpeter	Henshaw (1667)
1674	common salt (NaCl)	Borrichius (1674)
1675	animacula (infusoria)[d]	Leeuwenhoek (1677)
1686	organic matter in hail	Anonymous (1693)
1700	nitrate[a]	Ramazzini (1700)
1720s	organic matter, nitrate (in thundershower)[a], and NaCl[b]	Boerhaave (1735)
1749–1751	lime[c], NaCl[b], organic matter, and nitrate[a]	Marggraf (1753, 1786)
1790s	HCl in rainwater and gutter corrosion	Paping (1796)
1813	acidic rain from thunderstorm[f], $CaCO_3$	Stark (1813)
1821–1823	ammonium, iron, magnesium, manganese, potassium, Pyrrhin[e]	Zimmermann (1824)
1822	NaCl in rainwater is from sea salt	Dalton (1824)
1824	phosphoric acid in rain and fog	Witting (1825)
1825	carbonate, sulfate	Brandes (1826b)
1826–1827	nitric acid in rain from thunderstorm	Liebig (1827)
1827	aluminum	Neljubin (1827)
1834	acidic rain in London (sulfuric acid)	d'Arcet (1834)
1837	Pyrrhin = humus	Lampadius (1837)
1839	organic nitrogen in rain	Girardin (1839)
1841–1842	semi-quantitative and deposition estimate: NaCl, Ca, K, sulfate, carbonate	Bertels (1842)
1842	use of the term "acid rain" (*pluie acide*)	Ducros (1845)
1849–1850	iodine, bromine, hydrogen sulfide in rainwater	Marchand (1852)
1850	quantitative determination of iodine	Chatin (1851)
1851	first determination of chloride and sulfate	Pierre (1851)
1851	first determination of ammonium and nitrate	Barral (1852)
1854	rainfall and rainwater concentration	Boussingault (1853, 1854)
1860	phosphate in rainwater	Barral (1860)
1853–1916	long-term monitoring Rothamsted	Miller (1905a,b, 1913)
1863	hydrogen peroxide (H_2O_2) in rainwater	Meissner (1863b)
1869	nitrite (and H_2O_2) in rain (Tiflis)	Struve (1869)
1869–1870	acid rain due to SO_2 emission	Smith (1872)
1869–1870	quantitative determination of organic N in rain	Frankland (1874)
1876–1907	long-term monitoring Montsouris	Lévy (1903a,b)
1881	detection of alcohol in rainwater	Müntz (1881)

Table 2.1: (continued).

year	first findings ("highlights")	author
1884	detection of acetic acid in rainwater	Chairy (1884)
1889	detection of nitrite in rainwater	Failyer and Willard (1889)
1890	determination of nitrite in rainwater	Passerini (1891)
1910–1915	determination of sulfite in rainwater	Stoklasa (1923)
1912–1914	meteorology of rain chemistry	Anderson (1915)
1909–1911	first Russian rain chemistry network	Kossovich (1913)
1919–1921	first US rain chemistry network	MacIntire and Young (1923)
1932	detection of formaldehyde in rainwater	Dhar (1932)
1932	first pH measurement in rainwater	Cauer (1936)
1934	HCl from volcano in rainwater	Bottini (1935)
1935	first rain chemistry network in Japan	Miyake (1939)
1938–1939	nitrite in lightning twice increased	Ogura (1940)
1940	first determination of hydrogen carbonate	Usov (1940)
1948	first entitling "precipitation chemistry"	Sugawara (1948)
1947–1950	first North European rain chemistry network	Eriksson (1954)
1949	H_2O_2 determination in rainwater	Matsui (1949)
1958–1961	first USSR rain chemistry network	Drozdova et al. (1964)

[a] Detected as saltpeter
[b] Detected as common salt
[c] $CaCO_3$
[d] *Infusoria* is a collective term for minute aquatic creatures like ciliates, euglenoids, protozoa, unicellular algae, and small invertebrates
[e] Organic matter; the Webster American Dictionary of the English Language (1828) defined Pyrrhin as "A vegeto-animal substance, detected in rainwater by M. Brandes". However, it was first detected and named by Zimmermann (1824) and characterized as "organic matter derived from vegetation and animals" by Vogel (1828); Brandes (1826b) also found it in rainwater.
[f] Not related due to HNO_3 by *Stark* but "carbonic acid under the influence of electricity".
[g] Living dates unknown, flourished 1115–1130; Greek Muslim scientist, astronomer, physicist, biologist, alchemist, mathematician, and philosopher from Merv, then in the Khorasan province of Persia but now in Turkmenistan, who made important contributions to physics and astronomy. *Al-Khazini* is better known for his contributions to physics in his treatise *The Book of the Balance of Wisdom*, completed in 1121, which remained an important part of Islamic physics. See Hall (1973).

made it probable, according to his manner, that sourish particles of nitre ["*säuerliche Salpetertheilchen*"] floated about in the common atmosphere (cited after Gmelin 1801, p. 196 and 1798, p. 4). However, *Thomas Henshaw* (1618–1700)[310] likely first detected nitrate in atmospheric water (Henshaw 1667, pp. 264–265):

310 English lawyer, and scientific writer. While not a published alchemist, he was a significant figure in English alchemical work from the 1650s onward; in 1637 he became tutor there to *John Evelyn*.

> Salt-peter; though it be likely, that the Air is everywhere full of a volatile kind of Nitre, which is frequently to be seen coagulated into fine white Salt, like Flower of Wheat (but by the very taste may be easily known to be Peter) sticking to the sides of Plastered-walls, and in Brick-walls to the Mortar between the Bricks, (in dry weather, or where the wall is defended from the rain) for Lime doth strongly attract it; though Dew and Rain do conveigh much of it to the Earth. And the Clouds seem to be spread out before the face of the Sun, either to imbibe some part of his influence, or to have a Salt generated in them, for to advance the fertility of the Earth, and certainly they return not without a blessing; for I have more than once extracted Salt peter out of Rain and Dew, but from the latter more plentifully.

Sampling and treatment of precipitation were first demonstrated by *Ole Borch* (1626–1690), Latinized to *Olaus Borrichius*, Danish alchemist, in 1674:[311] "100 pounds of snow, or rain or hail water, evaporated in a glass vessel, transmuted into dusty earth, partly seemed to consist from common salt".[312] It looks to me as the first evidence for sea salt (NaCl) in precipitation. However, for *Borrichius* (he referred to *Dickinson*[313] and even *Newton*[314]) it was the confirmation of the ancient hypothesis of water convertibility (Kopp 1931); see below. The Swedish physician and alchemist *Urban Hjärne* (1641–1724) wrote:[315]

> Quod ante Clarissimus inter Danos Chymicus, Excellens et egregious Vir Doct. Olavus Borrichius in suo Hertete Aegyptico curiosi ac diligenter tentavit, salem ac pinguedinem seu oleum extrahere ex aqua pluvia et nivali milique apud eundem Virum Anno 1674 videre contigit; idem ipsum in nostro Laboratario, non sine jucundidate invenimus praceipue ex aqua nivis Martio mense hic et initio Aprilis decidua, quae agris terrisque manifestam et germinandi vim et succum conferre creditor [...] [It is a fact that first among the Danes the very famous chemist, an excellent and outstanding man, Doctor Olaf Borrichius, in his Egyptian Hermetic, carefully and diligently tried to extract salt and fat and oil from rain and snow water, and it happened to me to see the same man in the year 1674; not without pleasure we found the very same thing in our laboratory, especially from snow water fallen here in the month of March and the beginning of April, which is believed to bestow on the fields and lands the evident power of sprouting...].

311 Borricius, O. (1674) Hermetis, Aegyptiorum et Chemicorum sapienta ab Hermanni Conringii anidmadversionibus vindicate. Sumptibus Petri Hauboldi, Reg. Acad. Bibl., 448 pp. *Hermann Conring* (1606–1681), a Danish physician wrote "De hermetica Aegyptiorum vetere et Paracelsicorum nova medicina" (1648) 276 pp., were he affronted *Paracelsus*.

312 Cited after Kopp (1931) Vol. 3, p. 255 (also cited by Mellor, I, p. 81).

313 *Edmund Dickinson* (1624–1707) was the physician of King James II, and a much-respected person at that time who wrote "Epistola ad Theodorum Mundanum philosophum adeptum" (1686). He distilled water hundred times and found always earth after evaporation, what were interpreted as transmutation of water into earth.

314 Newton: *Optice*, London (1701) in which he expounded his corpuscular theory of light but speculated on alchemical transmutation.

315 In: Urbani Hierne Actorum chemicorum Holmiensium: Hoc est, Parasceve, sive, Praeparatio ad tentamina, in reg. laboratorio Holmiensi peracta, cum annotationibus Joh. Gotschalk Wallerii. 2 Vol. in 1, 1753, Stockholm, 292 pp. (cited after Klein et al. 1988, p. 929). This volume was posthumously edited by the Swedish chemist *Johan Gottschalk Wallerius*. See also Wallerius (1750, 1770).

Bernardino Ramazzini (1633–1714), an Italian physician in Padua, called the "father of occupational medicine", was likely the first who found nitrate in snow water (cited after Kopp 1931, p. 229), Ramazzini (1691, 1700, 1717).

In April 1675, Dutch naturalist *Antoni van Leeuwenhoek* (1632–1723) placed under a simple microscope a glass of rainwater, and perceived a multitude of variously shaped bodies; further experiments showed the presence of similar bodies in various animal and vegetable infusions (*New York Times*, 1886, July 18).[316] Leeuwenhoek (1677, p. 821) writes (see Dobell 1932 for more information on *Leeuwenhoek*'s "little animals"):

> In the year 1675 I discover'd living creatures in Rain water, which has stood but few days in a new earthen pot, glased blew within.
> The 26. May, it rained hard; the rain growing less, I caused some of that Rain-water, running down from the house-top, to be gather'd in a clean Glass, after it had been washed two or three times with water. And in this I observ'd some few very littly living creatures, and seeing them, I thought they might were produced in the leaden-gutters in some water, that had there remain'd before. Obser. III. On the same day, the Rain continuing, I took a great Porcelain-dish, and exposed it to the free Air upon a wooden vessel, about a foot and half high, that so no earthen parts, from the falling of the Rain-water upon the place, might be spatter'd or dashed into the said dish. With the first water that fell into the dish, I washed it very clean, and then flung the water away, and receiv'd fresh into it, but could discern no living creatures therin; only I saw many irregular terrestrial parts in the same. The 30. Of May, after I had, ever since the 26th, observ'd every day twice or thrice the same Rain-water, I now discover'd some, yet very few, exceeding little Animals, which were very clear. (Leeuwenhoek 1677, p. 823)

An unknown person (Anonymous 1693) collected in 1683 hailstones in Lisle (French Flanders)[317] and writes "One among the rest was observed to contain a dark brown matter in the middle thereof; and being thrown into the Fire, it gave a very great report" (Fig. 2.8) – the first report on organic matter in precipitation.

Boerhaave (1735, p. 349), who not only found rainwater to be rotting (putrefying) but also he had never observed it "to grow acid", observed under the microscope (citing also *Leeuwenhoek*) "principal native plants" and concluded that "these seeds fell into the water from the air", and, moreover, "eggs of minute animals" (later termed *infusoria* and *animacula*). Chemical treatment of rainwater and dew was carried out by distillation, the standard operation of the alchemists. We can conclude from these experiments on some ingredients in such water; indeed, Boerhaave (1735) writes:[318]

---

316 In 1668 *Francesco Redi* (1628–1697) confidently denied the ancient philosophers believe that decaying animal or vegetable matter, under the influence of heat, with sufficient air and moisture could beget anew (*de novo*) living thing.

317 Or Ryssel, today Lille, a city at the northern tip of France in French Flanders (West Flemish *Rysel*, Dutch *Rijsel*).

318 Sapo's (soaps today) = *Seiffen* (in German) = *savons* (in French); at that time a term for vegetable and animalic "oil-like solid" substances [huile fixed] containing potassium [*savons de potasse – pottaschige Seiffe* – potash soap] or calcium [*savon calcaire – kalkerdige Seiffe* – limy soap]. Earth's =

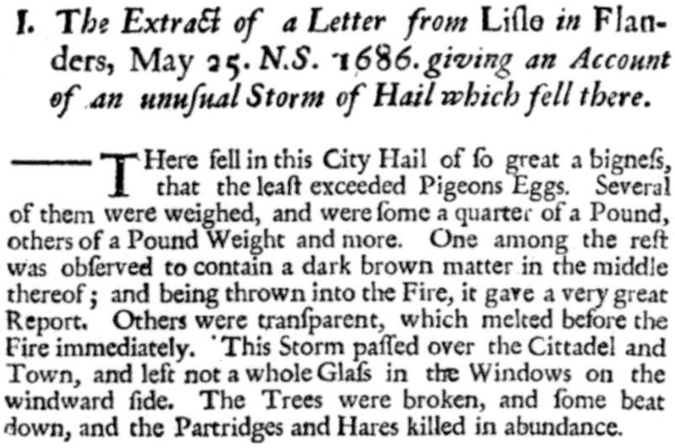

I. *The Extract of a Letter from* Liſle *in* Flanders, May 25. *N.S.* 1686. *giving an Account of an unuſual Storm of Hail which fell there.*

—— THere fell in this City Hail of ſo great a bigneſs, that the leaſt exceeded Pigeons Eggs. Several of them were weighed, and were ſome a quarter of a Pound, others of a Pound Weight and more. One among the reſt was obſerved to contain a dark brown matter in the middle thereof; and being thrown into the Fire, it gave a very great Report. Others were tranſparent, which melted before the Fire immediately. This Storm paſſed over the Cittadel and Town, and left not a whole Glaſs in the Windows on the windward ſide. The Trees were broken, and ſome beat down, and the Partridges and Hares killed in abundance.

Figure 2.8: The first written evidence of organic matter in precipitation; *Phil. Trans.* **17** (1693) 858.

> [. . .] Rain-water; and this certainly one may properly call the *Lixivium*[319] of the Atmosphere, in which are contained all the *Species* of Corpuscles, that were floating about in it. Of what kind, now, these are, and what great variety there is of them, we have already treated particularly in our History of Air: Let these things, therefore, be consulted from p. 282 to 292. There it appears, that in this Air is dispersed every kind of volatile Bodies. But Bodies are either spontaneously volatile, or become so by Fire, Fermentation, Putrefaction, Mixture, Separation, or Effervescence: And hence Salts, Spirits, Oil, *Sapo's*, Earths, and even Metals themselves, are found there (Boerhaave 1735, p. 347).

A very nice shortened German translation is found in Anonymous (1755, p. 231–232):

> Unter diesen ist das Regenwasser welches mit Recht eine Lauge unsers' Dunstkreises genennet werden ist, in welcher alle Arten der Körper enthalten sind, die sich in der Luft aufhalten; nehmlich, Salze, Geiste, Öle, Seiffen, Erde; ja selbst Metalle [Among them is rain water, rightly called leach of our atmosphere, that contains all kinds of bodies occurring in the air: namely salts, spirits, oils, soaps, earths, and even metals].

Boerhaave (1735) wrote in his "Elements of Chemistry" a section "Of rain water" (Boerhaave 1735, pp. 348–350) and it is sure that he made many own experiments with collected rainwater from different seasons and water from snow fallen. However, no description of how rainwater collection is given; we can only assume that such a "simple procedure" was known to any (al)chemist and not worth for description. It is very worth, to cite some more phrases from Boerhaave (1735):

Erden = *terre* (see also footnote 331 on p. 156): before 1800, five different "*Erden*" (systematically named only in German) were known, *Kieselerde* = la silice = silica (SiO_2). *Thonerde* (= *Alaunerde*) = *l'alumine* = earth of alum (Al_2O_3), *Kalkerde* = *terre calcaire* (= *chaux*) = quicklime (CaO), *Bittererde* (= Talkerde) = *talc terre* = magnesia (MgO), and *Schwererde* = *baryte* – barytes (BaO). However, at that time it was not fully clear whether "Erden" were oxides or also salts, namely carbonates in relevant minerals (derived from the oxides and acids) of the elements. Carbonate was also termed "*Kalke* = *chaux* = calces" (see footnotes 193 and 194 on p. 91), namely $CaCO_3$ (*Kreide* = *chalk* = lime).

319 Lixivium is the solution, containing alkaline salts, obtained by leaching wood ashes with water (also termed lye).

> [...] Rain Water, which comes from the upper regions, when after a long drought there happens Thunder, with very large Showers; for if the Rain-water is catch'd at such a time, it is found to abound within great variety of Bodies, as the Chemists have often taken notice (Boerhaave 1735, p. 319).
> [...] Rain-water being digested for a considerable time, putrefies, stinks, and if it is then distilled, yields oily Spirits, that are in some measure inflammable. And being digested, putrefied, distilled, and concentrated, it has assorted a very fragrant Spirit, which very gentle dissolves the Body of Gold, without any effervescence (Boerhaave 1735, p. 349).
> We have likewise observed, that when it has been exceeding hot and dry for a long time, and there succeeds great Thunder, with large showers if Rain, the water that then falls, if it is collected in pure Vessels, produce a Froth, which seems truly to contain something of a very fine kind of nitrous Salt (Boerhaave 1735, p. 350).
> The most impure Rain-water, now, of all, is that which falls in very hot and windy weather (p. 350).
> [...] especially if at the same time [when rainwater falls] that Air is foggy, thick, and stinking, so as to affect the Nose with a disagreeable Smell [...] (p. 350).

In modern terms, *Boerhaave* found in rainwater organic matter and chloride, and he is the first who associated nitric acid with thundershower.[320] At all, *Boerhaave* characterized rainwater qualitatively already in all features more than 100 years before systematic rainwater studies:
- the principal chemical composition,
- the origin of the chemical substances found in air and rainwater,
- the causes of variations in chemical composition by season and location.

We can state that it was "in fashion" to distill rainwater, as mentioned to look for the philosopher's stone. Hence it is remarkable that some chemists described the residue as different salts. Other (al)chemists have seen in it the transmutation of water into the earth. *Lomonossov*, who was obviously familiar with all these works, wrote in his "introduction into true physical chemistry" (Lomonossov 1752/1951, pp. 518–519):

> Aquarum plures dari species, varietate corporum, quibus impraegnata sunt, distinctas, experientia docuit. Alia enim indoles in pluvia, alia in fluviatili, alia in fontana deprehenditur. Pluvia, cum ex alto per athmosphaeram cadit, obvios vapores sulphureos et salinos in se recipit [Experiments show that several kinds of water exist that differ in its bodies. Rainwater represents one nature, river water another and well water a third. When rain falls through the atmosphere, sulphureous vapours and salts are taken up; translation from Russian by DM[321]].

First *Marggraf* (see below) is the first who describes sampling and treatment in detail; he cites the (obviously ancient) use of linen spread among piles and loaded with a poor

320 From the second cited phrase of *Boerhaave* we can assume that in fresh water from thunderstorm was free nitric acid that combined with sea salt to a week solution of mixed HNO_3 and HCl (aqua regia).
321 Опыт показывает, что существует несколько видов воды, различающихся находящимися в них телами. У дождевой воды наблюдаются одни свойства, у речной – другие, у родниковой – третьи. Когда дождь с высоты падает через атмосферу, то принимает в себя встречающиеся сернистые и соляные пары.

stone or a glass ball where a vessel, located down, to collect the rainwater. However, he rejects this method due to contamination, and he also correctly stated that rain from roofs is polluted because of the dust and smoke. The following phrases suggest that he carried out several experiments with different methods of rainwater collection in a modern understanding of "quality assurance" (Marggraf 1786, p. 69):

> Die Leinwand selbst bleibt wegen der noch anklebenden seifenartigen Theile, und wegen ihrer eigenen Bestandteile verdächtig, und auch die töpfernen Gefäße sind unsicher. Sind sie mit Bleyglasur überzogen, so springt in starker Kälte die Glasur sehr leicht ab, und das Wasser löset einen Theil der Töpfererde auf; sind sie mit Kochsalze gebrannt, so sind sie ebenfalls verdächtig [The linen itself stays because of ist adherent soap species suspicious, and because of ist own components, and the pottery vessels are suspicious too. Are they covered with lead glaze, so in strong cold the glaze spalls easily, and the water dissolves a part of the potter's earth; are they burnt with common salt, they are suspicious too].

Moreover, *Marggraf* notes (Marggraf 1786, p. 70) "I pass over the different methods to collect the rainwater [...]", which can be interpreted such that in older literature a description was found. *Andreas Sigismund Marggraf* (1709–1782), German apothecary and chemist in Berlin (Fig. 2.9),[322] was the first who collected and analyzed rainwater and snow (and water from the Spree River and local well water) from only analytical interest between 1749 and 1751 in Berlin.[323] *Marggraf* was interested in the chemical characterization of water because it became a trend to drink natural water and he argued that the water with the least pollution ("foreign bodies") is rain and snow water falling from the sky (Marggraf 1753, 1786); Fig. 2.10. It is worth describing his approach, translated from the German publication but shortened.

> From December 1749 until March 1750 he (with the help of an assistant) collected about 100 L rain water (this season he chooses because of less insects and dust in air) in a garden of suburb Berlin using glass bowls, 1 foot in height and 1.5 foot in diameter, cleaned with distilled water and being exposed only after half day rainfall (so that the air was already cleaned by the rain). At the end of the rainfall, the collected water he decanted into glass flasks (containing about 36 ounces), which also were cleaned before with distilled water. The glass flasks were stored in a basement. In same kind he collected in winter 1751 more than 100 L snow water.
> After collection of all rain, he filled the water to 3/4 in a retort which had a volume of 12 L and begun a gentle distillation (one drop after the other into the condenser). After a few days from 8–9 L remained about 2 L in the retort, becoming turbid. The distilled water was stored. Afterwards the retort was again filled with the collected rain water and the gentle distillation in the same procedure was repeated until all 100 L rain water were concentrated to a last rest of about 3 L; the same

---

322 In 1747, he announced his discovery of sugar in beets and devised a method using alcohol to extract it.

323 First published in French: *Examen chymique de l'eau*. In: *Historie de l'Académie Royale des Sciences et des elles-Lettres de Berlin* 1751. Haude et Spencer, Berlin (1753) pp. 131–157. After *Marggraf*'s death published in German: "4. Abhandlung Marggraf chemische Untersuchung des gemeinen Wassers". In: Physikalische und medicinische Abhandlungen der Königlichen Academie der Wissenschaften zu Berlin. Vierter Band, Ettinger, Gotha (1786) pp. 69–95.

operation was carried out later with snow water. The remaining 3 L water were then further distilled until 16 ounces and finally evaporated to 6–8 ounces. This water was then filtered and the filtrate residue extracted with hot pure water to solve all salty substances. This last filtrate was mixed with the first filtrate. After gentle drying the filtrate residue, about 100 grain white, slightly yellow earth remained, behaved like quicklime. To the filtrate he added 25–30 drops of a solution of salt tartaric and vaporized the solution until some 4 ounces remained. Precipitated white earth was filtered. Finally, the filtrate crystallized into oblongness, pike-like crystals, similar to them of saltpeter, and finally other crystals, entirely alike common salt. Both salts weighted only a few grains and were brownish, as a proof that rain water also contains sticky and oily components. Leftover 25–30 L of rain water has distilled in the same method up to a few ounces. By adding different metallic solutions in nitric acid (Ag, Hg, and Pb) a white precipitate evolves. Most precipitate come from silver, showing acido salis (HCl).
A few liters of rain water were exposed from May to September to sunlight, but the bottle was closed with clean blotting paper to allow air access but not insects and dust; the bottle neck was protected by another glass to avoid rain access. After a month bubble evolved and the water become muddy. In contrast, no putrefaction evolved with distilled rain water.

Figure 2.9: Portrait of *Andreas Sigismund Marggraf* (frontispiece from "*Neues allgemeines Journal der Chemie*", ed. A. F. Gehlen, Vol. 2, 1804).

At the end, he found nothing else as already known by *Boerhaave*: nitrate (saltpeter), calcium (lime), sodium, and chloride (common salt), and organic substrate ("sticky and oily brown remains"). He assumed that its origin was from salty and earthy components in the air. In snow, he found more hydrochloric acid and less nitric acid[324] and lime, and vice versa in rain. He also found the rainwater to be rotting because of the organic matter (he termed it "*schleimigt-öhlige Theilchen*" and "*particules visqueuses*

324 Note at that time with hydrochloric acid and nitric acid in rain, snow and dew chloride and nitrate is meant.

69

4. Abhandlung.

Marggraf chemische Untersuchung des gemeinen Wassers.

§. 1.

Will man gewisse natürliche Körper chemisch untersuchen, so muß man ohne Zweifel die reinsten Individua wählen, die die wenigsten fremdartigen Theile bey sich führen. Da nun unter den verschiedenen Arten des vom Himmel niederfallenden Wassers das Regenwasser und Schneewasser das reinste ist, so habe ich geglaubt mit der Untersuchung derselben die Untersuchung der verschiedenen Arten des Wassers anfangen zu müssen.

§. 2.

Sehr wenig rein ist das Regenwasser und Schneewasser, das von Dächern und Dachrinnen herunterfällt, weil die Dachziegel aus Thon und Eisenerde bestehen, und sich viel Staub und Rauch an dieselben ansetzt. Eben so unsicher ist die Methode reine Leinwand auf Pfälen auszuspannen, in die Mitte einen reinen Stein, oder eine reine schwehre Glaskugel zu legen, und so das Wasser in einem untergesetzten Gefäße aufzufangen. Die Leinwand selbst bleibt wegen der noch anklebenden seifenartigen Theile, und wegen ihrer eigenen Bestandtheile verdächtig, und auch die töpfernen Gefäße sind unsicher. Sind sie mit Bleyglasur überzogen, so springt in starker Kälte die Glasur sehr leicht ab, und das Wasser löset einen Theil der Töpfererde auf; sind sie mit Kochsalze gebrannt,

E 3 so

Figure 2.10: First page of *Marggraf*'s publication on studying common waters (Marggraf 1786, p. 69). Remarkable § 2 on description of the procedure of precipitation sampling and precautions.

& huileuses", respectively – mucoid-oily particles). Since this observation, the putrefaction to which rainwater is subject, shows that some organic matter is present (see on p. 161 the discussion concerned "*Pyrrhin*"). Figure 2.11 shows the Table of the chemical composition of different waters from the paper written in French (Marggraf 1753); what is follows the Table 2.2 including the terms from the paper written in German (1786).

It is obvious that during the long-time distillation most of the lime ($CaCO_3$) is originated from the dissolution of calcium oxide from the glass vessel; it corresponds to about 18 mg L^{-1} Ca that we can compare with a value typically to be 2 mg L^{-1} Ca, found in the 1980s in Berlin. By contrast, the value for nitrate (about 0.5–1.5 mg L^{-1}) corresponds to values found in the nineteenth century (see Table 2.7).

pour la page 157.

TABLE SYNOPTIQUE

EAUX TRE'S PURES QUI TOMBENT DE L'AIR.

Cent meſures dites Quartes. { d'eau de pluye contiennent dr. I. gr. LX. de terre calcaire ſubtile, & quelques grains d'acide nitreux & de ſel commun.
d'eau de neige - - - dr. I. d'une terre calcaire très déliée, & quelques grains d'un acide de ſel, comme imprégné d'une certaine vapeur nitreuſe.

EAUX PLUS IMPURES.

		d'une terre ſubtile calcaire.	*d'une terre gypſeuſe.*	*de ſel commun.*	*de Nitre pur.*	*de ſel moyen ſemblable au ſel d'Egre.*	*de la dernière leſſive des Ouvriers en nitre.*	*de leſſive alcaline.*
Cent meſures.	de l'eau de la pompe du vieux Château.	onc. I. dr. II. gr. XV.	gr. VII.	- - dr II. gr.XXX	gr. VIII. à X.	- - -	- gr. XXX.	- -
	de l'eau de la pompe du nouveau Château.	- dr. VII. gr. XX.	- dr. III gr. XXX.	- - dr. III.	- - - dr. I. gr. XXX.	- - -	- gr. XL.	- .
	de l'eau de la pompe de la ruë Electorale.	onc. I. - - - - -	- gr. XXX.	dr. I. gr. XXX.	- - dr. I gr. XXX	- - -	- - -	gr. XXX.
	de l'eau de la pompe de la Maiſon de M. *Marggraf*	onc. I. dr. II - - -	- dr. II.	- - -	onc. I. dr. I. gr. XV. mêlé avec un prode ſel commun.	- - -	- dr. IV.	- -
	de l'eau de la riviere de Sprée.	- - dr. IV. gr. XXVII	- -	- - gr. XXX. approchant du nitre.	- - -	- - -	- - -	très peu.
	de l'eau d'une ſource ſituée près d'un Moulin à papier.	- - - dr. V. gr. XXX.	- gr. IV.	- - -	- - -	- gr XXX.	- - -	- -
	de l'eau d'une ſource de Potzdam.	- - dr. VI. gr. XXIV.	- -	- dr. II. & quelques grains.	- - -	quelques grains.	- - -	- -

NB. *onc.* ſignifie une once, contenant VIII. dragmes
dr. - - une dragme, qui contient LX
gr. - - grains

Figure 2.11: Table with rainwater chemical composition (among other natural waters) on p. 157 in "*Examen Chymique de l'eau*" by Marggraf (1753).

Table 2.2: Composition of 100 Quartier (about 115 L) of rain and snow water after *Marggraf*. 1 *Quartier* (= 36 *Unzen*) = 1.145 L, 1 *Quent* = 31.1 g, and 1 *Gran* (grain) = 0.063 g (Prussian measures).

German	French	formula	rainwater	snow water
Zarte Kalkerde	terre calcaire subtile	$CaCO_3$	1 Quent and 40 Gran	1 Quent
Salpetersäure	d'acid nitreux	HNO_3	few Gran	few Gran
Kochsalzsäure	sel commun.	HCl	very little	few Gran

We have to consider while interpreting such "analysis" that the original chemical composition of freshly falling rainwater is different because of a) sampling and storage of rainwater over a long time (weeks and months) and b) the severe treatment through long-time distillation. Hence, we cannot wonder that no ammonium was detected (which should be much higher than nitrate at that time) because ammonium is unstable and will be lost in rainwater due to biological activities, but will not affect any other component[325] (Koch et al. 1986). Moreover, during the distillation process, ammonia will degas from ammonium because of $NH_4^+ + CO_3^{2-} \rightarrow NH_3 \uparrow + HCO_3^-$. Furthermore, it is interesting that no sulfate (easily to detect in form of gypsum crystals as done 100 years later by *Smith*) was found, a sign that no SO_2 emission appeared. In Berlin in the eighteenth century, stove heating and manufacturing was based exclusively on wood; still, around 1800 coal (imported in small amounts from England) did play no role; industrialization began stepwise after 1830. Before the construction of sewage systems,[326] the putrefaction of organic waste led to emissions of ammonia (including amines), methane, nitrogen oxides (gaining nitric acid after oxidation), and hydrogen sulfide in towns. We can conclude that the air at that time was full of carbonate of calcium (from soil dust) and ammonium[327] (from putrefaction of organic residues) and sea salt (NaCl) from marine air. Gaseous nitric acid and ammonia in the air form ammonium nitrate via gas-to-particle formation. It is not surprising that sulfate in rainwater in high concentration was first detected in England (after 1850).

Next to this investigation, *Torbern Olof Bergman* (1735–1784)[328] in 1774 used evaporation and precipitation to carry out a detailed examination of different natural waters and developed a list of the substances that he had identified in seawater. He introduced the technique of weighing the precipitated salts to determine their concentrations. He writes (Bergman 1788, p. 118):[329]

> Snow-water contains a small quantity of salited lime, together with some slight vestiges of nitrous acid; this water, when newly melted, is totally void of air and the aerial acid; substances which are found, in greater or less quantity, in almost waters: – and hence it is, perhaps, that snow-water is noxious to animals. Rain-water is generally contaminated with the same substances as the former; but in greater quantity: it is obvious that these waters, while suspended in air, must collect and absorb the various heterogeneous matters with which the atmosphere abounds, and

325 It could be assumed that through nitrification ammonium is converted into nitrate; own (unpublished) experiments in the 1980s have shown that the nitrate concentration not rise while that of ammonium decreased.

326 London 1842–1865, Paris 1850 and Berlin 1878.

327 100 years later Smith (1879) wrote that the air of towns is full of ammonia.

328 Swedish chemist and mineralogist in Uppsala; he greatly contributed to the advancement of quantitative analysis, and he developed a mineral classification scheme based on chemical characteristics and appearance; finally, he is noted for his sponsorship of *Carl Wilhelm Scheele*.

329 This term does not mean HNO_2 at that time but the "acid forms from nitrous gases" (detected as nitrate).

> therefore can never be obtained pure. Immediately after long continue rain or snow, these waters are found least loaded with heterogeneous matter.

Thus he "re-discovered" what *Marggraf* already found (it is likely that *Bergman* did not know the *Marggraf* papers from 1753 and 1786).

With his studies to prove that water is not transmuted into the earth (see p. 143), *Lavoisier* also investigated rainwater in August 1768, which he collected in a glazed vessel ["*grands vases de verre & de fayence émailée*"] at Le Bourget, that was clear of houses and tress (Meldrum 1933). His aim was not primarily to analyze rainwater but to study the "transmutation" (Lavoisier 1770, p. 90):[330]

> Mon objet étant d'examiner la nature de l'eau, de déterminer si elle contenoit de la terre, enfin de conflator ce qu'opéroit sur elle une suite de distallations successivers, [. . .]

Next, the rainwater was distilled in a glass alembic, the residue was weighed and the distillate was compared with the standard – it was water, from the Seine, once distilled. He found 2 Livres (1 Livres corresponds to 1 Pound ≈ 454 g) rainwater to be heavier than the standard (distilled Seine water) by 1.07 Grains (1 French Grain = 53.1 mg). The residue after once distillation of rainwater *Lavoisier* estimated to be 0.390 Grains per Livre (ibid., p. 106). *Lavoisier* also analyzed the residue in rainwater and found (Lavoisier 1770, p. 93–94):[331]

- *trois dixièmes de grain d'une terre insipide* & *très légere* [0.3 Grain tasteless and light earth]
- *un ouzième de grain de sol marin* [1/12 Grain sea salt],
- *partie à base d'alkali* [traces of alkali metals],
- *partie à base terreuse* [traces of alkaline earth metals]

It is obvious that the following authors cite the figures erroneously. Gairdner (1832, p. 113) writes (without citation) that "Lavoisier found it [rainwater] to contain no more than 0.397 of insoluble earth, and 1.31 soluble salts" (without giving the dimensional unit). Osann (1839, p. 114) writes

> Nach Lavosier's Untersuchungen enthält ein Pfund Regenwasser nur 0.30 Gran Erde und 0.99 Gran Salz [After *Lavoisier*'s investigations, one pound rain water contains only 0.30 grains of earth and 0.99 grains of salts].

---

330 See also: Lavoisier (1768) Expériences et analysis de diverses eaux, p. 30v, Archives de l'Académie des Sciences (cited after Meldrum 1833, p. 400).

331 *Erden, Terrae, Terres*, Earth: ancient term for simple, solid, tasteless and water-insoluble matter. The British chemist *Humphry Davy* demonstrated that they all consist from metals and oxygen; (water-soluble) alkaline earth are oxides of Ca, Ba, Mg, Na, K etc. Today the terms rare earth (metals) and alkali earth metals (Li, Na, K, Ru, Ca, Fr) and earth alkaline metals (Be, Mg, Ca, Sr, Ba, Ra).

At the end of the eighteenth century, *Bernard Joseph Paping* (physician in Amsterdam) presented a remarkable dissertation under the supervision of *Peter van Driessen* (1753–1828), professor for chemistry at Groningen. In that time, rain gutters were made from lead, and rainwater collected from roofs were used as potable water; however, for a long time already diseases from lead poisoning were observed. *Paping* (and *Driessen*)[332] concluded on the existence of free hydrochloric acid (HCl) in air, scavenged by rain and subsequent dissolution lead (Paping 1796, p. 25, cited after Driessen (1822)):

> Novimus, quam saepe tristi hoc malo afficiantur harum praesertim Urbium incolae, quarum atmosphaera aquae marinae effluviis est replete; ita ut vix ulla datur urbs, in qua frequentius aqua saturnine veneno incommodum creat, quam Amstelodami; videtur nempe salis marini acidum ex Muriate magnesiae, seu magnesia salita, solis calore elevari, et aqua pluviali commixtum, multum ad plumbi corrosionem contribuere [We know how often townspeople are unpleasant affected through atmospheric effluvia from sea water; so that barely a town like Amsterdam, which is frequently inconvenient influenced through poisoning water of Saturn; it seems that sea salt acid from chloride of magnesium, magnesia saline at high solar heat, and incorporated with rainwater, contributes significant to corrosion of lead].

2.3.1.2 Rainwater studies 1813–1863: an analytical approach

It is remarkable, that scientists from many disciplines were interested in the chemical composition of rain and snow water with the beginning nineteenth century, and, consequently, reviews on original studies can be found in several books after 1850. This interest comes first concerning water quality, then by the developing agricultural chemistry and later air pollution. In chapter VI in the book of *Joachim Isidore Pierre* (1812–1881), French agricultural chemist in Caen (Pierre 1859, pp. 31–44), almost completely included in the book of Smith (1872, pp. 232–244), an extensive review on precipitation chemical studies is given. Other good summaries of rain (and other natural) water studies are presented by Liebig and Kopp[333] (1853, pp. 705–709), Moleschott (1859, pp. 387–390, annex pp. 202–204), and Ludwig[334] (1862, pp. 1–29). Before 1850, all the early rainwater studies were based on evaporation as already done in the eighteenth century and earlier by alchemists. However, with the fast-increasing knowledge of chemical compounds and detection reactions for qualitative analytical chemistry, more and more substances could be detected and determinate in rainwa-

332 This is *Jan Constantin Driessen* (1790–1824) physician and chemist; it is unknown whether he was familiar with *Peter van Driessen*.

333 *Hermann Franz Moritz Kopp* (1817–1892) was a German chemist, *Liebig*'s student and professor in Heidelberg since 1843; known for his books on history of chemistry 1843–1847 (Kopp 1931) and Kopp (1869, 1886).

334 *Friedrich Hermann Ludwig* (1819–1873) was a German chemist and apothecary in Jena; assistant and later successor of *Ferdinand Wackenroder* (1798–1854).

ter; water analysis and agricultural chemistry stimulated the development of specific quantitative methods for ammonia and nitrate after 1840.

The remarkable first study in the nineteenth century was carried out by *William Stark* (1788–1863),[335] a chemist in Norwich (Stark 1813) who studied atmospheric acidity, using logwood instead of litmus, writing:

> [...] finding it to be more powerfully acted on both by alkalis and the acids.
> I was making some experiments on logwood during a violent thunder-storm, and exposed a few grains of it to the action of rain which fell at that time; the colour extracted was an orange red, the same as it produced by distilled water slightly acidified. This singular effect was so unexpected, that I doubted the accuracy of the experiment: a short time afterwards I repeated it, under similar circumstances, and found the same result.

There is no doubt that *Stark* find the rainwater a little acidic due to nitric acid, produced during the thunderstorm, but he does not recognize that it is due to HNO_3 (although its formation was known since *Cavendish* through electric discharges in the air); he writes further:

> On exposing logwood to the action of rain which fell when there had been *no thunder* for several weeks, a colour inclining to *lilac* was extracted; and water caught at the same time, at the height of 300 feet (on the top of the cathedral), produced also the lilac shade. These facts evidently indicated either the presence of lime, or some kind of alkaline matter.
> [...] my ingenious friend Mr. Robert Higgin,[336] he suggested the probability of its being caused by the presence of ammonia [...]

To proof this idea, *Stark* passed a stream of muriatic acid (HCl) into a receiver while distilling a quantity of rainwater; "[...] but no dense clouds were produced".[337] Further, he studied the precipitate found after boiling the water and concluded on the presence of super-carbonate[338] of lime ($CaCO_3$) and hence "[...] *lime* in the atmosphere" which he believed "is exhaled from the earth's surface [...] which came down with the rain".

---

335 *Stark* was one of the first dyers of fabrics of Norwich manufacture, "particularly of the colour termed Turkey red, the manufacturers in the North sending large quantities of goods for dyeing." He devoted much of his time to the prosecution of scientific studies, and was a Fellow of the Geological Society. Belonged to the Norwich Philosophical Society, "Mr. Stark contributed many papers at its meetings, in which he bore a distinguished part" (*Norfolks Annals*, chronological record of remarkable events of the nineteenth century by C. Mackie, Vol. III, 1901).

336 *Robert Higgin* (unknown living dates), of the City of Norwich, Shawl Manufacturer, for his Invention and Discovery of a new or improved Method of Consuming or Destroying Smoke (*The London Journal of Arts and Sciences*; containing reports of all new patents, Vol. IX, London 1823, p. 237); the method is based on the principle of a thermal post-combustion.

337 In presence of gaseous ammonia, particulate NH_4Cl should be formed. But because of the rainwater evaporation, all ammonia exhaled before "analysis" according to $NH_4^+ + CO_3^{2-} \rightarrow NH_3 + CO_2 + H_2O$ and $2HCO_3^- \rightarrow CO_2 + CO_3^{2-} + H_2O$.

338 The term "supercarbonate" was first used by *Thomas Beddoes* (1760–1808) in following sense: carbonate to a high degree with a high proportion of carbon dioxide.

To proof which acid he found in rainwater from the thunderstorm, *Stark* boiled it, and it became more acidic (that can be explained through NH_3 degassing and remaining nitric acid) and added "a small quantity of lime water" (CaOH), gaining a precipitate ($CaCO_3$). From further boiling the acidified rainwater and observing "a considerable quantity of gas to be expelled" he concluded on carbonic acid (CO_2); it is clear that simply the equilibrium between dissolved CO_2 with carbonate and bicarbonate shifted to CO_2 which degasses. *Stark*'s paper was never cited concerning his investigation of acidic rain from thunderstorms but only due to the finding of lime (what was found already by *Marggraf*).

Berzelius (1820, p. 402) included in his first textbook a short paragraph on rainwater and snow and writes:[339]

> Gewöhnlich ist das Regen- und Schneewasser mit atmosphärischer Luft, ein wenig Salpetersäure und, wie behauptet wird, von einer äußerst geringen Menge salzsaurem Kalke verunreiniget. Was inzwischen den letztern betrifft, so ist dessen Anwesenheit wenig wahrscheinlich; denn dieses Salz ist völlig feuerbeständig, und kommt, so viel wir jetzt wissen, nicht in Gasgestalt vor. Salpetersäure hingegen wird in geringer Menge bei Verbrennungen gebildet, und muss sich sonach immer in der Atmosphäre aufhalten [...] Die meisten Unreinlichkeiten welche das Wasser bei sich führt, sind mechanisch beigemengt, und bestehen aus zusammengeschlemmtem Staube, der in der Luft herum fliegt und vom Regen oder Schnee mit herabgespült wird [Normally rain and snow water is polluted with atmospheric air, a little nitric acid, and, as it is claimed, with a very small amount of hydrochloric lime. What concerns the latter now, so it is presence unlikely; since this salt is full fire-resistant and does not exist, as we know, in gaseous kind. Nitric acid, however, is gained in little amount during combustions, and thus must be always present in the atmosphere [...] Most impurities which are together with the water, are mechanically admixed, and consist from dust feasted together, which flies around and rinses down by rain or snow].

At the time of the original Swedish edition (1808), *Berzelius* could know only three rain and snow water studies, those by *Boerhaave*, *Marggraf*, and *Bergman*, and it was known from them that lime ($CaCO_3$), common salt (NaCl), and nitric acid (HNO_3) are permanent constituents. Thus, it is surprising that *Berzelius* doubt the presence of calcium (and) chloride in precipitation with the argument that the salt is *not* volatile, and states a phrase afterward that dust is suspended in air and *mechanically admixed* to precipitation (which not excluded the presence of lime and common salt in the air). On the contrary, the statement that nitric acid (a synonym in that time for nitrous oxides NO_x) is formed during combustions looks like today's modern knowl-

339 The first German translation from the second Swedish edition. However, in the first Swedish edition (*Lärbok I Kemien*, Stockholm, 1808, 483 pp.), the Swedish text (on p. 214) fits exact the German translation: Regn- och snövatten äro vanligen kemiskt orenade af atmos. Luft, litet salpetersyra och, som det pastas, as en ytterst ringa portion saltsyrad kalk. Hvod della sednare beträffar så är likväl dess närvaro mindre samolik, ty detta salt är aldeles eldsast och kan, etter hvad vi hittils veta, icke finnas I gasform. Salpetersyra deremot bildas till en ganfka ringa qvantitet vid förbränningar, och mäste sä1edes altid finnas I atmosferen.

edge, however, at that time formation of nitrous gases (NO_x/NO_y) was only known from *Cavendish*'s experiment with electric sparks. Lampadius (1820, p. 256), who first cited *Berzelius*' phrase on hydrochloric lime, writes that he often did observe its presence in precipitation and that it is rarely absent and strongest in precipitation from westerly directions (note, he lived in Freiberg near Dresden, rather far from the sea). *Lampadius* writes that he collected precipitation on free areas using clean glass containers, added some drops of an $AgNO_3$ solution, and put the vessel in the day- or sunlight, just gaining a brown-reddish color (see discussion concerned "*Pyrrhin*" on p. 161) like from the solution of $CaCl_2$ in water. *Lampadius* (Lampadius 1820) found that the black earth in the surroundings contains "*salzsauren Kalk*", but notes that most probably

> [...] das stark bewegte Meer Wasserstaub in die Luft sende, so fragt es sich warum vorzüglich der salzsaure Kalk und nicht vorwaltend salzsaures Natron sich in dem atmosphärischen Wasser vorfindet? [... the heavily moved sea sends water dust into air, and it wonders why prevailing hydrochloric lime and not hydrochloric sodium is present in atmospheric water?].

Notably, around 1820 several so-called "colored rains" (Chapter 3.2.2) have been investigated. It follows an example, the "red rain" fallen on 2 November 1819 at Blankenberge in Flanders (today Belgium) and analyzed by the chemist *van Stroop* from Brügge, finding a dissolution of "*rothem salzsauren Kobalt*" [red cobalt chloride].[340]

Gerardus Johannes Mulder (1802–1880), chemist and physician in Utrecht, analyzed many natural waters including rainwater in the early 1820s (for its use as potable water) and detected NaCl and dissolved CO_2. He describes how pollutants in the air affect the rainwater (Mulder 1825, p. 99):

> Nunquam aquam pluviae tam fatuam potavi, quam in nonnullis Privinciae Hollandiae locis, cujas rei causam puto impuricatem aëris, combinari cum principiis, quae ab aquis salsis, utique materiis vetetabilitus commixtis expulse, aër absorpsit rain
> [Rainwater can never be potable, and at some Dutch locations, in my opinion due to air pollution, combined with causes of salty water, surely dispersed mixed vegetable matter, the air takes up].

Wilhelm Ludwig Zimmermann (1782–1825), professor for mineralogy and chemistry in Giessen and predecessor of *Liebig*, must be seen as the first, who studied the (qualitative) chemical composition of rain (including snow and fog water) very intensively and extremely carefully. He collected rainwater using vessels made from platinum from

340 Published in several journals by different authors: Wurzer (1819) Sympathetische Tinte vom Himmel ergossen. *Gilb. Ann.* **63**, 230–231. *See*: Meyer and Stoop (1819). *Ferdinand Wurzer* (1765–1844) was a German physician and chemist, founder of the subject chemistry 1805 at University Marburg. Nothing is known upon *Meyer* and *van Stroop*. *Gilbert* ironically comments *Wurzer*'s note that hopefully *van Stroop* not mixed the rain before investigation with sympathetic ink (cobalt chloride).

280 events at 5 different sites and altitudes in Giessen between May 1821 and December 1823 (Zimmermann 1824). The total amount dissolved in rainwater he weighted to be 10–45 mg L^{-1}. He is the first who detected ammonium, iron, and manganese as permanent components in rainwater; furthermore, he found organic matter, hydrochloric salts (HCl), H_2CO_3, K, Ca and Mg, and occasionally Ni. The organic matter he termed *Pyrrhin* (pyrrhine, from πνρροξ – red; reddens a solution of silver), and believed it is gained from atmospheric volatile organic compounds that exhaled from the earth's surface:

> [...] daß das Pyrrhin als eine von der organischen Oberfläche der Erde mannigfaltigen, verflüchtigten Stoffen entstandene Substanz anzusehen ist [... that pyrrhine is a substance gained from various volatile matter from the organic earth's surface].

Chladni (1821, p. 356) refers *Zimmermann* who analyzed a "blood rain" fallen on 3 May 1821, in Gießen, containing a red-brown precipitate, detecting iron oxide, carbon, chromium, silica, lime but no nickel.[341]

Ernst Julius Heinrich Witting (1795–1861),[342] apothecary in Höxter (between Paderborn and Göttingen in North Rhine-Westphalia), a scholar of *Arend Joachim Friedrich Wiegmann* (1770–1853), chemist and apothecary in Braunschweig, studied rain, snow, hail and dew since 1823 (Witting 1823, 1824, 1825). He detected free HCl (but not in all events) and bonded (to Ca) and organic matter (in all events and also air), termed "*Kohlenhydrogen*" and "*Kohlen-Hydrogengas*", namely in dew but not in snow.[343] This was confirmed by Wiegmann (1824) who collected snow, rain, and dew in 1823 and found sulfate ("*Schwefelsäure*"), chloride ("*Salzsäure*"), calcium ("*Kalkerde*"), and carbonate ("*Kohlensäure*"). Daubeny (1836, pp. 2), quoting *Zimmermann* and *Witting*, comments:

> Upon the whole then it may be observed [...] the one relating to the existence of organic matter in atmospheric water is best substantiated; and this has more lately been ascribed by the distinguished Ehrenberg to the ova of a particular class of Infusoria [...] fill the atmosphere, and produce the pyrrhine observed by chemists.

Daubeny finally writes (Daubeny 1836, p. 3):

> To conclude my account of the foreign bodies met with in meteoric water, I may mention, that the fact of carburetted hydrogen having been detected in the water of rain, snow, and hail, is more

---

341 Published in (anonymously): Merkwürdige Naturerscheinung. *Allgemeiner Anzeiger der Deutschen* **62** (1821) 1483.

342 *Witting* wrote several books on toxicology and chemistry.

343 Osann (1839, p. 114) writes that *Kastner* and *Witting* could not detect traces of iron rain but in dew meteoric iron and nickel, citing *Kastners Archiv*, Vol. 5. p. 190. However, on that page is the article Witting (1825) and he neither found iron nor nickel; *Kastner* never published an article on rainwater analysis.

> credible, inasmuch as Boussingault has found this same gas in the atmosphere surrounding large cities.

John Dalton (1766–1844) found in rain in Manchester collected with a rain gauge (written at that time *gage*) after a severe storm on 5 December 1823 muriatic acid (as common salt) and later (on 24 February and 4 March 1824: 0.55 and 0.27 grain in 7500 of water, and between nineth and sixteenth fallen in ordinary circumstances traces of sulfuric acid and scarcely any of muriatic acid) also traces of sulfuric acid (Dalton 1824). He cites *John Blackwell*[344] who parallel examinates (the location is not communicated) rainwater and confirmed the amount of salt but who did not find sulfuric acid. *Dalton* was able to compare different samples of rainwater and to conclude that the muriatic soda (NaCl) determined was caused by sea spray (in grains per 7500 waters; in parenthesis concentration in ppm):

1.	After the 5th December from tub of the Society's premise	3	(400)
2.	From rain gauge just after the storm	0.86	(115)
3.	From the lead cistern, after the storm	0.81	(108)
4.	From the lead cistern, taken before the storm	0.13	(17)
5.	From the rain gauge a rainy subsequent day after the storm	0.13	(17)

Hence, he found rainwater after the storm to have 20 or 30 times more salt that had the water collected previously the storm. *Dalton* also collected rainwater in a clean glass vessel in the town, giving no traces of sulfuric but generally a slight one of muriatic acid. He argues (Dalton 1824, p. 370).

> It is owing, I apprehend, as was observed, to the muriate of ammonia [NH_4Cl], sublimed by the combustion of coals, and mingled with the atmosphere, which is carried down by the rain.

However, it is also likely that the determination method was not sensitive enough to find sulfate from sulfur dioxide pollution in the town (that also would be valid concerning ammonium). On the contrary, he believes that the sulfate found outside the town could be from the "earthenware bottle of the gage" (Dalton 1824, p. 370) – but is also probable that he detected sea salt sulfate.

Helmts (unknown given name and living dates), an apothecary in Achim near Bremen, found free HCl in rainwater, namely on the shores of the North Sea, but never metallic contaminants (Helmts 1826); his short paper is nicely entitled "*Fremde Beimischungen im Regenwasser*" [foreign admixtures in rainwater].[345] *Wiegmann* also

344 There is several *John Blackwell*, but likely the English civil engineer (ca. 1775–1840).
345 The term "*fremde* [or *fremdartige*] *Beimischungen*" [foreign admixtures] is synonym with the modern term *contaminants*; used also by Witting (1825) and Mons (1827, 1828).

studied rainwater collected in summer 1825 in the presence of dry fog (Chapter 3.2.4), namely from moorland burning and detected phosphoric acid, sulfuric acid, pyrrhine, and lime (Wiegmann 1826); however, during rainwater investigation over the previous four years, he was unable to detect phosphoric acid.

Simon Rudolph Brandes (1795–1842), a German apothecary in Salzuflen,[346] collected from January until December 1825 very careful daily rainwater (n = 92) which he investigated after sampling with the reagents, common at that time (AuCl, PtCl, $AgNO_3$, NH_4HS, $NH_4HC_2O_4$, KOH, $NH_4H_2PO_4$, $Ba(NO_3)_2$, $Pb(NO_3)_2$, $Ca(OH)_2$, $K_4Fe(CN)_6$)[347] and he describes color, taste and smell of each rainwater sample (Brandes 1826b). After qualitative detection of the compounds in each sample (not in each sample, but in total he found ammonia, sodium, potassium in traces, calcium, magnesium, iron, manganese, carbonate, chloride, sulfate, likely nitrate, and organic material) he mixed the rainwater samples to monthly samples for evaporation and determination the residue. Thus, he found the residue between 0.8 and 6.5 ppm (means of 2.6); from the 12 values the mean residue amounts in summer (April–September) 1.6 0.8–2.8) and in winter 3.6 (2.1–6.5) ppm. The total amount of the residue (2.75 gran) he again analyzed; it is worth presenting the list of the chemical composition:

Harz	resin
Pyrrhin	animal-vegetable matter
Mucus	sticky, gelatinous material
Salzsaure Bittererde	$MgCl_2$
Schwefelsaure Bittererde	$MgSO_4$
Kohlensaure Bittererde	$MgCO_3$
Salzsaures Natron	NaCl
Schwefelsaurer Kalk	$CaSO_4$
Kohlensaurer Kalk	$CaCO_3$
Salzsaures Kali	KCl
Eisenoxyd	FeO
Manganoxyd	MnO
Ammoniaksalz (salpetersaures?)	NH_4NO_3

346 *Brandes* published the famous "*Repertorium für die Chemie als Wissenschaft und Kunst*" [Repertory for Chemistry as Science and Art], where only 3 volumes appear (from A until Berzelit), Hannover (1826–1831). He should not mistake with the German meteorologist *Heinrich Wilhelm Brandes* (1777–1834). Brandes (1826c) describes on more than 300 pages the knowledge of chemical substances at that time; including sub-chapters on "*Meteorwasser*" (pp. 58–64) [meteoric water] and "*fremde Beimischungen der Luft*" (pp. 69–70) [foreign admixtures of the air].

347 In that time, not all formulas had been known yet, the name of the reagents in German as listed: salzsaures Gold, salzsaures Platin, salpetersaures Silber, hydrothionsaures Ammoniak, oxalsaures Ammoniak, Aetzkalilösung, phosphorsaures Ammoniak, salpetersaures Baryt, salpetersaures Blei, Kalkwasser, eisenblaues Kali).

The organic matter he distinguished in one soluble in water and another, “animal, soluble in potash”. Because he also noted the rainfall (totally 638 mm),[348] he calculated the total deposit to be 16.5 kg ha^{-1} yr^{-1}. *Brandes* is not only the first who conducted a nearly complete chemical analysis but he also discusses the origin of the substances and reviews previous articles on this subject. He remarks that for studying the pollutants in the air (he termed it “foreign bodies” the investigation of rainwater (he termed it “meteoric water”) is necessary; he writes:

> Die Gegenwart fremder Stoffe in der Atmosphäre kann uns bei Stoffen, welche einer luft- oder Dampfform fähig sind, nicht befremden. Was aber die der Verflüchtigung, wenigstens bei gewöhnlicher Temperatur, unfähigen Körper anlangt: so kann niemand leugnen, dass mehrere, wie z. B. das Kochsalz im Meerwasser, bei der Verdunstung des Wassers im aufgelösten Zustande, mechanisch mit fortgeführt, andere aber als höchst feiner Staub von der Luft aufgenommen und schwebend erhalten, oder von dem Wasserdunste der Luft aufgelöst werden [The presence of foreign bodies in the atmosphere cannot alienate us for substances which are capable to be in a kind of air or vapor. But what about bodies which are unable to evaporate, at least at common temperature: nobody can negate that several bodies, e. g., common salt in sea water, mechanically continued, other as very fine dust gathered by the air and held suspended or dissolved by the mist[349]].

Liebig collected 77 rainwater samples in Giessen from 1826 to 1827 (Liebig 1827); he always found NaCl, and in 17 samples collected during the thunder, nitric acid bonded to ammonium and calcium, and from the other 60 rains only 2 contained nitrate using the method of evaporation (citing *Zimmermann*, who was his predecessor, and *Brandes*). Liebig (1865) wrote, newer studies have shown that nitric acid is always contained in rain and often more in weight than ammonia, and consequently, that nitric acid is a permanent companion of ammonia in air.

Lampadius had already written in his book (Fig. 2.12, that different atmospheric waters still need more detailed investigations of their composition (Lampadius 1806); he cites *Marggraf* and (wrongly) *Priestley* who should had found nitric acid and lime in rainwater. Many years later (Lampadius 1834, 1835, 1837, 1838a,b), he studied the chemical composition of mineral and atmospheric waters; however, he contributed

---

348 *Brandes* used old measures like Parisien and Prussean Linien (lines), *Zoll* (inches), *Fuß* (foot), *Unzen* (ounce), *Pfund* (pound) und *Meilen* (miles) and presents the deposit as 9282545600 *Gran* = 19588220 *Unzen* = 1250166.6 *Pfund* per square *Meile*. He notes that 1 *Meile* = 24000 *Fuß* and it follows 1 *Zoll* = 12 *Linien* but other “recalculations” remain not clear. Here I present the deposit from the residue given 0.0000026 parts in 1 part meteoric water and the rainfall height given in 24.4 Prussian *Zoll* (= 283.7 Parisian *Linien* = 295.1 Prussian *Linien*).

349 I translated the German “*Wasserdunst*” as “mist” but it also includes fog and clouds (not rain in a narrow sense) but all suspended water droplets, gaining from condensation of water vapor (like *steam*), Thus, *Brandes* describes clearly (here in modern terms) the origin of atmospheric volatile trace substances from exhalations, sea salt mechanically from the sea, soil dust suspended in air and scavenged by atmospheric water. Note that oxygen, nitrogen, carbon dioxide and water vapor were considered as the “essential elements” of the atmosphere; we would say composing the clean air.

Systematischer Grundriß
der
Atmosphärologie,
von
Wilhelm August Lampadius,
Professor der Chemie und Hüttenkunde an der Freyberger
Bergacademie, Oberhüttenamtsassessor, mehrerer
gelehrten Gesellschaften Mitglied.
Freyberg, 1806.
bey Craz und Gerlach.

Figure 2.12: Cover of "*Systematischer Grundriß der Atmosphärologie*" [Systematic Treatise of Atmospherology] by *Wilhelm August Lampadius* (1806).

nothing essential news to that what other authors found before. Lampadius (1837) stated that pyrrhine is nothing else than humus from soil dust.[350] In snow from "east"

---

350 This was an essential new statement and can be seen in view of humic-like substances (HULIS) found in rain in the 1990s.

he did not find chlorine and sulfate but pyrrhine and in snow from "west" no pyrrhine but salts. Furthermore, he estimated (very incorrectly) the quantity of atmospheric gases in rainwater. In 1838, he detected hydrochloric acid and nitric acid in rain from a thunderstorm (Lampadius 1838a,b). However, some years before, *Heinrich August Vogel* (1778–1867), professor for chemistry in Munich, collected rainwater to study pyrrhine and stated that pyrrhine is not a "curious" ["*eigentümliche*"] substance and all organic matter derived from vegetation and animals shows the "pyrrhine reaction" (Vogel 1828). *Arend Friedrich August Wiegmann* (1802–1841),[351] a German Botanist, repeated *Vogel*'s experiments concern the extraction of wood in distilled water: after adding some drops of a solution of $AgNO_3$, no turbidity was gained but exposing the extract to sunlight soon later a dark ruby precipitate obtained that was again fully discolored through chlorine (Wiegmann 1829).[352] On the contrary, the wood extract gained light yellow after several days[353] and get discolored with a few drops of HCl solution and finally, after adding of iron sulfate solution a yellow precipitate is obtained from which *Wiegmann* concluded on the presence of humic acids.

The French chemist *Jean Pierre Joseph d'Arcet* (1777–1844),[354] traveling in 1831 to London, was the first who detected in fresh rainfall (collected with porcelain bowls) sulfurous and sulfuric acid, carbonate and an extractive, likely organic matter; blue litmus paper turns red (d'Arcet 1834, Anonymous 1835a).

Franz Josef Dominicus Bohlig (1815–1874), an apothecary in Mutterstadt (close to Mannheim) collected a sample of rain in Würzburg and estimated qualitatively carbonate, sulfate, calcium, chloride and traces of sodium, potassium, iron, and manganese (Bohlig 1834).

Hail[355] studies were popular in that time, e. g., *Harting*'s[356] book was reviewed by Ludwig (1856) with the interesting finding that hailstones always contain in the middle a white opaque nucleus (often several) and air bubbles; in the molten water were found earthy parts (volcanic ash), grains of sand, ammonia, diatoms, pyrite, and or-

351 Son of *Arend Joachim Friedrich Wiegmann* in Braunschweig.

352 Assuming organic molecules having reducing and/or chromophoric properties, metallic (black) silver gained, namely in sunlight due to photosensitization ($Ag^+ + e^- = Ag$) and (for example) organic bond hydroxyl converts in a keto group that is colored. In presence of chloride, white AgCl precipitates; thus, mixtures of colors may appear. Chlorine oxidizes Ag back to Ag^+.

353 Typically for dissolved humic substances.

354 Not to confuse with his father *Jean d'Arcet* (1724–1801); to him were *Leblanc* and *Pelletier* as students.

355 It is worth to note that the particle of hail (German *Hagel*, French *grêle*) is named hailstone (German *Hagelkorn*, in older German *Steine*, *Kiese* and *Schlossen* after *Muncke*), characterizing the core as solid stony matter, obviously known since ancient times. In French, hailstone is *glaçon*, i. e., an icy cube, only indicating solid water.

356 *Pieter Harting* (1812–1885) was a Dutch chemist and botanist in Utrecht. He wrote "*Skizzen aus der Natur*" [Sketches from Nature], translated from Dutch by J. E. A. Martin. Engelmann, Leipzig, 1854 and 1856.

ganic matter (naming the authors *Eversmann, Mène, Muncke, Pictet, Waller*).[357] *Georg Wilhelm Muncke* wrote the first review on hail (Muncke 1829), followed by *Julius Ludwig Ideler* (1809–1842), German philologist and naturalist in Berlin (Ideler 1829). *Marc-Auguste Pictet* (1752–1825), Swiss experimental natural philosopher in Geneve, detected in hailstones from hail (Pictet 1822), fallen in June 1821 in Ireland, pyrite (FeS_2). *Alexander Eduard Friedrich Eversmann* (1794–1860), German biologist (professor for zoology and botany at the University Kazan in Russia since 1828), also found metal sulfides (and feldspar) in hail and characterized the hailstone as crystal (Eversmann 1824). The English physician *Augustus Waller* (see p. 132 on his observation that fog droplets are not vesicles) investigated droplets of rain, fog, crystals of snow, and hail fallen in the surrounding of London in 1846 by microscope and found many organic bodies, mainly from vegetation (Waller 1847b). Muncke (1829, p. 39) cites falling hail in 1755 near the volcano Katlegiaa on Iceland that contained sand or volcanic ash.[358]

The Russian physician and pharmacist *Alexander Petrowitsch Neljubin* [Александр Петрович Нелюбин] (1785–1858), who worked and analyzed many mineral waters at the Royal Medico-Surgical Academy in St. Petersburg, is likely the first who carefully characterized physical and chemical the crystalline residues of hails (ice nuclei termed today), collected 1825 in the Orenburg Governorate (today Orenburg Oblast in Russia); he found the following chemical composition (in %); Neljubin (1827):

rothes Eisenoxyd (Fe_2O_3)	71
Manganoxyd (MnO_2)	7.5
Kieselerde (SiO_2)	7.5
Talkerde (MgO)	6.25
Thonerde (Al_2O_3)	3.75
Schwefel und Verlust (sulfur and loss of ignition)	5.0

Jean Pierre Louis Girardin (1803–1884), a French agricultural chemist, found a small quantity nitric acid and "*une matière organique azotèe assez abondante*" [pretty much organic bound nitrogen], termed "*miasmes*", and traces of sulfuric acid and ammonia as well calcium in hail collected in Paris (Girardin 1839). The French chemist *Ch. Mène*

357 Today's view on so-termed embryos, aerosol particles that can be efficient ice nucleators, is that soil dust particles (minerals like clays and namely potassium feldspar) provide the substrate for attachment of ice nucleating bacteria or biological particles (see Michaud et al. 2014) and references therein). The chemical composition of hail is similar to that of rainwater but in higher concentration (Samiha et al. 2018).

358 Eggert Olafson, und des Biorne Rovelsons Reise durch Island, veranstaltet von der K. Soc. D. Wissenschaften in Kopenhagen. Heiniche und Faber, Kopenhagen und Leipzig, 2. Vol. 1774 + 1775 (328 + 224 pp.). Original: Olafsen's og Land-Physici B. Povelsen's reise igjennem Island, Soroe (1772). Olafson, E. and Bjarni Pàlsson Travels in Iskland, performed by order of his Danish Majesty. Translated from the Danish. P. Philipps, London, 2 Vols.

(unknown living date) collected on 5 May 1851, in Paris 800 g hail on linen and gained after evaporation 2.78 g salt having the property of ammonium chloride, he also observed in the last moment of evaporation a black carbonaceous substance (i. e., organic matter), Mène (1851a).

The German physician *Emil Osann* (1787–1842), brother of the chemist *Gottfried Wilhelm Osann*, was the first who reviewed the chemical composition of meteoric water in view of differences to spring and other telluric[359] water (Osann 1839), citing *Marggraf*, *Zimmermann*, *Lampadius*, *Witting* and *Liebig* (but not *Bergman*, *Stark*, *Mulder*, *Dalton*, *Helmts* and *Bohlig*; cf. Table 2.4). He explains the chemical composition of precipitation, writing (Osann 1839, p. 128):

> Erzeugt in der Atmosphäre, abhängig von ihr und eben deshalb einem häufigen Wechsel unterworfen, werden die Mischungsverhältnisse des Meteorwassers zunächst bedingt durch die eigenthümlichen electrischen Prozesse unseres Dunstkreises und die gleichzeitigen Rückwirkungen der organischen Natur, mit welcher letztere in unmittelbarer Beziehung und einem steten Wechselverhältnis steht [Generated in the atmosphere, depending on it, and that is why the mixing ratios of meteoric water are attributable to the peculiar electric processes in our atmosphere and the simultaneous reaction of organic nature with them the latter is in direct relation and in a permanent interrelationship].

However, these relationships were already generally described by *Boerhaave* (Chapter 2.3.1.1, namely pp. 148–150) and citing (Osann 1839, p. 130):

> Sehr wahr nannte [...] Boerhaave [das Regenwasser] "eine Lauge des Dunstkreises" [Truly ... Boerhaave termed (rainwater) "a leach of the atmosphere"].

This expression ("*Lauge*", "*Lixivium*") was used as todays modern term "cleanser" of the atmosphere. Thus, precipitation was regarded as the ultimate removal pathway of all "impurities" (air pollutants) from the atmosphere. Half a century later, the atmosphere was still seen as the medium for dissipation and dilution of air pollutants.

Hippolyte Ducros (1805–1879), pharmacist in Nimes (France), studied the composition of hail, which fell in a severe thunderstorm in April 1842 (Ducros 1845) and found free nitric acid.[360] He entitled his paper "*observations d'une pluie acide*" [observation of an acid rain]; this is the first use of the term "acid rain" in literature, according to my investigations (Fig. 2.13). So far it was attributed in literature to *Smith* (1872); cf. footnote 51 on p. 14. However, *Lampadius* (see p. 278) reports on "acid dew" in India, sampling es medicine in the eighteenth century; which would be the oldest report on atmospheric acidity.

359 River and lakes.

360 *Ducros* believed to be detected potassium that is combined to nitrate; an unknown reviewer (*Repertorium für die Pharmacie*, Vol. 90, p. 36 – *Berichte über die neuen Fortschrittte der Chemie*) wrote that it is "more likely that the precipitate gained from tartrate is not potassium but ammonium tartrate".

***Observation d'une pluie acide*, par M. Ducros,**
pharmacien à Nîmes.

En mai, ou au commencement de juin 1842, éclata sur notre ville un orage, pendant lequel se succédèrent de violents coups de tonnerre ; il tomba une quantité considérable de grêle : ce fut cet orage qui ravagea les bois et les vignes dans la direction de Nîmes à Saint-Geniez de Melgoin. La quantité de grêle qui tomba en peu de temps fut assez grande pour recouvrir le sol de la cour de la maison que j'habite ainsi que la terrasse qui la couronne d'une couche d'au moins 4 ou 5 centimètres. La grosseur des grêlons ne dépassait guère celle des plus petits pois ordinaires, et leur forme était ovoïde ; ils paraissaient formés de plusieurs couches dont la première moins diaphane que les autres.

Figure 2.13: First page of "observation of an acid rain" (Ducros 1845, p. 273).

The first semi-quantitative analysis (by gravimetry) of rainwater and snow was carried out by *C. Bertels* (no living date and given name found), a chemist at the "*landwirthschaftliche Lehranstalt*" [agricultural school] in Regenwalde (Trans-Pomerania; today Resko in Poland),[361] who collected two samples per month over one year to estimate the deposition of mineral substances as a possible contribution to soils nutrition (Bertels 1842),[362] Table 2.3.

Eugène Marchand (1816–1895), a pharmacist from Fécamp in the Normandy, very clearly expresses the origin of iodine and bromine from sea water and their removal by atmospheric waters (Marchand 1852, p. 55):[363]

> L'origine de l'iode et du brome dans les eaux provient de l'enlèvement de ces principes aux eaux de la mer, par les vapeurs ou les particules aqueuses qui s'en échappent incessamment, et qui, transportées sur les continents, retombent à leur surface à l'état de pluie, de neige ou de grêle. Les eaux de pluie et de neige contiennent généralement une proportion appréciable d'iodures et de bromures.

361 He became known with his work "Chemische Unterschung des Guano" (*J. pr. Chem.* **28** (1843) 5–13).

362 Remarkable is the comment of the journal editor in a footnote on p. 94: "unfortunately that the author did not consider carbonated ammonia".

363 *Marchand* carried out his experiment 1849–1850 and send a letter to the French Academy in May, to claim the priority to him: Réclamation de priorité élevée à l'occasion des communications de M. *Chatin* et de M. *Barral*, sur la présence, dans l'air, de l'iode et de certains autres corps (*Compt. Rend.* 1852, p. 560). Boussingault (1856b, p. 243) cites *Marchand* to be the first who analyzed ammonium/ammonia in sea water about 8 km from the port of Fecamp with 0.57 mg L^{-1}. *Boussingault* is the first who concluded from this estimation that the sea is a source of ammonia to the atmosphere (*Mem. Acad. Med.* **19**, 1855).

Table 2.3: Chemical composition of snow and rain collected 1841–1842 in Regenwalde (Bertels 1842). 1 *Pfund* = 520,000 Mgr. (milligram), 1 Magdeburg Morgen = 180 *Quadratruthen* = 0.255321 ha (hectare); deposition recalculated from "*Pfund*" per "*Magdeburg Morgen*" and year; concentration recalculated from "Mgr. per 3 pounds of water".

Component	one event in April 1841		annual mean (n = 20)	
		concentration		deposition
	in Mgr.	in mg L^{-1}	in *Pfund*/*Morgen*/year	in kg $ha^{-1}yr^{-1}$
carbonated lime earth	10	6.4	31 7/10	4.2
carbonated talc	10	6.4	24 1/2	3.3
common salt	16	10.2	32 7/20	4.3
gypsum	10	6.4	24 13/20	3.3
sulfurous potash	7	4.6	20[a, b]	2.7[a, b]
organic bonded nitrogen	17	10.9	35 9/10	4.8
iron oxide	2	1.3	10 4/5	1.4
alum	4	2.6	13	1.7
silica	6	5.4	27	3.6
humic acids	4	2.6	8.6[c]	1.1[c]
total	86	55.1	215 9/10	28.7

[a]Deposition calculated as lime acid potassium ($KHCO_3$) by *Bertels*
[b]Original no means given; calculated from n = 9
[c]Original no means given; calculated from n = 16

Marchand also found in snow and rain in Fécamp: NaCl, $NaHCO_3$, NH_4NO_3, Na_2SO_4, $CaSO_4$, CaO, $MgSO_4$, organic matter, and traces of H_2S. French botanist *Gaspard Adolphe Chatin* (1813–1901) carried out measurements in 1850 in Paris (Chatin 1851) and reported on the presence of iodine in air (0.1–0.5 ng m^{-3}) and rain and snow (0.025–0.05 mg L^{-1}). *Sebastiano de Luca* (1820–1880),[364] sampled in Pisa in 1860 snow and rain at different heights (4, 18, and 54 m) at the Tower of Pisa, detecting (qualitatively) chlorine, iodine, bromine, sulfate, phosphate, and organic substances, namely organic nitrogen (de Luca 1888), but iodine and phosphate only at the lowest level; Schroeder (1873, p. 72) correctly concluded that soil dust, containing phosphate, could not reach the higher levels.[365] Interesting that de Luca (1854), after discussing numerous reagents for iodine detection, did not find iodine in air, rain, and snow in Paris (between November 1853 and February 1854). In the same journal, Chatin (1854) responds to *de Luca* that "no evidence is no evidence for absence". There was a curiosity of history to claim the priority of "non-presence of iodine in the atmosphere, in

364 Italian chair professor for chemistry at University Pisa (1857–1862) and afterwards at Napoli where he founded a new institute of chemistry. Close collaboration with *F. Arago*, *J.-B. Dumas* and *M. Berthelot*.

365 From today's knowledge we must correct this conclusion to that effect of declining dust concentration with altitude and insufficient analytical detection level.

rainwater and snow" by the Scottish chemist *Stevenson Macadam* (1829–1901), who extensively communicated to be the first (in 1852) who did *not* find iodine (Macadam 1858). Chatin (1855) communicated that *Martin* found in rain at Nice 0.004 mg L^{-1} and that *Renaud*, a physician on the frigate Érigone, found 0.05 mg L^{-1} iodine in the rain of French Guyana; Chatin (1859) conducted measurements of iodine in several environments.

The priority in the discovery of phosphate, however, in rainwater holds *Jean-Augustin Barral* (1819–1884), French chemist and agronomist in Paris,[366] who already in 1852–1853 is believed to detect phosphate in rainwater (and reported it to the French Academy), but it remained unsure whether it was a component of rainfall or the one of the sampling container (Barral 1860). *Barral* collected in Paris 1295 L and countryside 390 L and estimated the mineral residue to be 22.8 and 7.8 mg L^{-1}, respectively. The amount of phosphoric acid *Barral* determined in the range of 0.5–0.06 mg L^{-1} and discussed different sources of its origin, like mineral, biological material, hydrogen phosphide (PH_3) that was known from putrefaction. *Stéphane Robinet* (1799–1869), a chemist in Paris, determined between March 1863 and September 1863 at different sites in Paris the total hardness to be on average 3.27° (French degree, corresponding to 32.7 mg L^{-1} $CaCO_3$) and the $CaSO_4$ content to be larger than 20 mg L^{-1} (Robinet 1863).

In the 1850s, the main composition of rainwater and the principal meteorological influences on the rainwater concentration was known. However, monitoring of rainwater including all cations and anions being important began only after the 1950s (Chapter 2.3.5). In the review on "*Regen- und Schneewasser*" (Moleschott 1859, pp. 387–390), a Table presents the compounds found in rainwater and the researchers; the following substances were found (in modern terms): ammonium, nitrate, sulfate, carbonate, sodium, potassium, magnesium, iron, alumina, silica, humic acids, and organic matter (Table 2.4). Not listed in Table 2.4 but cited in Moleschott (1859) are iodine, bromine, manganese, HCl, and H_2S, occasionally found in rain.

The first exact determination of the composition of air absorbed by rainwater was conducted by *Friedrich Moritz Baumert* (1818–1865)[367] who found in the air of freshly fallen rainwater (11.4 °C) according to *Bunsen*'s method 64.37 % nitrogen, 33.76 % oxygen, and 1.77 % carbon dioxide. This corresponds to the following composition of the air, based on the absorption theory by *Bunsen*: 79.80 % nitrogen, 19.86 % oxygen, and 0.34 % carbon dioxide (Baumert 1853a, p. 17). Bunsen (1855, p. 48) calculated[368] the amount of air absorbed in pure water (like rainwater), using mean mixing ratios given by *Saussure* (414 ppm for CO_2); in % at different temperatures (in °C in parenthesis):

---

366 Editor of the *Journal of Practical Agriculture*. He made two balloon rises in 1850 for air sampling.

367 Studied first medicine and later chemistry at *Liebig*, *Redtenbacher* and *Bunsen*. Lecturer in Breslau and professor in Bonn.

368 Sometimes confused in literature to be measured by *Bunsen*.

2.92 (0), 2.68 (5), 2.46 (10), 2.26 (15), and 2.14 (20). The enrichment of CO_2 in rainwater is by a factor of 33 comparing to atmospheric air, but *Bunsen* also calculated the total deposition of carbonic acid by rainfall (2.569 g m^{-2} yr^{-1}) to be insignificant for the vegetation.[369] The French chemist *Eugène-Melchior Péligot* (1811–1890) analyzed the air in rainwater in Paris and found 2.4 % CO_2, referring to the value of 2.46 % for 10 °C derived by *Bunsen* and corresponding to 400 ppm CO_2 in air (Péligot 1857). Schulze (1871) found in rainwater at Rostock 1.073 mg L^{-1} carbonic acid (see also pp. 413–414).[370]

Table 2.4: Chemical composition of rainwater (concentration *c* in mg L^{-1}; gases in permille), summarized after Moleschott (1859).

substance (name in German)	formula	*c*	authors
Sauerstoff	O_2	97	Lampadius
Stickstoff	N_2	60–205	Lampadius, Barral,
Kohlensäure	CO_2	8–69	Lampadius, Baumhauer[g]
Ammoniak[a]	NH_3	0.3–24	Barral, Marchand, Boussingault, Bineau, Filhol
Salpetersäure	HNO_3	19	Barral
doppeltkohlensaures Ammoniak	NH_4HCO_3	17	Marchand
Chlornatrium[b]	NaCl	1–20	Dalton, Bertels, Mulder, Barral, Marchand, Meyrac, Filhol
kohlensaurer Kalk[c]	$CaCO_3$	7	Bertels
Bittererde[d]	MgO	2	Barral
Schwefelsaures Natron[e]	Na_2SO_4	10	Marchand, Barral
Eisenoxyd	Fe_2O_3	0.03–2	Bertels, Barral
Thonerde	Al_2O_3	3	Bertels
Kieselerde	SiO_2	6	Bertels
kohlensaures Kali	K_2CO_3	2	Bertels
organische Stoffe[f]	–	0.1–25	Bertels, Barral, Marchand

[a]Not gaseous; as ammonium additional NH_4HCO_3
[b]And additional KCl, $CaCl_2$, $MgCl_2$
[c]And additional $MgCO_3$, CaOH (lime)
[d]And additional $MgSO_4$
[e]And additional $CaSO_4$, $MgSO_4$
[f]Including nitrogen containing organic matter (*Bertels*)
[g]*Eduard Henrich von Baumhauer* (1820–1885) was a Dutch analytical chemist in Utrecht; cited in Mulder (1844, p. 719)
[h]First exact estimates were made by *Baumert* and *Bunsen* (see text)

369 This value is amazingly agreeing with the estimate by Möller (2014, p. 327) to be 0.53 g m^{-2} yr^{-1} (as carbon) at 400 ppm CO_2 for the continents.

370 The solubility of gases in water is (among others) defined by the absorption coefficient α that denotes the volume of gas absorbed by one volume of water at a pressure of 1 atm). It amounts (in parentheses temperature in °C): 1.713 (0), 1.424 (5), 1.194 (10), 1.050 (15), and 0.902 (20). Thus, *Schulze*'s estimation is rather correct, corresponding to a temperature between 10 and 15° C.

2.3.2 Rainwater analysis for understanding agricultural chemistry (1851–1919)

2.3.2.1 The role of agriculture experimental stations

The technical and scientific progress[371] together with a growing population in industrial countries in the middle of the nineteenth century was inevitably connected with an increase in agriculture and air pollution. The German population grew from 40.9 million in 1870 to 49.5 million in 1890; virtually all population growth occurred in towns and cities.

It is generally accepted by historians that *Justus von Liebig*, whose (1840) classic, "*Die organische Chemie in ihrer Anwendung auf Agricultur und Physiologie*" [Organic Chemistry in its Application to Agriculture and Physiology], created a feverish interest in agricultural chemistry both in Europe and America. In 1848, *Liebig* loudly proclaimed that since atmospheric ammonia always provided plants with the adequate nitrogenous matter, farmers needed only to be concerned with non-nitrogenous mineral fertilizers (Vogel 1883).[372] This (erroneously) hypothesis was accepted because of his great reputation. *Liebig*'s incorrect belief that all nitrogen assimilated by plants came from precipitation inspired further research. On the contrary, he correctly writes that rotting plants and animal manure convert the bonded nitrogen into ammonia (respectively *ammonium* in modern terms), from which a small percentage further converts into nitric acid, and that plants in his nutrition process also use nitric acid (in modern terms *nitrate*), and that also carbonic acid forms while putrefaction, hence soil air is richer in carbonic acid (in modern terms *carbon dioxide*) than the atmospheric air. Plant roots take up nutrients dissolved in soil water (that comes from rain);[373] the solubility depends on the amount of carbonic acid and ammonia in soils. These statements result in numerous investigations of ammonia, nitric acid, and carbon dioxide in rainwater, soil water, and air in the following decades of the nineteenth century, almost all carried out in newly founded agricultural experiment stations.

In 1860, there were only three serious European establishments doing scientific research in agriculture. In France was the private farm of *Boussingault* (Fig. 2.14) at

371 After the middle of the nineteenth century, Germany became worldwide the leading country in all scientific and technical disciplines (until end of the 1920s). Several important scientists from Great Britain, France, the USA and other countries were students at German universities or visitors.

372 *Liebig* was the first who described scientifically the role of minerals as plant nutrients ("*Mineraltheorie*"). Before 1840 it was an empirical experience of farmers that dust and ash improve plant growth. But only in 1870 *Liebig* claimed that he was wrong with the assumption that plants uptake nutrients from a solution gaining by rainwater to the soil. Only through intensive experiments the role of soil adsorption of ions by the soil (dependent from the soil type) were detected. First produced chemical fertilizers were unsuccessful because the ions (calcium, potassium, phosphate etc.) became insoluble through wrong operations in the production process.

373 Already Paracelsus (1577) stated that rainwater carries nutrient to the plants; Saussure (1804, p. 244) communicated that organic matter (remains from plants and animals), suspended in the air, deliver nutrient to the plant as well the same substances which are loaded by rainwater.

Figure 2.14: Portrait of *Jean-Baptiste Boussingault* (after a copperplate print in the possession of the author).

Bechelbronn in Alsace, founded in 1836. In 1847, *Boussingault* moved from the residence Bechelbronn at a distance of 5 km to Liebfrauenberg, where he later collected rainwater, fog, and dew for chemical analysis; at this new site was not a farm but an excellent vegetable garden, the soil of which was frequently referred to his later work on nitrification. The Franco-Prussian war of 1870–1871 removed these sites from the French influence. In England in 1843, *Sir John Bennet Lawes* (1814–1900),[374] with the help of *Sir Joseph Henry Gilbert* (1817–1901),[375] established the oldest agricultural experiment station in the world which is still in operation, the Rothamsted Experimental Station, located at Harpenden near St. Albans in Hertfordshire, 20 miles northwest of London. The first private establishment in Germany was founded in 1847 by *Ernst Theodor Stöckhardt* (1816–1896), a cousin of *Julius Adolph Stöckhardt* (from Tharandt), at the manor Brösa near Bautzen with *Emil von Wolff* (1818–1896)[376] as assistant and teacher, who became later the first director at Möckern. Möckern in Sax-

374 Owner of the Rothamsted estate and lord of the manor of Rothamsted; English entrepreneur and agricultural scientist (not graduated); he founded an experimental farm at Rothamsted, where he developed a superphosphate that would mark the beginnings of the chemical fertilizer industry.

375 British chemist who studied in Glasgow (*Thomas Thomson*), London (*Graham*) and Giessen (*Liebig*). *Lawes* invited him to join the Rothamsted station in 1843. Fellow of the Royal Societey in 1860, President of the Chemical Society (1882–3) and 1884–1890 professor for rural economy at Oxford.

376 German agricultural chemist, who teaches in Brösa at a private agricultural institute under *Stöckhardt* (1847–1854), was founding director of the Möckern Station and director of the Agriculture Academy at Hohenheim (1885–1894) and professor at university Stuttgart-Hohenheim.

ony near Leipzig was founded as the first state-supported agriculture experiment station in the world ["*Landwirtschaftliche Versuchstation*"] in 1852 by *Heinrich Wilhelm Leberecht Crusius* (1790–1858) on the suggestion by *Julius Adolph Stöckhardt. Crusius*, lord of the manor at Sahlis und Rüdigsdorf and agronomist, was one of the first farmers, who employed an agricultural chemist. Directors at Möckern (in parenthesis period of heading): *Emil Theodor von Wolff* (1852–1854), *Heinrich Ritthausen* (1854–1855), *Wilhelm Knop* (1856–1866), *Gustav Kühn* (1867–1892) and *Oskar Kellner* (1892–1911). Since 1934, the station focused on animal nutrition and obtained in 1952 the additional name "Oskar Kellner"; today, Möckern is a part of Leipzig. Nearly all station directors were chemists, not farmers, and all would have agreed with *Wilhelm Knop*, who declared:[377]

> Jedermann kann voraussehen, dass in Zukunft die Nothwendigkeit, Kunstdünger aller Art zum Betriebe der Landwirthschaft zu Hülfe nehmen, immer mehr hervortreten muß [Everyone can see that in the future, the necessity of artificial fertilizers of all types in the agricultural economy will become increasingly prominent].

Julius Neßler (1827–1905), a former industrial chemist and founder of the agricultural experiment station in Karlsruhe, developed in his dissertation in 1856 at University Freiburg a new method for determination of ammonia and ammonium still today accepted for wastewater analysis, known as *Nessler*'s reagent. German experiment station scientists were among world leaders in the field of research of nitrogenous matter in plants and animals.

In 1870, 11 stations existed, in 1873 already 33 (19 in Prussia and 7 Saxony) did and in 1904 the maximum of 78 stations in Germany (in France 58, USA 54, Russia 42, Austria-Hungary 38, Sweden 22, Italy 19, Great Britain 12, Norway and Holland each 6). The German stations became the model for American agricultural experiment stations; the first founded in 1875 (Finlay 1988). In practically every case, foreign station founders and promoters prefaced their speeches and writings with a tribute to *Justus von Liebig* or the Möckern station, to a large number of German stations, and to the high degree of support the German stations enjoyed. Yet despite the rhetoric, there was, in fact, no "model" of the German experiment stations. In the 1870s and 1880s, German agricultural experiment stations were characterized by their diversity, and no single model applied to all (Finlay 1992). In the 1870s, Germany employed nearly one hundred agricultural scientists, it was the home of the only significant professional organization of agricultural scientists, and German was the language of the field's leading books and periodicals.

Several German agricultural chemists were hired by the agricultural organization of other countries; in following some examples. In 1856, the Swedish Academy

377 Knop, W. (1861) Ueber die Frage: was nutzt dem Landwirthe die chemische Analyse der Dünger? Und Mittheilung verschiedener auf der Versuchsstation Möckern ausgeführter Analysen von Düngerstoffen. *Amtsblatt für die landwirtschaftliche Vereine des Kônigreich Sachsens* **9**, 2–3, 10–12, 28–29.

of Agriculture named a new agricultural chemist for its Stockholm laboratory, the German *Carl Alexander Müller* (1828–1906) from the experiment station in Chemnitz. In 1859, for example, the Baltic German *Jacob Johnson* (1806–1865), born in Estonia and graduated from Jena, later deputy chairman of the "*Kaiserliche freie ökonomische Gesellschaft zu Sankt Petersburg*"[378] and editor of the "*Mittheilungen*" (1844–1864), edited by the society, closed a brief article[379] on agricultural experiment stations with the concise appeal (cited after Finlay 1992, p. 322):

> Wann werden wir in Rußland unsere erste landwirthschaftliche Station haben?! [When will we in Russia have our first experiment station?!].

Another Baltic German, *George Thoms* (1853–1902)[380] who completed his advanced studies under the German chemists *Bunsen* and *Kekulé*, was named the director of the "*landwirthschaftlich-chemische Versuchsstation am Polytechnikum zu Riga*" in 1872. Later, *Johann Karl Woldemar von Knieriem* (1849–1935), a scholar of the agricultural chemist *Adolf Mayer* at the University Heidelberg, as well accepted a position as the agricultural professor at Riga Polytechnic in 1880. Beginning in 1856, *Louis Mulder* (1828–1902; sun of *Gerardus Johannes Mulder*) of the agricultural school in Deventer (Netherland) headed a committee hoping to fund an experimental facility, and vigorously praised German stations. The ministry in Netherland hired *Adolf Eduard Mayer* (1843–1942), an agricultural chemist at Heidelberg, to establish and direct a new station at Wageningen. In Belgium, *Phocas Lejeune* (1823–1881), director of the Gembloux agricultural school, to tour German stations, established an experiment station at the school, and hired *Arthur Petermann* (see p. 191), to be its director.

Even nations that lacked frequent or direct contacts with German universities and the German-language agricultural literature respected and admired the German experiment stations. As a consequence, German agricultural scientists migrated into many corners of the world. For instance, in 1872, a Greek agricultural society hired *Julius Neßler*. In 1877, a Spanish agricultural society engaged the chemist *Otto Wolffenstein* (1848–1879)[381] to establish a fertilizer control station in Valencia, where he became in 1879 the director of the "*Estación Agronòmica de la Societat*". The Portuguese government hired four German agricultural scientists. Brazil's government also pursued the Austrian agricultural scientists, *Franz Wilhelm Dafert* (1863–1933),

378 Royal Free Economic Society at Sankt Petersburg (1766–1919); Kuum (2018).

379 According to Kuum (2018) it is: Johnson, J. (1859) Ein kleiner Beitrag zur Kenntniss der Wirthschaftlichen Verhältnisse in Esth- und Livland. *Mittheilungen der Kaiserlichen freien ökonomischen Gesellschaft zu St. Petersburg*, pp. 401–424.

380 See for example: G. Thoms "Die landwirthschaftlich-chemische Versuchstation am Polytechnikum in Riga". *Rigaische Zeitung*, 26.2.1874.

381 Born in Berlin (died in Spain) as the brother of the well-known later chemist *Richard Wolffenstein* (1846–1926); chemist at the Berlin University. Dissertation "*Ueber das Sortieren von Saatgut*", Göttingen 1875, 45 pp.

a former assistant at the Munich and Bonn experiment stations, was chosen. Another striking example of German agricultural science's impact abroad was invitation of *Oskar Kellner* (see p. 191) to Japan, to direct its agricultural chemistry laboratory near Tokyo; when he was hired in 1880, *Kellner* was an assistant at Hohenheim's experiment station.

The impact of German stations on France, England, and Scotland was somewhat different. As mentioned above, those nations boasted several of Europe's leading agricultural chemists and agricultural science institutions earlier in the century in many ways. Italian experiment stations came closest to the German type. Though they were funded entirely by the state government, unlike German stations, there were several similarities. Most notably, Italian stations imitated the Germans' emphasis on specialized research projects. *Alfonso Cossa* (1833–1902), the Italian stations' principal founder, was well known among German agricultural scientists. In 1857, as a young chemist, *Cossa* translated two of *Liebig*'s agricultural chemistry texts into Italian; he subsequently found jobs as a chemistry instructor at the University of Pavia and at a technical school in Udine. Whereas French chemist *Adolphe Wurtz* (1817–1884) begins his book (Wurtz 1869, p. 1) with the remarkable statement "*La chimie est une science française*", *Cossa* stated that agricultural science was clearly a "German science" (cited after Finlay 1992, p. 349).

2.3.2.2 The "nitrogen period"

As already mentioned, ammonia in rain was first detected[382] by *Zimmermann*, the processor of *Liebig*, who soon later also found nitrate and ammonium in rain. Deposition studies (precipitation chemistry) began more systematically after *Liebig*'s statement in his book that plants assimilate their nitrogen from ammonia that is washed down from the air in the rain, and this ammonia, he added, is constantly renewed by the putrefaction of animal and vegetable material occurring everywhere on the earth's surface (Liebig 1840, p. 70):

> [...] conveyed to their roots in the state of ammonia, derived from the putrefaction of animal matter.

These views on ammonia were generally accepted until about 1855[383]when Boussingault (1856a) showed conclusively there was not enough ammonia in the rain to satisfy the nitrogen needs of plants. *Boussingault* asks, where does the nitrogen come from?

382 Miller (1905a) wrongly writes that "the first determinations of ammonia seem to be those by Payen in 1845–6" citing *de Gasparin* (Cours d'Agriculture, 2nd. Ed., Paris), but this work appeared in 1844 (!).
383 The question to which percentage atmospheric nitrogen deposition contributes to the plant nitrogen budget, remains with the Rothamsted Experiment Station in England up to 1916 and was renewed in the 1980s due to increasing emissions of NH_3 and NO_x and subsequent nitrogen deposition in Europe (see e. g., Möller and Schieferdecker 1982, 1989).

He was the first to start several careful experiments on nitrogen fixation and found it "reasonable that the green parts of plants are able to assimilate the nitrogen of the atmosphere" (cited after Aulie 1970, p. 446). Boussingault (1838, p. 366) writes:

> Indeed, nitrogen can enter directly into the organization of the plant, if the green parts are able to fix it. This element can enter the plant by aerated rain water, which is absorbed by the roots. Finally, it is possible, as some physicians think, that it exists in the air as an infinitely small percentage of ammoniacial vapor.

Boussingault (1845, p. 494) still speculates (a few years later he determined nitrogen in different waters and air):

> The most reasonable supposition in the actual state if science, is to consider the ammoniacial vapors diffused through the atmosphere as the prime source of the azotique principle of vegetable, and then through them to animals; a consequence of which hypothesis would be to assume with Liebig, that carbonate of ammonia existed in the atmosphere before the appearance of living things upon the face of the earth [...]
> Now nitrate of ammonia is one of the constant ingredients in the rain of thunder-storms.
> [...] the light of flashing, as the mean by which the azote of the atmosphere is made free for assimilation by organized beings.

The French agronomist *Adrien Étienne Pierre de Gasparin* (1783–1861) entitled chapter II of his book "*Cours d'agriculture*", "*De différentes substances variables qui font accidentellement partie* de l'*air*" and writes (Gasparin 1844, p. 34), note the wrong spelling of *Marggraf*:

> [...] l'existence et la formation de l'ammoniaque et de l'acide nitri que, trouvés dans les eaux de pluie par Magraft, Bergman, et dont Liebig, d'après eux, a constaté la presence [...]

The problem of analyzing ammonia in air (and rainwater) was solved only in 1847 by *Eugene Melchior Péligot*, whose procedure for measuring quantities of combined nitrogen in solution using titration against a known acid (Péligot 1847), *Boussingault* readily adapted. Berthelot and André (1886) recommend carrying out nitrogen analysis soon after sampling because organisms in rainwater consume ammonia and nitric acid.

No wonder that such ideas (French first edition 1844[384]) proposed other scientists (*Gräger*, *Kemp*, *Fresenius*) to determine ammonia in the air between 1845 and 1848 (Chapter 4.6.1.2). Soon later, several French scientists (*Barral*, *Bineau*, *Boussingault*, *Martin*)[385] started rain and deposition studies, almost because of agricultural chemical interest, hence, focused on ammonia and nitric acid. *Boussingault* analyzed together with the French analytical chemist *Anselme Payen* (1795–1871) nitrogen as am-

384 Èconomie rurale considerée dans ses rapports avec la Chimie et la météorologie. Paris 1884, Vol. 2, pp. 247 and 697.

385 *Martin* worked at the Observatoire de Marseille; no given name, no data of living I found.

monium and nitrate in biomass, soils, and waters;[386] he also developed new analytical methods for nitrogen, used by other chemists too.

Joachim Isidore Pierre (1812–1881) was a professor for agricultural chemistry in Caen and published a book[387] (Pierre 1859); unfortunately he not analyzed ammonia[388] in rain collected 1851 in Caen (Pierre 1851), but chloride and sulfate (as Na, K, Mg, Ca), given as deposition (here recalculated in kg ha^{-1}): 17.5 (Na), 29.7 (Cl), 19.2 (sulfate), 7.9 (K), 3.3 (Mg), and 1.9 (Ca), showing typical marine influence, the Na/Cl ratio corresponds closely to that in seawater and the other cations would fit stoichiometrically sulfate.

The first *quantitative* determination of ammonium and nitrate in rain must be denoted to *Jean-Augustin Barral* in Paris at the observatory in 1851 (Table 2.5) who also determined other components and present a value for the total iodine deposition[389] from January to June 1851 to be 0.15 g ha^{-1}. The solid residue in rain he found in Paris was larger (22 mg L^{-1}) than at the rural site (7 mg L^{-1}).

Armand Bineau (1812–1861), professor of chemistry at the University of Lyon, found the following characteristics (Bineau 1852, 1855a): a) in the surrounding of Lyon the ammonium concentration found in rain is between 1/4 and 1/8 less than that in the town, b) but that of nitrate is less (up to a factor of 8) in the environ-

Table 2.5: Results of the first quantitative rainwater analysis in Paris by Barral (1852); monthly rainwater collection and analysis (in mg L^{-1}).

component	first site 1851[c]	second site 1851[d]	1852[e]
nitrogen	6.397	7.939	4.355 (2.012–11.131)
ammonia	3.334	2.769	3.717 (1.135–9.646)
nitric acid	14.069	21.800	6.209 (1.837–11.774)
chlorine	2.801	1.946	3.597 (1.899–7.606)
calcium[a]	6.220	5.397	–
magnesium[b]	2.100	2.300	–

[a] Given as *chaux* (CaO)
[b] Given as *magnésie* (MgO)
[c] Means from July to December (observatory site Montsouris)
[d] Means from August to December (second station at the Paris court)
[e] Means from both sites including variation of the monthly means (January to June)

386 They published their results in *Annales de Chimie et de Physique* and in *Comptes rendus hebdomadaires des séances de l'Académie des sciences*.

387 He wrote the book *Chimie Agricole* (Pierre 1859) including chapter VI on several substances in air and rainwater (pp. 31–44) and chapter VII on the importance of atmospheric constituent for the agriculture (pp. 44–47).

388 He determined ammonia in air in 1851 (Chapter 4.6.1.2).

389 *Barral* estimated the chloride deposition to be 9.4 kg ha^{-1} from which a I/Cl ratio of $1.6 \cdot 10^{-5}$ follows, very close to the modern sea water ratio ($1.1 \cdot 10^{-5}$).

ment of Lyon, and c) ammonia has a maximum in winter and minimum in summer and vice versa for nitric acid. Only the points that ammonia/ammonium in town is much larger can be explained by the decomposition of waste and sewage (see also *Smith*'s finding) and that the amount of nitric acid is larger in summer than in winter is understandable.

Victor Meyrac (1792–1875), French pharmacist in Dax (Gascogne), found qualitatively ammonia and organic substance in rainwater (and in fog and dew) as well NaCl in large quantities (up to 20 mg L^{-1}) and iodine on small traces (Meyrac 1852). Martin (1853) collected on May 27, 1853 in Marseille 14 L rainwater and found ammonia in a concentration (Table 2.7) according to *Boussingault* and *Barral*. Chloride he estimated to be 4.29 mg L^{-1} (corresponding to 7.06 mg NaCl) and HNO_3 in traces.

Boussingault (1853, 1854)[390] found already characteristics of rainwater chemistry (concerning ammonia, but today we know that they are valid also for other species except for sea salt components): concentration is much higher at the beginning of a shower than at the end and after several days without rain, the proportion of ammonia rose at the begun of the first shower; unfortunately, he only estimated ammonia and not nitrate. As an example, on 28 August, he found in rain at Liebfrauenberg in 7 consecutive samples each of 1 liter, the following series of decreasing concentrations of ammonia (in mg L^{-1}): 1.15; 0.77, 0.61; 0.23; 0.14; 0.08; 0.10, and in the eighth sample (3.6 L) 0.03 with a means to be 0.30 mg L^{-1} and total 10.6 L of rainwater (Boussingault 1853). He explains correctly the observation that at the beginning of a rain event the concentration is found to be higher than at the end through scavenging of particulate and gaseous substances, which accumulate in air before the rain (Boussingault 1856c, p. 265). He is also the first, who careful listed the rainfall amount for each sample collected and calculated the total amount of ammonia to give a weighted-mean value, which is considerably less than the arithmetic means from all single samples (Liebfrauenberg 26 May 26–18 October 1853, $n = 64$):[391] 0.99 mg L^{-1} arithmetic means[392] versus 0.47 mg L^{-1} from total rainfall and total ammonia collected. Further-

390 *Boussingault*, who presented his result on May 9, 1853, to the Academy of Sciences in Paris (1854, 1856b) confirmed the finding of *Barral* and *Bineau*; the latter already published his results in 1851 in *Annales de la Societe Linneenne de Lyon*. The method for ammonia determination and analysis of different waters is described in Boussingault (1853, 1856b). The results on ammonia in rainwater, dew and fog also appear in German (Boussingault 1856b,c). In later editions (after 1854) of the English edition of his famous book on "Rural Economy, in its relations with Chemistry, Physics, and Meteorology or, Chemistry applied to Agriculture" (first edition 1845) no data on rain, dew and fog have been included.

391 *Boussingault* continued the series until November 16, but from October 26 he collected fog deposition with his rain gauge (he termed it "*pluie de la broullard*"). In contrast to sheet and porcelain bowls used before as rain sampler he used a linen similar the *Marggraf* (see p. 152) to get a large surface (4.9 m^2) and to collect also spare rain. The linen he stretched only just before rain to avoid contamination. Parallel he used a udometer (rain gauge) to measure the rainfall.

392 In Boussingault (1853) he cites 0.898 mg L^{-1} and total 1750 L of rainwater; he also gives slightly different values for the first two classes of rainfall amount: 3.11 and 1.21 mg L^{-1}.

more, he found that the ammonia content depends on the rainfall amount (Boussingault 1854, p. 147); he classified the mean ammonia concentration into 7 classes of rainfall in steps of 5 mm (the classes between 5.01 mm and 31.0 mm show only insignificant variation and no trend):

rainfall intervals (in mm)	ammonia concentration (in mg L^{-1})
0–0.5	2.94
0.51–1.0	1.37
1.01–5.0	0.70
5.01–31.0	0.41 (0.36–0.43)

Boussingault confirmed *Barral*'s values on ammonia in rain found in Paris with two estimates in spring 1853 (4.34 and 3.0 mg L^{-1}, respectively) and conducted the Liebfrauenberg series (see p. 180) whereas *Houzeau* (at that time his assistant) collected and analyzed rainwater in Paris (at the Royal Conservatory of Arts and Crafts). Although he found in Paris less ammonia than it was determined by *Barral* (between May and July 1.6–2.0 mg L^{-1}), the mean value at Liebfrauenberg was to be determined significantly less (1 mg L^{-1}). A few rain events (n = 7) at Liebfrauenberg *Boussingault* were also found to be rich in ammonia: 3.4 (2.1–5.2) mg L^{-1}, which he characterized either as short showers or rain with very low rainfall. The much higher ammonia concentration in Paris (or generally speaking in towns) was not surprising to him (Boussingault 1856b, p. 237); he writes sarcastically (translated from the German):

> In relation to exhalations, Paris can be compared with a large dunghill.

Boussingault (1856b, p. 244) also collected freshly fallen snow in March 1854 and found 0.70 mg L^{-1}, in snow from his patio 1.78 mg L^{-1}, and 36 h later from snow in his garden 10.34 mg L^{-1}; he concluded on ammonia exhalations from the soil. Motivated by these discrepancies, Vogel (1872) studied in Munich in Winter 1871–1872 the content of ammonia in snow and found no traces in freshly fallen snow, however, after 24 h of exposition of the sampling bowl he found ammonia and concluded:

> In solchen Fall ist aber der Ammoniakgehalt kein ursprünglicher im Schnee, sondern durch die Atmosphäre zugeführt [In such case the ammonia content in snow is not originally but supplied by the atmosphere].

Thus, the enrichment in the sampling gauge is due to dry deposition of gaseous (and particulate) ammonia – an unknown process at that time but in contrast to *Boussingault* correctly concluded by *August Vogel* from his observations.

Boussingault is the first who stated (without experimental proof because of the low concentration; he only analyzed glue and glairing dissolved in water to prove his analytical method) that organic matter in natural waters contains nitrogen (Boussingault 1856b, p. 240). *Boussingault* published only 1882 on nitrogen in rain and snow at high altitude, collected by *Aimé Civilale* (1821–1893), French photographer (Boussingault 1882); he found in rain at St. Bernhard (2600 m) 1.10 mg L^{-1} ammonium and 0.30 mg L^{-1} nitric acid, whereas in snow only traces, but a fresh snow shower at Cirque Comboë (2100 m) to be acid (0.66 and 0.30 mg L^{-1} nitric acid and ammonia, respectively).

Henry Bence Jones (1813–1873), an English physician and chemist, was the first who stated from a few qualitative analyses of rainwater collected in London and far from the city[393] that nitric acid was always present and not only while thunderstorm[394] as pointed out by *Liebig* (Bence-Jones 1851); he writes (Bence-Jones 1851, p. 408):

> Should nitric acid found to be present in the air at all times and in all places, its importance to the growth of plants will not be less than that of ammonia which was detected there.

Boussingault who refers *Bence-Jones*,[395] speculates the presence of ammonium nitrate in the air in dust (Boussingault 1856c, p. 262).

Jean Pierre Bernard Édouard Filhol (1814–1883), Professor of Chemistry at the University of Toulouse, collected from January until June 1855 rainwater in Toulouse; apart from nitrogen (Table 2.7) he found 26.54 mg L^{-1} solid residue and referred it to 2.85 mg NaCl (Filhol 1855). *Pierre Adolphe Bobierre* (1823–1881), chemist and pharmacist, professor at "*l'École préparatoire des Sciences et des Lettres de Nantes*", studied from January until December 1863 ammonia, nitrate, and chloride at two different altitudes. The monthly means of ammonia and nitrate varied largely (0.7–15.0 and 1.6–16.0 mg L^{-1}, respectively). As *Bineau*, he found ammonia at minimum and nitrate at maximum in summer, and vice versa. The much higher ammonia concentration on the ground compared to the altitude of 47 m (at the observatory) *Bobierre* explained through putrefaction[396] (Bobierre 1864).

The "French period of rainwater nitrogen" was concentrated between 1851 and 1855, showing finally that with rainwater not enough nitrogen will be supplied to plants. *Jens Jakob Berzelius* (1779–1848) comments in 1848 that (cited after Aulie 1970, p. 453):

---

393 In Kingston in Surrey, Melburg in Dorsetshire, Clonakilty in the country Cork (in January; December in London).

394 Nitrate was first detected by *Marggraf* in Berlin 1851 in rain collected over several month.

395 Cited as *Henry Ben=Jones* (or *Henri Ben-Jones* in Boussingault 1854, p. 150), who found always ammonium nitrate ["*salpetersaures Ammoniak*"] in rainwater; however, *Bence Jones* (or *Bence-Jones*) only detected nitric acid in rain.

396 This is hard to understand because putrefaction should be highest in the warm season; but because HNO_3 seems to be in summer (by photochemical oxidation) much larger then in winter, it will combine with gaseous NH_3 to particulate NH_4NO_3. Nitrate in rainwater shows no strong difference between the altitudes.

> Boussingault covers the same field as *Liebig* but *Boussingault* takes the hard tiresome way of answering every question by one or more experiments. He gives his answers not so quickly but they are reliable.

Müntz (see pp. 58 and 189), an assistant of *Boussingault*, succeeded him as director of the chemical laboratories in the "*Institut National Agronomique*" (1887–1914) and carried out later (1883–1877) rainwater sampling at tropical sites and in the Antarctica (Müntz and Laine 1911). In view of *Liebig*'s finding, it is a bit surprising that in Germany only a few researchers investigated nitrogen in rainwater 10 years after the "French period". The agricultural chemists *Wolf*[397] and *Wilhelm Knop* analyzed only ammonia in few rainwater samples at Möckern (in 1860), only cited in the book "*Der Kreislauf des Stoffs. Ein Lehrbuch der Agricultur-Chemie*" [The cycle of matter. A textbook of agricultural chemistry], Knop (1868).

By order of the Prussian Ministry for Agriculture ["*Preussisches Ministerium für landwirthschaftliche Angelegenheiten*"] all agricultural academies and research stations (Insterburg, Ida-Marienhütte, Lauersfort, Kuschen, Regenwalde, Proskau, Eldena, Waldau)[398] have started in 1864 a program to study the amount of nitrogen that comes per annum with precipitation to earth. The results were compiled in many tables by the "*Centralkommission für das agrikulturchemische Versuchswesen*" [Central Commission for agricultural chemistry experimentation] and published.[399]

At this point, it is worth citing *Georg Meissner*[400] who detected nitric acid (and hydrogen peroxide) in rainwater (Meissner 1863b). He noted that nitric acid is always

---

397 Knop (1868) wrote *Dr. W. Wolf*, his assistant (not to confuse with *Emil von Wolff* (1818–1896), German agricultural chemist and director of the Agricultural Academy Hohenheim). Nothing is known on W. Wolf.

398 "*Königliche Staats- und landwirtschaftliche Akademie*" Eldena (founded 1835) as institute belong the University Greifswald, 1902 associated with the research station Köslin; "*Landwirtschaftliche Akademie*" Waldau at Königsberg (1862–1867); "*Versuchsstation*" Lauersdorf near Crefeld (founded 1864), transferred 1902 to Bonn. The German agricultural chemist *Eduard Heiden* (1835–1888) worked as assistant 1855–1862 at Eldena, 1862–1867 at Waldau and as station's head from 1868 until his death in Pommritz (Saxony). History of other research stations see footnotes 402 and 405. At the stations Regenwalde and Dahme also CO_2 measurements were carried out (Chapter 4.3.1.3). Rainwater investigation only 1864 in Eldena, Lauersfort and Waldau, 1864–1865 in Kuschen, Insterburg and Proskau, and 1864—1867 in Regenwalde and Ida-Marienhütte.

399 Annalen der Landwirthschaft in den Königlich Preußischen Staaten. Herausgegeben vom Präsidium des Königl.- Landes-Oekonomie-Kollegiums und redigiert von dem General-Sekretair desselben C. v. Salviati. Fünfundzwanzigster Jahrgang. Achtundvierzigster Band. Verlag von Wiegandt u. Hempel, Berlin, 1866. In abstracted form with many tables the results can be found (online) in "*Jahresbericht über die Fortschritte auf dem Gesammtgebiete der Agricultur-Chemie*" (Ed. A. Hilger), Berlin, Paul Parey", chapter "*Die Luft*" (rapporteur Th. Dietrich) Vol. 9 (1868) 67–71 and Vol. 11/12 (1871) 150–154.

400 Known for his experiments on ozone and antozone, and adopting *Schönbein*'s view that in thunderstorm antozone, hence hydrogen peroxide, should be formed also concluded that nitric acid in thunderstorm is gained according to *Cavendish*'s experiments. See also p. 483 on *Meissner*.

associated with ammonia, but the latter is produced on the earth's surface. It is notable that shortly before his new measurement, he was unable to detect nitric acid in rainwater but Meissner (1863a, p. 137) argued that nitrous acid is fast oxidized to nitric acid by hydrogen peroxide. The German agronomist *Ewald Wollny* (1846–1901)[401] studied ammonia and nitric acid at different German stations 1864–1866[402] (Table 2.7). At one of these stations, the "*chemisch-physikalische Station für landwirthschaftliche Zwecke*" Insterburg [chemico-physical experiment station], Dr. *Pincus* (1819–1890)[403] became the head (1858–1869) and carried out rainwater analysis, published in curious journals (Pincus and Rollig 1867).[404]

401 Born in Berlin, studied 1866–1868 at the "*Landwirtschaftliche Akademie*" in Proskau and later Halle; 1880 he became full professor at the Technical University Munich. He edited the journal "*Forschungen auf dem Gebiet der Agricultur-Physik*" also cited in literature as "*Wollny`s Forschungen*" (in 20 Volumes 1878–1898).

402 Agriculture Experimental Station Dahme (founded 1857) near Berlin; Regenwalde (founded 1863) in former West Pomerania, today Resko in Poland; Proskau (founded 1857) as part of the "*Königliches pomologisches Institut*" in former Upper Silesia, today Prószków in Poland; Insterburg (founded 1857) in former East Prussia, today Tschernjachowsk (Черняховск) in Russia; Kuschen (founded 1861) and 1877 transferred to the new station Posen; Kuschen was 1911 integrated to Schmiegel, today Šmiegel near Poznan (Poland). See Nobbe (1877).

403 In Chemisches Zentralblatt 105 (1934) Band 1, Nr. 14, p. 2081, the following phrase is given: *Pincus*, Ein Brief Justus von Liebig's. Brief Liebig's an den Vater des Verfassers (damals Arzt in Insterburg), nach welchem Liebig noch 1860 um die Anerkennung seiner Anschauungen zu kämpfen hatte [letter Justus von Liebig's. Letter Liebig's to the father of the author (at that time physician in Insterburg), according to which Liebig still in 1860 had to fight for the recognition of his views]. However, at that time only one physician with the name *Pincus* in Insterberg can be identified: *Salomon Pincus* (1819–1890), Kreisphysikus [country physician] in Insterburg 1859, and later (1875) professor for forensic medicine at the University of Königsberg; one of his sons, *Oscar Pincus* (born 1859 in Insterburg) became also physician (PhD 1887 in Königsberg), married in Posen. It is known that *Salomon Pincus* corresponded with *Liebig*, *Kirchhof* and *Bunsen*; he also dealt lifelong with other issues than medicine; he published several papers on physics and chemistry (e. g., Ann. Phys. **135** (1868) 157; J. Prakt. Chem. **78** (1859) 112–115). Thus, it is likely that the local authorities in Insterburg commissioned him with the direction of the newly founded research station. On that occasion he holds in 1857 an introductory lecture: Pincus (1858) "Denkschrift zum Organisationsplan einer chemisch-physikalischen Versuchsstation für landwirthschaftliche Zwecke" [Memorandum to the organization plan of a chemical-physical experiment station for agricultural purpose]. *Zeitschrift des Landwirthschaftlichen Provincial-Vereins für die Mark Brandenburg und Niederlausitz* **14**, 191–204 (online available). The station was founded by *landwirtschaftlicher Zentralverein für Lithauen und Masuren* [Agricultural National Association for Litunia and Masuria], whose activities were at first limited on the investigation of fertilizers.

404 Pincus (1861) Agriculturchemische und chemische Untersuchungen und Versuche, ausgeführt bei der landwirthschaftlichen chemisch-physikalischen Versuchstation zu Insterburg. Bericht. Hrsg. von dem Curatorium, Gumbinnen and Berlin 1861. Gumbinnen was a town in East Prussia near Königsberg; today Гусев [Gusev] in Russia near Kaliningrad. Nothing is known on *J. Rollig*.

Paul Bretschneider (living dates unknown), director (1858–1878) of the agricultural experiment station Ida-Marienhütte,[405] investigated over 7 years rainfall and its nitrogen content (Bretschneider 1872); he only presents the total nitrogen deposition to be 14.5 kg ha^{-1} yr^{-1}, but total mineral solids in rain and organic matter to be 13.2 and 8.2 mg L^{-1}, respectively. However, he analyzed ammonium (vaporization with sulfuric acid and determination with brominated Javanese lye[406] in an azotometer) and nitrate after *Schlösing*'s method (Table 2.7); and he used a rain gauge (ombrometer) with 30 square feet, rainwater was collected after each rainfall, filtrated and weighted and stored with further rainfalls for monthly collection and analysis. In Anonymous (1871b) beside Bretschneider, Dr. A. Beyer (unknown given name and living data) is referred as observer of rainwater at Regenwalde, the report cites quite high values, also from Ida-Marienhütte (Table 2.7); it is impossible to assess the trueness of such data because of missing original literature. Extensively and detailed described are the sampling and analytical procedure from all experimental stations in the Annual Report of the Central Commission for Agriculture since 1864; only at Regenwalde and Ida-Marienhütte rainwater sampling was continued until 1867 (Eichhorn 1868).[407] Finally, in the nineteenth century *Reinhold Heinrich* (1845–1917), German professor for agricultural chemistry and physiology at University Rostock and director of the agricultural research station (1875–1909), investigated ammonia in rainfall (Heinrich 1881). *Heinrich* (like *Schloesing*) was also interested in studying the absorption of atmospheric ammonia by soils, and hence he conducted an experiment lasting over two years to measure the absorption of ammonia from the air over a glass vessel (78.5 cm^2 surface) filled with a hydrochloric solution (20 %). Based on monthly data he found 30.6 kg NH_3-N ha^{-1} yr^{-1} and from parallel rainwater sampling a total deposition by rain of 7 kg NH_3-N ha^{-1} yr^{-1} – this is the first study on parallel dry and wet deposition that also shows that soil absorption is more important than rainfall deposition.

However, after pioneering work in France and Germany, many more countries began more or less extensive to determine nitrogen in rainwater at agriculture stations (Table 2.7). In Italy, more sporadic investigations of rain were carried out; Mosello et al. (1992) give a summary of papers;[408] like *Alfonso Cossa* (1833–1902), a chemist in Pavia, Udine and at the agricultural experiment station Torino (Cossa 1865), *Giuseppe Bel-*

---

405 Agricultural research station in Laasan (today Łażany) near Saarau (today Żarów) not very far from Breslau (Wroclaw), founded in 1857 and 1877 transferred together with the station Blomberg into the new station Breslau. *Bretschneider* left the station in 1878 and found employment with a superphosphate manufacturer in Stettin (Finlay 1992); later he became the director of the station Pommritz, Saxony (1888–1895).

406 *Javellische Lauge* (German), *eau de Javelle* (French): aqueous solution of sodium hypochlorite.

407 *Carl Herrmann Alexander Eichhorn* (1816–?) was a teacher for physics and chemistry in Möglin and since 1853 in Bonn-Poppelsdorf (agricultural colleges).

408 In the nineteenth century, 7 researcher are cited, 9 authors on 7 sites between 1900 and 1945, and 1984–1988 40 authors and more than 100 sampling sites were operating in Italy.

lucci (1844–1921), a chemist in Perugia, but more prominent as paleo-ethnologist (Bellucci 1888), and the pharmacologist *Piero Giacosa* (1853–1928) in Turin (Giacosa 1895). The chemist *Edmund Mach* (1846–1901) was the first director of the agricultural college ["*Landwirtschaftsschule*"] in San Michele all'Adige (Tirol), founded in 1874. *Mach* collected from March 1885 until February 1886 ($n = 105$) 1114.3 mm rainfall, containing 13.03 kg NH_3-N and 7.44 kg HNO_3-N; he found more nitric acid in summer than in winter (Mach 1888).

Christian Frederik August Tuxen (1850–1903), a Dansk agricultural chemist in Copenhagen, collected rainwater for 5 years (1880–1885) on the campus of the Royal Veterinary and Agricultural University ("*Landbohøjskole*"); Table 2.7 (Tuxen 1890).

In New Zealand, the first rainwater analysis was carried out by *George Gray* (1849–1934)[409] between 1884 and 1888 and later again 1907–1909 (Gray 1888, 1910).

Hans Anton Rosing (1828–1867), Norwegian chemist at the agricultural high school ["*landbrukshøyskolen*"] in Ås, who met *Stöckhardt* in Tharandt and *Boussingault* in Paris; *Georg Reinholdt Wankel* (1843–1907), landowner and politician, published posthumously the result of rainwater investigation in Ås, Norway (1864–1865), Rosing and Wankel (1870).

Joseph Henry Gilbert (1817–1901), with a doctor degree from *Liebig*'s laboratory (later Sir), and Sir *John Bennett Lawes* (1814–1900) began their joint career in 1843 by stirring up a vigor controversy with *Liebig* over the validity of his new "mineral theory". In so doing, they firmly established in 1843 the agricultural research site Rothamsted Experiment Station (Russell 1942), the first long-term rain sampling monitoring, whose work continues to this day (Lawes et al. 1881, 1882, Brimblecombe and Pitman 1980).[410] The heated exchange of opinions in the literature between Rothamsted (*Lawes*) and Giessen (*Liebig*) had lasted for twelve years (1843–1855), also termed "nitrogen controversy". The work in Rothamsted had shown conclusively that *Liebig*'s "mineral theory" simply would not work, and that *Boussingault*'s earlier, more balanced views on minerals and nitrogen were essentially correct (Aulie 1970, p. 454).

Early ammonia analysis in rainwater was carried out occasionally in 1853–1856 and 1855–1856 for both ammonia and nitric acid by *John Thomas Way* (unknown dates of life),[411] a British chemist, who also determined together with *Barral* ammonia and

409 Born near Southampton (Britain), where he afterwards became teacher of chemistry at the Hartley Institution. He held this position till 1874 when he moved to New Zealand and become first Professor of Chemistry and Physics at Canterbury College, Christchurch.

410 First determination of ammonia in rainwater at Rothamsted were made 1853–1854 (Lawes and Gilbert 1854). Further determination, both of ammoniacial and nitric nitrogen were made by Way (1856) in 1855 and 1856; ammonia was determinate in monthly samples between 1877 and 1888, with some interruptions. Since January 1888, ammonia and since September 1886, nitric acid uninterruptedly. Chloride was determined each month since June, 1877, and sulfuric acid from 1881 to 1887 (Warington 1887).

411 1847–1848 he was a consulting chemist at the Royal Agricultural Society of England and 1865 he became a member of the Commission of River Pollution.

nitric acid in rain (as total amount in yearly rainfall) in Paris (1855–1856), Way (1855, 1856). Way (1856, p. 818) writes:

> Nitric acid was found in the rain of each month of the year, and consequently, if it be the result of electrical action, such action must continuously in exercise and not continued to special seasons.

No further analysis was made until 1877 when monthly determinations began. The British water chemist Sir *Edward Frankland* (1825–1899)[412] collected 71 samples from rain events 1869–1870 and found an average from 34 samples (after omitting those with pollution) 0.14 (0.01–0.44 mg L^{-1} NH_3, which agrees with *Way*'s results (cited by Warington 1889). König (1893 and 1904, pp. 1378–1381) cites results, attributed to *Frankland, Denison*, and *Morton* from 72 rainwater samples collected in England (he does not mention Rothamsted), worth to give here (in mg L^{-1}): 39.5 (residue), 6.3 (Cl), 0.50 (NH_3-N), 0.07 (NO_3-N), 0.99 (organic carbon), 0.22 (organic nitrogen), 0.71 (total nitrogen); concentration variations for chloride 0–16.5 and the residue 0.2–85.8; most remarkable is the first quantitative statement of the amount of organic bounded nitrogen in rain, contributing to 1/3 to total nitrogen.

In 1877, a large rainfall gauge was established in Rothamsted and permanent determinations of ammonium and chloride began, 1877 regularly done by the agricultural chemist *Robert Warington* (1838–1807), who also developed new methods for the analysis of ammonium and nitrate and analyzed the rainwater (Warington 1887, 1889), see Table 2.7; he left Rothamsted in 1890. From January 1888, the chemist *N. H. J. Miller* (?–1917)[413] carried out the analysis until his death (Russell 1917, 1942). *Miller* also analyzed (1881–1887) chloride with a means 2.01 mg L^{-1} (summer 1.23 and winter 2.85) and sulfate (1881–1886) 2.52 mg SO_3 L^{-1} (summer 2.67 and winter 2.39). Miller (1905a, p. 290) grouped the chloride values according to the amount of rainfall (between <1 and >4 inches); from the data follows for summer and winter an exponential decrease ($r^2 = 0.97$). *Miller* wrote the first review on precipitation chemistry citing 26 authors and 19 sites or stations in Europe between 1865 and 1900: 5 in Germany, 3 in France, 4 in Italy and 1 in Austria, and 14 overseas (Pretoria, New Zealand, USA, and 10 at tropical sites) with a range of ammonia concentration (in mg L^{-1}) 0.4–3.2 (tropical rain 0.08–1.55) and nitrate 0.06–1.73 (tropical rain 0.00–0.58); Miller (1905a, p. 286).

An interesting finding from Rothamsted is that the deposition of ammonia remains constant between 1853 and 1980 (about 3 kg ha^{-1}) whereas that of nitrate rose from around 1 kg ha^{-1} in the 1880s to around 2 kg ha^{-1} end of 1970s); note between 1917

412 *Frankland* was a member of the River's Pollution Commission together with the Scottish agriculturist *John Chalmers Morton* (1821–1888) and Sir *William Denison* (1804–1870). *Frankland* became the first Professor for Chemistry in Manchester at Owens College in 1851.

413 Russell (1917) wrote in *Nature* a necrology on *N. H. J. Miller* but not tells the date of birth and for what *N. H. J.* stands. My personal inquiry to the Rothamsted stations also gave no result.

and 1955 no rainwater chemistry monitoring were carried out at Rothamsted (Brimblecombe and Pitman 1980).

Besides the long-term rainwater monitoring at Rothamsted, a second monitoring site was established in Paris. *Albert Lévy* (1844–1907),[414] chief-chemist at the observatory Montsouris (Paris),[415] carried out over 30 years of analysis of rainwater and air components[416] (Lévy 1903a,b). Rainwater sampling was started in 1881; the station is also worldwide known for first quantitative monitoring of ozone (Chapter 4.4.5.7), carbon dioxide, and ammonia (Chapters 4.3.1.3 and 4.6.1.2). Lévy (1882) presents a list, comparing the deposition of total nitrogen (in kg ha^{-1}):

414 There is a (most frequent used) variation of the name: *Albert-Lévy*. *Directeur du service chimique à l'Observatoire de Montsouris* (in 1894). He also used the pseudonym *A. Bertalisse* (Georg D'Heylli (1887) Dictionaire de Pseudonymes. E. Dentu, Paris, 2e èd.; Slatkine Reprint 1971, Genéve). *Lévy* wrote: L'argot de l'X: 150 ans de la vie à l'École polytechnique (1883), Mes souvenirs / par A. Bertalisse (1890) and Histoire de l'air, G. Baillière, Paris (1879) 184 pp. and many more books.

415 The Paris Observatory was founded in 1671. In France, the first systematic analyses of atmospheric air are those of the Montsouris Observatory (founded 1871, see Anonymous (1881b)) since 1876 (ozone 1876–1883, ammonia and organic nitrogen 1876–1882, and carbon dioxide 1876–1910). *Lévy* also reports on nitrogen in snow, fog and dew. The organization of the observatory is described in *Nature* (55, 1897, 340) as: "Montsouris Meteorological Observatory was established by M. Duruy in 1871, by the influence of Dumas then president of the Municipal Council of the City of Paris. The work carried out in it is divided into three principal sections, which, while including purely scientific researches, take in also subjects relating to the climatology and hygiene of Paris. The first branch of the Observatory's work belongs to physics and meteorology, among the subjects included in this section being: atmospheric electricity, the usual meteorological observations, the influence of smoke and vapours upon atmospheric variations, &c. The chemical work done at the Observatory refers to the variations in the composition of air in different parts of Paris, analysis of water, variations in the composition of sewage waters and of the Seine at different points, and methods of filtration. To the micro-graphic section of the Observatory is entrusted not only bacteriological statistics, and the determination of the meteorological conditions which affect the abundance of micro-organisms in air, soil, and water, but also the examination of the specific characters of bacteria. A bacteriological laboratory for the diagnosis of diphtheria, and of other diseases of which the active principles are known, was joined to the micrographic section of the Observatory at the beginning of last year, and has been found of great service to the public and to medical practitioners."

416 *Edme Hippolyte Marié-Davy* (1820–1893), "Professeur de Physique à la Faculté des Sciences de Montpellier and vice-director of Montsouris Observatory, has organized, partly at the expense of the French Government, partly at the expense of the City of Paris, a chemical and microscopical laboratory for the analysis of all the matters in suspension in the air of Paris, both quantitatively and qualitatively. A certain quantity of air is constantly aspired by an aspirator in continued operation. The ozone acting on iodide of potassium and starch liberates iodine. The quantity of ozone liberated is measured by a titrated solution of arsenite of sodium. The matters in suspension are collected on a glass plate, and the crop is placed under the object-glass of a powerful microscope magnifying 1,000 times. The principal forms are drawn and plates are executed and published monthly in the *Annuaire de l'Observatoire Municipal de Paris, di Observatoire de Montsouris*. The analysis of rainwater is conducted on the same principles, and the results of chemical analysis are calculated and compared with the wind and other atmospheric circumstances" (Fonvielle 1876).

Paris (Montsouris) (1877–1881)	13.86 (10.91 as NH_3-N and 3.85 as HNO_3-N)
Rothamsted (1881–1882)	6.24–8.85
Vallombrosa (1872–1875)	11.63
Florence (1870–1872)	11.03–14.9
Regenwalde (1864–1867)	11.63–18.41
Dahme (1865)	7.46
Insterburg (1864–1866)	6.15–7.63

Intensive rainwater sampling at tropical sites was done by the French agricultural chemist *Charles Achille Müntz* and his coworkers[417] (Müntz and Marcano 1889, Marcano et al. 1891) concerning the question of whether thunderstorm originated fixed atmospheric nitrogen; as seen from the data, nitrogen concentrations in tropical rain are significantly less (Müntz 1892, p. 186):

> Je ne puis donc que maintenir mes conclusions au sujet de la richesse plus grande des eaux de pluie dans la station d'observation du Venezuela.

Jean Auguste Houzeau (1829–1911),[418] who started his carrier as an assistant of *Boussingault*, found that ammonia in freshly collected rainwater under sunlight and heat fast decompose and recommended to protect samples (Houzeau 1883). Miller (1905a, p. 287) concludes that tropical rain does not supply to the soil an essential greater amount of nitrogen. Other studies of tropical rainwater had been carried out for 5 years at Barbados (Harrison and Williams 1897) and British Guiana for 20 years (1890–1910)[419] and Ceylon (1898–1899),[420] East Java (1891),[421] and Mauritius (1895).[422] Clarke (1920, p. 52) summarizes that in most cases ammonium is in excess over nitric acid; but in the tropics the reverse seems to be true.

417 I like to name here *Vicente Marcano* (1848–1891) who was a pioneer chemist of Venezuela.

418 French chemist, since 1852 he worked at the agronomic institute in Versailles and at *Conservatoire National des Arts et Métiers* in Paris; 1858 director of the "*station agronomique*" and professor for chemistry in Rouen.

419 Cited by Miller (1905a) as pers. communication by *John Burchmore Harrison* (1856–1926), the pioneer (geologist and colonial analytical chemist) in research on tropical weathering (on Barbados) in the period 1897–1910; likely *John William Williams* (unknown living data), British chemist. See also Carrol (1962). Rainwater chemical monitoring was obviously continued at the Botanic Garden on Barbados at least until 1909, reported by *Experimental Station Record* **20** (1908–09) p. 518 from "Composition of Barbados rainfall" in "*Repertorium of Agricultural Work of Barbados*, Imp. Dept. Agr., West Indies, 1906–08, pp. 3 ff). From given values (1906–1908) for deposition in lbs/acre and rainfall in inches, the concentrations here are recalculated (in mg L^{-1}): 151 for chloride and 12.7 for nitrogen.

420 Miller (1905a) cites M. K. Bamber (1900) Report on Ceylon Tea Soils, Colombo, a paper not publicly available. *Montague Kelway Bamber* (1866–1924) was an agricultural chemist at the Royal Botanic Gardens at Peradeniya in Ceylon and wrote several books on tea plantation. He was a Governmental agricultural chemist at the Tea Research Institute in Ceylon (today Sri Lanka) at Thalawakele.

421 Miller (1905a) cites Th. Marr (1891).

422 Miller (1905a) cites P. Bonâme (1895), but see Bonâme (1896).

Table 2.6: Determinations of nitrite and nitrate in rainwater; values in mg L^{-1} (as N).

site	nitrite	nitrate	author[a]
Tbilisi	0.34–2.8	–	Struve (1871)
St. Chamas[a]	0.133 (0.0–0.316)	0.162 (0.001–0.716)	Chabrier (1871)
Algier[k]	0.07–0.14	–	Chairy (1884)
Scandicci[d]	0.0045 (0.0–0.0171)	0.255 (0.033–0.729)	Passerini (1891)[b]
Catania (Sicily)[e]	0.0006 (0.0–0.0027)	0.161 (0.035–0.244)	Basile (1895)[b]
Ploty (Ukraine)[f]	0.011 (0.0–0.025)	0.178 (0.002–0.178)	Welbel (1903)
Buenos Aires	0.014 (±0.023)	0.88 (±0.38)	Anonymous (1907)
Tomsk	0.03 (0.01–0.08)	0.41 (0.1–2.0)	Bykov and Karpov (1911)
Melbourne[g]	0.021 (±0.02)	0.108 (±0.013)	Anderson (1915)
Canterburry[h]	0.008 (±0.009)	0.072 (±0.06)	Anderson (1915)
Tokyo (1937)	0.0067	0.028–0.240	Miyake (1965)
Bombay (1938)	0.00062 (0.0–0.02)	0.133 (0.08–0.24)	Narayanaswami (1939)
Frankfurt (1957)	0.004	0.06	Georgii (1965)
Wahnsdorf (1958)	0.013	0.41	Mrose (1961)
Seehausen[i]	0.05	0.71	Marquardt, pers. Commun.

[a] 12 samples collected between January and July 1870
[b] Cited after Miller (1905a)
[d] Near Florence; ammonia: 0.614
[e] Ammonia: 0.327
[f] Ammonia: 0.854
[g] Standard variation based on given quarterly means (November 1912–February 1914)
[h] Standard variation based on given monthly means (all means from daily means); the values for February 2014 I have omitted because they are extreme due to low rainfall (0.1 inches): 0.132 (NO_3) and 0.0064 (NO_2) in mg L^{-1} as nitrogen.
[i] Precipitation-weighted mean 1982–1989 (cf. Table 2.21)
[k] Given as 0.14–0.29 nitrogen mg L^{-1} as ammonium nitrite (i. e., 50 % as nitrite).
[m] The ratio nitrite to nitrate in rainwater ranges from 0.018 to 0.062, which is very close to the ratio of gaseous HNO_2 to gaseous NO_2 of 0.02–0.08 we found 100 years later (see Möller 2020, p. 559).

Early determinations of nitrite (beside H_2O_2) in rainwater (Table 2.6) are very remarkable. Struve (1869) was the first (in Tbilisi) soon followed by Failyer and Willard (1889) who detected qualitatively nitrite in rainwater at the Kansas State Agricultural College Experimental Station[423] (data on ammonia and nitric acid from 1886–1889 are published also by Failyer (1889–1890)). They showed that in winter only a small number of samples (28 %) contained nitrite, whilst in summer 89 % of the samples gave a positive signal.[424] The first published values by Chabrier (1871) are not reliable (Table 2.6).

423 Founded 1887 in Manhattan (Kansas, USA); The chemist *George Henry Failyer* (1849–1945) was the first head of the Department of Chemistry (Today Kansas State University) but the analysis was done mostly by the chemist *Julius Terrass Willard* (1862–1950), later head of the experimental station (1900–1906) and professor at the university.

424 Today we know that nitrous acid is more abundant in summer due to photochemical formation.

Another first quantitative estimation of nitrite in rainwater (together with hydrogen peroxide, see Chapter 4.5.2.2) was done already in 1869 by Struve (1872, p. 28) to be 1.15 mg L^{-1} (and a maximum of 9.34 g L^{-1}). *Struve* refers *Goppelsroeder* who also accepts the presence of nitrous acid while analyzing nitric acid, whereas *Hugo Trommsdorff* (1838–1918)[425] neglects it. Chairy (1884) determined in two rainwater samples ammonium nitrite; also, likely too high.

Thus, the first quantitative determination of nitrite must be attributed to *Napolenoe Pio Passerini* (1862–1951), a Italian agronomist, who studied precipitation chemistry at the meteorological station of the Agronomical School of Scandicci between 1888 and 1890 (Passerini 1891, 1893).

The nineteenth century ends with rainwater analysis by a few important agricultural chemists. *Oskar Kellner* (1851–1911), a German chemist, came to Japan in 1881 and conducted first precipitation chemistry measurements from 1883 to 1885 (today he is termed the "father" of Japanese agricultural chemistry).[426] The work of *Kellner* was continued by his students 1931–1934 (Kurashige 1934, Kurashige and Kagei 1935). *Julius Stoklasa* (1857–1932)[427] first studied nitrogen in rainwater and snow in Bohemia 1877–1878 and 1883–1868 (Stoklasa 1882, 1887). He found 1878 in snow ten times more NH_3, but the same amount of HNO_3 and no difference between summer and winter values for NH_3 but two times more HNO_3 in summer and generally more HNO_3 than NH_3 in rainwater.[428] The agricultural chemist *Arthur Julius Petermann* (1845–1902),[429]

---

425 German chemist in a famous dynasty of pharmacists, who studied the chemistry of waters; Son *of Christian Wilhelm Hermann Trommsdorff* (1811–1884) who begun in 1797 with the production of pharmaceuticals and founded in 1837 a chemical manufactory (1837) in Erfurt, sold in 1885 to a consortium (but still works under this name). His father *Johann Bartholomäus Trommsdorff* (1770–1837) took over the "*Schwanen-Ring-Apotheke*" in Erfurt from his father *Wilhelm Bernhard Trommsdorff* (1738–1782).

426 *Kellner* is named in line with *Liebig* and *Hermann Hellriegel* (1831–1895) as pioneer in agricultural chemistry. He was an assistant of *Wolff* at Hohenheim. In 1880 he was termed at Tokyo as professor and stayed in Japan 12 years. In 1893 he became the director of the experimental station Möckern until his death in 1911.

427 Czech agricultural chemist; studied in Leipzig, Vienna, Leipzig and Prague. 1901–1927 full professor at Technical University Prague and 1899–1918 director of the State Chemical-Physiological Experimental Station Prague.

428 The relative enrichment of NH_3 comparing to HNO_3 in snow (and fog water, studied by *Stoklasa*, see p. 255) can be explained through the occurrence of NH_3 mainly particulate and HNO_3 gaseous.

429 *Petermann*, born in Dresden (Germany) and graduated from Göttingen (student of *Wilhelm Henneberg* and in contact with *Liebig*), started his career as an assistant at the agricultural station Pommritz in Saxony (first founded in 1852 at the manor Weidlitz and 1864 transferred to Pommritz, and still in operation), afterwards in Nancy (France) were *Louis Nicolas Grandeau* (1834–1911) was the director, followed by a brief stint as director of the Prilep station in Moravia, former Austria, founded by *Carl Max Graf Seilern* (1825–1905). In 1871, the "Association for the Founding of Agricultural Experiment Stations" was founded in Belgium, supported by subscriptions and private donations. *Grandeau*, who was an enthusiast of *Liebig*, invited *Petermann* to work at "*Rijkslandbouwinstituut Staatsschool voor landbouwwetenschappen*" [State Agronomic Institute] at Gembloux (founded in 1869) where he ar-together with his assistant *Jean Graftiau* (1858–?) and later chief chemist, carried out rainwater (and CO_2 measurements) at the agricultural research station based in Gembloux (Belgium); Petermann and Graftiau (1893). The analysis of rainwater concerning plant nutrients flux has continued in the following twentieth century (and – that must be stated – never ended until present).[430]

The twentieth century begins with rainwater sampling in Bologna by Casali (1901); Miller (1905a), Drischel (1940), and Eriksson (1952) cited him but any reference within the text is missing. *Adolfo Casali* (1834–1905)[431] made a very remarkable note (cited after Mosello et al. 1992) that human activity caused an increase in ammonia deposition and he surmised that this could be an environmental problem.

The first precipitation chemistry measurement in Hungary (ammonia and nitrate) was carried out by the Hungarian apothecary and polymath *Endre Kazay* (1876–1923) at the Ógyalla Observatory[432] in 1902 (Kazay 1904) (Table 2.7), continued 1966 by *Ernő Mészáros* and later *László Horváth* (Mészáros 1965, 1966, 1973, 1974, Horváth 1978, 1983, Horváth and Mészáros 1984).[433]

Maurice Aufray (unknown living dates), director of the Agricultural Laboratory of the Institute of Hygiene and Bacteriology at Tonkin (Province of Phú Thọ, North Vietnam) collected in Hanoi from April 1902 until September 1909 rainwater (Table 2.7). *John Walter Leather* (1860–1934) was an agricultural chemist who carried out rainwater analysis in Cawnpore (1904–1905, $n = 18$) and Dehra Dun (1904, $n = 21$), India (Leather 1906); *Leather* was the first Imperial Agricultural Chemist at the Imperial Agricultural Research Institute in Pusa, Bihar since 1892 until his retirement.

---

rived in 1872 and became the first director of the "*Station agricole de l'Association pour la fondation de stations agricoles en Belgique*", renamed in "*Station expérimentale de l'Etat* 1883–1888 and "*Station agronomique de l'Etat*" from 1888.

430 I'm rather sure that all persons who carried out rainwater analysis in the nineteenth century (75 authors, see Table 2.12 on p. 233) were mentioned; between 1900 and 1950, 73 researchers are listed, but there are several studies carried out in the USA with limited precipitation chemistry, like eight authors who determinate only sulfate deposit at 70 sites between 1910 and 1947 in the USA and 12 researchers who studied nitrogen in rainwater between 1912 and 1932 at Mt. Vernon, Iowa (cited by Eriksson 1952). For the twentieth century, only researchers until end of World War II can be listed relatively completely, almost based on the papers by Drischel (1940) and Eriksson (1952).

431 Professor of chemistry at the regional institute of technology [*Regio Istituto Tecnico*] and director of the laboratory of agriculture chemistry [*Laboratorio di Chimica Agraria*] in Bologna.

432 A seismological, meteorological and astronomical observatory in the former Hungarian city of Ógyalla. Since 1948, the city belongs to Slovakia and is now known as Hurbanovo.

433 *Ernő Mészáros* (born 1935) is a Hungarian meteorologist, professor and member of the Hungarian academy of sciences, who was director of the Institute of Atmospheric Physics (belong the Hungarian Meteorological Service) in Budapest 1976–1990 and afterwards head of research group of the Academy of sciences at Veszprém until 2005. *László Horváth* (born 1950) studied chemistry and worked in the Institute of Atmospheric Physics (Budapest) since 1975 in precipitation chemistry with specials interest in deposition of nitrogen compounds.

Table 2.7: Determinations of ammonia and nitric acid in rainwater before 1950; values in mg L^{-1}. It is not always clear to which formula the analytical value for "nitric acid" is related, HNO_3, NO_3 and N_2O_5, and also NO_5 is used; however, the variation of the factor giving elemental N is small (0.23–0.26) and surely less than the analytical error.

site	year	ammonia	nitric acid	author[a]
Paris	1851	2.8–3.3	14.1–21.8	Barral (1852)[b]
Paris	1852	3.7 (1.1–9.6)	6.2 (1.8–11.8)	Barral (1852)[b]
Lyon	1852	4.4	–	Bineau (1852)
Lyon	1853	6.8 (2.1–30)[c]	0.3–2	Bineau (1855a)
Lyon, env.	1852–1853	0.3–3.1	0–3.2	Bineau (1855a)
Paris	1853	1.77 (1.56–2.0)[d]	–	Boussingault (1854)
Liebfrauenberg	1853	0.99 (0.0–5.4)[e]	–	Boussingault (1854)
Marseille	1853	3.144	(traces)	Martin (1853)
Rothamsted	1853–1854	0.89	–	Lawes and Gilbert (1854)
Toulouse	1855	0.65 (0.44–0.83)	0.5[f]	Filhol (1855)
Rothamsted	1855–1856	1.3 ± 0.4[g]	0.47 ± 0.2[g]	Way (1856)
Rothamsted	1855–1856	0.98 (0.64–1.72)[g]	0.53 (0.3–1.3)	Warington (1887)
St. Bernhard	1859	1.10	0.30	Boussingault (1882)
Möckern	1860	1.0 (0.3–3.0)[h]	–	Knop and Wolf (1861)
Nantes	1863	0.997 / 5.939[i]	7.360 /5.683[i]	Bobierre (1864)
Jena[j]	1863	3.247	0.526	Reichardt (1863)
Insterburg[k]	1864–1867	2.297	1.49	Pincus and Rollig (1867)
Insterburg[k]	1864–1865	0.55	0.30	Wollny (1896)
Insterburg[k]	1865–1866	0.76	0.49	Wollny (1896)
Dahme[k]	1865	1.42	0.30	Wollny (1896)
Regenwalde[k]	1864–1865	2.03	0.80	Wollny (1896)
Regenwalde[k]	1865–1866	1.88	0.48	Wollny (1896)
Regenwalde[k]	1866–1867	2.28	0.56	Wollny (1896)
Proskau[k]	1864–1865	3.21	1.73	Wollny (1896)
Kuschen[k]	1864–1865	0.54	0.16	Wollny (1896)
Kuschen[k]	1865–1866	0.44	0.16	Wollny (1896)
Manchester	1869–1870	6.47	1.03	Smith (1872)
Florence	1869–1875	1.00	0.57	[m]
Basel	1870–1871	–	3.9 (0.1– 6.2)[l]	Goppelsroeder (1871)
England[n]	1869–1870	1.07	0.75	Smith (1872)
Scotland[n]	1869–1870	0.305	0.372	Smith (1872)
Scotland coast	1869–1870	0.738	0.432	Smith (1872)
Rothamsted	1869–1870	0.34	–	Frankland (1874)
London	1870	10.57	8.12	Smith (1872)
Vallombrosa (I)	1872–1875	0.62	0.25	[m]
Vallombrosa (I)	1873	0.82[o]	0.92[o]	Bechi (1873)
Florence	1873	1.25[o] / 1.004[p]	1.62[o] / 0.358[p]	Bechi (1873)
Munich	1875	3.8 (2.3–5.0)	2.5 (0.6–8.6)	Eugling (1878)[r]
Saint-Louis[q]	1875	1.1 (0–2.7)	–	Louvet (1876)
Tisis (Austria)	1876–1877	2.8 (1.2–6.2)	1.8 (0.9– 2.8)	Eugling (1878)[r]
Montsouris[s]	1875–1881	1.59 (1.20–1.98)	0.33 (0.2–1.6)	Lévy (1882)

Table 2.7: (continued).

site	year	ammonia	nitric acid	author[a]
Montsouris	1876–1890	2.2	large	Lévy (1891)
Rothamsted	1877–1880	0.54 (0.16–1.04)[t]	–	Miller (1905a)
Liebwerda[u]	1877	1.5 (0.1–8.8)	4.8 (1.6–9.8)	Stoklasa (1882)
Copenhagen	1880–1890	2.1 (1.3–2.8)	0.5 (0.2–1.2)	Tuxen (1890)
Rostock	1880–1881	0.892	–	Heinrich (1881)
Montsouris	1881–1882	1.26	2.93	Lévy (1903b)[v]
Montsouris	1881–1890	1.85	0.71	Lévy (1891)
Rothamsted	1881–1886	0.36 (0.33–0.40)[w]	–	Miller (1905a)
Pic du Midi	1882	0.34–0.80[x]	–	Müntz and Aubin (1882a)
Peček, Bohemia	1883–1886	1.53	2.21	Stoklasa (1887)
Tokyo	1883–1884	0.154	0.328[y]	Kellner et al. (1886)
Caracas	1883–1885	–	2.23 (0.2–16.2)[z]	Müntz and Marcano (1889)
Lincoln (N. Z.)	1884–1888	0.076	0.169	Gray (1888)
Barbados	1885–1890	0.08	1.60	Harrison and Williams (1897)
St. Michele	1885–1886	1.65 (0.86–3.41)	2.85 (0.92–5.45)	Mach (1888)
Réunion	1886–1887	–	2.67 (0.4–12.5)[aa]	Müntz and Marcano (1889)
Rothamsted[bb]	1886–1888	0.52 (0.29–1.24)[cc]	0.62 (0.4–1.1)[dd]	Warington (1889)
Rothamsted[ee]	1889–1903	0.45 (0.34–0.57)	0.19 (0.16–0.23)	Miller (1905b)
Rothamsted	1888–1916	0.405	0.20	Russell and Richards (1919)
Caracas	1889–1890	1.55 (0.37–4.01)[ff]	–	Marcano et al. (1891)
Manhattan (Kansas)	1889	0.393	0.154	Failyer and Willard (1889)
British Guiana	1890–1900	0.055	0.078	Harrison and Williams (1897)
British Guiana	1904–1909	0.018 (±0.014)[gg]	0.044 (±0.024)[gg]	Harrison (1911)
Calcutta	1891	0.172	0.115	Miller (1905a)
Gembloux	1893	1.14	0.35	Petermann and Graftiau (1893)
Ceylon	1898–1899	0.196	0.069	Bamber (1900)
Mississippi	1894–1895	0.235	0.074	[oo]
Ploty (Russia)	1902–1903	0.85	0.06	Welbel (1903)
Ógyalla (Hungary)	1902	1.99[hh]	0.74[hh]	Kazay (1904)
Hanoi	1902–1909	0.571	0.451	Aufray (1909)
Odessa	1903	0.8[ii]	0.8[ii]	Posnjakov (1904)
Dehra Dun, India	1904	0.104 (0.04–0.82)	0.070 (0.01–0.73)	Leather (1906)
Cawnpore, India	1904–1905	0.221 (0.04–1.24)	0.068 (0.03–0.84)	Leather (1906)
Garforth[jj]	1906–1909	1.04	0.31	Crowther and Ruston (1911)

Table 2.7: (continued).

site	year	ammonia	nitric acid	author[a]
Buenos Aires[kk]	1907	3.18 (±0.75)	0.88 (±0.38)	Anonymous (1907)
Lincoln (N. Z.)[ll]	1907–1909	0.13 (0.02–0.36)	0.18 (0.04–0.76)	Gray (1910)
Uithuizermeeden[mm]	1905–1907	0.86 (±0.35)	0.40 (±0.19)	Hudig (1910)
Uithuizermeeden[mm]	1908–1910	0.58 (±0.31)	0.18 (±0.09)	Hudig and Welt (1911)
Hebrides[nn]	1907–1913	0.04–0.15	0.03–0.14	Miller (1913)
Brisbane[pp]	1908	0.22	0.19	Brünnich (1909)
Ottawa[qq]	1908–1924	0.525	0.278	Shutt and Hedley (1925)
Leeds	1909	1.98	0.20	Cohen and Ruston (1925)
Bloemfontein	1909	0.986	0.323	Juritz (1910)
Flahult, Sweden	1909	0.450	0.177	Feilitzen and Lugner (1910)
Tomsk (Russia)	1910	0.94 (0.55–2.20)	0.41 (0.1–2.0)	Bykov and Karpov (1911)
Uithuizermeeden	1910–1912	0.72 (0.28–1.26)	0.24 (0.15–0.32)	Hudig (1912)
Iceland	1911–1913	0.09	0.03	Miller (1913)
Montevideo	1912	0.24	0.27	Schröder (1915)
Mt. Vernon[rr]	1912–1913	0.766	0.152	Wiesner (1914)
Ithaca, N. Y.	1915–1920	1.73[ss]	0.01[ss]	Wilson (1921)
Geneve, N. Y.	1919–1928	0.99	0.12	Collison and Mensching (1932)
Alfred, N. Y.	1923–1925	1.67	0.03	Wilson (1926)
Kentucky	1922–1923	1.19	0.74	Freeman (1924)
Denmark[tt]	1925–1926	0.71–1.07	0.35–0.60	Hansen (1931)
Trondheim	1928–1929	0.132	0.018	Braadlie (1930)
Weihenstephan	1928–1935	1.97	–	Scharrer and Schropp (1940)
Goodwell (Texas)	1930	0.283	0.129	Finnell and Houghton (1931)
Chicago	1931–1933	0.744	0.042	Ives et al. (1936)
Allahabad (India)	1932	0.31 (0.10–1.02)	0.90 (0.43–2.56)	Dhar and Ram (1933a)
Tokyo	1937	0.58	0.06	Miyake (1965)
Kobe	1937	0.28	–	Miyake (1965)
Hamamatsu	1937	0.19	–	Miyake (1965)
Oberschreiberhau[uu]	1937	0.76	0.47	Drischel (1940)
temperate zones[vv]	1909–1936	0.45 (0.04–0.83)	1.02 (0.1–3.0)	Drischel (1940)
tropical zones[ww]	1900–1933	0.26 (0.04–0.69)	1.15 (0.3–3.4)	Drischel (1940
Tashkent	1945–1947	0.6	0.1	Kudrin (1948)
Sweden	1947–1950[xx]			Torstensson (1954)

[a]Not all authors are additional referred within the text.
[b]He also estimated Cl, Ca and Mg (see text). Monthly samples June–December 1851 from two sites and means from two sites of the observatory January–June 1852; given as NH_3 and HNO_3. Wrong cited in *Annalen der Pharmacie* **123** (1853) 193–194 as (in mg L^{-1}) total N (8.36), NH_3 (3.61) and NO_5 (19.09).
[c]Winter 16.3, spring 12.1, summer 4 and autumn 6.8.

Table 2.7: (continued).

[d] $n = 4$ (May–July), carried out by *Houzeau*; one further event on January 3, 1854 with 3.08 mg L^{-1}.
[e] May–November 1853, $n = 64$ (arithmetic mean); because the total amount of rain fall is given for each event and the sum, the weighted-mean can be calculated to be 0.47 mg L^{-1}.
[f] Original given as 3.0 mg $NaNO_3$.
[g] Values are originally given in grains per Imperial gallon
[h] $n = 9$ (April–September) research station Möckern near Leipzig; study together with W. Wolf (cited after Knop 1868, Vol 2, p. 77).
[i] At two different altitudes: 154 feet and ground-based (means of January to December 1863): he found NaCl to be 14.09 and 13.80, respectively.
[j] One event with hail June 26, 1863; no nitrous acid detected. In snow (one event November 27, 1862) he determined 3.33 mg L^{-1} ammonia and nitric acid as sum.
[k] See footnote 402 on p. 184 for the stations; note that *Wollny* was the reporteur and not operator. Eriksson (1952) cited partly wrong data and "Kushen" instead of Kuschen and did not regard that the original data are not given in nitrogen-N but as NH_3 and HNO_3. In *Jahresbericht über die Fortschritte auf dem Gesammtgebiete der Agricultur-Chemie* **10** (1868) 58–62, (see also: *Annalen der Landwirthschaft* **50**, p. 249) the following results on ammonia (in NH_3) and nitric acid (in NO_5) in mg L^{-1} are cited: Regenwalde (1856) 2.42 / 2.49, Ida-Marienhütte (1865) 2.72 / 0.85, Ida-Marienhütte (1866–1867) 2.297 / 0.455, Dahme (1865–1866) 1.72 / 1.16, Kuschen (1856) 0.56 / 0.72, Kuschen (1866) 0.74 / 0.82, Insterburg (1865) 1.06 / 1.63.
[l] Arithmetic mean from cited single event samples ($n = 95$) collected at the Botanical Garden October 1870 – July 1871 (monthly means 0.7–4.9 mg L^{-1}) without three extreme values (100, 123 and 136 mg L^{-1} May, March and January, respectively), given as N_2O_5. Eriksson (1952) cites *Goppelsroeder* with a (wrong) mean of 0.46 mg L^{-1} NO_3-N; Drischel (1940) does not cite him. It is worth to note that Miller (1905a) lists *Goppelsroeder* in the bibliography but there is no citation in the text of the paper.
[m] Values cited by Miller (1905a) referring *Bechi* (*Saggi di esperienze agrarie* 1870–1876) and given as NH_3 and N_2O_5 (here recalculated as HNO_3); Eriksson (1952) cites also *Bechi* but mentioned that he used the data given by Miller (1905a, 1913) but he cited wrongly "Miller (1902)" and made the mistake to present the original values in NH_3-N and HNO_3-N. This is nice example how errors appear by wrong citations and not looking into original publications. *Emilio Bechi* (1820–1900), professor for chemistry at Florence; studied in Gießen at *Liebig*.
[n] Several sites inland.
[o] Recalculated from original values given as total amount of rain fall (7645 m^3 and 12915 m^3) and ammonia (9560 g and 10592 g) and nitric acid (12394 g and 11866 g) for Florence and Vallombrosa, respectively (*Berichte dtsch. Chem. Ges.* **9** (1876) 346).
[p] Other values given by Drischel (1940), citing *Berichte dtsch. Chem. Ges.* (1873) **6**, 1203.
[q] City in NW of Senegal; *Louvet* also observed ozone (see Table 4.18); apart from ammonium he determined NaCl: 69 (7–233) mg L^{-1}.
[r] *Carl Friedrich Wilhelm Eugling* (unknown living dates), assistant of the agriculture research station Tisis (Vorarlberg) in Austria. He wrote "*Handbuch für die praktische Käserei*" [handbook of practical cheesery] (Heinsius, Leipzig, 1901, 281 pp.), still published in new editions. The study aimed to show that nitrogen quantity in rain is larger in cities than at rural sites.
[s] Ammonia in dew ($n = 6$) 4.13 (3.01–4.86) mg L^{-1} in October 1880; in fog 2.75 mg L^{-1}, in snow ($n = 4$) 2.14 (1.40–2.77) 2.75 mg L^{-1}; nitrate in dew in September 1880 ($n = 6$) 2.1 (1.6–2.5) mg L^{-1} more than ammonium in September 1.28 (0.8–1.4) mg L^{-1}.
[t] Cited after Miller (1905a); October 1877–December 1880 (arithmetic mean from 28 monthly means).

Table 2.7: (continued).

[u]The analysis were carried out by Dr. *Ullika* in Libverda (Bohemia); listed are the values from 1877 (n = 52/50), May-October; in 1878 (September-December, n = 13) in rainwater 1.2 (0.2–4.3) NH_3 and 1.9 (0.1–2.7) HNO_3 and in snow (n = 5) 17.2 (6.0–23.6) NH_3 and 1.8 (0.6–4.0) HNO_3. In 1877, additional fog water was analyzed: NH_3 (n = 34) 18 (8–40) and HNO_3 (n = 19) 9 (5–21); all data calculated from published single values.
[v]n = 13; in snow (n = 7) 0.06–0.14 mg L^{-1} and in fog (n = 5) 0.19–0.64 mg L^{-1}; gaseous NH_3 (n = 8) 0.19–3.03 ppb. Mean chloride concentration in rain 0.34 mg L^{-1}.
[w]Cited after Miller (1905a); arithmetic means from 4 annual means including yearly variation.
[x]In snow, less ammonia was found: 0.06–0.14 mg L^{-1} (n = 7).
[y]November 1883 – October 1884; summer means 0.227 and winter means 0.562 due to large difference in rainfall.
[z]07/1883–07/1884 and 01–12/1885 (n = 121). Eriksson (1952) and Drischel (1940) cite 0.55 and 0.578 mg L^{-1} (as NO_3-N), respectively, but Müntz and Marcano (1889) give 2.23 mg L^{-1} (as NO_3).
[aa]Müntz and Marcano (1889) note that 19 samples were collected by *M. Raimbault*, director of the agronomic station at Réunion.
[bb]Mean from monthly rainfalls.
[cc]Values are given originally in nitrogen, recalculated here in NH_3.
[dd]Values are given originally in nitrogen, recalculated here in NO_3.
[ee]Arithmetic means including monthly variation.
[ff]Cited by Drischel (1940) and Eriksson (1952) wrongly as NH_3-N.
[gg]Arithmetic means from the given annual means (n = 6).
[hh]January–December 1902 (n = 36, 10-day sampling), nitrate given as N_2O_5. Horváth (1983) corrected these values by a factor of 1.09 and 0.65 for ammonium and nitrate, respectively, because of parallel analytical determination of the visual colorimetric (1904) and spectrophotometric (1980) method. Whereas the ammonium concentration in rainwater did not change (only concerning the seasonal cycle), the nitrate value increased by a factor of 8 between 1902 and 1980 (Horváth 1983).
[ii]Eriksson (1952) cites values being (in mg L^{-1}) 0.49 and 0.18 for NH_3 and HNO_3, obviously recalculated in nitrogen.
[jj] Garforth is small country-site town, 5 km eastward from Leeds.
[kk]Arithmetic mean including standard variation from (n = 21) single values. For the year 1906, Eriksson (1952) cites (in mg L^{-1}): 2.96 for NH_3 and 0.18 for HNO_3.
[ll]For 1907–1908: 0.17 (0.024–0.36) NH_3 and 0.24 (0.054–0.76) HNO_3; for 1908–1909: 0.094 (0.02–0.22) NH_3 and 0.117 (0.04–0.65) HNO_3.
[mm]Means from cited single values; 1906–1907 n = 9 (single events) and 1908–1910 n = 40 (monthly means).
[nn]Laudale (1907), Barra Head (1910–1913), Butt of Lewis (1908–1913); for values see Eriksson (1952).
[oo]Miller (1905a) cites (it is *Tracy*) "S. M. Tracey: Ammonia in rain-Water (*Eight Ann. Rep. Miss. Agr. Tech. Coll. Exp. Stat.* 1895) and Perkins, Tenth Ann. Rep., 1897", Drischel (1940) cites "Hutchinson: *10th ann. Rept. Mississippi Agr. Mech. Coll. Exp. Sta.* 1897", Eriksson (1952) cites "Tracy, S. M., 1895, Ammonia in rainwater. *Miss. Agr. Exp. Sta. 8th Ann. Rpt.*, p. 102". See also Table 2.12 on p. 233.
[pp]Similar values *Brünnich* found in Roma, a city 500 km westward of Brisbane (in mg L^{-1}): 0.42 for NH_3 and 0.21 for HNO_3.
[qq]This value is an average from 16 years with yearly about 150 analyzed rain samples.
[rr]At this site several investigators collected rainwater between 1911 and 1932 and determined nitrogen (see Eriksson 1952 for full bibliography): Knight (1911/12), Freeman (1924), Knox (1915), Artis (1916), Peck (1917), Trieschmann (1919), Shafer (1922), Woelk (1923), Ribble and Bowman (1926), and Williams and Beddow (1932).

Table 2.7: (continued).

[ss]Recalculated from period mean given as the total rainfall (29.32 inches per annum and acre) and total deposit (in pounds per annum and acre): 11.5 (NH_3-N) and 1.01 (HNO_3-N).
[tt]Askor (1922–1927), Blangsted, Spansbjerg and Hornum (1925–1926); for values see Eriksson (1952).
[uu]Today Szklarska Poręba in Poland; $n = 24$ (nitrate) and 29 (ammonia). *Drischel* also determined nitrite to be 0.10 mg L^{-1} ($n = 24$).
[vv]Arithmetic mean from 12 stations (USA, Australia, New Zealand, Europe), data from Drischel (1940).
[ww]Arithmetic mean from 7 stations (India, Hanoi, Brit. Guiana), data from Drischel (1940).
[xx]See p. 238.

Compared to Rothamsted, *Leather* found significantly smaller concentrations both of ammonia and nitric acid in rain whereas the ratio NH_3/NO_3 was similar and the total deposit also similar (due to much more rainfall). He also confirmed previous observations that an event with small rainfall contains significantly more nitrogen than at heavy rainfall and that within periods without rainfall nitrogen compounds are accumulated in the air; thus, he concluded that ammonia and nitrate are "washed out of the atmosphere by the falling rain and not obtained principally from that stratum in which the rain is formed" (in modern terms the latter means in-cloud scavenging).

Charles Frederick Juritz (1864–1945), an analytical chemist and pioneer working on the agricultural soils of Cape Colony (South Africa) studied 1909 nitrogen in Bloemfontain (Juritz 1910). Professor *Patrick Carmody* (1855–1927), Director of Agriculture and Government Analyst of Trinidad, analyzed rainwater (Carmody 1902).[434] *Miller* not only studied the nitrogen content at Rothamsted but also at other sites, like in Calcutta and Madras where he analyzed rainwater 1904–1905 (Miller 1905a), at Vifelstadir (Iceland) in 1911/1912, and at three sites at the Hebrides between 1907 and 1913 that showed low levels of ammonia at remote coastal sites (Miller 1913). *Carl Axel Hjalmar von Feilitzen* (1870–1928), Swedish agricultural chemist and son of *Carl Henrik Jobst von Feilitzen* (1840-–1901), who founded 1890 the agricultural experiment station Flahult[435] (Småland), analyzed together with *Ivar Henrik Lugner* (1871–?) rainwater (Feilitzen and Lugner 1910).

The Russian agronomist *Benzion Movscha-Morduchovich Welbel*[436] studied nitrogen deposition and drainage at the first Russian agricultural station in Ploty

434 *Carmody* also conducted a chemical and mineralogical examination of volcanic dust from the eruptions in St. Vincent and Martinique in May 102.

435 The station Flahult was part of the "*Svenska Mosskulturföreningen*" [Swedish Moor Culture Association] founded 1886 by *Carl Henrik Jobst von Feilitzen* (1840–1901), the father of *Carl Axel Hjalmar von Feilitzen*, who was also the director of the Jönköping chemical station.

436 *Вельбель, Бенцион Мовша-Мордухович* (unknown living dates; he belonged to the Jewish community in Odessa).

(1902–1905).[437] After this first Russian[438] rainwater investigation, Posnjakov (1904) studied between January and December 1903 rain and snow ($n = 83$) and dew and fog ($n = 43$) in Odessa[439] concerning the nitrogen and chloride content (5.3 mg L^{-1} Cl); the ammonia concentration in the rain he found to be dependent on wind direction, i. e., air mass characteristics (in mg L^{-1}): 0.7 (sea), 1.1 (country), and 1.4 (town); snow contains only about 50 % of the amount. Bykov and Karpov (1911) studied (it was a diploma work in the department of civil engineering) at the meteorological station of the Tomsk University the amount of NH_3, HNO_2, and HNO_3 in daily rainfall between February and May 1910 ($n = 38$), see Table 2.6 and 2.7; it was the first publication, expressing with the phrase "in dependence on meteorological factors" in the title the focus on atmospheric processes. They found as other researchers before that heavy rainfall contains less nitrogen, the first rainfall contains more than at the end of the rainfall and that rain in town is richer in ammonia and nitrous acid, but contains less nitric acid then suburb. They did not find a systematic relation between wind speed and the amount of nitrogen in rainfall but the highest amounts in rain from the south and southwest south directions.

Then, several decades later in the Soviet Union, an intensive investigation began on rain and cloud chemistry, no longer focused on nitrogen but predominantly on chloride and sulfate; the last paper that exclusively deals with nitrogen in rainwater published by Kudrin (1948).

Joost Hudig (1880–1967)[440] made for several years (1906–1912) observations on the composition of rainwater, collected by Mr. *H. Welt*, at the experimental station ["*Rijkslandbouwproefstation*"] Uithuizermeeden, a village in the northern part of the Province of Groningen (Hudig 1910, 1912, Hudig and Welt 1911). Historical interesting is the remark that the values for nitrate are not correct because of the improper analytical

---

437 Ploty [Плоть] is a village in the former Gouv. Kamenez-Podolsk, now Ukraine. Cited in literature as "*Landwirthschaftliche Versuchsstation des Fürsten Pawel Peter Trabezkoj*" [*Павел Петрович Трубецкой*] (1834–1914); сельскохозяйственная опытная станция Плоть Павла Петровича Трубецкого [Plotyanska agricultural research station of prince *Paul Trubetskoy*], founded in 1894. See also: Trubetzkoj, P. P. [Трубецкой] (1901) Химические исследование за 1899 г Плотянской агролаборато [Chemical investigation of the Ploty agrochemical laboratory for 1899]. *Ежемес. Метеор. Бюлл.* ГФО (Monthly Meteorological Bulletin].

438 To be correct, the first chemical analysis of hail (see p. 167) was carried out by Neljubin (1827) and that of rainwater (for hydrogen peroxide only) was carried by the German-Russian *Heinrich Struve* in 1869 in Tiflis (Chapter 4.5.2).

439 Магнитно-метеорологической обсерватории Новороссийского университета в Одессе [Magneto-meteorological Observatory of the University Odessa], founded in 1894; first head was the Russian meteorologist *Александр Викентьевич Клоссовский* (1846–1917) [*Alexander Vikentievich Klossowski*].

440 Chemical engineer and specialized in pedology who worked at the "*Rijkslandbouwproefstation*" [State Agricultural Laboratory] in Groningen, of which he also became the director. In 1929 he was appointed professor of "chemistry and the science of fertilization" at what is now the Agricultural University of Wageningen (Netherland), where he worked until his retirement in 1949.

method according to *Schloesing*: "*De bepaling van het nitraatstikstof in den aanvang verricht, kan niet op grote betrouwbaarheid bogen*" (cited after Buijsman 2011, p. 25).

First rainwater analysis in Australia after 1900 were made by *Johannes Christian Brünnich* (1861–1933), agricultural chemist in Brisbane[441] (Brünnich 1909).[442]

The Municipal Chemical Laboratory in Buenos Aires (no name of the reporter given) analyzed rainwater (1906–1907, likely in 1890–1908 according to the yearbooks) and gases (NH_3, organic nitrogen, O_3, CO_2) in the air (NH_3 and CO_2 until 1913 and likely longer). Besides ammonia and nitric acid (Table 2.7), this laboratory additional analyzed organic nitrogen and nitrous acid (in mg L^{-1}): 0.93 ± 0.29 (organic N) and 0.014 ± 0.023 (HNO_2); the HNO_2/HNO_3 ratio (from single values) amounts 0.017 ± 0.028. *Johannes Schröder* (1879–1942), a German chemist from Göttingen, professor for agriculture chemistry in Montevideo (1907–1920), first collected rainwater in Uruguay (Schröder 1915).

After several decades of rainwater studies on the nitrogen amount that deposits to soils, the "nitrogen period" ends, mostly by the findings from the Rothamsted station. Miller (1913, pp. 211–212) writes:

> We have evidence that some soils are more or less continuously giving into the air small portions of the ammonia produced from organic residues. Some soils may be expected to lose more ammonia than is returned in the rain, whilst others may gain in this manner more than they lose. It is evident from all these results that as a source of combined nitrogen, the rain is of no great importance to crops – an average of wheat or barley will contain eight-time nitrogen.

Finally, Russell (1919, p. 182) concluded:

> It is evident that the quantities of nitrogenous compounds present in rain water are too insignificant to exert any appreciable effect.

---

441 Born at Görz, Austria-Hungary (now Gorizia, Italy), went with his family to Stäfa (canton Zurich), Switzerland in 1874 and won entry to the Federal Polytechnic School [*Eidgenössische Technische Hochschule*] at Zurich. There he studied chemistry under *Viktor Meyer* and *Georg Lunge* and was briefly *Meyer*'s personal assistant; he migrated early in 1885 and became government agricultural chemist for the Queensland Department of Agriculture 1897–1931 (after T. J. Beckmann, Australian Dictionary of Biography, Vol. 7, 1979).

442 Eriksson (1952) cites *Brünnish* (wrong spelling), giving the values listed here in Table 2.7, and lists under references with the same paper title "Weedon, T. (1909) Fertilizing value of rainwater. *Annual Report of the Department of Agriculture and Stock*, Queensland (1908/09), pp. 59, 60, 77, 78" (after *Experimental Station Record* **20** (1910) 518). *Thornhill Weedon* (1849–1918), Queensland's first Government Statistician (1904–1914), reports on nitrogen-nitrate and -ammonia with total deposition of 3–4 lb. per acre and year (the paper is not available). Apart from that, Drischel (1940) lists under references *Brünnich* (1908/09) and *Weedon* (1908/09) not giving the papers titles, but cites values (in mg L^{-1} as N) 0.080 (NH_3) and 0.104 (nitrate) for "Cairus" (what is Cairns in North Queensland) concerns *Weedon*. Drischel (1940) not refers *Brünnich* in his text.

Only very few chemists of that period analyzed other compounds than nitrogenous. Reasons are adduced for supposing that ammonia arises from several sources. The sea, the soil, and city pollution may all contribute. Neither the sea nor city pollution seems to be able to account for all the phenomena: the soil is indicated as an important source by the fact that the ammonia content is high during periods of high biochemical activity in the soil and low during periods of low biochemical active (Russell and Richards 1919). *Eric Hannaford Richards* (1878–1939) and *Sir Edward John Russell* (1872–1965),[443] both researchers at Rothamsted (1913–1919 and 1907–1943, respectively) saw a routine analysis of rainwater more as a survey of air pollution than an evaluation of nutrient load (Russell and Richards 1919). However, the "nitrogen story" from the 1850s remained actual until the present. Galloway et al. (2004) concluded that global and regional nitrogen budgets were increasingly influenced by anthropogenic activities and estimated that global atmospheric nitrogen emissions (NO_x and NH_3) increased from 23 Tg N yr^{-1} in 1860 to 93 Tg N yr^{-1} in the early 1990s to 189 Tg N yr^{-1} in 2050.

2.3.3 Rainwater analysis for understanding cycling salts (1877–1956)

Chloride was regarded in relation to the salinization of agricultural soils namely in cases of limited run-off. With the insight that condensation nuclei in air are essential for cloud and subsequent rain the formation, and the idea that sea salt (namely NaCl) could provide such nuclei, rainwater and later fog and clouds were investigated concerned chloride until about 1950. Chloride was ubiquitously recognized from sea salt in air and rain;[444] several investigators analyzed chlorine apart from nitrogen (for example *Pierre, Barral, Filhol, Bobierre, Miller*) and already identified its source from the sea. Barral (1852) was the first who found NaCl in rainwater collected at two different heights at the Paris Observatory at 50 m larger (14.09 mg L^{-1}) than at 7 m (13.80 mg L^{-1}) above ground. Smith (1872) first identified Cl found in rainwater with industrial emission as HCl from potash industry and coal combustion. Since the late nineteenth century, chloride was measured by titration with silver nitrate using potassium chromate as an indicator (*Mohr*'s method) and sulfate through the formation of barium sulfate turbidity (Collins and Williams 1933).

The idea that the chloride content of continental water can be explained by airborn sea salt is often credited to *František Pošepný* (1836–1895), a world-renowned

---

443 British chemist who was appointed at the Rothamsted station in 1907 where he became its director (1913–1943), succeeding *Alfred Daniel Hall*.

444 Gautier (1899) determined the amount of sea salt in the air at the lighthouse Roche-Douvres, 19 miles from the coast at the Channel, in October 1898 to be 22 mg m^{-3} as NaCl. Chairy (1884) found at Algiers 17–53 mg m^{-3} NaCl in air.

Czech scientist, working in geology and related fields, who presented in 1877 his hypothesis on the origin of continental chloride (Pošepný 1877); see also p. 57 with another citation on salt cycling:

> Die Chlorverbindungen stammen zwar aus dem Meere, doch hat sich an ihrem Transport die Atmosphäre betheiligt. Durch den Wellenschlag in die kleinsten Theilchen zerschlagenes Meerwasser wird bei der Verdampfung in kleinen Mengen mit fortgerissen, gelangt bei dem Niederschlag dieser Dämpfe auf das Festland und wurde hier in sämmtlichen Quellen, Flüssen und Seen, wenn danach gesucht wurde, auch aufgefunden. Der Chlornatriumgehalt wurde ferner auch in dem atmosphärischen Niederschlage selbst nach gewiesen, die vollständigste Untersuchungsreihe liegt über das im Jahre 1863 in Nancy gefallene Regenwasser vor und ergiebt den ansehnlichen Gehalt von 14 Gramm in einem Cubikmeter. Die Salzlagerstätten repräsentiren gewissermaassen meteorologische Daten über die Beschaffenheit des Klimas früherer Formationsalter. [Although chlorine compounds come from the sea, the transport through the atmosphere is included. Through the action of waves sea water in its smallest particles gets carried away and arrives by precipitation on the continent, and will be detected, if you are looking for it, in all wells, rivers and lakes. The content of sodium chloride was detected in atmospheric precipitations; the most complete investigation conducted in 1863 in Nancy gives the considerable amount of 14 gram per cubic meter. The salt deposits represent virtually the meteorological data on the nature of the climate in earlier formation ages].

However, the German naturalist *Heinrich Friedrich Link* (1767–1851) already stated (but not so clear) a few decades earlier (Link 1826, p. 307) that common salt, found in precipitation in a large distance from the sea, is derived from "evaporation" of seawater:[445]

> Da sich überall Kochsalz im Regenwasser, im Schneewasser, sogar in bedeutenden Entfernungen vom Meere findet, so ist es nicht schwer, den Ursprung desselben abzuleiten [. . .] Es ist also höchst wahrscheinlich, daß durch die Verdampfung aus dem Meere das Kochsalz empor gehoben wird, wodurch zugleich wahrscheinlich würde, daß die Wasserdämpfe aus dem Meer sich weit über das feste Land verbreiten [As everywhere common salt is found in rainwater, in snow, even in far distances from the sea, so it is not hard to derive the origin of it [. . .] It is very likely that through evaporation from the sea common salt is lifted up, whereby it would be likely that the vapors from the sea diffuse far on the continent].

Nevertheless, Lampadius (1820) was the first who expressed the origin of chlorine in rainwater from the sea (see p. 160). The British physician *Meredith Gairdner* (1809–1837) included in his book on mineral and thermal springs a short chapter on "Atmospheric Water: Rain and Dew, Ice and Snow" (Gairdner 1832, p. 115; without reference, only mentioned *Zimmermann*, *Witting*, and *Brandes* who never stated some on the origin of chloride, whereas the effect of electric discharges in air is known since *Priestley*'s and *Cavendish*'s experiments):

---

445 In a footnote on p. 307 *Link* remarks that he collected at Breslau snow at sites far away from buildings and found always common salt in it.

> The idea was hazarded, that muriatic acid [HCl] might derive its origin from the volatilization of a portion of the muriated contained in the sea water; and that nitric acid might be formed by electric processes and decomposition going on in the higher regions of the atmosphere.

In his investigation of rain during a severe storm on 5 December 1823 (see p. 162) *John Dalton* writes (Dalton 1824):

> It is well known that the spray of the sea during a storm, is often carried inland by the wind to a considerable distance. (Dalton 1824, p. 324)
> We may now speculate upon the causes [...] it must be brought by the violent impetuosity of the wind acting upon the spray from the tumultuous waves of the ocean. (Dalton 1824, p. 335)

In an appendix to the first article *Dalton* (Dalton 1824, p. 371–372) first expresses that not only seawater is carried from the ocean but it evaporates and leaves salty particles in the air that are moved to land and deposited by rain:

> Now if the spray from the foamy waves be liable to be wafted away to considerable distances by the breeze, even in the driest state of the atmosphere there must particles of salt floating about it; for, the *water* resolving into *steam* will desert the salt; and this being found in infinitely fine particles, will be so much be carried even farther then the sea-water, in consequence of the particles being of superior tenuity.

Dalton (Dalton 1824, p. 366) cited *Leuwenhoeck* at Delft in Holland, about 10 miles from the sea who studied rain on 8 December 1703:[446]

> Supposing this might be the sea-water which the storm had not only dashed against our windows, but spread also over the whole country, I viewed the particles with my microscope, and found they had the figure of our common salt, but very small, because the water was little from whence those small particles proceeded; and where the glass, there were indeed a great number of salt particles, but so exceedingly fine that they almost escaped the sight through a very good microscope.

Tissandier communicated in his book "*Les poussières de l'air*" (1877) that *Dominique François Jean Arago* (1786–1853), French physicist and astronomer, described the mechanism of sea spray formation that is transported by the wind to the land ("*Arago a décrit le mécanisme de cette action en parlant des poussières d'eau que la brise soulève sur les rivages océaniques*") Boussingault (1854, p. 151) cited *Arago* who termed it "*poussières de l'Océan*" [sea dust].[447]

446 *Dalton* (1824, p. 364) also cites *John Fuller* on a storm on 27th Nov. 1703 "[...] the sea water was blown thus far, or that during the tempest the rain was salt": A strange effect of the late great storm in Sussex. In a letter dated 6 December 1703. In *Philosophical Transactions* (1704) Vol. 5, p. 19. *John Fuller* (1680–1745), British landowner in Sussex.

447 [...] *ces molécules liquides qu'Arago considérait comme les poussières de l'Océan*. In the German version (Boussingault 1856c, p. 263) it is named "*Meeresstaub*". Unfortunately, neither *Tissandier* nor *Boussingault* give a reference; and I was unable to find it in several books and papers by *Arago*.

Between about 1920 and 1940, atmospheric sea salt (chloride and iodine) was studied concerning balneology, namely in Germany and the Soviet Union. In the period after 1950, chloride became again in the interest of geochemical concern: it is the ion that dominates most discussion of "cyclic salts". Unquestionably, there is a chloride cycle in which chloride from the sea is incorporated in masses of moving air than in precipitation then runoff and groundwater, and finally in water discharged from the continents back to the sea (Feth 1981). Brierly (1970) cites 157 articles related to rainwater, 48 articles concern fog and cloud water, and 28 on salts as nuclei and aerosols (extended by Brierly 1979).

Between 1870 and 1900, chlorides were determined on the basis of six-month rainwater collection (October-March, April-September), in 1870–1879, by Sir *Arthur Herbert Church* (1834–1915), 1879–1881 by *Edward William Prévot* (1851–1920) and since 1881 by *Eduard Kinch* (1848–1920), all professors for agricultural chemistry at the Royal Agricultural College in Cirencester (founded in 1845) (Kinch 1887). *Kinch* notes that the amount of chloride increase with winds and storm from the Bristol Channel; the value is similar to that found by *Miller* at Rothamsted and by *Smith* at different British locations: 3.2 (1.15–10.4) mg L^{-1}.

In a letter to the journal "*Nature*", headed "Salt Rain and Dew", Draper (1883), citing the "School Geography" by *James Clyde* (Edinburgh, Oliver & Boyd, 1870) and the paragraph headed "Russian Lakes", is given the following remarkable statement: "In the south-east region, not only lakes but the very rain and dew likewise are salt, a phenomenon common to all the shores of the Caspian and sea of Aral". Gillman (1883) sent a reply to *Draper's* note, after having a correspondence with his father-in-law, *Alexander Petzholdt* (1810–1889), a German professor for agronomy at the Dorpat University (1846–1872), who made a special study of South Russia and answered "that it is a fact long known to chemists that the aqueous vapor in the atmosphere due to the evaporation of sea and salt-lake waters invariantly contains chloride of sodium, which is precipitated to the ground by rain and dew". Chairy (1884)[448] studied rainwater collected at the meteorological observatory in Algiers and found 17–53 mg L^{-1} NaCl (and 1–2 mg m^{-3}) in air.

Paulinus Jorissen (1869–1959), an agricultural chemist at Leiden, collected in 1905–1906 rainwater at the monitoring station of the meteorological institute in Den Helder. He only analyzed chloride and explained its large amount by proximity to the Sea (Jorissen 1906); Buisjman (2010, p. 9) remarks that *Jorissen* did not explain the reason of this study.

There are several Russian (and accordingly Soviet) studies on chloride and sulfate in rainwater, namely concerning the sea; see Maksimovich (1955), who gives a survey.

448 "*Professeur de physique* (2e *chaire*) *au lycée d'Alger*" (given name and living dates unknown). He also found iron in rainwater to be in the range 0.7–7.8 mg L^{-1} and acetic acid 0.14–0.29 mg L^{-1} but no iodine and hydrogen peroxide.

Welbel (1903, 1903–06, 1906), as already mentioned concerned the nitrogen content (Table 2.7) also determined sulfate and chloride in rainwater at Ploty. Posnjakov (1904) estimated Cl in rainwater at the Black Sea coast in Odessa to be on average 12.4 mg L^{-1} and in two fog events in November and December 1904, having 443 and 230 mg L^{-1}, respectively. Pirovarov (1906) studied the sulfur and chlorine deposit by rainwater in Odessa and found 1.53 mg L^{-1} as SO_3 and 4.8 mg L^{-1} Cl (recalculated from values given for rainfall and deposition; 17 pounds per acre and year by Clarke 1911). *Kossovich* proposed the chemical analysis of rainwater at seven stations in Russia in 1909–1910, whereby the analysis was carried out by *Jan Vityn* (1885–1951)[449] in the petrological laboratory of the Forest Institute in Petersburg [Санкт-Петербургский лесной институт], founded in 1877; from the results Kossovich (1913) concluded on the atmospheric transport of salts from seas to steppe soils. Vityn (1911) investigated, in 1909–1910, the chloride and sulfate content in rain in Petrograd and at the Schotilovski Station in the Tula Rayon and determined (in mg L^{-1}) 4.2 (1.5–19.3) and 15.5 (2.1–39.5) in Petrograd for chloride and sulfate (n = 18) and 4.6 (0.4–20.1) and 4.9 (1.2–9.3) for the Tula site for chloride and sulfate (n = 12); the industrial city influence on sulfate content is clearly seen.

Burkser and Fedorova (1949) studied rainwater chemistry on the coast of the Black Sea and Sea of Azov in 1947 and found (n = 21) that the main ionic composition is given by Ca-HCO_3-SO_4; concentration ranges of (in mg L^{-1}) 12–63 (HCO_3), 2–15 (SO_4), 3–17 (Ca), 0.5–4 (Na) and 1–4 (Cl as well K). The Na/Cl ratio amount 0.71 ± 0.50 (n = 20; calculated from original single values by DM). Blinov (1950) studied, in 1947–1948, rain and atmospheric aerosol at different distances from the Caspian Sea and found 14–56 mg L^{-1} sulfate (n = 5) and 23–68 mg L^{-1} Cl (n = 16) and a remarkable constant ratio sulfate to chloride being 0.607 ± 0.007 (n = 5, calculated from single original values by DM).

Peter Samsonovich Kossovich [Пётр Самсонович Коссович] (1862–1915), the most important agronomist and soil scientist in Russia at the end of the nineteenth century, who founded in 1900 the Russian "*Journal of Experimental Agronomy*", published in 1913 a treatise on the cycle of sulfur and chlorine in a similar approach as 46 years later by Eriksson (1959, 1960); *Eriksson* does not cite *Kossovich*. Kossovich (1913) spends chapter two on the chemical composition of rain collected at eight sites of the European part of Russia between 1909 and 1911 (in the area of Petersburg, Tula, Smolensk and Samara) and discussed the data in comparison with the results from other countries, summarized by Miller (1905a). He made the following conclusions:

– The amount of Cl in precipitation varies considerable between 0.42 and 58.1 mg L^{-1} in individual rainfalls but as annual mean only between 1.46 and 9.72 mg L^{-1}. Rain in inner parts of Russia contains Cl between 2 and 3 mg L^{-1}.

449 *Jānis Vītiņš* (also written *Janis Vitins* or *Wityn*) and *Ян Яковлевич Витынь*, Lithunian-Russian soil scientist.

- The chlorine amount in precipitation enlarges with approaching the sea but also increases in areas of with soils rich of salts.
- The amount of chlorine is larger at rainfall of low intensity and vice versa.
- The amount of sulfate in rainwater is between 0.28 and 14.17 mg L^{-1} (as SO_3); in rain close to large towns, railroads and industries the sulfate content is larger than 10 and at rural sites between 2 and 8 mg L^{-1}.

Kossovich furthermore discusses the amount of chloride and sulfate in rivers, groundwater, and soils concerning precipitation and evaporation, depending on sites and interannual variations, and concludes that the permanent input of sulfur from the atmosphere is the precondition for plant growth. Finally, he concluded from budgets including the amount of sulfur which is released by harvesting and leaching that with time a deficit of sulfur gains. At sites close to towns, however, he found sulfur deposition from the atmosphere to be sufficient. *Kossovich* describes the cycle of chlorine mainly as a mechanical transfer between sea, land, and atmosphere, whereas the sulfur cycle he has seen more complicated because of the permanent transformation between organic and inorganic matter including processes of oxidation and reduction with participation of microorganisms.

Several authors found a strong coastal gradient in the chloride concentration (it was already around the middle of the nineteenth century known that Cl concentrations are much higher at coastal sites compared to sites in distance). Miller (1913) determined 759.6 mg L^{-1} at 50 m from the coast of Barrahead (Hebrides). Bordas and Desfemmes (1928) studied the amount of chloride in rainwater at several distances from the French Atlantic coast (in mg L^{-1}): 95 (0 km), 33 (10 km), and 5 (50 km). Farcy (1931)[450] likely was the first to characterize quantitative the gradient found in the Bretagne (in mg L^{-1}) in rain in distance from the coast: 77.08 (30 m), 19.49 (2 km), 15.41 (25 km), and 13.48 (60 km). The gradient is principally not different from the chloride amount found by Leeflang (1938) in Leiden (Netherland). Collins and Williams (1933) concluded from several rainwater analysis in the USA the following decreases in chloride concentration at distance from the coast (in mg L^{-1}): 6 near the coast, which decreases to 1 between 50 and 100 miles inland and further to 0.2 beyond about 300 miles; the sulfate concentration at rural sites amounts 1–2 and in towns 6–10 mg L^{-1}. Teakle (1937) reported data for NaC1 in rainfall at Merredin and Salmon Gums Research Station (located in the south of Western Australia) between 1933 and 1936 in the range of 18–36 kg ha^{-1} and concluded that cyclic accumulation is an adequate explanation for the present levels of soil salt.

Cauer (1938), who was interested in an establishment of a chemical bioclimatology, summarized systematically the amount of chloride in rainwater with increasing

450 French chemist who was secretary of "*Société des Experts Chimistes de France*" in the early 1920s (nothing known on given name and living dates).

distance from sea coast (in mg L^{-1}): 70–100 (30–100 m), 15–30 (1–2 km), 6–13 (5–10 km), 4–9 (50 km), 3–5 (100 km), 1.2 (500 km), 0.5–1.5 (1000 km), and <0.3 (2000 km); rainwater collected in large towns and industrial areas contains 4–9 mg L^{-1} chloride whereas sulfate amounts (in mg L^{-1}) 1–4 at rural sites, 4–10 in large towns and 20–50 in industrial areas. Drischel (1940) found in 1937–1938 at Bad Reinerz (n = 33) 1.3 mg L^{-1} SO_4-S and (n = 95) 0.96 mg L^{-1} Cl and at Oberschreiberhau (n = 47) 1.0 mg L^{-1} SO_4-S and (n = 52) 1.3 mg L^{-1} Cl (mean pH 4.1–4.5).

Menzl (1948)[451] studied the chloride content in rain at Mt. Donnersberg (today Milešovka in Bohemia, Czech Republic, 836 m) and found its significant increase when the air was passing before a smelting works.

2.3.4 Rainwater analysis for understanding air pollution

> Für die nähere eingehende Analyse der Luft hat die Analyse atmosphärischer Niederschläge hohe Bedeutung. Bedenken wir z. B., welche großen Mengen von sogenannter normaler Luft wir in Arbeit nehmen müssen, um darin Wasserstoffperoxyd, salpetrige Säure, Salpetersäure and Ammoniak, sowie organische Stoffe nachweisen zu können, während es uns möglich wird, mit kleinen Mengen Regenwasser oder Schnee, im Nothfall mit 100 Cubikcentimetern derselben, alle diese Stoffe nachzuweisen, ja sogar Salpetersäure quantitativ zu bestimmen [For a more in-depth analysis of the air, the analysis of atmospheric precipitations has a high importance. For an example, let us consider which large amounts of so-termed common air we have to take to detect therein hydrogen peroxide, nitrous acid, nitric acid and ammonia as well as organic substances, whereas it is possible for us, with a small amount of rainwater or snow, in case of need with 100 cubic centimeters of it, to detect all of these substances, yes, even quantitatively nitric acid]. Goppelsroeder (1871, pp. 140–141).

Sulfur as "symbol" for air pollution and acidity was first recognized by *Robert Angus Smith* (Chapter 2.3.4.1). Deposits of sulfur (and calcium, magnesium, potassium) have also been studied in the first half of the twentieth century as essential plant nutrients.

There is one exceptional rainwater study under the aspect of smoke pollution parallel to the studies of *Smith* conducted by *Julius von Schroeder* (1843–1895)[452] in Tharandt (Saxony), who asked (Schroeder 1873) whether the influence of smoke from hard-coal fired locomotives influences the chemical composition of rainwater and whether the mineral deposition is sufficient as nutrition for forest soils. 1870–1871 he collected rainwater in Tharandt at the forest institute, situated in a small valley where the railroad goes, and at Grillenburg, a hillside location surrounded by forest, 10 km away from Tharandt. Tharandt is more surrounded by fields, which explains the large

451 *Oswald Adam Menzl* (1903–?) was a German meteorologist at the weather station in Kinding (Bavaria).

452 *Theodor Julius Reinhold von Schroeder* was born in Dorpat (today Tartu in Estonia) where he studied chemistry (1861–66), and continued in Heidelberg; since about 1870 in Tharandt, he became 1883 professor for agricultural chemistry; known for his investigations of smoke damages.

quantity of mineral matter (soils dust blown up from the farmlands and accumulated in the valley) in rain; the larger quantities of soot, sulfuric acid, and iron *Schroeder* explains by the railroad smoke: in mg L^{-1} (the column on the right shows values found more than 100 years later at the same institute by *Herbert Lux*, see Möller and Lux (1992, p. 247), 10-fold higher for chloride, sulfate and calcium due to coal-fired power plants in Bohemia):

	Tharandt	Grillenburg	Tharandt 1985–1989
combustible matter (soot and organic)	5.8	4.2	–
total minerals	13.3	2.9	–
potassium	0.53	0.54	0.5
sodium	0.74	0.85	0.8
chloride	0.42	0.17	3.3
calcium	0.61	0.77	5.6
sulfate	1.84	0.86	12.2
phosphate	0.23	0.10	–
iron(III)oxide	1.79	0.37	0.7

2.3.4.1 *Robert Angus Smith*: air and rain (1845–1879)

Before 1850 only qualitative rainwater studies were carried out, where those of *Zimmermann*, *Brandes* and *Bohlig* were from mere analytical interest (between 1821 and 1834). Many authors of the nineteenth century (and the beginning of twentieth) were only interested in the amount of nitrogen (Table 2.7); hence, it is worth noting (cf. Table 2.1 on p. 146) here that Pierre (1851) was the first who determined also chloride and sulfate,[453] and Na, K, Mg, and Ca (all the so-termed major ions in rainwater). Barral (1852), Meyrac (1852), Martin (1853) and Filhol (1855) determined apart from nitrogen also chloride (as sea salt), and Barral (1852) Ca and Mg.

A more systematic examination of chloride and sulfate in rainwater began in 1869–1870 with *Smith* (Fig. 2.15) at different locations in England and Scotland and later at Rothamsted (Miller 1905a,b). *Robert Angus Smith* (1817–1884), born at Pollokshaws (suburb Glasgow), was educated at the University of Glasgow in preparation for ministry in the Church of Scotland but left before graduating. He worked as a personal tutor and, accompanying his family to Giessen in 1839, he stayed on in Germany to study chemistry under *Justus von Liebig*, earning a Ph. D. in 1841. On returning to England the same year, he was attracted to Manchester to join the chemical laboratory of *Lyon Playfair* (1818–1898) at the Royal Manchester Institution. Here, he became

453 Sulfate in rainwater was first detected by Brandes (1826a) and semiquantitative estimated by Bertels (1842).

Figure 2.15: Photography of *Robert Angus Smith* (1817–1884). Reproduced courtesy of the library of the Royal Society of Chemistry.

involved in some of the environmental issues of the world's first industrial city and in many others issues concerned sewage, water pollution, and analytics (Smith 1845, 1852, 1859b, 1872, 1877, 1879). After *Playfair*'s departure in 1845, *Smith* worked as an independent analytical chemist. When the Alkali Inspectorate was established by the Alkali Act in 1863, *Smith* became inspector-general of alkali-works and he held the post until his death.[454]

> This appointment led him to continue those laborious researches into the composition of the air which he had already carried on for seventeen years; and the publication, in 1872, of that

454 This Act, which was an early example of environmental regulation, set limits on hydrochloric acid gas emissions from alkali plants, and it established a system of inspectors to monitor and enforce the legislation.

> most valuable work, *Air and Rain*, is the outcome of these researches [. . .] his valuable papers on chemical and physical subjects, not immediately connected with sanitation, indicate the vigour and wide range of his working power. A careful observer and close reasoner, he was, above all things, a practical man [. . .].[455]

As a scientist, *Smith* was far ahead of his time both in his research and in the ideas behind it. *Smith* began to collect and analyze rainwater in Manchester in 1851 (Smith 1852, p. 216) and concluded this exceptional phrase:

> We may therefore find easily three kinds of air – that with carbonate of ammonia in the field at a distance; that with sulphate of ammonia in the suburbans, and that with sulphuric acid, or acid sulphate, in the town (see also Smith 1872, p. 230).

First rainwater analysis *Smith* carried out in Manchester on only three days in June 1851 for hydrochloric acid and sulfuric acid[456] (values in mg L^{-1}): 86 (57–128) for HCl and 145 (128–154) for H_2SO_4. These values are considerably higher than those from 1870 determined in Manchester (Table 2.8) in a range (as monthly means in mg L^{-1}): 3–10 (HCl) and 21–105 (H_2SO_4) giving doubt on the trueness of these early estimates. Smith (1852, p. 213) writes that the presence of free sulfuric acid in the air "explains the fading of colors in prints and dyed goods, the rusting of metals, and the rotting of blinds." Between 1869 and 1870 *Smith* carried out a large amount of rainwater analysis collected at many sites in England and Scotland from the coast, rural country, and town (Tables 2.8 and 2.9). Figure 2.16 shows the residual crystals under the microscope after evaporation of rainwater; already many years before there had been a method for detection of the chemical composition (likely no "modern" analytical chemist is able to identify a compound from its crystalline shape).

> In the rain of Manchester there is confusion of crystals; although sulphates are prominent. I do not see crystals of common salt. It would appear that decomposition takes place [. . .] when sulphuric acid passes into the atmosphere in towns it probably seizes part of the sodium of the common salt in the same way, and so liberates hydrochloric acid – an alkali-work continually in action (Smith 1872, p. 382)

According to the subtitle of *Smith* monograph (1872) "The Beginning of a Chemical Climatology", he came to many important conclusions, to constitute him as the *first atmospheric chemist*. *Smith* also studied the impact of air pollution (namely SO_2) on plants and humans; the last one (Fig. 2.17) not even according to today's regulations (further reading: Eyler (1980)). From his famous book (Smith 1872) I give some citations in the following.

455 The British Medical Journal (1884) p. 976.

456 Furthermore, he estimated organic matter between 100 and 900 mg L^{-1}.

Table 2.8: Rainwater analysis by *Robert Angus Smith* (values in mg L^{-1}); means from several sites and dates. The rain has "been collected by several gentlemen in various parts of the country"; a description on how to observe quality whilst rainwater collection Smith (1872, pp. 248–250) provided.

substance	Manchester 1869/1870	Darmstadt 1869	London 1869/1870	Glasgow 1870	Liverpool 1870	sea-coast England	inland England
ammonia	6.47	1.91	10.57	9.10	5.38	1.90	1.07
chloride[a]	5.83	0.97	10.46	8.97	10.16	56.15	3.99
sulfate[b]	44.82	29.17	57.67	70.19	20.49	5.88	5.52
nitrate	1.03	2.89	8.12	2.436	0.58	0.37	0.749
org. N[c]	0.25	0.12	0.33	0.30	0.16	0.40	0.375
acidity[d]	12.02			15.13		none	none

[a] Hydrochloric acid (HCl)
[b] "Sulphuric acid (sulphates)" given as "sulphuric anhydride" (SO_3)
[c] "Albuminoid ammonia" (organic bound nitrogen)
[d] "Free acidity" given as SO_3 (note *Smith* also gave "total acidity")

Table 2.9: Rainwater analysis by *Robert Angus Smith* in Scotland in 1869–1870 (values in mg L^{-1}); means from several sites; "towns without Glasgow" (consider also footnotes in Table 2.8).

substance	east-coast	west-coast	inland	towns
ammonia	0.992	0.484	0.305	1.164
chloride	12.91	12.28	3.37	5.86
sulfate	7.66	3.61	2.06	16.50
nitrate	0.476	0.305	0.372	1.164
organic N	0.106	0.105	0.039	0.212
free acidity	2.44	0.14	0.314	3.16

On acidity and sulfate:

> All the rain was found to contain sulphuric acid in proportion as it approached the town, and with the increase of acid the increase also of organic matter (Smith 1872, p. 225).
> I do not mean to say that all the rain is acid; it is often found with so much ammonia in it as to overcome the acidity; but in general, I think, the acid prevails in the town (Smith 1872, p. 227).
> When the sulphuric acid increases more rapidly than the ammonia the rain becomes acid (Smith 1872, p. 245).
> Acidity is caused almost entirely by sulphuric acid, [...] (Smith 1872, p. 276).
> As rule rain is not acid far from towns. If it is acid, artificial circumstances must be suspected (Smith 1872, p. 266).
> The word *sulphuric* is used instead of *sulphureous* acid, although it is the second which is most perceived in towns. It does, however, pass rapidly into sulphuric, and the two are found mixed in the rain (Smith 1872, p. 273).
> I was led to attribute this effect [...] on stones, bricks, mortar etc.] to the slow, but constant, action of acid rain (Smith 1872, p. 444).

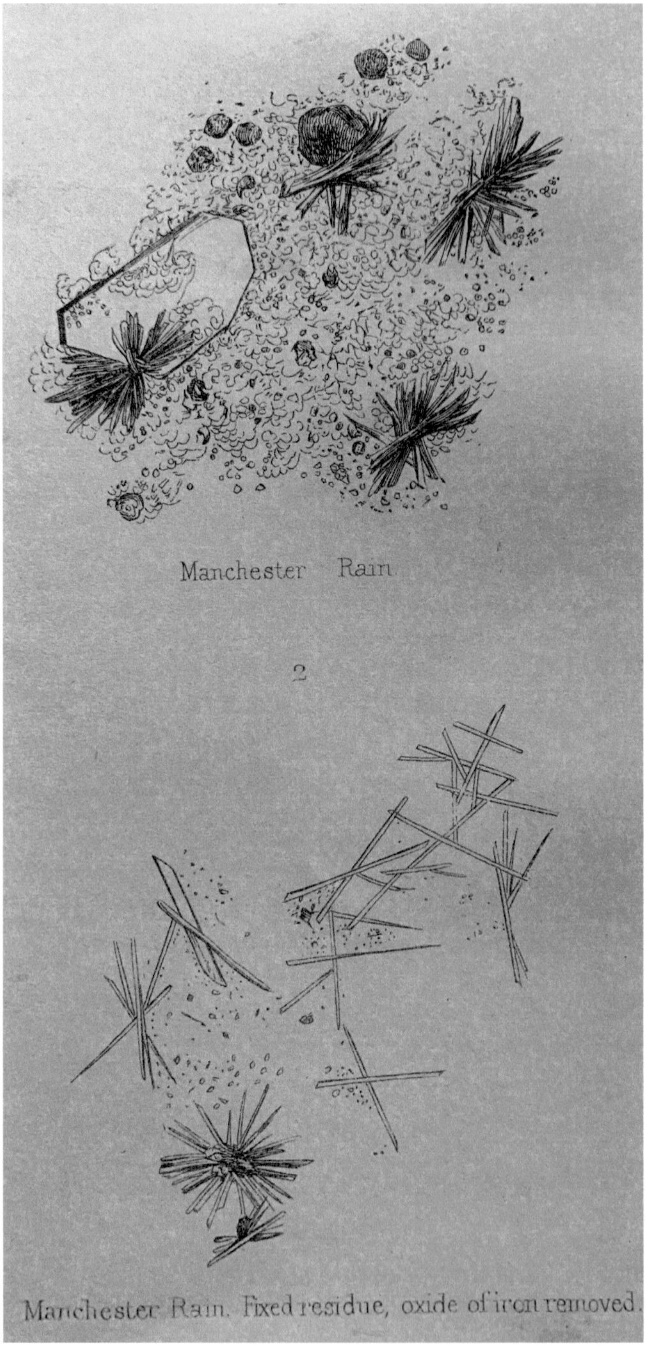

Figure 2.16: Residual from evaporated rainwater; Smith (1872, p. 382).

Figure 2.17: A lead chamber constructed by *Robert Angus Smith* as part of his experimental research into air quality (from Smith 1872, p. 131).

On ammonia:

> Ammonical salts increase in the rain as towns increase (Smith 1872, p. 246).
> The ammonia is one of the sewages of the air; it is a result of decomposition (Smith 1872, p. 277).

On hydrochloric acid and chloride:

> The chlorides become, therefore, a test both of sea-water and of coal-burning, especially as connected with certain manufactures. When the sea is the source, the rain will not be acid, but when the source is coal acidity will arise from the sulphuric acids (Smith 1872, p. 263 f).
> [...] the amount of chlorides depends on two causes mainly, these are (1st) the distance from the sea and (2nd) the distance from manufactories (Smith 1872, p. 272).

On organic matter:

> The organic matter found in the rain seems to be in perfect solution, [...] (Smith 1872, p. 226).

2.3.4.2 The "sulfur and nitrogen period" (1869–1947)

Soon after *Smith*'s finding of polluted rainwater, the long-range transport of smoke was observed by *Waldemar Christofer Brøgger* (1851–1940), Norwegian geologist, as "*smudsig snefeld*" [dirty snowfall] in Norway and attributed to either a large town or an industrial district in Great Britain (Brøgger 1881). *Karl Freytag* (1831–1908)[457] found in rain collected in 1869 near a zinc work (at Borbeck near Essen, Germany) 85 mg L^{-1} sulfuric acid, 46 times more than found in rainwater in Tharandt (Schroeder 1873), a site where forest damages by smoke were described by Stöckhardt (1871). The American chemist *Charles Mabery* (1850–1927) studied in 1894 the air pollution in Cleveland (USA) and found in freshly fallen snow from five different locations (in mg L^{-1}) 42–111 soot, 7–21 sulfuric acid and 0.00–0.12 ammonia (Mabery 1895). Kossovich (1913) found in rain close to large towns, railroads, and industries in Russia the sulfate content is much more than far from towns. Gray (1910) studied in two one-year periods (1908–1909) the chemical composition of rainfall in Lincoln, New Zealand and found (in mg L^{-1}, the original spelling used): 23.4 (11–108) total dissolved solids, 2.37 volatile compounds, 5.58 chlorine, 5.14 sodium oxide, 2.86 sulfuric anhydride, 1.74 carbonic acid, 1.60 silica, 0.72 potash oxide, and 0.63 magnesia oxide (Table 2.7 for nitrogen in 1907–1909 and 1884–1888). At the same time, he found in rain at the Campbell Islands (in ppm) 172.5 total solids, containing 77.5 chloride and each 0.084 HNO_3-N and NH_3-N.

The English chemists *Charles Crowther* (1876–1964) and *Arthur Gough Ruston* (living dates unknown), Lecturer in Agricultural Economic at University Leeds, stated that

457 Cited after Schroeder (1873): *Mittheilungen der landwirtschaftlichen Akademie Poppelsdorf*, Vol. 2 (1869); *Freytag* was head of the "*Versuchsfeld der Akademie Poppelsdorf*" [trial field of the Academy Poppelsdorf] and from 1870 head of the trial field in Halle and became professor for animal science.

the acidity decreases as they move further from the town (Leeds) (Crowther and Ruston 1911). Crowther and Steuart (1913) collected rain between July 1911 and July 1912 at 14 sites at distance between 3 and 7 miles so arranged as to give a complete circle of three miles from the center of Leeds and additional stations at 5 and 7 miles in directions between east and northeast (prevalent westerly winds). They analyzed sulfur, chlorine, and total nitrogen and described the plant damages. The samples were collected with a copper funnel, 37.5 cm in diameter, and 291 samples were dealt with. Unfortunately, the copper funnel neutralized much of the free acid as shown with parallel sampling at Garforth station with a copper and a glass funnel; in the following the values are listed (in mg L^{-1}):

	mean values	minimum to maximum
Cl	0.5–10.9	(1.0–58.7)
SO_3	17.0–61.0	(3.7–128.5)
Total N	0.86–1.52	(0.31–4.92)
Total suspended matter	9.6–40.4	(0.4–228.2)

In Germany, a remarkable study was done by Küppers (1918), a teacher for chemistry at the mining school in Bochum, on the amount of sulfuric acid in snow and rain.[458] He first mentions that sulfurous acid cannot be found in snow and rain due to its fast oxidation to sulfuric acid. In freshly fallen snow he found 6–9 mg L^{-1} (SO_3) but at sites close to industrial areas 15–20 mg L^{-1}; after 11–12 days without new snowfall the amount of SO_3 increased 3–5 fold. Similar results *Küppers* found for rainwater: an increase by 1.2 mg SO_3 per day and 100 cm^3 areas. He correctly concluded (without knowing dry deposition) that the permanent smoke led to the increase of sulfate on exposed surfaces.[459] Thus, estimation of sulfate in freshly fallen snow was considered as the simplest method for an assessment of the degree of air pollution. Lehmann (1921, p. 653) determined in three days old snow at the botanical garden in Würzburg 7.9 ppm and in the city center 39.1 ppm SO_3. Guth (1919, p. 465) found in two days old snow in Saarbrücken 76 and 96 mg L^{-1} SO_3. The absorption of SO_2 (and/or sulfate deposition) onto snow cover were found also by Stoklasa (1923, p. 408). The SO_3 amount in fresh fallen snow on January 16, 1914, in Prague was determined to be 14.18 mg L^{-1} and increased with time:

458 He cites two former works by *Sendtner* (1887: *Bayer. Industrie- und Gewerbeblatt*) and *Lehmann* (1910: *Die Luft*, Vol. 2, Springer, Berlin, p. 398) who got similar results for Munich and Würzburg, respectively. *Stöckhardt* (1881: *Thar. Forstl. Jahrb.* **21**, 229) has allow to act sulfurous acid (SO_2) on snow and found that it is no longer detectable after 6 hours. *Rudolf Sendtner* (1853–1933) was a German food chemist in Munich.

459 Similar, absorption of ammonia on snow, likely originated from coal combustion in winter, was studied by Müller (1888); see p. 677.

24 hours later	24.19
48 hours later	34.80
72 hours later	50.94
96 hours later	58.49

Further exposure did not show a more systematic increase in the SO_3 content (between 60 and 70 mg L^{-1}); likely due to the limitation of SO_2 absorption with increasing acidity of the snow. *Stoklasa* and his coworker (Drs. *Nekola*, *Šebor*, *Zdobnický*, and *Horák*) collected, in 1910–1915, precipitation in Prague and determined sulfate and sulfite. This was likely the first analysis of sulfite in rainwater. Stoklasa (1923, p. 406–407) found the sulfite (and sulfate) concentration in winter generally to be higher than that in summer; the following values are recalculated as sulfur (in mg L^{-1}):[460]

Season	sulfite	total sulfur	% of sulfite
January–March	12.6	36.5	35
April–June	6.0	15.1	40
July–September	4.3	7.4	28
October–December	18.4	36.6	50

Stoklasa also collected rainwater at rural sites in winter 1914 and determined sulfate (total sulfur) within the range of 0.2–1.0 mg L^{-1} which shows the large SO_2 pollution in towns at that time (Chapter 4.7.2.1). To get an idea on the SO_2 level in the air of Prague – and its relation to the rainwater composition – *Stoklasa* and coworkers measured in 1912–1915 in winter 3–7 ppm and in summer about 2 ppm (Stoklasa 1923, p. 251); this is about 50 times the SO_2 concentration in air of Dresden in the early 1970s and more than 1000 times of the present SO_2 concentration found in Berlin. *Wilhelm Liesegang* (1894–1953),[461] a German chemist and mostly known for his dust sampling studies

460 The percentage of S(IV) in relation to total sulfur corresponds to own results (Möller 2003, p. 375) obtained at simultaneous rainwater collection in November 1997 at ground and the Frohnau tower at 323 m altitude (NW Berlin); we found the percentage in the height to be 22 % and at ground 35 %, clearly showing the importance of sub-cloud scavenging of SO_2 (total sulfur in rainwater 1.0 mg L^{-1} thus 35 time less than in winter in Prague some 80 years before); in cloud water at Mt. Brocken we found a significant less percentage of free sulfite to be on average only 4 %.

461 German chemist, begun his carrier 1924 in the field of air hygiene in the worldwide oldest environmental institute, the "*Preußische Landesanstalt für Wasser-, Boden- und Lufthygiene*" [Prussian State Institute for Water, Soil and Air Hygiene], founded 1901 in Berlin-Dahlem as "*Königliche Versuchs- und Prüfungsanstalt für Wasserversorgung und Abwasserbeseitigung*" [Royal Experimental Institute for Water Supply and Sewages Disposal]. *Liesegang* became 1932 appointed professor and 1935–1945 director of the department for air hygiene (see also Liesegang 1927, 1932 and 1935 for his views on chemistry and air hygiene). This institute became 1934 part of "*Reichsgesundheitsamt*" [Reich Health Office] with *Hans Lehmann* (1893–1938) as first head (another important air hygienist was *Arnold Heller*

Table 2.10: Sulfate in rainwater (in mg L^{-1} SO_3) and as monthly dust deposition (g m^{-2} SO_3) at different sited in Germany 1932–1934.

location	year	winter	summer	dust deposit
Essen	26.8	45.3	16.5	6.12
Dortmund	19.5	32.8	12.0	6.10
Berlin (city center)	20.8	35.0	9.7	3.98
Berlin-Dahlem	18.5	41.8	6.1	3.40
Müncheberg[a]	8.3	11.6	6.1	3.36

[a] 50 km east of Berlin

Table 2.11: Content of rainwater (in mg L^{-1}) as annual mean 1937/38 (Nehls 1939).

location	sulfate (SO_3)	chloride (Cl)	ammonia (NH_3)
Hürth	158.4	11.8	0.5
Essen	31.6	9.8	1.6
Dortmund	30.1	7.8	0.8
Berlin (city center)	23.7	6.7	0.7
Berlin-Dahlem	15.8	5.7	0.8
Müncheberg	11.6	4.3	0.3

collected rainwater (monthly samples) at different sites in Germany (Tables 2.10 and 2.11), and found in the snow (in mg L^{-1} SO_3), (Liesegang 1927/29, 1934/36):

Berlin-Dahlem, after 9 hours	3.66
Berlin-Dahlem, after 105 hours	4.20
Berlin-Dahlem, after 175 hours	5.66
Berlin, Potsdamer Platz, after 50 hours	10.83
Berlin, Stettiner Bahnhof, after 52 hours	10.01

Remarkable is the much higher sulfate value in winter in towns due to domestic coal heating; sulfate in Essen and Dortmund is higher in summer than in Berlin due to large industrial complexes; furthermore, *Liesegang* determined chloride (3.6–8.0 mg L^{-1}) and ammonium (0.5–1.0 mg L^{-1}). This network was in operation until 1939 (Nehls 1939);[462] a sampling station at Hürth (1200 m lee side of the power station, Fig. 2.18) was included in 1934, indicated the highest pollution and deposition in

(?–1981) and 1952 as "*Institute für Wasser-, Boden- und Lufthygiene*" ("WaBoLu") part of the "*Bundesgesundheitsamt*" [Federal Health Agency], and was 1994 finally integrated as a Department into the "*Umweltbundesamt*" [Federal Environmental Agency: UBA]. The chemist *Erwin Lahmann* (1925–2001) headed the department air hygiene (1956–1990).

462 *H. Nehls* (unknown given name and living dates) was technician in the department of *Liesegang* in Berlin.

Figure 2.18: Panorama of the chemical manufacturing complex ["*Chemiepark*"] Knapsack and power station Goldenberg (the largest power plant in Europe with 530 MW based on lignite) at Huerth near Cologne, Germany; about 1936 (source: InfraServ Knapsack – Archive).

Germany at that time; whereas between 1933 and 1939 the total deposition of dissolved and undissolved substances doubled at Hürth, at the other station no trend was observed.

Between 1910 and 1950, several studies were carried out in the USA[463] at about 60 sites only related to sulfur deposition. MacIntire and Young (1923) made the first detailed study on precipitation chemistry in the USA (10 sites between 1915 and 1921) and demonstrated the local influence of towns on rural precipitation, like Smith (1872) in Great Britain and Kossovich (1913) in Russia before, they also found that calcium and potassium in rain is twofold higher in a town (Knoxville) comparing to rainfall 2.4 km from the city center. Wilson (1926) found large differences in rainwater sulfur between summer (3.4 mg L^{-1}) and winter (15 mg L^{-1}). Eaton and Eaton (1926) found in rainwater (1921–1922) in Chicago 5–6 times more sulfate than determined at rural sites in the USA. It was found that a moderate rainfall of about 0.5 inches removes practically all sulfur from the atmosphere; if the rain continues to a total of 2 inches, the latter 1.5 inches serves only to dilute the concentrations of sulfur. Alway et al. (1937) note that in certain earlier studies the recording of excessively high values of sulfate may be

463 Hart and Peterson (1911), Stewart (1920), Erdman (1923), Johnson (1924), Ellett and Hill (1929), Fraps (1930), Volk et al. (1945), and Bertramson et al. (1950). *Robert* Stewart (1877–?) was the Associate Chief of soil analysis at the Illinois Agricultural Experiment Station; *Lewis Wilson Erdman* (unknown living dates) was a soil scientist at the Iowa State College, all other scientists are mentioned in the text Chapter 2.3.4.1 or in Table 2.12.

attributed to the fact that rainwater was collected in containers made from corrosive metal; such metals react with atmospheric sulfur (i. e., SO_2) and result in errors up to six-fold. Alway et al. (1937) found in Minneapolis (but argued that SO_2 was additional absorbed into the gauge) and at Copperhill (copper mining 1850–1987) sulfate also more than 10-fold in excess; they showed the effect of densely populated areas on the sulfur content in rainfall and also found that moist soils absorb sulfur dioxide from the atmosphere. Harper (1942)[464] presents an excellent review on American sulfur deposition studies (1918–1936) and from Stillwater (Oklahoma), a rural site, 1927–1942 (mean of 1.0–1.7 mg L^{-1} varying between the seasons). From all studies cited by Harper (1942) at 22 rural sites, a mean sulfur deposit of 35 (13–77) pounds per annum and acre follows.

There is an exceptional early study on the air chemistry of nitrogen acids that was no longer under the view of agricultural chemistry. *Valentine George Anderson* (1885–1969), a chemical analyst at the chemistry department of the Working Men's College at Melbourne, first stated (Anderson 1915, p. 101):

> Investigations of the amounts of oxidized nitrogen in rain-water have hitherto been planned and carried out almost entirely from the standpoint of the agricultural chemist, and principally with a view to finding the monthly and yearly amounts of combined nitrogen per acre, carried by the rain to growing crops and pasture-land.

Anderson kept in view other objectives of his research. First, he considered the relations between the amount of rainfall and the meteorological conditions, second, discussed the atmospheric formation of oxidized nitrogen and third studied the effect of city air (Melbourne center) concerning a remote site (Canterbury). It is the first rainwater investigation using modern methods, like glass funnels with a storage vessel, daily cleaning of the gauge and sample collection and as far as possible daily chemical analysis of nitrate and nitrite (describing the analytical procedure). Remarkable are *Anderson*'s conclusions that a) daily rainwater collection (against the practice of agricultural chemists in collecting only at the end of each month) has the advantage that little change can occur due to contamination by dust particles, bacterial action (nitrification and denitrification), and absorption of nitrogen compounds from the air, and b) daily variations in the amount of nitrogen may be directly compared with the daily change in weather and any parallelism noted. The investigation covers 16 months (1912–1914). Thus, he classified the samples according to air masses, in relation to the rainfall and season. He found the concentration of nitric acid to be highest in summer and lowest in winter and inversely that of nitrous acid (Table 2.6). What concerns the origin of the acids, *Anderson* believes that NO_2 (he termed it nitrogen peroxide), mainly produced in air from lightning, reacts with the rainwater under the well-known (very slow) reaction gaining equimolar HNO_3 and HNO_2. Because he did not find an influence of

464 *Horace J. Harper* (1896–1961) was a soil scientist at the Oklahoma State University.

local thunderstorm on the oxidized nitrogen, he assumes that there is a large-scale atmospheric formation and distribution of the precursor NO_2. Interesting is the discussion to the paper, where most respectable scientists like *Russell* from Rothamsted as well *Wyndham Dunstan* (1861–1949) from the Imperial Institute London and *William Napier Shaw* criticize the missing ammonia values. *David Wilson-Barker* (1858–1941), president of the Royal Meteorological Society, said to the *Anderson* paper that "the chemical processes of the atmosphere are not often the subject of so much attention" and emphasizes that "professional chemists as well as meteorologists might be well employed on similar investigations". Finally, *Miller* from Rothamsted argued that a seasonal cycle derived from only such short period should not be overestimated; from over 30 years of rainwater analysis at Rothamsted there were different cycles but at all, the amount of nitrate was the same in winter and in summer. Miller (1905b, p. 23) further states that

> [...] as regards ammonia the question will remain an agricultural one (point of view) as long as the relations of atmospheric ammonia and soil nitrogen are not completely understood. At the same time, it seems likely that the more frequent estimation of even one constituent of rainwater – chlorides, for instance – might be of value to meteorologists.

Unfortunately, *Anderson* had never continued his research on atmospheric nitrogen chemistry and rainwater.[465]

Several rainwater monitoring stations existed in the first half of the twentieth century from pollution point of view (using so-termed deposition gauges, e. g., Ashworth 1933). From 1920 to 1945, only a few studies on nitrogen deposition were carried out; more studies began on sulfate, chloride and progressively on other compounds detected in rainwater. The following authors are worth mentioning. In the USA: Wiesner (1914), Wilson (1921), and Finnell and Houghton (1931); *Houghton* we meet later again with first fog water studies (Chapter 2.4.3),[466] and in Canada:[467] the agricultural chemist *Frank Thomas Shutt* (1859–1940); Shutt (1914), Shutt and Dorrance (1917), Shutt and Hedley (1925). *Benjamin Dunbar Wilson* (?–1940) was an agricultural chemist at the Department of Agronomy of the Cornell University (1918–1940)

465 *Anderson* began practice on his own account in 1910 as associate to *David Avery* (1871–1956), and was his partner from 1915 in the firm known as Avery & Anderson, analysts. In 1913–15 the University of Melbourne granted him government scholarships totaling £250 to undertake research from his own laboratories into nitrogen in rain-water. The work was partly sponsored by the British Association for the Advancement of Science; it was reported to and by them and published, with high praise, in the *Journal of the Royal Meteorological Society* in 1915 (Radford, J. T. (1979) Australian Dictionary of Biography, Vol. 7).

466 *Henry Howard Finnell* (1894–1960) was an agronomist and erosion specialist who pioneered methods to combat soil erosion during the Dust Bowl that afflicted North America in the 1930s.

467 The Dominion Experimental Farm System was inaugurated in 1886 in Canada based on the models from Rothamsted, Möckern and Cirencester.

and published several papers on nitrogen and sulfur in rainwater; later he became a specialist in organic chemistry of soils (Wilson 1921, 1923, 1926). The chemist *Victor G. Heller* (1886–?), head of the Department of Chemistry, Oklahoma Agricultural Experiment Station of the Agricultural and Mechanical College (later State University Oklahoma), conducted in 1935–1936 at 8 stations rainwater sampling by the help of local observers and analyzed ammonium, nitrite, nitrate, chloride, and sulfate (Heller 1938); notable is his conclusion that "sulfates are surprisingly high [...] suggesting contamination from fumes of burning oil, gas or coal". Hansen (1926) in Denmark, Braadlie (1930) in Norway, and Scharrer and Schropp (1940), and Drischel (1940) in Germany must be named.

Drischel (1940)[468] published a remarkable survey on rainwater analysis and present own results on chloride, sulfate, ammonium, and nitrate (including pH measurements) in rainwater collected in 1937–1938 at Bad Reinerz (today Duszniki-Zdrój in Poland) and Oberschreiberhau (today Szklarska Poręba Górna in Poland), former Lower Silesia (today Dolny Śląsk). Worth citing his averaged values from previous studies (in mg L^{-1} as N) given by Drischel (1940) and a comparison with values from the 1980s (Table 2.18; being around 3 mg L^{-1}) as N for NH_3 (first column) and HNO_3 (second column) as well:

Midlatitudes: rural sites	0.4–0.5	0.2–0.3
Midlatitudes: towns	1–2	0.05–0.1
Midlatitudes (mean from 34 stations)	0.603	0.272
Tropical sites (mean from 14 stations)	0.274	0.410

Ottaviano Bottini (1905–1969), professor for chemistry at the University Napoli, studied rainwater composition around the Mont Vesuvius during major eruptions in 1934 and 1936 and found pH between 3 and 4 with a minimum of 2.8 due to free HCl (Bottini 1935, 1939). Between 1901 and 1936 several investigations of rainwater in Italy have been conducted: in Firence (1901) by Passerini and D'Achiardi (1901), in Pisa (1903–1914) by Di Vestea (1916), in Napoli (1927) by Majo (1927) and 1934–1936 by Bottini (1935, 1939), in Udine (1928–1931) by Comel (1933),[469] and in Rome (1929–1931) by Marimpietri and Simoncelli (1932) were carried out. The French biochemist *Gabriel Emile Bertrand* (1867–1962) studied the content of sulfur, magnesium, potassium, and calcium in rainwater and published 14 papers between 1930 and 1947 (Bertrand 1935, see also Eriksson 1952).

468 See p. 4 for biography.

469 *Alfonso Di Vestea* (184–1938) was an Italian physician; *Alvise Comel* (1902–1988) was an Italian agricultural chemist. *Ester Majo* (unknown living dates), Italian physicist; she was Director of "*Istituto di Fisica Terrestre Della R. Università di Napoli*", Italy (the institute operated a meteorological station since 1872).

2.3.4.3 Rainwater studies in Japan

A complete history on precipitation chemistry in Japan is given by Wilkening (2004). Japan, similar to Germany, but because of other reasons, began after 1880 a rapid industrial development without any experience and knowledge concerning pollution control. The earliest record after the first rainwater examinations made by *Kellner* in Japan, but now obviously from air pollution point of view date from the year 1912 in Osaka, briefly summarized by Akimoto (1929), was a certain *Hirayama*, who analyzed rainfall for acidity, chloride, ammonia, and sulfate. *Minoru Akimoto* of the Osaka Public Health Laboratory conducted rainwater analysis at eleven sites around Osaka in 1926–1928, which he describes as "smoke city" (Wilkening (2004, p. 97). The meteorologist *Yasuo Matsudaira* and his colleague *Takeo Katō* from the Kobe Meteorological and Oceanographic Observatory, enthusiastically took up the first rainwater analysis by *Okada Takematsu*, director of the Tokyo Central Meteorological Observatory (today Meteorological Agency) and later director of the Kobe Observatory (Wilkening 2004, p. 98). *Takematsu* was the first meteorologist, who has seen the importance and necessity of conducting chemical analysis along with traditional meteorological measurements. Due to his urging, a Chemical Analysis Laboratory of the Central Meteorological Observatory (CMO) in Tokyo was established in 1935, and the geochemist *Yasuo Miyake* (1908–1990)[470] was appointed a chief chemist. After the war, the CMO was reorganized in the Geochemical Laboratory of the Meteorological Research institute and *Miyake* was appointed the director; he retired in 1969. *Miyake* first reviewed methods of precipitation chemistry and launched in 1935 the first network in Japan (Tokyo, Hamamatsu, Utsunomiya, Miyako, Chichijma, and Kobe). The first comparative analysis of urban precipitation chemistry in Japan was published by Miyake (1939), who presents data sets for three cities:[471] chemical composition of rainwater in Tokyo in 1936 (1.79, 0.58, 0.0067; 4.1), Kobe (2.33, 0.28, 0.004; 5.2) and Hamamatsu (2.39, 0.19, 0.0027; 5.6); in parentheses mean values in mg L^{-1} in the order chlorine, ammonium, nitrite, and pH. It is seen that pH is lowest in Tokyo; sulfate in rainwater at Tokyo (1936) was estimated to be 5.2 and 8.1 mg L^{-1} for the summer and winter season, respectively. *Miyake* furthermore determinates ammonia, nitrate, and nitrite and found in summer ammonia twice as much but in winter twice as much nitrate. The amount of ammonium and nitrite increases and the pH decreases in proportion to the pop-

470 After graduation from Tokyo Imperial University in 1931 in chemistry, *Miyake* began his professional career at the Hokkaido University as research assistant (1931–1935). He moved to the Central Meteorological Observatory as a chief chemist (1935–1946), and became a pioneer of marine chemistry in Japan. He served as Director of the Geochemical Laboratory at the Meteorological Research Institute (MRI) during 1946–1969. In 1957 he was appointed as Professor of Inorganic Chemistry at Tokyo Kyoiku University. His life-time works were compiled in three volumes of "Geochemical Studies of the Ocean and the Atmosphere" in 1978, 1979 and 1990.

471 According to Wilkening (2004), some data were earlier published by *Yasuo Matsudaira* and *Takeo Katō* (1933) for Kobe and by *Mitsuo Shimizu* (1936) for Hamamatsu.

ulation of the three cities, *Miyake* notes. *Miyake* published in 1937 and 1938 papers on air pollution in Tokyo (all in Japanese). Soon later, *Yutaka Ogura* analyzed rainfall at the meteorological station in Utsunomiya in 1938–1939 and found the interesting result that the nitrite concentration in rainwater during lightning storms were about twice that of ordinary rain (Ogura 1940). In 1939, *Kenzō Hirasawa* analyzed rainwater in Miyako, a village in northern Japan (Hirasawa 1941). *Hideo Matsui*,[472] who took over *Miyake*'s work, studied air pollution in Tokyo and rainfall in Maebashi (Matsui 1939). Between 1941 and 1944, he investigated at six different sites and distances from the sea rainwater for Na, K, Mg, Ca, Cl, and SO_4. Thus, he found that sodium and chloride at all sites corresponds to sea salt origin (magnesium only close to the sea) whereas calcium, potassium, and sulfate are everywhere in excess to sea salt components. The paper by Matsui (1942) is the first known extended rainwater analysis (Na, K, Mg, Ca, Mn, Ti, Al, Fe, Cu, Si, Cl, and S; unfortunately, no nitrate and ammonium determinate), showing the large depletion of chloride comparing to seawater in Tokyo but not suburbs, and concluded on the following mineral composition: $MgSO_4$, $MgCl_2$, KCl, and $CaSO_4$. *Miyake* further found that there is a marked correlation between chloride content and wind velocity, the sulfuric acid content in rainwater is larger in winter than in summer, the nitrite content in rain is very small when compared with that in air, and in the yearly variation of ammonia, the maximum exists in June, which is in good agreement with that in air (Miyake 1939, 1965).

After the war in Japan, only the Kobe monitoring was continued (since 1935) until 1961 (Ishikawa and Hara 1997). The “tradition” in rainwater chemistry in Japan was continued by *Ken Sugawara* (1899–1982) at the Nagoya University, the first chemist in Japan who studied the chemical composition of natural waters, but now from the point of “geochemistry of the atmosphere”. He first entitled an article “precipitation chemistry” (Sugawara 1948). Sugawara et al. (1949, 1953) studied in 1946–1957 in the Nagoya area the rainwater composition and have shown that NaCl is of sea salt origin, the sulfate content is small and alkali and earth alkali metals are surprising high (in mg L^{-1}): 3.4 (Ca), 4.8 (Na), 5.8 (Cl), 2.6 (K + Mg), and 1.2 (S). *Ushio Takeuchi* and *Zenichi Nakazawa* studied the amount of chloride in rainfall at Nagano City 1947–1949 depending on the incoming air masses (Takeuchi 1949, Takeuchi and Nakazawa 1951). Matsui (1949) was the first who determined hydrogen peroxide in atmospheric precipitation; between April 1948 and March 1949, he collected precipitation at the Kanagawa Prefecture and determined it as mean 0.52 (0.08–0.86) mg L^{-1}, in rain more than in snow and with a maximum in June and a minimum in January (H_2O_2 concentration parallel to the UV radiation), from which he concluded on photochemical formation.

472 It is notable that his name is wrong written as *Matui* in his papers in the Journal of the Meteorological Society of Japan.

Likely *Toshiichi Okita*, radio chemist from the Japanese Institute of Public Health, must be termed as the first Japanese atmospheric chemist. He studied in the 1950s also chloride content concerning air trajectories and in single raindrops and (Okita 1955, Okita and Tsuji 1958), and published several papers (1963–1968) on the conversion of SO_2 into sulfate. It is an irony of past that despite serious air pollution problems were described in Japan already around 1960, the first National Acid Deposition Monitoring Network (adapted from European and American networks) was established only in 1983; a committee on acid rain was headed by *Okita*. At present, *Hiroshi Hara* (born in 1946) is the most known exponent from the "third generation" after *Miyake*, the "grandfather" in Japanese rainwater chemistry.

2.3.4.4 Rainwater studies in Russia and the Soviet Union

In Chapter 2.3.3 on "salt cycling" several Russian studies were cited (Welbel 1903–06, 1906, Posnjakov 1904, Pirovarov 1906, Vityn 1911, Kossovich 1913, Burkser and Fedorova 1949, Blinov 1950). Vernadsky (1934b) in his great work "History of Minerals of the Earth's Crust" paid attention to the chemical composition of precipitation from meteorological and geological point of view. Thus, at the end of the 1930s, the Leningrad Institute for Experimental Meteorology [Ленинградский институт экспериментальной метеорологии (ЛИЭМ)][473] began to work on the development of analytical methods for water samples with a small volume (50–100 cm^3) based on colorimetry with the aim of cloud water collection. Testing was carried out in Leningrad and Tosno (50 km away) in preparation for the Elbrus Experiment (1939–1940). This was the beginning of extensive investigations of cloud physics and dynamics, including weather modification in the USSR, also by aircraft experiments (1946–1947) first conducted by *Vasili Aleksandrovich Zaizev* [Василий Александрович Зайцев] from the Main Geophysical Observatory [ГГО] and *Ivan Ivanovich Gaivoronsky* [Иван Иванович Гайворонский] (1910–1975) from the Central Aerological Observatory [ЦАО], first founded in 1941 in Pavlovski near Moscow and after the war established in Dolgoprudny. The Institute for Experimental Meteorology [Институт экспериментальной метеорологии (ИЭМ)] was established in 1968 in Obninsk near Moscow (Selezneva 1967). *Evgenia Semenovna Selezneva* [Евгения Семеновна Селезнева] (1905–1997), the first woman in geophysics and meteorology

473 The first meteorological institute in Russia was founded in 1849 in Sankt Petersburg as Physical Main Observatory [Главная физическая обсерватория (ГФО)], headed by *Adolph Theodor Kupffer* [Адольф Яковлевич Купфер] (1799–1865), renamed in 1924 to Main Geophysical Observatory (MGO) [Главная геофизическая обсерваторая (ГГО)] and since 1929 the leading scientific institute belong the Hydrometeorological Service of Russia. End of 1941 the former Institute for Experimental Meteorology was integrated as a department into the MGO (during the German blockade of Leningrad 1942–1944, the MGO was evacuated to Sverdlovsk). In 1949 the institute yield a name affix after the Russian climatologist *Alexander Iwanowitsch Voeikov* [Александр Иванович Вое́йков] (1842–1916).

in Russia, pioneer in meteorological interpretations of chemical analysis from rainwater at LIEM and later MGO.[474] After the war, *Valentina Ivanovna Drozdova* [Валентина Ивановна Дроздова], *Olga Petrovna Petrenchuk* [Ольга Петровна Петренчук] (1929–1983), *Rimma Fedorovna Lavrinenko* [Римма Федоровна Лавриненко] (born in 1936) and later *Peter Filippovich Svistov* [Петр Филиппович Свистов] (born in 1935) enlarged the team on aerosol, rain, and cloud water chemistry (Selezneva 1945a,b, 1947a,b, 1967, 1968, Selezneva et al. 1963, Selezneva and Drozdova 1966, 1974, Selezneva and Petrenchuk 1971, Drozdova 1962, Drozdova et al. 1962, 1964, Petrenchuk 1979a,b, Petrenchuk and Selezneva 1962, Petrenchuk and Drozdova 1966, Lavrinenko 1968).[475] It is worth mentioning the studies on acidity and alkalinity in precipitation due to sulfate and carbonate compounds (Selezneva and Lavrinenko 1971, Lavrinenko 1979, Petrenchuk and Nesterova 1979).

The first rainwater monitoring network in the USSR began with 13 stations, located mainly in the European part (1958–1961) within the scope of the International Geophysical Year (IGY) in 1957–1958 (summarized by Gushchin and Selezneva 1961, Zagar and Cholodova 1961, Drozdova et al. 1964). Since that time the Soviet program has expanded (Drozdova and Makhonko (1970)) and continues until present times, including 21 stations in the polar region and one drifting station in 2007 at the North Pole (Svistov et al. 2018a,b).[476] The team (consisting of about 20 scientists) from the MGO in Leningrad together with regional hydrochemical laboratories and local teams in Tbilisi, Kyiv, Alma-Ata, and Vladivostok, supervised more than 300 stations. The analytical program includes from the beginning 11 main components (sulfate, chloride, nitrate, bicarbonate, ammonium, sodium, potassium, calcium, magnesium, pH, and conductivity); quality control maintained by annual interlaboratory and intercomparison experiments. I do not know any other long-term monitoring program that includes bicarbonate in precipitation, an essential component for alkalinity, and thus for assessment of the capacity for neutralization of acidity due to sulfur dioxide emission. Since 1972 it is partly (with 11 stations) under the guidance of the WMO-BAPMoN network, Artz and Lavrinenko (1993). At the end of the 1980s, the network impended to collapse and only the activities of the head of State Hydrometeorological Service,

---

474 Just after the war, *Selezneva* published studies on cloud structure (1945b, 1947) and condensation nuclei in air (1945a).

475 They also published investigations on technical aspects of rainwater sampling (Selezneva 1968, Selezneva and Petrenchuk 1971, Selezneva and Drozdova 1974) and aircraft sounding (Aleksandrov and Petrenchuk 1959).

476 The results from the USSR network were published by the Main Geophysical Observatory in following: Ежемесячные данные по химическому составу атмосферных осадков за 1962–1965г. г. [Monthly data on chemical composition atmospheric precipitation for 1962–1965], Leningrad [since 1991 Sankt Petersburg], Ртп. ГГО (1970) 58 pp. With the same title and publisher: 1966–1970 (1985, 45 pp.), 1971–1976 (1986, 38 pp.), 1976–1980 (1986, 54 pp.), 1981–1985 (1989, 196 pp.), 1958–1961 (1991, 41 pp.), 1986–1990 (1994, 155 pp.), 1991–1995 (1998, 65 pp.), 1996–2000 (2006, 226 pp.), 2001–2005 (2010, 128 pp.), 2006–2010 (2014, 68 pp.).

Juri Antonovich Israel [Юрий Антóниевич Израэль] (1930–2014) who emphasized its role in air pollution and climate change monitoring, even on global scale, the network survived, and measurement of the chemical composition of precipitation is now regarded as the most simple and economic method in air pollution observation. From the results over more than 60 years, some characteristics were derived (Svistov et al. 2018a):

Mineralization 1–650 mg L^{-1}
Acidity in terms of pH 3.2–8.5
Percentage of anthropogenic pollution 85 %
Percentage of water quality like distilled (0.6–1.0 mg L^{-1} and pH ≈ 5.6) 1 %
Largest acidity (pH ≈ 3.2) with lowest (1.4–6 mg L^{-1}) in Arctics

The latter characteristic is an important finding that under atmospheric conditions missing any acid neutralization potential (e. g., from soil dust), already very small quantities of atmospheric acids, like hydrochloric, nitric, and sulfuric acid, resulting in acid rainwater. Many results from different regions within the network were published in Russian in reports, institutes journals, and proceedings (Ryaboshapko et al. (1994) and literature therein).

However, some rainwater chemical studies were conducted before establishing the network (see also p. 205). Usov (1940) collected (1936–1937) at four stations in the lower Volga region Nizhny Povolzhye [Нижнее Поволжье] rainwater and gives the mean values (in mg L^{-1}): 62.0 (HCO_3), 44.0 (SO_4), 36.5 (Cl), 35.0 (K + Na), 25.5 (Ca), and 3.9 (Mg); bicarbonate measurements are very rare and the data suggest that apart from sea salt (NaCl), $CaCO_3$ derived from soils was originally the main part of particulate matter that combined with SO_2 partly to sulfate in atmospheric water. Zhavoronkina (1955) studied the content of chloride and sulfur in rainfall along a trajectory between the Black Sea and the Barents Sea concerning sea salt distribution. Denisov (1956) studied rainfall in the mountains of the Tien Shan (border of Kazakhstan to China) and Denisov and Bugaev (1956) in northeastern parts of Ukraine.

2.3.4.5 Rainwater studies in India in the 1930s

In India, two exceptional scientists, *Nil Ratan Dhar* (1892–1896)[477] and *Atma Ram* (1908–1983),[478] investigated rainwater for nitrogen compounds (Dhar and Ram 1933a)

477 Regarded as "founder" of physical chemistry in India. After stays in England, Switzerland and France he returned to Allahabad and became professor for chemistry in 1919. In the late 1920s he again stayed in Europe and met among others *Baur, Haber, Nernst, Freundlich, Bonhoeffer, Hahn, Ostwald*, and *Weigert*. He is one of the first who studied photochemical reactions under environmental conditions; later he studied nitrogen fixation in soils and organic agricultural chemistry.

478 *Ram* made his PhD in the field of photochemistry under *Dhar's* supervision in 1935; later he became a leading chemist in ceramic chemistry in India.

and were the first who determined formaldehyde in rain (Dhar 1932), dew (Dhar 1933, Dhar and Ram 1933c), and in the air (Dhar and Ram 1933b,e); furthermore they first studied aqueous-phase photochemistry (Dhar and Ram 1932b, 1933c), the formation of OH in water by photolysis under sunlight (Dhar 1934), the photoreduction of CO_2 to HCHO in rainwater (Dhar 1934), the photooxidation of nitrite to nitrate in rainwater (Dhar et al. 1934), however, with some speculation and nowadays no longer supported results. Dhar and Ram (1933a) also discuss the origin of nitric nitrogen (ammonia emission from soils due to putrefaction was long known before). First, they state that thunderstorms cannot explain the permanent presence of nitric nitrogen (nitrate) in the air. To explain why more nitrate than ammonia was found (it should be noted that they collected only 13 rainwater samples between the end of August and the beginning of October 1932) they believe that a) ammonia is photo-oxidized under the action of sunlight and b) the slow combination of nitrogen and oxygen forming oxides of nitrogen in the presence of the ultraviolet light of the sun. Today, however, we know that atmospheric ammonia oxidation is negligible and the second process only occurs during lightning. On the contrary, Dhar and Ram (1933a) emphasize by experiments that in tropical soils both nitrification and ammonification are greater than in non-tropical climates. However, at that time it has been not yet known that during these processes, soils emit NO. *Dhar*'s works on the photochemical formation of formaldehyde from carbon dioxide and water in the presence of sunlight and photosensitizers (Dhar and Ram 1932a, Dhar 1934), and the hypothesis that it is produced in the atmosphere in upper layers via H_2O photolysis and subsequent CO_2 reduction by H atom (Dhar and Ram 1933b) was ingenious but wrong as we know today. It was concluded that formaldehyde can be washed out by rainwater, which was found in summer 1932 ($n = 40$) in Allahabad in the range of 0.015–0.12 mg L^{-1} (Dhar 1932, Dhar and Ram 1933c,d). Their experimental findings that the amount increases if the rain is preceded by bright sunny days and decreases when the days are cloudy and there are frequent showers, supported their idea on photochemical formation of HCHO in air.

A study by Narayanaswami (1939) is worth to mention because he determined nitrate and nitrite (no ammonium but chloride) in rainwater collected in 1938 at the most southern coastal site of Bombay. He found that thundershowers contain more nitrate and nitrite than at other time; however, the variation is not very large and the nitrite values seem to be much too low (in mg L^{-1} as N): nitrate 0.133 (0.08–0.24), and nitrite 0.00062 (0–0.02).

2.3.4.6 On the pH of rainwater (1936–1960)

Smith (1872, p. 276) stated that "acidity is caused almost entirely by sulphuric acid, which may come from coal [...]" but is limited on town. Unfortunately, *Smith* does not report how he determined the "free acidity"; its dimension is given in ppm of SO_3. From the ionic balance calculated from the concentrations given in Table 2.8 and 2.9 (see Table XIII on p. 285 in Smith 1872), it follows an acidity equivalent in terms of H^+

up to 550 μeq L^{-1}, corresponding to pH about 3.2. This calculated ionic balance is twice as much than that given "free acidity" that results in pH between 3.5 and 3.8 for British towns and 4.5 for Darmstadt in Germany, and about 5.5 for inland Scotland and none for other country sites; we can only speculate that *Smith* determined it by titration. The most remarkable conclusion is that the pH in rainwater obviously did not change in Central European towns for the next 100 years until the 1980s.

The first characterizing of rainwater to be "acidic" (detected likely by taste) is likely from Paping (1796) in Amsterdam due to free hydrochloric acid. Stark (1813) observed during a thunderstorm in Norwich rain a little acid (by litmus paper) but did not recognize that it could be due to the formation of nitric acid. The first communication that rainwater from a thunderstorm is acid due to free nitric acid is by Ducros (1845) who also coined first the term "acid rain" ["*pluie acide*"].

Before the 1950s, only very few pH measurements were conducted, the first in Germany in 1932 by Cauer (1936) and then in Tokyo in 1936 by Miyake (1939), pH was 4.1 (in contrast to values of 5.2 at Kobe and 5.6 at Hamamatsu). Cauer (1956, p. 506) refers Bergerhoff (1928, pp. 81–82) who already did try in 1927 to measure the pH in precipitation using test-paper; he found that in the highly polluted Ruhr area the pH value became higher after storing samples, explained by *Cauer* due to the dissolution of alkaline dust particles. Cauer (1936) first measured pH in a few rainfalls ($n = 5$) in the High Tatra in 1932 (5.2 in rain, 4.8 in snow, and 6.2 in rime) and between June and September 1936 at Bad Reinerz (three value to be 4.2, 4.3 and 4.7), Cauer (1937b); see Fig. 2.19. Ernst (1938),[479] who entitled his paper "On pH measuring in precipitation", collected rainwater and measured only pH[480] at three sites in Silesia, Bad Reinerz ($n = 47$), pH = 3.5–6.4, Oberschreiberhau ($n = 80$), pH = 3.1–6.1, and Neue Schlesische Baude[481] ($n = 9$), pH = 4.4–5.2, and confirmed the values already published by *Drischel*; 79 % of the pH values were in the range 3–5 and 15 % < 4. Cauer (1949a, p. 232, 1951a, p. 1131) summarizes the mean pH values of the aerosol ("*Nebelkerne*") found at different locations (as pH-controlling agent *Cauer* indicates ammonia):

3.4 Nebelhorn (Alps, >1000 m)
4.0 High Tatra (>1000 m)
4.7 Oberschreiberhau (750 m)
6.2 Berlin

479 Unfortunately, nothing is known on Dr. *Ernst*. He was the head of "*Forschungsstelle für Feinchemie der Luft und Geophysik*" [research institute for fine chemistry of the air and geophysics] in Reinerz (Vogt 1939, p. 49); in this brochure also a paper by Ernst "in press" entitled "*Über den H^2O^2-Gehalt der Niederschläge*" [on the amount of H_2O_2 in precipitation] is mentioned, but obviously never published.
480 In that time, it was already common to use the hydrogen or glass electrode (Poggendorff compensation method) to measure pH; however, *Ernst* used the colorimetric method using indicators, namely the spotting technique, which has an error of ±0.2 pH units. *Miyake* used the electrochemical method.
481 Now: Schronisko na Hali Szrenickiej (Karkonosze Mountains) in Poland, 1195 m a. s. l.

Figure 2.19: Forschungsstelle Oberschreiberhau, Feinchemische Luftuntersuchungen [Research Institute Oberschreiberhau, fine-chemical air examinations], source: Vogt (1939, p. 40). The person in front, bent over the desk, is Dr. Cauer, and the standing person is Dr. Ernst.

Ernst concluded correctly that flue ash particles containing alkali and earth-alkali metals (like Ca and Mg) increase the pH. Quitmann and Cauer (1939) measured pH in two rainfalls in January 1939 in Berlin, and it was 4.4 and 4.9. Drischel (1940) who also studied the chemical composition of rainwater measured pH in Silesia (former Germany) in 1937 (May–August; $n = 29$), pH = 4.28 (3.51–6.37) and August 1938 ($n = 8$) to be 4.0 (3.6–4.7).[482] Cauer (1948a) reports summarizing that he conducted within the past 10 years a large number of pH determinations of fog particles;[483] he presents mean pH from the following three sites: 4.0 at Tatra Mountains, 1000 m a. s. l. ($n = 53$), 4.7 at Riesengebirge Mountains 750 m a. s. l. ($n = 332$), and 6.2 at Berlin ($n = 250$). Cauer (1938, p. 411) communicated that the pH of rainwater at the coast of the Bretagne was always in the range of 5–6, and that of condensed water (using his dew-cup) constantly around 5.3 (it seems from the text that he expected a weak alkaline pH due to the sea-water alkalinity).

After the war, *Helmut Landsberg* was the first who conducted preliminary pH measurement in August 1949 at Washington, D. C., continued near Boston over a year

482 *Drischel* calculated wrongly the mean pH (4.54 in 1937) as arithmetic mean from single pH; however, because *Drischel* listed the rainfall, we can also calculate the precipitation weighted pH to be 4.43 for the 1937 data set.

483 He writes in German "*wässrige Schwebestoffe der Luft, der Nebelkerne*" [aqueous suspended matter in the air, the fog condensation nuclei]; Cauer (1948a) not reports on the sampling procedure but he used the "condensation dew-cup" (see Fig. 2.20 on p. 257), whereby aerosol particles and soluble gases were collected, representing the total "air pollution" (Cauer 1956).

(1952–1953) with observations of pH values of individual raindrops and snowflakes (83 precipitation events), that showed acidic values ranging from 3 to 5.5 with a mean of 4 (Landsberg 1954); in 39 dew samples, he found same pH values. Remarkably, he found a tendency for smaller drops to have lower values of pH than the larger ones.

Hermann Harrassowitz (1885–1956)[484] collected since summer 1952 (until 1956) about 500 samples of rain, snow, and dew at Giessen[485] and found pH values between 3.1 and 8.5 with 4.2–4.5 as the most frequent value (Harrassowitz 1956); he is likely the first who pointed out that acid rain (using the German term "*saurer Niederschlag*") will acidify soils and negatively impact waters. He also refers the French physician *Charles-Noël Martin* (1923–2005) who notes that atomic bomb tests produce nitric acid and thus lead to acid rain (Martin 1954).

Cauer (1956) reports detailed on pH values of "potential aerosols" (see p. 228) and precipitation; sampling using his condensation dew-cup he conducted at more than 20 different sites from town to high mountains, principally confirming his reported values from 1948 (mean pH value ranging from 3.4 to 6.5).[486] Cauer (1951b) named it "*Aerosoltröpfchen*" [aerosol droplets] or "*feinste Schwebestofftröpfchen der Luft*" [finest suspended droplets in air], best named in today's termination haze particles, and reported on pH down to 2.7 at Mt. Nebelhorn near Oberstdorf. He also named it "*stark saurer Höhendunst*" [strong acid airborne particles], likely due to nitric acid and other still unknown strong acids. For 13 sites he presents mean pH values of precipitation, citing those from his former co-workers *Drischel* and *Ernst* and own results, not published elsewhere (Tatra, Zugspitze, Nebelhorn, Oberstdorf, Pic du Midi, Berlin, Bochum, Bretagne (France), and Island at North Sea (all his measurement sites between the late 1930s and 1950s), showing mean values ranging from 4.1 to 4.8 at high mountains, 3.0 to 6.6 in towns and industrial areas, and at coastal sites 4.8 to 5.4 (Cauer 1956, pp. 502–504). Cauer (1956, p. 477) who also conducted many measurements using his condensation-cup in different indoor environments, regarded the pH value as an important chemical-bioclimatic parameter, dividing it in three categories of solutions: pH < 4.2 bactericide, pH 4.2–5.5 virulence paralyzing, and pH 5.5–7 virulence promoting. Cauer (1956, p. 504) comments that the pH values from the Swedish network (Egnér and Eriksson 1955) of 43 stations in 1954 were obtained only after the monthly sampling do not reflect the "true" mean pH values which could be obtained from individual samples with immediately pH determination. Thus, he visionary proposed (Cauer 1956, p. 506) to combine rain gauges with automatic pH recorders. In

484 German geologist and soil scientist; professor at Giessen since 1920 (forcible retired in 1934 by the Nazi's). He announced a further publication "Niederschlags-pH" that was never realized due to his soon death.

485 Thus, he is in historic line with Zimmermann (1824) and Liebig (1827).

486 Unfortunately, he nothing writes on the sampling periods, the number of samples and the averaging of pH.

contrast to his belief in 1935 that the pH value can be used as an "absolute indicator" for the air quality, he now considers the pH value as an "excellent reference indicator", and that chemical investigations of air and dust are essential. *Cauer* first correctly discusses (Cauer 1956, p. 480–481) that the pH is a result of alkaline and acid atmospheric substances (forming mixed "neutral" aerosol particles), and that alkali and earth-alkali metals from industrial dust buffer acid gases (namely sulfuric acid) but due to its limited dispersion, in more remote areas the pH is lower. Vice versa, at industrial sites with the combustion of coal with high sulfur content but missing alkaline dust emission, low pH values must be expected.

Helmut Berg (1908–1960)[487] collected atmospheric precipitation in and near Cologne from autumn 1957 to summer 1958 at 5 sites and measured (only) pH, ranging from 3.4 to 9.9; neither a diurnal nor an annual variation was to be found, and no dependency from air masses. 80–90 % of the samples have a median pH between 6.9 and 7.5 (Berg 1959).[488] In Vienna, *Ferdinand Steinhauser* (1905–1991)[489] found in Vienna from March 1956 to December 1957 the pH values in 114 rain falls between 3.4 and 7.4, and most values between 5–6 (Steinhauser 1959a; this paper cites all single estimates including rainfall amount). In precipitation at Wahnsdorf near Dresden, Mrose (1961) measured between April 1958 and January 1959 pH 4.3 without variation.

First pH values in rainwater are available from the Scandinavian network (1954–1956) with a mean of 5.5, varying from 4.8–5.0 along the northeastern shore up to 6.3–6.9 at the elevated interior of northern Sweden; the acidity is being greatest in winter and the alkalinity greatest in late spring (Carrol 1962). Gorham (1955) found rainwater pH between 4.0 and 5.8 (mean 4.45) at the Lakes District in England. Visser (1961) found in Uganda (1958–1959, $n = 78$) a mean pH of 7.8 (5.7–9.8). These results suppose the role of suspended soils particles in the air as emphasized by Junge and Werby (1958) and Barrett and Brodin (1955).

Long before recognizing the "acid rain period" (about 1970–1990, Chapter 2.3.5.2) as a consequence of regional sulfur dioxide pollution in Europe, America, and Japan, some scientists like Gorham (1955) and Barrett and Brodin (1955) recognized the acid-

---

487 German meteorologist, 1938 head of the Cologne weather aviation office ("*Reichswetterdienst*") and docent at university, since 1947 professor and head of the meteorological institute at the university Cologne.

488 *Berg* correctly notes that an arithmetic mean cannot be derived (this is only possible when the rainfall amount is known for each sample of precipitation). Thus, it is a mystery why Mrose (1966) cites *Berg* with pH values around 5.5 (from the frequency distribution given by *Berg*, and assuming equal amounts of rainfall for each sample, for Cologne a mean pH of 3.6 would be derived).

489 Austrian climatologist and meteorologist, director of "*Zentralanstalt für Meteorologie und Geodynamik* (ZAMG)" in Vienna (1953–1976). The measurements were carried out by the chemist *Kurt Chaluppa* (unknown living dates) who also measured after 1967 SO_2 and NO_2 at the "*Zentralanstalt*".

ity in Scandinavian precipitation.[490] They considered pH of 5.7 as the neutral point for atmospheric water (resulting from the equilibrium with atmospheric CO_2 at about 300 ppm), not in a chemical sense (that would be a pH of 7.0), but as a reference point from which changes that may take place by the addition of cations and anions have to be discussed. Pioneering work in understanding the causes of precipitation acidity made *Eville Gorham*[491] who studied the chemical composition of rainwater between June 1952 and May 1954 at Kentville, Nova Scotia, in a predominantly agricultural area (Herman and Gorham 1957) and the English Lake District between January 1955 and January 1956 based on daily samples (Gorham 1958a). He separated samples according to the pH into classes <5.7 and >6.2 and found the sulfur to ammonium-nitrogen ratio to be 3.3 and 1.4, respectively, far in excess of the stoichiometric ratio (1.1). Based on precipitations samples collected at British towns (Sheffield, Manchester, Salfort) he concluded that hydrochloric acid is responsible for the low pH values in spite of large sulfate deposits (Gorham 1958b). His arguments were experimentally supported some 25 years later by numerous other investigators. *Gorham* writes (Gorham 1958b, p. 276):

> [...] sulphur dioxide oxidizes rather slow, and that only the portion which has both oxidized and reacted with basic substances is trapped into the deposition gauges. In this connection it may be remarked that there is a strong correlation between sulphate and calcium [...] Ammonia is also known to be correlated with sulphate in condensation nuclei produced by fuel combustion (Junge 1954), and in acid rains of pH less than 5.7 (Herman and Gorham 1957), so that it may well account for some of the residual sulphate.

Some decades later he emphasizes again the neutralization of acid rain (Gorham 1994).

2.3.5 Precipitation chemistry after 1945

Finally, more than 150 scientists, who carried out for different reasons chemical rainwater analysis, are listed in Table 2.12 to whom we must pay respect for pioneering work; only for a very few of them, it was the main point of their general scientific work. Table 2.13 lists countries and researcheres of the first rainwater examination before 1950. The number of researchers after 1950 will be uncountable, but the main

490 Many scientists studied the acidity as budgets between acid and bases and estimated the "natural" acidity as a reference for acidification by humans (e. g., Möller and Zierath 1986, Möller and Horváth 1989).

491 *Gorham* received a BSc degree in biology and an MSc degree in zoology in Nova Scotia, Canady. In 1947 he began at the University College, London, his doctoral work studying mineral content of plants in the Lake District and became interested in the acidification of ecosystems. as graduated biologist was interested in the pure science of the chemistry of rain, bog, and lake waters.

Table 2.12: Scientists, who investigated the chemical composition of rainwater, snow, and hail before 1950; in chronological order with the year(s) and site(s) of measurements.

Author	nationality	year(s)	site(s)
Boerhaave	Dutch	1720s	Leiden
Marggraf	German	1749–1751	Berlin
Lavoisier	France	1748	La Bourget (Paris)
Bergman	Swedish	1774	Uppsala
Stark	British	1813	Norwich
Mulder	Dutch	1820s	Utrecht
Zimmermann	German	1821–1823	Giessen
Witting	German	1823, 1824	Höxter (near Göttingen)
Dalton	British	1824	Manchester
Brandes	German	1825	Salzuflen (near Bielefeld)
Helmts	German	1826	Achim (near Bremen)
Link	German	1826	Breslau (now Poland)
Liebig	German	1826–1827	Giessen
Lampadius	German	1834(–1837)	Freiberg (Saxony)
d'Arcet	French	1834	London
Bohlig	German	1834	Mutterstadt (near Ludwigshafen)
Girardin	French	1839	Paris
Bertels	German	1841–1842	Regenwalde (now Poland)
Ducros	French	1842	Nimes
Marchand	French	1849–1850	Fécamp
Chatin	French	1850	Paris
Barral	French	1851	Paris
Meyrac	French	1852	Dax (Gascogne)
Bineau	French	1852–1853	Lyon
Boussingault	French	1853/1859	Paris, Liebfrauenberg, St. Bernhard
Lawes and Gilbert	British	1853/1854	Rothamsted
Martin	French	1853	Marseille
Filhol	French	1855	Toulouse
Way	British	1855/56	Paris and Rothamsted
Warington	British	1855–1888	Rothamsted
de Luca	Italian	1860	Pisa
Knop and Wolf	German	1860	Möckern (Leipzig)
Pincus and Rollig	German	1860–1866	Insterburg (now Russia)
Rosing	Norwegian	1860–1865	Ås, Norway
Bobierre	French	1863	Nantes
Meissner	German	1863	Göttingen
Reichardt	German	1863	Jena
Beyer	German	1864–1867	Regenwalde (now Poland)
Wollny	German	1864–1866	5 stations (almost now Poland)
Cossa	Italian	1865	Pavia
Bretschneider	German	1865–1870	Ida-Marienhütte (now Poland)
Struve	German	1869	Tiflis
Smith	British	1869–1870	many stations in Great Britain
Frankland	British	1869–1870	Rothamsted
Goppelsroeder	Swiss	1869–1871	Basel

Table 2.12: (continued).

Author	nationality	year(s)	site(s)
Schroeder	German	1870–1871	Tharandt
Church	British	1870–1879	Cirencester
Chabrier	French	1871	St. Chamas (France)
Vogel	German	1871/1872	Munich
Bechi	Italian	1872–1875	Florence, Vallombrosa
Schöne	German	1874	Moscow
Louvet	French	1875	Saint-Louis
Eugling	German	1875–1877	Munich, Tisis
Miller	British	1877–1904	Rothamsted, Calcutta (1891)
Lévy	French	1875–1890	Paris
Stoklasa	Czech	1877, 1883–1886	Bohemia
Prévot	British	1879–1881	Cirencester
Heinrich	German	1880–1881	Rostock
Tuxen	Dansk	1880–1886	Copenhagen
Kinch	British	1881–1900	Cirencester
Müntz and Aubin	French	1882–1885	Pic du Midi, Caracas, Reunion
Kellner	German	1883–1884	Tokyo
Marcano	Venezuela	1889–1890	Caracas
Gray	New Zealand	1884(–1909)	Lincoln (New Zealand)
Mach	Italian	1885–1886	S. Michele (Tirol)
Harrison and Williams	USA	1885–1890	Barbados, British Guiana
Bellucci	Italian	1886	Perugia
Passerini	Italian	1888–1900	Florence
Russell and Richards	British	1888–1916	Rothamsted
Failyer and Willard	USA	1889	Kansas
Bamber	British	1889–1899	Ceylon
Marr[a]	Dutch	1891	East Java
Petermann and Graftiau	Belgium	1892	Gembloux
Erwin[a, f]	USA	1891–1893	Salt Lake City
Mabery	USA	1894	Cleveland
Hutchinson[b]	USA	1894–1895	Mississippi
Tracy[b]	USA	1895	Missouri
Basile[a]	Italian	1895	Catania
Bonâme[c]	French	1895	Mauritius
Giacosa	Italian	1895	Mount Rosa
Casali	Italian	1900	Bologna
Kasay	Hungarian	1902	Ógyalla (now Slovakia)
Carmody	British	1902	Trinidad
Aufray	French	1902–1909	Hanoi
Welbel	Russian	1902–1905	Ploty (now Ukraine)
Di Vestea	Italian	1903–1914	Pisa
Pirovarov	Russian	1903	Odessa
Posnjakov	Russian	1904	Odessa
Leather	British	1904–1906	Cawnpore, Dehra Dun (India)
Jorissen	Dutch	1905–1906	Den Helder
Unknown	Argentina	1906–1908	Buenos Aires

Table 2.12: (continued).

Author	nationality	year(s)	site(s)
Russell	British	1907–1919	Rothamsted
Brünnich	Australian	1908	Brisbane
Gray	New Zealand	1908–1909	Lincoln
Shutt and Hedley	Canadian	1908–1924	Ottawa
Cohen and Ruston	British	1909	Leeds
Juritz	South African	1909	Cape Colony, South Africa
Feilitzen and Lugner	Danish	1909	Flahult
Vityn[a]	Russian	1909–1910	Petrograd and 4 other sites
Kossovich	Russian	1909–1911	Petrograd and 7 other sites
Hudig and Welt	Dutch	1905–1912	Uithuizermeeden
Bykov and Karpov	Russian	1910	Tomsk
Anderson	Australian	1912–1914	Melbourne
Schröder	German	1912	Montevideo
Stewart	USA	1912–1919	Urbana, Illinois
Richards	British	1913–1919	Rothamsted
Wiesner[a]	USA	1914	Mt. Vernon, Iowa
Wilson	USA	1915–1920	Ithaca, New York
MacIntire	USA	1919–1921	10 sites in Tennessee
Collison and Mensching[a]	USA	1919–1928	Geneva, New York
Johnson[a]	USA	1921–1922	7 sites in Kentucky
Eaton[a]	USA	1921–1922	Chicago
Erdman	USA	1921–1922	Ames, Indiana
Freeman[a]	USA	1922–1923	6 sites in Kentucky
Ilkov[a, d]	Bulgarian	1922–1925	Bulgaria
Hansen	Danish	1922–1926	Askov, Hornum, Spangsbjerg
Ellett and Hill[a]	USA	1923–1928	Blackburg, Virginia
Fraps[e]	USA	1924–1927	10 sites in Texas
Majo	Italian	1927	Napoli
Liesegang	German	1927, 1932–1934	Berlin and 3 other sites
Harper	USA	1927–1942	Stillwater, Oklahoma
Braadlie	Norwegian	1928–1929	Trondhejm
Comel	Italian	1928–1930	Udine
Scharrer and Schropp	German	1928–1935	Weihenstephan
Marimpietri and Simoncelli[a]	Italian	1929–1931	Rome
Farcy	French	1930	Bretagne
Bertrand	French	1930–1947	different sites in France
Collins and Williams[a]	USA	1931–1932	9 sites in USA
Finnell and Houghton	USA	1931	Goodwell, Oklahoma
Ives[a]	USA	1931–1933	Chicago
Cauer	German	1932	High Tatra (now Poland)
Dhar and Ram	Indian	1933	Bombay
Bottini	Italian	1934, 1936	Mt. Vesuvian
Alway (et al.)[a]	USA	1936	4 sites in Minnesota
Usov	Russian	1936–1937	Lower Volga region
Drischel	German	1937–1938	Lower Silesia (now Poland)
Ernst	German	1938	Lower Silesia (now Poland)

Table 2.12: (continued).

Author	nationality	year(s)	site(s)
Narayanaswami[a]	Indian	1938	Bombay
Akimoto	Japanese	1926–1928	Osaka
Matsudaira	Japanese	1931–1933	Kobe
Kurashige[a]	Japanese	1931–1933	Tokyo
Miyake	Japanese	1937...1957	different sites in Japan
Hirasawa	Japanese	1939	Miyako, Japan
Matsui	Japanese	1939, 1949	Maebashi, Japan
Leeflang[a]	Dutch	1938	Leiden
Ogura	Japanese	1938–1939	Utsunomiya, Japan
Sugawara	Japanese	1946–1947	different sites in Japan
Burkser	Ukrainian	1947	Black Sea and Azov Sea
Kudrin[a]	Russian	1945–1947	Tashkent
Egnér and Eriksson	Swedish	1947–1950	different sites in Sweden
Blinov[a]	Russian	1950	Caspian Sea

[a]Unknown biographic data
[b]*Samuel Mills Tracy* (1847–1920) was the director and *W. L. Hutchinson* (unknown living dates) the chemist (and 1903 director) at the Mississippi Agricultural Experiment Station, established in 1888 at the Agricultural and Mechanical College of the State of Mississippi that was established in 1878.
[c]*Philipp Bonâme* (1831–1919) was a French agricultural chemist, Director of the Agricultural Station at Mauritius.
[d]*Ilkov* studied over three years the nitrogen deposition (in kg ha^{-1} yr^{-1}) as ammonia (11.86) and nitric acid (0.69).
[e]Fraps (1930, p. 10): 8–15 pounds sulfur per acre deposition by precipitation. *George Stronach Fraps* (1876–?) was the chief chemist at Texas Agricultural Experiment Station.
[f]Erwin (1893).

purpose of studying rainwater changed to an understanding of atmospheric chemical processes. With an increasing interest in understanding atmospheric physics and subsequent weather forecasts in the 1930s (and attempts for weather modification), the chemical nature of condensation nuclei came into the focus of atmospheric research. However, as in the first decades of the twentieth century, many studies considered the ion supply to soils from the atmosphere, it remained also the main issue just after the ending of World War II. Precipitation chemistry research contributes to the global understanding of many major contemporary environmental issues, e. g., air pollution, atmospheric chemistry, acidification and eutrophication of ecosystems, and climate change. Roughly we can characterize the precipitation chemistry[492] research after World War II by three periods,

492 In chemistry, precipitation means a precipitate, i. e., a solid deposition within a solution. Therefore, it is termed in literature almost *rainwater chemistry*. Here, the term precipitation is used in its meteorological sense. The often-used term "chemistry of acid rain" is not scientifically because acidity is already a chemical property of rainwater.

Table 2.13: First rainwater study by country between 1800 and 1950 (and additional before 1800).

year	country	author
1720s	Holland	Boerhaave (1735)
1749	Germany	Marggraf (1753)
1768	France	Lavoisier (1770)
1774	Sweden	Bergman (1788)
1813	Great Britain	Stark (1813)
1827	Russia	Neljubin (1827)
1839	France	Girardin (1839)
1860	Italy	de Luca (1888)
1860	Norway	Rosing and Wankel (1870)
1870	Switzerland	Goppelsroeder (1871)
1877	Bohemia (former Austria, now Czech R.)	Stoklasa (1882)
1880	Denmark	Tuxen (1890)
1884	Algeria	Chairy (1884)
1884	New Zealand	Gray (1888)
1883	Japan	Kellner et al. (1886)
1883	Venezuela	Müntz and Marcano (1889)
1885	Barbados	Harrison and Williams (1897)
1886	Reunion	Müntz and Marcano (1889)
1889	USA	Failyer and Willard (1889)
1890	British Guiana	Harrison and Williams (1897)
1891	Indonesia	Marr (1891, 1893, 1898)
1892	India	Miller (1905a)
1893	Belgium	Petermann and Graftiau (1893)
1895	Mauritius	Bonâme (1896)
1898	Ceylon	Bamber (1900)
1902	Vietnam	Aufray (1909)
1902	Hungary	Kazay (1904)
1902	Trinidad	Carmody (1902)
1906	Argentina	Anonymous (1907)
1907	Australia (Queensland)	Weedon (1909)
1908	Canada	Shutt and Hedley (1925)
1909	Sweden	Feilitzen and Lugner (1910)
1909	South Africa	Juritz (1910)
1909	Uruguay	Schröder (1915)
1910	Antarctica	Müntz and Laine (1911)
1911	Iceland	Miller (1913)
1912	Australia	Anderson (1915)
1924	Palestine (later Israel)	Menchikowsky (1924)[a]
1925	Bulgaria	Ilkov (1925)

[a] *Felix Menchikowsky*, Chemist, born 1887 in Russia and worked since 1922 at the Agricultural Experimental Station at Tel-Aviv (directed by *Otto Warburg*). Cited by Eriksson (1960); *Menchikowsky* determined the chloride deposition (131 kg ha^{-1} yr^{-1} and 1/3 of this amount further inland)

- The "global cycle period" (about 1947–1980): quantifying global biogeochemical cycles, namely that of sulfur and nitrogen by using experimental data from deposition (and calculated emissions data; note, modeling of global cycles using chemistry-transport-models started end of the 1980s),
- the "acid rain period" (about 1970–1990): recognizing acidification of lakes and soils with subsequent ecological effects by limnologists and soils scientists, and new-type forest decline by foresters initiated an intensification of rainwater chemistry and deposition studies, and
- the "air pollution control period" (about 1980–2000): the beginning abatement of SO_2 emissions as a political consequence of "acid rain" led to a significant extension of networks and the attempt to construct relationships between emission and deposition (to regard whether there is a linear or non-linear relation with the never realized aim of feedbacks for an "intelligent" air pollution control).

In other words, we can also term the whole era as "sulfur dioxide period", closing the historic cycle back to *Angus Smith* 100 years ago, who identified sulfate in rainwater as the cause of acidity and coal combustion as the source of sulfur dioxide.

2.3.5.1 The "global cycle period" (1947–1980)

The first network for precipitation chemistry was established in Sweden in 1947 (carried out between October and September 1950), initiated by *Gunnar Torstensson* (1888–1981), agronomist, who studied in Leipzig and Jena and the organic chemist *Hans Gabriel Egnér* (1896–1989), both at the Royal Agricultural College at Uppsala, and *Anders Knutsson Ångström* (1888–1981), meteorologist and director of the Swedish Institute of Meteorology and Hydrography. These scientists also worked out the sampling method (Egnér et al. 1947). *Erik Eriksson* (1917–2019), *Arne Emanuelsson* (1923–2000) and *Linus Vilhelm Högberg* (1887–1959) from the Meteorological Institute at Stockholm carried out in the following years the interpretations (Ångström and Högberg 1952, Emanuelsson et al. 1954, Eriksson 1954, Egnér and Eriksson 1955). At the beginning, still atmospheric nitrogen deposition for farming was in the focus (Torstensson 1954)[493] and most data are given as deposition and not rainwater concentration. Ångström and Högberg (1952) concluded from observations that

> [...] the almost constant ratio of about 2:1, in which the NH_4-N, and NO_3-N respectively, are in the precipitation, seems to suggest a common origin – perhaps some kind of nuclei, in which they enter in an almost fixed ratio.

The network was expanded to the whole Scandinavia, including 5 sites in Britain and 3 in Germany in 1952–1953 (totally 53 stations) and followed by analogous activities in other European countries. *Gunnar Brodin* (unknow ling data) from the Royal Agri-

493 He gave to my knowledge the last citation of *Marggraf*'s early finding that nitrate is in rain but wrongly written as "Markgraf" and without giving bibliographic data.

cultural College at Uppsala reported in 1955–1957 regularly on the results.[494] Thus, in 1955, the first "European Air Chemistry Network (EACN)" was established, after *Carl-Gustaf Rossby* (1888–1957), American meteorologist of Swedish origin had highlighted the need for atmospheric composition measurements in addition to conventional meteorological measurements to understand the atmospheric circulation of chemical substances. Monthly measurements of precipitation constituents were made at 49 sites across Scandinavia (41 in Sweden and 3 in Denmark) and Britain (5 stations). The network in Sweden has been re-arranged several times; no single site was used over the decades until now; thus 357 sites existed since 1955 stabilized to 40–60 since 2002 (Ferm et al. 2019). Eriksson (1954) concluded that the interpretation of chemical data on precipitation is a rather complex problem (see also Munn and Rodhe (1971) concerned interpretation of monthly rainwater samples):

> Until more direct evidence is available, it is a matter of intuition and, perhaps, taste in which way they are interpreted.

Mordy (1953) studied total nitrogen, sodium and chloride in a few rain events on Hawaii, found them to be in good agreement with the figures elsewhere in ocean climate, citing Eriksson (1952). An inverse relationship between nitrogen content and rainfall intensity exists in these data but not for sodium and chloride. The network growth to about 160 sites with monthly sampling. In many cases, the data show large discrepancies in ion balance or poor correlation between hydrogen ion concentrations determined from pH measurements and calculated from ion balance considerations. In view of these uncertainties, interpretation of the changes in precipitation chemistry should proceed with care (Kallend et al. 1983). Later it became clear that the major part of the variance was explained by local rather than regional influences, and that monthly sampling does not reflect "true" wet deposition and thus rainwater chemistry.

Before around 1965, less was known on the atmospheric chemistry of effluents from fuel combustion (Leighton 1961, Junge 1963a,b), and investigation of rain chemistry gave some ideas namely on the role of the aqueous phase for the oxidation of SO_2 to sulfate. Gambell and Fisher (1964) write that the "measurements support the hypothesis by Junge and Ryan (1958) that much of the sulfate in rainwater is formed from oxidation of SO_2 in cloud droplets".

Interesting is the comparison of values at Flahult gained in 1909 (Feilitzen and Lugner 1910) for ammonia and nitric acid with those gained in 1951–1954 (Egnér and Brodin, unpublished, cited by Tamm 1958), and those in 1985 (maximum) and 2015

494 Current data on the chemical composition of air and precipitation. *Tellus* **7** (1955) 395–400, 522–527, **8** (1956) 112–113, 283–285, 411–412, 513–517, **9** (1957) 140–143, 250–254, 423–427. See also: Egnér, H. and E. Eriksson: *Tellus* **7** (1955) 134–139, 266–271. See further: Oddie, B. C. V. (Meteorological Office, Bracknell): *Tellus* **10** (1958) 172–175, 281–286, 500–508, **11** (1959) 139–145, 259–165, 366–372.

(after Ferm et al. (2019), based on deposition values – no concentration values given – normalized to be one in 1955); in mg L^{-1} as nitrogen:

NH_3	0.4 (1909)	1.2 (1951–1954)	2.8 (1985)	2.0 (2015)
HNO_3	0.2 (1909)	0.8 (1951–1954)	3.2 (1985)	2.0 (2015)

It is interesting to compare this trend with that found in Hungary between 1902 (Kazay 1904), around 1980 (Horváth 1983), and around 2010 (Móring and Horváth 2014), on the basis of deposition values (in mg N L^{-1}):

NH_3	1.7 (1902)	1.5 (1970–1980)	0.6 (2010)
HNO_3	0.1 (1902)	0.8 (1970–1980)	0.5 (2010)

It is not the aim here to discuss these trends; rainwater concentrations reflect in a complex way regional (country) air pollution (in other words, emissions) as well as long-range transport of air masses, but also local processes.

Adopting the Swedish rain chemistry network, further countries established networks, such in the USA, where 1955–1956 at 62 stations rainwater was collected only during rainfall to avoid the influence of dry fallout, and further collected to monthly samples for analysis and finally published as maps concerned the distribution of sea salt (Junge and Gustafson 1957) and nitrogen compounds (Junge 1958a). *Junge*'s investigations were accompanied by Eriksson (1957) who concluded that Hawaiian rainwater does not differ basically from rainwater collected near the Swedish west coast. From this huge data set, Junge and Gustafson (1957) concluded that far enough from the coast sea salt particles must be uniformly distributed in the troposphere and rainfall not greatly depletes airborne sea salt particles. Junge and Werby (1958) discuss the distribution of Na, Cl, K, Ca, sulfate concerning sea salt origin and continental sources. They found the Cl/Na ratio at inland USA to be 0.5–0.9 and hence significantly lower than in seawater (to be 1.8 as mass ratio). Despite already published ideas on Cl depletion, they preference soil as the source for Na "enrichment" as discussed for K and Ca, and not Cl depletion. They believe (mistakenly, based on today's knowledge) "that there is little evidence that decomposition of sea-salt particles is of importance in the chemistry of rain water" (Junge and Werby 1958, p. 422). From the homogeneous potassium contribution and the very small concentration found in inland rainwater they concluded that K occurs in soils primarily as silicates and will be converted into soluble salts predominantly in moist areas by different weathering processes; on the contrary, $CaCO_3$ as a major constituent of soils in arid areas, is enriched in such areas and transported with air masses to other areas, they concluded.

In the late 1960s, the "Background Air Pollution Monitoring Network (BAPMoN)" was established. It focused on precipitation chemistry, aerosol, and carbon dioxide measurements, included regional and background stations, and had the "World Data Centre for Precipitation Chemistry (WDCPC)" center established in the USA, which receives and archives precipitation chemistry data and complementary information from stations around the world. During the 1970s among other important atmospheric issues (ozone layer and global warming), acidification of lakes and forests in large parts of North America and Europe, caused principally by the conversion of sulfur dioxide into sulfuric acid by precipitation processes, was addressed.

In West Germany, *Hans-Walter Georgii* conducted the first rainwater studies (together with measurements of NH_3, SO_2, and NO_2 in gas and aerosol phase) in Frankfurt (1956–1957, $n = 89$ and 1960–1961, $n = 138$) and parallel at the Taunus Observatory at the Kleiner Feldberg (825 m a. s. l.: 1957, $n = 66$ and 1969, $n = 91$), in Langen, a small city 20 km south of Frankfurt, (1960–1961), $n = 50$) as well at the Zugspitze in summer 1957 and 1958 (2966 m a. s. l., $n = 13$); Georgii (1960, 1965). Unfortunately, no pH measurements were carried out. *Georgii* was the first who studied experimentally the washout processes; he found no significant difference in rainwater concentrations (ammonium, nitrate, sulfate, chloride, sodium, calcium) between Frankfurt and Langen but 20–40 % less at the mountain sites. Interestingly he found that the ammonium to sulfate ratio amounts 0.32–0.37 (theoretically 0.37 for stoichiometric ammonium sulfate) at the mountain sites whereas in Frankfurt it amounts only 0.25 due to sulfate excess from SO_2 sources.

In East Germany (GDR) at the Meteorological Observatory Wahnsdorf at the beginning 1950s, the chemical analysis of rainwater was seen as a method for monitoring air pollution, namely in a network with a central analytical laboratory using proofed methods; thus methods for determination of turbidity, pH, chloride, sulfate, ammonium, nitrite, and total organic substance were developed and adapted, respectively (Teichert and Greifenhagen 1954). As part of the International Geophysical Year (IGY), the GDR National Committee initiated a network of 14 stations for monthly collection of precipitation (Arkona, Boltenhagen, Mt. Brocken, Mt. Fichtelberg, Gardelegen, Gera, Görlitz, Halle-Kröllwitz, Kaltennordheim, Karl-Marx-Stadt [now Chemnitz], Lindenberg, Neustrelitz, Potsdam, and Wahnsdorf) between 1957 and 1959. The chemical analysis (sulfate, chloride, nitrate, ammonium, Na, Ca, and K) was carried out at the Meteorological Observatory Wahnsdorf near Dresden, first by *Friedrich Teichert* (p. 33–34 and footnote 47 on p. 11) and his assistant *Greifenhagen* (unknown living dates) and continued in 1957 by *Helmut Mrose* and others (Mrose 1961); at Wahnsdorf, individual rain events were collected and analyzed between January 1958 and August 1959, and additional nitrite and pH. Mean values in mg L^{-1}: 9.5 (sulfate), 1.9 (chloride), 1.8 (nitrate), 0.044 (nitrite), 2.1 (ammonium), and 3.1 (calcium), pH = 4.3. Mrose (1966) continued rainwater analysis at the meteorological observatory Wahnsdorf until 1964 (Table 2.17)[495] and published the first frequency distribution of pH values ($n = 206$) with a mean of 4.5 (3.4–6.0).

495 The long-term rainwater chemistry monitoring in the GDR begun in 1978 in Neuglobsow within the scope of EMEP and in 1988 a network of 32 stations was in operation under the guidance of the

Likewise, in the scope of the IGY, in the Czech Republic a network involving 22 stations operated for 18 months began in 1958 (Macků et al. 1959, see also Moldan and Veselý 1987);[496] unfortunately, without determination of sulfate. Compared with the date from 1982–1984 (Moldan and Veselý 1987), chloride concentration and pH (4.7–5.5) were significantly higher and nitrate (2.5–3.6 mg L^{-1}) lower.

Having a view on the trends of rainwater constituents in Europe over the last 100–150 years (Table 2.14), we state that the sulfate concentration increased from the beginning of the nineteenth century until 1980 by a factor of 3 and decreased relative to the 1980 period by a factor of 6 until now, hence now about on a 50 % level of the nineteenth century value (remember that the nineteenth century already was the coal combustion period with sulfur pollution). In contrast, nitrate increased by about one order of magnitude from 1900 and is not well abated until now. This reflects the twentieth century as the mobile fuel-based engine period. Remarkably, ammonium has not changed over the 1900–1980 period (for more information and discussion see Sutton et al. 2008). The pH (in terms of acidity) also remains about constant and only increased slightly after 1990 due to the almost complete SO_2 abatement.

Table 2.14: Changing chemical composition in rainwater (European average) by factors, relative to 1 (1900); before 1980 citations see in Möller (2003).

compound	~1900	1900 → 1960	1960 → 1980	1980 → 2000[b]
nitrate	1	4–6	6–12	~constant
sulfate	1	~2	~3	≤0.5
ammonium	1	~constant	~constant	~0.8
pH[a]	1	1	1	~1.2

[a]pH remains probably constant (4.0–4.5) between 1860 and 1980 due to neutralization of acidic by alkaline compounds and increased (in other terms, the acidity decreased) between 1990 and 2000 to about 5.0–5.5

[b]Since 2000 no more change observed

As seen from Table 2.15, rainwater chemical composition reflects the extreme city air pollution in the nineteenth century. The London values also show the high importance of NH_3 emissions at that time from missing wastewater management. As a result, the pH was not different from the late twentieth century values. Table 2.16 shows some rainwater analysis from different parts of the world between 1980 and 2000.

Meteorological Service (stepwise build-up since 1985). Zierath (1981) studied rainwater chemistry at 8 sites 1976–1979 and the station Seehausen was in operation 1982–2002.

496 *Bedřich Moldan* (born 1935) is a Czech geochemist and ecologist; in 1990 he became the first Minister of Environment of the Czech Republic.

Table 2.15: Historical comparison of rainwater chemical composition.

Site	pH	SO_4^{2-}	NO_3^-	NH_4^+	Cl^-	sum[a]	Δ[b]	author
		in mg L^{-1}				in μeq L^{-1}		
London 1870	4.8	57.7	8.9	10.6	10.5	1029	1014	Smith (1872)
Beijing 1981	6.5	16.6	5.0	4.0	2.1	254	254	Zhao et al. (1988)
Seehausen 1985	4.3	13.4	4.3	2.7	2.9	274	229	Möller (2003)

[a] $[SO_4^{2-}]+[NO_3^-]+[Cl^-]-[NH_4^+]$
[b] $[SO_4^{2-}]+[NO_3^-]+[Cl^-]-[NH_4^+]-[H^+]$ = missing neutralizing cations like Ca^{2+}

Table 2.16: Selected results of chemical composition of precipitation in different regions of the world (in μeq L^{-1}, rounded).

Site	H^+	Na^+	K^+	NH_4^+	Ca^{2+}	Mg^{2+}	Cl^-	NO_3^-	SO_4^{2-}
China (1981–1984)[a]	0.2	44	21	160	460	–	39	34	162
India (1988–1996)[b]	0.1	18	8	40	56	–	32	18	36
Cape Grim (1977–1985)[c]	1	1167	30	2	78	247	1372	3	152
Central Australia (1980–1984)[d]	17	4	1	3	2	1	8	4	4
Lancaster (NW England)[e]	34	57	4	55	15	15	79	28	89
Central Bohemia (1980s)[f]	42	7	3	62	30	–	–	51	116
Hungary (1980s)[g]	32	23	9	61	85	16	26	41	119
Seehausen (1983–1989)[o]	44	33	6	85	59	12	63	45	150
North Sweden (1983–1987)[h]	23	4	2	13	3	2	5	11	33
South Sweden (1983–1987)[k]	46	24	2	36	10	7	30	36	67
Seehausen (1996–2002)[o]	14	20	2	52	16	5	25	39	40
Melpitz (2000)[m]	13	8	1	44	13	3	8	31	35
Neuglobsow (2000)[n]	13	14	2	39	18	4	14	31	33

[a] Suburb of Beijing (Zhao et al. 1988)
[b] Suburb in semiarid area (Kumar et al. 2002)
[c] Tasmania, south Pacific air (Ayers and Ivey 1988)
[d] Katherine, annual rainfall 75–136 mm (Likens et al. 1987)
[e] Harrison and Pio (1983)
[f] Hradec, mean of 483 samples (Moldan et al. 1988)
[g] Means from 6 stations (Horváth and Mészáros 1984)
[h] Means from 10 stations (Granat 1988)
[k] Means from 20 stations (Granat 1988)
[m] Rural station near Leipzig, weekly wet-only samples (UBA)
[n] Background station 60 km north of Berlin, weekly wet-only samples (UBA)
[o] Meteorological station about 40 km west of Salzwedel, 100 km north-westerly of Magdeburg and 150 km northeast of Mt. Brocken

2.3.5.2 Period of "acid rain" (1970–1990)

One of the most serious air pollution problems between about 1970 and 1990 was the acidification of large parts in Europe, Northern America, and Asia, called the era of "acid rain". In Sweden, negative effects of anthropogenic acid deposition on surface waters were first reported in 1967 and as early as 1976, long-term time series were being used to assess the impacts of acid deposition on surface waters (Odén 1968, 1976). Between 1961 and 1963, *Svante Odén*,[497] as a visitor to the International Meteorological Institute of Stockholm University, analyzed precipitation chemistry data collected by *Erik Eriksson* and his colleagues and he noticed a systematic change in the chemical climate with increasing amounts of sulfur and a lowering of the pH level. Under *Bert Bolin*'s inspiring leadership, a group of scientists, including among others *Odén*, *Carl Olof Tamm*,[498] *Lennart Granat*[499] and *Henning Rodhe*,[500] prepared the first comprehensive, integrated assessment of the acidification problem. These results were first presented publicly at a conference by the Swedish Natural Science Research Council in 1967. In 1972, a group of scientists including the American limnologist *Gene Likens* (born 1935) "re"-discovered the rain that was deposited at White Mountains of New Hampshire was acidic (Likens et al. 1972, 1979). Likens and Borman (1974) wrote:

> At present, acid rain or snow is falling on most of the northeastern United States. The annual acidity value averages about pH 4, but values between pH 2.1 and 5 were recorded for individual storms. The acidity of precipitation in this region apparently increased about 20 years ago.

The acid rain period – one of the most serious environmental issues of the twentieth century – finally was the trigger for the first large-scale measures in pollution control in the 1980s.[501] Likens and Borman (1974) further wrote:

> Only some of the ecological and economic effects of this widespread introduction of strong acids into natural systems are known at present, but clearly they must be considered in proposals for new energy sources and in the development of air quality emission standards.

497 (1924–1986), Swedish soil scientist and chemist.

498 (1919–2007), Swedish soil scientist.

499 (born in 1949), Swedish meteorologist at the Stockholm University; studied precipitation chemistry.

500 (born in 1941), Swedish meteorologist at the Stockholm University; world-leading atmospheric chemist.

501 The Convention on Long-range Transboundary Air pollution (CLRTAP) was signed in November 1979 (ratification in 1982), after considering the international problem that pollutants may originate in one country, with consequent effects felt in another country (but without binding effects). In the USA, the "Acid Precipitation Act of 1980" was signed. It is interesting to note that 15 years after air pollution abatement in central Europe, acid-sensitive Swedish lakes show slow recovery from historic acidification as shown from long-term (1987–2012) water quality monitoring (Futter et al. 2014). Overall, strong acid anion concentrations declined, primarily as a result of declines in sulfate. Chloride is now the dominant anion in many acid-sensitive lakes. Base cation concentrations have declined less rapidly than strong acid anion concentrations, leading to an increase in charge balance acid neutralizing capacity.

Likens et al. (1979) write, that "the principal cause is the release of sulfur and nitrogen oxides by burning of the fossil fuels". Galloway et al. (1978) report on the establishment of a research program on acid rain in the USA.

At the end of the twentieth century, the emission of sulfur dioxide was fully under control in Europe and Northern America; but that of nitrogen oxides only partly. Reviews of the early history of research on acid deposition were written by Cowling (1982), Gorham et al. (1984), Gorham (1989, 1998), Brimblecombe (1991),[502] Menz and Seip (2004),[503] and specifically in Scandinavia by Tamm (1995).

Whereas Smith (1872) used the term *acidity* without further explanation, first Granat (1972) defined the terms *acidity* and *alkalinity* based on the ion balance, simply $[H^+]$ = sum of anions – sum of cations, but without H^+. Granat (1972, p. 551) writes:

> However, there is organic material and thus presumably also weak acids and bases. Of course there is also insoluble material of inorganic nature. Some of these compounds may very well take part in acid-base reactions.
> If bicarbonate is present (in detectable quantities) alkalinity is numerically equal to the amount of bicarbonate. Otherwise the alkalinity is the titrated amount of acid-with a negative sign.

The mathematical description by *Granat*, however, is based on the assumption that the "water comes to equilibrium with the carbon dioxide pressure of the atmosphere": a pH = 5.60 (acidity = alkalinity = 0) represents the reference point for *neutral* atmospheric water, i. e., water in equilibrium with atmospheric CO_2 (at 340 ppm). However, as shown by later researchers – a wrong assumption. Natural emission of gaseous and particulate species leads to a natural acidity or alkalinity, which varies in time and space (Charlson and Rodhe 1982, Möller and Zierath 1986, Möller and Horváth 1989). Galloway et al. (1982) show that in the case of missing strong acids, weak organic acids (mainly from natural sources) contribute significantly to acidity. Zhao and Sun (1986)[504] write:

> It is a fact that alkaline soil with a pH of around 7 to 8 spreads over a vast area in the northern part of China, and thus airborne particles possess a pH similar to that of the soil itself [. . .] Ammonia is released into the air in large quantities from the alkaline soils [. . .] Atmospheric buffering capacity is therefore considered very strong in the north.

---

502 *Peter Brimblecombe* (born 1948 in Australia), studied in Auckland, New Zealand where his PhD concerned atmospheric chemistry of sulfur dioxide. 1974–2013 Lecturer and then Professor in Atmospheric Chemistry at the School of Environmental Sciences, University of East Anglia, Britain.

503 *Hans Martin Seip* (born 1937) is a Norwegian chemist, who began 1970 with investigation of acid precipitation (since 1988 cooperation with *Dianwu Zhao* from China).

504 *Dianwu Zhao* (born in the early 1940s) was the leading Chinese scientist in acid rain and air pollution research at the Chinese Academy of Sciences.

In China, air pollution and acid rain came into the research focus in the late 1970s, and since 1981 a network for precipitation chemistry is in operation. In the north of China, rainwater pH > 6 and in the south, pH < 5 was measured in that time. Thus, despite naturally alkaline rainwater in northeast China, "acid rain" was observed with pH > 5.6 as pointed out by Möller (1998, 1999b).

Werner Stumm[505] introduced (Stumm and Morgan 1981, Stumm et al. 1983) the acidity as a *base neutralizing capacity*, corresponding to the equivalent of all acids in solution, titrated to a given reference point (Zobrist 1987), and a corresponding alkalinity as an *acid neutralizing capacity*. This conception has been further developed by Facchini and Fuzzi (1993)[506] who define *atmospheric acidity* in the sense of a multiphase concept as the *base neutralizing capacity* per unit of volume of the atmospheric system. They stated that "in analogy with solution chemistry the addition of acidic or basic components to an atmospheric system can be viewed as a titration of the global system". Möller (1999b) introduced the atmospheric *acidifying capacity* as the sum of the *potential* acidity in gas and aerosols as well as the acidity in the aqueous phase. Nevertheless, from today's knowledge, it should be pointed out that any definition of atmospheric acidity was unhelpful in the understanding of *acidification*. *Svante Odén* expressed the phenomenon with the excellent phrase that *acidification* is used to describe a process by which a given environment is made more acidic (Odén 1976).

Brosset (1976) defined "black episodes" where dark, airborne particles containing ammonium sulfate and nitrate are transported together with increased concentrations of SO_2 and NO_y from the European continent to Sweden. In contrast, he defined "white episodes" which occur during dry sunny weather when colorless airborne particles that consist mainly from ammonium sulfate are transported over the North Sea. Later Rahn et al. (1982) clearly stated that the black episodes are from Central Europe whereas white episodes came from eastern Europe and west USSR, but they differ only by time or the extent of aging in the air. In the northwest part of the USSR, sulfate in precipitation increased (like in Scandinavia) by a factor of 2 between 1962 and 1976 whereas nitrate and ammonium did not change, but also pH remained between 5 and 6 in obviously buffered or neutralized by alkaline species solutions (Petrenchuk and Nesterova 1979).

However, from the *scientific* point of view, the huge number of studies after 1980 did not result in new insights of *acid precipitation* and atmospheric chemistry. They manifested that the acid-base status of precipitation is a result of a balance between acidifying compounds, mainly oxides of sulfur and nitrogen, and alkaline compounds, mainly ammonia and alkaline material in windblown soil dust. Monitoring became a governmental task; nevertheless, it remains a data basis for scientific

505 (1924–1999), Swiss chemist, leading in aquatic chemistry; 1954–1969 in USA and 1970–1992 at ETH Zürich. *Jürg Zobrist* (born 1947), Swiss chemist in water research.

506 *Maria Christina Fachini* (born 1960) and Sandro Fuzzi (born in 1951), Italian atmospheric chemists (they are married to each other) at the Institute of Atmospheric Sciences and Climate in Bologna.

interpretations regarding the different sampling procedures (exposure time and sampler type).

On the contrary, it was recognized that *acid deposition* largely as a result of SO_2 dry deposition and that due to increasing SO_2 emissions in the 1960s (and further on in eastern Europe until the 1980s) long-range transport led to acid deposition also in remote areas. Furthermore, from theoretical considerations and beginning modeling of aqueous-phase chemistry in the 1980s, it became clear that formation of sulfate and hence acidity in droplets of clouds and rain was limited with increasing gas-phase SO_2 concentration (Granat 1978). This did occur in Sweden, where the observed acid rain was believed to result from the transport of pollutant emissions from Great Britain and central Europe. Also, the Canadian acid rain was partially due to the transport of pollutants from the United States (Wisniewski and Keitz 1983). Cogbill (1976) first reviewed the history of precipitation acidity in the USA, referring papers by MacIntire and Young (1923), Ellett and Hill (1929), and Collison and Mensching (1932) to deduce on rainwater pH, indicating low acidity (pH 6–7), but over northeast North America since 1955 pH below 4.5 was measured.

From acidity examinations in rainwater since Smith (1872) until the early 1950s (Chapter 2.3.4.6), it is evident that the pH of rainwater did not change in areas near town and industrial zones, being around 4.5. In Germany, systematic pH measurement began in 1976 at the Meteorological Observatory Hamburg (Winkler 1977).[507] Winkler (1980, 1982) concluded, on the basis of the very few measurements known since the end of the 1930s, that the rainwater pH in Central Europe shows no trend, discussing that the acid formation capacity is exhausted. Indeed, this "trend" remained until the middle of the 1990s, and with the significant control of SO_2 emissions, the pH rose to around 5 ends of the 1990s in Europe and Northern America (Möller 2020).

Using historical values of rainwater composition (Tables from 2.7 to 2.9), hydrogen ion concentration is estimated further on the basis of the electroneutrality condition and simplified through consideration of sulfate, nitrate, and ammonium only, i. e., neglecting neutralizing soil dust[508] and assuming sea salt to be neutral, and, furthermore, neglecting bicarbonate, which is appropriate for pH < 5 but not for pH > 8, we can roughly estimate the pH:

507 *Peter Winkler* (born in 1943) German meteorologist, initiating air chemical research at the Meteorological Observatories Hamburg (1974–1992) and Hohenpeißenberg (1993–2005).

508 This is a drastic assumption because calcium is a proxy for acid neutralization, either from industrial dust emission or soil dust; Mészáros (1965) found in rainwater (n = 15) 1993–1964 in Budapest (in mg L^{-1}) mean values for total calcium 2.47 (0.34–2.8) and calcium bicarbonate 1.30 (0.34–2.8) and in air (n = 13, in µg m^{-3}) soluble calcium 0.020–0.026 and insoluble calcium 0.030–0.056, more soluble in winter and more insoluble in summer.

– rural sites in France and Germany (1850–1880)	pH 8–9
– towns in France and Germany (before 1880)	pH 4.7
– towns in Great Britain with coal combustion around 1870	pH 3.2
– rural sites in Great Britain around 1870	pH 4.2

The early nineteenth century (namely before sewage management) was characterized by a large ratio of ammonia (from human and animal excrements) to nitrate (from nitrification in soils and waters); from German rural sites (1860–1870) it follows NH_4/NO_3 = 2.6 (±1.2) in contrast to a value of 0.27 in the 1980s. Moreover, 85 % of ammonia was gaseous in the air and only 15 % in particulate matter (as NH_4NO_3). On the contrary, bicarbonate in rainwater served as the buffer; sulfate concentrations in the range of 2–4 mg L^{-1} in rainwater corresponds to pH of 4.5–5; this sulfate level was typical before the 1950s for rural sites in the USA, British inland, and central Europe (note that in British towns of the nineteenth century sulfate was in the range 20–70 mg L^{-1}).

2.3.5.3 Period of "air pollution control": the East German case (1980–2000)[509]

Since many decades it was one of the most important tasks in atmospheric chemistry to construct the relationship between the emissions of a substance (pollutant) and its concentrations (and deposition) concerning the distance from the source, first by measurements and later by modeling (beginning with simple Lagrangian dispersion and later followed by Eulerian chemistry-transport models). This forms crucial information for the design of air pollution abatement and – inversely – assessment for efficiency of these measures. However, European policy in the 1980s was more or less "linear"; depending on the economic capacity (or political trading) of different countries, reduction targets were set up (for example 30 % reductions of SO_2 emissions). Despite a vast number of publications, there is no clear relationship between long-term climate goals (concentrations, radiative forcing, or temperature change) and short/medium-term emission targets (Vuuren and Riahi 2011). The relationships between the reduction of SO_2 emissions and changing sulfate concentrations were simpler, a top theme after the 1980s (Leck and Rodhe 1989, Möller et al. 1996, Acker et al. 1998a, Möller 1999b, Lajtha and Jones 2013, Ferm et al. 2019). Whereas in many parts of the world (Western Europe, Northern America, Japan), significant reduction of SO_2 emissions occurred over a long time (since about 1975 continuously over some 20 years), in Eastern Germany[510] a drop-down of >90 % SO_2 emission occurred in connection with the German reunification over only 7–8 years (1991–1998).

509 This chapter is dedicated to my colleagues who died too soon: *Wolfgang Rolle* (1934–1996), *Wolfgang Marquardt* (1933–2002), and *Eberhard Schaller* (1950–2015).

510 The former German Democratic Republic (GDR). Parallel to this development, in the neighboring Czech Republic, namely the northern part close to Eastern Germany (Saxony), SO_2 emissions went down. Much slower was the SO_2 reduction trend in Poland.

The "political mission" of establishing a measurement site for wet deposition at Seehausen[511] was to prove that air masses transported from the GDR are less acidic due to the neutralization potential of flue ash from lignite-fired power stations, or, in other words, to verify that the GDR was not responsible for acidification of lakes in Sweden. Indeed, the "acid rain research program" established in 1982 in the GDR[512] (Marquardt and Ihle 1988, Marquardt et al. 1996, 2001) did show that precipitation, evolved from clouds transported over the GDR was less acidic and rich in calcium, a typical constituent of flue ash from lignite-fired power station, hence proxy for neutralization or buffering, respectively. However, rainwater acidity was only a very little less acidic in eastern air masses (pH = 4.4) compared to that in western air masses (pH = 4.3) before 1990; after 1990, acidity in "eastern" rainwater increased a bit and decreased in "western" rainwater. Only locally, close to polluted sites in GDR, acidity was significantly less (pH > 5). Most acid deposition, however, was due to SO_2 dry deposition. Without any doubts, the GDR was an exporter of atmospheric acidity.

Nevertheless, several studies and later monitoring of rainwater chemistry were carried out in the GDR, at the beginning of 1954, with few measurements by Teichert and Greifenhagen (1954) and Mrose (1961, 1966). A unique historic data set was created by *Reinhard Zierath* (1950–2013)[513] from 8 different sites in the GDR (1976–1979), Zierath (1981). Other earlier rainwater measurements are listed in Table 2.17. The See-

511 52°06′N, 12°24′E, 21 m a. s. l., mean precipitation 556 mm; about 40 km west of Salzwedel, 100 km north-westerly of Magdeburg and 150 km northeast of Mt. Brocken. Established by the former Institute for Energetics (Leipzig), continued from 1992 by the Institute for Tropospheric Research (Leipzig) and since 1996 by the Chair for Atmospheric Chemistry and Air Pollution Control (Cottbus) until ending the monitoring in 2002. The chemical analysis was carried out from 1982–1995 by Dr. *Erika Brüggemann* (Leipzig) and from 1996 to 2002 by Dr. *Renate Auel* (Berlin) with highest quality standards. No change of any sampling procedure and data classification was made over the whole period. Wet-only sampling every 4 hours (best for correlation with synoptic air changes).

512 Since 1982 all data on air pollution in the GDR became secrecy by law; the terms "*Waldsterben*" and "*saurer Regen*" [acid rain] were officially tabu. *Detlev Möller*, together with *Marquardt* and *Rolle*, hold 1984 in Leipzig the lecture "*Saure Niederschläge – Bildung und Einschätzung*" [Acid precipitation – formation and assessment] at 16th *Lufthygienisches Kolloquium*, published in a curious journal (Marquardt et al. 1984). Möller (1977) published still an own estimate of GDR's SO_2 emission (4.9 million tons); but he got a call from the Ministry of Health expressing official disapproval. In 1981, *Möller* published in the popular journal "*Wissenschaft und Fortschritt*" [science and progress] an article "*Schwefel in der Atmosphäre*" (Vol. 31, p. 438) but he was obligated from the editor to delete all absolute values (only given in %). In December 1982, the "*Zeitschrift für Chemie*" (a serious scientific journal, where *Möller* published since 1970) rejected an article on rainfall acidity he submitted.

513 German chemist at the Institute for Geography and Geoecology (belongs to the AdW) in Leipzig (1974–1992) who "spend" the time of fall of Berlin wall in autumn 1889 at the GDR Antarctic station. *Zierath* was an excellent scientist, who not found a scientific job after liquidation ["*Abwicklung*"] of the AdW end of 1992.

hausen measurement series, including 3668 samples (4-h sampling, to each sample a trajectory) over 20 years is surely the most sophisticated ever conducted in the world. Finally, the Meteorological Service built up a large network from 32 stations operating since 1988; the background station Neuglobsow belonged to EMEP earlier (1978) in operation (see for more details Möller 2020).

Table 2.17: Rainwater investigation in the former GDR (Möller and Lux 1992), concentrations in µeq L^{-1}. Not listed are measurements which were made in forests from the forestry institutes in Tharandt (*Herbert Lux*, born 1937) and Eberswalde (*Klaus Westendorff*, born 1949 and *Karl-Hermann Simon*, 1930–2011), see Möller and Lux (1992) for more details.

pH	Cl^-	SO_4^{2-}	NO_3^-	NH_4^+	Na^+	K^+	Ca^{2+}	Mg^{2+}	
–	54	198	29	117	26	–	156	52	Wahnsdorf (1958–1959)[a]
4.2	84	190	50	105	43	10	168	86	GDR mean (1976–1979)[b]
4.8	–	600	144	119	–	–	506	–	Cottbus (1982–1987)[c]
4.4	87	186	53	119	45	6	118	90	Seehausen (1982–1988)[d]
4.5	94	176	48	99	59	12	82	33	Greifswald (1984–1988)[d]
4.5	54	233	66	2	65	13	175	37	Oberbärenburg (1985–1989)[e]
4.4	93	254	66	78	35	13	280	44	Tharandt (1985–1989)[e]
4.7	61	292	71	127	46	10	176	33	GDR mean (1988–1989)[f]
5.2	60	740	58	121	39	14	523	75	Leipzig (1988–1989)[f]
4.9	50	352	59	73	38	11	226	38	Cottbus (1988–1989)[f]
4.4	47	316	50	127	22	14	194	44	Wiesenburg (1988–1989)[d]

[a]Mrose (1966)
[b]9 stations, pseudo-wet-only, Zierath (1981)
[c]bulk samples, *Heinz Jursch* (Hygiene Institute Cottbus), person. communication (1990); the concentrations seem to be too large for rainwater (exceptional ammonia), likely influenced by dust precipitation
[d]Marquardt, pers. Communication (1990), wet-only sampling
[e]Erzgebirge, wet-only sampling (*Herbert Lux*), $n = 133$ (Oberbärenburg), $n = 223$ (Tharandt); the low ammonia concentration at Oberbärenburg cannot be explained.
[f]32 stations, weekly bulk samples, Meteorological Service of the GDR

According to the huge air pollution in a few larger cities surrounded by industrial complexes, sulfate (SO_2 precursor), calcium and magnesium (dust precursors) were extremely enlarged compared to the background. Except for Bitterfeld, nitrate and ammonium show no significant regional differences. Today, no significant differences in rainwater chemical composition are found between the cities listed in Table 2.17 and the background. The decline after 1990 is between a factor of 2 (ammonium) and 40–100 (sulfate and calcium) at heavily polluted sites and 4–5 for the background, respectively. For more information see Möller (2020).

From the Seehausen monitoring of precipitation chemistry, it was seen that the variation between single samples (or events, but due to 4-h sampling time, longer lasting rain events were characterized by several samples) is huge, the maximum concentration is 10–20 times larger than the mean and the minimum goes towards zero. The reasons for large variations in concentration are manifold: rainfall rate (determining droplet sizes and amount), air mass origin, time with and without rain before sampling, and local contamination. The longer the dry period persists before the rain event, the higher the ionic concentration (up to 2–3 times) due to atmospheric accumulation. Conversely, the longer the rain event persists, the smaller (by a factor 0.3–0.8) the ionic concentration due to wet removal with the exception of sea salt NaCl and hydrogen ions. Thus, a larger number of samples are required to concluded about air pollution; moreover, the concentrations should be precipitation-weighted and the samples should be classified according to entry sectors, based on back trajectory calculations. Moreover, timely changes due to seasonal and interannual variations occur because of variations of meteorological conditions, air mass characteristics and air chemistry. The observers who were involved in this rainwater chemistry monitoring had no sufficient knowledge of the studies carried out before 1950, and it is no use arguing whether they had modified the program concerning the scientific progress in the background of the *history*, now in our hands.

2.3.5.4 An outlook

In 1989, the network BAPMoN was consolidated into the current WMO "Global Atmosphere Watch (GAW)" program. In Europe, the "Co-operative Programme for Monitoring and Evaluation of the Long-range Transmission of Air Pollutants in Europe (EMEP)" started in 1979 under the "Convention on Long-range Transboundary Air Pollution (LRTAP)" establishing a measurement network including rainwater (135 sites in 2015), Slanina et al. (1979). Data from GAW stations were obtained from the regional measurement programs to which they contributed, like EMEP and the "Acid Deposition Monitoring Network in East Asia (EANET)", and from national networks like the "Canadian Air and Precipitation Monitoring Network (CAPMoN)" and the "United States National Atmospheric Deposition Program (NADP)".

Currently, GAW coordinates activities and data from 31 Global stations, more than 400 Regional stations, and around 100 contributing stations. Measurements of common chemicals found in precipitation were performed by national and regional monitoring networks for many years in North America and Europe and increasingly in East Asia and Africa as well as at some GAW global and regional stations. Among the commonly measured ions (referred to as major ions) are sulfate, nitrate, ammonium, chloride, sodium, potassium, calcium, magnesium, phosphate, fluoride, pH, and acidity.

Standard protocols and data quality objectives described in the "Manual for the GAW Precipitation Chemistry Programme: Guidelines, Data Quality Objectives, and Standard Operating Procedures as well as the GAW Interlaboratory Comparison

Studies", have helped reduce uncertainties and inconsistencies in sampling and analytical methods by improving the quality, comparability, and representativeness of these measurements.[514] Available regionally-representative high-quality precipitation chemistry and precipitation depth measurements from monitoring stations around the world, in combination with global model-ensemble results, formed the basis for a scientific assessment entitled "A global assessment of precipitation chemistry and deposition of sulfur, nitrogen, sea salt, base cations, organic acids, acidity and pH, and phosphorus" (Vet et al. 2014). In 1995, the first global precipitation chemistry assessment was released as a World Meteorological Organization publication (Whelpdale and Kaiser 1996). Since the 1990s, basically at each site of the world rainwater was collected and analyzed; the number of publications is uncountable.

2.4 Fog and cloud water studies

Clouds – together with fog and precipitation – are surely the most fascinating weather phenomena of our atmosphere being observed by humans since antique eras (cf. Figs. 2.4 and 2.5). A phenomenological understanding of the water cycle was first described by *Aristoteles*. The ancient views of air and water before exploring the chemical composition of the atmosphere and the theories on condensation and evaporation were described by Möller (2020, pp. 44–66).

In contrast to precipitation, it is much more difficult to collect water from fog and clouds; due to missing appropriate sampling techniques (from the analytical point of view), low occurrence of fog and laborious sampling on mountains, there are only very few studies before the 1980s. The US American atmospheric chemist *Stephen E. Schwartz* (born in 1941) presents a fair definition of cloud chemistry (Schwartz 2003, p. 331): The term *cloud chemistry* is considered here to comprise both cloud composition and reactions that take place in clouds. But this is valid too for fog and precipitation chemistry.

The differences between *haze*, *mist*, *fog* and *cloud* were explained in Chapters 1.3.4 and 2.2.2. and cited "a fog, as a celebrated naturalist said, is a cloud in which one is, and a cloud is a fog in which is not". Even today among scientists is disagreement whether a cloud at a mountain summit is a cloud or a mountain fog. If the "cloud" is passing (in other words not being orographic) over the summit and its bottom is not very far below the summit, it is a *cloud*. Without any doubt there is a *cloud* far above the ground – collecting cloud droplets is possible only by airborne platforms (aircraft, balloon, etc.). Without any doubt, there is also *fog* in a plain with contact to the ground or only a few meters above. Monitoring of cloud chemical composition

514 The group of the author (DM) participated very successful within this program from 1993 until 2008 with respect to our precipitation chemistry monitoring Seehausen and Mt. Brocken cloud chemistry program.

is only possible with ground-based stations; (exceptional) aircraft collection began only after 1950. Thus, we do not separate between fog and cloud before 1950 and the first cloud that was investigated by *Boussingault* at Liebfrauenberg was likely a fog. Town fog[515] will be treated in Chapter 3.1.2. Research goals and aims of investigation concerned fog and cloud can be classified as follows:

- analytical interest together with rainwater and dew studies (nineteenth century),
- understanding condensation nuclei and rain the formation (1925–1955),
- understanding physics and dynamics for fog dissipation and cloud seeding,
- air pollution impact research, namely on forest decline,
- understanding aqueous-phase chemistry,
- understanding multiphase-chemistry (gas-liquid interaction),
- monitoring with respect to climate change.

The investigations were conducted by

- field measurement campaigns (between days and weeks),
- occasional measurements over a limited period (between month and years),
- monitoring (over many years),
- laboratory studies on kinetic and mechanisms of aqueous-phase chemistry,
- modeling aqueous-phase and multiphase chemistry (cloud chemistry modeling).

The experiments can be separated into ground-based experiments (fog at lowlands and clouds on mountain summits) and air-borne experiment (aircrafts and other air platforms).

2.4.1 Fog water studies before 1900

A scientific understanding of fog and cloud formation did not exist before discovering the role of condensation nuclei after 1870 (cf. Chapter 3.3). Prior to that, condensation was linked correctly with heat exchange and (incorrectly and mysterious) with “aqueous electricity”. However, Lampadius (1806, pp. 125–129) describes already rather well phenomena of formation and dissipation of fog. He also defined clouds as fog above our heads.[516] Furthermore, he considered fog still as a suspension of vesicles which include an electric fluid, and notes that “Fogs sometimes show a smell, or are escaping kinds of gases only mixing with the fog?”, and writes (Lampadius 1806, p. 125):

---

515 The London fog was studied and chemically analyzed first by Cohen (1896).

516 Wenn sie über unsern Köpfen schweben, nennen wir sie Wolken (p. 125). [...] Die Nebel der Gebirge sind wahre Wolken (p. 126) [...] Wolken sind gar nichts anders als hohe Nebel (p. 129).

> Die Nebel zeigen zuweilen einen Geruch, der in einigen Fällen jenem der brennbaren Luft, in anderen aber sich dem Geruch der elektrischen Materie nähert. Diese Erscheinung kann uns vielleicht auf einen chemischen Prozeß der Wasserzerlegung bey genauerer Beobachtung führen, oder vermengen sich bloß aufsteigende Gasarten mit dem Nebel? [Fogs at time show an odor which in some cases is close to those of inflammable air, in other cases come near to the odor of electric matter. This phenomenon may lead us only to a chemical process of water decomposition more precisely observed, or are only combining rising kinds of gases with the fog?]

Similar, Prout (1834, p. 315)[517] wrote that "fogs were sometimes observed of a strong odour, apparently the result of an admixture of foreign bodies". However, likely the first written statement comes from *Boyle* (1692, p. 37):

> And you may have observed, as well as I, that Fogs, some of which I have known to be very lasting, and to have a large Spread, did require no tender Nostrils to perceive them to stink.

Besides *moist fog*, *dry fogs*[518] and *stinking mist* (see Chapter 3.2.4) were distinguished. British physicist *Geoffrey Ingram Taylor* (1886–1975) separated between "clouds of smoke" or "smoky fogs", where the small particles of which these fogs are made are small solid particles of which smoke consists under atmospheric conditions which prevent them from dispersing and on the other hand fogs made by drops of water (Taylor 1917). *Jean-Baptiste Ferdinand Antoine Joseph van Mons* (1765–1842), a Belgian chemist and naturalist, describes the different kinds of fog (Mons 1817, 1827, 1828). *Mons*, who cites *Sebald Justinus Brugmans* (1763–1819), a Dutch botanist and physician in Leiden, who claimed the presence of sulfurous acid in stinking mist, investigated the fog that occurred on 12 June 1825, to search after foreign bodies. Unfortunately, *Mons* describes through all over 40 pages in philosophical manner different fogs, but writes so short on collecting the fog that it is impossible to assess the usefulness of the method:

> He brought the air impregnated with the fog by using a forcing pump and a bellows into three large glass vessels equipped with necks and valves, from which one contains distilled water, another one very diluted lime solution and a third rectified spirit of wine. The through flow was continued about two hours [translated from the German version, Mons 1828, pp. 47–48 and 66–67]

Mons notices no smell and no reaction with reagents (silver nitrate and lead acetate). *Mons* further noted that fogs could be very different in chemical composition and mentioned that *Witting* observed free acid (see Witting 1826, 1827).[519] Moreover, *Mons*,

517 *William Prout* (1785–1850) spent his life as a practicing physician in London and occupied himself with chemical research in biology and physics. He is known for his idea that atomic weight of every element is a multiple of that of hydrogen.

518 Expressed in the 1950s by *wet fog* and *dry smoke*.

519 *Witting* observed yellow leaves on crops namely wheat as a result of a fog from May 26, 1824. The yellow powder he investigated and concluded on free acids, being likely a mixture of phosphorus, hydrochloric and sulfuric acid, partly bonded on lime.

draw attention to the fact that rainwater contains a lot of different foreign bodies (Mons 1828, pp. 68–69)

> Das Meteorwasser enthält eine Menge fremder Beimischungen [. . .] es ist daher, meines Erachtens, zu vermuten, dass es Nebel geben kann und giebt, welche mancherlei Beimischungen enthalten [Rainwater contains many foreign additions [. . .] thus in my opinion, it is to assume that there are fogs, which contain various additions].

Brugmans (1783) describes the "*zwavelagtigen nevel*" [sulfureous fog] and plant damages; it was caused by the Laki eruption on Iceland, beginning on 8 June1783, and lasting over eight months.

Some scientists, who investigated rainwater, also mentioned that they analyzed fog (Zimmermann 1824, Meyrac 1852, Posnjakov 1904) but any description of the kind of fog collection is missing, and we must assume that it was collected with the same rain gauge (bowls) as mentioned by *Boussingault*, who used stretched linen with a large surface; he termed it "*pluie la brouillard*" [rain of fog]. Boussingault (1854, 1856c) was the first explicitly gave analytical values for fog water composition, but he only analyzed ammonia. *Boussingault* conducted chemical analyses of cloud (or fog) water in 1853 at Mt. Liebfrauenberg (Alsace) in October and November 1853 and in January 1854 in Paris. He found ammonium (as ammonia) to be 4.6 (2.6–7.2) mg L^{-1} at the mountain ($n = 6$) and one extreme event at Mt. Liebfrauenberg (November 14–15, 1853) had 49.7 mg L^{-1}, showing an alkaline reaction. In Paris, January 18–23 1854 he found in fog 137.85 mg L^{-1} ($n = 2$), corresponding to 0.64 g ammonium carbonate; the fog water was clear but a bit brown, what he concluded from fumes suspended in the air. The fog water also shows an alkaline reaction; a large amount of ammonia (compare his comment to the air pollution in Paris on p. 181) could explain why in some circumstances fog has a penetrating smell, he states. Stoklasa (1882) apart from rainwater also investigated fog water (likely with the same rain gauge) in summer 1877 at Libverda (Bohemia) and found 18 ± 9 mg L^{-1} NH_3 ($n = 34$) and 4.8 ± 2.4 mg L^{-1} HNO_3 ($n = 19$); thus, *Stoklasa* is the first who determined nitric acid in fog. Müntz and Aubin (1882a) found in 5 samples of fog water at Pic du Midi 0.19–0.65 mg L^{-1} NH_3, considerably less than *Boussingault* at Liebfrauenberg. There are no other fog/cloud studies in the nineteenth century known except for town fog (Chapter 3.1.2).

2.4.2 Fog and cloud water studies: understanding condensation nuclei (1925–1955)

Since *Aitken* (Chapter 2.2.3) first proposed that sea salt and hence NaCl particles act as condensation nuclei, investigation of droplet formation and interaction with condensation nuclei came into an interest in the first half of the twentieth century for understanding cloud droplet and subsequent rain the formation. Swedish *Hilding Köh-*

ler (1888–1982),[520] professor of Meteorology at the University of Uppsala, studied fog chemistry by sampling rime deposits[521] at the Haldde Observatory (Finnmark, 904 m a. s. l.; in operation 1899–1926) and determining sodium chloride (mean of 3.59 up to maximum of 450 mg L^{-1}) in the same ratio (14.2–14.8) to magnesium like in seawater (Köhler 1925). This was the first "proof" that sea salt acts as condensation nuclei. These investigations were continued by Albrecht (1925), Köhler (1929) at Mt. Sonnblick in Austria (3106 m a. s. l.) and Lipp (1932) at Mt. Zugspitze (2962 m a. s. l.). Collecting rime deposits was surely a simple method in contrast to collect liquid cloud droplets. Köhler (1929) found at the Sonnblick chloride in rime much smaller in the range of 0.01–0.06 mg L^{-1}. Lauscher and Schwarz (1932) collected rime at Sonnblick between August 1930 and September 1931 ($n = 15$) with a mean of 0.21 (0.04–0.55) mg L^{-1}. At the same time, Lipp (1932) found at Zugspitze similar values ($n = 12$) for chloride to be 0.55 (0.005–0.590) and for magnesium 0.89 (0.070–0.210); the ratio Cl/Mg to be 0.76 (0.2–10.7) shows that on continental mountains other condensation nuclei exist than close to the sea.

Original for sampling and chemical analysis of condensation nuclei, Quitmann and Cauer (1939) developed the "*Cauersche Kondenskugel*" [Cauer's condensation dew-cup], used by *Cauer* until the middle of the 1950s (Cauer and Cauer 1941, Cauer 1956, best described in Quitmann 1940); Fig. 2.20. However, already Mrose (1940) criticized this method because on the cooled wall almost all water by sublimation together with gaseous species is collected. However, Junge (1953, p. 20) noted that *Cauer*'s cup in a first approximation reflects the natural condensation process in clouds; joint experiments together with *Cauer* have shown that only a small proportion of the aerosol is captured by the dew but a substantial amount of gaseous substances deposit. Cauer (1948b, 1956)[522] used this sampling method later for the determination of "potential aerosols", i. e., existing nuclei (he termed it "*Nebelkerne*" in German) and gaseous precursors like HCl, NH_3, etc.; this principle of artificial dew formation had been already used by *Emil Schöne* for determination of H_2O_2 in the air (Schöne 1874), see p. 515–516.

520 Swedish theoretical meteorologist at University Uppsala (also written *Koehler* but *Kohler* is wrong).

521 *Rime* (in German: *Nebelfrost* = *Raureif* = *Raufrost* = *Raueis*) occurs when supercooled water droplets (at a temperature lower than 0° C) in fog or cloud come in contact with a surface that is also at a temperature below freezing; the droplets are so small that they freeze almost immediately upon contact with the object. Rime is common on windward upper slopes of mountains that are enveloped by supercooled clouds. Frost (= hoarfrost; in German: *Reif*) is – similar to dew – resublimation (i. e., direct crystallization) of atmospheric moisture on the ground and on exposed objects. The German Weather Service distinguished between *Raureif*, *Raufrost* and *Raueis* with respect to humidity, temperature and wind speed (in English all terms = rime); *Nebelfrostablagerung* = rime deposit.

522 In several publications *Cauer* argued that the "atmospheric" pH (namely that of aerosol) impacts human health, such as the skin (Cauer 1957).

Figure 2.20: The condensation dew-cup by *Hans Cauer* ["*Cauersche Kondenskugel*"]. Left: condensation in action; photo by courtesy of *Grete Cauer*. Right: The complete experimental set-up with *Aßmann* psychometer (right), thermohydrograph (below in the background), and funnel with sampling glass (below), Source. Cauer (1951a, p. 1128).

Helmut Mrose, who also was the first in Europe who collected fog and cloud water (Chapter 2.4.3) was also the first who analyzed rime on compounds owing to artificial air pollution; before, only sea-salt compounds were seen as essential condensation nuclei. Mrose (1940) collected in the surrounding of Jena (Thuringia) in winter 1939–1940 samples of rime water ($n = 9$) and found them almost black due to soot, containing much of sulfate (12–387 mg L^{-1} as SO_3), nitrate (4.2–14.2 mg L^{-1} as N_2O_5), chloride (2.1–12.9 mg L^{-1}), and ammonium (0.4 –12.9 mg L^{-1} as NH_3). In the filter residual, he determined iron, calcium, silica, and aluminum and concluded that all compounds are originated from coal combustion and industrial dust emissions. *Robert M. Cunningham* (1919–2008), an American cloud physicist and specialized in the study of fog, found sea fog in relation to the evaporation of seawater droplets (Cunningham 1941); see further on condensation nuclei in Chapter 3.3.2. *William H. Radford* (1909–1966) developed the first active fog sampler[523] and reported on 8 years fog wa-

523 The principle is analogous to the Caltech active cloud water collector (US Patent 4697462 from 1986) by B. C. Daube Jr., R. C. Flagan and M. R. Hoffmann. A "rotating-arm collector (RAL)" was patented by R. J. Pilie and E. J. Mack in 1974 (US 3889532) and later modified by Jacob et al. (1984).

ter sampling using a passive collector in Round Hill (USA); Radford (1938). Houghton and Radford (1938a, p. 14) wrote:[524]

> Chemical analyses of numerous samples of fog water indicate that although it always contains dissolved salts, the concentration is usually so low that, for the purposes of fog dissipation, fog particles may be regarded as practically pure water.

Unfortunately, Radford (1938) only published the total ionic content to be 70 (8–480) mg L^{-1}; he wrote (Radford 1938, p. 30):

> The principal constituents appear to be chlorides and sulphates. The purposes, methods and results of the analysis of these and other fog water samples will be fully described in a forthcoming paper.

However, likely due to the war, no further publication did appear. Only *Radfords* coauthor *Henry Garrett Houghton* (1905–1987) published in 1955 new results on chemical investigations from 56 fog samples collected at the NE coast (Brooklyn, Kent Island, Round Hill, and Nantucket Island) and 35 cloud water samples from Mt. Washington with the aim for a better understanding of CCN; hence, only chlorides (1–98 mg L^{-1}), sulfate (6.5–52 mg L^{-1}) and pH (3.5–7.4) were estimated, concluding that "the active nuclei of condensation are composed primarily of sulfates and chloride" (unfortunately, ammonium and nitrate have not been determined).

In the same context of studying condensation nuclei, the well-known Japanese geochemist *Yasuo Miyake* (1908–1990) analyzed in the 1940s cloud water at Mt. Kirigamine (1930 m a. s. l.), Prefecture Nagano, rime at Mt. Fuji (3780 m a. s. l.) and Mt. Oiwake (1200 m a. s. l.) as well as sea fog at Mt. Sharidake (1200 m a. s. l.) at Hokkaido on Na, K, Mg, Ca, chloride and sulfate (Miyake 1948, 1965, 1978).[525] This is the first comprehensive fog, rime, and cloud chemistry investigation worldwide (Table 2.21).

Pioneering work in the field of cloud physics, dynamics, and condensation nuclei was conducted in Russia at the Central Aerological Observatory soon after the war. Different techniques for sampling and measurements were developed, namely cloud water impactors by *Aleksandr Khristoforovich Khrgian* [Александр Христофорович Хргиан] (1910–1993),[526] *Solomon Moiseevich Shmeter* [Соломон Моисеевич Шметер]

---

524 *Radford* and *Houghton*, both electrical engineers, worked at the Massachusetts Institute of Technology (MIT), were a member of the research staff at MIT's Round Hill Research Station from 1928 to 1938, where her work on the physical properties of fog (Houghton 1931, 1932) led to the development of one of the first practical methods of dissipating fog (Houghton and Radford 1938b). An excellent early fog climatology of the USA is given by Stone (1936).

525 This textbook was published first in Japanese in 1954. Within the group of *Miyake* many analyses of rain, rime and seawater were carried out between 1935 and 1939; however, in the English edition (1965) no original papers are cited, also not by Miyake (1978).

526 Soviet meteorologist was one of the world-leading cloud physicists; he worked at the CAO (Central Aerological Observatory) and the Moscow University.

(1920–2005), *Aleksandr Moiseevich Borovikov* [Александр Моисеевич Боровиков] (1917–1975), andaerosol particles by *Rostislav Ivanowitsch Grabowski* [Ростислав Иванович Грабовский] (1914–?) using different aircraft (of type LI 1, IL 28, PO 2) from the airbase Dolgoprudny near Moscow (1949–1950) and analyzed only for chloride (Grabowski 1951a, Shmeter 1952, Khrgian 1952, Khrgian and Mazin 1952, Khrgian et al. 1961).[527] At the MGO in Leningrad, pioneering work in developing instruments for the determination of liquid water contents and droplets sizes in fogs and clouds were done by *Vasili Aleksandrovich Zaizev* [Василий Александрович Зайцев) and *Aleksei Aleksandrovich Ledokhovich* [Алексей Александрович Ледохович][528] (Zaizev 1948a,b, 1950, Zaizev and Ledokhovich 1960, 1962). Sophisticated aircraft samplers for supercooled droplets were further developed at the end of the 1950s (Aleksandrov and Petrenchuk 1959, 1962).

Takeshi Ohtake (1926–2017), a Japanese meteorologist, collected cloud drops at different heights at Mt. Zao (1956–1957) in a clever experiment, using collodion films for droplet sampling with simultaneous microphotographs to get the original drop sizes, whereas the droplets were evaporated and the residue examined by electron microscopy. The results are given for cloud nuclei classification as sea salt (mean 25 %), combustion (mean 71 %) and soil matter (mean 2 %) depending on the altitude. The chlorinity increases significantly with height.

2.4.3 Fog and cloud water studies: understanding air pollution (1955–1970)

In the 1950s, only two studies existed about fog and cloud chemistry, that by Houghton (1955), see above, and another by *Helmut Mrose*, German meteorologist at the Wahnsdorf Observatory, who is the second scientist in the world after *Miyake* who collected fog and cloud water (since 1954) in Germany for extensive chemical analysis at four mountain stations (Großer Inselsberg, Geisingberg, Mt. Brocken, and Collmberg) and at the Baltic Sea in Kap Arkona (Mrose 1955, 1958, 1961, 1966), see Table 2.18. Mrose (1955, p. 38) writes that between 1950 and 1955 at Mt. Inselsberg 118 fog or cloud samples were analyzed on sulfate, showing a strong dependency on the wind direction (in mg L^{-1}): lowest from SE (15.5), largest between NE an E (47.7 and 57.8), and all other directions between 22.6 and 34; the range of concentrations concerned other substances he noted to be 1–10 for chloride, 3–6 for ammonium and 2–15 for nitrate.

---

527 Addional scientist must be named in relation to cloud and aerosol sampling and chemical analysis: B. B. Манцевич [Manzevich], Юрий А. Гильгнер [Juri Gilgner], Владислав Евгеньевич Минервин [Vladislav Evgenjevich Minervin] (1920–2002), B. C. Хахалин [Chachalin], Соломон Моисеевич Шметер [Solomon Moiseevich Shmeter] (1920–2005), Александр Моисеевич Боровиков [Aleksandr Moiseevich Borovikov] (1917–1975), Илья Павлович Мазин [Ilja Pavlovich Mazin]. The two papers are cited by Petrenchuk and Drozdova (1966) in Latin and different name spelling.

528 Romanization of Russian names is made by the British Standard if the name is not found romanized in international literature.

Table 2.18: Early estimates of fog and cloud chemical composition (in mg L^{-1}). A: Boussingault (1853), B: Köhler (1925), C: Lauscher and Schwarz (1932), D: Lipp (1932), E: Miyake (1948), F: Houghton (1955), G: Mrose (1966), H: Petrenchuk and Drozdova (1966), I: Okita (1968).

site	type	Cl^-	SO_4^{2-}	NO_3^-	NH_4^+	Na^+	K^+	Mg^{2+}	Ca^{2+}	pH	author
Liebfrauenberg[a]	fog	–	–	–	4.6	–	–	–	–	–	A
Paris[b]	fog	–	–		138	–	–	–	–	–	A
Haldde[c]	rime	3.59	–	–	–	–	–	–	–	–	B
Sonnblick[d]	rime	0.21									C
Zugspitze[e]	rime	0.55	–	–	–	–	–	0.89	–	–	D
Mt. Kirigamiru[f]	rime	4.8	13.6	–	–	2.4	1.3	3.1	3.5		E
Mt. Fuji[g]	rime	0.18	0.26	–	–	0.13	–	0.21	0.03		E
Mt. Sharidaku[h]	cloud	4.5	12.6	–	–	3.8	–	1.8	1.7		E
Round Hill[i]	fog	34.8	11.4	–	–	–	–	–	–	–	F
Mt. Washington[j]	cloud	0.14	7.1	–	–	–	–	–	–	4.5	F
Collmberg[k]	fog	20.7	158.5	11.8	35.7	–	–	–	63.6	4.2	G
Gr. Inselsberg[l]	cloud	8.3	63.0	9	10.8	–	–	–	–	–	G
Brocken[m]	cloud	7.3	37.2	13.9	12.1	19	3.4	–	4.4	5.1	G
Kap Arcona[n]	fog	61.7	89.3	28.1	39.7	34.5	9.5	–	15.0	3.9	G
North ETU[o]	cloud	6.3	0.6	0.1	0.6	2.4	1.7	1.0	0.4	5.3	H
North-west ETU[o]	cloud	2.5	6.0	0.5	0.9	1.2	0.8	0.2	0.6	4.7	H
South-west ETU[o]	cloud	2.1	8.9	0.8	1.8	0.7	0.6	0.5	1.0	4.8	H
Mt. Norikura[p]	cloud	4.6	26.2	3.4	3.3	2.2	2.8	–	–	3.8	I
Mt. Shiobara[q]	fog	17	60	8	7	11	7	–	–	4.9	I
Mt. Tsukuba[r]	cloud	24	64	1	11	7	6	–	–	–	I

[a] 1853, France, Alsace (309 m a. s. l.)
[b] One event
[c] 1925, in Finnmark (904 m a. s. l.)
[d] 1931, Austria (3106 m a. s. l.)
[e] 1931–1932, Germany (2962 m a. s. l.)
[f] 1940s, Japan, Hokkaido (1200 m a. s. l.)
[g] 1940s, Japan (3780 m a. s. l.)
[h] 1940s, Japan, Hokkaido (1200 m a. s. l.)
[i] 1954, USA, Virginia
[j] 1954, USA (1917 m a. s. l.)
[k] 1954–1960, Germany, between Dresden and Leipzig (370 m a. s. l.), $n = 12$
[l] 1950–1955 ($n = 118$), 1956–1960 ($n = 30$), Germany, Thuringia (916 m a. s. l.)
[m] 1954–1960, Germany, Harz (1152 m a. s. l.), $n = 18$
[n] 1954–1960 Germany, Island Rügen (Baltic Sea), $n = 42$
[o] 1961–1964, aircraft sampling above the European territory of the USSR ($n = 16$, 29, and 46)
[p] 1963, Japan (3026 m a. s. l.), $n = 10$, nitrite = 0.04 (0.00–0.10)
[q] 1963, Japan, foothill of mountains, $n = 2$, nitrite = 2.0 (1.9 and 2.2)
[e] 1963, Japan (875 m a. s. l.), $n = 5$, nitrite = 0.4 (0.08–1.05), LWC = 0.19 (0.09–0.27) g L^{-1}

Whereas *Houghton* has already used a more sophisticated active sampler, but determinates only chloride and sulfate, *Mrose* determines chloride, sulfate, nitrate, and ammonium in 1954 (later additional sodium, potassium and calcium), but used a simple passive sampler: linen was clamped in a wooden frame where the running water was collected in a bottle. Mrose (1958) writes that chemical analysis of fog water is an excellent method for the determination of trace substances in the air due to the large enrichment from the gas to the water phase but the amount of liquid water content must be known; he determined it psychrometric. The first passive sampler was constructed by the German meteorologist *Johannes Grunow* (1902–1971) in form of a string-collector (Grunow 1955), but was used only for estimation of the fog water deposition in comparison with rainfall. Cauer (1956, p. 502) began cloud water sampling at two mountain sites (Rodskopf and Vogelsberg), but due to the low number of samples he only notes that pH values were in the range of samples collected with his condensation cup; notable is that he had personal contact with *Mrose* concerns his fog water sampling at Inselsberg (Thuringia).

In the early 1960s, Okita (1968) collected fog and cloud water using an active string collector at three sites in Japan (Table 2.18); he was the first who determine nitrite, the ratio nitrite/nitrate amounts 0.02 (0.004–0.05) at Mt. Norikura (3026 m) and 0.4–0.5 (0.04–1.3) at the other lower sites. The latter value is much larger than found in rainwater (Table 2.6).

A unique cloud water sampling program using aircraft was launched in 1961–1964 over the European part of the USSR, collecting 150 samples over three regions on the routes Leningrad – Archangelsk – Narjan – Mar – Pechora and Leningrad – Minsk – Kyiv – Dnepropetrovsk. (Drozdova et al. 1962, Petrenchuk and Drozdova 1966). Together with cloud chemical composition, the rainwater chemical composition of the regions from the network are presented in comparison. They found the Cl/Na ratio to be significantly larger (2.1–3.0) compared to that in rainwater (0.8–1.5); however, they assume a chlorine emission from an industrial source, whereas the data show that the ratio increases with increasing sulfate, thus suggesting Cl loss due to acidification. Clearly is seen the increase of the total ionic concentration when approaching towns and industrial areas. Furthermore, as a remarkable result, the influence of cloud types have been found; St (stratus) and Sc (stratocumulus) having small droplets show large concentrations of sulfate, nitrate, ammonium, a low pH (4.5) but much less bicarbonate comparing to frontal Ns (nimbostratus) clouds (whereas the content of Na, K, Mg, Ca and Cl is not different and pH is 5.2). In that time, nothing was known on chemical cloud processing, but the data suggest that scavenging of the gaseous precursors SO_2, NH_3 and HNO_3 is more important for smaller droplets whereas the crustal elements in particulate matter, activated by sulfate, ammonium and nitrate provide the nucleation mode.

An interesting study conducted Oddie (1962), who collected precipitation samples by an aircraft over Southern England at heights up to 2500 m, i. e., in the cloud level with the aim to obtain samples free from contamination either from industrial sources or from sea spray. However, like the results (much later presented to the public) by *Petrenchuk* and *Drozdova*, most of the samples still showed clear evidence of industrial pollution. It is notable that in that time little was known on the relations between cloud type, droplet sizes, air mass characteristics and cloud water chemical composition, as the following conclusion by *Oddie* express: It is suggested that these differences must arise from some reaction, occurring either at the sea surface or in the atmosphere itself, but there is insufficient evidence to determine its nature. A next cloud chemistry study was conducted end of 1967 by Lazrus et al. (1970) at the top of Pico del Oeste (1020 m) in Puerto Rico with the aim to understand the mechanisms of sea salt fractionation; cloud water was collected with a large passive aluminum screen ($n = 8$) and analyzed for Cl, Na, Ca, Mg, and sulfate. They found no chemical fractionation of sea salt and neither enrichment nor depletion of chlorine. The chemical concentration in fog water is almost significant higher compared to that in cloud water (Tables 2.19 and 2.20) for different physico-chemical reasons, namely, local pollution and emission sources are reflected by fog water chemical composition, for example, chloride and sodium close to the ocean, ammonium at agricultural sites, and nitrate and organic substances in areas with heavy traffic, such as California, where numerous investigations have been conducted in the 1970s (Chapter 2.4.4).

Moreover, high acidity (pH < 3) and likely phytotoxic substances found in fog, led at the beginning of the 1980s to the idea that forest damages observed at lower mountain sites could be caused by fog or cloud. Conversely, observation of forest declines stimulated fog research in the 1980s.

Table 2.19: Chemical composition (in μeq L^{-1}) of fog water at different regions of the world; arithmetic mean if not other noted.

Site/region	Cl^-	NO_3^-	SO_4^{2-}	Na^+	NH_4^+	Ca^{2+}	Mg^{2+}	H^+	pH
Brooklyn, N. Y. (USA)[a]	28	–	312	–	–	–	–	20	4.7
Nantucket Island (USA)[a]	2500	–	1061	–	–	–	–	–	–
Kap Arkona (Rügen)[b]	1738	153	1860	1500	1653	750	–	158	3.8
Collmberg (315 m)[b]	583	190	3302	–	2100	3176	–	63	4.2
Zinnwald (877 m)[m]	450	680	1960	–	1500	–	–	5	5.3
Taunus[c]	642	1540	1210	–	–	–	–	1450	2.8
Central California[d]	223	1110	292	139	1580	84	27	479	3.3
Po valley, North Italy[e]	174	775	820	37	1606	52	17	292	3.7
Jan Joaquin valley, California[f]	572	1425	493	316	750	74	89	1105	3.0
Jan Joaquin valley, Bakersfield[g]	47	850	1160	19	1440	47	6	60	4.2
Bern-Belpmoos (Switzerland)[h]	605	578	1300	54	1534	309	58	3	5.5

[a]11 and 5 samples, respectively (Houghton 1955)
[b]28–42 samples, around 1960 (Mrose 1966)
[c](Schrimpff 1983)
[d]1980–1982, 10 samples, 150 km north of Los Angeles (Brewer et al. 1983)
[e]November 1984 and 1985, 169 samples (Fuzzi 1988)
[f]1981–1983, 53 samples from 11 stations (Jacob et al. 1985)
[g]California, December 1982 until January 1983, 108 samples (Jacob et al. 1984)
[h]Median, agricultural area 515 m a. s. l., October 1983 until March 1985, 27 samples (Fuhrer 1986)
[m]105 samples from 35 events May–November 1987 (Zier 1992)

Table 2.20: Chemical composition (in µmol L^{-1}) of fog water in California (USA), 108 samples from December 1982 until January 1983, after Jacob et al. (1984).

substance	concentration	substance	concentration
TOC	4000	V	55
S(IV)	515	Cu	34
Fe	400	Mn	14
Pb	330	sum of organics	4165
HCHO	165	sum of inorganics	3487
Ni	61	sum of metals	894

2.4.4 Fog and cloud water studies: understanding chemistry (1976–1990)

After 1980, the number of fog and cloud chemistry studies rose by several reasons: air pollution assessment, chemical processing, high-altitude deposition, and cloud-aerosol-interaction. Namely the so-termed new-type forest decline initiated the first long-term cloud chemistry monitoring in Northern USA and several case studies in Germany and Italy. Fog and clouds in mountains upslope and forces clouds which than intercept with the vegetation. Vegetative interception is a source of water and nutrients for the ecosystem but also acidification.

Mack et al. (1974, 1977) started with the investigation of coastal fog in California; it was systematically continued by Waldman et al. (1982, 1985), Brewer et al. (1983), Munger et al. (1983), Hoffmann 1984, Jacob et al. (1984, 1985), and Fuzzi et al. 1984) in the Los Angeles area (Pasadena, Lennox, and Bakersfield). A large number of compounds were analyzed: hydrogen ion, sodium, potassium, calcium, magnesium, ammonium, nitrate, sulfate, chloride, and often fluoride, iron, lead and nickel, and organic compounds.

Waldman et al. (1982) observed six fog events during 1981 and 1982 in the Los Angeles basin; the so-termed Caltech rotating arm collector (RAC) was used for sampling (first design by *Mack* and *Pilie*), Fig. 2.21. The reported pH values were in the range of 2.2–4.0 and *Waldman* et al. concluded that the dominant processes controlling the fog composition are the preexisting aerosol and the scavenging of gaseous nitric acid. This work is also the first that reports on formaldehyde (HCHO) in fog water (3–14 mg L^{-1}).

Brewer et al. (1983) used two different simple passive samplers, made as a polyethylene sheet and the other as a polypropylene mesh and collected 20 mist and 10 fog samples between 1980 and 1982 at 16 locations. They do not define the difference between mist and fog, but adopting the meteorological definition that mist droplets are smaller (because of higher visibility), the somewhat higher concentrations (namely for ammonium) in mist comparing to fog are well explained. Median pH amounts 3.3 (fog) and 3.6 (mist). The aqueous-phase concentrations reflect the air pollution which is dominated by NO_2 ($\geq$0.365 ppm) in contrast to SO_2 ($\leq$0.1 ppm).

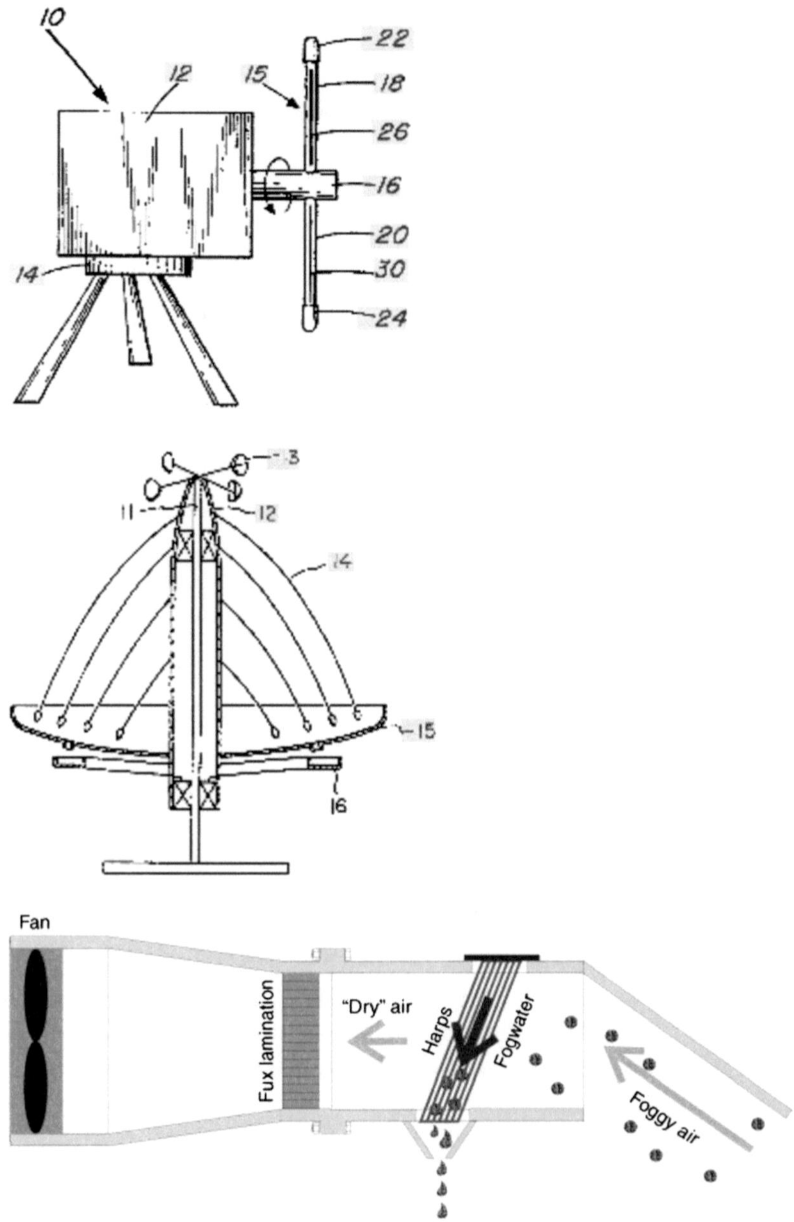

Figure 2.21: Historic cloud water samplers. A) Rotating slotted stainless-steel tube by *Roland J. Pilie* and *Eugene J. Mack* 1975 (USP 3,889,532). B) windmill with flexible rods by *Youshio Usui* 1994 (USP 5,275,643). C) Caltech Active Strand Cloud Water Collector (CASCC), Daube et al. (1987).

The work by Munger et al. (1983) was the first experimental study in the background of topical results from laboratory and modeling studies on the conversion of SO_2 to sulfate. Eight samples were collected using the rotating arm collector. The concentrations correspond to the previous studies, ranging over two orders of magnitude (sulfite, Cu and Mn were additionally included to the compounds listed above). This paper first measured sulfite at high concentrations (200–400 µeq L^{-1}), accounting for 10–20 % of the sulfate concentration and discusses the adduct formation between sulfite and formaldehyde. Formaldehyde and other carbonyls were just determined in air (Grosjean 1982) and fog (Grosjean and Whright 1983).

Jacob et al. (1985) conducted a more extensive study between 1982 and 1983 including nine sites ($n = 64$). A problem at that time was the absence of a widely accepted method for LWC measurement (the PVM-100 was later introduced by Gerber 1984); the LWC was estimated from the collection rate of the rotating arm collector assuming 60 % efficiency to allow (for the first time) to present weighted-mean values of the fog water composition. Nitrate in the mM range and pH < 3 were observed (Table 2.20). Kawamura and Kaplan (1984) first detected organic acids in fog (and rainwater), and Munger et al. (1984, 1986) carbonyl sulfonates in fog. Glotfelty et al. (1987) detected pesticides in fog.

In 1976, the Atmospheric Science Research Center (ASRC) has begun a routine program of continuous cloud water collection and chemical analysis at Whiteface Mountain in Adironbacks in the northeast of New York State (Castillo 1979, Falconer and Falconer 1980). It was primarily focused on studying the reasons for new-type forest decline.[529] This worldwide first cloud chemistry monitoring is still in operation (with several interruptions), and results in several important findings, which were the basis for the first establishment of chemical cloud climatology. Mostly stratus type clouds were observed during this program. However, it was also shown that a greater database and more detailed information (e. g., determination of exact cloud base height for each event) are required to arrive at a climatology that is truly representative of a region (Castillo and Jiusto 1983, Fuzzi et al. 1984, Castillo et al. 1985, Weathers et al. 1988, Mohnen 1989, Mohnen and Kadlecek 1989, DeFelice and Saxena 1991, Mohnen and Vong 1993). Since 1986, Mt. Whiteface cloud water sampling was included in a larger project, the "Mountain Cloud Chemistry Program (MCCP)", which includes four other mountain sites (Mitchell, Moosilanke, Senandoah, and Whitetop) between 1986 and

529 Also termed "*Waldsterben*" according to German. Observations of unexpected foliar diseases, needle yellowing and needle loss on trees were made in West Germany in the 1970s, later named "new-type" forest decline. Smoke damages [*Rauchschäden*] were first identified by Stöckhardt (1850) and since that time systematically investigated in Saxony (see Chapter 4.7.2.2). End of the nineteenth century, the terms acidity [*Säuregehalt*] of air, smoke, and precipitation, acid fogs [*saure Nebel*] were used (in parenthesis the German original word) which was very likely unknown before 1985 to modern researchers (Wislicenus et al. 1914), Waldman et al. (1982), Hileman (1983) and Hoffmann (1984) rediscovered "acid fog".

1989 (Mueller and Weatherford 1988, Saxena et al. 1989, Saxena and Lin 1990, Li and Aneja 1992). Additionally, at three mountain sites in Quebec, Canada, "The Chemistry of High Elevation Fog (CHEF) Project" was carried out from 1985 until 1991 (Schemenauer 1986, Schemenauer et al. 1988, 1995). The whole program was focused on cloud water deposition to forests (Baumgardner et al. 2003). Thus Castillo et al. (1985) collected over 50 cloud water samples during five comprehensive case studies concerned stratiform clouds chemistry at Whiteface Mountain, New York. The water samples were analyzed for pH, conductivity and ions of sodium, potassium, magnesium, calcium, ammonium, sulfate, chloride, and nitrate. Trajectory analyses and cloud condensation nucleus concentrations confirmed that the air masses in all five of these cases represented continental air that was relatively clean (low aerosol concentration) for the northeastern United States. However, a mean pH = 3.6 for all collected cloud samples was not different from the fog studied in the highly polluted air of California. The low pH values were related to a normal background of nitrate ions found in the rural continental air masses plus sulfate ions largely from the industrial emissions of the Midwestern United States. This example shows very nicely that not the total ionic content determines the pH but the acid-base budget. Summarized data on the Mountain Cloud Chemistry Program can be found in Saxena and Lin (1990) and Kim and Aneja (1992).

Additionally, airplane sampling of cloud water began; Mohnen (1980) developed at the end of 1970s a special sampler; an excellent overview on aircraft sampler is given by Straub and Collet (2004). In that time (ending 1970s), it became clear that conversion of SO_2 into sulfate occurs preferably in clouds (Möller 1980); thus, Hegg and Hobbs (1981, 1982) conducted aircraft measurements and found strong evidence for sulfate production in some of the clouds. Between 1982 and 1983, Daum et al. (1984) collected 88 samples of stratiform cloud water above the eastern United States and measured ammonium, nitrate, sulfate both in cloud water and in aerosol as well as parallel in gas phase NO_x, H_2O_2, and O_3. As expected from theory, a strong inverse relationship was observed between cloud water H_2O_2 (not measured but calculated from the *Henry* equilibrium)[530] and interstitial SO_2.

Hegg and Hobbs (1986) studied, in 1984, the in-situ formation of sulfate and nitrate while collecting samples in cloud layer by aircraft in and around cumuliform clouds in the western portion of Washington State. Analyses of chemical measurements indicate that for the 13 cases studied, the mean sulfate production led to $0.9\,\mu g\ m^{-3}$ with an uncertainty of $0.5\,\mu g\ m^{-3}$. No nitrate production in the clouds was detected and the nitrite concentrations in the cloud water (up to 0.19 ppm) appeared to be greater[531] than can be explained by the absorption of NO_x and HNO_2.

---

530 *William Henry*.

531 In that time, photocatalytic HNO_2 formation from NO_2 was still unknown, but these data support our own findings (Acker et al. 2008a).

Isaac and Daum (1987) conducted aircraft measurements in January and February 1984 in Ontario, Canada, and found a mean cloud water pH of 3.6 and an equivalent ratio of nitrate to sulfate of 0.7 and 1.4 for cloud water and ground-based rainwater; this shows also the importance of sulfate production in clouds and the scavenging of nitric acid by precipitation.

As Mrose (1966) first stated, that chemical cloud water monitoring is a simple way to get insights into non-local air pollution (in contrast to fog chemistry), Ogren and Rodhe (1986) first studied, in 1984, systematically cloud water chemistry in relation to sectors in southern Sweden; they found that clouds connected with southern air masses, mostly from East Germany and Poland, contain much more sulfate (33.6 mg L^{-1}) and nitrate (4.2 mg L^{-1}) than from all other sectors (2.2 mg L^{-1} sulfate and 0.4 mg L^{-1} nitrate).

In Europe, a long-term fog chemistry program was started in 1982 and lasted over fifteen years in the Po Valley at San Pietro Capofiume in Northern Italy by *Sandro Fuzzi* and co-workers (Fuzzi 1988, Fuzzi et al. 1983, 1992, 1996, 2002, Facchini et al. 1992). Systematic studies[532] around the world began in the 1980s: in Austria (*Hans Puxbaum*), in Japan (*Manabu Igawa* and others), and in Germany (*Hans-Walter Georgii, Wolfgang Jaeschke, Otto Klemm*); intensive cloud and fog studies also were started in Taiwan and China after 1990. Research on forest damages and acid deposition started in Germany at the beginning of 1980s mainly at mountain sites (Wisniewski 1982, Hileman 1983, Schrimpff 1983). Besides air pollution research, the very early understanding of nutrient budgets again came into focus (Fuhrer 1986, Zobrist et al. 1985, Turner 1991); it is remarkable that the nitrogen load was similar to that 159 years ago found by *Baussingault* (38 mg L^{-1} and 63 mg L^{-1}). A leading institute in Germany for rain, fog and cloud water research was under the leadership of *Hans-Walter Georgii* at the Meteorological Institute (1965–1992), University Frankfurt/Main including the Taunus Observatory at Kleiner Feldberg. First cloud water sampling was conducted in the early 1980s at Kleiner Feldberg (Schmitt 1987), analyzing Na, K, Mg, Ca (in 10-fold higher concentrations than in rainwater), and Al, Fe, Cd, Mn and Cr in the range 0.01–0.4 mg L^{-1}. Schell et al. (1992) and Winkler (1992) constructed fog water collectors and organized intercomparison campaigns. In 1990, the first European cloud chemistry field campaign was carried out at Kleiner Feldberg under the participation of scientists from Italy, Austria, Sweden, and Britain (Wobrock et al. 1994, Fuzzi 1995).[533]

It is notable that fog (and cloud) water sampling in Eastern Germany, which was begun by *Mrose* in the middle of 1950s, was continued in the "Erzgebirge" (Saxony) by *Manfred Zier* (born 1937) from the Observatory Wahnsdorf in the late 1980s

532 Some of them (and other not cited) were focused on fogs and/or mixed mountain clouds.

533 My co-worker *Wolfgang Wieprecht* took part while staying some month there.

(Zier 1992)[534] and again in the late 1990s by *Frank Zimmermann* from the Mining Academy [*Bergakademie*] Freiberg (Zimmermann and Zimmermann 2002, Lange et al. 2003). It is important to note (several authors "forget" it or do not consider it) that the concentration of fog water constituents is not only several times higher than that in rainwater but also compared to ("true") cloud water what it follows for ammonium and nitrate (in mg L^{-1} as N); these dates also illustrate that between the 1950s and end of 1980 there was no change and with the drastic emissions abatement at the beginning 1990s the concentrations in fog and cloud water declined, as later shown within the Mt. Brocken cloud chemistry program:

year	type	ammonium-N	nitrate-N	author
1950s	fog	28–36	12–40	Mrose (1966)
1950s	cloud	12	14	Mrose (1966)
1987	fog	27	42	Zier (1992)
1990s	fog	5	14	Lange et al. (2003)
1990s	cloud	5	5	Möller et al. (1996)

The number of cloud chemical measurements is rather limited; about at 50 sites cloud water was sampled, almost only for short time (Table 2.21 for selected examples). Smallest ever measured concentrations were found at Åreskutan from northwestern (polar) air masses, likely representing North Atlantic CCN background. At this station, when southern (continental) air masses arrive, concentrations increase by more than one order of magnitude. Local pollution (e. g., at Kleiner Feldberg) is responsible not only for higher sulfate and nitrate but also for ammonium, which is originated from agricultural activities.

2.4.5 Fog and cloud water studies after 1990: an outlook

In Europe, an intensive phase in organizing cloud field experiments was realized within EUROTRAC[535] (as GCE project) between 1989 and 1993, and later in the 1990s, in Asia (Japan, China, Taiwan), long-term cloud chemistry programs were established which partly are going on. 13 research groups from 7 European countries participated

534 During 35 fog episodes in 1987, 105 fog water samples were gained by a passive fog water collector at the weather station Zinnwald on the ridge of the Erzgebirge.

535 European scientific research program within the EUREKA framework. It was accepted as a EUREKA project in 1985 and, after a two-year definition phase, began work in 1988. The project finished at the end of 1995. A second phase of the project runs from 1996 to the end of 2002. EUREKA was established 1985 as a European research initiative, still working as an open platform for international cooperation in innovation, also associating non-EU countries.

Table 2.21: Chemical composition of cloud water at different sites in the world (µeq L^{-1}); A – altitude (a. s. l.) in m.

site /area	A	Cl^-	NO_3^-	SO_4^{2-}	Na^+	NH_4^+	K^+	Ca^{2+}	Mg^{2+}	H^+
Mt. Rigi (Switzerland)[a]	1620	53	490	300	34	660	8	46	12	23
Seeboden (Switzerland)[a]	1030	77	440	380	20	920	15	88	21	5
Mt. Wilson (LA, USA)[b]	750	190	1486	859	241	580	21	139	90	1184
Whiteface Mt. (USA)[c]	1483	–	73	410	–	157	–	–	–	32
Whiteface Mt. (USA)[d]	1483	31	110	140	11	89	20	17	6	280
Mt. Brocken[e] (1993/95)	1142	73	236	213	303	72	6	57	22	103
Mt. Brocken[f] (1996/02)	1142	92	235	176	279	96	4	38	22	61
Kleiner Feldberg[h]	826	242	779	409	107	1543	22	56	–	316
Åreskutan (Sweden)[k]	1250	5	9	32	6	10	<0.8	0.7	0.8	35
Åreskutan (Sweden)[m]	1250	19	68	700	33	–	6	22	12	4
Åreskutan (Sweden)[n]	1250	0.8	2	6	<0.4	1	<0.8	0,3	<0.2	13
Great Dun Fell[o]	847	238	238	472	246	306		37	67	321
Mt. Srenica[p]	1362	46	502	447	58		471	88	29	354

[a]Winter 1990 until spring 1991 (Collett et al. 1993)
[b]120 samples between 1982 and 1983 (Waldman et al. 1985)
[c]1986–1988 (Mohnen and Vong 1993)
[d]28 samples from stratus clouds 1976 (Castillo et al. 1985)
[e]3476 samples 1993–1995; note that Mrose (1966) found much higher concentrations (Table 2.18); this led to the assumption that *Mrose* sampled cloud water near the cloud base
[f]8086 samples 1996–2002
[g]655 samples from aircraft between 1960 and 1970 (Petrenchuk 1979b)
[h]1983–1986 (*Wolfgang Jaeschke*, pers. Commun.)
[k]80 samples summer 1983/1984 (Ogren and Rodhe 1986), only sectors NE + W
[m]4 samples from south (polluted air)
[n]41 samples from NW (clean air)
[o]Middle England, 4 cloud events, 30 samples, LWC weighted mean (May 1993); Acker et al. (1999a)
[p]Krkonoše, Southwest Poland, 6 cloud events, 63 samples, LWC weighted mean (September/October 1993); Acker et al. (1999a)

at 3 large field campaigns. It was the aim of GCE (**G**round based **C**loud **E**xperiment) to characterize in principle the interaction between droplets (clouds and fog) with the gaseous and solid particulate phase. Using different measurement sites, the scavenging and aqueous-phase transformation were studied in different polluted air masses and cloud types as well fog (Po valley campaign in 1989 in an agricultural area; Kleiner Feldberg, near Frankfurt am Main campaign in 1990 in the polluted urban area; Great Dun Fell experiment 1993 is mostly marine influenced orographic clouds). During the Kleiner Feldberg experiment mainly the formation of cloud drops, the time-depending evolution of chemical species in the multiphase system and studying of different sampling systems were done (Wobrock et al. 1994). Both experiments (Po valley and Kleiner Feldberg) were running in autumn and can be chemically characterized by oxidant limitation. In contrast, the Great Dun Fell experiment took place in spring with low pollution but even high oxidant concentrations (Choularton et al. 1997, Colvile et al. 1997, Swietlicki et al. 1997, Wells et al. 1997, Cape et al. 1997, Lüttke et al. 1997, Sedlak et al. 1997, Bower et al. 1997, Pahl 1996, Pahl et al. 1997, Laj et al. 1997).

Continuous monitoring of cloud and rain samples (time resolution weekly) in three sites in Britain over one year has allowed consideration of the impact of the enhancement of wet deposition of pollutants by orographic effects (Fowler et al. 1995), especially the scavenging of cap cloud droplets by rain falling from above (the seeder-feeder effects).

Vertical distributions of aerosol particles and CCN in clean air around the British Isles were obtained during several flights through research fields of small cumulus clouds over the sea; results are given in Raga and Jonas (1995). Also, as a result of research flights within the CLEOPATRA[536] project the distribution of trace substances inside and outside of clouds is determined (Preiss et al. 1994). Cloud events with high pollutant loading were observed during the EASE[537] field campaigns in the so-termed black triangle region at Mt. Szrenica in the Sudeten Mountains (Poland) in 1995 and 1996 (ApSimon 1997, Acker et al. 1999a, Kmiec et al. 1998). The idea from the Great Dun Fell experiment to study cap-cloud chemistry was continued in the joint project FEBUKO,[538] where several campaigns at Mt. Schmücke (Thuringian Forest, Germany) were carried out (e. g., Acker et al. 2003, Herrmann 2005). The focus was on the microphysics and chemistry of different types of aerosols, the role of aerosol chemical composition for cloud formation as well as the chemical transformation, where the behavior of organic substances was of special interest (e. g., organic acids, peroxides, organic carbon, soot).

In the 1990s, the number of studies increased that was focused on special questions (SOA formation, photochemistry, namely photocatalysis) and substances (nitrite, H_2O_2, trace metals, and organic substances). As early fog and cloud studies were focused on the nature of condensation nuclei, new techniques now allow complex measurement programs to study the interaction between aerosols, clouds, air pollution and climate. Besides only very few long-term monitoring programs worldwide, many case studies were carried out, in the 1990s mainly in Europe and more recently in Asia. Interesting that after 1990 again in Japan extensive cloud research programs started (Minami and Ishizaka 1996). The longest (since 1988) and still working monitoring occurs at Mt. Oyama (Igawa et al. 1998, 2001, Igawa 2016). Other sites in Japan were studied either by campaigns or only in selected years, like Mt. Fuji (Watanabe et al. 2006), Mt. Norikura (Watanabe et al. 1999), and Mt. Rokko (Aikawa et al. 2001).

536 "Cloud and Cloud Transport Experiment" organized by "*Deutsches Zentrum für Luft- und Raumfahrt*" (DLR) Oberpfaffenhofen.

537 "Emission Abatement Strategies and the Environment" (EU project).

538 Project "Field Investigations of Budgets and Conversions of Particle Phase Organics in Tropospheric Cloud Processes" organized by IfT.

After the year 2000, cloud (and fog) chemistry monitoring began in China (Sun et al. 2010, Lu et al. 2010 and see Watanabe et al. 2011 for citations) namely at Mt. Tai (Guo et al. 2012, Li et al. 2017). A long-term monitoring started in 1996 at Mt. Bamboo in Taiwan (Lin 2015, Simon et al. 2016), still ongoing.

Besides regional air pollution detection, ecosystem deposition and local budgets are of interest (Igawa et al. 1998, Klemm and Wrzesinsky 2007). In the late 1980s and the 1990s, a "renaissance" of fog water (from a hydrological point of view) and nutrient deposition to agricultural and namely forestry areas (Dasch 1988, Schemenauer et al. 1995, Thalmann et al. 2002, Bridges et al. 2002, Burkard et al. 2003, Hůnová et al. 2004, 2011).

The Mt. Brocken[539] cloud chemistry program (1992–2009) was the first continuous long-time mountain site cloud monitoring[540] in Europe (1142 m a. s. l., Harz Mountains, 51.80 °N, 10.67 °E) located at the former border between East and West Germany

539 Mt. Brocken, has a long tradition in cloud and weather research. Several famous personages climbed Mt. Brocken, for example in June 1658 *Otto von Guericke* (1602–1686) who was also known for his studies on the change of air pressure with altitude, climbed Mt. Brocken with his glass barometer, however his attendant who carried the instrument, "stumbled, shattered the glass and frustrated the project" (Guericke 1672, Book III, section 30 "Experiments which show how air pressure varies at different altitudes"). *Jean-André Deluc* presented at the end of eighteenth century (Deluc 1787) the most advanced theory on the reason why cloud drops breeze in air. He climbed Mt. Brocken on October 30, 1776 with his newly constructed hygrometer but the weather was so "bad" that the instrument "was covered with icicles" and came "down to the point of extreme humidity" (Deluc 1777, p. 425). In 1777, 1783, and 1784 *Johann Wolfgang von Goethe* (inspired to write the drama "*Faust*"), and the poet *Heinrich Heine* who wrote later "*Die Harzreise*"; first evidently climbing was by the scientists and cartographer *Tilemann Stella* in 1561. *Claus Eduard Nehse* (with a successor until 1859) began in 1840 with simple meteorological observations. *Adolph Richard Aßmann* (see also p. 132), observed cloud droplets in November 1884 at Mt. Brocken using the best available microscope at different altitudes within the cloud. After evaporation of the droplets he was unable to see any residual and concluded that the nuclei (CCN) must be smaller than 1 µm (see also p. 325). Instantly *Aßmann* proposed (later supported by *Gustav Hellmann*) to continue the measurements; however only in 1895 a wooden building was constructed as first observatory equipped with all modern instruments. A larger stony house followed in 1913 that was replaced (250 m south) in 1939 by a new tower building still existing (but fully reconstructed). Mt. Brocken was closed to the public between 1961 and 1989 as part of the "*Mauer*" borderline between West and East Germany.

540 Together with *Uwe Feister* I climbed Mt. Brocken in April 1990, claiming a wooden house some 100 west of the observatory (used by GDR secret service before 1990) to establish a station for cloud sampling and air pollution monitoring, operated until 2010; it was teared down in 2014 to renature the site. Interestingly to note that our first idea was to establish a measurement site for ozone within the Tropospheric Ozone (TOR) project of EUROTRAC, however, despite the Brocken site would have been the most representative TOR station, it was rejected by officials with the argument that there were already some mountains sites in West Germany (Schauinsland, Zugspitze, and Wank) – from today's knowledge surely due to aspects of competition. Nevertheless, O_3 measurements were conducted together with the cloud chemical program giving unique scientific insights into multiphase chemistry of ozone (see p. 592).

to study the role of clouds in the atmospheric budget of air pollutants, Fig. 2.22. Experiences from the *Mountain Cloud Chemistry Project* in the eastern United States (Mohnen and Vong 1993) have been adopted; for almost 18 years we analyzed 23,743 cloud water samples based on 1-h collection (for details see Möller et al. 1993, Acker et al. 1996, 1998b, 2002).[541] A full set of meteorological data and gases (SO_2, NO, NO_2, O_3) were continuously measured. LWC measurements and cloud water sampling were carried out only during frost-free period, as a rule between the middle/end of April and middle/end of November. The monitoring program was accompanied by extended field campaigns and case studies; several other compounds were measured during campaigns, like organic compounds (Lüttke et al. 1997, 1999), heavy metals (Plessow et al. 2001), HNO_2 (Acker et al. 1999b, 2000, 2001, 2008a), sulfite (Tian et al. 1999), macromolecular organic substances (Feng and Möller 2004) and hydrogen peroxide (unpublished); see also Möller (2020) for details.

The large differences we found in the chemical composition of cloud water from event to event, much more than it was known from precipitation chemistry studies, are caused principally by three different reasons (as already stated by Petrenchuk 1979b):
- differences in the air pollution situation,
- large scale transport characteristics and
- cloud dynamic and microphysical behavior.

This program was also aimed to establish *cloud chemistry climatology*; however, open questions remained and missing information (like monitoring of atmospheric aerosol[542] and cloud droplet size distribution). Long-term cloud water sampling and analyzing is very laborious and costly. Despite meanwhile existing longer time-series in the world, the Mt. Brocken program remains unique for Europe, most valuable for covering the period of significant air quality changes in central Europe in the early 1990s (Table 2.22).

The only other European stations including cloud chemistry on an occasional basis are the Puy de Dôme Microphysics and Chemistry Station in France and the High-Altitude Research Station Jungfraujoch, located in the Swiss Alps in the center of Europe; both stations are more focused on aerosol physics and chemistry.

541 The data set is available in the World Data Center PANGAEA, Data Publisher for Earth & Environmental Science. https://doi.pangaea.de/10.1594/PANGAEA.909620

542 *Möller* asked *Heintzenberg* (IfT) in 1997 to participated at Mt. Brocken with an aerosol measurement program – but he was obviously not interested.

Figure 2.22: The Mt. Brocken Cloud Chemistry Monitoring Station (Harz Mountains, Germany), middle. Above: The automatic wet-only cloud water collector – string collector after *Mohnen* and *Kadlecek* and meteorological instruments (left: *Gerber* LWC sensor, right: ultrasonic wind sensor) under riming in winter. Below: ISCO sampler for cloud water (left) and the station in winter (right).

Table 2.22: Comparison of cloud (Mt. Brocken) and rainwater (Seehausen) chemical composition; weighted means (in µeq L^{-1}). LWC in mg m^{-3} and yearly rainfall amount R (in L m^{-2}) as mean over the period; n – number of samples.

period	n	LWC/R	Cl^-	NO_3^-	SO_4^{2-}	NH_4^+	Na^+	K^+	Ca^{2+}	Mg^{2+}	H^+
cloud water											
1993–1995	3476	331	73	236	213	303	72	6	57	22	103
1996–2002	8086	362	92	235	176	279	96	4	38	22	61
rainwater											
1993–1995	643	529	39	42	60	53	26	4	24	7	38
1996–2002	1388	475	25	40	40	52	20	2	16	5	14

Here I would like to name (without making claim of being complete) leading researchers who significantly contributed within the last 25 years to experimental[543] fog and cloud chemistry research (alphabetically listed): *Cort Anastasio* (University of California), *Viney Aneja* (North Carolina State University), *Tom Choularton* (University of Manchester), *Jeff Collett* (Colorado State University), *Werner Eugster* (ETH Zürich), *Bruce C. Faust* (California Institute of Technology), *Jaroslav Fišák* (Institute of Hydrodynamics Prag), *Sandro Fuzzi* (CNR Bologna), *Hartmut Herrmann* (TROPOS Leipzig), *Manabu Igawa* (Kanagawa University, Japan), *Wolfgang Jaeschke* (University Frankfurt/Main), *Otto Klemm* (University Münster), *Paolo Laj* (CNRS-LGGE, France), *Neng-Huei (George) Lin* (National Central University Taiwan), *John Ogren* (Colorado State University), *Hiroshi Okochi* (Waseda University Tokyo), *Bob Schemenauer* (Canada), *Mieczysław Sobik* (University Wroclaw, Poland), *Tao Wang* (Hong Kong Polytechnic University) *Koichi Watanabe* (Toyama Prefectural University, Japan).

The pioneers of fog cloud chemistry research in the USA from the 1930s are *William H. Radford* and *Henry Garrett Houghton*; from the early 1980s, I will name *Daniel Jacob, Jed Waldman, William Munger, Michael Hoffmann, Volker Mohnen, Peter Victor Hobbs, Dean A. Hegg, Peter H. Daum, Raymond A. Castillo, Vin Saxena*, and others. For Japan, the pioneers are *Yasuo Miyake* and *Takeshi Ohtake*. And last but not least, in Europe (Germany) *Helmut Mrose*. What was a long time not known in the western world (and likely will make first public with this book) are the Soviet scientists from the 1950s and 1960s, who published almost all results of investigations and instrumental development in Russia in Institute journals (and, if cited than in English translation which makes it impossible to find the original paper): *Rostislav Ivanovich Grabovski, Solomon Moiseevich Shmeter, Alexander Moiseevich Borovnikov* concern more phys-

543 Another big issue since the beginning 1980s is cloud chemistry modeling but in this book out of the focus. Furthermore, many researchers (not listed here) are more interested on cloud – aerosol interaction. The relevant literature is almost not cited here.

ical aspects, *Valentina Ivanovna Drozdova* and *Olga Petrovna Petrenchchuk* concern chemical aspects.

2.5 Dew water studies

> Dew, far from being a pure or unadulterated water, is in reality a more mixed mixture than most others (*Thomas Henshaw* 1665).
> A *dew* too collected [...] has yielded by distillation a liquid [...] an alcaline lixivium, or aqua fortis: and this liquor was inflammable like spirits of wine, [...] But the nature of *dew* is likewise surprisingly various, according to the different dispositions of the weather, and according to the various and successive changes of the meteors in the air [...] The principal part, therefore, of *dew* is water; the rest, on account of their manifold variety, cannot possibly be described (George Shelby Howard (1788, p. 685–686) The New Royal Cyclopaedia, and Encyclopaedia, Vol. II. London, Alex. Hogg).[544]

Dew is water that has condensed directly onto a surface where the temperature is at or below the ambient dew-point temperature; note that dew appears as droplets on surfaces (no water film). Hence dew (and frost) is not termed to belong atmospheric water but was in ancient times seen as "heavenly". The chemical substances measured in dew water come from the following sources:

- dissolution of substances being on the surface and due to reactions with the surface,
- exhalations (namely from leaves) from the surface,
- sedimentation of particles onto the surface,
- dry deposition of molecules (gases) and small particles, and
- aqueous-phase chemical reaction in dew droplets

The importance of dew concerning the ecosystem and the atmospheric chemistry includes:

- water and nutrient supply to the ecosystem (plants and animals),
- plant damages and surface corrosion (acid dew) at high air pollution, namely due to SO_2 absorption,
- interfacial chemistry, e. g., decomposition of ammonium nitrite and photosensitized formation of HNO_2 and of oxidants.

544 Parts of this text are equally worded in "The Complete Dictionary of Arts and Sciences, Vol. 1" under "dew" by Temple Henry Croker, Thomas William and Samuel Clark (London 1766), and referred to *Boerhaave*'s "Elements of Chemistry".

During the spring of 1664 and 1665, the Royal Society of London showed considerable interest in May-dew,[545] (Henshaw 1665) carrying out systematic experiments with it (Taylor 1994). *Thomas Henshaw* made several experiments reported on his observations:[546]

> He did at last extract about two ounces of a fine small *white Salt*, which, looked on through a good *Microscope*, seemed to have Sides and Angles in the same number and figure, as *Rochpeeter* (Henshaw 1667, p. 36).

Already alchemists found that dew circumstantially provides "mysterious" chemical features, it is likely that the alchemists found nitre (nitrate)[547] in dew because *Mosheim*[548] says (cited after Waite 1887, p. 6):

> Dew, the most powerful dissolvent of gold which is to be found among natural and non-corrosive substances, is nothing else but light coagulated and digested in its own vessel during a suitable period it is the true menstruum of the Red Dragon, i. e., of gold, the true matter of the Philosophers.

The German *Johann Heinrich Zedler* (1706–1754) edited an encyclopedia in 64 volumes and cites (Zedler 1735, p. 374–375) *Ramazzinus*[549] (De constitutione anni 1690 in Miscell. Nat. Cur. Dec. II, An. IX, Adpend. P. 37) under the keyword *Honigthau* [honeydew]:[550]

545 The dew of May, esp. that of the morning of the first day of May, to which magical properties were attributed and which is said to whiten linen, and to enable a face washed with it to keep its beauty. Not to confuse with honeydew.

546 *Rochpeeter* (in the 1809 edition written as *rockpetre*, (rock salt) = NaCl (rocky salt). Note that the printed papers have been written by unknown persons (reporters) after the lecture. For example, the citation at the beginning of this chapter is not printed in the *Transaction* (1667) but only in *Transaction* (1808, p. 13). Thus, the reporter made a footnote (1809) writing "The most important part of the paper is the concluding paragraph, demonstrating that a great quantity of saline matter is contained in dew". *Henshaw* who also set large quantities of dew to putrefy, found larvae which developed into insects.

547 This quotation by *Moosheim* tells us that dew contains chloride (sea salt) and nitric acid, likely generated through photocatalytic processes. Nitrate (and nitrite) in dew were found in quantities much higher than expected only by dry deposition. Acker et al. (2008a) studied systematically formation of nitric and nitrous acid in dew and explains it by heterogeneous formation due to reaction of NO_2 with water surfaces (Acker and Möller 2007).

548 *Johann Lorenz von Mosheim* (1693–1755) Institutionum historiae ecclesiasticae libri IV, appeared in 1726.

549 *Bernardino Ramazzini* (1633–1714): Dn. Bernardini Ramazzini In Mutinensi Lycaeo Med. Professoris De Constitutione Anni M. DC. LXXXX. Ac De Rurali Epidemia, Quae Mutinensis Agri, & vicinarum Regionum Colonos graviter afflixit, Dissertatio, Ubi quoque Rubignis natura disquiritur, quae fruges, & fructus vitiando aliquam caritatem Annonae intulit. In: Miscellanea curiosa sive ephemeridum medico-physicarum Germanicarum Academiae Caesareo-Leopoldinae Naturae (1690) pp. 15–56.

550 Today, *honeydew* is defined as a sweet sticky substance deposited on leaves by insects and/or a substance produced by the leaves of some plants. *Guttation* is the exudation of drops of sap on the tips

> [. . .] eine Art eines verschlimmerten Thaus, und theils von einer lixivalischen alcalischen, Theils und meist von scharffer, sauerer und zerfressender Beschaffenheit, so wie der ordentliche Thau, aus welchem man so gar einen geist der metalle auflöset, zubereiten könne; wie z. B. des Herrn *Cnoeffelii Menstr. Pro paranda Tinct. Corall*, zu welchem, wenn sich salpetrichte und saure Cörpergen gesellen, solche in der luft zusammen jähren, und alsdenn herab fielen, der Honig-Thau erzeuget würde [. . . a kind of impaired dew, and partly of lixivalic alkaline, partly and mostly of hot, acid and corroding property, as the proper dew, from which one can prepare a spirit that dissolves metals; as e. g., Mr. *Cnoeffelii Menstr. Pro paranda Tinct. Corall*, to which, if nitrous and acid bodies combine in air and afterwards fall down, the honeydew could be generated].

Andreas Cnöffel [or *Andrzej Cnoeffel*] (1601–1658), Polish physician in Marienburg, writes on "*Tinctura nitri corallata*" (Cnöffel 1676, p. 49), "*Rosa triplicate*" (Cnöffel 1676, p. 47), and "*liquor Alcahest*" (Cnöffel 1676, pp. 47–49), obscure solution or solvents.[551] *Johann Samuel Trommsdorff* (1676–1713), German theologist and philosopher, wrote that honeydew is no dew but caused by the plant itself (Trommsdorff 1699). In contrast, *Leeuwenhoek* believed that honeydew (rubigo in Latin, *Honig-dauw* in Dutch) falls from the sky and consists of small worms and their eggs (Leeuwenhoek 1697, pp. 148–157). *Boyle* writes (likely observations already long known)[552] on decolorization of linen at remote sites, obviously due to air chemical operations (Boyle 1692, p. 50):

> In some Places also, which are judged likely to afford subterraneal Steams, guesses may be made, whether this or that kind of Salt ascends into the Air, by spreading upon the Ground, in Places free from Dirt and Dusts, large Pieces of clean and white Linen Cloth, that has no Relish of Sope or Lees; and observing, after they have lain a competent while, whether, and how they are discoloured, and what kind of Saltness, if any, is to be found in the Moisture imbibed by them, from the ascending Steams and falling Dew.

In Chapter 2.1, some alchemistic insights on dew are presented; but around that era, *Robert Boyle*, who remained an alchemist believing the transmutation of metals to be

---

or edges of leaves of some vascular plants. it is not to be confused with dew, which condenses from the atmosphere onto the plant surface.

551 Also termed *ignis-aqua*; hypothetical "universal solvent", able to dissolve every other substance, including gold. First described by *Paracelsus* who believed that alkahest was, in fact, the philosopher's stone. *Menstruum* is the general meaning for solvent in alchemy. It remains unclear what *nitri corallata* (the species name, *corollata*, is Latin meaning "like a corolla", referring to the appendages of the nectar glands that resemble a normal corolla; spurge, having milky often poisonous juice; *nitri* = salpeter), *Rosa triplicate* (triplicate dew) means.

552 Textiles have long been whitened by grass bleaching (spreading the cloth upon the grass for several month), a method virtually monopolized by the Dutch from the time of the Crusades to the eighteenth century. They developed a technique in which goods were alternately soaked in alkaline solutions and grassed, or crafted, a procedure in which they are exposed to air and sunlight; the goods were then treated with sour milk to remove excess alkali. Now, we explain this phenomenon by surface photo-catalytic oxidant formation, i. e., radicals (OH) either via HNO_2 formation or direct electron transfer onto oxygen.

a possibility (More 1941), but is largely regarded today as the first modern chemist, wrote in his "history of air" (1692) some remarkable phrases, which shows that the salinity of dew was recognized in the seventeenth century (Boyle 1692, p. 58):

> And since Dew is made of Steams of the Terrestrial Globe, which whilst they retain that Form, and were not yet convened into Drops, did swim to and fro in the Air, and made part of it; the Phenomena that show the Power of Dew in working on solid Bodies, may help to manifest how copiously the Air may be impregnated with subtle saline Parts. [. . .] the Subtilty of Brasilian Dew, and its Power to rust Metals [. . .] in several Places the Portugals kept their great Guns cased over, that the Dew might not fall upon them, and by its Corrosiveness so rust them [. . .] And when I demanded whether he tasted the Dew, to observe the Saltness of it? He replied, that he had not, but that he had in divers Places observ'd, that it left the Grass, &c. that it had rested on, cover'd over with a pure white Salt, as if it had been a hoar Frost.

According to today's knowledge, it is sure that the "white salt" was NaCl from sea salt deposit and the corrosive properties of the dew were likely due to HCl formation during evaporating dew containing magnesium chloride. The first statement I found in the literature that dew contains apart from water traces of atmospheric origin (and not alchemistic elements) is given by *Lampadius* (on basis of his experiment from 1796) who said that dew consists os pure water and some carbonic acid (2 %) but more than in rainwater (Lampadius 1806, p. 121). He also said that dew may contain substances emitted from plants ("*Ausdünstungsmaterien*" – effluvia), and thus used before as medicine; bleaching properties of dew are known for centuries and were used for cleansing of clothes. The dew sampler used in the nineteenth century (Fig. 2.23) was very similar to that used by alchemists (Fig. 2.1 on p. 114). Lampadius (1806, p. 121) quotes in the paragraph on dew in a footnote:

> In den Edinburger medicinischen Commentarien 1 B. S. 215 findet sich eine Nachricht von einem in Indien im September und October fallenden sauern Thau, welchen man auf Musselin auffange und als stärkendes Arzneymittel gebrauche [In the Edinburgh medical commentaries Vol. 1 p. 215 a message is found on acid dew, falling in September and October in India, sampling on muslin, and using as medicament][553]

Probably the first chemical study of dew was conducted by the French chemist *Jean-Sébastien-Eugène Julia de Fontenelle* (1790–1842)[554] who collected 4 L of dew (assuming that the evaporations of swamps should condense together with dew) in 1819 in

553 Medicinische Commentarien von einer Gesellschaft der Aerzte zu Edinburgh. Altenburg. Vols. 1–10 (1774–90); ser. 2, vols. 1–10 (1791–97); ser. 3, vol. 1 (1799); translation from English. Original: Medical and philosophical commentaries by a society of physicians in Edinburgh, Murray, Vol. 1–2 (1774). Unfortunately, I was unable to find the quotation neither in available German nor in English editions.

554 It is historically remarkable that the price essay (on the influence of swamp air on human health announced by the Academy of Lyon) by *Julia de Fontenelle* over 155 pages on swamp air and heavy criticized in a review by Rust (1824) because of its incompleteness and superficiality.

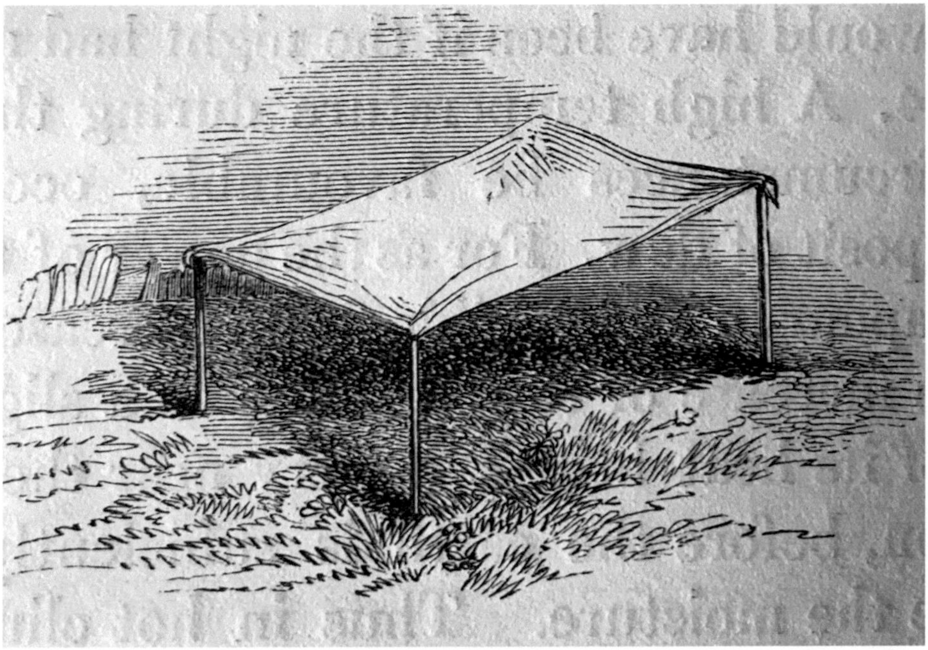

Figure 2.23: Dew sampler used by *Wells*. From "The dew-drop and the mist or, an account of the nature, properties, dangers, and use of dew and mist in various parts of the world" (Tomlinson 1847), p. 28, published in London (115 pp.), Anonymous (1847), but likely by *Charles Tomlinson*, who published 20 years later a very similar enlarged title: "The Dew-Drop and the Mist: An Account of the Phenomena and Properties of Atmosphere" (Tomlinson 1847).

the marshes of Cercle, France (Fontenelle 1823), and analyzed the dew together with the French chemist *Louis-Nicolas Vauquelin* (1763–1829). They found (apart from dissolved oxygen and carbon dioxide) organic matter, chloride, sodium, sulfate, calcium, and carbonate (Fontenelle 1823), and found in the air the same substances except for organic matter and mentioned that the swamp air has an own odor, its cause not was proved by the dew analysis (cited after Rust 1824). On the contrary, Pierre (1859, p. 41) and Smith (1872, p. 241) quotes *Fontanelle*: "this water was inodorous, without color, and clean; in a short time, it deposited small flakes of nitrogenous matter" (without doubt organic nitrogen). Gairdner (1832, p. 114) writes (without giving a reference):

> Dew much purer than rain-water; but immediately on its deposition, it becomes impregnated with extractive and salinic matter, which often from abundant efflorescences on the surface of the ground in those countries where the fall of dew is most considerable.

Witting (1823, 1824) collected dew directly from plants and found carbonate, chloride, and organic matter; he concluded that exhalations from the plant cause the dew water composition.

Some of the scientists who studied rainfall in relation to agricultural chemistry also investigated fog (Chapter 2.4.5) and dew, like Meyrac (1852), Boussingault (1854), Knop (1868), Lawes et al. (1881, 1882), Lévy (1882), Posnjakov (1904), and Leather (1906). The first quantitative estimates of ammonia in dew (he termed it "*pluie la rosèe*" [rain of dew]) is known from Boussingault (1854) at Mt. Liebfrauenberg who found in dew at this mountain site similar mean values like in fog (n = 5, 18 August–28 September 1853): 5.6 (3.1–6.2) mg L^{-1}; one event (not included in the mean above) on 24–25 September had only 1.06 mg L^{-1} but after a rainy day. Not much later, the German agricultural chemists, *Wolf* and *Knop*, carried out measurements in 1860 in Möckern, near Leipzig. While *Boussingault* collected dew using a sponge, *Wolf* and *Knop* collected dew with a glass cup from grass leaves before sunrise (in Knop 1868, annex pp. 77–78) and found 1–6 mg L^{-1} ammonia. Lévy (1882) found at Montsouris (Paris) in September–October 1880 a mean of 6 events to be 4.13 (3.01–4.90) ammonium-N mg L^{-1} and 2.9 (1.6–2.6) nitrate-N. This is the first determination of nitrate in dew. At Rothamsted, Lawes et al. (1881, p. 316) state that "most of the very small deposits represents dew", but only one of these deposits is especially noticed, namely that of September 1881, where it measured in 0.007 inches precipitation 5.48 ppm ammonia. Leather (1906) collected 1904–1905 dew (n = 10) in India using a large rain gauge to study the nitrogen content and determined 1.9 (0.8–2.6) mg L^{-1} ammonia and 1.7 (0.5–4.1) mg L^{-1} nitric acid. *Leather* cites Welbel (1903) who found at the agricultural experiment station Ploty in 1902 a means of 5 mg L^{-1} ammonia in dew.

The first dew chemistry study in India was conducted by Dhar (1932) and Dhar and Ram (1933c); they found that dew contains formaldehyde (approximately 0.15 mg L^{-1}) and that the concentration is appreciably greater than that in rainwater. However, their speculation on a photochemical formation from organic substances in aqueous solution is obsolete (see also p. 227).

Yaalon and Ganor (1968) collected in Jerusalem 35 dew samples using a glass plate during the dry summer of 1967; 21 samples sufficed for a complete chemical analysis like in rainwater normally carried out; the median pH amounts were 7.7. The concentrations of soluble salts found in dew were usually within the range of concentrations found in bulk precipitation but richer in calcium, magnesium, chloride, and bicarbonate. They also found that in precipitation the variations are much larger and conclude that dew condensations represent the dry fallout of sea salt and dust (calcareous dust from the surrounding). Brimblecombe and Todd (1977) collected 122 dew drops with plastic pipettes from 12 different surfaces at the University campus at Norwich and analyzed single droplets on Na and K only. The mean pH from mixed droplets was 5.7. They found a high molar K/Na ratio (about 1) in dew from natural surfaces, and, in the case of washing grass blades, a ratio to be 13 and could be explained by potassium release from the soil and plants. When exposing plastic "grass", the ratio amounts 0.35, not far from that (0.15) found from glass plates by Yaalon and Ganor (1968) in contrast to rainwater being 0.05 according to Junge (1963a). Thus, the paper by Brimblecombe and Todd (1977) is the first that emphasize the importance of biological contribution to dew water. Despite they note that dew may act as a sink for SO_2 with subsequent oxidation of S(IV) to sulfate in dewdrops, there was not yet experimental evidence found. Soon later Brimblecombe (1978) calculated that the transfer of SO_2 into leaf wetness ("dew") is rapid and should remain in equilibrium with the atmosphere.

A few years later, Smith and Friedman (1982) and Cadle and Groblicki (1983) studied experimentally dew composition at different sites in the USA[555] under the aspect of SO_2 dry deposition. Mulawa et al. (1986) collected natural dew in 1982 from a Teflon surface suburb of Detroit and found that concentrations of calcium and chloride were 29 and 16 times, respectively, larger in dew than that in rainwater whereas the enrichment for sodium, potassium, nitrate, and sulfate were around 4; the median pH was 6.5. Artificial dew samples were generated in 1984 to measure dry deposition rates of several species and compare them to a dry surface deposition; they found deposition rates were 2–20 times greater to the artificial dew than to the dry surface, were in the upper range ammonium, sulfite and sulfate were found. Schroder et al. (1989) found the chemical composition of natural dew is similar to rainwater with the exception that dew has much higher concentrations of calcium and chloride and much lower acidity. Pierson et al. (1986) conducted experiments in August 1983 at a mountain site (838 m) in Pennsylvania to quantify the role of dew in acid deposition. They found the dew (collected on Teflon frame) to be acid (mean pH around 4) and atmospheric HNO_3 as well as SO_2 were chiefly responsible for dew nitrate and sulfate. Remarkable are the measurements of nitrite (in the range of 0.2–2.0 $\mu eq\ L^{-1}$) and the (molar) nitrite/nitrate ratio to be 0.04 (0.006–0.2). Pierson et al. (1986) derived averaged deposition velocities to be significantly less (due to nighttime stable conditions) than in summer daytime. Thus, deposition onto dew was addressed to be unimportant comparing to removal by fog and rain. Wesely et al. (1990) conducted experiments to examine the chemical properties of dew (and guttation) that affect the dry deposition of SO_2 to prairie grasses; chemical analyses showed that aqueous concentrations of sulfite, sulfate, and several other ions in the leaf wetness were often highly comparable to the values typically found in rain, while ammonium was not very abundant.

In Poland, a first dew monitoring program was performed to outline the chemical composition of dew water in a meteorological context; dew samples were collected from eight measurement stations from August 2004 to November 2006 (Polkowska et al. 2008).

In the last two decades dew has also been studied not only concerning ecology (dew as source of moisture for plants, biological crusts, insects, and small animals, e. g., Jacobs et al. (2000), but also concerning its use as potable water (Beysens et al.

555 These papers are cited by a few researchers but not publicly available: Smith and Friedman (1982) The chemistry of dew as influenced by dry deposition: results of Sterling, Virginia, and Campaign, Illinois, experiments. Work Pa. 82WOO141, MITRE Corp., McLean, Va.; see Cadle and Groblicki (1983) The composition of dew in an urban area. In: The meteorology of acid deposition (ed. P. J. Samson), Transaction of an APCA Specialty Conference, pp. 17–29.

2006, Muselli et al. 2006), and, finally, from air chemical point of view, now also termed interfacial chemistry (Rubio et al. 2006, Acker et al. 2008a). Beysens (2018, p. 213) presents a Table compiling 30 studies on chemical composition (more or less completeness) of dew water in the period 1986–2017 in 13 countries, ranging from pH 4.0 to 7.4 and concentrations ranging over two orders of magnitude. The French physicist *Daniel Beysens* (born 1945) closed the "historical cycle" to Wells (1814) with the description of dew formation, and its physical and chemical properties (Beysens 1995, 2018). From the dew chemistry research, we can learn that a) artificial collector surfaces (like Teflon, glass, etc.) can neither represent the natural dew formation nor natural dry deposition and interfacial chemistry and b) natural surfaces give less insight in atmospheric chemistry but more in specific surface chemistry. Besides dew, the wetness of leaves and other surfaces due to rainfall and fog and cloud interception were considered. According to their observations (Acker et al. 2005, 2008a), dew remains on grass leaves, namely at mountain sites, until noon and later, thus providing at daytime and under sunlight a site for interfacial chemistry.

3 Investigation of particulate matter in the air

> [...] Air, as a compound, in which very great and heavy Bodies are capable of swimming, [...] as Dust (Boerhaave 1735, p. 252).

The English word[556] *dust* is derived from Proto-Germanic "*dunstaz*" and also the source for Old High German "*tunst*" [storm, breath] and German "*Dunst*" [haze, mist], Swedish "*dust*", Danish "*dyst*", and Dutch "*duist*".[557] The German word for dust is "*Staub*", which is derived from Old High German "*stoup*" and Middle Low German "*stof*" (Danish and Norwegian "*støv*") and modern German "*Stoff*" [matter, substance, material]. Thus, dust, haze, mist, fog, fume, and smoke [Latin fumus = smoke] are terms that are physically very closely related to a "cloud of particles" – or in modern terms *particulate matter*, whereas *atmospheric aerosol* denotes the colloidal system air – particles. The term *dust* includes solid particles (but excluding solid hydrometeors) from nanosized (clusters of atoms and/or molecules)[558] to giant particles (there is no upper size limit, but particles larger than 100 μm are very rare). Thus, dust particles not only cover a sizes range over several orders of magnitude but also show a large variation in chemical composition, structure, and shape. There are several physical criteria to distinguish dust (and surely more):

- volatile – nonvolatile,
- soluble – insoluble,
- visible – invisible,
- crystalline – amorphous,
- mineral – biological (vegetable / animal),
- inorganic – organic,
- suspended – sedimentation.

From the point of origin, we can separate among[559]

- soil dust,
- sea salt (cycling salts),
- volcanic dust (ash),

556 Dust: dried earth reduced to powder; other dry material reduced to powder.

557 Already in ancient time smoking caves (in German "*Dunsthöhle*") were known to be killing due to gases (CO_2, H_2S, SO_2); *Plinius* said "Mortiferum Spiritum exhalantia" (Landener 1856). There were the believe that these caves were the entrance to the underworld with his evil spirits and demons.

558 So-termed nanoparticles, with the exception of primary soot particles from combustion, are secondary produced particles in gas-to-particle reactions from gaseous precursors like HCl, SO_2, SO_3, HNO_3, NH_3, H_2O, halogenides and organic compounds. They grow very fast and have a changing chemical composition. Artificial nanoparticles are uniform in size and chemical composition, chemical inactive and its biological and chemical implications in nature are still widely unknown.

559 The English word "ash" (powdery remains of fire) comes from Proto-Germanic "*askon*" (source also of Old Norse and Swedish "*aska*"), Old High German "*asca*", German "*Asche*".

https://doi.org/10.1515/9783110732467-003

- combustion dust (smoke, fume, soot, ash),
- extraterrestrial dust (meteors),
- secondary aerosol.

Constantine Samuel Rafinesque (1783–1840),[560] is to my knowledge, the first who wrote (thoughts) on atmospheric dust in a scientific style (Rafinesque 1819):

> Atmospheric dust [...] existing every where in the lower strata of our atmosphere. This dust is invisible, [...] The size of the particles is very unequal, and their shape dissimilar; [...]
> Its chemical analysis has never been attempted by chemists; but the earthy sediment which is the result of its accumulated deposition, proves that it is a compound of earthy particles in a peculiar state of aggregation, and in which alumine appears to preponderate, rather than calcareous or silicious earth or oxides.
> Its formation is sometimes very rapid, and its accumulation very thick in the lower strata of our atmosphere, but the intensity is variable. Whenever rain or snow falls, thus dust is precipitated on the ground by it; but a small share is still left, or soon after formed.

Scientific understanding of dust in the atmosphere began with the definition of a colloid in 1861 by *Thomas Graham* (1805–1869).[561] As well as the terms *sol* and *gel*, *Graham* introduced the terms *colloid* and *colloidal condition of matter* (Graham 1861). Over 100 years ago, it had already become clear that the sciences to study colloids – chemistry and physics – must deal with the colloidal state and not only with the colloid, the dispersed phase itself (Ostwald 1909).[562] The colloidal system was characterized as multiphase or heterogenic. For colloids like "gas and solid", cigarette smoke, atmospheric dust, and for "gas and liquid", the atmospheric fog was given as an example (Ostwald 1909). Besides fine dust particles, soot and coarse particles were known; at the end of the nineteenth century, it was known that the coarse fraction was soil dust and organic matter (with the latter constituting up to one-third of the total). *August Schmauß* (1877–1954)[563] introduced the term *aerosol* in 1920 (Schmauß 1920, Schmauß and Wigand 1929). Schmauß and Wigand (1929) considered the three categories of dispersed particles behind atmospheric aerosols: ions, dust, and hydrometeors.

560 French polymath born near Constantinople and self-educated in France. He traveled as a young man in the United States, ultimately settling in Ohio in 1815.

561 Scottish chemist in London who is best-remembered today for his pioneering work in dialysis and the diffusion of gases.

562 (*Carl Wilhelm*) *Wolfgang Ostwald* (1883–1943) was a German chemist and founder of colloid chemistry; sun of *Friedrich Wilhelm Ostwald* (1853–1931), who got the Nobel prize for research in physical chemistry (catalysis and equilibrium).

563 German physicist and meteorologist in Munich. 1910–1948 director of "*Bayrische Meteorologische Centralanstalt*" (1879 founded and 1917 renamed in "*Bayrische Landeswetterwarte*"), 1921 professor for meteorology at the University Munich and head of "*Institut für Meteorologie und Klimatologie*" (1935 established as "*Meteorologisches Institut*" at the university) at "*Forstliche Versuchsanstalt*" (founded 1881 and moved to Freising/Weihenstephan in 1992); 1923–1945 head of the "*Deutsche Meteorologische Gesellschaft*", member of several academies.

3.1 Dust, soot, and air pollution in a historical view

> I observed some very small Corpuscles floating in it [air], glissening in the Sun, and by the variations of their little Surfaces sparkling with wonderful coruscation (Boerhaave 1735, p. 249).[564]

The Roman poet and philosopher *Titus Lucretius Carus* (about 99–55 BC) saw dust floating in the light (De Rerum Natura):[565]

> The sun's light and the rays, let in, pour down across dark halls of houses: thou wilt see the many mites in many a manner mixed [...]

There is no doubt that dust from fire, sand storm, and volcanic eruption is known since ancient times.

3.1.1 Air pollution

Air pollution has been recognized from the earliest of time, with perceptible odor an issue of ancient Egypt and smoke of concern to Roman administrators and lawyer, Brimblecombe (2004) writes. Remarkably, the mixing of air (as it was still regarded as a uniform body) and water with pollutants (accurately referred to as "foreign bodies" in the old terminology) was known since *Aristoteles*. The perception of bad air is likely as old as the human settlement. The great Jewish naturalist, philosopher and physician, *Moses Maimonides* (1135–1204), is the first who describes after the ancient world the air pollution in towns and its effect on humans (Rosner 1987). From the fourteenth century, the first regulations of air pollution and avoiding infections are known. *Hermann Boerhaave* described in his "*Elements of Chemistry*" polluted air with some imposing phrases (Boerhaave 1735);

> [...] the Air-sphere, and which on account of the very large quantity of Vapours exhaled into it, has hitherto called the Atmosphere. (p. 252).
> [...] Soot, collected at the very top of a Chimney, from the volatile Smoke of a burnt Vegetable [...] which floats at liberty through the Air [...] and widely disperse it through the Air (p. 284).
> Excrementitious matter will be always dispersed into the Air [...] as for Urine, how quickly does that spontaneously become perfectly volatile, and exhale (p. 285).
> The Vapours about Mines, which are often so fatal, that no living Creature can breathe in them safety [...] Arsenicks, Orpiments, Cobalts, Sulphur of Antimony, Bismuth, Zincq, and other Bodies furnish the matter of these Vapours. (p. 287)

---

564 These particles were termed in German with the beautiful poetic term "*Sonnenstäubchen*" (not translatable; it corresponds to *solar mote*). The dictionary Lloyd and Noehden (1836) sets "*Sonnenstäubchen*" = atom; Ebers (1799, p. 279) writes: "*Sonnenstaub, Sonnenstäubchen*" – an Atome; the small or fine Dust one sees flying in the Air, in the Rays or Beams of the Sun.

565 On the Nature of Things. Translated by W. E. Leonhard (2004) Dover, New York, p. 37 (first published 1921).

[...] every such chemical Operation infects the very Air [...] mixtures of Oil of Vitriol, [...] with Nitre, Sea-Salt, or Sal-Gem,[566] converts in an instant those very fixed Salts into Fumes, [...] with which the Air is in a short time so strongly impregnated as to carry those Salts to great distances all around (p. 287).

John Evelyn wrote the first book on air pollution "Fumifugium: or, The Inconveniencies of the AER and Smoake of London Dissipated" (1661)[567] (italics set by *Evelyn*):

[...] in Clouds of Smoake and Sulphur, so full of Stink and Darkness [...] (Evelyn 1661, p. 8)
[...] It is this horrid Smoake which obscures our Churches, and makes our Palaces look old, which our Clothes, and corrupts the Waters, so as the very Rain, and refreshing Dew which fall in the several Seasons, precipitate this impure vapour, which, with its black and tenacious quality, spots and contaminates whatever is exposed to it (Evelyn 1661, p. 20).
[...] the *Aer* with so dark and thick a Fog, as I were hardly able to pass through it, for the extraordinary stench and *halitus* it send forth; [...] (Evelyn 1661, p. 22)
[...] *Arsenical* vapour, as well as *Sulphur*, breathing sometimes from this intemperate use of *Sea-Coale...* (Evelyn 1661, p. 27).
[...] our *London* Fires, there results a great quantity of volatile Salts, which being very sharp and dissipated by the Smoake, doth infect the *Aer*, and so incorporate with it, that the very Bodies of those corrosive particles [...] (Evelyn 1661, pp. 28–29).

Evelyn's remark (Evelyn 1661, p. 28) concerns volatile salts and their corrosive effects after distribution in air, which may be the first evidence for gaseous (and, consequently, dissolved) HCl in town air. His expression "[...] traveler [...] sooner smells, than sees the city [...]" (Evelyn 1661, p. 19) gives us an idea of the air pollution level. In an English cook book[568] an appendix "Remarks on Kitchen-poisons [...] and the Nature and properties of Water" is written, citing *Boerhaave* and describing the London rainwater "quality" (Mason 1780, appendix p. 36):

The rain, which falls through the smoke of large towns, is rendered foul and black; whose especially if it be collected, as it generally is, from the roofs of houses; when it brings with it a great many particles of soot, which give a very disagreeable taste and color.

We know from Boyle (1692), who also cited observations from travelers, that a qualitative understanding of *different* dust (mineral, saline, animal, vegetable, metallic, sooty) began very likely in the Middle Age; *Boyle* speaks on different sorts of *aerial*

566 Sal gemmae (old) = halitite or rocky salt (NaCl).
567 First published by W. Godbid "for G. Bedel and Th. Collins" (London 1661); reprinted in 1772 "for B. White" with a preface by S. Pegge, and in 1825 in the edition of Evelyn's Miscellaneous Writings ed. by W. Upcott. Here I used a reprint in original form in 100 numbered Copies (Swan Press in Baskerville), no year but before 1955 (No. 10), 49 pp. Further reading: Jenner (1995).
568 First edition 1773; in this edition the full name *Mrs Charlotte Mason* is printed, but remains a mysterious and elusive figure; the book's author was described as a professed housekeeper, who had upwards of thirty years' experience of families of the first fashion.

salts originated from "subterraneal Steams and Menstruments" and names *nitre* (saltpeter, i. e., nitrate) and *lime* (calcium carbonate) as well as sea salt:

> I think it not impossible, that the Air may, at least in some Times and Places, be impregnated with Corpuscles of a saline Nature, whether simple, or compounded, or of both kinds, not easily reducible to any of the sorts (Boyle 1692, p. 47).
> We guess the Air to be impregnated with Nitre, we may expose Lime to it, or some other Body that we think disposed to imbibe or retain such a Saltness (Boyle 1692, p. 49).
> [...] though the Air hereabout (viz. Oakly in Buckingham-shire) is very damp about the End of Autumn, and Beginning of Winter. So that I conceive this Salt either proceeds from the Sea-Vapours near the Sea-Coasts, or else from the Dissolution of this Esurine Salt in the Air, upon the burning of Sea or other mineral Coles (Boyle 1692, p. 56).

Berlin physician *Ludwig Formey* (1766–1823) wrote (Formey 1796, p. 9):

> Die Strassen in Berlin sind breit, gerade und geräumig, und gewähren nicht allein dem Auge einen schönen Anblick, sondern tragen zur Gesundheit der Einwohner viel bey, indem der Wind allerwärts streichen, die Luft erneuern und von den Ausdünstungen und Unreinigkeiten befreien kann [The streets of Berlin are wide, straight and spacious, and do not only grant the eye alone beautiful view, but provide a lot to the health of the inhabitants because the wind can everywhere flow, renew the air, and lost them from effluvia and impurities].

Local air pollution, namely urban pollution in towns and cities must be regarded at the end of the nineteenth century tremendously, likely by about two orders of magnitude higher in concentration than at present concerns the "classical" pollutants (soot, dust, SO_2). Strong-smelling substances (likely organic sulfides), ammonia (NH_3), and organic amines must be dominant in cities before the establishment of sewerage systems in the second half of the nineteenth century. *The Great Stink* or *The Big Stink* was a time in the summer of 1858 during which the smell of untreated sewage was very strong in central London.[569]

Before the discovery of chemical air composition, it was long known that air contains *foreign bodies* whereas air was regarded as a gas, named "common air". *Christlieb Ehregott Gellert* (1713–1795)[570] wrote in his book *Metallurgic Chymistry*

569 For London smoke history see Brimblecombe (1977, 1987); see also Davenport and Morgis (1954), a bibliography on air pollution with almost 4000 citations.

570 *Gellert* studied at Meissen and Leipzig. From 1736 through 1737, he worked as an instructor at the Gymnasium in St. Petersburg, then he became adjunct at the "*Akademie der Wissenschaften*" [the Royal Academy of Sciences at St. Petersburg] until 1746. In 1747 he settled in the mining town of Freiberg he held several positions related to the mining and metallurgical industry. At the founding of the "*Berg-Akademie*" in 1765, he was appointed professor of metallurgy and chemistry. *Gellert* introduced the process of cold extraction of precious metals by the use of mercury, a works (the largest of the time) being built in at Halsbruck in 1790. His famous work "*Anfangsgründe zur Metallurgischen Chimie*" (Gellert 1750) was published in an enlarged 2nd edition in 1776; translated into English (1776), French (1758), Italian (1790). The English edition is based on the German 2nd ed. From 1776, but reads easier than the German.

(Gellert 1776) a chapter on air (pp. 80–90) which represents the knowledge of the middle of the nineteenth century. *Gellert* writes (Gellert 1776, p. 80): "Since therefore no chemical operation may be done without fire, it follows that they can neither be performed without air [...]." He emphasizes further that no fire can exist without air, and nobody known could live and grow without air. *Gellert* describes (Gellert 1776, pp. 81–84) air as a fluid body where the air particles are very small but larger than fire particles. Water evaporates from the whole earth daily in immense quantities and falls again in dews, rains, and snow. Further, apart from the old language and expressions, the text reads like a modern textbook (*italic* by *Gellert*):

> [...] all that immense number of volatile substances, which continuously, from the whole surface of this globe, rise up in the air [...], falls down by itself, or united with other substances. (Gellert 1776, p. 85)
>
> As then soot is nothing else but a collected smoke of combustible, vegetable matter, &c. and of this immense quantities are daily raised up into the air, the existence of earth in the air is rendered beyond all doubt. [...] dust and sand from great wilderness, the ashes from volcanos &c., which are frequently known to be carried away to the distance of a hundred miles. (Gellert 1776, p. 86)
>
> [...] all *odoriferous vegetables* emit these their essential spirits or vapours into the air [...] in so great a quantity, that often mariners and travellers by sea, have the fragrant smell in the open sea, discovered the vicinity of land long before it could reach the eye. An amazing quantity of *vineous spirits* are continually generated by fermentation and over the surface of the earth [...] All *natural acids*, *bitter* and *alkaline salts* of vegetables evaporate at last in the air [...] (Gellert 1776, p. 86)
>
> It has evern been observed, that entire particles of vegetables, in some kind of seeds, were taken up by air, and carried to immense distances: from which event, among other errors of that kind, those sulphurous rains, bloody and like waters, have taken rise. (Gellert 1776, pp. 86–87).

Further, *Gellert* mentioned animals as a source of air emissions, urine, and dung, which evaporate into the air. On p. 88 he writes the following headline on the left margin:[571] "Air contains even fossil bodies, sulphur, nitre, acids of sulphur, of nitre, if common salt, even fixed salts". The text continues:[572]

> *Sulphur* is burnt up daily in great smelting-houses, by the roasting and smelting of minerals, and all this is dispersed into the air [...] The existence of *nitrous acid* in the air [...] Further, when considered that these two acids, the *sulphureous* and *nitrous*, being actually in air, are both of a stronger kind than that of common salt, and that continually such immense quantities of common salt are exposed to the air, it is plain that a great quantity of this *saline acid* must thereby be freed

571 Moreover, the other margins of that chapter (pp. 88–90) contain the following phrases: Air contains water and earthy particles. Air contains vegetable, spirituous, oily, saline particles. Air is full of animal particles, and the seed of insects. Air contains metals.

572 Concerns HCl degassing from sea salt see Chapter 3.4.1. This is the oldest citation I found on this topic and the first exactly describing a chemical process in air.

> from its fixed earth by the two former stronger acids, and thence carried off into the air. [...] so does nature work the same process likewise,[573] and perhaps in many other ways (p. 88).
> The existence of *mercury* in the air is daily experiences by miners with the loss of their health, in those mines where ores are worked by amalgamating mills [...] that metals may be dissolved by air alone, and thence carried off with the air [...] (Gellert 1776, p. 89).

Finally, *Gellert* summarizes the effects of air (Gellert 1776, p. 90):

> Since so many particles of almost every kind of bodies are actually contained in the air, and as they are therein in a continual motion, they can unite and mix themselves with the air in various way and manners, and produce by that means those sudden and astonishing effects therein. And as not every part of the globe produces the same kind of bodies, nor the same quantity of each, the bodies within the airy region must likewise differ in their kind and quantity, consequently the phenomena of air cannot be the same everywhere alike.

The problem of dust collection is well described by *Edward Elway Free* (1883–1939),[574] a physicist in the U. S. Bureau of Soils (he also first demonstrated the significance of dust inputs in soil genesis and reviewed much of the early literature); Free (1911, p. 114) writes:

> The collection and examination of atmospheric dust is matter of peculiar difficulty. If the air be passed through water, as, for instance, in a gas-absorption bulb, the dust is completely collected but its properties are often changed by the contact with water, and it is difficult to remove the water without modifying the nature of dust. The method of collecting dust on plates smeared with Vaseline or glycerin is open to similar objections. Filters of cotton-wool stop most of the dust, but is cannot afterwards be separated from the material of the filter. Gun-cotton may be used instead of ordinary cotton, and then dissolved in ether, but this will mean the loss of the ether-soluble constituents of the dust itself [...] funnels built on the raingauge principle and arranged to face of wind, collect the dust in excellent condition, but do not collect all. The finest materials are blown clear through and escape.

So-termed dust precipitation (deposition) became largely in use after 1920 in air pollution monitoring; it is collected by an open gauge (for example, the German *Bergerhoff* sampler)[575] and given in mass per square and time (e. g., in mg m^{-2} d^{-1}) included almost PM > 10 μm (diameter) due to sedimentation. In England, according to a proposal

573 *Gellert* meant here "evaporation, and inspissation, be entirely destroyed and rendered so volatile, as to fly off into the air by themselves".

574 *Edward Elway Free* is not well known but produced some excellent work during this time. *Free* majored in Chemistry at Cornell University and graduated in 1906. He worked as a chemist with the University of Arizona from 1906–1907, as a physical scientist with the Bureau of Soils from 1907–1912, and as a private consultant in chemistry and physics after 1912. In 1917 *Free* earned a PhD from Johns Hopkins University. He continued to run his consulting company until his death (Brevik 2004).

575 Bergerhoff, H. (1956) Staubpegelzonen nach Sedimentationsmessungen der Landesanstalt für Bodennutzungsschutz des Landes Nordrhein-Westfalen, Bochum. *Heinz Bergerhoff* (living dates unknown).

of Cohen (1896), the Department of Scientific and Industrial Research (the investigation of atmospheric pollution, yearly reports since 1914),[576] earthenware vessels with a diameter of 30 cm have been used as "standard gauge" (in 1936 115 gauges were in operation; Lehmann and Heller (1940, p. 588). In the founding decree of the Prussian Government concerned the establishment of "*Landesanstalt für Wasser-, Boden- und Lufthygiene*" (footnote 461 on p. 216), it was written (Heller 1974, p. 35):

> Die Rauchfahne von Berlin dient bei klarem, windstillem Wetter den Luftschiffern auf 100 km und mehr Kilometern als Wegweiser [The plume of the city Berlin serves as a guide for airships in 100 km or more in clear, windless weather].

3.1.2 Smoke and fog: smog and the soot plague

> Busy factories meant smoky chimneys; so, therefore, smoky chimneys were needed that the factories were busy. Smoke was the symbol of industry, the emblem of a new wealth. The effect became the cause. What was wanted was more smoke (Marsh 1947, pp. 149–150).

John Evelyn says, speaking of fog during the Great Frost of London in January 1684 (Bray 1827, p. 109):

> London by reason of the excessive coldness of the aire hindering the ascent of the smoke, was so fill'd with the fuliginous steame of the sea-coale that hardly could one see crosse the streetes, and this filling the lungs with its grosse particles exceedingly obstructed the breast so as one could hardly breath.

Muncke writes in Gehler's (1833, Vol. VII. p. 52):

> [. . .] der nebelartige Rauch über großen Städten selten und namentlich über London nie fehlt [. . .] Diese Ansicht wurde bei mir hauptsächlich hervorgerufen, als ich in der Nähe von Birmingham von einem einzigen Standpuncte aus 95 hoch hervorragende Kamine zählte, die vielen niedrigen nicht mitgerechnet, aus deren jedem eine schwarze Rauchsäule emporstieg, so dass alle vereinigt die ganze unübersehbare Fläche mit einer undurchsichtigen Rauchwolke überdeckten [. . .] [. . . the foggy fume never is absent above large towns and namely London [. . .] This view was mainly caused with me when I have termed from a single point near Birmingham 95 highly outstanding chimneys, not calling the many low, from each one a black column of smoke rose, so that all unified covered the whole area with a non-transparent plume].[577]

The need for strategic approaches to urban air pollution grew in late-eighteenth-century Britain in cities such as Manchester, where the steam machine had been

576 *G. M. B. Dobson* was the chairman of the Atmospheric Pollution Research Committee in the 1930s.
577 The German terms "*Rauchsäule*" [column of smoke] and "*Rauchwolke*" [cloud of smoke] were likely first used here by *Muncke*. The English term *plume* (see footnote 239 on p. 112) was not yet used at that time as a "trail of smoke" (best corresponds to the German "*Rauchfahne*"), surely derived from observation of "smoking" driving steamers and locomotives.

adopted with great enthusiasm. In the 1880s, most strikingly, in his influential publication London Fogs (1880), *Francis Albert Rollo Russell* (1849–1914)[578] charged that the smoke issuing (Russell 1880):[579]

> A London fog is brown, reddish-yellow, or greenish, darkens more than a white fog, has a smoky, or sulphurous smell (p. 6),
> In winter more than a million chimneys breathe forth simultaneously smoke, soot, sulphurous acid, vapour of water, and carbonic acid gas, and the whole town fumes like a vast crater, at the bottom of which its unhappy citizens must creep and live as best they can.
> We should thus expect to find that the darkness and severity of yellow fogs are in the proportion of the size of towns, their conditions being similar with regard to the use of coal. And this is strictly the case. Liverpool, Birmingham, Manchester, and Glasgow, consume large quantities of coal, and their fogs resemble those of London (p. 14).

The terms (wet and dry) fog, mist, smoke, and dust have been used for several decades by the investigators in a different sense. Aitken (1880) writes:

> Dust, fogs, and clouds seem to have but little connection with each other, and we might think they could be better treated of under two separate and distinct heads.

Russell (1881) comments:

> Where the fog extended high the smoke mixed with it and produced a yellow fog, but where it remained low smoke escaped into the upper air and drifted away, leaving a white fog below, [...]

Further in response, Aitken (1881, p. 312) writes:

> It is condensed vapour which forms the fog and dust simply determines whether it will condense in fine- or coarse-grained particles.

Russell (1888) explains the formation of London smog:

> [...] produced by the mechanical combination of particles of water with particles of coal or soot, and require for their fullest development the following conditions. A still air, a temperature lowest at or near ground, [...] absence of clouds overhead, and free radiation into space.

578 Meteorologist, civil servant and author, most noted for his studies of London fog and smog, as well as the worldwide effects of the 1883 volcanic explosion on the island of Krakatoa in Indonesia. Third son of Lord *John Russell*, later the first *Earl Russell* (1792–1878). Not to confuse with the meteorologist *William James Russell* (1830–1909) who also studied London's fog.

579 See also: Horsfall, Th. C. (1893) The nuisance of smoke from domestic fires, and methods of abating it. Manchester and Salford Noxious Vapors Abatement Association, No. 3, 20 pp. *Thomas Coglan Horsfall* (1841–1932) was a noted philanthropist, town planner, writer and founder of the Manchester Art Museum, Ancoats Hall. *Horsfall* considers injurious effects of smoke and noticed that smoke from domestic fires is greater in quantity and more injurious than smoke from industrial furnaces, and domestic smoke is preventable by improved devices and appropriate methods of heating and cooking.

Soon later Russell (1889) writes:

> Unlike fog, haze commonly occurs in this country when the lower air is in a state of unusual dryness (p. 60).
> Is it not possible that condensation to a slight degree may occur upon some minute crystalline particles, at temperatures above the dew-point? (p. 63)

Besides *Russell*, *Bailey* and later *Cohen* must be named in smoke and fog research. However, urban air pollution monitoring networks did not really begin until the development of deposit gauges, inspired by the work of *The Lancet* in 1910 (Mosley 2009). In the years after 1891, *George Herbert Bailey* (1859–1924)[580] put attention on town pollution (Bailey 1893, p. 81):

> Pollution of the atmosphere by organic matter and by gaseous emanations must be regarded as one of the most fertile sources of diseases and of the lowering of tones which are invariably associated with overcrowded districts.

Smoke and fog as contemporaneous phenomena were scientifically described by *Julius Berend Cohen*, professor for organic chemistry in Leeds (Fig. 3.1). He writes (Cohen 1896, p. 369):

> Town fog is mist made white by Nature and painted any tint from yellow to black by her children; born of the air of particles of pure and transparent water, it is contaminated by man with every imaginable abomination. That is town fog.

Cohen conducted laboratory experiments and concluded (Cohen 1896, p. 371): "The more dust particles there are, the thicker the fog"; carbonic acid (CO_2) and sulfurous acid (SO_2) were observed to increase rapidly during fog, and, "[. . .] although I have no determinations of soot to record, the fact that it increases also is sufficiently evident." Fog water particles become coated with a film of sooty oil. Consequently, fog persists longer than under clean conditions. Giberne (1890) denotes this phenomenon as "*London fog*"; however, it was described earlier for Paris and Amsterdam (Anonymous 1835, Flammarion 1874). The term "London fog" is likely coined by *James Main* (unknown living data); Main (1832):

> Fogs are happen everywhere [. . .] But fogs are more dense about London [. . .] by the clouds of smoke from every fire [. . .] Beside fogs, we have also mists and haze [. . .]

Russell (1895, p. 234) used the word combination "smoky fog" and writes:

> Town fogs contain an excess of chlorides and sulphates, and about double the normal, or more, of organic matter and ammonia salts.

580 Chemist at the Owens College (today University of Manchester) who studied under *Roscoe*, and later obtained his PhD at *Bunsen* in Heidelberg.

Figure 3.1: Industrial city (Sheffield) in rest and work; after Marsh (1947, p. 113).

Table 3.1: Composition of soot deposit in 1891, in % (Cohen and Ruston 1925, p. 73).

site	soot	hydrocarbons	organic bases	SO_3	HCl	NH_3	mineral matter	water (by difference)
Chelsea	39.0	14.3	2.0	4.3	1.4	1.4	33.8	5.8
Kew	42.5	4.8	–	4.0	0.8	0.8	41.5	5.6

During the last fortnight of February 1891, the weight of the foggy soot deposit in Kew (just outside London) was 0.84 g m^{-2} (Table 3.1). Fig. 3.2 illustrates samples collected in Leeds.

In Manchester 1892, SO_2 concentrations[581] were measured that were by a factor of 100–700 larger than that today (in mg m^{-3}): 0.2–0.4 (September), and 1–2 (December) and 3–6 in the presence of fog (Cohen and Ruston 1925, p. 72). Of 39 fog events in London between 1 September 1902 and 31 March 1903, eight were described as "smoke fogs" (Cohen and Ruston 1925, p. 68). Another interesting figure is given for 1883–1885, citing *Russell* (on the impurities in London air, The Monthly Weather Report of the Meteorological Service, 1885) for "the solid organic impurities in the air" (in mg m^{-3} recalculated):

Fine weather	0.12
Dull weather	0.36
Foggy weather (no dense fog)	0.86

The concentration of sulfuric acid (SO_3) in Manchester in autumn 1909 was estimated (in parenthesis the suburban value) to be 1.7 (0.7) mg m^{-3} without fog and 11.7 (3.8) mg m^{-3} with fog. The CO_2 concentration was several times larger than that in the background (300 ppm). The dustfall was estimated to be 700–800 mg m^{-2} d^{-1} (in Leeds "only" 314 mg m^{-2} d^{-1}),[582] whereas in Leeds about 15 % of the dust was "nasty, sticky oil" (Cohen 1896); *Cohen* writes:

> Were the soot pure carbon it would be comparatively harmless. Unfortunately, this is not the case.

Cohen and Ruston (1911) describe the composition and properties of soot. *Cohen* also determined in Leeds the particle number using the *Aitken* counter in the range of 40,000–260,000. Friese (1909)[583] measured in Dresden dust in the air to be 1300 (100–2700) mg m^{-3} and in Summer 700 (100–1600) mg m^{-3} (what was about 50 times

581 Given as sulphureous (old spelling) acid and value as milligrams of SO_3 per cubic feet (recalculated as SO_2 in mg m^{-3}).

582 Dustfall (soot) in Hamburg, December 1903–January 1904 (n = 32) 41 (3–300) mg m^{-2} d^{-1} (Liefmann 1908).

583 *Walther Friese* (1879–1954) was a German chemist who worked lifelong in the Public Health sector in Dresden. *Renk* and *Hempel* were the reviewers of his dissertation (1908).

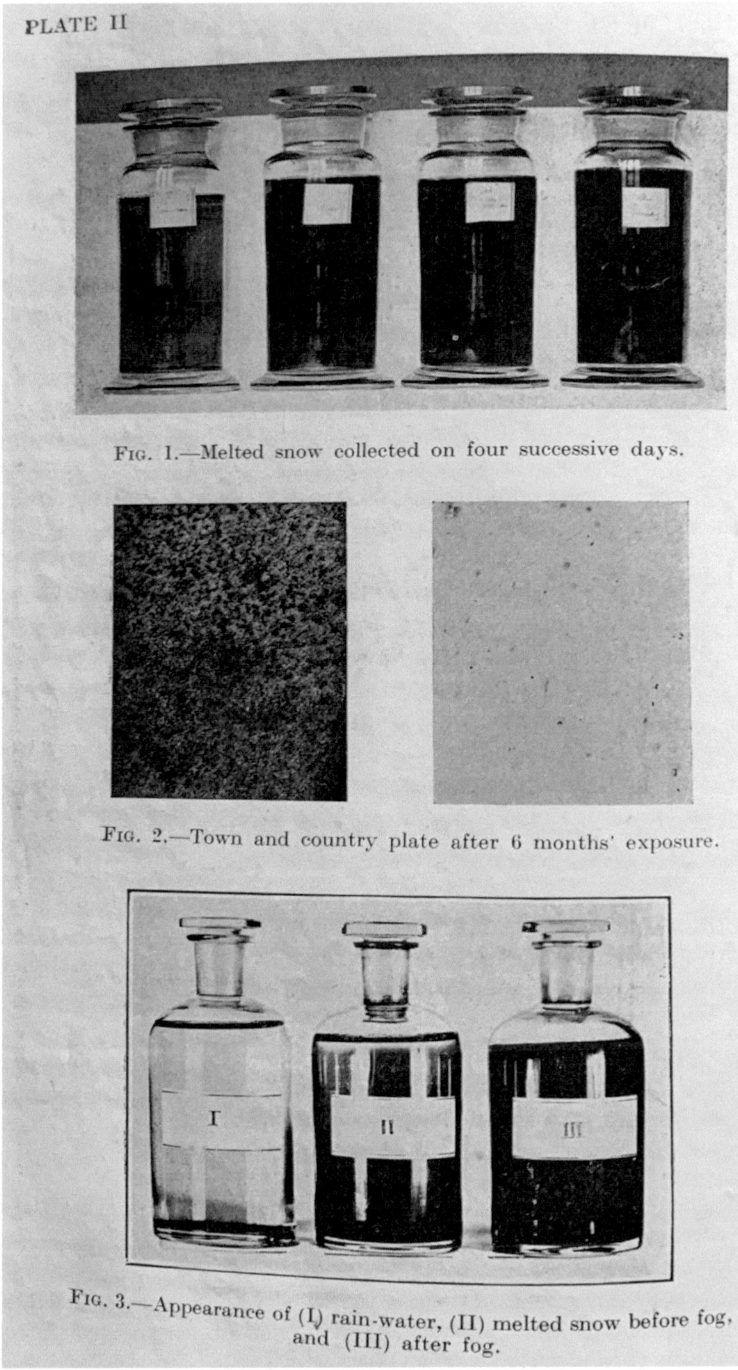

Figure 3.2: Appearance of different rainwater and snow samples, collected in 1909 in Leeds, Great Britain (after Cohen and Ruston 1925, p. 14).

higher than today) and found in rainwater total mass of 60 (20–124) mg L^{-1}; of these 16 mg L^{-1} suspended and 34 mg L^{-1} dissolved. Mabery (1895) determined soot in 1894 in Cleveland to be 71.5 (42–111) mg L^{-1} in 14 (1.5–40) mg m^{-3} in freshly fallen snow and the air,[584] respectively, and writes (Mabery 1895, p. 112):

> When moistened with water the soot [collected on the windows] gave a strong acid reaction on test paper, and analysis showed the presence of sulphuric acid [1.91 %].

Despite recognizing that the atmosphere is contaminated, Mabery (1895, p. 122) "accepted" the condition to be "normal":

> From the results of this examination, it is evident that a city atmosphere, contaminated by the universal consumption of bituminous coal, where no efforts are made to prevent the escape of soot, soon reaches a stage in which it is destructive to property and not conductive to health. In this respect the atmosphere of Cleveland is, doubtless, no worse that of other cities, and, perhaps, in a better condition than some that use the same fuel. Under the usual conditions of life in cities, sanitary regulations require careful attention and constant supervision.

The terms "smoake" and "clouds" in *Evelyn*'s booklet (only once he used the term *fog*) mean surely what we now understand behind *smog*. Coinage of the artificial word "smog" is generally attributed to *Harold (Henry) Antoine Des Voeux* (1861–1942)[585] in his paper (1905), "Fog and Smoke" for a meeting of the Public Health Congress to avoid that the *London fog* is considered as natural phenomenon beyond human control.[586] The 26 July 1905 edition of the London newspaper *Daily Graphic* quoted *Des Voeux*:

> He said it required no science to see that there was something produced in great cities which was not found in the country, and that was smoky fog, or what was known as 'smog'.

The following day the newspaper stated that "Dr. Des Voeux did a public service in coining a new word for the London fog". Already in a letter to the Times (27 December 1904, p. 11), *Des Voeux* wrote that he wished to call London's fog problem by the name

---

584 In determining soot in the atmosphere, air was drawn through a combustion tube filled partly with ignited asbestos and the carbon was determined by ignition and absorption of the CO_2. *Mareby* notes "that the quantity of organic dust in the suspension is very much larger than is represented by the values given above".

585 French physician, living in London. Honorable treasure of the Coal Smoke Abatement Society (formed in 1882) and later President of the National Smoke Abatement Society (Marsh 1947).

586 However, this is predated by a *Los Angeles Times* article of January 19, 1893, in which the word is attributed to an unidentified "witty English writer": "The fact that the death rate of London has recently almost doubled, going to over thirty in the thousand, is sufficient attestation of the evil effects of the dense, black fog which hung over that city for six consecutive days not long ago. These visitations, which a witty English writer once designated by the name "smog", represent a condition of the atmosphere when it is saturated with moisture and charged with soot and the fumes of sulfur and carbonic acid gas from the chimneys and smokestacks of the great city".

"smog" to show that it consisted "more of smoke than of true fog." *Des Voeux* believed that ordinary fogs were caused by the condensation moisture and that nothing could be done to eliminate them. Smog, on the contrary, was a different matter entirely. "Fog", *Des Voeux* wrote, was "incurable and infrequent" while "smog" was "curable and frequent" (cited after Clay and Troesken 2010, pp. 6–7, see also Thorsheim 2006, p. 30).

Shaw (1906) discuss the relation between fog and coal smoke, and presents a speculation (which is from today´s view a remarkable feedback of air pollution onto the weather and climate, resp.) that (meteorological) London early morning fog´s are often near the ground and have little vertical depth; over the top of it is a layer of warmer smoky air which seriously interfere with the effect of the sun´s substantial power of dissipating early morning fog.

The Committee for the Investigation of Atmospheric Pollution, appointed at the International Smoke Abatement Conference and Exhibition held in London, March 1912 was headed by *Napier Shaw*, director of the Meteorological Service; the secretary was *John Switzer Owens* (1871–1942).[587] There were published 23 annual reports between 1915 and 1937. Dust monitoring began in England by the standard filter method introduced by *Owens* in 1913 and improved in 1913–1914 with automatic changing filter paper, that was compared against a standard "scale of shades". Another "standard" was produced by weighting the filters according to the scale of 25 units (Waller 1964, Spurny 1998). *Owens* also introduced a "jet dust counter", where a very small jet of air is made upon a glass surface, and examined microscopically (Owens 1922, 1924); he nicely notes the object having in view to provide some simple and effective methods making investigation (Owens 1922, p. 18):

> [. . .] for examining the quantity and nature of suspended dust, and its relation to visibility, fumes from industrial processes, dust produced by mines, fogs such as are experienced in our larger cities, and generally in the hope of throwing some light on problems which are concerned with the presence of fine suspended impurities in the air.

The problem was named in Germany "*Rauch- und Rußplage*" [smoke and soot plague] in the books devoted to public health and town planning (Liefmann 1908, Reich 1917).[588] Anonymous (1895) writes

---

587 He was an engineer and hold a medical degree; technical adviser to the London Smoke Abatement Society and since 1912 had been Superintendent of Observations on Atmospheric Pollution in the Department of Scientific and Industrial Research. He was joint author with Sir *Napier Shaw* of a book on "The Smoke Problem of Great Cities" published in 1926, and also presented papers to the Institution on "The Smoke of Cities" in 1936, and "Identification of the Source of Deposited Matter" in 1939.

588 *Max Harry Liefmann* (1877–1915) was a German physician who habilitated in Halle on "*Über den Nachweis von Ruß in der Luft*" [On the detection of soot in air] in 1907; since 1909 head of the department bacteriology at the Rudolf Virchow hospital in Berlin, fallen in the war in Lithuania. *Albert Johann Friedrich Reich* (1859–?) was a German civil engineer and architect in Berlin.

> Gegen die Rußplage wird seit Jahrzehnten mit verschiedenen Mitteln angekämpft [Since decades is fought by various means against the soot plague]

and describes an apparatus for the removal of dust and soot from chimneys, "*Löfflers selbstthätiger Ruß- und Funkenfänger*" [Löffler's, automatic soot and spark collector], based on a simple impactor principles.[589] In Germany, the first city which established a "*Kommission zur Bekämpfung des Rauches*" [commission for abatement of smoke][590] was Königsberg in Prussia, but not because of immense smoke problems as because the municipality was interested in the hygienic questions. Ascher (1907, p. 126)[591] reports on estimations of soot and sulfuric acid on snow deposits (in mg L^{-1}):

location	soot, total	soot, inorganic	SO_3
Alniken (Baltic sea side)	4.0	3.5	0.52
Volksgarten (near gasworks)	51.0	32.0	8.4
Pathological Institute	49.5	25.0	5.5
Mitteltragheim (city center)	99.0	54.5	3.6

Remarkably, the term "smog" is not used in the book "The Smoke Problem of Great Cities" by Shaw and Owens (1925) and in *Marsh*'s book "Smoke" (Marsh 1947).[592] Shaw and Owens (1925, p. 16) write:

> We propose to restrict the name *fog* to a cloud of water globules, whether fouled by smoke or not and to use the word *haze* for a cloud which is not necessary dependent upon water globules.

Without using the term *atmospheric chemistry*, between about 1850 and 1950 (and later of course), it was *air pollution chemistry* research as notified by the British geographer *Eva Germaine Rimington Taylor* (1879–1966); Taylor (1942). Lawther (1959, p. 89) cites

589 *B. Löffler* (unknown living data) was a German manufacturer in Frankfurt/Main; see: Ein Ruß- und Funkenfänger. *Deutsche Gerber-Zeitung* **54** (1890); *Illustrierte schweizerische Handwerker-Zeitung* **11** (1895) 358; *Polytechnisches Journal* (*Dingler's J.*) **291** (1894) 96. The apparatus was produced by industrialist *Paul Lechler* (1849–1925) in Stuttgart, who was socially very engaged.

590 Members of the Commission: Dr. *Ascher*, "*Kreisassistenzsarzt*" [district resident], *Korbert*, director of the gasworks, *Robin*, chief engineer of the Steam Boiler Inspection Association [*Dampfkesselrevisionsverein*], and Dr. *Hurdelbrink*, city chemist. Erster Bericht der Kommission zur Bekämpfung des Rauches in Königsberg i. Pr. *Schriften der physikalisch-ökonomischen Gesellschaft in Königsberg* **48** (1907) 121–154. Zweiter Bericht [...] *Deutsche Vierteljahreschrift für öffentliche Gesundheitspflege* **41** (1909) 369–379. See also Ascher (1907), Hurdelbrink (1907, 1909). *Franz Hurdelbrink* (unknown living data).

591 *Ludwig Ascher* (1865–1942) was a German medical doctor who worked practical in different German towns but spend his life concerns public health and hygienic issues; he passed in the Ghetto Litzmannstadt.

592 *Arnold Marsh* (? –1968) was the secretary and later director of National Society for Clean Air.

observations made by *Owens* in 1924, that smoke and fog are not so closely connected as the word “smog” would make us believe. Concerning the “relationship between fog and smoke”, *Marsh* wrote that fog was a natural phenomenon, whereas smoke, passing through a fog, could not dissipate as it could in non-foggy weather because of the absence of air currents, and the “clean natural fog gradually becomes more and more impregnated with smoke”(this is not correct, because smoke particles act as condensation nuclei, and thus “clean natural fog” could not possibly have appeared in cities like London in those years).

As to the time when the air pollution problems associated with the combustion of fossil fuels were solved, by the end of twentieth century (still unsolved is the climate change problem due to carbon dioxide), sulfur dioxide and soot (smoke) were the key air pollutants since centuries. Coal is used in cities regularly since the beginning of middle age; we can term that beginning of (and not ending period) the *coal age*.

Obviously, the word “smog” was widely used only after World War II, in connection with the photochemical air pollution in Los Angeles and the “winter” smog like in London in 1952. Thus the term “smog” goes far beyond the initial sense in combining “*smoke and fog*”, denoting a complex air pollution situation (Green et al. 1964, Leighton 1961). The largest smog event was recorded in London in 1952, initiating research of the oxidation of SO_2 and the dispersion of air pollutants. Wilkins (1954) writes:

> In some areas [of the British Isles during 5th to 9th December 1952] there was no fog, or at the most a patchy white mist, while in many industrial areas the fog became grimy smog. In London and the Thames Valley, fog or smog covered upwards 1,000 square miles.

Leighton (1961, pp. 1–2) writes in the first modern textbook on “photochemistry of air pollution”:

> Yet the word “smog” has quite a different meaning to a resident of London than it does to a resident of Los Angeles... London smog generally occurs at high relative humidities accompanied by fog; Los Angeles smog occurs at low relative humidities under a clear sky. London smog is accompanied by radiation or surface inversion, Los Angeles smog by subsidence or overheated inversion. To chemical agents, London smog is reducing; Los Angeles smog is oxidizing. The immediate effect of London smog to humans is bronchial irritation; in Los Angeles it is eye irritation. In both cases visibility is reduced, but this is general more severs in London.

The smoky skyline of towns and industrial areas remained unchanged until the 1970s (Fig. 3.3). Only with recognizing regional (acid rain period) and global pollution, abatement strategies were implemented and additional exhausting coal deposits in Europe, “historical smoky sceneries” have changed stepwise into “green landscapes” in the 1990s.

Figure 3.3: Town and industrial pollution views. From top to bottom: *L'atmosphère des villes industrielles...* (after Flammarion 1888, p. 97); the industrial skyline of Dortmund (Ruhr area, Germany) from a Postcard about 1912; Ostrava in the Czech Republic about 1963.

3.2 On dust

The president of the British meteorological society, *William Marcet* writes (Marcet 1890, pp. 73–74):

> The infinitely small particles of matter we call *dust*, [. . .] Dust is concerned in many meteorological phenomena, such as fogs, as it is generally that fogs are due the deposit of moisture on atmospheric motes.
> I the vegetable and mineral wo crumbles into *dust*, on the other hand it is highly probable that dust was the original state of matter before the Earth and heavenly bodies were formed.

Alfred Ditte (1843–1908), French chemist at Sorbonne in Paris; writes (Ditte 1904, p. 235):

> The earth's atmosphere contains an enormous quantity of dust, particles of which float in the air for varying periods of time.

3.2.1 *Ehrenberg*'s red dust and other dust fall

William Phipps Blake (1827–1910) was an American geologist, who first considered rock corrosion as a source of atmospheric dust (Blake 1855). However, all dust falls until the middle of the nineteenth century were seen as either of the local origin or cosmic (meteoric) one. Prestel (1858) first gives evidence for long-range transport of smoke (dry fog) from moorland fires (see next chapter). Another early evidence of long-range transport of air pollution is likely due to dirty snowfall in Norway observed by Brøgger (1881), see p. 214. Colored rainfall and dust fall (Chapter 3.2.2) were regarded largely *heavenly* (in earlier time religious) and later of cosmic origin. Humboldt (1849, p. 5) accepted the finding by *Christian Gottlieb Ehrenberg* (1795–1867), one of the founders of modern soils science, who was likely the first investigator of atmospheric dust concerning microorganisms, and concluded on a systematic relation between dust events and their causes. Ehrenberg (1830, p. 485) writes:

> Die Atmosphäre erfüllt sich mit rothem Staube, wobei zufällig eintretender Regen als Blutregen, und in dessen Folge rothe Färbung der Flüsse und stagnierenden Gewässer erscheint [The atmosphere meets with red dust, where accidental rain appears as blood rain, and as a result appears the red color of rivers and stagnant waters].

In contrast, the French astronomer and physicist *François Arago* writes "that the careful observation of dust falls led to the conclusion that they do not differ from common falls of meteoric stones" (Arago 1857, p. 206). *Ehrenberg* investigated 12 samples of "*Staubmeteore*" [dust meteors],[593] collected (partly by *Darwin*) between

593 Ehrenberg (1872, p. 12) states that it was inccorect to name this dust particles meteors (making the confusion with true meteors); remember that in ancient time all atmospheric phenomena were named meteoric, and also rainwater as *meteoric water*, still in use.

1803 and 1847 over the Atlantic, at the Cape Verde Islands, Malta, Genoa, Lyon, and Tirol (Blume and Bölter 2011), having all colors between yellowish and reddish, and being identical with "*unserem gewöhnlichen Luft- und Gewitterstaub*" [our common air and thunderstorm dust], (Ehrenberg 1849, p. 18). From his mineralogical, biological and partly chemical investigations he concluded (Ehrenberg 1849, pp. 22–23) from "*gleichmäßigen Mischung in so großen Mengen und bei so großen Raum und Zeit-Unterschieden*" [consistent mixing in so large amounts and at so large difference in time and space]:

> Sie scheint einem constanten Verhältnisse, einen constanten, schwebenden, sich lange und immer von Neuem mischenden Staubnebel angehören zu müssen, welchen ein zufällig dazu tretender Orkan in beliebige Richtung verbreiten kann [it seems to be in a constant proportion, in a constant, suspended, and for a long time and renewed dust fog, which is dispersed into any direction by an accidentally occurring storm].

Ehrenberg rejected the already existing idea that this dust is originated from the Sahara[594] because of the distribution of altogether 141 species (no longer living) of diatoms and earthy freshwater, terrestrial and maritime animals he found. *Ehrenberg* (1849, p. 58) writes:

> [...] ein unabsehbar großes Staubnebeldepot in den oberen Schichten der Erdatmosphäre in über 14,000 Fuß Höhe [...] die optischen Verhältnisse beeinflussen müsse [... an immeasurable large depot of dust fog in the upper layers of the earth's atmosphere in an altitude of more than 14,000 feet [...] must influence the optical properties].

Ehrenberg denotes this phenomenon "*Passatstaub*" [dust with the trade wind] and assumed the origin from South America. Ehrenberg (1872, p. 263), who termed these organic atmospheric species "*organische Atmosphärilien*", listed 148 shapes, which distribute to the different parts of the world as follows (number in parenthesis): Australia (24), Asia (53), Africa (32), America (78), Europe (60), and the atmosphere (12). Some of them occur in all parts of the world, but most forms are of specific origin. Thus, it became possible to identify the origin of dust falls by analyzing the distribution of species found in the dust. Because of the large-scale and long-range dispersion, he writes (Ehrenberg 1849, p. 44):

> [...] dass derselbe nur an Dunst-Nebel (Wolken) gebunden gewesen, da in den Zwischenländern kein Staubfall beobachtet worden ist [that the same is bonded to haze-fog (clouds) because between the countries no dust fall were observed].

594 In a letter *Darwin*'s to *Ehrenberg* from March 23, 1845, *Darwin* doubts this estimation (cited after Blume and Bölter 2011). Today we know that such events are caused by Sahara dust storms. Gorbushina et al. (2007) investigated some of *Ehrenberg*'s dust samples that were stored in the Museum of National Science ["*Naturkundemuseum*"] in Berlin and found bacteria and spores from fungi which germinated; the main source was identified as the Bodélé depression in northern Chad.

Ehrenberg (1872, pp. 15–59) describes 198 "events" of "*Passatstaub*" and blood rain between 1154 BC and 1870 and established a "climatology", distinguishing different types per month. There are interesting results of the analysis of rainwater collected at Catania on 23 March 1869; Ehrenberg (1872, pp. 62–63) cites *Orazio Silvestri* (1835–1890), an Italian geologist and volcanologist at the University of Catania; in mg L^{-1} (note that this rainwater concentration is about 100-fold higher than in "normal" rainwater):

Al_2O_3	910
$CaCO_3$	289
SiO_2	121
Fe_2O_3	252
NaCl	216
organic nitrogen	540

Further examination has shown that the organic nitrogen was completely from biogenic matter; microscopic observations resulted in living infusoria, germs, spores, plant debris, etc.; *Ehrenberg* later estimated 107 species in the seven dust samples (having a yellow-reddish color) collected on 24 March 1769, in Catania (Ehrenberg 1872, p. 65). He still doubts that the Sahara is a source of the dust, but concluded from many dust sample investigations that the dust fall contains a local biological and geological component and that the dust comes from high altitudes, transported over long distances and time, likely mixing with dust from different regions (Ehrenberg 1872, p. 82). Thus, from the organic species, occurring in changing proportions, but always mixtures of several geographic origins, he concluded that suspended dust particles in higher atmospheric layers permanently exist and mixe according to the atmospheric currents (by this explanation he avoided his earlier fixation on South American dust sources only). Chemical analysis has shown a chemical composition typical for soil dust[595] (in %), see Ludwig (1862, pp. 28–29):

SiO_2	40–73
Al_2O_3	6–17
Fe_2O_3	8–15
Ca	4–10
K	3
Mg	2
Na	2

595 Due to weathering, however, primary iron containing minerals (e. g., amphiboles and pyroxenes) of volcanic rocks are not identified in mineral dust (Shi et al. 2012). Thus, this composition is typical for volcanic dust (see also below citing *Nordenskiöld* who believed it to be of cosmic origin).

He estimated the presence of volcanic dust not only by the black color but mainly by missing organic matter. From the studies of numerous dust samples, which *Ehrenberg* received from different sites, he also concluded on different soil dust characteristics. *Ehrenberg* must be named as the discoverer of *bioaerosols*, "*das unsichtbare organische Leben des Luftkreises*" [the invisible organic life of the atmosphere] (Ehrenberg 1872, p. 100).

Hellmann (1879) describes 65 dust falls between 1858 and 1871, observed from ships in the "*Dunkelmeer*" [dark sea][596] and analyzed the events from a meteorological point of view and finally concluded that *Ehrenberg* is mistaken with his assumption, red dust is from South America (which he does not exclude to be as minor part) and permanent in the air, but dominant from the western Sahara, strongly varying from year to year and dominant in the period December to May (80 %). *Hellmann* further states that *Ehrenberg* confused dust fall (sedimentation dust) with dust-clouds or dry-haze or fog (suspended particulate matter).[597] *Hellmann* cites a note of *Roussin*[598] in 1817: "the thick fog or haze prevails almost all the year on the coast of northwest Africa". Furthermore, *Hellmann* critical comments on *Ehrenberg*'s microscopic observations (Hellmann 1879, p. 401):

> Der herabsinkende Staub oder Sand hat, mag er herkommen, woher er will, auf seinem Wege vom Anfangs- bis Endpunkte der Luftreise die mannigfachsten Veränderungen erlitten, da er den verschiedenartigsten meteorischen und mechanischen Einflüssen ausgesetzt gewesen ist, und kann daher unmöglich in derselben Beschaffenheit und Zusammensetzung am Orte seines endlichen Niederschlags zur Erdoberfläche wie dem seiner Erhebung von derselben befunden werden [the falling dust or sand, wherever it comes, suffered manifold changes on his way from the beginning to the ending of the air journey because it is impacted by different meteoric and mechanical influences, and thus cannot be in the same condition and composition at the site of deposition as on the site of its source]

Similar, Tissandier (1877, p. 87) writes that a large variety of fine and coarse dust particles, including plant debris and organic matter, in large amounts will be moved by the wind, like from the Sahara and the Gobi Desert. Boussingault (1856c, p. 263–264) describes in the air (and higher atmospheric layers) permanent suspended dust from

596 *Hellmann* states that *Edrisi* (1099–1166), Arabian cartograph from Ceuta (*Abu Abd Allah Muhammad ibn Muhammad ibn Abd Allah ibn Idris al-Idrisi*), first describes the "*Dunkelmeer*" (mare tenebrosum). Humboldt (1836, pp. 65–65) termed the sea "*finsteres Meer*" and cites *Edrisi* who said that the "mare tenebrosum" is so named because of difficult navigation, namely permanent heavy storms. *Humboldt*, however, mentioned that older writers stated that this sea is windless. Further, *Humboldt* cites *Johann Philipp Kurzmann* (?–1794) who explains in his "Commentatio de Africa Geographi Nubiensis" (1791, p. 8) that the name of this sea is because of a cloud, resting above the surface of the sea.

597 From British logbook entries *Hellmann* notes the terms dust (37), sand (8), sand or dust (3), powder (2), fog (6), cloud-dust (4), and African dust (3); in parenthesis the number (2 unclassified).

598 *Albin Reine Roussin* (1781–1854) was a French admiral and statesman.

volcanoes, by wind blow from the sea and land surface, and removed by rain, and cites *Bergman* who named them "*Unrath der Atmosphäre*" [garbage of the atmosphere].[599]

Adolf Erik Nordenskiöld (1832–1901)[600] is the first who find evidence that dust is part of cosmic origin (Nordenskiöld 1874), later confirmed by *Gaston Tissandier* (1843–1899).[601,602] *Pietro Tacchini* (1838–1905), an Italian astronomer, director (1859) of the Astronomical Observatory of Palermo observed[603] in February 1879 at different sites in Italy dust deposits (also together with rain) containing black magnetic iron balls having diameters between 4 and 41 µm. *Nordenskiöld* collected snow at different locations in Stockholm, northern Finland, and Greenland in 1871 and 1872 and found always small black and grey magnetic grains, containing metallic iron, cobalt, phosphor, and likely nickel, typical for meteoric origin. He describes this dust to be similar to that he found in 1870 on ice in Greenland, which he termed *Kryokonit*. *Nordenskiöld* speculates that it is either of volcanic or of cosmic origin, but he excludes mineral soil dust. Remarkable is the content of black carbon and humic-like matter ["*einen humusartigen organischen Stoff*", Nordenskiöld 1874, p. 163]. He reports on the chemical analysis of Kryokonit, carried out by *Gustav Lindström* (1829–1901), a Swedish paleontologist (Nordenskiöld 1874, p. 161), in % (here named with the formula and not in historic terms):[604]

599 *Tobern Olof Bergman* is meant; no original citation found; not in Bergman (1788).

600 Finland-Swedish aristocrat (Baron), geologist, mineralogist and Arctic explorer; son of *Nils Gustaf Nordenskiöld* (1792–1866), studied at *Berzelius* in Stockholm and is regarded as founder of mineralogy in Finland.

601 French chemist, meteorologist, aviator (balloon rises) and editor in Paris.

602 Ditte (1904, p. 240) referred beside *Ehrenberg*, *Nordenskiöld* and *Arago* concerns extra-terrestrial origin of atmospheric dust *Daubrée* and Baron *de Reichenbach*; *Gabriel Auguste Daubrée* (1814–1896), French geologist who studied meteorites, namely that of Orgueil. *Karl* (*Carl*) *Ludwig Friedrich* (since 1839 Freiherr *von*) *Reichenbach* (1788–1869) who was a notable German industrialist and scientist. Ditte (Ditte 1904, p. 248) refers the combustion of coal as source of atmospheric iron.

603 Sui granuli di ferro trovati nella polvere di scirocco del febbraio 1879. *Annali delle'ufficiao centrale die meteorologia Italiana*, Ser. II., Vol. 1 (1879); Sur des particules ferrugineuses observées dans la poussière amenée par un coup de vent de siroco en divers points de l'Italie. *Compt. Rend.* **88** (1879) 613.

604 From today's knowledge both dust types were surely mixed. The presence of coal-type organic matter could also be from carbonaceous chondrites or large forest fires in Northern America. The fact that these particles also found on snow in Stockholm *Nordenskiöld* attributed to smoke from combustion. Back snow in Norway Brøgger (1881) believed to come from Great Britain (see p. 214).

SiO_2	62.0
Al_2O_3	15.0
CaO	5.1
FeO	4.6
Na_2O	4.0
MgO	3.0
K_2O	2.0
Fe_2O_3	0.7
P_2O_5	0.1
MnO	0.1
H_2O and org.	3.0

Tissandier is the first who collected street dust in Paris at 3 m height in 1870 and 1872 using an aspirator to study the amount in the air, the particle size, chemical composition, and the dust fall (Tissandier 1874, 1877):

> [...] déterminer la proportion des corpuscules solides contenus dans un volume d'air connu, et de rechercher la composition chimique des poussières aériennes.

Tissandier found that the dust quantity in the air is much higher after several days without rain (in mg L^{-1}):

23.0	after 8 days dry weather (July 1870)
6.0	after a rainy day (July 1872)
6.0–8.0	normal weather conditions (summer 1870 and November 1872)

The diameter of particles derived from wood and coal was estimated to be 100 µm and that of mineral origin 1–10 µm (he termed the particles "*corpuscules aériens*"). He also determined the dust precipitation in Paris and suburb (at Saint-Mandé), in mg m^{-2} per 12 h: 1.5–3.5 and 1.0–2.5, resp., i. e., only insignificant larger in Paris. The dust collected with the aspirator consists of 25–34 % organic and 66–75 % inorganic matter, containing chloride, sulfate, traces of nitrate and mostly iron, calcium, and silica. Dust from a 60-m height tower of Notre-Dame, i. e., from the area that has not been entered for several years,, consisted of 32.35 % organic and 67.7 % inorganic matter; the latter consists of (in %):

9.220	water soluble matter (ammonium nitrate, chlorides and sulfates of alkali and earth-alkali metals)
6.120	Fe_2O_3 (in hydrochloric acid soluble matter)
15.940	$CaCO_3$ (in hydrochloric acid soluble matter)
2.121	$MgCO_3$, traces of phosphate, Al, etc. (in hydrochloric acid soluble matter)
34.334	in acid insoluble matter (silicates)

Because of the large quantity of dust in the air, *Tissandier* believed that it plays an important role in the atmosphere, but he also emphasized that the dust amount must largely vary depending on the meteorological and local conditions. The suspended dust amount found in 1870–1872 in Paris is about 200-fold of that found in Berlin shortly after the year 2000, but the principal chemical composition is the same (Möller 2020, p. 566). These values (9 % of 7 mg m^{-3}) are consistent with the air-borne fraction of chloride, ammonium, and sulfate, collected in London in 1869 by Smith (1872) to be around 2 mg m^{-3} (sulfate dominates with 85 %). Boussingault (1856c, p. 263) already concluded that the non-volatile ammonium nitrate, like sodium chloride and iodide, and altogether all soluble and non-volatile substances must be components of atmospheric dust. Ditte (1904, p. 247) writes concluding,

> [. . .] that the air ordinarily contains only a small number of metals – sodium, calcium, magnesium, aluminum, and more especially nickel, cobalt, and iron. [. . .] The proportion of solid matter in the air does not appear great enough to be of significance in the physiography of the earth, but almost a third of it is composed of organic matter containing living germs.

Hundred years before, Saussure (1804, p. 279) already stated that the atmosphere contains all elements suspended, mostly from ashes that consist of alkaline salts of potassium and sodium, phosphoric acid salts of lime and magnesia, oxides of iron and manganese.

3.2.2 On falling stones and colored precipitation

> Man hat seit einiger Zeit von so manchem sonderbaren Regen und Stoffen, welche im Regen und ohne Regen vom Himmel hernieder gefallen sind gelesen, z. B. Blut, Milch, Fleisch, Staub, Salpeter, Schwefel, Steine, Papier, Haare, blaue Seide (!?), Gallerte und bituminöse Materie [One has read since some time upon some curious rain and substances which fall from the sky with or without rain, e. g., blood, milk, meat, dust, saltpeter, sulfur, stones, paper, hairs, blue silk (!?), gelatinous and bituminous matter]. (Anonymous 1821, p. 3375)
> Noch merkwürdiger sind jedoch, wie mir deucht, die Frosch-, Kröten-, Schlangen-, Fisch- und Insecten-Regen [Still more curious, however, it seems to me, are rains of frogs, toads, snails, fishes and insects]. (Anonymous 1821, p. 3376)

The observation that substances other than water (so-termed foreign bodies) were connected with precipitation was known since *Aristoteles*. Different colors and odors were described, mostly in connection with fog and mist. Showers of blood, earth, sulfur, plants, frogs, and various kinds of animals as well colored snow and clouds was described in the literature (first in *Homer*). All these phenomena, connected with blooms of insects, pollen, and dust, were first identified by the French naturalists *Nicolas Claude Fabri de Peiresc* (1580–1637) and later by *René-Antoine Ferchault de Réaumur* (1683–1757). The Danish *Ole Worm* (1588–1655), termed *Olaus Wormius*, states that on 16 May 1646, there fell a heavy shower in Copenhagen and contained dust exactly like sulfur.[605] *Simon Pauli* the younger (1603–1680), a German physician living in Denmark, states that on 19 May 1665, there was a tempest in Norway with dust so like sulfur that, when thrown into a fire, it produced the same smell, and, that, when mixed with spirits of turpentine, it produced a liquor the odor was just like that of balm of sulfur; the close neighborhood to Iceland volcanoes is sufficient to explain this occurrence (cited after Flammarion 1874). *Honoré Fabri* (1608–1688), French Jesuit and naturalist, wrote (Fabri 1670, p. 414, cited after Camuffo 1992)

> Sometimes rain of a different colour and smell falls, both because of mixing of different air masses and because of different maturation. Sometime red rain can be seen; but that colour was due the inclusion of certain red lime corpuscles which, in the air sometimes seem to condense by accident. Although there is nothing strange in this, the common people thing that it sometimes rains blood.

Ernst Florens Friedrich Chladni (1756–1827), a German physicist, published a catalogue on atmospheric falls in chronological order (Chladni 1826).[606] In his subsection "Falls of Dust and soft Substances, dry and moist" *Chladni* writes: "In the red and black dust,

---

605 Note that "sulfur" at that time was a synonym for odor and fire, inflammable things.

606 In 1794, *Chladni* published "Ueber den Ursprung der von Pallas gefundenen und anderer ihr ähnlicher Eisenmassen und über einige damit in Verbindung stehende Naturerscheinungen" [On the origin of the iron masses found by Pallas and others similar to it, and on some associated natural phenomena] in which he proposed that meteorites have an extraterrestrial origin. This was a controversial statement at that time, since meteorites were thought to be of volcanic origin. This book made *Chladni* one of the founders of modern meteorite research. *Chladni* published a large number of contributions to "falling meteoric stones": Ueber den Ursprung der von Pallas gefundenen und anderer ihr ähnlicher Eisenmassen, und über einige damit in Verbindung stehende Naturerscheinungen. Johann Friedrich Hartknoch, Riga, 1794, 63 pp. Ueber Feuer-Meteore, und über die mit denselben herabgefallenen Massen. Nebst zehn Steindrucktafeln und deren Erklärung von Carl von Schreibers. 2 Vol., J. G. Heubner, Wien 1819, 434 pp. Chronologisches Verzeichniss der Meteor-Stein- und Eisenmassen. *Allgemeine nordische Annalen der Chemie* **2** (1813) 409–424. Chronologisches Verzeichniss der mit einem Feuermeteor niedergefallenen Stein- und Eisenmassen, nebst einigen Bemerkungen. *Annalen der Physik* **15** (1803) 307–328. Furthermore in *Annalen der Physik*: Beiträge zu den Nachrichten von Meteorsteinen. **29** (1808) 375–383. Neues Verzeichniss der herabgefallenen Stein- und Eisenmassen, in chronologischer Ordnung. **50** (1815) 225–256. Erste Fortsetzung des Verzeichnisses der bisher bekannt gewordenen herabgefallenen Stein- und Eisenmassen. **53** (1816) 369–393. Ueber einige vom Himmel gefallene Materien, die von den gewöhnlichen Meteorsteinen verschieden sind in St. 7.9.1815, nebst Beiträgen zur Geschichte der Meteorsteine, und einigen diesen Gegenstand betreffenden Bemerkungen. **55** (1817) 249–276. Ueber Unabhängigkeit der Meteorsteinfälle und der Feuerkugeln von Jahres- und Tageszeiten, von den Himmelsgegenden, von der geographischen Lage, vom Wetter und von bestimmten Perioden. **57** (1817) 121–143. Zweite Fortsetzung des Verzeichnisses der vom Himmel gefallenen Massen. **54** (1816) 329–357. Dritte Fortsetzung des Verzeichnisses und der Geschichte der vom Himmel gefallenen Massen. **56** (1817) 375–385. Vierte Fortsetzung des Verzeichnisses der vom Himmel gefallenen Massen. **60** (1818) 238–254. Fünfte Fortsetzung des Verzeichnisses der vom Himmel gefallenen Massen. **63** (1819) 17–54. Neue Beiträge zur Kenntniß der Feuermeteore und der herabgefallenen Massen. **68** (1821) 329–370. Neue Beiträge zur Kenntniß der Feuermeteore und der herabgefallenen Massen. Zweite Lieferung. **71** (1822) 359–386. Neue Beiträge zur Kenntniß der Feuermeteore und der

the oxide of iron is probably the principal coloring matter. In the black dust, there is no doubt carbon will also be found". The fewest events were related to meteoric falls[607] or volcanic eruptions, most were (soil and combustion) dust events, occasionally associated with rain and snow.[608] Nevertheless, this text suggests that the persons *must* have collected rainwater for "chemical consideration"; remember that rain collection for the estimation of the rainfall amount was done hundreds of years before. Ehrenberg (1830, p. 488) writes:

> Der rothe Regen zu Brüssel im Jahr 1646 schon mit Destillation des aufgefangenen Wassers geprüft [...] Dr. Thomas Ran machte chemische Versuche mit dem heutigen Regenwasser zu Ulm vom 15.11.1755 [The red rain at Brussels in year 1646 already proofed by distillation of the collected water [...] Dr. Thomas Ran carried out chemical experiments with today rainwater at Ulm of November 11, 1755].

The matter (*foreign bodies*) was considered suspended in the atmosphere in a state of the mixture and "those which pervade the atmosphere in a state of solution" (Prout 1834, p. 347). Its nature was found to be of vegetable origin, meteoric stones or aerolites, as formed from volcanoes and soils, but also meteorites. Red rain was explained by sand blown by the sirocco, from the Sahara to Europe, and by sands and organic material from other terrestrial sources[609] (Chapter 3.2.2).

---

herabgefallenen Massen. Dritte Lieferung. **75** (1823) 229–257. Neue Beiträge zur Kenntniß der Feuermeteore und der herabgefallenen Massen. Vierte Lieferung. **78** (1824) 151–168. Neue Beiträge zur Kenntniß der Feuermeteore und der herabgefallenen Massen. Fünfte Lieferung. **82** (1826) 21–35 and 161–183. Neue Beiträge zur Kenntniß der Feuermeteore und der herabgefallenen Massen. Sechste Lieferung. **84** (1826) 45–60. And finally: E. F. F. Chladni's Beschreibung seiner Sammlung vom Himmel herabgefallener Massen. Nebst einigen allgemeinen Bemerkungen. *Archiv für gesammte Naturlehre* **4** (1825) 200–240.

607 Chladni (1826) cites a work (translated by Jean-Pierre Abel-Rémusat) of Ma Touan Lin (1245–1322), a Chinese author, who describes 96 instances of falling stones, which "are always in a state of ignition at the moment they reach the ground; their external surface is black; and some of sound when struck like metallic substances. The name by which they termed them signifies falling stars changed into stones". Thus the Chinese already assumed extraterrestrial origin of such "stones" 500 years before *Chladni*.

608 Biomass burning aerosols mainly consist of black carbon (BC) and organic aerosols (OAs), and some of OAs are brown carbon (BrC). Both BCs and tar balls (a class of BrC) appear brownish at small particle sizes and blackish at large sizes. At realistic size distributions, BCs look more blackish than tar balls, but still exhibit some brown color. However, when the absorptance of aerosol layer at green wavelength becomes larger than approximately 0.8, all biomass burning aerosols look blackish (Liu et al. 2016).

609 Some citations are given here from *Charles Fort*: Bailey, J. W. (1842) Yellow showers of pollen. *American Journal of Science* **42**, 195–197. Moore, J. W. (1879) Pollen showers. *Meteorological Magazine* **14**, 96. Thorpe, T. E. (1903) Red rain and the dust storm of February 22. *Nature* **68**, 53–54. Noble, A. (1904) Dust in the atmosphere during 1902–3. *Monthly Weather Review* **32**, 364–365.

Tizenhauz[610] investigated particles fallen in April 1846 near Wilna (now Vilnius in Lithuania) while a thunderstorm, likely together with hail. The particles (up to 5 Gran, i. e., 0.3 g) were organic, and *Tizenhauz* writes that it is originated from (Tizenhauz 1848, p. 365)

> [. . .] balsamischen Pflanzen-Aushauchungen, die, in den oberen Regionen angehäuft, und, durch Wirkung der Electricität metamorphosiert, in Form von Hagel auf die Erde niederfielen [. . . balsamic plant exhalations, accumulated in upper regions, and, metamorphized by the electricity, fallen down as hail tot he earth].

This is likely the first indication for ice nuclei of agricultural soil origin (Tobo et al. (2014) for recent studies).

In chapter III of "The Book of the Damned" by *Charles Fort* (1874–1932)[611] many citations can be found; remarkable are the observations on black rain and snow, i. e., precipitation colored by soot from combustion and slag from coal processing

> Such a rain as that which fell in Ireland, May 14, 1849, described in the *Annals of Scientific Discovery*, 1850, and the *Annual Register*, 1849. It fell upon a district of 400 square miles, and was the color of ink, and of a fetid odor and very disagreeable taste. The rain at Castlecommon, Ireland, April 30, 1887 – "thick black rain" (*Amer. Met. Jour*. 4, 193). A black rain fell in Ireland, Oct. 8 and 9, 1907 (*Symons' Met. Mag*., 43, 2). It left a "most peculiar and disagreeable smell in the air." The orthodox explanation of this rain occurs in *Nature*, March 2, 1908: cloud of soot that had come from South Wales, crossing the Irish Channel and all of Ireland. So the black rain of Ireland, of March, 1898: ascribed in *Symons' Met. Mag*., 33–40, to clouds of soot from the manufacturing towns of North England and South Scotland. Black rain in Switzerland, Jan. 20, 1911. Switzerland is so remote, and so ill at ease is the conventional explanation here, that *Nature*, 85, 451, says of this rain that in certain conditions of weather, snow may take on an appearance of blackness that is quite deceptive [and many more examples].

Josef Redtenbacher (1810–1870)[612] analyzed "*Meteorsteine*" [aerolites] that fall in 1841 together with a short shower at the village of Iván near Sopron (Hungary). He describes them as black-brown stones having a dimension between 1 and 5 mm and being a mixture of silica (sand), clay, iron oxide, and traces of potassium, manganese, calcium and magnesium, likely derived from mold (Redtenbacher 1842). *Chlani*'s publications are a unique source of past precipitation events (rain, snow, dust); however,

610 Likely Count [Graf] *Konstantin Tizenhauz* (or *Konstanty Tyzenhauz* (1782–1853) [Константин Тизенгауз], Polish-Lithuania-Russian naturalist and ornithologist.

611 Publisher: Boni and Liversight, New York (1919) 298 pp. This book deals with various types of anomalous phenomena including strange falls of both organic and inorganic materials from the sky. *Fort* does not purport to explain the phenomena as a whole, simply stating the facts as they are, and letting readers to make their own conclusions. In my opinion, it is not always easy to decide whether the "facts" are true or wrong without reading the cited original paper (some citations are incomplete and wrong).

612 Austrian chemist, professor in Prague and Vienna.

he was primarily interested in meteorites, but writes that (Chladni 1821, p. 345) they always contain nickel. Thus, deposits, that show no nickel, could be attributed to earth-derived organic and mineral matter; Chladni (1821, pp. 347–359) reports on several precipitation events and some analytical results.

Grotthuß (1821, pp. 367–370) analyzed a black papery sample found in his father's collection, named "*Himmels-Papier*" [sky paper], deposited in January 1686 at the village of Rauden in Kurland (Courland, now in Latvia), combustible giving a sulfur-like odor, and detected iron, silica, aluminum, nickel, lime, and carbon. Finally, he stated that the meteoric origin is without doubt, but of "*vegetabilischen Ursprung*" [vegetable origin] (Grotthuß 1821, p. 369).[613] This appears to us as a contradiction with the assumption of meteoric origin. Hence, Chladni (1821, p. 358) concluded that "organic matter is delivered from space", and quoted professors[614] *Steinmann* from Prague and *Ficinus* from Dresden who analyzed "red snow dust", fallen at alp Anceindax (near Bex in Switzerland). Besides iron, manganese, silica, lime and sulfur they detected soot and organic matter (Chladni 1821, pp. 356–357). It was *Steinmann* who first suggested that meteorites contain organic matter (Chladni 1821, p. 358). However, when reading *Chaldni*'s publications and description of "colored precipitations" that were (at least partly) analyzed, he was obviously so fascinated by extraterrestrial events that he called all observations belonged to meteorites,[615] but today's readers would name it terrestrial phenomena. Nevertheless, without having serious evidence, *Chladni* adapted enthusiastically the idea of cosmic origin of organic matter (Chladni 1821, p. 358–159), writing concerning the general formation of organic substances

> [...] sind wir wohl berechtigt anzunehmen, daß auch in andern Gegenden des Weltalls die Natur nicht unvollkommener oder unkräftiger seyn wird, als bei uns, und daß dieses vielmehr eine allgemeine Naturerscheinung seyn möge [... we are probably justified to assume that also in other areas of the cosmos the nature is not less imperfect or powerful than with us].

The following phrase is likely one of the first written descriptions of chemical processing in the atmosphere (Anonymously 1821, pp. 3375–3376)[616]

> Einige Naturkundige lassen sie in unserer Atmosphäre, andere im großen Weltraume entstehen; alle aber stimmen darin überein, daß sie durch einen chemisch-electrischen Luftprozeß ihr Daseyn erhalten müssen, und aus seinen luftartigen Stoffen sich erzeugen, welche der Aether

613 See *Boerhaave*'s comment on vegetable matter in air on p. 54.

614 *Josef* (also *Joseph*) *Johann Steinmann* (1779–1833) was an Austrian pharmacist and chemist (also student at *Hermstädt* in Berlin), since 1812 at the Prague Polytechnical Institute, where in 1817 he was appointed a professor for chemistry. *Heinrich David August Ficinus* (1782–1857) was German pharmacist and chemist, professor in Dresden.

615 The black rain event from Montreal (see below on p. 314) he comments that the soot was not from forest fire but from the explosion of a meteorite!

616 This publication is signed with "B" (an unknown author), and represent an interesting sketch upon colored precipitations.

oder die Luft enthalten [Some naturalists let them originate in our atmosphere, others in the large cosmic space, but all agree that they preserve their existence through chemical-electric atmospheric process, and are generated from aerial substances that the ether or the air contains].

3.2.3 On organic particulate matter

The observation of organic matter in the atmosphere is surely made a long time before, as reported from colored precipitation (Chapter 3.2.2) and found in rain by Boerhaave (1735) and concluded by Marggraf (1753) from rainwater investigation. Saussure (1804, p. 244) correctly assumed that the air contains organic matter remained from plants and animals, dissipated and turned back to plants which take up them after dissolution. Wiegmann (1824, p. 208) speculated that honeydew, a sweet sticky substance found on plants (see also p. 166) could also be of atmospheric origin; the following phrase reads like the formation of secondary organic aerosol (SOA):

> [...] klebrig süße Materie in der Luft erzeugen kann, da die Urstoffe derselben ja alle in der Atmosphäre befindlich sind, und es nur auf die physischen Ursachen und Verhältnisse ankommt, daß sie als sogenannter Honigthau niederfallen [...] sticky sweet substance might be generated in the air because all her parent material[617] are available in the atmosphere and it depends only from the physical causes and relations that the so-termed honeydew falls down].

Wiegmann was emphasized from the rainwater investigation by *Zimmermann* who found "organic matter" that he called "*Pyrrhin*" (see p. 161) but which also has been regarded from lower organisms (Chapter 3.2.2). Hence, Wiegmann (1824, p. 206) referred Märklin (1823, p. 37)[618] on the removal of (translated from German and shortened):

> With every rainstorm, the substances of organisms, that have been evaporated with the water, will condense with the clouds. What happens on our Earth very slowly, originates here, probably under the influence of electricity suddenly; and in this moment they fall down with the rain.

Thus, the wonderful statement of the formation of organic matter in the air by *Wiegmann* was clearly stimulated by *Märklin*'s phrase.[619] Brandes (1826c, p. 69) writes that this substance behaved as nitrogen-containing organic matter.

In the middle of the nineteenth century, microorganisms, fossils, and plant debris were found in dust. *Robert Angus Smith* first paid attention to airborne organic matter

---

617 In modern term: precursors.

618 Märklin, G. F. (1823) Ueber die Urformen der niedern Organismen [archetypes of lower organisms]. Heidelberg. *Georg Friedrich Märklin* (1761–1823) was a German pharmacist and botanist in Wiesloch (about 30 km south of Frankfurt/Main).

619 Wiegmann (1824, p. 205) referred his teacher *Medicinal-Rath* [medical council] *Krüger* from Pyrmont who 1806/07 already has dealt with this issue, but does not know whether he published his observations; *Johann Friedrich Krüger* (1771–1843) was physician and pharmacist.

(Smith 1845, 1859a). In the 1850s, *Smith* frequently mentioned that much of the organic matter was "organized" and "are the cause of some hitherto entirely mysterious phenomena, putrefaction included" (Smith 1872, p. 388). Using the sampling method "air-washing", i. e., a bottle containing pure water is filled several times with air and shaken, Smith (1872, pp. 487–490) got a clouded liquid (collected in total 2496 L of air) and requested *John Benjamin Dancer* (1812–1887)[620] to exanimate it by the microscope, and reports on the following findings (today termed biological aerosol):

- fungoid matter: spores or sporidiae, 100th of an inch in diameter with an average number of 100 and spores varied from 10,000th to 50,000th of an inch with a number of 250,000 in a single drop,
- vegetable tissues, where larger particles had evidently been partially burnt, and were brown in color (likely burnt wood),
- fragments of vegetation, doubtless weather-worn, and a few hairs of leaves of plants and fibres,
- several long elliptical bodies, similar to the pollen,
- hairs of animals.

Louis Pasteur sampled air at Arbois (France) to investigate the hypothesis of so-termed "spontaneous" generation; his experimental method was to draw air through gun-cotton, dissolve the latter in ether or alcohol and collect the residue of particles for staining and microscopic examination. He found many different "germs" in collected dust, which were able to germinate with different substrates.[621] *Pasteur* also studied air from different sites (rural and urban) and different altitudes and found a decrease in air-borne germs with height (up to 2000 m a. s. l. at Mt. Montauvert); Pasteur (1862).[622] However, systematic studies of biological particles and dust (termed "*matière organisée*" and "*poussière organisée*") in the size range of 2–100 µm at Montsouris began with *Pierre Miquel* (1850–1922),[623] also called "father of aerobiology" (he termed this new science *Atmospheric Micrography*), Miquel (1878, 1883). *Miquel* refers *Ehrenberg* (Chapter 2.2.2) and *Gaultier de Claubry*[624] who studied already biological dust. Before *Pasteur's* findings, it was believed that air contains

---

620 British scientific instrument maker and inventor of microphotography.

621 Today, the word "germ" refers to micro-organisms (bacteria, fungi, micro-algae).

622 *Pasteur*'s original papers: Mémoire sur les corpuscules organisés qui existent dans l'atmosphère, examen de la doctrine des générations spontanées. *Ann. Sci. Nat.* **16** (1861) 5–98 and *Ann. Chim. Phys.* **64** (1862) 5–110.

623 *Docteur ès-sciences physiques* (Paris, 1877) *et docteur en médecine. Chef de service micrographique à l'observatoire de Montsouris, puis directeur du laboratoire de bactériologie de la ville de Paris.* He wrote: *Les Organismes vivants de l'Atmosphère* (Paris: Gauthier-Villars, 1883, 310 pp.); see also review by Varigny, H. de (1883) Les Organismes vivants de l'Atmosphère. *Nature* **28,** 76–77.

624 *Henri-François Gaultier de Claubry* (1792–1878), French chemist and toxicologist, who studied miasmas and conducted a series of experiments to verify the correctness of *Gay-Lussac*'s hypothesis to use (1824) an instrument (chlorometer) for cleaning of air from organic impurities (Claubry 1827).

"miasmas", foul-smelling gases, transferring diseases. It is notable that *William James Morton* (1845–1920), the USA physician, at the end of the nineteens century still writes on "miasmatic emanations" (Morton 1894, p. 355). Fig. 3.4 shows biological particles, sampled in the air of Paris by Lévy (1879).

3.2.4 On dry fog

The German physicist *Georg Wilhelm Muncke* (1772–1847) describes in Gehler (1833, pp. 38) under the keyword "*Nebel*" [fog] a type that does not consist of water vesicles but of fine suspended particles from the smoke of combusted matter, best termed "*trockner Nebel*" [*Brouillard sec*; dry fog];[625] included them he also termed "*gemischte Nebel*" [mixed fog] and smoke in towns, and that they consist generally from

> [...] verdampften oder verflüchtigten, größentheils trockenen, Substanzen bestehen [...] evaporated or volatilized, mostly dry substances [...].

Muncke correctly writes that forest fires can result in large-scale, smelling smoke (black rain and fog) and thus recognized in larger distances from the fire. Another cause, *Muncke* writes, maybe the electricity (a widely investigated phenomenon since the end of the eighteens century) to which all kinds of miraculous effects were attributed. Although *Muncke* cites many more speculations on the role of "electricity" for dry fog, finally he doubts this cause and termed volcanic eruptions, like that one in 1783 on Iceland, which led to the most famous dry fog. Finally, *Muncke* termed apart from fires and volcanic eruptions the third type of dry fog, "*nebelartiger Rauch*" [foggy smoke] above large cities like London (Chapter 3.1.2). Anonymous (1822) reports on dense black rain falls on 9 November 1819, in Montreal; the medical doctor *Martin Paine* collected them and send a bottle to the Lyceum in New York for analysis – as only foreign body soot has been found, originated by forest fires south of Ohio. On 16 November 1819, black dust was deposited onto snow in Broughton (Ohio, 1000 km SW from Montreal).

The thick dry fog was observed in Germany after the eruption of Kötlugia (Iceland) in 1721 and both in Germany and in the rest of Europe after the eruption of 1755. "The great dry fog of 1783, which covered all Europe and persisted for a month, causing all kinds of unusual meteorological phenomena" Free (1911) writes, according to

---

625 Also termed dry haze (in French *brume rousse*). *Muncke* also cites the German (historical) terms *Höhrauch, Höhenrauch, Heerrauch, Haarrauch, Landrauch, Sonnenrauch, Heiderauh*. This phenomenon was known since centuries especially in the Netherlands and northern Germany due to fire management of peatlands, heathlands [*Heide*] and moorlands. A nice historic article on this issue: Plagge (1840) Der Haar- oder Heiderauch auch Moorrauch und Höherauch. Deutsche Vierteljahres Schrift. Erstes Heft, J. G. Cotta, Stuttgart und Tübingen, pp. 191–217. *Hellmann* used the term "*trockener Nebel*".

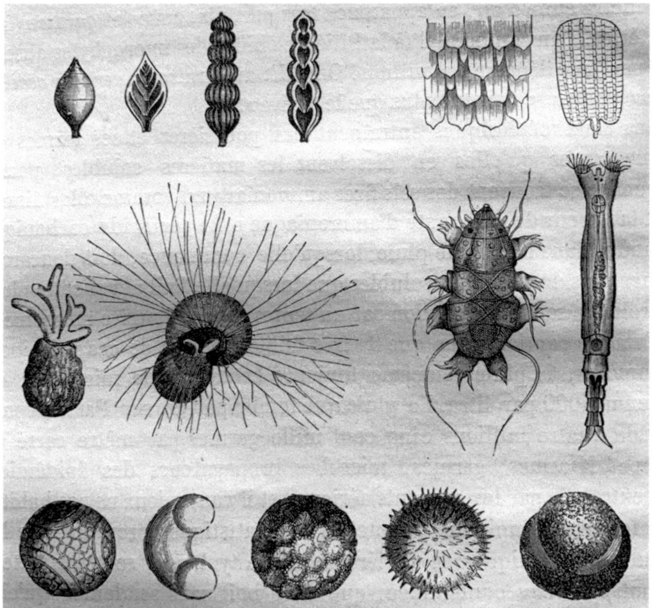

FIG. 48. — Ce que nous respirons. Corpuscules en suspension dans l'air.

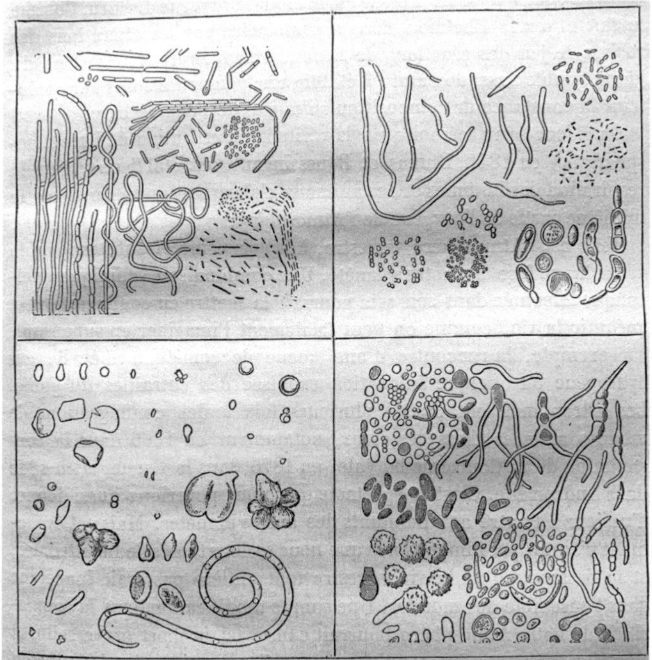

FIG. 49. — Ce que nous respirons. Microbes, bacilles et bactéries.

Figure 3.4: Different biological particles collected in the air of Paris (Flammarion 1888, pp. 100 and 101).

Stothers (1996) the first scientifically studied volcanic eruption (Laki in Iceland, beginning on 5 June 1783); see also Sigurdsson (1982). However, at the time of the eruption, it was not recognized that such "fogs" were from volcanic plumes; the German mathematician, astronomer, and physicist in Jena, *Johann Ernst Basilius Wiedeburg* (1733–1789), was one of the first who wrote on the "dry fog" of 1783 (Wiedeburg 1784) and estimated a connection with the Calabrian earthquakes in 1783, referring *Torbern Bergman* (Wiedeburg 1784, p. 77) who regarded such fogs in connection with earthquakes. *Wiedeburg* termed it "*allgemeiner Nebel*" [general fog] that lasted 5–6 weeks over the whole Europe and resulted in a decrease in visibilities to 20–30 steps, furthermore, in dry air and little rain, and, finally, in damages of the vegetation and breathing difficulties. *Wiedeburg* collected dew with porcelain bowls and characterized it after evaporation (Wiedeburg 1784, p. 70):

> Dieser zusammengezogene Niederschlag hatte einen zusammenziehenden Geschmack: rauchte auf Kohlen mit einem ekligen Geruch, und beraubte die Luft ihrer Feder⸗Kraft [This precipitation had a stringend taste: fumed on coals with a nasty smell, and robbed the air of her elastic force]. [. . .] daß er ein trockner, irrdischer und vorzüglich sulphurischer Nebel war, bei welchem die Luft weder sigirt, noch auch dephlogisticirt war [. . . a dry, telluric and above all sulphureous fog, at which the air is neither converted nor dephlogisticated], Wiedeburg 1784, p. 82].[626]

This fog was named "*trockner*" or "*dürrer Nebel*" [dry or arid fog] to separate this phenomenon from the common wet fog in autumn and winter. *Wiedeburg* further writes that there are different types of dry fog according to "*trocknen Dunstheilgen*" [dry haze particle] like soil, salt, sulfur, and other minerals – and even electric – particles. He recognized the most significant difference to "*Höherauch*" (from peat burning] which is limited in its dispersion whereas the fog in 1783 covered Europe and even the world. Hence, he concluded the origin with the earthquakes in Calabrian (between February and March 1783). The term "*brouillard électrique*" [electric fog] was introduced in 1784 by Chevalier *Robert de Lamanon* (1752–1787); see Demarée and Ogilvie (2017) for a list of historic publications. *Carl Hoyer* (1775–?), "*Prorector des Gymnasiums*" [high school prorector] in Minden, adopted this view ["*elektrischer Nebel*"] and cites himself that he already in 1819 (without knowing *Lamanon*'s article) (Hoyer 1846, p. 303) in "*Mindener Sonntagsblatt*" No. 24

> [. . .] den Höhenrauch für ein zersetztes (oxydiertes??) Gewitter erklärt [. . . dry haze explained as disintegrated (oxidized??) thunderstorm].

---

626 *Sigirt* from *sigiren*, an alchemistic term, nowhere explained but depicted transformation in sense of transmutation. Note that *sulphureous* was not specifically sulfurous acid (SO_2) as we expect it in connection with volcanic activities but also terming a fiery property. See also Brugmans (1783) and on p. 255.

Further, *Hoyer* separates between "*Höhenrauch*" [dry haze] and "*Moorrauch*" (moor smoke] and writes (Hoyer 1846, p. 303):

> Der unangenehme Geruch beim Höhenrauch ist kein Torfgeruch, sondern ein elektrisch-phosphorischer [...] [The unpleasant smell of dry haze is no smell of peat but an electric-phosphoric ...].

Today, we cannot imagine how observers were able to smell dry haze being similar to that odor of an electrostatic generator, likely similar to *Marum*'s "odor of electrical matter" (ozone), see Chapter 4.4.1. On the contrary, *Hoyer*'s speculation (Hoyer 1846, p. 308) that lightning can cause decomposition of gases and produce fog, appears from today's perspective as an early idea of gas-to-particle conversion

Ernst Witting, who was cited concerning rainwater sampling and analysis on p. 161 (Witting 1825), also studied dry fog from moorland burning using the "bottle shake method"[627] and detected phosphoric acid and hydrochloric acid, and speculated its formation under the influence of atmospheric electricity, namely through "*zersetzte Gewitter*" [disintegrated thunderstorm]; he wrote that, unfortunately, he did not collect rainwater during the event (Witting 1824, 1826). *Rudolph Brandes* wrote in a postscript to this paper by *Witting* (*Brandes* was the editor of this journal) that he does not agree with the conclusions of his friend *Witting* but with *Finke* that this kind of dry fog is simply caused by moorland burning (Brandes 1826a). Remarkable is the critical comment by *Brandes* (Brandes 1826a, p. 192):[628]

> Auch kann ich mir keine klare Vorstellung machen von dem, was man sich unter zersetzten Gewittern zu denken hat [Even I have no clear idea what one can imagine behind disintegrated thunderstorms].

Leonhard Ludwig Finke (1747–1837) was a German physician known for his contributions to medical geography, who wrote an essay on dry fog based on long-term observations (Finke 1820); see also Hellmann (1883, p. 507) for literature on "*Moorrauch*" [moor smoke]. *Cotte*[629] (in Gehler (1833), Vol. VII, p. 45) suggest that

---

627 A bottle is cleaned with distilled water, 1/8 of the volume remained in it and the bottle trapped until sampling of air, which is dissolved in the water while agitating.

628 *Brandes* (without knowing the processes and mechanisms) writes a remarkable phrase (Brandes 1826a, p. 192), based on observations, on cloud dissipation by dry fog: *Ich habe mehrmals beobachtet, daß Heerrauch kam, wenn Gewitterwolken am Himmel standen, oder eine diesen ähnliche Wolkenformation, und daß mit dem Erscheinen des Rauches diese Wolken schwanden, und kein Gewitter entstand* [I have several time observed that dry fog appeared when thunderclouds where in the sky, or a similar one cloud formation, and that with the appearance of the smoke these clouds disappeared, and no thunderstorm developed].

629 It will be meant the French priest *Louis Cotte* (1740–1815), who carried out many atmospheric observations at Montmoreney near Paris and wrote the book "*Traité de Météorologies*" (L'Imprimerie Royal, Paris, 1774).

> [...] es seyen mineralische Ausdünstungen, begleitet von elektrischer Materie [...] [... there are mineral effluvia, accompanied by electric matter ...].

In modern terms, this phrase could be interpreted as "secondary photochemical particle formation".[630] Wiegmann (1827) separates between volcanic dust (also termed "*Sonnenrauch*")[631] and such from fires, having different chemical characteristics. As *Witting*, *Wiegmann* "analyzed" these fumes and detected phosphoric acid, hydrochloric acid bond with calcium, and organic matter (pyrrhine-like he writes) and also sulfate (he cites his rainwater investigations). *Karl Ernst Adolf von Hoff* (1771–1837), a German naturalist and geologist in Gotha, first mentioned the smell of such dry fog to be "*bituminös*" [bituminous] and "*Braunkohlengeruch*" [lignite coal smell]; this observation is likely also the first indication of long-range transportation of dust events (Hoff 1827), 20 years later confirmed by *Prestel*. *Johann Schön* (1771–1839),[632] the first who observed that together with dry fog the atmospheric electricity "*vorzüglich thätig sey*" [is excellently active] and that such fog either dissipates with drizzle or disperses with the wind. *Christian* (or *Christoph*) *Heinrich Veltmann* (unknown living data), a pharmacist in Osnabrück, used the term "*Moordampf*" [moor vapor], which shows the long-lasting dispute whether this phenomenon was (wet) fog, cloud, or smoke (Veltmann 1827). There are numerous publications on dry fog, namely in the German literature (not further worth citing). *Michael August Friedrich Prestel* (1809–1880), a German high school teacher and meteorologist, writes (Prestel 1858, p. 106):

> Sollten die Franzosen und Engländer für die Sache keinen eigenen Ausdruck haben, so macht der Name Höhrauch, für uns Deutsche wenigstens, den Namen trockner Nebel ganz entbehrlich [Should the French and Englishmen no have an own term for this case, so does the name height smoke, for us Germans at least, the name dry fog completely dispensable]

Prestel (who also studied the ozone climatology, Chapter 4.4.5.3) first concluded that moorland burning is the cause of this smoke, transported over large distances; he reported that the smoke, first observed at Emden on 6 May 1857 (and likely began with burning a few days before) was observed on May 18 in Vienna and May 19 in Cracow. He proposed as special terms "*Moorrauch, Moordampf, Heiderauch*"; in Dutch "*veenrock*". Prestel (1868, p. 365) writes very clear the different causes and the impact:

---

630 Dry fog has also been observed in connection with lightning [... *auch Höhenrauch im Zusammenhang mit Gewittern beobachtet*]; today we know that through lightning NO_x and O_3 form and in the presence of organic compounds secondary organic aerosol is gained.

631 Likely derived from *Sonnenstäubchen*, the term for very fine dry dust particles (cf. footnote 564 on p. 287. Ebers (1799, p. 280) writes: *Sonnenrauch*, a dry Fog; Exhalations in the dry Summer-Days that are not so wet or moist as in Autumn. The latter definition, however, is not exclusively related to volcanic dust.

632 Professor for mathematics in Würzburg; observed and recorded from 1811 until his death weather phenomena. *Archiv für die gesammte Naturlehre* **8** (1827) 475, cited after Wiegman (1827, p. 491).

> Jeder in Folge von Moor- oder Waldbränden oder bei Ausbrüchen von Vulcanen in die Atmosphäre übergegangenen Rauch verursacht eine Trübung der Luft [Each as a result of moor or forest fires or volcanic eruptions into the atmosphere escaped smoke causes a turbidity of air].

He discusses meteorological processes, namely the temperature gradient, for rising to higher altitudes and hence realizing long-range transport. This was confirmed by *Domenico Ragona* (1822–1894), the director of the observatory in Modena, and *Angelo Secci* (1818–1878), head of the Vatican observatory, who observed in large parts of Europe dry fog events.[633] *Johann Friedrich Dellmann* finally disproved the idea of "*zersetzte Gewitter*" and stated that "*Höhenrauch*" is not fog but *smoke*. The observed red sky (this observation was already known before 1800) is never connected with fog and clouds but only with smoke particles ["*Rauchtheilchen*"], which take up water molecules and thus becomes heavier and finally sink, writes Dellmann (1869); this is the first description of haze formation by activation according to the modern theory (see also p. 131 on nucleation theory). Already in 1852 (Dellmann 1869), he observed that the dust blown from fields and streets into the atmosphere decreases the normally positive electric state of the atmosphere or even turned it into a positive state, but smoke causes the contrary. *Dellmann* writes (Dellmann (1869, p. 522):

> Die Austrocknung der Luft durch Höhenrauch und Staub muss gedacht werden wie die Niederschlagung beider durch Regen, nur mit dem Unterschied, dass bei letzterm Vorgange die Wassertheile Tropfen sind, bei ersterm Dampfmoleküle [the drying up of air through high smoke and dust must be imagined as precipitating them through rain, but with the difference that in the latter process the water bodies are droplets, and in the first vapor molecules].

Free (1911) refers dry fog to *haze*, as an unusually large quantity of dust in the air, and this will happen whenever meteorological conditions are like to cause the accumulation, or whenever a large quantity of dust is rapidly supplied by fires, volcanic eruptions, etc.

The term *dry fog* is nowadays no longer in use; at the beginning of this chapter, it was stated that the terms *dust*, *haze*, *mist*, *fog*, *fume*, and *smoke* were often synonymously used but are now rather well distinguished by definitions of Meteorological Organizations. Today we call *dry fog* atmospheric dust (or aerosol) and separate according to the origin between soil dust (Chapter 3.2.3), sea salt aerosol, smoke from biomass burning, and fossil fuel combustion (also termed *brown clouds*), and secondary inorganic and organic aerosol from gas-to-particle formation (homogeneous nucleation). In German, "*Staubwolke*" is an appropriate term, but in English used only as an interplanetary dust cloud.

633 Kleine Mitteilungen, in *Zeitschrift der Österreichischen Gesellschaft für Meteorologie* **4** (1879) 379–382, 465, 533.

3.3 Condensation nuclei, ions and dust particles

What we today name atmospheric particulate matter or aerosol particles, ranging in its size from molecule clusters of about 0.001 μm to giant dust particles up to 100 μm, and which include mineral, biological, inorganic, and organic particles of different origin, was termed and classified differently over the time. The term "particle" (from Latin particula, diminutive of pars) concerning atmospheric relevance was to my knowledge not used before the ending eighteenth century, and then only for visible particles,[634] which were mostly further named *meteors* (later vesicles and droplets, and finally hydrometeors), *corpuscles* and *bodies*. Crucial was the point of view:
- dust (fume, smoke) as air pollutant,
- nucleus (introduced by *Aitken* in view of condensation),
- ions (introduced as carrier of atmospheric electricity).[635]

Methods for measurement of dust and nuclei are described by Grundmann (1942) and for methods before 1960 by Spurny (1998), and deposition gauges by Ashworth (1933). It is also worth to refer to the dissertation of the Swiss chemist *Hans Forster* (1902–?), who writes some 100 pages on the history of aerosol and nuclei research (Forster 1940). Furthermore, Schmauß and Wigand (1929), Burckhardt and Flohn (1939) refer almost all aerosol measurements including a large bibliography.

3.3.1 Dust particles and haze formation: investigations before 1900

We have referred already *Robert Boyle* in this Volume (p. 78 and 287) with the presence and formation of "Corpuscles" in atmospheric air, but the following phrase reads like new-particle formation (NPF), Boyle (1690, p. 64):

> [. . .] two sorts of Particles convene, there results from their Coalitions certain Corpuscles of a new nature.

The following phrase by Boyle (1692, p. 46) very clearly shows the observation of gaining atmospheric particles from volatile gases like HCl, HNO_3, and NH_3, escaping from its aqueous solutions:

634 In solutions such as rainwater, *Leeuwenhoek* already observed living bodies in the middle of the 17th century.

635 The German physicists *Julius Elster* (1854–1920) and *Hans Geitel* (1855–1923), pioneers in investigation of atmospheric electricity, introduced the term "ion" in analogy of the already known ion hypothesis in the electrolysis, and they concluded that the positively charges particles are ions (Elster and Geitel 1899); the negatively charged particles they did not term "electron", although it was already discovered by the Irish physicist *George Johnstone Stoney* (1826–1911); Stoney (1894).

> I shall now add, that besides the more simple Salts hitherto enumerated, it is not unlikely, that in some Places the Air may sometimes contain compounded Salts. For I have elsewhere shewn, that some sorts of saline Spirits, meeting one another in the Air, may there convene. And I elsewhere teach how to order a couple of Liquors so, that one will never of it self afford any thing in a dry form; and yet the spirituous Effluvia of this Liquor, meeting with those of the other, will exhibite a volatile and saline Body in a dry form; though the Liquors themselves being mingled, will not afford any such Substance.

This process of homogeneous nucleation, nowadays also termed *gas-to-particle conversion* (or new-particle formation, *Charles Darwin* describes in his book "The Voyage of the Beagle" in 1839) as a phenomenon we now call *blue haze* (Went 1955, 1960); Darwin (1839, p. 33):

> During this day I was particularly struck with a remark of Humboldt's, who often alludes to the thin vapour, which, without changing the transparency of the air, renders its tints more harmonious, and softens its effects. This is an appearance, which I have never observed in the temperate zones. The atmosphere, seen through a short space of half or three-quarters of a mile, was perfectly lucid, but at a greater distance all colours were blended into a most beautiful haze, of a pale French grey, mingled with a little blue. The condition of the atmosphere between the morning and about noon, when the effect was most evident, had undergone little change, excepting in its dryness. In the interval, the difference between the dew point and temperature had increased from 7.5 degrees to 17 degrees.

However, 150 years before, *Robert Boyle* had already described such haze (Boyle 1692, pp. 54–55):

> Those adventurous Navigators that have made Voyages for Discovery in unknown Seas, when they first discerned something of obscure near the Horizon, at a great Distance off, have often doubted, whether, what they had so imperfect a Sight of, were a Cloud, or an Island, or a Mountain: But though usually it were more likely to be the former...

Whereas *Darwin* first described in 1839 the phenomenon of *blue vapor*.[636] The French physicist, *Jean François Auguste Morren* (1804–1870), professor of physics at Marseille, was fascinated by *Tyndall*'s pioneering work on radiant heat, first linked it with atmospheric photochemical processes. Morren (1869, 1870) observed that SO_2 (in a gas mixture of $5\,O_2 + 2\,N_2 + 3\,SO_2$) under the influence of intensive solar light forms "fog" (he proposed the reaction $3\,SO_2 = 2\,SO_3 + S$; see further on p. 724. To me, the visioneering conclusion (obviously overseen by laboratory photochemists and unknown to atmospheric chemists) is given by *Morren* and concerns the role of solar radiation on the formation of "*vapeur bleuâtre*" [blue vapor] in the atmosphere namely near mountains, although the causes are still unknown:

636 The Dutch biologist *Frits Warmolt Went* (1903–1990), working at CALTECH, postulated the importance of biogenic volatile organic compounds emitted by forests for atmospheric new particle formation which was highly influential in the field of atmospheric chemistry in his 1960 article "Blue Haze in the Atmosphere" (Nature **187**, 641–643).

> [. . .] circonstances qui me sont encore inconnues, que, dans l'air atmosphérique, les de lumière donnaient naissance, dans le premier surtout et dans le voisinage du sommet, à un nuage bleuâtre très-sensible, bien que d'une légèreté et d'une délicatesse excessives. [. . .] belle vapeur bleuâtre qui, sous l'action de la lumière d'un beau soleil et d'un ciel pur, baigne et entoure le pied des montagnes dans les vallées des Alpes ou des Pyrénées (Morren 1870, p. 348).

Christian Friedrich Schönbein was the first who observed in 1811 (Schönbein 1840, p. 631) another kind of blue haze as a result of lightning stroke into a church, first recognizing a "sulphureous odor" which he later believed to be ozone (see p. 311):[637]

> Das ganze Schiff der Kirche war mit einem bläulichen Dunst erfüllt [The whole nave of the church was filled with a bluish haze].

James Glaisher, Superintendent of the Meteorological Department at Greenwich Observatory, observed in 1866 "a dense blue mist", which was "apparent on all sides, and extended fully to the tops of the trees, though not there so easy to distinguish". Against a background of distant trees, it resembled "thin smoke from a wood fire". Glaisher (1866) observed the blue mist nearly every day from the end of July until early October 1866. On 8 October (mainly fine), there was a "thin blue mist during the morning, which disappeared as the day advanced". This appears to be *Glaisher*'s final sighting of the blue mist. No further mention of it appears in the Greenwich Weather Record for 1866.[638]

The first direct observations of fine dust particles dispersed in the air were made in 1870 by *John Tyndall* (1820–1893) (Tyndall 1870) and in 1880 by *John Aitken* (1839–1919) (Aitken 1881, 1883).[639] The first phrase in the paper by Tyndall (1870, p. 339) reads as follows:

---

637 The flash produces O_3 but also NO, both reaction to NO_2 and father to NO_3 which with always present organic substances, further reaction to HNO_3 occurs; ammonia from putrefaction also was at that time in the air, thus forming submicron particles of ammonium nitrate. In principle, *Schönbein* concluded on this general process – of course without understanding the mechanism – because *Cavendish* had shown in 1784 (see p. 96) that nitric acid forms during electric dischargers and *Scheeele* in 1786 found ammonium chloride on the cork of a bottle containing hydrochloric acid (see p. 671).

638 *Glaisher* evidently thought that the blue mist had some connection with cholera, although he preserved scientific caution about it. He said, "Whatever may be its nature, the fact is very remarkable, that since the cholera period of 1854 this phenomenon has not been observed till the present time". See Mukharji (2012) for the history of "cholera clouds". A bibliography is found in Marriott (1904).

639 *Cargill Gilston Knott* (1856–1922), assistant at the Edinburgh University 1879–1883, edited a Volume with collected papers by *Aitken*. In citing *Aitken*'s papers on "dust, fogs and clouds", there is some confusion (the paper list given at Wikipedia, No. 10–13, is incorrect). The original first paper (Nature 1881) is reprinted entitled "On dust, fogs and cloud" in *Van Nostrand's Engineering Magazine* 24, 1881, 308–319. In Knott (1923) the papers appear as "On dust, fogs and cloud", Trans R.S.E XXX, 1880–01, pp. 34–64 and "Dust, fogs and clouds", Nature, XXII, 1881, pp. 65–67.

SOLAR light in passing through a dark room reveals its track by illuminating the dust floating in the air.

Tyndall further writes that he was troubled by the appearance of floating dust, which, though invisible in diffuse daylight, once revealed by a powerful condensed beam. In several experiments, including absorption by sulfuric acid, potash, and burning the passing air, he detected organic matter. The following phrase – in modern terms – is the discovery of secondary organic aerosol, the blue haze (Tyndall 1870, p. 340):

When the air was sent too rapidly through the flame, a fine blue cloud was found in the experimental tube. This was the *smoke* of the organic particles.

Aitken also found that the presence of fine particles is necessary for the formation of rain (Chapter 2.2.3). *Aitken* constructed an "apparatus for enabling us to count the number of particles of dust in the atmosphere" (later termed *Aitken* dust counter or *Aitken* nucleus counter),[640] Aitken (1888, 1889a,b, 1890, 1892, 1894). Aitken (1889a, p. 1) writes:

The solid matter floating in our atmosphere is every day becoming of greater and greater interest as we are gradually realizing the important part it plays in the economy of nature, whether viewed as to its physical, physiological, or meteorological aspects. One fundamental point on which we have at present very little information of anything like a definite character, is as to the number of solid particles present in our atmosphere. We know that they are very numerous, and it seems probable that the number varies under different conditions of weather; but what numbers of particles are really present under any conditions, and how the number varies, we have at present very little idea.

He counted instrumentally between 43 (at Kingairloch in 1891) and 42,500 (at Hyéres in 1892) dust-particles in a cubic centimeter of the air (Aitken 1894); he observed, however, that there was the least dust, the air was very clear; whereas with the maximum of dust, there was a very thick haze. *Lord Rayleigh* had shown that these fine dust particles, by their scattering action on the small waves of light at the violet end of the spectrum, are the cause of blue sky.

In some experiments with light, rich in ultra-violet rays, admitted through a quartz window into a vessel containing moist dust-free air, Lenard and Wolf (1889)[641] found that if the air was allowed to expand after being exposed for some min-

640 A sample of air is mixed in an expandable chamber with a larger volume of dust-free air containing water vapor. Upon sudden expansion, the air in the chamber cools adiabatically below its dewpoint, and droplets form with the dust particles as nuclei. A portion of these droplets settle on a ruled plate in the instrument and are counted with the aid of a microscope. Free (1911) comments that the number of dust particles (with no regard to size or weight) may be fair accuracy determined by this condensation method.

641 *Philipp Lenard* (1862–1947) German physicist, Nobel price 1905; anti-Semite and later Nazi. The *Lenard* effect is referred to two phenomena, (1) ionization of air through UV, and (2) spray electri-

utes to the light, a fog was produced showing that nuclei of some kind had been produced by the action of the ultra-violet rays. They regarded their experiments as proving that the ultra-violet rays caused the posterior surface of the quartz to disintegrate, the small particles are thrown off constituting the nuclei on which condensation took place (cited after Wilson 1899, p. 412). *Charles Thomson Wilson* (1869–1959),[642] a Scottish physicist and meteorologist, repeated and extended the experiments by *Lenard* and *Wolf* (Wilson 1899); he found that "the nuclei produced cannot have arisen from the disintegration of the quartz" (Wilson 1899, p. 417), and "that the nuclei are from throughout the volume of the most air, and not at the surface of the vessel containing it" (Wilson 1899, p. 417). Furthermore, *Wilson* proved that "no condensation could be produced in pure steam even with prolonged exposure (Wilson 1899, p. 422), and "no fog could be obtained in dry air" (Wilson 1899, p. 423). Thus "[...] both air and water vapor are necessary for the production of the ultraviolet light fog; the air doesn't need to be saturated" (Wilson 1899, p. 423). *Wilson* also speculated on the nature of the nuclei. He detected no fog in pure hydrogen gas but in pure oxygen ("with sufficient strong ultra-violet rays"; Wilson 1899 p. 424) and in carbon dioxide (but at stronger UV irradiation than in the case of O_2), hence concluding that the presence of oxygen and water vapor is necessary. His general explanation of droplet formation in "dust-free" air is based on the idea of photochemical formation of water-soluble compounds that lower the vapor pressure of minute water droplets like HNO_3 and H_2O_2; he writes (Wilson 1899, p. 428):

> The view here taken is then, that under the action of the ultra-violet light small drops of water combine with the oxygen in contact with them, and in consequence of the lowering of the equilibrium vapour pressure by the dissolved H_2O_2 they are able to grow, when similar drops of pure water would evaporate.
> Growth of the nuclei into visible particles [...] that [...] some compound is formed in solution in each drop.
> [...] combination of oxygen and nitrogen with the water of the incipient drop to form nitric acid as the most likely reaction which could accounts for the phenomena.
> When the clouds are produced in oxygen, however, the only possible combination which can account for the phenomena is that of oxygen and water to form hydrogen peroxide.
> The formation of ozone would not enable us to explain the production of the clouds, and indeed although clouds are very easily produced in ozonized oxygen it is, as the experiments of Meissner and others shew, only as a consequence of reactions, by which some of the ozone is destroyed.

---

fication or waterfall effect (separation of electric charges accompanying the aerodynamic breakup of water drops). *Maximilian Franz Joseph Cornelius Wolf* (1863–1932), German astronomer in Heidelberg.

642 He first studied biology in Manchester, and moved in 1887 to Cambridge where he became interested in physics and chemistry (graduated in 1892). He became particularly interested in meteorology, and in 1893 he began to study clouds and their properties. The invention of the cloud chamber was by far *Wilson*'s signature accomplishment, earning him the Nobel Prize for Physics in 1927.

However, *Wilson* did not exclude impurities; the following phrase reads like a definition of the term *gas-to-particle conversion* (Wilson 1899, p. 425):

> The small quantity of water in these clouds makes it very difficult to exclude the possibility of their formation being due to the presence of traces of some vapour, which might become oxidised under the influence of the ultra-violet rays.

Finally, from experiments made with sunlight, he writes:

> It is plain, therefore, that sunlight, unlike the light from the other sources tried, contains nucleus-producing rays which can penetrate glass. (Wilson 1899, p. 430)
> In the atmosphere, according to this view they would persist for an almost indefinite time, and might finally become large enough to act like "dust" particles in helping condensation. (Wilson 1899, p. 431)
> The cloud or nucleus-producing effect of ultra-violet rays has obviously bearings on other meteorological phenomena. The nuclei which enable clouds to form may in many cases arise from this source. The upper clouds especially may owe their formation in this way to the action of sunlight. It is possible, too, that owing to the action of the ultra-violet rays, sunlight may sometimes cause clouds to persist in unsaturated air. (Wilson 1899, p. 431)

Russell and Watson (1891)[643] emphasize the role of dust particles as condensation nuclei

> Sulphur burnt in the air is a most active fog-producer, so is salt. Many hygroscopic bodies form nuclei having so great an affinity for water that they can cause its condensation from an unsaturated atmosphere. At the same time non-hygroscopic bodies, like magnesia and many others, are powerful fog-producers.

Before *Wilson* constructed his *cloud chamber* and intensively continued his studies on fog formation, some German scientists conducted similar experiments, worth refer here. *Erich Barkow* (1882–1923), a German meteorologist and polar explorer, studied fog formation using *Kiessling*'s fog chamber[644] caused by ozone formation and first used the German term "*blauer Nebel*" [blue fog] (Barkow 1907), although *Pringal* soon later describes this fog soon later as "*feiner Dunst*" [fine haze]. The German physicist *Erich Julius Otto Pringal* (1882–1915) found that ozone in pure air hardly forms fogs, but in most in the presence of nitrous gases; however, he further assumed as other researchers that ozone oxidizes nitrogen in presence of humidity to nitrous gases (Pringal 1908).

---

643 *William James Russell* (1830–1909) was an English chemist and Fellow of the Royal Society. He made an extended study of the formation of London's fog. Not to confuse with the meteorologist *Francis Albert Rollo Russell* (1849–1943), who also studied London's fog.

644 The German teacher and physicist *Johann Kiessling* (1839–1905) studied fog formation (Kiessling 1885).

A remarkable paper presents the British meteorologist *Ernest David Fridlander* (1870–1960) with the ending of nineteenth century on atmospheric dust observations taken on a round-the-world voyage with *Aitken*'s Pocket Dust Counter in places such as the Atlantic and Pacific Oceans, New Zealand, California, the Indian Ocean, and Switzerland (Fridlander 1896).

3.3.2 Understanding the nature of condensation nuclei: research 1900–1960

At the beginning of twentieth century, all small dust particles were regarded as ions, since Elster and Geitel (1899) discovered them as carriers of electrical current in atmospheric air. Further pioneers in atmospheric ion studies were Ebert (1901) and Langevin (1905).[645] Large ions (electric charged molecular clusters) were first observed in air by Langevin (1905) and later Pollock (1909)[646] as being produced photochemically from atmospheric trace gases as first proposed by Lenard (1900) and directly from combustion processes (Wigand 1913).[647] *Wigand* counted the number with *Aitken*'s apparatus at a balloon ride on 5 January 1913, from the city Bitterfeld, crossing Mt. Brocken to Lalendorf near Güstrow (Mecklenburg); during the ascent, the number concentration increased from 150.000 at ground level at 80 m to 10.850 (1250 m), 3.400 (3000 m) and less than 25 (7000 m), finally at the madding site at 50 m (remote areas) to 10.000.

Later studies (Pollock 1915a,b, Wigand 1919, 1930) showed that different classes of larger ions exist in the air under ordinary conditions. Whereas small ions (0.1–1 nm) have been regarded as charged air molecules, the ion spectrum ranging from 1–100 nm (subdivided in middle and large ions) was identical with *Aitken* nuclei (also called initial nuclei) and the uncharged were simply termed nuclei, further subdivided into coagulations nuclei (or condensation nuclei) and dispersion nuclei (soil dust and sea salt particles).[648]

It is a clear statement by Aitken (1881) that without condensation nuclei (CCN) no fog, no cloud, and hence no rain forms. As a source of such particles, he named sea salt and smoke from combustion. From the microscopic investigation of evaporating cloud droplets under the microscope, Aßmann (1885) concluded that these nuclei must be smaller than 1 μm. Wigand (1913) doubts that condensation occurs on solid particles and Lenard (1936) suggested that nuclei may be formed from "hygroscopic"

645 *Hermann Ebert* (1861–1913) was a German experimental physicist in Leipzig, Kiel and Munich. *Paul Langevin* (1872–1946), French physicist in Paris; in 1926 he became director of the *École de Physique et Chimie*.

646 *James Arthur Pollock* (1865–1922) was an Irish-born physicist, active in Australia.

647 *Albert Wigand* (1882–1932) was a German meteorologist; supported the Nazi development.

648 Air ions larger than 1.6 nm in mass diameter are defined as charged aerosol particles. The early research of air ions is well summarized by Israel (1970) and Flagan (1998).

gases under the influence of UV light (cited after Quitmann and Cauer 1939). Boylan (1927)[649] writes:

> In discussing observations upon atmospheric pollution, considerable confusion has caused by failure to distinguish between the fine dust suspended in the atmosphere, [. . .], and nuclei the of cloudy condensation, [. . .] This confusion is chiefly due to Aitken himself, who usually referred to the condensation nuclei as "dust" and to his different instruments as "dust counters".

See also the book by Whytlaw-Gray and Patterson (1932) on aerosol research. Likely the first monograph on aerosols (in terms of a general treatment) was written by *William Edward Gibbs* (1890–1934), Ramsay professor of chemical engineering at University College London (Gibbs 1924).

The statement of *Albert Wigand* that electric variations caused by ions can influence human health, and the fact that *Carl Dorno* agreed with this opinion, stimulated the beginning of bioclimatic nuclei investigations in the late 1920s. In the 1930s, Germany became leading in condensation nuclei and atmospheric aerosol research in the focus of bioclimatology; there were established several observational sites:
- *Observatorium* Wahnsdorf,
- *Agrarmeteorologische Forschungsstelle* [research center] Münchehof,
- *Bioklimatologische Forschungsstellen*: Braunlage, Bad Elster, Friedrichroda, Wyk auf Föhr,
- *Bergobservatorien* [Mountain observatories]: Mt. Brocken, Feldberg, Kalkmit, Schneekoppe, Zugspitze.

Another motivation for aerosol research was the need for an understanding of cloud and rain formation (Chapter 2.2.3, p. 129 and Chapter 2.2.5, p. 135), namely the nature of condensation nuclei.

The ancestors in investigations fog and haze formation in the nineteenth century were named in the previous chapter: Darwin (1839), Schönbein (1840), Glaisher (1845, 1866), Meissner (1869), Morren (1870), Kiessling (1885), Lenard and Wolf (1889). Wilson (1899, 1912) stimulated much research in the early twentieth century: Barkow (1907), Pringal (1908), Bieber (1911), Rothmund (1917).[650]

After *Aitken* estimated the particle number with his "dust counter", likely the British meteorologist *Henry Lionel Wright* (unknown living data) observed in 1928–1930 at the Kew Observatory smoke particles (range of number concentration 60–2400 cm^{-3}) and condensation nuclei (range of number concentration 15000–

649 *Boylan* worked at the University College in Dublin, likely as co-worker of *James Nolan* (unknown given name and living data).

650 The knowledge from rainwater studies that sea salt is omnipresent even far from the sea, and *Köhler*'s finding in 1925 that rime consists mainly from NaCl, the "sea salt hypothesis" initiated much research for the next 30 years (see Chapter 2.4.2).

55000 cm^{-3}), showing a strong seasonal variation (minimum in summer and maximum in winter) and concluded on the sources of CCN, a) from the same sources as smoke particles, and b) from other sources, e. g., smokeless fuel and sea spray (Wright 1932). Concerning the discovery of large ion by *Langevin*, Nolan and Guerrini (1936, p. 1175) write:

> It was natural, then, to assume that the large ions were the condensation nuclei, familiar owing to the work of Aitken, which had acquired positive and negative charges, and that the condensation nuclei were the stable uniformly-sized drops, the theoretical possibility of which had been indicated.

The general discussion to the contribution by the more theoretical work on the mass and size of the atmospheric condensation nuclei offers the knowledge on condensation nuclei and fine particles of the mid-1930s (Nolan and Guerrini 1936, pp. 1179–1181):[651]

> Professor R. Whytlaw-Gray (leeds) said: These experiments appear to show that particles of various kinds and of a considerable size range are present in town air. Of these a certain proportion are amicroscopic[652] and a much smaller proportion of microscopic dimensions.
> Mr. A. G. Grant (Darlington) said: Has any attempt been made to define the properties of a nucleus? Previous papers have referred to hygroscopic nuclei, for example, sea-salt, and it has often been suggested that organic spores and particles of free carbon (in industrial fogs) may form nuclei.
> Mr. J. H. Coste (Teddington) said: There seems no reason, in the nature of things, why hygroscopic nuclei should, in the air of inhabited districts, be all of the same kind or size. Wright and I have found various ions (in the chemical sense) in the-condensates from London air, with some evidence of selective condensation.
> Dr. J. J. Nolan (Dublin), in reply, said: We have not been concerned in this work very directly with the nature of the condensation nuclei, but rather with an essay at the determination of the sizes of the bodies- whatever they may be-which are counted by the Aitken instrument.

Thus, it was only agreed that the size spectrum of aerosol particles covers a large range and that they are of different origin – but nothing was known on their chemical nature. Between 1930 and the early 1950s, a large number of studies concerns the number of nuclei and ions (with respect to atmospheric electricity), resp., in the atmosphere at many different locations conducted. This was possible especially by an improved Aitken counter, the *Scholz Kernzähler* [nuclei counter], and the ZEISS konimeter (see p. 39); see Spurny (1998) for instruments used before 1960. A few significant German researchers[653] should be named: Scholz (1931), Kähler (1932), Israel and Schulz (1932),

651 *John James Nolan* (1888–1952) was an Irish physicist. *John Henry Coste* (1871–1949) was a British chemist in air pollution studies. *Robert Whytlaw-Gray* (1877– 1958) was an English chemist. Nothing is known on *Grant*.

652 sub-microscopic is meant.

653 *Joachim Scholz* (1903–1937) was a physicist at the Observatory Potsdam, participated 1932 at the Soviet polar expedition. *Karl Wilhelm Kähler* (1880–1966) studied physics and chemistry in Kiel and

Landsberg (1934, 1938), Junge (1935, 1936), Schulz (1939), Findeisen (1938, 1939), Burckhardt and Flohn (1939), Lettau (1939) Effenberger (1940).

Junge first estimated the range of CCN between 0.001 and 7 µm, *Findeisen* between 0.007 and 0.1 µm and soon later (Findeisen 1939) between 0.01 and 0.1 µm. *Effenberger* found at the geophysical observatory Collmberg (University Leipzig) in 1939 a diurnal cycle of the number of condensation nuclei which was interpreted by Mrose (1954) due to formation by solar radiation.

Although indirect ideas arose many decades before that the nuclei consist from NaCl from sea salt and sulfates from combustion, such as Coulier (1875) who observed that smoke and flame gas increase the number of nuclei, and also Aitken (1911, 1912, 1917) observed the formation of nuclei from atmospheric photochemical action, namely the oxidation of SO_2 to sulfuric acid, it was only after World War II that *Christian Junge* found that aerosol particles smaller 0.8 µm mainly consists from ammonium sulfate and not chloride (Junge 1953).

The British meteorologist *George Clarke Simpson* (1878–1965) wrote a paper to controvert the idea current among "most meteorologists" at that time that the nuclei on which cloud particles form are composed of sea-salt derived from sea spray; he discusses that the nuclei are more likely to be of the nature of acid than of salt but the data do not rule out sea salt as an important source of nuclei nor do they rule out sulfuric acid, ammonium sulfate, ammonium bisulfate, nitrite or nitrate (Simpson 1941b). Simpson (1941a) names nitrous acid as probably the most important source of atmospheric nuclei. An anonymous reviewer (S. H. in *Bull. Am. Met. Soc.* **24** (1943) 123–124) comments that ammonium nitrite and nitrate might be more likely sources of nuclei than nitrous acid.

Jakow Iljitsch Frenkel [Яков Ильич Френкель] (1894–1952), world-famous Soviet physicist, stated (but still confused) that condensation nuclei are formed by solar radiation with participation of ozone (Frenkel 1953):

> Вопрос о происхождении ядер конденсации водяного пара в атмосфере, вызывающих конденсацию при ничтожном пересыщении, а иногда и в ненасыщенном паре, остается до сих пор неясным. [...] Ядра конденсации водяного пара возникают под влиянием ультрафиолетового излучения Солнца над слоем озона, и далее, под влиянием своего веса, опуска ются в нижние слои атмосферы [The question on formation of condensation

made his habilitation 1924 in Berlin; professor 1929 at the Berlin University and at the Observatory Potsdam (1907–1945). *Heinz Lettau* (1909–2005) was a German geophysicist and meteorologist in Königsberg, Potsdam at the geodetic institute (1931–1933) and Leipzig (since 1934 as assistant of the geophysicist and meteorologist *Ludwig Weickmann* (1882–1961)), during the war chief-meteorologist of Generalfeldmarschall *Erwin Rommel* and after war interned at Camp Luston as a prisoner. Since 1957 at the University of Wisconsin and later Professor of Meteorology. *Ernst Friedrich Effenberger* (1916–1989) was a German physician and physicist; 1964 professor for occupational medicine and industrial hygiene at the University Hamburg. *Israel* see p. 39; *Schulz* see p. 36; *Burckhardt* and *Flohn* see p. 10; *Junge* see pp. 39 f; *Findeisen* see p. 140.

> nuclei for water vapor in the atmosphere, caused by low supersaturation and even under undersaturation, is until now unclear, [...] Condensation nuclei for water vapor form under the influence of ultra-violet solar radiation above the ozone layer, and further transfer down into lower layers of the atmosphere through by sedimentation].

Before the war, only some chemical analysis of fog droplets and rime (*Cauer, Houghton, Mrose*) were carried out which showed that sulfate and likely also nitrate could be important constituents of condensation nuclei. However, Cauer (1949a, p. 229) writes (translated from German, but not stated in the English version Cauer 1951a):

> Contrary to the opinion often expressed, in Europe sulfuric acid from exhausts of chimneys provided the vast amount of condensation nuclei, far away from industrial areas it was found that nuclei from sea water according the spray theory are dominant.

With *Junge*'s fundamental contributions to the atmospheric aerosol research since the beginning 1950s (Junge 1951, and following years until 1963), the modern period in understanding atmospheric particles began, and with further significance in the 1970s, and is not ending. Continuous monitoring of aerosol particles began in Germany at the meteorological observatory Hamburg in 1976, and of a few chemical constituents using a cascade impactor with three sizes range in 1980 and 1985 (Kaminsky and Winkler 1994). Although Junge (1956, p. 136) writes the important need for measurement, it was not before the 1980s, that techniques were developed to meet this challenge (italics by *Junge*):

> Unfortunately, almost all previous work in air chemistry (e. g., Cauer 1951a, and Egnér and Eriksson 1955) measures a mixture of particles and gases. *A real step forward in the understanding of the basic processes in air chemistry can be gained only if aerosols and gases are measured simultaneously but separately, and if the aerosols, in turn, are separated according to size.*

It is interesting to note, that Junge (1956) is often not correctly referred by succeeding investigators – likely because of *Junge*'s large reputation. First, *Junge not* state that chloride is lost in aerosol; he writes that the chloride-to-sodium ratio (in giant particles, i. e., >0.8 μm) equals that of seawater. From the aerosol composition of "large" particles (<0.8 μm), however it can be seen that the Na/Cl ratio is definitively smaller than that in sea water, with a distinct difference between samples from land breeze (0.81) and those from sea breeze (1.24), but it is rather surprising that *Junge* did not comment it. Second, *Junge* has cited concerning the reaction of NO_2 with NaCl (see below with respect to chlorine loss), what not is correct. Junge (1956), first presenting conclusions that are no longer supported nowadays

> These data show that the continent cannot be the source of the nitrate (p. 133)
> Cities and industrial areas do not represent great sources of NO_3" (p. 134)

further writes (p. 134); note that NO_3 means the nitrate ion:

> We believe that the nitrate is formed by oxidation of NO_2 to NO_3 when sea-spray aerosols mix with continental air masses in coastal areas.

There is no doubt concerns *Junge*'s experimental values, but with the very little knowledge on sources and chemical reactions in the atmosphere in the 1950s, he was unable to conclude today's view.[654]

3.4 On sea salt

It must have been early observed that the atmosphere in the vicinity of the sea frequently becomes impregnated with saline materials, writes *John Brodhead Beck* (1794–1851), a New York physician (Beck 1819). One of the first written notes is from Borrichius (1674), see p. 147. The British botanist *Richard Anthony Salisbury* (1761–1829, born *Markham*) found on the 14th of January, 1803, after a large storm salt deposited in his garden; he writes remarkable observations (Salisbury 1807):

> The effect of the *Salt Wind* was still more and more evident as the sea approaches (p. 288). [...] were wind thus loaded with the spray of salt-water to blow [...] whenever the wind is strong enough to turn over the tops of the waves into what the sailors call *White Caps* (p. 289).

Beck, asking "in what way does the salt exist in the atmosphere in that storm" writes that "the spray of the sea drive onwards by the force of the wind", according to the opinions of *Joseph Banks*[655] and *Humphrey Davy*.[656] Further *Beck* (Beck 1819, p. 390) communicates a view (which is not different from today's view), referring that it is maintained by Dr. *Mitchill*[657] "muriate of soda is continually rising into the atmosphere from the surface of the ocean, and that the air, in all mar-

---

654 Worth to note is that in the smaller particles, the content of nitrate is only 1/10 of that of sulfate, whereas in the larger particles both substances are about equal. Clearly seen in his data is some maritime sulfate, later called excess-sulfate. The principle of electroneutrality is fulfilled for the smaller particles, suggesting that ammonium-sulfate-nitrate is originated by gas-to-particle conversion; in the larger particle fraction, there are missing cations from continental sources in the balance, and particulate sulfate and nitrate are of continental origin, namely man-made SO_2 and NO_x emission, chemical transferred by cloud processing.

655 The reference given by *Beck* (*Transactions of the Linnean Society* Volume **8**, 289), however, concerns Salisbury (1807) who refers Sir *Joseph Banks* (1743–1820), an English naturalist and botanist, who obviously observed plan damages due salt deposits. It remains unclear, whether the interesting statement in *Salisbury*'s paper is his own or that of *Banks*.

656 The reference *Agricultural Experiment*, p. 339 is given.

657 *Samuel Latham Mitchill* (see pp. 65, 70 and 688) is meant; unfortunately, *Beck* provides no reference.

itime situations, is thus constantly more or less impregnated with salt". Finally, *Beck* writes:[658]

> The most striking fact in support of this doctrine is the existence of muriate of soda in the rain and snow which fall in the vicinity of the ocean.

Beck's friend Dr. *John Torrey*[659] investigated the last snow which fell, and proved "that a *free acid* does not exist in snow water, but that the muriate exists in it combined with an alkali, which is most probably soda".

Although sodium chloride was detected in early rainwater investigations, and Pošepný (1877) clearly describes the origin of NaCl in air from evaporating seawater (pp. 57 and 202), well-known German pharmacist *Herrmann Emil Schelenz* (1848–1922)[660] denied the presence of NaCl in the air at sea-side resort St. Peter (North Sea) after experiments with aspiration of air through silver nitrate solution and no detection of any precipitate (Schelenz 1886).[661] *Schelenz* (obviously not knowing the above cited works) begins his paper with the comment that there exists the view that sea air is salty, but that he found no data in chemical literature supporting this opinion. However, he refers Dr. *Scheby-Buch*[662] who wrote in "*Allgemeine medicinische Zeitung*" and in "*Bäder-Almanach*" that (quoted by Schelenz (1886) p. 1016):

> Der Seewind nimmt die verdunstende salzgeschwängerten Wassertheile mit und trägt sie ans Land [The sea brise takes the evaporating salty water particles and carried them on land]. [...] das Salz [...] soll aber auch rein mechanisch etwa in der Art, wie es bei den Zerstäubungsapparaten geschieht, aus der Gischt der Brandungen und Wogenkämmen durch den Wind in die Luft geführt werden [... but salt ... should also carried into air purely mechanical by the wind, something like in nebulizers, from the spray of the surf and combs].

658 See Chapters 2.3.3 and 2.3.5.1 for measurements of the content of chloride in fog and cloud water and in rainwater.

659 *John Torrey* (1796–1873) was an American botanist, chemist, and physician.

660 Known for his "*Geschichte der Pharmacie*" (Springer, Berlin, 1904, 925 pp.). *Schelenz* also investigated the air by aspiration concerns organic matter (not found) and germs (14 in air from land and only 1 in air from sea); on his ozonometric investigations see p. 533.

661 *Schelenz* collected 350 L air at the beach and 1000 L in 250 m distance through a "*Kugelrohr*" [ball tube], unfortunately without calling the volume of the absorption solution. According to Lunge (1890) we can assume it to be in the range 50–100 mL. From modern measurements at the North Sea coast (station Westerland) we know an average PM10 concentration of 20 $\mu g\ m^{-3}$, i. e., the chloride content may be less than 3 $\mu g\ m^{-3}$. Hence, *Schelenz* collected only 1–3 µg chloride, and assuming an absorption volume of only 20 mL, it would not precipitate as AgCl (solubility corresponds to about 40 µg in 10 mL). Another example (see also p. 696) for "no evidence is no evidence of absence".

662 *Oskar Louis Scheby-Buch* (1846–?) was a German medical doctor; unfortunately, *Schelenz* not gives details on publication. According to *Schelenz*, *Scheby-Buch* also wrote on sea salt in "*Bäder-Almanach*" (first edition 1882, and continuously edited), but publication not found.

The last phrase reds like the spray water theory by Melander (1897) and the ideas by *Blanchard* and *Woodcock* in the early 1950s (see below).

In connection with studying the chloride content in rainfall (Grabowski 1951a), atmospheric chloride aerosol as nuclei (Grabowski 1951b, 1952, 1953a, 1956) and chloride washout (Grabowski 1953b), *Grabowski* (we met him before concerns aircraft cloud water studies, p. 259) developed the "marine hypothesis" [морская гипотеза] – similar and independent from the German "*Spritzwasserhypothese*" [splash water hypothesis] by Cauer (1938), but remember that Pošepný (1877) first published such hypothesis – and long before "western" scientists (Woodcock et al. (1953) and reaffirmed by Duce (1979). Grabowski (1952) found that sea salt escapes from the sea surface at wind speeds of 10–16 m s^{-1} with an annual global formation of $27 \cdot 10^{12}$ kg, corresponding to $15 \cdot 10^{12}$ kg chloride. On the other hand, he estimated the global deposition of sea salt by precipitation to be only $2.5 \cdot 10^{12}$ kg.[663] Furthermore, *Grabowski* concluded that sea salt plays an important role in the budget of cloud condensation nuclei. *Grabowski*'s papers remain unknown by western scientists.

3.4.1 Sea salt and condensation nuclei

The origin of salt particulates in the atmosphere was the subject of considerable debate, Wright (1940) and Simpson (1941a,b). *Wright* sought to meet one of *Simpson*'s arguments by invoking the work of the Finnish physicist *Gustaf Melander* (1861–1938), who had suggested there was a production of salt nuclei by evaporation of salt solutions (Melander 1897). A first study on the chloride content in atmospheric aerosol was conducted by the American meteorologist and climatologist *Woodrow Cooper Jacobs* (1908–1990) who found that "the most outstanding relationship is that of chloride content to wind velocity" (Jacobs 1937, p. 148). He further writes (p. 150):[664]

> We would not expect the droplets of sea water produced to be of uniform size but to vary from that of the smallest liquid drops possible to those of such large size that they immediately fall back into the sea.

Findeisen (1937) asked whether condensation nuclei generate at the sea surface. Cauer (1938, p. 411) cites *Leo Schulz* (oral communication) who did not find more nuclei in the surf zones than on land, and argues that *Melander*'s "*Spritzwassertheorie*" [spray water theory] may explain the origin of alkali and earth alkali metals (Mg, Na, K) but

---

663 Later global sea salt production estimates are ranging $(1–30) \cdot 10^{12}$ kg, with the most recent estimate of $(5 \pm 4) \cdot 10^{12}$ kg (Lewis and Schwartz 2004); Möller (1990) estimated $(6.81 \pm 3.41) \cdot 10^{12}$ kg).

664 The mechanics of bubble burst and the formation of mist during effervescence was first studied by Stuhlman (1932).

not that of volatile substances, namely the halogens (see also next chapter). Cauer (1951a, p. 1133) argued, citing the results by Köhler (1937) who found in dew and hoarfrost at the North Sea relatively small chloride values, but still assumed NaCl nuclei, as well his own results from Oberschreiberhau, that hydrochloric acid nuclei "could have originated from a few molecules of HCl with eight times as many molecules of water".

The western pioneers in the investigation of sea salt particles, their possible role as condensation nuclei and the mechanism of sea spray formation were *Alfred H. Woodcock* (1905–2005) and *Duncan C. Blanchard* (1924–2016)[665] (Woodcock 1950, 1952, 1953, Woodcock and Gifford 1949, Woodcock et al. 1953, Blanchard and Woodcock 1957). First sampling and measurement of the size distribution of sea salt particles Woodcock and Gifford (1949) conducted.

However, with *Junge*'s finding that *Aitken* particles (thus, CCN) consist mainly of ammonium and sulfate, and the early measurement result that sea particles are belong the giant particles, the hypothesis that sea salt acts as CCN, was almost rejected but was further controversially discussed. On the other hand, based on *Junge*'s results that sea salt is found at altitudes of 800 m, Jaenicke (1982) concluded on the existence of sub-μ sea salt particles. Early in the 1950s, Metnieks (1958)[666] already evaluated sea salt particles down to 0.1 μm on the west coast of Ireland. On the other hand, Junge (1963, p. 160–162) clearly finished the discussion on the role of sea salt particles as CCN, writing:

> The long-debated question of the role of sea-salt particles as atmospheric condensation nuclei has now been settled. Then total number of sea-salt particles is far too low to be of any importance fir the formation of cloud droplets. [...] Despite of their low numbers, the sea-salt particles may be important for the formation of a few large drops in clouds and thus for the initiation of rain.

Ogiwara and Okita (1952)[667] studied the residual of cloud droplets by electron microscopy and found on average 30 % NaCl but concluded from the data that the greater part of cloud and fog nuclei are composed of hygroscopic and non-hygroscopic substances which originate in combustion and that sea spray is not a significant source of condensations nuclei. Isono (1957) again considered sea salt as being CCN. From theoretical and experimental work, Lodge et al. (1954) showed conclusively that there is no basis for accepting the *Melander* hypothesis as an explanation of the origin of condensation nuclei in the atmosphere. For the much smaller and more numerous CCN,

665 American oceanographers, cloud physicists and friend at Woods Hole Oceanographic Institution, MA. See also Blanchard (1984).

666 *Arvids Leons Metnieks* (?–2017) was an Irish physicist, working together with *Pollak*.

667 *Sekiji Ogiwara* and *Toshiichi Okita* (see also p. 223) from Tohoku University, Sendai, Japan (unknown living data).

Twomey (1968, 1971)[668] demonstrated that the major constituent of sea salt (NaCl) is only a minor component of CCN number concentrations even over the ocean.

Finally, recent findings by O'Dowd and Smith (1993), Jennings (2000),[669] Geever et al. (2005), and Clarke et al. (2006) renew the old idea that sea salt particles well occur also in the submicron range and thus a larger proportion of CCN originate at the ocean surface; see also O'Dowd and de Leeuw (2007) for a survey on marine aerosols.

3.4.2 Sea salt and chlorine degassing

The production of sea salt (salis marini), namely at the coasts of Spain, is 1000-years old operation. There cannot be any doubt that such as dew and rainwater, seawater, and sea salt were treated by early alchemists in her search for materia prima (see Chapter 2.1). Simple heating[670] of sea salt or mixing with oleo vitrioli (sulfuric acid) forms HCl, known as spiritus salis marini for many centuries. Treatment of sea salt with spiritus nitri (HNO_3) gains aqua regia,[671] known by *Geber* (as a synonym for Arabic alchemists) in the eighth century, and to *Raymund Lull*, *Basilius Valentinus*, and *Paracelsus* (Gmelin 1799). Hydrochloric acid, namely from "decomposing" sea salt seems to be known since ancient times; in Plinius XXXI, 6, cited after Lersch (1863, p. 64) who notes that the latter is related to free hydrochloric acid, occurring in the sea air is written:

> Nec decolour species æris, argentive [. . .] Argentum medicates aquis inficitur, atque etiam afflatus salso, sicut in mediterraneis Hispania [Or discoloring air constituents [. . .] Silver, on mixing with mephitic water, and even through salt emanations, particularly in the Mediterranean Spain].

The first phrase is surely related to hydrogen sulfide, making silver black The name *marine acid* (or marine acid air),[672] clearly shows us that it was gained from sea salt by the distillation of seawater and treatment of sea salt. *Robert Hooper* (1773–1835), a British physician, known as a medical writer, describes in his "*Lexicon Medicum*" (Hooper 1845) the way in gaining HCl from sea salt:

---

668 *Sean Twomey* (1927–2012), Irish born physicist who worked at CSIRO (Australia) before coming to the University of Arizona in 1973.

669 *Stephan G. Jennings* (born 1944) and *Colin O'Dowd* (born 1966), both atmospheric physicists at National University of Ireland, Galway.

670 According to $MgCl_2 + H_2O = MgOHCl + HCl$.

671 A mixture of concentrated nitric acid and concentrated hydrochloric acid. In German *Königswasser*, in Swedish *Kungsvatten*, in French *eau régale*. Note that *aqua fortis* = nitric acid (in ancient German *Scheidewasser*).

672 Other historic names: spirit if salt (in German *Salzgeist*), acidum salis, acidum muriaticum, spiritus salis marini, spriritus acidus salis marine. *Gay-Lussac* called it hydrochloric acid.

> It was prepared by the older chemists in a very rude manner, and was called by them spirit of the salt.

Common salt or sea salt was mixed with dried clay and water, put into a retort; distillation being the proceeded upon, the muriatic acid came over. Another method consists by decomposing the salt by oil of vitriol. *Hooper* gives the explanation of the mechanism, referring *Davy*:

> [...] the water of the oil of vitriol is first decomposed, its oxygen unites to the sodium to form soda, which is seized on by the sulphuric acid, while the chlorine combines with the hydrogen of the water, and exhales in the form of muriatic acid gas.

Interesting that he also presents the "true" explanation, however, rejecting the argument that "neither muriatic acid nor soda exists in common salt":

> When common salt is decomposed by oil of vitriol, it was usual to explain the phenomenon by saying, that the acid, by its superior affinity, aided by heat, expelled the gas, and united to the soda.

The German physician and chemist in Erfurt, *Hieronymus von Ludolf* (1708–1764), on the contrary, writes correctly (Ludolf 1752, p. 627)

> Daß das Küchen- oder Meersalz ein Körper sey, welcher aus acido salis und einem alcalischen fixen Salz zusammen gesetzet [That the table salt or sea salt a body is which is composed from acido salis and a fixed salt].

Christlieb Ehregott Gellert writes (Gellert 1750, p. 82; Gellert 1776, p. 112):

> Wenn man ferner betrachtet, daß diese beyden sauren Geister [sulfuric and nitric acid is meant] in der Luft schweben, daß sie mächtiger sind, als der Kochsalzgeist, daß eine Menge Kochsalz beständig in freyer Luft lieget, so wird man sich vorstellen können, daß dieser Kochsalzgeist durch jene von seiner feuerbeständigen Erde losgemachet wird, und in die Luft steiget (see English version on p. 288).

In the following phrase, *Gellert* repeatedly emphasizes the importance of natural chemical processes (Gellert 1750, p. 82):

> Dieses kann die Kunst und ohne Zweifel eben so wohl die Natur, ja vielleicht noch auf verschiedene andere Arten.
> As commonly and surely then as this effect is obtained by artificial preparations, so does nature work the same process likewise, and perhaps in many other various ways (English edition 1776, p. 88).

This is the first written phrase (as far as I know) about the release of hydrochloric acid as a result of the action of atmospheric acids on table salt in the air. There can be no doubt that this was not based on *Gellert*'s observations but on knowledge transferred from alchemistic times. Boerhaave (1762) writes (translated from German):

> It is curious that, if the oil of vitriol solely is done upon the most-fixed sea salt, thereby a very volatile spirit form (pp. 39–40).

Further, he writes his observation when mixing sea salt with oil of vitriol (pp. 37–38):

> At the moment when both are blended, a volatile white haze rises in the air. You have to be careful of this: it suffocates man, and can suddenly spoil the lungs, that there cannot be helped.

Although Paping (1796) in Amsterdam (see p. 157) had already concluded the presence of muriatic acid (HCl) in the air, at the beginning of nineteenth century, some confusion arose, and *Gellert*'s knowledge obviously was forgotten. A nice paper (Anonymous 1823) summarized the results "from several papers by Hermstadt, Vogel, Pfaff, &c.".[673] Hermbstädt (1821) state that the air of the Baltic Sea contains a substance that he called "*färbendes Wesen*" [coloring being], proposing "*Schwefelwasserstoffgas*" or "*Phoshorwasserstoffgas*" (sulphuretted or phosphuretted hydrogen, i. e., H_2S or PH_3); Pfaff (1822) rejected *Hermbstädt*'s view, showing that steam (even from pure water) passing through a solution of "nitrate of silver" ($AgNO_3$) acts deoxidizing, and boiling seawater volatiles HCl that precipitates silver, thus, giving colors between yellow and dark-brown. Pfaff (1822, p. 72) found that boiling seawater releases free HCl, and clearly states that the sea air contains either HCl or NaCl (Pfaff 1822, p. 316). Vogel (1822) investigated the air in Heiligendamm near Bad Doberan and concluded that it contains either gaseous or saline chloride; he also found that from boiling seawater, excluding the mechanical transfer of water or salt particles, muriatic acid is gained. *Vogel* was supported by the Court pharmacist *Krüger* in Rostock[674] who conducted his own examinations and concluded that the air contains "*Wasserstoffgas*" [hydrogen]; Krüger (1822). In Anonymous (1823, p. 29) is stated:

> From all these observations and experiments, the following are the results: that the air near the sea-shore [. . .] contains generally muriatic acid; its quantity is increased by dry season, and ceases to exist in rainy weather.

Witting (1823, p. 225) had no doubt on the existence of free hydrochloric acid in the atmosphere but he (and referring the attempts of other chemists) had no idea which process could release HCl from NaCl; because higher temperatures (such as solar radiation) no freed HCl; *Witting* attributed electricity as a possible cause, such as already speculated by Vogel (1820, p. 98) who argued that free HCl in the air is unlikely asking the question:

---

673 *Sigismund Friedrich Hermbstädt* (1760–1833) was a German pharmacist and chemist in Berlin. *Heinrich August Vogel* (see p. 166). *Christoph Heinrich Pfaff* (1773–1852) was a German physician, chemist and physicist in Kiel (given name not *Christian* as printed in *Pogg. Anal.*).
674 *Friedrich Wilhelm Krüger* (1782–1866).

> [...] warum die Salzsäure sich nicht gleich wieder mit dem frei gewordenen Natron verbände, oder in der großen Menge Wasser aufgelöst bliebe, anstatt sich mit der ihr weniger verwandten Luft zu vereinigen [why muriatic acid not suddenly combines with the free natron, or remains dissolved in the huge quantity of water instead to combine with the less related air].

Finally, Brandes (1826c, p. 69) writes, that experiments together with his brother and *Herrn* [Mr.] *Reimann* has shown that from a solution of "*salzsaurer Bittererde*" ($MgCl_2$) a part of the acid (HCl) escapes at a temperature above about 110 °C; *Karl Ludwig Reimann* (1804–1871) was chemist and pharmacist in Jena, Heidelberg and Pforzheim.

Concerning the dominant idea in the 1930s that sea salt particles are effective condensation nuclei, *Hans Cauer* first found that iodine in rainwater is enhanced against chlorine compared to its ratio in seawater, due to the emission by seaweed burning; *laminaria* and *fucus* (Cauer 1938). From laboratory experiments (see below) he concluded that ozone containing air, passing the seawater surface, oxidizes chloride to hypochlorous acid, which further reacts with chloride in a reduction-oxidation reaction to chlorine that degasses. The chlorine in the air instantly converts into hydrochloric acid (he does not explain in which way); the HCl molecules and molecule complexes are very hygroscopic and dissolve in fast growing droplets, he writes. This rather crude description we could likely understand in modern terms as homogeneous nucleation. However, historically remarkable is his further comment that from the hydrochloric acid in these droplets (under the involvement of nitrous gases) volatile chlorine escapes. This idea he obtained earlier (Cauer 1935b, p. 404) during investigations of a sampling and determination method of total chloride in the air using aspirators, but first cited himself in Cauer (1949a, p. 229). *Cauer* passed in three experiments chlorine-free air (7 m^{-3} in 2 h), containing different amounts of ozone through diluted seawater (1:2000) having pH = 8.1, and determined the chlorine loss to be 18.8 % (common air), 42.3 % (weakly ozone-enriched air), and 55.3 % (strong ozone-enriched air), Cauer (1949a, p. 414). He explained the chlorine loss (Cauer 1949a, p. 404) that the "normal" equilibrium $Cl^- \rightleftarrows ClO^-$ shifts to the right side by oxidants, such as ozone. *Cauer* further argued that with the reaction $Cl^- + ClO^-$ molecular chlorine is gained, which escapes, and likely chlorate (which is missing in the chemical analysis of chloride). This mechanisms Yeatts and Taube (1949)[675] studied kinetically (without referring *Cauer* or any relation to seawater chemistry): $O_3(g) + Cl^-(aq) \rightarrow O_2(g) + ClO^-$ (slow), and $ClO^- + 2H^+ + Cl^- \rightarrow Cl_2 + H_2O$ (rapid). Besides ozone, the reaction of H_2O_2 with chloride (see older literature quoted in Gmelin 1927) was first described via HX + H_2O_2 gaining X_2 (X = Cl and Br) by Connick (1947).

Cauer (1949a, p. 229 and 1951, p. 1131) for the first time communicated that the ratio between chloride and cations (Mg, K, Na) was nowhere (even on the high seas, as determined during submarine voyages) near the ratio found in seawater. Thus,

---

675 *Henry Taube* (1915–2005) was a US American chemist; awarded in 1983 with the Nobel Prize for his work in the mechanisms of electron-transfer reactions.

he correctly concluded that sea salts are not transported chemically unchanged in droplets of seawater according to the spray theory. Cauer and Cauer (1942) were also the first who put attention to the role of gaseous chlorine compounds in the atmospheric chlorine budget; they proposed the release of chlorine (note not HCl as we today know) by the reaction between ozone and sea salt particles. In fog, studied at Oberschreiberhau, Cauer and Cauer (1941) already found evidence on participation of gaseous chlorine compounds (see also Junge 1952, p. 23). Today we know that chlorine in particulate matter is generally in the form of unreactive chloride; however, a variety of heterogeneous and multiphase reaction processes can lead to the conversion of particulate chloride into gas phase, reactive chlorine species. These processes include acid displacement and reactions of N_2O_5, ozone, and other species with chloride-containing aerosol.

As for many conclusions in atmospheric chemistry, rainwater analyses first have shown strong deviations of the Cl/Na ratio in rainwater over Scandinavia (Rossby and Egnér 1955). Junge (1956, p. 130) found that the Cl/Na ratio in giant particles in Florida equals to that of the seawater. From his Round Hill measurement, however, *Junge* concluded that over the continent independently sources of sodium and chloride exist. The problem of the very different experimental results of investigators (Hitchcock et al. 1980 for references) concerns the variation of ionic ratios in marine aerosol is best characterized by Chesselet et al. (1972), who found that the measured Cl/Na ratios in marine aerosol resemble the seawater values, but also minute amounts of terrigenous dust in northern hemisphere marine air in the same range of concentrations as sea salt particles and contain sodium and potassium of primary continental origin. Thus, mixing of particles of different origin and chemical composition, including non-adequate sampling techniques, did not provide a clear picture about Cl loss, although Junge (1956) showed the Cl loss in fine marine particulate matter. For example, Martens et al. (1973) clearly determined a significant Cl deficit in the coastal aerosol.[676]

On the basis of his aerosol chemical analysis at different locations in the early 1950s, Junge (1956, p. 134) writes that nitrate (never nitrite) is concentrated predominantly in giant particles (Junge 1956, p. 133), and writes further (Junge 1956, p. 134):

> Nitrate is formed by oxidation of NO_2 to NO_3 when sea-spray aerosols mix with continental air masses in coastal areas.

Motivated by *Junge*'s findings, Robbins et al. (1959) studied in the laboratory the reaction between NO_2 and NaCl particles and propose as the first the acid displacement according to $HNO_3 + NaCl = NaNO_3 + HCl$, and assume as the precursor reaction the hydrolysis of NO_2 in the gas phase ($3\,NO_2 + H_2O = 2\,HNO_3 + NO$) and due to small humidity of sea salt particles, not $2\,NO_2 + H_2O$ (liquid) $= HNO_2 + HNO_3$. Thus, what

676 Many investigations on the Cl loss and deficit begun only in the 1990s; see Möller 2019, pp. 429–430). See also Möller (1990) for references in the 1980s.

today is accepted, that first nitric acid is formed in the atmosphere from NO_x (by still unknown mechanisms in that time), was already suggested correctly. Eriksson (1959, 1960) argued that the reaction between O_3 and NaCl as proposed by *Cauer*, is too slow having atmospheric evidence. *Erik Eriksson* believed that the gaseous chlorine in the marine atmosphere was HCI released from sea-salt particles but suggested that it was hygroscopic SO_3 formed in the atmosphere by the oxidation of SO_2, which dissolves in the particles, forms H_2SO_4, lowers the pH, and releases HCl to the atmosphere. Junge (1963a, p. 168) comments:

> Summarizing, we can say that the observations show a release of Cl from sea-salt particles, but that the mechanism of this process is not quite clear.

Available evidence reviewed by *Robert Duce* suggests that the release of gaseous HCI from sea-salt particles is the major source of gaseous Cl in the marine atmosphere (Duce 1969). Many years later Brimblecombe and Clegg (1988) describes the acid replacement and argued that only strong acids (H_2SO_4 and HNO_3) can support the reaction (s stands for solid) $NaCl(s) + H^+(p) \rightarrow Na^+(s) + HCl(g)$.

Several field studies between 1996 and 2004 (Möller 2010, p. 429) have shown the importance of HNO_3 in the replacement of chloride and the release of HCI from sea spray. Finlayson-Pitts et al. (2003)[677] proposed a mechanism (recrystallization) based on laboratory studies.

Besides HNO_3, other oxidized nitrogen compounds have been found to react with NaCl under the release of reactive chlorine compounds. In contrast to Robbins et al. (1959), Schroeder and Urone (1974) found nitrosyl chloride (NOCl) in the NO_2 + NaCl reaction. Finlayson-Pitts (1983) also investigated the reactions of atmospheric NaCl particles with gaseous NO_2 gaining NOCl and proposed the mechanism $NO_2 + NaCl = NO_2Cl^- + Na^+$, followed by $NO_2Cl^- + NO_2 = NOCl + NO_3^-$. Finlayson-Pitts et al. (1989) report experimental evidence that gaseous chlorine nitrate ($ClONO_2$) and dinitrogen pentoxide (N_2O_5) also react with NaCl particles at 298 K, to form Cl_2 and $ClNO_2$ respectively. Zetzsch and Behnke (1992) observed that nitryl chloride ($ClNO_2$) is produced at high yields in the reaction of N_2O_5 with wet NaCl aerosol.

However, none of these researchers obviously knew that already in the early nineteenth century *Edmund Davy* (1785–1857)[678] studied the reaction between HNO_3 and powdered NaCl; Davy (1831) writes:

> In the mutual decomposition of chloride of sodium and nitric acid, the products appear to be chloro-nitrous and chlorine gases, and nitrate of soda.

Davy further investigated the "new" gas, and writes:

---

677 *Barbara J. Finlayson-Pitts* (born 1948) is Canadian-American atmospheric chemist. She married *James Pitts* (1921–2014) in 1970.

678 Irish chemist; cousin of *Humphry Davy*.

> By passing chloro-nitrous gas through water an acid is obtained, which appears to resemble very closely the common solvent od gold, or aqua regia, otherwise called nitro-muriatic acid.

Beckham et al. (1951, p. 319) write:

> Nitrosyl chloride, although accurately identified only during the first half of the nineteenth century, has been in the hands of alchemists and chemists for many centuries as one of the components of aqua regia.

Next to *Davy*, *Corneille Jean Koene* (1809–1883)[679] studied the chemistry of aqua regia (Koene 1844) and named the acid *hyponitric acid* (*Gay-Lussac*'s "*acide pernitreux*" and *Schönbein*'s "*Untersalpetersäure*"); not to be confused with today's hyponitrous acid $H_2N_2O_2$. The French chemist *Alexandre Edouard Baudrimont* (1806–1880) repeated *Davy*'s experiments (Baudrimont 1846) and believed that the gas has the formula NO_2Cl_3 (in nitric acid, NO_5, three equivalents of oxygen are replaced by chlorine), and named it "*acide chloronitrique*" (or "*chloroazotique*"); its aqueous solution he considered as aqua regia. Gay-Lussac (1848) who first stated that aqua regia consists of 1 part HNO_3 and 3 parts HCl (Gay-Lussac 1848, p. 208), evolved this gas by heating of aqua regia and presents the formula NO_2Cl_2, derived from his "*acide hyponitrique*" (NO_4) by exchange of two equivalents of oxygen (he named it "*acide hypochloronitrique*"); when heating of aqua regia, *Gay-Lussac* believed that a compound with the composition NO_2Cl is gained (named "*acide chloronitreux*", derived from "*acide nitreux*" NO_3, i. e., nitrous acid), and that aqua regia is a mixture of different acids. Gay-Lussac (1848, p. 622) writes:

> [...] vapeur, [...], mélange du chlore avec le nitreux, soit de l'eau régale ordinaire ou d'un mélange de sel marin et d'acide nitrique concentré.

Today we know that the equation $HNO_3 + 3\,HCl = NOCl + Cl_2 + 2\,H_2O$ is valid and that all these "compounds" were nothing else than nitrosyl chloride (NOCl), containing free chlorine in changing amounts, as stated by the German industrial chemist *Johannes Wilhelm Goldschmidt* (1861–1923) who was still a student of chemistry in the time of his investigation (Goldschmidt 1880); he named *Gay-Lussac*'s gas in German "*Unterchlorsalpetersäure*". However, *Goldschmidt* did not know the short communication by Tilden (1874)[680] who studied the formation of NOCl from aqua regia, characterized it as orange-yellow gas and deep orange liquid, boiling at −8 °C, and who doubt the existence of *Gay-Lussac*'s $NOCl_2$, writing that "dichloride was probably a solution of chlorine in the monochloride"; *Tilden* also presented for the first the above chemical equation. In that time, the constitution and names of different oxides and acids

679 Dutch-Belgian chemist in Brussels. In 1855 *Koene* introduced as the first the idea that the amount of CO_2 in the atmosphere is not constant.

680 *Sir William Augustus Tilden* (1842–1926) was a famous British chemist.

of nitrogen were confusing (Chapter 4.6.2.1), and finally led to the statement that *Gay-Lussac* already studied nitrosyl chloride (what is true), but without giving its name and the formula NOCl.[681] Müller (1862)[682] studied the properties of "*Untersalpetersäure*" (giving the formula NO_2 in contrast to *Gay-Lussac*, and nitrogen peroxide called in English) and first cites Weltzien (1860a):

> Da es durchaus nothwendig geworden ist, die Radicale der salpetrigen und Salpetersäure zu benennen, so schlage ich für das erstere (NΘ) den Namen "Nitroxyl", für das der zweiten ($N\Theta^2$) "Nitroxyd" vor. Demnach wäre die von Gay-Lussac dargestellte Verbindung NΘCl und die von R. Müller erhaltene Nitroxydchlorür [Since it has become quite necessary, to name the radicals of the nitrous and nitric acid, I propose for the first (NΘ) the name "Nitroxyl", for the latter ($N\Theta^2$) "Nitroxyd". According to this, the compound prepared by Gay-Lussac would be NΘCl, and that prepared by R. Müller Nitroxydchlorür].

According to Beckham et al. (1951) all investigators in the early nineteenth century made experiments with impure nitrosyl chloride that had been evolved from aqua regia, i. e., a mixture of nitrosyl chloride and chlorine. Müller (1862, p. 6) stated that most likely a compound with the formula NOCl is gained, but he also found evidence for a compound hitherto unknown NO_2Cl (now as nitroxyl chloride known). Finally, in discussing methods of formation of chlorine from the reaction between HCl and HNO_3, Lunge and Pelet (1895) critically reviewed previous works, also investigated the possible formation of nitryl chloride ["*Nitrylchlorid*"] NO_2Cl but never detected it.[683] *Lunge* and *Pelet* investigated the reaction between HCl (or NaCl + H_2SO_4) and HNO_3 in solution and proposed the following processes: $HCl + HNO_3 \rightleftarrows ClNO_2 + H_2O$, $ClNO_2 + NO \rightarrow ClNO + NO_2$, and $ClNO_2 + 2\,HCl \rightarrow ClNO + Cl_2 + H_2O$.

3.4.3 Sea salt and nitrate deposits in Northern Chile

This chapter deals with the question of the origin of nitrate of soda of Northern Chile, in the desert of Atacama. Among the many theories (geological such as volcanic and rock weathering, biological such as the decay of vegetation, nitrification of ammonia from guano, soil bacterial fixation of atmospheric nitrogen; see Ericksen 1963, 1981,

---

681 It seems that in citing the history of nitrosyl chloride, several errors appeared. Gmelin (1927, p. 424) and Mellor VIII, p. 612) refer wrongly Gay-Lussac (1848) with the equation $2\,NO + Cl_2 = 2\,NOCl$, later called by the French chemist *Eugène Wourtzel* (unknown living data) "*réaction de Gay-Lussac*" (Wourtzel 1912, p. 345): "*synthése du gaz chlorure de nitrosyle NOCl pare union directe des gaz NO et* Cl^2".

682 *Richard Müller* (unknown living data) was a German chemist and assistant of *Weltzien* at Karlsruhe.

683 NO_2Cl (also named nitroxyl chloride) was first gained by Hasenbach (1871, p. 11) in the reaction $NO_2 + Cl_2$ and named "*Chlorid der Salpetersäure*". *Hasenbach* was an unknow chemist in the laboratory of *Kolbe* in Leipzig.

1983), the "atmospheric theory of nitric acid formation" is only referred here. However, Diaz (1994) writes:[684]

> Whitehead (1920) stated: "Many quite divergent hypotheses on the origin of the deposits have been suggested with the universal acceptance of none". This statement still reflects the opinion of many geologists.

The French geographer and cartographer *Pierre Joseph Aimé Pissis* (1812–1889) writes (Pissis 1878, p. 15 and 16):

> The constant association of common salt with nitrate seems at first sight to point to a probable marine origin of those minerals; [...] The formation of nitric acid, [...], of transforming atmospheric nitrogen into nitric acid in the presence of other oxidizable matter.

Pissis further comments that "marine origin" excludes sedimentary action in water or by the concentration of a saline solution by evaporation because of total missing of limestone and marine shells. Ericksen (1981) noted that *Pissis* arguments are as valid today as they were a century ago. *Ericksen* further writes on the origin of nitrates that chemical weathering of feldspathic igneous rocks with atmospheric nitric acid form by either chemical reactions between atmospheric nitrogen and ozone or electrical discharge in dense fogs, citing succeeding investigators such as Sundt (1921), Wetzel (1928) and Knoche (1930).[685] However, *Ericksen* was wrong with the note that *Plissis* (such as *Wetzel*) suggested electric discharges as the cause of atmospheric nitric acid. The presence of atmospheric HNO_3 *Plissis* writes, is "according to the results of the experiments of Cloes" (it is meant *François Stanislas Cloez* with his paper in 1861). *Wetzel* suggested that the nitrate ion is gained by electrical discharge in the dense winter fogs (camanchaca), and whereas *Sundt* proposed diverse sources for nitrate and ammonia in the atmosphere, *Knoche* suggested formation by reaction of atmospheric nitrogen with ozone. Ericksen (1981) also first found that in addition to the nitrate in high concentration, perchlorate in the desert of Atacama is the only known natural source. *George Edward Ericksen* was the scientist who doubtless much contributed to an understanding of the saline deposits, and, who changed his opinion on its origin several times. Finally (Ericksen 1983) believes that two main sources must be considered, volcanic origin and sea salt deposits in combination with atmospheric chemical processes. To me, the idea that chloride replacement occurs in sea salt because of different NO_y compounds as discussed above (NO_2, N_2O_5, HNO_3), originated by lightning, which remains as saline nitrate, is not only fascinating but most likely compared to other proposals (what not exclude additional other sources). Atmospheric chemistry

684 *George Edward Ericksen* (1920–1996) and *Walter* Lucius *Whitehead* (1891–1969) were American geologists.

685 *Lorenzo Sundt* (1839–1933), born in Norway was an engineer and geologist in Chile. *Walter Wetzel* (1887–1978) was a German geologist. *Walter Knoche* (1881–1945) was a German-Chilean meteorologist.

is also the only explanation for the occurrence of perchlorate.[686] The spatial distribution of nitrate deposits on the plateau close to the sea in a band up to about 30 km wide and about 700 km along the Chilean coast is frequently affected by sea spray rich fog which forms a cloud bank. These clouds provide chemical processing under strong solar radiation, frequent lighting, hence sources of NO and O_3. Because of missing rain, the droplets evaporate, and the salty particles deposit, stabilizing in windblown surface salt, and then washed to greater depths by the infrequent rains, accumulating over geologic time periods.

Further to me, it is rather surprising that *Hans Cauer* did not combine his findings of strong acidity of nuclei or haze particles in remote air of high altitudes due to nitrous and/or nitric acid with his idea of chlorine depletion by ozone; he only refers *Wetzel* with his theory[687] of the formation of saltpeter in the Chilean Andes concerns the occurrence of nitric acid (Cauer 1949a, p. 232). In the end, a curious idea is cited here by *Henry M. Papée* (unknown living date, chemist in Canada), who concluded from analyzing *Junge*'s data on rainwater and aerosol composition, and referring *Cauer*'s idea on chloride liberation by the reaction with ozone, that (Papée 1959, p. 201) "this again supports the hypothesis that nitrate formation on sodium chloride particles is to be associated with the radiochemical excitation mechanisms".

686 Chloride and HCl react with ozone and generate as final products in contact with water (see Möller 2019, p. 378), first found by Allmand and Spinks (1929) in the gas-phase reaction between chlorine and O_3 after addition of moisture. See also Catling et al. (2010).

687 Wetzel (1928) proposed that passing air with dry fog that contains nitric acid (gained from lightning) reacts with the mountain rocks and sodium nitrate is transported by the very rare rain event to the desert and deposited there.

4 Investigation of gases in the air

Besides the major atmospheric gases – nitrogen, oxygen, and carbon dioxide – several gases were known from the experiments by *Priestley* and *Scheele* in the early 1870s (like hydrochloric acid, ammonia, nitrous oxides, sulfur dioxide) but not yet identified in the air. From alchemical experiments, these and other gases were known under archaic names (Chapter 1.3.3) but not its constitution, however, some of its main features.

Already in the first half of the nineteenth century, other gaseous substances were supposed and later detected in the air. In the second half of the nineteenth century, two "natural" constituents of the atmosphere were discovered, ozone and hydrogen peroxide, and its oxidizing capacity was recognized. Since the concentration of almost all trace gases are orders of magnitude smaller than those of the main gases, only with the development of analytical techniques in the late nineteenth century they were proved to be present in the air. Nevertheless, from chemical analysis of rainwater constituents (Chapter 2.3) it was already in the eighteenth century concluded on "foreign substances" in air.

The best review on the constitution of the atmosphere before 1865 is given by *Charles Mayer Wetherill* (1825–1871)[688] in a report to the US Government concerning warming and ventilation of the Capitol (Wetherill 1865) that includes chapters on measurement methods (11 pp.), eudiometric measurements in chronological order (11 pp.), analysis determining the carbonic acid in the atmosphere (15 pp.), miscellaneous substances in the atmosphere like ammonia, organic matter, and dust (5 pp.), and ozone (6 pp.). Despite referring all these substances, only a few were detected and determined in the atmosphere (almost in rainwater) – even less gaseous substances were determined but almost in much higher quantities in the air.

There are without any doubt many more published works on trace gas investigations in the nineteenth century that are not cited in this volume; many of them are forgotten, others are well-known for other research but in printed matters not simply available; for example, Leeds (1878), Kapustin (1879), Anonymous (1881a), Bujwid (1881), Long (1881–1882), Uffelmann (1888), Nékám (1890), Geralevich (1891), Linow (1892), and Russell (1892).[689]

688 American chemist; in 1862, he was appointed the first head of the Chemical Division in the newly organized U. S. Department of Agriculture.

689 *Odo Bujwid* (1857–1942) was a Polish bacteriologist in Warsaw who studied at *Robert Koch* and *Louis Pasteur*. *Julius August Christian Uffelmann* (1837–1894) was professor for medicine and hygiene at University of Rostock. *Carl Linow* (1865–?) studied medicine in Rostock and worked as physician in Dresden. *William James Russell* (footnote 643 on p. 325). *Lajos Alexander Nékám* (1868–1957) was Hungarian dermatologist. Other researcher are unknown.

https://doi.org/10.1515/9783110732467-004

4.1 A brief history of the discovery of gases in air

> Before the modern discoveries in chemistry, the atmosphere was considered as *one* simple elastic fluid, *sui genesis*, containing in it, by some means or other, certain foreign substances not essentially but accidentally mixed with it. (Dalton 1802, p. 545)
> [...] it now appears there are at least *four* distinct elastic fluids found in every portion of atmospheric air subject to examination. (Dalton 1802, p. 546)

In the eighteenth century, the interest in natural processes generally expanded. Travelers and biologists were interested in describing the climate and its relation to culture and biota. In the late eighteenth century, chemists began to understand the transformation between solid, liquid, and gaseous matter. A fundamental interest in biological processes, like plant growth, nutrition, and respiration among others, stimulated the study of the water cycle and the gas exchange between plants and air. Nitrogen (N_2), oxygen (O_2), water vapor (H_2O), carbon dioxide (CO_2), and rare gases are the permanent basic gases in the air. Thus, basic air composition was discovered in the eighteenth century. However, the role of water in its changing phases (condensation and evaporation) was fully understood only in the nineteenth century (Chapter 2.2.3). The assimilation of gases and the uptake of nitrogen dissolved in water by plants and the decomposition of dead biomass as a source of gases led to the first understanding of matter cycles by early agricultural chemists (Knop 1868). Finally, *Lord Rayleigh* (1842–1919)[690] and *Sir William Ramsay* (1852–1916)[691] identified the novel gases in the air between 1882 and 1898. At the end of the nineteenth century, many trace species were known in the air (NH_3, HCl, NO_x, HNO_x, SO_2, H_2S, O_3, H_2O_2, CH_4, H_2), with a few determinations in gaseous air and more systematically in rainwater (namely ammonium, nitrate, chloride, and sulfate). It is important to note that all trace species mentioned and discovered or assumed to be in the air were believed to be of natural origin or, in other words, substances with a specific function in nature (at that time yet unknown). *Angus Smith* was the first who considered sulfur and chlorine also to be of anthropogenic origin from coal combustion and chemical processing (Smith 1872). Table 4.1 summarizes the historic milestones.

The discovery of main and trace substances in the air can be termed as the beginning (or roots) of atmospheric chemistry. However, accepting that "atmospheric chemistry" is the discipline that deals with substances in the atmosphere, "air chemistry" (Gmelin 1798) termed "*Luftchemie*" in that sense for the first, see p. 2, this discipline is much older. The 150-years period of 1749–1898 we can now term the era of the discovery of main and trace substances in air and hydrometeors; beginning in 1749 with *Marggraf*'s rainwater study, in 1752 with *Black* who was the first to detect CO_2

690 *Rayleigh*, Lord (*John William Strutt*, British mathematician and physicist. Professor in Cambridge (1879) and London, Nobel prize in 1904 (together with *Ramsay*).

691 Scottish chemist who studied in Glasgow and Tübingen; professor in Glasgow (1880) and London (1887).

Table 4.1: Milestones in the investigation of the atmosphere air and discovery of gases.

air as an "element"	Anaximenes (around 600 BC)
first "bioclimatology"	Hippokrates (ca. 460–ca. 377 BC)
first phenomenology of air	Aristoteles (384–322 BC)
air as body	Heron (10–75)
air as "chaos" and "kinds of air"	Paracelsus (1493–1541)
weight of air	Galileo (1564–1642)
term "gas"	Helmont (1577–1644)
air pressure: vacuum and barometer	Torricelli (1608–1647)[a]
pressure decline with altitude	Pascal (1623–1662)[b]
combining pressure, volume and temperature	Boyle (1626–1691)
discovery of CO_2 in link with plant growth and combustion	Mayow (1643–1679), Hales (1677–1761), Black (1728–1799), Senebier (1742–1809)
discovery of N_2 in air	Rutherford (1749–1819)
discovery of O_2 in air	Scheele (1742–1786), Priestley (1733–1804)
first air analysis (N_2 and O_2)	Lavoisier (1743–1794), Cavendish (1731–1819), Gay-Lussac (1778–1850), Humboldt (1769–1859)
discovery of O_3 in air	Schönbein (1799–1868), Houzeau (1829–1911), Andrews (1813–1885)
"nutrients" from air	Liebig (1803–1873)
plant injuries from SO_2	Stöckhardt (1809–1886)
chemical climatology[c]	Smith (1817–1884)
sprouts in air	Pasteur (1822–1895)
discovery of rare gases in air	Ramsay (1852–1916), Rayleigh (1842–1919)
first observation of ultrafine particles in air	Aitken (1839–1919)

[a] An Italian physicist and mathematician in Pisa; best known for his invention of the barometer.
[b] *Blaise Pascal* was a French scientist in Rouen, Clermont-Ferrand and Paris.
[c] "Three zones of air pollution; fields and open country with carbonate and ammonia, ammonium sulfate in suburbs, and acid sulfate and sulfuric acid in town".

in the air of Edinburgh and ending in 1898 with *Ramsay*'s discovery of He, Ne, and Xe in air. The next 50 years – except for stratospheric O_3 studies and beginning accurate tropospheric O_3 measurements – were stagnant in further air chemical studies. Air pollution remained a local problem until the 1960s. It was not before the end of the 1970s that scientists recognized the occurrence of global changes. The techniques available to measure trace species, however, were still rather limited.

In Chapter 1.3.3 on airs and gases in alchemy, we mentioned that *van Helmont* was the first who distinguished different airs or gases, being related to atmospheric air (gas ventosum – common air) but recognized not identical due to different properties. *Stephen Hales* measured the amount of "Airs" produced by the destructive distillation of a variety of animal, vegetable, and mineral substances (Hales 1727),[692] but

692 Chapter VI: A Specimen of an attempt to analyse the Air by chymio-statical Experiments, when shew, in how great a proportion Air is wrought into the composition of Animal, Vegetable and Min-he had no conception of a multiplicity of chemically distinct airs (McEvoy 2014). *Hales* invented the "pneumatic trough" (Fig. 4.1) as a way of using the upward displacement of water to isolate and collect such an insensible and rarefied substance as 'the Air'; he focused solely on measuring the amount of "Air" produced by a chemical substance (like burning brimstone[693] or nitre, heating "*sal armoniack*", the sea salt mist of calx bone; Hales 1727, p. 177) and developed no techniques for its manipulation.

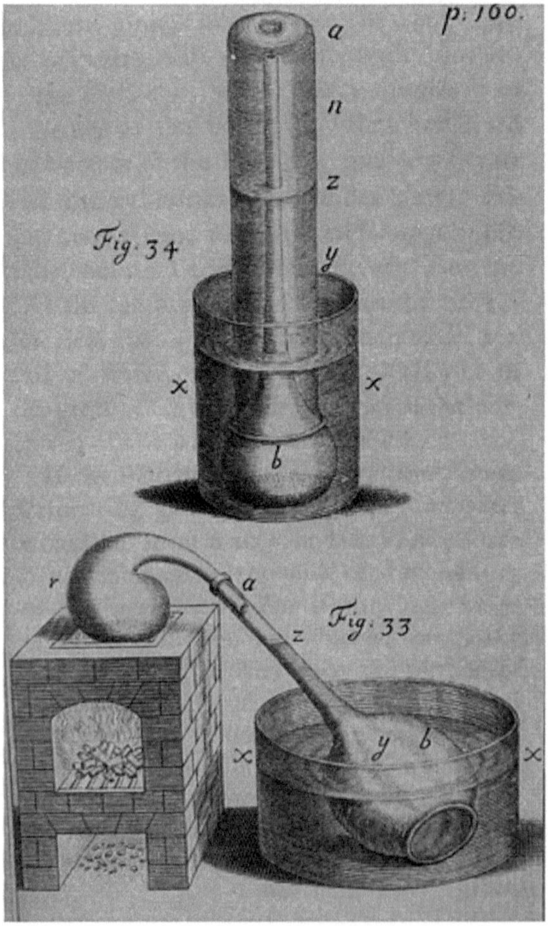

Figure 4.1: Instruments for "measurements of new generated Airs"; above the "pneumatic trough" (Hales 1727, p. 160).

eral substances: And withal, how readily it resumes its elastick State, when in the dissolution of those substances it is disengaged from them (pp. 155–316). Stephen Hales (1677–1761) was an English physiologist, chemist and inventor in Cambridge; he studied the role of air and water in the maintenance of both plant and animal life.

693 Formerly "the mineral sulfur", now restricted to biblical usage. Old English brynstān; related to Old Norse brennistein and brennusteinn (from Proto-Germanic brennan "to burn"); not to confuse with German Bernstein (amber).

Carbon dioxide is the first gas discovered in the air; the following scientists must be named: *John Mayow* (1643–1679)[694] and *Joseph Black* (1728–1799)[695] as well *Stephen Hales* (1677–1761), who early in the eighteen's century began his essential studies on air, absorption of water by plants and its transpiration to the atmosphere. Furthermore, Mayow (1674) concluded from his experiments on respiration of animals that there is a constituent of the air absolutely necessary for life (oxygen), which he termed "*spiritus nitro-aereus*" and another not supporting life (nitrogen); however, he was still unable to characterize these gases (see p. 90). Partington (1956, p. 409) writes:

> Mayow uses two names for the active part of air, *igneo-aerial particles* and *nitro-aerial particles* or *nitro-aerial spirit*, and he uses the name *sulphureous particles* or *salino-sulphureous particles* for combustible matter.

Partington (1956, p. 412) further quoted Mayow (1674, pp. 114, 300 and 302):

> The nitro-aerial particles are "truly elementary (revera Elementares esse)".
> The nitro-aerial particles are the pure and vital part of the air (particulae aeris vitals, aeris autem puri vitalisque), equal in effects to a great quantity of common air.

Joseph Black published the first account of air different from ordinary air (Black 1756). *Black* used the term *fixed air*, which *Hales* used to denote a physical state of 'the Air', to name this new chemical species. *Henry Cavendish* clarified the disciplinary boundaries of pneumatic chemistry when he dealt with more than one kind of air – *fixed air* and *inflammable air* (hydrogen)[696] – and focused on determining their characteristic properties, rather than their role in a specific chemical reaction or medical procedure (Cavendish 1766).

An early reference to *Priestley*'s interest in the air is contained in his "History and Present State of Electricity, with Original Experiments" (London 1767), in which he wrote (Priestley 1767, p. 598):

> Having read, and finding by my own experiments, that a candle would not burn in air that had passed through a charcoal fire, or through the lungs of animals, or in any of that air which the chymists call *mephitic*; I was considering what kind of change underwent, by passing through

694 An English chemist, physician, and physiologist who is remembered today for conducting early research into respiration and the nature of air. Date of born not exact known (between 1640 and 1643); he studied in Oxford and worked as physician in summer in bath and in winter in London. Member of the Royal Society in 1678.

695 A Scottish chemist and physician (born in France). He was professor of chemistry at Glasgow (1756) and from 1766 at Edinburgh; he is best known for his theories of latent heat and specific heat.

696 *Armand Gautier* was the first in 1898 to proclaim the presence of free hydrogen, and detected it in 1900 to be permanent present (to be 15 ppmv) in atmospheric air (Gautier 1898, 1900, 1901a,b,c,e). This was verified in 1902 by *Rayleigh*'s spectroscopic studies in air.

> the fire, or the lungs &c., and whether it was not possible to restore it to its original state, by some operation and mixture. For this purpose, I gave a great degree of intestine motion to it; I threw a quantity of electric matter from the point of a conductor into it, and performed various other operations upon it, but without any effect.

Joseph Priestley (1843, pp. 147–148) adopted *Black*'s usage of *fixed air*, distinguishing the other kinds of air by their properties, or another periphrasis, without mentioning any 'hypotheses' about their internal chemical compositions (McEvoy 2014). *Priestley* obtained *inflammable air* from metal-acid solutions in the manner described by *Cavendish*; he also procured less pure air samples (probably, mixtures of hydrogen and carbon monoxide) by heating other inflammable like coals, vegetables, and animal substances. He expanded on *Cavendish*'s account of the airs "sensible properties", stressing, in particular, its peculiar inflammability. Like *Cavendish*, he obtained from a dilute solution of copper in *spirit of salt* (HCl), not inflammable air as expected, but the highly soluble *marine acid air* (HCl), which he collected over quicksilver (Hg) instead of water. Repeating *Cavendish*'s experiments on metals – except lead and tin – dissolved in *spirit of nitre* (HNO_3), *Priestley* obtained *nitrous air* (NO), which played a prominent role in his subsequent pneumatic inquiries (Priestley 1774; cited after McEvoy 2014).

In the second half of the eighteenth century, the air consisted of two different constituents, maintaining respiration and combustion (O_2) and not maintaining it (N_2). The discovery of oxygen is credited to *Carl Wilhelm Scheele* (1742–1786)[697] and *Joseph Priestley* (1733–1804).[698]

The discovery of nitrogen is generally credited to *Daniel Rutherford* (1749–1819)[699] whose "Dissertatio Inauguralis de Aero Fixo Dicto, aut Mephitico" (On Air said to be Fixed or Mephitic) was published in Edinburgh in 1772. Seventeen years after *Black*'s dissertation on "fixed air" (CO_2), just before *Scheele*'s (and *Priestley*'s) discovery of "good air" (O_2), *Rutherford* conducted experiments where he removed oxygen from the air through burning substances (i. e., charcoal) and afterward carbon dioxide by ab-

---

697 A German pharmacist, born in Stralsund/Pomerania (Sweden at that time), druggist in Gothenburg (1757), Stockholm (1768), Uppsala (1770) and Köping (1775). In Uppsala he met the well-known chemist *Tobern Bergman* from the University to patronize him; member of the Royal Academy of Sweden (1775). Scheele (1777, p. 2) presents also a nice definition of chemistry: Die Körper in ihre Bestandtheile zu zerlegen, deren Eigenschaften zu entdecken, und sie auf verschiedene Art zusammen zu setzen, ist der Gegenstand und Hauptzweck der ganzen Chemie [To dissect the bodies into their component parts, to discover their properties, and put them together in different ways, is the object and main purpose of all chemistry].

698 An English theologist, teacher of languages (German, French, Italian, Latin, Greek, Arabian, Syrian, Chaldean, Hebraic), self-educated scientist (he only listened some lectures), and politician; due to his sympathy with the French revolution he moved to Philadelphia/USA in 1794.

699 A Scottish chemist and physician in Edinburgh who is most famous for the isolation of nitrogen in 1772.

sorption with lime; the rest (nitrogen) he denoted as "phlogisticated air", even though it was not flammable.

Scheele carried out between 1768 and 1773 a series of experiments to study air; before spring 1770 he found that 1/3 of the air (the oxygen) is taken up by a solution of copper oxide in ammonia (Nordenskiöld 1892, p. 60), and in 1771–1772, he gained and described oxygen, and it is very likely that he also before *Rutherford* was familiar with the remaining gas, nitrogen. *Priestley* wrote in 1771 about the goodness of air (air quality in modern terms) and noted that injured or green plants can restore exhausted air. He observed in his work "Air in which Candles or a Brimstone have burned" on the purification of air by plants that green plants give off oxygen (without identifying it – eight years later the Dutch physician and botanist *Jan Ingenhousz* (1730–1799) discovered that plants release O_2 under light) and thus improve it. However, he did not recognize the causes of this phenomenon. Priestley (1772, p. 22) writes:

> I were so happy as by accident to have hit upon a method of restoring air which was injured by the burning candles, and to have discovered at least one of the restauratives which Nature employs fir this purpose. It is vegetation. This restoration of vitiated air, I conjecture, is effected by plants imbibing the phlogistic matter with which it is overloaded by the burning of inflammable bodies.

This notice was undoubtedly the first written idea used by other researchers much later that green pants and forests remove impurities and that plants evolve "ozonized air" (see pp. 474–176, 532, and 569–570).

In 1772 *Priestley* started his studies on air using mercury for locking gases. After a break in 1776 he systematically began to investigate different "kinds of air": nitrous (salpetric) air (NO_x), acid (muriatic) air (HCl), and alkaline air (NH_3). He stated that these "kinds of air" are not simple modifications of ordinary (atmospheric) air. He published his observations in a book entitled *Observations on Different Kinds of Air;* Priestley (1772, 1775). By heating red mercury oxide, *Priestley* produced dephlostigated air (O_2) on 1 August 1774, and informed in October 1774 in Paris *Lavoisier* and some other French chemist on his discovery.[700] *Scheele* and *Priestley* worked independently, the discovery of oxygen, one of the most important steps in experimental chemical research, comes to both together. However, the honor of recognizing the significance of this discovery and using it for the chemical theory comes to *Lavoisier* (Nordenskiöld 1892, p. 408).

700 *Scheele* wrote a letter to *Lavoisier* (on September 9, 1774) informing him that he was already busy since some years with researches on air and fire [... *quelques années, de experiences de plusier sortes d'air*, ...] but that he was unable to gain common air from fixed air according to a prescription by *Priestley* (Nordenskiöld 1892, p. 406–407). *Priestley* characterized the new gas as part of common air only in march 1775.

On the basis of these findings, *Antoine-Laurent de Lavoisier* (1743–1794)[701] recognized the role of oxygen in combustion and established modern chemistry as a scientific discipline; he also gave the names of the main air constituents, "*oxygéne*" and "*azote*". Lavoisier presented his "*Traité élémentaire de Chimie, présenté dans un ordré nouveau et d'après les découvertes modernes*" (Paris 1789), soon later translated by *Robert Kerr*[702] into the "Elements of Chemistry in a new systematic order, containing all the modern discoveries" (Edinburgh 1790).[703]

Another remarkable scientist was *Henry Cavendish* (1731–1810),[704] who never rejected the phlogiston theory and did not publish his results on air studies until 1783. *Cavendish* was the first to study flammable air (hydrogen) in different mixtures with common air to investigate its explosion (1766). He separated hydrogen from other gases and showed that it burned to water. In connection with *Lavoisier*'s discovery of the role of airy oxygen (1777) it became clear that water is a chemical compound (Lavoisier 1783). Already in 1772, he privately told *Priestley* about his experiments with "mephitic air" (nitrogen); thus, it seems likely that *Cavendish* already knew before *Rutherford* about "inflammable air" (N_2). *Cavendish* had already tested to find whether airy nitrogen is a uniform matter and found a small residue (noble gases). He did not conclude, however, that these remains are an element (argon). In 1781, he realized that water is produced in a reaction of hydrogen ("flammable air") with oxygen ("vital air") and soon, he noted that there are also acidic substances not containing any oxygen. In 1781 he also sampled atmospheric air at different sites and analyzed it gravimetrically after sorption of water-soluble gases (CO_2, NH_3, and water vapor). In explosions, in which *Cavendish* used electric sparks, he found "[...] liquor in the globe [...]; it consisted of water united to a small quantity of nitrous acid" (Cavendish 1784). This statement is most remarkable to me; it forms the first evidence of HNO_x formation under atmospheric conditions by lightning (note that separation between nitric and nitrous acid was not yet known; "nitrous" was a term for oxidized nitrogen).

---

701 *Antoine-Laurent de Lavoisier*, a French chemist; 1764 onward at the Academy of Sciences, 1775 commissioner of the Royal Gunpowder Administration and residence in the Paris Arsenal; opening of a laboratory in the Arsenal; 1785–1789 collaboration with *Claude Berthollet*, *Antoine de Fourcroy* and *Guyton de Morveau*; 1789 co-editor of the newly founded *Annales de Chimie*. 8 May 1794 executed on the guillotine.

702 *Robert Kerr* (1755–1813), a Scottish physician and translator.

703 Reprint by Dover Publ. New York, 1965.

704 British chemist and physicist in London; he is considered to be one of the so-termed pneumatic chemists of the eighteenth and nineteenth centuries. Around 1783 *Cavendish* began to work closely with *Charles Blagden*, an association that helped *Blagden* enter fully into London's scientific society. In return, *Blagden* helped to keep the world at a distance from *Cavendish*. *Cavendish* published no books and only a few papers, but he achieved much. In 1790 he met *Humboldt* who soon accepted *Cavendish*'s chemical findings.

When reading these old papers in the light of what we know today, it is often difficult, if not impossible, to understand what the scientists meant by different terms; confusion also results from attributing the same term to different substances (we may only conclude that in those days such distinctions were not always possible): phlogisticated air for both N_2 and H_2, acid air for both CO_2 and O_2. Kopp (1869) accepted that phlogiston[705] was actually hydrogen. Cavendish (1784, p. 389) and Watt (1784) defined correctly:

water = dephlogisticated air [oxygen] + phlogiston [hydrogen]
dephlogisticated water = dephlogisticated air [oxygen]
dephlogisticated air [oxygen] = water deprived of its phlogiston

Wilhelm Ostwald explains in annotations to *Scheele*'s book (Ostwald 1894, p. 109):

phlogisticated = deprived of oxygen or united with hydrogen
dephlogisticated = oxidized or deprived of its hydrogen [phlogiston]

Thus, phlogisticated means reduced and dephlogisticated oxidized. Scheele (1777), however, defines unknowable (if we assume that phlogiston was actually hydrogen):

phlogiston = *Feuerluft* [oxygen] – *fixe Luft* [carbon dioxide]

Phlogiston has always been regarded as a substance (matter) and not simply "fire" in sense of heat.[706] However, *Scheele* (1777) writes that heat consists of "*Feuerluft*" [oxygen] combined with phlogiston. Phlogiston, writes *Scheele*, is a true element and a very simple principle but does not exist for itself; it can be transferred from one body into another by certain substances, combined with the exchange of heat and or light (Scheele 1777, p. 61). In all these combinations phlogiston does not change itself and can be eliminated from the last compound anew. *Scheele* cites the French chemist *Antoine Baumé* (1728–1804), who believes that phlogiston is elementary fire (or heat) combined with gentle earth-like "*Kieselerde*" (silica) and that the carbonaceous residue of distilled oils is nearly pure phlogiston (see next page the discussion phlogiston = carbon). When such gentle coal is burned, the residue is negligible and it should not be clear how such less earth could absorb so much heat. The following phrase in this context by *Scheele* is probably the first argument of mass conservation (Scheele 1777, p. 61):

705 Phlogiston = fire element, from Ancient Greek φλογιστόν (phlogistón), neuter of φλογιστός (phlogistós, "burnt up, inflammable"), from φλογίζω (phlogízō, "to set fire to"), from φλόξ (phlóx, "flame").
706 Note that the law of mass conservation was discovered only about 1785 by *Antoine Laurent Lavoisier*, and that heat, light, electricity etc. were regarded at that time as ponderable substances.

> Denn was hier nach dem Verbrennen fehlet, ist das Gewicht der Hitze. Allein, wieget denn die Luftsäure [CO_2] nichts, die sich in so grosser Menge von dieser Kohle unter dem Verbrennen scheidet? [Because what here is missing after combustion, is the weight of heat. Alone, weighs the aerial acid nothing, which gained in such a large amount from the coal while burning?]

Scheele was the first who distinguished between heat from fire and heat from light (which he termed "*strahlende Hitze*" – radiative heat) and found that light rays (e. g., sunlight) convert only into heat when they fall on a body (Scheele 1777, pp. 46–59).[707] From *Scheele*'s "definition", regarding this equation as a mass budget, phlogiston would be "negative" (minus-sign) carbon (C). It is close to *Lavoisier*'s "definition" of fixed air (Lavoisier 1774, p. 419), changing the minus-sign into plus-sign in *Scheele*'s equation:

fixed air [carbon dioxide] = common air + phlogiston

Priestley (1772, p. 22) writes concerns recovery of air (see p. 351) that plants imbibe [assimilate] "the phlogistic matter", which results from burning matter in the common air; in other words, "the phlogistic matter" was carbon dioxide, assimilated by plants and "turned" into oxygen. According to the phlogiston theory (remember that *Becher*'s idea comprised those combustible substances contain ignitable matter, the "terra pinguis", later termed by *Stahl* phlogiston). one can "define":

fixed air [carbon dioxide] = phlogistic matter = phlogiston + [oxygen]

Consequently, *Priestley* termed his later discovery (1775) "dephlogisticated air" on the theory that it supported combustion so well because it had no phlogiston in it, and hence could absorb the maximum amount of phlogiston during burning. *Lavoisier* already understood clearly in 1772 what means the reduction of calces (e. g., HgO) with charcoal but not yet the intrinsic link with oxidation (of charcoal with oxygen into carbon dioxide). Thus, charcoal (carbon) contains phlogiston. However, finally, it remains unclear what phlogiston means; it is the mass difference (plus- or minus-sign) in redox processes including solid and gaseous matter concerning the solid or gaseous matter, respectively. Table 4.2 and 4.3 lists the discoveries of chemical air constituent.

---

707 The influence of light onto horn silver [AgCl], first observed by *Johann Heinrich Schulze* in 1717 (cf. p. 98), was studied by *Scheele* who identified the "black colour" as elemental silver and that violet light acts more intensive on silver chloride then red light, arguing that violet light particles contain less phlogiston: "*Die violetten Strahlen reducieren das Hornsilber eher als die andern*" (Scheele 1777, p. 58); This is very likely the first statement of *photochemistry*.

Table 4.2: Phlogisticated and dephlogisticated substances. Note that nitrous air = NO, and marine air = HCl.

Archaic term	substance	formula	authorship
phlogisticated acid of vitriol	sulfur dioxide	SO_2	Stahl
phlogisticated nitrous air	nitrous acid	HNO_2	Bergman, Cavendish
phlogisticated air	nitrogen	N_2	Rutherford, Priestley
dephlogisticated air[b]	oxygen	O_2	Priestley
dephlogisticated nitrous air	dinitrogen monoxide	N_2O	Priestley (1789)[a]
dephlogisticated nitrous air	nitric acid	HNO_3	Bergman
dephlogisticated marine air	chlorine	Cl_2	Scheele
(pure) phlogiston	hydrogen	H_2	Cavendish, Kirwan (1787)

[a] *Priestley* first termed this gas nitrous air, diminished.
[b] As originally thought to be air deprived of phlogiston by *Priestley*.

Table 4.3: Discovery of atmospheric relevant trace gases.

Year	gas	explorer	in air
1755	CO_2	Helmont and Black	yes
1766 / 1901	H_2	Cavendish / Gautier	no / yes
1772	N_2	Rutherford	yes
1773 / 1774	O_2	Scheele[b] / Priestley	yes
1774	Cl_2	Scheele	no
1776	SO_2[a]	Priestley	no
1776 / 1786	NH_3	Priestley /Scheele	no / yes
1776	HCl	Priestley	no
1777 / 1796	H_2S[a]	Scheele / Berthollet	no
1841	O_3	Schönbein	yes
1854	SO_2	Thomson	yes
1863 / 1872	H_2O_2	Meißner / Schöne	yes
1883	Ar	Rayleigh	yes
1884	Ra	Rayleigh and Ramsey	yes
1898	Kr, Ne, Xe	Ramsey and Travers	yes

[a] This gas was already known in the Middle Ages
[b] It is impossible to fix the exact year of the discovery of oxygen by *Scheele* (before 1773 but not before 1771)

4.2 The main gases of the air

4.2.1 Nitrogen

Mayow (1674) found that from common air after combustion in a closed volume, the remaining "air" (mostly nitrogen) does not support life and further combustion. This observation was also made in 1710 by *Francis Hawksbee* the Elder (1660–1713), a scholar of *Boyle*. It was *Scheele* who likely isolated nitrogen from the air first around 1771 (but published only in 1777); independent from *Scheele* and *Rutherford*, *Cavendish* separated nitrogen from air (Weeks 1932).[708] *Daniel Rutherford* (1749–1819) a pupil of *Black*, isolated nitrogen from an air volume by oxygen consumption through combustion and removal of carbon dioxide by alkali (Rutherford 1772).[709] It was *Black* who gave *Rutherford* the problem of studying the properties of this residual "air". Since the residual gas did not support life, he termed it "noxious", or "injured" air. He did not recognize that this gas is the constituent of the atmosphere after the removal of oxygen and carbon dioxide. He thought that "noxious air" was atmospheric air that had taken up phlogiston from the substance that had been burned (Weeks 1956, p. 221). Priestley termed this gas in 1775 as phlogisticated air and by *Scheele* in 1777 as "*verdorbene Luft*" [*foul* or *corrupted air*]. *Jean-Antoine Chaptal* (1656–1831) termed it "*nitrogène*" after finding its relation to nitric acid.[710] *Lavoisier* termed it (Lavoisier 1789, p. 52–53)[711]

> [...] azote from the Greek privative particle ἀ and ζή, vita; hence the name of the noxious part of atmospheric air is azotic gas [...] it is proved to compose a part of the nitric acid, which gives as good reason to have termed it *nitrigin*.

Over the next decades, there were a lot of confusion and misunderstandings concerning the chemical nature of nitrogen. *Bergman* and *Scheele* considered nitrogen as nitric acid, being gaseous through uptake of phlogiston. *Lavoisier* assigned nitrogen to the elements in his antiphlogistic nomenclature (1787). Girtanner (1792) describes nitrogen gas to nitrogen with caloric ("*Salpeterstoffgas* = *Salpeterstoff* + *Wärmestoff*"). Without knowing at that time that nitrogen exists free only as molecule (N_2), this "definition" expresses that the atmospheric nitrogen (N_2) was not identified with the "element" nitrogen (N). Hence *Berzelius* termed it in 1820 *nitric radical*. He separated between *azote* (Az) and *nitric* (N), Berzelius (1813a). Berzelius (1814), starting from N (*nitricum*), denoted related compounds as follows (Crosland 1962, p. 274); oxygen was represented by a dot to other electronegative elements:

> Nitrogenium $\dot{N}$ (= NO) (modern notation N_2), Oxidum nitrosum $\ddot{N}$ (= NO^2) (modern notation NO), and Ammoniacum $\dot{N}H^6$ (= NOH^6) (modern notation NH_3)

Some chemists long disputed the elementary nature of nitrogen although Lavoisier (1789, p. 196–197) clearly describes the formation of nitrogen from ammonia:

---

708 *Mary Elvira Weeks* (1892–1975) was an American chemist and historian of science.

709 It is certain that *Cavendish* carried out experiments with air and found nitrogen before *Rutherford* but he told private on these finding in 1772 to *Priestley*. He published first results from his experiment on air in 1783 (Ramsay 1896).

710 Nitrogène, nitrigen and nitrogen are derived from the archaic name for potassium nitrate (KNO_3). *Nitrium* (Latin) and *nitrèn* (French), *nitre* (English) as well as saltpetre (British) or saltpeter (American) and *Salpeter* (German).

711 It must be noted that *Lavoisier* (and other chemists as well as naming in other languages) termed the "noxious part of atmospheric air [...] azotic gas".

> The hydrogen of the ammonia combines with the oxygen of the oxide,[712] and forms water, whilst the azote being left free escapes in form of gas [. . .] Mr Cavendish first observed it in nitrous gas and acid, and Mr Berthollet in ammoniac and the prussic acid. As no evidence if its decomposition has hitherto appeared, we are fully entitled to consider azote as a simple elementary substance.

Cavendish began first experiments with common air under the influence of electric discharges in 1777; his observation of products from the reaction between nitrogen and oxygen must be honored to be the first laboratory study in atmospheric chemistry.

The following nineteenth century was full of quantitative measurements of some atmospheric trace species having low quality and speculations based on limited observations. *David Low* (1786–1859), a Scottish agriculturalist, believed that nitrogen must consist of carbon and oxygen (Low 1848) because of its close relation to organic matter. Later (Low 1856, p. 204) he wrote that "nitrogen was seen to be a simple combination of hydrogen and carbon, namely HC that is a common element". A French chemist, *Auguste Laurent* (1807–1853), who was a student of *Jean-Baptiste Dumas*, has shown that nitrogen consists of diatomic molecules (Laurent 1846); the formula N_2 finally was derived from the gas density;[713] *Lavoisier* already estimated the density of nitrogen to be 0.4444 grain per cubic inch (1.450 kg/m^3), not very much different from today's estimate (1.250 kg/m^3). Note that all density measurements before *Rayleigh's* discovery of argon included nitrogen plus argon (1.2572 kg/m^3).

4.2.2 Oxygen

The English physicist and polymath *Robert Hooke* (1635–1703), a contemporary of *Boyle*, in his famous book "Micrographia", published in 1665,[714] thought that air contains a substance that exists in solid form in saltpeter.[715] *John Mayow* explained combustion by saying that air contains a "*spiritus nitro-aereus*" (p. 90), a gas that is consumed in respiration and burning, with the result that substances no longer burn in the air that is left (cited after Weeks 1956, p. 210). The first person to prepare oxygen by heating the saltpeter was *Ole Borch*, but he did not know how to collect it.[716] He stated in 1678 that it did not burn but that it made charcoal burn very vigorously (Weeks 1956, p. 211). *Stephen Hales* communicates in his "*Vegetable Statics*" (London 1727) that lead, calcinated with the minimum, releases a great portion of "air" (O_2).

---

712 In experiments when ammonia is heated together with metal oxides.

713 $M = (\rho/p)RT$.

714 See: Alembic Club Reprint No. 5 (1905) Extracts from Micrographia. Univ. of Chicago Press, pp. 43–47.

715 Remember; saltpeter (American) and saltpetre or salpetre (British).

716 See: Jorgensen, S. M. (1909) Die Entdeckung des Sauerstoffs (translated from Danish by Ortwed and Speler). F. Enke, Stuttgart, pp. 12–14.

Mayow states that the increase of weight during calcination is due to air absorption (O_2); however, these explanations did not find advocates (Kopp 1931, p. 140).

The facts on the air chemical composition were expressed most clearly by *Scheele* in his booklet "*Chemische Abhandlung von der Luft und dem Feuer*" [Treatise on Air and Fire], which was published in 1777 (Scheele 1777),[717] Fig. 4.2. *Carl Wilhem Scheele*[718] (Fig. 4.3) submitted his manuscript to the bookseller *Swederus*[719] at the end of 1775, however, to the great annoyance of *Scheele*, the edition of the book was delayed until August 1777 (after the comment by *Ostwald* in the reprint from 1894). From laboratory scripts and letters (Nordenskiöld 1892), as known, *Scheele* discovered oxygen before 1773 and, therefore, before *Priestley* did in 1774 (Priestley 1775). *Scheele* carried out the experiments on "Air and Fire" between 1768 and 1773; he left Malmö in 1768 for Stockholm and moved in 1770 to Uppsala (and in autumn 1775 for Köping). *Scheele*'s first phrase in his book reads:

> Die Untersuchung der Luft ist jetziger Zeit ein wichtiger Gegenstand der Chemie [The researches on air are at present an important object of chemistry].[720]

When *Scheele* started his experiment on air (1768), he described the existing knowledge on air (Scheele 1777, p. 8):

> Die Luft ist dasjenige flüssige unsichtbare Wesen, welches wir beständig einathmen, den Erdboden allenthalben umgiebt, sehr elastisch ist, und eine Schwere besitzt. Sie ist beständig mit einer erstaunlichen Menge von allerley Ausdünstungen angefüllet [...] Unter diesen fremden Theilchen haben die Wasserdünste beständig das Übergewicht. Es ist aber die Luft auch noch mit einem andern elastischen luftähnlichen Körper vermischt, welcher in vielen Eigenschaften von selbiger abweicht und von dem Herrn Professor *Bergman* Luftsäure, und zwar aus gutem Grunde genennt wird. Sie hat ihr Daseyn von denen durch die Fäulung oder Verbrennung zerstöhrten organisirten Körperchen [Air is that invisible fluid, which is constantly breathed by us, which everywhere surrounds the globe, and which is possessed of elasticity and weight. It is constantly filled with a great quantity of various effluvia [...] Among these heterogeneous[721] particles, wa-

717 For English edition see Scheele (1780); French edition, translation by Baron de Dietrich (1781): Traité chimique de l' air et de feu. Paris, 268 pp. A second German edition, extended by reports from *Leonhard*, *Kirwan* and *Priestley*: Scheele (1782). Reprint of the English edition by Alembic Club 1901. Reprint of the 2nd German edition in 1894, edited by *Wilhelm Ostwald* (Ostwald 1894). Note the German text is identical in all editions, and the English text is very close to German original; all page references here (Scheele 1777) are based on the edition Ostwald (1894).

718 *Marggraf* and *Scheele* (both German pharmacists) were regarded as the greatest experimental chemists in the 18th century who discovered many elements and compounds from vegetation, animals and minerals.

719 Magnus Swederus (1748–1836). In his "Vorbericht" [advance report] to the first German edition in 1777, Bergmann notes regretting that this work was ready two years before printing, and that Priestey, without knowing Scheele's work, even before described "the different new properties of air".

720 This and the following translations into English after the Edition "Chemical observations and experiments on air and fire" (London 1780).

721 This translation is incorrect: The German "*fremde Theilchen*" corresponds to "foreign particles".

Carl Wilhelm Scheele's
d. Königl. Schwed. Acad. d. Wissenschaft. Mitgliedes,
Chemische Abhandlung
von der
Luft und dem Feuer.
Nebst einem Vorbericht
von
Torbern Bergman,
Chem. und Pharm. Prof. und Ritter; verschied.
Societ. Mitglied.

Upsala und Leipzig,
Verlegt von Magn. Swederus, Buchhändler;
zu finden bey S. L. Crusius.
1777.

Figure 4.2: Cover of "*Chemische Abhandlung von der Luft und dem Feuer*" (Treatise on Air and Fire) by *Carl Wilhelm Scheele* (1777).

tery vapors are always most prevalent. But it is mixed with another elastic fluid similar to air, and which is in many properties different from it, termed properly by Professor Bergman Aërious Acid. It owes its existence to the destruction of organic bodies by putrefaction and combustion].

Scheele used very simple facilities (Fig. 4.4): "*Kolben, Retorten, Bouteillen, Gläser und Ochsenblasen*" [pistons, retorts, bottles, glasses and Ochs bladders], Scheele 1777, p. 26. In many experiments consuming oxygen from air by different chemicals,[722] *Scheele* found values between ¼ and ⅓, significantly more than the 21 % oxygen content in air. However, we have to appraise it concerning the simple experiments (*Scheele* observed gas loss through the Ochs bladders[723]). The "lost air" (oxygen)

722 The First, he used liver of Sulphur ["*Schwefelleber*"], a mixture of potassium sulfide, polysulfides, thiosulfate and sulfate, gaining from heating of potash and sulfur at 250 °C in the absences of air.
723 *Wilhelm Ostwald* comment it as one of the first observations of gas diffusion through membranes.

Figure 4.3: Portrait of *Carl Wilhem Scheele* (after a copperplate print in the author's possession).

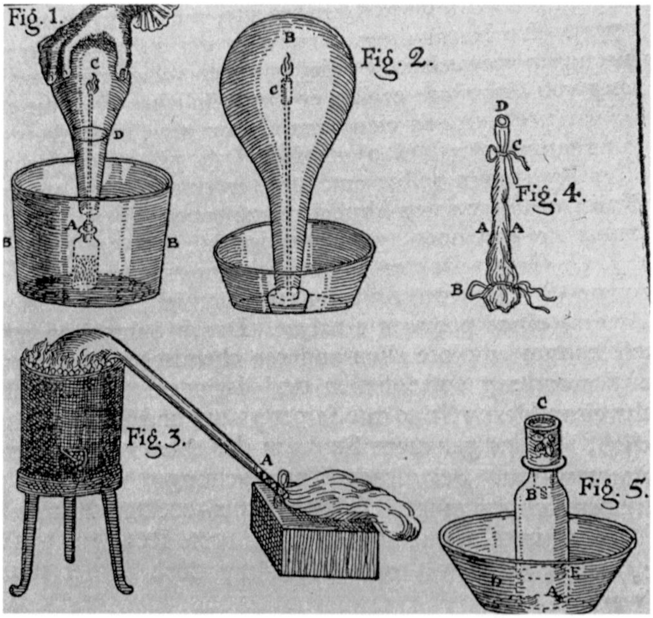

Figure 4.4: Laboratory instruments used by *Scheele* (from Bugge 1929, p. 286; note original in Scheele (1777) between pages 154 and 155). 1 – Flame of hydrogen, locked through water. 2 – Candle light, locked through lime water. 3 – Retort on heater with bladder receiver for gas collection. 4 – Animal bladder for collection of gases. 5 – Breathing insect, locked through lime water.

he could not regenerate chemical[724] but concluded early in his experimental work (Scheele 1777, p. 15):

> [...], dass die Luft aus zwey von einander unterschiedenen Flüssigkeiten bestehe, von welchen die eine die Eigenschaft, das Phlogiston anzuziehen, gar nicht äussere, die andere aber zur solchen Attraction eigentlich aufgeleget ist, und welche zwischen dem dritten und vierten Theil von der ganzen Luftmasse ausmacht [... thus much I see from the abovementioned Experiments, that Air is composed of two different fluids, the one of which attracts not the phlogiston, and the other has the quality of attracting it, and will this latter fluid make between a third and a fourth of the whole bulk of the air].

The experiments were based on consuming and producing gases by different reactions and measuring the change of volumes and weight the new "airs". Carbon dioxide was removed by absorption with alkali oxides. The gases were weighted (oxygen found to be lighter and nitrogen to be heavier than air); oxygen was characterized by supporting candles and animals. *Scheele* also found that animals consume oxygen and exhale carbon dioxide, whereas plants consume and produce both gases.

The first note on the isolation of oxygen (first termed "aer vitriolicus" – "*Vitriolluft*") is found in laboratory scripts from 1771–1772 (Nordenskiöld 1892, p. 458) through heating (thermal decomposition) of mercury oxide and carbonate; later he also used magnesium nitrate, silver carbonate, manganese oxide, and others. *Scheele* writes in his laboratory scripts:

> Der *mercurius praecipitatus ex solutione* in *acido nitri cum alkali fixo*, giebt *per dest.* der Blase *aerem vitriolicum*, in welchem 1/3 *aer fixus*, ein wenig gelbes Sublimat. Dann *mercurium vivum*, in welchem mit starkem Feuer doch mehr röthliches Sublimat, doch sehr wenig, und dann *mercur. vivus* folgt [The *mercurius praecipitatus ex solutione* in *acido nitri cum alkali fixo*, gives *aerem vitriolicum*, in which 1/3 *aer fixus*, a bit yellow sublimate. Then *mercurium vivum*, in which under strong heat more reddish sublimate, but very little, and then follows *mercur. vivus*].

During distillation of fuming nitric acid, *Scheele* observed that at the end of the operation, a light in the receiver burnt lighter than in common air;[725] he collected this "air" [$NO_2 + O_2$] and filled a glass, containing water, with this air (NO_2 dissolved in water and released "pure" oxygen) and put a small burning candle in it, and concluded (Nordenskiöld 1892, p. 25–26).

> Da diese Luft nothwendig zur Entstehung des Feuers erfordert wird, und etwa den dritten Theil in unserer allgemeinen ausmachet, so werde sie der Kürze halber inskünftige die Feuerluft nennen; die andere Luft aber, welche zur feurigen Erscheinung gar nicht dienlich ist, und welche in unserer Luft etwa zwey Drittheile ausmachet, will in der Folge mit dem bereits bekannten Namen, verdorbene Luft belegen [Since this air is absolutely necessary for the generation of fire, and makes about one-third of our common air, I shall henceforth for shortness sake call

724 We know that the oxygen combined (chemical) irreversible with reduced sulfur to sulfate.

725 Thermal decomposition of nitric acid proceeds via $4\,HNO_3 = 4\,NO_2 + O_2 + 2\,H_2O$.

> it *empyreal air*, [literally *fire-air*:] the air which is unserviceable fort the fiery phenomenon, and which makes almost two-third of common air, I shall for the future call *foul air* [literally *corrupted air*].

Scheele characterized "*Feuerluft*" (oxygen) in many experiments and also found that water absorbed relatively more oxygen from common air (Scheele 1777, p. 91):

> Diese in Wasser aufgelöste Feuerluft muss denen Wasserthieren eben so unentbehrlich seyn, als denen, so auf der Erde Leben: Sie müssen solche in ihren Leib einziehen und entweder als Luftsäure oder verdorbene Luft verändern [. . .] als Luftsäure bleibt sie nicht bei freyer Luft beym Wasser, und die verdorbene Luft kann sich gar nicht damit verbinden, als dann das Wasser wieder im Stande von neuem Feuerluft aufzulösen und solches den Thieren zuzuführen [This fiery air must be essential for the aquatic animals, as it is fort those who stay on land: they resorb it in her body and change it either into aerial acid or corrupted air [. . .] as aerial acid it cannot stand in free air by the water, and the corrupted air cannot combine with the water, as the water anew absorbs fiery air to deliver it to the animals].

In contrast to *Scheele*, the date of the discovery of oxygen by *Priestley* is known (Fig. 4.5): 1 August 1774; Priestley (1775, p. 8) writes (note *mercurius calcinatus* = HgO):

> With this apparatus, after a variety of other experiments, an account of which will be found in its proper place, on the 1st of August, 1774, I endeavoured to extract air from *mercurius calcinatus per se*; and I presently found that, by means of this lens, air was expelled from it very readily. Having got about three or four times as much as the bulk of my materials, I admitted water to it, and found that it was not imbibed by it. But what surprized me more than I can well express, was, that a candle burned in this air with a remarkably vigorous flame, very much like that enlarged flame with which a candle burns in nitrous air, exposed to iron or liver of sulphur; but as I had go nothing like this remarkable appearance from any kind of air besides this particular modification of nitrous air, and I knew no nitrous acid was used in the preparation of *mercurius calcinatus*, I was utterly at a loss how to account for it.

When *Priestley* (Fig. 4.6) repeated this experiment in Paris in October 1774 together with *Lavoisier* and other respectable French scientists, he had "no idea, at that time, to what these remarkable facts would" (Priestley 1775, p. 12). Because *Priestley* (Chapter 4.6.2.1) has recently "discovered" nitrous oxide he was slow in recognizing that he had "discovered" oxygen; he confused oxygen with nitrous oxide (Smith 1972, p. 301). *Lavoisier* (who likely already received the letter from *Scheele*, footnote 700 on p. 351), and who not yet solved the question which "air" (gas) is involved in calcination, repeated *Priestley*'s experiments in spring 1775. Still, on the read to French Academy on 26 April 1775, he neither mentioned *Priestley* nor *Scheele*, so that the audience must have believed that all was discovered only by *Lavoisier*; even later (in his "*Traité élémentaire de chimie*") he writes that oxygen was about at the same time discovered by him, *Priestley* and *Scheele*. *Priestley* was upset on it and claimed in 1780 to be the first to discover oxygen; he meant that Scheele's discovery was indeed independent of his, but later, he believed (Bugge 1929, p. 270).

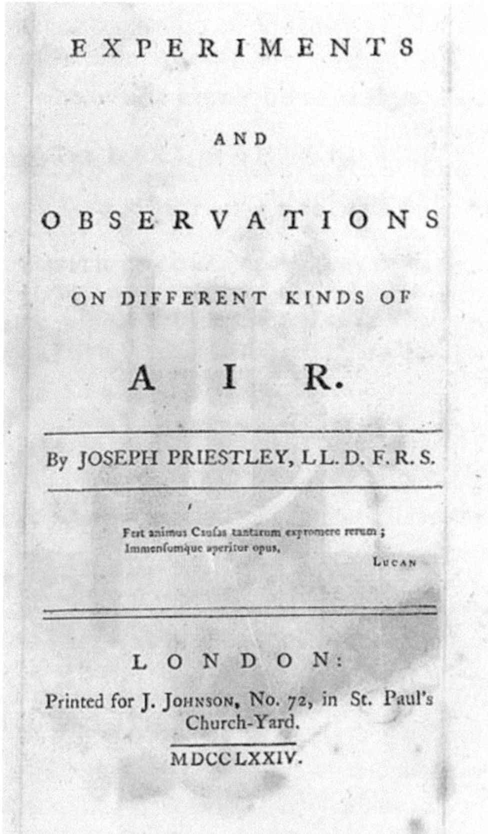

EXPERIMENTS

AND

OBSERVATIONS

ON DIFFERENT KINDS OF

AIR.

By JOSEPH PRIESTLEY, LL.D. F.R.S.

Fert animus Cauſas tantarum expromere rerum;
Immenſumque aperitur opus.

LUCAN

LONDON:

Printed for J. JOHNSON, No. 72, in St. Paul's Church-Yard.

MDCCLXXIV.

Figure 4.5: Cover of *Experiments and Observations on Different Kinds of Air* by Joseph Priestly (1775).

Figure 4.6: Portrait of *Joseph Priestley* (after a copperplate print in possession of the author).

Priestley who lived in Leeds near a brewery, in 1768 began to do experiments with fixed air (Partington 1933) and examine all the "airs" that might be released from different substances in the early 1770s. His researches on gases are collected in six volumes of "Experiments and Observations on Different Kinds of Air" (1774–1786),[726] abridged in three volumes (Priestley 1790).[727] *Priestly* wrote in a letter dated 31 May 1774, to *Caleb Rotheram* (1756–1796), Reverend in Kenton (Rutt 1831, p. 233):

> I do not expect that the account of my experiments on air will give you so much satisfaction as you expect, owing to the great difference between *seeing* and *reading*. I have not yet found any person, though ever so good a philosopher, and who has read my papers ever so carefully, but it surprised to see me actually make the experiments. But the instructions I have given are sufficient to enable any person to do every thing after me.

Priestley (and *Cavendish* too) was not able to draw correct conclusions from his experiments; this was reserved for *Lavoisier* (Partington 1933, p. 350). *Priestley* (as well as *Scheele* and *Cavendish*) remained proponents of the phlogiston theory until her death. Thus Priestley (1790, p. 9) writes:

> The *azote* in the new nomenclature is not expressive of any thing particular to what I have termed phlogisticated air; and the term *vital*, does not sufficient distinguished dephlogisticated from *common*, or *atmospherical* air.

Priestley also developed the "nitrous air test", which tested for the "goodness of air": using a "pneumatic trough" (Fig. 4.7), he would mix nitrous air with a test sample, over water or mercury, and measure the decrease in volume – the principle of audiometry (Chapter 4.2.5.1).

Antoine-Lauren Lavoisier (Figs. 4.8 and 4.9) began his series of experiments on combustion and calcination late in 1772 with the idea to "repeat everything with new precautions, in order to connect what we know about the air which becomes fixed in or disengaged from bodies with other knowledge about these bodies" (Holmes 1998, p. 15). In chapter XV "Dr. Priestley's inquires with different kinds of air" (Lavoisier 1776, pp. 120–121), *Lavoisier* writes that[728]

---

726 The French edition (1775–1777) in three volumes: Experiences et observations sur different especes d'air (translated by *Gibelin*), Paris; German edition (1778–1780) in three volumes: *Versuche und Beobachtungen über verschiedene Gattungen der Luft* (translated by Chr. Ludwig), Wien and Leipzig.

727 The first the results appeared between 1772 and 1789 in several papers in *Philosophical Transactions of the Royal Society*. He wrote beside phlogisticated (N_2) and dephlogisticated air (O_2) on fixed air (CO_2), vitriolic air (SO_2), nitrous air (NO, but often together with NO_2), marine (acid) air (HCl), inflammable air (H_2, mixture with CO, or CH_4 or HC), alkaline air (NH_3), and on fluor acid air (SiF_4). He also isolated nitrous air, dismissed (N_2O).

728 It is more than regrettable that *Lavoisier* did not know the works of *Scheele* before 1781 (after the French edition of *Scheele*'s book), or in other words, that *Scheele* did not publish earlier his (in my opinion more elaborate, and more interesting than *Priestley*'s) works.

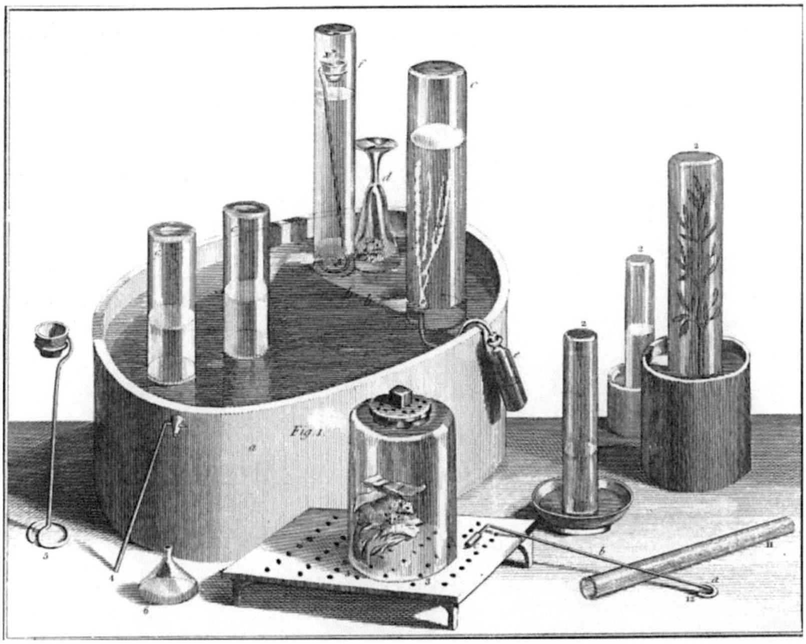

Figure 4.7: Laboratory instruments used by *Priestley*. This is plate 4 of Vol. 1 (frontispiece) *Experiments and Observations on Different Kinds of Airs* (1774) and shows the apparatus associated with the pneumatic trough that *Priestley* had adopted from earlier pneumatic researchers.

Figure 4.8: Portrait of *Lavoisier* (after a copperplate print in possession of the author).

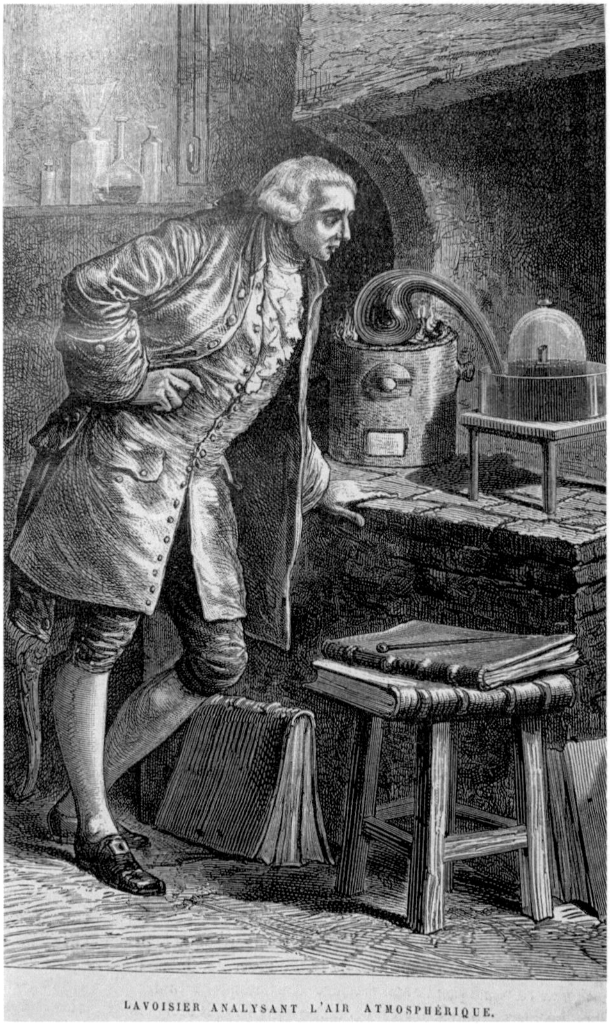

Figure 4.9: Contemporary illustration: *Lavoisier* analyses atmospheric air (from Flammarion 1888, p. 81).

> [...] this work must be regarded as the most elaborate, and most interesting, of any which has appeared since that of Dr. Hales, on the fixation and separation of air.

However, in the whole "*Opuscules physiques et chimiques*" (appeared in French in Paris 1775), *Lavoisier* widely discusses and repeats experiments on fixed air, nitrous air, etc. only in the Appendix (pp. 407 ff). On a read to the Royal Academy on April 26th (1775) *Lavoisier* asked (Lavoisier 1776, p. 407–408):

> Do there exist different species of air? [...] are the different kinds of air which nature affronts us, or which we are able to form, substances distinct of themselves or modifications of atmospheric air?

Lavoisier made two remarkable experiments concerns the reduction of "*mercurius praecipitus per se*"[729] (and repeated *Scheele*'s experiment without knowing from it) first by charcoal to elementary mercury and characterized the air in the receiver as fixed or mephitic air [CO_2] (Lavoisier 1776, p. 414). In a second experiment he carried out the reduction of HgO only by strong heating the retort and found the received air very different from that being CO_2 (Lavoisier 1776, p. 417):

> [...] candles and burning bodies were *not* extinguishing by it, but burned with an enlarged flame in a very remarkable manner; the light they gave was much greater and clearer than in common air [...] that this air was not only more respirable, more combustible, and consequently more pure even than the air in which we live.

From these experiments *Lavoisier* concluded (Lavoisier 1776, p. 418–419); Fig. 4.10:

> As common air is changed into fixed air when combined with charcoal, it should seem natural to conclude that fixed air is merely a combination of common air and phlogiston. This is Dr. Priestley's opinion.

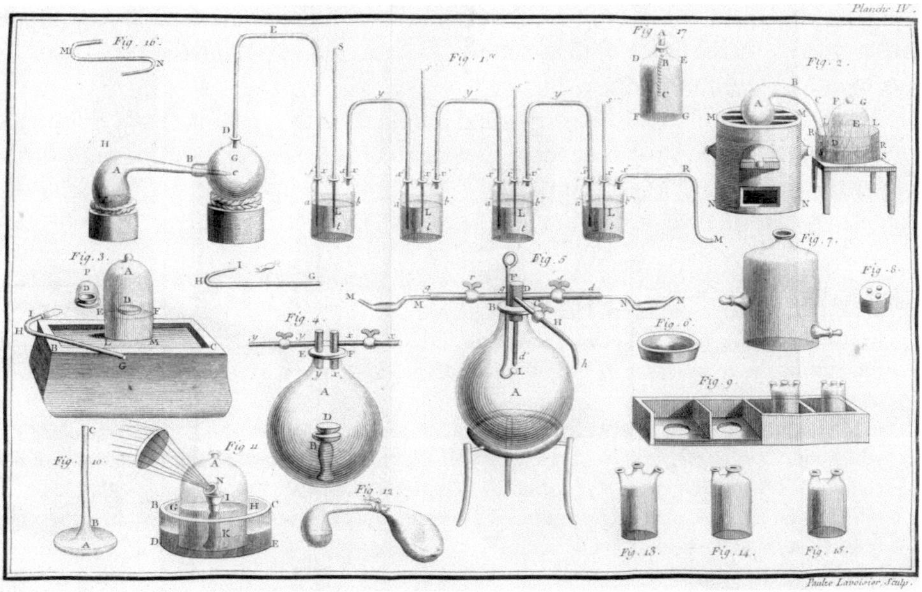

Figure 4.10: Laboratory instruments used by *Lavoisier* (from a plate in the French edition "*Traité élémentaire de Chimie, présenté dans un ordré nouveau et d'après les découvertes modernes*", Paris 1789).

729 It is HgO; there were many names for "red mercury oxide"; Anthon (1833, p. 179).

However, the translator, *Thomas Henry* (1734–1816), added a footnote, writing that "Dr. Priestley has, certainly never delivered such an opinion as M. Lavoisier here ascribes to him". At the very end of his publication, *Lavoisier* writes (Lavoisier 1776, pp. 427–428):

> From nitre [KNO_3] heated in glass vessels Priestley procured very pure dephlogisticated air. A mixture of equal parts of brimstone [sulfur] and nitre yielded air which was highly nitrous, and as the produce of nitre and charcoal is nitrous air, he concluded that the kind of air in also produced in the explosion of gun-powder.

Already Scheele (1777, p. 32) describes heating of "*Salpeter*" (KNO_3) as the best method to yield pure "*Feuerluft*".[730] About *Lavoisier*'s "*Opuscules physique et chemique*" the Scottish chemist *Andrew Norman Meldrum* (1876–1934) wrote in 1930 that "the *Opuscules*, at the first reading, is a disappointing work. One expects great things of Lavoisier" (Meldrum 1930, p. 32). The French historian *Maurice Daumas* (1910–1984) wrote in 1941 that "the *Opuscules* is not a work of combat. It contains no demonstration, no discovery, no conclusion" (Daumas 1941, pp. 99–100). The British Historian *Alfred Robert Hall* (1910–2006) wrote in 1954 that *Lavoisier* was "less the author of new experiments than the first to realize their full significance" (Hall 1954, p. 332). The American historian of science, *Frederic Lawrence Holmes* (1931–2003) wrote on *Lavoisier* that "By the end of 1773 its shape was clear, but its potential was only beginning to unfold" (Holmes 1998, p. 142).

In the period until 1777, *Lavoisier* had ample time to repeat some of *Priestley*'s latest experiments and perform some new ones of his own. In his "*Mémoire sur la combustion en general*" (Lavoisier 1780),[731] he stated the phenomena (cited after the English translation by Leicester and Klickstein 1952):

> Combustion may take place in only a single variety of air: that which Mr. Priestley has named *dephlogisticated air* and which I name here *pure air*.
> In all combustion, pure air in which the combustion takes place is destroyed or decomposed and the burning body increases in weight exactly in proportion to the quantity of air destroyed or decomposed.
> In all combustion the body of which is burned changes into acid by the addition of the substance which increases its weight. Thus for example, if sulfur is burned under a bell, the product of the combustion is vitriolic acid; if phosphorus be burned, the product of the combustion is phosphoric acid; if a carbonaceous substance be burned, the product of the combustion is fixed air, formerly termed the acid of chalk, etc.

730 Today we know that thermal decomposition of potassium nitrate led to potassium nitrite (KNO_2) and oxygen, however, the mixtures of KNO_3 with sulfur and charcoal produce a variety of gases, depending on the level of reduction of nitrate-nitrogen and oxidation of sulfur and carbon (SO_2, SO_3, N_2, NO, NO_2, O_2), not yet understood at that time.

731 Read to the French Academy of Science September 5, 1777; published in 1780 in the *Mémoires* for 1777.

Lavoisier was struck by the fact that the combustion products of such nonmetals as sulfur, phosphorus, charcoal, and nitrogen were acidic (Lavoisier (1781). He held that all acids contained "pure air" [oxygen] and that oxygen was, therefore, the acidifying principle [*principe oxygéne*]. In "*Reflexions sur le Phlogistique*" (Lavoisier 1786),[732] *Lavoisier* showed the phlogiston theory to be inconsistent with observation. In this paper *Lavoisier* termed his principle of combustion and acidity "*oxygine*"; he later re-spelled it "*oxygène*".[733] He writes (cited after Best 2015, p. 141):

> There can be no doubt that the substance that is combined with the metal to form the metallic calx is vital air, the oxygen principle.

For *Lavoisier*, "the oxygen principle" is just one component of that gas (pure air), along with the matter of heat. He believed that this element (oxygen, denoted as dephlogiston) is an immanent part of acids[734] and this gave him the name "*oxygéne*" (from Greek οξνς – acid).[735] Where phlogiston had formerly been released, oxygen was now absorbed. The metallic calces were afterward termed oxides. By 1785 his new theory of combustion was gaining support, and the campaign to reconstruct chemistry according to its precepts began.

Thus, *Lavoisier* built the fundament of modern chemistry,[736] finalized the often termed "Chemical Revolution" of the eighteenth century. The "oxygen theory", explaining the partitioning of oxygen between oxides and (oxo) acids (and its salts) was the key for a final understanding of combustion, respiration, and assimilation, and finally, matter cycling between the biosphere and atmosphere 100 years later. Mechanistic and kinetic understanding of combustion involving oxygen radicals needed additional 50 years, and knowledge of atmospheric oxygen chemistry about 200 years after *Lavoisier*.

---

732 *Lavoisier* read this paper at the Academy over two nights, starting 28th June and concluding 13th July 1783 (Best 2015). The original French text first appeared 1786 in *Mémoires de l'Académie Royale des Sciences année 1783*, pp. 505–38. The edition now most often cited appears in Jean-Baptiste Dumas (ed.), *Œuvres de Lavoisier*, vol. II – Mémoires de chimie et de physique, 623–655. Imprimerie Impériale, Paris (1862). English text with comments see Best (2015, 2016).

733 Lavoisier, A. L.: Mémoire sur l'affinité du principe oxygine avec les différentes substances auxquelles il est susceptible de s'unir. In: Dumas, J.-B. (ed.) Œuvres de Lavoisier, vol. II – Mémoires de Chimie et de Physique. pp. 546–556. Imprimerie Impériale, Paris ([1782] 1862). Originally published in: Mém. Acad. R. Sci. Paris année 1782, 530–540 (1785).

734 This wrong statement created for long time confusion in understanding the nature of muriatic acid (HCl), not containing oxygen.

735 Interestingly that in all languages the word for oxygen (if not directly transferred like to French, English, Greek, Spanish, Italian, Danish, Norwegian, Hungarian, Turkish, Albanian, Azerbaijani, etc. using prefixes like *oxy*, *oxi*, *oksi*) is constructed as "acid matter" using the word for acid (prefix *kis* in Slavic languages, *sauer* in German, *zuur* in Dutch, *syre* in Swedish). Thus, German *Sauerstoff* = acid matter.

736 Already soon later Séguin (1790, p. 467) writes "*M. Lavoisier, créateur de la Chimie moderne*".

4.2.3 Water

The observation that uncorrupted water (materia prima) only comes from the atmosphere (atmospheric water) certainly promoted experiments to derive the philosopher's stone from it (see also Chapter 2.1). Despite much progress in science at the beginning of the seventeenth century, the belief of convertibility between air and water, and water and soil (and vice versa) was widely accepted until the chemical composition of the air, and the structure of water was discovered by *Scheele*, *Priestley*, *Cavendish*, *Watt*, *Lavoisier*, and others[737] after 1770.

The debate about who actually discovered the chemical composition of water (H_2O) was termed the "water controversy" in the nineteenth century. Concerning the discovery of the chemical composition of water, three scientists must be regarded as candidates (Kopp 1869): *Cavendish* (Fig. 4.11), who was probably the first who carried out experiments in summer 1781 (Cavendish 1784, p. 134) to gain water by combining phlogiston (hydrogen) and dephlogisticated air (oxygen).[738] *Cavendish* mixed in a globe inflammable (hydrogen) with dephlogisticated air which then was "fired by electricity", i. e., he studied the explosion of hydrogen and oxygen. In several experiments, the mixture was more or less phlogisticated (nitrogen) and dephlogisticated, resp., i. e., with changing ratios between nitrogen and oxygen. Consequently, he obtained "liquor" being not acid (water) and with different degree of acidity (nitrous and nitric acid) (Cavendish 1784, p. 139). He concludes that dephlogisticated air (oxygen) is "dephlogisticated water, or water deprived of its phlogiston; or, in other words that water consists of dephlogisticated air united to phlogiston; and that inflammable air [hydrogen] is either pure phlogiston".[739] *Cavendish* mentioned his first results already 1781 to *Priestley*, and, in summer 1883 reported a friend of him to *Lavoisier* "that dephlogisticated air is only water deprived of phlogiston" (Cavendish 1784, p. 134).

Cavendish only announced his results to the Royal Society of London in January, 1784. However, when *Priestley* learned of *Cavendish*'s experiments in early 1783,[740] he repeated them (Priestley 1783, 1785), and quickly communicated his findings to his

737 It must be mentioned *Gaspard Monge* (1746–1818) who independently from *Lavoisier* carried out in 1783 similar experiments to gain water.

738 Ramsay (1896, p. 123) writes "[. . .] his experiments were in many cases not published until long after they had been made. He appears to have carried on his work for his own information, and to were indifferent to the impression which his labors made on his fellow-man". *Cavendish* was very rich due to heritages and consumed only a small percentage of it.

739 Two reaction chains occur parallel, $H_2 + O_2$ (= H_2O) and $N_2 + O_2$ (= $NO + NO_2 + HNO_2 + HNO_3$). The problem in understanding *Cavendish*'s (and others) papers result from different meaning of phlogiston; sometimes phlogiston is equal to hydrogen and other times to any unknown principle or matter (fire?).

740 *Priestley* writes that the experiments of sparkling inflammable air and dephlogisticated air to get water were originally performed by *Cavendish* (Schofield 1964).

Figure 4.11: *Henry Cavendish*; contemporary illustration, unknown author (Source: *The Popular Science Monthly* **59** (1901) 432).

friend *James Watt* (1736–1819). *Watt* then wrote to *Jean Andre Deluc* and *Joseph Black*, suggesting that[741]

> [. . .] water is composed of dephlogisticated air and phlogiston deprived of part of their latent or elementary heat [. . .] My assertion was simply that air was water deprived of its phlogiston and united to heat.

Thus, *Watt* formulated the composition of water in 1783 in a similar way to *Cavendish* (Watt 1784);[742] and finally, *Lavoisier*, who, in 1783, made the first public announcement that water consisted of inflammable air (hydrogen) and dephlogisticated air (oxygen), Lavoisier (1783):

741 This phrase is rather confusing; *Watt* and *Cavendish* remained until her dead proponents of the phlogiston theory.

742 In this paper, *Watt* summarized the experiments made by *Priestley* and *Cavendish* as well being repeated later in Paris. There is no evidence that *Watt* made own experiments; he notes that before he was believing that air was a modification of water (although *Boyle* already stated more than 100 years ago that water is not air). It is worth to note that *Watt* noted in this paper "that Mr Cavendish was the first who discovered that the combustion of dephlogisticated and inflammable air produces moisture".

> L'eau n'est pas une substance simple [...] Elle est susceptible de décomposition et de recomposition.

Watt claimed the priority only after *Lavoisier* announced "his" experiments (he was "forced" by *Deluc*). In June 1783, Sir *Charles Brian Blagden* (1748–1820)[743] visited Paris and told *Lavoisier* about *Cavendish*'s experiments. *Lavoisier* had been searching for the acid which he thought must be formed by the union of hydrogen with oxygen, the "principle of acidity". On 24 June, both together repeated *Cavendish*'s experiments, and the next day *Lavoisier* sent a memoir to the French Academy of Science, claiming the discovery of the composition of water. *Lavoisier* states "that we may conclude that water is not a simple substance and that it is composed, weight for weight, of inflammable air and vital air" (Partington 1928). *Lavoisier* is definitively not the discoverer of the composition of water but he was the first to give the correct explanation of them (1783).

Priestley had burned inflammable air (hydrogen) with dephlogisticated air (oxygen) before 1775. While *Alessandro Volta* (1745–1822) had exploded such mixtures with the electric spark before 1776, *Pierre-Joseph Macquer* (1718–1784) in 1777 was probably the first to note that water was the result of burning these two gases (Edelstein 1948).[744] In winter 1782 *Priestley* began experiments in an attempt to convert water into a permanent gas (Priestley 1783, 1785, 1788); the results were communicated from time to time to *Watt*, with whom *Priestley* had been friendly. *Watt* did not carry out many experiments with gases. Although *Priestley* observed the appearance of moisture several times, he neither realized that this water was the product of the union of the two gases.[745] *Cavendish* and *Watt* remained on friendly terms. Instead, the dominant question is why this discovery remained a subject of controversy among prominent figures in Victorian science, some 50–70 years after *Cavendish*'s experiments (Miller 2004).

743 He studied medicine in Edinburgh, worked in Gloucester and moved in 1772 to London. *Blagden* served as a medical officer in the Army (1776–1780). In June 1783, *Blagden*, then assistant to *Cavendish*, visited *Lavoisier* in Paris and described how *Cavendish* had created water by burning "inflammable air". He became in May 1784 one of the two Secretaries of the Royal Society in London.

744 Citing *George Wilson* (1818–1859), chemist at University Edinburgh, who wrote "The Life of the Honorable Henry Cavendish", London, 1851, p. 23.

745 The formation of "phlogiston" (hydrogen) from common air passing burning charcoal, was almost carbon monoxide and not hydrogen (*Priestley* observed black soot beside moisture); hence he produced carbon dioxide (Schofield 1964). In a letter to *Deluc* from December 27, 1783, *Priestley* writes that "no facts of my discovery proves that water consist of pure air and phlogiston [...] and it was Mr Cavendish who first found water on decomposing dephlogisticated and inflammable air. In my opinion Mr Watt first entertained the idea [on the composition of water] [...] and the experiments of Mr Cavendish proved the justness of it, tho Mr Cavendish had not that idea itself".

According to *Christoph Girtanner* (1760–1800), a Swiss chemist, who wrote the first German anti-phlogistic textbook in chemistry (Girtanner 1792), water consist of 0.06 hydrogen and 0.88 oxygen (0.11 and 0.88 would be correct to gain the mass ratio 1:8). The weight of oxygen could be calculated because of earlier works by *Humboldt* and *Gay-Lussac* in 1805, who found that water consisted of only two elements, hydrogen and oxygen (they found in many experiments that two volume parts of hydrogen combine with one volume part of oxygen to form water), Humboldt and Gay-Lussac (1805).[746] Humboldt and Gay-Lussac (1805, p. 76) cite (not giving a reference) *Fourcroy*, *Vauquelin* and *Séguin* who synthesized water from hydrogen and oxygen and found that 100 mass parts of water contain 85.662 parts oxygen and 14.338 parts hydrogen. Still, these figures must be corrected due keeping the gases on water and taking into account the different absorption capacity to the 87.41 oxygen and 12.59 hydrogen. This mass ratio (1:6.9) is already closer than that found be *Girtanner*. Given the correct volume ratio H:O to be 2:1 found by *Humboldt* and *Gay-Lussac* one has to assume that the hydrogen gas was not pure, containing gases that were negligible in relation to the whole volume but not concerning the mass.

Lacking any knowledge about how many atoms of hydrogen and oxygen combine in a molecule of water, *Jean-Baptiste André Dumas* (1800–1884) again had to make some assumptions in 1842. He assumed that nature is basically very simple and, therefore, one atom of hydrogen combines with only one atom of oxygen (this led to the formula HO for water, used for many years). Using this hypothesis and the fact that hydrogen was assigned a weight of one unit, it follows that oxygen is eight times heavier than hydrogen. In 1842, *Dumas* determined water composition by weight and found that two parts (by weight) of hydrogen combined with 15.96 parts of oxygen (giving the formula H_2O). Later experiments changed this ratio to 2:15.88.

However, the Proceedings of the Royal Society in London accepted still in 1843 a paper where "the author is of the opinion that the evidence on which the modern theory of the composition of water is founded, is fallacious; and believing water to be a simple body, he conceives that it forms hydrogen by combining with the electric fluid" (Stevenson 1843).

The Scottish chemist *Alexander Scott* (1853–1947) carried out between 1887 and 1893 studies on the composition of water and found that 1 volume of oxygen combines with 2.00285 volumes of hydrogen (Scott 1893), whereas Burt and Edgar (1916) found it to be 2.00288 from which they concluded on the atomic weight of hydrogen to be 1.00772 (today's estimate is 1.00794).

In 1895, the American chemist *Edward Williams Morley* (1838–1923) introduced a new value for the atomic weight ratio of oxygen to hydrogen, providing the most pre-

746 The original paper appears in French: Gay-Lussac, J. L. (1809) Mémoires sur la combinatiion des substancces gazeuse, les unes avec les autres. *Mémoires de Physique et de Chimie de la Societé d'Arcueil* **2** (1808) 207–234 and 252–153; and in German: Gay-Lussac, J. L. (1810) Ueber die Verbindung gasförmiger Körper eines mit dem andern. *Annalen der Physik* **36**, 6–36.

cise determination of the atomic weight of oxygen at the time (Morley 1895). His work on the atomic weight of oxygen covered a period of eleven years. Much time was spent in the calibration of instruments and improving the measurement accuracy. The final result was the density of oxygen 1.42900 ± 0.000034 g L^{-1} and hydrogen 0.089873 ± 0.0000027 g L^{-1}. The ratio of the hydrogen and oxygen combining volumes he found to be 2.00269, and thus it follows for the atomic weight of oxygen 15.877 (currently accepted 15.874).

Sidney Edelstein (1912–1994), the American research chemist, makes a persuasive case for *Watt*'s priority (Edelstein 1948) whereas the US historian *Robert Schofield* (1923–2011) and the British chemist and historian *James Riddick Partington* (1886–1965) argues that to *Cavendish* the first claim must be given (Partington 1928, Schofield 1964); myself (DM) come round to the latter opinion. An excellent short history on the composition of water is written by Wisniak (2004).

4.2.4 Argon and the other novel gases

In a certain sense, the discovery of argon must be attributed also to *Cavendish*. *Cavendish*'s investigations of phlogisticated air (nitrogen) did show that a small part of this "air" could not be oxidized into nitric air (by oxygen and electric sparks); and after removal of excess oxygen and water, he stated that 1/120 part of the total air remains (Bauer 1895, Rayleigh and Ramsay 1896).[747] This was 100 years before *Rayleigh* identified it to be argon.[748] *Lord Rayleigh* was the first who observed (between 1882 and 1892) that oxygen and other gases produced from different sources always showed the same density but not airy nitrogen (Rayleigh and Ramsay 1896). It was shown that nitrogen extracted from chemical compounds is about one-half percent lighter than "atmospheric nitrogen" (Rayleigh and Ramsay 1896, p. 1). The difference of about 11 mg was already far away from measurement errors. In his address on the occasion of receiving the Nobel Prize (1904) *Rayleigh* explained how he made his discovery, showing the (from today's point of view) simple but accurate experiments and conclusions:

> The subject of the densities of gases has engaged a large part of my attention for over 20 years. [...] Turning my attention to nitrogen, I made a series of determinations [...] Air bubbled through liquid ammonia is passed through a tube containing copper at a red heat where the oxygen of the air is consumed by the hydrogen of the ammonia, the excess of the ammonia being subsequently removed with sulfuric acid. [...] Having obtained a series of concordant observations on gas thus

747 *Alexander Emil Anton Bauer* (1836–1921) was an Austrian chemist.

748 "Although Cavendish was satisfied with his result, and does not describe whether the small residue was genuine, our experiments about to be related render it not improbable that his residue was really of a different kind from the main bulk of the "phlogisticated air", and contained the gas now termed argon" (Rayleigh and Ramsay 1896, pp. 8–9).

> prepared I was at first disposed to consider the work on nitrogen as finished. [...] Afterwards, however [...] I fell back upon the more orthodox procedure according to which, ammonia being dispensed with, air passes directly over red hot copper. Again, a good agreement with itself resulted, but to my surprise and disgust the densities of the two methods differed by a thousandth part – a difference small in itself but entirely beyond experimental errors. [...] It is a good rule in experimental work to seek to magnify a discrepancy when it first appears rather than to follow the natural instinct to trying to get quit of it. What was the difference between the two kinds of nitrogen? The one was wholly derived from air; the other partially, to the extent of about one-fifth part, from ammonia. The most promising course for magnifying the discrepancy appeared to be the substitution of oxygen for air in the ammonia method so that all the nitrogen should in that case be derived from ammonia. Success was at once attained, the nitrogen from the ammonia being now 1/200 parts lighter than that from air. [...] Among the explanations which suggested themselves is the presence of a gas heavier than nitrogen in air [...]

This new gas was identified by *Ramsay* in 1894, who made spectroscopic studies, as an element and named argon (Ar), derived from Greek αργόν = slack (*Ramsay* was awarded the Nobel Prize together with *Rayleigh* in 1904). While investigating for the presence of argon in a uranium-bearing mineral, he instead discovered helium, which since 1868 had been known to exist, but only in the Sun. This second discovery led him to suggest the existence of a new group of elements in the periodic table. *Ramsay* and his co-workers quickly (1898) isolated neon (Ne), krypton (Kr), and xenon (Xe) from the earth's atmosphere (Rayleigh 1901, Ramsay 1907). *Ramsay* and *Morris William Travers* (1872–1961) determined the content of Kr and Xe still rather inaccurate (0.05 to 1.0 and 0.0059 to 0.05 ppm, resp.);[749] all mixing ratios are volume-based.

Herbert Edmeston Watson (1886–1980)[750] first determined with a single analysis the content of Ne in air to be 18.2 ppm (Watson 1910) and Glueckauf (1946) with three analyses on average 18.21 ± 0.04 ppm. Krypton and Xenon were first precisely measured by *Charles Moureu* (1863–1929) and *Adolphe Lepape* (1900–1930) to be 1.0 ± 0.1 ppm (Kr) and 0.09 ± 0.02 ppm (Xe), and by *Gerhard Damköhler* (1908–1944)[751] to be 1.08 ± 0.1 ppm (Kr) and 0.08 ± 0.03 (Xe); Moureu and Lepape (1926),[752] Damköhler (1935). Even though vast quantities of helium constantly escape from the earth's crust, its content in air determined Paneth and Glückauf (1937, 1946) with 5.239 ±

---

749 Proc. Roy. Soc. **63** (1898) 404, **71** (1903) 426, **80** (1908) 599.

750 Professor of Chemical Engineering University College London; invented the neon glow lamp in 1911.

751 A German chemist at Göttingen, known for his research in chemical engineering; he passed by suicide.

752 *Moureu* was a French organic chemist and pharmacist, President of the *Sociéte Chimique de France*, who studied together with *Robert Biquard* (1895–1906) and since 1907 with *Lepape* for many years' rare gases of the air. *Lepape* was a physicist associated with the French institute of hydrology and climatology. He worked also on radioactivity. See also: Moureu, C. and A. Lepape (1914) Les gaz rares des grisous. H. Dunod et E. Pinat, Paris, 110 pp. The values published in 1926 were estimated before 1923: *Moureu* hold on June 14th, 1923 a lecture at the Chemical Society in London on these results (Moureu 1923).

0.002 ppm and concluded that turbulence quickly eliminates any nonuniformity resulting from localized He discharge. Glueckauf and Kitt (1956) measured the content (using a method of distillation and low-temperature gas-chromatography for the isolation of Kr and Xe) to be 1.39 ± 0.01 ppm (Kr) and 0.086 ± 0.001 (Xe). Argon was examined in air by *Henri Moissan* (1852–1907)[753] with an average (after Paneth 1937) of 0.9343 ± 0.0006 ppm (Moissan 1903). Interestingly that the content of argon decades later was determined much higher (1.16 ppm).[754] Today's accepted composition of air by rare gases is (Mackenzie and Mackenzie 1995); in ppm(v) – but these values are identical (only Xe was given as 0.00001) with those, quoted by Moureu (1923)!

Argon	0.9323
Neon	0.0018
Helium	0.0005
Krypton	0.0001
Xenon	0.000009

4.2.5 Air analysis

> The *azotic* atmosphere [...] supports the mercury in the barometer at a medium nearly 21.2 inches; [...] The *oxygeneous* atmosphere [...] 7.8 inches. The *aqueous* vapour is variable [...] of 0.6 and from that to one inches [...] The *carbonic acid* atmosphere has not perhaps been accurately ascertained in quantity; it is from everywhere in small proportion [...]; its pressure may probably amount to half an inch of mercury. (Dalton 1802, p. 547)

The precondition for any quantitative analysis is the knowledge of stoichiometry and the atomic mass of elements. First gas analysis (Chapter 4.2.5.2), at the end of the eighteenth century (and lasted until the middle of the nineteenth century), was based on volumetry, i. e., the gas (almost oxygen) undergoes a specific reaction or will be absorbed by a specific substance, and the change in volumes is measured, termed gasometry and absorptiometry (Bunsen 1857). Physical methods like spectroscopy, introduced by Kirchhoff and Bunsen (1860), were used systematically for analysis of atmospheric air constituent not before the 1950s (Thomas 1991); the only exception is ozone (Chapter 4.4.7) Forerunner of colorimetry, introduced after the work "*Einführung in die höhere Optik*" [introduction into higher optics] by *August Beer* (1825–1863) in 1853, who introduced the absorption coefficient, were at the beginning of the nineteenth century color reactions, giving a relative measure for the amount of a substance like the *Schönbein* paper for ozone (Chapter 4.4.5.2).

---

753 French chemist who won the 1906 Nobel Prize in Chemistry for his work in isolating fluorine from its compounds.

754 US National Bureau of Standards in 1968 (Taylor, J. K., ed. (1969) Microchemical Analysis chapter. Technical Note 505, Washington, 126 pp.)

After realizing the constancy of the main components of ordinary air, oxygen and nitrogen, air or gas analysis was focused on air in caverns, volcanoes, mines, basements, technical gases, etc., and some atmospheric constituents like carbon dioxide, ozone, and ammonia (Bunsen 1857, Hempel 1885, 1890a,b, 1913, Wanklyn and Cooper 1891). Air sampling was done simply by bottles and tubes (with small volumes between 50 and 500 mL) using barrier liquids, absorbing solutions, or even the empty device. The air was sucked in via a thin tube through breathing or (later) a manual pump (aspirator). Gases, soluble in water or specific absorbent solutions (CO_2, HCl, NH_3, HNO_3, SO_2) and transformed into ions (carbonate, chloride, ammonium, nitrate, sulfate) were analyzed by standard methods (titrimetry and gravimetry, having an error of >1 mg L^{-1}), developed already early in the nineteenth century for natural waters.

To my knowledge, the first extended chemical investigation of atmospheric air was conducted 1854 by *August Vogel* (1817–1889)[755] during the cholera epidemic August 10–20, 1854 in Munich, determining oxygen, nitrogen, carbon dioxide, and hydrocarbons in terms of carbon and hydrogen (Vogel 1854), Fig. 4.12.

Remarkable is the early attempt by *Hugo Reinsch* (1809–1884)[756] who first unsuccessful collected air using *Woulff*'s bottles (aspirator method) and afterward constructed "air absorbing" large linen roofs (26 square feet), mounted on 4 stakes, per-

Zur bequemern Uebersicht finden sich die an den 11 Versuchstagen gefundenen Zahlenwerthe in einer Tabelle zusammengestellt.

Datum	In 100 Volum. Luft				In 100 Gew.-Th.
	Stickstoff	Sauerstoff.	Kohlensäure	Wasserstoff	Kohlenstoff
10. August	79,8	20,2	0,0370	0,00221	0,00102
11. „	78,8	21,2	0,0315	0,00141	0,00116
12. „	79,1	20,9	0,0298	0,00130	0,00107
13. „	78,9	21,1	0,0301	0,00221	0,00099
14. „	79,5	20,5	0,0326	0,00069	0,00110
15. „	79,6	20,4	0,0404	0,00130	0,00102
16. „	79,3	20,7	0,0317	0,00166	0,00118
17. „	79,7	20,3	0,0321	0,00178	0,00121
18. „	78,0	22,0	0,0401	0,00082	0,00089
19. „	79,4	20,6	0,0382	0,00101	0,00100
20. „	79,6	20,4	0,0366	0,00097	0,00096

Vergleicht man mit diesen Zahlen die von Boussingault aufgefundenen Resultate, welcher in 100 Vol. der Luft von Paris von 0,005 bis 0,013 Wasserstoff und 0,022 Kohlenstoff in 100 Gew.-Th. nachgewiesen, so ergibt sich nach den hier angestellten Versuchen der Wasserstoffgehalt und Kohlenstoffgehalt bedeutend geringer, als in dem Boussingault'schen Versuche.

Figure 4.12: Results of the air analysis by Vogel (1854) in Munich.

755 Professor for agricultural chemistry at the University Munich; scholar of *Liebig*. Son of *Heinrich August von Vogel* (1778–1869), professor for chemistry in Munich.

756 *Edgar Hugo Emil Reinsch* (but in literature quoted only as *H. Reinsch*) was a German pharmacist and later technical chemist, teacher and Rector at Zweibrücken and Erlangen; wrote the book "*Grundriss der Chemie für technische Lehranstalten*" [fundamentals of chemistry for technical schools], Bassermann & Malty, Mannheim, 1854, 316 pp.

manent wetted and sprinkled by aqueous solutions over 12 hours per day for a 14-day experiment. A first absorber was sprinkled with diluted hydrochloric acid, a second with sodium hydroxide solution, and a third with distilled water. The linen was impregnated with gypsum (Reinsch 1865); later he suggested installing the linen vertically instead of horizontally.[757] In the solution from the first (acid) absorber roof, he found Al, Na, K, Ca, Mg, H_2SO_4, Pb, Zn, Cu, Fe, Mn, and SiO_2. In the solution from the second (alkaline) absorber roof, he found H_2CO_3, H_2SO_4, H_3PO_4, HCl, FeO, MnO, and SiO_2, and from the third roof mainly ammonium sulfate. A few years later (Reinsch 1869), he collected large amounts of air over four weeks using four bubblers inline (water + ammonia, water, water + HCl, and water) and concluded on the permanent presence of H_2SO_4, H_3PO_4, HNO_3, and HCl in the air.

The most extensive early and complete history of investigation of the composition of atmospheric air is written by *Emil von Wolff*, entitled (Wolff 1847) "*Zusammensetzung der atmosphärischen Luft*" [composition of atmospheric air] (pp. 3–64):

I *Wesentliche Bestandtheile* [essential constituents]
  A. *Stickstoff and Sauerstoff* [nitrogen and oxygen] (pp. 3–22)
  B. *Kohlensäure* [carbonic acid] (pp. 22–32)
II *Bestandtheile, deren Gegenwart zufällig zu sein scheint und die nur in geringer Menge auftreten* [constituents, which presence seems to be accidental, and which occur only in very small amount]
  A. *Kohlenwasserstoff, Schwefelwasserstoff, Pyrrhin u. s. w.* [hydrocarbon, hydrogen sulfide, pyrrhine etc.] (pp. 32–36)
  B. *Miasmen oder organisch fein zertheilte Körperchen in der atmosphärischen Luft* [miasma or subtly dispersed organic bodies] (pp. 36–41)
III *Atmosphärisches Wasser* [atmospheric water]
  A. *Gewöhnliche Bestandtheile* [common constituents] (pp. 41–46)
  B. *Ungewöhnliche Beimengungen* [uncommon admixtures] (pp. 47–51)
IV *Zusammensetzung der Luft, die man in dem Wasser aufgelöst findet* [composition of Air, dissolved in water] (pp. 51–61)
Rückblick [review] (pp. 61–64)

Dumas and Boussingault (1841) write (translation from German):

> All previous investigations show that atmospheric air contains under the different circumstances about 21 % oxygen and 79 % nitrogen, and thus carbonic acid and water vapor deducted, has a constant composition (Dumas and Boussingault 1841, p. 405).
> The vegetation, this is true, feeds carbonic acid, and exhalates oxygen for that the animals it convert again into carbonic acid (Dumas and Boussingault 1841, p. 407).

757 To my knowledge, no further experiments were carried out to collect atmospheric trace species; however, this principle (using large pieces of vertical canvas) was used in the twentieth century for fog water collection (Schemenauer and Cereceda 1994).

Hundred years later, *Eugen Glueckauf* (1906–1981),[758] German-born British scientist and expert for nuclear power and micro-gas analysis of atmospheric gases wrote: "The composition of the atmosphere" (Glueckauf 1951). The German chemist *Albert Ladenburg* (1842–1911)[759] included in his "*Handwörterbuch der Chemie*" [concise dictionary of chemistry] a chapter on the atmosphere (pp. 63–103) that includes a detailed description of determination methods (pp. 71–94) of oxygen, ozone, hydrogen peroxide, carbonic acid, water vapor, carbon monoxide, hydrocarbons, hydrogen sulfide, sulfurous acid, ammonia, nitrous and nitric acid, common salt, boric acid, ammonium chloride, iodine, organic compounds, and atmospheric dust; this shows, how important at that time "atmospheric chemistry" or "chemistry of the atmosphere" (in terms of chemical composition) were seen for general chemists.

4.2.5.1 Determination the goodness of air: eudiometry

For centuries medical doctors speculated that contaminated air ("bad" air) brought diseases. Even in the nineteenth century one of the proponents of the great Victorian sanitary works, *Edwin Chadwick* (1800–1890), wrote: "all smell is disease" (Finer 1952). For chemists of the nineteenth century and earlier, there were three "sensors" in the determination of chemical substances: smell, look, and taste.[760] From historic reports, we know that it stunk often in cities before the installation of sewerage. Moreover, factories were very near to residential areas and waste of all kind were exhausted into air or passed into rivers. Thus, the smell was omnipresent.

The percentage of nitrogen and oxygen in the air was only roughly estimated, and there was the belief that the concentration of oxygen varies and is lower in polluted air and higher in remote air. Even carbon dioxide was regarded in a few percent range and first removed (by alkalis) in all analyses of atmospheric air. The accuracy of the first air analysis was very low, resulting in oxygen values between 18 % and 27 %. The remaining gas was attributed to nitrogen and noxious gases.[761]

---

758 Name before 1947 *Glückauf* written. He graduated in Berlin and left Germany for England where he became an assistant of professor *Paneth*; *Glueckauf* contributed in the fields of atomic physics, radio chemistry and electrolyte solution chemistry.

759 *Ladenburg* studied chemistry in Heidelberg at *Bunsen* and *Kirchoff*, in Bonn at *Kekulé*, and in Paris at *Wurtz* and *Friedel*; habilitation 1868 in Heidelberg, professor 1874 in Kiel and 1889–1909 in Breslau.

760 This was still valid for (very experimentally working) chemists in the twentieth century. Many compounds have a characteristic smell, color or crystalline structure (Fig. 2.16) that helps for their identification.

761 Today we know that the oxygen concentration only in flue and process gas can be significantly lower (typically 5 % in flue gas).

In 1771, *Joseph Priestley* made a remarkable observation:[762] when nitrous air[763] mixed with ordinary air in the presence of water, there was a startling one-fifth contraction in the volume. *Priestley* believed he had found a way to measure the "goodness" of the air by keeping careful track of the volumes of gas he was mixing (Priestley 1799). For these experiments he used his own redesigned pneumatic trough (Fig. 4.7) in which mercury instead of water would trap gases that were usually soluble in water – this was the prototype of a eudiometer. The Italian physicist Abbé *Felice Fontana* (1730–1805), who was intrigued by *Priestley*'s idea to measure the "goodness" of air, constructed 1774 in Pisa several instruments to measure "*salubrità dell' aria*" [health of air].[764] Soon later, *Marsilio Landriano* (1751–1815), an Italian chemist, physicist, and meteorologist, believed that measuring the goodness of air might explain the origin of disease; he repeated *Priestley*'s experiments and designed a compact system[765] and termed it eudiometer[766] (Fig. 4.13); "*Richerche sisiche intorno alla salubrità dell' aria*" (Milano 1775).[767] After making several measurements in Italy, he sends his eudiometer as a gift to *Priestley* (Busch 1807, p. 264)

A eudiometer is a graduated glass tube (burette) used in the study and volumetric analysis of gas reactions (Fig. 4.13). *Landriano's* and *Fontana*'s "eudiometer of nitrous gas" is based on the gas reaction observed by *Priestley*, giving largely uncertain values because of the uncontrolled stoichiometry of a) the formation "nitrous gases" and b) the reaction of nitrous gases with oxygen. The procedure is in short, that nitrous gas, produced from copper wires and nitric acid, is mixed in a well-known volume with common air resulting in a gas volume that is smaller than the original volumes of nitrous air (NO_x) and common air because of the reaction between NO_x and oxygen. The volume difference (measured in millimeters of 100 or ratios) is a measure for oxygen and hence the "goodness" of air.

762 I am following here the article "Eudiometer" by Busch (1807, pp. 262–278) concerns the discovery of different eudiometers.

763 It remains unclear, what "nitrous air" denotes (Table 1.3): NO, NO_2 and/or HNO_3 (and HNO_2 because $NO + NO_2 = N_2O_3$ which reacts with water to HNO_2). *Ingenhousz* (and also described by *Scherer*) denotes it "*Salpersäureluft*". In that time, it was not exact clear which chemical compound is behind the terms (Table 2.2). Nitrous oxide (today N_2O) was almost NO [*Salpetergas*] and nitrous acid (today HNO_2) NO_2 [*salpetrige Säure*].

764 An Italian physicist who discovered the water gas shift reaction in 1780. This instrument (later also termed eudiometer) is described in Ingenhousz (1779). Fontana, F. (1775) *Descrizione, e usi di alcuni strumenti per misurare la salubrità dell'aria*. Firenze, per Gaetano Cambiagi Stampatore Granducale, 42 pp.

765 Andrea Sella (University College London, Britain): https://www.chemistryworld.com/opinion/landrianis-eudiometer/8200.article

766 The name comes from the Greek εὐδία meaning clear sky. An excellent and extensive history on eudiometry is given by Grapi (2019). See also Busch (1807, pp. 262–278) on eudiometer.

767 German translation: Landriani, M. (1778) Untersuchung der Gesundheit der Luft. K. A. Serini, Basel, 119 pp.

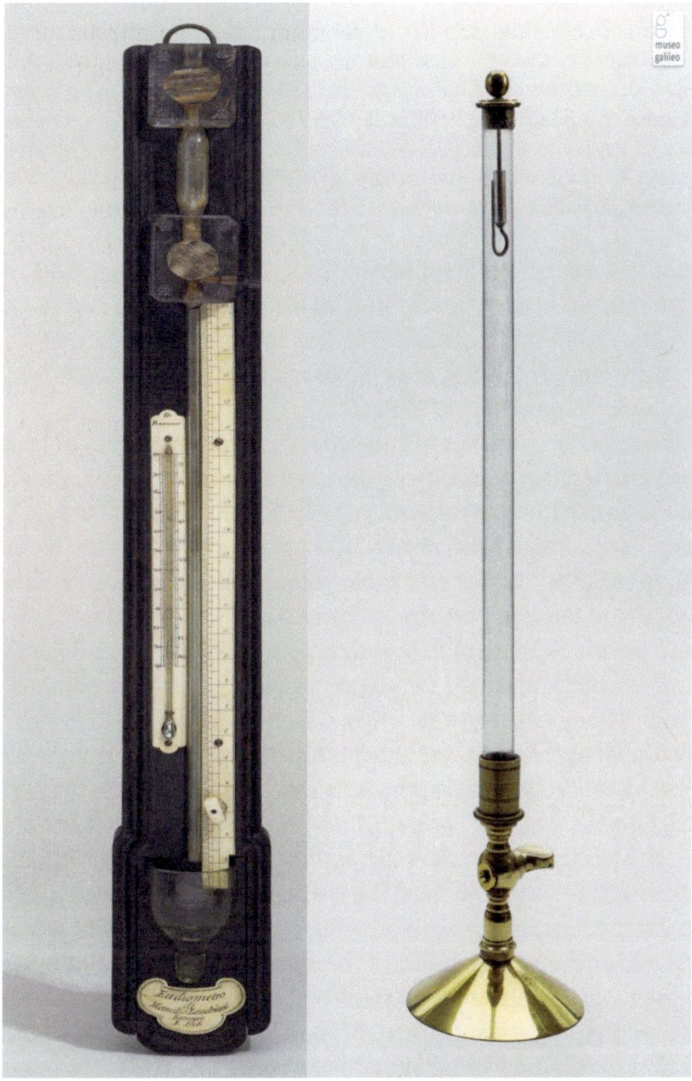

Figure 4.13: Eudiometer. Left the glass "*Eudiometro di Landriani*" (Inv. 1371, 1776, dimension 700 × 75 mm) and right "*Eudiometro di Volta per la detonazione dei gas*" (Inv. 1627, ca. 1790, dimension 490 × 20 *mm*), Museo Galileo, Florence https://catalogo.museogalileo.it/multimedia/Eudiometria.html

Landriani's friend *Alessandro Volta* (1745–1827) went one step further (Osman 1958), based on *Cavendish*'s experiments (Cavendish 1784). He equipped a eudiometer with spark wires to study the hydrogen-oxygen reaction; hydrogen was produced from zinc and hydrochloric acid. The accuracy of this device was already 0.1 % in oxygen. It is worth citing (Thomson 1807, p. 62) the process of examination:

> When 100 measures of hydrogen are mixed with 200, or any greater bulk of oxygen, the diminution of the bulk after detonation is always 146 measures [...] Hence the method of using this eudiometer is very simple: mix together equal bulks of the air to be examinated, and of hydrogen gas, ascertain the diminution of bulk after combustion, divide it by three, the quotient represents the number of measures of oxygen in the air [...] 200 measures of air and much of hydrogen amounting in 126 measures of bulk diminution. Hence 126/3 = 42, the quantity of oxygen in 200 measures, hence 100 parts of air contain 21 of oxygen.

Volta constructed several fore-runner (inflammable-air gun); his work beginned in 1775 and finished with continuous improvements in 1790 (Osman 1958). He had close contact with *Senebier*[768] who used this instrument for his own research. *Cavendish*, who as cited in the literature (Busch 1807, p. 264) improved *Landriani*'s eudiometer, using mainly a spark eudiometer according to *Volta*.[769]

Scheele developed in 1781 a "eudiometer of sulphurets", where a mixture of iron filings and sulfur, formed into a paste with water were used to absorb oxygen, which was significantly improved around 1800 by *Antonio de Marti y Franquès* (1750–1832), a Catalonian chemist and naturalist, who measured the oxygen content in the air to be 21–23 % (Marti 1805). *Marti* used bottles and tubes, filled with a solution of lime sulfur[770] (old term: sulphuret of lime; in German: "*Schwefelkalk*"), measuring the absorption of air and found mostly a figure of 0.21 with an error of 0.01; he also made measurements at "polluted" sites and stated that these gases, emanated by combustion and respiration cannot change the oxygen content of air by 1 %. *Marti* also comments that there is no prospect by using an eudiometer the detect the oxygen content with higher accuracy. There is a need for new methods to determine changes in the permille range. The Scottish chemist *Thomas Charles Hope* (1766–1844) changed *Marti*'s method around 1800 only by using two connected bottles (Hope 1805). The British chemist *Sir Humphrey Davy* (1778–1829) criticized the untrustworthiness of *Fontana*'s eudiometer and developed the "nitrous oxide eudiometer" using solutions of green hydrochloric iron ($FeCl_2$) and sulfuric iron ($FeSO_4$)[771] (Davy 1801, 1805); he confirmed the constancy of the oxygen values measured by *Marti* and concluded that health effects are not caused by changing oxygen ratios but through foreign substances in the air. The German pastor and naturalist *Johann Friedrich Salomon Luz* (1744–1827) improved *Fontana*'s eudiometer, namely for using it on field sites (Fig. 4.14).

---

768 *Jean Senebier* (1742–1809) was a Genevan Calvinist pastor and naturalist.

769 There is evidence that *Cavendish*'s eudiometer shown in the Chemistry Department of the Manchester University (and referred by historian like *Partington* and *Holmyard*) was made not before the death of *Cavendish* (Farrar 1963). This eudiometer was used as its emblem by the Cavendish Society in the early nineteenth century.

770 Mixture of calcium polysulfides formed by reacting calcium hydroxide with sulfur; contains traces of calcium thiosulfate.

771 In German: Salpetergas-Eudiometer.

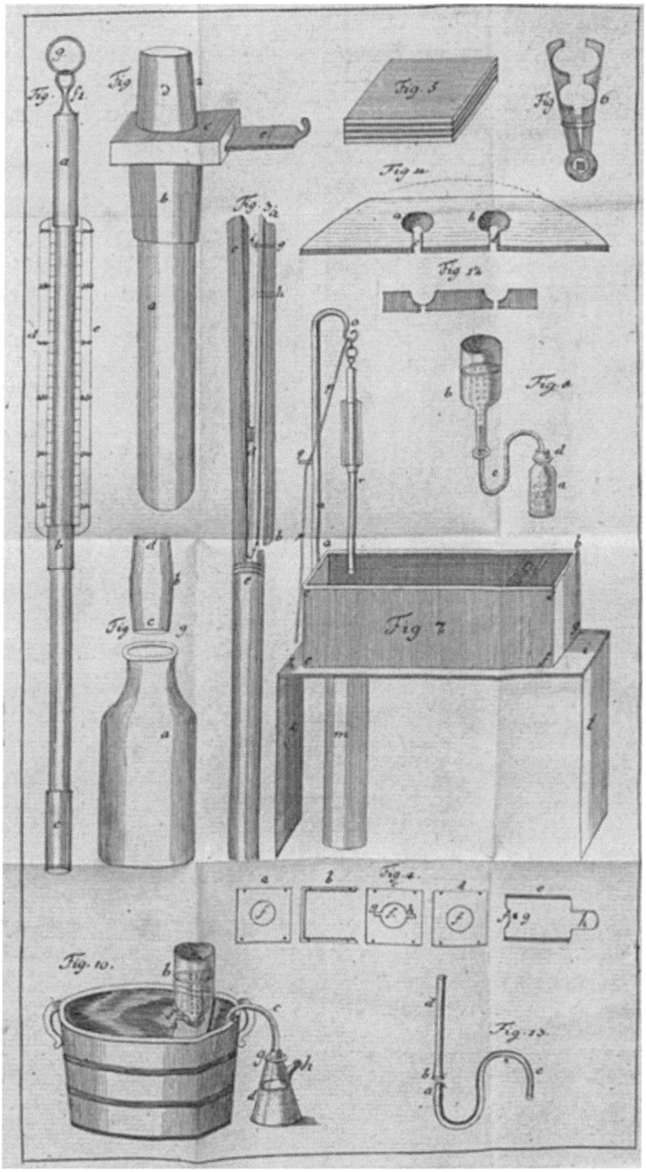

Figure 4.14: *Fontana*'s eudiometer improved by Luz (1807, p. 86).

A "eudiometer of phosphorus" was first proposed by *Franz Carl Achard* (1753–1821) in 1784 and soon later considerably improved by *Henri Paul Irénée Reboul* (1763–1839). *Reboul* carried out 800 examinations and found oxygen to be 18–20 % in air.[772] The Italian chemist *Giovanni Antonio Giobert* (1761–1834), in Riga the German scientist

772 *Annales de Chimie* (1793) **13**, 633.

Table 4.4: Invention of eudiometers.

eudiometer type	inventor	year
nitrous gas (reaction with oxygen)	Fontana	1774
nitrous gas (reaction with oxygen)[a]	Landriano	1775
electric spark (reaction of oxygen with hydrogen)	Volta	1776
sulphurets: alkaline sulfides (absorption of oxygen)[b]	Scheele	1781
phosphorus (reaction with oxygen)[c]	Achard	1784
nitrous oxide in sulphureous iron solution (absorption of oxygen)	Davy	1801
platinum sponge and hydrogen (reaction of oxygen with hydrogen)	Döbereiner	1817

[a]Further improved by *Cavallo* and *Magellan* in 1777
[b]Improved by *de Marti* and *Hope* around 1800
[c]Improved by *Reboul, Parrot, Giobert* and namely *Berthollet*

Georg Friedrich Parrot (1767–1852), and *Berthollet* further changed and used *Achard*'s eudiometer (Table 4.4).

Afterwards, many persons improved and constructed different eudiometers using different "reactions". In the preface to "*Recherches chimique sur la Végétation*", *Nicolas Théodore de Saussure* (1767–1845)[773] writes (cited after Hill 2013, p. ix):

> For my eudiometric tests I used either sulfide of potassium or phosphorus. This procedure allows me to achieve in my analysis an accuracy that the nitrous gas eudiometer used by my predecessors could not attain.
> In this work, when I have given readings of the phosphorus eudiometer, they have always been cleared of the error that nitrogen gas can introduce through the expansion it undergoes in dissolving the phosphorus [...] In the first moments of the disappearance of the oxygen gas, the expansion of the nitrogen gas is undetectable...
> For these experiments I used rapid combustion and the bent tube indicated by Giobert (*Analyse des Eaux de Vaudier*). I tilt the eudiometer when the phosphorus is melting, so that the phosphorus flows and spreads over the full length of the tube. With this procedure, the analysis of air is completed in less than half an hour, and at this time needs no correction.

Johann Friedrich Salomon Luz constructed and built numerous instruments for use in experimental physics; he describes on 47 pages (Luz 1807) in detail the construction of a eudiometer and writes (Luz 1807, p. 12):

> Das Herzstück bey einem Eudiometer ist eine Glasröhre, die ohngefähr 18 bis 20 pariser Zoll lang, ½ Zoll weit, von gleicher Weite und innen matt geschliffen ist [The centerpiece of the eudiometer is a glas tube, about 18 to 20 Parisian inches long, ½ inch wide, of the same width and inside tarnished].[774]

773 Swiss chemist and botanist, sun of Horace-Bénédict de Saussure (1740–1799) in Geneve.
774 This description is exactly that of a *denuder*, used 200 years later first by Ferm (1979) for absorption of ammonia and separation them from ammonium in particulate matter; later also used for HCl, HNO_2 and HNO_3.

Soon later other researchers (*Ingenhousz*, *Lavoisier*, *Magellan*,[775] *Saussure* (*Jr.*), *Scheele*, *Senebier*, *Cavallo*,[776] and others[777]) used such eudiometers, which were further improved, to estimate the "goodness of air" in many European towns, mountains, and on the sea as described in 1785 by *Johann Baptist Andreas Ritter von Scherer* (1755–1844) in his book "*Geschichte der Luftgüteprüfungslehre für Aerzte und Naturfreunde*" [History of air quality testing teaching for doctors and nature lover].[778] Scherer (1785, p. 27), in comparing different eudiometers, writes very clear the sources of errors to be

> imperfection of the air meter (eudiometer),
> the lack of experience with the device and
> random causes.

Furthermore (Scherer 1785, pp. 101–102) he describes results, which are correct but cannot be derived from the use of eudiometers, because a) the differences of the oxygen values measured did not reflect the reality and b) even in case of correct oxygen estimation the "bad" air was on concentration orders of magnitude smaller than the oxygen content:

> sea air is cleaner,
> country air is cleaner than city air,
> air in winter at cold weather is most clean,
> in summer, when a lot of rain is, the air is not so clean,
> marsh and morass areas are in winter not so harmful like in summer.

These conclusions are obviously based on smell and observation that higher temperature results in more putrefaction and rain wash air. On the other hand, Scherer (1785, p. 98) writes curiously that

> Die Luft in Oesterreich, Baiern and Schwaben war weniger dick als in der Schweiz [. . .] die Luft wäre in Italien noch dicker [The air in Austria, Bavaria and in Swabia was less fat than in Switzerland [. . .] air in Italy would be even thicker].

775 *Jean Hyacinthe de Magellan* (1723–1790) was a Portuguese natural philosopher.

776 *Tiberius Cavallo* (1749–1809) was an Italian physician and natural scientist birth but lived in London.

777 Beside the persons named here, Gmelin (1799, pp. 330–333) cites the following names: *J. H. v. Mahelhaen* 1777/1780, *le Roux des Tillet* 1777, *Servière* 1777, *J. G. Stegmann* 1778, *Gerardin* 1778, *Gatton* 1779, *Pickel* 1781, *Kratzenstein* 1781, *Carnus* 1782, *Cr. Viborg* 1783, *Wilcke* 1783, *Marwan* 1786, *Gruber* 1787, *Chaptal* 1787, *Gr. V. Sternberg* 1787, *Jurin* and *Gattoni* 1785, *Guyton* 1788, and (no year but name given) *Gren*, *Marq. De Bréze*, *von Breda*, *Späth*. He refers to Gmelin, J. F. (1794) *Progr. de aëris vitiosi exploratione* (Göttingen), where a detailed description is given. Translated in "Scherf's Beiträge zum Archiv", Vol. VIII (1789); *Beiträge zum Archiv der medicinischen Polizei und Volksarzneikunde*, Leipzig, ed. *Johann Christian Scherf* (1750–1818), a German physician.

778 Wien 1885 in two Volumes (214 and 218 pp.).

Scherer (Scherer 1785, p. 73) recommended a daily examination of "*die Reinheit des Luftmeeres*" [the goodness of the ocean of air] and to log "eudiometric tables" together with the values of temperature and pressure.[779]

4.2.5.2 Analysis of oxygen

Scheele determined the oxygen content in the air of Stockholm in 1779 to be between 9/33 and 10/33 (27–30 %) and *Lavoisier* found in Paris ¼ (25 %) in 1780. First Cavendish (1783) analyzed 1781 the composition of air using a "eudiometer of nitrous gas" for the first time, very close to the correct values (Table 4.5). *John Dalton* who determinates the composition of atmospheric air in 1802, gave the following composition in % (Dalton 1805, p. 257):

N_2	75.55
O_2	23.32
H_2O	1.03
CO_2	0.1

Gay-Lussac collaborated with *Humboldt* in using different eudiometers to examine the air; with the *Volta* eudiometer, they confirmed the values found by *Cavendish*, Table 4.5 (Gay-Lusssac 1805, Humboldt and Gay-Lussac 1805). On September 16, 1804, *Gay-Lussac* went up alone with a hydrogen-filled balloon up to 7016 m above Paris (Gay-Lussac 1804, 1805a,b). Besides measurements of magnetism, pressure, temperature, and humidity, he collected two air samples at altitudes of 6561 and 6636 m. He writes (Gay-Lussac 1804):

> When I arrived at Paris, my first care was to analyse the air I had brought back. All the experiments were made at the polytechnic School, and the inspection of Messr. Thénard and Gresset and found it precisely the same as the air at the earth's surface to be 0.2149 of oxygen.[780]

Humboldt and Gay-Lussac (1805, p. 88) refer (without giving a citation) *Armand Jean François Séguin* (1767–1835) who measured the oxygen content in halls of hospitals in Paris – despite its unbearable smell – to be the same like in the free atmosphere. *Séguin* used *Volta*'s electric eudiometer which he demonstrated in Paris in 1786. Humboldt and Gay-Lussac (1805, p. 92) write:

> Die atmosphärische Luft enthält dem Volumen nach nur 0.21 Sauerstoffgas, und variirt in ihrer Zusammensetzung nicht [The atmospheric air contains in volume only 0.21 oxygen, and does not vary in its composition].

779 This is surely the first idea of registration a "chemical weather". A curious instrument called "Chemical Weather Glass" describe Negretti and Zambra (1864, p. 137) referring *FitzRoy* who used it since 1825 but called it "camphor glass" (see FitzRoy 1863, pp. 439–441).

780 At École Polytechnique around 1800 were *Jean-Ch.-Alex Gresset* and *Alex-Jos.-Marie Gresset*.

Table 4.5: Historic data on air composition (in Vol-%).

Substance	Cavendish (1783)[e]	Gay-Lussac and Humboldt (1802–1805)[f]	Benedict (1912)[h]	Krogh (1919)[b]	Cadle and Johnston (1952)	present[d] (2018)
N_2	79.16[a]	79.2	79.016[a]	79.022[a]	79.02	79.014[a, c]
O_2	20.84	20.8	20.954	20.948	20.95	20.95[g]
CO_2	–	–	0.030[b]	0.030	0.030	0.0410

[a]Within these figures' novel gases is (Ar at present 0.934).
[b]*Benedict* himself gives the average of CO_2 as 0.031 %; corrected to 0.30 by *Krogh* due to 0.002 % CO formation while oxygen is absorbed by potassium pyrogallate.
[c]Ar amounts 0.9340 %.
[d]Earth Fact Sheet reference (https://nssdc.gsfc.nasa.gov/planetary/factsheet/earthfact.html).
[e]After Humboldt (1799) and Ramsay (1896, p. 125), *Cavendish* carried out the analysis 1781 in London and Kensington.
[f]Both scientists collected air samples from the different places, including balloon ascends and mountain tops; this average is taken from Humboldt (1850–52, p. 311).
[g]20.9392 (Tohjima et al. 2005), continuously decreasing.
[h]At the Nutrition Laboratory of the Carnegie Institution of Washington.

Humboldt and *Gay-Lussac* estimated the error of many measurements using the eudiometer by *Volta* to be only 0.001 to 0.003. Marti (1805), who used *Scheele*'s eudiometer,[781] also found the oxygen content to be not varying but a bit less accurate (error 0.01) between 0.21 and 0.23. Because *Gay-Lussac*'s summary and conclusions concerns oxygen measurements in the early nineteenth century are so clear, I will cite them here (Gay-Lussac 1805c):

> Saussure Jr., found[782] also that air collected on the Col-du-Geant contained, within a hundredth of part, as much oxygen as that of the plain, and his father confirmed the presence of carbonic acid on the summit of Mont Blanc. Besides, the experiments of Messrs. Cavendish, Macartney, Berthollet, and Davy, have confirmed the identity of the composition of the atmosphere over all surfaces of the earth to greatest heights to which it is possible to attain [...] I think, also, I have proved that the prosperous of oxygen and azote, which constitute the atmosphere, do not sensible vary in very extensive limits. There still remain a great many things to be cleared up in regard to the atmosphere.

Robert Bunsen (1811–1899)[783] showed in 1846 in Marburg that the oxygen content in air varies slightly between 20.840 and 20.970 % with an average of 20.924 % (measurement error was 0.03 %), Bunsen (1857).

781 Using bottles, filled with 1.6 to 6 ounces liquid sulfur lime and 0.25 to 1.25 ounce measures atmospheric air, standing without shaking.

782 With an "nitrous gas eudiometer" (Saussure 1804). Likely *George Lord Macartney* (1737–1806) who was ambassador in China.

783 German chemist, professor in Marburg, Kassel, Breslau and Heidelberg; he investigated emission spectra of heated elements, and with *Kirchhoff* he discovered the elements Cs and Ru; devel-

In the early nineteenth century, some people believed that epidemics like cholera are connected with changing air composition. *Julia de Fontenelle* was the first who shows from 50 analysis at 20 different places in Paris in April 1832 as all researchers before, a constant composition of 21 % oxygen and 79 % nitrogen (Fontenelle 1832), and later also by Baumgartner (1832)[784] in Vienna, Laskowsky (1850)[785] in Moscow, and Vogel (1854) in Munich.

Many well-known chemists analyzed the oxygen content of air (the difference was attributed to nitrogen) in the early 1800s like *Dumas* and *Baussingault* at Paris, *Marignac* at Geneve, *Brunner* at Bern, *Verver* at Groningen, and *Bernhard Carl Lewy* (1817–1863)[786] at different sites between around 20.5 % and 21.6 % based on the method[787] by *Dumas* and *Baussingault* (Lewy 1843).[788] Dumas and Boussingault (1841) write that famous chemists like *Prout*, *Döbereiner*, *Falkner*, and *Thomson*[789] regard oxygen and nitrogen in a fixed constant ratio in air (between 20:80 and 21:79). In contrast, *Dalton* considered air as a variable mixture of oxygen and nitrogen, richer in oxygen in inhabited regions and less in oxygen at higher altitudes. Brunner (1853) in Bern improved his method (developed in 1833) and found a mean of 20.93 % ($n = 7$; 20.83–21.07 %).

Henri Victor Regnault (1810–1878)[790] began in December 1847, upon the air of Paris, and December 1848 with the most careful oxygen measurement so far, having an error of only ±0.01 % absolute (Table 4.6). This measurement series was combined with the first "international measurement campaign", using uniform sampling tubes

oped several gas-analytical methods and was pioneering in photochemistry. *Gustav Robert Kirchhoff* (1824–1887) was a German physicist in Berlin, Breslau and Heidelberg, who contributed to the fundamental understanding of electrical circuits, spectroscopy, and the emission of black-body radiation by heated objects.

784 *Andreas Freiherr von Baumgartner* (1793–1865) was an Austrian Naturalist and Statesman.

785 *Nikolai Erastowitsch Laskowski* [Николай Эрастович Лясковский] (1816–1871) was a Russian chemist who studied at *Liebig* 1844 in Gießen, and also in Berlin and Paris. Since 1855 Professor in Moscow.

786 Also written as *Bergnart* and *Karl* (but *Levy* is a wrong spelling); a Danish chemist from Copenhagen who gained PhD in Berlin 1839, and became assistant of *Dumas* and *Boussingault* at Paris (1841) and 1847 professor for chemistry at Bogóta (New Granada).

787 It is based on weighting (i. e., values are cited originally as mass-% – about 23 % O_2): air is sucked through a red-heated tube filled with copper (oxygen is absorbed and the difference of the weight before and after absorption is estimated) into an evacuated bottle, where pure nitrogen is sampled and weighted.

788 In nearly all cases, *Lewy* seems to have found that an increased amount of carbonic acid in the air was accompanied by an increase in the percentage of oxygen (cited after Frankland 1861).

789 *Johann Wolfgang Döbereiner* (1780–1849) was a German chemist; *Johann Ludwig Falkner* (1787–1832) was a Swiss chemist. *Prout* see footnote 517 on p. 254 and *Thomson* footnote 375 on p. 174.

790 A French chemist and physicist, best known for his careful measurements of the thermal properties of gases.

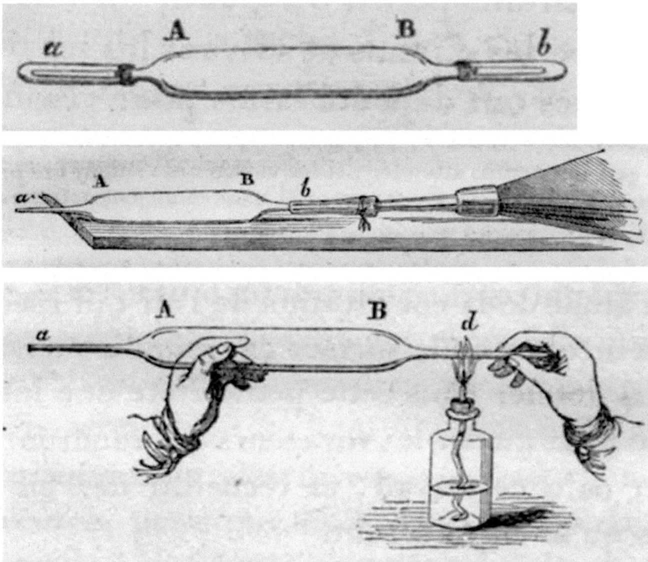

Figure 4.15: Determination of oxygen in atmospheric air: sampling tube and procedure (after Regnault (1852).

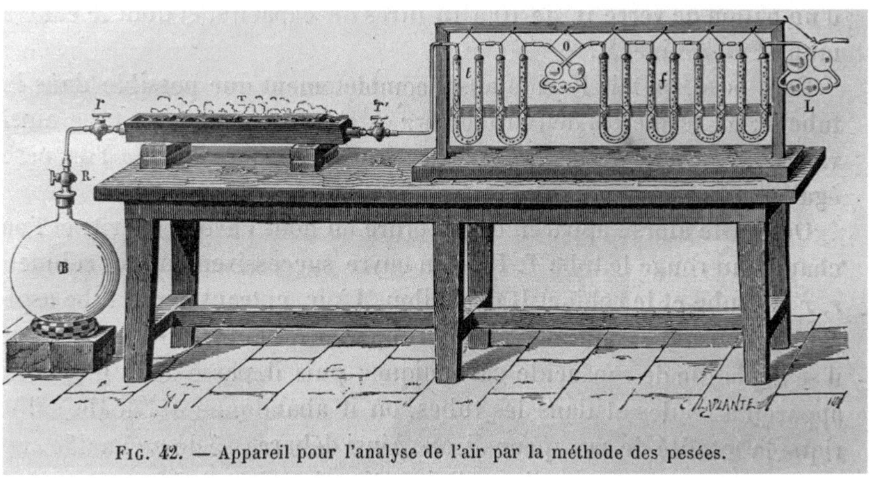

Figure 4.16: Apparatus for analyzing air by the method of weight (after Flammarion 1888, p. 85).

(Fig. 4.15) and protocols at 10 sites and onboard of three frigates, totally 125 analyses (and additional 113 in Paris, Table 4.6); all analyses were carried out in Paris. Fig. 4.16 shows the laboratory equipment for air analysis.[791]

791 This method, originally developed by *Dumas* and *Boussingault*, is much more accurate than methods measuring volumes. The apparatus used is composed of a tube which brings in the air from outside

Table 4.6: The first international measurement campaign for the determination of oxygen in the air (Regnault 1852).

site	*n*	period	oxygen in %	operator
Montpellier	7	1848	20.929–20.968	Marié-Davy
Lyon	3	1848	20.918–20.966	James de Bellecroix[a]
Normandie	1	1848	20.952	Izarn[b]
Berlin	32	1848–1849	20.961 (20.903–20.998)	Gustav Magnus[c]
Madrid	10	1848	20.956 (20.916–20.982)	Zarco de Valle[d]
Genève	15	1848–1849	20.903–20.993	Emilie Plantamour[e]
Mt. Salève	3	1848	20.928–20.963	Rochette[f]
Mt. Buet	2	1848	20.930, 20.981	Soret
Chamonix	1	1848	20.962	Soret
roads of Toulon	13	1851	20.943 (20.854–20.982)[l]	d' Elissalde[g]
Atlantic	5	1848	20.920–20.965	Castagnet[f]
Ecuador	2	1848/1849	20.960, 20.988	Ancelle[h]
Africa to India	10	1848–1850	20.387–20.936	Clèrin[k]
Ecuador (Luna)	1	1852	21.015	Fourrichon[g]
South Atlantic	3	1852	20.950–20.963	Fourrichong
Arctic Ocean	17	1848–1849	20.85–20.94	Franklin[g]

[a]Professor of chemistry in Lyon at l'École La Martinière (unknown living dates)
[b]Likely a descendant of *Joseph Izarn* (1766–1848), French physician and professor of physics in Paris
[c](1802–1870), notable chemist and physicist in Berlin
[d]*Manuel Remón Zarco de Valle* (1785–1866) was a Spain general and president of the Royal Academy of Sciences in Madrid
[e](1815–1882), Swiss astronomer in Geneve
[f]Unknown
[g]Captain of the frigate
[h]*Sébastien Wisse Ancelle* (1810–1863) was a French engineer and explorer (acquainted with *Humboldt*)
[k]Commandant la station des Indes
[l]Three samples additional in the port of Algiers (only 20.4 %) and near Minorca

of the room where the operation is proceeding; secondly of a set of *Liebig* balls (L) containing a concentrated solution of caustic potash; thirdly, of a tube (f) in the shape of the letter U several times repeated, and filled with fragments of caustic potash; fourthly, of a second set of balls (O), containing concentrated sulphuric acid: fifthly, of second set of U tubes (l), filled with pumice-stone steeped in concentrated sulphuric acid; sixthly, of a straight tube (T) of hard glass. This tube is filled with copper fillings, and laid upon a long iron furnace, so that it can be heated throughout its whole length, and is moreover furnished at its extremities with two taps (r and r'), which admit of its being emptied: seventhly, of a glass globe (B), holding from two to three gallons, and the neck of which is fitted with a tap (R). A complete vacuum is made in Tube T and glass ball B, thus emptied of air, and weighted. The air, entering from right, and passing the subsections, is removed from water vapor and carbonic acid, and losses the oxygen in the red-hot copper tube T, and then passes into the empty glass ball as pure nitrogen. The increase of weight gives the weight of oxygen and nitrogen. Text, shortened, from Flammarion (1874, pp. 59–61).

A large number of analyses by *Lewy* demonstrate that the air near the surface of the sea contains about the same proportion of carbonic acid as that resting upon the land and that the sea air is richer both in oxygen and carbonic acid by day than by night; a fact which he explains by assuming that dissolved air is liberated during the day from the heated surface-layer of the ocean, such dissolved air being, as is well-known, much richer than atmospheric air in the two gases just named (cited after Frankland 1861, p. 23). The British chemist Sir *Edward Frankland* (1825–1899) describes the "correlation between atmospheric oxygen and carbonic acid, for when the one increases the other diminishes" (Frankland 1861, p. 30).

Measurements in the second half of the nineteenth century became more accurate with values between 20.91 % and 20.96 % (cited by Krogh 1919).[792] Smith (1872) carried out in the 1860s many air analyses in England and found on average 20.94 % and no differences between remote, polluted, laboratory, and mountain air. He writes (Smith 1872, p. 33):

> We are exposed to currents of good air in the worst, and of bad air in best atmosphere in towns like Manchester.

Hempel (1885, p. 268) states that the natural variation of atmospheric oxygen seems to be in the order of 0.5 % based on his own analysis and that of *Jolly*, *Morley*, and *Vogler*,[793] but the causes were unknown. *Johann Philipp Gustav von Jolly* (1809–1884), a German physicist, believed in 1879 that oxygen varies naturally by 0.49 % (Jolly 1879). Today we know that the natural variation[794] is less than 0.01 %, a figure that was far away from the accuracy of oxygen measurements in the nineteenth century; from the mean figures listed in Table 4.7, it follows a mean of 20.94 ± 0.03 %. The German agricultural chemist, a founder of forest soils science in Munich, *Ernst Ebermayer*

---

792 The number of publications with the title "on the composition of air" (in English, French and German) is vast. Here are only citations from German journals (year, volume, pages): *Archiv der Pharmacie* **4** (1823) 215–229, **52** (1835) 132, **126** (1853) 149–156, **180** (1867) 107–108; *Journal für Praktische Chemie* **23** (1841) 237–243, 24, 66–91, **26** (1842), 294–297, 297–298, **27** (1842), 215–227, **30** (1843), 207–227, **32** (1844), 444–448, **52** (1851) 278–279; *Annalen der Physik und Chemie* 107 (1834) 148–158, **112** (1835) 436–456, 456–458, **129** (1841) 392–408, **242** (1879) 520–544; *Annalen der Chemie und Pharmacie* **68** (1848) 221–223, **78** (1851) 123–124, **80**, 227–229, **84** (1852) 207–209; *Archiv der Pharmacie* **4** (1823) 215–229, **51** (1835) 132, **126** (1853) 141–156, **160** (1862) 65–66, **180** (1867) 107–108; *Berichte der Bunsengesellschaft* **11** (1879) 1696–1698, **18** (1885) 267–282, **20** (1887) 991–999, 1864–1873.

793 *Christian August Vogler* (1841–1925) was a German geodesist. See: *Meteorologische Zeitschrift* (1882) p. 175; *Der Naturforscher* (1882) p. 245.

794 Today we know that variations in O_2 are strongly anticorrelated with variations in CO_2. The CO_2 cycle is known to be caused primarily by seasonal uptake and release of CO_2 by land plants and soils. In addition, the amount of O_2 that is dissolved in the upper ocean varies seasonally as the water warms and cools. Furthermore, the net production of O_2 in the upper 100 m of the ocean and the rate of vertical mixing undergo large seasonal variation. The seasonal O_2 variation varies between about 20 and 65 per meg, corresponding to (see p. 394) less than 0.0065 % (Keeling and Shertz 1992, Keeling et al. 1998).

Table 4.7: Percentage of oxygen in air freed from carbon dioxide and water vapor (cited after Clarke 1908, Blanchard and Pickering 1927). Mean values (min-max).

year	*n*	mixing ratio (%)	locality of samples	author
1846	28[b]	20.994 (20.840–20.970)	Heidelberg	Bunsen[g]
1852	113[a, b]	20.960 (20.913–20.999)	Paris	Regnault (1852)
1864	32[b]	20.943 (20.78–21.02)	Manchester	Smith (1872)[g]
1865	169[b]	20.952	several Britain sites	Smith (1872)
1881	778[b]	20.955	Cleveland (USA)	Morley[h]
1881	45[b]	20.933 (20.90–20.95)	Cleveland (USA)	Morley[g]
1883	20	20.864 (20.72–20.97)	Cape Horn	Müntz and Aubin (1884a, 1886)
1883	8[c]	20.93 (20.86–21.01)	Dresden[g]	Hempel (1885)
1884	94[c]	20.930 (20.877–20.971)	Dresden	Hempel (1885)
1886	45[e]	20.922 (20.901–20.939)	near Bonn	Kreusler (1887)[f]
1912	255[c]	20.951	Boston	Benedict (1912)[h]
1921	974	20.937		Boothby and Sandiford[d, h]

[a]*Regnault* did not measure CO_2; the figure must be corrected to 20.956 (Blanchard and Pickering 1927)
[b]Hydrogen eudiometer
[c]Potassium pyrogallate method
[d]*Walter M. Boothby* (1880–1953), US anesthetist; *Irene Sandiford* (1899–?) US biochemist.
[e]*Jolly*'s copper eudiometer, described in *Thiel's Landwirthschaftliche Jahresbücher* **14** (1885) pp. 333 ff.
[f]*Ulrich Kreusler* (1844–1921) was a German agricultural chemist at Poppelsdorf (Bonn)
[g]Simultaneously (same day and time) oxygen was determined to be 20.93 (20.80–21.09) crossing the Atlantic from Liverpool to New York (July 22–30, 1883) by *Ernst Bessel Hagen* (1851–1923) was a German physicist (Hempel 1885, p. 278)
[h]Data cited from Blanchard and Pickering (1927)

(1829–1908), who also presents a nice review on earlier oxygen measurements, and who obviously was motivated through the recent ideas on the variation of the oxygen content (in parenthesis in % mean; minimum-maximum), carried out measurements at several German forestry sites with different altitude inside (20.78, 20.61–20.94) and outside (20.83; 20.72–21.00) the forest; he stated that the variation is due to the measurement uncertainty and that there is no difference (Ebermayer 1886). Clarke (1908) writes that the variation in oxygen content (Table 4.7) is doubtless due to different methods of determination.

Sophisticated methods for air and gas analyses (Haldane 1918, Hempel 1913) were developed by *John Scott Haldane* (1860–1936), a British physiologist, and the German chemist *Walter Matthias Hempel* (1851–1916).[795] Hempel (1885) conducted – together

795 Founder of the chemical institute at the Dresden Technical University and professor 1889–1912, scholar of *Bunsen* and pioneer in gas analysis and electrolysis.

with colleagues in Tromsø (*Sparre Schneider*),[796] Para/Brasilia (*Fritz Pusinelli*),[797] Cleveland (*Morley*), and Bonn (*Kreusler*) – simultaneous sampling and analysis (the samples from Norway and Brazil were sent to Dresden for analysis[798]) between 4 and 15 May 1884. The results that did show no correlation to any meteorological parameter were (in %):

Tromsø	20.92
Dresden	20.93
Bonn	20.92
Para	20.89
Cleveland	20.93

Sixty years after *Renault*, in a survey involving many hundreds of precision analyses, *Francis Gano Benedict* (1870–1957) proved conclusively that during about two years, which involved a great variety of weather conditions, no variation occurred in the O_2 content of the air (standard variation ±0.006 %); Benedict (1912). Similar precision was obtained by *August Krogh* (1874–1949)[799] with a standard variation of ±0.005 % (Krogh 1919). This conclusion about constant oxygen content in the atmosphere was reinforced by Machta and Hughes (1970) many years later, although both studies were limited by the precision of the analytical techniques employed.

Friedrich Adolf Paneth (1887–1958)[800] wrote that despite innumerable analyses of the oxygen content in air the second decimal of this constituent is still not exactly known (Paneth 1937); he recommended an absolute value of 20.946 ± 0.002 % as the most likely figure for oxygen in the uncontaminated atmosphere, in combination with an average CO_2 value of 0.0033 %).

Besides nitrogen and oxygen, at the end of the nineteenth century, the concentrations of carbon dioxide (0.003 %), argon (0.93 %), and that of hydrogen (<0.0006 %) were well known. Other gases like ozone, hydrogen peroxide, ammonia, nitric acid, sulfur dioxide, and others as well as dust were detected (Chapter 4.1). For the analysis of N_2 no direct precision method was discovered, writes Glueckauf (1951). However, as the sum of nitrogen and rare gases (termed “atmospheric nitrogen” by *Krogh*) has a constant of at least ±0.002 %, nitrogen too must be constant to the same degree.

796 *Jacob Sparre Schneider* (1853–1918) was a Norwegian zoologist.

797 Unknown; family *Pusinelli* has its origin from Italian merchant in the thirteenth century and where also well-known wine dealer in Dresden in the 19th century.

798 Carried out by *Hempel*'s assistants *Oettel* (see p. 425) and the unknown *Schumann*.

799 A Danish physiologist; discovered how capillaries regulate oxygen and was awarded with the Nobel Prize in Physiology or Medicine in 1920.

800 An Austrian-born British chemist. Fleeing the Nazis, he escaped to Britain. He became a naturalized British citizen in 1939. After the war, *Paneth* returned to Germany to become director of the Max Planck Institute for Chemistry in 1953.

Since the finding by Keeling (1988a,b) using a new interferometric technique, who observed strong anti-correlation between atmospheric O_2 and concurrent measurements of CO_2 caused by fossil fuel combustion processes, and recognized the potential of concurrent atmospheric O_2 and CO_2 measurements in studying the global carbon cycle, a global oxygen network was stepwise established. Since this first study, high-precision atmospheric measurements of O_2 and CO_2 were made all over the world from a variety of platforms, including ships. To avoid O_2 variations due to photosynthesis and respiration (anti-correlation with CO_2), a conservative tracer termed "atmospheric potential oxygen" (APO)[801] was introduced (Stephens et al. 1998): $APO = O_2 + 1.1\, CO_2$. However, the oxygen content is measured as $\delta(O_2/N_2)$, where APO varies not more than about 5 per meg (1 per meg equals 0.0001 %).

4.3 Carbon compounds in air

4.3.1 Carbon dioxide

4.3.1.1 Discovery of carbon dioxide

Carbon dioxide was the first gas to be distinguished from common air, perhaps because it is so intimately connected with the cycles of plant and animal life. The discovery of carbon dioxide must be attributed to several scientists, *Jan van Helmont*, *John Mayow*, *Stephen Hales*, and *Joseph Black* (see also pp. 76 and 349 ff).

As mentioned before, *Helmont* already found about 1630 that the gas bubbling in a brewery during fermentation is the same as that obtained by burning charcoal, as both gases turned limewater milky – a test which is used even today to identify carbon dioxide; he termed it *spiritus sylvestre* [wild gas]; see Table 1.4 on p. 71 for the different name of CO_2. Hence, he is credited with its discovery. He was acquainted also with the air of the celebrated *Grotto del Cane* near Naples (CO_2 from volcanic exhalations),[802] as well as with the exhalations in mines, against the pernicious effects of which cautions had been given long before his time by *Andreas Libavius* and *George Agricola* (cited after Gmelin 1801, p. 193–194).

John Mayow was the first who prove that carbon dioxide is present in the atmosphere. He showed that air contains a gas that is a special agent for combustion and respiration and is fixed from calcified metals (i. e., metal oxides). He also showed that

801 Specifically, when fossil fuels are combusted and release 1 mole of CO_2, 1.4 moles of O_2 will be removed from the atmosphere. Similarly, when a terrestrial ecosystem removes 1 mole of CO_2 from the atmosphere, it also releases 1.1 moles of O_2 (Battle et al. 2006). Battle et al. (2006) define $APO = \delta(O_2/N_2) + 1.1 \cdot 4.8\,[CO_2]$ where the value $4.8 = 1/0.209$ derives from the O_2 content in air and 1.1 corresponds to the terrestrial CO_2/O_2 stoichiometry.

802 According to Kopp (1931), already *Plinius* mentioned the exhalation of lethal vapors (*spiritus lethales*) from caverns in the first century of our era.

the new gas extinguished a flame, that it could not support life, and that it was present in gas exhaled from the lung. *Stephen Hales*, who early in the eighteenth century began his important studies of the elasticity of air, pointed out that various gases, or "airs" as he termed them, were contained in many solid substances. His many careful measurements (published in *Vegetable Statics, or an account of some statical experiments on the sap in vegetables*, London 1727) of the absorption of water and its transpiration to the atmosphere were the basis for the understanding that air and light are necessary for the nutrition of green plants.

The careful studies of *Hales* were continued by his younger colleague, *Joseph Black*, a Scottish chemist, whose experiments concerning the weights of gases and other chemicals were the first steps in quantitative chemistry. *Black* was the first who found it in the air of Edinburgh, probably between the dates 1752 and 1754 (Black 1754). He writes (Black 1803, p. 74):

> [...] From this it was evident that the sort of air with which the lime is disposed to unite in a particular species which is in a small quantity only with the air of the atmosphere. To this particular species I gave the name of fixed air.

Black also identified carbon dioxide in exhaled breath, determined that the gas is heavier than air, and characterized its chemical behavior as that of a weak acid. *Priestley* made some observations on the gas in 1767 (Mellor VI, p. 1). Bergman (1788) called it acidum aëreum (aerial acid, "*Luftsäure*" in German). Bucquet (1778) called the gas "*acide à la craie*", and Keir (1777)[803] *calcareous gas*.

Gmelin (1801, pp. 199–200) cites many more scientists who contributed to our knowledge of carbonic acid.[804] Fixed air was renamed "*acide crayeux aériforme*" [aerial chalk acid] by *Lavoisier* in 1781. In 1787, *Horace Bénédict de Saussure* (*Saussure* the elder) employed lime water as a test for CO_2 in the air when he was at the summit of Mont Blanc (Saussure 1796). First quantitative attempts to determine CO_2 in atmospheric air were done by *Alexander von Humboldt* in 1797 (Humboldt 1798, 1799, 1800) and by the English chemist *John Dalton* in 1802 (Dalton 1805), however, only showing that its level is less than 0.1 %. *Dalton* guessed in 1803 that the molecule contains one carbon atom and two oxygen atoms; this was later proved to be correct.

Théodore de Saussure (*Saussure* the younger) was the first who carried out systematic determinations of carbon dioxide at several places. *Humboldt*, who named as minor species "carbonic acid gas", citing *Boussingault* and Lewy, mentioned in a footnote (Humboldt 1850–52, p. 311) that "the proportion of carbonic acid in the at-

---

803 *Jean-Baptiste Michel Bucquet* (1746–1780) was a French physician and chemist. *James Keir* (1735–1820) was a Scottish chemist and geologist.

804 Macbridge (1764), Jacquin (1769), Well (1771), Henry (1773), Smedt (1773), Fontana (1774), Venel (1774), Schinz (1778), Bergmann (1779), Köstlin (1780), Meyer (1781), and Laugier (1786); see references in Gmelin (1801).

mosphere [...] varied only between 0.00028 and 0.00031 in volume" [280–310 ppm]. Remarkably, many authors agreed with 0.03 % (according to 300 ppm) as "natural reference value" (Schloesing 1880, Blochmann 1886, Renk 1886, Letts and Blake 1900, Brown and Escombe 1905a, Friedheim 1907, Lode 1911, Benedict 1912, Krogh 1919, Loewy 1924, Quinn and Jones 1936, D'Ans and Lax 1943, Remy 1965). Petermann and Graftiau (1891, pp. 29–30) write:

> [...] acide carbonique de l'atmosphère est en général d'une grade constance [...] se rapproche très sensiblement de 3 litres par 10,000 litres d'air à 0 ° et 760 millimètres de pression.

Schloesing (1880) also determined the content of carbon dioxide and carbonate in seawater and discusses that between the sea and the atmosphere an equilibrium cannot be reached because there is a continuous change between them – when the atmospheric CO_2 decreases, the sea degases CO_2 and vice versa such the sea controls the air. *John Dalton* writes in an appendix "On the gradual deterioration of the Atmosphere, by respiration and combustion" to his article "On respiration and animal heat" (Dalton 1813, p. 41):

> [...] calculating the weight of the atmosphere at the rate of 15 lbs. upon a square inch, for such a number of miles we obtain 12 trillion of lbs. avoirdupois; – calculating also the quantity of carbonic acid which 1000 millions of men (the supposed population of the earth) would expire in the space of 6000 years, at 3 lbd. Per day, we shall find it to be 6 thousand billion of lbs. or just 1/2000 part of the whole atmosphere: now, supposing this doubled, to allow for the quantity of acid which may be supposed to the generated by combustion, we shall then have 1/1000 part of the atmosphere to be carbonic acid, which agrees with experiments as to the quantity now actually found in it.

Hadfield (1842, p. 11), a pupil of *Dalton*, cited this phrase shortened and a bit clearer:

> Mr. Dalton had made the calculation to shew that the quantity of acid gas at present found in the atmosphere may be no more than the natural produce of 6000 years, and that the supposed decomposition of it may not be necessary.

This is the first consideration of a global carbon budget based on own experimental data, although the carbon dioxide concentration determinate by *Dalton* was more than three-times larger than realistic.[805]

805 The other data (mass of the atmosphere and human average CO_2 exhalation per day) are very close to today's estimations of $5.2 \cdot 10^{21}$ g and 2.3 pound d^{-1}, resp. Without mention it, we can assume that *Dalton* considered the vegetation neutral in relation to CO_2 (1779 *Ingenhousz* published that plant consume CO_2 at daytime and release it at night); however, human's CO_2 exhalation is also "atmospheric neutral"; volcanic emission of CO_2 was not yet known in *Dalton*'s time. Based on *Dalton*'s assumption for the yearly CO_2 release by man-made combustion, today's per capita CO_2 emission is 200 time larger.

4.3.1.2 Methods for determination of carbon dioxide in the air

> Die Kohlensäurebestimmung in der atmosphärischen Luft ist an sich eine so schwierige Aufgabe, dass man darauf gefasst sein muss, ziemlich grosse Apparate oder beschwerliche Zurüstungen und Arbeitsmethoden anzuwenden [The determination of carbonic acid in air is a difficult task that you have to be prepared, to apply rather large instruments or troublesome preparations and methods] (Pettersson 1886, p. 477).

The basic principle of determination of carbon dioxide was introduced by *Théodore de Saussure*, on the basis of carbon dioxide absorption by baryta water (an aqueous solution of barium hydroxide). A large glass globe with a volume of about 20 L will be filled with the ambient air to be investigated. Then a quantity of baryta water will be added, more than sufficient to neutralize the carbonic acid in that volume of air. The bottle was closed and either several hours agitated or left several days with occasional shaking. The carbonic barite ($BaCO_3$) so formed will be collected, that which is sucked on the wall dissolved with hydrochloric acid and the collected solutions precipitated with sulfuric acid baking soda (Na_2SO_4) and finally weighted ($BaSO_4$). *Dalton* used a globe one-fifth part of the size and lime water instead of baryta; after shaking with the air, and afterward sulfuric acid for ascertaining the amount combined. It is worth citing the description of his first determination of carbon dioxide in the air (Dalton 1805, p. 254):

> I have found, that if a glass vessel filled with 102.400 grains of rain water be emptied in the open air, and 125 grains of strong lime water be poured in, and the mouth then closed; by sufficient time and agitation, the whole of the lime water is just saturated, by the acid gas it finds in that volume of air. But 125 grains of the lime water used require 70 grains measures of carbonic acid gas to saturate it, therefore, the 102,400 grains measure of common air contain 70 of carbonic acid; or 1/1460 of the whole. – The weight of the carbonic acid atmosphere then is to that of the whole compound as 1:1460; but the weight of carbonic acid gas in a given portion of air at the earth's atmosphere, is nearly 1/1000 of the whole; because the specific gravity of the gas is 1½ that of common air.

The Swiss chemist *Carl Emanuel Brunner* (1796–1867) in Bern changed this method two years later by passing about 25 L air through several tubes; the first is filled with pumice stone soaked in sulfuric acid, the second and third are filled pumice stone soaked in caustic potash solution (KOH) and finally through two further "sulfuric acid tubes" that hold humidity. Increase in the weight of potash and subsequently in sulfuric acid tube corresponds to carbon dioxide (pressure and temperature were registered). Many modifications were made (apparatus like bottles or tubes with an aspirator, instead of titration, gravimetric estimation, other titrating acids like HCl solution, different indicators for the neutralization point, etc.).

The most known method in the nineteenth century was developed in 1858 by *Max Joseph von Pettenkofer* (1818–1901),[806] the "*Flaschenmethode*" (Fig. 4.17 and 4.18) and

806 Bavarian chemist and hygienist; in 1865 he became also professor of hygiene in Munich; founder of the first Institute for Hygiene, today termed "*Max von Pettenkofer-Institut für Hygiene*".

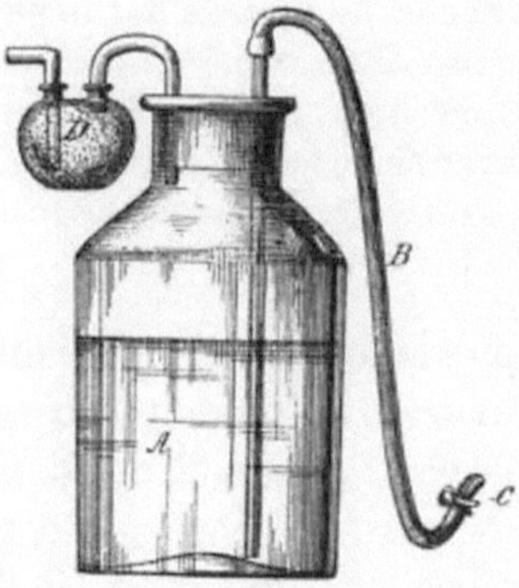

FIG. 19.

A, flask containing the barium water.
B, glass syphon for drawing off the solution.
C, rubber tube, closed by pinch cock, into which the tip of pipette may be inserted for drawing off the solution.
D, absorption bottle containing broken pumice stone saturated with strong solution of caustic soda. It robs the air passing into the flask of its carbonic acid.

Figure 4.17: *Pettenkofer*'s method (Billings 1893, p. 186)[807]; is worth noting that *Pettenkofer* does not present a figure in his paper (Pettenkofer 1862).

the "aspiration method" (Pettenkofer 1862), principally characterized that after precipitating of $BaCO_3$ (an excess of baryta must be used, and well shaken with the air), an extracted part of the clear solution is titrated by oxalic acid (to see how much of the baryta is left uncombined), and the point of neutralization is found by putting a drop of the liquid on a piece of turmeric paper). It is historically interesting to mention *Pettenkofer*'s motivation for developing a method added to the numerous methods that "we hold from recognized researchers like *Saussure*, *Brunner*, *Dumas*, *Boussingault*, *Watson*, *Regnault*, *Hlasiwetz* and *Gilm* and others, and which altogether yielded agreeing results" (Pettenkofer 1862, pp. 165–166): he set himself the task to study the ventilation in housing better than before and hence he needed a method to determine carbon dioxide in short time intervals (from 15 to 30 min) and more accurate than before. This new task, studying *air hygiene*[808] (in German "*Lufthygiene*"), for the first indoor, and later also in ambient air, remained for the next 100 years the main purpose for examination of carbon dioxide in air (note that *Pettenkofer* was the first hygienist in Germany, and likely worldwide). However, this method also remained for the next hundred years the basic principle (many details changed by other researchers) for investigation atmospheric air. Besides hygienic aspects and ventilation, carbon dioxide

807 *John Shaw Billings* (1838–1913) was a US American surgeon.

808 From middle of the nineteenth century a large number of studies deal with CO_2 in air of buildings and rooms, barn air, soil air, in mines, breezing air of animals and humans, and CO_2 in natural waters.

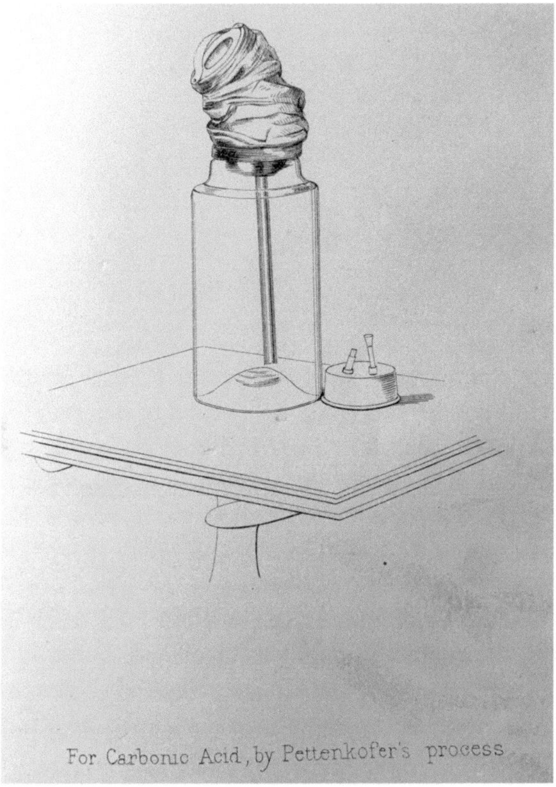

Figure 4.18: For carbonic acid, by *Pettenkofer*'s process, used by Smith (1872, p. 451).

measurements were conducted also from agricultural (namely plant growth) and medical (human physiology) aspects.

Important progress was the use of phenolphthalein as an indicator for the saturation point of the lime-water solution by carbonic acid from the volume of air (Blochmann 1884), however, first only for simple tests for carbonic acid in houses.[809]

809 A bottle of about 500 mL volume is filled with the air to be investigated and afterwards poured 5 mL saturated lime water with 3 drops of phenolphthalein dissolved in alcohol, the bottle closed with a cork stopper and 2–3 min agitated. Afterwards the bottle will be opened and through ambient air is sucked in until the red solution become colorless. The accuracy is given by the multiple of the bottle volume (not better than 10 %). *Blochmann* added the following reference: *Es sei mir gestattet darauf hinzuweisen, dass Mechanikus und Optikus J. C. Schlösser in Königsberg i. Pr. den completen Apparat: Versuchsflasche, Messgläschen, Tropfenzähler für die Phenolphtaleïnlösung, Vorrathsflasche für das gesättigte Kalkwasser u. s. w. nebst den für einige hundert Analysen ausreichenden Reagentien und specielle Gebrauchsanweisung in solidem, polirtem Holzkasten für 10 Mark liefert* [Let me point this out that the mechanic and optician J. C. Schlösser in Königsberg (in Prussia) delivers the complete instrument: test bottle, burettes, drop counter for phenolphthalein solution, storage bottle for saturated lime water etc. along with sufficient reagents for some hundreds analysis and a special manual in a solid, polishes wooden box for 10 Mark].

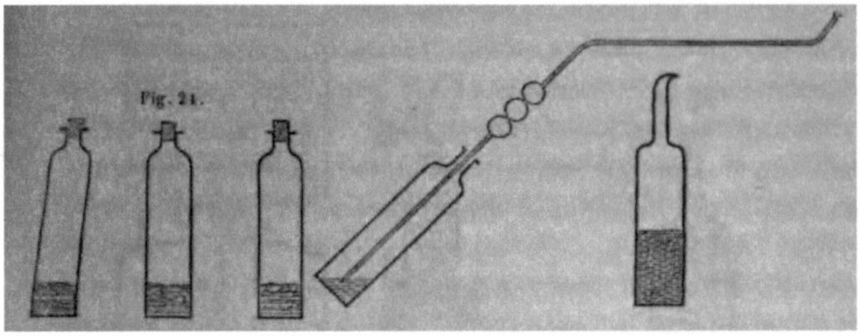

Figure 4.19: *Reid*'s carbonometer for testing of carbonic acid in the air (Reid 1844, p. 67).

This so-termed "minimetric method" was introduced by *Robert Angus Smith* in connection with the Royal Mines Commission to find a simple method of determining the value of the air in mines, based on the qualitative observation of beginning turbidity through precipitating lime (Smith 1872, pp. 192–205). *Smith* cites *Boswell Reid* (1805–1863)[810] who proposed in 1835 (Reid 1844) a test for carbonic acid levels in the air of the Palace of Westminster where air was passed through lime water in a phial, termed "carbonometer" (Fig. 4.19).

Robert Angus Smith was the first who used *Pettenkofer*'s method for systematic measurements of CO_2 (Chapter 4.3.1.3) but he modified it by using another bottle with a different shape and very wide mouth (to allow easy cleaning) and instead of blowing air into the vessel, he draws it out with a flexible bellows-pump; the advantage, *Smith* has seen, was the simple procedure.

The "minimetric" test method was later introduced by *Georg Lunge* (1839–1923)[811] to German researchers (Lunge 1877), and modified by several persons (Hesse 1879, Münnich 1880, Ludwig 1882, Blochmann 1884, Cordeiro 1885, Lunge and Zeckendorf 1888, Fuchs 1889, Anonymous 1890, Lebedinzeff 1891, Schultz 1892, Wolpert 1892), Teich 1893, Haldane 1895.[812] For all these instruments that were constructed for determination of carbon dioxide, almost not applicable for atmospheric but indoor air, the main point is the experience of the investigator – having less practice, the largest er-

810 *David Boswell Reid*, British physician and chemist, known for his studies on ventilation and regarded as "father of air-conditioning".

811 German technical chemist in Breslau, Zürich and England.

812 *Walther Hesse* (1846–1911) was a German physician in Schwarzenberg and Dresden who worked together with *Robert Koch* and is well-known for his studies on industrial hygiene. *Frederick Joaquium Barbosa Cordeiro* (1859–?) was an author of several book on physics, meteorology and the atmosphere. *Arsenius Lebedinzeff* (rare *Lebedinzew*) [Арсений Лебединцев] (1863–1920) was a Russian chemist at the Novorossiysk University [Новороссийский университет], today University Odessa. *Hermann Schultz* (unknown living dates), a physician in Mecklenburg-Schwerin. *Heinrich Wolpert* (1866–1924), student of *Rubner* and lecturer for hygiene at the Berlin University, son of *Adolf Worpert* (1832–1907), German professor for hygiene in Nuremberg who also constructed an instrument (D. R. P. Nr. 39382 vom 5. September 1886). *Gottlob Fuchs* (unknown living data) was a German physician, born in Wüstebriese (Silesia, today Brzezmierz in Poland). Nothing is known on *Teich*.

rors can occur as a comment by Lunge and Marchlewski (1891).[813] Many publications in the nineteenth century concerned methods and determination of CO_2 in rooms of schools and hospitals; not cited here.

Important progress was made by *Sven Otto Pettersson* (1848–1941) who constructed an instrument where water vapor was first removed from the air through phosphoric anhydride (P_2O_5) and afterwards carbon dioxide absorbed by soda lime (NaOH); the reduction in volume corresponds to the quantity of CO_2 (Pettersson 1886). *Pettersson* and *Palmqvist* soon later improved the method so that within the 15 min-period measurement, faster but in the same quality as *Pettenkofer*'s method, became possible (Pettersson and Palmqvist 1887). The handling of *Pettersson-Palmqvist*'s method was simplified by Troili-Petersson (1897), Bleier (1898), *Haldane* (see Swaab 1904), and Anderson (1913). The Swedish engineer *Klas Sondén* (1859–1940) further improved *Pettersson*'s apparatus (Sondén 1887), and finally was constructed for the measurements by Benedict (1912), Fig. 4.20. A good survey on the different methods is given by Bitter (1890) and Rauch (1920).

The first physical method for the determination of carbon dioxide was developed by *Heinrich Heine* (unknown living data) in 1881 (his dissertation) at the Physical Laboratory of the University Giessen at *Wilhelm Conrad Röntgen* (1845–1923) who soon before studied heat absorption by gases and found a (nonlinear) relationship between the proportions of two gases and the pressure, respectively, volume change when heating the gas. *Heine* used an apparatus shown schematically in Fig. 4.21 and estimated curves for mixtures of CO_2 (down to 0.012 %) and air. Thus, he was always able to estimate the unknown concentration of CO_2 in air samples collected in the suburb of Giessen.

Blochmann (1886), Letts and Blake (1900, 1901), and the Scottish chemist *James Walker* (1863–1935) who succeeded *James Alexander Brown* in Edinburgh in 1908 (Walker 1900) found that the ordinary *Pettenkofer* process gives systematically too high values due to additional absorption of CO_2 by the baryta water (e. g., when handling solutions in laboratory air). Letts and Blake (1901) made careful simultaneous sampling and analysis by the *Pettenkofer*'s and its own new process, showing that the *Pettenkofer* ordinary process results in 25–35 % higher values and such explaining the difference between "4" and "3" parts of 10,000. *Edmund Albert Letts* (1852–1918),[814] professor of chemistry in Queen's College, Belfast, and *Robert Frederick Blake* (1867–1944) together published two remarkable papers (Letts and Blake 1900, 1901), giving a new accurate CO_2 determining method and the most extensive critical overview on CO_2 observations in the nineteenth century, citing 198 authors. They write (Letts and Blake 1900, p. 172):

813 *Leon Paweł Teodor Marchlewski* (1869–1946) was a Polish biochemist, assistant of *Lunge* in Zürich and since 1903 professor in Krakow.

814 Further reading on *Letts* see Burns and Walker (2015).

COMPOSITION OF THE ATMOSPHERE

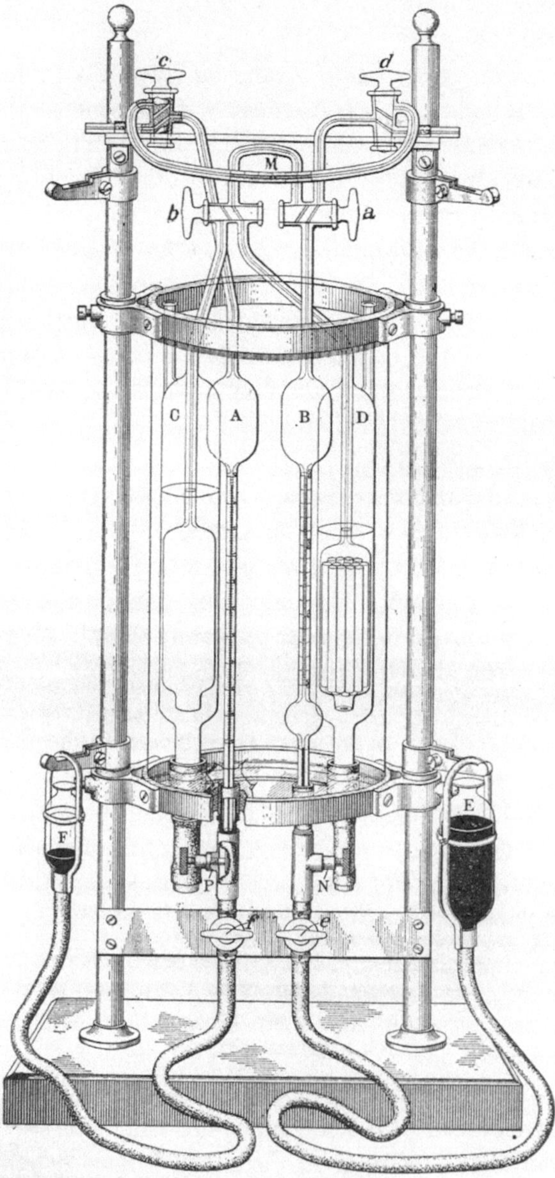

FIG. 1.—Sondén air-analysis apparatus.

The two pipettes *A* and *B*, and the reagent containers *C* and *D*, are immersed in water in the glass tank. Stop-cocks *a*, *b*, *c*, and *d*, permit intercommunication of all parts. Cabron-dioxide percentages are read directly on *A*, and oxygen percentages directly on *B*. Mercury reservoirs *F* and *E* are connected with pipettes *A* and *B* respectively. The manometer *M* aids in securing equal tension in *A* and *B*.

Figure 4.20: Apparatus for analysis of atmospheric aid, devised by *Klas Sondén* (Benedict 1912, p. 74).

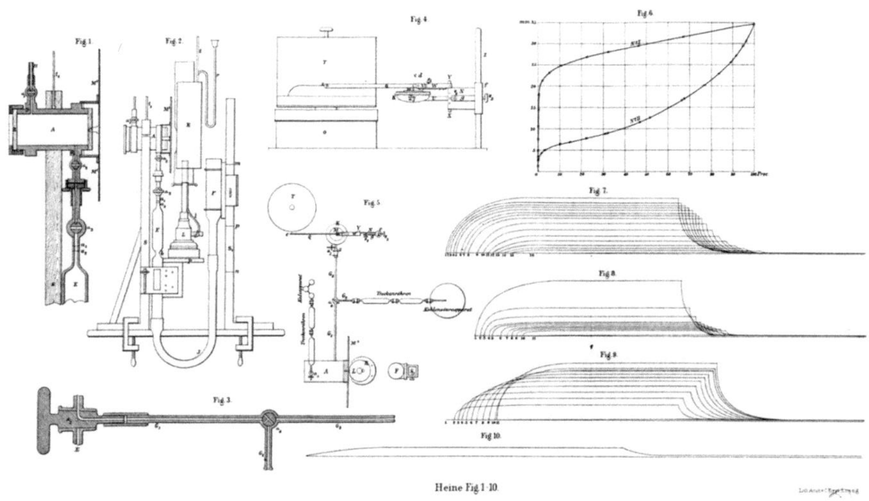

Figure 4.21: Apparatus for determination of gases due to absorption of heat (Heine 1882). It consists of (1) horizontal positioned tube for intake of gas and heat absorption (Figs. 1, 2, and 5), (2) combined with other glass tubes to mix different gases like CO_2 and air (Figs. 1 and 2), the *Knoll*'s pantograph to measure the pressure increase due to heat absorption (Figs. 4 and 5) and finally the *Bunsen* burner as a heat source with the chimney (Figs. 2 and 5). Figs. 6 to 10 absorptions curves (dependence of pressure change from CO_2 concentration in air), shown for several ranges of CO_2 concentration.

> De Saussure whose researches extended over the period 1809–1830, and who made some hundreds of observations [...] but it is interesting to notice, as Blochmann has pointed out, that his figures almost steadily decreased as his work progressed. [...] until about the year 1870, the normal amount of atmospheric carbonic anhydride appears to were taken as 4 in 10,000 vols.

What is the "error" of such analytical figures?[815] *Letts* and *Blake* further write:

> These variations, if actually small, are *relatively* large, and correspond with fluctuations of at least 10 per cent of the total quantity, and according to some observers to a much higher proportion. [...] and that very few of these were tested as regards their degree of absolute accuracy; and also,

815 It is known from chemical practice that a precision (it is the standard deviation from several measurements of a sample with the same concentration) to be ±3 % is excellent. However, a high precision does not mean automatically that the value is "true"; the trueness is the degree of match with the "true value". A high precise measurement can be far away from the true value, as found in inter-comparisons for rainwater analysis; today's wet chemical analysis of ions has a precision of 0.02 mg Ca L^{-1} and an inter-network bias of 15 % for calcium. However, when the analysis has a good accuracy, the mean from many measurements can fit the "true" value. Assuming this bias being similar also for barium; however, later we discuss that not the (excellent) precision of chemical analysis (already likely provided in nineteenth century) is important but the errors while CO_2 sampling/absorbing (incompleteness) and alkaline titration (additional CO_2 absorption) are essential. Such precision (0.02 mg Ba L^{-1}) would result only in 1 % relative error when sampling CO_2 from 6 L air.

> that the observers have collected their samples at different heights above the ground [. . .] (Letts and Blake 1900, p. 174)
> Müntz and Aubin are an exception to this rule (Letts and Blake 1900, p. 436).

It is clear to any chemist that many sources of errors can appear in the procedure and a long experience or practice by the operating person is necessary to get high accuracy. Fresenius (1875) states that for air analysis precision of ±0.002 % absolute CO_2 level (according to ±20 ppm) is excellent and Haldane (1918) gives ±10 ppm as precision (3–7 % relatively). Hempel (1913) first cites results from inter-comparisons; the accuracy is not better than ±100 ppm for three different methods (50–100 ppm for ambient air analysis and 100–200 ppm for indoor CO_2 analysis) and ± (30–50) ppm by comparing *Pettersson*'s and *Pettenkofer*'s method.

Today it is obvious that measurements are corrected to standard conditions and related to dry air. In many older papers (before the 1870s), this is not clearly expressed. When discussing historical values, we have to take into account that CO_2 in laboratory air can go up to 0.2 % (Hempel 1913), and therefore any not careful analytical operation might lead to higher analytical figures. The next fact which must be taken into account when comparing different historical CO_2 values is the town influence, which was estimated between 30 and 200 ppm higher than in the rural surroundings (Renk 1886, Rubner 1907, Lode 1911)[816] very close to today's values of city dome CO_2.[817] When wind speed comes larger than 3 m s^{-1}, ventilation transforms the CO_2 city dome into a plume according to the prevailing wind direction. Thus, in evaluating historical CO_2 concentrations, gained from sampling at different times of the day and the year as well as at different levels above ground, and namely in construction of mean values, one has to consider the following variations that are caused by biospheric and man-made activities:[818]

Diurnal variation	<5–30 ppm
Seasonal variation	5–20 ppm
Inter-annual variation, north hemisphere	10–20 ppm
Inter-annual variation, south hemisphere	2–3 ppm
North hemisphere minus south hemisphere	5 ppm (around 1980; increasing)
City plume pollution effect	<30 to >100 ppm

Accepting ±10 ppm CO_2 as the precision of single measurement, it becomes clear that many of the above listed natural variations can hardly be reflected by nineteenth-century measurements; in the following text, given standard variation had been calculated from cited individual values in original publications. The most serious effect that

816 *Alois Lode* (1866–1950) was an Austrian hygienist in Innsbruck.

817 Modern measurements show differences from 15 to 550 ppm (Möller 2019).

818 These figures are only for illustration; they depend strongly from the location and are changing with time. There are many papers in this issue; some general reference can be found in Möller (2019).

influenced historic measurements after 1870 was local air pollution; hence measurements in cities and close to cities cannot represent background (remote air) figures; see p. 422 the discussion concerns “pre-industrial” CO_2 levels.

In past, often only one value (and almost daytime) per day or only a few values per month were produced. According to the time of day, such figures cannot reflect a daily mean; only a large number of measurements made at different times, can reflect a mean daytime value. The majority of CO_2 measurements were carried out at business times; hence, a “corrected” daily mean could be around or higher than 10 ppm. The diurnal and seasonal cycles appear as a sinusoidal curves. The diurnal amplitude depends strongly on the local situation. Hence, the difference between continental daytime-means and nocturnal means could range from 10 to30 ppm.[819] Principally this was known at the end of the nineteenth century by a few more intense studies, but extensive sampling over longer periods was impossible due to a long time for sampling and single analysis (several hours) and hence large manpower needed. Only very few of the historical measurement series were able to reflect a seasonal cycle; accuracy with 10–20 ppm and a too small number of measurements mask this natural variation.

However, the season influence on CO_2 already was stated by *Saussure*, who found it in summer higher than in winter, supported by Fittbogen and Hässelbarth (1879)[820] who found a minimum in December whereas Petermann and Graftiou (1891) found a very small variation of only a few ppm. *Edme Hippolyte Marié-Davy* (1820–1893), a meteorologist at the Montsouris Observatory in Paris (see also p. 188), found a CO_2 maximum in December (Marié-Davy 1880).

More CO_2 is found in the ground air from lower than from higher levels; the gradient has a definite connection with the season and rainfall has a marked influence, Letts and Blake (1900, p. 216) summarize, citing *Jozef von Fodor* (1843–1902), professor for hygiene at the University Budapest, who found soil CO_2 emission to be 0.175 L m^{-2} d^{-1} (Fodor 1879):

> [...] the fluctuations in the amount of atmospheric carbonic anhydride are mainly due to the absorbing action of soil on the one hand, and the evolution of ground air on the other.

Summarizing the discussed points that affect the CO_2 level, it is no longer surprising that historically published values after 1870 between 270 and 360 can be seen as no principally erroneously but not representing a “true” mean (climatological) CO_2 figure. Furthermore, it is senseless to construct (as done by some “climate sceptics”) time

---

819 Today, CO_2 is measured continuously, “automatically” giving a mean daily value, resulting in representative monthly means and subsequent annual mean figures (when the time gaps of measurements are not too large).

820 *Paul Hässelbarth* (1850–?) was a German chemist at the agricultural research station Dahme. *J. Fittbogen* (unknown living data) was a German agricultural chemist, head of the station Dahme 1874–1885 (successor of *Hermann Hellriegel*, the first director). They measured 1875–1876 at Dahme a mean CO_2 level of 334 (270–417) ppm ($n = 347$).

series by combining values from different authors because of methodical errors by different authors and/or methods, timely influences (almost diurnal cycle), and different local influences. However, there are a few exceptional scientific serious measurements in the late nineteenth century, which can be used to construct pre-twentieth century background values.

4.3.1.3 Carbon dioxide measurements in the nineteenth century

Wolff (1847) cites *Bergman* who estimated in 1774 carbonic acid in air to be 0.06; however, it is more probable that *Scheele* conducted the measurements (0.06 corresponds to the value 1/16 cited below). There were the belief at the beginning that carbonic acid is found only in the lower parts of the atmosphere. Only with the measurements by *Saussure* (Montblanc) and *Humboldt* who analyzed air collected by *Garnerin*[821] in 1798 from a balloon in Paris at an altitude of 650 Toisen (about 1267 m) one concluded that carbonic acid is generally distributed in the atmosphere. *Theodore de Saussure* is the first chemist who began in 1809 to study systematical the variation of carbon dioxide in the air in the surrounding of Geneve (Saussure 1816, 1830). He cited the first chemist who developed analytical methods for CO_2 measurements, *Fourcroy*, *Dalton*, and *Thènard*, and describes the procedures in detail (Saussure 1830). Kämtz (1836, pp. 29–33) communicated in his textbook on meteorology the following scientists who determinate carbon dioxide in air (in parenthesis the mixing ratio):[822]

Scheele (1/16), *Priestley* (1/52), *Gehler* (1/10)
Girtanner (1/100)
Fontana, who denied the existence of CO_2 in the air
Humboldt (0.5–1.5 %)
Configliachi (max. 0.08 %)
Parrot (0.57 %)
Dalton (0.1 %); see pp. 386 and 395
Thènard (1812) (1/1761 = 0.057 %)
Saussure (1809–1810) (0.0425–0.0779 %)

Saussure (1830) found as mean from all his determination 415 ppm (370–620 ppm), and summarized the following phenomena:

the air above wet soils (due to rain) and lakes contains less CO_2,
the air above frozen or dry soils contains more CO_2,

821 *André-Jacques Garnerin* (1769–1823) was a French balloonist and first skydiver.
822 *Pietro Configliachi* (1779–1844) was a physicist in Padua, successor of *Volta* in 1804; he also determinates around 1808 oxygen to in the range 0.209 to 0.211 at different sites (Configliachi 1811, p. 143 and 144). *Parrot* (see p. 384) made in Russia (Riga) together with his friend *David H. Grindel* (1776–1836), pharmacist, and *Johann David Sand* (1748–?), later teacher for sciences at cathedral school, the first measurements of air composition in Russia (Parrot (1801); see also Stradins and Arons (1992, p. 181)). No original reference I found on *Scheele*, *Priestley* and *Gehler*.

the air contains in the country at night more CO_2 than in the city and than at daytime,
seasonal variation is similar at all sites,
on mountains, the air contains more CO_2 than in the plane,

There is a group of famous scientists (*Louis Jacques Thénard*, *Theodore de Saussure*, *Jean Baptiste Boussingault*, *Hugo von Gilm*) who measured CO_2 levels around 400 ppm (and more) between 1815 and 1850 at different sites in Central Europe; see Table 4.8. All later researchers agreed that such values were much too high. Blochmann (1886, p. 41)[823] cites mean values from several observers:

1830–1856 (n = 322)	360 ppm (6 observers)
1856–1886 (n > 4000)	310 ppm (21 observers)

Smith (1872, pp. 42–73) also presents an extended review on CO_2 measurements since *Saussure*, writing at the beginning:

> I am disposed to think that there may be a little excess in his results; but this will not affect the comparative amounts found at different times, and which form the most interesting part of the enquiry.

John Shaw Billings (1838–1913), an American librarian, building designer, and surgeon, wrote 500 pages on "ventilation and heating" that includes a chapter on "carbon dioxide" giving another excellent survey including data since *Saussure* (Billings 1893, pp. 61–84). Another review on carbon dioxide measurements is given by Wollny (1885) namely concern its variations. However, the best review on early measurements of carbon dioxide is given by Wolff (1847, pp. 22–32).

William Hadfield (living date unknown), a pupil of *Dalton*, adopted *Dalton*'s method and determined between December 1828 and 1830 (n = 9) at Cornbrook (neighbourhood of Manchester) on average 800 ppm carbonic acid (Hadfield 1842), be much too high as noted by Smith (1872, p. 450). Most remarkable, however, is *Hadfield*'s motivation for such investigation (Hadfield 1842, p. 10–11):

> The gas is formed in immense quantities in large towns from the combustion of fuel, [. . .] There are no experiments tending to shew whether the quantity of Carbonic acid gas in the atmosphere is upon the increase or decrease.

Henry Hough Watson (1810–1886), an English chemist in Bolton near Manchester, measured 530 (420–862) ppm in the city 1832–1833 (n = 19) and at a higher located rural site 414 (361–474) ppm 1833–1834 (n = 12); Watson (1850) gives a difference of about 100 ppm between city and surrounding. This is the first evidence (although the free-air level is much too high) for the influence of large CO_2 sources (big cities, manufactures)

823 *Georg Rudolf Reinhart Blochmann* (1848–1920) was a German chemist at Königsberg, student of *Bunsen*.

Table 4.8: Researchers who measured carbon dioxide in the nineteenth century: Footnotes are given for persons not referred in the text.

Observer	place	date
Parrot (and Grindel)	Riga	about 1800
Dalton	Manchester	1802
Configliachi	Alpes	1808
Saussure	Geneva	1809–1815
Thénard[a]	Paris	1812
Saussure	Chambeisy	1827–1829
Brunner[b]	Bern	1832
Watson	Manchester and surrounding	1832–1833
Emmett	Barbados	1836
Boussingault and Lewy	Paris, Andilly	1839–1843
Verver	Groningen	1840
Hadfield	Cornbrook (Manchester)	1842
Marchand	Halle	1845
Lewy	Atlantic Ocean	1847
Mene	Paris	1851
A. and H. Schlagintweit	Käruthen, Monte Rosa (Switzerland)	1853
Vogel	Munich	1854
Hlasiwetz	Innsbruck	1855–1856
Gilm	Innsbruck	1856–1857
Pettenkofer	Munich	1858
Regnault	Paris	1859
de Luna[d]	Madrid	1859
Frankland	Mont Blanc	1861
Schulze	Rostock	1863–1871
Roscoe	Manchester	1864
Smith	Manchester, London, Scotland	1864–1969
Thorpe	Irish Sea, Atlantic, South America	1865–1866
Storer and Pearson[c]	Boston	1870
Hill[c]	Cambridge	1870
Weaver	Leicester	1872
Henneberg[e]	Weende (near Göttingen)	1872
Risler[f]	Calèves (Switzerland)	1872–1873
Reiset	Diepe (Belgium)	1872–1873
Truchot[g]	Clermont-Ferrand, Puy de Dôme, Pic du Sancy	1873
Zittel[h]	Libyan Desert	1873
Farsky	Tabor (Bohemia)	1874–1875
Fittbogen and Hässelbarth	Dahme (Prussia)	1874–1875
Tissandier	Balloon ascension	1875
Claesson	Lund	1876
Marié-Davy	Montsouris (Paris)	1876–1879
Hesse[j]	Munich	1877
Lewy	Montsouris	1877–1883
Armstrong	Leeds	1879
Wolffhügel[k]	Munic	1879

Table 4.8: (continued).

Observer	place	date
Macagno[m]	Palermo	1879
Fodor	Budapest, Klausenberg	1879
Armstrong	Leeds	1879
Müntz and Aubin	Paris, Cape Horn, Atlantic, America	1880–1883
Dumas	Paris	1882
Heine[n]	Giessen	1882
Reichardt[o]	Jena	1882–1884
Ebermayer	Bavarian Highlands	1883–1884
Lévy and Miquel	Montsouris	1881–1890
Palmqvist	North Atlantic and Scandinavia	1883
Hempel and Oettel	Dresden	1884
Blochmann	Königsberg	1885
Selander	Stockholm	1885–1886
Spring and Roland[p]	Lüttich	1885
Carnelley and Mackey[q]	Dundee, Perth	1886
Van Nuys and Adams	Bloomington (USA)	1886
Marcet and Landriset	Mont Blanc	1885, 1886
Feldt	Dorpat (today Tartu in Estonia)	1887
Lorenz[r]	Sonnblick	1887
Carnelley and Wilson[s]	Scotch Moorlands	1887–1888
Heimann	Dorpat (today Tartu in Estonia)	1888
Frey	Dorpat (today Tartu in Estonia	1889
Roster[t]	Florence	1889
Petermann and Graftiau	Gembloux (Belgium)	1889–1891
Lebedinzeff	Odessa	1890
Puchner[u]	Bavarian forests	1891
Andrée	balloon flight, Sweden	1894
Williams	Sheffield, center and suburb	1896
Letts and Blake	Belfast	1897
Thierry	Mont Blanc	1898
Brown and Escomte	Kew (Britain)	1898–1901

[a]Thénard (1827)

[b]Brunner (1832) improved *Thénard*'s method using baryta in bottles.

[c]Storer and Pearson (1871). *Francis Humphreys Storer* (1832–1914) was a professor of agricultural chemistry at Harvard University. *Henry Barker Hill* (1849–1903) was assistant and 1884 director and professor of Chemistry at the Laboratory of Harvard University (Anonymous 1871a); *Pearson* found from 21 observations in Boston in spring 385 ppm and Hill in winter 1879 from 11 observations of the outer air of Cambridge an average of 337 ppm.

[d]*Ramon Torres Muñoz de Luna* (1822–1890) Spain chemist; measured high value (0.3–0.6 %); Luna (1860).

[e]Henneberg (1873); *Johann Wilhelm Julius Henneberg* (1825–1890) was a German animal nutritionist and physiologist at Agriculture Research Station Weende near Göttingen.

[f]Risler (1882); *Charles Eugene Risler* (1828–1905), *directeur de l'Institut national agronomique de Paris*.

[g]Truchot (1873a) *Paul Truchot* (unknown living dates) was a chemist in Paris and author of books.

Table 4.8: (continued).

[h]Air collected in tubes that were melted and analyzed later in Munich. Zittel, K. A. (1875) *Zeitschrift der Österreichischen Gesellschaft für Meteorologie* **10**, p. 175. See also: Pettenkofer, M. von (1874) Ueber den Kohlensäuregehalt der Luft in der libyschen Wüste über und unter dem Boden. *Sitzungsberichte der Kaiserlichen Akademie der Wissenschaften, Math.-Naturwiss. Classe*, IV. Bd., pp. 339–351.
[j]*Ludwig Otto Hesse* (1811–1874) was a German physicist and chemist in Halle, Heidelberg, and Munich; almost teaching. Cited by Billings (1893) but no reference found.
[k]*Gustav Wolffhügel* (1845–1899), German hygienist; worked at the *Hygiene-Institut* of the Munich University at *Pettenkofer*; Fodor (1881) cites *Wolffhügel* who carried out 205 determinations from December 1774 to July 1875 with an average to be 376 ppm (no individual reference found).
[m]Average 358 (50–760) ppm; this large variation suggests analytical problems. *Macagno* also determined oxygen, ammonia, and organic matter in air (see p. 674).
[n]Heine (1882).
[o]Reichardt (1884); *Eduard Reichardt* (1827–1891) was a German agriculture chemist at the University of Jena who investigated the soil absorption of CO_2. In 11 determinations he found an average of 299 ppm (290–310).
[p]Spring and Roland (1885).
[q]Carnelley and Mackie (1886); *Thomas Carnelley* (1854–1890) was a chemist in Manchester; *William Mackie* (1856–1932) was a Scottish physician and public health specialist.
[r]Two samples: 205 and 236 ppm.
[s]Carnelley and Wilson (1888).
[t]Roster (1889). *Giorgio Roster* (1843–1927) was an Italian physician in Firenze.
[u]Puchner (1892). *Heinrich Ruprecht Puchner* (1865–1939) was a German soil scientist in Munich; he carried out about 1700 determinations within a range of 0–1000 ppm; Baumann (1893, p. 13–14) criticized the result to be totally wrong and discusses in detail the different sources of errors.
[v]*Richard Weaver* (unknown living dates) who measured CO_2 in buildings (1000 to 3000 ppm) and the influence of ventilation, also determined the suburb value of Leicester to be 460 ppm, surely influenced by city CO_2 (Weaver 1872).

on elevated concentrations, measured at a level, strongly depending on meteorological parameters, which influence dispersion. An example for much lower estimates is the British *Anthony Emmett* (1790–1872), a soldier in the Royal Engineer, who determined CO_2 while crossing the Atlantic and 1836 on Barbados between 125 and 250 ppm (Emmett 1837).

According to *Heinrich Hermann Hlasiwetz* (1825–1875), an Austrian chemist in Innsbruck, all chemists used before the middle of the nineteenth century *Brunner*'s method often without mentioning it. Hlasiwetz (1856) used it to determine carbon dioxide in the air of Innsbruck (May 1855 – February 1856) with large variation from 23 to 830 ppm ($n = 33$) and discussed that such values cannot reflect the true CO_2 level. He discusses several sources of errors and did try to eliminate them, but without success as the result show, citing the German chemist *Karl Friedrich Mohr* (Hlasiwetz 1856, p. 194):

> Alle diese Apparate haben den gemeinschaftlichen, nicht zu beseitigenden Nachtheil, dass sie grosse äussere Glasflächen haben, deren hygroskopischer Zustand gar nicht mit Sicherheit in Anschlag gebracht werden kann [All these instruments have to common disadvantage that can-

not be eliminated, that they have large outer glass surfaces, whose hygroscopic state cannot be taken into account].

Richard Felix Marchand (1813–1850), professor for chemistry in Halle, was the first who estimated a mean from 150 observations (at or before 1845) in Halle significant less than the "4" figure to be 310 ppm (Marchand 1848, p. 24); his finding is only cited by very few following scientists (this paper deals with respiration and without any given details he mentioned this value; Marchand (1850, p. 30) in an extended paper on eudiometry and detailed information of the instrument and the use of it also mentioned as a reference that the air in Halle contains 20.920 % oxygen and 525 ppm(m) – corresponding to 270 ppm(v) – carbon dioxide).

Despite likely much too high values measured by *Bernhard Lewy*, it is worth citing the mean values from eleven different places in New-Granada to be 400.8 ppm and for different seasons found in Bogota, based on a large number of measurements (Lewy 1851):

382.2 ppm (rainy season) oxygen content 20.99542 %
457.3 ppm (dry season) oxygen content 21.02195 %

He also found a large difference in carbon dioxide at night (334.6 ppm) and daytime (542.0 ppm) above the Atlantic Ocean. Extreme CO_2 values *Lewy* found during forest clearing by burning being up to ten times larger (up to 490 ppm with reduced oxygen to 20.33075 % compared to 21.01425 under "normal" conditions). The general higher carbon dioxide values he explains by permanent volcanic emissions.

The British chemist *Henry Enfield Roscoe* (1833–1915), stimulated by the measurements of *Smith* in Manchester and *Lewy* in Paris, conducted some measurements in the city and suburb (Stretford, 4 miles westward) of Manchester 1863–1864, based on *Pettenkofer*'s method (Roscoe 1864); he writes:

> The first and most important conclusion to which these experiments lead is that the amount of carbonic acid contained in Manchester town air differs but very slightly (if at all) from that contained in the air of the neighbouring country.

He found (in ppm at Stretford / Manchester): 385 / 390 (Feb. 3, 1863) and 277 / 280 (March 19, 1863) and on average in Manchester ($n = 46$) 392 (280–560) ppm, whereas the highest value did occur during dense fog.[824] In London on 27 February 1857, *Roscoe* determined 370 ppm. The conclusion he draws is very remarkable for that time in "air pollution science":

> Hence we may conclude that the combustion of coal and respiration of animals exert no appreciate the quantity of carbonic acid contained in the town air of Manchester collected in an open

824 In rooms (chemical laboratory, Chemical Theatre, bedroom) he determined between 740 and 950 ppm and in large crowded meeting room (with 1000 person) 3600 ppm.

> situation; gaseous diffusion and the great motion of the atmosphere serving completely to disperse the millions of tons of this gas which every year evolved by the above-mentioned causes in this neighborhood.

The longest time-series of CO_2 measurements were conducted by the well-known Montsouris Observatory in Paris (1876–1910), Fig. 4.22. Stanhill (1982) discusses possible reasons for this high variability of annual, seasonal, and daily variations; although no direct proof of the reliability of the series is available, *Stanhill* believes a precision of measurement better than 2 % and proposes that a major and variable non-fossil fuel source of atmospheric CO_2 was active during the last quarter of the nineteenth century. However, the large timely variations (but not periodically as we know it today) of the single measurements can be interpreted as the scattering of

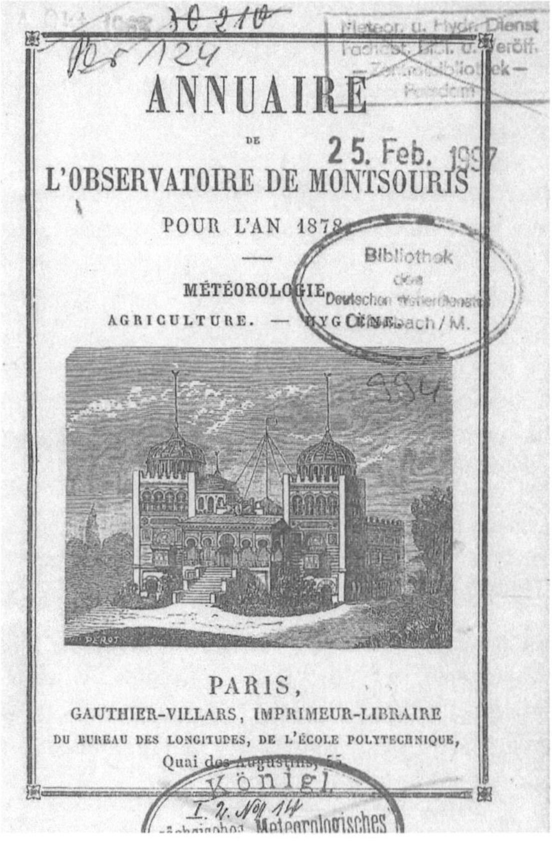

Figure 4.22: The building of the Montsouris Observatory: Front cover of "*Annuaire de l'observatoire de Montsouris pour l'an 1878*". Note the interesting stamps of different libraries, which hold this volume: "*Königlich Sächsisches Institut*" (Chemnitz, Nov 2, 1914), "*Zentralbibliothek des Meteorologischen und Hydrologischen Dienstes*" (Potsdam, October 4, 1963). "*Bibliothek des Deutschen Wetterdienstes*" (Offenbach/Main, 25 February 1997).

Table 4.9: Monthly means at Montsouris station (unknown number of single measurements) after Marié-Davy (1880), in ppm. Mean from all monthly data: 312 ± 39 ppm and from all annual averages: 319 ± 25 ppm.

	01	02	03	04	05	06	07	08	09	10	11	12
1876	–	–	–	269	249	256	261	–	–	313	307	280
1877	280	282	276	270	278	280	277	267	280	269	308	344
1878	333	335	322	331	359	351	342	350	347	353	354	355
1879	356	357	357	358	356	356	346	333	330	304	255	244

the analytical figure by errors and likely predominant by the different times of the single measurements and the subsequent average. Thus, the inter-annual (5–15 ppm) and monthly variations (>10 ppm) cannot be explained by natural physical reasons (as Waterman (1983) writes in response to the *Stanhill* paper) and furthermore, the marked rise in 1890 (by 27 ppm) suggests a change in the sampling and/or analytical procedure. Here are the characteristics of two periods (period average based on mean annual values, standard variation, min-max):

1876–1889 ($n = 9$) 283 ± 8 (270–294)
1890–1910 ($n = 20$) 311 ± 6 (300–325)

The obvious problems with the Montsouris analysis – mentioned already by Reiset (1880) – is illustrated by the data published by Marié-Davy (1880) despite his explanation of different air mass influences (without doubt, the Montsouris station north of Paris in a park – today within Paris – is much less suitable than the Dieppe station for background estimates). Table 4.9 shows the monthly means from April 1876 until December 1879; the standard variation of total monthly means ranges 35–56 ppm and that for annual means ranges 12–41 (monthly minimum-maximum 244–359 ppm). These data illustrate the above-discussed uncertainties. Lévy and Miquel (1891) give a mean from the Montsouris series (1881–1890) to be 287 ± 8 ppm (277–302) based on annual means whereas the mean from all monthly means amounts 293 ± 2.5 ppm (289–297) showing very small variations only.

Franz Ferdinand Schulze (1815–1873), professor for chemistry, physics, and pharmacy since 1850 at the University Rostock, proposed in 1863 to conduct daily measurements of carbon dioxide over a longer duration of time, together with meteorological observations, initially at agricultural research stations. *Schulze* carried out in Rostock from November 1863 to December 1864, but doubted by himself on the values (communicated in the preface in Schulze (1871)):[825]

825 The data he published together with weather observations by *Herrmann Karsten* (1809–1877) in *Die Landwirthschaftlichen Versuchs-Stationen* **9** (1869) 217–227.

1863–1864 ($n = 481$) 364 ± 21 (264–466) ppm

Thus, *Schulze* improved *Pettenkofer*'s method and tested it to conduct more than 1,600 estimations between October 1868 and July 1871 in Rostock from a balcony of the museum close to the chemical laboratory of the university (Schulze 1871):[826]

1868–1871 ($n = 1034$) 292 ± 12 (225–340) ppm

In discussion of his results, the most extensive series of observations at that time, *Schulze* first remarks the remarkable finding that the carbon dioxide content is substantially smaller than from most previous observers, citing *Saussure*, *Boussingault*, *Lewy*, and *Schlagintweit*. He cites *Thorpe* who found similarly "low" values above the sea but he did not cite *Marchand* who was the first to detect values close to 300 ppm at Halle in 1845. In contrast to *Saussure*, he does not find any regularity in the timely variation. Remarkable is *Schulze*'s observation that maritime air masses contain less carbon dioxide than continental; he concluded that the sea permanently absorbs carbon dioxide from the atmosphere in contrast to the land that dominates with sources of carbon dioxide. This observation, *Schulze* communicated, was done before *Thorpe*'s finding and independent of it. From the determination of carbonic acid in rainwater he concluded that rain cannot reduce the atmospheric content of carbon dioxide measurable. *Schulze* also determined the content of carbonic acid in Baltic seawater[827] by degassing it with hydrogen to be 3–5 mg L^{-1} (but also after several days degassing he found still carbonic acid in the water) and through cooking of the seawater 11.83 on mg L^{-1} average.[828] Because this quantum is much larger than expected from pure physical absorption (according to *Bunsen*'s experiments, *Schulze* notes), he assumed the participation of "weak chemical reactions" (electrolytic dissociation was still unknown at that time).

826 It is notable that *Schulze* is cited by Claesson (1876, p. 176) with more than 1600 determinations and a mean of 292 ppm, referring *Chemisches Central-Blatt* 1872 (Dritte Folge. III. Jahrgang, p. 243) where no data are given and *Chemisches Central-Blatt* 1875, where nothing on *Schulze* is communicated. Billings (1893) also cites a mean value of 292 ppm. Callendar (1958) cites $n = 1034$ and 292 ppm as average value.

827 1.073 mg L^{-1} in rainwater, and in distilled water, shake with atmospheric air, 0.831 mg L^{-1} on average. *Schulze* notes that he was unable to remove all carbonic acid from the rainwater. Total dissolved inorganic carbon (DIC), the sum of CO_2(aq), hydrogen carbonate HCO_3^- and carbonate CO_3^{2-}, where the absorption as well the species distribution strongly depend from the pH (see Fig. 4.6 on p. 213 in Möller 2019). Verma (2005) determined in rainwater (pH 5.48) 6.45 mg L^{-1} DIC (97 % H_2CO_3 or CO_2(aq), resp.), given as H_2CO_3. At pH = 4.5, rainwater contains about 6 mg L^{-1}, calculated from the equilibrium with atmospheric CO_2 (20 °C and 360 ppm CO_2), Möller (2019, p. 252).

828 However, in seawater the amount of DIC is on average 145 mg L^{-1} (as H_2CO_3, but chemically to 88 % as HCO_3^-); Möller (2019, p. 23 and 226). In Baltic seawater (Thomas 1997), DIC amounts only 16.2 to 19.2 mg C L^{-1}, corresponding to 90 mg L^{-1}, several times more than *Schulze* determinates.

Similar values were found in Gembloux, a rural site in Belgium, by Petermann and Graftiau (1893) for the period 1889–1891 with a mean of 294 ± 18 ppm from 525 day-time measurements. *Johan Peter Claesson* (1848–1937), a Swedish chemist at Lund and Stockholm (other writing *Clasen* and *Klasen*), developed a method, using crystalline barium hydroxide (*Barythydrat*) as an absorbent in tubes, combined with tubes filled with calcium chloride to fix water vapor and shows the general application also for atmospheric air but not citing the individual values (Claesson 1876):

1876 (n = 31) 279 (237–327)

He writes (Claesson 1876, p. 176):

> Kohlensäurebestimmungen aus der Luft sind somit sehr leicht ausführbar und können vielleicht in die meteorologischen Institute eingeführt werden. Man braucht ja nur zwei Wägungen am Tag zu machen [Determinations of carbonic acid in the air are therefore very simple to carry out and could be possibly introduced in meteorological institutes. One must do only two weightings daily].

As examples of other (unbelievable) high values owing to systematic errors, the following figures are given, by Gilm (1857)[829] from Innsbruck

November 1856 – March 1857 (n = 18) 415 ± 23 (381–458) ppm,

and by Farsky (1877)[830] from Tabor, Bohemia

1874–1875 (n = 295) 342 ± 8 (328–362) ppm.

Farsky (1877) denotes himself as a newcomer ("*Als Neuling dieser Art von Untersuchungen* [. . .]"), giving us an example that handling and long-term experiences are needed for "correct" CO_2 atmospheric air analysis as Blochmann (1886) emphasizes. Thorpe (1867, p. 152) writes:

> These old measurements, scarcely to be trusted as regards quantity, owing to the inaccuracy nature of the method employed.

Boussingault and Lewy (1844) conducted in Paris CO_2 first measurements in two time series (1839–1840 and 1843, latter simultaneous with that in Andilly, a village 240 km east of Paris) that are not unrealistic when comparing with later measurement by Müntz and Aubin (1882b):

829 *Hugo von Gilm* (1831–1906) was an Austrian chemist.

830 *Franz* (*František*) *Farsky* (1846–1927) was a Czech chemist.

October/November 1839 and February/March 1840 (n = 19)	365 (305–494)
September/October, Paris 1843 (n = 100)	319
September/October, Andilly 1843 (n = 94)	299

Robert Angus Smith obtained in 1864 (Smith 1872, pp. 49–54), in ppm:

Suburbs of Manchester, February–April (n = 14)	369 (291–467)
Thereof westerly wind (n = 5)	300 (291–313)
Thereof easterly wind (n = 9)	408 (340–467)
Manchester, February–April (n = 74)	442 (311–805)
Thereof during fogs	679
Thereof in streets in usual weather	403
London, on the River Thames, April 1864 (n = 8)	343 (298–354)
London, in the Open Places (n = 5)	302 (285–334)
London, in the Streets (n = 10)	380 (337–428)
London, November 1869 (n = 35)	439 (345–518)
London, Metropolitan Railway tunnels, November 1869 (n = 6)	1452 (780–3380)
Scotland, August 1865, different locations (n = 14)	341 (315–350)
Glasgow, winter 1869 (n = 21)	502 (337–618)

Smith also measured in Scotland at different elevated places, subdivided into below 1000, 1000–2000, 2000–3000, and above 3000 feet, as well as in forests, not finding any significant difference with means between 332 and 337 ppm.

Gaston Tissandier, together with his brother *Albert Tissandier* (1839–1906), conducted several balloon flights to study meteorological conditions (Tissandier 1878, 1879); on the journey March 23–24, 1875 from Paris to Arcachon (near Bordeaux) they sampled carbon dioxide and determined it to be 240 ppm at an altitude of 890 m and to be 300 ppm at 1000 m (Figs. 4.23 and 4.24).

Salomon August Andrée (1854–1897), a Swedish engineer and polar researcher, also carried out balloon flights (1892–1894) and found no changes with altitude up to 4290 m (n = 19, 305–358 ppm); the analysis was made by *Palmqvist* in Stockholm (Andrée 1894).

There are several historic measurements with averages significantly less than 290 ppm. Remarkable measurements were done 1878–1889 in Dorpat (today Tartu, Estonia) belong the Medical Faculty in three doctor theses: *Victor Richard von Feldt* (1860–1917), *Jacob Heimann* (1862–?), and *Eugen von Frey* (1862–?). The supervisor of all these doctorands was *Georg Dragendorff* (1836–1898), a German chemist and pharmacist who studied at *Franz Schulze* in Rostock, who summarized the results in a report to the "*Naturforscher-Gesellschaft*" in 1889 (Dragendorff 1892) with a total mean of 265 ppm (n = 1534), very close to the value found by *Heine* with infrared absorption to be 262 ppm. In Dorpat, the method by *Pettenkofer* was used, found a mean of

Figure 4.23: *Gaston Tissandier* and *Albert Tissandier*; contemporary illustration (etching by Auguste Boulard, 1883; Bibliothèque Nationale Paris).

Dorpat, February–May 1887 (n = 377), Feldt (1887)	266 (185–361)
Dorpat, (n = 601), Heimann (1888)	269 (182–375)
Dorpat, (n = 556), Frey (1889)	262 (189–336)

Variation among monthly means was small within the range 250–283 ppm. When reading these works, the reader gets a strong feeling of its seriousness and that the measurements are done very accurately. The interesting finding by Heimann (1888) is the day-night difference to be on average 28 ppm: day-time average 258 ppm (n = 379) and night-time average 286 ppm (n = 222). Similarly "low" values found Heine (1882) in Giessen, determined by a physical method during three days (7–10 July 1881):

Giessen, July 1881 (n = 50)   261 ± 28 (200–340)

This large variety of individual values (about ±80 ppm) shows an error of 30 %, however. It does not mean that the mean value is incorrect and its accuracy may be ±10 ppm. Unrealistic low levels found Truchot (1873a) at Puy-de-Dôme (1446 m) with 203 ppm and Pic du Sancy (1884 m) with 172 ppm, whereas his determination in Clermont-Ferrand (all in August 1873) with 313 ppm seems to be reliable.

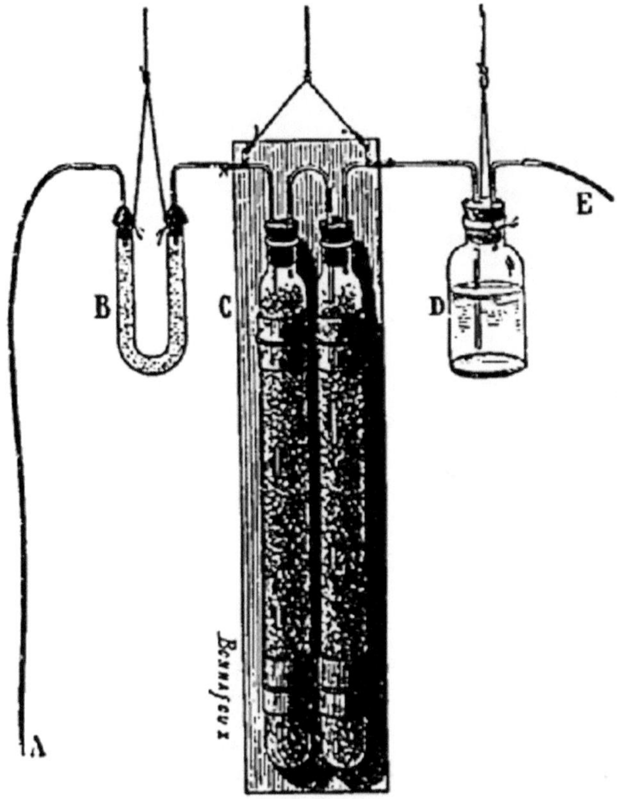

Fig. 1. — Appareil de MM. Hervé-Mangon et Gaston Tissandier pour doser l'acide carbonique de l'air, tel qu'il était disposé à bord du ballon *le Zénith*.

A. Entrée de l'air extérieur. — B. Tube à coton destiné à intercepter les poussières. — C, Tubes remplis de pierre ponce imbibée de potasse exempte de carbonate. — D. Flacon renfermant de l'eau de baryte. — Tube communiquant avec un aspirateur à retournement.

Figure 4.24: Apparatus to sample carbon dioxide on the balloon journey (Tissandier 1878, p. 331).

I fully agree with the statement by From and Keeling (1986) that analysis by French chemist *Jules Reiset* (1818–1896) shows the highest level in analytical science for that time and seems to be most reliable. Reiset (1879) carried out the first series at a rural station 8 km from Dieppe (September 1872 until August 1873) with analysis at day- and night-time (n = 92) giving a mean of 294 ppm; additional parallel measurements above different crops and under free field conditions show differences up to 10 ppm less above vegetation and near a herd of sheep's (300 head) the CO_2 increases to 318 ppm. In a later measurement campaign from June until November 1879, using improved sampling, described in very detail, Reiset (1880, 1882) estimates a mean of 298 ppm (n = 91), including daytime (mean 289 ppm) and nighttime measurement

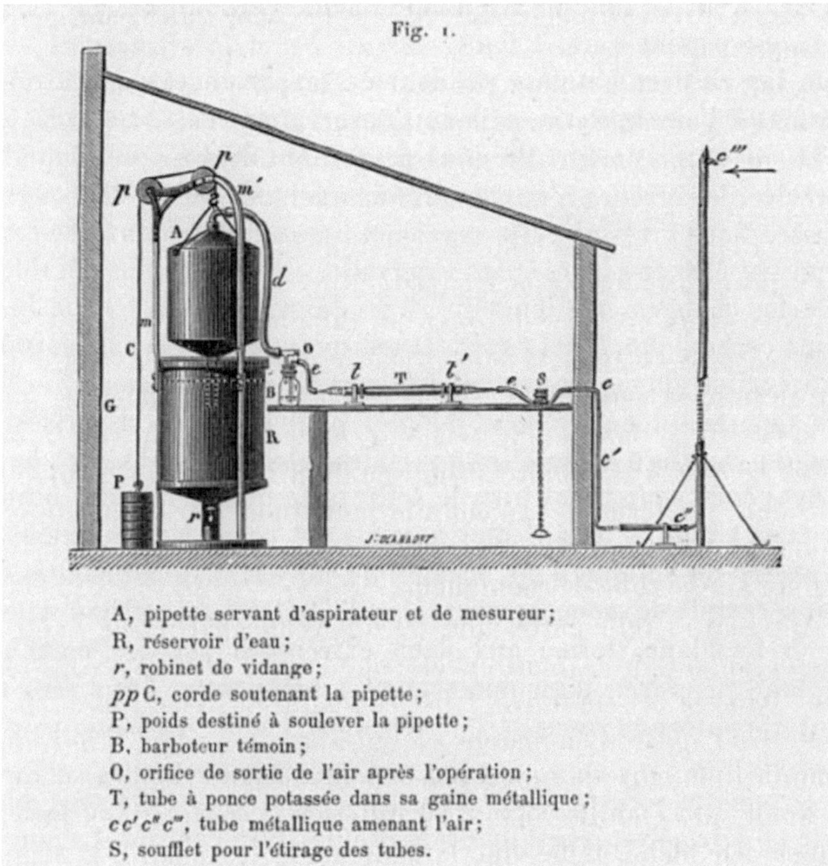

Figure 4.25: Equipment for carbon dioxide measurements at Cape Horn 1882–1882 (Müntz and Aubin 1886, p. A24).

(mean 308 ppm), unfortunately not mention the number of analysis. While fog, a mean of 317 ppm (up to a maximum of 342 ppm in intense fog) he found. All data were related by *Reiset* to standard conditions (dry air, 0 °C and 760 Torr).

Besides *Reiset*, Müntz and Aubin (1882b, 1884a, 1886)[831] made very careful measurements at different sites (Fig. 4.25). These data are of high quality and may well be the only nineteenth century data available which are unequivocally free from local or regional pollution effects (Wigley 1983); in ppm:

831 In difference to *Müntz* and *Schloesing*, I have not found any data on *E. Aubin* who was a chemist and is also known for the *Méthode de Schloesing at Aubin* for nitrogen determination.

Paris, December 1880 – March 1881 ($n = 21$)	339 ± 26 (301–422)
Paris, April 1881 – June 1881 ($n = 8$)	293 ± 6 (288–306)
Vincennes, suburb of Paris, in April and May 1881 ($n = 26$)	284 ± 13 (270–317)
rural site in 1881 (*institute agronomique*) ($n = 11$)	299 ± 16 (273–329)
Pic du Midi (2887 m a.s.l) in August 1881 ($n = 14$)	267 ± 9 (252–294)
Haiti, Pétioville, 1882 ($n = 8$)	276 ± 20 (252–314)
Patagonia, Santa Cruz, 1882 ($n = 9$)	267 ± 13 (258–197)
Chile, Cerro Negro 1882, ($n = 5$)	270 ± 9 (259–282)
Cape Horn, Bay Orange, October 1882 – July 1883 ($n = 40$)	258 ± 12 (231–185)
Martinique, 1882 ($n = 7$)	280 ± 16 (171–308)
Florida, Saint Augustin, 1882 ($n = 7$)	291 ± 8 (281–305)
all last seven remote location ($n = 91$)	265 ± 12 (231–314)

They also refer mean data for daytime and nocturnal measurements, in ppm:

Cape Horn	262 / 270
Florida	270 / 295
Martinique	290 / 295
Mexico	266 / 286
Santa Cruz	266 / 267
Chile	266 / 282

Furthermore, they present simultaneous data from Paris and Joinville, a small city about 180 km ESE from Paris; the mean difference amounts 20 (4–32) ppm, clearly showing the increased CO_2 level at Paris:

June 1, 1881	299 / 274
June 11, 1881	306 / 272
June 15, 1881	289 / 285
June 28, 1881	296 / 279

From the measurements conducted in Paris, a pronounced seasonal difference of 46 ppm results. The results of *Müntz* and *Aubin* are of invaluable value for the construction of pre-twentieth century CO_2 levels (in ppm):

large cities	>300
clean rural sites	<290
remote sites	<280

Dumas (1882) states as "*la grande moyennes*" 294–310 ppm CO_2 as "normal value" (background in today's sense). Petermann and Graftiau (1891) communicate a typical value around 300 ppm, but most of his cited values are between 290 and 295 ppm.

More findings around "300" can be added; however, there is no doubt that these figures do not represent the "background" but a "town background":

Halle, Germany (*Marchand*), 1845	310
Atlantic Ocean (*Thorpe*), 1865	295
Rostock (*Schulze*), 1868–1871	292
Leeds, Britain (*Armstrong*), 1879	313
Belfast, Ireland (*Letts* and *Blake*), 1897	300
Stockholm (*Selander*), 1885–1886	303
Kew, Britain (*Brown* and *Escombe*), 1898–1901	294

Selander (1887)[832] conducted 1885–1886 CO_2 measurements on the island Vaxholm near Stockholm, and found a remarkable dependency from the wind direction (quoted after Andrée 1894, p. 367):

Wind from land	310 ($n = 4$)
Wind from sea	291 ($n = 4$)

During the Swedish Greenland expedition 1888, *Laura Auguste Palmqvist* (1857–1932), Swedish chemists in Gothenburg, analyzed air samples in flasks from several localities in Norway, Sweden, Greenland, and surrounding seas; *Otto Pettersson* (1848–1941), professor of chemistry at Stockholm Högskola (*Experimentalfältet*), had initiated the measurements and supervised *Palmqvist*'s analyses (Palmqvist 1892). *Palmqvist* asked, "What is the atmosphere's normal content of carbonic acid, and what variations is this content subordinate to?" (Palmqvist 1892, cited after Bohn (2011, p. 168)). The high quality of measurement is expressed by *Palmqvist*'s comment:

> Die Differenz zwischen zwei Analysen derselben Luft übersteigt selten ein Hunderttausendstel des ganzen Volumens [The difference between two analyses of the same air rarely exceeds one hundred thousandth of the whole volume] (Palmqvist 1892, cited after Sondén and Tigerstedt 1895, p. 20).

It is worth to refer the work of *William Marcet* (1828–1900), President of the Royal Meteorological Society, and *Auguste Landriset* (unknown living data), a chemical assistant at the University of Geneva, who conducted at the Lake of Geneva and the summit of the Dole (1674 m) simultaneous determination in August 1885 and 1886. They found at the ground site 350 and the summit 362 ppm as average; but the remarkable finding was that "when the summit is in a cloud, the air in the cloud contains a smaller portion of carbonic acid than it would hold in fine clear weather" (Marcet and Landriset 1887, p. 172). In a discussion to this paper, *William James Russell* "was surprised to see that

832 *Nils Edvard Selander* (1846–1931) was a Swedish bacteriologist and regimental doctor.

there was so large an amount of carbonic acid present in the air of the neighbourhood of Geneva".

Baumann (1893) states in an excellent overview that the mean CO_2 content of the atmosphere varies only small; however, according to *Baumann*, the absolute minimum and maximum from all these measurements was 260 ppm and 354 ppm, respectively, referring the different localities as discussed above. Brown and Escombe (1905b)[833] estimated in Kew (Britain) from 1898–1901 ($n = 94$) a mean of 294 ppm CO_2 (243–360, where only 9 values were larger than 320 ppm); they give an error of only ±1 % (±3 ppm) which is surprisingly low.

William Carleton Williams (1850–1904), professor for chemistry at the University College Sheffield, gives a Table with the results of 32 authors from 1802 to 1892 (Williams 1897) and conducted own measurements in Sheffield, confirming the meanwhile known differences between the city center (390 (280–622) ppm, $n = 20$) and suburb (327 (216–514) ppm, $n = 142$). However, most remarkable is his finding that the CO_2 quantity depends on the air temperature: 406 ppm below 0 °C, 332 ppm 0–5 °C, 322 ppm 5–10 °C, and 298 ppm 10–15 °C. The mean for February until April (in summer no measurements were done) is 293 (259–313) ppm; thus, obviously heating caused CO_2 quantities larger than 290 ppm.

From all measurements listed in Tables 4.10 and 4.11 (but excluding the very low values below 250 ppm), having a mean below 300 ppm, it follows a total mean of 281 ± 15 (259–300) ppm, a representative for background places around the world in the second half of the nineteenth century. Callendar (1958) derived from the most reliable data sets (*Reiset*, *Müntz*, and *Aubin*, *Letts* and *Blake*, *Brown* and *Escombe*) an average of 288.5 ppm, and from a larger data set (18) 291.5 ppm, and he concluded that the data show no significant trend between 1870 and 1900.

On the other side, in large cities, in the winter time, an average of 400 ppm (varying from 300 to 800 ppm) was typical, whereas, at the place in the suburbs of cities and industrial sites, the typical mean CO_2 concentration was around 300 ppm.

The knowledge of the so-termed *pre-industrial* CO_2 level is important for two reasons, first, we need this value to estimate how much CO_2 humans were added to the atmosphere since the beginning of the Industrial Revolution, and second, this value is a reference for all atropogenic impacts which depend on the actual CO_2 level, e. g., climate forcing and oceanic acidification. First pre-industrial levels were given by Callendar (1938, 1949, 1958) and From and Keeling (1986), derived from historical measurements to be 290 or 292 ppm, respectively. This "reference" value was estimated by several authors to be in the range of 243–290 ppm (Holmén 1992 and

833 *Horace Tabberer Brown* (1848–1925) was a British chemist who worked since 1866 at the Worthington brewery in Burton-on-Trent. *Fergusson Escombe* (1872–1935), botanist. *Brown* and *Escombe* are well-known for her works on gases (namely carbon dioxide) and liquids in relation to photosynthesis and plant growth.

Table 4.10: Historic measurements of carbon dioxide in the nineteenth century (concentration in ppm); *n* number of samples or measurements. Note the scatter of individual measurements (in parenthesis min–max).

site or station	year	concentration	*n*	reference
Chambeisy	1827–1829	425 (315–574)	104	Saussure (1830)
Manchester	1832–1833	530 (420–862)	19	Watson (1850)
rural sites thereof	1833–1834	413 (361–474)	12	Watson (1850)
Paris	1839–1840	400	142	Dumas and Boussingault (1841)
Groningen	1840	419 (351–508)	90	Verver (1840)[r]
New Granada	1847	401 (361–504)[o]	large	Lewy (1851)
Halle, Germany	1845	310	150	Marchand (1848)
Alps	1849	580	?	Schlagintweit (1854)
Monte Rosa	1852	590–950	?	Schlagintweit (1854)
Paris	1851	232 (132–420)	12	Mène (1851b)[q]
München	1854	346 (298–404)	11	Vogel (1854)
Innsbruck	1856–1857	415 ± 23 (381–458)	18	Gilm (1857)
Mont Blanc	1860	420–1110	10	Frankland (1861)
Rostock	1863–1864	360 ± 21 (320–405)[c]	481	Schulze (1871)
Rostock	1868–1871	291 ± 12 (272–323)	1034	Schulze (1871)
Irish Sea	1865	308 (266–322)	26	Thorpe (1867)[k]
Atlantic Ocean	1865	295 (266–336)	51	Thorpe (1867)[k]
Para, Brasilia	1866	328 (307–348)	34	Thorpe (1867)[k]
Weende	1872	320 (303–338)	17	Henneberg (1873)
Dieppe[f]	1872–1873	294	92	Reiset (1879)
Dieppe[f]	1879	298[e]	91	Reiset (1880)
Tabor[h]	1874–1875	342 ± 8 (328–362)	295	Farsky (1877)
Lund	1875	259 (237–327)[b]	31	Claesson (1876)
Leeds, Britain	1879	313	115	Armstrong (1879)[m]
Montsouris	1877–1881	302 (297–309)	large	Lévy (1882)
Montsouris	1881–1890	287 ± 8 (277–302)[a]	large	Lévy and Miquel (1891)
Montsouris	1881–1890	293 ± 2.5 (289–297)[b]	large	Lévy and Miquel (1891)
Paris	1880–1881	288–422[p]	large	Müntz and Aubin (1881a)
Vincennes[i]	1880–1881	288 (270–299)	large	Müntz and Aubin (1881a)
Vincennes[i]	1881	284 ± 13 (270–317)	37	Müntz and Aubin (1882b)
Paris	1881/1882	319 ± 26 (301–422)	20	Müntz and Aubin (1882b)
Paris	1882	293 ± 6 (288–306)	8	Müntz and Aubin (1882b)
France, rural site	1881	300 ± 15 (273–329)	8	Müntz and Aubin (1882b)
Pic du Midi	1881	286 ± 9 (269–301)	14	Müntz and Aubin (1881b)
Giessen, suburb	1881	261 ± 28 (200–340)	50	Heine (1882)
Stockholm	1883	320 (250–390)	?	Palmqvist (1892)
Dresden	1884[n]	408 ± 45 (320–550)	51	Hempel (1885)
Stockholm	1885–1886	303 (224–481)	?	Selander (1888)
Bloomington	1886	282 (273–290)	18	Van Nuys and Adams (1887)
Gembloux	1889–1891	294 ± 8	525	Petermann and Graftiau (1893)
Dorpat[g]	1887	266 (185–361)	377	Feldt (1887)
Dorpat	1888	262 (189–336)	556	Frey (1889)
Dorpat	1889	269 (182–375)[d]	601	Heimann (1888)
Odessa	1890	304 (288–334)	7	Lebedinzeff (1891)

Table 4.10: (continued).

site or station	year	concentration	*n*	reference
Belfast	1897	289	67	Letts and Blake (1900, 1901)
Kew, Britain	1898–1901	294 (243–360)	94	Brown and Escombe (1905b)

[a] Based on annual means
[b] Means from all monthly means
[c] Doubt by the author himself
[d] Variation among monthly means is 250–283 ppm; day-night difference to be on average 28 ppm: day-time average 258 ppm ($n = 379$) and night-time average 286 ppm ($n = 222$)
[e] Daytime means 289 ppm and nighttime mean 308 ppm
[f] All data were related to standard conditions (dry air, 0 °C and 760 Torr)
[g] Today Tartu, Estonia
[h] Bohemia
[i] Suburb of Paris (*institute agronomique*; *le Plateau de Gravelle, bois de Vincennes*)
[k] Sir *Thomas Edward Thorpe* (1845–1925) was a British chemist
[m] *George Frederick Armstrong* (1842–1900) was an English sanitary engineer; daytime mean ($n = 53$) 296 ppm and night-time mean ($n = 63$) 330 ppm.
[n] Carried out by *Felix Oettel* (1862–?) assistant of *Hempel* at the university Dresden together with oxygen determinations, November 6 to December 24, 1884
[o] Means from 11 stations over about one year
[p] December 1880 to May 1881 (at *Conservatoire des Arts et Métiers*); maximum CO_2 levels at overcast and calm weather (322–422 ppm) and minimum at clean and windy days (289–310 ppm), which characterizes the local accumulation and dispersion, respectively
[q] French chemist (unknown given name and living dates).
[r] *Bernardus Verver* (unknown living dates) was a Dutch physician in Groningen. He wrote his 164-pages "Dissertatio politico-medica inauguralis, qua inquiritur: num publicae sanitati nocere possint venena metallica, quibus conserantur agri, ad occidenda animalia nociva" in 1841 (note that most dissertation in the nineteenth century was not longer than 20 pp.) *Verver* also determined oxygen ($n = 45$) to be 20.864 %.

citations therein, p. 242). Wigley (1983) concludes, mainly based on the data from *Müntz* and *Aubin*, that 260–270 ppm is more likely level than 290–300 ppm. Today's most accepted *pre-industrial* value is 280 ± 10 ppm whereas mainly data from bubbles in ice cores were employed (Neftel et al. 1985, Friedli et al. 1986, Etheridge et al. 1996) and from historical ocean CO_2 data, a smaller value of 268 ± 13 ppm was derived (Chen and Poisson 1984). From the ice core, the AD 1010–1850 average amounts to 280.9 ± 2.6 ppm and the 1850–1900-year average was 290.6 ± 3.8 ppm (further reading Möller 2019).

4.3.1.4 The twentieth century rise of carbon dioxide

The nineteenth century ends with the experimental finding that the background CO_2 concentration lies between 260 and 290 ppm (well supported by ice core data gained in the late 20th century), the CO_2 level outside towns is around 300 ppm, and the air of towns contains significantly more than 300 ppm CO_2. Furthermore, that the CO_2 quan-

Table 4.11: Historic values of carbon dioxide less than 300 ppm in the nineteenth century (extracted from Table 4.10).

site or station	year	concentration	*n*	reference
Atlantic Ocean	1866	295	51	Thorpe (1867)
Rostock	1868–1871	292 ± 12 (225–346)	1034	Schulze (1871)
Dieppe	1872–1873	294	92	Reiset (1879)
Pic du Sancy	1873	172	1	Truchot (1873a)
Lund	1875	259 (237–327)	31	Claesson (1876)
balloon ascends	1875	240, 300	2	Tissandier (1878)
Dieppe	1879	298	91	Reiset (1880)
Montsouris	1881–1890	287 ± 8 (277–302)	large	Lévy and Miquel (1891)
Pic du Midi	1881	286 ± 9 (269–301)	14	Müntz and Aubin (1881b)
Giessen, suburb	1881	261 ± 28 (200–340)	50	Heine (1882)
Vincennes	1881	284 ± 13 (270–317)	37	Müntz and Aubin (1882b)
overseas	1882–1883	265 ± 12 (231–314)	91	Müntz and Aubin (1886)
Jena	1883	299 (290–300)	11	Reichardt (1884)
Bloomington	1886	282 (273–290)	18	Van Nuys and Adams (1887)
Sonnblick	1887	220	2	Lorenz (1887)
Dorpat	1887	266 (185–361)	377	Feldt (1887)
Dorpat	1888	262 (189–336)	556	Frey (1889)
Dorpat	1889	269 (182–375)	601	Heimann (1888)
Mont Blanc	1898	269	1	Thierry (1899)
Sheffield	1896	293 (259–313)	large	Williams (1897)
Belfast	1897	289	67	Letts and Blake (1900, 1901)
Kew, Britain	1898–1901	294 (243–360)	94	Brown and Escombe (1905b)

tity in air shows typical seasonal and diurnal variations but no trend. Callendar (1958) proposed a base value of 290 ppm for the year 1900. The second half of the twenteith century, with the Mauna Loa record, also known as the *Keeling* curve, certainly the best-known icon illustrating the impact of humanity on the planet as a whole, will herald a new era, *climate change*.

The measurement series carried out in Boston by *Francis Gano Benedict* (1870–1957)[834] between April 1909 and January 1912 is probably the longest and most precise of that time. Benedict (1912) tried carefully to avoid town influences, and by separating the values shown in Table 4.12 it is possible to derive a seasonal amplitude of about 9 ppm and a town contribution (adding to the rural background) of about 45 ppm; the most likely background mean is 307 ppm.[835] Krogh (1919) states from

834 American nutritionist (director of the nutrition institute in Boston), who developed a calorimeter and a spirometer, used to determine oxygen consumption and measure metabolic rate; known for his precise CO_2 and O_2 measurements.

835 To give an example, how different values become when averaging, Callendar (1958) cites *Benedict*'s analysis as mean of 317.5 ppm ($n = 645$), much higher than shown in Table 4.12 for all cases but correct *Krogh*'s analysis to be 300 ($n = 40$).

Table 4.12: Statistics of the CO_2 measurements by Benedict (1912) carried out in Boston at the Nutrition Institute from April 1909 to January 1912 (assessed imprecision larger than 10 ppm).

	all data	**summer**[a]	**winter**[b]	**all <320 ppm**	**all >320 ppm**
mean (ppm)	307	302	311	298	343
standard deviation	24	19	26	15	21
min	210	260	260	210	330
max	460	460	430	320	460
n	670	313	356	534	123

[a] April-September
[b] October-March

Benedict's analysis the background averages for CO_2 of 0.03 % (0.001 % accuracy) are identical to his own measurements (adding 10–70 ppm by town influence from Copenhagen). Thus, in the early twentieth century a continental background CO_2 level of around 300 ppm seems reliable, i. e., already showing an increase by 20 ppm from the time before 1880 (Table 4.13).

August Krogh was the first to use a new gas analysis apparatus in Copenhagen to show that oxygen varies inversely with carbon dioxide (however, the data do not provide a stoichiometric balance because of still missing precision). He wrote that "there is usually an oxygen deficit corresponding to the increase in carbon dioxide" (Krogh 1919, p. 12), but the air is rapidly renewed by atmospheric ventilation and that urban CO_2 emission contributed less than 0.001 % to the ventilation.

Guy Stewart Callendar (1898–1962)[836] presented in 1938 the hypothesis that fossil fuel combustion had contributed to an increase in the atmospheric content of CO_2, which, through its effect on "sky radiation" had caused mean temperatures to increase; he stated (Callendar 1938):

> By fuel combustion man has added about 150,000 million tons of carbon dioxide to the air during the past half century. The author estimates from the best available data that approximately three quarters of this has remained in the atmosphere.

Callendar (1940) used measurements by the Finnish chemist *Kurt Buch* (1881–1967), because they were among the most recent values of atmospheric CO_2 content, and stated that "The modern measurements in Table II show an increase over those taken in the nineteenth century of approximately 30 ppm of CO_2". Bray (1959) writes "The magnitude of the increase for weighted yearly non-urban data was 25 ppm (from 294 to 319) for the quarters 1857–1881 to 1932–1956". British limnologist and ecologist *George Evelyn Hutchinson* stated that "There can be little doubt that during the first half of the

836 A leading British steam and combustion engineer; known for propounding the theory that linked rising carbon dioxide concentrations in the atmosphere to global temperature, but he thought this warming would be beneficial.

Table 4.13: Measurements of carbon dioxide in the twentieth century (concentration in ppm); *n* number of samples or measurements. Note, the scatter of individual measurements (in parenthesis min–max).[837]

site or station	year	concentration	*n*	reference
Boston, USA	1909–1912	307 (210–460)	645	Benedict (1912)
Copenhagen	1917	302	40	Krogh (1919)
near Stockholm	1920–1923	313[a]	20	Lundegårdh (1924)[j]
High mountains	1925	293	9	Reinau (1930)[g]
USA[b]	1930's	310	?	Callendar (1938)
North Atlantic	1932 and 1935	319	81	Buch (1939)
North Finland	1934–1935	321	95	Buch (1939)
Boston, USA	1930–1932	310 (260–350)	291	Carpenter (1937)
Durham, USA	1934–1935	310 (270–350)	75	Carpenter (1937)
Baltimore, USA	1933–1936	310 (280–600)	790	Carpenter (1937)
rural Scotland	1936	324	153	Haldane (1936)
urban site	1936	399	500	Duerst (1939)[k]
urban site	1939	442	730	Kreutz (1941)[h]
Bavaria, rural site	1951	321[d]	235	Huber (1952)[i]
Scandinavia	1954	329 (319–347)	594	Fonselius et al. (1956)
Mauna Loa	1959	316	large	Keeling et al. (1976)
Mauna Loa	1970	326	large	Keeling et al. (1976)
remote sites[e]	1981–1990	347 (345–350)[e]	large	Keeling and Whorf (2004)
Mauna Loa	2019	411	large	[f]

[a] At costal site 304 ppm
[b] Rural site means
[c] Means from 10 sites in Scandinavia
[d] Daytime means 306 ppm and nighttime mean 410
[e] Means of 9 remote stations worldwide (in parenthesis variation of station mean)
[f] Dr. *Pieter Tans*, NOAA/ESRL (www.esrl.noaa.gov/gmd/ccgg/trends/) and Dr. Ralph Keeling, Scripps Institution of Oceanography (scrippsco2.ucsd.edu/).
[g] *Erich Hellmut Reinau* (1884–?) was a German agriculture chemist.
[h] *Wilhelm Kreutz* (1900–1964) was a German agricultural meteorologist.
[i] *Bruno Huber* (1899–1969) was a German botanist in Tharandt and Munich.
[j] *Henrik Gunnar Lundegårdh* (1888–1969 was a Swedish botanist.
[k] *Johann Ulrich Duerst* (1876–1950) was a German agronomist at the Bern University (Switzerland).

twentieth century the mean CO_2, the content of the air in north temperate latitudes has increased" (Hutchinson 1954).

Buch also took an interest in *Callendar*'s work. Based on his measurements (Buch 1939), *Kurt Buch* suggested that increasing industrial combustion had upset a possible previous equilibrium and that CO_2 was now moving from the atmosphere to the sea (Buch 1945, 1948). On a conference in atmospheric chemistry held in Stockholm in

837 Effenberger (1951) and Stepanova (1952) give a bibliography on carbon dioxide in the atmosphere.

1954, opened by *Carl-Gustaf Rossby*, director of the Institute of Meteorology at Stockholm Högskola, *Buch* spoke on the subject (Eriksson 1954). In order to examine the question of an increase in the level of CO_2 in the atmosphere, *Buch* suggested making some new measurements. *Erik Eriksson* at the institute was making plans for places at which to take samples (Bohn 2011, p. 170), and at the end of 1954, 13 stations in Scandinavia were in operation, gaining 2–3 samples monthly.[838] On the basis of the results of this network (closed in 1956), Fonselius et al. (1956, p. 178)[839] wrote that

> [. . .] the difference between Buch's and our values suggest an increase of the CO_2 content in Scandinavia, but it is impossible to say at present whether this increase is just a fluctuation in the regional CO_2- climate or if it represents a steady increase since 1935.

In that time, it remained an open discussion whether atmospheric CO_2 increases or the differences in measured quantities are caused by natural variation, and what the CO_2 absorption capacity of the sea is.

Slocum (1955) doubts *Callendars* finding of increasing CO_2 in the atmosphere due to the combustion of fossil fuels, arguing that comparison of measurements made in the nineteenth century and recent years do not demonstrate that there was a significant increase in atmospheric CO_2. *Slocum* concluded that the mean value for the nineteenth century is 335 ppm (Slocum 1955, p. 229) and refers to German values of more than 400 ppm from the 1930s (see Table 4.13). However, the decision, which of the older values are reliable, *Callendar* made by deliberate rejection criteria (Callendar 1858, p. 244). Fonselius et al. (1956) conducted in 1954 at 10 sites in Scandinavia CO_2 measurement using *Krogh*'s technique and get a mean of 329 ppm (319–347 sites mean, $n = 504$); they wrote (Callendar 1958, p. 176):

> In spite of Callendar's careful selection of values it remains a fact that the samples were taken with different techniques and that the analyses were made in different ways. Some samples were taken in few minutes in gas burettes, others during many hours or days by slow absorption. No attention was paid to the geographical distribution of the sampling places and to the seasons. If we are to detect trends, we need samples taken by the same technique, at the same hour of the day and at same place for several years, and the analyses have to be carried out with identical techniques.

838 The method of analyzing the air samples was a modification of Danish physiologists *August Krogh* and *Poul Brandt Rehberg*'s microtitration technique from 1929; *Rehberg* (1895–1989) was a Danish physiologist at Copenhagen.

839 *Stig Fonselius* (1921–2003) from the Swedish Fishery Board of Sweden, *Folke Koroleff* (1918–2003) from the Institute of Marine Research, Helsinki), and the Swedish scientist *Karl Erik Wärme* (1920–1970). *Giles Slocum* (unknown living data) was meteorologist at U. S. Weather Bureau.

Revelle and Suess (1957)[840] wrote:

> Thus human beings are now carrying out a large scale geophysical experiment of a kind be reproduced in the future. Within a few centuries we are returning to the atmosphere and oceans the concentrated organic carbon stored in sedimentary rocks over hundreds of millions of years. (p. 19)
> In contemplating the probably large increase in CO_2, production by fossil fuel combustion in coming decades we conclude that a total increase of 20 to 40 % in atmospheric CO_2, can be anticipated. (p. 26)

When *Charles David Keeling* (1928–2005)[841] started his CO_2 measurements in the 1950s, he almost always derived the same value of 310 ppm (Keeling et al. 1976). Previous measurements of CO_2 in the atmosphere did not show this constancy, but these measurements had been made by wet chemical methods, which were considerably less accurate than the dry manometric method he deployed. He concluded that a) the earth system might behave with surprising regularity and b) there was a need to make highly accurate measurements to reveal that regularity. Since 1958, CO_2 concentrations at the Mauna Loa Observatory were obtained using a non-dispersive, dual detector, infrared gas analyzer. Air samples are obtained from air intakes at the top of four 7 m towers and one 27 m tower. By the early 1970s, this curve was gaining serious attention and played a key role in launching a research program into the effect of rising CO_2 on the climate.

4.3.1.5 Carbon dioxide and climate

The carbon dioxide theory emerged as a consequence of the experimental work of *John Tyndall*, who wrote that slight changes in the amount of any of the radiative active constituents of the atmosphere – water vapor, carbon dioxide, ozone, or hydrocarbons – may have been produced (Tyndall 1861)

> [...] all the mutations of climate which the researches of geologists reveal [...] they constitute true causes, the extent alone of the operation remaining doubtful of climate which the researches of geologists reveal [...].

Luigi de Marchi (1857–1936) first explains glacial periods through changes of atmospheric properties like transmission (Marchi 1895). The American geologist *Thomas Chrowder Chamberlin* (1843–1928) developed at length the idea that climate changes

840 *Roger Randall Dougan Revelle* (1909–1991) was an US oceanographer and climatologist, and was among the early scientists to study anthropogenic global warming; the *Revelle* factor that describes the solubility of CO_2 in sea water was named after him. *Hans Eduard Suess* (1909–1993) was an Austrian born American physical chemist and nuclear physicist.

841 American chemist, most known for his CO_2 measurement at Mauna Loa, Hawaii; affiliated with Scripps Institution of Oceanography, University of California, San Diego, from 1956 until his death in 2005. His son *Ralph F. Keeling* (born 1959) continues CO_2 research.

could result from changes in the concentration of atmospheric carbon dioxide (Chamberlin 1899). At the same time, *Svante August Arrhenius* (1859–1927)[842] first established the carbon dioxide theory. He wrote (Arrhenius 1896):

> The air retains heat (light or dark) in two different ways. On the one hand, the heat suffers a selective diffusion on its passage through the air; on the other hand, some of the atmospheric gases absorb considerable quantities of heat. These two actions are very different [...] (Arrhenius 1896, p. 238)
> The selective absorption of the atmosphere is, according to the researches of *Tyndall*, *Lecher* and *Pernter*, *Röntgen*, *Heine*, *Langley*, *Ångström*, *Paschen*, and others, of a wholly different kind. It is not exerted by the chief mass of the air, but in a high degree by aqueous vapour and carbonic acid, which are present in the air in small quantities. Further, this absorption is not continuous over the whole spectrum, but nearly insensible in the light part of it, and chiefly limited to the long-waved part, where it manifests itself in very well-defined absorption-bands, which fall off rapidly on both sides [...] (Arrhenius 1896, p. 239)
> [...] that there exists as yet no satisfactory hypothesis that could explain how the climate conditions for an Ice Age could be realized in so short a time as that which has elapsed from the days of the glacial epoch... a preliminary estimate of the probable effect of a variation of the atmospheric carbonic acid on the belief that one might in this way probably find an explanation for temperature variations of 5–10 °C [...] (Arrhenius 1896, p. 267)
> One may now ask how much the carbonic acid must vary according to our figures, in order that the temperature should attain the same values as in the Tertiary and Ice Ages respectively. A simple calculation shows that the temperature in the arctic regions would rise about 8 ° to 9 °C, if the carbonic acid increased to 2.5 or 3 times its present value. In order to get the temperature of the ice age between the 40th and 50th parallels, the carbonic acid in the air should sink to 0.62–0.55 of its present value (lowering of temperature 4–5 °C.) (Arrhenius 1896, p. 268).

This idea could only answer the riddle of the ice ages, however, if such large changes in atmospheric composition were really possible. For that question, *Arrhenius* turned to a colleague, *Arvid Gustaf Högbom* (1857–1940), a Swedish professor of mineralogy at Uppsala University. It happened that *Högbom* had compiled estimates for how carbon dioxide cycles through natural geochemical processes, including emission from volcanoes, uptake by the oceans, and so forth. Arrhenius (1896) cited Högbom (1894) in the following:

> The world's present production of coal reaches in round numbers 500 million of tons per annum [...] (Arrhenius 1896, p. 270)
> [...] that the most important of all the processes by means of which carbonic acid was removed from the atmosphere in all times, namely the chemical weathering of siliceous minerals, is of the same order of magnitude as a process of contrary effect, which is caused by the industrial development of our time, and which must be conceived of as being of a temporary nature [...] (Arrhenius 1896, p. 271)

842 Swedish scientist in Stockholm, originally a physicist, but often referred to as a chemist, and one of the founders of the science of physical chemistry.

Carbonic acid is supplied to the atmosphere by the following processes:
1. volcanic exhalations and geological phenomena connected therewith,
2. combustion of carbonaceous meteorites in the higher regions of the atmosphere,
3. combustion and decay of organic bodies,
4. decomposition of carbonates, and
5. liberation of carbonic acid mechanically enclosed in minerals on their fracture or decomposition.

The carbonic acid of the air is consumed chiefly by the following processes:
6. formation of carbonates from silicates on weathering, and
7. the consumption of carbonic acid by vegetative processes.

The ocean, too, plays an important role as a regulator of the quantity of carbonic acid in the air by means of the absorptive power of its water, which gives off carbonic acid as its temperature rises and absorbs it as it cools. The processes named under (4) and (5) are of little significance, so that they may be omitted. So too the processes (3) and (7), for the circulation of matter in the organic world goes on so rapidly that their variations cannot have any sensible influence. From this, we must accept periods in which great quantities of organisms were stored up in sedimentary formations and thus subtracted from the circulation, or in which such stored-up products were, as now, introduced anew into the circulation. The source of carbonic acid named in (2) is wholly incalculable. Thus the processes (1), (2), and (6) chiefly remain as balancing each other. As the enormous quantities of carbonic acid (representing a pressure of many atmospheres) that are now fixed in the limestone of the earth crust cannot be conceived to have existed in the air but as an insignificant fraction of the whole at any one time since organic life appeared on the globe, and since therefore the consumption through weathering and formation of carbonates must be compensated using continuous supply, we must regard volcanic exhalations as the chief source of carbonic acid for the atmosphere. But this source has not flowed regularly and uniformly. Just as single volcanoes have their periods of variation with alternating relative rest and intense activity, in the same manner, the globe as a whole seems in certain geological epochs to have exhibited a more violent and general volcanic activity, whilst other epochs were marked by a comparative quiescence of the volcanic forces. It seems therefore probable that the quantity of carbonic acid in the air has undergone nearly simultaneous variations, or at least that this factor has had an important influence (Arrhenius 1896, p. 272).

Wilhelm Pfeffer (1897, p. 331)[843] writes:

[...] das eine geringe und unschätzbare Zunahme der Kohlensäure in der Luft die photosynthetische Production erheblich steigert und dann ein schnelles Wachstum der Pflanze gestattet. Durch solche vermehrte Tätigkeit wird also selbstregulierend auf Erhaltung eines constanten Kohlensäuregehalts in der Luft hingearbeitet und wenn, wie wahrscheinlich, in früheren Erdperioden die Luft einmal mehr Kohlensäure als derzeit enthielt, so war damit ein das Wachstum begünstigender Faktor gegeben [... that a small and uncountable increases of carbonic acid in the air enhances the photosynthetic production and allows a fast growth of plants. Such increased activity results self-controlling in maintaining of a constant amount of carbonic acid in the air, and, if likely, in earlier earth's periods the air contained more carbonic acid then today, it was likely provided a beneficial factor for plant growth].

843 *Wilhelm Friedrich Philipp Pfeffer* (1845–1920) was a German botanist.

Twenty years later, Alt (1916)[844] noted that since Dufour (1870)[845]

> [...] the question of climate change in historic time is still completely open and the statement of the majority of meteorologists that the climate does not change is neither proved nor rejected.

The carbon dioxide theory of climate change was in deep eclipse in 1938 when *Guy Stewart Callendar* reviewed it and placed it on a firm scientific basis. Callendar (1938, 1940, 1949) believed that nearly all the carbon dioxide produced by fossil fuel combustion has remained in the atmosphere, and he suggested that the increase in atmospheric carbon dioxide may account for the observed slight rise of average temperature in northern latitudes during recent decades. He thus revived the hypothesis of Chamberlin (1899) and Arrhenius (1903) that climatic changes may be related to fluctuations in the carbon dioxide content of the air. Callendar (1938, 1949) documented a significant upward trend in temperatures for the first four decades of the twentieth century and noted the systematic retreat of glaciers. He compiled estimates of rising concentrations of atmospheric CO_2 since pre-industrial times and linked the rise of CO_2 to the combustion of fossil fuel. Finally, he synthesized information newly available concerning the infrared absorption bands of trace atmospheric constituents and linked increased sky radiation from increased CO_2 concentrations to the rising temperature trend. Today this is termed the *Callendar Effect*. The next great contributor to the carbon dioxide theory was *Gilbert Norman Plass* (1920–2004)[846] who established connections between the physics of infrared absorption by gases, the geochemistry of the carbon cycle, feedback loops in the climate system, and computer modeling (Plass 1953, 1956a,b). Using recent measurements of the influence of the 15-µm CO_2 absorption band, he calculated a 3.6 °C-increase in the surface temperature for the doubling of atmospheric carbon dioxide and a 3.8 °C-decrease if the concentration were halved. Contrary to the assumptions of many scientists at the time, the effect of water vapor absorption did not mask the carbon dioxide effect by any means. He used these results to argue for the applicability of the carbon dioxide theory of climate change for geological epochs and in recent decades (Fleming 2010).

4.3.2 Carbon monoxide (CO)

In the last year of his life (1800), the Scottish physician and chemist *William Cruickshank* (?–1811)[847] prepared carbon monoxide by passing carbon dioxide over heated

844 *Eugen Johann Alt* (1878–1936) was a German meteorologist and climatologist; during 1921–1934, director of "*Sächsische Landeswetterwarte*" [Saxony Weather Station] in Wahnsdorf near Dresden.

845 *Louis Dufour* (1832–1892), professeur de physique à l'Académie de Lausanne.

846 Canadian physicist from Harvard and Princeton University, later professor at Texas A&M University.

847 Not to confuse with the chemist and anatomist *William Cumberland Cruikshank* (1745–1800).

iron. He described it as "heavy, inflammable air" and concluded that it must be a compound containing carbon and oxygen, and termed it "*gaseous oxide of carbone*" (in French "*gaz carbonés combustible*" and in German "*Kohlenoxyd*"). Saussure (1802) did show that CO_2 is decomposed under electric discharges to CO and that CO_2 with hydrogen gives water and CO. Carbon dioxide was for a long time considered to be the only oxide of carbon, and the lower oxide of carbon being combustible was for a long time confused with inflammable air – hydrogen (Mellor, V, p. 907). *Friedrich Hoffmann* (1660–1724) German physicist in Halle,[848] wrote on the toxic vapors which emanate from burning charcoal (Hoffmann 1719), but he did not recognize this gas. The French physician *Joseph-Marie-François de Lassone* (1717–1788) and the French chemists *Morveau*, *Macquer*, *Fourcroy*, and *Lavoisier* describe the formation of a combustible gas by different treatment of charcoal between 1778 and 1793. *Priestley* get this inflammable air when passing steam over red-hot charcoal, and the American chemist *James Woodhouse* (1770–1809) obtained it by heating the charcoal with oxides; *Woodhouse* concluded that the gas is a compound of hydrogen and carbon. Finally, *Dalton* expresses in 1808, that "carbonic oxide is a binary compound consisting of one atom of charcoal and one of oxygen [...]" (citations after Mellor, V, p. 908).

Theodore de Saussure was the first who speculated on the presence of CO in atmospheric air (Saussure 1830, p. 431) after eudiometric analysis of air freed from CO_2. However, he rightly concluded that it can be the sum of hydrogen, carbonic oxide, and hydrocarbons [*présence d'un gaz combustible dans l'air*]; but he preferred to have been detected carbonic oxide (0.94 parts in 2000 parts of CO_2 free air, i. e., 470 ppm). Boussingault (1834) carried out in Paris similar investigations but believed that the "combustible carbon" exists as hydrocarbons [*l'hydrogène carboné*]; he estimated (n = 11) 80 (50–120) ppm. Verver (1840), citing Boussingault (1834), also conducted determinations of hydrogen and (organic) carbon at Groningen in July 1838 (n = 16) to be 148 (92–220) ppm and 144 (104–173) ppm, respectively (note, it corresponds to the sum of combustible gases). Boussingault (1863) observed that small quantities of CO are formed at ordinary temperatures by the oxidation of organic substances by oxygen in air.[849] Müntz and Aubin (1884b) estimated in Paris in autumn 1882 "*gaz carbonés combustible*" in the range 3–10 ppm whereas at the rural site Vincennes "only" 2–4.7 ppm.

Pettenkofer (1851) was the first who studied the presence of CO in rooms heated by iron ovens and found the concentration too small to be of importance. An excellent review on the health problems from indoor carbonic oxide is given by *Ferdinand Fischer* (Fischer 1880),[850] whereas the German physician *Max von Gruber* (1853–1927) found that below 0.05 % no toxic effects are found (by self-experiments); Gruber (1881). Determinations of CO in atmospheric air are basically absent in the nineteenth century;

848 Inventor of *Hoffmann*'s drop (spirit of ether).

849 Wilks (1959) found that green plants grown in a closed, illuminated system liberate small quantities of carbon monoxide.

850 *Heinrich August Wilhelm Ferdinand Fischer* (1842–1916) was a German technological chemist.

different proposed analytical methods (detection limit 0.002 %) based on iodometry and the reaction with $PdCl_2$ were developed end of the nineteenth century, particularly for determinations in the blood of animals and humans, and under indoor conditions. Kostin (1901) presents the most comprehensive overview on methods for the determination of CO in air and blood; see also Komar (1927). The interest in CO was only focused on its toxicity (fully described until 1930) and thus on measurements in business premises, school classes, and private rooms using gas lightning.

Süpfle et al. (1933)[851] carried out CO measurements in the streets of Dresden, and found concentrations between 3 and 44 ppm, depending on the traffic, whereas in parks far from streets less than 1 ppm (or "zero", i. e., below the detection limit); the results agree with those by Connolly et al. (1928) from American cities. These values (and also the historic Paris data) agree well with results from CO measurements in different large American cities end of the 1960s with nocturnal minimum values of 1–4 ppm and daytime maximum values of 5–15 ppm (Butcher and Charlson 1972, p. 140).

Since the 1940s, CO was observed spectroscopically in the earth's atmosphere (Adel 1952 and literature therein). Until the end of the 1960s, CO was believed to be inert in the troposphere. *Hermann Thiele* (1866–1916), a German chemist in Dresden, studied gas reactions under UV radiation, like $H + O_2$ and $CO + O_2$, and the influence of humidity on the process (Thiele 1909). Concerns the latter reaction, under dry air conditions, ozone formation was dominant and in moist air, oxidation of CO to CO_2. Without knowing the mechanisms, this finding is the first indication of the role of water in photochemistry; under dry air conditions, only $O + O_2 \leftrightarrow O_3$ occurred ($CO + O$ is unimportant, see below), and under moist conditions O (more specifically $O(^1D)$) reacts with H_2O gaining OH, which further oxidizes CO to CO_2 (see below).

The reaction $CO + O$ was found to be too slow by Jackson (1934) compared with $O + O_2$ to be of importance. After the finding of Weinstock (1969) that the atmospheric lifetime of CO is rather short, Heicklen et al. (1971) and Westberg et al. (1971) first proposed the reaction $CO + OH \rightarrow CO_2 + H$ as the main tropospheric sink[852] and emphasized its role in photochemical smog chemistry through the subsequent radical reactions $H + O_2 \rightarrow HO_2$ and $HO_2 + NO \rightarrow OH + NO_2$, providing radical chains and tropospheric net ozone formation (see also Chapter 4.4.8.2).

4.3.3 Organic matter and compounds

Let us first address what "organic" means in chemistry because in the colloquial language it is equated with "life". The term derives from όργανον (Greek) and organum

851 *Karl Süpfle* (1880–1942) was a German hygienist in Munich (at *Gruber*) and since 1927 in Dresden.
852 This reaction was considered already earlier in combustion processes (Bonhoeffer and Haber 1928, p. 282, Jost 1939/46) and the stratosphere (Bates and Witherspoon 1952).

(Latin) which means tool, device, and voice and was first used in the seventeenth century in French with various derived forms and enlarged senses (*organe, organiser, organisme, organisation, organique*). "Organic" compounds were known for centuries but were not distinguished from "inorganic" matter. Systematic classification of chemistry was done into mineral, vegetable, and animal according to its origin by the French chemist *Nicolas Lémery* (1645–1715) who wrote "*Cours de chymie*" in 1675 (cited after Kopp 1931).

First *Lavoisier* found systematically that vegetable matter is composed of C, H, and O and that in animal matter additionally N and P are present. To *Lavoisier*, the quantitative analysis of organic substances by combustion and product analysis must be credited (Meyer 1914, p. 361). This elementary analysis based on gravimetry was further developed in the early nineteenth century (*Berthollet*, *Gay-Lussac* and *Thénard*, *Berzelius*, *Döbereiner*, *Prout*) and finally perfected by *Liebig*; at the beginning twentieth century the Slovenian and Austrian chemist and physician *Fritz Pregl* (1869–1930) developed the micro-elemental analysis that remained until the 1960s the fundamental organic analysis; IR (infrared spectroscopy), NMR (nuclear magnetic resonance) and MS (mass spectroscopy) for structural analysis developed in the early twentieth century but commercial instruments were available not before the mid-1950s. The ultimate analytical techniques like GC (gas chromatography) and LC (liquid chromatography) were available since around 1960 but only in the 1970s sophisticated instruments and detectors became widely used.

According to Walden (1941), the first use of the term "organic chemistry" is now attributed to the Swedish chemist *Jöns Jacob Berzelius* who termed it "*organisk kemie*" in a book published in 1806. After *Lavoisier*'s revolutionary book (Lavoisier 1789), "*Traité élémentaire de chimie*" (1789), *Berzelius* wrote the first textbook "*Lärbok I kemien*" (1817–1830) in six volumes (Berzelius 1820). It was soon published in French (1829) still with the now traditional title, "*Traité de chimie minerale, vegetale et animale*" in 8 volumes with the subtitle "*Chimie organique*" (*2ème partie* – 3 volumes). In Germany, the third volume of 1819; of the first "*Handbuch der theoretischen Chemie*" [handbook of theoretical chemistry] by *Leopold Gmelin* is subtitled "*Dritter Band welcher die Lehren von den organischen Verbindungen enthält*" [third volume that contains the sciences of organic compounds]; later rearranged into separate volumes of inorganic and organic chemistry (from 1848). It was believed since that time that organic matter (also termed "organized bodies") could not be synthesized from its elements and that a special force, the *vital force*, is needed for its production. The oldest known and simplest hydrocarbon, methane, was known for centuries as marsh gas but not recognized as an organic compound; its formula was established only in 1835. Gmelin (1819, pp. 936–937) writes that all organic compounds consist of at least three substances [*Stoff* in German] (not yet using the term *element*), namely carbon, hydrogen, and oxygen. Further, he writes (Gmelin 1819, p. 938) that organic compounds are either solid or liquid but never gaseous (if not some miasma). Gmelin (1817, pp. 173–179) discusses compounds of carbon with hydrogen as inorganic compounds; he describes "*olefiant gas*" (ethene) and "*Kohlenwasserstoffgas*" (methane), and states that further compounds of hydrocarbons [*Kohlenwasserstoffe*] are unknown. As further compounds of carbon *Gmelin* (Gmelin 1817, p. 179) list "*Schwefelalkohol*" (ethyl mercaptan C_2H_5SH), cyan (CN) and prussic acid (HCN), compounds of carbon with some metals, namely iron to carbon metals, and finally remarks that carbon is an essential constituent of all organic compounds. Gmelin (1819) subdivides organic compounds only into two classes, in organic acids (and describes 28 acids) and organic oxides, describing 38 substances – almost composed of naturally occurring substances; only "*Weingeist*" (ethanol) and "*Aether*" (diethyl ether) were known as molecular substances.

However, organic molecules can be produced by processes not involving life as first shown by *Friedrich Wöhler* (1800–1882)[853] who destroyed the theory of vital forces by the synthesis of urea in 1828; an event generally seen as a turning point. *Liebig* ("Organic chemistry in its application to agriculture and physiology", 1840) defined the task of organic chemistry as follows:

> The object of organic chemistry is to discover the chemical conditions which are essential to the life and perfect development of animals and vegetables, and, generally, to investigate all those processes of organic nature which are due to the operation of chemical laws. (The first phrase in Part I: "Of the chemical processes in the nutrition of vegetables".)

Liebig's definition is still very much focused on the original idea that all "organic" is definitely equal to "life", although *Wöhler* already rejected the "vital force" in association with organic compounds. Gmelin (1848) wrote that carbon is the only element never missing and hence it is the only essential constituent in an organic compound.

At that time, the *isomerism*, i. e., substances having the same chemical composition but different properties (because of different chemical structure, which was found much later) was widely proved and accepted. In the middle of the nineteenth century, the *theory of types* was introduced for the classification of organic compounds (by *Charles Frédéric Gerhardt,*[854] *Charles Adolphe Wurtz, August Wilhelm von Hofmann,*[855] *and Alexander William Williamson*) and further developed by *August Kekulé.*[856] *Hermann Kolbe* renewed the radical theory in the 1860s. At about the same time the beginnings of the *structural theory* proceeded (*August Kekulé, Sir Edward*

853 German chemist in Göttingen; he is regarded as a pioneer in organic chemistry as a result of him (accidentally) synthesizing urea but also the first to isolate several chemical elements (Al, Yt, Be, Ti).
854 *Charles Frédéric Gerhardt* (1826–1856), an Alsatian chemist in Strasbourg.
855 *August Wilhelm von Hofmann* (1818–1892), a German chemist in Gießen (at *Liebig*), London and Berlin.
856 *August Kekulé* (1829–1896), a German chemist in London and Bonn.

Frankland,[857] *Arthur Rudolf Hantzsch*,[858] *Wilhelm Traube*,[859] and *Johannes Adolf Wislicenus*[860]).

4.3.3.1 Methane

"Hydrocarbon" ("*carbure d'hydrogèn*" in French and "*Kohlenwasserstoff*" in German), not yet specified as methane was known from marshes and swamps (termed swamp gas)[861] and many natural gas sources (from which it was already sometimes used as fuel), and first described by *Alexandre Volta*, who distinguished for the first among different inflammable gases like hydrogen, carbonic oxide and methane (Volta 1778). *Claude-Louis Berthollet* made the most thorough and penetrating analysis of heavy inflammable air (methane) of anyone up to this date and showed that it was composed of inflammable air from metals (hydrogen) and carbon, presented to the *Académie* in December of 1785 but published only in 1788 (Berthollet 1788). The formula was given as C^2H^4 or C^2H^2,H^2 (Gmelin 1848).[862] This gas [methane CH_4] was feared by coal miners, who termed it *firedamp*[863] because it caused dangerous explosions. Methane is also the first atmospheric trace gas that was identified from its source (swamps, marshes), long before it was detected in the atmosphere.

On p. 433 it was mentioned that *Boussingault* was the first who made determinations in the air of Paris of combustible gases containing hydrogen. By air combustion

857 *Edward Frankland* (1825–1899), a British chemist.

858 *Arthur Rudolf Hantzsch* (1857–1935), a German chemist in Berlin, Würzburg and since 1902 in Leipzig.

859 *Wilhelm Traube* (1866–1942), a German chemist; sun of *Moritz Traube*.

860 *Johannes Adolf Wislicenus* (1835–1902), a German chemist; father of *Hans Adolf Wislicenus*. 1853–1856 USA (Harvard and New York), Halle, Zürich, Würzburg, and Leipzig.

861 In German *Sumpfgas* and *Grubengas* (also: *schwere brennbare Luft*, *Kohlenwasserstoffgas*, *gekohltes Wasserstoffgas*). In Latin gas hydrogenium carbonatum, and in French *gas inflammable de marais*, *gas hydrogène carburé*.

862 It is interesting that the "*Stammkern*" [translated in the English edition as *primary nucleus*] *Methylen* C^2H^2 (not to confuse with ethyne; today methylene is the term for the CH_2 group) was termed *methylène* (in French "*methylene*", in German "*Methylengas*"), gained by Dumas and Peligot (1835) in connection with "*esprit de bois*" [wooden spirit; in historic German "*Holzgeist*": methanol CH_3OH], having the old formula $C^2H^4O^2$ (= C^2H^4,H^2O^2), and *hydrate de methylène* [methyl ether; in historic German "*Holzäther*" C_2H_6O or CH_3OCH_3). Hence *methylene* (its existence was never confirmed as compound and likely confused with methane) corresponds to the methyl radical CH_3 and the CH_2 unit in aliphatic hydrocarbons, respectively. Dumas and Peligot (1835, p. 67) note "*sur la théorie du methylene*" that in analogy to the formula of ammonia (Az^2H^6, today NH_3) there exist a "*radical inconnu*" C^4H^6 (= CH_3). Gmelin (1867) changed the term in *Methyl* and the formula in C^2H^3 = C^2H^3,H. Wöhler (1863) presents the formula CH^2 for methane and CH for ethene. Roscoe and Schorlemmer (1871) present the correct formula CH_4 for methane, C_2H_2 for ethine and C_2H_4 for ethene. They further write that coal gas (in German "*Leuchtgas*") consists from hydrogen, methane, carbon monoxide and traces of propene (still with wrong formula C_2H_6), ethene (C_2H_4) and vapors of benzene (C_6H_6).

863 In German "*Grubengas*".

he determined the amount of water gained; in 11 analyses he obtained a volume ratio of hydrogen in air of $8(5–13)\cdot 10^{-5}$ (this value is about 2 orders of magnitude higher than it would be expected for CH_4 in that period derived from ice core record to be about 800 ppb) and states that it could be equivalent to the sum of free hydrogen, hydrogen sulfide, and hydrocarbons but "*il est plus probable que ce principe est en grande partie de l'hydrogène carboné*" (Boussingault 1834, p. 172). The presence of hydrocarbons in atmospheric air was no surprise to *Boussingault* who argued that assimilation and putrefaction of plants [*planteur exhalent*] occurs everywhere. In 1898, *Armand Émile Justin Gautier* (1837–1920), Professor for Chemistry at the university in Paris, reported the presence of "*gaz de marais*" (methane) using a similar measurement technique (Gautier 1898, p. 1684):

> [...] apparent entre les nombres si satisfaisants que nous avons obtenus, et les idées l'on s'était faites de l'existence du méthane dans l'atmosphère.

Gautier found by successive air combustion (Fig. 4.26) in five experiments in 100 L of air (recalculated at 0 °C and 750 Torr) 3.96 mg hydrogen and 12.45 mg carbon (corresponds to $22 \cdot 10^{-5}$ CH_4 volume ratio), i. e., a ratio C/H = 3.1, and writes (Gautier 1898, p. 1683) "*autour du rapport théorique* C/H = 3 *qui caractérise le gaz de marais (formène)*". Thus, to *Gautier* must be credited the first evidence of the presence of methane in the air. It is worth citing here in French original the findings for the city air of Paris (Gautier 1901e, p. 96); note that the splitting of hydrocarbon *Gautier* made by the speculative assumption of a mixture $C_6H_6 + 7\,CH_4$ in air to fit the experimental mean of C/H = 13/34 as an atomic ratio or as mass ratio C/H = 156/34 = 4.6 (in cm^3 per 100 L of air at 0 °C and 769 Torr):

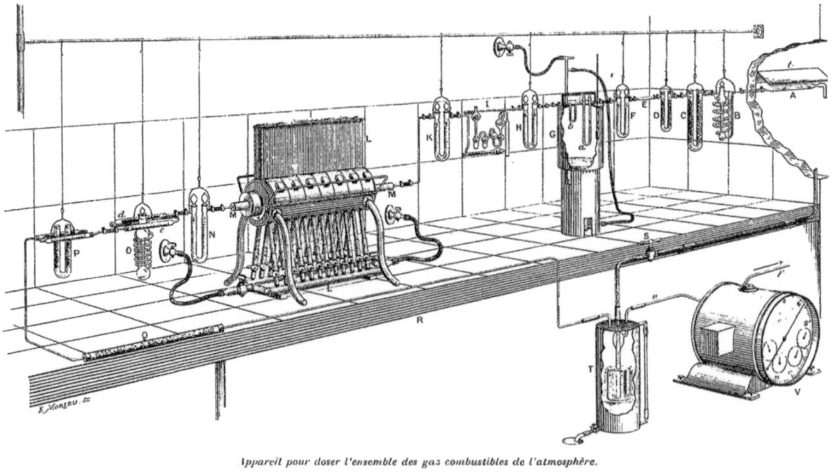

Appareil pour doser l'ensemble des gaz combustibles de l'atmosphère.
(Toute la partie GHIK n'est indispensable que lorsqu'on veut doser séparément l'oxyde de carbone.)

Figure 4.26: Apparatus for determination of hydrocarbons in the air through combustion used by Gautier (1898, p. 1679).

Hydrogène libre aérien	19.4
Formène	12.1
C^6H^6 ou vapeurs analogues trés carburées	1.7
CO moyen avec traces d'hydrocarbures en C^nH^{2n-2} et C^nH^{2n}	0.2

In modern literature, the presence of methane in the earth's atmosphere is attributed to the discovery in 1948 by the Belgian solar spectroscopist scientist *Marcel Victor Migeotte* (1912–1992) who began in 1947 at the Ohio State University to investigate the atmospheric transmission in infrared and to identify atmospheric constituents (Migeotte 1948), using a spectrometer designed and built by *Robert Hamilton Noble*[864] in 1946, and later continued at the observatory Jungfraujoch in 1950 where he also detected carbon monoxide together with the Belgian astronomer *Lucien Neven* (1913–1987); Migeotte and Neven (1950). McMath et al. (1948) are the first who computed the abundance of methane in the atmosphere from IR measurement to be 1.18 ppm by mass. Glueckauf (1951) postulated (by re-calculation) that the CH_4 content by *McMath* et al. maybe 1.2 ppm by volume instead of 2.2 ppm (or 1.18 ppm by mass).

The first direct determination of methane in atmospheric air by gas chromatography begun in 1965 by Ortman (1966); Bainbridge and Heidt (1966) conducted the first vertical profile measurements using aircraft and balloon. Between 1966 and 1980 many scientists carried out CH_4 measurements at different places. Graedel (1978) reports a global background methane value in the troposphere of about 1.4 ppm, whereas concentrations as high as 6 ppm were measured in urban areas.

Reinhold Albert Rasmussen (1936–2019) and *Mohammad Aslam Khan Khalil* (born 1950) discovered the increase of methane in the earth's atmosphere in 1981 (Rasmussen and Khalil 1981);[865] this discovery was only possible due to the beginning systematic methane measurements, gaining seasonal cycles and understanding atmospheric distribution; pioneers in this research were *Dieter Ehhalt*,[866] *Paul Fraser* and *Paul Steele*,[867] *Hanwant B. Singh*,[868] and *Peter Fabian*[869] (Ehhalt 1967, Singh et al.

864 Unknown living dates, Doctor thesis 1946 under supervision of the American physicist *Harald Herborg Nielsen* (1903–1973); later professor for optical sciences at the University Arizona.

865 An American Botanist and Atmospheric Chemist and Pakistani American physicist, both Oregon Research Center.

866 A German physicist (born 1935), University Heidelberg, 1964–1967 and 1969–1974 National Center of Atmospheric Research (NCAR), USA, 1974–2000 director of Institute for Atmospheric Chemistry (Nuclear Research Facility Jülich).

867 *Paul Fraser* (born 1949), working 1974–2014 at CSIRO Marine and Atmospheric Research, Australia (the "man" of Cape Grim Baseline Station, together with *Paul A. Steele, Paul B. Krummel, Ray L. Langenfelds* and *David M. Etheridge*).

868 An Indian-American chemist, Ph. D. from the University of Pittsburgh (1972), NASA Ames Research Center, retired in 2018.

869 (1937–2014), German physicist and meteorologist ("atmospheric chemist") at University Munich since 1988, known for stratospheric ozone research.

1979, Fabian et al. 1979, 1983, Fraser et al. 1981, Steele et al. 1987) and many others (e. g. Cavanaugh et al. 1969, Stephens and Burleson 1969, Swinnerton et al. 1969).

The oxidation of methane via $CH_4 + O_2 = CO_2 (+H_2O)$ was already known to *Dalton*, *Henry*,[870] *Davy*, and *Dumas*. The oxidation of hydrocarbons in combustion processes was studied in the early 1930s, namely the role of radicals in the thermal decomposition of organic compounds. Some investigators have believed that radicals started a chain, others have thought that radicals simply recombined to give final products obtained, and still, others have believed that the radicals are due to unimportant reactions, the main reaction taking place through rearrangement and simultaneous breaking the bond in the organic compound (Rice 1935, p. 36). Frear (1934) found that the thermal decomposition of methane proceeds via a chain mechanism and Pease and Munro (1934) detected organic peroxides, aldehydes, and alcohols in the products of pentane oxidation. The reaction $CH_4 + OH \rightarrow CH_3 + H_2O$ with the fast subsequent reaction $CH_3 + O_2 \rightarrow CH_2O + OH$ was first described in 1942 by the Belgian chemist *Adolphe Van Tiggelen* (1914–1969) who made important contributions to the understanding of flame processes (Tiggelen 1942). The oxidation of CH_4 in the atmosphere became clear after recognizing the role of OH radicals and providing first reaction cycles by Levy (1971), McConnell et al. (1971), Weinstock and Niki (1972), and Crutzen (1972); *Dieter Ehhalt* provided the first global methane cycle and estimated the residence time to be 5 years (later corrected to about 10 years) based on the kinetic data by Greiner (1970).

4.3.3.2 Other hydrocarbons and organic compounds

Wetherill (1865, p. 204) reports in the Congressional Serial Set of the U. S. Governmental Office (note wrong spelling of *Thomson*):

> Dr. Thompson appears to be the first recognized the importance of organic matter as a constituent of the air of towns. He found that the air of London, when passed through oil of vitriol, communicated a dark tinge to it, and if large quantities of air were passed though distilled water the inevitable result was the formation of fungi.

Richard Dundas Thomson (1810–1864), a British chemist and pioneer in public sanitation, already attempted in 1849 to investigate air passed through absorbent porous bodies to find any condensing vapor or solid particles; but the result was entirely negative. Thus, during the cholera epidemic in London in 1854, he used aspirators (Thomson 1855). When passing the air through distilled water, he also found the air in rooms and external to be acid (p. 711) but the atmosphere of a sewer to be alkaline; he also found "siliceous particles mixed with fuliginous matter" (Thomson 1855, p. 127)[871] and writes (Thomson 1855, p. 129–130):

870 *William Henry.*

871 Fuligineous = sooty, or in modern terms carbonaceous.

> [...] for while all the other atmosphere investigated were highly acid, the air in the sewer was strongly alkaline, and the amount of animal life in the sewer was found greatly to predominate over that which was detected in the atmosphere above ground. It is obvious, then, that, in addition to the noxious gases, like sulphuretted hydrogen, carbonic acid, carburetted hydrogen &c., which may be evolved in various proportions from the decomposing organic matter in the sewer fluids, there exist in the sewer atmosphere living beings of vegetable and animal origin, or at least the sporukles and germs of these organisms exist there.
> Simultaneous observations on the meteorological condition of the atmosphere are likewise of importance, as bearing on the power on its mass, to obviate any stagnant tendency of the air and to prevent the accumulation of extraneous and morbid influences.

The last phrase is likely the first written recognition on the importance of dispersion of air pollutants concerning atmospheric mixing ratios and thus air quality threshold values.

What we can see is that according to the ancient view of chemistry (vegetable, animal, and mineral), the term "organic matter" includes biological (organized matter) and chemical matter. Smith (1872, p. 480) writes:

> The question of organic matter in the air and organised matter are somewhat mixed together in their origin, and are only now beginning to be treated separately. The real question in late days is not the presence of organic matter or even organized bodies; we require to discover now the quantity, character, and functions.

Organic matter in atmospheric air was first detected in hail (Anonymous 1693; see p. 149), in rain by *Boerhaave* in the 1720s (p. 149) and by *Marggraf* in 1749 (pp. 152 ff), in the air by *Witting* and *Wiegmann* in the 1820s (pp. 161 and 166); many more scientists detected organic matter before 1850 in precipitation (Chapter 2.3.1.2). See also in this volume the discussion on pyrrhine (pp. 161–166) and miasma (pp. 314, 456, 497, 501, 518). Wetherill (1865, p. 204) further reports in his chapter "on organic matter in the air"

> Dr. R. Angus Smith tested the air for organic matter by ascerting the amount of air necessary to discharge the color of a solution of permanganate of potassium, of which the strength was obtained by decomposing a known weight of sugar or of oxalic acid. Supposing the sugar and the organic matter in the air to be composed by exactly the same amount of permanganate, which will not involve a great error, he obtained the proportions of organic matter existing in the air.

However, *Wetherill* made a large error citing the values from Smith (1859a) denoting the proportions as "grains of organic matter in 100 cubic inches of air"; *Smith* (Smith 1859a, p. 178) writes "relative quantities of organic and other oxidizable matter in the air", he notes that "the sulphureous acid takes the oxygen of the chameleon [permanganate], and an apparently large amount of organic matter appears". The values correspond to "grains of solution of permanganate used decolorized by 100 cubic inches of air" (see also Smith 1872, pp. 399–409, where somewhat different values are cited).[872] Smith (1859a, p. 178) determined the relative amount of organic matter (including other oxidizable substances; so-called "permanganate values") by air-washing:

54.5	Manchester ($n = 238$)
29.2	London ($n = 6$), experiments during warm weather
12.3	London, after a thunder-storm
18.1	Near Milan, moist fields
3.3	open sea (German Sea, 60 miles from Yarmouth)
2.8	forest at Chamonix
1.4	Lake Lucerne

Stöckhardt (1846, p. 85) writes in a paragraph "minor constituents of air" of his textbook:

> Dass die Luft noch andere Beimengungen enthalten müsse, wenn man bedenkt, dass Alles, was auf unserer Erde sich verflüchtigt oder verstäubt, von ihr aufgenommen wird. Die von den Gewürzinseln kommende Luft riecht noch in einer Entfernung von 8 bis 10 Meilen nach Zimt und Nelken [That the air also contains other foreign ingredients is not strange, since it is the constant receptacle of volatile substances and dust. The air coming from the Spicy islands, even at the distance of eight or ten miles, is impregnated with the odor of cinnamon and cloves], English version from Stöckhardt (1851, p. 90).

The oldest known organic compounds are ethanol (wine spirit) from alcoholic fermentation and wooden vinegar from the dry distillation of wood, containing different organic liquids (first recognized by *Boyle* in 1661, however, was identified only after 1839 by *Berzelius*). The oldest known organic acid is vinegar (acetic acid), described by *Plinius*[873] but confused with other "vegetable" acids (like tartaric acid, malic acid, tannic acid, gallic acid) still in the eighteenth century. Formic acid was known already in the fifteenth century but also confused with vinegar; only in the late eighteenth century, the different characteristics of organic acids was recognized (Kopp 1931, IV, p. 331). Flammable gaseous hydrocarbons were long confused with carbon monoxide

872 600 grains of it (the solution of chameleon – i. e., permanganate) are required to decompose 5 grains of a standard solution of oxalic acid. The standard solution of oxalic acid is so made that 1000 grains neutralize 1 grain of carbonate of soda. A thousand grains contain therefore 1.184 grain of crystalline oxalic acid (Smith 1872, p. 176). Thus it should follow 1/1000 of the "permanganate value" corresponds to about 12 $\mu g\, m^{-3}$ of carbon in air. Assuming that at that time at remote sites (like lake Lucerne) CH_4 represents almost the atmospheric burden with organic compounds, but will not be oxidized with this "solution of chameleon", the values are much too high (10–500 ppm) to represent even the sum of oxidizable matter in air (VOC and SO_2); however, the relative variation among the different sites of investigation is very feasible.

873 *Gaius Plinius Secundus Maior* or *Pliny the Elder* (~23–79) was a Roman author, a naturalist and natural philosopher known for his encyclopedic *Naturalis Historia*.

and hydrogen. Swamp gas (CH_4) was mentioned by *Plinius* as effluvia from different sites of the earth, and *Basilius Valentinus* refers to fiery appearances in mines; in the seventeenth and eighteenth centuries, many reports on firedamp appeared but without any discussion on its nature (Kopp 1931, III, p. 297). Another simple hydrocarbon, ethene (old: ethylene) appears to be discovered by *Johann Joachim Becher*, who obtained it by heating ethanol with sulfuric acid; he mentioned the gas in his "*Physica Subterranea*" (1669). This hydrocarbon is the first to be studied in 1795 by four Dutch chemists, *Johann Rudolph Deimann, Adrien Paets van Troostwyck, Anthoni Lauwerenburgh*, and *Nicolas Bondt*, who found that it differed from hydrogen gas and that it contains both carbon and hydrogen. This group of chemists also discovered that ethylene can combine with chlorine, gaining an oily liquid (1,2-dichlorethen; the composition was only known after 1831), termed "*Dutch liquid* or *oil of the Dutch chemists*" [*Flüssigkeit der holländischen Chemiker* in German]. Hence, the gas (ethene C_2H_4) was termed "*olievormend gas*" (English "*olefiant gas*", in French "*gaz olefiant*" and in German "*ölbildendes Gas*"). It is said that the old name *olefin* (now alkene) for unsaturated hydrocarbons that contain a carbon–carbon double bond is derived from this archaic term. Both gases (CH_4 and C_2H_2) *Carl Jacob Löwig* (1803–1890)[874] denotes "more or less as inorganic compounds" (Löwig 1832, p. 33). Still, in 1867, only methane and ethene were known as gaseous hydrocarbons (Gmelin 1867). In 1871, additional ethyne, all with the correct formula (Roscoe and Schorlemmer 1871). However, *Charles Adolphe Wurtz* gained in 1855 "mixed" radicals between methyl (C^2H^3), ethyl (C^4H^5), and butyl (C^8H^9, termed "valyl" by *Kolbe*). The German chemist *Adolph Wilhelm Herrmann Kolbe* (1818–1884)[875] describes in his textbook (Kolbe 1854) *butyl* as liquid (C^8H^9), the gas "*Butylengas*" (C^8H^8) and "*Butylenwasserstoffgas*" (C^8H^9H).[876] In the 1890s, namely due to chemical analysis of natural oil and gas, a modern understanding of alkanes (paraffines) with the general formula C_nH_{2n+2} was achieved (Brühl 1897).

Similar to *Boussingault* and later *Gautier*, *August Vogel* conducted during the cholera epidemics in Munich 1854 eleven determinations of carbon and hydrogen; after drying of the air and removal of carbon dioxide and ammonia, the air passed through a red-hot tube, the gained CO_2 was absorbed by potash, H_2O, and calcium chloride and the weight increase recalculated as carbon and hydrogen per 10.000 mass parts of air (Vogel 1854): 0.105 (0.089–0.118) for carbon, 0.140 (0.069–0.221) for hydrogen and 0.90 (0.46–1.48) for the C/H ratio. Remarkably, the carbon amount does not very much in contrast to that of hydrogen; furthermore, it is remarkable that the amount of "hydrocarbon" in air is 100 times less than estimated by *Gautier* in Paris about a half of century later. Whereas *Vogel*'s carbon amount corresponds to about

---

874 German chemist, discovered bromine independently of *Antoine Jérôme Balard*.

875 Successor of *Bunsen* in Marburg 1851.

876 In today's understanding it is butane C_4H_{10} and butene (or butylene with the older name) as collective term for the alkenes C_4H_6.

1 ppm methane (and thus being more reliable than *Gautier*'s estimate); however, the hydrogen amount is too large to represent hydrocarbons and could be thus interpreted as free hydrogen in air (but much larger with a value of 8 ppmv). Coming back to *Gautier*'s determination of "methane" in Paris, he made comparative experiments with purified air containing traces of methane, giving the quotient C/H as 2.4, indicating that, under these conditions, the hydrogen burns faster than the carbon (Gautier 1900). Gautier (1901a) found that the atmosphere of a forest in July contains only half as much carbon as the air of Paris and analyses of the air of Mt. Canigou, Pyrenees, indicated a quotient C/H being only 0.33; the hydrogen present is, therefore, almost entirely uncombined. Examinations of the air at the Roches-Douvres lighthouse, 40 kilometers from the coast of France showed the almost complete absence of hydrocarbons, whilst free hydrogen was present in the volume proportion of $19.45 \cdot 10^{-5}$ (this value is about 25 times higher than we know it from first reliable measurements in the 1920s). *Gautier* concluded that air normally contains about 1/5000 of its volume of free hydrogen together with variable quantities of hydrocarbons, the latter being due to the action of vegetable and animal life, to industrial operations, and others. However, today we know that the amount of methane in the air does not vary compared with non-methane hydrocarbons. Hence, another conclusion of Gautier (1900) "that it is still an open question as to whether the combustible gases of the atmosphere consist wholly of methane or a mixture of this substance with hydrogen and hydrocarbons richer in carbon than the paraffins" is correct. Thus, a reinterpretation of the "methane value" in the air of Paris (about 2 ppmv) could lead us to the conclusion that the amount of non-methane hydrocarbons (NMVOC) of city air in 1900 was significant (around 600 ppmv). *Armand Gautier* mentioned in an article on "the smoke of Paris" (Gautier 1901d) that it contains hydrogen sulfide, phenol, acetylene, and diverse hydrocarbons.

The Swiss hygienist *Friedrich Huldreich Erismann* (1842–1915),[877] who founded 1884 in Moscow the first institute of hygiene (after the Munich model of *Pettenkofer*'s institute), was likely the first who studied the (indoor) pollution by hydrocarbons from different light sources, like petroleum, rape oil, town gas, and candles, showing that air pollution by petroleum is lowest and highest due to candles (Erismann 1876, quoted after Bebber 1895, p. 51). The Swiss botanist *Carl Eduard Cramer* (1831–1901)

877 *Erismann* was one of the first who emphasized the importance of interlinked consideration of air, water and soil *hygiene* (or in other terms *quality* or *pollution*). However, air quality was almost studied indoor and the quality of water was in the foreground. The problems in "*Außenlufthygiene*" [untranslatable: open (or external) air hygiene; note that the terms *indoor* and *outdoor* in conjunction with air pollution became in use only after World War II] were the establish the relation between damage and air pollution, mainly because of meteorological processes. Beyreis et al. (1955) state that no other country then the USA established in short time big cities and enormous industry in the first half of the 20th century resulting in huge air pollution; thus, no wonder that with the beginning 1950s intensive smog research begun.

continued *Erismann*'s investigations to obtain absolute values of the combustion products (Cramer 1890); apart from CO_2 he determined (by elementary analysis) incomplete combustion products of different illumination materials as carbon and hydrogen per gram of fuel carbon. 30 years after *Erismann*, *Heinrich Wolpert* (without citing *Cramer*) but describing *Gautier*'s experimental arrangement in detail and some improvements, studied external air and air from soil, respiration, and in rooms (Wolpert 1905). In the air of Berlin *Wolpert* determined gaseous combustible hydrocarbons to be 15 ppmv (given as 0.015 volume permille), varying in the range of 6–25 ppmv, or as 4.4 (1.9–7.2) % of CO_2 (corresponding to 343 ppmv), and thus considerably less than *Gautier*'s value in Paris. Finally, *Wolpert* remarks that his experiments made more evidence on the presence of combustible gaseous carbon compounds than *Gautier*'s.

At the end of the nineteenth century, the effects of air on human health (under indoor conditions) became into focus; apart from carbonic acid and microorganism, the smell was caused by organic compounds (Haldane and Smith 1892). In Germany, the hygienists and physicians *Max Rubner* (1854–1932),[878] *Karl Bernhard Lehmann* (1858–1940), and *Carl Flügge* (1847–1923) might be considered as the founders of air hygiene; note that *the* founder in Germany and for the world was *Pettenkofer* (p. 398). However, they were more interested in indoor and manufacturing pollution than in outdoor, i. e., atmospheric air pollution.

Müntz (1881) detected alcohol in rain and snow water in Paris and concluded that it occurs in the atmosphere as a product of decomposition of organic matter from soils. *Henry Henriet*, together with *Albert Lévy* (Henriet 1904), found while studying the composition of the air an "energetically reducing gas" which he qualitatively identified as formaldehyde (HCHO)[879] and then determined in the air at Montsouris in 1903 to be in the range of 1–5 ppm (20–60 mg m^{-3}) and mentioned its extreme amount in comparison to that of ozone (10–30 μg m^{-3}); the value for HCHO is orders of magnitude too high.

The problems in the quantitative determination of the organic matter in the nineteenth century are summarized and described by the American bacteriologist *David Hendricks Bergey* (1860–1937) who refers first different methods published by Moss (1871), Smith (1872), Fodor (1881), Nékám (1890), Wanklyn and Chapman (1891),[880] Lehmann (1890), Uffelmann (1890), Remsen (1881), and others (Bergey 1896). Methods are distinguished in those to determine organic nitrogen as "albuminoid nitrogen"

---

878 *Rubner* became the successor of *Robert Koch* at the Berlin University (1891–1909).

879 The Indian chemists and *Nil Ratan Dhar* and *Atma Ram* detected formaldehyde in rain and dew (pp. 226 and 280) and speculated (Dhar and Ram 1933e) on its existence in the upper atmosphere due to photochemical processes involving H_2O and CO_2 (equilibrium $HCHO \rightleftarrows CO + H_2$).

880 *James Alfred Wanklyn* (1834–1906), Irish-American analytical chemist; *Ernest Theophron Chapman* (1845–1872), US American analytical chemist.

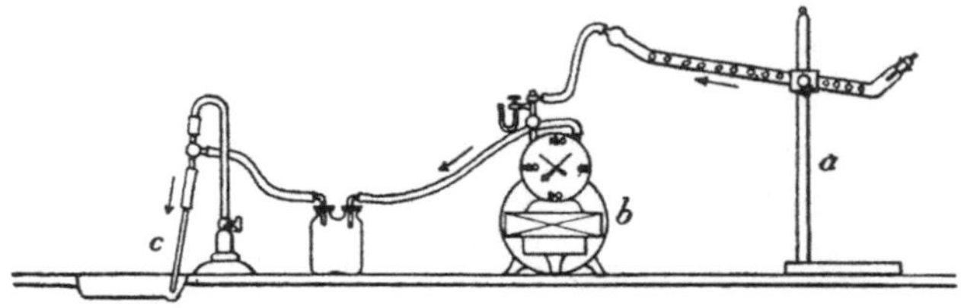

Fig. 1.—Apparatus used with Methods I, V, and VI. This apparatus consists of: *a*—a stative which supports the Pettenkofer absorption tube containing the absorbent material; *b*—the gas-meter which measures the volume of air aspirated through the apparatus; and *c*—a Chapman water pump, by means of which a slow, steady current of air is maintained through the apparatus.

Figure 4.27: Apparatus for the sampling of organic matter in the air through absorption (aspiration method) used by *David Hendricks Bergey* (Bergey 1896, p. 26).

(parallel to free ammonia) and others to determine the total organic matter as "oxidizable substances". Organic nitrogen in air was estimated by aspiration of the air through cotton-wool, or gun-cotton, asbestos, or pumice-stone, and then subjected the absorbent to distillation where the organic matter is converted into ammonia and estimated with *Nessler*'s reagent, or by direct nesslerization of the absorbent used to abstract the organic matter from the air (Bergey 1896, p. 1). In the second group of methods the organic matter (note, the total oxidizable matter) by boiling with a dilute solution of permanganate or potash and titration with oxalic acid; all these methods are colorimetric, and first developed for water analysis. The amount of oxygen consumed for a given volume of air corresponds to the value for "total oxidizable matter"; *Bergey* thus also notes that sulphuretted hydrogen (H_2S) and sulphureous acid gas (SO_2), which he determined iodometric using the *Pettenkofer* tube as sum within the range 24–74 mg m^{-3} as sulfur, cannot be excluded in the values for organic matter; he determined in external air free ammonia and albuminoid ammonia in the range of 1–14 and 1–100 mg m^{-3} as ammonia (it is much higher in rooms). Finally, *Bergey* tested several methods (Fig. 4.27) and compared them with the conclusion that *Remsen*'s method is the "most reliable method"; however, the marked variation in the total amount of organic matter found in simultaneous determinations with the methods shows that no reliance can be placed on the results. Nevertheless, he concluded that the amount of albuminoid nitrogen is larger than free ammonia in most analysis and that room-air contains more organic matter than external air, and that there is no relation between gaseous and dust-like organic matter but that the gaseous organic matter forms an exceedingly small proportion of the total organic matter.

Chapman (1870) found that the "air contains suspended nitrogenous organic matter" (collected through cotton-wool) as well as "volatile organic bases" (when the air is conducted into water). *Ira Remsen* (1846–1817), US American analytical chemist, used a modified *Chapman*'s method based on pumice stone absorption for organic nitrogen (Remsen 1881). *Remsen* comments to the method by *William Amphlett Moss* (unknown living data), US American apothecary (Moss 1871), that very little is new. *Cornelius Benjamin Fox* (1839–1922)[881] who also wrote the first monograph on ozone, prefers a method what he calls the "pulverization of water method" (Fox 1886); using *Bergson*'s spray producer (developed as asthma inhalator)[882] the air is washed of all its impurities as it passes through the fine spray formed in the cylinder by the spray-producer (Waldenburg 1864, p. 88).[883] Finally, *Karl Bernhard Lehmann* (Lehmann 1890) recommended *Uffelmann*'s method (Uffelmann 1888).

Thus, organic air pollutants were not specified and summarized under "putrefying gases" where it remained unclear whether the pestilential smell was of hygienic importance. Blücher (1900, p. 22) also cites smelling hydrocarbons as vapors from crude oil products like gasoline, benzene, petroleum. The German meteorologist *Wilhelm Jacob van Bebber* (1841–1909) used in his textbook (Bebber 1895) "*Hygienische Meteorologie*" [hygienic meteorology] the classification according to the physiological effects, introduced by Renk (1886) and termed hydrocarbons belong indifferent substances but having characteristic smells, citing carbon disulfide (this compound he termed noxious), naphthalene, mercaptan, being in so small concentrations that chemical analysis is impossible.

Hydrocarbons and other organic substances have received considerable attention as air pollutants not before 1950 after recognizing that "they may participate in reactions in the atmosphere which produce objectionable intermediate compounds and products" (Cadle and Magill 1956, p. 3–13).[884] There were two reasons to keep organics not within the focus of further interest after the few investigations at the ending nineteenth century, first missing analytical methods to specify organic compounds in the air at low concentrations and secondly, the overwhelming air pollution by sulfur dioxide and dust, resulting not only in "permanent" London smog but also in some deathly smog catastrophes like the fog of 1930 in the Meuse River Valley area of Belgium, 1948 in Donara (Pennsylvania) due to Zinc Works,[885] in 1950 in Poza Rica, Mex-

881 A British physician in Scarborough who studied in London, Edinburgh and Paris; after his early retirement he wrote numerous articles on public health.

882 *Joseph Bergson* (1812–1902) was a German medical doctor at Berlin who introduced in 1862 his *Bergson*'sche *Apparat* (Hydroconium) and *Bergson* tube in English.

883 This sampling device, 100 years later "re"-constructed by Cofer et al. (1985), and termed "mist chamber" (or *Cofer* scrubber) was widely used for absorption of soluble gases, namely for sampling water-soluble organics in air. This system (also used in my group in the 1990's) we connected with an automatic sampler and analysis by ion chromatography combined with automatic sampling technique. It is still used for sampling of organics (Spaulding et al. 2002).

884 *Paul L. Magill* (unknown living data) was an US chemist and air pollution researcher, namely Los Angeles (Magill 1949, Magill et al. 1956). The earliest articles on air pollution ("smog") in Southern California were published in conjunction with the first National Air Pollution Symposium held in Pasadena, 10–11 November 1949.

885 Following the deadly smog, President *Truman* convened the first national air pollution conference in 1950. Congress didn't pass its first Clean Air Act until 1963, but progress continued steadily

ico, and 1952 the infamous London Fog. However, despite successful research on photochemical smog in the Los Angeles area (Cadle and Magill 1951, Haagen-Smit 1952), concentrations of hydrocarbons in the atmosphere of most cities have not been measured, probably largely due to the difficulties of making such determinations. Benzene and other aromatic hydrocarbons from the use of solvents and also automobile-exhaust were found locally in the air at relatively high concentrations; in city air, the concentration of aromatic hydrocarbons is probably generally less than 0.1 ppm (Cadle and Magill 1956). *Cadle* and *Magill* further write those aldehydes and ketones are introduced into city air from automobile-exhaust and incinerator smoke, but also evidently produced by the oxidation of hydrocarbons in the atmosphere; maximum concentrations ranged from 0.12 ppm (Baltimore) to 1.0 ppm (Los Angeles), however, the differences may represent differences in analytical techniques rather than the significant difference between the cities.[886] In the Los Angeles atmosphere, at least, less than half of these substances are formaldehyde. Cadle and Johnston (1952) reported on the presence of formic acid in Los Angeles smog. Organic halides have also been identified in Los Angeles smog, and they are probably present in the air of other cities with concentrations generally below 0.1 ppm (Cadle and Magill 1956).

As for methane, many organic substances which evaporate into the air were "detected" by source identification. However, despite the smell of vegetation was known for centuries (alchemists extracted oil from plants), but several terpenes were identified only at the beginning of the twentieth century and its biosynthesis was discovered in 1964. In 1960, attention was paid to the role of terpenes in the troposphere by the Dutch-US-American botanist *Frits Warmolt Went* (1903–1990), director of the Missouri Botanical Garden (Went 1960). This paper, full of visionary hypothesis which was proved decades later, offered several ideas on the formation and role of *blue hazes* on the basis of observations. Citing *Tyndall*'s experiments (pp. 321–223) and that of *Haagen-Smit* (p. 13) who mixed dilute petrol or olefin vapors with low concentrations of ozone observing a blue cloud formed. Obviously, *Went* made "experiments", "dropping a handful of crushed pine needles, or any other aromatic plant material, into a container with ozone" (Went 1960, p. 642), a blue haze develops and writes:

> A pine forest produces quantities of volatile terpenes, such as pinene, which give the forest its piney smell [...] The amount of these volatile organic compounds produced yearly by all land-plants and their decomposition products is of the order of 10^8 tons.

---

after that, with President *Nixon* creating the Environmental Protection Agency in 1970, the same year that Congress passed a more comprehensive Clean Air Act.

886 This explanation also can be used to explain the difference (by a factor of 10) between *Gautier*'s estimate in Paris and *Wolpert*'s estimate in Berlin; however, the historical values for "combustible gases" (and it must include free hydrogen, carbon monoxide and methane beside nonmethane hydrocarbons) is two orders of magnitude larger than in US cities in the 1950s – not unreliable when considering that "classical" pollutants (dust and sulfur dioxide) were also higher by about a factor of 100 around 1900 in large cities.

From the observation that the smell during the day disappears at a short distance from the forest, *Went* concluded on fast oxidation which results in condensation of terpenes to macromolecules. He further concluded that the sub-micron (he writes *submicroscopic*) blue haze particles must grow into µm size particles, losing the blue colors and providing condensation nuclei, so finally, they can be precipitated by rain and snow. *Went* not only suggested that the blue hazes and gained veil clouds play a very important part in the heat balance of the Earth but also (Went 1960, p. 643)

> [. . .] that volatization of terpenes and other plant products results in the production of, first, blue haze, then veil clouds, and finally, bituminous materials which, when returned to the Earth by rain and snow, produce the source materials for petroleum formation.

Due to insufficient analytical facilities, first estimates of monoterpene emissions from forests and atmospheric measurements are known not before 1980 (Isidorov et al. 1985, Isidorov 1990). Biogenic VOC emissions research in North America reached a period of peak activity in the mid-1970s to early 1980s because of speculation that these compounds had a role in urban ozone formation (Altshuller 1983). These efforts included the systematic enclosure survey of North American vegetation species and above-canopy flux measurements of emissions from entire ecosystems (Rasmussen 1972, Zimmermann 1979, Winer et al. 1982, Arnts et al. 1982). *Thomas E. Graedel* (born 1939), US American physicist, professor emeritus of industrial ecology and chemical engineering, listed organic compounds which were identified in ambient or source-related air (Graedel 1978):

- Well over 500 hydrocarbons, including more than 100 different alkanes (with a mean concentration of nonmethane hydrocarbons of 1370 µg m^{-3}; maximum concentration of ethane 95 ppb), about 100 alkenes (maximum concentration of propene 52 ppb) and alkynes (at a few ppb), 21 terpenes (typical values near forests 0.1–1 ppb), more than 40 cyclic hydrocarbons (2 to 50 ppb), 67 aromatic compounds derived from benzene (toluene up to 129 ppb, benzene is the next most abundant, followed by the xylene and ethyl benzene; most compounds less than 1 ppb but 19 sometimes more), more than 20 aromatic compounds derived from naphthalene (at ambient concentrations of a few ng m^{-3}), and more than 130 polynuclear aromatic hydrocarbons.
- Carbonyl compounds, including 30 aliphatic aldehydes, 15 olefinic aldehydes, 5 cyclic aldehydes, and 33 aromatic aldehydes; more than 25 aliphatic ketones, 6 olefinic ketones, about 40 cyclic ketones, and more than 60 aromatic ketones.
- Oxygenated organic compounds: more than 80 organic acids (data are available for the concentration of some 30 organic acids in atmospheric aerosol up to a few µg m^{-3}), more than 100 esters (1–100 ppb), nearly 150 alcohols (methanol up to 100 ppb, phenol at a few ppb), only two organic peroxides (unknown concentration), about 80 ethers.

- Nitrogen-containing organic compounds. 15 nitriles, 43 amines, 32 nitro compounds, 60 heterocyclic nitrogen compounds.
- Sulfur-containing compounds: a dozen different mercaptans, 11 organic sulfides, 12 heterocyclic sulfur compounds, and 4 compounds of thioacids and thiocyanates.
- Organic halogenated compounds: more than 40 halogenated alkanes and alkenes, nearly a dozen halogenated aromatic compounds, and more than two dozen chlorinated pesticides.
- A few organometallic compounds.

The Soviet-Russian chemist *Valerij Alekseevich Isidorov* [Валерий Алексеевич Исидоров] (born 1942) from Leningrad (St. Petersburg) describes in his book (Isidorov 2001) about 500 organic compounds concerning their source and ambient concentrations detected and determined since the 1980s; he also carried out many measurements. The general processes of hydrocarbon oxidation and the formation and participation of radicals were already studied and known in the 1930s in combustion processes (Chapter 1.3.5.2), and continued – namely using advanced measurement techniques to estimate reaction rate constants – between 1950 and 1970 (Heicklen 1976, pp. 239–273).

4.4 Ozone (O_3)

> Many hypotheses have been set forth respecting this mysterious something, ozone (Mackereth 1868, p. 213).

Much is written already on the history of ozone.[887] After *Schönbein*, the first monograph about ozone was written by *Guiseppe Belluci* (Bellucci 1869). Fox (1873) writes in the preface of his monograph: “The importance of the subject of Ozone is so great, and the amount which was written respecting it is so large [. . .]”. Whereas *Fox*’s book only deals with ozone and antozone in the atmosphere, and Harries (1916)[888] only describes reactions between ozone and organic compounds (ozonolysis), further mono-

887 The most comprehensive history on ozone (not only on atmospheric chemistry) is written by the Israel chemist *Mordecal B. Rubin* (1926–2012) in a series of eight papers (here cited only those to be relevant to atmospheric chemistry: Rubin 2001, 2002, 2003, 2007, 2009, Braslavsky and Rubin 2011). Books on the history of ozone: Anonymous (1913), Schmidt (1988a,b), Lemmerich (1990), Mégie (1991), Singh and Fabian (2003), Bojkov (2012). Books on *Schönbein*: Kahlbaum (1899), Kahlbaum et al. (1901), Schmidt (1988a), Nolte (1999). Collected volumes of *Schönbein*’s publications: Schönbein (1867b), Schütt (1996). Note books on stratospheric ozone, the ozone layer and hole are not listed here.

888 *Carl Dietrich Harries* (1866–1923) was a German chemist who worked mainly on rubber and ozone (ozonolysis is also termed *Harries* reaction); 1890–1904 at Berlin University (assistant of *Emil Fischer*), 1900 professor and 1904–1916 at University Kiel, since 1916 leading position at Siemens & Halske; he was married with *Hertha von Siemens* (1870–1939), daughter of *Werner von Siemens*.

graphs (Fonrobert 1916, Rideal 1920, Moeller 1921, Vosmaer 1916)[889] are mainly focused on the industrial production and application of ozone. Thus Vosmaer (1916, p. 1–2) writes:

> The early history of ozone is of little interest [. . .] In our days of excessive literature there is no time for the consideration of obsolete speculations about the nature of a substance; it is quite enough to read established facts; hence we may pass over those ancient publications and speculations about ozone being a nitrogen or a hydrogen compound, etc. [. . .]
> For those interested in the history we refer to the work of C. Engler, published in 1870, and called: "Historisch-Kritische Studien über das Ozon." (note, the reference by *Vosmaer*, 1870 is wrong; it is 1879)

Indeed, Engler (1879) describes excellent the long way from *Schönbein*'s first observation and speculations on the nature of ozone until recognizing its constitution. Further, Andrews (1874) wrote a review on the early history of ozone. A review on the history of ozone can be found in Gmelin (1907, pp. 124–148; 1952, pp. 5628–534), on its occurrence in Gmelin (1958, pp. 101–111) and the chemistry in Gmelin (1960, pp. 984–1184). In Gmelin (1852), ozone is not mentioned yet.

The American chemist *Albert Ripley Leeds* (1843–1902)[890] wrote a comprehensive history on ozone and hydrogen peroxide from the beginning until 1882 (Leeds 1879b, 1883), but the most important thing of both publications are two tables with the literature on ozone and hydrogen peroxide (on the formation and properties, analytical methods, atmospheric studies, and action on organic substances, plants, and animals) citing 183 authors with 573 references (without the title but including a short description of the subject; note that at that time the same subject was published in several journals, e. g., the first *Schönbein* paper on the "electric odor" 1840 was published in 9 different publication organs). *Leeds* Table includes 36 authors with 48 references on observation of ozone in the atmosphere. However, there are many important references of ozone observations in the atmosphere missing: e. g., Schlotfeldt (1849), Gräger (1852), Smallwood (1854, 1857), Glaisher (1855), Schiefferdecker (1855), Reslhuber (1855, 1856), Boehm (1858), Prestel (1865, 1866), Möhl and Dietrich (1868), Mitchell

889 *Ewald Fonrobert* (1887–1964) was German chemist, who made his dissertation 1913 at University Kiel (co-worker of *Carl Dietrich Harries*) and later in Wiesbaden; known for his research on rubber and oil fuel. Sir *Eric Keightley Rideal* (1890–1974) was an English physical chemist who gained PhD 1912 at the University Bonn (under *Richard Anschütz*), 1919–1910 visiting professor at the University of Illinois at Urbana, 1920–1946 in Cambridge (Trinity Hall), becoming Professor of Colloid Science in 1930; 1946–1955 in London. *Alexander Vosmaer* (1866–1944) was a Dutch chemical and electrical engineer and director of the electrotechnical laboratory in Harleem and of the Ozon-Maatschappij system [Society for treatment of water by ozone] in Amsterdam. *Max Moeller* (?–1945) was a German physical chemist at Siemens & Halske in Berlin (date of birth unknown); later director of "*Wernerwerke für Messtechnik*" (killed 1945 by soviet soldiers).

890 US chemist, Prof. of Chemistry at Haverford College from 1867 to 1871 and from 1871 professor at Stevens Institute of Technology, Hoboken. Vice President of the American Chemical Society.

(1869), Lévy (1878a,b,c), Wolffhügel (1875), and he included also curious references like Montani (1865),[891] and some not to be find in the given literature.

Furthermore, Day (1868) and unknown Dr. *Schürmann* published an excellent review on ozone, namely including the early history and chemistry (Schürmann 1875). Another, but more popular article on ozone was written by *Otto Eduard Vincenz Ule* (1820–1876) who was a German writer, known for his popularization of natural sciences (Ule 1863). *Abraham Lissauer* (1832–1906), a German physician in Danzig, wrote also on the relation between ozone and human health (Lissauer 1864). It follows a list of authors having most references about ozone (name, period of publications, number of publications):

Schönbein	1840–1870	131
Houzeau	1854–1873	34
Andrews	1855–1872	21
Soret	1853–1871	21
Osann (G. W.)	1847–1866	20
Bérigny	1855–1867	13
Williamson	1845–1853	5
Moffat	1861–1872	5

Gustav Dachauer (living dates unknown)[892] was a Chemist in Munich who wrote the first review on ozone research in the period of 1840–1863 (Dachauer 1864), presenting summaries of publications on the nature of ozone (in parenthesis number of references) from *Schönbein* (62), *Fischer* (9), *Marignac* (1), *Williamson* (3), *Marchand* (2), *Osann* (11), *Fremy* and *Bequerel* (1), *Baumert* (2), *Houzeau* (5), *Andrews* (8)), *His* (1),[893] *Neumann* (1), *Cloez* (1), *Andrews* and *Tait* (2), *Clausius* (1), *Gorup-Besanez* (1), *Schrötter* (1), *Weltzien* 81), *Böttger* (1), and *Kolbe* (1).

It lies in the aim of the volume in your hand that "my" history of ozone is within the frame of *atmospheric* ozone, and I try to refer not only well-known authors but also less known or even forgotten, giving my respect to her contributions, even for small or wrong findings. The following citations show that the importance of (atmospheric) ozone was well recognized in the ninetheenth century but the increase

891 This short communication informs on ozonometry at Constantinople during the cholera 1865 without giving any results; it was likely the Italian cleric *Francesco Fabi Montani* (1802–?).

892 Doctoral Thesis 1858 in Göttingen "Über den Caprylaldehyde"; he wrote some books on chemistry (*Chemisches Taschen-Wörterbuch*, 1863; *Haupt-Grundlehren der Chemie zur Einführung in diese Wissenschaft für angehende Chemiker, Mediciner, Pharmaceuten und Techniker leicht fasslich dargestellt*, 1863, *Kosmetische Receptirkunst für Ärzte und Apotheker*, 1864).

893 His, W. (1856) Ueber die Beziehungen des Blutes zum erregten Sauerstoff. *Virchows Archiv für Pathologische Anatomie und Physiologie und für Klinische Medizin* **10**, 483–499 (wrong cited as *Hiss* by *Dachauer*). *Wilhelm His* (1831–1904) was a German-Swiss physician.

of tropospheric ozone and the decrease of stratospheric ozone was pointed out not before the late 1970s.

Several books and essays (no periodicals are listed here) have been published before 1951: Schönbein (1844, 1849c), Wolf (1855), Boeckel (1856), Scoutetten (1856), Smallwood (1857), Floderus (1859), Meissner (1863a, 1869), Dachauer (1864), Fox (1873), Lender (1872c), Engler (1879), Engler and Weissberg (1904), Le Chatelier (1913), Fonrobert (1916), Harries (1916), Vosmaer (1916), Rideal (1920), Moeller (1921), Wildt (1934), Fabry (1950), Prokofewa (1951).[894] The following historical phrases should illustrate the diverse interest of chemists in understanding the role of ozone in the atmosphere – as an intrinsic component of air and not the foreign body – and thus in the balance of nature.

> To the philosophers, the physician, the meteorologist, and the chemist, there is perhaps no subject more attractive than that of ozone (Fox 1873, p. 1).
> By the agency of electricity, – probably, too, by the influence of light, – the oxygen in the air undergoes a peculiar change, by which it is rendered far more energetic than it is in its ordinary state. This is the condition to which the name of *ozone* was applied. Now, this ozone, or this peculiar oxygen, always exists in the air we breathe; but its quantity is subject to great and rapid variations (Hunt 1854, p. 217).[895]
> It is probable, as we have already had occasion to remark, that ozone may be the active agent in removing from the atmosphere those organic poisons to which many forms of pestilence are traceable; and it is a curious fact, that a low electrical intensity, and a consequent deficiency of atmospheric ozone, marks the prevalence of cholera, and an excess distinguishes the reign of influenza (Hunt 1854, p. 300).
> Ozons, jenes oxydierenden Stoff, über dessen Nature man ungeachtet der schönen Arbeiten der HH Schönbein, Marignac und De la Rive, Frémy und E. Becquerel so wenig übereinstimmt [Ozone, this oxidizing matter, on that nature despite the fine works of Messrs Schönbein, Marignac and De la Rive, Frémy and E. Becquerel, one agrees so little] (Houzeau 1855).
> Es giebt wohl keine Körper, welche in der Literatur der Chemie in den letzten Jahrzehnten so häufig auftreten, als das Ozon und das Wasserstoffperoxyd [There are probably none bodies, which occur so often in the chemical literature of the last decades like the ozone and the hydrogen peroxide] (Weltzien 1866).
> Der electrische Sauerstoff, welcher Stoff und Kraft zugleich ist, zeigt die Einfachheit des irdischen Haushalts, weil er in der grossen Natur grosse Aufgaben zu lösen hat [The electricized oxygen,

894 *Manfred Mustafa Floderus* (1832–1909) was a Swedish teacher of natural sciences at *gamla skola* [old school] in Upsala. On *Carl Engler* see p. 470. On the German chemist *J. Weissberg* no information is available. *Henry Le Chatelier* (1850–1936) was a French chemist and physicist who is most known for "The Equilibrium Law" also termed *Chatelier's* principle. This volume was prepared by *Chatelier* in cooperation (editing committee) *H. Abraham*, *H. Gautier*, and *J. Lemoine*. *Rupert Wildt* (1905–1976) was a German-American astronomer; he was born in Munich, studied in Berlin (1923–1928) and worked at the observatories in Bonn and Göttingen, emigrated to the USA in 1935 and worked at observatories in Pasadena and Princeton, discovered in 1939 the energy transport within Sun's photosphere. *I. A. Prokofewa* [И. А. Прокофьева] (unknown living data) was working at the astronomic observatory in Pulkovo in Leningrad (now Saint Petersburg); nothing more is known about her.
895 *Robert Hunt* (1807–1887) was a British mineralogist and amateur poet.

> which is together matter and power, shows the simplicity of the earth's economy, because it has to solve great tasks within the great nature] (Lender 1872a).
> Though it was known for more than a century that air and oxygen acquire a peculiar odor when exposed to the action of electric sparks, and though Schönbein ascertained nearly half a century ago that this odor is due to a distinct form of matter, now termed ozone, which is produced by the electrolysis of dilute sulphuric acid, by the action of electric discharge in air, and as a product of the slow oxidation of phosphorus, chemists are still trying to learn the exact conditions of the formation of this substance, and still investigating some of its simplest reactions; whilst inventors are but beginning the work of making it useful to man (Anonymous: The production and uses of ozone. *Nature* **58** (1898) 416–418).

It was recognized to be a substance of unusual properties that has never been isolated in the pure state. The given names (before accepting *ozone*) show the relation to oxygen: electricized oxygen, allotropic oxygen, nascent oxygen, active oxygen, excited oxygen (from the German "*erregter Sauerstoff*"). The bleaching, antiseptic properties, and its action as deodorizer, disinfectant, and a germicide were soon known and used in the nineteenth century. The English chemist *William Mogford Hamlet* (1850–1931 writes (Hamlet 1881, pp. 330–331)

> [...] it is active oxygen which must be regarded as the greatest known enemy to bacteria life, whether from the action of chlorine, nitric oxide, ozone, or peroxide of hydrogen. It also follows that these bodies are the bed disinfectants.

An extensive paper on ozone and its uses in medicine was published by *Willian James Morton* (p. 314); he writes, citing *Moffat*, *Boeckel*, and others (Morton 1894, p. 357):

> They claim that the spread of epidemic diseases is caused by the absence of or a diminution of the normal quantity of ozone in the atmosphere, and is coincident with the appearance of cholera, etc. Others, however, explain that the absence of ozone is due to the increased number of germs which absorb the ozone, arguing that their opponents mistake the cause for the effect.

After *Schönbein* discovered ozone in the atmosphere, the nineteenth century boomed in ozone research and observation. Beyond question, this volume comprises more historic references on atmospheric ozone than each other previous publications. However, I do not doubt that less than 10 % of nineteenth-century ozone observers are included here; probably more than a thousand observers produced millions of ozone data between 1850 and 1900. On the other hand, I am very sure that I have listed *all* researchers who have significantly contributed to an understanding of atmospheric ozone before 1960.

With the industrial and scientific development in most European countries and Northern America since the mid-nineteenth century, an innumerable number of societies, institutions, observatories, research stations, governmental boards, etc. arose, which almost all produced reports, proceedings, journals, and other volumes; many of them were digitalized but also many were lost, are forgotten or unknown nowadays. From the individuals who carried out the investigations and observations, only a small percentage remained in our mind, are known by name and her living data. Thus, many assistants, laboratory workers, and temporary workers were lost in the darkness of the past, but, sometimes (e. g., by Fox 1873) you can read the name but no more.

A huge bibliographic source under the keyword "Ozone" is the "Index-catalogue of the Library of the Surgeon-General's Office, United States Army. Authors and subjects. Government Printing Office, Washington", published in three series between 1880 and 1932: series 1, Vol. 10 (1889) pp. 338–341; series 2, Vol. 12. (1907) pp. 388–390, and series 3, Vol. 9 (1929) pp. 544–545. This work comprises about 250 medical journals from all over the world, however, at a first glance I have seen many missing medical journals from Germany and France. On the other hand, many cited sources cannot be found either because of the transcription in Latin characters, namely from the Cyrillic alphabet[896] or because the abbreviation of the bibliographic source is unknown today or the reference does not exist in public archives.[897]

The history of atmospheric ozone chemistry is inextricably linked with the chemistry of atmospheric oxygen, and the chemistry of oxygen cannot be separated from the chemistry of oxides and oxoacids (namely those of carbon, nitrogen, and sulfur); however, only the presence of water vapor makes the chemistry complex, providing oxo-hydrogen compounds (HO_x) in atmospheric air, namely the hydroxyl radical, OH, as the general oxidizer. Therefore, O_2 and H_2O in combination with solar radiation create the never-ending source of atmospheric chemistry.

4.4.1 Early history and discovery of ozone

Before discovering ozone in 1839 by the German chemist *Christian Friedrich Schönbein* (1799–1868)[898] in Basel while conducting electrolysis experiments with water (Schönbein 1840, 1843), other scientists observed, that oxygen gas, when electrical sparks

---

896 For example: Strzhizhovski, A. (1876) Reszultati nabludeniee nad ozonom [результаты наблюдение над озоном – results upon ozone observation]. *Zdorovje* [*Здорóвье* – Health] (St. Petersburg) **2**, 22. It is likely Стржижовский Антон Станиславович (1836–?) of polish origin, written *Strzyżowski*. Furthermore, one can found quotations of known and unknown persons but with references not available, e. g., Heidenreich (1850) Ozon und Katarh. *Neue medicinisch-chirurgische Zeitung* **7**, 3 (*Friedrich Wilhelm Heidenreich* (1798–1857) was a German medical doctor and naturalist), Bona H. (Henri) (1864) De l'ozone. Paris (likely dissertation – not found), Desplats, V. (1857) De l'ozone. Dissertation, Faculté de médecine de Paris, 44 pp. (*Victor Desplats*, medical doctor, unknown livings data; nothing is known on *Bona*). Between 1850 and 1890 numerous dissertations about ozone were made by candidates of medical doctor (several are cited in this volume).

897 Here are examples for periodicals not to be found: Bergheim, F. (1857) Das Ozon der Luft. *Kosmos. Zeitschrift für angewandte Naturwissenschaften* **1**; Lobuto, J. G. (1879) Ozonometria en al afio de 1873, seguidaensde al 31 de mayo. *Observador méd. México 1876–79*, **5**, 190–195.

898 German-Swiss chemist, born in Metzingen (Germany). 1812–1819 he worked as trainee in a pharmaceutical factory in Böblingen and begun to study chemistry in 1820 in Erlangen and later in Paris (Sorbonne); worked as teacher in England, 1828 professor at university Basel and 1828 full profeshad been passed through it, acquired a peculiar smell. At different places of this volume, we mentioned that people since ancient times did smell the smoke, fog, miasmas, and other pollution. The "sulphurous smell" [Θεείωι means divine and θήϊον – sulfurize, cleansing with sulfur but also divine] in association with thunderbolt was described by *Homer* in Aeolic dialect (Beard 1874, p. 459, Cunliffe 1963, p. 187, Leaf 2010, p. 172).[899] On the other hand, the volcanic origin of sulfur and the disinfecting power of sulfurous fumes were known in the antique.

Lionel Wafer (1640–1705), a Welsh explorer, wrote "[...] violent Thunder and Lightning: During which time, the Air has often a faint Sulphureous Smell, where pent up among the Wood" (Wafer 1699). The odor of thunderbolt is described by Gehler (1825, p. 1030) as "*Schießpulver- oder Schwefelgeruch*" [gunpowder or sulfurous smell], however, he further writes that this smell cannot result from the bodies affected by the flash, because it is always the same smell independent form the kind of bodies, and in the air, as we know this mixture until now, there is nothing sulphureous.[900] Fox (1873, p. 143) writes that "the odor of ozone is so powerful, that air containing only one-millionth of it is said to have a decided smell of the gas". The odor threshold of ozone is about 40 μg m^{-3} (20 ppb) but ranges between 7 and 250 ppb, thus in clean air it could be recognized.[901] In Gehler (1825, p. 1002)

sor for chemistry. *Schönbein*, most known for the discovery of ozone, was interested in many fields of chemistry.

899 Mohr (1854) was obviously the first who provided the information on "sulfurous" odor when lightning strikes in the epics of Ilias (VLLL, 135. XIV, 415) and of Odysee (XII, 417. XIV, 307).

900 It is said that the smell of burnt gunpowder is similar to that of ozone or "electricity"; combustion of gunpowder consists of many differing complex reactions, but SO_2, NO_x and organic compounds, not fully oxidized to sulfate, nitrogen and CO_2, creating a sweet-sharp smell. I remember it when I "experimented" as "young chemist" in my childhood with mixtures of charcoal, sulfur and potassium nitrate. Note that "sulphureous" in historic literature (today's spelling sulphurous in English and sulfurous in American) is used in sense of sulfurlike, sulfur-containing, and sulfury; in still older alchemistic literatur also as expression for stinking and/or burning.

901 From four olfactory detection studies carried out between 1939 and 2007, an ozone detection limit was derived to be 15–20 ppb (see Cain et al. 2007). However, these estimations were based on filtered air containing only ozone. I never heard in the last 40 years that any person dealing with ambient ozone measurements (also much higher than 100 ppb) was able to detect ozone in the air by his nose. Note that chemists used for centuries smell, tasting and appearance, until nowadays, for identification of chemical substances. The ambient air, in cities and industrial areas, namely before the 1990s, was odorous due to a mixture of different air pollutants from chemical processing, traffic exhaust gases, and house fire. Moreover, the air pollution was to be seen in kinds of smoke, fumes, hazes, dust and soot deposition. On the other hand, people living permanent in "bad air" acclimatize and become unable to recognize the odor. In autumn 1990 at a meeting in Leipzig with air chemists from West Germany, they recognized the typical "east-odor", whereas I said, I cannot do it. But when working as student in the large historical laboratory of the Chemical Institute of the Berlin University, that has a permanent "chemical background smell", I did recognize when another student opened a bottle containing potassium cyanide several meters in distance. Thus, olfaction is much more sensitive whilst significant short-time changing chemical composition in the air. When I had my practicum as student *deLuc*[902] (see Deluc 1787) is cited who linked the flash with chemical processes that immediately produce "electric matter". Ludwig (1871, p. 230) wrote that the ozone is particularly characterized through its strange odor, according to *Wöhler* it remembers on that of phosphorus.

Dutch *Martinus van Marum* (1750–1837),[903] subjecting different kinds of air to electrical discharges with his newly constructed electrifying machine in 1785, found that the volume of pure oxygen reduced by 1/20 but the "air" did not show a reaction on lime water or litmus; he wrote (van Marum 1786, p. 25):

> Indem wir nun diese Luft aus einem Gefäße in das andere brachten, bemerkten wir, daß sie einen starken Geruch angenommen, welcher nun sehr deutlich der der electrischen Materie eigene Geruch zu sein schien, allein er war viel stärker, als wir ihn je vorher empfunden hatten [By doing this air from one device into another, we mentioned, that they picked up a strong odor, which seemed to be very clearly that of the electric machine own odor, but much stronger that we have it ever before perceived].

Thus, *van Marum* reported the odor of ozone but he failed to identify it as a unique form of oxygen. The Scottish chemist *William Cruickshank* (1745–1800) also mentioned in his last life year during electrolysis of sulfuric acid (note that oxygenated muriatic acid is chlorine) "a peculiar smell, somewhat like that of ox. Mur. acid when very dilute" (Cruickshank 1801, p. 261); he further writes (note the silver wire is the anode):

> [. . .] from the silver wire a small quantity of gas was likewise disengaged, but in a short time the fluid in this leg of the tube lost its transparency, and became milky and opake [. . .] The obtained gas during this process was not particularly examinate, it was only found not to be flammable.

He was referred later only with his observation of a "chlorine-like odor"; however, this full citation here (for the first time in historic literature) clearly shows, that he recognized this smell like *gas*, as *Schönbein* in 1839. Only due to his death, he was unable to examine it in more detail. From his experiments, *Cruickshank* made the following notable remark (Cruickshank 1801, p. 257) which can be regarded as the beginning of *electrochemistry*, or, even interlinked oxidation and reduction processes (redox):

> [. . .] simplest mode of explanation would be, to suppose that the galvanic influence (whatever it may be) is capable of existing in two states, that is, in an oxygenated and deoxygenated state.

in Bitterfeld in 1967, I did taste the air to be acidic (Bitterfeld was known in that time for its horrible air pollution). However, I remember the "electric odor" due to sunlamps in my childhood (which were a strong source of ozone) and from first heavy photocopying machines.

902 Wrong written, sometimes *de Luc*.

903 Dutch physician, inventor, scientist and teacher, known for his "Large electricity machine". He introduced modern chemistry in the Netherlands after the theories of *Lavoisier*.

Beard (1874, p. 459) and Bridge (1907, p. 357),[904] the latter copied his text passage completely from *Beard*, wrote that *Tiberius Cavallo*, a prominent scientist in the history of electricity, soon after *Cruickshank*'s observation, reported that "electrified air" had a purifying effect, and that *Cavallo* spoke of this condition of the air as the "aura electrica". However, in Cavallo (1803), he spoke in another relation on "electrical aura" (Cavallo 1803, p. 37) and at no place on the "purifying effect of electrified air"; in contrast, he noted that "the electric spark taken repeatedly in the common air, diminishes a little it's purity" due to the formation of nitrous acid as shown before by *Cavendish*. *Cavallo* wrote the chapter "Atmospheric Electricity" (Cavallo 1803, pp. 436–450). But it is worth referring *Cavallo* (Cavallo 1803, pp. 445–446) with his "most plausible" explanation of the source of atmospheric electricity; this phrase led surely some of the later ozone observers between 1850 and 1880 to the belief that this process is the source of ozone in the air:

> [. . .] to derive it from the evaporation of water, and from the condensation of vapours.

Bridge (Beard 1874) further writes:

> In 1826 Dr. John Davy (Lecture on Agricultural Chemistry) recognized the existence of this principle in the atmosphere, and published a formula for the preparation of chemical tests to be used in its detection, resembling that afterwards adopted by later investigators.

Whereas *Beard* (Bridge 1907) writes:

> In 1826 D. John Davy, in a measure anticipating Schönbein, recognized this peculiarity of the atmosphere, and devised tests for detecting it.

John Benjamin Dancer (1812–1887), a British scientific instrument maker, first referring that "In 1826 John Davy believed this principle existed in the atmosphere, and proposed test for its detection" (Dancer 1876, p. 122), conducted 1838 in Liverpool experiments with the electric decomposition of water, and "noticed a white cloud, or vapor, escaped from the bubbles of the mixed gas [. . .] in the jar' (Dancer 1876, p. 122–123). The prevailing opinion that it was caused by impurities from the sulfuric acid, he excluded in further experiments, however, he did not continue the research. Notably, Mellor (I, p. 877) wrote that *Cavallo*'s "electrified air as the *aura electrica*" is referred in Cavallo (1782), i. e., many years before *Cruickshank*'s observation. However, in this volume, the latter term is not mentioned and also nothing can be found on its "purifying action". *Mellor* further writes that "H. Davy, in his *Lectures on Agricultural*

---

904 *George Miller Beard* (1839–1883) was an American neurologist and *James Howard Bridge* (1858–1939) was a British author. A similar text concerning *Cavallo* and *Davy* wrote Bartlett (1883, p. 2). *Bridge* wrongly noted the year (1783 instead of 1785) of *van Marums*'s observations.

Chemistry (London 6, 1826) says that in 1826 Dr. John Davy [...]". Thus, obviously, *Mellor* copied these wrong citations.[905]

Schönbein reported on March 13, 1839, to the local "*Naturforschende Gesellschaft*" [Society for Natural Sciences] in Basel (founded in 1817) on an odor at the positive electrode which was the same as the odor produced by an arc between electrode:

> Herr Schönbein macht die Gesellschaft auf die merkwürdige und bisher noch nicht beobachtete Thatsache aufmerksam, dass bei der Electrolyse von Wasser an der positive Electrode ein Geruch entwickelt wird, auffallend ähnlich demjenigen, den man beim Ausströmen gewöhnlicher Electricität aus Spitzen wahrnimmt [Mr. Schönbein calls the Society's attention to the noteworthy new observation that a smell develops at the positive electrode during electrolysis of water which is strikingly similar to that obtained by the flow of electricity across electrode], cited after Rubin (2001).

As early as 1840 *Schönbein* proposed the name ozone; in an annotation (Schönbein 1840, p. 635) he writes:

> [...] den Vorschlag, das riechende Princip *Ozon* zu nennen, wenn es sich bei ferneren Untersuchungen entweder als ein elementarer oder zusammengesetzer Salzbildner verhalten sollte [... the proposal, to name this smelling principle *ozone*, if in further studies it behaves either as elementary or composed salifying agent].

Schönbein names this gas "ozone" after the Greek word όζειν (to smell). His friend *Karl Friedrich Mohr* (1806–1879), a pharmacist in Coblenz, explains in a wonderful poetically communication (Mohr 1854) the word as onomatopoeia "oz–zo", derived from όζω (I smell) and όδμή or οσμή (odor).[906]

---

905 *John Davy* (1790–1868) an army doctor and self-educated chemist, was the brother of *Humphry Davy* (1778–1829), who hold lectures from 1802 until 1812 before the Board of Agriculture, published in 1813 as "Elements of Agricultural Chemistry" (Longmann, London); *John Davy* produced nine volumes on the collected works of his brother from 1836 to 1840. I found nothing in the works of *John* or *Humphry Davy* in this field of "electrified air"; *Humphry* included in his book a chapter "on atmospheric composition". Burns (1997, p. 182) wrote first on this mystery regarding *Davy*, and also searched the publication of *Humphry Davy* and *John Davy*, and not found this statement. *Burns*, however, did not known on the paper by *Beard*, some that this is the source of the error.

906 From "ozone" soon later derived the verb *ozonise* (American *ozonize* and German *ozonisieren*, but never *ozonieren*) and the noun *ozonation* (also referred to as *ozonisation* and American *ozonization*); German *Ozonierung* or *Ozonisierung* (French *ozonisation*). In the late nineteenth century, particularly "ozonation" was used (with respect to air and water treatment). Not often the verb *ozonate* (from noun *ozonation*, but not to confuse with substances called *ozonates*) is used, and never in the nineteenth century. It is meant a) convert of oxygen into ozone (ozonized air) and b) treat of substances with ozone (first used for water treatment in the late nineteenth century) and later in the twentieth century (not before 1950) of organic substances (formation of ozonides). Furthermore, *ozoniferous* (ozone containing) was used by Fox (1873, p. 28), *deozonisation* (Fox 1873, p. 95) and *ozoniser* (device to produce ozone) *ozonograph*, *ozonometer* (instrument to measure ozone).

In a letter sent in 1840 to *François Arago* and submitted to the French Academy of Sciences, *Schönbein* suggests that ozone could belong to the chemical group of chlorine or bromine. The Swiss physicist *Auguste Arthur de la Rive* (1801–1873) argued that this smell is derived from lost metallic particles (de la Rive 1841, pp. 402–405). Schönbein (1858a), who never identified the constitution of ozone, proposed the existence of another form of reactive oxygen species which he termed *antozone*. He believed that ozone and antozone were both formed under ozone-producing conditions (quiet electric discharges of air) and react together to give oxygen[907] hence explaining the low yield of ozone by its destruction by antozone. He first believed that this new substance is bonded together with hydrogen, and a body similar to bromine and chlorine, and later identified with hyponitrous or nitrous acid, after revising this view, he stated that ozone is a higher compound of oxygen with hydrogen, but different from *Thénard*'s hydrogen peroxide (see Table 4.15). An amusing dispute arose between *Schönbein* and *Fischer*[908] who argued in 1845 that the three methods gave three different substances: the odor from arcing air was the odor of nitrous or nitric acid, the odor from electrolysis was due to hydrogen peroxide, and the odor from the reaction of phosphorus[909] was simply phosphoric acid (Fischer 1845, Schönbein 1845b). Fellenberg and Rivier (1845)[910] have been thrown some doubt on the whole theory of ozone by finding nitric acid, or some other acid or oxide of nitrogen that were produced in experiments pass-

907 Later described as $O_3 + H_2O_2 \rightarrow H_2O + O_2$.

908 *Nathaniel Wolfgang Fischer* (1782–1850) was professor for chemistry in Breslau.

909 Phosphorus (white) was first produced as an element (but not recognized as an element) by the German alchemist *Hennig Brand* (c. 1630–1692) in Hamburg in 1669 from the heating of distillated urine remaining with sand. *Bernhardt Siegfried Albinus* (1697–1770) isolated phosphorus from the charcoal of mustard plants and cress, confirmed by *Andreas Sigismund Marggraf* (1709–1782) in 1743 and likely *Scheele* in 1769 found it from bone (Kopp 1931). *Lavoisier* recognized P as an element. Natural sources of phosphuretted hydrogen (phosphine PH_3) were identified as sewage sludge, swamps and human flatus. In the early nineteenth century phosphine (which is spontaneously inflammable) was also known from cemeteries where it sometimes burned with blue flames. In continental rainwater, Lewis et al. (1985) found phosphate in such concentrations that it could not be explained by the scavenging of particulate phosphorus and concluded that it was a terrestrial source of a volatile P compound. First *Glindemann* and coworkers (Glindemann et al. 2004, 2005 and citations therein) detected monophosphane PH_3 (formerly phosphine, also termed hydrogen phosphide) in air in a concentration range of pg m^{-3} to ng m^{-3}. Close to identified emission sources (paddy fields, water reservoirs and animal slurry) reported concentrations are significant higher. Glindemann et al. (2003) found for the first time PH_3 in remote air samples (low ng m^{-3} range) in the high troposphere of the north Atlantic. For century's *ignis fatuus*, a phosphorescent light seen over marshy ground and around graveyards at night (ghostly lights) is now known to be caused by the spontaneous combustion of gases emitted by decomposing organic matter where phosphanes would act as a "chemical match".

910 *Ludwig Rudolf von Fellenberg* (1809–1878) was a student of *Liebig* 1841 in Gießen, took over the paper mill of his father, became professor of mineralogy and chemistry at Lausanne, later established a private laboratory in Bern.

ing of electricity through moist air and the oxidation of phosphorus under the same circumstances.

Schönbein also ascribes the smell occurring together with lightning strokes, since antiquity described as "sulfureous", to this new substance ozone. In a letter to *Justus von Liebig* from 5.9.1853, *Schönbein* emphasized the role of ozone in the earth's atmosphere (Kahlbaum and Thon 1900, p. 10):

> Geneigt zu glauben, das atm. Ozon spiele im Haushalte der Erde eine wichtige Rolle, halte ich es fuer wünschenswerth, dass möglichst zahlreiche, sowie grosse Zeiträume als bedeutende Laenderstrecken umfassende, untereinander vergleichbare Beobachtungen ueber die Veränderungen des Ozongehaltes der Atmosphäre angestellt werden [...] [Inclined to believe that atmospheric ozone plays an important role in the earth's budget, I think it is desirable to carry out as many as possible as well as big periods and comprehensive major routes comprising observations on the variation of the atmospheric amount of ozone, comparable with each other [...].

Robert Angus Smith writes (Smith 1873, p. 5):

> The formation of ozone in the upper atmosphere does not give us more oxygen; we obtain only a more active condition of that element. The formation of ozone at the surface, and of nitrous gas also, by evaporation, affect, in conjunction with the elimination of oxygen by organism, the supply of that which may be removed.

This is the first written statement of ozone formation in the upper atmosphere; the wrong belief that ozone is also formed by plants and soil emanations was under long discussion in the nineteenth century (not highlighted in this volume). However, *Smith* writes in his "Air and Rain" (Smith 1872, p. X):

> Some people will look for ozone in this volume, but nothing appears regarding it: I have not been able to make the projected trials; and this work does not attempt to be historic. Many other things are left out in this *beginning*.

The nineteenth-century hype of atmospheric ozone paper tests lasted only a few decades after *Schönbein*'s discovery. In his 1906 presidential address, "the meteorology of daily life", *Richard Bentley* (1854–1936)[911] pointed under the subtitle "the state of air" (Bentley 1906, pp. 83–84): "The observations of ozone are now nearly discontinued, but have given place to far more interesting ones of the ionization of the air".

911 He was formerly head of the famous publishing house of Richard Bentley & Son. He was President of the Royal Meteorological Society in 1905–6, Fellow of the Society of Antiquaries and of the Royal Geographical, Linnean, Statistical and Royal Botanic Societies.

4.4.2 On the nature of ozone

4.4.2.1 Constitution

In 1845, *Auguste de la Rive* communicates in a postscript "*Sur l'ozone*" that together with the Swiss chemist *Jean-Charles de Marignac* (1817–1894) they produced ozone from pure and dry oxygen, and suggested that ozone is a form of oxygen [*l'ozone ne provient que de l'oxygène*], de la Rive (1845, p. 1291). This was the first correct view on the nature of ozone. *Schönbein*, on a read to the Basel Society on July 14, 1847 (Schönbein 1849b,c), however, rejected the view of *de la Rive* and *Berzelius* that ozone is simply oxygen; he derived for the peroxides of hydrogen [in German at that time called "*Superoxide*"], manganese, etc., the formula

$$\mathrm{HO} + \mathring{\mathrm{O}}, \quad \mathrm{MnO} + \mathring{\mathrm{O}} \quad \text{etc.}$$

Furthermore, he holds chlorine, bromine, and iodine as normal peroxides with the formula

$$\mathrm{ClO} + \mathring{\mathrm{O}} \quad \text{etc,}$$

and ozone he regarded as

$$\mathrm{HO} + \mathrm{X}\mathring{\mathrm{O}}$$

Schönbein (1851a, p. 387) noted, based on his newest experiments, that he now agreed with *de la Rive* that ozone does not contain hydrogen. The US American chemist *Thomas Sterry Hunt* (1826–1892), professor at MIT, was the first who proposed speculative in 1848 the formula O_3 (cited after Rubin 2001). Frémy and Becquerel (1852) first termed ozone "*l'oxygène électrisé*" to express "*l'idee de la transformation de l'oxygène en un autre corps*" [transformation of oxygen into another body].

The French physicist *Alexandre Edmond Becquerel* (1820–1891) was the first who studied extensively the properties of ozone after *Schönbein* (Becquerel 1850); he found it to be insoluble in water, that it decomposes at 250 °C, it impacts at higher concentrations the respiratory tract, and it reacts in the presence of water with chlorine and bromine.

Moritz Baumert conducted in the Heidelberg laboratory at *Bunsen* careful experiments (Baumert 1853b) and believed that "ozone" derived by electrolysis is different from that gained by electric sparks; he proposed the formula HO_3 and writes that it is a kind of acid, and termed it "*hydrogenige Säure*" (analogous to "*salpetrige Säure*" = nitrous acid)[912] but does not form salts with bases, decomposes, and is chemically

912 And analogous to "*schweflige Säure*" (sulfurous acid]; in German (like in English), the words arose from *hydrogen* + *ige*, *salpetr* + *ige*, and *schwefel* + *ige*. Hence in English, it could name (hydr + ous)

like that of hydrogen peroxide. This was according to the historic view that oxygen was acidic, and hence ozone too. Its chemical properties he describes like hydrogen peroxide. Remarkable is his oxidation series of hydrogen (Baumert 1853b, p. 48):

HO water
HO^2 hydrogen peroxide
HO^3 ozone

Baumert's view on the compositions is curious because it was known since 1842 that water consists of two atoms of hydrogen and one atom of oxygen (H_2O), and Clausius (1858) in discussing the nature of ozone, that oxygen exists as a molecule, composed of two atoms oxygen (OO).[913] *Baumert*'s (Baumert 1853b, p. 49) following phrase reads like the presence of hydroxyl radicals (OH) in his experiments, gaining either from O_3 or H_2O_2:

> Das Ozon giebt endlich bei Gegenwart leicht oxydierbarer Stoffe seinen Sauerstoff an diese ab, während sich Wasser abscheidet [the ozone finally transfers its oxygen in the presence of easily oxidable substances to them while releasing water].

hydrous acid. In English, *hydrogenous* means "containing hydrogen" (the term *hydrogeneous acid* in unknown but hydrogeneous ion = hydrogen ion). Furthermore, in English, the term *hydrogenic* means "formed by the agency of water" (hydr + genic], see: *Merriam-Webster.com Dictionary*, sometimes wrong translated into German as "*wasserstoffhaltig*" (containing hydrogen). The term *hydrous acid* appeared in English only from the translation of the German term "*Wassersäure*" (Sertürner 1818); he defines *hydrates* as a combination of an acid with water (as an indifferent substance) and termed it *hydrous acid* in the sense of a "water containing acid". *Friedrich Wilhelm Adam Sertürner* (1783–1841) was a German apothecary and the discoverer of morphine. Rose (1831, p. 398) adopted this definition: hydrous acid as a (weak) combination of an acid with water, where water supplies the place of a base. *Hydrous* means in English "containing water"; synonyms: hydrated (thus *hydrate* as a substance).

913 Clausius (1857) in his famous paper on heat (the fundament of the first law of thermodynamics) on p. 369 notes that oxygen must be a diatomic molecule (based on the principles of volume ratios when mixing gases). *Clausius*, who believed to be the first who expressed the idea of a diatomic oxygen molecules, however, refers in a footnote (Clausius 1858, p. 645) comments by *Verdet* and *Marignac* to the French translation of his work on heat, that already *Dumas*, *Laurent* and *Gerhardt* expressed the idea that "in simple gases molecules consist from several atoms". *Clausius* further refers that *Gerhardt* in Vol. 4 of his textbook (*Traité de chimie organique*, Paris 1856) from other, pure chemical reasons, that free hydrogen as hydrogen hydride (H, H) and chlorine chloride (Cl, Cl) has to be seen, and that oxygen is composed from several (at least two) atoms (p. 612 in Traité...). *Dalton*'s axiom ("atom principle") did not allow for like atoms to combine in diatomic molecules, like oxygen atom O to oxygen molecule OO. *Berzelius* also did not favor combination of like atoms. First *Laurent* and *Gerhardt* after the death of *Berzelius* in 1842 developed and published the idea that elementary gases like nitrogen and oxygen consist of diatomic molecules (Laurent 1846). Finally at the Karlsruhe Congress of chemists in 1860 it was generally accepted that certain elements, like hydrogen, nitrogen, and oxygen, were composed of diatomic molecules and not individual atoms. *Marcel Émile Verdét* (1824–1866) was a French chemist.

Baumert's phrase, written as a (incorrect) chemical equation, HO_3 (ozone) = HO (water) + O (oxygen), however, according to today's knowledge, the first idea is that ozone (O_3) is able to transfer an oxygen atom (O) onto oxidable substance (e. g., sulfite, see Chapter 4.7.2.3), and that ozone decomposes in a multistep chain (in gas and aqueous phase)[914] to the hydroxyl radical (OH), which as oxidizer finally turns into water (H_2O). Andrew (1856, p. 2) cites "Baumert has concluded it is tetroxide of hydrogen, HO_3".

Thomas Andrews (1813–1885), a professor for chemistry at the Queen's College in Belfast, summarized the findings from the experiments of *Marignac*, *de la Rive*, *Berzelius*, *Williamson*,[915] *Fremy* and *Becquerel*, and *Baumert*, that (Andrews 1855)

- the substances comprehended under the name of ozone are not identical,
- the ozone obtained by the action of electric spark on oxygen gas is oxygen itself in an altered or allotropic state,
- the ozone obtained by the electrolytic decomposition of water is an oxide of hydrogen, having the formula HO_3.

Andrew further concluded (Andrews 1855, p. 340):

> [. . .] that ozone, from whatever source derived, is one and the same substance, and is not a compound body, but oxygen in an altered or allotropic condition.

Conducting excellent experiments, *Andrew* clearly shows that no gaseous compound having the composition HO_3 is formed during the electrolysis of water, and that ozone, from whatever source derived, is one and the same body, having identical properties and the same constitution.[916] Baumert (1856, 1857) however, in response to *Andrew*, remained on his idea that "the electrolytic ozone is a higher oxidation state of hydrogen". Lichtenstein (1860b, p. 594) used for the first the modern formula writing O_3H. Heldt (1861, pp. 16, 48, 63) regarded ozone as volatile (or *gaseous*) hydrogen peroxide, but different from the *liquid* (*Thénard*'s) hydrogen peroxide.

914 Today this compound is assumed as intermediate in aqueous-phase ozone decomposition via ozonide: $O_3 \xrightarrow{e^-} O_3^- \xrightarrow{H^+} HO_3 \rightarrow HO + O_2$.

915 The British chemist *Alexander William Williamson* (1824–1904) get similar results like *Baumert*, i. e., production of water, when ozone obtained by electrolysis was decomposed by being passed over heated copper (Williamson 1843). Berzelius (1847) discusses the formation and constitution of ozone. *Berzelius* writes in a letter to Schönbein dated March 12, 1847: "Ihre Entdeckung von Ozon ist aus diesen Gesichtpunkte eine der schönsten die je gemacht worden sind"; Kahlbaum et al. (1901, p. 46).

916 Osann (1864) also strongly argued against the existence of HO_3. It is historically interesting that he speaks on "*Ozon-Sauerstoff*" [ozone-oxygen] and "*Ozon-Wasserstoff*" [ozone-hydrogen], Osann (1860), meaning ozone and hydrogen peroxide. Nasse (1869, p. 393) obviously misinterpreted "*Osann*'s *Ozonwasserstoff*" in form of an allotropic hydrogen similar to oxygen, that does not exist *Nasse* argued. However, "*Ozonwasserstoff*" = H_2O_2 explains the observations. Of course, at that time nothing was known on OH and HO_2 as further candidates of oxygen-hydrogen compounds; however, the oxidizing and reducing capacity of H_2O_2 was recognized.

Auguste Houzeau first rejected the idea that ozone consists from hydrogen and oxygen by careful experiments: "*n'est pas un oxide d'hydrogène*" (Houzeau 1855, 1856, 1865). This was soon later supported by *Andrews*, studying the decomposition and density of ozone together with *Peter Guthrie Tait* (1831–1901), and not detecting water in (Andrews 1856, Andrews and Tait 1858, 1860), and confirmed later by Soret (1863) and von Babo (1865). The English chemist *William Odling* (1829–1921)[917] suggested in his "*Manual of Chemistry*" (Odling 1861) that the formula of ozone is O_3 based on the volumetric experiments by *Andrews* and *Peter G. Tait*. In the years 1863–1864, the Swiss chemist *Jacques-Louis Soret* (1827–1890) in Basel determined the density of ozone gas using *Graham*'s law of diffusion, and found the ratio 3 to 2 for ozone to oxygen and thus confirmed the formula O_3 (Soret 1864, 1865, 1868).

30 years later *Albert Ladenburg,*[918] who showed respect to *Soret*'s experiments but because he used a very dilute (5 %) mixture of ozone in oxygen, *Ladenburg* gets nearly pure ozone from liquidized ozonized air after evaporation of oxygen (as bluish-black opaque liquid), and estimated the density ratio ozone to oxygen to be 1.456; the theory required 1.5 but because of the difficult experiment and the resulting errors he regarded it as confirmation of the formula O_3 (Ladenburg 1898a). In a later publication (Ladenburg 1898b), he communicated that he made an error, which however does not influence the result essentially, because he assumed a priory that 253.06 parts of iodine correspond to 48 parts of ozone, i. e., making circular reasoning. Soon later two papers appeared attacking *Ladenburg*. *Wilhelm Staedel* (1857–1919), a German chemist in Tübingen and Darmstadt, first attacked *Ladenburg* with discussing the stoichiometry of the iodometric titration; he argued that the ozone molecule, expresses as $O_2 + nO$, i. e., one molecule of ozone can set free nI_2 molecules; however, *Ladenburg* set $n = 1$, therefore "*damit wird vorausgesetzt, was bewiesen werden soll*" [so it is assumed, what is to be proven] writes *Staedel* very elegant (Staedel 1898). Furthermore, *Max Gröger* (1857–1939), an Austrian chemist in Brünn (today Brno in Slovakia) and Vienna, indicated a mistake in *Ladenburg*'s experiment, noting that it is impossible to determine the molecular weight but only the difference in the density between oxygen and ozone, independent from the size of the ozone molecule (Gröger 1898), giving as an example two different stoichiometry: $O_3 + 2\,HI = O_2 + H_2O + I_2$ and $O_4 + 4\,HI = O_2 + 2\,H_2O + 2\,J_2$. A reply followed by Ladenburg (1899) rejected all arguments by *Staedel* and *Gröger*. In a subsequent communication, Ladenburg (1900) again advocated his determination, and finally, he conducted a similar experiment like *Soret* (Ladenburg 1901) to estimate the molecular weight of ozone in a simple experiment by weighting oxygen-ozone mix-

917 English chemist who contributed to the development of the periodic table.

918 *Ladenburg* was the first who get nearly pure gaseous ozone through evaporation of oxygen from liquified oxygen-ozone mixtures; he studied the properties often together with his assistant *Curt Gustav Adolf Krügel* (1876–?), who made his dissertation in Breslau 1897.

tures and estimation the volume of ozone through absorption by turpentine oil to be an average 47.78 (45.3–50.5; $n = 5$).[919]

The German physicist *Rudolf Julius Emanuel Clausius* (1822–1888), who discovered the second main of thermodynamics, explains the formation of ozone from oxygen through electric discharge by the dissociation of both atoms of the molecule [*Trennung beider Atome eines Moleküls*], and the observation that only a small amount of ozone is gained, by recombination of single atoms [*Wiederverbindung der einzelnen Atome*]; hence only a small part remains separated and forms ozone [*kleiner Theil bleibt vereinzelt und bildet Ozon*] (Clausius 1858). Curious is that *Clausius* explains the decomposition of ozone (remember he considered ozone as oxygen atom O) at higher temperatures by the "argument" that high temperature can results not only in combination between oxygen and hydrogen but also in the combination of segregated oxygen atoms (Clausius 1858, p. 648). According to the oxidation capacity of ozone, he compared it with oxygen in *status nascens*,[920] but the latter is considered additional in an "electric state". Thus, oxygen in *status nascens* might still exceed the action of ozone in some cases. Interesting that *Clausius* at the end of his paper (Clausius 1858, p. 651), referring the density estimations of ozone recently done by *Andrew* and *Tait*, notes that his hypothesis on the nature of ozone is in contradiction, and hence withdraw his "ozone hypothesis" (Clausius 1858, p. 652). *Clausius*, referring his treatise of 1858, that meanwhile new results were published by *Meissner*, *Andrew* and *Thait*, *von Babo*, and *Soret*, which confirm his view principally, summarizes it in the following two phrases (Clausius 1864a, p. 353):

> 1) Gewöhnlicher Sauerstoff besteht aus gepaarten, activer Sauerstoff aus ungepaarten Atomen [Ordinary oxygen consist from paired, active oxygen from unpaired atoms].
> 2) Die beiden Atome, welche ein Molecül gewöhnlichen Sauerstoffs bilden, befinden sich in entgegengesetzten electrischen Zuständen [The two atoms, from which one moelcule of ordinary oxygen consist, are in opposed electric states].

Another idea on the constitution of ozone *Carl Weltzien* (1813–1870)[921] published (Weltzien 1860a); he correctly considered ozone to be composed of several atoms of oxygen, but he regarded oxygen to be O and hence ozone to be OO as "*verdoppelter Sauerstoff*" [doubled oxygen] (Weltzien 1860a, p. 128). Weltzien (1860a, p. 125) regarded HO_2 as hydrogen peroxide (like *Baumert*), but HO_3 as "*hydrogenige Säure*" [hydrous acid] and not ozone, and *Thénard*'s hydrogen peroxide as H_2O_4 (an adduct of two HO_2, written as H, H (O_4).

---

919 See also footnote 925 on p. 468 concern thermal decomposition of ozone.

920 Weltzien (1860a, p. 124) comments that *Schönbein*, but also many other chemists, talk on *status nascens*. However, this is grammatically wrong because it means "arising state" [*entstehender Zustand*], but one would talk on the "state of formation" [*Zustand des Entstehens*], thus *status nascendi* would be more correct.

921 Since 1843 professor for chemistry at polytechnic in Karlsruhe.

Lambert Heinrich Clemens Karl Freiherr von Babo (1818–1890), who studied medicine in Heidelberg and Munich, but begun a carrier in chemistry at *Liebig* in Gießen and since 1844 at the University in Freiburg (Breisgau), constructed an apparatus to produce ozone (Babo 1861) and conducted experiments to prove that ozone does not contain hydrogen (the formulas HO, HO^2 and HO^3 were discussed before) and that oxygen can be converted completely into ozone (Babo 1863). *Babo* also stated that the oxidizing properties of ozone are only observed in the presence of not bonded water (Babo 1865).

With *Soret*'s and *Ladenburg*'s estimates of the ratio of the density between ozone and oxygen, it was quantitatively established that ozone is an allotropic form of oxygen: OOO or O_3 (oxygen trioxide). Traube (1893, p. 1481) proposed a triangle structure of the ozone molecule ∇ with a double bond between two oxygen atoms –O=O–. *Aleksander Wolkowicz* (1867–?)[922] considered the structure of ozone to be O=O=O and in analogy to SO_2 and SeO_2 as OO_2 as anhydride of the unknown acid H_2OO_3, indicated that the sulfuric acid H_2SO_3 is also not known molecular gaseous, and named it "*oxygenige Säure*" [oxygenous acid] further arguing that corresponding salts are known as tetroxides like $(KO)_2O{=}O$ (K_2O_4)[923] (Wolkowicz 1894); see p. 637 on ozone acid. *Gilbert Newton Lewis* (1875–1946) and *Linus Pauling* (1901–1994) assumed a linear structure O=O=O like that of SO_2 (Lewis 1923), based on calculated formation enthalpy, and stated that a triangle structure is not possible (Pauling 1932). An isosceles triangle with an angle ahead to be 19.5° proposed Hettner et al. (1935). *Michael James Stuart Dewar* (1918–1997) proposed a π-complex structure for ozone which requires the molecule to be an acute-angled triangle in shape; the resonance behavior does not show unpaired electrons and no characteristics of a radical (Dewar 1948). The bonding in ozone is discussed until present (Laing 1995), however, the physical and chemical properties are well-known today for an understanding of atmospheric chemistry.

4.4.2.2 Properties

Soret also conducted several experiments on the consumption of ozone in different chemical reactions. It seems that *Clausius* was annoying on his ozone hypothesis and "happy" that *Soret* in the most loyal way recognized [*dabei in loyalster Weise*

922 He was a chemist of Polish origin in the laboratory of the Swiss institute for testing of building materials [*Eidgenössische Anstalt zur Prüfung von Baumaterialien*] in Zürich.

923 Potassium tetroxide is not mentioned in modern textbooks of inorganic chemistry. E. W. Neumann (*J. Chem. Phys.* **2** (1934) 341) proposes the formula KO_2 containing the ion O_2^-, whereas W. Klemm and Sodomann (*Z. anorg. allg. Chem.* **225** (1935) 273) proposed K_2O_2 with the ion O_4^{2-}. W. Kassatochkin and W. Katow (*J. Chem. Phys.* **4** (1936) 458) confirmed from X-ray investigations *Neumann*'s view as KO_2. Today KO_2 is named potassium superoxide that is derived from the hydroperoxyl radical HO_2, whereas potassium peroxide K_2O_2 with the ion $^-O{=}O^-$ is derived from hydrogen peroxide.

anerkannt] that through this new view his hypothesis is not rejected but only slightly modified [*nicht umgestoßen, sondern nur leicht modificiert werde*];[924] Clausius (1869, p. 104).

Berthelot (1876) was the first who studied the thermal formation of ozone and who measured calorimetric the formation enthalpy of ozone very close to today's value (34.4 kcal per 24 g of ozone, still assuming that 1 equivalent of oxygen corresponds to 8 g). The thermal instability of ozone was one of the first properties observed by *Schönbein* and many other researchers. Berthelot (1878a) first studied the decomposition at 12 °C in a glass vessel and found that the concentration was reduced to 55 % of its original value after five days. The Dutch chemist *Eduard Mulder* (1832–1924; son of *Gerardus Johannes Mulder*) in Utrecht studied the rate of the thermal decomposition depending on the concentration of ozone, on pressure, and temperature, based on the pressure change during the decomposition (Mulder 1885). Remsen (1883) suggested that ozone decomposes in one molecule and one atom of oxygen. Thus, it was shown the inverse reaction of ozone formation: $O_3 = O_2 + O$ and $2\,O_3 = 3\,O_2$.[925]

The British chemists *David Leonard Chapman* (1869–1958), who was a pioneer in studying the rate of explosion in gases, studied the decomposition of ozone and the effect of various gases, founding that nitrogen dioxide and chlorine accelerated the rate of decomposition (Chapman and Jones 1910).

Paul Hautefeuille (1836–1902)[926] and *Joseph Chappuis* (1854–1934)[927] investigated the conditions of ozone formation during silent discharges, the properties and liquification of ozone (Hautefeuille and Chappuis 1884). *Ernst Herrmann Riesenfeld* (1877–1957)[928] describes pure ozone (the boiling point at −112.3 °C) as a nearly

924 This more than kindly statement by *Soret* can only be explained by the large reputation of *Clausius*; from today's perspective it must be said that *Clausius* was wrong, and his distinction between ozone, seen as freed oxygen atom from oxygen molecule, and oxygen as atom in *status nascens*, gained from electrolysis of water, did not sound convincing at that time. With the knowledge of *Soret*'s density estimate of ozone *Clausius* should have withdrawn also his publication.

925 The thermal decomposition of ozone at ambient temperatures (half-life amounts some days) in atmospheric air would result in a steady state like the photolysis due to the inverse reaction $O + O_2 = O_3$. However, at solid surfaces, ozone adsorbs and decompose; the adsorbed oxygen atom O(ads) can react with the surface material. This was already observed by *Schönbein* and others that rubber tubes used in experiments were soon destroyed. On water surfaces, many reactions occur which decompose ozone.

926 A French mineralogist and chemist in Paris. From 1870 to 1885, he served as co-director of the chemical laboratory at the *École Normale Supérieure* and in 1885 he was appointed professor of mineralogy at the Faculty of Sciences in Paris. He is well-known for mineralogical syntheses.

927 *Louis Philibert Claude James Chappuis* was a French chemist and physicist, who first worked as teacher and who became in 1881 professor for physics at *École Centrale des Arts et Manufactures*. He earned his doctorates in 1882 on "*Étude spectroscopique sur l'ozone*" (Paris).

928 A German chemist, one of the pioneers in electrochemistry, who studied extensively the properties of ozone; moved in 1933 to Sweden.

black-blue-colored liquid, which solidified at −249.6 °C to black crystalline matter (Riesenfeld 1927).

Basically, the ozone chemistry is well described phenomenologically.[929] Many more chemists addressed ozone around the middle of the nineteenth century (it was *en vogue*) without contributing significant news. Here I like to refer the well-known *Gottfried Wilhelm Osann* (1797–1866), a German chemist in Dorpat, Erlangen, Jena and professor for chemistry and physics in Würzburg, who published 29 articles on this subject between 1848 and 1863. *Osann*, who advocated *Schönbein*'s view but doubt soon that ozone represents an unknown oxidation state of hydrogen (Osann 1851a,b). Osann (1848) realized that the gas gained from the slow combustion of phosphorus[930] is identical to that from water electrolysis,[931] i. e., active oxygen (ozone), that he termed "*Ozon-Sauerstoff*" [ozone-oxygen], explaining (Osann 1851b, p. 537):

> Ich glaube, daß diese Bezeichnung besser ist, als Ozon, weil hiermit sogleich gesagt ist, daß es sich um einen Körper handelt, welcher in seinen Eigenschaften denen des Sauerstoffs zunächst zu stellen ist [I believe that this meaning is better, than ozone, because it is instantly expressed that it is a matter of a body which is close to oxygen in his properties].

Thomas Shapter (1809–1902), an English doctor of medicine, "through the intervention of Drs. Moffat and Drew"[932] made in summer 1854 observations with *Schönbein*'s

---

929 Peroxide of manganese was at that time the name for manganese dioxide (MnO_2). The well-known property of MnO_2 and CuO (and other oxides) to decompose O_3 goes via a catalytic surface mechanism, which were understood 150 years later. Li et al. (1998) suggest the peroxide intermediate formation (S – surface of the catalyst): $S + O_3 = S{-}O + O_2$, $S{-}O + O_3 = S{-}O{-}O + O_2$, and $S{-}O{-}O = S + O_2$ (slow). There is no direct reaction between O_3 and H_2O_2 in the gas phase. Indirectly, O_3 reacts with OH (which can be produced from slow H_2O_2 photolysis, but rather from O_3 photolysis itself) in air. In aqueous phase, H_2O_2 and O_3 turn into OH radicals by separate reactions, thus itself converting into H_2O and O_3, respectively. Finally, the oxidizing capacity of O_3 (with the exception of a direct reaction with olefins) results from the OH radical, gained in subsequent reactions.

930 The slow combustion of elemental phosphorus at air ("glow" reaction), finally gaining P_4O_{10} (P_2O_5) proceeds via $P_4 + O_2 \rightarrow P_4O + O$, $P_4 + O + M \rightarrow P_4O + M$, $P_4O_n + O_2 \rightarrow P_4O_{n+1} + O$, and $O + O_2 \rightarrow O_3$ (Dainton and Kimberley 1950).

931 Ozone gas evolves by a six-electron reaction at a voltage higher than 1.511 V, accompanied by oxygen evolution according to $3\,H_2O \rightarrow O_3 + 6\,H^+ + 6\,e^-$ ($E° = 1.511$ V); see Christensen et al. (2013) for details. The model of elementary steps was provided by Da Silva et al. (2003) and includes the rate determining step $OH^- \rightarrow (OH)_{ads} + e^-$. The hydroxyl radical further anodic oxidizes $(OH)_{ads} \rightarrow (O)_{ads} + H^+ + e^-$. From the oxygen atom dioxygen evolves and it combines then with monooxygen to ozone. The OH radical can dimerizes to H_2O_2. Hence, according to the experimental conditions, hydrogen peroxide remains dissolved in solution whereas oxygen and ozone can escape gaseous. However, H_2O_2 can also be oxidized at the anode to $HO_2 + H^+ + e^-$ and the peroxo radical further into dioxygen, thus gaining hydrogen peroxide only as intermediate.

932 *John Drew* (1809–1857), British astronomer and meteorologist in Southampton, visited *Schönbein* in Basel, who was interested that after first observations in Germany also in England ozone observations were conducted, and who was also familiar with *Faraday* (*Schönbein* and *Faraday* corresponded paper (protected from direct sunlight and rain) and stated, for the first time using the term *oxidizer* (Shapter 1854), that ozone is

> [...] a most ready and powerful oxydiser, and in a great number of cases acts like Thénard's peroxide of hydrogen, or chlorine, or bromine.

From alternating coloring and decoloring periods of the test-paper *Shapter* concluded "that there is developed in the atmosphere, at certain periods, a deoxidizing property, as well as, at other periods, an oxidizing property, writing

> There is, therefore ample ground for concluding that, if this double property of the atmosphere do not belong to that allotropic form of oxygen to which the term ozone was applied, there must exist either some other form of this same agent, or some other and distinct agent to which the phenomena of deoxidizing belong.

Cornelius Fox used the term "air-purifiers" which included ozone, peroxide of hydrogen, and nitrous acid (Fox 1873, pp. 299 and 300, 1876, p. 272), and "purifying agents" Fox 1873, p. 275). He writes (Fox 1873, p. 99):

> It is nature's great deodorizing and purifying principle that oxidizes the emanations from decomposing animal and vegetable substances with which the air is constantly being contaminated.

German chemist *Carl Oswald Viktor Engler* (1842–1925)[933] in Halle, coined the term "air cleanser" for oxidative substances like ozone and hydrogen peroxide – hundred years before Crutzen (1986) coined the phrase "detergent of the atmosphere" to describe this important cleansing role of OH (O_3 and H_2O_2 are the OH precursors); Engler (1879, p. 65) writes in his book "*Historisch-kritische Studien über das Ozon*" [historic-critical studies on ozone]:

> [...] dass in unserer Atmosphäre und auf der Oberfläche der Erde eine grosse Zahl von Bedingungen für die Bildung derjenigen Atmosphärilien – Ozon, Wasserstoffsuperoxyd, salpetrige Säure und Salpetersäure – gegeben ist, die wir unter der Bezeichnung "Luftreiniger" zusammenfassen [... that in our atmosphere and on the earth's surface a lot of conditions exist for the formation of such substances – ozone, hydrogen peroxide, nitrous and nitric acid – which we summarize under the term "air cleanser"].

Peter (1849), in his "remarks on ozone" refers:

---

about the nature of ozone), wrote on July 23, 1853 in a letter the British Medical Association "to undertake daily observations for a year, forwarding the results to me every three month, for transmitting to Bâle" (Anonymous 1853).

933 A German chemist, studied in Freiburg (promotion at *Carl Weltzien* in Freiburg 1862) and private lecturer, professor for chemistry in Halle 1872–1876, and since 1876 in Karlsruhe; "father" of petrochemistry (after 1885).

> A few years since a mysterious notice, by Leuch, appeared in the *Journal fur prak. Chem.*, vol. cxxv, p. 191, of a new bleaching agent, which was described as at ones quick and efficient, as well as very economical. It is believed that the agent proposed was ozoned air.

It is reported (Anonymous 1845), quoting "*Die Allgemeine Polytechnische Zeitung von Leuchs*" that[934]

> [. . .] Bleichen durch blosse atmosphärische Luft bewirkt, die in einen andern (elektrischen?) Zustand versetzt ist. [. . . bleaching is achieved by mere atmospheric air which is transferred in another (electric?) state].

Another very short piece of information, also likely by *Leuchs* (Anonymous 1835b) entitled "upon bleaching in the sun", reports that decoloring is due to the "decomposing impact of the light", whereas the intensity of solar radiation and heat accelerate the bleaching. Furthermore, the grass on which the items lie seems under the influence of light oxygen gas to release.[935]

4.4.2.3 Sources

The well-known Scottish organic chemist *Alexander Crum Brown* (1838–1921), who never studied atmospheric ozone to my knowledge, wrote a short paper (Brown 1869), summarizing the knowledge in an excellent scientific style:

> Andrew and Tait established that ozone is merely a form of oxygen, and that oxygen can be transformed into ozone, and ozone into oxygen, without the production of any other substance. Soret's experiments seem further to prove that ozone is denser than oxygen, in the proportion of 3 to 2. Common oxygen can be transformed into ozone by means of the silent electric discharge, oxygen is ozonized in the course of many slow processes of oxidation, and particularly by the slow oxidation of certain volatile oils. These processes undoubtedly occur to a great extent in nature.

However, the formation of ozone while slow oxidation of oils like turpentine was already doubted by the English chemist *Charles Thomas Kingzett* (1852–1935) who writes

934 It is *Johann Carl Leuchs* (1797–1877) who was a technical writer in Nuremberg, and editor of *Allgemeine Polytechnische Zeitung und Handlungs-Zeitung* (1845) No. 18, p. 77 (text is identical with that in *J. pr. Chem.*). Interestingly, that nothing is told on the technical measure, but that (translated from German) "information on the kind of bleaching takes place if 100 enterprisers are found who each pay 100 Gulden Augsb. Cour or 68 Preuss. Thaler [. . .] Registrations are to be made to C. Leuchs & Comp. Nürnberg"].

935 Because of the historic remarkable notice, it is here given in its original text in German: Beim Bleichen an der Sonne ist es die zersetzende Einwirkung des Lichtes, welches die Entfärbung bewirkt. Mit der Stärke des Sonnenscheines und der Wärme wird das Bleichen daher beschleunigt. Auch scheint der Rasen auf den man die Zeuge legt, beizutragen, weil die Pflanzen, so lange sie dem Licht ausgesetzt sind, Sauerstoffgas entbinden. Man könnte das Bleichen beschleunigen, wenn man durch weiß angestrichene, oder noch besser mit Spiegel belegte Wände an der der Sonne entgegengesetzten Seite, Licht und Wärme verstärkte.

(Kingzett 1874) "It is generally stated in books that the oxidation of oil of turpentine is attended with the production of ozone". He further writes (Kingzett 1874):

> [...] that the oxidising properties of such oil must be regarded as direct evidence that the oxygen is united with a hydrocarbon, forming a true definite chemical compound subject to the ordinary laws. (p. 512)
> When oil of turpentine is exposed to air or oxygen in the presence of moisture, it oxidises, producing an agent which resembles ozone and peroxide of hydrogen, [...] (p. 520)

Kingzett (1880, p. 793), referring his own former investigations, concluded:

> [...] have shown that ozone is not produced in the operation, but that peroxide of hydrogen results as a secondary product of the action of water upon a peroxidised derivative of turpentine.

Otto Wolffenstein found experimentally that the oil of turpentine absorbs oxygen and does not contain ozone (Wolffenstein 1870).

What is follows is a brief history of the formation of "oxidizing principles" while combustion processes. *Th. V. Saussure* was the first (Saussure 1802) who notified (while eudiometric experiments) that during combustion of hydrogen in atmospheric air nitric acid (and ammonia) is formed. Schönbein (1845a) showed that during the combustion of hydrocarbons an "*oxydierende Materie*" [oxidizing matter] is gained, but he was unable to decide whether it is nitric acid or another substance. Schönbein (1861c, p. 221) finally could show that the slow combustion of phosphorus in atmospheric moist air is a complex process, not only gaining ozone but additional forming no less than six different compounds (in parenthesis his old formulas): phosphorous acid (PO_3), phosphoric acid (PO_5), nitrous acid (NO_3), nitric acid (NO_5), ammonia (NH_3) and hydrogen peroxide (HO_2). *Rudolf Böttger* reported on a meeting hold in Speyer in September 1861 to German naturalists and physicians on combustion of hydrogen in presence of air; the water thus generated was neutral in its reaction but reacts with potassium iodide under precipitation of iodine (i. e., is oxidizing), but also decolors a solution of potassium permanganate, i. e., shows reducing properties (Böttger 1862). *Böttger* did not believe that it is hydrogen peroxide, but that ammonium nitrite is obtained. *Schönbein* who was present at the lecture proposed the formation of hydrogen peroxide, and because the oxidizing principle remained after boiling of the water, also traces of ammonium nitrite gained (today's reader should know that in the air at that time always much higher concentrations of ammonia were present then nowadays, also in the laboratory). Meissner (1863a, p. 283), however, believed that H_2O_2 is produced in these experiments. *Salomon Pincus* (1819–1890), a district physician in East Prussia, was not only interested in rainwater chemistry and agriculture (p. 184) but also in air chemistry; he found in experiments made in 1867[936] that a

936 The first report was published in 1867 by *Pincus*: *Agriculturchemische und chemische Untersuchungen und Versuche, ausgeführt bei der landwirthschaftlichen chemisch-physikalischen Ver-*

flame of purified hydrogen and oxygen gave a condensate with a very strong smell of ozone (Pincus 1871). *Oscar Loew*[937] also has definitively shown the formation of ozone by rapid combustion; while blowing a strong current of air through a tube into the flame of a *Bunsen* burner; in the air collected he identified ozone by its smell and by the common tests (Loew 1870b). *Loew* further wrote the equation 3([OO]) = 2([OO]O) and noted that

> [. . .] in every combustion even the most rapid and energetics, an intermediate decomposition of the molecule of common oxygen must take place if the single atoms will enter into combination with the elements, and that ozone or antozone would be detected in a flame if the high temperature would not destroy it again as quickly as it is formed.

This was also found by *Carl Than* (1834–1908)[938] who notes that ozone is only formed in fast combustion flames at the moment when air is in contact with the outer edge of the flame because in the inner hot zone it is immediately decomposed (Than 1870). *Heinrich Struve* finally burnt hydrogen into a long-extended funnel and detected ozone, and in the condensed water hydrogen peroxide and ammonium nitrite (Struve 1871, p, 296). *Struve* further notices that due to the current decomposition of ozone and hydrogen peroxide, these compounds exist often for only short time and only ammonium nitrite as salt remains. Furthermore, Böttger (1873a) found that in the oxyhydrogen reaction ozone is gained. Böttger (1873b) found in experiments with electric discharges in a glass ball filled with absolute dry atmospheric air after a few minutes a yellow vapor, smelling like nitrous acid, and when the atmosphere is wet due to wetted inner walls and some distilled water at the bottom, no vapors but the odor of ozone and in the liquid water hyponitrous acid.[939] Nagy-Ilosva (1889), in his paper first presenting the results of previous investigations, discusses the different methods for detection of ozone and hydrogen peroxide, and then conducted combustion experiments with different flames, denied its formation and found only higher oxides of nitrogen. However, he notified that ozone is completely destroyed at 240 °C and hydrogen peroxide at the higher temperature. Riesenfeld (1927) detected ozone in flame

---

suchsstation zu Insterburg. Bericht V, Gumbinen (1867). Gumbinen (also written Gumbinnen) was a town and Governmental District in East Prussia; today Гусев [Gussew] in Oblast Kaliningrad (Russia).

937 *Oskar* (also *Oscar*) *Loew* (1844–1941) was a German agricultural chemist, the last student of *Liebig*. He studied *at Carl Wilhelm von Nägeli* in Munich and *Hermann Kolbe* in Leipzig. 1867–1876 he worked in New York and Washington, returned 1877 to Leipzig and later Munich where he became professor in 1886; 1893–1907 in Tokyo succeeding *Oskar Keller*, and back in Munich 1910.

938 *Károly Antal Than de Apát*, also termed as *Carl von Than*, was a Hungarian chemist who discovered carbonyl sulfide (COS), see p. 709.

939 Note that in an atmosphere containing no water, no nitrous acid can be gained, only NO and NO_2 (and NO_3 + N_2O_5) and O_3; the latter is to a large extent consumed in oxidation of primarily produced NO. In presence of water, HNO_2 and HNO_3 (not $H_2N_2O_2$ – hyponitrous acid) is gained which dissolve in the liquid water. The odor of ozone could not be recognized due to the dominant odor of nitrous gases.

gases, namely oxygen-hydrogen combustion, but in the case of great quenching, the formation of hydrogen peroxide was dominant; he concluded that the production of ozone is not thermal but chemical in subsequent reaction (today we know that the starting phase consists in the thermal O_2 dissociation).

The source of energy to produce ozone was either electric discharge in the laboratory or in the atmosphere during a thunderstorm; the energy for ozone decomposition was regarded generally as heat (radiation energy to photolyze the molecules were regarded only in the late nineteenth century, see p. 478). Brown (1869), however, was wrong with the statement that ozone is gained from oxygen in slow oxidation processes, later termed *autoxidation*, discovered by Traube (1882) as a source of hydrogen peroxide (see next Chapter). This can be attributed to the close affinity between O_3 and H_2O_2, confusing by the same detection in air with the *Schönbein* paper and similar oxidative behavior, not knowing at that time that both compounds produce OH radicals, the "true" air cleanser. Brown (1869, p. 254) writes (italics by *Brown*):

> (1) Ozone is converted into common oxygen by heat. (2) Ozone is converted into common oxygen by contact with various substances which do not themselves undergo change in the process; such substances are peroxide of manganese [. . .], etc. (3) Ozone acts upon a small class of substances as a *reducing* or *deoxidizing* agent; thus it reduces peroxide of hydrogen to water, being itself, at the same time, converted into common oxygen. (4) Ozone acts on most substances as an *oxidizing* agent. Thus it corrodes organic matter, bleaches indigo, converts sulphuret of lead into sulphate of lead, etc.

Anton Schrötter (1802–1875), an Austrian chemist and mineralogist, describes a fluorspar from Wölsendorf (Bavaria), having a peculiar odor, what has been already described years before by others as "hypochloric acid", and now "identified" as ozone because of precipitating iodine from potassium iodide (Schrötter 1860). However, this "*Stinkspat*" (decaying fluorspar) liberates free fluorine after hit with a hammer. Schönbein (1858b) first studied this mineral without referring the odor similar to ozone. Schönbein (1861b, p. 97), stimulated by *Schrötter*'s publication, writes that "*in diesem Mineral freies Antozon oder positiv activer Sauerstoff enthalten sei*" [this mineral contains free antozone or positive active oxygen]; when rubbing it with water, he detected that hydrogen peroxide gains[940] and presents the equation $\oplus + HO = HO_2$. *Osann*, on a read to the "*Physikalisch-Medicinische Gesellschaft*" [Societas Physico-Medica] in Würzburg, communicates that several observers during the lightning on 6 January 1865, into the tower of the church "*Neubaukirche*" recognized a smell which was explained by ozone, whereas the spreading fog was due to antozone (Osann 1866).

Another curious finding, based on ozonometry, was, that plants exhale ozone, together with oxygen which they give out during the daytime (Scoutetten 1856,

940 Today it is known that free fluorine combines with water in different way, like $F_2 + H_2O = 2\,HF + O$ (with subsequent O_3 formation) and $F_2 + 2\,H_2O = 2\,HF + H_2O_2$); Cady (1935).

Kosmann 1862, Daubeny 1867).[941] However, *François Stanislas Cloez* (1817–1883)[942] found that *Schönbein*'s ozonometer is blued under several conditions, namely under moist air and the influence of volatile plant exhalations; he speaks on water decomposition and the formation of "*l'oxygène naissant*" [nascent oxygen][943] under the influence of light Cloez (1856). He concluded from his observations, that the *Schönbein* paper "for detecting the presence and estimating the amount of ozone in the air is entirely without value".

Lawes et al. (1863), investigating the sources of nitrogen of vegetation, noticed that "in gas exhausted from plants, no ozone was detected (Lawes et al. 1863, p. 144), and "ozone was observed most readily in the vicinity of such plants as emit freely essential oils" (Lawes et al. 1863, p. 152). *Andrés Poey*, making experiments in Havana, also denied the formation of ozone by plants (Poey 1863b).

Most clearly the Italian neurologist, physiologist, and anthropologist *Paolo Mantegazza* (1831–1910) noticed (Mantegazza 1870) that many essential oils when in contact with the oxygen of the atmosphere in the presence of sunlight, develop very large quantities of ozone (cited after Anonymous 1872). *Mantegazza* conducted the experiments in closed vessels and observed that "the action is strongest in direct sunlight, far less so in suffused daylight, and very weak or at an end in the dark". Today we know that this observation was not ozone but hydrogen peroxide due to photocatalysis.

Whereas atmospheric electricity after *Schönbein*'s first suggestion was generally accepted (*Houzeau*, *Roscoe*, and others), the other proposed sources had protagonists and opponents contemporary. However, because of seldom thunderstorms but the permanent presence of ozone in the air, an explanation was found best expressed by Dancer (1876, p. 126):

---

941 *Constant Philippe Kosmann* (unknown living data, French chemist in Strasburg) conducted 253 experiments in the Botanical Garden at Strasburg with the *Schönbein* paper placed on growing plants which are more deeply affected in a given time, then others left in contact with the open air at a distance from all vegetation. *Schönbein* in a letter to *Ebermayer* on June 7, 1867, expressed his believing that conifers because of its content of oil of turpentine should be more important as source of ozone than deciduous trees (Ebermayer 1873b, p. 241). Bellucci (1874) noticed that green plants do not exhale ozone.

942 A French chemist, known for his pioneering role in analytical chemistry. In 1864, *Cloez* was the first scientist to examine a carbonaceous chondrite, the Orgueil meteorite, after it had fallen in France. *Cloez* said that its content "would seem to in indicate the existence of organized substances in celestial bodies".

943 In that time, it was surely unclear that nascent oxygen is atomic oxygen (O). However, in an oxygen containing medium (water and air) O combines immediately with O_2 to O_3. Hubbs (1930) investigated the toxic property of nascent oxygen, gained from ozone (O_3), sodium hypochlorite (NaOCl) and hydrogen dioxide (H_2O_2) and found no difference: the toxic oxygen species (ROS) are the radicals OH, HO_2 and O_2^-, only became known later. The believing of *Cloez* "that emanations escaping from plants impart to the air the power of affecting the test-paper in the same manner as ozone, without, first ozonizing the atmospheric oxygen" (citation by Rogers 1858, p. 36) could be interpreted as the reaction between terpenes with ozone producing radicals like OH and HO_2.

> Probably the generation of ozone is constantly taking place in the upper regions of the atmosphere; the formation of clouds, rain, hail, &s., developing free electricity under conditions highly favorable for the conversion of oxygen into ozone.

Houzeau wrote [correctly] in 1872 that lightning in atmospheric air and electric sparks in the air of the laboratory produce nitrous acid [note that this term denoted nitrous gases NO_x] and not ozone; ozone is gained only in chemically pure oxygen by discharges. Thus, only the slow compensation of different electricity, constantly occurring between clouds and the earth's surface, is the source of ozone.

John Benjamin Dancer further argued that conversion of water into aqueous vapor from the water surface (the sea, lakes, reservoirs, etc.) always develop "free electricity and manifestation of ozone" (Dancer 1876, p. 127); he also believed "that the oxygen which is evolved by plants is in an ozonized condition, we should have an additional source for the production of this purifying agent". These conclusions were based on the experiments of his friend *Joseph Baxendell* (1815–1887), an English meteorologist and astronomer (Baxendell 1869, 1876). Protagonists of the "water evaporation" theory were *Lender, Gorup-Besanez, Baxendell*, and others).

William Stanley Jevons (1835–1882), an important English economist and philosopher, who first studied chemistry, spend some years in Australia, and in 1866 he became a professor in Manchester, wrote in a response to *Baxendell*'s paper (Baxendell 1869) to my knowledge for the first time on the role of vertical transport processes for air pollutions studies (Jevons 1869, p. 33):

> The quantity of ozone which reaches the surface will depend on three circumstances:
> 1. The thickness of the current of air touching the surface.
> 2. The proportion of ozone existing therein.
> 3. The degree in which the current is rendered uniform by constant mixture.

He further noticed from *Glaisher*'s balloon observations that the atmosphere usually consists of several strata of air that are separated by distinct boundaries. From own observations both in England and Australia he noticed "that smoke from a great town or from extensive bush fires rises only to a definite height". Then he explained the causes of diurnal variation (Jevons 1869, p. 34):

> At night the air tends to lie in a stagnant layer which can often be detected by the mist or smoke which it contains. This layer will be rapidly exhausted of ozone, and will be filled with organic exhalations from the earth [...] In the day time, on the contrary, the circulation or convection [...] the air is thus successively brought to the surface and organic substances are carried off and oxidated.

Thomas Mackereth (1828–1896), Reverend and schoolmaster in Salford, interested in astronomy and observer of the Royal Meteorologic Society in Eccles, a close friend of *Baxendell* and *Dancer*, observed for many years' atmospheric ozone (Mackereth 1867, 1868), but his pioneering view, based on own observations and correlations is the connection of atmospheric ozone with solar radiation and the number of groups of solar spots observed in each year by *Schwabe*[944] (Mackereth 1868):[945]

> Here it is obvious that the amounts of ozone have decreased each year with the amount of solar radiation, and pretty nearly too as the group of solar spots have decreased (p. 218).
> [...] that ozone is the result of an activity produced in oxygen by the direct action of solar radiation (p. 219).

Despite *Clausius* already in 1858 published the idea that for the formation of ozone the dissociation of the oxygen molecule O_2 must be seen as the first step, and *Soret* in 1864 the formula O_3 definitively confirmed, the largely unknown Austrian medical doctor *Johann Hammerschmied* (unknown living data),[946] on a lecture in April 1873 proposed his "molecular" theory on ozone formation, arguing that it is supported by the experimental finding from observations that condensation of water vapor is the source of ozone. It is worth citing the phrase to see that curious imaginations on chemistry still existed apart from already published scientifically sound ideas (Hammerschmied 1873, pp. 453):

> Indem nun die Wassermoleküle und kleinsten Wassertröpfchen auf einander oder gegen einander stürzen, auf ein Attraktionscentrum (Molekülgruppe) hinfallen, (Tyndall) drängen sie, reissen sie die zwischen ihnen liegenden Stickstoff- und Sauerstoffatome von oder auseinander, befreien sie dieselben aus den Fesseln ihrer gegenseitigen Anziehung, mit einem Worte: machen sie frei [...] sie wurden aus dem passiven Zustande in den aktiven überführt. Diesen aktiv gewordenen Stickstoff- und Sauerstoffatomen der Luft stehen die Wassermoleküle (HO_2) gegenüber [...] [While now the water molecules and water droplets on each other or against each other tumble, fall on a center of attraction (molecular center), they split up the nitrogen and oxygen atoms, lying between them, liberate them out of the shackles of her mutual attraction, with this word: make them free [...] (Tyndall) they were I from a passive into an active state. These activated nitrogen and oxygen atoms of the air are confronted with water molecules (HO_2) [...]

In short words, *Hammerschmied* also assumed the dissociation of oxygen (and nitrogen), but what now follows in his text, is bizarrely chemistry. His verbal description is

---

944 *Samuel Heinrich Schwabe* (1789–1875) was a German astronomer remembered for his work on sunspots.

945 This does not support a photochemical ozone formation near the earth's surface; stratospheric ozone transport down to the surface could be correlated with the maximum intensity of the sun's light in April, and, on the other hand, less radiation in Manchester is due to air pollution (resulting in ozone removal) and in valleys due to more ozone dry deposition comparing to hill sites.

946 *Hammerschmied* wrote "*System der Medicin. In drei Bänden*" (Wien 1863). From his experiments with heat and absorption, Tyndall (1863, pp. 84–86) writes "I hold that ozone is produced by the packing of the atoms of elementary oxygen into oscillating groups" hence, like compounds, encountering resistance while moving through the aether. He rejected the proposal that it was a hydrogen compound, because no water vapour was detected when ozone was decomposed while heating. *Tyndall* remained aloof from the disputed areas, writes McCabe (2012); it seems that *Hammerschmied* was the only person who cited Tyndall (even Rubin (2001)) did not cite him).

here written in equations, using his "formulas" (he argued that only activated nitrogen can react with water and not activated oxygen because the water already contains oxygen):

$$N + 3\,HO_2\ (\text{water}) = NH_3 + 6\,O$$
$$N + 4\,O = NO_4\ (\text{nitrous acid})$$
$$N + 5\,O = NO_5\ (\text{nitric acid})$$
$$O + \text{oxygen} = \text{ozone}$$

Finally, he concluded (Hammerschmied 1873, p. 455) that the condensation of water vapor results in the formation of ammonia, nitric acid, and ozone, explaining the permanent presence of these compounds in the air (namely ammonium and nitrate in rainwater). The generally observed spring maximum in the amount of atmospheric ozone *Hammerschmied* (Hammerschmied 1873, pp. 448–449) ascribes to another source than electricity, "*chemischen Prozesses der erwachenden Natur*" [chemical processes of the awakening nature], however, he notes that growing plants provide no ozone. The view, however, without any understanding of the mechanisms of chemical reactions, that ozone is ubiquitarian, led already Lichtenstein (1862, p. 16) to the statement that "[...] *jeder Chemismus ist mit Ozonentwicklung begleitet*" [... each chemistry is associated with the formation of ozone]; see also pp. 525–526.

The first idea that not electricity but solar radiation is the sole cause of "activation of oxygen" comes from the German chemist *Casimir Wurster* (1854–1913), known for *Wurster*'s reagent (tetramethylparaphenylene diamine) that blues in the presence of oxidizing substances like ozone and hydrogen peroxide. He published an article on his "ozone" observations based on his nose to smell the odor and then using a very sensitive test-paper impregnated with his reagent. He found (Wurster 1886b) that the "formation of ozone" is highest "*durch Sonnenstrahlen immer in den obersten Nebelschichten der Wolke*" [through solar radiation always in the upper fog layers of a cloud] (Wurster 1886b, p. 3213). *Wurster* referred *Schöne* who concluded that the formation of hydrogen peroxide in air is in relation with the sun (see p. 641). *Wurster*'s test-paper could not distinguish between O_3 and H_2O_2, hence he wrote "activated oxygen". He could not know that activated oxygen also includes other reactive oxygen species (ROS) like the superoxide anion, hydroxyl radical, and hydroperoxyl radical, which would all show a "*Wurster*'s reaction". Many observations of *Wurster* we can interpret on the basis of our today's knowledge as the photochemical formation of hydrogen peroxide and not ozone. He writes (Wurster 1886b):

> Die Activierung des Sauerstoffs und die Bildung seiner Dauerform Ozon oder Wasserstoffsuperoxyd erfolgt nur dann in reichlichem Masse, wenn Sauerstoff, Wasser und Sonnenstrahlen zusammentreffen, wie dies Schönbein schon nachgewiesen hat und wie diese Thatsache ja seit Jahrtausenden zum Bleichen benutzt wird [The activation of oxygen and the formation of its permanent form ozone or hydrogen peroxide proceeds only effectively when oxygen, water and

> solar radiation meet, as already shown by Schönbein, and used since thousands of years for bleaching] (p. 3212).
> Die Elektricität ist nicht die Ursache der Ozonbildung, sondern es erscheint bei der Zersetzung des Ozons in gewöhnlichen Sauerstoff ein Theil der chemischen Energie des Ozons in Form von electrischer Spannung [The electricity is not the cause of ozone formation, but it seems that in the decomposition of ozone into common oxygen a part of the chemical energy appears in the form of electric current] (p. 3217)

Helmholtz (1887) who studied the interaction between a vapor yet and the flame of a *Bunsen* burner, is the first who believed to has been detected monoatomic oxygen or dissociated oxygen in or near the surrounding of the flame (and termed it antozone) by *Wurster*'s test-paper, and stated that in the flame apart from the combustion products also occur very oxidizing substances. He found that with increasing temperature the dissociation of O_2 increases. *Helmholtz*'s statement (Helmholtz 1887, p. 13, translation from German) that "in the flame molecules of each kind exist, unburned and burned, new formed and dissociated, saturated and unsaturated, heavy vibrating and low temperate, everywhere and permanent mixed". These phrases can be seen as a first idea on radical chemistry in flames. Table 4.14 summarized the most important notations on ozone in the nineteenth century.

Based on two findings by *Schönbein* and his great reputation, that ozone formation is first accompanied with lighting and thus electric discharges in the atmosphere, and second with slow combustion processes (what later was termed autoxidation), several hypotheses developed in the course of ozone observations in the ninetheenth century:

- Because lightning is rare and occurs not evenly distributed over the year, the general atmospheric electricity, and its sources, was seen as the sole cause of ozone formation (electrized oxygen); this was accepted by all researchers (only after around 1870 first ideas arose on oxygen dissociation through solar radiations).
- Thus, all phenomena with water spraying and evaporation (waterfall, fountains, sea surface) were regarded as a source of ozone by a few researchers (*Dancer*, *Zittel*, *Lender*, *Gorup-Besandez*, *Baxendell*, *Bellucci*).
- With an increasing understanding of assimilation and the oxygen release by green plants, some researchers believed that plants also emit ozone (*Scoutetten*, *Kosmann*, *Daubeny*, *Mantegaza*).
- Only Lichtenstein (1862) concluded from experiments that humans emit ozone via the skin and respiratory air.

Despite the first ideas in the later nineteenth century that it was not electricity that caused the formation of ozone but solar radiation it was only in 1900 that it became evident that the oxygen photolysis needs wavelengths unavailable in the troposphere. Until the early 1950s, it was believed that tropospheric ozone arises only as a result of the downward transport of stratospheric ozone.

Table 4.14: Pioneering steps in the history of ozone research.

Year	investigator	notation
1785	Van Marum	odor of "electric matter" whilst electric discharges; also observed whilst electrolysis of sulfuric acid together with gas formation
1801	Cruickshank	"chlorine-like" like smell whilst electrolysis of sulfuric acid, and being a gas
1839	Schönbein	discovery of a gas, termed "*riechendes Prinzip*"
1840	Schönbein	proposed the name "*Ozon*" [ozone]
1845	de la Rive	ozone is a form of oxygen
1848	Hunt	proposed speculative O_3 as formula
1851	Osann	ozone is not an unknown oxidation state of hydrogen
1854	Shapter	first use of the term "oxidizer"
1856	Andrew	the substances comprehended under name ozone are not identical
1856	Plesse and Pierre	first reliable ozone determination using an aspirator (KI method)
1858	Clausius	dissociation of oxygen (O_2) as the first step in ozone formation
1858	Houzeau	first reliable ozone determination in the atmosphere
1861	Babo	experimental confirmation that ozone does not contain hydrogen
1861	Odling	proposed the formula O_3 based on the previous experiments
1864	Soret	confirmation of the formula O_3 through density measurements
1868	Mackereth	activity of solar radiation onto oxygen results in ozone formation
1875	Lévy	first long-time measurements using as novel method
1881	Hartley	solar radiation cut-off at 300 nm likely due to O_3 absorption
1882	Chappuis	found first absorption lines in infrared and visible spectrum
1883	Schöne	first spectroscopic detection in the air
1886	Wurster	stated that only solar radiation causes ozonation
1893	Schöne	systematic spectroscopic determination of ozone
1900	Lenard	showed experimentally that ultra-violet light causes ozonation
1903	Meyer	first absorption spectrum, confirmation of *Hartley*'s hypothesis
1904	Warburg	formation of ozone is a photochemical process
1906	Regener	photochemical decomposition of ozone
1909	Pring	first chemical ozone determination up to 20 km height
1913	Fabry and Buisson	detection of the ozone layer
1920	Fabry and Buisson	first determination of total ozone
1924	Dobson	first spectrometer for total ozone determination
1926	Dobson	first network from 6 stations in Europe
1928	Götz and Dobson	have found the average height of the ozone layer
1930	Chapman	first theory on the ozone equilibrium ("dry" mechanism)
1934	Regener	ozone determination by balloons up to 40 km height
1935	Cauer	first wet-chemical ozone analyzer
1949	Haagen-Smit	ozone formation in Los Angeles smog observed
1950	Bates and Nicolet	O_x-HO_x-mechanism ("wet" mechanism)
1964	Hampson	proposes that water vapor from rockets destroys ozone
1971	Crutzen	proposes that nitrogen oxides destroy stratospheric ozone
1974	Stolarski / Cicerone	propose that chlorine destroy stratospheric ozone
1974	Molina/Rowland	propose that chlorine from CFCs destroys stratospheric ozone
1985	Farman	discovered the "Antarctic Ozone Hole"
1986	Solomon	first proposal on heterogeneous ozone decomposition
1986	Crutzen and Ehhalt	propose the formation of nitric acids clouds

Table 4.15: Different "activated" oxygen species in the history of ozone.

Observer	designation	formula	structure
Thénard	peroxide of hydrogen	$HO + O$	
Schönbein	ordinary inactive oxygen	O	$\oplus\ominus$
Schönbein	negative active oxygen (ozone)[a]	O^-	$\ominus\oplus\ominus$
Schönbein	positive active oxygen (antozone)	O^+	$\oplus\ominus\oplus$
Schönbein	ozonide (ozonized oxygen)	$\overline{O}$ or $\mathring{O}$	$\ominus$
Schönbein	antozonide	O^+	$\oplus$
Schönbein	hydrogen peroxide[b]	HO^2	$HO + \oplus$
Horn	jodosmone (positive electricity)	?	
Baumert	ozone (from electrolysis)[b]	HO^3	
Clausius	ozone	O	
Weltzien	ozone	OO	
Soret	ozone	OOO (O_3)	
Heldt	hydrogen peroxide	$H + O_2$	
Heldt	ozone (volatile hydrogen peroxide)	HO_3	$HO_2 + O$
Meissner	electricized oxygen	?	
Meissner	atmizone[c]	?	
Engler and Nasse	antozone (= hydrogen peroxide)	H_2O_2	
Helmholtz	monoatomic oxygen (= antozone)	O	

[a] Consistent with both positive and negative active ones

[b] Note that the formula HO was seen for water

[c] Heldt (1861, p. 46); first view as peroxo compound and not oxidized water HO + O; Heldt (1861, p. 48) found that ozone and hydrogen decompose together to water and oxygen, writing as a phrase, here in his formulas presented: HO_2 (hydrogen peroxide) + HO_3 (ozone) = 2 HO (water) + HO_2; the latter decomposes further to water and oxygen ($2\ HO_2 = 2\ HO + O_2$)

[d] First named "*nebelbildender Sauerstoff*" [fog-forming oxygen] and soon later by *Meissner* identified as antozone (Meissner 1863a, p. 35 and 102), from Greek ατμίζω = fume

4.4.3 Ozone and antozone

The apparent contradiction that water, even with the strongest oxidant, may not be oxidized to hydrogen peroxide (which formula was described as HO^2), was explained by *Schönbein* ("antozone theory") in that oxygen exists in three different modifications: negative active (ozone O^-), positive active (antozone O^+) and ordinary inactive oxygen (O_2), consistent with both positive and negative active ones (Table 4.15). The corresponding oxygen compounds he termed ozonides and antozonides. He claimed the credit of having isolated antozone in 1861 by treating finely powdered peroxide of barium with cold monohydrated sulfuric acid (today we know that hydrogen peroxide gained and not ozone). Schönbein (1852, p. 43) writes:[947]

947 Note that HO_2 was seen as formula for hydrogen peroxide and HO for water. Nevertheless, this phrase reads like the reaction of the hydroperoxyl radical (HO_2) in its transformation to the hydroxyl radical OH: $HO_2 + X = HO + XO$ ($X = O_3$. NO, NO_2 etc.). Later it has been found that hydrogen peroxide (H_2O_2) does not loose atomic oxygen (O).

> Das oxidirende Vermögen von HO^2 [...] sucht man gewöhnlich durch die Annahme zu erklären, dass ein Theil des in dieser Verbindung enthaltenen Sauerstoffs nur locker an das Radical gebunden sei [...] [The oxidizing property of HO^2 [...] one tries to explain normally by the assumption that a part of oxygen in this compound is only loosely attracted to the radical...].

Schönbein did try to show in many experiments (and subsequent papers) that always when ordinary oxygen (O_2) transfers to active oxygen both modifications formed, i. e., ozone and antozone or ozonide and antozonide, respectively. Of the antozonides, he termed hydrogen peroxides (with the formula HO^2) and superoxide of barium (and others).[948] However, never was recognized antozone simultaneously with the formation of ozone. Osann (1855) contributed to another confusion in the discussion whether ozone contains hydrogen; first, he confirmed the observation that through electrolysis of acid solution (but interestingly not alkaline) ozone is gained (Osann 1855, p. 313),

> [...] dafs Sauerstoffgas durch blofses Elektrisiren in die Modification des Ozon-Sauerstoffgases übergeführt werden kann [... that oxygen gas can be transformed into the modification of ozone-oxygen gas due to simple electrifying]

and he rejected the assumption that at the oxygen pole hydrogen in higher oxidation state is formed. However, he found in his experiments "that the electrolytic generated hydrogen gas has a larger reducing capacity than that gained in ordinary matter" (Osann 1855, p. 315, translated from the German), which he termed "*Ozon-Wasserstoff*" [ozone-hydrogen], believing it being an allotropic modification of hydrogen, similar to that of ozone-oxygen; see Jensen (1990) for the history of the nascent state.

The English chemist Sir *Benjamin Collins Brodie* (1817–1880) studied intensively the chemistry of peroxides and showed that there is no difference in the general chemical action of oxygen bonded in ozonides or antozonides (Brodie 1862), later confirmed by the *Carl Weltzien*. Weltzien (1860a) was the first who fight against the antozone theory and prove that hydrogen peroxide also has reductant properties. Further, *von Babo*

---

948 Wurtz (1868, *Dictionnaire de Chimie pure et appliquée*. Vol. 2, p. 721) discovered "ozonides" in reaction between ozone and strong alkali solutions as yellow-brown precipitate which immediately decomposed with absence of ozone. The chemical constitution remained unclear; Baeyer and Villiger (1902) explained this compound as salt of the hypothetical "*Ozonsäure*" H_2O_4 (p. 637). The Soviet chemist *Isaak Abramovich Kazarnovski* [Исаак Абрамович Казарновский] (1890–1981) who studied in Zürich, recognized in 1949 the constitution MO_3 and coined the term ozonide (Kazarnovski et al. 1949); In 1969 *Kazarnovski* enlightened the mechanism of the formation of hydrogen via peroxide hydroxyl radicals and the ozonide ion (O_3^-) while decomposing ozone in solution (Kazarnovski et al. 1969). Organic ozonides first studied 1898 by the French chemist *Marius Paul Antoine Gabriel Otto* (1870–1939) in reaction between ozone and organic compounds. *Carl Dietrich Harries* discovered in 1902 the ozonolysis, but the mechanism was enlightened only 1952 by the German chemist *Rudolf Criegee* (1902–1975).

also denied the existence of two electric-polarized modifications of oxygen; he writes (Babo 1863, p. 290):

> So wahrscheinlich es die schönen Untersuchungen Schönbein's und Meißner's machen, dass neben Ozon auch stets sogenanntes Antozon gebildet wird, [...] durch die Electricität allein kein Körper von den Eigenschaften des sogenannten Antozons aus reinem Sauerstoff entstehen kann [So presumably the beautiful experiments Schönbein's and Meißner's do it, that apart from ozone always the so-termed antozone forms, [...] through electricity solely no body with the properties of the so-termed antozone can gained].

However, *Schönbein* stuck to his point of view and found supporters, such as *Georg Meissner* (1829–1905),[949] who carried out experiments (Meissner 1863a) to remove the ozone from the gas mixture after its formation and to show the properties of the remaining asserted antozone, supposing that antozone forms fog in contact with water. Meissner (1869, p. 3) writes:

> [...] unter der Einwirkung elektrischer Spannung neben dem Ozon ein zweiter Zustand des Sauerstoffs oder eine zweite Sauerstoffmodification entsteht, welche vor Allem dadurch ausgezeichnet und characterisiert ist, dass sie, nachdem das Ozon durch gewisse oxydierbare Substanzen absorbiert ist, den Wasserdampf ohne Mithülfe einer Temperaturerniedrigung zu Nebelbläschen zu condensieren vermag [... under the impact of electric discharge apart from ozone a second state of oxygen is gained, which is especially characterized through the capability – after absorption of the ozone by certain oxidizable substances – to condense the water vapor without decrease of the temperature].

Thus *Meissner* concluded that ozone does not favor the formation of fog but "antozone". *Meissner* also states that "electricized oxygen" can oxidize free nitrogen (N_2), but ozone does not.[950] Finally, he defined antozone as this modification of oxygen, which is able to oxidize water into hydrogen peroxide.[951] *Meissner* also believed a short time that the third modification of oxygen exists, which he termed "*Atmizon*".[952] Passing atmizone-containing air through water, the solution is later proved as diluted hydrogen peroxide (Meissner 1863a, p. 95); *Meissner* stated that atmizone (or later equated with antozone) oxidizes water. Whereas *Schönbein* recognized the white vapor forming above phosphorus as "*salpetrigsaures Ammoniak*" (NH_4NO_2), confirmed by Osann (1865, p. 58), who explained "[...] *seine Bestandtheilen nach als eine in Salz metamorphirte atmosphärische Luft* [...]" [... in its constituents a salt, metamorphized atmospheric air...], *Meissner* believed it to be antozone.[953] *Werner Schmid*

949 A German anatomist and physiologist; University Professor at Basel (from 1855), Freiburg (from 1857) and Göttingen (from 1860 to 1901).

950 This is known from oxygen atoms according $N_2 + O = N + NO$ (slow reaction).

951 This is known from $O(^1D) + H_2O = 2\,OH$ with subsequent formation of H_2O_2 via HO_2 radicals.

952 I found this term in no English written publication, hence I will "translate" it into *atmizone*.

953 This view was based on *Schönbein*'s believing that ammonium nitrite is generated from atmospheric air and water by the aid of heat.

(unknown living dates) from Wiesbaden (probably working in the Chemical Institute Fresenius) on the initiative of *Fresenius*, conducted several careful experiments with phosphorus and found the formation of ozone and hydrogen peroxide only in moist air (Schmid 1866). Finally, Bieber (1911, 1912) conducted interesting experiments on the condensation of water vapor in the presence of ozone to characterize the chemical nature of the fog (blue haze in today's termination). *Willie Bieber* (unknown living dates) interpreted *Meissner*'s antozone as free oxygen atoms, which appear during the formation and decomposition of ozone.[954] Prestel (1868, p. 327), however, writes:

> [...] dass das Ozon, wie bei bekannten Experimenten unter der Glasglocke, so auch in der Atmosphäre nebelartige Trübungen verursachen kann [... that ozone, as known from experiments under the bell jar, also may causes nebulous turbidity].

Rothmund (1917), investigating *Meissner*'s fog, concluded after experiments, passing dried ozone through solutions of potassium iodide or sulfuric acid, that the fog is gained by the reaction between O_3 and iodide in solution and the freed iodine, transferred into the gas phase, further oxidizes to iodic acid[955] which forms with the water vapor fog droplets.

Finally, Weltzien (1866) believed that the so-termed antozone is simply the "peroxide of hydrogen" (H_2O_2), soon confirmed by Engler and Nasse (1870). Engler (1879, p. 18)[956] writes:

> [...] dass das sogenannte Antozon nur dann entsteht, wenn Ozon in Gegenwart von Wasser zerstört wird, so dass die Annahme nahe lag, das Antozon sei [...] weiter nichts als Wasserstoffsuperoxyd [... that the so-termed antozone only be formed if ozone in the presence of water decomposes, so that the assumption was likely that antozone [...] is nothing else than peroxide of hydrogen].

The physician *Otto Johann Friedrich Nasse* (1839–1903),[957] who cooperated with the chemist *Carl Engler* in Halle, writes that together with the formation of ozone frequently hydrogen peroxide appears, and in some cases it is found exclusively, namely, when ozonation and oxidation follow quickly after this, and thus the ozone is no

954 The observation of fog formation (not always) of almost all scientists of the nineteenth century during experiments with ozone can be also interpreted due to homogeneous nucleation. Depending from the experimental condition, during the formation of ozone by electric discharges, also condensation nuclei can be produced from trace substances containing in the air or not completely removed by air washing such as NH_3, SO_2, NO_x and VOC and thus forming of haze or (at high humidity) fog.

955 More likely is the formation of oxides of iodine acting as condensation nuclei.

956 This small booklet "*Historisch-kritische Studien über das Ozon*" (Halle 1879) is to my mind the best history on all aspects of ozone and includes all literature concerning discovery and nature of ozone, formation, production. properties, detection and determination, ozone in the atmosphere, its sanitary role and technical use.

957 A famous physician in third generation, born in Marburg, 1866–1880 private lecturer and professor in Halle, 1880–1888 in Rostock, and finally moved to Freiburg in 1902.

longer detectable (Nasse 1869, p. 392).[958] Nasse (1870, p. 205) clearly expresses the elementary step in oxidations by the ozone molecule:

> Wenn Ozon Oxydationen ausführt, so kann dieses wohl niemals (wenigstens kennen wir keinen solchen Fall) so geschehen, dass das ganze unversehrte Ozonmolekül eingeht, sondern stets muss eine Zerstörung der Ozonmoleküle voraufgehen [...] ein Atom abspalten, während die beiden anderen sich zu einem Molekül gewöhnlichen Sauerstoffs zusammensetzen [...] Die sogenannten Ozonreagentien sind Stoffe, welche zuerst das Ozonmolekül spalten [...] So hat man es denn im Grunde nur mit Sauerstoff im status nascenc zu thun [...] Ueber den Ursprung dieser freien Sauerstoffatome können [...] die Ozonreagentien ohne Weiteres gar keinen Aufschluss geben [When ozone performs oxidations, this can never be (at least we do not know such cases) so happen that the whole unbrokenly ozone molecule reacts, but always a decay of the ozone molecule must precede [...] one atom dissociates while the other two combine to an ordinary molecule of oxygen [...] The so-termed ozone reagents are compounds that first dissociate the ozone molecule [...] Thus basically one has to do only with oxygen in status nascence [...] On the origin of this free oxygen atom [...] the ozone reagents cannot simply give an answer].

Later (at beginning of the 1880s), *Moritz Traube* (1826–1894) proved the antozone theory to be wrong and that H_2O_2 is not "oxidized water" but a reduced form of the oxygen molecule (Traube 1882, 1887).

4.4.4 Understanding atmospheric chemistry of oxygen compounds

> For tropospheric chemistry it can be said that ozone is the fuel and the hydroxyl radical is the propellor (Seinfeld 1986, p. 137).

Francis Edward Blacet (1899–1990) from the University of California and Los Angeles first made the following significant statement for the atmospheric chemistry in the troposphere (Blacet 1952):

> Reactions requiring radiant energy below 2900 A cannot be significant. This, of course, rules out photochemical ozone formation from diatomic oxygen absorption and the photo-chemical synthesis of oxides of nitrogen from their elements. In fact, it eliminates all of the high-altitude reactions. Consequently, at low altitudes photochemical processes of significance can only occur when they involve one or more substances not normally regarded as being present in pure air (p. 1340).
> The photochemical reactions which do occur in the immediate environment are initiated as a result of sunlight absorption by pollutants produced primarily in the activities of man. Many of these may be of minor importance (p. 1341).

958 The chemical knowledge of *Nasse* is amazing for today's readers. This cited paper is based on a lecture *Nasse* read to the "*naturforschende Gesellschaft zu Halle*" [Society of natural scientists]; see also *Abhandlungen der Naturforschenden Gesellschaft zu Halle* **11** (1869 27–28. *Nasse* argued theoretical that monovalent elements possess no allotropic states, like hydrogen which becomes saturated in the H–H molecule.

The discovery of atmospheric ozone led soon later to the view that the earth's atmosphere is oxidizing, and ozone directly reacts with foreign substances, *pollutants*, thus *cleaning* the atmosphere. This view remained until about 1950. Between 1950 and 1970 the view dominated that atomic oxygen (O), a product of the ozone photolysis, oxidizes the pollutant. Only after 1970, the primary role of the hydroxyl radical (OH) has been recognized as *cleansing species*. Another reactive oxygen species, hydrogen peroxide (H_2O_2), discovered long before ozone, was with certainty also detected in the air before ozone, but not seen before the 1950s as an important oxidizing species in the atmosphere (see Chapter 4.5.3 for understanding the atmospheric chemistry of H_2O_2). Without knowing chemical mechanisms, at the end of the nineteenth century the germicidal properties of O_3 and H_2O_2 have been used in many (also industrial) applications, but an understanding of the chemistry behind developed only after 1980. Although *Christian Junge* already in the late 1950s put attention on the role of aqueous-phase oxidation of sulfur dioxide in air, exclusively the atmospheric gas-phase chemistry has been studied until about 1980. The view that atmospheric oxygen chemistry can only be understood as multiphase chemistry, i. e., interlinked gas, aqueous and particulate chemistry was not before the mid-1980s.

However, long before understanding the atmospheric significance of these reactive oxygen species (later abbreviated as ROS), their chemistry in gases and aqueous solutions has been studied. Most thermal chemical reactions and radicals such as H, O, OH and HO_2 we today consider after primary photochemistry in the atmosphere, have been postulated in studying combustion chemistry in the 1920s and 1930s, and are connected with *Bonhoeffer*, *Harteck*, *Hinselwood*, *Bates*, *Joost*, *Lewis*, *von Elbe* and many more. However, direct proof of oxygen radicals succeeded after World War II.

Understanding the photochemistry of dioxygen (O_2) and ozone began in 1900, and the first model of stratospheric chemistry was presented in 1930 (Chapter 4.4.8.1). Photochemistry has many pioneers (Chapter 1.3.5.3 for the roots before 1900), but *Leighton* and *Blacet* must be named as *the* pioneers of atmospheric photochemistry. The necessary presence of nitrogen oxides and volatile organic compounds in air to gain photochemically surface-near ozone has been recognized in the early 1950s. Despite all main reactions that are involved in this mechanism being known until the early 1970s, a full understanding of photochemical ozone formation in the troposphere has been only reached the end of the 1990s (Chapter 4.4.8.6). Furthermore, new marine atmospheric ozone removal chemistry and biogeochemical loops only recently have been discovered and still are under the investigation.

4.4.4.1 Gas-phase chemistry

The chemistry of radicals such as O, H, OH, and HO_2 have been rather well understood around 1940 from laboratory investigations, and the first values of activations energies and reactions rate have been estimated (e. g., von Elbe and Lewis 1942).[959] However, indeed only very few of these reactions are relevant for tropospheric chemistry, and several re-investigations of mechanisms and re-estimations of kinetic parameters in the following decades (particularly about 1960–1990) have been conducted to create kinetic models for atmospheric chemistry modeling.

In the gas phase (water vapor at high temperature or during electric discharges and in hydrogen-oxygen flames), evidence of OH radicals was spectroscopically found by the American physicist *William Weldon Watson* (1899–1992)[960] (Watson 1924), and later by the German chemist *Karl-Friedrich Bonhoeffer* (1899–1957). Bonhoeffer (1928) found evidence for the equilibrium $2\,OH + H_2 = 2\,H_2O$ in thermic water dissociation, further investigated by Bonhoeffer and Reichardt (1928). Bonhoeffer and Haber (1928) finally proposed a chain mechanism involving OH and H for the oxygen-hydrogen reaction ($H_2O \rightleftarrows H + OH$) in flames and suggest the (later in atmospheric chemistry) important reactions $H_2O_2 = 2\,OH$ and $2\,H + O_2 = 2\,OH$. The latter reaction was earlier proposed in the form $O_2 + H \rightarrow O_2H$ (as today accepted in atmospheric chemistry) by Marshall (1926).

In the atmosphere, "pure" gas-phase chemistry occurred, but in reaction chambers and vessels, the wall presents a surface for heterogeneous chemistry, making experimental investigations very complicated. *Abraham Lincoln Marshall* (1897–1974), American photochemist, writes in 1926 very clearly on the problems of studying gas-phase reactions, but also on the importance of heterogeneous chemistry, studied under atmospheric conditions, not before the 1980s (Marshall 1926, p. 34):

> Kinetic studies of gas reactions have played a very important role in the development of theories concerning the mechanism of chemical reactions. There are, however, very few such reactions occurring in the gas phase free from disturbance due to the walls of the containing vessel. Bodenstein and his co-workers have made many comprehensive studies regarding the behavior of thermal gas reactions and have accumulated a large amount of valuable data. In all of this work remarkably few truly homogeneous reactions were found, by far the greater number taking place at some solid surface [...]
> The studies on heterogeneous gas reactions have thrown a great deal of light on the influence of physical condition and chemical nature of the surface on the course of the reaction but have yielded practically no information on the mechanism by which the reaction proceeds at the surface [...]
> Photochemical studies of gas reactions have given us considerable additional information on reaction mechanism in the gas phase. The number of such reactions available for exact quantitative study has in the past been limited, however, by the small absorption coefficient of many reactants

959 *Guenther von Elbe* (1903–1988) was a German-American chemist and *Bernard Lewis* (1899–1993), born in London and immigrated to USA as child; both were well-known combustion chemists.

960 The first water spectrum was gathered by *George Downing Liveing* (1827–1924) and *James Dewar* (1842–1923) who found that no other element than O and H participate in the spectrum but either oxygen nor hydrogen alone (Liveing and Dewar 1880).

in the wave length range available. Other complications were secondary reactions induced in the primary products [...]

The German physical chemists *Wilhelm Groth* (1904–1977) and *Hans Eduard Suess* (1909–1993),[961] studying the photochemical decomposition of carbon dioxide and water vapor using *Harteck*'s xenon lamp (Groth and Suess 1938), found that the primary products (CO, H, and OH) further react to gain aldehydes; they refer Dhar and Ram (1933c; see p. 226) who detected formaldehyde in rainwater. *Groth* and *Suess* discuss that this process could be a source of free oxygen and organic compounds in the early atmosphere, and thus providing a precondition for the formation of organic life. The *Miller-Urey* experiment in 1953 (see Möller 2020, p. 305) showed that under UV radiation organic molecules can be formed in an atmosphere containing NH_3 and CH_4. *Suess*, who emigrated in 1950 in the USA, stayed first until 1951 at *Harold Urey* at the Chicago University

Detection of OH was carried out since the 1930s based on UV absorption, however, until the 1960s with detection limits of about 1 ppm (10^5 times higher than OH concentrations later detected in atmospheric air). The existence of the hydroperoxyl radical HO_2 (former name perhydroxyl) was first proposed by the English chemist *Hugh Stott Taylor* in the photochemical reaction between hydrogen and oxygen at room temperature where H atoms were gained from H_2 in a mercury arc (Taylor 1926). Thus, he termed this first process (hydrogen photolysis) *photosensitization*, followed by radical chain reactions, proposing (Taylor 1926, p. 566)[962]

$$
\begin{aligned}
H + O_2 &= HO_2 \\
H + O_2 &= OH + O \\
HO_2 + H_2 &= H_2O_2 + H \\
OH + H_2 &= H_2O + O \\
H_2O_2 &= H_2O + O \\
O + H_2 &= H_2O
\end{aligned}
$$

One of the key reactions in atmospheric chemistry, $O(^1D) + H_2O = 2\,OH$, was first proposed by Taube (1957)[963] while studying the decomposition of O_3 in solution by UV (254 nm) and the mechanism to explain the final formation of H_2O_2 (no formation at

961 Grandson of the geologist *Eduard Suess*, who introduced the term biosphere.

962 In this paper *Taylor* also studied the reaction $H_2 + CO$, and identified among the reaction products HCHO, CH_4 and solid polymerized products; this is surely the first evidence of photochemical formation of secondary organic aerosol (SOA); remarkable that *Taylor* proposed HCO (formyl radical) as the first intermediate.

963 *Henry Taube* see footnote 675 on p. 338.

IR decomposition). This reaction does not proceed with $O(^3P)$ according to Harteck and Kopsch (1931).[964] McGrath and Norrish (1957) found the first evidence for this reaction to be rapid in gas phase through their experiments on the flash photolysis of moist ozone. Leighton (1961, p. 148) first recognized that this reaction is of particular interest in polluted air.

The hydroperoxide radical HO_2 was detected by mass spectroscopy in 1953 by Foner and Hudson (1953). Many of the huge number of investigated and hypothesized elementary reactions were considered in smog chemistry after 1950 but later neglected due to the unimportance of most radical-radical reactions in atmospheric air compared to radical-molecule reactions; only in the stratosphere (because of the small concentration of molecules containing other atoms than H and O) some of them are crucial (Chapter 4.4.4.1).

The formation of ozone as a result of oxygen photolysis became clear around 1920, but the presence of ozone in the lower atmosphere was only explained by downward vertical transport. An understanding of photochemical ozone formation in the troposphere became evident in the 1950s but was not fully understood before the 1980s (Chapter 4.4.8.6). The Dutch chemist *Arie Jan Haagen-Smit* (1900–1977) from Utrecht came in 1936 to the California Institute of Technology, first working on plant chemistry. He started his air pollution research in 1948 when Southern California residents suffered stinging eyes and respiratory irritation from smog. In his pioneering paper "*Chemistry and Physiology of Los Angeles Smog*" (Haagen-Smit 1952) he attributed the decrease in visibility, crop damages, eye irritation, objectionable odor, and rubber deterioration caused by nitrogen oxides and hydrocarbons from automotive exhausts which produce local ozone, resulting in chain mechanism including radicals[965] and formation of aldehydes and peroxides. In a further series of studies between 1953 and 1956, *Haagen-Smit* and co-workers (Haagen-Smit and Fox 1956) conducted laboratory experiments with mixtures of nitrogen dioxide and several hydrocarbons and their derivatives to estimate the ozone-forming potential of organic compounds. Haagen-Smit (1952, p. 1346) writes:

> The oxidant is represented by a complex of factors consisting of oxygen, atomic oxygen formed by the photochemical dissociation of nitrogen dioxide, and ozone produced in the photochemical oxidation of organic material.

---

964 *Paul Harteck* (1902–1985) was an Austrian physical chemist, doctorate 1926 under *Max Bodenstein*, 1926–1927 he was the assistant of *Arnold Eucken* in Breslau and 1928–1933 of *Fritz Haber* in Berlin, 1933/1934 with *Ernest Rutherford* (1871–1937) in Cambridge and since 1934 professor at University Hamburg.

965 A survey of the "free" radical literature published to 1959 including approximately 2200 references and a brief description of the work was made by Lavin et al. (1961). Note that according to newest IUPAC definition, the term "free" radical is obsolete, only "radical".

Lewis H. Rogers (1910–2000), chief chemist (1954–1958) of the Southern California Air Pollution Formation, coined the term *photochemical smog* (Rogers 1958) and summarized the key reactions (based on the compilation by Johnston 1957);[966] the reaction $NO + O_3 = NO_2 + O_2 + 48$ kcal was first established by Johnston and Crosby (1951):

$$NO_2 + h\nu \rightarrow NO + O$$
$$O + O_2 \rightarrow O_3$$
$$NO + O_3 \rightarrow NO_2 + O_2$$
$$O_3 + \text{olefins} \rightarrow \text{radicals}$$
$$R + O_2 \rightarrow RO_2$$
$$RO + NO \rightarrow RONO$$
$$RO_2 + NO \rightarrow RO_2NO \rightarrow ?$$

Compilation of such reaction schemes was based on results from earlier researches on photochemistry, namely organic compounds which were photolyzed in the gas phase. Berthelot and Gaudechon (1910b)[967] found equal volumes of CO and ethane upon UV photolysis of acetone. The British chemist *Ronald George Wreyford* Norrish (1897–1978) suggested a free-radical mechanism for carbonyl photoreaction in the 1930s (Norrish 1934, Roth 2001). *Philip Albert Leighton*, who became the first leading atmospheric photochemist after the war, began to study aldehyde photolysis in the 1930s (Leighton and Blacet 1932).

The photodissociation of NO_2 is the only definitively established process for the formation of ozone in the troposphere (Warneck 1988, p. 183) and was already studied by Norrish (1929) who proposed the mechanism with the collision between a photoactive and an inactive NO_2 molecule (which was still named nitrogen peroxide):

$$NO_2 + h\nu \rightarrow NO_2{}^* \quad (\lambda = 546\,\mu m)$$
$$NO_2{}^* + NO_2 \rightarrow 2\,NO + O_2$$

Holmes and Daniels (1934) first estimated quantum yield and absorption coefficients of the NO_2 photolysis and also studied the photolysis of N_2O_4 and N_2O_5. The reaction $NO_2 + h\nu \rightarrow NO + O(^3P)$ was first formulated by Leighton (1961, p. 57). *Richard Allan Graham* first studied the photochemistry of NO_3 and the kinetics in the N_2O_5-O_3 in 1975 (PhD thesis, University of California); Graham and Johnston (1978). Pioneers in gas phase photochemistry, namely in the study of smog and air pollution in Los Angeles County (chronological order by year of birth):

966 *Harold Sledge Johnston* (1920–2012) was a US chemist, known for studies on gas phase reaction kinetics.

967 *Henry Gaudechon* (?–1934) was a French chemist.

Philipp Albert Leighton (1897–1983), Stanford University
Francis Edward Blacet (1899–1990), University of California and Los Angeles (UCLA)
Arie Jan Haagen-Smit (1900–1977), Dutch origin, California Institute of Technology (Caltech)
Richard D. Cadle (1915–2010), Stanford Research Institute in Palo Alto and moved in 1963 to National Center for Atmospheric Research (NCAR)[968]
David H. Volman (1916–2000), University of California Davis (UC)
Harold Sledge Johnston (1920–2012), University of California Berkeley
James N. Pitts Jr. (1921–2014), University of California Riverside (UCR)
Jack G. Calvert (1923–2016), Ohio State University

In the 1950s, photochemical modeling of stratospheric chemistry first implicated the strong radical oxidants O and OH, generated from photolysis of O_3 and H_2O, in the oxidation of CO and CH_4 (Bates and Witherspoon 1952). The importance of photochemically generated radicals in the chain oxidation of hydrocarbons leading to urban O_3 smog was also recognized in the 1950s (Leighton 1961). Smog models of that time hypothesized that O atoms produced in urban air from the photolysis of NO_2 and O_3 would provide the main pathway for hydrocarbon oxidation (Altshuller and Bufalini 1965, 1971). This mechanism was thought unimportant outside of urban areas because of low O_3 and NO_2 concentrations, and transport to the stratosphere was viewed as necessary for the oxidation of CO, CH_4, and other gases present in the global troposphere (Cadle and Allen 1970). Long atmospheric lifetimes for these gases were implied because of the 10-year residence time of air in the troposphere. This view of a chemically inert troposphere was first challenged by Weinstock (1969), who found from ^{14}CO measurements that the atmospheric lifetime of CO is only ~0.1 years, requiring a dominant sink in the troposphere. Levy (1971) then presented photochemical model calculations for the unpolluted troposphere showing that high concentrations of OH could be generated from photolysis of O_3 in the presence of water vapor and account for the missing sink of CO in the Weinstock (1969) analysis. The chemical mechanisms responsible for the presence of hydroxyl radicals in the troposphere were further refined by McConnell et al. (1971), Wofsy et al. (1972), Crutzen (1973), and Warneck (1974). Further work in the early 1970s confirmed the importance of tropospheric oxidation by OH as the main sink of CO and CH_4 (McConnell et al. 1971, Weinstock and Niki 1972, Levy 1973) and further showed that OH, not O, is the main oxidant of hydrocarbons in urban air (Heicklen et al. 1971, Kerr et al. 1972, Demerjian et al. 1974).[969]

968 In the year 1959 in the USA the establishment of *National Atmospheric Program* and the founding of the *National Center for Atmospheric Research* (NCAR) in Boulder (Colorado) opened in 1960.

969 *Kenneth L. Demerjian* (born 1947) US American chemist, University at Albany-SUNY. *Julian Heicklen* (born 1932), US American chemist at Pennsylvania State University. *J. Alistair Kerr* (born 1934), British chemist, Univ. of Birmingham, 1994 ETH Zürich. *Hiram Levy II* (?), US American chemist at NOAA. *John C. McConnell* (1945–2013), Irish-born Canadian chemist and physicist, York University Toronto. *Hiromi Niki* (1937–1995), York University Toronto, Canada. *Bernard Weinstock* (1917–1981), US American chemist, student of *Harold Urey*, since 1961. *Steven C. Wofsy* (born 1945), US American chemist at Harvard University, 1961 at Ford Motor Company.

Calvert et al. (1972) recognized that the photolysis of formaldehyde is an important source for hydrogen atoms which further react with oxygen to gain the HO_2 radical. Considerable evidence over the past three decades supports the view that tropospheric OH is the main oxidant for non-radical species in the atmosphere.

4.4.4.2 Aqueous-phase chemistry

After the discovery of ozone, many reactions in aqueous solutions and later in organic solvents have been studied; this chapter deals only with chemical reactions that are relevant for the atmosphere, although its importance often has been recognized many decades after its discovery. As presented in Chapter 4.4.3, the close chemical relationship between ozone and hydrogen peroxide ("ozone and antozone") was soon found and has been associated with O_2 and H_2O (Chapter 4.5.3.2 for aqueous-phase H_2O_2 chemistry).

Schönbein (1844, pp. 62, 94, 110) found that ozone is removed from air after bubbling through alkaline solutions. This was proven quantitatively by Soret (1864) but Cossa (1867) found that O_3 will not be destroyed in pure KOH solution free of any organic substance. However, the role of certain dissolved organic compounds (DOC) working as photocatalyzers became only clear in the 1980s. Thus, *Cossa*'s remarkable early observations support the essential role of water-dissolved organic compounds.

The German chemist *Casimir Wurster*, who also contributed with first ideas to the formation of ozone in air (p. 478), ascribed the yellowing of groundwood pulp containing paper under the influence of sunlight through "activation of oxygen" and oxidation of the gum in the paper (Wurster 1886c); this is the first reference on the mechanism later termed photocatalytic oxidation (formation of superoxide and subsequent oxygen radicals). Chapman and Jones (1911) found that ozone when passed into strong alkalis forms unstable colored ozonates MHO_4, and proposed the equilibrium[970] $2\,OH^- + 2\,O_3 \rightleftarrows 2\,HO_4^- \rightarrow H_2O + O_2 + 2\,e^-$, to be assumed that ozone is slightly acidic. A mechanism was not known before Weiss (1935) first proposed the following three reactions, based on the observation that ozone decay is effective only in alkaline solution: $O_3 + OH^- \rightarrow O_2^- + HO_2$, $O_3 + HO_2 \rightarrow 2\,O_2 + OH$, and $O_3 + OH \rightarrow O_2 + HO_2$. In the aqueous phase, existence of OH radical was first proposed by Haber and Willstädter (1931)[971] but definitively detected only in 1965 (Thomas 1965). Another spontaneous radical formation in alkaline solution was suggested (Schroeter 1963, Walling 1957): $O_2 + OH^- \rightleftarrows OH + O_2^-$.

970 This is very close to the modern view: $O_3 \xrightarrow{e^-} O_3^- \xrightarrow{H^+} HO_3 \rightarrow OH + O_2$.

971 *Fritz Haber* (1868–1934), German chemist, awarded with the Nobel Prize for Chemistry in 1918. *Richard Martin Willstätter* (1872–1942), German chemist, awarded with the Nobel Prize for Chemistry in 1915.

The ozonation of drinking water was widely used after recognition of the germicidal properties of ozone, first shown in 1859 by *Gorup-Besanez*[972] and later applied to treat organically polluted water in 1885[973] by *Friedrich Emich* (1860–1940).[974] After manufacturing of ozone generators, the first water treatment plant was established in 1892 in Martinikenfelde,[975] near Berlin, by the firm *Siemens* and *Halske*. Many studies have since been carried out at this plant, supported by the German government, to investigate the efficiency of ozonation as a basis for developing larger water treatment plants (Fonrobert 1916).

In fact, the roots of aqueous-phase atmospheric chemistry lie in the academic aqueous radical research between 1930 and 1950 as well as in understanding chemical water treatment processes. The "aquatic surface chemistry" (Stumm 1987) mechanism, now often termed as interfacial and photocatalytic chemistry, was not understood before the beginning of the 1980s. Thus, Yeatts and Taube (1949), in studying the reaction between chloride and ozone (p. 338), proposed the reaction $Cl^- + O_3 + H_2O \rightarrow Cl + OH + HO^- + O_2$ as a one-electron oxidation-reduction process between O_3 and Cl^-; however, also noted that atomic Cl initiates a rather pronounced catalytic decomposition of ozone also in water solution.

In 1967, Japanese chemist *Akira Fujishima* (born 1942) discovered photocatalytic water decomposition (water photolysis by sunlight conditions) on TiO_2 surfaces, later termed the *Honda-Fujishima* effect (Fujishima 1972). However, the phenomenon that finely divided ZnO and TiO_2 are capable of acting as photosensitizers for several chemical reactions was known long before (Chapter 4.5.3.2), obviously unknown to *Fujishima*.[976] With time, the meaning of photosensitization changed towards reactions involving a *photocatalyst* (chromophoric substances[977] and semiconductors) similar to *Baur*'s conception of molecular electrolysis (Kavarnos and Turro 1987, Hoffmann

---

972 Gorup-Besanez (1859) studied the reaction of ozone with several cyanides and organic substances in aqueous solution; cyanide reacts very fast, some organic compounds (almost natural substances) also effective but other not.

973 Monatshefte für Chemie 6 (1885) p. 77 (cited after Fonrobert 1916). Emich, F. (1885) Zur Selbstreinigung natürlicher Wässer. *Sitzungsberichte der Kaiserlichen Akademie der Wissenschaften in Wien – mathematisch-naturwissenschaftliche Classe*. 91, 67–84. Self-cleaning of river water under the influence of organisms was shown first in 1869 by *Carl Alexander Müller* (1828–1906), professor for agricultural chemistry in Chemnitz and later in Stockholm (*Landwirthschaftliche Versuchsstationen* 1873, Vol. 73, p. 263. *Müller* used the term "Selbstreinigung" [self-cleaning] for the first time.

974 Austrian chemist at Technical University Graz; one of the founders of microchemistry.

975 This former village is now an urban region at the northwestern edge of downtown Berlin.

976 I met *Fujishima* in Berlin in 2004 where he holds a lecture and had a discussion with him. He was obviously surprised about my question on the role of water films or humidity on the photocatalytic effect.

977 The Russian-German chemist *Otto Nikolaus Witt* [Отто Николаус Витт] (1853–1915) coined the term *chromophore* for unsaturated organic compounds containing groups like >C=C<, >C=O, –N=O, and –N=N– though to be essential for color (Witt 1876).

1990, Hoffmann et al. 1995, Faust and Allen 1992, Faust et al. 1993, Anastasio et al. 1997): $W + h\nu \rightarrow W^* \xrightarrow{A} (W^+ - A^-) \rightarrow W^+ + A + e^-$ products. Hence, the theory of photoinduced release of electrons goes back to the conception of "molecular electrolysis" by *Emil Baur* in 1918 (see pp. 102–103).

The idea that an electron transfer occurs onto oxygen had been proposed by Evans and Uri (1949) and was accepted by Leighton (1961) as a mechanism likely important for particulate matter: $O_2(ads) \xrightarrow{e^-} O_2^-(ads)$. The hydrated electron e^-_{aq}, also written as H_2O^- or simple e^-, and first postulated by the radiation chemists *Gabriel Stein* (1920–1976) in 1952 (Stein 1952) and characterized in 1962 by recording its absorption spectrum (Stein 1968, Hart and Anbar 1970, Hughes and Lobb 1976). The solvated electron[978] and especially the hydrated electron has since been found to be an extremely important reactive species. It is the simplest electron donor and all of its reactions are in essence electron transfer reactions (Logan 1967). The aqueous chemistry of the hydrated electron is described by Hart and Anbar (1970), Hughes and Lobb (1976), and Buxton (1982). Photoinduced electron transfer was demonstrated in many molecules where the donor and acceptor are linked together intramolecularly (Kavarnos and Turro 1987).

4.4.5 Studies of atmospheric ozone in the nineteenth century

4.4.5.1 Introduction

Christian Friedrich Schönbein (Fig. 4.28), who discovered this new gas while performing different experiments in the laboratory, that he termed ozone, but never elucidated its chemical constitution, but was the first to get evidence for its presence in atmospheric air (Schönbein 1845a). However, already in 1840, he assumed that "this smelling substance which the thunderbolt releases, the same is which through the electricity of our machines and while electrolysis of water releases" (Schönbein 1840, p. 632); he further writes (Schönbein 1840, p. 634):

> [...] und so dürfte dieser Körper in dem Haushalte der belebten, wie unbelebten Natur eine große Rolle spielen. [... and so should this body play an eminent role in the economy of animate and unanimate nature].

978 Many papers refer to the first observation of a solvated electron in 1864 by Weyl (*Pogg. Ann.* **123**, 350) who studied the dissolution of some alkali and earth alkali metals in liquid ammonia and who observed the characteristic deep blue color of the solution (owing to the light absorption of solvated electrons) and formation of hydrogen, termed "hydrogen ammonia". However, Kraus (1908) first stated the existence of "an electron, surrounded by an envelope of solvent molecules". Liquid ammonia dissociates similar to H_2O: $2NH_3 \rightleftarrows NH_2^- + NH_4^+$. The reaction of potassium with liquid ammonia is thereby very similar to that of sodium with water: $K + NH_3(l) \rightarrow KNH_3 \rightarrow KNH_2 + H$. Amides ($-NH_2$) are well-known compounds; today, we know that K reacts with NH_4^+ via electron transfer, forming the ammonium radical NH_4, which decomposes into $NH_3 + H$.

Figure 4.28: *Christian Friedrich Schönbein*. Source: Wikimedia Commons.

In retrospect to his childhood, he reports on a lightning strike in the church of his birthplace (Metzingen in Württemberg) in 1811 where he perceived the "sulphureous" odor (Schönbein 1840, p. 631). The first phrase of his 1840 paper (Fig. 4.29) reads (translation from German): "Carefully conducted and often repeated experiments have put me beyond doubt, that starch paste mixed with iodide of potassium (totally free of iodate of potassium), becomes soon blue in external air". The fact that his test-paper (later termed *Schönbein* paper and simply ozonometer)[979] and "*unaufhörlich stattfindenden electrischen Entladungen in der Atmosphäre*" [of ceaseless occurring electric discharges in the atmosphere] led him to the following conclusion (Schönbein 1840, p. 162):

> [. . .] machen es daher wenn nicht gewiß, doch höchst wahrscheinlich, daß das fragliche in der Atmosphäre vorhandene oxydierende Princip nichts anderes ist, als eben das auf electrischen Wege entstandene Ozon [. . . hence does make it not sure, but very likely, that the questionable in the atmosphere existing oxidizing principle represents nothing else than the ozone gaining electrically].

979 The reagent was later called "*starch*" [starch iodide pastes]. See also footnote 990 on p. 500 and 1006 on p. 512 concerns the chemical mechanism.

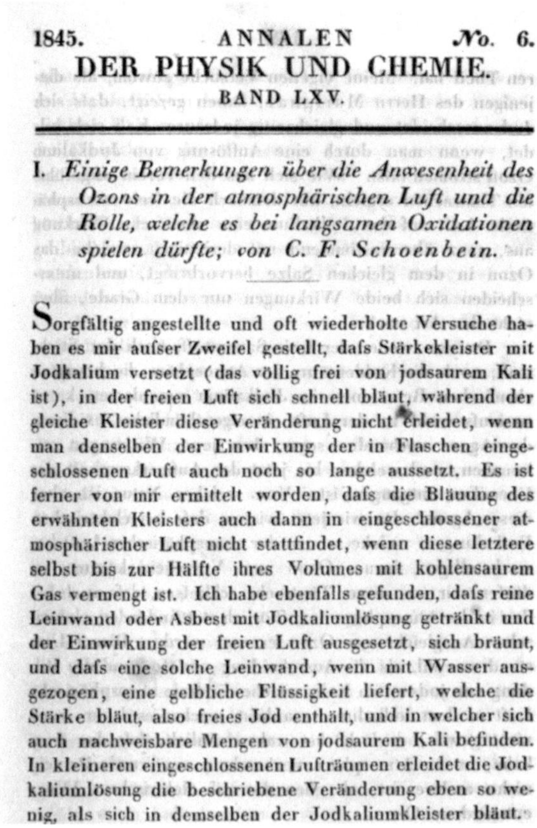

1845. ANNALEN №. 6.
DER PHYSIK UND CHEMIE.
BAND LXV.

I. *Einige Bemerkungen über die Anwesenheit des Ozons in der atmosphärischen Luft und die Rolle, welche es bei langsamen Oxidationen spielen dürfte; von C. F. Schoenbein.*

Sorgfältig angestellte und oft wiederholte Versuche haben es mir aufser Zweifel gestellt, dafs Stärkekleister mit Jodkalium versetzt (das völlig frei von jodsaurem Kali ist), in der freien Luft sich schnell bläut, während der gleiche Kleister diese Veränderung nicht erleidet, wenn man denselben der Einwirkung der in Flaschen eingeschlossenen Luft auch noch so lange aussetzt. Es ist ferner von mir ermittelt worden, dafs die Bläuung des erwähnten Kleisters auch dann in eingeschlossener atmosphärischer Luft nicht stattfindet, wenn diese letztere selbst bis zur Hälfte ihres Volumes mit kohlensaurem Gas vermengt ist. Ich habe ebenfalls gefunden, dafs reine Leinwand oder Asbest mit Jodkaliumlösung getränkt und der Einwirkung der freien Luft ausgesetzt, sich bräunt, und dafs eine solche Leinwand, wenn mit Wasser ausgezogen, eine gelbliche Flüssigkeit liefert, welche die Stärke bläut, also freies Jod enthält, und in welcher sich auch nachweisbare Mengen von jodsaurem Kali befinden. In kleineren eingeschlossenen Lufträumen erleidet die Jodkaliumlösung die beschriebene Veränderung eben so wenig, als sich in demselben der Jodkaliumkleister bläut.

Figure 4.29: First page of the first communication of *Christian Friedrich Schönbein* on the presence of ozone in atmospheric air (Schönbein 1840, p. 616): *Some remarks on the presence of ozone in atmospheric air and the role that it should plays during slow oxidations.*

On a read to the Society in Basel on 10 July 1844, *Schönbein* showed that water, shakes some time with ozone, immediately blues his test-paper. He also presented water, falling while a strong thunderstorm, which also showed all reactions of that treated with ozone. While shaking both waters with easily oxidable substances, both waters lost their property to impact the starch-iodide paper (Schönbein 1847b). From a modern view, this is the first evidence of hydrogen peroxide in rainwater. His next "observation", read on March 17, 1847, was the beginning of the numerous following studies with the erroneous conclusion that snowfall, due to the atmospheric electricity, is in relation to large amounts of ozone in the air, and that ozone is a disease-causing agent (Schönbein 1849a):

> [...] theilt seine im Laufe des Winters gemachten Beobachtungen über den wechselnden Gehalt der atm. Luft an Ozon mit und findet wiederholt bestätigt, dass bei Schneefällen der Jodkaliumkleister sehr rasch und stark gebläut werde und somit unter diesen Umständen auch die Luft reichlich Ozon enthalte. Der Referent leitet diese Thatsache aus dem Umstande ab, dass bei Schneefall bedeutende electrische Ausgleichungen in der Atmosphäre stattfinden. Das gleichzeitige Auftreten catarralischer Zustände mit einem verhältnismässig starken Ozongehalt der

> Luft wird zu Gunsten der Vermuthung hervorgehoben, dass Ozon eine krankmachende Materie sei und namentlich catarrhische Uebel verursache [... communicates his observations made during the winter on the changing amount of atmospheric ozone, and find it repeatedly confirmed that while snow fall the potassium iodide starch is soon and strong blued, thus under these circumstances the air contains abundant ozone. The speaker derives this fact from the view that during snow fall important electric equalization occurs in the atmosphere. The simultaneous occurrence of catarrhal conditions with a relatively large amount of ozone in the air led him to the assumption to regard ozone as a disease-causing agent, namely causing catarrh].

The nineteenth century, in the background of developing different scientific disciplines, techniques, and methods for investigations, and the widespread enthusiasm in educated circles, for the following about 40 years after *Schönbein*'s discovery of ozone in 1841 a period of ozone hype lasted. The early idea that ozone could play a fundamental role in the economy of nature led first physicians to believe that atmospheric (and soon later artificial produced) ozone is an important factor for human health and diseases. This was supported by the already long-existing belief that miasma, atmospheric disease-causing matter, will be decomposed through ozone which was early recognized as having oxidizing properties. Schönbein (1851a, p. 391) understood behind miasmas gaseous or vaporous substances in atmospheric air (in modern terms *pollutants*) which create physiological impacts. He writes (Schönbein 1851a):

> Es ist wohl bekannt, dass die Chemie eine Anzahl gasförmiger Substanzen unorganischer Art zu erzeugen vermag, welche in beinahe unwiegbarer Menge der atmosphärischen Luft beigegeben, diese wahrhaft vergiften und zur Unterhaltung des thierischen Lebens völlig untauglich machen [It is well known that chemistry might produce many gaseous inorganic substances, which in nearly imponderable amounts added to the air, it truly contaminates and make it disable for supporting animal life].

He termed to those miasmas or miasmatic substances (we would name it pollutants) compounds of hydrogen with sulfur, phosphorus, arsenic, and antimony, but also ozone, chlorine, bromine and iodine, and the large number of organic materials, which namely is released through rotting of organic matter. *Schönbein* further writes (Schönbein (1851a, p. 393)) that nature itself has taken precaution through the permanent formation of ozone which decays these miasmas.[980] Hence, it is not new when Scoutetten (1856, p. 188) writes:

> Les miasmes oxydables sont détruits par l'ozone, mais celui-ci à son tour est détruit par les miasmes.

Therefore, it is not surprising that practical physicians, which had at that time a much better knowledge in chemistry than at present, began with the first observations of

980 Schönbein (1851a, pp. 393–398) reports on own experiments with rotting meat under the influence of artificial produced ozone that show disappearance of smell as long as ozone is presence.

ozone in the air, followed by meteorologists at many stations in the emerging meteorological networks in the mid-nineteenth century. Soon curious hypotheses arose on the natural sources of ozone, its benefit and disadvantage for human health, and its use for disinfection.

Based on *Schönbein*'s first ozone observations, the German anatomist *Johann Alexander Ecker* (1816–1887) in Basel stated that ozone is likely the cause of diseases (Ecker 1847), and this paper stimulated many physicians to observe ozone in relations to epidemics. The virtual (positive) importance of atmospheric ozone for human health is discussed to these days; the German physician *Franz Xaver Herman Horn* (1815–?) in Munich is likely the first who makes propaganda for ozone as therapy (Horn 1856, 1857), making a relationship between odors (pleasant and noxious) and air electricity, defining that positive electricity (ozone) smells noxious and free negative one (jodosmone) pleasant. What "*Jodosmon*" (German written) stands for, remains unclear; *Horn* denotes it as electric nitrogen[981] and created the German verb "*jodosmonisirt*" ["jodosmonized" in contrast to ozonized].

The toxicity of ozone concerning animals was first studied by *Valentine Schwarzenbach* (1830–1890)[982] who exposed rabbits and other animals to (obviously strong) ozonized air who torturous died within a few hours (Schwarzenbach 1850); assuming that the atmospheric air contains "only" 10^{-5} parts of ozone (this is 100 times more than under smog events), sensible persons could get the respiratory problem, he concluded.

The German medical doctor *Constantin Lender* (1828–1888)[983] applied in 1870 ozone (Lender 1870) and recommended "*Ozonwasser*" [ozone water] as therapy (Lender 1872b,c). However, already Horn (1857, p. 10) produced ozone water and recommend it for therapeutic use.[984] *Wilhelm Waldmann* (1845–1905) fighted against *Lender*'s theory, but proposed the oxygen therapy and states that ozone water contains, citing the chemist *Böttcher* (it is *Böttger*), nitrous acid and/or according to *Cremer* (it is *Kremer*), hydrogen peroxide (Waldmann 1872a).[985] Waldmann (1872b) refers *William Thierry Preyer* (1841–1897), an English-born physiologist who worked

---

981 Horn (1856, p. 26) writes that "*die Luftströmungen von Osten positive, und die von Westen negatice Electricität freimachen*" [air currents from east gain positive, and those from west negative electricity]. Thus, ozone and jodosmone was seen as different states of atmospheric electricity (electrotherapy).
982 Medical doctor who also studied chemistry; 1872 professor for chemistry at University Bern. These investigations he carried out still as student. With the middle of the 19th century, the number of studies concerns impacts of ozone on animals, plants, and materials increases (not subject of this treatise).
983 *Carl Friedrich Constantin Lender* was a German physician in Berlin since 1866 who worked every summer since 1867 in Kissingen and is known for his ozone therapy; member of Leopoldina 1886.
984 Horn (1857, pp. 15–16) explains the positive impact of ozone on human health first chemical due to its oxidation capacity onto the blood and secondly "dynamic-electric" onto the neural system.
985 It was understandable that *Lender* as "inventor" and profiteer of the ozone water complained against these findings based on the review given by *Carius*, referring and using obscure arguments (Lender 1872b).

in Jena, Germany, and who investigated *Lender*'s ozone water in 1870 with the result that no hydrogen peroxide, nitrous or nitric acid was present, but ozone, which, however, could not be stable for a longer time. *Rudolf Christian Böttger* (1806–1881) was a German chemist in Frankfurt/Main who found no ozone but nitrous acid in the water and declared it as deception (Böttger 1871). Dr. *Albert Kremer* (unknown living data) of Stettin (today Szczecin, Poland) detected even no traces of ozone in that water but traces of hydrogen peroxide (Kremer 1872). *Georg Ludwig Carius* (1829–1875), German chemist and director of the chemical institute in Marburg, also investigated ozone water (produced by the *Krebs, Kroll u. Comp.* in Berlin) and found its properties like a solution of ozone in water made in the laboratory. He prepared also fresh ozone water and found that the dissolved ozone is soon lost when storing in air, but he detected no hydrogen peroxide or nitrous acid, which would, if present, be oxidized by ozone to nitric acid (Carius 1872, 1873a).

The German chemist *Carl Friedrich Rammelsberg* (1795–1864) in Berlin, found that ozone water only shows a very weak "ozone reaction" and contains 16 ppm chloride (Rammelsberg 1873) and referring *Behrens* and *Jacobsen*,[986] who detected in the same ozone water hypochlorous acid (HOCl) and termed it a diluted solution of hypochlorous acid (Behrens 1873).[987] *Carius* afterward studied the absorption of ozone in water and found that water dissolves ozone without chemical change and that therefore rainwater surely absorbs atmospheric ozone (Carius 1873b, p. 809). Schöne (1873) studied the absorption of ozone produced from oxygen-free of nitrogen, and found a maximum of 0.0189 g ozone per liter of water; it decomposes in water under the formation of oxygen but not hydrogen peroxide. *Schöne* communicates that he did not investigate the commercial ozone water but agrees with *Waldmann, Behrens, Jacobsen, Böttger*, and *Rammelsberg* that it does not contain ozone due to its limited stability; furthermore, he rejected *Carius*' idea that rainwater takes up ozone. The first who studied the action of ozone on the water was Andrews (1855, p. 340), writing:

> It is not absorbed by water, but when sufficiently diluted with other gases, is destroyed by agitation with a large quantum of water; it is also, contrary to the common statements, destroyed by being agitated with lime-water and baryta-water.

Besides *Schönbein*'s, other "ozone meters" were developed by the German chemist *Gottfried Wilhelm Osann* (Osann 1853), the French physician *Adolphe Bérigny* (p. 552 ff), who also conducted measurement at different places in France and abroad

986 *Theodor Heinrich Behrens* (1842–1905) and *Oscar Georg Friedrich Jacobsen* (1840–1889) were German chemists in Kiel; *Jacobsen* is known for his sea water analysis during the Pommerania expedition 1871 and 1872 in the North Sea and is considered as the first German marine chemist.

987 Today we know that ozone decays in water, primarily into OH radicals and oxygen. In pure water, H_2O_2 would be gained. In potable and natural water, the decomposition depends from the traces; O_3 reacts with chloride gaining hypochlorite (OCl^-) and oxygen, with nitrite to nitrate etc.

(Bérigny 1857, 1858a), and the British physician *Thomas Barbour Moffat* (Moffat 1861), and modified by many others later. In an extended chapter "how is ozone observed in the atmosphere" Fox (1873, pp. 167–301) first lists the substances that were employed in the detection of ozone (iodide of potassium, red litmus, and iodide of potassium, pure silver, copper and acetic acid, iodide of potassium and starch, oxide of thallium, sulphate of manganese, sulphide of lead, resin of guaiacum, indigo, fungi), and describes different ozone-boxes (*Moffat*, *Clark*, *Lowe*, *Smyth*, *Houzeau*, *Festing*)[988] and the self-recording instrument by *Cann-Lippincott*. Poey (1865)[989] introduced an instrument, which he termed an *ozonograph* and *actiongraph*, intended to register half-hourly the atmospheric ozone and the action of light. It consists of drums to carry the paper, and clockwork to move them, with the arrangement to admit air and light.

The most commonly used method was based on iodide of potassium and starch, a rudimentary method based on paper strips impregnated with potassium iodide solution (KI) and starch ($((C_6H_{10}O_5)_n$-$(H_2O))$ in distilled water.[990] There are several versions of the method, which vary in the concentrations of starch, potassium iodide, and paper used. The precision of the measurements varies considerably as a result of the different concentrations used. Among the different types of paper available, that produced by *Jame de Sedan* is the most widely used and is recognized as that which allows the greatest precision (Bojkov 1986). Fox (1873, pp. 169–171) refers apart from *Schönbein*'s first paper, those of *Lowe*, *Negretti* and *Zambra*, *Moffat*, *Jame* de Sedan, *Day*, *Polli*, *Osann*.[991] *Schönbein*, *Huizinga*, and *Bérigny* have used paper dipped in a so-

988 Sir *James Clark* (1788–1870), British physician to Prince *Augustus Morton Festing* (1837–1887), deputy paymaster and fellow of the British Meteorological Society in 1868 (Festing 1869). All other persons will be name below in the text. On Clark's ozone cage see Negretti and Zambra (1864, p. 135).

989 *Andrés Poey y Aguirre* (1825–1919), founder (1858) and director of physical-meteorological Observatory in Havana, sun of the naturalist and zoologist *Felipe Poey* (1799–1891) in Havana, later lived in Paris; Poey (1865, 1867). Also written *Poëy*.

990 Starch contains different polysaccharides in its structure, is insoluble in cold water, but swells up in hot water to a thick and viscous substance, forming a colloidal mixture known as a starch paste. When adding acid to water containing starch, a hydrolysis process begins with a reduction of the molecular mass of the substance and the formation of "soluble" starch. The test-paper, exposed to atmospheric air, colorize in presence of ozone (and other oxidizing substances) to brown and blue colors due to formation of iodine; see footnote 1006 on p. 512 for description of the chemical mechanism.

991 *Henry Negretti* (1818–1879) and *Joseph Zambra* (1822–1897) formed a partnership in 1850, the "Negretti and Zambra company" that produced scientific instruments in London. *Jame* (de Sedan) is referred by several authors, first by Bérigny (1857) as "Mons Jame de Sedan" who tested it to be the "best" paper (*réactif de Jame de Sedan*). Borius (1879) refers "*papier ozonométrique de Jame* (*de Sedan*)". No author gives a reference or his full name. I found him in Thiriet (1886, p. 21): *Édouard Jame* (1815–1883) was a botanist, meteorologist and pharmacist. He first studied pharmacy at "*collège de Charleville*", where he established a pharmacy. In 1840 he moved to Sedan, and became a member of the Medical

lution of oxide of thallium, which becomes changed by ozone into brown peroxide;[992] however, *Fox*'s experiments with this substance were unsatisfactory (Fox 1873, p. 171). Many early observers (for example, *Reslhuber* and *Boehm*) found *Schönbein*'s paper unsound; nevertheless, first modest and later slashing criticisms did not prevent other ozone workers to continue his observations or establishing the observational sites.

At the end of his scientific career, Schönbein (1867a) argued that despite the uncertainty of his potassium iodide paper, there cannot be any doubt on the presence of ozone in atmospheric air; other substances also being sensitive to this paper like chlorine and bromine, he rejected to be free in air, and nitrous acid (nitric acid does not show a reaction) he assumed to be present in air only bonded (ammonium nitrite);[993] he regarded atmospheric ozone as the agent to clean the air [*das atmosphärisches Ozon schon in dieser Beziehung zur Reinhaltung der Luft diene*]; (Schönbein 1867a, p. 329).[994] The early finding of the oxidizing properties of ozone by *Schönbein* and following researchers, and the authority of Pettenkofer (1855) who concluded that developing miasma consumes atmospheric ozone, based on first observations by several physicians, that during epidemics the amount of ozone is low, stimulated around 1855 a large number of ozone observations (Table 4.16).

From the many observations made with the original *Schönbein* paper, only a few are referred here; the results are always given in numbers of the *Schönbein* color scale but from today's perspective without value. Moreover, many relationships between "ozone", meteorological elements, diurnal and seasonal variation, locations, etc. are contractionary and questionable. However, the authority of *Schönbein* was so great[995] that nobody doubts him, and constructed arguments – based on observations, which are not caused by ozone (for example, the influence of humidity on the test-paper) – that evaporation and condensation and all kinds of precipitation (namely snow, but not during fog) form atmospheric electricity and thus ozone. On the other hand, there were early results based on careful observations that could have caused to turn the ozone research in another, more scientific direction, avoiding the huge number of following observations.

Jury (until his death). In 1858 he became a member of the "*Société météorologique de France*". *Fox* likely means *Henry Day* (s. p. 572).

992 First shown by Schönbein (1864b, p. 263).

993 It is worth to note that *Schönbein* writes (Schönbein 1867a, p. 323) "that likely no chemist investigated rainwater, namely from thunderstorm more frequent than I have done it, but never I found such water, that only weekly bleuing this paste or redden litmus paper" (translated from German).

994 The German word "*Reinhaltung*" is untranslatable [keeping it clean] (also to French); "*Reinhaltung der Luft*" or "*Luftreinhaltung*" is the modern political term for "air pollution control". In my studies of the historic literature, it seems that *Schönbein* was the first to use this German word. In Grimm's dictionary (*Deutsches Wörterbuch*) this word is missing but "*Reinerhaltung*" [prevention in sense of preclusion] is mentioned.

995 This is historical identical with *Liebig*'s statement of fertilizing plants through atmospheric ammonia (see Chapter 2.3.2), resulting in many rainwater studies.

Table 4.16: Observations of ozone in atmospheric air by ozonometry in the nineteenth century. Note, authors, not listed in the annotation, are mentioned within the text.

year[a]	location	author	remarks
1840	Basel	Schönbein	first "evidence" of ozone in air
1841	Brussels	Quetelet	together with electrometer
1846	Roggendorf	Spengler	influence and ozone
1846–1849	Frankfurt/Main	Clemens	ozone near a chemical factory
1847–1848	Schorndorf, Germany	Faber	daily observations
1847–1851	Quedlinburg	Brenner	observations
1848	Oschersleben	Schlotfeldt	cross-sensitivity to coal burning
1848/1849	Heilbronn et al.[b]	Betz et al.[b]	measurement campaign ("tests")
1848–1861	Harwarden, U. K.	Moffat	with meteorology and diseases
1849	Lexington	Peter	no relation to cholera
1850	Stuttgart	Plieninger	daily three observations
1850–1851	Mühlhausen	Gräger	higher at night
1850–1862	St. Martin, Melbourne	Smallwood	observation with meteorology
1850–1883	Urbino	Serpieri	monitoring
1852 ff	Toronto	Kingston	observations
1852–1853	Frankfurt/Main	Mettenheimer	daily three observations
1852–1853	Königsberg	Schiefferdecker	systematic study at 12 sites
1852–1853	Lyon	Bineau	low in town
1852–1854	Erfurt	Heydloff	daily three observations
1852–1859	Danzig	Schaper	monitoring
1853	Berlin	Schultz	observation during cholera
1853	Würzburg	Osann	each snowflake shows reaction
1853	Kiel	von Maack	analysis of error sources
1853	Edinburgh	unknown	reported by Wells (1854)
1853–1854	London	Glaisher	17 stations during cholera
1853–1854	München	Seitz	no relation with illnesses
1853–1855	Bern	Wolf	first systematic study
1853–1855	Nancy	Simonin	daily two observations
1853/1856	Milan	Strambio	daily two observations
1853–1866	Kremsmünster	Reslhuber	daily two observations
1853–1873	Krakow	Karliński[c]	observation with meteorology
1854	Bridgewater	Haviland	at different distance increasing
1854	Thann, France	Conraux	observations with illness
1854–1856	Bedford, U.K.	Barker	Schönbein and Moffat test-paper
1854–1859	Erlangen	Küttlinger	observations
1854–1864	Strasbourg	Boeckel, Th.	observation day and night, daily
1854–1866	Prague	Boehm et al.[d]	daily two observations
1854–1874	Klagenfurt	Prettner	daily observations
1854–1920	Wien	Flechner, Haller et al.	Central Meteorological Institute
1855–1859	Algiers	Bertherand	medical meteor. observations
1855	Amiens	Decharmes	150 observations at different places
1855	Metz	Scoutetten	higher with rain
1855	Paris	Silbermann	no ozone while cholera
1855	Lemberg	Pless and Pierre	first quantitative determination
1855	Arnstadt (Thuringia)	Niebergall	systematic observations

Table 4.16: (continued).

year[a]	location	author	remarks
1855	Knypshausen	Toel	systematic observations
1855	Dresden	Flemming	systematic observation
1855	Cüstrin	Lindner	observation in September
1855	Alger	Mitchell	aspirator testing
1855	Texas	Pope[e]	observations
1855 ff	French "network"	Bérigny et al.	observations
1855–1856	Norfolk, Portsmouth	Webster	with fever epidemic
1855–1856	Karlsruhe	Weber	systematic observations
1855–1856	Constantinople	Grellois	systematic observations
1855–1861	Exeter	Shapter	at three locations
1855–1862	Kirchdorf (Austria)	Schiedermayer	monitoring network
1855–1866	Neudorf (Strasbourg)	Robert	during cholera epidemic
1855–1866	Salzburg	Woldřich	monitoring network
1855–1868	Ofen (part of Budapest)	Frenreisz and Schenzl	monitoring network
1855–1873	Versailles	Bérigny	daily
1856	Ottweiler	Zimmermann	observations in January
1856	Clausthal	Schoof	observations
1856	Paris	Cloez	ozonometry without value
1856	near Paris	Houzeau	semi-quantitative attempt
1856	Crim	Bérigny et al.[f]	observations
1856–1857	Boston	Rogers	observation with meteorology
1856–1863	Lisbon [g]	unknown	daily night and day
1857–1863	Hobart Town	Abbott	observations
1856–1865	Oxford	Main	observations
1856–1870	Greenwich	Glaisher et al.[h]	monitoring at Royal Observatory
1857	Kuttenberg near Prague	Hartmann	observations
1857	Neusohl[k]	Zenger	quantitative determination
1857	London	Evans	observations
1857	Thionville	Grellois	observations
1857–1859	on sea	Wüllerstorf	observations
1857–1860	Melbourne	Smyth, R. B.	observations
1857–1873	Emden	Prestel	ozonometric wind rose
1857–1875	Scotland	Mitchell (et al.)	28 stations
1858–1865	Mannheim	Weber	systematic daily observations
1858–1866	Berlin	Lichtenstein	monitoring
1859	Dresden	Sachse	observations
1859	Kussouli, India	Ireland	in different altitudes
1859–1867	Koblenz	Schaper	monitoring
1860	Berlin	Heldt	testing
1860	Calais, lighthouse	Bérigny	observations
1860	Hobart, Australia	Hall	observations
1860ff	Salford	Mackereth	observations
1860–1861	Rio de Janeiro	Tschudi and Chomet	observations
1860–1863	Sydney	unknown[m]	observations
1861	Pisa	Silvestri	observations
1861–1863	Napoli	Palmieri	observations

Table 4.16: (continued).

year[a]	location	author	remarks
1861–1870	Rouen	Houzeau	systematic observation
1862	Greenwich	Glaisher	balloon ascent, testing
1862	Plauen (Germany)	Pfaff	observations
1862	Bonn, Wiesbaden	Ireland	observations
1862/1883	Havana	Poey	observations
1862–1864	Highland, Illinois	Bandelier	systematic observations
1862–1868	Brisbane	MacDonnell, Bliss	observations
1863	Hannover	Begemann	observations
1863	Eaux-Bonnes (Pyrenees)	Pietra Santra	observations
1863ff	Edinburgh	Mitchell	use of aspirator
1863–1865	Melbourne	Ellery	observations
1863–1867	Bombay	Cook	observations
1863–1868	Kremsier (now CZ)	Gstein	monitoring network
1864	Iceland	Jacolot	ship expedition, daily observation
1864 (?)	Washington	Wetherill	observations (no ozone at night)
1864–1865	Torquay and Oxford	Daubeny	plants as source of ozone
1864–1865	Zwickau	Meier	day value larger than nighttime
1864–1865	Leipzig et al.[n]	Bruhns et al.[n]	observations
1864–1866	Geelong, Australia	Day	observations
1865	St. Louis, USA	Wislizenus	not in the city but in the country
1865	Clausthal (Harz Mt.)	Schoot	daily observation
1865	Havana	Bérigny	observations
1865/1868	Banbridge	Smyth, J.	test of new ozonometer
1866	Phares du Touquet	Sainte-Claire Deville	with meteor. observations
1866	Groningen	Huizinga	testing
1866–1867	Kassel and suburb	Möhl / Dietrich	values in town much lower
1866–1867	Herrieden	Heidenschreider	observation
1866–1867	Paris	[o]	network of 18 stations
1866–1869	Colmar	Kampmann et al.	daily observations at 4 sites
1867	Paris and suburb	Pieraggi	two sites compared
1867	Moncalieri	Denza	August-September, electricity
1867	Modena	Ragona	influence of temperature
1867	Mexico	Bérigny	observations
1867/68	Old Trafford	Vernon	ozone paper comparison
1867–1877	world-wide	Mulvany	as surgeon afloat
1868	Bavaria	Ebermayer	6 forest stations
1868/1869	Paris and suburb	Lamy	thallium paper tests
1868–1870	Gotha (Saxony)	Luedicke	daily observation day and night
1868–1875	Manchester	Baxendell	occasional studies
1869	Over (near Bristol)	Lippincott	diurnal variation
1870	Petersburg	Dohrandt	testing
1870 (?)	Scarborough	Fox	observations
1871	Kissingen	Lender	in thorn house much ozone
1871	Driburg	Hüller	July – October observation
1871/1872	Arctic region	Meyer	Polaris expedition
1871/1872	Marienbad et al.[p]	Lender et al.[p]	observation

Table 4.16: (continued).

year[a]	location	author	remarks
1871–1903	Michigan	Kedzie et al.	network
1872–1873	Sassnitz (Baltic Sea)	Lender	observation
1872/1873	Fuscher Bad	Haller	on mountains is more ozone
? –1873	Eccles, Britain	Mackereth et al.[q]	several years observation at 3 sites
1873	Hildburghausen	Lösecke	one year observation
1873	Langdale, Ceylon	Heelis	observation
1873	Helgoland	Zimmermann	April/May, 28 days observations
1873–1874	Saint-Louis (Senegal)	Borius	with meteor. observations
1873–1877	St. Honorine-du-Fay[r]	unknown	with meteor. observations
1874	Libyan Desert	Zittel	at different places
1874 (?)	Mexico	Rocha	testing ozone recorder
1874	Munich	Wolffhügel	testing
1875	Saint-Louis (Senegal)	Louvet	with meteor. observations
1875–1904	Krasno	Friesenhof	agrometeorological observations
1876	Libyan Desert	Ascherson	at different places
1876	forest near Paris	Kunze	in, out and above the forest
1876–1877	Colorado	Ambrook et al.[s]	three different locations
1876–1877	Recife[t]	unknown	with meteor. observations
1877–1878	Otisville (USA)	Nicholson	observations
1877–1907	Paris	Lévy	Montsouris series; quantitative
1878	Menton (France)	Daremberg	at sea in distance from land
1879	Moscow	Schöne	comparison KI and TlOH papers
1881	New York	Gratacap	inverse correlation with wind
1883	at sea	Wragge	low values on sea to Australia
1885	Izera Mt. (now Poland)	Rudeck	testing new ozonometer
1885	at sea around UK	Black	observations
1886	St. Peter, North Sea	Schelenz	more ozone from sea
1886–1892	Belgium	Bastelaer	network
1888	Blue Hill Observatory	Rotch and Pickering	with meteor. observations
1894/1895	Mont Blanc	Thierry	observations

[a] Year(s) of observation or – when this information is not given – year of publication

[b] In Tübingen *Schlossberger*, *Griesinger*, *Fallati*, and *Baur, in* Stuttgart *Renz*, in Mergentheim *Krauß* and *Wrede*, in Calw *Müller*, and *Höring in Heilbronn*.

[c] With assistants *Jan Kowalczyk* and *Daniel Wierzbicki*

[d] *Jenner*, *Andres*, and *Lippich*

[e] *John Pope* (1822–1892) was a U. S. military officer; appointed in 1851 as the Chief Topographical Engineer of the Department of New Mexico, who made several expeditions (he reported to the US Congress and was in contact with the Smithsonian Institute). Gaillard (1864, pp. 234–236) reports on his Tables giving "the registers of the barometer, thermometer, hygrometer, and ozonometer together". Observations were made at Indianola and the Gulf of Mexico. *Pope* writes (Gaillard 1864, p. 236) "ozone increases in quantity, rapidly and regularly, in receding from the low lands which border the Gulf of Mexico, and is greatest on the summit of the table lands, in the interior. On the low lands, animal and vegetation decomposition is rapid; on the table lands, extremely slow, and with little escape of offensive gases." *Gaillard* correctly concluded from *Pope*'s observations that there is no relation between malaria and ozone, and that obviously more moisture not led to more ozone, writing "... for it is claimed by most observers that moisture e is eminently to the production of ozone."

Table 4.16: (continued).

[f] *Scrive*, *Leroy*, and *Méry*
[g] Probably *João Carlos de Brito Capelo* (1831–1901), Portuguese meteorologist at "*Observatório Astronómico de Lisboa*", reported in: Annales Do Observatorio Do Infante D. Luiz em Lisboa, 1856 a 1865, Vol. 1, 131 pp. on p. LXIX, and also in Vol. 2 (1870) on p. 144, for the year 1870.
[h] *Nash*, *Lippincott*
[k] In former Austria (Hungary), today Banská Bystrica (Slovakia).
[m] Unknown, cited after Baxendell (1869)
[n] Gotha, Gross-Breitenbach, Greiz, Werningshausen and Königsstein by the observers *Luedicke*, *Bertram*, *Oettler*, and *Beck*.
[o] Carried out by municipal body of Paris (*Bulletin hebdomadaire de statistique municipale de la Ville de Paris*, 1868); cited after Fox (1873, p. 106).
[p] Carlsbad, Mentone, Meran and Wiesbaden with the observers *Wernigh*, *Barsdorf*, *Friedrich*, *Bernsdorf*, *Stiege*, *Schulze*, and others
[q] *Samuel Marshall* in Kendall, *Heap* in Royton near Oldham, and *Thompson*, in Wilmslow
[r] City in Normandy; *Annuaire de la Société Météorologique de France* **22** (1874) 22, **23** (1875) 45–54, **24** (1867) 70-94, **27** (1879) 54–80.
[s] Sergeant *James A. Barnich* and Sergeant *M. Hobbs*
[t] Capital of Pernambuco, Brasilia; *Annuaire de la Société Météorologique de France* **26** (1878) 58, 72–102.

Likely the earliest observation of ozone was made by *Adolphe Jacques Quetelet* (1796–1874), a Belgian astronomer, at the Royal Observatory in Brussels on a day in August 1841. *Quetelet* was interested in the comparison between atmospheric ozone and electricity, using *Peltier*'s electrometer and ozone by test-paper on an hourly basis in August 1842 at the Royal Observatory in Brussels but published 20 years later (Quetelet 1864). A maximum of electricity he found to be at 6–9 AM and 8–10 PM and a minimum at 2–4 PM, corresponding to humidity; his diurnal development of ozone shows the same variation as observed by *Lippincott* in 1857 at Greenwich. *Francesco Denza* (1834–1894), an Italian Padre and professor of meteorology, found at the meteorological station Moncalieri that there are no relations between ozone and atmospheric electricity (Denza 1868). *Smallwood* (see p. 562) also did not find a pronounced relationship between electricity and ozone. A few observers found that a higher altitude (even a few dozen meters above the ground, almost on cathedrals) the amount of ozone in the air is higher; Bérigny at Versailles (1856), Decharmes (1856) in Amiens, Scoutetten (1856) in Metz, Haller (1874) in Vienna. *Böckel*, but also *Decharmes* state that a sudden decrease in temperature favors the formation of ozone in the air.[996]

996 These observations are correct and caused through removal of ozone at the ground, and transport of (stratospheric) ozone down from upper air layers, and of different amount of ozone in different air masses (cold air from upper layers; today often warm air is connected with higher ozone levels due to tropospheric photochemical formation).

Another belief, hold for more than 100 years, on "ozone rich forests" was already proved to be false by Ebermayer (1873b) who found ozone above the canopy and close to the forest but not inside the forest. A similar observation was also made by Friesenhof (1904). Table 4.16 lists (I hope most important) scientists who conducted ozonometric observations in the external atmosphere (many more than quoted by Fox 1873 and Leeds 1879b, 1883); Table 4.17 lists first ozone investigation by countries.

Cark Flügge, who included in his textbook on hygienic investigations methods (Flügge 1881) a chapter on the determination of ozone, writes (Flügge 1881, p. 107) that "at some meteorological stations ozone was observed since more than 20 years without finding something else than [...]", and calling the main feature (which already found the first observers), and citing *Mackereth, Boehm, Prestel, Karliński, Zittel*, and *Lender*.[997]

At the hygienic institute in Munich each year (since 1876) a practical course was offered by *Max Pettenkofer*, that also included "*Ozon: Methode der Ozonoskopie*" (quoted after *Breslauer Aerztliche Zeitschrift* **2** (1880) 202); the German term "*Ozonoskopie*" [ozonoscopy] was used only from very few scientists at that time. The "science of ozone measurement" was named ozonometry (*Ozonometrie* in German, *l'ozonométrie* and – very seldom – *d'ozonologie* in French), and the "instrument for ozone detection" was named ozonometer and – very seldom – ozonograph, or ozone meter (*Ozonometer* in German).

It is beyond the scope of this volume to refer all investigations that were conducted on ozone in the nineteenth century, namely, a presentation on the different ozone papers, ozone meters, and ozonoscopes, etc. I refer the interested reader to Fox (1873) and Engler (1879).

4.4.5.2 On the criticisms of the ozonometry

The test-papers were unable to differentiate between oxidants such as ozone, hydrogen peroxide, and HNO_2 (the latter reacts only in acid medium according to Abeledo and Kolthoff 1932).[998] From the early ozone observations (note that the term "amount of ozone" corresponds to the "ozone reaction", i. e., the scale value of the test-paper), and the majority of observers, the following relationships were derived:

---

997 *Lender*, known for curious theories on ozone, lamented in 1873 "enough ozone observations are still missing" (Lender 1873); he summarizes the "observations": annual periodicity with ozone maximum at equinoxes and minimum at solstices, thus causal relationship with air currents; the "ozone reaction" is highest at night in winter and at day in summer, and proportional to the wind speed; largest amounts of ozone are transported by sea winds.

998 The elementary steps are: $HNO_2 + H^+ \rightleftarrows H_2NO_2^+ \xrightarrow[(-H_2O)]{} NO^+ \xrightarrow[(-NO)]{I^-} I$. A negative signal comes from iodine reducing compounds like SO_2 ($HSO_3^- + I \rightarrow HSO_3$ (in multisteps to sulfate) $+ I^-$); however, SO_2 was measurable only locally near emissions source (ore mining and coal combustion) in the nineteenth century.

Table 4.17: First ozonometric observations by country.

Year	country	observer
1840	Switzerland	Schönbein
1846	Germany	Spengler
1848	England (United Kingdom)	Moffat
1849	USA	Peter
1850	Canada	Smallwood
1853	France	Bineau
1853	Poland (former Austria)	Karliński
1853	Austria	Reslhuber
1853	Italy	Strambio
1854	Czechia (former Austria)	Boehm
1855	Ukraine (former Austria)	Pierre
1855	Algeria	Bertherand
1855	Turkey	Grellois
1856	Portugal	unknown
1857	Australia	Abbott
1858	Russia	Bérigny
1859	India	Ireland
1860	Brasilia	Tschudi
1862	Cuba	Poey
1864	Iceland	Jacolot
1867	Mexico	Bérigny
1871/72	Arctic region	Meyer
1873	Ceylon	Hellis
1874	Libya	Zittel
1874	Senegal	Louvet
1875	Mexico	Rocha
1886	Holland	Huizinga
1888	Belgium	Bastelaer

- increase of the amount of ozone with humidity,
- maximum amount of ozone on days with precipitation,
- minimum amount of ozone during hazy and foggy days,
- maximum amount of ozone in winter and minimum in summer,
- less ozone in large towns than in small towns and maximum on rural sites.

Two other phenomena, observed while exposing the test-papers, the dependency of the "ozone reaction" from wind forces and decolorization under the direct influence of sunlight, were explained by the increase of the amount of ozone transported to the passive sampler with increasing wind force, and due to evaporation of iodine, respectively (hence, the use of ozone-boxes was strongly recommended, but not used by all observers).

The period 1850–1880 of uncounted ozone measurements using the *Schönbein* paper has often been criticized and it is likely that apart from ozone, also nitrous acid[999] and mostly hydrogen peroxide were also responsible for the blue coloring of the paper. The objection to *Schönbein*'s ozonometer (starch-iodide paper) and *Houzeau*'s ozonometer (potassium iodide on red litmus paper) lies in the fact that their materials are hygroscopic, and their indications vary widely with the moisture of the air; the intensive "ozone reaction" near places with intensive water evaporation, e. g., waterfalls were wrongly interpreted as an ozone source. Andrews (1867) writes

> It was assumed for many years, chiefly on the authority of Schönbein, that the body in the atmosphere which colours iodide-of-potassium paper is identical with ozone; but this identity has of late been called in question.

Andrews was aware that chlorine and nitric acid also color the test-paper, however, he conducted laboratory experiments and got the evidence that ozone was indeed present in atmospheric air. Atmospheric air, giving a distinct colorization of the test-paper, was passed through a tube containing pellets of manganese dioxide (it was known that ozone decomposes), and not the slightest coloration occurred. In the next experiment, he heated atmospheric air to 260 °C in a glass vessel and passed it afterwards through a tube cooling down, and no reaction occurred in the test-paper. The same effect was found with an artificial atmosphere of ozone. On the other hand, when small quantities of chlorine or nitric acid, largely diluted with air were drawn through the apparatus, the test-paper was equally affected, whether the glass globe was heated or not. From these experiments, *Andrews* concluded that "the body in the atmosphere, which decomposes iodide of potassium, is identical with ozone".

Petrus Heinrich Karl von Maack (1806–1873),[1000] a German physician in Kiel, known for lectures and publications on the history, geography, and geology of the State Schleswig-Holstein, published a very remarkable article on numerous causes of errors by using *Schönbein*'s paper; from systematic investigations conducted in summer 1853 in Kiel, he concluded (Maack 1856, p. 29)

> [...] dass demselben alle wissenschaftliche Brauchbarkeit abgesprochen werden muss [that all scientific usefulness must be denied],

and, some phrases later, the test-papers are as slow-acting ozonometer completely useless. It is amazing to see from our today's critical view concerns scientific correctness and quality assurance, that *Maack* as a largely unknown medical doctor was able

---

999 Before 1850 this acid was wrong attributed to be NO_2 or N_2O_4; note that sulfurous acid was SO_2 and no difference with the acid anhydride was made (e. g., $SO_2 + H_2O$).

1000 Referred in "*Adressbuch der Stadt Kiel*" (1852) as P. C. H. van Maack, Dr. med et chirurg. His living data are referred in: Index-catalogue of the Library of the Surgeon-General's Office, United States Army. Authors and subject. Vol. VIII, Washington, Gov. printing, p. 467. His dissertation, written in Latin, 1834, 38 pp.

to study and analyze the limitations of this "ozonometer" – whereas *Schönbein* as a celebrated chemist was obviously unable to recognize or accept it. Many physicians and natural scientists conducted "ozone observations" for the following 2–3 decades, having no value; however, *Maack*'s article, printed in a curious local journal was disregarded or remained unknown (only referred by *Mettenheimer*, see p. 521), and thus not recognized, but would avoid so many meaningless later "studies" if it would become known, and, what is even more important, also accepted. What we have seen is that also later critical voices by well-known researchers did not prevent further "ozonometric studies" until the early twentieth century. It is worth summarizing *Maack*'s observations made by *Schönbein*'s original paper (Schönbein 1850) produced by bookbinder *Bürgy* from Basel:

- he soon found the principal problems of this passive samples like protection against precipitation due to decoloring (washing out), dependency from the wind speed due to more "ozone reaction" at same ozone amount in air (i. e., *Maack* recognized that the color depends from the dose and not concentration);
- thus, he recommended (and constructed) a wooden box (long before other scientists published her ozone-box) to avoid such influences;
- furthermore, he criticizes the scale of colors not showing linearity to the amount of ozone, and the inhomogeneity of the paper, resulting in subjective scale estimations by different observers;
- he proposed a different preparation and storage of the paper and recommended never to touch it with the fingers, and,
- most important, he found a decoloring with the time of exposure and subsequent storage of the ozonized paper as well as with increasing temperature due to evaporation of iodine and,
- a further chemical reaction of the precipitated iodine by ozone, gaining "*Jodozone*" (likely iodate) which decays into iodic acid and iodine, thus at all
- the "ozone values" depend on temperature and humidity, despite other reacting substances present in the air also resulting in the reaction.

Summarizing he writes, when *Schönbein* found in summer less ozone than in winter, this can perhaps be true, but the difference of the seasonal values depends on the location and time.

Wilhelm Heldt (1823–1865),[1001] a German chemist in Berlin, conducted various tests and interesting experiments with the starch-iodide paper; he found that carbon dioxide, hydrogen sulfide, and methane do not affect it, but sulfur dioxide[1002] does little, and nitrous acid does quickly (Heldt 1861). Remarkable is his observation that

1001 A student of *Liebig* in Gießen (1842), doctorate in Berlin 1846 (worked on citric acid and santonin); later papers dealt with bleaching, cement, metallurgy, etc.

1002 He made an important note concerning atmospheric oxidation of sulfuric acid, i. e., SO_2 (Heldt 1861, p. 23, translated from German): In towns, where much of sulphureous soft and hard coal, cokes the paper closed in a glass tube for 14 days and exposed to sunlight colorizes, but not the paper held in the dark. According to today's knowledge, it was photosensitization from starch, oxygen, and moisture. Decolorization, i. e., reduction of precipitated iodine, *Heldt* found due to several substances (he named them iodine-separating substances) from wood smoke, ammonia, hydrogen sulfide, sulfuric acid. Finally, he concluded (Heldt 1861, p. 25):

> Aus allen angeführten Tatsachen wird man so viel entnehmen, daß die Beobachtungen des Reagenspapiers in freier Luft in Summa etwas ganz Anderes ausdrücken, als was man bisher in dieselben hineinlegen wollen, und daß die Schlüsse, zu welchen die Reactionen des Jodkaliumpapiers Veranlassung gaben, völlig irrthümlich sind [From all these facts one can learn that the observations by the test-paper in free air express at all something completely different than that what has been interpreted, and that the conclusions drawn from the reactions of the starch-iodide paper, are completely in error].

The French chemist *Edmond Frémy* (1814–1894)[1003] writes (Frémy 1865, p. 940):

> Without denying the importance of results, provided by the papers of Schönbein and Houzeau, I don't think that these agents demonstrate with sufficient certainty the presence of atmospheric ozone.

Additionally, he made in a lecture the following remarks (Frémy 1866, cited after Fox 1873, p. 42):

> When it is known that the air contains Oxygen, and that atmosphere is constantly traversed by electric discharges, it is difficult not to admit the formation of Ozone in the air. But can this body remain in it? Has anyone as yet clearly proved its presence in the air? The electric discharges, which transform Oxygen into Ozone, ought necessarily to effect also the combination of Oxygen with Nitrogen and produce Nitric Acid, which, in truth, is invariably found in thunderstorms.

Similarly communicates *Giovanni Cantoni* (1818–1897), an Italian physicist and director of the Italian meteorological service from 1865–1878, and further noted that all observations must be done under the same conditions for them to be of scientific value (Cantoni 1866). Payen (1866) found that even atmospheric carbonic acid freed iodine, but this was shown to be wrong by *Schöne* later. *Dirk Huizinga* (1840–1903)[1004] tested all known reagents on ozone until that time (1867) and concluded that none is suitable for identification of ozone (Huizinga 1867). He found that potassium iodide is also sen-

---

etc. is burnt, one could assume that the released sulfurous acid would results in iodine-separating. However, this is not the case because it oxidizes fast to sulfuric acid, which returns back to the earth's surface through dew, rain or snow. One cannot be surprised if rainwater and snowflakes, falling onto white starch-iodide paper, leave brown spots, as all acids etc., escaped into air, constantly return to the earth through these intermediaries.

1003 In 1831, he entered the laboratory of *Gay-Lussac* and succeeded him in the chemistry chair at the Museum of Natural History, Paris, in 1850, of which he became the director (1879–1891).

1004 Dutch professor for physiology at Groningen; 1867 dissertation entitles "*Enige opmerkingen over ozone*" [some remarks about ozone]; father of the famous historian *Johan Huizinga*.

sitive against hydrogen peroxide but not thallium paper such as Schöne (1880c) did, and further that the *Schönbein* paper did not show any difference in color between day and night. Concerns the thallium paper, *Huizinga* (Huizinga 1867, p. 202) found that nitrous acid decolorizes the brown precipitate, hence *Huizinga* concluded from the weaker color at night, a) decrease of ozone, b) increase of nitrous acid, or c) both causes together on the other hand. It is clear that *Huizinga* was wrong in his view that hydrogen peroxide has no effect on the thallium paper. On the other hand, we know today that ozone decreases at night and nitrous acid increase, but historically *Huizinga* was right by wrong observations. The problems of detection of ozone in the air, Brown (1869) describes clearly:

> [. . .] (a) by oxidation of iodide of potassium, (b) by oxidation of oxide of thallium, (c) by oxidation of sulphate of manganese, which is of brown colour. (c) is very characteristic, converting sulphate of manganese into peroxide of manganese [MnO_2]; no other substance which can be supposed to exist in the air is known to have this action.

Brown then explains that the brown color is hard to differentiate concerns O_3 levels and that the thallium paper[1005] is sensitive against CO_2, gaining brown thallium carbonate; the first methods result in violet-blue colors due to the formation of free iodine and thus giving relative quantities of ozone in air, but are sensitive concerns humidity and all other oxidizing substances.[1006] Thus Fox (1873, p. 41) writes:

> Does the atmosphere contain ozone? Frankland would reply "Not proven", whilst Frémy asserts that the evidence in favour of its presence in the air is of most doubtful character.

The Russian-German chemist and meteorologist *Ferdinand (Bogdanovich) Dohrandt* [Фердинанд Богданович Дорандт] (1846–1878),[1007] mostly known for his meteoro-

1005 Thallium-I-oxide Tl_2O, in German "*Thalliumoxydul*" (historic term).

1006 The mechanism of the reactions in determination of ozone via reaction with potassium iodide was not known before 1952, and described in detail in the 1970s. Under neutral conditions, iodine is formed (which can evaporate under dry conditions), which can react with the base (OH^-) formed in this reaction gaining iodate ion. This reaction is reverse upon acidification. Iodate also can be formed in reaction between iodide and ozone. Furthermore, under acidic conditions the production of hydrogen peroxide has also been considered (Burns 1997, and citations therein). The fast initial reaction is $I^- + O_3 \rightarrow I + O_3^-$ (ozonide decays into hydroxyl ion and the radical), but also direct oxidation to iodate occurs very fast: $I^- + O_3 \xrightarrow[-O_2]{} IO^- \xrightarrow[-O_2]{O_3} IO_3^-$. The oxidation of iodide by hydrogen peroxide is also fast: $I^- + H_2O_2 \rightarrow I + OH + OH^-$. The first investigation of the reaction mechanism between iodide and hydrogen peroxide was made by the Austrian-American chemist *Hermann Alfred Liebhafsky* (1905–1982), supposing a series of observable and short-living intermediates (Liebhafsky 1932); see also for a complex of equilibria, oxidation and reduction reactions Schmitz (2001).

1007 Studied chemistry in Dorpat and Heidelberg, after his return to St. Petersburg he exclusively investigated meteorological problems; he became in 1878 the director of the physical observatory in Tiflis [Tbilisi]. Due to his early death in his age of only 32 years, the atmospheric science of the 19th century lost without any doubt a great researcher.

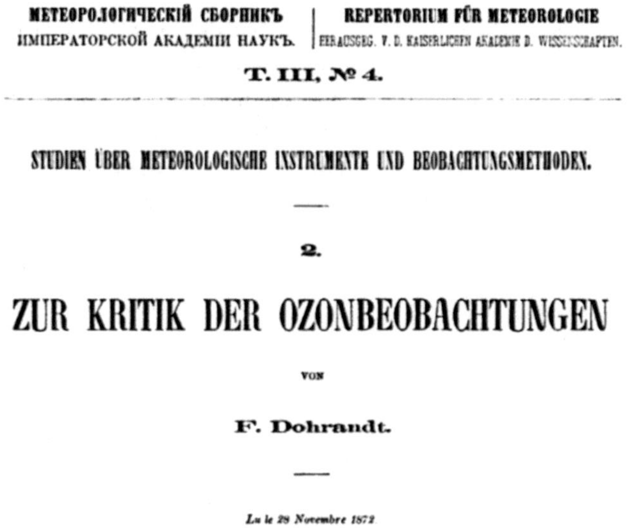

МЕТЕОРОЛОГИЧЕСКІЙ СБОРНИКЪ ИМПЕРАТОРСКОЙ АКАДЕМИИ НАУКЪ. | REPERTORIUM FÜR METEOROLOGIE HERAUSGEG. V. D. KAISERLICHEN AKADEMIE D. WISSENSCHAFTEN.

T. III, № 4.

STUDIEN ÜBER METEOROLOGISCHE INSTRUMENTE UND BEOBACHTUNGSMETHODEN.

2.

ZUR KRITIK DER OZONBEOBACHTUNGEN

VON

F. Dohrandt.

Lu le 28 Novembre 1872.

Figure 4.30: Front cover of *Dobrandt*'s "Critique of the ozone observations", submitted 1872 to "*Repertorium für Meteorologie*", St. Petersburg.

logical and magnetic measurements during the Amu-Darja expedition 1874–1875 in Russia, wrote, in my opinion, the best article on the problems of ozone observations at that time, entitled "*Zur Kritik der Ozonbeobachtungen*" (Dohrandt 1874), see Fig. 4.30.

Dohrandt (1874, p. 7) writes that physicians and meteorologists conducted in good confidence atmospheric ozone observation for many years, and although this confidence would be unsettled through some not irrelevant disadvantages, they continued his observations and derived not always very consistent regularities. In his article, *Dohrandt* not only criticizes the inadequacy of the observation method but also the locations of observations;[1008] he correctly states that (I refer here the German original text because of the importance of this statement concerns understanding the variation of surface-near ozone concentrations); Dohrandt 1874, pp. 11–12:

> Weil also bei den ozonometrischen Bestimmungen immer nur die Differenz des in der freien Atmosphäre enthaltenen Ozons und des durch Oxydation zerstörten Ozons erhalten wird und eben wegen dieses überwiegend localen Characters aller anzustellenden Messungen kann den bisherigen Ozonbeobachtungen keinerlei Bedeutung für die allgemeine Physik der freien Atmosphäre zugesprochen werden [Because ozonometric determinations always only measure the difference between the ozone which is present in the free atmosphere, and the ozone which is decomposed through oxidation, and because of this mainly local characteristics of this measurements, one cannot attribute any importance to the previous ozone observations to the general physics of the free atmosphere]

1008 He not only provides the exact bibliography of all referred authors, made own tests, and compared different results on their consistency, in other words, he applied scientific rules which were by far not self-evident at that time. His article is under: https://books.google.de/books?id=tVJYAAAAYAAJ&printsec=frontcover&hl=de&source=gbs_ge_summary_r&cad=0#v=onepage&q&f=false

Dohrandt argued similar to *Maahs* (without knowing his paper) concerns the periodicity due to the volatility of iodine (humidity and temperature variation), pointed out the difference between the "ozone reaction" and "the amount of ozone" (nonlinearity), and finally, communicate that ozone measurements because of meteorological parameter collections, can only be carried out according to the aspirator method by *Pless* and *Pierre*. From a large number of ozone observations, he finally summarizes the following notes with certainty:

- ozone is abundant in atmosphere,
- with increasing elevation ozone also increases,
- the hitherto used ozonometers do not measure the amount of ozone,
- observed variations of the amount of surface-near ozone in relation with meteorological variation do not allow conclusions on the cause of ozone formation,
- atmospheric ozone observations cannot be seen as compiling of meteorological data but can perhaps use for sanitary assessment, and this is the only reason to continue ozone observations.

Another Russian-German chemist, *Heinrich Struve*, who came from rainwater investigations to the conclusion that ozone and hydrogen peroxide always are together present in the atmosphere, notified (Struve 1869, p. 318) that "all ozonometric determinations can only be accepted as a result of a double reaction", and that science aims to find new methods, not only to separate both bodies but also the determine them quantitatively. On the other hand, *Struve* wrote that a future task of science is studying the occurrence of these atmospheric substances depending on other natural processes. However, this is not the task of chemists alone but chemists and meteorologists must go hand in hand. Unfortunately, *Struve* further wrote, that many meteorologists namely Germans not just do not deal with ozone in the atmosphere but even ignore it, and state that this issue should be in the hand of physicists.

Matthias Eugen Oskar Liebreich (1839–1908), physician, chemist, pharmacologist and toxicologist, in 1872 successor of the chemist *Eilhard Mitscherlich* (1794–1863) on the chair for pharmacology at the Berlin Charité, writes (Liebreich 1880, p. 318):

> Der erste Fehler beginnt damit, dass man überhaupt Ozonmessungen vornimmt und veröffentlicht, und dass Balneologen von einer ozonreichen Luft sprechen und einen besonderen Werth darauf legen. Es hat mich immer gewundert, dass sich nicht Jemand gefunden hat, welcher gegen solche Dinge ankämpfte, die das Publikum zum Unglauben und Aberglauben verführen müssen. [The first mistake starts with that one at all conduct measurements of ozone and published them, and those balneologists talk on air rich of ozone, and put a special emphasis on it. It always surprised me that nobody finds to fight against such things, which must mislead the public to disbelief and superstitiousness].

Liebreich not only vehement rejected ozone "measurements" by the test-paper but also its therapeutical use. In the same way argued *Joseph Maria Pernter* (1848–1908), Austrian physicist and weather observer, who refers *Schöne* that the presence of ozone is not proved but that of hydrogen peroxide (Pernter 1881). The finding by *Schöne* (and many other observers) that the "amount of ozone" (that is the blueing of the test-paper) correlates with the relative humidity (hence, he termed it "rough hygroscope"), *Pernter* proved by analyzing the ozone observations at Vienna of the last five years. He found no relation to the humidity but he did not conclude that *Schöne* and others were to be wrong, but to the contrary, that all previous ozone observations have no value (Pernter 1881, p. 396). Moreover, he found it incredible, namely after *Dohrandt*'s critics, that such observations were continued, as useless ballast entrained in meteorological journals and published.

Fox (1873, pp. 182–184) refers phrases on the "chaotic state of ozonometry" by *Richardson*, *Wetherill*, *Boehm*, *Lowe*, Dr. *Scoresby-Jackson*, *Davies*, Prof. *Parkes*, Admiral *Fitzroy*, and Dr. *Henry* of the Smithsonian Institution.[1009] Fox (1876, p. 272) writes:

> It was found that test thus exposed rather acted as anemometer than ozonometer, on account of the continued changes in the forces of the wind and the ever varying quantities of air to which they were in this way exposed. Of course the higher the wind the deeper the coloration, for the simple reason that more air passes over the test paper.

Hermann Emil Schöne (1838–1896),[1010] who proved that hydrogen peroxide is a "normal constituent of atmospheric air" (Schöne 1874, 1878), presented a criticism of "ev-

1009 *Robert Edmund Scoresby-Jackson* (1833–1867) was a short-lived but influential British physician and historian. Likely *Herbert Davies* (1818–1885), physician in London. *Edmund Alexander* Parkes (1819–1876) was an English physician. Vice-Admiral *Robert Fitz Roy* (1805–1865) was an English officer of the Royal Navy and a scientist; he achieved lasting fame as the captain of HMS *Beagle* during *Charles Darwin*'s famous voyage. *Joseph Henry* (1797–1878), US American physicist.

1010 *Hermann Emil Schöne* [he changed his name according to the Russian tradition in *Шене, Эмиль Богданович*] was a German chemist, born in Halberstadt, who studied in Halle, Berlin and Göttingen and moved to Moscow in 1863 on the recommendation of *Rammelsberg* and *Rose*. From 1864 he worked as assistant of the chemist *Pavel Antonovich Iljenkov* [Павел Антонович Ильенков] (1821–1877) in the chemical laboratory, organizing the establishment of the Petrowski Agricultural and Forest Academy [Петровская земледельческая и лесная академия], founded in 1865, where he became in 1875 a professor for inorganic and analytical chemistry. Petrowskoje Rasumowskoje [Петровское-Разумовское] was a village near Moscow (today Metro station Петровско-Разумовская in the northern part; *Kirill Grigorjewich Rasumowski* [Разумóвский] (1744–1803) was a count, the last hetman of the Saporosch's [Запорожцы] Cossacks, 1746–1776 president of the Russian Academy of Science in St. Petersburg, and co-founder of the Moscow University). *Schöne* deceased in Moscow. He was a member of the "*Deutsche Chemische Gesellschaft*" [German Chemical Society] and read many results of his studies on atmospheric ozone and hydrogen peroxide at the Society's meetings in Berlin. *Schöne* is also known for his works on soil chemistry. He is forgotten in Germany but not in Russia. The Petrowski Agricultural and Forest Academy is today part of the Russian State Agricultural University K. I. Timirjasev [Российский государственный аграрный университет – Московская сельскохозяйственная академия имени К. А. Тимирязева (МСХА)]; *Kliment Arkadjewitsch Timirjasew* (1843–1920) was biologist and physiologist who became professor at the Petrowski Academy in 1871 and at the Moscow University in 1877. The building, were *Schöne* worked, still exist. An extended biography on *Schöne* is written in Russian by Smarygin (2018).

Figure 4.31: Photo of *Emil Schöne*. Source: website of the Russian State Agricultural University https://www.timacad.ru/news/oni-byli-pervymi-18o-let-so-dnia-rozhdeniia-emilia-shione.

idences for the presence of atmospheric ozone" (Schöne 1880b), based on his previous works that hydrogen peroxide "catalytic decomposes in a solution of potassium iodide into water and oxygen whereby small amounts of iodine and hydroxide remain (Schöne 1878, 1880a), Fig. 4.31.

Schöne (1880b, p. 1504) criticized *Schönbein*'s claim that hydrogen peroxide[1011] does not affect test-papers based on potassium iodide, manganese sulfate, or thallium oxide, and shows that all these papers also have been colored through hydrogen peroxide. Further *Schöne* refers *Houzeau* who indeed found that precipitation of iodine from potassium iodide does not occur through free halogens and acids, even not through nitrous acid, but that hydrogen peroxide causes the release of iodine. Finally, citing *Andrews* as the third proponent of the presence of atmospheric ozone, *Schöne* notified that *Andrews* fully ignored atmospheric hydrogen peroxide, and his experiments (see p. 509) can be also explained with the presence of atmospheric hydrogen peroxide.

Schöne also discusses the smell which has been observed after the lightning strikes by *Schönbein* and others to be like that of ozone. *Schöne*, knowing the odor

1011 Schönbein (1864c, p. 289) proposed a reagent to detect H_2O_2 (in solution) based on a solution of salts from Ni, Co, Pb or Bi salts which is dropped in the solution to be tested, adding some drops of a KOH solution, afterwards some drops of diluted starch with potassium iodide and finally acetic acid or diluted sulfuric acid, gaining promptly a blueing. Soon before Schönbein (1864b, p. 113) communicated that he has shown that acidified iodide starch pastes together with iron-II-oxide [*Eisenoxydul*] is very sensitive on traces of hydrogen peroxide; another, but not so sensitive reagent he found in a solution of indigo, decolored through hydrothionic acid [*Wasserstoffschwefel* or *hydrothionige Säure*] which blues in presence of H_2O_2. *Schönbein* used the formula HO_2 for hydrogen peroxide. Hydrothionic acid (acidum hydrothionosum) was assumed to be H_2S_5 (polysulfane), discovered by *Scheele* as a yellow oily liquid, which decays to H_2S and sulfur (Liebig 1843, p. 275). The effect of Fe(II) on hydrogen peroxide (many years later named *Fenton* reaction) consists in formation of OH radicals and thus an increase of the oxidation capacity of hydrogen peroxide.

of ozone well, describes lightning odor like burning gunpowder, arguing that electric sparks produce nitrous acid but not ozone in air (Schöne 1880b, p. 1506).[1012] Nonetheless, *Schöne* did not exclude the presence of ozone in atmospheric air, despite evidence of the presence of atmospheric hydrogen peroxide, writing (Schöne 1880b, p. 1507):

> Die Abwesenheit stichhaltiger Beweise schliesst selbstverständlich die Möglichkeit des Vorkommens des Ozons in der atmosphärischen Luft keineswegs aus [The absence of substantial evidences does naturally not exclude the possibility of the presence of ozone in atmospheric air].

In a subsequent paper, Schöne (1880c) discusses that the so-called test-paper "ozone reaction", even if ozone is present in atmospheric air, could be a result of several oxidizing bodies, hydrogen peroxide and ozone, and likely others still unknown. But even if one is satisfied with the estimation of the total amount of oxidizing bodies in air, the potassium iodide papers both *Schönbein*'s and *Houzeau's* have defects, making them even for "*Bestimmungen der rohsten Art unbrauchbar*" [determinations of the rawest kind unusable]; (Schöne 1880c, p. 1509). Referring *Dobrindt, Schöne* further discusses the influence of atmospheric humidity and the hygroscopicity of the used materials on the determination. He states that several workers found that ozone does not impact completely dry papers, but that at the same amount of ozone the degree of blueing increases with the increase of humidity, which can be found in laboratory experiment. *Schöne* emphasizes that also hydrogen peroxide needs the presence of water to impact potassium iodide. In simultaneous observations, using the *Schönbein* ozonometer and a psychrometer, he found that the curves of the ozonometer were parallel with

1012 Many researchers noted the formation of nitrogen oxides in preparation of ozone by silent discharges on air (Böttger 1858, 1873b, Houzeau 1871, Boillot 1873, Carius 1874a, Hautefeuille and Chappuis 1881, Shenstone and Evans 1898), and in phosphorus generation of ozone from air (Berthelot 1877a, Leeds 1879a,b). The smell of NO is described to be sharp sweet-smelling, whereas NO_2 has a strong harsh odor (also described as pungent acrid), and the odor of O_3 as sharp reminiscent; surely to distinguish by a chemist who was familiar with the pure gases but not for non-chemists. It should be noted that under conditions of simultaneous formation of NO and O_3, NO immediately is oxidized by O_3, and the odor of the mixture depends from the mixing ratios. NO_2 is a reddish-brown gas at high concentrations which has a threshold of odor of 5–10 ppm, that of O_3 is about by a factor of 100 smaller. Therefore, it is very likely that in all these experiments only the odor of NO_2 has been recognized. Götz (1934, p. 25) notes "[...] *den so häufigen starken Geruch von Ozone (oder Stickoxiden?) in solch großen Höhen*" [... the common strong smell of ozone (or nitrogen oxides?) at such great heights]. An interesting remark made Cauer (1936, p. 21) who did not smell ozone during measurements in the High Tatra (1000 m altitude) even at largest values up to 100 $\mu g\ m^{-3}$, but a similar concentration of ozone when produced in the laboratory through a quartz lamp or an ozonisator, proposing that perhaps the simultaneous presence of nitrogen oxides plays a role and/or the sudden increase of the amount of ozone only can be recognized as odor. Reynolds (1930b) reports on observations that with indoor globular lightning a strong peculiar smell and grey or reddish smoke occurred; Chameides et al. (1977) first report on NOx formation with atmospheric lightning.

the relative humidity, marking that *Schönbein*'s ozonometer is at most a raw hygrometer. Finally, *Schöne* proposed to use thallium hydroxide TlOH (in old German "*Thalliumoxydulhydrat*") because these papers do not need the presence of water and are therefore independent of the atmospheric humidity. *Schöne* presents monthly mean data from January until December 1879 of parallel determinations using the thallium paper and the iodide-starch paper together with observation of humidity, cloudiness, precipitation, and wind strength at the agriculture research station Petrowskoje Rasumowskoje near Moscow. Comparing the results, low conformity is found, and on the whole, both papers show opposed results. From the thallium paper observation, *Schöne* found that the intensity of coloration is

- stronger at day than at night,
- stronger with southern air, namely in winter,
- less with increasing cloudiness and stronger rain,

and notifies that these results are the same as his observations of atmospheric hydrogen peroxide (pp. 427 ff, Chapter 4.5.2.2). Therefore, it is not necessary, *Schöne* concluded, to assume another oxidizing body than hydrogen peroxide. However, scientifically appropriate he notes that further investigations are needed to get evidence for the presence of ozone in the atmosphere. According to today's knowledge, photochemically produced near-surface ozone would result in similar dependencies.[1013]

4.4.5.3 Observations of atmospheric ozone in Germany

The first published observations of ozone after *Schönbein* were done by the German physician and balneologist *Ludwig Spengler* (1818–1866) from Bad Ems, who lived a short time in 1846 in the small village Roggendorf near Gardeleben in Mecklenburg (Spengler 1849, 1857). He found that the occurrence of influenza is combined with a larger ozone reaction (December 1846 and January 1847). Hence, he termed medical practitioners to make ozone observations everywhere, and wrote the wonderful phrase (Spengler 1849, p. 73):

> Die Atmosphäre ist der Sammelplatz einer grossen Anzahl der verschiedenartigsten gas- und dampfförmigen Materialien, allein noch sind wir nicht im Stande, mehr als Stick-, Sauerstoff, Wasserdampf und Kohlensäure nachzuweisen [The atmosphere is the collection site of a large number of various gaseous and vaporous matter, but we are not yet able to detect more than oxygen, nitrogen, water vapor and carbonic acid].

The next ozone observation carried out by the practical physician (*Johannes*) *Theodor Clemens* (1824–?) in Frankfurt am Main near a chemical factory between 1846 and

1013 It remains a speculation to assume a certain amount of surface-near photochemical ozone formation in populated areas after the mid-19th century with increasing manufacturing and industrialization (Chapter 4.4.8.6).

1849 (Clemens 1849); he found no ozone in the factory, arguing that exhaust gases like hydrogen sulfide and oil-forming gases (light hydrocarbons) consume ozone; like *Schönbein* (and many other later) he found ozone highest at snowfall. In a later study between 1850 and 1851 (Clemens 1853), he investigated the relation between gaseous emanations from waters (ponds, privies, river Main, and others) and the amount of ozone; he also investigated the quality of water and its exchange with the bottom and surface air. *Clemens* found the varying amounts of ozone above the water surfaces, arguing that different processes of consuming and forming ozone cause these changes (Clemens 1853, p. 28–29). According to his former view on ozone consuming gases (Clemens 1849), he notes that on the one hand miasma from swamps destroy ozone but on the other hand "*eine eigenthümliche Ozonisirung des von manchen Wasserflächen reichlich entbundenen Oxygens*" [a specific ozonation of the abundant escaped oxygen from some water surfaces] occurs (Clemens 1853, p. 70). Generally, he found at night less ozone (Clemens 1853, p. 90), and concluded summarizing that standing waters could be grouped in those (swamps) that release ozonized oxygen under the influence of sunlight, and those, which under all circumstances, remove the atmospheric ozone.

The physician *Wilhelm Eberhard Faber* (1787–1872) in Schorndorf (near Stuttgart) who visited together with Dr. *Müller* from Calw (see p. 521) in late 1847 *Schönbein* in Basel to learn his view on ozone. *Faber* found it very interesting (remember that *Schönbein* advocated somebody to conduct ozone observations with his paper) that he decided to carry out between December 1847 and December 1848 such observations (Faber 1849); he communicates his experiences concerning preparation of the test-papers and found at days with precipitation a pronounced increase in the "ozone reaction" and less ozone on days with increased temperature.

The practical doctor *August Rudolf Brenner* (1821–1884) in Quedlinburg, who spend many years (1854–1875) in Russia and became a professor for electrotherapy in Leipzig after his return to Germany in 1877, conducted ozone observations since early 1848 over 15 months (Brenner 1849). Brenner (1849, p. 6) describes the role of the atmosphere in a wonderful phrase:

> Die Atmosphäre ist der zeitweise Träger von Stoffen, welche aus dem anorganischen in das organische Reich, aus Organismen einer Art in solche anderer Art übergehen; sie ist ferner selbstthätige Erzeugerin von Stoffen und Vorgängen, welche von wesentlichstem Einfluss auf die verschiedenen Reiche der Natur sind, und im Fall sie örtlich beschränkt sind, oft höchst auffallende Erscheinungen in denjenigen anorganischen und organischen Körpern hervorbringen, welche sich vermöge ihres örtlichen Vorkommens und Aufenthalts ihren Einflüssen nicht zu entziehen vermögen [The atmosphere is the intermittent carrier of substances, which transfer from the inorganic into the organic kingdom, from organisms of one kind into those of another kind; it is further the producer of substances and processes, which are of important influence onto different parts of nature, and, in case they are localized, often generate highly conspicuous phenomena in those inorganic and organic bodies, which cannot evade these impacts due to its local occurrence and residence].

Such as *Spengler*, *Brenner* believed that influenza occurred with larger ozone values in the air; the coincidence of diseases of the respiratory system in the cold season with the highest "ozone reaction" was referred to larger amounts of ozone in the air (irritations through inhalation of air containing large ozone concentrations was recognized by *Schönbein*)[1014] but the misinterpretation of the colored test-paper at cold, wet and stormy weather was not yet seen. *Brenner* continued his "observations" (Brenner 1851), notified that coloring of the test-paper often goes together with "*einem grossen Wassergehalt der Atmosphäre*" [a large amount of water in the atmosphere], argued that carbonic acid in the air does not result in an "ozone reaction", and communicated that he was able to forecast catarrhs with his ozonometer a few days before, and thus writing "[...] *um so wahrscheinlicher wird mir die Schönbein'sche Hypothese*" [... the more likely Schönbein's hypothesis will be for me] (Brenner 1849, p. 17).

The German apothecary *Schlotfeldt* (living dates unknown) in Oschersleben, after listening to a lecture of the young physician *Brenner* from Quedlinburg on *Schönbein*'s ozone, was very emphasized by his idea on the occurrence of influenza with increasing ozone levels and continued observations with *Schönbein*'s ozonometer (Schlotfeldt 1849). The only interesting fact he reports, that *Schönbein*'s paper also colors brown when it hangs over a bottle with nitric acid, and that this paper, hanging over hydrogen sulfide water, lost the brown color. *Schlotfeldt* refers to the permanent use of bituminous coal in his surrounding area (Harz Mountains) and the formation of hydrogen sulfide from its combustion (however, it is sulfur dioxide, much later known for interference in iodometric determinations). At that time, interest in ozonometric observations was aroused mainly from the idea that ozone – because of its oxidizing properties – could reduce "miasma" (poisoning organic substances) and thus be "healthy" to people.

The first person who noted that *Schönbein*'s paper is useless, and who doubt that atmospheric ozone causes influenza and respiratory diseases,[1015] was *Philipp Friedrich Betz* (1819–1903), a German practical doctor in Tübingen, who experimented with *Schönbein*'s paper in winter 1847–48 (Betz 1848); and he termed up colleagues to prove this finding. Thus, in January 1849 *Betz* organized experiments to investigate *Schönbein*'s paper for ozone observations, together with[1016] Prof. Dr. *Schlossberger*,

1014 Clearly expressed by Schönbein (1851a, p. 388) that pure ozone act as a strong poison, but also at low concentrations in the air after longer exposition it might result in harmful effects to the organism. Further *Schönbein* emphasized the (questionable due to the effect of humidity on test-paper) observation that the maximum amount of ozone occurs in winter and the minimum in summer.

1015 In *Archiv physiologische Heilkunde* **7** (1848) and *Zeitschrift für rationelle Medicin* **6** (1848). It is remarkable that Betz (1848) wrote *Schönlein* instead of *Schönbein*. Betz (1848) communicated that *Schönbein* attacked him with arguments which probably harm him themselves.

1016 *Julius Eugen Schlossberger* (1819–1860) was a German biochemist and physician. *Wilhelm Griesinger* (1817–1868) was a German physician and neurologist, since 1864 at the Charité in Berlin and best known for the first establishment of a neurological station. *Karl Nicolaus Fallati* (1803–1868),

Prof. Dr. *Griesinger*, Dr. *Fallati* and Dr. *Baur* in Tübingen, Dr. *Renz* in Stuttgart, Dr. *Krauß* and pharmacist *Wrede* in Mergentheim, Dr. *Müller* in Calw and Dr. *Höring* in Heilbronn (Betz 1849); this is the worldwide first "measurement campaign" (*Schlossberger* prepared the test-papers), and from the results, *Betz* draws guidelines for further observations. This was also the first investigation (noted by all observers) that the paper in rooms never shows a "reaction". *Griesinger* and *Baur* indicated that the test-paper during snowfall do not color and that snow water has no other colors than rainwater. *Betz* was already in contact with the German paleontologist and naturalist *Heinrich Theodor Plieninger* (1795–1879) who conducted in Stuttgart since January 1850 daily three observations with two different test-papers (Plieninger 1850). However, *Plieninger* was not satisfied with the paper obtained from *Betz* and writes (Plieninger 1850, p. 171):

> Aus allem dem wird hervorgehen, dass diese Beobachtungen des atmosphärischen Ozon, (oder welch anderer Stoff es sein möge, der auf dem Papierstreifen reagiert), noch im Keime liegen und erst auf dem Wege der Beobachtung selbst eine weitere Ausbildung erhalten müssen, wenn sie zuverlässige Resultate liefern sollen [From all this will emerge that these observations of atmospheric ozone (or whatever a substance it is which reacts on the paper) are still at the beginning, and must obtain a further training through observations to get reliable results].

The German apothecary in Mühlhausen (Thuringia) *Nicolaus Gräger* [also written *Nikolaus Graeger*] (1806–1873), who also measured ammonia in air, "measured" (from December 1850 until June 1851) daily at three different times ozone using the *Schönbein* paper; the only systematic variation he found that the ozone amount in air is largest at night (Gräger 1852). Interestingly that he finally comments that nitrous acid also shows this "ozone reaction" and other substances, which also react (from laboratory studies), are unlikely to be present in the atmosphere.

Carl Friedrich Christian von Mettenheimer (1824–1898), a famous physician in Frankfurt am Main (1849–1861) and afterward in Schwerin, conducted daily three times ozone observations in 1851–1852 in Frankfurt am Main, but communicated only 9 years later (Mettenheimer 1863); long-winded he represents all values in detail. However, *Mettenheimer* was obviously the only person who not only referred *Maack*'s paper on the criticism of *Schönbein*'s paper but also discussed it in detail over several pages (Mettenheimer 1863, pp. 496–499), agreeing and confirming *Maack*'s arguments also based on own observations (it remains unclear from his observations in 1851–1852, i. e., before *Maack*'s paper appeared, or at the time of writing his paper).

German physician in Tübingen and Wildbach (Black Forest). *Ferdinand Christian Baur* (1792–1860) was a German theologist. *Friedrich Krauß* [or *Krauss*, wrongly written by *Betz* as *Kraus*] (1803–1885), German physician. *Josef Johann Nepomuk Wrede* (1786–1863), apothecary in Mergentheim. *Karl Müller* (1803–1877), physician in Calw. *Franz Jacob Höring* (1802–1891) physician in Heilbronn and Mergentheim.

The German physician *Reinhold Heydloff* (?–1879) conducted 1853–1854 in Erfurt daily three ozone observations, published all his data and draw the following conclusions (Heydloff 1853–54): the amount of ozone increases from summer to winter, the daytime maximum occurs at night, it increases with winds from South through West to North and decreases with winds from East to South, increases with precipitation and equal to zero during fog.

August Wilhelm Ferdinand Schultz (1805–1890), a famous physician of Berlin, conducted ozonometric observations between August and November 1853 (Schultz 1854) and concluded that the amount of ozone was significant during the period of cholera epidemics, and about zero in November when the number of diseased persons rapidly decreased; it is remarkable that he did not make any conclusion on a relation between ozone and cholera (he writes that "other observer like to communicate explanations of these facts"). Furthermore, *Schultz* noted introducing to his paper correctly that *Schönbein*'s ozonometer is like older hygrometers (of that time), i. e., it presents no real values and is not comparable with ozonometers used by other observers. The Austrian epidemiologist *Anton Drasche* (1826–1904) of Vienna, strongly doubt that the diminution of the amount of ozone in air is in relation to the cause of cholera (Drasche 1860, p. 145); he also refers *Schultz* (wrongly cited *Schulz*) of Berlin with increasing ozone values during the cholera epidemics in 1853, and similar observation by *Voltolini* in Silesia and *de Wette* of Basel (not giving bibliographies).[1017] *Drasche* further refers *Carl Haller* with ozone observations in Vienna in autumn 1854 and from all 6 Austrian ozonometric stations, showing large variations but no relationship between ozone and the epidemics, namely the ozone values for Vienna in 1854 and 1855.

Friedrich Wilhelm Beneke (1824–1882), a medical doctor and one of the founders of German "*Seebäderwesen*" [seaside resorts], communicates on ozone observations (Beneke 1858) by *Niebergall* and *Toel* (see p. 523) and named "*Ozon als den Reiniger der Luft*" [ozone being the cleaner of the air], (Beneke 1858, pp. 225 and 266). Beneke (1855, p. 490), however, communicates that the physiological effects of seaside resorts are not yet understood aside from the freeing of the common professional and daily life, the absence of pollutants, the larger amounts of humidity and ozone.

Carl Gottfried Matthäus August Niebergall (1811–1890) observed ozone in Arnstadt (Thuringia); Niebergall (1858). The German physician *Toel* (unknown living data), who succeeded in 1820 *Friedrich Wilhelm von Halem* as district doctor in Knipshausen (today Knyphausen) near Oldenburg, conducted ozone observations in 1855 simultaneous at two sites, direct above a swamp and at 12 feet above the earth (Beneke 1858, pp. 257–268).

1017 *Friedrich Eduard Rudolf Voltolini* (1819–1889) was physician at Falkenberg 1854–1860 and then in Breslau. *Ludwig de Wette* (1812–1887) was born in Berlin, the son of the well-known theologian *Wilhelm Martin Leberecht de Wette*, and physician of Basel between 1851 and 1884; see: De Wette, L. (1856) Bericht an Löbl. Cholera-Commission über den Verlauf der Cholera in Basel im Jahre 1855. Schweighauser, Basel, 44 pp.

Franz Seitz (1811–1892), a German professor of medicine at University of München, conducted ozone observations in 1853–1855 in Munich according to the *Schönbein's* paper and found no relation between the amount of ozone in the air and any illness[1018] (Seitz 1865, pp. 359–362).

Ferdinand Flemming junior (1825–?), a practical physician in Dresden, with colleagues systematically studied the variations of ozone in Dresden and at other locations in Saxony and Prussia in 1855 (Flemming 1855). They concluded that a cholera epidemic occurs only some time (6–14 days) later after ozone absence the air,[1019] and with increasing ozone the epidemics decreases. *Flemming* found the amount of ozone to be highest with winds from SW to W, with increasing temperature, and in the afternoon between 1 and 3 pm. *Flemming* made an urgent request to his colleagues to resume ozone observations, namely four times daily, and also to note temperature and wind direction. Another observation in Dresden was made by *Karl Traugott Sachse* (1815–1863), a teacher for mathematics and natural sciences at the "*Kreuzschule*" [Protestant Evangelical School of the Holy Cross] in Dresden in 1859 (Sachse 1861). Stimulated by *Flemming*, in Ottweiler (a small town near Saarbrücken) the physician *H. Zimmermann* (unknown living dates) conducted in January 1856 ozone observations (Zimmermann 1856). Another physician, Dr. *Lindner* at Cüstrin (today Kostrzyn nad Odrą in Poland) "measured" ozone during a cholera epidemic in September 1855 (Lindner 1855). The medical doctor *Carl Wilhelm Ludwig Schaper* (unknown living dates), born in Danzig, conducted ozonometric observations in 1852–1859 in Danzig (today Gdańsk in Poland) and in 1859–1867 in Koblenz, presenting (according to the meteorological recommendation by *Dove*) a statistics of the "ozone values" and writes poetically (Schaper 1867, p. 170):

> [...] dass das Ozon eine Verwendung in der Oekonomie unseres Erdkörpers finden müsse [...] durch die nachgewiesene überaus grosse Regelmässigkeit, mit welcher dasselbe nach Jahr, Tag, Nacht, Oertlichkeit aufgefunden oder vermisst wird. [... that ozone must find an application within the economy of our Earth due to proven regularity with which it is found or missed to year, day, night, and locality].

Schaper found a pronounced seasonal variation (maximum in May) at both sites but always in Danzig higher *Schönbein*-values; the higher nocturnal values, however, show us the limitation of representative ozone determination. The German physician *Schiefferdecker* and the Austrian naturalist *Reslhuber* (Chapter 4.4.5.4) were the first who studied systematical different factors that influence the ozone reaction, and

1018 German original: *Es war durchaus keine Beziehung zwischen irgend einer Krankheit und dem Ozongehalte der Luft aufgefunden* (p. 362).

1019 It is interesting to note that *Flemming* refers a verbal information by Professor *Seitz* in Munich that 14 days before outbreak of epidemic in 1854 permanent absence of atmospheric ozone was observed (Flemming 1855, p. 788). This an early nice example how short-time observations led to wrong conclusions, or data are misinterpreted.

agreed in almost all conclusions; all later observations contributed nothing else more. However, it was in vogue to carry out at many different places in Europe and overseas for the next 20 years ozonometric observations (Table 4.16), and it was the belief that – likely because of a stepwise improving understanding of what climate means – that the locality is essential for the amount of ozone in the air. *Wilhelm Friedrich Schiefferdecker* (1818–1989)[1020] conducted ozone observations in 1852–1853 in Königsberg (today Kaliningrad in Russia) and draw the following conclusions (Schiefferdecker 1855):

- *Schönbein*'s method is unreliable; wind and humidity show a large cross-sensitivity,
- the amount of ozone at different locations in the town varies considerably so that a single measurement in a town makes no sense,
- the amount of ozone outside the city is rather constant and larger than a suburb,[1021]
- the amount of ozone is at night and in winter larger,[1022] and
- is larger on days with snowfall than on days with rainfall, and on those again larger than on nice days,[1023]
- increases sometimes during a thunderstorm,
- humidity in air favors the "ozone reaction",[1024]
- there is no relation between the amount of ozone and the occurrence of illnesses.

The latter result was the reason to terminate the observation after this year (Schiefferdecker 1855, p. 214). *Carl Julius Adelberg Küttlinger* (1817–1868), a district physician in Erlangen, conducted ozone observation during 1854–1859 and almost confirmed *Reslhuber*'s and *Schiefferdecker*'s findings (Küttlinger 1859). "*Oberlehrer*" [senior teacher] *Christian Ludwig Schoof* (1810–1891) from Clausthal measured in 1856 (cited by Prestel 1866 but wrong written as *Schoot*); Schoof (1860).

1020 Physician in Königsberg; he was member of the new medical society "*Verein für wissenschaftliche Heilkunde*" [Society for Scientific Medicine]; *Helmholtz* served as first director and encouraged *Schiefferdecker* to carry out ozone measurements. Stimulated by *Schönbein*'s publication, the Society decided in 1852 to establish a "*Section für Ozonometrie*" [department of ozonometry] with *Schiefferdecker* as head. At 12 stations urban and suburban (with 9 local observers) from June 1852 until May 1853 daily two observations with 12 hours exposure were conducted. Remarkable of *Schiefferdecker*'s paper is that on 25 pages he presents all single data in Tables and Diagrams, showing the timely variation.

1021 The first published statement that the amount of ozone is outside towns larger than inside towns is given by Ecker (1847).

1022 This observation can be explained by the fact that ozone at daytime and in summer is photochemically destroyed (note that photochemical ozone production was yet unimportant).

1023 This observation can be explained similarly (depression of photochemical activity) and, as cited in the text, by the wet removal of particulate matter that provides a surface for catalytic ozone removal (but much less effective than photochemical ozone removal under low NO_x conditions).

1024 This is surely the reaction of O_3 on the wetted paper surface, giving radicals and hydrogen peroxide which results in a stronger "*Schönbein* reaction"; moreover, absorption of H_2O_2 is favored.

Eduard Weber (1811–1871), a practical doctor and later army physician, conducted ozonometric observations (daily twice, protected against sun and precipitation, and together with meteorological parameters) in Karlsruhe between October 1855 and September 1856, and in Mannheim from 1856 until 1866 (Weber 1858, 1867). In contrast to other observers, *Weber* found in Mannheim a distinct annual variation, with a maximum in July and a minimum in December, and the highest values at wind from south to west and lowest at wind from north to northeast (these observations correspond to modern ozone observations). From his observations made in Karlsruhe, he found that the amount of precipitation did not correlate with the ozone reaction but that it is the highest at the beginning, namely at snowfall.

Eduard Lichtenstein (1818–?), a practical doctor and geophysicist in Berlin, member of the Leopoldina Academy since 1856, conducted regularly ozone observations since 1858 until 1876. He reports (Lichtenstein 1876, p. 126) that together with *Dove* and *Arndt* they have observed ozone at the Berlin Meteorological Station since 1866.[1025] *Lichtenstein* first used *Schönbein*'s ozonometer and since 1874 *Houzeau*'s method. Since 1866 he exposed the ozone test-paper together with blue litmus paper and concluded if the litmus reddens, and although the ozone paper shows a reaction, that nitrous acid is (he used the formula NO^3) in air and not ozone. He excluded that the reddening is caused by carbonic acid due to the finding that in closed rooms, containing much more CO_2 than in external air, the litmus never reddens. He coined the term "*Ozonologie*" [ozone science],[1026] defining this "science" (Lichtenstein 1876, p. 127)

> Aufinden einer Gesetzlichkeit im Auftreten des atmosphärischen Ozongehalts und Differenzierung je nach geographischen und meteorologischen Verhältnissen [Detection of regularities in the occurrence of the amount of atmospheric ozone and distinction depending on geographic and meteorological situations].

Lichtenstein (1860a) conducted at his residence in Grabow, province Posen in Prussia (today Grabów nad Prosną in Poland) in October 1858 a few ozone observations while observing aurora and concluded from less amount of ozone at the northern side of his

1025 *Heinrich-Wilhelm Dove* (1803–1879) was a famous German meteorologist who became in 1848 the director of the (*Royal*, later *Prussian*) Meteorological Institute in Berlin, founded in 1847 on the initiative of *Humboldt*. Before 1866 he was the only employee when he gets the High School professor *Johann Albert Arndt* (1811–1882) who became the director after *Dove*'s death in 1882.

1026 This term is also used in French (*d'ozonologie*) as the science of ozone. Curiously enough, that the Austrian physician Dr. *Thiele* used "*Ozonologie*" as synonym for ozone therapy (published at the Ozone Congress 2019 held at Baden near Vienna). The editors of "*Wiener Medicinische Wochenschrift*", Vol. 4, 1854, p. 29, write unflattering in response to the secure cancer cures, offered by the Italian physician *Landolfi*: *Wien, wohin er sich begibt, ist besserer Boden als Prag für alle Ausgeburten von Schein, Lug und Trug*, [. . .]. *Ozonologen* [. . .] [Vienna, where he is going, is better soil than Prague for all spawns of appearance, lies and deception, … ozonologists. . .]. *Nicola Landolfi* (unknown living data) was a Neapolitan surgeon who travelled in 1853 through Germany to cure people of cancer by *Landolfi*'s etching paste.

house than from the southern one, that atmospheric ozone is in connection with aurora. He holds this belief, but wrote that the amount of ozone is larger with northern air masses because they pass the sea, and evaporating of salty water is connected with electricity, and argued (Lichtenstein 1876, p. 123) that due to relationship between electricity (and its inverse correlation with magnetism) and ozone, the amount of ozone at the equator must be higher than that at northern latitudes. Lichtenstein (1860b) published 27 questions to be answered by ozonometric observers; no. 25:

> Sollte man nicht durch allgemeine und fortgesetzte Beobachtungen isozonische Curven auf unserem Planeten oder doch zunächst in Europa bezeichnen, so wie es isotherische und isochimetrische Curven giebt? [Should one not through general and continued observations isozonic curves on our planet, but first in Europe construct as there are isotheric and isochemic curves?][1027]

As another curious "result" by Lichtenstein (1862) in ozonometric experiments with healthy and ill persons in the Berlin Charité he "found" that the human skin emits ozone and the respiratory air contains ozone; as "reference" he notes that he and *Scoutetten* in Metz as well *Bérigny* in Versailles never found ozone in closed rooms.

Michael August Friedrich Prestel (1869–1880), director of the "*Naturforschende Gesellschaft zu Emden*" [Natural Research Society of Emden] systematically investigated the ozone concentration in Emden between 1857 and 1874 (Prestel 1865, 1874b); he found the same annual variation as *Karliński* at Krakau. Remarkable is *Prestel*'s "ozonometric wind rose" which we now call *air pollution climatology*[1028] (Figs. 4.32 to 4.34). Prestel (1872) wrote that (cited after Lender 1874, p. 28)

> [. . .] ist die Bestimmung der Ozonreaktion der Luft, [. . .] ein wichtiger Gegenstand der meteorologischen Statistik und Feststellung des periodischen und nichtperiodischen Auftretens des Ozons wesentliches Moment für die Climatologie [. . . the determination of the ozone reaction in air [. . .] is an important issue of meteorological statistics and an identifying of periodic and non-periodic occurrence of ozone is an essential factor of climatology].

Prestel (1866) compared his long-term means with data (Table 4.16) from Highland, USA (1862–1864) and Clausthal, Harz Mt. (1865) and found that at all three sites are of the same annual cycle (maximum March–April and minimum November–January), whereas at Clausthal the highest ozone values with the lowest daily variation and at Highland the lowest amount of ozone in the air was observed. Prestel (1874a) describes

1027 Isotheric means a geoisotherm of equal mean summer temperature; the term "*isochimetrisch*" is unknown (misprint?); "*isochemisch*" means invariable chemical composition.

1028 In German, "*Immissionsklimatologie*". The term "*Immission*" – not known in English (only *emission* despite "immission" is used in official translations from German into English)– is deduced in analogy to emission (entry of matter into the atmosphere), however, it does not describe the discharge (which is referred to as deposition), but the concentration of a matter at the effective location, a somewhat spongy definition.

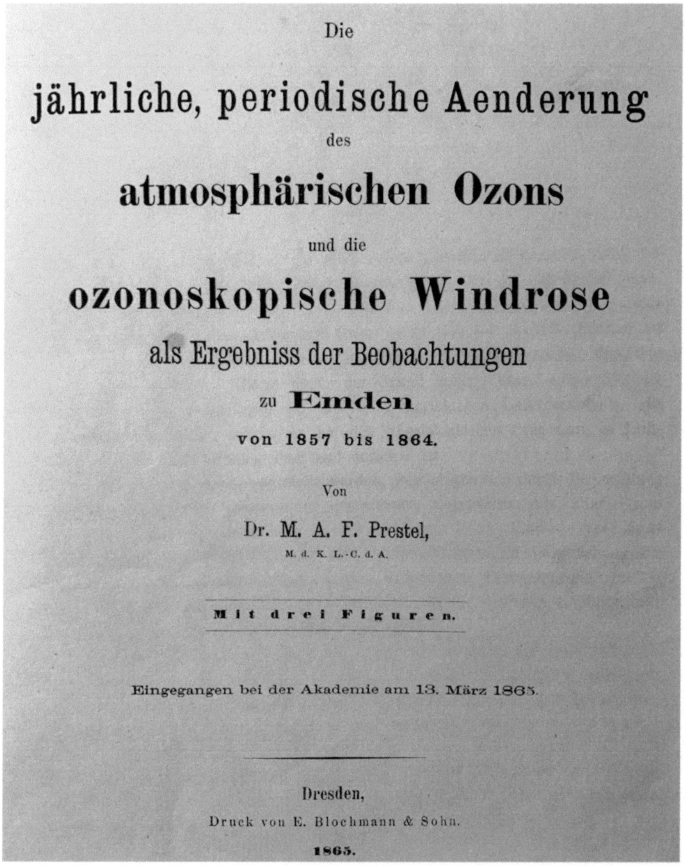

Die

jährliche, periodische Aenderung

des

atmosphärischen Ozons

und die

ozonoskopische Windrose

als Ergebniss der Beobachtungen

zu Emden

von 1857 bis 1864.

Von

Dr. M. A. F. Prestel,

M. d. K. L.-C. d. A.

Mit drei Figuren.

Eingegangen bei der Akademie am 13. März 1865.

Dresden,

Druck von E. Blochmann & Sohn.

1865.

Figure 4.32: Cover of the publication by *Michael August Friedrich Prestel* "Annual periodical variation of atmospheric ozone and the ozonoscopic wind rose as result of observations at Emden from 1857 upon 1864" (Prestel 1865).

the insufficient properties of test-papers and recommends how the paper must be prepared.

In Plauen (Saxony) between May and September 1861, *Emil Richard Pfaff* (1827–1871), "*Bezirksarzt*" [district physician] in Plauen/Saxony, observed ozone and recordered temperature, wind, and illnesses (Pfaff 1862). He found the air often free of ozone, and when approaching a thunderstorm, *Schönbein*'s ozonometer showed a strong reaction; thus, he concluded that thunderstorms entail ozone and take it away. Similarly he observed a changing amount of ozone with a weather change. *Anton Heidenschreider* (1826–1870), a German physician and meteorologist at Herrieden (the town in Bavaria near Nuremberg) made observations in 1866–1867 (Heidenschreider 1868).

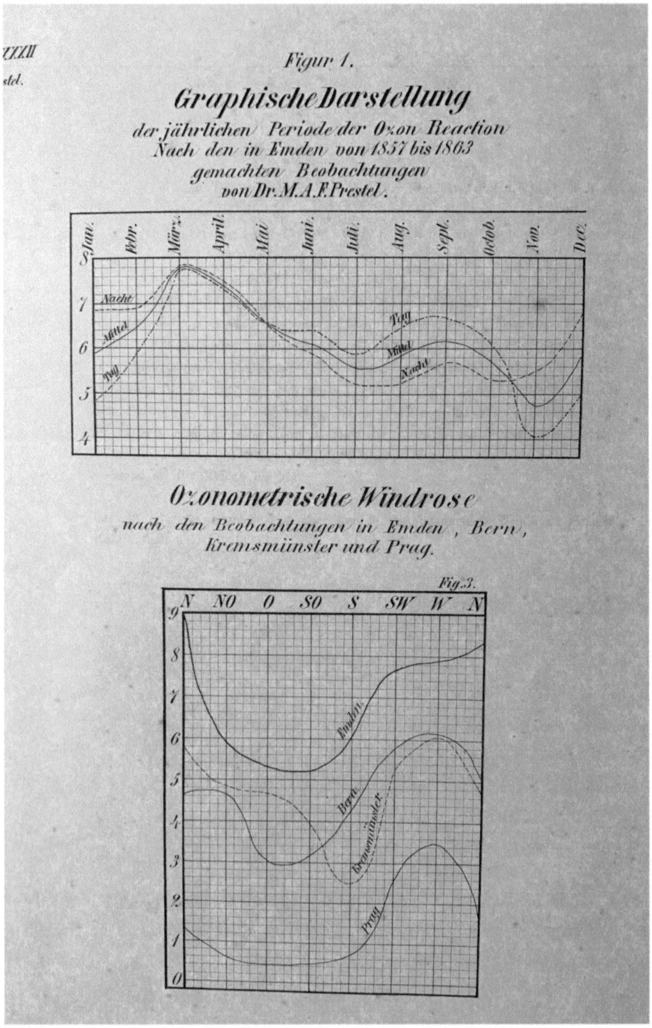

Figure 4.33: Annual variation of the “ozone reaction” and ozonometric wind rose according to observations in Emden, Bern, Kremsmünster, and Prague (Prestel 1865).

Carl Begemann (1815–1885), graduated pharmacist, a teacher for physical science at the “*Königlische Tierarzneischule*” [Royal Veterinary School] in Hannover and meteorological observer, doubted the existence of ozone in the atmosphere because of the many uncertainties when exposing test-papers, judging from already published papers and many of his own observations (Begemann 1864). Referring Martius (1862),[1029]

1029 *Theodor Martius* (1796–1863) was a German pharmacist. After the lightning strike on June 22, 1859 into a house in Erlangen, the owners told him that the brass chain of the watch, now black, was

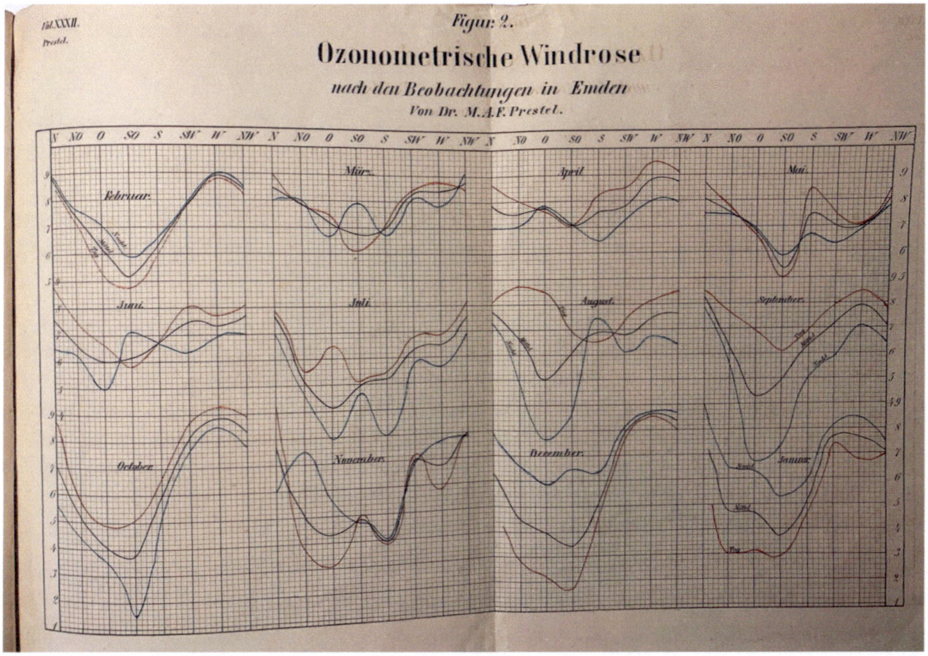

Figure 4.34: Monthly variation of "ozone" in dependency from wind direction at Emden (1857–1864); upper curve daytime, middle curve medium, and lower curve at night (Prestel 1865).

that the peculiar sulfureous smell which was often observed accompanying lightning strikes in rooms, could be ozone, *Begemann* (Begemann 1864, p. 9) states that also hydrogen peroxide could be formed.

Heinrich Möhl (1832–1903), a German mathematician and geologist, and *Theodor Dietrich* (p. 3) conducted observations using *Schönbein*'s paper in Kassel and a nearby rural site (Altmorschen) from March 1866 until May 1867 and found in the city air only one-third of the ozone amount of that in the country air (2.3 and 6.9 on *Schönbein*'s scale, respectively), explaining it by the consumption of ozone through gaseous emanations in the city (Möhl and Dietrich 1868).

Constantin Lender began in summer 1871 with ozone observations in Marienbad, Carlsbad, and Kissingen, supported by the gentlemens *Wernigh* from Berlin, *Barsdorf* and *Friedrich* from Liegnitz and his son, and in December at Mentone (Bretagne, France) with the help of *Graf von Bernsdorf* and Dr. *Stiege*, and other gentlemens (Lender 1872a). *Lender* also refers ozone observations from Meran and Wiesbaden, made in 1872 by "*Stadtgerichtsrath*" [city judge] *Max Schulze*. Lender

blank before; *Martius* investigated a part of it but find no sulfur but only the black oxide of copper gained from the stroke heat. This is surely the first direct experimental evidence that the "sulfureous smell" is not caused by any sulfur compound.

(1874, pp. 50–53)[1030] communicated on ozone observations in Sassnitz (Baltic Sea) from August 1872 until April 1873, and noted (Lender 1874, p. 53):

> In dem überaus ozonreichen Sassnitz genügt daher eine sechsstündige Beobachtungszeit [At the greatly ozone-rich Sassnitz a six-hour observation is sufficient].

Lender was well informed on the literature concerns "atmospheric chemistry" (ozone, rainwater composition, and chemical experiments) and wrote many remarkable phrases such as (Lender 1872a, p. 38):

> Der Vernichtungskampf, welchen das Ozon gegen die combustiblen Körper der Luft führt, geht in dem gasförmigen Medium der Atmosphäre nach und nach vor sich, geschieht nicht in explosiver Weise. Man darf aus der Gegenwart von Ozon nicht auf das Freisein seiner Luft von oxydablen Stoffen schliessen [The "battle of annihilation", that ozone is waged against combustible bodies, proceeds gradually in the gaseous medium of the atmosphere, and not explosive. One may not conclude from the presence of ozone on the absence of oxidizable substances in the air].

Lender introduced the poetic name "*Ozonräuber*" [ozone predator]. *Gustav Wolffhügel*, a hygienist at the *Pettenkofer* Institute in Munich, presents an excellent review on ozone studies at this time (Wolffhügel 1875), developed a sophisticated passive sampler, termed "*Ozonbox*" (Fig. 4.35) and found (like other scientists before, namely *Houzeau*) that the ozone amount in air is less in cities than that at rural sites, in rooms, it is undetectable (because it reacts with dust, indoor pollution, and the walls), and missing in ground air; he studied the ozone removal from the different surfaces in the laboratory.

Ernst Ebermayer conducted in 1868 at different locations in Bavaria ozone measurements: Rohrbrunn (Spessart), Altenfurth, Seeshaupt (Lake Starnberg), Johanniskreuz (Palatinate Forest), Ebrade (Steigerwald), Duschelberg (Bavarian Forest), and Achaffenburg; he also compared his results with ozone observations made in Leipzig and Zwickau, referring *Bruhns* (Ebermayer 1873a,b). *Ebermayer* was a protagonist of the "water evaporation theory", i. e., the sea (the North Sea and Baltic Sea) he regarded as a source.

In Saxony, six stations (from total of 21 meteorological stations) for ozone observation were operated in 1868–1870 (Table 4.18). *Otto Carl Friedrich Luedicke* (?–1875), a pharmacist and meteorological observer in Gotha, published the results (Luedicke (1871, 1875). *Luedicke* (also *Lüdicke*) analyzed the ozone data in relation to wind direction and power, and humidity; as other scientists, he found a strong connection between the amount of ozone in the air and the humidity.

1030 Nothing is found on the persons referred by *Lender*. *Bernsdorf* (almost *Bernsdorf* or *Bernsdorff* written) is an old Mecklenburg noble family; *W. Wernigh* (*Lender* called him his friend) was an engineer and constructor in Berlin.

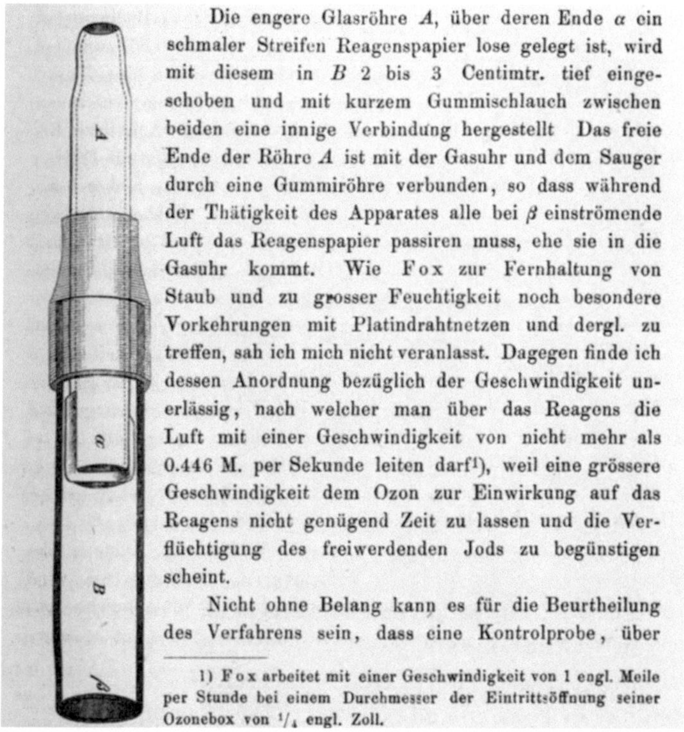

Die engere Glasröhre A, über deren Ende α ein schmaler Streifen Reagenspapier lose gelegt ist, wird mit diesem in B 2 bis 3 Centimtr. tief eingeschoben und mit kurzem Gummischlauch zwischen beiden eine innige Verbindung hergestellt Das freie Ende der Röhre A ist mit der Gasuhr und dem Sauger durch eine Gummiröhre verbunden, so dass während der Thätigkeit des Apparates alle bei β einströmende Luft das Reagenspapier passiren muss, ehe sie in die Gasuhr kommt. Wie Fox zur Fernhaltung von Staub und zu grosser Feuchtigkeit noch besondere Vorkehrungen mit Platindrahtnetzen und dergl. zu treffen, sah ich mich nicht veranlasst. Dagegen finde ich dessen Anordnung bezüglich der Geschwindigkeit unerlässig, nach welcher man über das Reagens die Luft mit einer Geschwindigkeit von nicht mehr als 0.446 M. per Sekunde leiten darf[1]), weil eine grössere Geschwindigkeit dem Ozon zur Einwirkung auf das Reagens nicht genügend Zeit zu lassen und die Verflüchtigung des freiwerdenden Jods zu begünstigen scheint.

Nicht ohne Belang kann es für die Beurtheilung des Verfahrens sein, dass eine Kontrolprobe, über

1) Fox arbeitet mit einer Geschwindigkeit von 1 engl. Meile per Stunde bei einem Durchmesser der Eintrittsöffnung seiner Ozonebox von 1/4 engl. Zoll.

Figure 4.35: "*Ozonbox*" as a passive sampler to protect the test-paper against light, rain and dust, and diffusion-controlled, using a *Bunsen* sucker and a gas meter. Air income down (β) and outcome aloft at "A"; diameter of glass tube "A" 8 mm and "B" 12.5 mm; the test-paper is fixed at "α" at the upper glass tube "A" (Wolffhügel 1875, p. 415).

Karl Christian Bruhns (1830–1881), a German astronomer at University of Leipzig, published 10 Volumes (issues I–XII) between 1864 and 1880 of the results of meteorological observations in the Kingdom of Saxony, including the results of ozone observations in Zwickau, Leipzig and Greiz (e. g., Bruhns 1867, 1870).

Karl Alfred Ritter von Zittel (1839 –1904), a German geologist in Vienna and Munich, who also measured the carbonic acid in the desert, found that the amount of ozone is strongest in the desert, less at the oasis and smallest in the Nil valley. Namely when dew appears, the "ozone value" increases even before sunrise; *Zittel* speculates that it could be possible that not only during water evaporation but also condensation oxygen can be transformed into ozone. It is surprising that in 1874 – after many publications concerning the uncertainties of the test-paper – *Zittel* (and other authors) still believed that different color deepening corresponded to different ozone levels in the air (Zittel 1874). *Paul Friedrich August Ascherson* (1834–1913), a German botanist in Berlin, continued *Zittel*'s ozone observations in the Libyan desert between February and May 1876 (Ascherson 1877). He also concluded (as *Zittel*) that dew formation,

Table 4.18: Ozone observation stations in Saxony.

Period	place	observer
01.01.1868–01.07.1870	Leipzig[b]	Prof. *Bruhns*
01.07.1868–31.12.1870	Gotha	*Luedicke*
01.12.1868–31.12.1870	Gross-Breitenbach (Thuringia)	major *Bertram*
01.01.1868–31.12.1870	Greiz[b]	Dr. *Oettler*[c]
01.07.1868–31.12.1870	Werningshausen (north of Erfurt)	pastor *Beck*[d]
01.01.1866–31.12.1870	Zwickau[a, b]	teacher *Meier*
1872–1875	"*Festung*" [castle] Königsstein[b]	?

[a]Ozone observations were carried out also in the years before (1864–1865) by an unknown observer (Bruhns 1867, pp. 75–77), likely by *Meier*.
[b]The observations were continued until 1875: *Ozonbeobachtungen in Leipzig, Greiz, Zwickau, Festung Königstein*; Bruhns (1877, pp. 56–59) and Bruhns (1880, pp. 56–58).
[c]Medical council (unknown given name and living dates) who supervised the meteorological station 1867–1877 (after Lehmann 1884, p. 172).
[d]Very likely *Johann Gottlieb Beck* (unknown living dates), cantor, and teacher.

according to the "theory" of water evaporation and condensation, is responsible for the strongest "ozone reaction" (Ascherson 1877, p. 88).[1031] *Oswald Hüller* (or *Oswaldus Hueller*), a German physician in Driburg near Paderborn (unknown living dates) measured ozone in 1871 (from July to October) and believed that forests and waters are sources of ozone (Hüller 1872). The German physician (balneologist) *Friedrich Zimmermannn* (1843–1910) carried out between April and May 1873 (28 days) ozone observations on the island of Helgoland (Zimmermann 1873). *August Georg von Lösecke* (1837–1912), a pharmacist in Hildburghausen, conducted ozone observations using *Lender*'s paper (Lösecke 1874). His conclusion that it would be desirable to carry out ozone observations at different locations and with the observation of meteorological, clearly shows that the literature on this issue was unknown to him (and many other observers), although he began his publication with the comment that in recent times multiple attentions was drawn to ozone.

The German forest scientist *Max Friedrich Kunze* (1838–1921), who studied in Tharandt (1857), Giessen and Leipzig and became 1870 professor for mathematics in Tharandt, and was the founder of "*forstliches Versuchswesen*" [forest experimentation in the field], measured ozone in June and July 1876 in the forests of Halatte (*Forêt d'Halatte*) in Picardy and Ermenonville (both north of Paris); he found less ozone in the forest than outside, and (only for coniferous forest) in 14 m height above the canopy more ozone (Kunze 1877).

Hermann Oeffinger (1842–1889), a district physician of Müllheim (Baden), communicated that he has established on 1 June 1874, numerous stations in the world that

1031 From today's knowledge we can assume that this finding was due to photocatalytic formation of hydrogen peroxide in dew after sunrise.

begin with ozone observations using the same method and expressed the request to friends of meteorology and natural science outside Europe to participate within this network (Oeffinger 1874); I have nothing found in the literature on "his network" and any answer to his request.

Herrmann Emil Schelenz (1848–1922), pharmacist in Rendsburg (Schleswig-Holstein) conducted ozonometric observations in St. Peter (North Sea coast) and found there "more" ozone than at Rendsburg, writing (Schelenz 1886, p. 1017)

> Ich möchte die Heilkraft der Seeluft auf ihre absolute Reinheit und ihren Reichtum von Ozon zurückführen [I like to attribute the healing power of sea air on its absolute purity and the abundance of ozone].

However, *Schelenz* adapted the "ozone influencing factors" like water evaporation as a source, and notified without knowing already published criticisms that on a hot day ozone is less and at night more.

The German pharmacist *Eugen Rudeck* (1854–?) from Breslau hold a lecture on ozone (Rudeck 1886), "nothing new communicating" as notified in "*Archiv für Pharmacie*" (**225**, 222), and as "highly interesting" notified in "*Jahresbericht über die Fortschritte der Pharmokognosie, Pharmacie und Toxikologie*" (**46**, 126), but presenting a new ozonometer (Fig. 4.36). *Rudeck* communicated that he conducted 800 measurements in the Izera Mountains, namely in Flinsberg (today Świeradów Zdrój in Poland), and found that the ozone content does not increase proportionally with the altitude. He believed the view that ozone is gained through electricity and friction due to moving water, and named turpentine oil, emanated from conifers, "*Ozonerreger*" [ozone cause] because it incorporates large amounts of oxygen, and writes (Rudeck 1886, p. 1225):

> Durch die Verharzung lagern sich die einzelnen ausgeschiedenen Sauerstoffatome an einander und gehen zu je drei in ein Molekel Ozon über [Through the resignification the individual discharged oxygen atoms attach and each three transfer in one molecule of ozone]

It is remarkable that 15 years later *Rudeck* remained his view on ozone formation by plants and, moreover, he communicated a curious view on ozone formation at haylofts: due to the agglomeration of oxygen in wet and warm layers of hay at lofts in combination with the water evaporation gained ozone due to friction and electricity (Rudeck 1902); now, in the year 1902, *Lenard* has found in 1900 that ozone is gained by ultra-violet light, and the idea that the slow oxidation of turpentine produces ozone was already rejected in the 1870s by *Kingzett* and *Wolffenstein*, he still named these oils "*Ozonträger*" [ozone carrier].

4.4.5.4 Observations of atmospheric ozone in Austria

In the nineteenth century, Austria (Empire of Austria 1804–1867 and Austro–Hungarian Monarchy 1867–1918) was a multinational state comprising modern-day

Lakmus. Roth.	Phenol-phtaleïn. Roth.	Curcuma. Braun.	Lakmus. Blau.
1. **Dunkelroth.** Sehr viel Säure.	1. **Sehr blassroth.** Sehr viel Ozon.	1. **Dunkelbraun.** Sehr viel Ozon.	1. **Dunkelblau.** Sehr viel Ozon.
2. **Roth.** Viel Säure.	2. **Hellroth.** Viel Ozon.	2. **Braun.** Viel Ozon.	2. **Blau.** Viel Ozon.
3. **Hellroth.** Säure.	3. **Roth.** Spuren Ozon.	3. **Hellbraun.** Spuren Ozon.	3. **Hellblau.** Ozon.
4. **Blassroth.** Spuren Säure.	4. **Roth** die Grundfarbe. Kein Ozon.	4. **Gelb** die Grundfarbe. Kein Ozon.	4. **Ganz schwach** blau-roth. Spuren Ozon.
No. I.	**No. II.**	**No. III.**	**No. IV.**

Ozon ist nicht nachweisbar, wenn Tab. I. zugegen ist.

Tab. IV. dient auch zur Messung alkal. Luftbestandtheile.

Figure 4.36: *Rudeck*'s ozonometer: the filter paper, impregnated with different indicators and subdivided into 16 squares, should indicate the presence and/or absence of acids, alkalis and ozone in air in four degrees: "*sehr viel*" [plenty of], "*viel*" [a lot], "*genügend*" [ample], "*mäßig*" [moderate] (Rudeck 1886).

Austria, Hungary, Czechia, Slovakia, parts of Poland, Ukraine, Italy, Slovenia, and Croatia. In the 1850s in Austria were carried out ozone observations at 21 stations which were stepwise implemented (Krakau, Senftenberg, Czaslau, Scheunitz, Neusohl, Szegedin, Wien, Kahlenberg, Mauer, Kaltenleutleben, Kirchdorf, Klagenfurt, Venedig, Salzburg, Bad Gastein, St. Maria, Stilfserjoch, Lemberg, Prag, Linz)[1032] – a few of them are reported in this volume; monthly means were published, and the Cracow values are given in day and nighttime means (Anonymous 1856).

Augustin Reslhuber (1808–1875), who first studied theology in Linz and then astronomy in Vienna, became 1841 professor for natural history at "*Stiftsgymnasium*" [collegial high school] in Kremsmünster and 1857 director of the observatory Kremsmünster, is the pioneer of ozone research in Austria. In August 1853, *Schönbein* traveled to Kremsmünster (near Linz in Upper Austria) and Vienna to agitate for

1032 Here listed with the German name; see text for actual names.

Table 4.19: Results from ozonometric observation in Kremsmünster in October 1853 – October 1854 (Reslhuber 1855, 1856).

weather condition	*n*	day mean	nighttime	daytime
wet days	72	6.83	7.17	7.17
dry days	41	1.72	3.96	2.84
bright days	102	3.71	4.23	3.18
bright days October–March	47	5.55	6.10	4.99
bright days April–September	55	2.50	3.35	1.64
days with snow fall	18	8.93	9.72	8.14
days with rime formation	1	10.75	11–12	10.0
days with thunderstorm	11	4.82	6.72	2.91
total mean	113	5.21	5.91	4.51

ozonometric observations based on his method, which have already begun in September 1853 at the observatory (Reslhuber 1855). The Central Meteorological Institute in Vienna [*k. k. Central-Anstalt für Meteorologie und Erdmagnetismus Wien*][1033] then organized a network of stations with local observers, who remain unfortunately almost unknown. Already in November 1854, *Reslhuber* reports to the Academy in Vienna on his observations from September 1853 until October 1854 (Reslhuber 1855), and later until December 1855 (Reslhuber 1856), comparing the results from Kremsmünster with other stations, namely Vienna. *Reslhuber* made day and night observations, studied the relations between the ozone reaction and pressure, temperature, humidity, wind (direction and power), cloud types, and bright days and days with precipitation, snow, rime, thunderstorm, presenting monthly means and annual means (Table 4.19). An important observation *Reslhuber* made by exposing the test-paper only four two hours: longer observational time does not intensify the blueing, but could result in discoloration. During experiments, he also found that with a longer exposition of test-papers in the air the discoloration is due to the evaporation of iodine. Another important observation he made with the two-hour exposed papers (which is confirmed by modern in-situ measurements of tropospheric ozone) is that "[...] *ein schneller Wechsel der Luft eine größere Ozonreaction zur Folge hat*" [... a fast change if air mass results in a larger ozone reaction]; Reslhuber (1856, p. 366). *Reslhuber* further noticed "*dass bei Gewittern sich der Ozongehalt nach der Menge und Art der sie begleitenden Niederschläge richte*" [that during thunderstorm the amount of ozone is based upon the amount and kind of precipitation]; (Reslhuber 1856, p. 366).[1034] The general variation of ozone he describes as follows (Reslhuber 1856, p. 356):

1033 Founded in 1851 in the district Wieden, moved in 1872 to Hohe Warte in Wien-Döbling; first director *Karl Kreil* (1798–1862) 1851–1876, followed by *Carl Jelinek* 1863–1876 and *Julius Hann* 1877–1897.

1034 Friesenhof (1904) noticed from long-term observations that local thunderstorm results in less ozone than steady rain and at lightning in distance the amount of ozone is very low.

> Überhaupt ist in den kälteren, feuchten, trüben Monaten der Ozongehalt am grössten, in den warmen, trockenen, heiteren Monaten am kleinsten [Generally, the amount of ozone is largest in the cooler, moister, cloudy month, in the warmer, dryer, bright month lowest].

To prove the idea that atmospheric electricity causes ozone, *Reslhuber* conducted a series of experiments in 1855 with simultaneous measurements of atmospheric electricity, but he found that not in all cases (indeed in most cases) there was no relation between electricity and amount of ozone (Reslhuber 1856, p. 375). The data (Table 4.19) clearly show that on days with a thunderstorm the mean ozone reaction is less than the annual mean. Also, other observers (e. g., *Ebermayer*) found that lightning does not influence the amount of ozone in the air.

As suggested by *Karl Kreil*, *Joseph Georg Boehm* (1807–1868)[1035] begun in January 1854 at the Prague observatory with ozone observation (Boehm 1858). Already in the first year, *Böhm* found the test-paper unreliable. To investigate whether the location has a specific influence on the amount of ozone, he organized another location, Baumgarten (today Stromovka park in Prague), where Dr. *Jenner* carried out the observations in 1857. Because he found the ozone reaction by a factor of two higher in the park, *Boehm* conducted in August and September in Königstadtl (today Městec Králové, about 50 km NE of Prague) observations whereas *Karliński* continued the observation in Prague. The "ozone value" was to be found by a factor of four higher at Königstadtl. Independent from *Boehm*, the pharmacist *Hartmann* of Kuttenberg (today Kutná Hora, about 60 km E of Prague) exercised ozone observations and, comparing the mean values from October to December 1857 with the values of Prague resulted in 20 times lower values at Prague. Finally, *Boehm* conducted from October 1857 until February 1858 with the help of *C. Andres* and *Lippich* apart from the observatory in Prague at Hradschin and St. Katharina measurements and found the amount of ozone at Hradschin (the castle) comparable with that detected at the observatory but at Katharina to be much higher. *Boehm* communicated two remarkable conclusions, first (Boehm 1858, p. 428) "how difficult it is to derive general regularities from short observational series and from individual sites of observation" (Fig. 4.37), and, secondly that the air pollution of Prague, located in a valley with low wind speeds, is the reason for low amounts of ozone; it is worth to cite him (Boehm 1858, p. 438):

> [. . .] den dicken Qualm [. . .] zu sehen und zu riechen, und man wird keinen Augenblick daran zweifeln, dass man es mit einem completen chemischen Laboratorium zu thun habe, in dem kaum eines der zahlreichen Reagentien fehlen dürfte [. . . thick smoke [. . .] to see and to smell, and nobody would not doubt for a moment that one has do with a complete chemical laboratory, in which barely one of the numerous reagents is missing].

1035 Austrian astronomer and mathematician who studied in Prague and worked as assistant 1833 at Vienna Observatory. 1839 full professor in Salzburg (1848 rector), 1852 director of the observatory in Prague and professor of astronomy at Charles University.

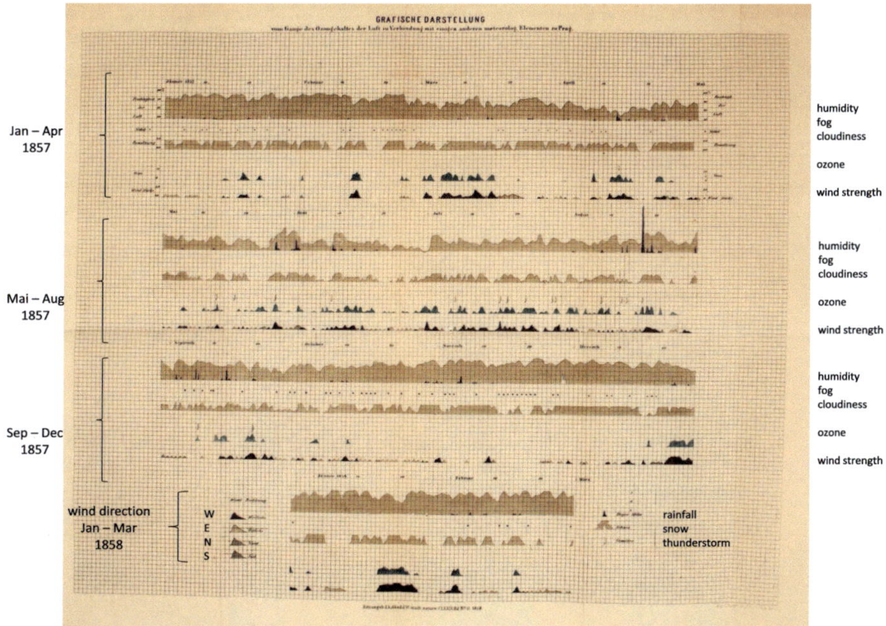

Figure 4.37: "*Grafische Darstellung vom Gange des Ozongehaltes der Luft in Verbindung mit einigen meteorologischen Elementen zu Prag*" [Gaphic representation of the history of the amount of ozone in air in combination with some meteorological elements]. This is the first printed diagram of timely variation of ozone (Boehm 1858).

Concerned the health issue of ozone, *Boehm* comments (translated from German) sardonically (Boehm 1858, p. 440) that he (if he would be a physician) "would not send his patients because of a little blue paper to this or that place". In line with the Austrian ozone network (Table 4.20), the Austrian meteorologist *Johann Prettner* (1812–1875) must be named who also was the founder of the Carinthian [Kärntner] weather observation network (Prettner 1874).[1036] The physician *Anton Emerich Flechner* (1807–1883) conducted in Wien 1855–1859 ozone observations in relation to illnesses (Flechner 1855–1860).

The Austrian naturalist *Franz Neumann* (1837–?) is the first (Neumann 1857)[1037] who presents not own data but an analysis of relations between the amount of ozone

1036 A necrology on *Prettner* see: Carinthia **65** (1875) 81–95 (including a list of *Prettner*'s published works). *Prettner* published regularly "*Meteorologische Beobachtungen zu Klagenfurt*" in "*Jahresberichte des naturhistorischen Landesmuseums in Kärnten*".

1037 According to the style of publications at that time, no references are given, and it is difficult to relate named authors to the data and the stations cited. *Neumann* refers Dr. *Wolf* in Bern, the Radcliffe observatory in Oxford and *Schiefferdecker* in Königsberg. He further refers *Quetelet* (Brussels), Dr. *Lamont* (Munich), and Dr. *Böckel* or *Boeckel* (Straßburg). He presents data from stations in Vienna, Prague, Cracow, Kremsmünster, Stanislau (Galicia in former Austria, today Iwano-Frankiwsk

Table 4.20: Stations in the Austria-Hungarian Empire (note that not all are listed and that observations at some stations could be extended over the year 1866 due to missing information). This network was built on the initiative of individuals but was soon coordinated by the Central Meteorological Service in Vienna [*k. k. Central-Anstalt für Meteorologie*] where all data were compiled.

Location	period	operator
Wien	1853–1920	*Boehm*, *Flechner* and others (see Lauscher 1983)
Kremsmünster	1854–1866	*Reslhuber*
Krakau	1853–1873	*Karliński*
Klagenfurt	1854–1874	*Prettner*
Prag[a]	1854–1866	*Boehm*
Lemberg	1855–1866	*Pierre*
Kirchdorf[b]	1855–1862	*Schiedermayr*[l]
Salzburg	1855–1866[b]	*Woldřich*
Ofen[c]	1856–1868	Dr. *Frenreisz* (1856–60) and Dr. *Schenzl* (1861–68)[m]
Brünn[d]	1858–1866	?
Kremsier[e]	1863–1868	Dr. *Gstein*[n]
Czaslau[f]	1855–?	?
Szegedin[g]	1855–?	?
Schemnitz[h]	1855–?	?
St. Maria[i]	1855–?	?
Altdala[j]	1898–1905	

[a]Between December 1857 and February 1858 at three places: Observatory, Hradschin, and St. Kathrina
[b]In Upper Austria at the river Krems
[c]Buda (today part of Budapest), further series 1871–1898
[d]Today Brno in Slovakia
[e]Today Kroměříž in Czech Republic
[f]See footnote 1037 on p. 538
[g]Today Szeged in Hungary
[h]Selmecbánya, today Banská Štiavnica in Slowakia
[i]Today St. Marien near Linz
[j]Ó-Gyalla (also written Ógyalla), today Hurbanova in Slovakia (near Bratislava): meteorological and terrestrial magnetic observatory 1898 (before astrophysical observatory)
[k]1855–57 and 1863 by *Schiedermayr* countryside, 1857–1862 in the city (St. Peter)
[l]*Karl Schiedermayr* (1818–1895) was an Austrian physician and botanist
[m]*Franz Frenreisz* (unknown living data) was a Hungarian physician; Dr. med in 1832. *Guido Schenzl* (1823–1890) was an Austrian naturalist, director of the "königlich ungarische Central-Anstalt für Meteorologie und Erdmanetismus" in Ofen
[n]Unknown

(Івано-Франківськ] in Ukraine), Lemberg (today Lwow), Czaslau (Čáslav in Bohemia), and St. Maria (liekely St. Marien today). Furthermore, he cites *Th. Brorsen* in Senftenberg (Austria) and Dr. *Gräger* (Mühlhausen). Annotations to stations and names not referred in the text of this volume: *Johann von Lamont* (1805–1879), professor for astronomy at Munich university; *Theodor Brorsen* (1819–1895), Danish astronomer (Senftenberg in Bohemia, today Žamberk in Czech Republic). Kremsmünster was operated by *Reslhuber*, Vienna by *Boehm*, Prague by *Pierre*, and Cracow by *Karliński*.

in the atmosphere and meteorological elements (air electricity, humidity, wind force, temperature, pressure) based on published and unpublished data in 1854 and 1855, and cited by Knop (1868, p. 83) who summarized, that 1) in January and February the amount of ozone is highest and in July and September lowest, 2) at low barometer reading the air ich richer in ozone, 3) at strong wind greater than at weak wind, and 4) corresponds the amount of ozone the atmospheric humidity and electricity.

The Austrian meteorologist *Karl* (*Carl*) *Jelinek* (1822–1876) reports several stations with ozonometric observations (21 stations) of the Austrian meteorological network which began in 1853 at Kremsmünster (*Reslhuber*), Jelinek (1866); Reslhuber (1856, p. 355) noticed that the Central Meteorological Institute in Vienna published the results of ozone observations from 20 stations ("*Übersicht der Witterung in Österreich im Jahr 1855*");[1038] Table 4.20, see also Table 4.16.

In Prag [Prague] and Ofen the lowest ozone values were observed among all Austrian stations; generally, day values are larger than nocturnal ozone. *Carl Fritsch* (1812–1879), a meteorologist and one of the founders of the meteorological society in Austria, wrote on the uncertainties of ozonometric observations within the Austrian network, citing *Jan Nepomuk Woldřich* (1834–1906), a Czech geologist, and observer in Salzburg, and *Prettner* who complain the uncertainty of the test-papers (Fritsch 1866).

One of the first long-term ozone measurement series in the world was established in Poland (Kraków) during 1854–1878. The Polish astronomer and meteorologist *Michal Franciszek Ignacy Karliński* [*Franz Michael Karlinski* written in German] (1830–1906), director of the Astronomical Observatory of the Jagiellonian University in Cracow, carried out observations from December 1853 until November 1874, presenting monthly, seasonal and total means (Karliński 1874); between 1855 and 1862 he also worked at the observatory in Prague. *Karliński*'s assistants were the astronomer *Jan Kowalczyk* (1833–1911) and the meteorologist *Daniel Wierzbicki* (1838–1901); Wierzbicki (1882). The data obtained by the method of *Schönbein*'s papers reveal that annually averaged concentrations of ground-level ozone in the second half of nineteenth century ranged between 11 and 24 ppb (Pawlak and Jarosławski 2014).

First quantitative estimations of ozone (with reliable values from today's view) in the air using ozonometer based on aspiration and titration of potassium iodide were conducted by Pless and Pierre (1856),[1039] who determined in Lemberg in 1855 in three experiments between 1.4 and 7.94 mg in 100.000 L of air (about 6–40 ppb),

1038 Jahrbücher der k. k. Zentralanstalt für Meteorologie und Erdmagnetismus, Bd. 1–8, 1848–1856. Jahrbücher der k. k. Zentralanstalt für Meteorologie und Erdmagnetismus, Neue Folge, Bd. 1–86: der ganzen Reihe Bd. 9–94. 1864–1949 (since 1904 Jahrbuch der k. k. Zentralanstalt für Meteorologie und Geodynamik).

1039 *Victor Pierre* (1819–1889), sometimes wrong written as *Viktor*; Austrian physician and physicist, professor of physics in Lemberg (today Lwiw [Львів] in Ukraine) 1851, at Prague 1857 and Vienna 1866 at "*Technische Hochschule*". Nothing is known on the co-author *Pless*.

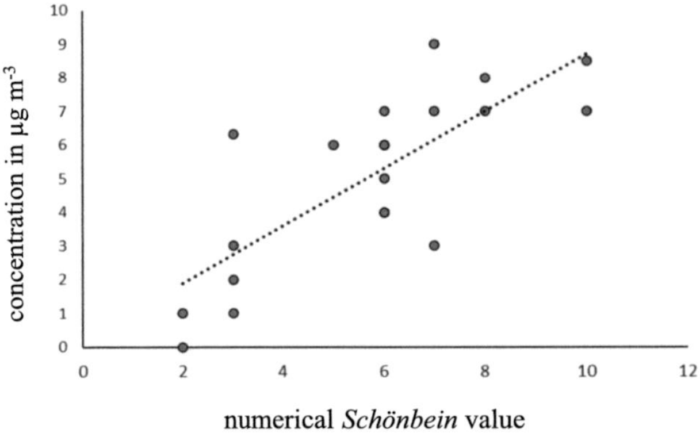

Figure 4.38: Correlation between ozonometric *Schönbein* paper values and simultaneously determined ozone using an aspirator and titration (data from Zenger 1858, p. 90). Note that at that time the formula of ozone (O_3) was not yet known and it was termed Ozone-Oxygen, i. e., the amount of "ozone" in the air is surely calculated as O_2. Furthermore, no information is given on the stoichiometry in the reaction between ozone and released iodine.

and the Czech-Austrian physicist *Karel Václav Zenger* [*Karl Wenzel* written in German] (1830–1908), who developed a new method for ozone determination, and found from 20 experiments (1000 L air sucked over 12 h daytime) in Neusohl a range of 0.002–0.01 (0.0058 ± 0.022) mg in 100 L of air (about 10–50 ppb); Zenger 1858). Using parallel *Schönbein*'s paper, *Zenger* compared the numerical values (scale up to 10) with the aspirator values (Fig. 4.38), showing a large scattering but a general trend of correlation.

Bernhard Freiherr von Wüllerstorf (1816–1883), Austrian admiral, who made a circumnavigation with the frigate Novara (Novara expedition 1857–1859) reports on a session of the Austrian Meteorological Society on March 1, 1867 (*Zeitschrift der Österreichischen Gesellschaft für Meteorologie* **2** (1867) 143) that he found everywhere ozone but at the crow's not more than aloft. *Carl* (*Karl*) *Haller* (1809–1887), physician in Vienna, conducted in the summers in Bad Fusch (today's name, near Mt. Grossglockner in Austria) 1872 and 1874 ozone observation and compared them with the result from the Central Meteorological Institute in Vienna (Haller 1874). He found the mountain air to be richer in ozone.

The Austrian physician *Joseph Schreiber* (1835–1908), who became in 1872 the second docent of climatology at the University Vienna (Faculty of Medicine), wrote e few remarkable phrases, being well worth citing here (Schreiber 1877):

> Das Vorhandensein von Ozon beweist jedoch indirekt, dass in solcher Atmosphäre keine organischen Stoffe vorhanden sein können [The presence of ozone proves indirectly that in such atmosphere no organic compounds are present] (p. 164).
> Das aus den Nadelhölzern ausschwitzende Terpentin besitzt nämlich unter allen Körpern am meisten die Eigenschaft, den Sauerstoff der Luft in Ozon umzuwandeln [The turpentine, evaporating from conifers, has he among all bodies most of all the property to convert oxygen of the air into ozone [p. 164].

> Sowie das Ozon auf chemischem Wege die Atmosphäre reinigt, so befreit der Regen (Schnee) die Luft auf mechanischem Wege von den organischen, gleichzeitig aber anorganischen Beimengungen [As ozone cleans the atmosphere in a chemical way, so rain (snow) cleans the air in a mechanical way from organic but also inorganic impurities] (p. 164).
> [...] die Meteorologische Central-Anstalt möge in Erwägung ziehen [...] dass an einzelnen Stationen auch Untersuchungen über den Gehalt der Luft an organischen Beimengungen und an Ozon angestellt werden [... the Central Meteorological Institute should consider [...] to carry out at selected stations observation on organic impurities and ozone] (p. 166).

Long-term ozone observations (1875–1904) were made by "*Freiherr*" [baron] *Gregor von Friesenhof* (1840–1913), who was a Carpathian German gentleman farmer at the village Alt-Krasno in district Bajmöez (former Hungary, today part of Krásno in Slowakia) in "Neutraer Camitate" (Nitriansky komitá, today in Slovakia). The "*agrarmeteorologisches Observatorium des Neutrathaler landwirthschaftlichen Vereines*" [agricultural-meteorological observatory of the Neutrathal agricultural society] in Krásno was 1875 founded and published monthly reports and annual "*Abhandlungen*" [treatises]; Friesenhof (1904).

4.4.5.5 Observations of atmospheric ozone in Switzerland and Italy

Johann Rudolf Wolf (1816–1893),[1040] Swiss astronomer, carried out between 1853 and 1855 daily observations in Bern (Wolf 1854, 1855, 1856); like many other observers, he found the "ozone reaction" much higher on days with rain, snow or thunder. In Wolf (1855) all daily values are listed together with those from Strasbourg (by *Boeckel*) and Cracow (by *Karliński*), and citing *Boeckel* that the "ozone reaction" is lowest during cholera epidemic and presenting an (obscure) relation between the amount of ozone in air and the mortality in Bern.

In Italy, the first who measured ozone, was Padre *Alessandro Serpieri* (1823–1885), an Italian scientist known for work in astronomy and seismology, who founded the meteorological observatory at Urbino in 1850 and measured ozone until 1883 (published in "*Bullettino dell'Osservatorio Meteorologico di Urbino*"). The Italian physician *Gaetano Strambio* (1820–1905) carried out ozone observations in Milano between July and November 1855 during an epidemic and showed that the ozone variation has no relation to the occurrence of illnesses (Strambio 1856). *Strambio* also wrote a nice essay on the early history of ozone, on the different ideas on its constitution (citing *Schönbein*, *Marignac*, *Baumert*, *de la Rive*, *Berzelius*, *Faraday* and *Houzeau*), and citing many observers (partly with bibliography) like *Spengler*, *Böckel*, *Wolf*, *Yung*, *Clement*, *Schwarzenbach*, *Bérigny*, *Richard*, *Silbermann* and *Simonin*.[1041]

1040 *Johann Rudolf Wolf* was professor of astronomy in Bern (1844) and Zürich (1855), and director of the Bern Observatory in 1847.

1041 *Clement* is mistaken; it is *Theodor Clemens*. *Yung* and *Richard* not identified.

Luigi Palmieri (1807–1896), Italian meteorologist, vulcanologist, and director of the meteorological observatory in Napoli, was famous for his scientific studies of the eruptions of Mount Vesuvius, conducted meteorological, electrical, and ozonometric observations since 1861 (Palmieri 1863, 1865, 1866). Furthermore, he was the first in 1882 to detect by spectral analysis the presence of helium on Earth in the lava of Mount Vesuvius. Another Italian geologist and volcanologist, *Orazio Silvestri*, conducted ozone observation in Pisa (Silvestri 1861). *Francesco Denza* founded the meteorological observatory in Moncalieri (near Turino) in 1866 and measured ozone in 1867 (Denza 1868). *Domenico Ragona* (1822–1894), director of the observatory in Modena made 1867 ozone observations (Ragona 1868).

According to Polli (1850), many ozonometric observations were conducted in Italy, and he wondered, why, using the same papers, the data show no consensus. I call here as early Italian ozone researchers the botanist *Giberto Scotti* (1818–1880) and *Giovanni Polli* (1812–1880), a Milanese physician who earned his laureate in medicine in 1837 and taught medicine and chemistry in various schools in Milan and was one of *Mantegazza*'s teachers (Scotti 1853, Polli 1850).

4.4.5.6 Observations of atmospheric ozone in the United Kingdom

The earliest ozonometric observation in the United Kingdom was made by the British physician *Thomas Barbour Moffat* (1813–1882) in 1848 in Harwarden, Flintshire, who continued the daily observations together with meteorological parameters and registration of different diseases until 1861 (Moffat 1853, 1854, 1861, 1862a,b, 1873). Since 1852, he often reported on "medical meteorology and ozone" to the Annual Meetings of the British Meteorological Society (published as reports before the first volume of *Proceedings of the British Meteorological Society* in 1862 did appear).[1042] *Moffat*'s observations (and as a medical doctor he also studied chemical aspects of the ozone test-paper and constructed the first ozone-box, providing the most reliable measurements at that time) are worldwide first and unique in its systematic approach and careful conclusions; and it is wondering, that many following observers conducted much less reliable measurements and draw wrong conclusions. In February

1042 In this volume further contributions to *ozone* are referred (here are only those of *Moffat* and *Barker* referred): *John Charlton Bloxam*: Ozone at Newport and Stapler (Sessions 1856–57); *H. S. Eaton*: Atmospheric conditions and ozone (Session 1858–59); *Smallwood*: why of observating ozone (Session 1859–60). The Ninth and Tenth Reports of the Council of the British Meteorological Society for the years 1859 and 1860, contains the scientific papers: (pp. 32–45) On some of the atmospheric conditions favourable to the development of ozone, as deduced from observations taken at Little Bridy, Dorset, from February 20th 1857 to February 20th 1859, by *Henry Storks Eaton* (1835–1909 was meteorologist and president of three Societies 1876–77). *Moffat*'s work on ozone and his observations at Harwarden were first referred to in *Parliamentary Papers* **57** (1854–55) pp. 71, 89, 102. Nothing is known on *Bloxom*.

1852,[1043] after four years of ozone observation, referring the British physician *William Addison* (1803–1881) who stated that "the changes in the weather are prejudicial to health, and that mean readings of meteorological instruments do not show the connection between the prevalence of diseases and atmospheric conditions" (Addison 1853), Moffat (1853) concluded:

> I feel confident that meteorological tables cannot be of any value, in a medical sense, unless they are formed on daily observations, [. . .], and that ozone never appears without a decrease in the reading of the barometer, and that its presence or absence in the air depends entirely on the direction of the wind.
> If the wind continues from the northern points [. . .] from N. W. to S. E. [. . .], the barometer readings will remain high, there will be few or no cases of diseases, and no ozone; but if the wind passes [. . .] southern [. . .] to points from N. W. to S. E., the reading of the barometer will decrease, there will be (diseases); there will be ozone, and general change in the weather [. . .]
> [. . .] that decreasing reading of the barometer will follow in a few hours, and ozone will be detected, or not, just as the northern or southern current prevails.

Moffat (1854) reports on the preparation of ozone test-paper and his chief conclusions:

> [. . .] owing to the action of light and the influence of atmospheric currents in producing decomposition of the iodide of potassium, it was necessary, in order to secure uniformity in results, to place the test paper in darkness, and to keep them protected from atmospheric currents.
> That ozone periods almost always commence with decreasing readings of the barometer, and increase of temperature; and terminate with increasing readings of the barometer, and decreases of temperature.

Moffat (1861, p. 89) noticed that "the diseases may be attributed rather to the vicissitudes of weather than to ozone", and:

> In conducting ozone observations, it must be borne in mind that light causes colorization of the test-papers, and that moisture, sulphuretted hydrogen and ammonia cause loss of colour.

Charles Leeson Prince (1821–1899)[1044] was a local doctor at Uckfield who constructed an ozone-box (Prince 1854). In connection with the cholera epidemic 1853–54, *James Glaisher* used *Moffat*'s and *Schönbein*'s paper for daily tests at 17 stations in London (Glaisher 1855). He remarks (Glaisher 1855, p. 71):

> It was found that the paper prepared by Dr. Moffat were more sensitive than those of Dr. Schönbein, and accordingly indicated the presence of ozone when none was indicated by those of Schönbein.

1043 He prepared a short paper for the spring meeting of the British Meteorological Society. Obviously, his reading to the Society contained much more information, as seen from another short communication (Moffat 1854).

1044 He was interested in astronomy and meteorology, giving meteorological observations for Uckfield for 1843–1870; in 1872, he retired from general practice and moved to Crowborough to create an observatory.

At the Royal Observatory Greenwich, the ozone observations were continued until 1870 by *Glaisher*, Superintendent, and his assistant *Thomas Downs* (1822–1904) until 1863, followed by *William Carpentier Nash* (1841–1926), published in "*Results of Magnetical and Meteorological Observations*" (*extracted from the Greenwich Observations*) in yearbooks from 1856 until 1870 (no further description); hourly observations from August 1856 to December 1857 made by Lippincott (1869), printed only 1856 in the *Greenwich Observations*, afterward monthly means; no description of the ozonometric method. In 1862, *Glaisher* used *Moffat*'s and *Lowe*'s test-paper on a balloon ascend to study the "ozonized condition of the atmosphere" and "to make known for the first time the results obtained" (*Proceedings of the British Meteorological Society*, **1** (1861–63) pp. 357–358), Fig. 4.39.

Figure 4.39: *James Glaisher* and *Henry Coxwell* (1819–1900) from the Greenwich Observatory took a balloon rise on 17 July 1862 to the highest altitude so far (36,752 Parisian feet, about 11,000 m) without oxygen and lost nearly her life, they measured ozone, temperature, and pressure; contemporary illustration after Flammarion (1888, p. 147). Lender (1874, pp. 44–47) describes the rise in detail (in German). See also Marriott (1904, p. 16).

Tomas Herbert Barker (1814–1865), a British physician at Bedford, who was familiar with *Moffat* and interested in medical meteorology, preferred daily observations instead of weekly; although occasionally, hourly observations would be desirable (Barker 1853). He published a famous intercomparison between *Schönbein*'s and *Moffat*'s ozone test-paper (Barker 1856), confirming *Glaisher*'s result; he found from 21 monthly observations (1854–1856) relative values of the amount of ozone (*Schönbein* / *Moffat*) as monthly means (17.33 / 41.13) and daily means (1.59 / 3.79).

Stimulated by *Schönbein*'s statement that ozone is the greatest natural disinfectant, was the British surgeon *Alfred Haviland* (1824–1903) at Bridewater who conducted in March 1854 simultaneous ozone observations (mean ozone level given in parenthesis) at different distances from the center of Bridgewater (0), 3/4 mile from Bridgewater (4.2) and two sites 3 1/4 miles away at a park (6.1) and a hill site (10.3); Haviland (1855, pp. 130–131). Local reports of epidemic and endemic diseases (Anonymous 1855, pp. 58, 302–303) note that *Haviland* "furnishes a continuation of his medico-meteorologic observations") and publish the ozone results from December 1854 until August 1855.

Dr. *Conway Evans* (c. 1830–1892), who was the Medical Officer of Health Strand District in London in 1855 (*Strand* was a local government district within the metropolitan area of London from 1855 to 1900), and editor of "The Health of The Metropolis" (Weekly Return of the Rise and Decline of Diseases in London) and author of "Reports On The Sanitary Conditions Of The Strand District", wrote in July 1857 (Evans 1857, cited after Fox 1873, p. 96, Evans and Pittard 1858):

> The N. E. wind reaching Hackney from the country is found to be highly charged with ozone, but on arriving at Fulham, after having crossed London, it appears to have lost almost all traces of this substance.

Thomas Shapter conducted ozone observations for 6 years at 3 localities in Exeter, Devon, and writes (Shapter 1862, p. 54):

> Perhaps the only features of any certainty to be noted here are, that in the city the amount developed is at its minimum; that distances from the city increases the amount; and that recent observations on the high ground in the neighbourhood (the Pennsylvania Hill) shows not only an increase but a full development to be almost constant.

Edward Joseph Lowe (1825–1900), a renowned English botanist, meteorologist, and astronomer, who founded in 1855 the Beeston Observatory (near Nottinghamshire), recommended that "the test should be in a dark box; and no box was former to answer"; he constructed "Lowe's Ozone box" (Lowe 1863, p. 534). *John Smyth* (1830–1914), a civil engineer at Milltown, Banbridge (Ireland) constructed in 1864 an ozone-box and tested it in the years 1865, 1866, and 1868, writing "By the advice of Dr. Andrew I have tried pure linen instead of paper and found it to work very well" (Smyth 1869, p. 380); Fig. 4.40.

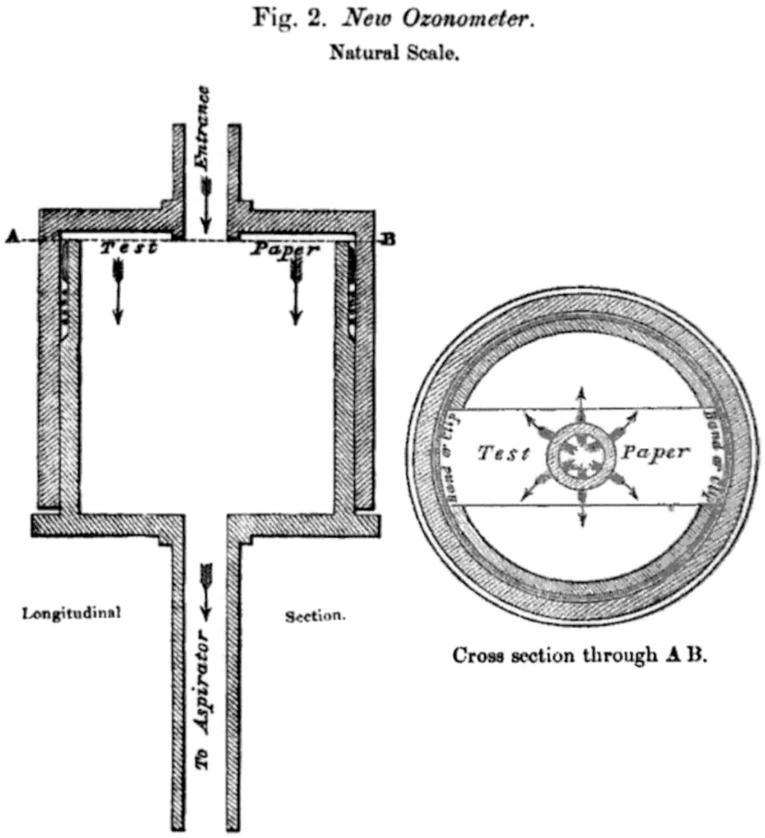

Figure 4.40: New ozonometer by *John Smyth* (using a colored ozone-scale), Smyth (1869).

Robert Angus Smith found that the air of Manchester did not react with ozonoscopic papers nor was there any reaction in the country when the air came from the city, which he attributes to the absorption of ozone by the products of combustion arising from the large amount of coal consumed in the city (Smith 1859a). In the vicinity of Manchester (e. g., Salford and Eccles), several ozone observations were made by *Baxendell* and *Mackereth*, likely in the period 1860–1865. Mackereth (1868) states[1045] that no ozone is found in air that passes over the town. In Salford, *Mackereth* "hardly any has been detected with the most sensitive tests". *Mackereth* who conducted for more than three years in Eccles (four and a half miles west of Manchester) where he lived, and found considerably less ozone than obtained by Mr. Vernon[1046] at Old Trafford, two miles

1045 He quoted observers *Samuel Marshall*, of Kendall (headmaster of Stramongate School), Mr. *Heap*, of Royton near Oldham, and Mr. *Thompson*, of Wilmslow.

1046 *George Venables Vernon* (1831–1878) was a meteorologist in Manchester who observed since 1849 at Old Trafford many meteorological parameters. Vernon (1868) compared unsized papers prepared by nearer to Manchester but less densely populated and "superior sanitary condition", concluding (Mackereth 1868, p. 213) "Hence it is clear that large towns either destroy ozone or prevent its manifestation". But ozone is detected in considerable quantities on the sea coast, and in rural valleys far removed from towns, he writes (Mackereth 1868, p. 214), but "far larger amounts are detected in elevated districts, as on hills and mountains.

Baxendell (1869) stated that from his "observations it seemed probable that the amount of ozone near the earth's surface depended upon the height at which clouds are formed in the atmosphere". *Baxendell* continued his observations and found that during fogs and haze were prevalent no ozone in the air, and "in clear states of the air the test-papers were more or less coloured". He further writes (Baxendell 1876, p. 114)

> Moreover, not only does haze prevent the coloration of the test papers, but it often rapidly bleaches those which had already been coloured by the action of ozone.

A similar observation was made by *Albert Lévy* (see p. 556). *Baxendell* concluded:

> It appeared, therefore, from these results that atmospheric ozone was absorbed or decomposed by haze or fog, and given off or produced when evaporation changed haze or fog into invisible aqueous vapour.

Baxendell's further discussion is a bit inconsistent, noting the low solubility of ozone in water, the different generation of "vesicles or globules" in fog, haze, and clouds, and from waterfalls or fountains, and "their effects upon the oxygen or ozone in the atmosphere might be very different". In a series of observations in 1872 on the lee side of the reservoirs of the Manchester Corporation Waterworks he observed "no effect in increasing the amount of ozone in the air passing over it" (Baxendell 1876, p. 115). Moreover, "experiments made at Southport on the action of spray from the sea led to no satisfactory result". In 1875, *Baxendell* carried out at the fountains at Arnfield and Hollingworth, and found in 17 experiments leeward significant more "resulting tints" than windward, concluding that "the spray from the fountain on evaporating gives off or produces atmospheric ozone, and in this respect is similar to ordinary fog or haze" (Baxendell 1876, p. 119). In some of the experiments the paper "became sensible damp", and did not show colorization, now concluding "that spray has, like for or haze, the power of absorbing or decomposing atmospheric ozone" (Baxendell 1876, p. 120). *Baxendell*'s subsequent statement that "the air has the power of condensing oxygen upon its surface ["water in the visible form of clouds, fogs, and haze"] into a thin film of ozone" reads like a modern view of photocatalytic conversion of oxygen into super oxide anion (and subsequent formation of hydrogen peroxide).

Moffat's method with *Moffat*'s ordinary (sized) papers, and found on average a mean ratio to be 2.22 between the old (unsized) and new (sized) paper.

Charles Daubeny (1795–1867), English chemist, botanist, and geologist, conducted 1864–1866 ozonometric observations during the winter at Torquay, and during the rest of the year at Oxford, and studied different plants, concluding that "green parts of plants during the day are constantly evolving ozone" (Daubeny 1867, p. 26).

As many observers (*Rogers*, *Möhl* and *Dietrich*, *Bineau*, *Conraux*, *Böhm*, *Shapter*, *Smith*, and others; see Table 4.17), *John William Tripe* (1821–1892), English physician and since 1871 President of the Royal Meteorological Society, observed the "deozonisation" of air passing over London (*Report of the Council of the British Meteorological Society* 1858, pp. 7 and 36); cited after Fox (1873, p. 95).

The Reverend *Robert Main* (1808–1878), an English astronomer, who became in 1860 the director of the renowned Radcliffe Observatory at Oxford University, conducted 1856–1865 ozone observations. The observatory was founded in 1772 and is also known for stratospheric ozone research by *Dobson* since 1922; results of all observations were published annually in *Astronomical and Meteorological Observations made at the Radcliffe Observatory, Oxford* (see also Baxendell (1869) for data).

In 1863, a "Committee on Ozone Observations" was founded in Edinburgh with *Arthur Crichton Mitchell* (1864–1952), Scottish physicist and meteorologist, who already used an aspirator in Alger 1855 for ozonometry, "with the view of eliciting information on that mysterious element of the atmosphere, *ozone*" (Mitchell 1869, p. 252, Anonymous 1873). *Mitchell* also reports (with illustrations) on *Mitchell*'s dry aspirator (1855), *Andrew*'s water aspirator, and *Smyth*'s box for the ozone paper (Mitchell 1869, pp. 256–258). Mitchell (1880) reports that 1857–1875 at 28 stations in Scotland ozone observations were carried out.

Hourly observations once a week were made by *Robert Cann Lippincott* (1814–1890), landowner and meteorologist at Over Court, an Elizabethan Manor house, in Over, a small village about a mile south west of Almondsbury village, near Bristol, owned by the Cann-Lippincott family. *Lippincott* was the first who studied the "influence of sun's hour-angle on the development of ozone" (in modern terms the diurnal cycle) using the hourly data from Greenwich in 1857 (Lippincott 1869), and noticed (Lippincott 1869, p. 160),

> I. That the *maximum* development of ozone takes place when the sun is near the *eastern horizon*; and that there is a tendency to a second maximum when the sun is near the western *horizon*.
>
> II. That the *minimum* development of ozone occurs when the sun is near the zenith; and that there is a tendency to a second *minimum* when the sun is near the *nadir*.

Lippincott (1871) also published hourly observations made daily from February to November 1869 at his observatory at Over Court, 6–7 miles from Bristol, without any further comments. Fox (1873, p. 64) noticed that they differ from those in 1857 (Fig. 4.41),

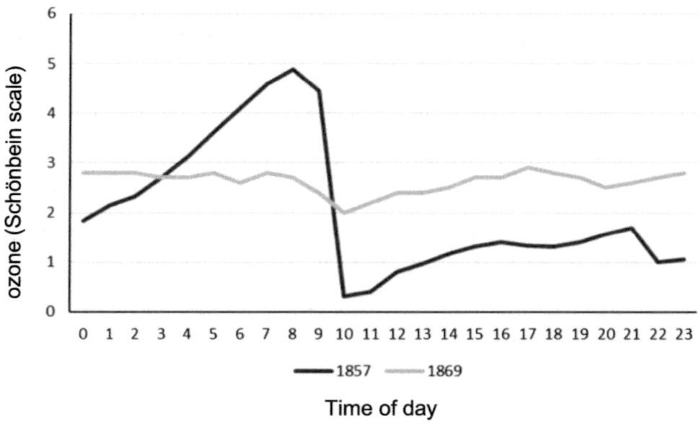

Figure 4.41: Mean diurnal variation of surface ozone observed hourly (once a week over the whole year 1857) by *Schönbein* test-paper; data from Lippincott (1869, p. 159). The diurnal variation shown for 1857 is "typical" for test-papers with humidity interference. Worldwide the first 1-hour-measurements of daily variation.

> [...] to in the fact that they were made daily instead of weekly, also display the *midday minimum* and the maximum at sunrise and sunset. They do *not*, however, show the existence of a *minimum when the sun is near the nadir*; but, on the contrary, they indicate a very decided excess.

These are the first measured diurnal variations of surface ozone, however, in 1857 almost due to the artefact increase in nocturnal ozone caused by the increasing humidity. The diurnal variation made in 1869 seems to be realistic, basically showing no diurnal slope. And thus representing that ozone at that time was exclusively of stratospheric origin. The fact that the 1869 value as daily mean (2.63) is larger than the value in 1857 (2.06) shows not an increase but the year-to-year variation in ozone (not really understood at that time).

Fox (1873, p. 65), however, is not surprised about the maximum towards the sunrise, citing the experience of *Osann*, and arguing that the atmospheric moisture is precipitated, based on the belief that this aqueous precipitation is attended by the development of ozone due to the electricity. *Fox* (Fox 1873, p. 71) "own observations at Scarborouh have not exhibited an increased amount of test-colorization, either during thunderstorm or auroral displays". Fox (1873) cites the following observers of ozone without giving a reference and the year of observation: Mr. *Harrison* of Weybridge Heath (p. 79), an anonymous observer at the Botanical Garden in Edinburgh (p. 79), Mr. *Sawyer* at Brighton (p. 79), Mr. *Matthewson* at Lerwick, Shetland (p. 80), Rev. J. *Slatter* of Streathey, Berkshire (p. 80), Mr. *Harris* at Worthing (p. 80),[1047] and Mr. *Crosthwaite* (p. 82). Fox (1873, p. 80) cites further stations without giving the name of the observer and year of observation: Eastbowner, Bath, Guernsey, Culloden, Island of Isle, Cockermouth.

1047 Most likely: *Frederick Ernest Sawyer* (1852–1891), and the surgeon *William Harris* (1808–1896).

Thomas Spencer Wells (1818–1897), British surgeon, reports on ozone observations in Edinburgh, October – December 1853 by an unknown observer (Wells 1854, pp. 839–841).

"The observations of ozone are now nearly discontinued, but have given place to far more interesting ones of the ionization of the air", writes Rev. *Fenwick William Stow* (1841–1904), who became in 1866 member of the Royal Meteorological Society, and conducted meteorological observations at the lighthouse near Whitby, and "found the sea air so abundantly loaded with the body or bodies which color the tests, as render it extremely difficult to estimate their highest tints" (cited after Fox 1873, p. 100).

John Mulvany (?–1927; lost at sea) was a medical doctor of Portsmouth and staff surgeon afloat, who made ozone observations between 1867 and 1877 around the world (Lake Erie, Mozambique, SE Africa, Persian Gulf, Zanzibar, Madagascar, Falkland Island), who concluded from his observations that "the action of rain and other conditions favorable to ozone is easily understandable" (Mulvany 1880, p. 188), and concluded on the source of ozone (Mulvany 1880, p. 190):

> It appears to be formed in the upper strata, and to be carried downwards by rain drops, whose office is vehicular. The spherules of water which constitute clouds and have their origin in radiation and condensation, have a similar office. Ozone does not appear to diffuse readily downwards [...]

Mulvany, who read his paper on 17 June 1880 to the Meteorological Society on the occasion of his election as a member, also concluded "no disease can be clearly traced to ozone met in the atmosphere"; striking in the discussion of his lecture is that honorable members[1048] like *Tripe*, *Whipple*, *Williams* and *Symons*, do not doubt on the vesicle hypothesis although it was already rejected by other scientists (see Chapter 2.2.4), and that *Williams* had noticed that the author had not refer whatever to *Schönbein*'s statement that ozone produced a form of influenza (he believed *Schönbein* to be correct). In a certain sense, *Mulvani*'s idea is close to that of Baxendell (1876, p. 121), who believed that ozone is formed in clouds, and "due to descending currents bringing air from the cloud regions to lower and warmer levels and thus causing the rapid evaporation of the condensed vapour which it contains".

In August 1885, meteorologist *W. J. Black* (unknown living dates) observed ozone at sea during a voyage on the ship *Ceylon* around the United Kingdom, using *Moffat*'s paper (Black 1885). Black (1886) found that ozone papers already previously colored and again exposed in London became all bleached to their original white; he asked: "as to the chemical changes that had taken place, and if these had been due to an an-

1048 *George Matthew Whipple* (1842–1893) was geophysicist at Kew Observatory. *Charles Theodore Williams* (1838–1912) was an English physician, known as a leading authority on pulmonary tuberculosis. *George James Symons* (1838–1900) was a British meteorologist who founded and managed the British Rainfall Organisation.

tozone causing a decomposition of the ingredients (iodide potassium) to their original constitution".

4.4.5.7 Observations of atmospheric ozone in France

In France, the first ozonometric observations using *Schönbein*'s papers were conducted 1852–1853 by *Armand Bineau* in Lyon, who detected it seldom in the air of the city, although be observed always in the air of the neighboring country (Bineau 1855a,b). He attributes the deficiency in the city to the action of organic substances in the air. Fox (1873, p. 96) writes[1049] that the "officials of the Imperial Observatory, MM. Fournet, Lambert, Rassinier, and Bineau, have all testified to its great scarcity in Lyons", hence it was generally named "the town without ozone".

Next ozone observation in France was made by *Jean Baptiste Simonin* (1785–1870), a French physician in Nancy,[1050] who conducted morning and evening sampling from April until December 1853 (Simonin 1854); he continued the observations until December 1854 (Simonin 1869). *Simonin* states that in contrast to the general observations of highest ozone values in the cold season, in Nancy it is not this case, and ozone is distributed with variations over all seasons and temperatures. During frequent fogs in autumn and winter, he found no ozone and concluded on its decomposition by the fog [... *ces broulliards qui anéantissent l'ozone*], (Simonin 1869, p. 230).

Thèodor Boeckel (or *Böckel*) (1802–1869), physician in Strasbourg, made daily two observations 1854–1865 (Boeckel 1865); his sun *Eugène Boeckel* (1831–1900), French surgeon, made his dissertation on ozone (Boeckel 1856).

Dr. *Billiard* (unknown living data), of Corbigny, is of the opinion, the diminution of ozone in the atmosphere is the first cause of cholera (Billiard 1854). *François Joseph Conraux* (1817–?), French physician in Thann (Haute-Rhine), noted in 1854 the disappearance of ozone from the air during a cholera epidemic whilst ozonoscopes

1049 It is "*Observatoire impérial de Paris*"; the "*Observatoire de Lyon*" was founded in 1878. *Joseph Fournet* (1801–1869) was a French geologist and meteorologist, professor at University Lyon. Likely *Charles Lambert* (1804–1864), French explorer and engineer. Nothing found on *Rassinier* concerns Lyon.

1050 The *Simonin* family at Nancy is famous for three generations of surgeons, having the same name: *Jean-Baptiste Simonin* (1750–1836), his son *Jean-Baptiste Simonin* (1785–1870), and the grand-son *Jean-Baptiste Edmond Simonin* (1812–1884) as "*Directeur de l'École de médecine et de pharmacie*". *Simonin* (1785–1870) is the senoir (Simonin père in French), who observed ozone in 1853 (Simonin 1854). Fox (1872, p. 133), who not provides the bibliography, refers *Simonin* senior who remarked the coincidence of the appearance of cholera at Nancy in 1854 with "no ozone period", but did not observe a similar agreement during the epidemic of 1855; this could be a mistake because *Simonin* observed ozone on 1853 and 1854 and wrote nothing on cholera and ozone. Strambio (1856, p. 226) also cites *Simonin* (senior) and presents the monthly mean ozone values from 1853. Furthermore, Gaillard (1864) refers "Simonin Sr and Simonin Jt, at Nancy" without any more information. *Simonin* present his meteorological observations (but no more on ozone) yearly until 1872 in the "*Mémoires de l'Académie de Stanislas*".

simultaneously colored at villages (Rodern and Leimbach) situated 3 km from the town, to which the disease had not spread; cited after (Pettenkofer (1855, pp. 364–365) who cites "*Beobachtungen über das Verschwinden des Ozon*" from "*Feuille d'annonces et avis divers de Thann et de Cernay*" 1855, p. 13). *Joseph Constantin Decharmes* (1815–1905), "*professeur de science physianes et naturelles au Lycèe Impérial d'Amiens*", conducted in 1855 extended ozone observations (Decharmes 1856), and *Aimè-Augustin Joseph Robert* (1813–1880), French physician, at the village Neudorf (Alsace), during a cholera epidemic in 1855, believed that a relation does exist (Robert 1855).

The French physician *Adolphe Bérigny* (1809–1885) first conducted at Versailles in August 1855 ozone observation using *Schönbein*'s paper and found it to be abundant whereas in Paris at the same time, where a cholera epidemic prevailed, *Johann Theobald Silbermann* (1806–1865)[1051] found no ozone (Bérigny 1855). Afterward *Bérigny* initiated a "network" of ozone observations in France and abroad with the help of local observers (Bérigny 1857, 1858a, 1870). Bérigny (1857) first compared different test-papers prepared by *Schönbein*, *Moffat*, and Drs. *Richard*, *Lerebours*,[1052] and *James* from Sedan. *Bérigny* conducted ozonometric measurements at different sites in France: at Sevastopol on the Crim (Bérigny 1858b,c), at Versailles (Bérigny 1855, 1856, 1859, 1870), at the lighthouse Calais (Bérigny 1860), and 1862 at Havana and 1865 in Mexico (Bérigny 1867). Bérigny (1856) refers persons who carried out ozonometric observations in Metz (architect *Noble*), professor *Pourriau* at the agricultural school in La Saulsaie (Ain), and other locations where observation were made (Verneville, Thionville, Talange, Sailly, Yutz, Ennery, Ventouse). Bérigny (1858c) reports on observations between May and June 1856 at three locations on Crime: Sevastopol by Dr. *Scrive*, Bataklava by Dr. *Leroy*, and monastery Saint-George by Dr. *Méry* (nothing found on these observers).

The longest monitoring (1855–1873) was carried out at Versailles, published in "*Annuaire de la Société Météorologique de France*" with the heading "*observations ozonométriques*" 1855–1858 in Vol. 18 (1870) 25–28, 1959–1865 in Vol. 19 (1981) 21–25, 1866–1871 in Vol. 21 (1873) 22–27, 1872 in Vol. 21 (1873) 18, and monthly and annual means 1859–1872 in Vol. 21 (1873) 28. On *Bérigny*'s proposal, *Urbain Jean Joseph Le Verrier* (1811–1877), astronomer, mathematician and directors of "*Observatoire de Paris*", established 24 observational stations in Paris (in each "*Département*").

Leonard Endymion Pieraggi (1830–1889), a French mining engineer, conducted at "*Bois de Colombes*" ("*plaine de Gennevilliers*", suburb of Paris) ozone is twice as much compared to Paris downtown (Pieraggi 1867). *Eugène Grellois* (1811–1887), a French

1051 Also *Jean-Thiébault*; French engineer and physicist at *Conservatoire National des Arts-et-Métiers* (University of *Grandes écoles*) and the Sorbonne.

1052 *Noël Marie Paymal Lerebours* (1807–1873) was a French optician and daguerreotypist. *Richard* not identified.

military physician, observed from July 1855 until March 1856 ozone and the meteorology at Constantinople (and found relations like *Boeckel* at Strasbourg and *Simonin* at Nancy) and state that during the earthquake the emanations disengaged, destroy the ozone present in the air (Grellois 1857a). In April/May 1857 he observed ozone in Thionville (Grellois 1857b).

Auguste Houzeau carried out the first atmospheric ozone measurements at "*l'ermitage*" in Montmorency near Paris from July 9 to August 8, 1856 (Houzeau 1857). He later developed a semiquantitative method, a paper involving a mixture of iodide of potassium and litmus together with a specific holder, where different colors correspond to certain ranges of the ozone concentration; additional he presented the first possibility to determine ozone quantitative by dimensional analysis of the precipitated iodine (there is a series of papers on the new method: Houzeau 1858, 1860 and 1863, and on measurement: 1865, 1868, 1872a, 1872b). Rather casually Houzeau (1872b, p. 46) communicated in a footnote the first quantitative estimate of the ozone mixing ratio in air (10 ppb if it is based on volume) without further comment:

> En 1857, j'avais trouve que cette proportion d'ozone s'élevait à 1/100000000 environ.

Such estimate was surely based on a "detection limit" ["*limite de leur sensitivité*"] of his paper to show the lowest degree of color, which he estimated to be 0.0002 to 0.0003 mg of ozone which corresponds to 1/261000 mass parts in the air (Houzeau 1872b, p. 20). Consequently, *Houzeau* notes that an exact determination of the quantity of ozone is very difficult due to its low concentration in the air; he presents a Table (Fig. 4.42) of observations made with his paper in Ècorcheboeuf in December 1865. Furthermore, he states that the amount of ozone depends on the height above the soil; without further comment *Houzeau* communicates (Houzeau 1872b, p. 29) that air at two meter above ground contains a maximum of 1/450000 (mass) or 1/700000 (volume) parts of ozone; the latter corresponds to 1400 ppb, unrealistic large from today's knowledge, however,

30 A. HOUZEAU.

		AIR CALME.	AIR TRÈS-AGITÉ.
Altération du papier vineux mi-ioduréà l'air libre après une exposition de........	2 heures.	Nulle.	Violet faible.
	4 id.	Violet.	Violet.
	6 id.	Bleu très-faible.	Bleu faible.
	8 id.	Id.	Bleu.
	10 id.	Bleu.	Très-bleu.
	12 id.	Id.	Id.
	24 heures.	Très-bleu.	Id.

Figure 4.42: Table showing the degree of blueing dependent on the time of exposure and the state of air agitation (Houzeau 1872b, p. 30).

the above figure of 10 ppb seems to be reliable for that time. It should be noted that with the exposure of a test-paper the amount of passing and acting air can only be very roughly estimated with the registration of the wind speed and the time of exposure.

Houzeau (after Fox 1873, p. 72) writes "[...] that ozone is more frequently present in the air during days of rain than during days of fine weather". Moreover, *Robert Joseph Henri Scoutetten* (1799–1871), a French professor and head of medicine at the military hospital in Metz, found that rain and fogs determine different effects, according to their conditions of production of "ozone reactions" (Scoutetten 1856):

> If rain follows a storm, and returns after a temporary reappearance of blue sky, the test-paper exhibits deep tints. If, on the contrary, the rain is fine and continuous, and the temperature is slightly elevated, there is little ozone.

Houzeau presents the first systematic (relative) measurements, carried out in Rouen 1861–1870, using his "*papier de tournesol vineaux mi-joduré*". As quantitative mean he termed the number of days with and without ozone in air; he separates between three levels of blueing ("*bleu faible*", "*bleu*", and "*très bleu*"), but for getting monthly means he made no difference and termed only the number of days with reaction.[1053] Only while specific meteorological situations (storms and thunderstorms) he considered the degree of blueing. The seasonal cycle of ozone (maximum in May and June, minimum in December and January), *Houzeau* contributed to causes of natural variations ["*réveil de la nature*"] and separated a very active and non-active season in link with the sun and vegetation, but also speculates that the less rainfall in summer may be a reason. Furthermore, he found that the atmospheric amount of ozone is largest when the air in Rouen comes from south to south-west and smallest from the west (by a factor of 3–4). *Houzeau* is also the first who found that the concentration of ozone is at night lower than daytime (most other observers found it vice versa). Based on his measurements in different cities of France, Houzeau (1872b, p. 26) writes the remarkable phrase:

> [...] evidence l'influence des localités sur la manifestation de l'activité chimiques de l'air.

The French scientist and writer *Louis Figuier* (1819–1894) wrote a popular article upon *Bérigny*'s and *Houzeau*'s investigations (Figuier 1866). At that time, nothing was known about the atmospheric formation reactions of O_3 and H_2O_2 and which chemical mechanisms existed between both species in the gas and aqueous phases. The formation of ozone, hydrogen peroxide, and nitric acid in air had been attributed

1053 There is a remarkable sophisticated comment to *Houzeau*'s averaging by Dohrandt (1874, p. 15). According to *Houzeau*, the lowest degree of blueing should correspond to 0.00025 mg ozone, fixed on the paper. Despite this exact number, this result for the amount of ozone in the air can only be of relative importance and there is no apparent reason why a day with a low – but still detectable – amount of ozone should be excluded from the ozone observations.

generally to "variations in the electrical condition of the atmosphere" (Fox 1873). Thus Houzeau (1872b, p. 62) concluded that ozone is a natural constituent of the air:

> [...] dans la nature et permettent à l'électricié atmosphérique d'ozoniser l'oxygène de l'air.

Concerns the origin of ozone in the atmosphere, *Houzeau* writes, referring *Marum*, *Fremy* and *Bequerel*, who gained ozone from pure oxygen through electric discharges, the atmospheric electricity is responsible ["*Évidemment l'ozone de l'air dérive, pour la plus grande partie du moins, de ce qu'on a appelé l'lectricité atmosphérique*"], Houzeau (1872a, p. 713). Further *Houzeau* writes, referring *Cavendish*, who gained nitrous acid in atmospheric air through electric discharges, but under atmospheric lightning conditions the electricity exchanges like in a capacitor between clouds and the earth forming ozone ["*condensateur analogue à mon tube ozoniseur*"], (Houzeau 1872a, p. 715):

> Il est fort probable que dans les orages de la première classe (négatifs au papier de tournesol mi-ioduré) l'éclair est surtout une étincelle directe qui nitrifie l'air sans l'ozoniser sensiblement, et que, dans ceux de la deuxième classe, l'éclair participle plus généralement des propriétés de l'étincelle de condensation elle produit beaucoup d'ozone et peu d'acide nitreux [It is very probable that in the first case the lash occurs as direct spark, loading the air with nitrous acid without remarkable ozonation; that in the other case the flash act as a condensed spark, generating much ozone and less nitrous acid].

Houzeau (like *Schiefferdecker* and *Reslhuber*) found that on rainy days the ozone level is larger than on "nice" days (38 days with ozone to 100 days with rain and 28 days with ozone to 100 "nice" days). Engler (1879, p. 51) presents a remarkable explanation; rain and snow wash out dust particles from the atmosphere and thus reduce ozone decomposing substances; as already stated by Fox (1873, p. 74):

> The washing out of the air of the impurities continually passing into it, in the oxidation of which the ozone was previously consumed.

French physicist *Albert Lévy* (1844–1907) and his co-worker *Paul Léon Félix Allaire* (unknown living data) developed in 1875 the chemical method suggested by Thénard (1872)[1054] to observe the abundance of ozone almost continuously from 1877 to 1910 at the municipal Observatory of Parc Montsouris in Paris (Lévy 1878a, 1907, Lévy et al. 1905); cf. Fig. 4.22 and Radau (1877)[1055] on the early history of air analysis at Montsouris. This newly aspirator method is based on bubbling the ozone containing air through a solution of potassium arsenite where the arsenite (AsO_3^{3-}) is transformed

1054 Baron *Arnould Paul Edmond Thénard* (1819–1884), son of *Louis Jacques Thénard*, the discoverer of hydrogen peroxide.

1055 *Jean Charles Rodolphe Radau* (1835–1911) was a German-French astronomer.

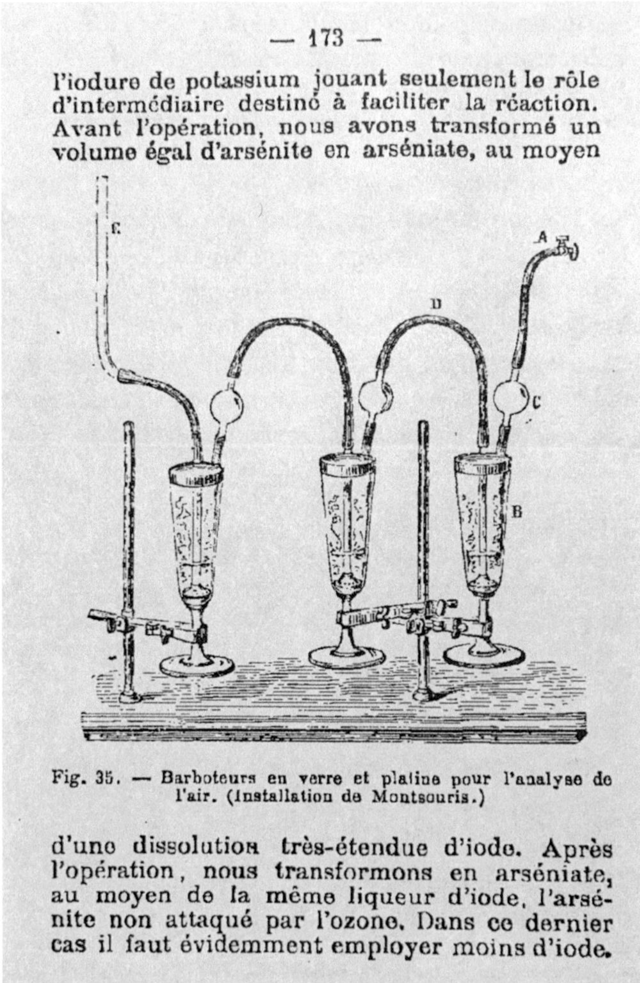

— 173 —

l'iodure de potassium jouant seulement le rôle d'intermédiaire destiné à faciliter la réaction. Avant l'opération, nous avons transformé un volume égal d'arsénite en arséniate, au moyen

Fig. 35. — Barboteurs en verre et platine pour l'analyse de l'air. (Installation de Montsouris.)

d'une dissolution très-étendue d'iode. Après l'opération, nous transformons en arséniate, au moyen de la même liqueur d'iode, l'arsénite non attaqué par l'ozone. Dans ce dernier cas il faut évidemment employer moins d'iode.

Figure 4.43: System of aspirators for air analysis at Montsouris (Lévy 1879, p. 173).

into arsenate (AsO_3^{4-}), and the not converted arsenite titrated after *Mohr*'s method with potassium iodide solution;[1056] Fig. 4.43.

Marié-Davy (1876) reports for the first time on reliable ozone concentrations measured in March 1876 in the air of Paris between 7.6 and 11.3 µg m^{-3}. Lèvy (1882) presents

1056 Volz and Kley (1988) reconstructed the type of bubbler used at Montsouris and, comparing the method with the modern ultraviolet absorption technique, found it to be reliable. They also found a negative interference concerns SO_2. Grouping the daily values (about 3000) into wind sectors, *Kley* and *Volz* estimated the ozone concentration from SE (rather unpolluted air) to be 8 ppb and in air passing over Paris to be 5–6 ppb.

Table 4.21: Monthly and annual mean concentration of ozone (in µg m^{-3}) in the air at Montsouris (Paris); data from Lauscher (1983, p. 76), s. d. – standard deviation. Mean concentration 1886–1899: 20.2 ± 3.4 µg m^{-3} (about 10 ppb). See also Fig. 4.45.

Year	J	F	M	A	M	J	J	A	S	O	N	D	year
1886	18	23	22	22	24	24	18	21	12	21	15	19	20
1887	17	18	22	24	21	18	22	22	15	17	16	26	20
1888	22	18	26	23	27	25	21	17	22	18	21	10	21
1889	10	12	15	14	15	22	22	15	13	12	9	10	14
1890	12	12	17	12	23	19	20	19	8	8	11	11	14
1891	11	7	17	19	25	25	21	17	18	18	19	16	18
1892	15	17	17	22	22	43	24	21	14	14	13	20	20
1893	23	23	28	41	35	21	28	23	20	15	12	15	24
1894	10	7	3	6	18	34	7	27	29	22	16	23	17
1895	24	34	33	31	34	10	29	32	33	12	12	20	25
1896	21	16	23	26	29	31	29	22	29	21	19	21	24
1897	18	24	32	29	29	35	37	31	18	17	17	18	25
1898	17	19	21	30	30	27	29	35	26	17	12	16	23
1899	19	19	20	26	19	14	15	14	14	11	13	13	16
mean	17	18	21	23	25	25	23	23	19	16	15	17	20
s. d.	0.5	0.7	0.8	0.9	0.6	0.9	0.7	0.6	0.8	0.4	0.4	0.5	0.4

Table 4.22: Monthly mean concentration of ozone (in µg m^{-3}) in the air at Montsouris (Paris) over 20 years (1877–1896); Anonymous (1898, p. 294).

	J	F	M	A	M	J	J	A	S	O	N	D	mean
mean	13.8	15.8	16.7	17.2	19.8	20.3	18.5	17.9	16.6	14.2	13.4	13.6	16.5
max	24	34	33	41	35	43	29	32	33	22	21	26	27
min	1	6	3	5	4	7	4	6	6	5	3	3	6

for the first five years of daily measurements the monthly mean ozone concentrations – for the first time in mg per 100 m^{-3}. The monthly means vary between 5 and 30 µg m^{-3} (highest in February and lowest in December) and the yearly means (1877–1881) from 6 to 19 (total mean of 11.5) µg m^{-3}; the overall mean corresponds to 6 ppb O_3 (Tables 4.21 and 4.22). Lévy found a good correlation between simultaneous determinations of ozone in the air by *Schönbein*'s paper and his arsenite method (Fig. 4.44). Volz and Kley (1988) as well as Anfossi and Sandroni (1997) reanalyzed the Paris data and present reconstructed annual means of being around 10 ppb; see Bojkov (1986, pp. 345–346)[1057] for more details on the Montsouris measurements. Lévy (1878c), who measured ozone daily at Montsouris by his quantitative method parallel with *Schön-*

1057 *Rumen D. Bojkov* (born 1931 in Bulgaria) made his PhD 1964 in physics and mathematics in Moscow, his habilitation 1971 in Rostock and worked many years for the WMO in Geneve and served as

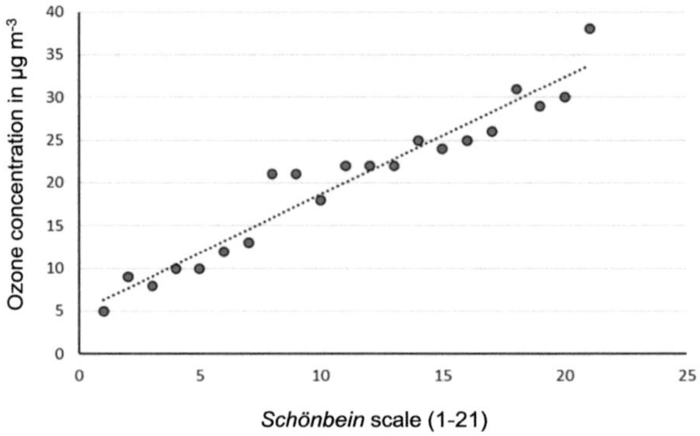

Figure 4.44: Correlation between simultaneous determinations of ozone in the air by *Schönbein*'s paper and aspirator arsenite method 1876–1878 ($n = 273$) at the Observatory Montsouris; data from Lévy (1878a, p. 502).

bein's paper, notes that despite some errors of the test-paper, the ozonometric observation can be continued with benefit. The occasional discoloring of the ozone-paper, namely when the air, arriving Montsouris, was passing over Paris (i. e., from North), and during the formation of fog, *Lévy* explains (Lévy 1878c, translated from German):

> It shows the presence of a body in the air, still unknown, which is able to decompose the compound between iodine and amylum, which has been formed previously under the influence of ozone.

The French meteorologist *Charles Joseph Sainte-Claire Deville* (1814–1876), one of the founder of the meteorological observatory Montsouris in 1871, argued that long-term observations are needed to get means due to the large variability of single values (Sainte-Claire Deville 1865b). Sainte-Claire Deville (1865a) presents daily values from August 1865 from Versaille and the Lighthouse du Touquet, giving means from three short periods, showing the correlation between ozone (in µg m^{-3}) and temperature:

Temperature (in °C)	ozone at Versailles	ozone at lighthouse Touquet
5 to 9	15.5	15.0
10 to 13	13.1	12.6
14 to 18	17.0	16.1

Prosper de Pietra Santa (1820–1898), French physician and founder of "*Société Française d'hygiène*", exposed seven test-papers of *James* (de Sedan) in the village

Secretary of the International Ozone Commission 1984–2000. He lives since some years in Radebeul (Germany).

Ozone contenu dans 100^{m3} d'air du parc de Montsouris, à Paris.

(Valeurs exprimées en milligrammes.)

MOIS.	1877	1878	1879	1880	1881	1882	1883	1884	1885	1886	1887	1888	1889	1890	1891	1892	1893	1894	1895	1896
Janvier . .	2,3	1,6	0,5	0,1	0,5	0,7	0,7	1,8	1,1	1,8	1,7	2,2	1,0	1,2	1,1	1,5	2,3	1,0	2,4	2,1
Février . .	3,0	1,6	1,3	0,7	1,0	0,9	0,6	2,0	1,6	2,3	1,8	1,8	1,2	1,2	0,7	1,7	2,3	0,7	3,4	1,6
Mars . . .	2,8	1,5	0,5	0,4	1,6	0,6	(1,2)	1,0	1,4	2,2	2,2	2,6	1,5	1,7	1,7	1,7	2,8	0,3	3,3	2,3
Avril . . .	2,0	1,4	0,7	0,8	0,5	0,5	1,5	1,5	1,5	—	2,4	2,3	1,4	1,2	1,9	2,2	4,1	0,6	3,1	2,6
Mai. . . .	1,5	2,2	0,9	0,4	0,9	0,9	0,7	2,1	2,6	2,4	2,1	2,7	1,5	2,3	2,5	2,2	3,5	(1,8)	3,4	2,9
Juin . . .	1,3	1,8	1,0	0,7	1,1	0,8	1,0	2,4	2,4	1,8	2,5	2,2	1,9	2,5	4,3	2,1	3,4	1,0	3,1	3,3
Juillet . . .	1,8	1,5	0,9	1,2	1,2	0,4	1,2	2,4	2,2	1,8	2,2	2,1	2,2	2,0	2,1	2,4	2,8	0,7	2,9	2,9
Août . . .	1,3	1,7	(1,6)	0,9	0,6	0,6	0,7	2,4	2,4	—	2,2	1,7	1,5	1,9	1,7	2,1	2,3	2,7	3,2	2,2
Septembre .	1,1	1,8	1,1	0,6	1,0	0,6	1,2	2,0	2,5	1,2	1,5	2,2	1,3	0,8	1,8	1,4	2,0	2,9	3,3	2,9
Octobre . .	1,9	1,5	0,6	0,5	1,0	0,6	1,3	1,4	1,7	2,1	1,7	1,8	1,2	0,8	1,8	1,4	1,5	2,2	1,2	2,1
Novembre .	2,2	0,6	0,3	0,6	1,7	1,0	1,8	0,5	1,7	1,5	1,6	2,1	0,9	1,1	1,9	1,3	1,2	1,6	1,2	1,9
Décembre. .	1,5	0,3	0,5	0,3	0,8	0,4	1,5	1,2	1,6	1,9	2,6	1,0	1,0	1,1	1,6	2,0	1,5	2,3	2,0	2,1
Moyenne. .	1,9	1,5	0,8	0,6	1,0	0,7	1,1	1,7	1,9	1,9	2,0	2,1	1,4	1,5	1,9	1,8	2,5	1,5	2,7	2,4

296

CIEL ET TERRE.

Figure 4.45: Copy of the Table of the 20 years ozone observations by *Albert Lévy* at Montsouris (1877–1896) given in mg per 100 m^3 (Lévy 1899, p. 296).

Eaux-Bonnes (Pyrenees, 747 m a. s. l.) for 12 hours at different locations indoor and outdoor, showing the "natural" variability in a range between 1 and 20 degrees (Pietra Santa 1864); remarkably, he published in large temporal distances (1878 and 1895) again on atmospheric ozone (Pietra Santa 1878–1895).

Friedrich Eduard (Frédéric) Kampmann (1797–1873), French-German pharmacist and naturalist in Colmar (Alsace), conducted at 4 stations daily ozone observations 1866–1869 (Kampmann 1868); he notes (the unknown) Messrs. *Hirn*, *Pin*, *Martz*, and *Giorgio* as stations assistants. French chemist *Claude Auguste Lamy* (1820–1878) studied the reliability of thallium test-paper at different locations in Paris and rural sites over a whole year (Lamy 1869).

Georges Daremberg (1850–1906), a French physician, found that ozone levels 150 m from the coast on sea at Menton (near Nice) were much lower than that on land and discusses the influence of wind, whereas Houzeau (1872b) argued that dry wind has an influence on the test-paper, writing "[...] as to storms which generated ozone and storms which did not"; Daremberg (1878); he also discusses the value of ozone test-papers (Daremberg 1879).

The French chemist *Maurice de Thierry* (1858–?),[1058] who also designed an absorption spectroscope (Thierry 1886), conducted 1894–1895 at several stations of the Mont Blanc massif at different altitudes for ozone measurements (Fig. 4.46), using

1058 Surprisingly little is known about *Thierry*: "*préparateur de manipulations chimiques à la Faculté de médecine de Paris; rédacteur en chef*" (Revue des sciences et des lettres). He wrote: *Atlas* de *Manipulations de Chimie. Métalloïdes. Avec une préface du Dr Armand Gautier*. Paul Rousseau, Paris, 1890.

Fig. 2. — M. Maurice de Thierry entre ses deux guides, en haut du mont Blanc, à la porte de l'observatoire de M. Janssen. (D'après une photographie instantanée.)

Figure 4.46: Mont Blanc Observatory *Janssen*: *Maurice de Thierry* between his two guides, at the top of Mont Blanc, at the door of the observatory. "There are a certain number of stations," *Maurice de Thierry* further wrote, commenting on his attempts to determine variations in the relative concentration of ozone and carbon dioxide with respect to altitude, "provides especially favorable conditions for the studies I have undertaken" (Thierry 1897, p. 460; see also Thierry 1899). Illustration is taken from Janssen (1895, p. 328); *Pierre Jules César Janssen* (1824–1907) was a French astronomer and director of the Physical Observatory of Meudon, near Paris Anonymous 1892).

Lévy's method (Thierry 1897), and compared his results with those from Montsouris (August–September 1994), in μg m^{-3}:

Montsouris	27 ± 7 (n = 21)
Chamonix (1050 m)	35–39
Grand-Mulets (3020 m)	94

4.4.5.8 Observations of atmospheric ozone in America

The medical doctor *Robert Peter* (1805–1894)[1059] is the first who carried out ozonometric observations in Lexington (USA) between April and September 1849 during a cholera epidemic, writing (Peter 1849, p. 156):

> The experiments prove that there is by no means a constant relation between the probable amount of ozone in the air, as indicated by the iodized paper, and the severity of the diseases as

1059 *Peter* was a native of Cornwall, England, who migrated to the United States in 1817. He received his MD from the Transylvania College of Medicine in 1834 and served many years at Transylvania as chairman of chemistry, pharmacy, dean, and librarian of the Medical School.

> determined by the number of deaths; and that, especially, the indications of ozone do not cease with cessation of cholera [...] The ozone theory of cholera, therefore, is not sustained.

In Canada, the first long-term monitoring (1850–1862) on the American continent was conducted by *Charles Smallwood* (1812–1873)[1060] at a fully equipped meteorological station, best described by himself (Smallwood 1854, p. 231):

> Ozonometer. – Obervations were carefully registered twice daily, for some years, of the amount of ozone present in the atmosphere; the slips of iodized paper are carefully preserved in a dark place after having been exposed to the atmosphere, shaded from the sun, and rain. As a general rule, rain or snow shows an increase, and so far as my own observations go, a high electric state of the atmosphere does not show an increase in the amount of ozone.

Smallwood (1858) describes the station at Saint-Martin (Isle-Jésus, now Laval), near Montréal, which he had established in 1841. Here he began the first long-term record of local weather conditions in Canada, testing his instruments and improvised recording devices – these even using photographic technology – three times daily (Fig. 4.47). An eighty-foot mast hosted an electrometer for measuring atmospheric electricity. The observatory was connected via the Montréal telegraph early on, to major cities in the US, linking him to American astronomers and meteorologists, with whom he corresponded extensively. In 1864, he moved his observatory from Saint-Martin to the McGill campus. Results of his observations (1850–1856) are summarized in Smallwood (1857).

Between 1856 and 1862, the monthly report on meteorology from the observatory St. Martin included the phrase "Ozone was in small [moderate, rather large, or large] quantity" (published in the *Canadian Journal, new series*, Vol. 1, no 1 (1856) until Vol. 6, no 39 (1862)). Smallwood (1857, p. 5) noted the importance of locating his (nearly 6000, over eight years) observations at a 3-mile distance from the river, to avoid the higher incidence of O_3 in localized mists, fog, and vapor; interestingly, he termed his "instrument" *Ozonometer* (he compared the *Moffat*'s with the *Schönbein* paper and found so little difference that he continued that of *Schönbein*'s; see p. 544 on the different results by *Glaisher* and *Barker*). Analyzing these with his meteorological records, he established that high ozone levels correlate well with days of high humidity and precipitation, rather than prevailing atmospheric electricity (and even the aurora borealis)

1060 Physician, professor of meteorology, and founder of McGill Observatory at Montreal. Born in Birmingham, England, *Smallwood* obtained his MD from University College, London before immigrating to Lower Canada at age 21 in 1833. He had been elected an honorary member of the Montreal Natural History Society in 1856, and was president in 1865. In 1863 all the apparatus of *Smallwood*'s observatory was moved from the frame building at Saint-Martin to a stone structure built for the purpose in the McGill University grounds. *Smallwood* was a member of four foreign scientific societies: La Société Météorologique de France, the Pulkovo astronomical observatory in Saint Petersburg, Académie Royale de Belgique, the National Institute of the United States, and the Academy of Natural Sciences (Philadelphia). Cited from: Dictionary of Canadian Biography, Vol. X.

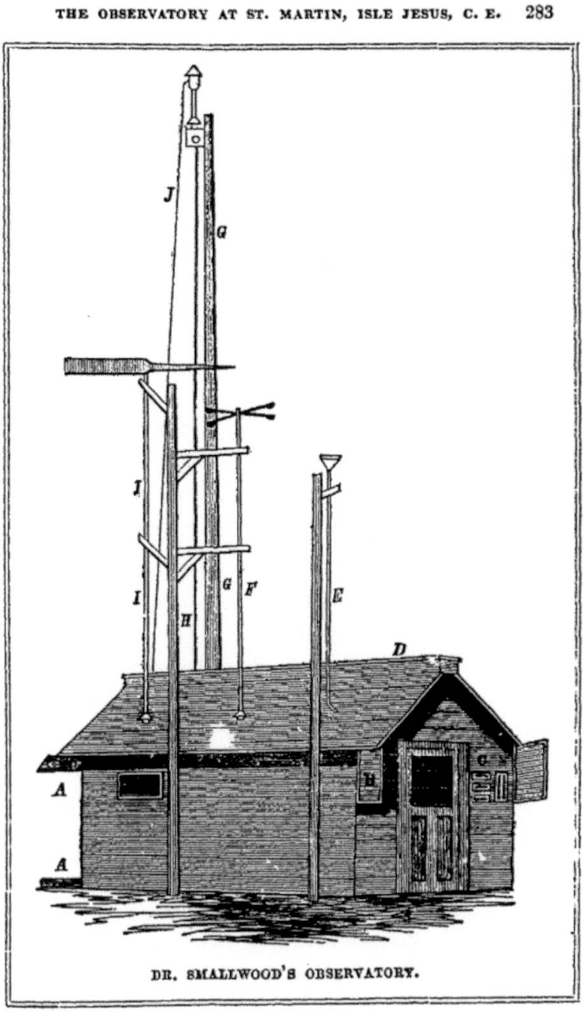

Figure 4.47: The Observatory of *Charles Smallwood* at St. Martin (Isle Jesus) near Montreal (from Smallwood 1858, p. 283).

as proposed by earlier researchers (Smallwood 1857, p. 8). The fundamental historical importance of *Charles Smallwood* is based on his idea, to combine chemical (ozone) with physical (temperature, pressure, wind, electricity, precipitation) monitoring at *one* meteorological station. Another remarkable observation Smallwood (1859, p. 266) made:[1061]

1061 Today it is known that clouds scatter solar radiation and can reduce as well as increase photodissociation; this is the fact that hydrogen peroxide was found to be increased just above the and close to clouds (Möller 2010, pp. 282 and 385).

> The amount of ozone last year has shown an increase on the usual average. Observations are now being taken here, intended to show effects of the different clouded rays of light on the Ozonometer.

George Templeman Kingston (1816–1886), a meteorologist and director of the observatory Toronto since 1852, published meteorological observations (monthly tabulations of the Saint-Martin observation) including “state of ozone” in *The Canadian Journal of Industry, Science, and Art* (bimonthly pub. By H. Scobie, for the Council of the Canadian Institute), Vol. 1 (1852) until Vol. 3 (1855). Dr. *Webster*, an American physician (unknown giving name and living data), carried out ozonometric observation in Norfolk and Portsmouth (for two years or more) and found in one year no ozone and the other ozone in abundance (cited in *Canadian Journal* **7** (1857) p. 72).

William Barbon Rogers (1804–1881), American geologist and MIT’s first president, stimulated by *Cloez*’s paper, conducted over two years experiments, but did not find any influence of different plant oils, exposed to sunlight near the test-paper (Rogers 1858); he found the “usual” reactions, in the night being stronger than at daytime and under cloudy weather stronger than under brilliant sunshine, and observed that the action of the city Boston deprived air passing over its with ozone. Bellucci (1874) also confirmed earlier work by Cloez (1856) that ozone is not produced together with the oxygen of growing plants; he states that the coloration of test-paper was not produced by ozone but to the complex and simultaneous action of humid oxygen and solar light.

Samuel Edwin Gaillard (1827–1885), an American medical educator and editor, wrote an essay on ozone and its relation to health and diseases (Gaillard 1864), noted (Gaillard 1864, p. 296):[1062]

> Ozone has received the most enthusiastic attention from the savants of Europe; the list of observers is along one, [. . .] Wolf, at Berne; Simonin Sr. and Simonin Jr, at Nancy; Boeckel Sr. and Boeckel Jr., at Strasbourg, Billiard, at Carligny; Robert, at Neudorf; Bérigny and Silbermann, at Paris, Seitz, Slechner, Schnepf, Abt, Eschach and Knolz, at Vienna; Pourian, at Saulsaic; Decharmes, at Amiens; Grellois, at Constantinople; and vary many others [. . .] But what shall we say of America? What can we say? Near Montreal, Canada east, Smallwood quietly but industriously pursues his labors. Capt. Pope, U. S. Army, has made many and most valuable observations. With two, or perhaps three other names, the American list ended.

Gaillard refers first *Billiard, Boeckel, Schnepf, Knolz, Robert*, and *Wolf* with the finding that an “invasion of cholera incised with the zero of the ozonometer” or “diminution of ozone as the cause of cholera”, but than that (Gaillard 1864, p. 269)

1062 Not *Slechner* but *Flechner*. *Bernard Schnepf* [or *Bernhard Schnepp*] (unknown living dates) was a German physician in Paris. I was unable to find any reference or information about *Abt, Eschach, Knolz*, and *Pourian*. It is amazing that *Gaillard* had access in Baltimore to publications, fully forgotten nowadays. On the other hand, he not cited many other observers, referred in this volume.

> A committee at Kœnigsburg [it is Königsberg/Prussia] could not discover that there was a direct relation between cholera and the amount of atmospheric ozone. How can this discrepancy be explained? Any one familiar with the subject of ozone can easily anticipate the reason.

Gaillard correctly writes that the "ozonometer used were not uniform", and, "when many hundreds of such observations are taken, and the numbers added, it will be seen at a glance how totally different must be the result, with each individual", and finally "[...] refer them to different ozonometers, all is confusion and obscurity".

Wetherill (1865) ascertained that the air of the public ground of Washington yielded at night an abundance of ozone, whilst that of the streets of the city, examined at the same time, indicated an absence of this gas. *Adolph Francis Alphonse Bandelier* (1840–1914), was a Swiss-born American archaeologist who studied ozone in Highland (Illinois) 1862–1864. Bandelier (1868) found that great humidity is generally accompanied by a considerable quantity of ozone in the air; daily maximum 2–5 pm and 12–2 am, minimum in the morning and evening.

Friedrich Adolph Wislizenus (1810–1889), American physicist and botanist of German origin, who immigrated 1834 in the USA, first to New York and since 1839 in St. Louis, was interested in atmospheric electricity and meteorological studies and "stated that he had often tried to find traces of ozone in the atmosphere of the city, without discovering any, but that, a few days since, he had made some experiments in the country with perfect success" (*The Transactions of the Academy of Sciences of St. Louis* 1861–1868, **2** (1868) p. 548).

Robert Clark Kedzie (1823–1902), professor of chemistry at the Michigan Agricultural College, was pioneering in establishing a statewide meteorological observation network since 1871 (Kedzie 1876); Linvill et al. (1980) have uncovered monthly average ozone values from 1871 through 1903 from approximately 20 stations (Fig. 4.48). Kedzie (1876, p. 139) writes (cf. with the statement of Dohrandt (1874) on p. 303):

> Ozone is found everywhere in the air that rests upon the earth's surface. It is found wherever the agents which cause its destruction are not present in sufficient amount to cause its entire removal. The ozone found in the air is always the difference between the amount of ozone formed and the portion destroyed.

A. W. Nicholson (unknown living dates) of the State Board of Health of Michigan, made at Otisville 1877–1878 ozone observation to study *Schönbein*'s paper (Nicholson 1881). Hatcher and Arny (1900) comment on this work with the phrase "it is, however, much to be regretted that the unreliable Schönbein paper was used in all his work, thereby vitiating much of his date". *Charles Ambrook* (1840–1911), who was the secretary of the Colorado State Board of Health, reports on ozone observations 1876–1877 in Colorado at Denver (taken by Sergeant *James A. Barnich*), Pike's Peak and Colorado Spring (taken by Sergeant *M. Hobbs*), and not presumed to deduce general laws from this 2-years observations, writing (Ambrook 1877, p. 108–109); Fig. 4.49:

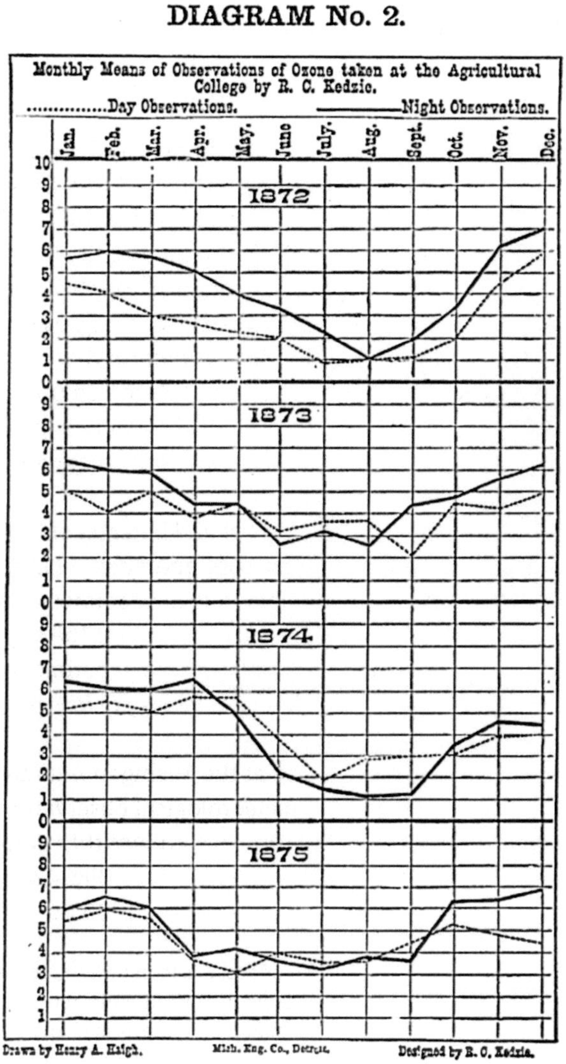

Figure 4.48: Annual variation of the amount of ozone in the years 1872 to 1875 at the Agricultural College of Michigan (Kedzie 1876, p. 141).

> A careful research of the literature at my command, has impresses me with the belief that the "Ghost that Schonbein raised," will not be so easily laid, for a more contradictory set of results from apparently equally competent observers, is hard to find, than is on record about ozone.

Ambrook refers Sergeant *Meyer*,[1063] U. S. signal service, on duty with the Polaris expedition to the Arctic regions in 1871–1872, who reports (Ambrook 1877, p. 109):

1063 The German-born Meteorologist *Frederick* (*Friedrich*) *Meyer* (unknown living dates) was together with the German zoologist *Emil Bessels* (1847–1888) on board the Polaris, described by Bessels (1879).

Figure 4.49: Photo of *Charles Ambrook* in 1899. Source: Carnegie Library for Local History, Boulder.

> Observations were made daily, no ozone was obtained on our voyage north, more or less noted at winter quarters, and some on our voyage south; at winter quarters, and during the winter greatest amount of ozone was observed with north-east winds, and especially with heavy snow-drifts. During the summer, ozone was very abundant, and decreasing in the fall.

Ambrook, citing observations by *Kedzie*, *Ebermayer*, and *Fox*, and noting very different results, states (Ambrook 1877, p. 109):

> It is hardly possible that these observers were mistaken in their conclusions, yet they are contradictory in the extreme; may there not be some factor necessary to the formation of ozone, whose presence, or absence has been overlooked? I believe there is, ...

Finally, *Ambrook* concluded "that ozone reaches Boulder in waves or currents" and argued that the production of ozone could be in a distance from the site of observation (Ambrook 1877, p. 110).

The US geologist *Louis Pope Gratacap* (1851–1917) found in January 1881 in New York a rather good inverse correlation between ozone and "weather" in terms of the storm (no ozone) and clear (ozone), presenting it in a graph (Gratacap 1881). *Albert Lawrence Rotch* (1861–1912), meteorologist and founder of the observatory, and the astronomer and physicist *Edward Charles Pickering* (1846–1919) observed ozone at the Blue Hill Meteorological Observatory, Massachusetts, U. S. A., in the year 1888 at the base (valley) and the summit (Rotch and Pickering 1889, pp. 149 and 153); both distinguished scientists made careful and very detailed meteorological observations but very rudimentary ozone observations.

4.4.5.9 Observations of atmospheric ozone in other countries

In the previous chapters, it was notified that several observers not only did observe ozone in her home countries but also around the world. In this chapter, further ozone observations in Australia, Senegal, Algeria, Brasilia, Mexico, Cuba, India, Ceylon, Denmark, Iceland, Holland, Belgium are noted.

Francis Abbott (1799–1883) was a watchmaker, amateur astronomer, and meteorologist at Hobart (Tasmania, Australia) who in 1854 began to record meteorological observations, and of ozone 1857–1863, which were published each month in the *Proceedings of the Royal Society of Tasmania*; his tables became the standard reference for the local climate, were sent to the United States Meteorological Department for incorporation in world charts and printed for many years by the Tasmanian government. *Edward Swarbreck Hall* (1805–1881), a medical practitioner at Hobart, refers *Abbott*'s ozonometric observatory and stated that he also found always high ozone values in contrast to English sites despite the presence of epidemics (Hall 1860). *Robert Brough Smyth* (1830–1889) of British origin, civil servant and mining engineer, arrived in Melbourne in 1852 and worked as a meteorologist in the Flagstaff Observatory, that was founded by *Georg von Neumayer* (1826–1909), a German explorer and polar researcher, who also was its director until 1864. *Smyth*, Secretary to the Board of Science of Victoria, wrote in an 1858 report (Smyth 1859), referring the Europeans experiences "that ozone is intimately connected with health and diseases" (cited after Proeschel 1860). *Robert Lewis John Ellery* (1827–1908), an English-Australian astronomer, one of the founders of the Royal Society of Victoria and its president from 1866 to 1884, conducted ozone observations in Melbourne 1863–1865 (Ellery 1866). In Sydney, ozone observations were carried out by unknown observers in 1860–1865 (cited after Baxendell 1869). *Frederick Proeschel* (1809–1870), map-maker and publisher of French-German origin, published an article in *The Sydney Morning Herald* (Fr 23, Nov 1860) entitled "Ozone in the Air", and referred the ozone observations done under *Neumayer* in Melbourne, but lamented that "is more an ozonoscope than ozonometer, as it does not indicate the percentage of the ozone compared to the bulk of the air". *Edmund MacDonnell* (1827–1897), a Government Meteorological Observer (1858–1887) at the meteorological station in Brisbane, measured (1862–1868) ozone according to a piece of anonymous information in "*Zeitschrift der Österreichischen Gesellschaft für Meteorologie*" (**4** (1867) 445–446), however, according to *Queensland Government Gazette*, the first report on ozone is in the year 1866 by the observer *John Bliss* (unknown living data). The last report on ozone measurements in Vol. 40 (1887), and monthly data are published. *John Day* (1816–1881), a British physician in Australia (born in Wales), known for studying ozone and hydrogen peroxide sanitary applications, made ozone observation in Geelong 1864–1866 (Day 1866); see also Due (1995) on *John Day*'s life.

The French physician *Alfred Borius* (1837–1885) studied the meteorology at Saint-Louis in Senegal in 1873–1874 and published the monthly means of ozone in a graph

(Borius 1874, p. 181, Borius 1875, p. 191). Borius (1875) wrote in a monograph a chapter on ozone (pp. 216–220). These observations were continued by French pharmacist *Albéric Louvet* (unknown living data) in Saint-Louis in 1875 (Louvet 1876); he also collected rainwater and analyzed it for ammonium and NaCl (Table 2.7). Borius (1879) found with parallel measurements using the Piche evaporometer and the ozone test-paper of *Jame* (de Sedan) in Senegal identical results.

Alphonse François Bertherand (1815–1888), French army doctor [*Médecin militaire*] at Algiers, founded in 1856 and edited until 1864 "*Gazette Medical* de l'*Algérie*", edited 1865–1889 by his brother *Èmile-Louis Bertherand* (1821–1890), and conducted ozone observation 1855–1859 in Algiers. The Swiss naturalist and traveler *Johann Jakob von Tschudi* (1818–1889) found, together with the French physician *Joseph Antoine Hector Chomet* (1808–1884) in Rio de Janeiro in 1860–1861, a correlation between illnesses and the amount of ozone (Tschudi 1862). The Mexican physician *Manuel Rocha* (unknown living dates) reported on atmospheric ozone and constructed an automatic ozone recorder based on *Houzeau*'s method (Rocha 1875), see Fig. 4.50.

Andrés Poey y Aguirre found that in the city of Havana the ozone reactions dismiss with the elevation, whilst in the neighboring country the reverse was observed, but he also found (as usual with the test-paper) more at night than daytime; interesting to note that he found in Havana no dependency from sun and humidity[1064] (Poey 1863a,b).

The British physician *Henry Cook* (1831–1927) registered ozone in the Bombay Presidency (India) since 1863 and found a distinct annual variation with a maximum in June/July and a minimum in February/March (Cook 1865, 1870). *Edward Heelis* (?–1882), who was working in the tea business on Ceylon, made observations at Langdale in 1873 (Heelis 1873–1876). *Clement Lindley Wragge* (1852–1922), a meteorologist in Adelaide (Australia), writes "During my voyage hither from London in the *Maranoa*, via the Canal, and calling at Malta, Aden, and Colombo, I was surprised at the low values for ozone as registered by *Moffat*'s tests" (Wragge 1884, p. 336. *William Wothesporn Ireland* (1832–1909), a Scotch physician in the East Indian Company, studied in 1859 in Kussouli, a former British hill resort station near Shimla (Hindustan, India), ozone at six locations of different altitudes, from 1000 feet to 6400 feet; he did not find what he expected, namely the greatest quantity of ozone at the highest elevation (obviously due to the local vegetation and climate), but an increase between 1000 and 4000 feet (Ireland 1862). *Ireland*, who wrote the paper in Bonn (Germany) in 1862 also communicated that he observed ozone in and near Wiesbaden in February 1862 and found it highest at hills outside to town.

The French physician *Alphonse Aristide Marie Jacolot* (1831–1889) conducted on the frigate Danaé during an Iceland expedition from April until October 1864 daily ozone observations (Jacolot 1865). Since 1886 ozone observations were conducted

1064 The whole year is sunshine and humidity at Havana.

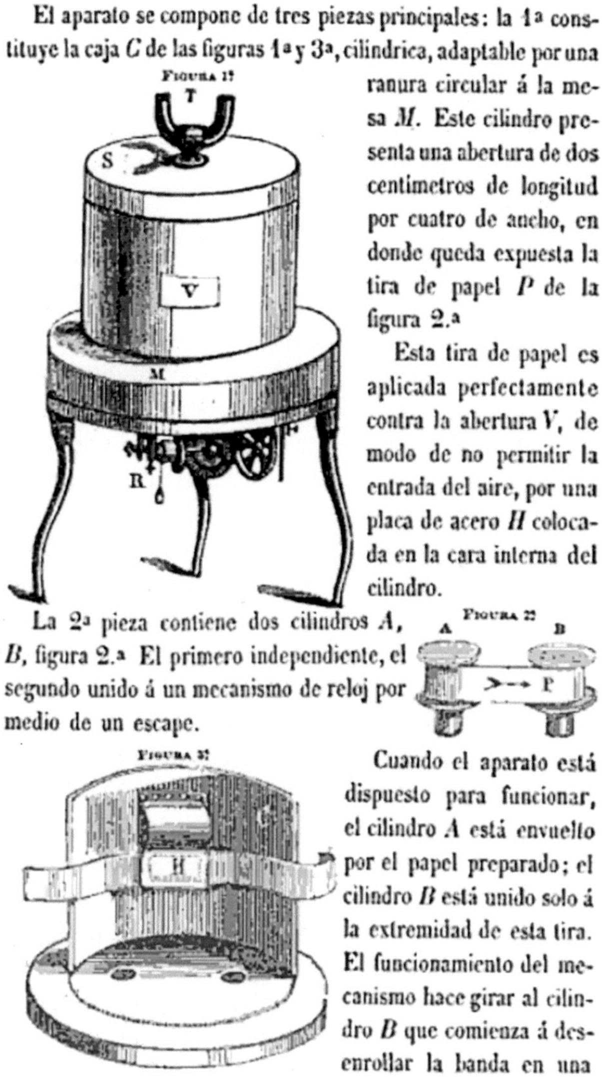
El aparato se compone de tres piezas principales: la 1ª constituye la caja *C* de las figuras 1ª y 3ª, cilindrica, adaptable por una ranura circular á la mesa *M*. Este cilindro presenta una abertura de dos centimetros de longitud por cuatro de ancho, en donde queda expuesta la tira de papel *P* de la figura 2.ª

Esta tira de papel es aplicada perfectamente contra la abertura *V*, de modo de no permitir la entrada del aire, por una placa de acero *H* colocada en la cara interna del cilindro.

La 2ª pieza contiene dos cilindros *A*, *B*, figura 2.ª El primero independiente, el segundo unido á un mecanismo de reloj por medio de un escape.

Cuando el aparato está dispuesto para funcionar, el cilindro *A* está envuelto por el papel preparado; el cilindro *B* está unido solo á la extremidad de esta tira. El funcionamiento del mecanismo hace girar al cilindro *B* que comienza á desenrollar la banda en una

Figure 4.50: Automatic ozone recorder constructed by Rocha (1875, p. 134).

at many stations in Belgium; *Désiré Alexandre Van Bastelaer* (1823–1907), Belgian pharmacist, botanist, and archeologist, was the head of the committee for ozonometry, founded in 1888, including 28 stations like Brussels, Gand, Liége, Mons, Brogs, Soignes and others (Bastelaer 1893).

4.4.6 Do ozonometric observations have value?

Over the whole period of ozonometric observations, many workers found the test-papers not suitable, beginning with first criticism by Schiefferdecker (1855) or even completely useless by von Maack (1856), Huizinga (1867), Dohrandt (1874), Schöne (1880b,c), and Pernter (1881). The most serious problem of the iodide-starch paper was its relation to moisture and humidity, hence the ozonometer was named by Schöne (1880c) as a hygrometer. A less significant problem was its cross-sensitivity to other oxidizing substances; halogens could be excluded due to their absence in atmospheric air and likewise nitrous acid (also in terms of NO_x) due to their very low concentration compared to that of surface-near ozone. Schöne (1873, 1874), however, doubt the presence of ozone in air due to non-evidence, but not exclude its presence; he favored the presence of hydrogen peroxide in atmospheric air. In Chapter 4.4.5.2 arguments are compiled that the test-papers, if any, detected the sum of ozone and hydrogen peroxide. Another problem was the reading of the coloration scale (between 10 and 20 degrees for different papers), almost resulting in subjective estimations. Houzeau (1872a) therefore reduced observations by test-papers on three degrees: weak, medium, and strong, and finally on "present " or "absent" only. Nevertheless, from many observations some findings are "true" according to today's knowledge:

- ozone is abundant in atmospheric air,
- the amount of ozone increases with height,
- the amount of ozone is significant less in towns, and
- the amount of ozone is largest in spring in its annual variation.

The most wrong findings were (note that the term "ozone" means "ozone reaction", i. e., the coloration of the test-paper:

- the different actions of water (precipitation, condensation / evaporation, sea spray, fountain spray, waterfalls etc.) are a source of ozone,
- plants are a source of ozone,
- ozone has a relation to diseases, namely influenza and cholera,
- the amount of ozone is larger at night than daytime, and
- the amount of ozone is larger in the cold season than in the warm season.

All the latter findings were largely based on the positive effect of moisture and the relative humidity in air on the test-paper. It must be noted, however, that several observers also found no relation between the amount ozone in air and the occurrence of diseases. Furthermore, some observers rejected the idea that plants emit ozone (it is notable to mention that upon present the belief lasts of "ozone-rich forests").[1065] Other

1065 Likely also due to the odor of "fresh air" (identified with the smell of ozone), but the truth is the absence of pollutants and the presence of a variety of pleasant smelling volatile organic compounds.

observers did not find more ozone at night and close to waters. The observation of a larger amount of ozone at strong wind is simply due to passing more air at the same time to the exposed test-paper, already recognized by some observers. Thus, nothing can be derived from the amount of ozone (i. e., concentration) in atmospheric air from these observations; the first reliable determinations were based on titration methods by *Houzeau* and particularly *Lévy*.

An important result of laboratory research was that ozone is slightly soluble in water and decomposes (Andrews 1855, Schöne 1873, Carius 1874a); furthermore, that during (not before and after) fog events the air is free of ozone (Heydloff 1853–54, Simonin 1869, Baxendell 1876, Lévy 1878c), where *Baxendell* already concluded that ozone is destroyed by the fog. Very early it was recognized from laboratory studies that ozone acts as an "air cleanser", and thus concluded that ozone decomposes foreign bodies such as miasma, etc. in the air. Notable is the finding from many investigators that the amount of ozone in towns is significantly less than at rural sites, and the given explanation, that the ozone is consumed by "pollutants". Another (correct) finding, that the amount of ozone very close to the earth's surface is less than a few meters above it, was not explicitly explained.

Today we know that dry deposition is an efficient removal pathway and can reduce surface-near ozone down to zero during nights with a low inversion layer; moreover, without further discussion, it is evident that in towns dry deposition is more efficient than on farmland. The answer to the question, which pollutants were the ozone consumers, remains speculative. Assuming that tropospheric photochemical ozone formation was negligible or insignificant, stratospheric ozone was permanent photochemically decomposed in the troposphere and finally absorbed by surfaces. But it was unknown that ozone dominantly decomposes photochemically in the gas phase, and does not directly react with organic compounds (except for olefins). On the contrary, in presence of hydrocarbons, the "photocatalytic" decay of ozone slows down because the reaction $OH + O_3$ goes in competition with OH + hydrocarbons. Chemical ozone consumers like SO_2 and NO_x were surely more abundant in towns than in rural sizes, but likely the smoke in towns provided a heterogeneous pathway of catalytic ozone destruction.

An important sink of atmospheric ozone goes via oxidation of nitrogen oxides $O_3 + NO_x$, (today named "ozone titration" in laboratory jargon) but the concentrations of NO and NO_2 were surely very low in the nineteenth century, but increased in the second half in towns because of combustion based on hard coal. With the latter, sulfur dioxide became a dominant pollutant, namely in British towns (already before 1850);

Fox (1873, p. 143) writes: "In the early morning, and during snowy weather, it is most striking, when we suddenly admit the country air into our bedrooms by opening the windows". Becker (1977, p. 8) refers that unfortunately one ascribes ozone in remote areas and "ozone-rich" forest a health-promoting affect.

SO_2 consumes ozone via the aqueous phase (clouds, fog, rain). The oxidation of nitrite in solution by ozone was already recognized by Schönbein (1847a) and that sulfite is fast oxidized in solution (Chapter 4.7.2.3) and precipitation (e. g., Küppers (1918). Furthermore, there is no doubt that the extreme dust and smoke (namely soot) pollution of the air in towns removed ozone much greater than today. It is no further need to comment Fox (1873, p. 7):

> Notwithstanding all obstacles, however, to the progress of truth, we are advancing steadily towards full daylight, and are apparently on the eve of great and important discoveries respecting these interesting and important substances.

4.4.7 Photophysics, photochemistry and formation of ozone 1878–1932

"The spectrum of ozone is exceedingly complex and was the subject of numerous investigations", Rideal (1920, p. 9) writes. The first idea to detect ozone in air spectroscopically was given by *Henry Day* (1814–1881),[1066] physician of Stafford, cited by Fox (1873, p. 173) who writes (not giving a reference):

> Dr. H. Day of Stafford has suggested that a delicate test for ozone might possible be found by observing the difference in the color of the electric spark passed through rarefied ozonised air, and that transmitted through ordinary air. He thought that rarefied air ozonized air might have some characteristic effects on the spectrum, and advocated the employment of the spectroscope in determining this point.

Day (1868, p. 81) who had "great objection to these test-papers" writes "bringing the spectroscope into use", and "that the spectrum would give a different band or bands with rarefied ozonized air from that or those afforded by common air under similar circumstances and conditions." The French physicist *Alfred Cornu* (1841–1902) found by measurements between June and August 1878 that the atmosphere absorbs all solar radiation below 294–195 nm (Cornu 1879a, p. 1104), soon later fixed to 293 nm (Cornu 1879b, p. 1286). *Joseph Chappuis* (who made the investigations together with *Hautefeuille*) first observed the blue color of ozone in sufficient thickness and found strong absorption between 606 and 613 nm, which could explain the blue color of the sky (Chappuis 1880). Soon later *Chappuis* found eleven lines lying in the region between 628.5 nm and 444 nm in the visible spectrum (Chappuis 1882a), for liquid ozone he presented two further bands (Chappuis 1882b); the bands between 430 nm and 750 nm were later named *Chappuis* bands.

1066 *Henry Day* received his medical education at Guy's Hospital and in Paris, qualifying in 1835. For some years he practiced in London but about 1850 moved to Stafford. He was president of the St. Andrews Medical Graduates' Association and was an office-holder at the Sanitary Institute's meeting at Stafford in 1878.

In the ultra-violet region, the British chemist *Walter Noel Hartley* (1845–1913)[1067] first estimated the ozone absorption band at 256 nm and proposed that likely atmospheric ozone is responsible for the total absorption of solar radiation below 295 nm as found by *Cornu* (Hartley 1880, 1881). *Hartley* found that the ultra-violet absorption of ozone corresponds to the total amount of ozone derived from earlier measurements, and also believed that the cause of the blue color of the sky is due to ozone.[1068] This band (255.3 nm with the maximum absorption) was later named *Hartley* band; this strong absorption is only exceeded by oxygen at 145 nm (*Schumann-Runge* continuum).[1069]

The first direct evidence of ozone in the atmosphere was provided by Schöne (1884) who measured the spectrum of the atmosphere in early winter 1883 near Moscow during intense frost to avoid overlapping with the bands of water vapor (599–610 nm). The absorption spectrum of ozone (the principal band between 595 and 613 nm) was found to be well following the description given by *Chappuis*. Besides the eleven bands found, *Schöne* detected another at 516 nm and, still subject to some doubt, at about 452 nm. Schöne (1894b) communicates that the spectrum of ozone consists of 13 more or less intensive bands. He found as result of his spectroscopic observations that the lower atmospheric layer contains less ozone in the morning than in the evening, the maximum occurs in February and March, afterward the ozone concentration declines to a minimum in July; during lightning and strong rain, the spectroscope never detected ozone.

Edgar Meyer (1879–1960), who gained a doctoral degree on UV absorption by ozone at *Emil Warburg* in Berlin, estimated the maximum absorption band at 257 nm and for the first time the absorption coefficient α between 180 and 300 nm (based on the modern equation $J = J_0 10^{-\alpha d}$, where d layer thickness), and could thus confirm

1067 British pioneer of spectroscopy, and the first person to establish a relationship between the wavelengths of spectral lines of the elements (1883). He published the following books concerns the atmosphere: Air and its Relations to Life – Being, with Some Additions, the Substance of a Course of Lectures Delivered in the Summer of 1874 at the Royal Institution of Great Britain (1876), Longmans, Green, London, 280 pp.; Water, Air and Disinfectants (1877), Society for promoting Christian knowledge, London, 128 pp.; A Course Quantitative Analysis for Students (1887), Macmillan & Co., London, 248 pp.

1068 It was *John Tyndall* who recognized, based on *Rayleigh*'s work on electro-magnetic waves, that the blue frequencies in the sunlight are 16 times more efficient scattered than the red frequencies; therefore, the cloudless sky appears blue.

1069 *Victor Schumann* (1841–1913, a German engineer and hobby physicist in Leipzig, discovered in 1893 the ultra-violet region below 183.5 nm (Schumann 1893, p. 449). *Carl Runge* (1856–1929) was a German mathematician and physicist. The lines in the region between 91 nm and 121 nm were discovered between 1906 and 1914 by *Theodore Lymann* (1874–1954); at the *Lymann*-alpha line (121.5 nm) photodissociation of H_2O occurs (in the mesosphere) into H + OH, where H escapes into interplanetary space and oxygen remains in the earth's atmosphere (intermediates are HO_2, H_2O_2 and O_3); Bates and Nicolet (1950), Nicolet (1950).

Hartley's hypotheses of the abrupt ending of the solar spectrum at 293 nm (Meyer 1903). *Meyer* found a minimum of ozone absorption at 205 nm and a maximum at 258 nm (*Hartley* at 256 nm). Using a mean atmospheric ozone concentration based on *Lévy*'s long-term quantitative ozone determinations, *Meyer* calculated from his absorption measurements a mean of 1.65 mg / 100 m^{-3} (corresponds to 7.7 ppb).

Knut Johan Ångström (1857–1910)[1070] studied the spectrum in fare infrared and found bands at 4.8 μm and 9.1–10.0 μm (Ångström 1904); he concluded on a significant amount of ozone in the atmosphere but that ozone does not have a large effect on solar absorption because of the primary absorption through carbon dioxide and water vapor between 0.7 and 4.35 μm.

Ladenburg and Lehmann (1906)[1071] noticed that all previous researchers used for their investigation about 8 % ozone gas mixtures with oxygen whereas they evaporated liquid ozone into a vacuum for the first time to obtain high percentage of gaseous ozone. They found 14 bands between 449 nm and 627 nm and determined the absorption in percentage for 49 wavelengths between 0.59 μm and 12.2 μm, thus generally confirming *Chappuis*'s and *Ångström*'s results; in the ultra-violet range, they detected 12 bands between 322 nm and 356.5 nm. The latter spectral region, which was identified later as bands by Fowler and Strutt (1917), and was named *Huggins* bands (300–3600 nm).[1072] Furthermore, they found five bands between 610 and 670 nm under experimental conditions which led them to conclude on another body, excluding nitrogen oxide (they used nitrogen-free oxygen for ozone production) *Ladenburg* and *Lehmann* believed that a higher oxygen compound than ozone was gained.[1073]

1070 Swedish physicist, son of *Anders Jonas Ångström* (1814–1874) and father of *Anders Knutsson Ångström* (1881–1981); all studied atmospheric radiation. The unit is called after *Anders Jonas Ångström*.

1071 *Erich Robert Ladenburg* (1878–1908), sun of *Albert Ladenburg*, an older brother of the famous physicist *Walter Rudolf Ladenburg* (1882–1952), was a German physicist and the first assistant of *Heinrich Rubens* (1865–1922) at the physical institute of the Technical University Berlin (TH); tragic death by drowning in 1908. *Erich Lehmann* (1878–?) was a German photochemist who studied at TH Berlin; 1904–1914 assistant, 1908–1933 private lecturer and later professor (photochemical laboratory of the TH). 1933 dismissed due to his Jewish origin with unknown fate.

1072 The British astronomers *William Huggins* (1824–1910) and his wife *Margaret Lindsay Huggins* (1848–1915) found six broad lines (319.9–333.8 nm) to be characteristic of Sirius, Vega, and other white stars, and which was identical as the continuation of the spectrum of hydrogen beyond H (Huggins and Huggins 1890).

1073 This observation stimulated *Carl Dietrich Harries*, who already observed in 1904 that in a reaction between high-concentrated ozonized air (e. g., 14 %) and organic compounds more oxygen than equivalent to O_3 is added, to assume a polymer of oxygen such as O_4 that decays into $O_2 + O + O$ (Harries 1911); he called it later (Harries 1912) *oxozon* [oxozone], today named tetraoxygen. Riesenfeld and Schwab (1922) isolated pure ozone and reported that there was no evidence at all for a higher molecular weight component such as O_4.

Johannes Nikolaus Stark (1874–1957)[1074] has shown that ozone molecules give rise to many bands lying between visible green and the ultra-violet 219 nm (Stark 1914). *Philipp Lenard* first studied the impact of ultra-violet light on gases (Lenard 1900) using quartz tubes in combination with a spectrometer and as a light source an inductorium consisting of several *Leyden* bottles;[1075] he found an absorption maximum in air to be 160–180 nm. His main research goal was studying the formation of condensation nuclei (Chapter 3.3.1) but also showed that ultra-violet light was an effective agent for ozonizing oxygen (Lenard 1900, p. 503). Referring *Schumann*'s investigations that hydrogen provides the most powerful source of ultra-violet light, he concluded (Lenard 1900, p. 504)

> Unmittelbare Wirkungen solcher Sonnenstrahlung dürften am ehesten in den oberen Schichten der Erdatmosphare zu suchen sein [Most likely a direct impact of such solar radiation should be in the upper layers of the earth's atmosphere].

Without further argumentation, Henriet and Bonyssy (1908) considered that ozone is formed by ultra-violet light in the higher atmosphere. Soon later *Eugen Goldstein* (1850–1939)[1076] investigated the formation of ozone under the influence of ultra-violet light using a *Geissler* tube (Goldstein 1903). These findings were also soon later confirmed by Aubel (1909), Johnson and McIntosh (1909), and Berthelot and Gaudechon (1910a, p. 1172). *Witalius Chlopin* (1890–1950)[1077] first studied the influence of ultra-violet light on humid air detected not only ozone but also hydrogen peroxide in significant quantity and in traces the anhydride of nitrous acid ($NO + NO_2$); Chlopin (1911).

Emil Warburg was the first who found that during the silent discharge not only ozone is gained (*ozonized air*) but also decomposed (*deozonised air*), and that it de-

1074 German physicist who was awarded with the Nobel Prize and who was a proponent of the Nazis and the "German Physics".

1075 The German *Ewald G. Kleist* (1700–1748) invented a precursor of a capacitor in 1745 in Cammin (today Kamien Pomorski in Poland), first named "*Verstärkungsflasche*" [amplification bottle]. *Pieter van Musschenbroek* (and others) used and improved this first electric condenser since 1746, and carried out various experiments in Leiden (Holland), later named "*bouteille de Leyden*" by the French *Jean Nollet* (1700–1770); in German "*Leydener Flasche*". Another instrument often used (e. g., by *Goldmann*) was the *Geissler* tube ["*Geißler'sche Röhre*"], an early gas discharge tube invented by the German glass blower *Johann Heinrich Wilhelm Geißler* (1814–1879).

1076 *Gotthilf-Eugen Goldstein* was a German who worked as experimental physicist and educations at the Berlin Observatory; he discovered the canal rays and is known for his research on gas discharges. It is a human (German) tragedy that both *Lenard* (anti-Semite and later Nazi) and *Goldstein* (Hebrew) showed that UV in the *Schumann* portion of the spectrum 120–180 nm exerted the maximum activity in this respect.

1077 *Witali Grigorjewitsch Chlopin* [Виталий Григорьевич Хлопин], in Germany his given name was *Witalius*, was a Russian-Soviet radiochemist in St. Petersburg/Leningrad who studied in Göttingen. Chlopin (1911) also proved that the *Schönbein* paper shows much more sensitivity to hydrogen peroxide than to ozone.

pends on the amperage (Warburg 1902). Two years later Warburg (1904) recognized that the process of ozonation cannot be seen as a process similar to electrolysis (comparing the electric charges in experimentally estimated coulomb compared between both processes) and that the formation of ozone must be interpreted as a photochemical process; he notified that during silent discharges both cathode rays and short-wave ultra-violet radiation occur, referring both *Lenard* and *Goldstein*. *Warburg*'s doctor student *Erich Regener* (1881–1958), German physicist, thus studied the oxygen-ozone equilibrium under the influence of UV radiation and noticed the fact that although light between 120 and 180 was a powerful ozonizing agent, yet light in the wavelength range of 230–290 nm (esp. 257 nm) exerted an equally effective catalytic decomposing effect (Regener 1906). *Regener* concluded the ozone production occurs at a wavelength lower than 193 nm based on the finding by Kreusler (1901)[1078] that dry and atmospheric air free of carbon dioxide absorbs only below 193 nm (*Regener* furthermore studied the photolysis of other gases like NH_3, NO, N_2O).

In the discussion of *Bodenstein*'s theory, *Fritz Weigert* (Bodenstein 1913, p. 849) proposed the photolysis of ozone in our present modern view; Warburg (1914) estimated the second reaction (O + O) based on energetic calculations to be unimportant:

$$O_3 = O_2 + O$$
$$O + O = O_2$$
$$O + O_2 = O_3$$

Oliver Reynold Wulf (1897–1987), an American chemist, physicist, and meteorologist, found evidence that the molecule O_4 is responsible for absorption by oxygen in the region 247 nm and, in addition, give a plausible explanation of *Warburg*'s results that ozone absorbs at 206 nm and 253 nm but not oxygen, as found shortly before (Wulf 1928). Lewis (1924) proposed O_4 as a metastable dimer of molecular oxygen. 80 years later the bands between 240 and 290 nm in the O_2 spectrum (now called *Wulf* bands) have been proposed to be due to collision-induced absorption (CIA) of molecular oxygen $[O_2–O_2]$.[1079]

1078 *Hans Kreusler* (unknown living dates) was a German physicist in Berlin.

1079 See Braslavsky and Rubin (2011) for references. The major constituent's oxygen and nitrogen possess no dipole moment so that their vibrational and rotational energy cannot be excited directly by absorption of radiation. However, dipole moments are induced during collision processes; This process was discovered by the Canadian physicist *Harry Lambert Welsh* (1910–1984) and co-workers (Crawford et al. 1949). Cacace et al. (2001) believed having detected tetraoxygen. O_4 might also make a fleeting appearance in atmospheric chemical reactions that are responsible for the phenomenon of "nightglow" on Earth and other planets. Wagner et al. (2002) observed O_4 spectroscopically in the atmosphere. What O_4 looks like remains probably a mystery. O_4 could be also be an intermediate in the thermal decay of O_3: $2O_3 \rightleftarrows O_6 \rightarrow O_4(\rightarrow 2O_2) + O_2$.

Sydney Chapman (1888–1970), who was a British physicist in Manchester and London, presented the first complete theory of stratospheric ozone; the main part of his communication was to explain the daily and seasonal variation of ozone as well the variation of the amount of ozone by latitude (Chapman 1930). He stated that the production of ozone is solely by ultra-violet radiation and already assumed that the O atom formed by dissociation of O_2 is likely in different electronic states but any knowledge of it was still missing.[1080] The transport to lower atmospheric layers *Chapman* explained through gravity. The ozone–oxygen cycle, later named *Chapman* cycle, includes the following reactions:

$$
\begin{aligned}
O + O &= O_2 \\
O + O_2 &= O_3 \\
O + O_3 &= 2\,O_2 \\
O_3 + h\nu &= 2\,O_2 \quad \text{(photochemical decay)} \\
O_3 + O_3 &= 3\,O_2 \quad \text{(thermal decay)}
\end{aligned}
$$

The photolysis of O_3 was kinetically first studied at yellow-red light (about 600 nm) by Kistiakowsky (1925),[1081] but interpreted by Schumacher (1932)[1082] in terms of

$$
\begin{aligned}
O_3 + h\nu &\rightarrow O(^3P) + O_2 \\
O + O_3 &\rightarrow 2\,O_2 \\
O + O_2(+M) &\rightarrow O_3(+M)
\end{aligned}
$$

Reinhard Mecke (1895–1969), a German physicist and a pioneer of infrared spectroscopy in Heidelberg, has stated that the amount of ozone found in the lower layers of the atmosphere may be partly due to some ozone molecules dropping down due to higher gravity from the higher parts of the atmosphere and thus surviving their decomposition by ultra-violet light. The mechanism for the formation of ozone *Mecke* describes as $2\,O_2 + h\nu$ (<202.5 nm) = $O_3 + O$, and the decomposition of ozone as $O_3 + h\nu$ (<265.5 nm) = $O_2 + O$ (Mecke 1931). *Mecke* also discusses for the first time that a three-body collision is essential while ozone formation is due to uptake of collisional

1080 The electronic states and photodissociation of O_2 was intensively studies after 1950; see Leighton (1961) and Heicklen (1976) for details; further on the history of radical chemistry Wallington et al. (2018).

1081 *George Bogdan Kistiakowsky* [Ukrainian: Георгій Богданович Кістяківський), who escaped after Bolshevist revolution to Germany where he obtained PhD at *Max Bodenstein*; 1926 immigrated to the USA; wrote e textbook on photochemistry (Kistiakowsky 1928) and contributed to the US atomic bomb research. Nephew of *Vladimir Alexandrovich Kistiakowsky* (1865–1952).

1082 *Hans-Joachim Schumacher* (1904–1985), German chemist, first at *Bodenstein* in Berlin, 1933–1945 professor in Frankfurt and since 1945 in Buenos Aires; he was a Nazi and denunciator.

energy of the $O{-}O_2$ body, and otherwise the reaction system would be returned to O and O_2. Harteck (1931), based on laboratory investigations and the proposed mechanism by *Chapman* and *Mecke*, discusses the fluctuations of ozone due to pressure and temperature dependence of these reactions; he argued that photolysis is practically not dependent on the temperature, the ozone formation strongly decreases with decreasing pressure and the reaction rate of the two-body reaction $O + O_3$ decreases with lowering the temperature, hence with increasing altitudes and in polar winter, resulting in long lifetimes of O atoms, gaining a reservoir.

Beretta and Schumacher (1932) studied the ozone decomposition in ultra-violet light (at 313 nm) and proposed the same ozone cycle but also the formation of excited oxygen atoms:

$$O_3 \rightarrow O(^1D) + O_2.$$

Schumacher (1932) continued these studies to the red-light spectrum. Eucken and Patat (1936)[1083] studied experimentally the temperature dependency of the reaction system proposed by *Schumacher* and *Mecke*, and confirmed the values for the reactions rates and activation energies as well *Harteck*'s view, and concluded from kinetic calculations that the formation of atmospheric ozone at 40 km height is significantly smaller than at 25 km height. Further progress in the understanding of stratospheric chemistry was made only after 1950 (Chapter 4.4.8). At the conference on ozone (Tharandt, April 1944), *Karl Adolf Erich Schröer* (1904–1945), a German professor for physical chemistry in Berlin, presented an extended kinetic-chemical model and calculated the vertical distribution of ozone, comparing it with the experimental results (Schröer 1944/1949). *Richard Ansel Craig* (1922–1978), who was a foremost American meteorologist, submitted his doctoral thesis on ozone observations at the MIT in 1948, and to advance the photochemical theory, but found later that many of his photochemical calculations were already the work of previous investigators, whose papers had not come into his attention due to the war (Craig 1950); however, this monographic work is a rich source of historic literature.

4.4.8 Atmospheric ozone research in the twentieth century

In the 1920s, an international community ozone research was developed by three leading scientists: *Charles Fabry* from France, *Gordon Dobson* from the United Kingdom,

1083 *Arnold Thomas Eucken* (1884–1950) was a German physicochemist; doctorate 1906 and habilitation 1911 at *Nernst* in Berlin, 1915 professor in Breslau and since 1930 in Göttingen. *Franz Xaver Maria Theresia Patat* (1906–1982) was an Austrian-German chemist who became in 1952 Professor for Technical Chemistry in Hannover.

and *Paul Götz* (1891–1954)[1084] from Switzerland. Götz (1951, p. 288) writes:

> It is fascinating to observe how ozone pulsates through the atmosphere like blood circulating in an organism. Ozone is created by radiation in high-altitude layers. Primarily at the shadow boundary of the polar night, and at the altitude of the warm layer, it produces the temperature contrasts a resulting polar vortex which enable it to sink as a secondary layer down to the theater of meteorological activity. And finally, diffusing to the vicinity of ground level, the ozone re-enters the oxygen metabolism.

In 1929, *Fabry* organized the first international ozone conference in Paris. However, investigation of ozone in the free troposphere remained of little interest, although it was known since the end of the nineteenth century that at higher altitudes (on mountain sites) the amount of ozone was found to be significantly larger than that at lowland stations. In the 1960s, it was assumed that transport of ozone from the stratosphere was thought of as the only major source of tropospheric ozone (Junge 1962); Danielson (1968) first investigated the transport processes down through the troposphere. Surface-near events with increased ozone concentrations, found in the 1930s, have been explained exclusively through the inhomogeneous distribution of ozone in the troposphere caused by different transport from the stratosphere. An understanding of the photochemical formation of ozone in the troposphere began only in the late 1970s (Fishman and Crutzen 1978, Crutzen 1988, Crutzen et al. 1999). At the beginning of tropospheric ozone research, its contribution to the "greenhouse" effect was the main issue (Fishman et al. 1979); only in the late 1980s, its role of *oxidant* (remember, it has been already recognized 130 years before – but without understanding the chemistry beyond) and harmful atmospheric agent (keywords: *Waldschaden*, human health) became into the focus of tropospheric ozone research. Still, in 1980 (Dütsch 1980, p. 26), the near-surface ozone was regarded exclusively from the stratosphere.

Unfortunately, it remains unclear even today whether the near-surface concentration of ozone has already increased since the early nineteenth century (such as proposed by Marenco et al. 1994); evident is an increase of tropospheric ozone worldwide with the beginning 1960s. However, like in Los Angeles in the late 1940s, it is highly likely that at many industrial sites and towns ozone was locally produced in presence of nitrogen oxides and organic compounds much earlier – and likely soon later again consumed by air pollutants.

1084 *Friedrich Wilhelm Paul Götz* was a German-Swiss who fell ill with tuberculosis in 1914 and spend some time in Davos, and decided to remain there in 1916; since 1919 he worked with *Carl Dorno* at the Observatory and founded in 1921 the "*Lichtklimatisches Observatorium*" Arosa [Light Climatic Observatory]: his successor became the physicist *Hans Ulrich Dütsch* (1918–2004). *Dütsch* wrote his dissertation under the supervision of *Götz* in 1946 "*Photochemische Theorie des atmosphärischen Ozons unter Berücksichtigung von Nichtgleichgewichtszuständen und Luftbewegungen*". In 1988, the observatory was integrated into the Swiss Meteorological Institute.

4.4.8.1 Total and stratospheric ozone

William Jackson Humphreys (1862–1949), an American physicist and atmospheric researcher, analyzing the available data on solar spectroscopy, recommended measuring the amount of ozone in the upper atmosphere and he concluded (Humphreys 1910, p. 108)

> [...] any proportionate decrease in the ultra-violet radiation, [...] leads to a diminished production of ozone in the upper atmosphere, [...] and a cooling off the earth and the lower atmosphere.

The French physicists *Charles Fabry* (1867–1945) and *Henri Buisson* (1873–1944) discovered the ozone layer[1085] during their investigations at the faculty of sciences of Marseilles in 1913. They determined the total amount of pure ozone to be 5 mm and assumed that it exists in the upper atmosphere (Fabry and Buisson 1913). The British astronomer *Alfred Fowler* (1868–1940) and physician *Robert John Strutt* (1875–1947)[1086] carried out the first undisputed spectroscopic measurements of total atmospheric ozone (Fowler and Strutt 1917). Strutt (1918) found that the lower atmosphere is much more transparent to ultra-violet radiation than the upper atmosphere, from which follows that the upper atmosphere contains more ozone than the lower one; the complete reduction of radiation corresponds to 0.26 mm pure ozone.

In 1924, *Gordon Miller Bourne Dobson* and *Douglas Neill Harrison* (1901–1987)[1087] designed an instrument for the determination of the ozone amount from photographic plates using a photometer, but with elaborate handling (Dobson and Harrison 1926). In 1931, *Dobson* developed a new instrument with two prisms (double monochromator) and photoelectric detection (sodium cell), which made the observations much simpler. It also became the standard instrument for measuring both total column ozone and the ozone profiles[1088] (Dobson 1931), further improved in the 1940s, but other spectrometers also have been constructed and used for observations (Brönnimann et al. 2003). The total column ozone (mass per square) is expressed in the *Dobson* unit (1 DU is defined as 0.01 mm thickness at STP).[1089]

1085 A recommended “brief history of stratospheric ozone research” is published by Müller (2009).

1086 Sun of *John William Strutt*, 3rd Baron *Rayleigh*; who became the fourth Baron *Rayleigh* in 1919.

1087 Nothing more is known about *Harrison* then he was the head of upper air instruments at Meteorological Office Development branch.

1088 The vertical distribution of ozone is derived by the Umkehr method (developed by Götz et al. 1934). Direct solar intensity is measured at two different wavelengths, one being more absorbed by ozone than the other. At sunrise and sunset, the intensities decrease at different rates. The ratio shows an inversion. This is called the Umkehreffekt and gives information about the vertical distribution of ozone in the atmosphere. An Umkehr measurement takes about three hours, and provides data up to an altitude of 48 km, with the most accurate information for altitudes above 30 km.

1089 For example, 300 DU of ozone brought down to the surface of the earth at 0 °C would occupy a layer only 3 mm thick. One DU corresponds to $2.69 \cdot 10^{16}$ ozone molecules cm^{-2}, or $2.69 \cdot 10^{20}$ m^{-2}. This is equivalent to 0.4462 millimoles of ozone m^{-2}.

After first observations (1924–1925) in Oxford, seven instruments were built and distributed throughout Europe (Oxford/Britain, Arosa/Switzerland,Lindenberg/Germany,[1090] Abisko/Sweden, Valentia/Ireland, Lerwick/Britain), and at Montezuma in Chile (Dobson 1930). After this period, the instruments were installed in California (Table Mountain), Egypt (Helwan), India (Kodaikanal), and New Zealand (Christchurch), and at Spitzbergen, where data were collected from August 1928 to November 1929. More than 6000 plates of the spectrographs have been developed and analyzed in Oxford (Bojkov 2012, p. 2).

In the 1930s, many more sites in Europe and outside of Europe were added to total ozone observations (Brönnimann et al. (2003) for details). Fabry and Buisson (1931) estimated the total amount of ozone in the atmosphere to be 0.18 mm, corresponding to a mean concentration of 43 μg m^{-3} (at 15 °C). *Frederick Eugene Fowle* (1869–1940),[1091] an American physicist, observed the amount of atmospheric ozone between 1921 and 1928 both in the northern (Table Mt., California) and southern (Montezuma, Chile) hemispheres, detecting maxima in the spring and minima in the autumn in both hemispheres (Fowle 1929). *Père Pierre Lejay* (1898–1958) was a French physician in Grenoble, who measured total ozone 1932–1936 at the meteorological and magnetic observatory zi-ka-wei (Chinese: 徐家汇, Romanized as Zikawei and Xujiahui, today part of Shanghai); Lejay (1937), Fig. 4.51. The Jesuit *Maurice Burgaud* (1884–1977), who began working at the observatory in 1924, was the magnetic station's final director and continued the ozone observations until 1942.[1092]

Götz and Dobson (1928) have found the average height of the ozone layer between 30 and 40 km above the earth's surface, and Götz (1931) reported on altitudes up to 50 km for the ozone layer (it means the maximum of ozone concentration). The knowledge of stratospheric ozone, Harteck (1931, p. 859) shortly summarize as follows: the amount of ozone increases with the latitude and is about twice as much than above the equator, in spring the amount of ozone is larger than in autumn, probably larger in the morning than in the evening, and especially high in the polar night. Soon later, using a new method, Götz et al. (1934) have reported that the average height of ozone in the atmosphere at Arosa, Switzerland, is about 20 km, much less than former estimates.

1090 The Aeronautical Observatory Lindenberg was founded by *Richard Aßmann* in 1905.

1091 Graduated in physics 1894 at MIT and afterwards Research Assistant at the Smithsonian Astrophysical Observatory. He was an editor of "Smithsonian Physical Tables". *Fowle* together with *Charles Greeley Abbot* (1874–1973), astronomer and director of the Astrophysical Observatory of the Smithsonian Institution, determined from 1900 to 1907 at first the sola constant at Washington to about 2.1 cal m^{-2} s^{-1} (Abbot and Fowle 1908).

1092 The Jesuits established some forty-seven meteorological stations, if one includes those in observatories that kept records for sufficiently long periods of time. Geographically, there were seventeen in Europe, six in the United States, fifteen in Central and South America, and nine in Asia, Africa, and Australia. Among them, the most important were the observatories of Belén (Cuba), Manila (Philippines), Zikawei (China), and Antananarivo (Madagascar), which were dedicated to observing, forecasting, and researching tropical hurricanes (Udías 2019).

Figure 4.51: The 'Siccawei' (or Zi-ca-wei) astronomical observatory was built in 1912 and at that time considered one of the first three in the world. Source: Raymond Vibien Family Album (https://www.virtualshanghai.net/Photos/Images?ID=283).

Dobson studied extensively the amount of ozone as a function of the latitude and the season of the year (Dobson 1930) and discussed it in relation to meteorology (Dobson and Meetham 1934, Meetham and Dobson 1935).[1093] Ladenburg (1935)[1094] discusses the problems of light absorption and distribution of atmospheric ozone.

Interestingly that already the first ozone measurements in Marseille and Arosa revealed irregular ozone fluctuations from day to day, explained by Dobson and Harrison (1926) as meteorological fluctuations. Kiepenheuer (1938)[1095] found such fluctuations during single measurements over 1–2 h during observations in summer 1937 at Jungfraujoch (3547 m) and writes (Kiepenheuer 1938, p. 356):

> Diese Schwankungen legen die Existenz von vorbeiziehenden Ozonwolken bzw. Ozonlöchern nahe [These fluctuations suggest the existence of drifting clouds of ozone and holes of ozone, respectively].

Notably, we (Fig. 4.52) found ozone fluctuations in the free troposphere on a time scale of a few minutes in layers close together.

1093 *Alfred Roger Meetham* (1910–?) was a British physicist working lifelong in the field of air pollution.

1094 *Walter Rudolf Ladenburg* (1882–1952) [also written *Rudolf Walter*], sun of *Alfred Ladenburg*, emigrated in 1932 to the USA and became professor for physics at Princeton University.

1095 *Karl-Otto Kiepenheuer* (1910–1975) was a German astronomer in Göttingen and since 1942 in Freiburg.

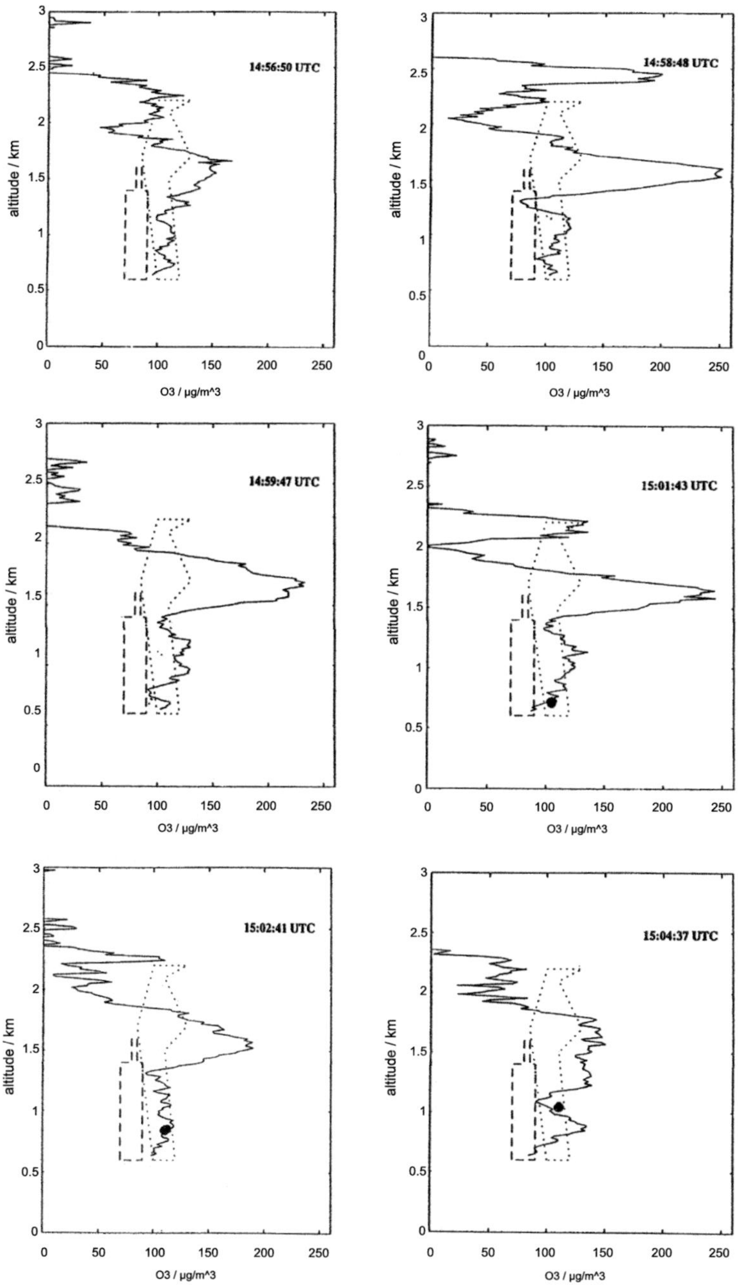

Figure 4.52: “Ozone clouds” at two layers (about 1500 m and 2500 m altitude) and its timely variation within about 8 min. Ozone profile measurements downtown Berlin (Alexanderplatz) August 25, 2000, around 15:00 UTC: a) solid line – BTU mobile lidar Elight M510 (in high time resolution; one profile every minute), b) dotted line – aircraft measurements (duration of one profile about 20 min). BTU lidar was operated by my assistant *Rolf Fabian* and the aircraft flew *Volker Mohnen*.

From a series of balloon flights in 1932 and 1933, *Erich Regener* (1881–1955) measured cosmic rays with a balloon-borne electrometer. The first successful balloon-borne measurements of the vertical distribution of ozone *Regener* conducted with his son *Victor Regener* (1913–2006):[1096] they constructed a small spectrograph, transported by two sounding balloons attaining heights of 21, 20, and 31 km respectively; at 30 km they estimated that 70 percent of the ozone layer was below the apparatus (Regener and Regener 1934). Such profile measurements were continued by *Regener* Jr. who also developed a new spectrograph for balloon ascents (Regener 1938b). *Victor Regener* continued sounding-balloon flights in New Mexico (USA) since 1959 (Regener 1951), showing that the vertical distribution of ozone between 10 and 30 km is subject to strong variation.

Such spectrographs also have been used for surface-near ozone measurements by *Götz* and *Regener* (Regener 1938a), These measurements were labor-intensive but have shown that the amount of ozone varies significantly through different vertical and horizontal mixing processes as explained by Glückauf (1944). Whereas the layer at an altitude of 20–30 km was characterized by a maximum ozone production (70 % of total ozone),[1097] the layers below 10 km have been seen as the removal zone through dust and oxidable substances (Götz 1951); the catalytic ozone decomposition (HO_x) process was not yet known. The first vertical distribution of ozone using German V2 rockets was obtained between 1946 and 1948 by photographing the ultraviolet spectrum of the Sun with small automatic spectrographs (Johnston et al. 1951). Rockets remain for many years the experimental tool for exploring ozone chemistry in the upper atmosphere; the first satellite measurements were made in the early 1960s. The prewar measurements from Arosa and Tromsö are doubtless to most important (Fig. 4.53).

In 1929, the first International Ozone Congress was held in Paris with 35 participants (Bojkov 2012); the second followed 1936 in Oxford with 58 participants; lectures were presented by *Dobson*, *Chapman*, *Götz*, *Regener*, *Wulf*, *Ladenburg*, and others.[1098]

1096 *Erich Rudolph Alexander Regener* became a Professor of Physics at Stuttgart University (1937–1945 released by the Nazi regime) in 1920; he founded 1936 a private research laboratory for the physics of the stratosphere [*Forschungsstelle für Physik der Stratosphäre*] in Friedrichshafen (Bodensee), which has been integrated 1938 into the "Kaiser Wilhelm Society", destroyed during allied bombings in 1944, and moved to Weißenau; in 1952 established as Max-Planck-Institute for Physics of the Stratosphere (with a new director in 1955 this institute moved to Katlenburg-Lindau/Harz and has been integrated in the existing Institute for Ionospheric Research). His son *Victor H. Regener* who earned his doctorate in physics in 1938 in Stuttgart, left afterwards Germany, first taking a two-year research position in Italy, and then to the USA (University of Chicago, and in 1946 University of New Mexico). See Carlson and Watson (2014) for more information upon the research *Regener*.

1097 From the profile shown in Fig. 4 in Regener and Regener (1934) it can be derived that 10 % of total ozone is located in the layer up to 10 km, 20 % between 10 and 19 km, and 38 % between 19 and 28 km.

1098 The presentations have been published in: Conference on atmospheric ozone held at Oxford, September 9th to 11th, 1936. *Quarterly Journal of the Royal Meteorological Society* **62**(S1), 1–76.

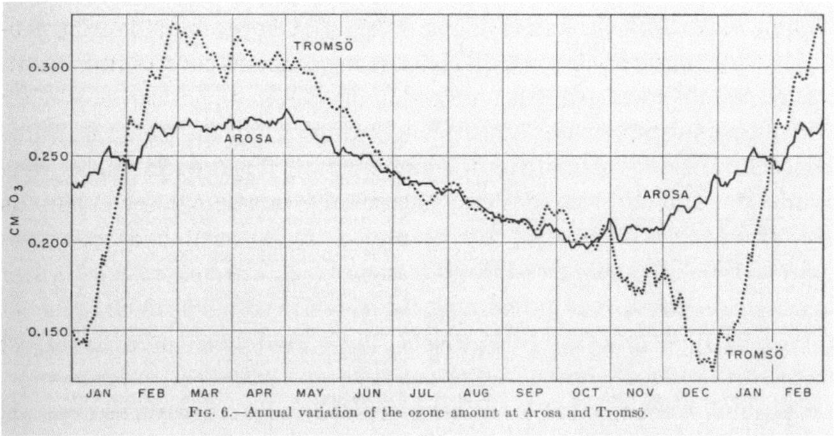

Figure 4.53: Annual variation of the ozone amount based on five-day averages of extensive measurements periods at Arosa (1926–1946) and Tromsø (1939–1948); Götz (1951).

Under different names, 35 ozone conferences and symposia were organized until 2019. In 1948, the International Ozone Commission (IO_3C) was founded, with *Dobson* as president. The network expanded around the world.

Shortly before the end of World War II, a third ozone symposium was held in Tharandt (17 and 18 March 1944), but only German researchers attended: *Regener*, *Ehmert*, *Kiepenheuer*, and seven other investigators,[1099] and *Götz* published only after the war (Ozon 1949). *Otto Hoelper* (1891–1944)[1100] presented a survey on ozone research, reporting that at the conference in Oxford in 1936, the establishment of a European network including 15 stations have been agreed, but due to the war, the first stations ran the measurements only in 1939, just No. 9 was received in Germany (Potsdam) a few days before the mobilization, and besides the measurements in Arosa, only the stations in Aarhus and Tromsø ran the operational measurements. Moreover, *Hoelper* reports that, at *Kiepenheuer's* suggestion, the construction of stronger and simpler instrument was planned and implemented until 1944 but only was used at Schauinsland from February 1943. This instrument was based on a spectrometer described by *Hoelper* in 1930. *Stranz*, who worked at the Observatory Collmberg and the geophysical institute of the University of Leipzig, reports on the development of an ozone radiosonde, which not has been realized due to the war. The first ozone radiosonde that also has been developed by the American physicists *William Weber Coblentz*

1099 *Julius Bartels* (1899–1964), *Rudolf Penndorf* (1911–?); *Hubert Dietrich Stranz* (1916–?); *Rudolf Karl Ludwig Ritschl* (1902–1989), professor for physics at the Berlin University and since 1967 head of the institute for optics and spectroscopy of the AdW; *Helmut Moser* (1903–1991), German physicist. Without lecture, *Mecke* participated.

1100 German meteorologist, director of the Meteorological Main Observatory Potsdam (1935–1944); Succeeded by *Harald Koschmieder* (1897–1966).

(1873–1962) and *Ralph Stair* (unknown living dates) (Coblentz and Stair 1939) and successfully used for stratospheric ozone measurements (Coblentz and Stair 1941); *Stair* conducted several measurements until end of the 1940s.

From the measurements before 1938 "it has confirmed that the advection of air with different ozone content seems to be of importance and that the relation between the ozone variations and the passage of cyclones was dependent on the season of the year" writes *Kaare Langlo* (1913–1985)[1101] in the introduction to his study on the causes of the regular and irregular ozone variations (Langlo 1952). The *Dobson* No. 8 which was first installed in Tromsø, was in 1939 replaced by No. 14, and then in operation at Dombaas[1102] (1940–1946) and Oslo (1946–1949). *Langlo* used these three Norwegian stations, and data from Arosa (1939–1949) as well zi-ka-wei (1932–1942) to describe the latitudinal distribution of ozone in the northern hemisphere, to compare the variation from year to year at several stations, and the mean annual variation of the amount of ozone (this publication is also a source of historic references on the subject). *Langlo* concluded that the horizontal transport (advection) of ozone in the atmosphere is due to the general circulation of the atmosphere and thus explains the seasonal changes in the amount of ozone. The rapid ozone variations from day to day is regarded as chiefly a dynamic problem by *Langlo*. Furthermore, he assumed that the chemical destroying of ozone in the troposphere and in the ozone incline region is appreciably greater in low latitudes than in high latitudes.

The instrument *Dobson* No. 9 arrived in Germany[1103] a few days before the beginning of World War II and was operated (1941–1945) in Potsdam by the meteorologist *Maria Dorfwirth* (unknown living dates); because of the unknown wedge con-

1101 Norwegian meteorologist who held leading positions at the secretariat of the WMO from 1952 to 1975 and played an important role in establishing international programs.

1102 Dombås in Norway.

1103 Gushchin (1995) reports that in 1946 the MGO Leningrad received the Dobson No. 9 from Germany as reparation payment, however, in miserable condition, and only 1952 it was used for comparison, and in 1955 it was fully reconstructed. However, this cannot be true, because the instrument Dobson No. 9 was at the Meteorological Observatory Potsdam until 1950, and moved in 1951 to the Observatory Wahnsdorf to determine the hitherto unknown wedge constant (Hinzpeter 1952). Feister (2009) reports that end of the 1950s No. 9 was send to Ealing-Beck Ltd. for modernization and became then No. 64, and was used together with a No. 71 at Potsdam. No. 64 was lent to the University Sofia 1963–1967 and operated by *Rumen Bojkov* (pers. information *Uwe Feister*, 2021). The reference status of No. 71 was transferred to the modernized Dobson instrument No. 64 in 1981. In 1992 Dobson No. 64 was moved to Hohenpeißenberg and operates as Regional Reference Instrument (the primary standard represents Dobson No. 83 in Boulder, USA). After closing the Potsdam Observatory in 2003, No. 71 was transferred to Lindenberg. At Hohenpeißenberg operates Dobson No. 104 since 1967, a Brewer No. 010 since 1983, and a lidar since 1987. Another Dobson instrument (No. 44) came about 1963 from Weißenau to the University Cologne (with *Helmut Zschörner*), never used there, moved 1998 to Hohenpeißenberg for refurbishment and is now in Armenia in operation (personal information *Adolf Ebel*, 2021); its origin is unknown, but according to its number it was built after war. *Adolf Ebel* (born 1933) was professor for meteorology and geophysics at the University of Cologne.

stant, it was used that from the No. 7 instrument (*Götz* in Arosa). *Hinzpeter*,[1104] who worked since 1947 at the Potsdam Observatory, has shown in 1951 that this constant was different after determination the hitherto unknown wedge constant at the Observatory Wahnsdorf, and thus all calculations have been performed again by *Dorfwirth* (Hinzpeter 1952). *Karl-Heinz Grasnick* (1916–2000), who returned to the Potsdam Observatory after Soviet war imprisonment in 1950, improved the *Hoelper* spectrometer, which had already been used to measure diffuse ultraviolet sky radiance at Potsdam by Schloemer (1938)[1105] in the 1930s, and resumed atmospheric column ozone measurements on the occasion of the IGY in 1957–58 (Grasnick 1963). In fact, Dobson No. 71 was ordered for that purpose but arrived at Potsdam only in 1959; from 1959 to 1961, it was operated by *Günther Skeib* (1919–2012) and *Christian Popp* (1928–1960)[1106] at the Soviet Antarctic station Mirny (Skeib and Popp 1961, Feister 2009). A regular daily column ozone recording started by the *Dobson* spectrometer No. 71 at the "*Meteorologische Hauptobservatorium*" (MHO) [Meteorological Main Observatory], belonging to the Meteorological Service of the GDR[1107] in Potsdam in January 1964. In 1974, the

1104 *Hans Georg Theodor Hinzpeter* (1921–1999) was a German meteorologist who studied at the Berlin University, worked at the Meteorological Main Observatory Potsdam and the Humboldt-University (1947–1958), became the director of the Observatory Wahnsdorf (1958–1961), was scientists at the University Kiel (1961–1970), and then Professor for Meteorology in Mainz (until 1975) and Hamburg. Historically remarkable, that *Hinzpeter* was at a conference in Vienna during building the Berlin Wall (1961), and decided (because his family was in that time in Freiburg) not to return to the GDR. *Hinzpeter* was head of the "founding committee" (1991–1992) of the Institute for Tropospheric Research (IfT, later abbreviated by TROPOS) in Leipzig where the author met him several times as "invited member" of this committee. Founding director became *Peter Warneck* and first director *Jost Heintzenberg* (born 1943); originally the author was "elected" as head and professor of the department atmospheric chemistry (this position received *Wolfgang Rolle*), but he managed to receive the position "head of the Branch for Air Chemistry" in Berlin belong the Fraunhofer Institute for Atmospheric Environmental Research (IFU), Garmisch-Partenkirchen. From today's fairness it is incredible that as author of this publication not *Maria Dorfwirth* was named; only she made the not simple measurements over the war years and the complicated calculations.

1105 *Wolfgang Schloemer* (?–1993) was a German meteorologist, dealing with atmospheric optics.

1106 From 2 to 3 August 1960 a fire broke out on the station, in which *Christian Popp* and other seven polar explorers died.

1107 Including the country weather services, in 1950 the uniform Meteorological Service of the GDR ("*Meteorologischer Dienst der DDR*" = MD) was founded. In the years from 1951 to 1964, the MD was called Meteorological and Hydrological Service with tasks corresponding to those of the Hydrometeorological Service in the USSR. The "Main Office for Climatology" (HAK = "*Hauptamt für Klimatologie*"), which was established in 1952, became the leading institution for climatology. Within the MD, the present Meteorological Main Observatory Potsdam (MHO), the Aerological Observatory Lindenberg and the Meteorological Observatory Wahnsdorf continued to be research centers. The Greifswald Observatory was closed in the second half of the 1950s, whereas the Observatory Kühlungsborn, which had specialized in the physics of the high atmosphere, was incorporated by *Ernst August Lauter* (1920–1984) into the Heinrich-Hertz Institute of the Academy of Sciences in 1968 (since 1969 with the common name "Central Institute for Solar-Terrestrial Physics").

MHO had become a regional ozone Centre of the Regional Association VI of the WMO; Feister (1985). The leading scientist at that time was *Grasnick*, followed by *Uwe Feister* (born 1950); Feister (1985, 2009). An improvement in the instrumentation for ozone measurements at Potsdam became available in May 1987, when a *Brewer* spectrophotometer[1108] of the type MKII (single monochromator, No. 030), which takes ozone measurements automatically several times per day, was installed. Its ozone data were compared with *Dobson* ozone as well as column ozone values derived from a (Soviet) filter instrument of the type M-124 (Feister 1994). At the beginning of the 1990s, after the Potsdam Observatory had become part of "*Deutscher Wetterdienst*", the responsibility of the Regional Ozone Center was transferred to the Meteorological Observatory Hohenpeißenberg, where column ozone measurements had been taken since 1967 with *Dobson* instrument No. 108; *Dobson* Regional Reference Instrument No 64 (the former No. 9 from 1939) was also moved from Potsdam to Hohenpeißenberg. In 2003, The Potsdam Observatory (founded in 1893) was closed by "*Deutscher Wetterdienst*", and the tasks of the WMO Regional Radiation Center as well as its staff were transferred to the Lindenberg Observatory.

Besides *Dobson*, *Brewer* and *Dütsch*, *Yasuo Miyake* and *Katsuko Saruhashi* from the Meteorological Institute Tokyo also carried out after the war theoretical studies on the photochemical formation of ozone in the upper atmosphere and the meridional and annual variation (Miyake 1978, pp. 660–676; note, the original publications appeared in 1951). Miyake and Kawamura (1954) conducted total ozone observations in Tokyo in 1951–1952 using a chromium phototube and glass filters and found a marked seasonal variation having the maximum value in late spring (0.28 cm) and the minimum in autumn (0.22 cm).

An important role in the early history of stratospheric ozone research played India (Mani 1990).[1109] It was the Indian physicist *Kalpathi Ramakrishnan Ramanathan* (1893–1984) and his co-workers (namely *V. R. Karandikar* and *M. W. Chiplonkar*) who were responsible for the pioneering investigations of atmospheric ozone in India and who carried out detailed measurements of total ozone and its vertical distribution in the atmosphere using Umkehr techniques during the forties and fifties and established the main features of the horizontal and vertical distribution of ozone over the tropics. The measurements of total ozone in India were first made at Kodaikanal during 1928–1929, followed by measurements in Bombay and Poona during 1936–1938 (Chiplonkar 1939), using the same photographic instrument. The first *Dobson* ozone spectrophotometer used for regular measurements of ozone was acquired by the India Meteorological Department (IMD) in 1940, Poona (Karandikar 1948). The extremely

1108 *Alan West Brewer* (1915–2007) was a Canadian-English physicist and climatologist.

1109 *Anna Modayil Mani* (1918–2001) was an Indian physicist and meteorologist, responsible for instrumentation of the network and ozone measurements.

important connection between atmospheric ozone and the general circulation of the atmosphere, announced by Ramanathan (1956) in his Presidential address to the International Association of Meteorology in Rome in 1954, was, in a sense, responsible for the establishment of a large number of ozone measuring stations in the world during the IGY and the theoretical studies, that followed, of the behavior and transport of ozone in the atmosphere. Regular measurements at a network of six stations in India were organized by *Ramanathan* during the International Geophysical Year. *Ramanathan* and his associates confirmed the fact that the day-to-day variations in the ozone amounts in the tropics are small and showed that the level of maximum ozone is higher (25–28 km) in the tropics than at higher latitudes. They also showed that in the tropics, unlike in other places, the vertical distribution does not change with the seasons.

The first total ozone measurements made in Antarctica were started by the Falkland Islands Dependencies Survey (which later became the British Antarctic Survey) at the Argentine Islands in early 1955. In September 1956, measurements started at Halley Bay as part of the IGY network. These data series lead to the discovery of the ozone hole by Farman et al. (1985). It is notable that the term "ozone hole" appeared for the first time in German as "*Ozonloch*", Paetzold and Zschörner (1955),[1110] while spectral observation of solar light at Mt. Zugspitze were explained by strong local variations in the vertical distribution of ozone. A key development in the history of atmospheric ozone research was the IGY in 1957.

Besides the Arosa series (Staehelin et al. 1998), at the German meteorological observatory Hohenpeißenberg a systematic ozone measurement program began by a Dobson-104 for total ozone and a Brewer-Mast sonde for vertical profile measurements in 1967, and today it belongs to the longest high-quality series. Surface-near ozone is measured wet-chemical since 1971 and with chemiluminescence since 1977, and lidar measurements of stratospheric ozone have been included in the program in 1987 (Claude 1996). Finally, it is to note that since the end of the 1950s, the additional regular ozonesonde measurements have been conducted at many locations worldwide; in Germany at the observatories Lindenberg (near Berlin) and Hohenpeißenberg (Bavaria).

The first theory of ozone formation and destruction was presented by Chapman (1930). Bates und Nicolet (1950)[1111] proposed the first cycle of ozone destruction in the

---

1110 *Helmut Zschörner* (1920–2005) studied physics in Berlin and Stuttgart and worked since 1951 in *Regener*'s institute in Weißenau in the working group of *Paetzold*. With this group he came 1963 to Cologne at the institute for Geophysics and Meteorology (he became later known for philosophic essays on the cosmic happenings).

1111 *Marcel Nicolet* (1912–1996) was a Belgian physicist and meteorologist and Sir *David Robert Bates* (1916–1994) was a Northern Irish mathematician and physicist; both scientists published together many articles in atmospheric ozone and photochemistry in the 1950s.

stratosphere (later named HO_x cycle):

$$OH + O_3 = HO_2 + O_2,$$
$$HO_2 + O = OH + O_2 \quad \text{and}$$
$$O_3 + h\nu = O + O_2,$$

resulting into the gross reaction $2\,O_3 + h\nu = 3\,O_2$. However, only through the implications of anthropogenic influences on the stratospheric ozone cycle by Crutzen (1971), Johnston (1971), Molina and Rowland (1974) as well as Stolarski and Cicerone (1974) ozone depleting processes have attracted increased attention.

Following the ideas of Bates and Nicolet (1950) that reactions of HO_x species (H, OH, and HO_2) may have considerable interferences with the odd oxygen (O and O_3), *John Hampson* (1923–?)[1112] suggested that HO_x-catalyzed reactions, due to rocket and high-flying aircraft water-vapor emissions, would lead ozone loss in the stratosphere (Hampson 1964), and was tested in model calculation by Hunt (1966), however, despite (also from today's knowledge) all essential reactions have been included, the reaction rates, based on laboratory measurement by Norrish and Wayne (1965),[1113] were still very uncertain. It is notable to explain that at that time the mechanism was named "dry" when only the *Chapman* reactions were involved and "wet" when including OH and HO_2, formed by photolysis of the water molecule. *Paul Crutzen* found that the HO_x mechanism could not explain the ozone observations (Crutzen 1969) and proposed a year later the NO_x-catalyzed mechanism (Crutzen 1970). Indeed, *Harald Johnston* raised the concern about depletion of the ozone layer from human sources that flying a fleet of supersonic transport (SST) planes in the stratosphere might produce enough NO_x (Johnston 1971).

1112 He grew up in England and Wales and served in the Royal Navy during World War Two, in several parts of the world. After the war he proceeded in a scientific career and became a military researcher for the Canadian government. He pointed out that in a nuclear war, vast amounts of nitrogen oxides would be produced (Hampson 1974). This could have a devastating effect on stratospheric ozone levels, resulting in intense doses of ultraviolet light at the earth's surface. He left Canada in 1969 to Britain but was invited back to Laval University to continue the work, but soon felt that he was being hindered in his work, and left again Canada. *Hampson* survived on very little money and he waged a lonely battle for his message to be heard where it counted (information by *Brian Martin*, Australia). Obviously frustrated he wrote in a 21-pages letter from 16 July 1981 to *Martin* as a last sequence: "Are there any jobs in Australia for errant Canadians? Particularly for one who would like to forestall nuclear disaster through failure to understood the implications of impact and who has the necessary contacts in the US and USSR to create a route to this end if only a single responsible politician would have the courage to give the support which would make this possible [...]". See also Crutzen and Birks (1982) on impacts of "nuclear war" on the atmosphere.

1113 *Richard Wayne* (born 1938) was a Professor of Chemistry at the Oxford University who wrote books on atmospheric chemistry and photochemistry.

A few years later, the American atmospheric scientists *Richard Stolarski* (born 1943) and *Ralph John Cicerone* (1943–2016) of the University of Michigan described a process by which chlorine from rocket exhausts could catalyze the destruction of large amounts of ozone in the stratosphere over a period lasting many decades (Stolarski and Cicerone 1974, Crutzen 1974b). In 1972 before the chlorine issue had come to the forefront of the field, *Stolarski* and *Cicerone* had a conversation with *Don Stedman*[1114] in which he said (Stolarski 2001): "Chlorine destroys ozone. Everybody knows that!" This was known since 1854.[1115] In quick succession, Solomon et al. (1986), McElroy et al. (1986), and Crutzen and Arnold (1986) published variants of the same theory. Independently and almost simultaneously, two University of California researchers, *Mario José Molina* (1943–2020) and *Frank Sherwood Rowland* (1927–2012) voiced similar concerns about chlorine-catalyzed ozone loss but suggested the existence of a much larger source of anthropogenic chlorine in the stratosphere (Molina and Rowland 1974).[1116]

Crutzen and *Arnold*[1117] pointed out that the application of fertilizers could damage the ozone layer due emission of nitrous oxide and its photodissociation in the stratosphere ($N_2O + h\nu = N_2 + O$) and reaction with $O(^1D)$, gaining NO. Chemical radicals X, such as nitric oxide (NO), chlorine (Cl), bromine (Br), hydrogen (H), or hydroxyl (OH), serve to catalyze the first reaction and make the net reaction $O + O_3 \rightarrow O_2 + O_2$:

$$X + O_3 \rightarrow XO + O_2$$
$$O + XO \rightarrow X + O_2$$

1114 *Donald H. Stedman* (1943–2016) was an American chemist, most known for his studies on motor vehicle emissions.

1115 Schönbein (1854, p. 285) first describes that ozone reacts with chlorine, iodine and bromine, gaining substances with great bleaching properties. Weigert (1907, p. 255) first reports on decomposition of ozone in presence of chlorine in visible light (UV was filtered); under strong radiation, ozone decays completely and chlorine remains unchanged. Weigert (1908) found the reaction rate to be independent from the amount of ozone and proposed for the first the formation of an intermediate that very quickly decays. From today's knowledge we interpret this process first as photochemical decay of O_3 and a subsequent reaction of $O(^3P)$ with Cl_2 under formation of ClO and Cl. The mechanisms under thermal decomposition have been described as a chain reaction with the start reaction $O_3 + Cl_2 = ClO + ClO_2$, followed by propagation $ClO_x + O_3 = ClO_y + z\,O_2$ and the termination $2\,ClO_x = 2\,Cl_2 + O_2$ (Bodenstein et al. 1929). Furthermore *Norrish*, *Kristianowsky*, *Bonhoeffer* and other scientists investigated the photochemical reactions in the 1930s.

1116 The Mexican-born *Molina* and the American *Rowland* were chemists at the University of California. They argued that many widely used industrial chlorofluorocarbons (CFCs) had the potential to migrate into the stratosphere, where they would eventually break down as a result of exposure to intense ultraviolet radiation and release significant quantities of chlorine. They were awarded with the Nobel Prize in chemistry along with *Crutzen* for their fundamental works on the destruction of the ozone layer.

1117 *Frank Arnold* (born 1943), professor emeritus for physics at the Heidelberg University.

Furthermore, the importance of heterogeneous reactions in the stratosphere have been already suggested in the mid-1970s (Cadle et al. 1975). However, heterogenous (later named multiphase) chemistry was long time regarded to be unimportant compared to the bulk gas-phase chemistry. Only the paper by the atmospheric chemist *Susan Solomon* (born 1956) prompted numerous laboratory studies (Solomon et al. 1986). At the same time, Crutzen and Ehhalt (1977) proposed that the nitric acid cloud formation in the cold Antarctic stratosphere could be the major cause for the springtime "ozone hole". See Bojkov (1995, 2012), Staehelin et al. (2009) and Müller (2009) for more about history. Schmidt (1988b) presents the milestones in ozone research and includes (very) short biographies.

Notably, the Soviet scientists in this field of research are widely unknown (except for *Gushchin* and *Khrgian*); hence it follows a very brief history of the Soviet ozone research (see also p. 599 on tropospheric ozone). In 1928, the first article on atmospheric ozone, stimulated by the question of the radiative balance of the atmosphere (Aderkas 1928),[1118] appeared in the Soviet Union. In 1933 (continued until 1935), the astronomer *N. P. Lugin* [Н. П. Лугин] was the first who measured total ozone at the astrophysical observatory in Kuchino [Кучино], a small village 8 km east of Moscow, using a quartz spectrometer, and estimated it to be 0.339 cm (Lugin 1936). At the same time, *Sergey Rodionov* and his co-workers began to investigate the ultra-violet spectrum of solar radiation in preparation of the Elbrus Expedition at the Leningrad University in 1934 and 1935 (Ambarzumjan 1934, Balakov et al. 1936); for example, in August 1935 at the Caucasus Mountain, total ozone was estimated to be 0.245 cm. In 1949, *Rodionov* and his co-workers developed a simple ozonometer based on glass filters at the Leningrad University (Rodionov et al. 1949). In the 1950s, three different methods for ozone observations have been developed at the MGO and CAO, spectrophotometer with prisms or diffraction grating, spectrographs, and electrophotometer with light filters (Gushchin 1960). Ozonometers M-83 (145 built) and M-124 (450 built) with broadband glass light filters were developed and introduced into production (Khrgian 1973, Gushchin 1995). The Soviet geophysicist *Gennadij Petrovich Gushchin*[1119] [Геннадий Петрович Гущин] (1920–2009) at the Main Geophysical Observatory (MGO) was a leading scientist and member of the International Ozone Committee, being involved in the IGY with 11 operational stations covering all regions of the USSR during 1957–1958 and afterward establishing the network, which in 1960–1970 consisted of 45 stations, for total ozone monitoring on the territory of the USSR, (Gushchin 1957, 1960, 1963,

1118 *Olga Jurevna Aderkas* [Ольга Юрьевна Адеркас] (1872–?) was a Russian/Soviet physicist at St. Petersburg in "лесной институт" [forest institute] at the department physics and meteorology.

1119 In the membership list of the International Ozone Commission his name (1957–1984) is (wrongly) written as *Genady Gushtin*, and in literature he is also (wrongly) cited as *Gustin*. The Cyrillic letter "щ" (Гущин) has different transcriptions according to different norms: ŝ (ISO), šč (DIN), shh (GOST), shch (BSI and ALA), and schtsch (Duden), the Russian sound corresponds best with "shch" and "schtsch".

1964 and Gushchin and Vinogradova 1983). In 1960, aircraft ozone measurement (using an aircraft version of the M-83, developed by *Gushchin* in 1957) began at the Central Aerological Observatory (CAO) near Moscow under the leadership of *A. S. Britaev* [Бритаев].[1120] In 1970, regular rocket observations began at CAO; the first attempts have been made in 1955 (for reviews including historical aspects see Levine (1966)). From the beginning of the 1970s, besides *Gushchin*, *Arkady Mordukhovich Shalamyansky*[1121] [Аркадий Мордухович Шаламянский] (1936–2021) became a leading scientist from the MGO for the total ozone network in the USSR. With the collapse of the USSR in the 1990s, the ozone network in Russia was also close to complete collapse, and *Shalamyansky* saved it, preventing the closure of many stations. Through his efforts, the laboratory in the MGO and the network have been running at a high level in recent years (on average 1 % deviation in comparison to Dobson-108 that has a deviation of 0.18 % to the WMO etalon). On the territory of Russia, 28 stations are in operation, and additional Tomsk (Antarctica) and stations from the former Soviet Republics, altogether about a third of the worldwide number of stations (Gushchin 1995, Shalamyansky et al. 2018). See also Dobson (1968) for the history of upper-atmosphere ozone research.

4.4.8.2 Troposphere: observations 1900–1944

With the beginning twentieth century, the view that ozone is formed through electric sparks in the atmosphere has been rejected, and (already in the late nineteenth century, ideas arose that solar radiation originates ozone) evidence was found that only ultra-violet light causes atmospheric ozone in upper layers (Chapter 4.4.5). Consequently, ozone research was focused on the stratosphere and total ozone estimations (Chapter 4.4.8.1), and tropospheric ozone measurements were sparse, surely also due to missing adequate chemical methods. Only a few locations, ozonometric observations using test-papers have been continued, such as in Vienna, Ó-Gyalla, and Krasno (Tables 4.16 and 4.24). The only reliable measurement series comes from Montsouris,[1122] Paris, based on the arsenite method.

1120 *Victor Amazaspovich Ambarzumjan* [Ви́ктор Амаза́спович Амбарцумя́н] (1908–1966) was an Armenian-Soviet astronomer. *Sergey Fedorovich Rodionov* [Сергей Федорович Родионов] (1907–1968) was a Soviet physicist in atmospheric optics at the Leningrad University; he participated at the Elbrus expedition in 1936. Nothing is known on *Lugin*, *Balakov* and *Britaev*.

1121 German transcription (that is closer to the Russian phonics): Schalamjanskij.

1122 After *Lévy*'s death in 1907, *Pierre Miquel* (wrong cited as *Miguel* by Kley et al. 1988) edited the ozone observations in *Annales de l'Observatoire Municipal* **9** (1908) 8, 242; **10** (1909) 6, 228; **11** (1910) 6–7. Note that the title of Annales Observatoire de Montsouris varies: 1877–1887: *Annuaire de l'Observatoire de Montsouris*. 1888–1899: *Annuaire de l'Observatoire Municipal de Paris, dit Observatoire de Montsouris* (*météorologie, chimie, micrographie, applications á l'hygiène / Ville de Paris*. 1890–1910: *Annales de l'Observatoire Municipal* (*Observatoire de Montsouris*). All at Gauthier-Villars, Paris.

Ozone was further only seen as a natural constituent of the atmosphere and the view, that it is beneficial to human health persisted until the middle of this century; *Erich Regener* stated "*Anwesenheit von Ozon ist ein Indikator für gute Luft*" [the presence of ozone is an indication of good air]; Regener (1946). The German-American geophysicist *Bernhard Haurwitz* (1905–1986)[1123] wrote (Haurwitz 1938, p. 417): "The importance of atmospheric ozone for meteorology consists in its high absorbing power for certain regions of the spectrum".

The observation in Vienna (1853–1920) is the longest series based on *Schönbein*'s paper; *Friedrich Lauscher* (1905–?), an Austrian geophysicist at University Vienna, recalculated the degrees of blueness from the original *Schönbein* scale (scaled in 10 units), used until 1872, and *Schönbein-Lender* scale (scaled in 14 units), used since 1873, using specific monthly factors (unfortunately not explaining how these were derived); additional he presents (absolute) ozone values for the years 1976–1980); Lauscher (1984). *Lauscher* notified no trend, and from the data, an average over the whole period of $25 \pm 5\,\mu g\,m^{-3}$ can be derived (monthly minimum $3\,\mu g\,m^{-3}$ in February 1914 and monthly maximum $62\,\mu g\,m^{-3}$ in June 1906). *Josef Bauer* (unknown living dates) discusses the Vienna values 1874–1910 (Bauer 1912) with respect to the wind speed and direction; the only remarkable result seems that air masses from the south, i. e., passing Vienna, show a significantly less amount of ozone. Furthermore, *Bauer* investigated meteorological influences which could support the hypothesis that ozone is delivered from upper parts of the atmosphere but not found evidence.

The first quantitative (but unreliable) determination of atmospheric ozone in the new century was made by Hatcher and Arny (1900)[1124] in the neighborhood of Covington (Lousiana). First, they refer, after "careful examination of the literature" (Hatcher and Arny 1900, p. 425):

> Houzeau states that the maximum of ozone in the atmosphere is 1 part to 450,000 by weight (or 0.28 mg to 100 litres air). Schöne (Brochure, Moscow, 1897) gives amount as varying from 1 to 10 mg to 100 litres air; while H. de Varigny says that the average is 1 mg and the maximum is 3ö mg to 100,000 litres, is at such variance with the other figures that we can only consider it as a typographical error.

1123 *Haurwitz* was graduated in Leipzig (diploma, dissertation and habilitation), spend three months at Oslo and was invited in 1932 by *Carl-Gustaf Rossby* to the MIT and the Blue Hill Observatory of Harvard for seventh month. In early 1933 *Haurwitz* accepted an invitation from the seismologist *Beno Gutenberg*, a former colleague in Germany, to visit the California Institute of Technology in Pasadena where he gave lectures on atmospheric dynamics where he met *Albert Einstein*. As *Hitler* was appointed chancellor of the German Reich, both *Haurwitz* and *Einstein* independently chose not to return to Germany. *Haurwitz* spend several years in Canada and returned to the United States in 1941.

1124 *Robert Anthony Hatcher* (1868–1944) and *Henry Vinecome Arny* (1868–1943) were both pharmacologists; *Arny* wrote the book "Principles of Pharmacy" (1917).

Houzeau, however, also gives a figure for the ozone concentration to be about only 10 ppb (p. 553) and Varigny (1896),[1125] who never observed ozone, gives an average of 1876–1881, equal to 11.5 µg m^{-3} (about 6 ppb); his citation is very likely referred to *Lévy*'s data. The *Schöne* reference (Brochure, Moscow, 1897) does not exist;[1126] furthermore, to my knowledge, *Schöne* never determined quantitatively the amount of ozone in the air. Thus, it was not a typographical error and *Varigny* (*Lévy*, resp.) and *Houzeau* agree with about 10 µg m^{-3} ozone. *Hatcher* and *Arny* used the aspirator method based on potassium arsenite and parallel the potassium iodide solution with subsequent titration; his values are unreliable large and vary more than would be expected (February and March 1900), in mg ozone per 100 L of air:

Arsenite (n = 9)	1.1 ± 1.3 (0.08–3.45)
Iodide (n = 5)	5.0 ± 5.9 (0.015–15.81)

At this point, it is necessary to spend some words on methods of analytical determination of ozone in the air. The German chemists in Berlin *Otto Liebknecht* (1876–1949)[1127] and *Walter Katz* (unknown living dates) wrote extended chapters on the analytical chemistry of ozone (and hydrogen peroxide), comprising a large number of chemical, electrochemical, and physical methods which have been developed in the first half of the twentieth century (Liebknecht and Katz 1953). However, almost all described methods were laboratory techniques. At the ozone conference held in Oxford in September 1936, the need for a chemical method of determining atmospheric ozone, and the objections against those applied so far, were discussed by *Guéron*, *Pettre*, *Dobson*, and *Chalonge* (Ozone 1936).[1128] In the following discussion, *Chapman* said: "It seems important to make new comparisons between the spectroscopic and the chemical methods for the measurement of low ozone". *Vassy*,[1129] reporting on comparisons between spectroscopic and the chemical methods at different locations, stated "[...] the lack of accuracy of the chemical method was clearly noticed." *Paneth* said; "It seems difficult but not impossible to increase the accuracy of the chemical methods so as to make it possible to measure the ozone content of a small amount of air. If this could be done

1125 *Henry Crosnier de Varigny* (1855–1934) was a French naturalist and journalist; the citation (Varigny 1896, p. 27) reads as follows: "This amount is on the average of 1 milligram per 100 cubic meters of air; 3 1/2 milligrams are a maximum."

1126 *Schöne* died already in 1896 and diseased 1–2 years before; his last publication was from 1894.

1127 Sun of the socialist leader *Wilhelm Liebknecht* and brother of *Karl Liebknecht*.

1128 *Jules Guéron* (1907–1990) was a French chemist and atomic scientist: *Marcel Prettre* (1905–1976) was a French physicochemist. *Daniel Chalonge* (1895–1977) was a French astronomer and astrophysicist, student of *Fabry*. Another important contributor to stratospheric ozone investigations was the French astronomer *Ernest Vigroux* (1902–?); he was still alive in 1999 when his "Necrologia: Daniel Barbier". *Ciel et Terre* **81**, 137 was published in Paris. See Vigroux (1953).

1129 *Étienne Vassy* (1905–1969) was a French geophysicist who constructed in 1958 an optical ozone sonde.

it would permit the determination, in a few hours, during an airplane flight, of the ozone content at different heights". *Van Everdingen*[1130] wondered whether the high cost of *Dobson*'s photoelectric instrument;[1131] he asked whether a simpler apparatus, based on the same principle, could not be devised, even if the accuracy were much reduced. *Dobson* pointed out "that the main cost of the photo-electric instrument is in the double spectroscope which is necessary to determine accurately the wave-lengths used. If this is not done, the observations might be so much in error as to be valueless" (Ozone 1936, pp. 13–14).

The chemist *Jules Guéron* and his wife *Geneviève* (1906–1995), maiden name *Bernheim*), a physicist, studied together with *Marcel Prettre* the reaction mechanism of the ozone potassium iodide reaction and concluded that the homogeneous reaction

$$O_2 + 2\,KI + H_2O = O_2 + J_2 + 2\,KOH$$

should be avoided because if very complex parallel and subsequent reactions that make the stoichiometry unsure and thus an accurate determination impossible, and that only the heterogeneous reaction

$$3\,O_3 + KI = KIO_3 + 3\,O_2$$

supplies absolute sure results (Guéron and Prettre 1935, Guéron et al. 1935).

The French chemist *Robert Lespieau* (1864–1947), in continuation of *Thierry*'s measurements in 1897 at Mont-Blanc, conducted 1900 and 1901 at three different altitudes, Glacier du Bossons (1250 m), Grands-Mulets (3050 m), and Sommet (4810 m) at all 11 determinations, showing on average 41, 33, and 26 ppb, but under standard conditions very little variation, and having an average of 48 $\mu g\ m^{-3}$ (Lespieau 1906). Henriet and Bonyssy (1908) made short time ozone observations (together with carbon dioxide) and considered that ozone is formed by ultra-violet light in the higher atmosphere.

In 1909 at the meteorological station of the University of Manchester, using kites, Hayhurst and Pring (1910)[1132] determined, the amount of ozone (and nitrogen dioxide) to be of the order of 25 $\mu g\ m^{-3}$ at different altitudes at ground level. Using a balloon up to a heigh of 10 miles, they found 130–400 $\mu g\ m^{-3}$ ozone. In 1913 at the Alps at two

1130 *Ewoud van Everdingen* (1873–1955) was a Dutch meteorologist.

1131 Dobson (1968) reports that No. 2 was built by R & J Beck (London) for £ 500, and No. 3 was built for only £ 100 to a purchaser in China; in 1936 further instrument were offered for £ 385 (Nichol 2018, p. 17); 1 £ of that time corresponds to a purchasing power of 85 Euro today. The price for the Soviet spectrometers M-124 (450 instruments built, and used in many countries until now) was cheaper by two orders comparing with Dobson or Brewer instruments, reports Gushchin (1995).

1132 *Walter Hayhurst* (unknown living data) and *John Norman Pring* (1884–?), both physicochemists at the electrochemical laboratory of the University of Manchester.

different altitudes, Pring (1914) determined much more ozone, on average from every three experiments: 2.5 ppm (2130 m) and 47 ppm (3580 m), and with free balloons, which rise to about 10 miles and then burst, they found (n = 10) on average 2.1 ppm (varying between 0.48 and 5.4 ppm at altitudes between 6.5 and 20 km), concluding that there is not a very large increase in the amount of ozone between altitudes of 4 and 20 km.

An interesting investigation was conducted by Usher and Rao (1917)[1133] in Bangalore, India, who were motivated by the frequent observation that in tropical climates rubber articles, cotton, and silk fabrics perish and certain coloring matters are fade much faster than in temperate latitudes. They proposed the hypothesis that the atmosphere in the tropics contains some chemically destructive substance that is either absent from or present in far smaller quantities in the atmosphere of higher latitudes that obviously suggests ozone. It is remarkable that they first present objections to the older methods of ozone determination. Referring Rothmund and Burgstaller (1913)[1134] who first showed that in the estimation of ozone, potassium iodide is untrustworthy on account of a secondary reaction, *Usher* and *Rao* suggest that all the recorded determinations based on the direct use of potassium iodide are unreliable for chemical reasons; they also reject the results found by *Pring* (see p. 597) for physical reasons (absorption of ozone by passage the air over a closely packed solid reagent or mechanically before passage through a liquid reagent). *Usher* and *Rao* used the reaction between ozone and alkali nitrite in an aqueous solution with subsequent colorimetric determination of the remaining nitrite from the standard solution, and taking two samples of air, in one bottle ozone was removed from the air through a passage of manganese dioxide; thus, they believed to estimate the sum of O_3 and NO_2 and in the other sample only NO_2, hence the analytical difference resulting in ozone. However, they found the error corresponds to 40 ppb, and believed it to be unimportant (likely expecting much higher value such as found by *Hayhurst* and *Pring*); no wonder (!) that *Usher* and *Rao* present from 14 complete determinations between July 1916, and January 1917: "The results are interesting chiefly because on no occasion was any ozone found". They further write (Usher and Rao 1917, p. 808) "On November 18th, and again on the 22nd, nitrogen peroxide was present to the extent of 1 part in 5 million and 1 in 4 million respectively, the weather having been thundery". The interpretation of the results, however, is historically interesting (Usher and Rao 1917, p. 809):[1135]

1133 English chemists *Francis Lawry Usher* (1885–1969) and Indian chemist *Basrur Sanjiva Rao* (1895–1975).

1134 *Viktor Rothmund* (1870–1927) was a German professor for physical chemistry at the German University in Prague; *Alexander Burgstaller* (unknown living data), Austrian chemist.

1135 Note that nitrogen peroxide was the name for N_2O_4 (later named nitrogen tetroxide), thus NO_2 in the atmosphere. A concentration of 250 ppb appears to be unbelievable in 1917 even in a tropical city).

> [. . .] there is some ground for the opinion that ozone and nitrogen peroxide never occur together in the atmosphere [. . .] ozone rapidly oxidises nitrogen peroxide [. . .] it not only helps to explain the surprisingly small quantities both of ozone and of nitrogen peroxide found, but may also be considered an important factor in the production of the nitric acid which is a normal constituent if air.

Furthermore, it is worth noting that *Usher* and *Rao* were the first who proposed (what is today a standard method) manganese dioxide for ozone removal (confirmed later by *Paneth* and *Edgar*).

The spectrographic instrument (see Chapter 4.4.8.1) which have been used for total ozone determination based on UV absorption (against the light of the Sun and the Moon), have also been applied for surface-near determinations at different altitude against an artificial light source in different distances (up to 6 km), but needed specialists for the measurement and analysis (see Table 4.23). In line with the quantitative determinations here the average from Montsourios 1875–1908 with 10–20 mg m^{-3} ozone must be referred, which corresponds to 0.1 to 0.2 mm ozone layer according to Lepape and Colange (1929), and who compared it with the total ozone measurements at Arosa in 1929 having an average of 2.5 mm (varying monthly 2.3–3.2 mm ozone at height altitudes (see also Table 4.23). Notable is a short notice by the German physician *Kurt Bieling* (1874–?) and the meteorologist *Irma Bleibaum* (1888–?) from Friedrichroda

Table 4.23: Amount of surface-near ozone (in ppb, at standard conditions), determined with spectroscopic methods (the values are equivalent to that given originally in 10^{-2} cm per 1 km air under standard conditions; Barbier et al. did measure the absorption on a distance of 400 m).

year	location	value	author
1929	Arosa (1900 m), spring	29	Götz and Ladenburg (1931)
1930	Province, summer	22	Fabry and Buisson (1931)
1932	Friedrichroda	10	Bieling and Bleibaum (1934)
1932	Chur (600 m), autumn	18	Götz and Maier-Leibnitz (1933)[a]
1933	Jungfraujoch (3600 m), summer	30	Chalonge et al. (1934)[b]
1933	Lauterbrunn (800 m), summer	17	Chalonge et al. (1934)
1934	Zürich (500 m), July	10	Götz et al. (1935)
1934	Arosa (1900 m)	27	Götz et al. (1935)
1934	Arosa (1900 m), April	24.9 ± 0.5	Barbier et al. (1935)[c]
1934/35	Abisko (Lapland), winter	25 ± 6 ($n = 21$)	Barbier et al. (1935)[c]
1935	Jungfraujoch (3600 m), September	18.8 ± 0.4 ($n = 7$)	Barbier et al. (1935)
1935	Jungfraujoch (3600 m), Febr-March	27.2 ± 6.1 ($n = 13$)	Barbier et al. (1935)
1937	Jungfraujoch (3600 m), September	18.4 ± 5.7 ($n = 7$)	Barbier et al. (1935)
1938	Jungfraujoch (3600 m), Febr-March	20.0 ± 2.4 ($n = 22$)	Barbier et al. (1935)
1938	Poona (India)	10–50	Chiplonkar (1939)

[a] *Heinz Maier-Leibnitz* (1911–2000) was a German atomic physicist
[b] *Daniel Barbier* (1907–1965) was a French astronomer
[c] See also Barbier and Chalonge (1939)

concerns the amount of ozone in a sanitarium; although it was a very preliminary work with the spectrographic-photometric method, the result looks reliable (note that *Bleibaum* was an experienced investigator in radiation measurements).

Alexandre Dauvillier (1882–1979), a French physicist who was primary laboratory assistant to *Maurice de Broglie* (1875–1960), determined ozone at Scoresby Sund (Greenland) from November 1932 until August 1933 with the wet-chemical arsenite method (n = 251) and found on average 50–60 µg m^{-3} with maxima between 100 and 570 µg m^{-3} (Dauvillier 1933).

Widely unknown are ozone measurements from the Soviet Union, carried out in Moscow and the Caucasus Mountains during the Elbrus expedition 1934/1935 by *Maria Aleksandrovna Konstantinova-Schlesinger* [Мария Александровна Константинова-Шлезингер] (1891–1987)[1136] who developed a new fluorescence method to determines ozone. Furthermore, aircraft measurements have been conducted (Konstantinova-Schlesinger 1936a,b, 1937a,b, 1938); note that the original data were given in 10^{-8} g L^{-1} at actual p, T conditions. Gmelin (1943) referred these values in 10^{-8} "Vol.-Tl." [volume parts] or 10^{-3} cm/km, respectively, erroneously assuming standard conditions. Möller (2003, p. 589) recalculated the initial concentrations into mixing ratios (ppb) according to the real p, T conditions, resulting in an approximately linear relationship between altitude and the mixing ratio:

	Schlesinger	Gmelin	Möller
Moscow (100 m)	<2.0	9.2	9
Elbrus (2200 m)	5.7	26.8	32
Elbrus (4300 m)	7.3	34	58
Aircraft (9620 m)	8.7	40.4	119
Aircraft (13000 m)	9.8	45.5	152
Aircraft (14000 m)	10.0	46.6	158

Konstantinova-Schlesinger used the fluorescence obtained by the action of ozone on 9,10-dihydroacridine that is oxidized to acridine and which fluoresces; only 2–3 liters of air must be passed through the solution. Parallel the same amount of air was passed

1136 Soviet chemist and professor (*Schlesinger* were her maiden name), known for developments of luminescence analysis (monograph in 1948: Люминесцентный анализ. М.; Л.: Изд-во АН СССР, 287 pp.). 1921–1931 she worked in the Institute for Physics and Biophysics and 1934–1964 in the Physical Institute of the Academy of Sciences in Moscow. The founder of luminescence analysis was the famous physicist *Sergey Ivanovich Vavilov* [Сергей Иванович Вавилов] (1891–1951) who encouraged *Konstantinova-Schlesinger* in 1934 to introduce it for *chemical* analysis. It is historically interesting that *Konstantinova-Schlesinger* reports on a talk between *Vavilov* and the Austrian polymer chemist *Hermann Franz Mark* (1895–1992) at a colloquium held at the Institute for Physics and Biophysics in 1935, that *Mark* did not believes that the luminescence analysis could be used in chemistry.

through a filter of manganese dioxide to remove the ozone and to control interferences by other substances. Her result at 9620 m altitude (40 ppb) agrees fully with results at 10 km heigh measured by Regener and Regener (1934), Götz et al. (1934), and Meetham and Dobson (1935) giving values between 40 and 60 ppb.

The Swiss chemists *Emile Briner* (1879–1965) and *Ernest Perrottet* (unknown living data) from the University Geneve determined ozone at different Mountain locations in Switzerland (Briner and Perrottet 1937), in ppb:

Geneve (400 m)	7
Zermatt (1650 m)	14
Rochers de Naye (2045 m)	17
Gornergrat (3200 m)	38

It is worth mention shortly the measurement method used by *Briner* and *Perrottet*, which is based on the oxidation of aldehydes to acids in organic solution in the presence of a peracid, and measurement the increase of acidity. Using butyl aldehyde and octane, they found detection limits below 10^{-9} (other investigators never used this method).

The chemist *Hans Cauer* was the first who developed a reliable and applicable chemical method for surface-near ozone determinations using a micro-aspirator (wash tubes with frit in sequence) containing an acidic potassium iodide solution at pH = 2.8, and *Griess* reagent or *Ilosvay*'s reagent, respectively, to detect the presence of nitrite. Laboratory investigations confirmed that no hydrogen peroxide is formed while passing the air through the solutions. If it is present in ambient air, its concentration is too small to be detected (Cauer 1935b).[1137] First measurements Cauer (1936) conducted in Weszterheim (today Tatranska Polianka in Slovakia), High Tatra (1000 m altitude) from June until October 1934 (n = 104 on 16 days, 5–7 values per day over the whole day distributed with a sampling time of 1–2 h). The ozone values vary between 6 and 101 µg m^{-3}, and the noon values between 15 and 57 µg m^{-3}, giving an average of 29 ± 11 µg m^{-3} (corresponding to about 15 ppb). Further estimations *Cauer* made at the Jungfraujoch in September 1932 (12 ± 5 µg m^{-3}) and in Bad Nauheim in April 1934 (12 ± 7 µg m^{-3}). *Cauer* compared his result with those found by *Götz*, *Mettham* and *Dobson*, and found them agreeing by the order (recalculated *Götz* values 13–22 µg m^{-3}). *Cauer* also noted that only in Bad Nauheim the daily variation of ozone shows a regularity. Cauer (1937b) continued ozone measurements at different locations of the *Glatzer Bergland* (today Góry Ziemi Kłodzkie in Poland), in Bad Reinerz between June and September 1936 with an average of about 25 µg m^{-3}, in Bad Kudowa (Kudowa-Zdrój),

1137 Cauer (1937a, p. 555) notes that his apparatus for ozone determination can be purchased by A. Pfeiffer, Wetzlar, "*Fabrik für wissenschaftliche* Geräte" [manufactory for scientific instruments]. *Arthur Pfeiffer* founded the manufactory in 1890, today still existing as Pfeiffer Vacuum Technology AG.

Bad Landeck (Lądek-Zdrój), Bad Altheide (Polanica-Zdrój), Bad Langenau (Długopole-Zdrój), Wölfelsgrund (Międzygórze), and Grunwald (Zieleniec). *Cauer* communicated that the daily variations in ozone are different, and "relatively normal" (maximum at noon and minimum at evening) deviation only shows Bad Langenau, Bad Altheide and Wölfelsgrund; lowest values have been found in Bad Landeck (less than 10 µg m^{-3}). Cauer (1937b, p. 560) refers *Drescher*,[1138] who measured 1882 in Bad Reinerz the daily ozone variation (no information on the method is given) to be similar like *Cauer* it found: maximum in the morning, and more ozone in winter than in the summer as found by *Cauer* at Grunwald and Bad Kudova at higher altitudes. At all, his discussion of the ozone variations clearly shows the still missing knowledge of surface-near ozone removal. In uplands, complex transport processes occur that also result in complex variations of (stratospheric) ozone, which cannot be reflected only with a few measurements.

Johannes Alois Karl (*Hans*) *Tichy* (1888–1970)[1139] conducted measurements in Schreiberhau of the ultra-violet radiation parallel to ozone determinations (using *Cauer*'s apparatus) intending to obtain relations between the diurnal variations of both parameters; he measured at 24 clear days between February 1937 and March 1938. According to today's observations, the parallelism between ozone and UV is fascinating (Fig. 4.54). *Tichy*, of course, does not state that surface ozone was gained due to ultra-violet radiation, knowing that the formation of ozone is associated with wavelength of about 187 nm. He states that the annual variation of surface-near ozone (based on the relatively few measurements by *Cauer*) seems to be similar to that of high-altitude ozone, thus linked with the origin and transport of air masses. Notable is his comment that at overcast days, the diurnal variation (based on *Cauer*'s observations) is contrary to that in Schreiberhau, showing a maximum in the morning and minima at noon and afternoon again an increase over the night.

1138 Drescher, W. (1883) Der Kurort Reinerz: seine Heilmittel und Indicationen in Verbindung mit statistisch-medicinischen Nachrichten über die Saison 1882 und der 15 Vorjahre. Selbstverlag des Verfassers, 289 pp. (available in the library of University Jena, signature: EHH Med.I,72e). *Wilhelm Drescher* (unknown living data) was balneologist and medical council in Reinerz.

1139 German physician and balneologist, who founded 1923 in Oberschreiberhau (now Szklarska Poręba Gorna in Poland) a "*Heilklimatische Forschungsstelle*" [Medical Climatic Research Center], since 1933 "*Bioklimatische Abteilung der Reichsanstalt Breslau in Oberschreiberhau*" (director *H. Vogt*) and who became in 1937 the head of "*Rheumaforschungsstelle der Reichanstalt für das Deutsche Bäderwesen in Breslau*" [Research Institute for Rheumatology of the Reich Institute for German Balneology] in Bad Warmbrunn (today Cieplice Śląskie-Zdrój in Poland, since 1975 district of Jelenia Góra, the former Hirschberg). Besides, *Tichy* was the head of chemical laboratory of the bioclimatic branch of the Reich Institute in Ober-Schreiberhau. After the war, *Tichy* became the most-known rheumatologist in the GDR; he founded the Institute for Rheumatology in Dresden.

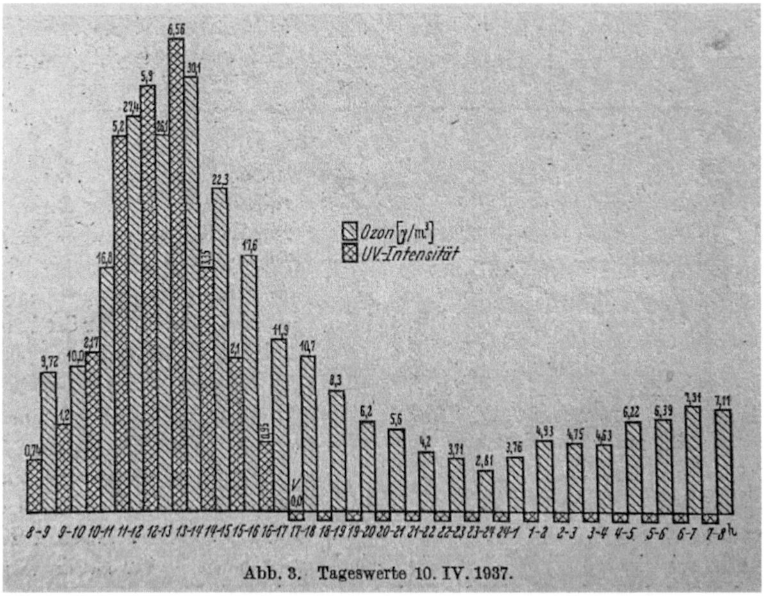

Figure 4.54: Diurnal variation of ozone (in µg m^{-3}) and UV intensity ("*klimatologisches UV-Dosimeter der I. G. Farben*") on 10 April 1937 at in 1-hour-measurements at Schreiberhau (700 m altitude); Tichy (1939, p. 127). Note that the dosimeter units are directly related in a known fashion to the so-called normal erythema dose (Landsberg 1937).

Another young scientist in the surroundings of *Cauer* was *Friedrich Gerhard Renger* (1918–2015)[1140] who conducted measurements of ozone using *Cauers*'s "*Waschrohr*" in October and November 1941 and May 1942 at Oberschreiberhau. The results he re-evaluated together with *Otto Lucke* (1908–1968)[1141] concern the relation between ozone and weather (Renger and Lucke 1953). In contrast to *Tichy*, *Renger* did not find a connection between UV and O_3. He also did not find any parallelism between O_3 and other meteorological parameters and air mass types.

Regener (1938a), who was interested in measuring the tropospheric vertical ozone variation, developed an elegant method where 25 liters of air were passed over a few drops of a solution of potassium iodide containing already thiosulfate, and afterward titration using a micro burette, he found in Friedrichshafen 21 ppb in the air. *R. Auer* (unknown living data), a co-worker of *Regener* at Friedrichshafen, who observed the

1140 *Renger* studied medicine in Leipzig and wrote his doctoral dissertation 1944 at University Breslau. Since 1948 he worked as physician at Charité Berlin, 1962 professor for internal medicine (1965–1969 in Dresden). Note, his similar career to the physician *Hans Drischel* from atmospheric chemistry in Breslau to medicine in Leipzig.

1141 German meteorologist and geophysicist; 1962–1968 director of the "*Institut für Meteorologie und Geophysik*" at Humboldt University in Berlin.

daily variation of ozone using the microchemical apparatus by *Regener*, and explained it by the transport from the stratosphere and its decomposition at the surface, writes (Auer 1939, p. 140):

> Der Ozongehalt der bodenhahen Luftschichten ist sehr gering und seine genaue experimentelle Bestimmung ziemlich schwierig, so daß eine Zeitlang die Troposphäre überhaupt frei von Ozon gehalten wurde [The amount of ozone in surface-near air layers is very small, and its exact experimental determination rather complicated, so that for a while the troposphere has been considered free of ozone].

Albert Ehmert (1910–1971),[1142] known for the development of ozone measurement techniques in the 1950s (Ehmert 1949), who continued (partly together with his wife *Hedwig*, a laboratory technician at the institute in Friedrichshafen) *Auer*'s ozone measurements with *Regener*'s method, which he improved and applied in aircraft measurements but also stationary to investigate diurnal variations (Ehmert 1941b, Ehmert and Ehmert 1941). This electrochemical method (Ehmert 1841a) was based on variations in the conductivity of ozone oxidized potassium iodide solutions. The diurnal variation was found similar to those measured by *Auer* and *Tichy* and well-known today in high pressure weather: increase of the ozone concentration in the course of the morning, maximum in the early afternoon, and rapid decrease in the evening. It was exclusively explained by vertical transport and surface ozone destruction – nothing was known on photochemical ozone formation in the troposphere. Daily ozone maxima were in the range 10–20 ppb. When the instrument was used in an aircraft,[1143] the titrations were conducted the latest 50 min after landing. In altitudes between 400 and 4000 m, variable ozone concentrations between 4 and 29 $\mu g\ m^{-3}$ were found and explained by the different past of the air bodies, either characterized ozone poor, due to photochemical decomposition, or ozone rich, because of fresh stratospheric ozone transport.

Cauer marked that his method determines the "*Gesamtoxidationswert*" [total oxidant value] in the atmosphere; he also defined the "*Gesamtreduktionswert*" [total reduction value], i. e., the sum of oxidizable substances. Thus he believed to determine (or better to say to validate) the air quality. During his ozone measurements (almost conducted in "clean air areas"), he excluded interferences through NO_2, and he also assumed that hydrogen peroxide is – if any – at very low concentrations (this is true from today's knowledge, and be within the error of ozone determinations). Nevertheless, Edgar and Paneth (1941, p. 519)[1144] write:

1142 German physician, student of *Regener* and since 1938 at the "*Forschungsstelle für Physik der Stratosphäre*" in Friedrichshafen and 1944 in Weißenau (see biography *Regener*); 1955 in Katlenburg-Lindau and 1965–1971 director of the Institute for Aeronomy there.

1143 The measurements were conducted within the research program of the "*Deutsche Forschungsanstalt für Segelflug*" [German Research Institute for Gliding] that was under the heading of *Walter Georgii* (1888–1968), father of *Hans-Walter Georgii*.

1144 Nothing is known on *Edgar*, he was a co-worker of *Fritz Paneth* at the Imperial College in London.

> Unfortunately, these numerous measurements are open to the criticism that the chemical reaction used is not specific for ozone, other oxidising agents having the same effect on potassium iodide; as the presence in air of at least one of them, nitrogen dioxide, is certain, and can amount to the same order of magnitude [...] the value of all these "quantitative" ozone records is extremely doubtful. It is true that it could be shown that part of the oxidizing property of air could be destroyed by reagents which decompose ozone; but since all attempts to identify it by specific reagents were unsuccessful some chemists were inclined altogether to deny the presence of ozone in the atmosphere.

Soon before, Paneth and Edgar (1938) argued that the spectroscopic measurements of ozone are too difficult and costly for routine measurements and developed a chemical method where ozone first is condensed from the passing air on silica gel at liquid air temperature and afterward re-distilled, whereas NO_2 is held back. The purified ozone is then determined by titration with potassium iodide. They determined both substances in the air of London at the Kew Observatory (South Kensington); further, ozone in the air of a seaside (Southport) was also measured between 10 am and 5 pm; results in ppb (Edgar and Paneth 1941, p. 526):

Kew Observatory, street level February-July 1938, $n = 21$)	12.1 ± 6.7 (5–26)
Kew Observatory, roof (June 1939, $n = 5$)	22.6 ± 15.3 (4–45)
Southport (July 1939, $n = 3$)	17–29

The concentration of NO_2 was measured in 1938 to be 6.8 ± 6.2 (0.5–20) ppb on street level ($n = 25$). They found in the air of London on average ($n = 14$) between February and May, 1938, 11 ppb ozone (and 4 ppb NO_2, varying from less than 0.5 to 13 ppb), but significantly more in May (22 ppb) than in the time before (9 ppb). This publication was commented by *William Colebrook Reynolds* (1870–1940),[1145] who found an average of 12 ppb ozone in spring, as determined by *Paneth* and *Edgar*. Like *Usher* and *Rao*, *Reynolds* used a differential method and notified that except for ozone the other gases are not normal constituents of country air (Reynolds 1938). Several years ago, Reynolds (1923) reported that he has "for some time been measuring the proportion of certain variable gaseous constituents in London and country air" and succeeded now to measure ozone; during severe thunderstorms and a few days after the storm, he found much more ozone (70–300 ppb) in the air than before (40–50 ppb), but no change in the content of other gases.[1146]

1145 British pharmaceutical chemist who also wrote on decomposition of nitric acid by light (Reynolds and Taylor 1912) and globular lightning (Reynolds 1930b). Reynolds (1938) communicates from 1923 to 1927 inclusive he "made almost continuous measurements of the chlorides, ammonia, nitrogen peroxide, sulfur dioxide, and ozone at Plaistow and the suburban village Upminster, simultaneously. *William Henry Taylor* (1866–1917) was American physician and chemist.

1146 The concentration of other gases did not vary over the period of thunderstorm; NO_2: 8–9 ppb at London and 2–3 at Upminster; SO_2: 50 ppb in London and 22 ppb in Upminster; NH_3: 5 ppb at both

Paneth and Glückauf (1941) developed a quick electrochemical method, noting that the method developed two years earlier is cumbersome. The new method was aimed for meteorological purposes, hence, very fast and continuous, and with automatic recording. This was the first record of ozone in high time resolution: 12 measurements within 2 hours and a changing ozone concentration between 1 and 36 ppb due to sudden changing wind directions from WNW (28–36 ppb) to NE (rapid fall to 1 ppb). Glückauf (1944) performed an excellent study to investigate the variations of ozone during depressions at Durham, England. First, he stated that due to a limited number of single measurements, "an unqualified average of the ozone content was to be used for the comparison" by *Auer*. Then again *Glückauf* supported *Auer*'s view that "there is apparently no daily variation in the ozone content of surface air, provided that the wind seed remains at a high speed". *Glückauf* used for comparisons "only maximum values on days with considerable turbulence", an idea, also proposed 50 years later by *Dieter Kley*, *Heiner Geiss* and *Volker Mohnen* (Kley et al. 1994, p. 153)[1147] but not citing *Glückauf*. It should be noticed that before the 1950s, the tropospheric chemistry of ozone was fully unknown. It was generally accepted that "ozone is a compound that is easily destroyed by contact with organic matter and especially by reducing agents (e. g., SO_2 from the combustion of coal) one cannot expect to find it in air which has been in contact with or near the ground for any great length of time, especially in the vicinity of human habitations" (Glückauf 1944, p. 13–14). Furthermore, that ozone is "destroyed" at the surface. However, the process of dry deposition was still unknown. Based on our today's knowledge, one cannot exclude that ozone was photochemically produced in the lower troposphere (see also p. 630), namely in largely habituated and industrialized areas such as in England. However, we learned only in the late 1980s and early 1990s that the photochemical *net* formation of ozone (Chapter 4.4.8.6 for details) is not a fast process, thus needs time and space (mesoscale), and hence that meteorological phenomena determine short-time ozone variations. These variations *Glückauf* explains, based on very few ozone observations with a time-resolution of only 1 h, could not have been done better today.[1148] The variation of the surface-near

---

locations (note that Upminster is east of London, thus receiving the city plume). In rainwater he determined 50 ppm nitric acid.

1147 *Dieter Kley* (born 1937) was a German chemist and director of "*Institut für Chemie der Kernforschungsanlage Institut 2, Chemie der belasteten Atmosphäre*" (ICH-2); the other institute was ICH-3: "*Atmosphärische Chemie*" (director *Dieter Ehhalt*, succeeded by *Andreas Wahner* in 2000). Both institutes were unified after *Kley*'s retirement in 2002 and renamed in Institute for Chemistry and Dynamics of the Geosphere, which again was renamed in 2000 in Institute for Energy and Climate Research IEK-8: Troposphere. *Heiner Geiss* (born 1946) was a German physician and co-worker of *Kley* in Jülich. *Volker Armin Mohnen* (born 1937) studied physics in Karlsruhe and Munich, obtained his doctorate 1966 there, and moved to the USA where he worked since 1967 at the University Albany (New York) at the Atmospheric Sciences Research Center; 1985 he became professor and director.

1148 The author, together with his co-workers and in research collaboration, begun in the early 1990s to study tropospheric ozone for some 15 years by monitoring (Mt. Brocken Cloud Chemistry Station) ozone content – and this is the important implication for today's researchers – due to exclusively stratospheric caused ozone, between less than 20 ppb and more than 60 ppb was due to its inhomogeneous distribution in the troposphere and as a result of advection and turbulent mixing. *Glückauf* describes three definite conditions, apparently without exception:

- a drop of the ozone content up to 300 miles in advance of warm fronts, with high values of ozone after their passage,
- exactly the same behavior shows an occluded front of the warm front type,
- a drop of the ozone content is found after the passage of backbent occlusions, persistent until the front has moved on up to 200 miles, after which the ozone returns to its normal concentration.

A complete absence of ozone appears when air masses have prolonged contact with the surface, and ozone replenishment from higher regions occurs. The term "normal concentration" used by *Glückauf* corresponds to "background concentration" used nowadays, but this is now the sum of stratospheric and "tropospheric" ozone.

4.4.8.3 Troposphere: after-war observations until about 1960

Soon after the war, a "remake" of *Schönbein*'s paper methods was introduced by Curry (1946), who believed that not only ozone is determined but also a hypothetic ozone modification with higher atomicity, termed "*Aran*" ($O_3{\cdot}O_x$). In the late 1940s at several sites in Switzerland and Bavaria, he carried out thousands of determinations of the "aran content" [*Arangehalt*] in the atmosphere; it was in the range of 10–60 µg m^{-3} (typically for ozone). In the 1950s, several leading German physicians accepted *Curry*'s aran hypothesis,[1149] i. e., that the oxidation capacity of the atmosphere influences the biological events of humans. Aran (i. e., the sum of oxidizing substances, later identified with ozone) has been regarded as an indicator for *biotropic* (unfavorable) weather conditions. In a book review, Flohn (1949)[1150] writes that this hypothesis undoubtedly

---

and a large number of field campaigns in Europe. Unfortunately, he had no knowledge on the history of ozone research, and, with his today's knowledge he would have design this research more specific and had avoided some needless studies (Möller 2000, 2002a, 2004, Möller et al. 1997a,b, 1999, Treffeisen et al. 2002, Thomasson et al. 2002).

1149 The *Time* (Monday, June 16, 1947) reports on the occasion of a lecture by *Curry* at the Congress of Allergology in Atlantic City, citing him: "Pessimists among them feared that the aran theory, whatever its merits, would prove a gold mine for quacks and medical faddists. And writes further: "Curry decided that the health-governing material in air is a mysterious gas he calls "aran." Aran's concentration in the air, Curry computed, varies with the time of day (it is low at night and high in midafternoon, and with the weather (low in warm south winds, high in cool north winds). He is pretty certain that varying the concentration of aran can increase or decrease inflammation, start bleeding, and produce all sorts of spasms". See p. 21 on Curry.

1150 *Hermann Flohn* (1912–1997) was a German meteorologist and climatologist; since 1935 he worked in the "*Reichswetterdienst*" (first at *Knoch*), 1938–39 he was the head of the bioclimatic research station gave new impetus to bioclimatology and serial measurements of atmospheric trace substances; *Fohn* refers *Cauer*, who already made precious contributions to this field. Mrose (1959b), however, writes that *Curry*'s theory meanwhile was no longer admitted but that it stimulated air chemical investigation after World War II.

Stimulated by *Curry*'s view, in the 1930s, Cauer (1948a) was obviously looking for explanations for the ozone variations he found and proposed the first idea on tropospheric formation of ozone. *Cauer* suggested (Cauer 1948a, p. 64–65) that the main source of ozone in the boundary layer is due to oxygen dissociation by electron impact ionization via quantum tunneling. This hypothesis was developed based on ozone observations in cooperation with *Amelung* in Königsstein (Taunus) between March and August 1948, finding ozone formation at ground lightning arresters, namely while passing of low clouds (Cauer 1948a).

Thus, the first half of the twentieth century ended with the open question of where does the surface-near ozone comes from: only from the stratosphere through (already rather well described) transport processes and additional or exclusively through (still unknown) formation processes in the boundary layer. The final answer will be obtained further some 40-years pass. Despite the observed phenomenon of photochemical smog, in the Los Angeles area at the end of the 1940s, and the first observation of increased ozone levels due to photochemical involving nitrogen dioxide and organic material (Haagen-Smit 1952), there was consensus among most scientists at that time that the source of surface ozone is of natural origin, as it is transported from its stratospheric source region to the lower troposphere by turbulent transport, and chemically destroyed at the earth's surface.

Renger and Lucke (1953, pp. 27–29), discussing *Renger*'s ozone measurements from 1941/42 (see p. 602), still refer "theories on the origin of surface-near ozone" that are curious,[1151] but finally adopted *Regener*'s theory of ozone transportation from high altitude and local phenomena of advection and removal. Still, in 1964, *Warmt* stated that there is no explanation for the high ozone levels found in Los Angeles, observed up to 0.5 mg m^{-3}, as reported by Haagen-Smit (1952), being as high as in the stratosphere. Junge (1963a, p. 49) writes:

> There is no evidence of tropospheric ozone sources, except polluted areas, e. g., Los Angeles, and the profiles seem to be in qualitative agreement with Regener's concept that tropospheric ozone is exclusively of stratospheric origin.

Bad Elster, and from 1939 in leading positions in Berlin and after the war at "*Deutscher Wetterdienst*" in West Germany; 1961–1977 professor and director of the new founded meteorological institute of the University Bonn.

1151 For example, *Cauer*'s ideas on the consumption of ozone while formation of condensation nuclei, and its release after dissipation of fog and clouds.

The first ozone measurements in the second half of the twentieth century have been conducted by *Ernst Friedrich Effenberger* (1916–1989)[1152] between May and September 1947 at Westerland (Sylt) using *Regener*'s method (Effenberger 1949); mean at 14 CET (n = 95) 45 (10–120) µg m^{-3}. The diurnal variation (between 8 and 18 CET measured only) shows mostly a distinct maximum at the afternoon and a minimum at the morning (average 11 µg m^{-3}), explained by turbulent mixing. Variations in the ozone maximum *Effenberger* explained by the progression of fronts and occlusions. He also found a relatively good agreement between *Regener*'s and *Cauer*'s instruments. *Uwe Karl Albert Werner Jessel* (1916–1979)[1153] continued these ozone measurements in 1951 (22 µg m^{-3}, n = 62, *Cauer*'s method) and 1952/54 (59 µg m^{-3}, n = 248, *Ehmert*'s method). Parallel measurements by Jessel (1952) results in 21 µg m^{-3} with *Cauer*'s and 56 µg m^{-3} with *Ehmert*'s method (Ehmert 1941a). However, Ehmert and Ehmert (1941) have shown that both ways provide the same absolute value on average. *Cauer* criticized in several publications *Effenberger* and *Jessel* not being consistent concerns sampling and measurement conditions, as later also notified by Warmbt (1964, p. 42), who furthermore comments curious chemical perception by *Jessel* (Jessel 1955). Notwithstanding, in 1960, *Warmbt* also found with parallel measurements that under the same sampling conditions *Ehmert* values are on average 50 % higher, but when taking into account a blanc correction factor for *Ehmert*'s solution, both methods agree.

Hans Cauer conducted in Königsstein (Taunus) in spring and early summer 1948 and in the late summer 1949 in Wyk (Island Föhr), and at Norderney (Cauer 1949a,b, 1950, 1951a,b); values in µg m^{-3}:

Taunus (n = 600)	20 (0–189)
Wyk (n = 310)	26 (4–86)
Norderney (n = 229)	28 (3–126)

Cauer had the erroneous thought (Cauer 1949b, 1951c) that not descend of ozone-rich air from the stratosphere but local formation through electric fields must be seen as the source of surface-near ozone (Cauer 1949b, p. 96). He argued that 5–8 eV (corresponding to 242 nm radiation) is necessary for dissociation of oxygen molecules and that an electric field strength of 5000 V cm^{-1} should be sufficient and gained near clouds and the earth's surface, citing experts in atmospheric electricity. Otherwise, the same energy also dissociates the ozone molecule and surface-near reducing substances (*Cauer* assumed almost the oxidation of NH_3 to NO_2) and also removes ozone, thus explaining

1152 German physician; assistant of *Pfleiderer* in the institute of bioclimatology and marine medicine of the University Kiel at Westerland (Sylt) and professor for occupational medicine 1964–1982 at University Hamburg.

1153 German physician, who worked 1949–1970 in the institute of bioclimatology and marine medicine of the University Kiel at Westerland (Sylt), dissertation 1953 and since 1964 professor.

the strong variations of the ozone content over time and short distances. He believed that "ozone schlieren" exist (today we would name it bubbles or plumes). On the other hand, *Cauer* clearly concluded from his measurements that any further ozone research at one site with a single ozone monitor would not get progress, but only ozone determinations with automatic instruments, combined with "accurate analytical" methods from time to time for calibration and control, together with chemical analysis of the aerosol, meteorological parameters and the "reducing value" within a network; furthermore, additional to the surface-near monitoring, aircraft measurements must be conducted. It is remarkable that he noted the need for about eight trained persons in a permanent contract. In this context, it is historically interesting to note that Cauer rejected Effenberger (1948) critique on his method but refused *Effenberger* and *Jessel* to be "not chemical trained persons".

Hans Ungeheuer (1908–1987)[1154] conducted between March and August 1949 half-hours ozone measurements (he still names it "*Aranmessungen*" [aran measurements]) with the "Curry-Dirnagel" instrument (Dirnagl 1949),[1155] an automatic instrument developed at the bioclimatic station Bad Tölz. Based on 7200 measurement values he found no relation with any of the meteorological elements with the exception of the wind speed and direction, explaining the diurnal ozone variation with the mountain-valley circulation and thus ozone as an indicator with the otherwise complicated separation of wind from mountain wind (Ungeheuer 1950). These instruments were later used by *Eugen Obenland* (1804–1975),[1156] who conducted ozone measurements at the weather station Oberstdorf (Bavaria) at the basis (810 m) and the nearby summit Nebelhorn (1932 m) in the summers of 1951 and 1952 (Obenland 1953). He found on average 20–25 $\mu g\ m^{-3}$ more ozone at the summit than at the basis station, and a linear relationship between relatively humidity and ozone content at both sites which he correctly explained through different mixing of dry and ozone-rich air from upper layers with wet and ozone-poor air from the valley. Very notable is his observation of significantly reduced ozone values when the summit was within clouds. *Obenland* refers *Friedrich Volz* (unknown living data),[1157] who investigated the decomposition of ozone at different surfaces (Volz 1952) and noted that the clouds on earth, comprising

---

1154 German meteorologist and head of "*Medizin-Meteorologische Versuchs- und Beratungsstelle*" [medical-meteorological research and information center] in Bad Tölz since 1948.

1155 *Karl Dirnagl* (1917–2004) was a German physicist and balneologist who worked in the American Bioclimatic Research Institute of *Manfred Curry* in Riederau and changed 1950 to the new founded "*Institut für Balneologie*" at the University Munich (he writes in his autobiography that *Curry*'s "research" was increasingly dowsing influenced; Dirnagl (2002, p. 16).

1156 German meteorologist (engineer) who worked 1928–1968 at the weather station Oberstdorf, since 1950 as head (1950–1968 "*Bioklimatische Forschungsstelle*").

1157 German physicist; he worked in the Institute for Meteorology and Geophysics at University Frankfurt and afterwards at University Mainz in the 1950s and in the 1960s at the branch of the Astronomical Institute of University Tübingen in Weißenau near Ravensburg (former *Regener*'s institute). Known for his work on optics of aerosol.

on average a hundred times larger surface than the Earth, thus would play an important role in surface-catalyzed ozone decomposition. *Volz* found by measurements at Arosa and Chur in 1951 that the ozone concentration decreased to 50 % in clouds, and therefore concluded that "there is no more any doubt on decomposition of ozone on water surfaces" (Volz 1952, p. 259). In a written comment to *Volz*, *Hans Cauer* (Volz 1952, p. 260–261), however, argued that ozone cannot simply be destroyed on surfaces but only decomposed by chemical reactions (he listed many reducing substances) and heat or radiation; further *Cauer* discusses that ozone is not consumed by the clouds of "*ausgewachsenen*" [fully grown] clouds that have relatively large droplets, even containing reducing substances, but through reducing matter in the moment of the heterogeneous nucleation process. Note that this termination was unknown at that time, *Cauer* expressed it a bit circumstantial:

> Viel stärker wirken sich die reduzierenden Stoffe dann aus, wenn sie bei hoher Feuchte noch gerade gasförmig vorliegen [. . .] Den Ozongehalt wesentlich mindern, [. . .] der Vorgang der Nebelbildung, der Kondensation und der Vorkondensation. Diese Vorgänge zerstören Ozon rasch bis auf geringe Spuren [Much stronger affect reducing substances if they are still present gaseous at high humidity [. . .] The content of ozone distinct reduce, [. . .] the process of fog formation, condensation, and pre-condensation. These processes quickly destroy ozone down to small traces].

Today we know that OH radicals (naturally almost gained via ozone photolysis) led to gas-to-particle formation and thus condensation nuclei and that clouds interrupt the photochemical net formation of ozone (through uptake of HO_2 radicals) but also effectively consume ozone via different aqueous-phase chemistry: Lelieveld and Crutzen (1990, 1991),[1158] Möller and Mauersberger (1992), Acker et al. (1995), Möller et al. (1997a,b, 1999). Whereas *Möller* and co-workers found the same ozone variations at Mt. Brocken in clouds in the early 1990s like *Volz* and *Obenland* at the beginning of the 1950s, and explained it through oxidation of dissolved sulfur dioxide, the atmospheric SO_2 content around 1950 at Mountains around Arosa and in south Bavaria was surely too low making this mechanism important. Fortunately, some other aqueous-phase chemical reactions decompose ozone (see Möller 2019, p. 300–304).

The German meteorologist *Gerhard Zimmermann* (1919–?) conducted in 1950 and 1951 ozone measurements in Bad Kissingen at several locations (Zimmermann 1953) as suggested by *Karl Heinrich Knoch* (1883–1972)[1159] aimed to study small-scale variations

1158 *Jos Lelieveld* (born 1955) is a Dutch chemist who was a student of *Paul Crutzen* and became his successor in 2000 as director of the Department for Atmospheric Chemistry of the MPI in Mainz.

1159 German climatologist ("father" of German climatology) who also was in close contact with *Hans Cauer*. Remember that in Bad Kissingen *Constantin Lender* observed ozone in 1871 with *Schönbein*'s paper. After War II, in Bad Kissingen the "*Zentralamt des Wetterdienstes in der US-Zone*" [Central Office of Weather Service in the US zone] was established for Bavaria, 1948 renamed in "*Deutscher Wetterdienst in der US-Zone*" [German Weather Service in the US zone] and 1952 integrated in the German Weather Service with Offenbach as head quarter.

from bioclimatic point of view in spa towns (such as *Cauer*'s investigations). Two interesting findings *Zimmerman* communicated; first, he could not confirm *Cauer*'s idea that ozone is produced near-surface through electric fields, and second, for the first time, the ozone concentration at a busy highway was found to be considerably lower (10–24 µg m^{-3}) comparing to the ozone content in the spa park, only located 12 m in the distance from the street (52–59 µg m^{-3}), and correctly denoted to the exhaust gases of car traffic. Another "modern" result was the explanation of the "normal" ozone value already a few meters away from the street, such as at other sites in the garden due to the "one meter wide and above head height lilac hedge", filtering the exhaust gases. The reduction of ozone (in modern chemical jargon "ozone titration" named) in streets with automobile traffic *Zimmermann* observed at any time and in any weather.

In May 1948, *Victor Regener* and his student *Irby Gerald Bowen* (unknown living dates) conducted ozone measurements with 3-h-intervals (Bowen 1949) at the campus of the University of New Mexico on a tower, 55 feet above ground level. They found distinct daily variation like Auer (1939), and after construction of new automatic ozone recorder (Bowen and Regener 1951) continued these measurements at different stations in 1950. Historically most interesting is her note that not photolysis and thermal decomposition but (1) oxidation of foreign particles suspended in the atmosphere, and (2) oxidation of subjects on the surface of the earth or catalytic destruction of the ozone molecule on these objects; liquid water in clouds and on the surface, as well as water vapor, are to be include destroying agents under (1) and (2).

The next ozone measurement was carried out by Götz and Volz (1951) at Arosa (1860 m) with *Ehmert*'s apparatus; they found a concise annual variation (Fig. 4.55), no diurnal variation but irregular small variation in connection with changing weather situations. The annual maximum of 50 µg m^{-3} at local pressure (608 mm Hg) corresponds to about 32 ppb at standard conditions, while the annual average (30 µg m^{-3}) corresponds to about 19 ppb. *Robert Henry Kay* (unknown living dates), working with *Dobson* in the Clarendon Laboratory in 1952, adapted a chemical technique for making in situ measurements of ozone over England using an aircraft, and in winter during 1952–1953, found from 13 flights an approximately linear increase (however, at individual flights strongly varying) from 15 to 27 ppb at 11.6 km height (Kay 1953). In contrast, in July 1955, Brewer (1955) found over Norway about a constant ozone content of 30 ppb up to 10 km, and then an increase. *Brewer* speculated that these differences might be due to the absorption of ozone in the intake tube or because of real geographic differences.

Ehmert (1952) notified that "for measurement of local concentrations of ozone in the atmosphere today chemical methods are in fact at an advantage compared to the physical methods", and thus to conduct large-scale monitoring; his method (Ehmert 1941a) he further improved and simplified. *Ehmert* reports (in parenthesis the name of the operator) on simultaneous measurements at Weißenau (*Ehmert*), Bad Tölz (*Ungeheuer*), Freiburg (*Person*), Tübingen (*Daubert*), and Arosa (*Volz*). From the low ozone values found in Tübingen, *Albert Ehmert* concluded (Ehmert 1952, p. 191):

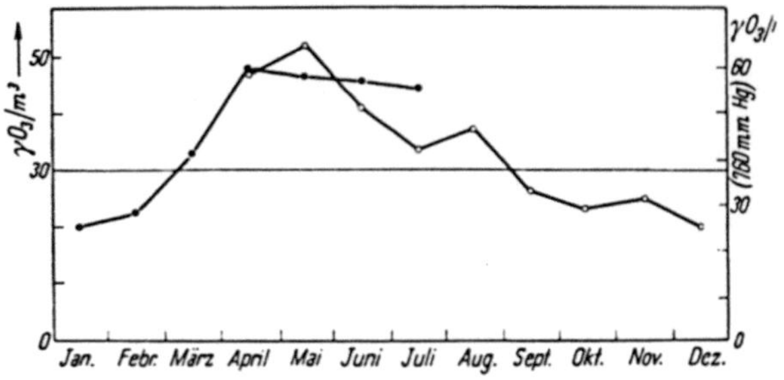

Abb. 4. Monatsmittel des Ozongehalts in Arosa
(• • • 1950; ∘∘∘ 1951).

Figure 4.55: Monthly means (based on hourly measurements at daytime) of ozone at Arosa (1950–1951) after Götz and Volz (1951, p. 638); note 1 $\gamma = 10^{-6}$ g (in that time often the unit "pphm" was used, meaning "parts per hundred million" $= 10^{-8} = 10$ ppb).

> Die Heizgase der Stadt liegen bei solchem Wetter als sichtbarer Dunst über der Stadt und zerstören das Ozon sehr wirksam [The fuel gases of the town are seen in such weather as visible haze above the city and destroy ozone very effectively].

Referring Auer (1939) and Regener (1948) and his previous works the view is again confirmed that ozone is quickly decomposed at the ground of the earth, and that ozone as rather a permanent constituent of air can provide meteorological information. *Ehmert* further stated (what has been neglected by almost all following observers):

> Allerdings sollte man nicht am Boden und in Städten, sondern in der freien Atmosphäre messen [However, one should not measure at ground and in towns but in the free atmosphere].

Erich Regener, based on results from previous measurements of ozone at ground and its vertical distribution, including calculations by his co-worker *Hans-Karl Paetzold* (1916–2002)[1160] based on the model by Schröer (1944/49), discussed extensively the variation of ozone in the troposphere and stratosphere, writing (Regener 1951, p. 173):

> The complexity of the problem of atmospheric ozone is exhibited in the fluctuations of the ozone concentration at all heights. In the layer of air immediately adjoining the ground, the ozone content always sinks to zero when the air stagnates, due to the effect of the ground, which is very destructive to ozone. In the troposphere, advection plays the main part in causing fluctuations. Widely varying vertical ozone distributions have been found recently by means of the spectrographic method in balloon ascents. To explain this, large-scale horizontal and vertical movement of air at great heights must be assumed.

1160 German geophysicist in Weißenau (scholar of *Regener*) until 1961 and afterwards professor at University Köln; he developed 1959 an ozone sonde (Paetzold 1953). His habilitation thesis in 1954 were entitled "*Über die Photochemie der Erdatmosphäre unter besonderer Berücksichtigung der Ozonschicht*" [On the photochemistry of the earth's atmosphere with special consideration of the ozone layer].

Paetzold (1952, 1953), using different methods, investigated mainly the vertical distribution of ozone in relation to chemistry and meteorology. Thus at the beginning of the 1950s, *Regener* together with his co-workers *Ehmert* and *Paetzold* (and his son *Victor Regener* in the USA) already described *all* processes (at least phenomenological) of the variation of the ozone content in space and time (Paetzold and Regener 1957).

The author of this volume (DM), who made with his co-workers and together with other scientific groups between 1991 and 2002 at different locations in Europe several so-called "intensive field campaigns" to study the photochemical formation of ozone, and conducted 18-year ozone monitoring at Mt. Brocken, must now state that (and I like to include millions of ozone measurements at thousands of sites in the world since the 1970s) the results are in no relation to the efforts and expenditures, and that a few of our scientific ancestors gained the fundamental understanding with relatively few measurements but careful observations and conclusion more than 50 years ago. Malley et al. (2016), who give a chronology of atmospheric composition monitoring networks since the 1800s, suggest, in the background of the unprecedented volume of data that is now collected across the world, more integrated analysis of data within a "chemical climatology" (referring *Angus Smith* but not *Hans Cauer*) framework that seeks to more directly link the impacts, state and drivers of atmospheric composition.

This chapter ends with referring the eminent Japanese Geochemist *Yasuo Miyake* and his co-workers and the ozone observations by the Austrian meteorologist *Ferdinand Steinhauser*, who were already motivated by the Los Angeles observations of the photochemical formation of ozone. However, the surface-near ozone content increase due to tropospheric photochemistry has been observed only in the 1970s after long-term monitoring (Chapter 4.4.8.4).

The Japanese meteorologist *Kyo Sekihara* (unknown living dates) from the Meteorological Research Institute Tokyo conducted with *Ehmert*'s method ozone observations in August 1953 at Mt. Fuji, and observed an anomalously increase before the occurrence of St. Elmo's fire reaching 90 $\mu g\ m^{-3}$ and a notable decrease to 10 $\mu g\ m^{-3}$ with the passage of a frontal zone, and "normal" values around 25 $\mu g\ m^{-3}$ (Sekihara 1954). Other early metropolitan ozone and NO_2 measurements were carried out by *Yasuo Miyake* and his co-workers *Kiyoshi Kawamura*, and *Sumiko Sakurai* (unknown living data) in Tokyo midtown and suburban between December 1957 and March 1959 using *Ehmert*'s method (Miyake et al. 1961). They also studied the interferences of the ozone measurements with SO_2 and NO_2, and introduced correction factors (also for O_3 interference in NO_2 determinations). Hence the values are likely uncertain, but they clearly found more NO_2 at midtown (30–80 $\mu g\ m^{-3}$, around by a factor of two) and also little but distinct more ozone than a suburb, concluding on photochemical ozone formation in the city such as in Los Angeles. The mean suburban ozone value amounted rather low, with about 20 $\mu g\ m^{-3}$ including monthly mean variations between 14 and 29 and daily variations between 0 and 80 $\mu g\ m^{-3}$. Miyake et al. (1962)

studied ozone and NO_2 at Mt. Norikura in Japan at the top (2770 m) and a base station (1450 m) for some days in late summer in 1959 and 1960; the ozone values at the top have been found higher than that at the base, but very different among the years (58/35 and 30/28 μg m^{-3} in 1959 and 1960, respectively), while the concentration of NO_2 was rather constant with 2.3–3.7 μg m^{-3} and such as measurement by Junge (1958b) at Mauna Kera, Hawaii (2900 m) to be 2.8 ± 0.4 μg m^{-3}.

Ferdinand Steinhauser reports on ozone observations in Vienna conducted between September 1957 and February 1959 with *Ehmert*'s method and 2-hourly sampling (Steinhauser 1959b), motivated by the smog formation in Los Angeles, and citing *Cauer*'s (wrong) idea that ozone in humid air cause fog formation, but obviously not knowing the works by *Teichert* and *Warmt* from Wahnsdorf. *Steinhauser* noted that atmospheric substances that react with free iodine, such as due to pollution from heating (without naming SO_2), result in "negative ozone values", often observed in winter. Thus, he noted that the averages represent "minimum values", and we have to consider that the annual mean of 21 μg m^{-3} is too low and the annual variation mostly – like at Wahnsdorf – determined by measurement artefacts (winter only 5–10 and summer 30–40 μg m^{-3}). The annual and daily variation *Steinhauser* explained through the seasonal dependency of vertical exchange.

4.4.8.4 The beginning of systematic surface-near monitoring

As mentioned on p. 33, the meteorological observatory Wahnsdorf pioneered air chemical research after World War II. According to Teichert and Warmbt (1955), bioclimatological aspects, namely the finding and hypotheses by *Cauer*, *Tichy*, *Effenberger*, *Curry*, and *Ungeheuer* and *Regener*'s hypothesis of stratospheric origin (Regener 1948, 1951, 1952) stimulated the beginning of tropospheric ozone research that resulted in the most extended time series in the world. Finally, Teichert (1952), who first tested the methods by *Regener*, *Cauer*, and *Paneth*'s silica gel method, suggested the potassium iodide method (*Cauer's* method) because of its simplicity, although the principally known disadvantages. Furthermore, Teichert (1953) reports on experiments with fluorescein because it is referred to be specific for ozone and shows no interference with NO_2 as found by Heller (1935)[1161] and also citing *Konstantinova-Schlesinger*; he found this method not applicable for routine measurements.

Since 1952, sampling over one hour was carried out (first with three times daily and stepwise enlarged). A pioneering investigation *Teichert* conducted with ozone measurements at the 80-m tower of the observatory Lindenberg and at the ground for several months; because of the later discussed too low ozone values determined at Wahnsdorf due to negative SO_2 interferences, these values are probably historically

1161 *Wilfried Heller* (1903–1982) was a German chemist who emigrated 1933 to Paris and 1937 to the USA where he became in 1953 professor for colloid chemistry. The French chemist *Louis Benoist* (unknown living dates) first introduced fluorescein for determination of ozone (Benoist 1919).

important reference data for the early 1950s surface ozone (in µg m^{-3}), the first column at the tower and second at ground (Teichert 1955):

August 1953 ($n = 54$)	37	32
June/July 1954 ($n = 230$)	30	26
September 1954 ($n = 161$)	23	22

More distinct are the differences between tower and ground in the morning while in the afternoon the difference is seen but much smaller:

	in the morning		in the afternoon	
August 1953	29	21	44	42
June/July 1954	27	22	32	30
September 1954	13	12	31	30

Junge (1963a, p. 51–52) found it worth reffering also the hourly data measured between 7 and 20 CET. *Teichert* (and *Junge* adopted this view) found *Regener*'s view on ozone transport from the upper atmosphere confirmed; *Junge* additionally stated "that fairly representative tropospheric ozone values can be obtained at the ground when the meteorological conditions are properly chosen". *Junge* (Junge 1963a, p. 44) concluded on an average (or let's say "reference") of 25 ppb[1162] from the available tropospheric ozone values in the 1950s. *Junge*'s statement was clearly not relevant to the Wahnsdorf data, where the ozone content was found to be much lower in 1952; average values ($n = 1092$; individual values range from 0 to 66 µg m^{-3}):[1163]

Winter (November–February)	8.1
Summer (June–August)	16.5

Teichert and *Warmt* concluded from his observations on the same causes of variation in the ozone content like Glückauf (1944), but obviously without knowing *Glückauf*'s

1162 In a recently published extended paper, Tarasick et al. (2019) report on ozone measurements since 1875, and choosing a large number of data sets, they conclude on practically not changing averages between 24 and 26 ppb for four periods (1896–1901, 1929–1934, 1938–1941, and 1951–1970) with rather small variation among the site (18–30). This paper is not yet "history" but I like to comment that their "calculated" change of ozone from the historical period to the modern period (Table 7, p. 25) seems to be questionable. However, this paper is very worth to read because all techniques and methods of ozone measurement are well described.

1163 It is notable that Teichert and Warmbt (1956, p. 265) stated the relatively high ozone values 1952/53 (about 12 µg m^{-3}) in Wahnsdorf, whereas since 1954 a remarkable decline was observed, and believed as cause the annual weather (however it was due to the increase of SO_2 emissions).

paper. Furthermore, they discuss the daily variation such as *Tichy* and *Effenberger*, and the annual variation similar to *Ehmert* and confirm *Regener*'s turbulence theory. Not yet, they discuss the interference with SO_2 and the pollution from Dresden they consider to be negligible as ozone consumer. However, from today's knowledge, it is likely that the annual variation was driven (as artefact due to "negative" ozone as called by *Steinhauser*) almost through SO_2 from coal heating in winter and manufacturing also in summer. It is worth noting that Junge (1962) notified a need for a global surface near ozone network to understand better the stratosphere – troposphere exchange; but such network was realized not before the late 1970s, and then from other reasons, tropospheric photochemistry and new types of forest damage.[1164] Warmt (1964) reports for the first time on the results from the ozone network;[1165] it follows a list of the average ozone concentration (in μg m^{-3}); annual variation in parenthesis):

Wahnsdorf (1952–1961)	9.7 (6.6–13.5)
Lindenberg (1956–1961)	21.2 (16.0–24.2)
Arkona (1956–1961, without 1960)	31.8 (29.2–38.7)
Boltenhagen (1956–1961)	28.4 (21.4–34.8)
Kaltennordheim (1955–1961, 487 m)	21.6 (20.6–25.3)
Fichtelberg (1955–1961, 1213 m)	36.0 (32.8–42.3)
Mt. Brocken (1955–1961, 1142 m)	35.9 (31.7–43.0)

Warmt adopted *Junge*'s view that the daily maximum values largely represent the satiation with negligible surface destruction potential (Junge 1962), as already noted by *Glückauf* (see p. 605). Thus Warmbt (1964, p. 53) communicated the absolute values of ozone maxima during 1956–1961, which amounts between 113 and 121 μg m^{-3} at mountain sites (Fichtelberg and Brocken) and all other stations between 95 and 101 μg m^{-3}. On the other hand, the mean daily maxima duirng 1960–1961 that were very close between 51 and 53 μg m^{-3} for the "remote" stations (Brocken, Fichtelberg, Arkona, Boltenhagen) correspond to *Junge*'s assumption of 25 ppb as a mean ozone content for the free troposphere in the 1950s. The data also show that there was no trend until 1961.

The average ozone concentration at Mt. Brocken 30 years later is rather exactly by a factor of two larger (76 μg m^{-3}, see Möller 2020, p. 432), an increase accepted by

1164 The terms "dieback, decline, forest dieback, stand level dieback, canopy level dieback, "*Waldsterben*" and "*Waldschäden*" have been used, more or less interchangeably, to describe this condition. In contrast to the "classical" damages caused by sulfur dioxide, ozone-caused stress to trees and forests have been named "novel forest decline", where the German term "*neuartige Waldschäden*" have been also used in English literature.

1165 Expanded 1968 with the station Großer Inselsberg (916 m), which was moved in 1979 to Mt. Schmücke (380 m), and the background station Neuglobsow in 1979.

most investigators at different locations in Central Europe, and, vice versa showing that the ozone measurements by *Warmbt* at "remote" locations were not unreliable. However, the data from Wahnsdorf (and likely at other central lowland stations) were highly affected by SO_2. Thus between 1952 and 1971, a "constant" O_3 concentration of $25 \pm 5\,\mu g\,m^{-3}$ was registered because of rising SO_2 concentrations. Only in 1967, when SO_2 concentrations remained constant, and in 1972, after introducing a pre-filter, the data can be considered reliable.[1166] Thus, almost the data series from Arkona (Island Rügen at the Baltic Sea) represents the long-term trend.

Warmt (1965) conducted the first-time ozone measurements over the ocean (Northern Atlantic, crossing Iceland and Greenland in July–August 1963) using *Ehmert*'s method (*Cauer*'s method was not applicable due to interferences with sea salt spray). He found a little daily ozone variation, varying between about $50\,\mu g\,m^{-3}$ at the eastern North Atlantic and $23\,\mu g\,m^{-3}$ west of Greenland, no decrease in ozone in fog, and concluded that the ozone content above the boundary layer amounts $50–60\,\mu g\,m^{-3}$ like Junge (1963a). Only a small layer of reduced ozone above the sea surface exists. Kolbig and Warmbt (1978)[1167] report on ozone measurements at the Soviet Antarctic station Mirny (February 1960–January 1962), with a minimum in summer ($34\,\mu g\,m^{-3}$) and a maximum in winter ($50\,\mu g\,m^{-3}$), similar to the results ($45\,\mu g\,m^{-3}$ annual mean) from measurements 1957–1958 at Little America in the Antarctics by Wexler et al. (1960).[1168]

Teichert and *Warmbt* conducted for the first time after war ozone monitoring at a mountain site (Fichtelberg) and found a distinct annual variation with two maxima, first April–June (about $50\,\mu g\,m^{-3}$) and in September (about $40\,\mu g\,m^{-3}$) for the years 1954 and 1955. The annual variation at Fichtelberg in 1955 was very close to that observed at Mt. Brocken, while at Kaltennnordheim practically no annual variation has been found. Teichert and Warmbt (1956) plotted the monthly ozone maxima and found between January and June a broad maximum of about $100\,\mu g\,m^{-3}$ for both mountains and about $50\,\mu g\,m^{-3}$ for the other locations and the lowest value ($60–70\,\mu g\,m^{-3}$ at the mountain sites) in December. This clearly shows with today's knowledge that the source of ozone was exclusively from the stratosphere[1169] (with a "normal" free tropo-

1166 An automatic instrument, the "Ozonograph" was developed by Mrose and Warmbt (1964) on the basis of the iodometric principle to take quasi-continuous measurements, and its new design, the "Ozonograph II" installed at the network sites in 1981/1982 to improve data coverage with 24 hours operation and stored 30-minute ozone averages. Due to the interfering effect of sulfur dioxide (SO_2) to the iodometric method, surface ozone measurements were affected at the sites with high ambient SO_2 concentrations, before a chromium trioxide filter was introduced into the network in 1971/1972 to remove the SO_2 effect (Warmbt et al. 1974).

1167 *Joachim Kolbig* (1933–1999) was a German meteorologist at the Observatory Potsdam, who was belong the pioneers of Antarctic research in the GDR. Beside surface measurements, also ozone sonde experiments were conducted.

1168 *Harry Wexler* (1911–1962) was an American meteorologist, the "father of weather satellites".

1169 An indication that ozone in the time before 1965 was originated from the upper atmosphere was also found by the positive correlation with artificial radioactivity due to nuclear tests (Warmbt 1966).

spheric ozone value of 50–60 µg m^{-3}), that there was no photochemical ozone formation (which would have a maximum in July), and that lower ozone values were due to ozone degradation in the boundary layer. In other words, when decades later, we were talking about the "ozone formation potential", in the time before air masses had a different "ozone destruction potential" – now resulting in a budget.

Based on measurements conducted at several heights above ground, *Warmbt* argued, such as *Regener* and *Ehmert* in the early 1950s, that the vertical profile results from ozone removal at ground and ozone transport downwards from the upper atmosphere (Fig. 4.56). Further, *Warmt* argued that the variation corresponds to profiles given by Junge (1962) when adopting an exchange coefficient to be 100 g cm^{-1} s^{-1}. The much higher ozone level at Arkona *Warmbt* is explained by a much smaller removal rate (0.1 cm s^{-1}) compared to it inland (1.0 cm s^{-1}) according to *Junge*. From today's knowledge, however, these values are questionable. The German stations, except for Fichtelberg and Arkona, were strongly influenced by pollution, and thus interferences with SO_2 in the measurements.

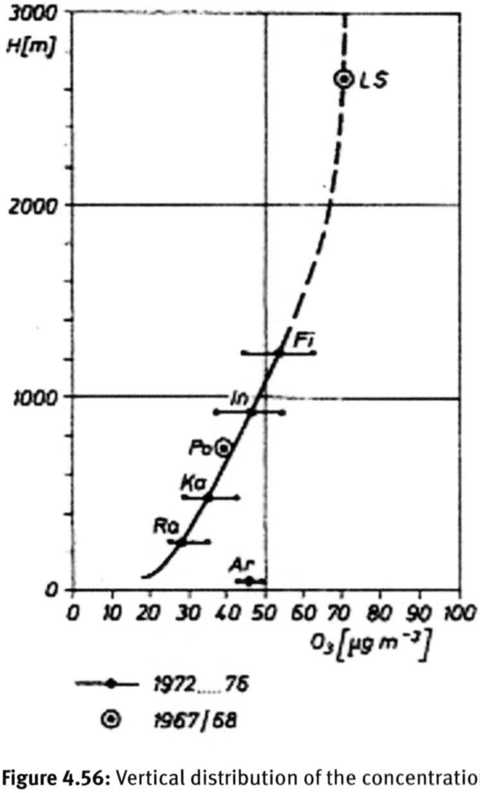

Figure 4.56: Vertical distribution of the concentration of ozone. Ar – Arkona (42 m), Ra – Radebeul (station Wahnsdorf, 246 m), Ka – Kaltennordheim (487 m), Po – Poprad (High Tatra, 707 m), In – Großer Inselsberg (914 m), Fi – Fichtelberg (1213 m), Lo – Lomnicky Štit (High Tatra, 2632 m), Warmbt (1980, p. 194).

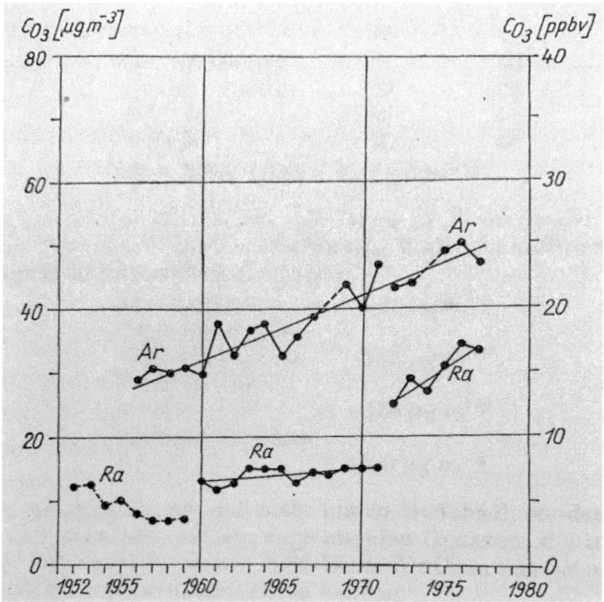

Figure 4.57: Annual mean ozone concentration at Radebeul-Wahnsdorf and Arkona (Warmbt 1979, p. 28): world-wide first published trend of rising tropospheric ozone.

Warmbt (1979) reported for the first time on the pronounced increase of the annual means of the ozone concentration from 30 µg m^{-3} (1956/57) to 48 µg m^{-3} (1976/77) with the following linear trend regressions for Arkona (1956–1977), in µg m^{-3} (x year with 1956 $x = 0$), Fig. 4.57:

Summer half year (April–September)	$[O_3] = 33.6 + 1.29 \cdot x$
Winter half year (October–March)	$[O_3] = 22.4 + 0.67 \cdot x$
Summer (June–August)	$[O_3] = 35.2 + 1.36 \cdot x$
Winter (December–February)	$[O_3] = 20.7 + 0.60 \cdot x$

The ozone content at the Baltic Sea in the mid-1950s (about 15 ppb) corresponds to many ozone values at different remote sites in Europe (cf. Table 4.24) since the beginning of the century, suggesting that this figure can be seen as the "normal" surface-near ozone concentration, largely unaffected by humans.[1170] The increase of ozone *Warmbt* contributed to the increasing pollution of the surface-near atmospheric layers from nitrogen oxides and reactive hydrocarbons, and noted that the "observed increase of ozone seems to be a large-scale rather than a local phenomenon". It is notable that at the beginning of the ozone monitoring, the "trend" at Wahnsdorf was

1170 This means that the "reference" value of 25 ppb by Tarasick et al. (2019) is too high.

"negative" due to the interferences with increasing atmospheric SO_2 concentrations in the 1950s. On the other hand, the jump in the ozone concentration since 1960 was due to a changing measurement method.

A similar observation so far was only observed in two short measurement series by Galbally (1972) and Atmannspacher et al. (1984). It is interesting to note that Atmannspacher (1976)[1171] reports on ozone maxima at the Hohenpeißenberg up to more than 800 µg m^{-3}, whereas Warmbt (1980) found in December 1967 at a Mountain (2632 m) in High Tatra a daily maximum of "only" 102 µg m^{-3}. *Ian Edward Galbally* (born 1945)[1172] reported on an increase of atmospheric ozone in Aspendale, Australia, from 35 µg m^{-3} in 1965 to 80 µg m^{-3} in 1971 (Galbally 1972). Based on the world's longest continuous ozone record from the Arkona-Zingst,[1173] we can conclude that from about 1955 in all industrialized areas of the world the tropospheric ozone has begun to rise (see also Feister and Warmbt 1987, Parrish et al. 2012). The overall percentage change until 1990 was 37.5 % (Feister 2009, p. 31).

Continuous records in southern Germany began at the rural hilltop site of Hohenpeißenberg in 1971 (Atmannspacher 1976), and the mountaintop site of Zugspitze (2670 m) in 1978 (Gilge et al. 2010),[1174] while continuous measurements began at the summit of Whiteface Mountain in upstate New York in 1973 (Oltmans et al. 2013). At Hohenpeißenberg, however, the long-term trend of increasing free tropospheric ozone could be observed by ozone sondes from 1967 to 1982 by about 50–70 %, and the increase of surface-near annual mean ozone from 26 µg m^{-3} in 1971 to 34 µg m^{-3} in 1982 (Atmannspacher et al. 1984). It is interesting to note that obviously, the increase in the free troposphere was much stronger than surface-near (there only 30 %). The long-term increase in surface ozone was observed in other networks as well (Hov et al. 1978, Logan 1985, Oltmans and Komhyr 1986, Staehelin et al. 1994) in Europe and North America, but did not occur at all remote sites with long-term observation records (Lefohn et al. 1992).[1175] These ozone increases coincided with rising European NO_x

1171 *Walter Attmannspacher* (1921–1991) was a German meteorologist at Hohenpeissenberg and director of the observatory 1968–1986. He constructed a wet-chemical ozone sonde for balloons.

1172 Retired atmospheric chemist at CSIRO, Division of Marine and Atmospheric Research, and Lead Scientist for the Reactive Gases Program at the Cape Grim Baseline Atmospheric Station in Tasmania, Australia.

1173 The measurements began at Cape Arkona in 1956 using a wet chemical method until 1990 (Warmbt 1964), after which continuous ozone measurements began at the nearby coastal site of Zingst, 60 km to the southwest. Combining the two sites produces a continuous record with more than 57 years of data (Parrish et al. 2012).

1174 *Stefan Gilge* (born 1962) is a German chemist, graduated from Technical University Aachen, who worked 1984–1994 at *Kley*'s Institute in Jülich, 1994–2015 as leading scientist at the Observatory Hohenpeißenberg and since 2015 as head of the Dept. Air Hygiene in "*Zentrum für medizinmeteorologische Forschung*" [Center for Medical Meteorological Research], Freiburg.

1175 *Allen S. Lefohn* (born about 1944) is an American physical chemist, most known for his contribution to air quality studies who was the President and Founder (1981) of A. S. L. & Associates in Helena, emissions that increased by a factor of 4.5 between 1955 and 1985 (Staehelin et al. 1994). An understanding of tropospheric net ozone formation in the presence of nitrogen oxides, carbon monoxide and hydrocarbons began in the mid-1970s (Chapter 4.4.8.6).

4.4.8.5 Tropospheric ozone research after 1970: an outlook

Such as Glückauf (1944) has already noted that ozone values due to "unqualified" averaging must be used very cautiously, Tarasick et al. (2019, p. 45) write that spatial representativeness is often the largest source of uncertainty in the use of ground-based data. The urban stations exhibit larger diurnal amplitude (due to strong night time ozone loss and strong daily photochemical production) while remote and high elevation stations show much flatter diurnal profiles; overall, the 60 available datasets during 1896–1975 indicate an ozone mole fraction in the well-mixed unpolluted boundary layer that lies in the range 22 to 26 ppb (Glückauf 1944, p. 46).

The key finding and discussion in tropospheric ozone research in the second half of the twentieth century concerns the photochemical formation of tropospheric ozone, the increase in the ozone content in the troposphere, and strategies to control tropospheric ozone. Almost all investigators in the 1990s agreed that the surface-near ozone concentration rose by a factor of two (and more) compared to the ozone content in the late nineteenth century. The "pre-industrial" reference value remains uncertain as discussed in the previous chapter; see also Cooper et al. (2014) for more information and discussion of trends, whereas Tarasick et al. (2019) believe that the tropospheric ozone content was about 25 ppb at the end of the nineteenth century and did not change until 1970, and increased only between about 35 and 75 %, more in the northern than in the southern latitudes.

Until the 1970s, wet-chemical methods (iodometry) remained the only routine procedure. Thereafter, both chemiluminescence and UV-absorption based methods have been used for the measurement of ozone in ambient air. Since about 2000, it has been generally accepted to use UV-photometry as the primary calibration method. When the tropospheric ozone research came into a renaissance in the 1980s, the reliability and relevance of the nineteenth century ozone data based on *Schönbein*'s paper was critically assessed by Bojkov (1986), Kley et al. (1988), Volz and Kley (1988), Anfossi et al. (1991), Varotsos and Cartalis (1991), Lefohn et al. (1992), Marenco et al. (1994), Pavelin et al. (1999), Weidinger et al. (2011); see Table 4.24.

Montana. *Øystein Hov* (born 1950) is a Norwegian meteorologist and atmospheric chemist at the Norwegian Institute for Air Research (NILU). *Jennifer Logan* (born likely 1950) is professor emeritus from the Harvard University. *Samuel J. Oltmans* (born 1946) is retired professor for atmospheric and environmental research of University of Colorado Boulder. *Johannes Staehelin* (born 1949) is a Swiss chemist at the "*Institut für Atmosphäre und Klima*" at ETH Zürich, most known for his works on ozone chemistry (also in aqueous phase) and ozone trends.

Table 4.24: "Pre-industrial" values of the ozone concentration derived from historical measurements and critical assessment. Except for Montsouris, at all other stations *Schönbein* (or *Lender*) papers have been used and humidity corrected by empirical relationships between modern ultraviolet absorption ozone measurements and reconstructed test papers.

Years	location	value (in ppb)	author
1850–1900	Europe	17–23	Bojkov (1986)
1853–1856	Szeged	11 (6–14)	Weidinger et al. (2011)
1870–1895	Buda	9 (7–11)	Weidinger et al. (2011)
1876–1880	Lansing, Michigan	35	Linvill et al. (1980)
1874–1895	Pic du Midi	10	Marenco et al. (1994)
1895–1909	Pic du Midi	14	Marenco et al. (1994)
1868–1893	Moncalieri	~8 (5–15)	Anfossi et al. (1991)
late 19th century	North America	19	Bojkov (1986)
1853–1920	Vienna	13	Lauscher (1984)
1891–1895	Vienna	9–14	Pavelin et al. (1999)
1891–1903	Montventoux	~20	Pavelin et al. (1999)
1876–1920	Montsouris (Paris)	11 (5–16)	Volz and Kley (1988)
1893–1905	Ó-Gyalla	13 (12–15)	Weidinger et al. (2011)
1901–1940	Athen	~20	Varotsos and Cartalis (1991)
1895–1901	Hiroshima	~16	Pavelin et al. (1999)
1895–1901	Tokyo	10–15	Pavelin et al. (1999)
1918–1929	Montevideo	10–13	Pavelin et al. (1999)
1898–1909	Mauritius	7–13	Pavelin et al. (1999)
1890–1895	Luanda	5	Pavelin et al. (1999)
1876–1879	Hobart	10–12	Pavelin et al. (1999)
1883–1907	Adelaide	11	Pavelin et al. (1999)

After recognizing the increase of surface-near ozone, a key question arose concerns the "pre-industrial level" which was not yet influenced by human influences. From all the ozone determinations made in the 1930s in Europe using spectroscopic and chemical methods a "typical" value is within 10–20 ppb; lower values seem to be found in towns and higher values at rural sites. Referring the "normal concentration" by Glückauf (1944) to be ~20 ppb, I agree with Bojkov (1986) that the "pre-industrial" value is also around 20 ppb; we will see that this value will remain the "normal concentration" for further two decades after World War II before rising due to net-ozone formation in the troposphere. The Montsouris data often has been regarded as the only reliable one from the nineteenth century; however, Tarasick et al. (2019, p. 13) discusses the interferences with SO_2, NH_3 and NO_2 and concludes that the average of 11 ppb as proposed by *Kley* and *Volz* is too low.

The French physicist *Alain Marenco* (*Université Paul Sabatier*, Toulouse) evaluated the ozone measurements from Pic du Midi, using the regressions based on comparing measurement at Montsouris (Marenco et al. 1994), and pointed out that the increase in ozone from 10 to 14 ppb from 1895 until 1909 coincided with the global rise in methane concentrations, while the ozone decrease at Montsouris may have been the result of increased emission of NO in Paris that destroyed ozone. However, this argumentation is not convincing, first because the tropospheric ozone chemistry is more complex, and second the question remains unanswered why at an altitude of 3000 m the ozone content is not significantly higher than at Montsouris in Paris. Staehelin et al. (1994) present ozone values at three different altitudes (800, 1850, and 3450 m) in the vicinity of Arosa from the 1930s and 1980s, showing a linear increase with the same slope but the different values at zero-altitude (mathematically), to be 12.5 ppb and 41.7 ppb, respectively (Möller 2003, p. 590). Möller (2003, p. 588) added more values into the *Marenco* diagram, suggesting that there was no exponential increase before the 1950s and that the Pic du Midi values are likely too low.

Several studies with atmospheric chemistry models confirmed that the long-term increase in ozone concentrations in the troposphere and near the ground in regions of the Northern hemisphere could have been the result of increasing photochemical ozone production due to increased emissions of nitrogen oxides and hydrocarbons by anthropogenic activities (e. g., Isaksen and Hov 1987, Crutzen 1988, Crutzen and Zimmermann 1991, Renner and Rolle 1989).[1176] Several North American (starting in the mid-1980s) and European (beginning in 1990) Ozone Research Programs, a combination of sophisticated monitoring, field campaigns, and modeling, led to a sufficient understanding of tropospheric ozone formation in different regions, under different time-spatial scales, and to abatement strategies but also to new research needs (e. g., investigation of natural organic emissions that were underestimated in the past, the importance of heterogeneous processes in oxidant chemistry and others); the recent history cannot be part of this volume. At the end of the twentieth century, the increase in tropospheric ozone content was stopped, and most important for local air quality, exceeded ozone values no longer or only seldom occurred.

Ozone in the center of the European continent typically experiences a summertime or broad spring/summer ozone peak, whereas sites located in less polluted regions of northern or western Europe experience a springtime peak (Scheel et al. 1997, Monks 2000, Wilson et al. 2012). Ozone at rural sites across the eastern United States typically peaked in summer during the 1990s (Cooper et al. 2012), while western US sites had a broader spring/summer peak. In contrast, the 2006–2010 time period shows that ozone in the eastern US has strongly decreased in summer while remaining constant in spring so that the summer maximum has been replaced by a broad spring/summer peak. This shift in the seasonal ozone cycle appears to be a response to emission reductions. Parrish et al. (2013) examined the seasonal ozone cycle at four rural sites in Europe and one rural site in the western United States. They discovered that not only is springtime ozone greater in recent years (2005–2010) than in earlier

1176 *Ivar S. A. Isaksen* (1937–2017) was a Norwegian meteorologist at the Institute of Geophysics, University of Oslo.

decades (the 1970s through the early 1990s), but the seasonal maximum now occurs earlier in the year.

Thus, it seems that the "history" of atmospheric ozone goes slowly back to its roots in the ninetheenth century due to the abatement of the "reactive" hydrocarbons in the atmosphere. However, the basic photochemical ozone formation in the free troposphere due to methane remains, and according to its increased atmospheric content, with a doubled source strength compared to the mid-nineteenth century. Nevertheless, academic research on atmospheric ozone is going on; see Prather et al. (2011).[1177] *Hans Cauer* would be delighted to know that some 80 years after his research on iodine (Cauer 1929, 1932) and ozone (Cauer 1936, 1937b, 1948a) a new generation of atmospheric chemists detected by combining measurements of iodine and sodium in the upper 130 m of an ice-core in Greenland by modeling and laboratory studies about the chemistry between ozone, iodine, and iodide, that the human-driven increase in tropospheric ozone has led to an amplification of the natural cycle of oceanic iodine emissions that has consequently decreased the lifetime of ozone in the marine atmosphere, thus closing a negative feedback loop (Prados-Roman et al. 2015, Cuevas et al. 2018). Finally, Yeung et al. (2019) conclude that the total tropospheric ozone burden probably increased by less than 40 % between 1850 and 2005, primarily near the surface, with most of the increase occurring between 1950 and 1980. A synchronous increase in oceanic emissions of iodine – the product of ozone-iodide reactions at the air-sea interface – seems to corroborate this timing, as well as the importance of halogen chemistry in late twentieth-century ozone budgets. Moreover, historical global emission estimates of ozone precursors, combined with current model chemical schemes, appear to capture the main features of the tropospheric ozone increase since 1850. Yeung et al. (2019) stress that these findings do not invalidate the late nineteenth-century surface ozone observations, which may simply not be representative of the tropospheric ozone burden at that time. To bridge the past with the future (see also final Chapter 5), the final phrase of the paper by Lamarque et al. (2005, p. 14) illustrates it best:

> The importance of chemistry in shaping the lifetime of greenhouse gases and aerosols will need to be investigated more, both for the recent past and the future, with the ultimate goal of incorporating and understanding the feedbacks between chemistry and climate.

4.4.8.6 Understanding tropospheric ozone formation and budget

Until the mid-80's, there has been considerable controversy regarding the relative importance of the two sources of ozone in the natural troposphere: injection from the stratosphere and tropospheric photochemical production (Finlayson-Pitts and Pitts 1986, p. 962).

1177 *Michael J. Prather* (born 1947), Professor of Earth System Science at University of California, who studied mathematics, physics and astronomy.

Cadle and Johnston (1952) found that 1000 ppb NO_2 produces about 100 ppb O_3, but on average 300 ppb ozone has been observed in Los Angeles (Neiburger 1959). Thus it does not explain the high ozone concentrations. Maximum measured ozone concentrations in towns have been reported by Bruckmann (1991):[1178] 1160 ppb Los Angeles 1970's years, 760 ppb Tokyo 1988, 543 ppb Mannheim 1976, 291 ppb Hamburg 1990. Although of some very rare observations of extreme surface-near ozone values up to several hundreds of ppb due to stratospheric intrusions found at Tromsø (Götz 1951) and Hohenpeißenberg (Attmansspacher 1976), daily ozone maxima due to transport from the stratosphere have been estimated to be between 10 and 60 ppb, far less than ozone concentrations often found in Los Angeles (500–800 ppb) and worldwide in large cities and industrialized areas (200–350 ppb) according to Becker (1977). The photolysis of NO_2, also when considering that the conversion $NO \rightarrow NO_2$ proceeds by HO_2 gained from photolysis of aldehydes, cannot explain the large concentrations of ozone observed in situations of photochemical smog. Although all important reactions of tropospheric chemistry were known at the beginning of the 1970s, an understanding of tropospheric net ozone production, including the NO-NO_2 and OH-HO_2 cycling, and the role of ozone precursors like CO, CH_4, and hydrocarbons became clear only after the mid-1970s by modeling of chemical systems (e. g., Crutzen 1974a, Chameides and Walker 1977, Stewardt et al. 1977, Chameides 1978).[1179] The chemistry of "photochemical smog" has been rather well understood in the 1970s from "smog chamber" investigation (Kerr et al. 1972, Demerjian et al. 1974, Schurath 1977). From the photo steady state of the reactions $NO_2 + \text{light} + O_2 = NO + O_3$ (1) and $NO + O_3 = NO_2 + O_2$ (2) it follows $[O_3] = k_1[NO_2]/k_2[NO]$, i. e., the higher the NO_2 concentration, the higher the ozone concentration, and vice versa, the higher the NO concentration, the lower the ozone concentration. However, it was further known in that time that the transfer $NO \rightarrow NO_2$ also goes via HO_2 and RO_2 radicals, which are gained from the oxidation of hydrocarbons (and CO),[1180] and thus the question arose, which of the so-called precursor, NO_x or VOC, should be controlled. The approach adopted by the EPA of the USA was to control hydrocarbon (namely NMHC) emission only (what has been found in the 1990s to be the right way), and reduction in NO_x would be only required to keep ambient NO_2 concentrations below the level of adverse health effects. The be-

---

1178 *Peter Bruckmann* (born 1948) was department head at "*Landesumweltamt*" [State Environment Agency] in North Rhine-Westphalia and a leading scientist in German air pollution control.

1179 *William L.* ("Bill") *Chameides* (born 1949) is an American atmospheric scientist who worked 25 years at the Georgia Institute of technology, most known for excellent works in modeling of atmospheric smog and aqueous-phase chemistry. *James C. G. Walker* (born about 1939) is an American geochemist, professor emeritus from the University of Michigan, and most known for his works on evolution of the atmosphere. *Richard W. Stewart*, who made his PhD in 1967 at the Columbia University and worked at NASA Goddard Space Flight Center.

1180 Another never considered ozone precursor was SO_2 (with turns $OH \rightarrow HO_2$); mentioned by Möller (2003, pp. 607–608) who learned this idea by a personal communication of *Wolfgang Rolle* about 1988.

lief of Japanese Authorities was that abatement programs should meet the stringent NO_2 standard, which would also contribute in a very high degree to the reduction of photochemical oxidants (OECD 1975).

The next ten years brought a rapid expanding network of ozone measurements in Europe and elsewhere, and understanding rose for the role of transport processes and meteorological conditions (e. g., Georgii et al. 1977, Fricke 1983).[1181] Furthermore, modeling studies showed that the rate-determining step in the formation of ozone is the reaction OH + VOC (e. g., Hough and Derwent 1987), and chemical mechanisms developed to describe the ozone formation (e. g., Derwent 1990). Furthermore, modeling studies have begun to explain the ozone trends by the increasing emissions of so-called precursors. It has been found that some NO_2 is removed from the boundary layer by dry deposition, up to more than 50 % of NO_2 in Europe may be converted into peroxyacetyl nitrate (PAN), and that transport of PAN, and therefore of NO_x plus anthropogenic carbon monoxide, methane and less reactive hydrocarbons are responsible for the observed ozone increase in the troposphere in the northern hemisphere (Guicherit 1988).[1182] Some PAN decomposes into NO_2 and peroxy radicals. This would suggest that the effect of anthropogenic NO_x emissions is felt only on a regional scale, i. e., several hundreds of kilometers.

There is, however, a general tendency for ozone concentrations to be lower in urban areas than in downwind rural areas because in urban areas, some ozone is removed by reaction with other pollutants, mostly nitric oxide. In polluted atmospheres, these reactions constitute a major sink for ozone. Since more than 90 % of the emissions of oxides of nitrogen occur as NO an amount of ozone equivalent to the flux of emitted NO will be converted to NO_2. This led to the defining of a parameter, oxidant (O_x) by Guicherit (1988) $O_x = O_3 + NO_2$ and noting that O_x is a conservative quantity over short times scale. Kley et a. (1994) compared O_3 and O_x concentrations in Cologne, a large city, with those from a rural site (Rothaargebirge, some 200 km downwind); in ppb:

	Cologne	rural site
O_3	9	22
O_x	30	31

Smog chamber experiments have been conducted with changing mixtures of nitrogen oxides and hydrocarbons to study ozone formation capacity of single hydrocarbons,

1181 *Wolfgang Fricke* (born 1948), German meteorologist, since 1990 at Offenbach who became in 2006 the director (until 2013) of the Hohenpeißenberg Observatory.

1182 *Robert Guicherit* (born about 1945) is a Dutch who has been working at TNO, The Hague (Netherlands).

and the effect of changing concentrations ratios on the reaction products (e. g., Schurath 1977, Becker 1977, 1991). At the end of the 1980s, it became clear that reduction of NMVOC led to reduction of peak concentrations of ozone and that solely reduction of NO_x results in a growth of ozone concentration in areas with the excess of NO_x, namely near the source area. Measurements have shown that in-situ ozone formation rates up to 50 ppb h^{-1} can be gained (Volz-Thomas et al. 1997),[1183] but under consideration of removal and transport processes, the ozone formation is on average "only" 10–15 ppb d^{-1} (Beck and Grennfelt 1994, Möller 2004 and literature therein). The conception of NO and VOC limitation, i. e., the function of the formation potential of ozone dependent from the precursor concentration developed by Sillman (1995).[1184] In the mid-90's, it was understood that tropospheric ozone involves three different sources:

- stratospheric ozone due to transport from the upper atmosphere,
- free tropospheric ozone gained in low specific formation rates from CO and CH_4 as precursors, and
- boundary layer ozone gained from NMVOC in high specific formations rates (called "hot" ozone by Möller 2003).

The source strength of ozone in the troposphere thus depends on stratosphere-to-troposphere transport (STT), i. e., the intrusion of ozone through tropospheric folding. There is most evidence that seasonal variation of STT is not strong. However, generally, a summer minimum and a spring maximum is observed. The amount of ozone observed near the ground depends on the occurrence and frequency of folding events and its depth. Folding occurs in the Northern Hemisphere in the mid-latitudes (30 °) with subsequent air mass transport to the north and at the North Pole with subsequent transport to the south. Monks (2000, p. 3546)[1185] writes:

> Ozone measurements carried out at the Arkona station (Baltic coast of the former East Germany) since 1956 were used to calculate the seasonal ozone cycle for the period 1956–1983. A clear broad spring-summer maximum is indicated, which peaks in May (Feister and Warmbt 1985, 1987).[1186] The Montsouris ozone time series (1876–1886) also exhibits a clear spring maximum characteristic of the seasonal ozone cycles inversely clean and remote atmospheres in the northern hemisphere (Volz and Kley 1988). Other historical ozone series show similar seasonal features to Montsouris, e. g., Athens (1901–1940) (Varotsos and Cartalis 1991, Cartalis and Varotsos 1994). In contrast, the Moncalierei (Northern Italy) ozone series (1868–1893), while similar in magnitude

1183 *Andreas Volz-Thomas* (born 1946), chemist at the Jülich institute for atmospheric chemistry and the co-worker of *Dieter Kley*.

1184 *Sanford Sillman* (born 1954), studied physics and mathematics; worked at the University of Michigan and is now emeritus.

1185 *Paul S. Monks* (born about 1969) is Professor of Atmospheric Chemistry and Earth Observation Science at the University Leicester.

1186 The reference Feister and Warmbt (1984) is quoted for 1980, but both is wrong: it is Feister and Warmbt (1985).

to Montsouris, does not show that same spring maximum, but rather a broad spring-summer maximum (Anfossi et al. 1991). A comparison of the Arkona and Montsouris time series (Janach 1989) suggested, that the hemispherical increase (approximate doubling) of ozone has only occurred over the last few decades.

Thus, in the seasonal cycle of tropospheric ozone, three different source characteristics overlay, the stratospheric cycle, the free tropospheric ozone production cycle, having an opposite direction to the stratospheric one, and the "hot" ozone cycle in the boundary layer, having a strong seasonal and diurnal variation parallel to the radiation intensity. This explains as well the often observed two maxima in the nineteenth century, which moved with increasing emissions of CO and CH_4 to a broad but shallow maximum in summer, and finally with the increasing NMVOC emissions after 1950 to a distinct summer maximum. Furthermore, it was recognized that the worldwide increase of the (large-scale) tropospheric ozone content was almost caused by the rise in emissions of CO and CH_4. However, the observation of short-time high ozone values (not only area-limited situations of photochemical smog) was exclusively due to the presence of NMVOC in air, having a large ozone formation potential. Moreover, in the United States and in the Mediterranean, biogenic organic emissions, underestimated mainly before the 1990s, contribute to extended ozone values. Thus the "hot" ozone consists of artificial and biogenic compartments.

The latter was studied by several field experiments in Northern America and Europe in the 1990s and later also in Asia. These experiments were aimed to investigate the development of chemical compounds in the plume of cities or industrial areas during transport into the surrounding rural area. The Berlin Ozone Experiment (BERLIOZ) between 5 July and 7 August 1998 near the city of Berlin was the most extensive field campaign ever initiated in Europe addressing the photochemical smog problem and funded with 35 million DM by the Federal Ministry (BMFT). For this purpose, in additional to 45 permanent monitoring stations under operations of Environmental Authorities, 10 surface sites were equipped (including 6 lidars and 5 aircrafts) along an axis extending from 50 km SSE to 70 km NNW of the city center of Berlin, 15 German universities and research institutes, various environmental agencies, the German Weather Service and several researchers from European countries, e. g., from France, Italy, Switzerland and Poland participated (Becker et al. 1999). The Photochemistry Experiment during BERLIOZ (PHOEBE) aimed at a quantitative understanding of the fast radical chemistry in rural and suburban air by simultaneous measurements of the major radicals (OH, HO_2, RO_2, and NO_3) and the chemical compounds and physical parameters that control the radical concentrations (Volz-Thomas et al. 2003. Furthermore, different models have been applied to study the ozone evolution (Becker et al. 2002).[1187]

1187 The author participated as local organizer (deputy of *Karl-Heinz Becker* who was the main coordinator) together with *Eberhard Reimer* (FU Berlin) who was responsible for forecasting of "smog"

One cannot state that we – 20 years later – understand every detail of tropospheric ozone formation, but obviously the formation of "hot" ozone is not only fairly well described but also largely under control due to the reduction of artificial NMVOC emissions.

Now, ending this chapter, we turn to the tropospheric ozone budget, i. e., balancing the sources with the sinks. Already in the nineteenth century, it was recognized that the local ozone content in air depends strongly on the removal process, which has been seen in that time almost in reactions between ozone and pollutants. There are chemical reactions that decompose ozone, most important the photolysis of O_3 gaining $O(^1D)$ with the subsequent reaction $O(^1D) + H_2O$ yielding OH radicals, known since the end of the 1960s. Further, depending on the content of oxidable substances (NO_x, VOC, CO, SO_2 etc.) which are in competing reactions to O_3, the reactions O_3 + HO_2 and O_3 + OH are significant contributors to ozone removal, namely in the remote atmosphere, also known for 50 years. Other minor contributions come from the reaction of O_3 + terpenes, with SO_2 via the aqueous phase, with NO_x and subsequent HNO_3 removal almost via wet deposition, and with halogen compounds.

The total strength of the global ozone sinks due to chemical loss[1188] has been estimated by models between 3470 and 4360 Tg yr^{-1} (see Hu et al. 2017 for the most recent model and literature therein). The global tropospheric chemical source strength has been estimated between 3420 and 4970 Tg yr^{-1}. Hu et al. (2017) note that estimates from pre-2000 modeling are lower than estimated from post-2000 modeling due to improved chemical mechanisms, better emission estimated, and better models. However, the amount of stratosphere-troposphere exchange was "estimated" as the difference between chemical sources and sinks (chemical loss + dry deposition), i. e., obviously, until the present, does not exist a direct way for estimation of the downward flux! This stratospheric ozone source strength thus amounts between 325 and 770 Tg yr^{-1}. According to the models, the tropospheric ozone burden shows no large scatter-

---

situations to activate sophisticated measurement systems from stand-by such as aircrafts, lidars etc. *Möller* and his coworkers operated a mobile lidar (cf. Fig. 4.52) and the station Eichstädt, NW at the edge of Berlin (see for example Fig. 7.41 in Möller 2003, p. 600, on simultaneous ozone developed at this station and Mt. Brocken, 250 km SE). Unfortunately, no "typical" weather situation (high pressure system with air masses from SE lasting 5–7 days) occurred thus the general aim of the project was not achieved and only results from selected days at different stations has been elaborated and published. *Möller* and his group participated at several complex measurement campaigns which had "wrong" weather comparing to the planned measurement strategy. Nobody never wrote on it – but always enough papers, graduations and thesis gained; this is science, not economy.

1188 Hu et al. (2017) presents for wet removal of ozone a value of 17 Tg yr^{-1}, mainly as HNO_3 (additional sulfate may be considered). A simple plausibility consideration, however, results in a much larger value of 70–100 Tg yr^{-1}: assuming a global NO_x emission of 25 Tg N yr^{-1} which is oxidized 100 % to NO_2 from which 50 % is removed by dry deposition and 50 % further oxidizes to HNO_3, this amount corresponds to 1.5–2.2 Teq O_3 (dependent from the oxidation pathway). There is no need for any comment concerns the reliability of such model estimates.

ing among the models with values between 337 and 363 Tg and the mean tropospheric ozone residence time ranging between 22 and 24 days. Assuming a tropospheric ozone burden of 350 Tg and a residence time of 23 days, it follows simply from the flux equation (residence time = mass/flux) a mean tropospheric ozone concentration of 60 µg m^{-3} which is very consistent with measurements. On the other hand, adopting the mean concentration and the volume of the troposphere, it follows 350 Tg as total content of ozone on the troposphere; furthermore, assuming a global mean dry deposition velocity of ozone of 1 cm s^{-1}, it follows a dry deposition flux of 950 Tg yr^{-1}. The different models over the last 20 years provide values between 770 and 1090 Tg yr^{-1}.

It seems that models, even sophisticated, did not contribute to much progress in understanding the global ozone budget for the last 40 years. Seiler and Fishman (1981) and Logan et al. (1981) stated that the "fuels" of tropospheric ozone are CO and CH_4; thus, its increase in the air results in rising ozone values. Möller (2003, p. 615) "calculated" the global tropospheric ozone production from different precursors (CO, CH_4, NMVOC, isoprene, and terpenes) based on its current emissions and using simple chemical stoichiometry and a few "logical" chemical assumptions to be 4000 Tg yr^{-1} in agreement with the models. At the end, one can note that chemistry is simple in a well-mixed bulk reactor – the only "disturbance" is due to inhomogeneous mixing, i. e., "meteorology".

To understand the O_3 chemistry in the remote atmosphere, which is characterized by the absence of NO_x, SO_2, and NMVOC, or at extremely low concentrations, the observations by Ayers et al. (1996) at Cape Grim baseline station during 1990–1991 are fundamental. Both the seasonal and diurnal cycles of peroxides concentration were anticorrelated with ozone concentration: peroxides have a maximum in summer and noon. Thus, the central processes were daytime and summertime photolytic destruction of ozone, transfer of reactive oxygen into the peroxides, and continuous heterogeneous removal of peroxides and ozone at the ocean surface.[1189]

Note that the tipping point between ozone decomposition and ozone formation is given by the NO concentration; an absence of CO and/or VOC (the "fuels" of tropospheric ozone formation) in the natural atmosphere can be excluded because of their biogenic sources.[1190] Finlayson-Pitts and Pitts (1986, p. 969) have given a simple criterion to distinguish between regimes of ozone production and decomposition by comparing the rates of the reaction $O_3 + HO_2$ with that of $O_3 + NO$: for concentrations of NO less than about 10 ppt the ozone decomposition dominates. Lightning (NO source strength about 6 Tg N yr^{-1} according to latest estimations) can only provide locally, and limited NO concentrations are larger. Biogenic soil NO emission (likely <10 Tg N yr^{-1} according to Weng et al. 2020) can also only result in regional ozone production

1189 Under steady-state it follows $[O_3]/[H_2O_2] = j(H_2O_2)/j(O_3) < 1$, j = photolysis rate.

1190 However, these sources (biomass burning for CO and wetlands for CH_4) are largely also due to human activities.

in the continental atmosphere. Thus, it cannot exclude that in populated areas in the ninetheenth century ozone was produced photochemically; the "fuels" (CO and CH_4) were present and without doubt NO_x at low concentration but sufficient to initiate the ozone formation cycle. However, due to missing (or only present at low concentration) reactive NMHC, surface-near photochemical ozone formation was not fast. On the other side, in the nineteenth century, large parts of the earth's atmosphere (oceans, deserts, arctic regions) were photochemical regimes of ozone decomposition.

Thus, in evaluating the ozonometric observations of the ninetheenth century, we can consider that both ozone and hydrogen peroxide were registered, but in different relations according to the season and the weather conditions. Finally, we can describe the other influences and observations:

- in winter, the stratospheric ozone input is stronger than in summer but much less photochemical ozone decomposition occurs (ozone maximum);
- in summer, the stratospheric ozone input is weaker than in winter but the photochemical ozone decomposition is very effective (ozone minimum);
- under "wet conditions" (high humidity, clouds, precipitation), the photocatalytic formation of hydrogen peroxide occurs, but much stronger in summer than in winter (maximum "ozone reaction").

It seems to be very likely that the ozonometric observations revealed ozone in winter, and in summer, ozone plus hydrogen peroxide. Those observations showing large "amounts of ozone" in connection with the appearance of atmospheric water (precipitation, dew, fountains, etc.) cannot be explained by atmospheric ozone, but through the presence of hydrogen peroxide. *Gottfried Wilhelm Osann* found that each snowflake falling on a paper coated with potassium iodide is followed by a "*Schönbein* reaction" (Osann 1853). *Otto Nasse* argued concerns the so-termed "ozone reaction" that due to the likely permanent presence of hydrogen peroxide or substances which easily dissociate under release of oxygen that (Nasse 1870, p. 210):

> [...] der sichere Beweis für Anwesenheit von Ozon ist, besonders wenn es sich um kleine Mengen handelt, ausserordentlich schwer [... the sure proof for the presence of ozone is extremely difficult, namely at low amounts].

On the other hand, the almost observed correlation of the "amount of ozone" with the atmospheric humidity was due to the sensitivity of the test-paper concerns humidity. Notwithstanding, the finding of low amounts of ozone in large towns can be explained by the air pollution, and that of low and even "zero" ozone during fogs either by aqueous-phase ozone removal[1191] and/or scavenging gaseous hydrogen peroxide, in other terms cleaning the air by oxidants.

1191 The decline of atmospheric ozone in clouds was first time experimentally shown by Acker et al. (1995), and soon before by modeling (Lelieveld and Crutzen 1991, Möller and Mauersberger 1992).

4.5 Hydrogen peroxide (H_2O_2)

One of the most interesting, mysterious, and important species in air is hydrogen peroxide.[1192] H_2O_2 is produced in the atmospheric gas-phase only by a single pathway, the HO_2 radical recombination, and thus closely linked with the photochemistry of ozone (other sources of HO_2 radicals are photolysis of aldehydes and ozonolysis of alkenes). Note that aqueous-phase and interfacial atmospheric chemistry provides more sources of HO_2 partly independent of the presence of ozone. The American atmospheric chemist *Gregory Kok* suggested that hydrogen peroxide is an index of the HO_2 radical concentration and would better represent the atmospheric oxidation capacity than ozone (Kok et al. 1978). However, early specification on ozone as the oxidant leader-species, the relatively complicated and man-power consuming measurement technique limits H_2O_2 measurements to very few sites and occasionally only. Moreover, from the last decade, nothing can be read on atmospheric hydrogen peroxide, and no further measurements are known; hydrogen peroxide does not deserve this disregard.

The first review on hydrogen peroxide, but only concerns its constitution and properties, was written by the German chemist *Julius Wilhelm Brühl* (1850–1911) who stated that its properties are still just as puzzling as at the time of its discovery by *Thénard*, and writing that it is a relatively weak oxidant but a strong reducing agent (Brühl 1897). Extended monographs on hydrogen peroxide wrote Machu (1937) and Kausch (1938), who also include a vast history, Schumb et al. (1955) and Ardon (1965); historic analytical methods for hydrogen peroxide Birckenbach (1909) published a book, and Liebknecht and Katz (1953) an extended book chapter. Modern review articles about the atmospheric chemistry of hydrogen peroxide are by Gunz and Hoffmann (1990a), Jackson and Hewitt (1999), Lee et al. (2000), Vione et al. (2003).[1193] See also Gmelin (1907, pp. 124–148; 1966, pp. 2097–1516), on OH (Gmelin 1969, pp. 1527–2600) and on HO_2 (Gmelin 1969, pp. 2600–2623).

4.5.1 Early history and discovery

Hydrogen peroxide (H_2O_2) was discovered by *Louis Jacques Thénard* (1777–1857)[1194] in 1818 while treating barium peroxide with sulfuric acid (Thénard 1818, 1819). He termed

1192 High-concentrated H_2O_2 was used namely in Germany during World War II as propellant in submarine and torpedo as well as rocket propulsion. From today's view it is a curious story that authorities in China asked the German chemist *Heinrich Bremer* (1920–2008) during a working stay end of the 1950s, whether H_2O_2 can be extracted from the air for industrial use (pers. communication).

1193 The article of *Vione* et al. is not recommended because of several unclear statements and confusing schemata.

1194 French chemist and professor at École Polytechnique in Paris; he published a textbook, and his "*Traité de chimie élémentaire, théorique et pratique*" (4 vols., Paris, 1813–1816), which served as a standard for a quarter of a century.

it "*l'eau oxygènée*" (oxygenated water), but he also already used the name "*peroxide d'hydrogéne*".[1195] *William Prout* (1785–1850)[1196] was the first to propose the existence of H_2O_2 in the atmosphere, and he termed it "*deutoxide d'hydrogène*",[1197] draw from such observations the following clear conclusion (Prout 1834, p. 570):[1198]

> [...] the bleaching qualities of dew, and of the air itself; as to the large proportion of oxygen sometimes contained in snow water and in rain water [...].

In his famous book "Chemistry, Meteorology and the Function of Digestion" he writes (Prout 1834, p. 569):

> [...] that a combination of water and oxygen is a frequent, if not a constant, ingredient in the atmosphere. This ingredient, which we suppose to be a vapour, and analogous to (we do not say identical with) the deutoxide of hydrogen,[1199] may be imagined to act as a foreign body, and thus to be the cause of numerous atmospheric phenomena, which at present are very little understood [...] The oxygen and vapour in this combination are so feebly associated, [...].

It is notable that *Prout*'s ideas were established before the discovery of ozone by *Schönbein* in 1839 (Schönbein 1844). The study of H_2O_2 in air was closely connected with studying the chemistry of O_3 in the nineteenth century (Engler 1879, Rubin 2001). Remarkably, the existence of H_2O_2 in air (as gas as well as dissolved in hydrometeors) was definitely established before 1880, but the existence of O_3 in the atmosphere was still discussed around 1880; definite proof of the existence of O_3 in the atmosphere was not before first spectrometric measurements with the ending ninetheenth century.

A hundred years before *Prout*, Boerhaave (1735, p. 350–351) writes on his investigations of water, in which Venice Soap (hard soap made from olive oil) is dissolved and exposed to sun light, some phrases which reads like first evidence for atmospheric hydrogen peroxide: [1200]

1195 It was first termed in German "*oxydiertes Wasser*" (oxidized water) and "*Sauerstoffwasser*" (oxygen water), and later "*Wasserstoffhyperoxyd*" (19th century), changed "*Wasserstoffsuperoxyd*" (before the 1960's) and finally to *Wasserstoffperoxid* (rare *Hydrogenperoxid*).

1196 English chemist, physician, and natural theologian; in 1815 he hypothesized that the atomic weight of every element is an integer multiple of that of hydrogen, suggesting that the hydrogen atom is the only truly fundamental particle.

1197 This term was introduced by the Scottish chemist *Thomas Thomson* (1773–1852). It was subsequent termed "peroxide of hydrogen".

1198 From our current knowledge on photosensitized formation of oxygen radicals in surface wetness this is a clear evidence for H_2O_2 formation.

1199 Hence, we can speculate that beside H_2O_2 also another strong oxidant (such as the hydroxyl radical) is meant; *Prout*'s first phrase upon the "oxidant power" of snow and rain clearly indicates the presence of H_2O_2.

1200 Metallic copper causes the explosive decomposition of hydrogen peroxide when the solution has a concentration of 30 percent or greater. *John Papiewski* https://sciencing.com/copper-explode-16298.html

> [...] poured upon wax, that is to be exposed to the Sun, or thrown upon Linnen that is to be whiten'd, it gives them an exquisite whiteness, [...]. [...] thrown into this Water, [...] Copper, Water and Vessel will be agitated with an incredible noise and fury not to be restrained, to the imminent danger of all around it. [...] This surprising property, now, I am of Opinion, cannot be understood, or explained, from any common principle of Bodies.

4.5.2 Studies of hydrogen peroxide in the ninetheenth century

4.5.2.1 On the nature of hydrogen peroxide

Schönbein (1858a), who never identified the constitution of ozone, proposed the existence of another form of reactive oxygen species, which he called antozone. He believed that ozone and antozone were formed under ozone-producing conditions (quiet electric discharges of air) and react together to give oxygen, hence explaining the low yield of ozone by its destruction by antozone.[1201] Meissner (1863a, p. 126) found hydrogen peroxide while passing electrified air through water, in fact, more when using pure oxygen than atmospheric air. Thus, he concluded (Meissner 1863a, p. 336) that in clouds, due to the action of electricity ozone and antozone are formed, and then the water from thundercloud should contain hydrogen peroxide. Independently Schönbein (1869, p. 273) argued that antozone, always gained together with ozone, combines with the water in the atmosphere to hydrogen peroxide.

From the first observations of hydrogen peroxide in air (in 1869), *Heinrich Wilhelm von Struve* (1822–1908)[1202] proposed in 1870 that hydrogen peroxide is produced during all burning processes in air (Struve 1871).[1203] Moreover, *Struve* concluded that ozone, hydrogen peroxide, and nitrous acid are gained simultaneously. *Eugen Frei-*

1201 The "reaction" $O_3 + H_2O_2 = 2O_2 + H_2O$ was often studied in the 19th century (see Rothmund and Burgstaller 1917 and literature therein); these authors found O_3 decomposition catalyzed by H_2O_2 in acid solution.

1202 Генрих Васильевич Струве: a German-Russian chemist, born in Dorpat (now Tartu in Estonia), worked 1849–1867 in Sankt Petersburg and from 1867 in Tiflis; in 1876, *Struve* became a member of the Russian Academy of Sciences in Petersburg. Son of *Friedrich Georg Wilhelm von Struve* (In Russian, his name is often given as *Vasilii Yakovlevich Struve* [Василий Яковлевич Струве]; the *Struve* family was a dynasty of five generations of astronomers from the eighteenth to 20th centuries. Members of the family were also prominent in chemistry, government and diplomacy.

1203 Over the following 150 years, formation of H_2O_2 in combustion processes was assumed and established not only during explosions, but also in the slow combustion of hydrocarbons (1955), in CH_4 burner flames (1959) and in combustion engines (1959) as described by Gmelin (1966). The only known paper by the atmospheric scientific community concerning H_2O_2 formation while (biomass) burning was published by Lee et al. (2000). Taken into account the well-known combustion chemistry (e. g., Warnatz et al. 1996), it is more than surprising that no modern studies are available on direct emission of H_2O_2 from different combustion sources like cars and power plants. Possible (and important) primary H_2O_2 sources (i. e., not via ozone photolysis) could influence our understanding of man-made changes of atmospheric chemistry.

herr von Gorup-Besanez (1817–1878),[1204] motivated by the finding of *Lender*[1205] that in thorn houses, the *Schönbein* paper shows the "ozone-reaction", studied systematically the influence of artificial springling in a botanical garden.[1206] He also found the "ozone-reaction" but excluded nitrous acid and hydrogen peroxide by parallel use of three different "detection papers" (potassium iodide on starch paper, litmus paper and thallium paper) and concluded that the general process of water evaporation causes the formation of ozone (Gorup-Besanez 1872).[1207] On the other hand, *Gorup-Besanez* supported *Struves* conclusion and writes (Gorup-Besanez 1872, p. 250):

> Nach Allem, was über das Vorkommen des Ozons, des Wasserstoffsuperoxyds und des salpetrigsauren Ammoniaks in der Atmosphäre bekannt ist, stellen diese Körper eine eng vcrbundene Trias dar [After all we know on the occurrenc of ozone, hydrogen peroxide and ammonium nitrite in the air, these bodies represent a closely linked triad].

Soon after this phrase, *Gorup-Besanez* expresses, but still very vague, the participation of oxo radicals ("oxygen activation and polarization"):[1208]

> In der That scheinen mir alle in der Luft möglichen Bildungsweisen dieser Körper immer wieder auf eine vorgängige Polarisation, auf ein Activwerden des Sauerstoffs, d. h. auf die Bildung von Ozon zurückzuführen, und was für die drei genannten Körper gilt, gilt auch für in der Luft etwa vorhandene freie salpetrige Säure, wie die einfachste Ueberlegung der directen Bildungsweisen gerade dieser letzteren darthut [Indeed, it seems to me that all possible formation ways of these bodies in air can be attributed to a previous polarization, an activation of oxygen, i. e., the formation of ozone, and what is valid for the three mentioned bodies, is also valid for free nitrous acid in air, as the simplest consideration concerning the direct formation of the latter shows], Gorup-Besanez 1872, p. 251.

The formation of hydrogen peroxide in autoxidation processes was proposed by Traube (1882) despite nothing being known on formation mechanisms. Years before, Schönbein (1866) detected the formation of hydrogen peroxide from essential oils and turpentine oil under sunlight. Before *Moritz Traube* has brought forward a number of experiments to show that hydrogen peroxide is reduced oxygen, chemists have

---

1204 An Austrian-German chemist who studied medicine (1836–1839) and chemistry (1839–1842) in Munich, Göttingen and Erlangen; since 1855 professor in Erlangen and 1850 member of Leopoldina.
1205 Lender (1878) laments that Gorup-Besanez (1872) does not refers him concerns his first identification the thorn house in Bad Kissingen as a strong source of ozone; *Lender* claims to be the first discoverer of this "ozone source".
1206 Bellucci (1876) also reports of an ozone odor associated with waterfalls.
1207 Today we know that this is not the case. See comments to the fact of water evaporation (e. g., in dew) on the formation of HNO_2 and the decomposition of ammonium nitrite (Chapter 4.4.2.3, pp. 472–473). The much stronger "ozone-reaction" can be caused by the transformation of interfacial ozone into oxyradicals and due to photocatalytic surface processes.
1208 Today we know that O_3 together with NO_x is within a photochemical loop and interlinked with H_2O_2 and HNO_2 via OH and HO_2 radicals.

looked upon hydrogen peroxide as oxidized water, from the act that it yields up one of its atoms of oxygen with great readiness, and therefore acts as an energetic oxidizing agent (Leeds 1883, p. 147); thus H–O–H + O = H–O–O–H. *Traube*, who rejected the antozone theory, was also the first (Traube 1887) who showed that water electrolysis in acid solution (never in alkaline solution) always first produces hydrogen peroxide when oxygen is bubbling around the cathode and then hydrogen. He explained it by a reaction between *in statu nascendi* hydrogen (H) and molecular oxygen[1209] and concluded the formula H–O=O–H ($2H + O_2$) in contrast to H–O–O–H for H_2O_2. Thus *Traube* (in several works between 1882 and 1893) definitively stated that hydrogen peroxide is not an oxidation product of water but reduced molecular oxygen; however, note that Heldt (1861) this already stated 20 years before; see footnote c in Table 4.15 on p. 481.

The Swiss chemist *Friedrich Goppelsroeder*, who did not doubt the presence of hydrogen peroxide in the air, but still accepted *Schönbein*'s view of ozone ("*Minus-Ozon*") and antozone ("*Plus-Ozon*"), and its permanent formation at the earth's surface in oxidation processes through "ozonisation" and "antozonisation" of oxygen, wrote poetically (Goppelsroeder 1871, p. 142):

> Das Minus-Ozon wirft sich auf den sauerstoffbegierigen Kohlenstoff und Wasserstoff der organischen Stoffe, verwandelt dieselben in Kohlensäure und Wasser, das plus oder Antozon tritt aber mit Wasser zusammen und bildet Wasserstoffsuperoxyd, das wegen seiner Verdampfbarkeit zum Theile, jedenfalls zum geringsten Theile, in die Atmosphäre gelangt [Minus-ozone throws himself on oxygen appetent carbon and hydrogen of organic substances, converts them into carbonic acid and water, whereas the plus or antozone joins with water and forms hydrogen peroxide, which due to is evaporability, partly, at least to a slightest extent, escapes to the atmosphere].

It is worth describing the following experiments by Osann (1865), who studied the gas gained from passing atmospheric air over (white) phosphorus. The observed white vapor was believed to be antozone according to *Meissner* or ammonium nitrite according to *Schönbein* (today we know that it is finally P_4O_{10}). He (correctly) found that it contains ozone (footnote 916 on p. 464), and by passing through a solution of "*salpetersaures Silberoxyd-Ammoniak*"[1210] he observed a black precipitate, identified as Ag_3O (in truth it is elemental silver), and assuming that it is produced by "deoxidation" (reduction) and not oxidation through ozone. Thus, he concludes with the formation of antozone, a reducing agent. In a logical following experiment *Osann* then removed ozone by absorption in a solution of "*pyrogallussaures Kali*" [potassium pyrogallate], found the air still enriched with the white vapor but no smell of ozone more. Finally, he collected the air in a *Woulff* bottle, containing some water, and again he obtained

1209 That could be interpreted as the first (not yet clear) idea on the reaction $H + O_2 \rightarrow HO_2$. However, we know that the anodic process goes via $OH^- \xrightarrow[-e^-]{} OH \xrightarrow{OH} H_2O_2 \xrightarrow[-(e^- + H^+)]{} HO_2 \xrightarrow[-(e^- + H^+)]{} O_2$.

1210 It is $[Ag(NH_3)_2]^+$, also called *Tollens*'reagent.

with the silver solution the black precipitate but now excluding the presence of ozone. The water from the *Woulff* bottle he named "*antozonehaltige Flüssigkeit*" [antozone-containing liquid] and proved it to by hydrogen peroxide by comparison the same experiments with hydrogen peroxide gained from potassium peroxide. Leeds (1879a, 1881) confirmed experimentally that both ozone and hydrogen peroxide are produced during the ozonation of purified air by moist phosphorus; along with these two bodies, and as a necessary part of the same series of reactions, incident originally to the setting free of nascent oxygen, a certain amount of nitrate of ammonia is invariably produced. Not very wrong, *Leeds* (Leeds 1879a, 1881, p. 8) communicated the equations $P_4 + 6O_2 = 2P_2O_5 + 2O$, $O + O_2 = O_3$, $H_2O + O = H_2O_2$, and $N_2 + 2H_2O + O = NH_4NO_3$. The latter $H_2O + O$ (but only via $O(^1D)$ and in the autoxidation of phosphorus $O(^3P)$ is gained only) is today regarded as the key reaction yielding OH radicals in atmospheric chemistry. The reaction $N_2 + O = NO + O$ is known today. However, the formation of ammoniacal salts only occurs due to the presence of gaseous NH_3 in the air. In laboratory experiments, this formation is a source of error not being inherent in the experiments themselves and is only remained to guard against it by very complete washing of the air drawn through the ozonator.

The formation of H_2O_2 while slow combustion of P_4 in moist air can now be explained via formation of phosphine PH_3 ($P + H_2O = PH_3$ + products) and the reaction $O + PH_3 = OH$ + products, followed by $OH + O = O_2 + H = HO_2$ ($+ HO_2 = H_2O_2 + O_2$) according to Davies and Thrush (1968).

Marcellin Pierre Eugène Berthelot (1827–1907),[1211] who studied the reaction between potassium permanganate and hydrogen peroxide, proposed as intermediate[1212] "*trioxide d'hydrogene*" HO^3 (today seen as H_2O_3); the formula of hydrogen peroxide (at that time in French named "*l'eau oxygéne*") he presents as HO^2 (Berthelot 1880). Some years later *Dmitrii Ivanovich Mendeleev* [Дмитрий Иванович Менделеев] (1834–1907)[1213] proposed the intermediate H_2O_4 (Mendeleev 1895, cited after Plesničar 2005); *Leopold Gräfenberg* (1878–1962)[1214] named it "*Ozonsäure*" [ozone acid] ($H_2O{\cdot}O_3$) (Gräfenberg 1902, 1903). Baeyer and Villiger (1902)[1215] concluded from the effect of ozone onto dry KOH, gaining an orange-brown colored salt, on the existence of potassium ozonate or tetroxide and thus concluded on ozone acid as hydrate of ozone:

1211 French organic and physical chemist, science historian, and government official. When he died in 1907, he was honored throughout the nation, with most French towns naming a street or a square after him.

1212 In analogy to the known trisulfide H_2S_3 (*Berthollet* writes HS^3).

1213 Russian chemist in Saint Petersburg; developed the modern periodic table.

1214 Born in Göttingen, he obtained a PhD degree based on experimental work "Contributions to the Knowledge of Ozone" done at the Göttinger Institute of *Walter Nernst*. Immigrated 1933 to Palestine.

1215 *Johann Friedrich Wilhelm Adolf* (since 1885 Ritter von) *Baeyer* (1835–1917) was a German chemist, professor for organic chemistry in Berlin (1866), Strasbourg (1866) and since 1875 in Munich; Nobel laureate in 1905. *Victor Villiger* (1868–1934) was a Swiss chemist, coworker of *Baeyer* and since 1905 at BASF.

$O_3 + H_2O = H_2O_4$. Bach (1902) termed it hydrotetroxide. Bach (1893) presented a curious idea on formation of H_2O_2 in the atmosphere, stimulated by *Schöne*'s finding (see p. 641) that obviously solar radiation is essential for its formation; he proposed in analogy to the CO_2 assimilation by plants the photocatalytic reduction of CO_2 in water, i. e., H_2CO_3, to formaldehyde and hypothetical peroxocarbonic acid HO–CO–O–OH [in old German "*Ueberkohlensäurehydrat*"], where the latter decays into CO_2 and H_2O_2 and the latter oxidizes into O_2 and H_2O.

Bach (1897) and Engler and Weissberg (1904) stated that the whole O_2 molecule is taken under the formation of a peroxide. Both statements, O_2 reduction and the O_2 transfer into H_2O_2, are amazingly correct. German chemist (Nobel prize in 1927) *Heinrich Otto Wieland* (1877–1957) and his student *Wilhelm Franke* (1903–?), after World War II professor for enzyme chemistry at the University Cologne, studied the kinetics of H_2O_2 formation in autoxidation processes (Wieland and Franke 1929). The Russian chemist *Valentin Alekseevich Kargin* [Валенти́н Алексе́евич Карги́н] (1907–1969) estimated the dissociation constant ($H_2O_2 \rightleftarrows H^+ + HO_2^-$), (Kargin 1929).

The bleaching and germicidal properties of H_2O_2 had been found very early and opened many industrial, wastewater treatment, and medical applications. However, unbalancing the oxidation potential could result in oxidative stress (Sies 1986). Thus Möller (1989) proposed increasing atmospheric H_2O_2 to be responsible for the new-type forest decline in Europe, which was first recognized in the second half of the 1970s.

4.5.2.2 Detection of hydrogen peroxide in the atmosphere

As mentioned on p. 631, *Osann* found in 1853 that each snowflake falling on a paper coated with potassium iodide is followed by a "*Schönbein* reaction". Which substance other than H_2O_2 could it be? This seems to be the first evidence for the presence of hydrogen peroxide in the atmosphere. It was soon known that the *Schönbein* paper (potassium iodide) also shows reactions with other oxidizing compounds in air, namely hydrogen peroxide (Houzeau 1868, Struve 1871). In contrast, thallium paper only shows coloration with hydrogen peroxide. It was also known that O_3 reacts very slowly with permanganic acid ($HMnO_4$) (in old German "*Üebermangansäure*") whereas H_2O_2 fast reacts, which was used for its analytical separation. Struve (1869, p. 319) stated that it remains a mystery why *Schönbein* never detected hydrogen peroxide in atmospheric precipitation in Basel (*Schönbein*'s "last news" with his finding was later published and only in 1871 known to *Struve*). Further, *Struve* (1869) noted that *Dumas* and namely *Frémy* assumed the presence H_2O_2 in the atmosphere for theoretical reasons, which surely *Houzeou* caused to his study. Thus Houzeau (1868), for the first searched the literature concerning proof whether H_2O_2 is present in the atmosphere or not, and did not find anything. During a starry night in June 1867, the test-paper showed "*grande avtivité chimique de l'air*", *Houzeau* collected 320 mL dew water that showed no reaction with respect to H_2O_2. As detection limit, he communicates 1/25000000, i. e., 0.04 mg L^{-1}. To prove whether the procedure of dew concentrate decomposes H_2O_2, he added 0.122 mg H_2O_2 to 100 cm^{-3} of remaining dew water, and after evaporation to 3 cubic centimeter he found a distinct reaction on H_2O_2. Therefore, *Houzeau* concluded on the absence of H_2O_2 in air.

Schönbein (1869) noted that he tried to detect hydrogen peroxide in rainwater for many years but was unsuccessful. However, he did not conclude on the absence of H_2O_2 in atmospheric air but explained it with the insufficient sensibility of his reagents and had no doubt on its permanent presence in air. On 21 June 1868, he detected it for his first time in rainwater from a strong thunderstorm, and further on 5 July, and all following events. *Schönbein*, knowing that H_2O_2 decomposes in water, was not surprised to find no H_2O_2 after some storage and a weak reaction in later rain. Further, he stated that the atmosphere contains changing quantities of H_2O_2 according to the (electrical) source strength and believed this source to be more important than the earth's surface with its oxidation processes.

Hagenbach[1216] communicated in a footnote (Schönbein 1869, p. 270)[1217] that *Houzeau* and *Schönbein* did not know *Meissner*'s work from 1863. Like *Schönbein*, *Georg Meissner* believed (Meissner 1863a, pp. 335–336) that H_2O_2 must be found in rainwater from thunderclouds, and in summer 1863 in Göttingen he detected H_2O_2 with no doubt in fresh fallen rainwater from thunder,[1218] and also in "common" rainwater, but in smaller amounts (Meissner 1863b). He further found that the shower contains more H_2O_2 than drizzle does. With respect to the known fact that H_2O_2 slowly decomposes in water, *Meissner* stated that it is necessary to detect it soon after collection, whereas the next day, no H_2O_2 in the rainwater is longer found, with the exception of a thunderstorm; he writes (Meissner 1863b, p. 267):

> [...] man kann Gewitterregenwasser geradezu wie eine reine verdünnte Wasserstoffsuperoxydlösung zu Versuchen benutzen [... one can use rainwater from thunderstorm really like a pure diluted solution of hydrogen peroxide for experiments].

Meissner further notified that he has no doubt to find H_2O_2 also on snow but experiments to conduct in next winter obviously he never realized (Fig. 4.58).

Heinrich Wilhelm von Struve, who conducted in 1869 investigations on the pollution of the river Kur in Tbilisi, also analyzed rain and snowfall concerns nitrite and nitrate. On the presence of an oxidizing agent, he concluded after his observation that

1216 *Eduard Hagenbach-Bischoff* (1833–1910) was a Swiss physicist in Basel.

1217 This is nice historic example (and well-known today) that pioneering scientific results published in curious journals (here "*Göttinger Nachrichten*") are hardly recognized by contemporaries. The language was not the problem – all scientists were able in that time to read German, English, French, and Latin.

1218 Zuo and Deng (1999) believed to be the first authors (!) to establish lightning induced H_2O_2 production in a Florida thunderstorm.

Ueber die Bestandtheile des Regenwassers.

Von G. Meißner.

Ich habe in meinen „Untersuchungen über den Sauerstoff" Seite 335 und 336 mit Bezug auf die im 1. Abschnitt des Buches mitgetheilten Versuche die Gründe entwickelt, aus denen ich schließen mußte, daß das aus Wolken herabfallende Wasser, besonders bei starken elektrischen Spannungen, wie beim Gewitter, Antozon, also Wasserstoffsuperoxyd enthalten würde, und ich durfte um so weniger bezweifeln, daß sich diese aus meinen Untersuchungen resultirende Forderung erfüllt finden würde, als jedes Regenwasser nach Liebig's Beobachtungen Salpetersäure enthält, diese aber nicht anders entstanden sein kann, als in der Weise, wie in Cavendish's bekanntem Versuch, die Analyse dieses Versuches aber ergibt, wie ich a. a. O. S. 129 und f. ausgeführt habe, daß zur Bildung von Salpetersäure aus dem Stickstoff der Luft die Verbindung mit Ozon und Antozon zugleich stattfinden muß. Während mir damals keine eigene Beobachtungen über die Bestandtheile des Regenwassers zu Gebote standen, habe ich im Laufe dieses Sommers diese Lücke ausgefüllt und in der That meine Vermuthung über die Gegenwart des Wasserstoffsuperoxyds in dem Wasser der Wolken vollkommen bestätigt gefunden.

Figure 4.58: First page of "*Ueber die Bestandtheile des Regenwassers*" [On the constituents of rainwater], Meissner (1863b). The first evidence of atmospheric hydrogen peroxide.

nitrite soon disappeared in the precipitation samples after collection. Now he studied systematically the precipitation and found H_2O_2 in 15 events between January and May 1869. Struve (1869) found (according to the strength of the detection reaction) that snow[1219] contains more H_2O_2 than rain does, but rain from thunderstorm contains more than common rain. Struve (1871) first communicated that he now became familiar with the work of Meissner (1863b) due to late access to literature in his laboratory in Tbilisi ("far in Asia" he writes) and apologizes for claiming the first finding of H_2O_2 in the atmosphere in 1869. From his investigations, he concluded on the close relation between ozone, hydrogen peroxide, and ammonium nitrite in the atmosphere where (Struve 1871, p. 294)

> [...] unter dem Einfluss electrischer Entladungen, vielleicht auch anderer Ursachen, Verbrennungs-Erscheinungen vor sich gehen, deren Resultate wir in der Luft selbst und in den atmo-

1219 However, *Schöne* found a few years later that snow contains less H_2O_2 than rain (as all modern investigations also did show).

sphärischen Niederschlägen nachweisen können [... under the influence of electric discharges, likely also other causes, combustion phenomena occur, and their results we detect in the air and atmospheric precipitation].

In the following publication, Struve (1872, p. 28) was able to determine in one rainwater sample 0.46 mg L^{-1} H_2O_2 and 1.15 mg L^{-1} HNO_2. The German chemist *Werner Schmid* (unknown living dates) found in rainwater in Breslau, showing hydrogen peroxide acid reaction (Schmid 1869). Goppelsroeder (1871, p. 143), who listed several analytical methods of doubtless detection of H_2O_2, detected it in freshly fallen rainwater and noted that negative observations might be explained through catalytic decomposition of H_2O_2

[...] dass gewisse unorganische und organische Substanzen das Wasserstoffsuperoxyd zu katalysieren vermögen [... that certain inorganic and organic substances catalyze hydrogen peroxide].

The German chemist *Emil Schöne* (p. 516 and Fig. 4.31) was the first to state that H_2O_2 is a permanent natural constituent of our atmosphere. His statement that the existence of H_2O_2 in air is due to the influence of solar radiation (onto water) was later proven through observations by Thiele (1907), Tian (1911), Chlopin (1911), and Kernbaum (1911). These authors' experiments show that the ordinary moist air of a room, after subjection for a few minutes to the action of ultra-violet light, shows the presence of ozone, hydrogen peroxide, and nitrogen trioxide. Hence the formation of H_2O_2 during electric discharges in air and water vapor was established at the beginning of the twentieth century, which explained its excess occurrence in rain from thunderstorm. However, the formation mechanisms were still unknown.

Emil Schöne was the first scientist who studied (1874–1875) systematical atmospheric H_2O_2 in detail in rain, snow, and air near Moscow at Petrowskoje Rasumowskoje, an agriculture research station (Schöne 1874, 1878, 1893, 1894a).[1220] *Schöne* also extensively studied analytical methods to determine H_2O_2 (Schöne 1875, 1880a) and investigated the reactions of hydrogen peroxide with several substances in a series of five articles 1878–1879.[1221] Beginning with referring previous researchers[1222]

1220 Furthermore, *Schöne* wrote the following publications in Russian on atmospheric ozone and hydrogen peroxide: "К вопросу нахождения в атмосферном воздухе и атмосферных осадких перекиси водорода" [On the question of the presence of hydrogen peroxide in atmospheric air and atmospheric precipitation]. Moscow, 1893, 18 pp. "О законностях в колебаниях количеств атмосферной перекиси водорода" [On regularities of the variation of the amount of atmospheric hydrogen peroxide]. Moscow, 1881, 49 pp. "Опытные исследования над перекисью водорода" [Experienced investigations on hydrogen peroxide]. Moscow, 1875, 162 pp.

1221 Experimentaluntersuchungen über das Wasserstoffhyperoxyd. *Annalen der Chemie* **192** (1878) 257–285, **193**, 241–297, **195** (1879) 228–252, **196**, 58–74, and 239–258.

1222 *Schöne* wrote wrongly *Schmidt* (instead of *Schmid*) what was likely the reason that Schmid (1869) was wrong referred by all following researcher as Schmidt (1869).

(*Meissner*, *Schönbein*, *Struve*, *Schmid*, and *Goppelsröder*) who detected H_2O_2 in rain-water (*Struve* also in snow) and *Houzeau* who was unable to detect it in rain, dew, or snow, Schöne (1874, p. 1693) first asked the questions to be answered with investigations

- it is present dissolved only in precipitation or also as vapor,
- what is the relation between its content, the meteorological phenomena, time of day and year,
- how it is related to ozone,
- how it is produced in the atmosphere,
- What it is impact after removal to the earth's surface on the geology and vegetation,
- if it exists as vapor, what it is impact on respiration,
- is there a hygienic relevance.

Schöne collected 387 samples (215 rain and 172 snow samples) between July 1874 and June 1875; additional, he observed meteorological data (in time intervals between 30 min and some hours), and used *Schönbein*'s ozonometer (since August 1874). He found only 86 snow and seven rain samples without detection of H_2O_2. However, he noted that likely in the 86 snow samples traces of H_2O_2 were present but so low that he did not include these samples in his quantitated elaboration. The rain samples without detection of H_2O_2 were either only analyzed more than 12 h after sampling or were events together with dense fog or dust, thus H_2O_2 may be decomposed, he notified. *Schöne*'s method for quantitative analysis of H_2O_2 is based on the slow precipitation of iodine from a neutral potassium iodide solution with colorimetric determination, described in Schöne (1875). His procedure seems to be very reliable, making H_2O_2 comparison standard solutions between 0.1 and 1.0 mg L^{-1} every two weeks new. *Schöne* comments that nitric acid and other oxidizing substances that immediately precipitate iodine do not disturb his determination. He also not claimed absolute trueness of the values but notified that the conclusions he has drawn could be seen to be reliable. As deposition gauge, he used glass funnels with a square of 0.4425 m^{-2}, and 1.8 m above ground fixed in a holder where the precipitation is collected in large glass bottles, permanently cleaned. Immediately after collection (3–5 hours and shorter), the sample was filtered and analyzed. The H_2O_2 concentrations determined to range between 0.04 and 1.0 mg L^{-1}. *Schöne* (to my knowledge) first used besides arithmetic averages also precipitation-weighted means. Snow he collected in large porcelain bowls. From his precipitation investigations *Schöne* noted the following remarkable observations (Schöne 1874, 1878); Fig. 4.59:

- in showers the H_2O_2 concentration was higher than in drizzle,
- in rain from southern air masses H_2O_2 was higher than in rain from polar air masses,

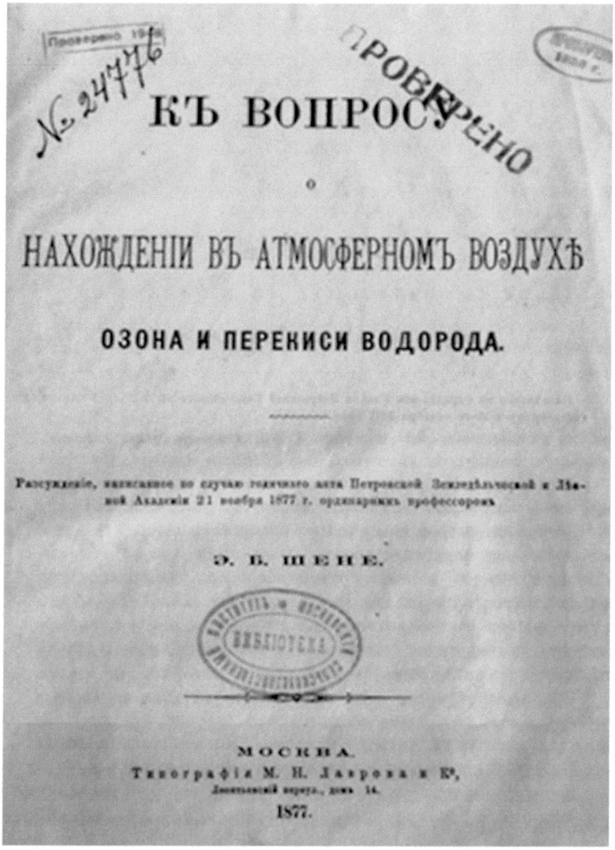

Figure 4.59: Cover of the publication by *Emil Schöne* "К вопросу нахождения в атмосферном воздухе озона и перекиси водорода. Рассуждение, написанное по случаю годичного акта Петровской земледельческой и лесной академии" [On the question of the presence of ozone and hydrogen peroxide in atmospheric air. Written on the occasion of the annual act of the Petrowski Agricultural and Forest Academy]; Tipografiya M. N. Lavrova, Moscow, 1877, 22 pp.) Source: website of the Russian State Agricultural University: https://www.timacad.ru/news/oni-byli-pervymi-180-let-so-dnia-rozhdeniia-emilia-shione.

- the H_2O_2 concentration shows a distinct seasonal variation with maximum in summer and minimum in winter, and
- in snow less H_2O_2 is found than in rain during the same season.

Fig. 4.60 shows the annual variation of the content of H_2O_2 in precipitation and the "ozone reaction" from *Schönbein*'s paper. Impressive is the anti-cyclic (wrong as discussed in Chapter 4.4.5.2) variation of ozone. It is remarkable that *Schöne* not concluded that the very low H_2O_2 content in winter is due to the occurrence of snow and not rain (he found in hail in summer a high H_2O_2 content), but due to the season. On

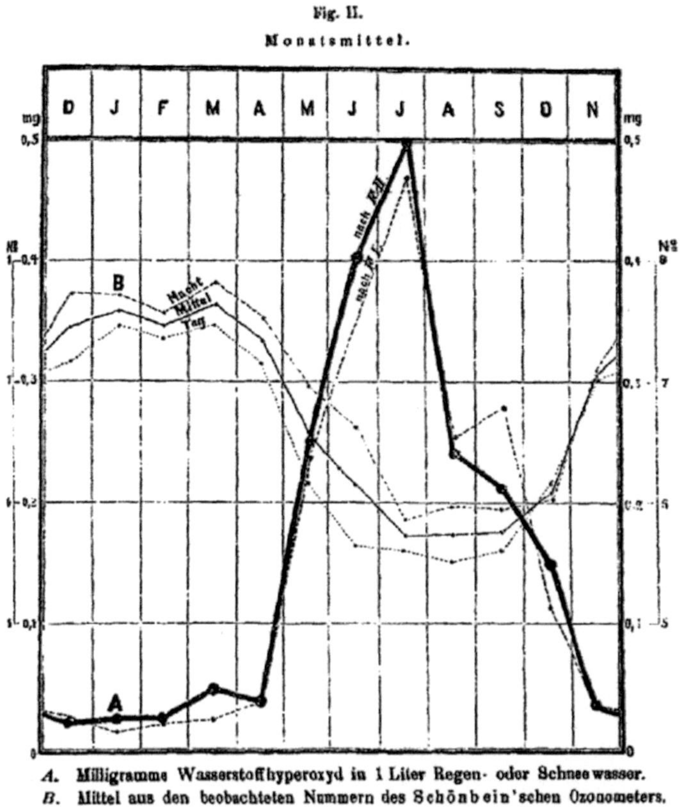

Figure 4.60: Monthly means (arithmetic means dotted line and weighted means solid line) of H_2O_2 in precipitation (mg L^{-1}) and of ozone from *Schönbein*'s paper (curves at night, mean and day from top to bottom); Schöne 1878, p. 489.

the other hand, he notified the role of the phase transfer, i. e., that rain scavenge gases but snow less or not.

In dew[1223] and rime, he never detected H_2O_2. However, to investigate whether H_2O_2 occurs vaporous in air he used artificial dew samples, with ice filled large beakers placed in a porcelain bowl, and the condensed water from the outer wall of the beaker sampled. He named this sampler "*Condensationsapparat*" [condensation apparatus], principally a method which was used 60 years later by *Hans Cauer* (Fig. 2.20 on p. 257). Between July and October, 47 daytime and 16 nocturnal samples he collected. In nighttime samples, he found only traces of H_2O_2 around 0.05 mg L^{-1}, and the maximum afternoon up to 0.4 mg L^{-1}, showing a diurnal variation as known today. Furthermore,

1223 Schöne (1878, p. 875–876) argued that dew formation occurred during the night when the atmospheric H_2O_2 content is naturally very low and additionally, emanated substances from the ground may decompose it.

he found that the content of H_2O_2 decreases in artificial dew from July to October: 0.40, 0.35, 0.15, 0.09 (monthly means in mg L^{-1}). The content of H_2O_2 in artificial dew (equivalent to the atmospheric content of gaseous H_2O_2) increases

- with increasing temperature and
- less cloudiness.

Rain and fog decrease the amount of H_2O_2 in the artificial dew immediately below the detection limit; he writes in modern terms as nowadays (Schöne 1874, p. 1707):

> Der Regen wäscht eben aus der Luft das in ihr dampfförmig enthaltene Wassserstoffhyperoxyd aus [The rain washes out the hydrogen peroxide that is present in the air as vapor].

Like *Cauer* he calculated the content of gaseous H_2O_2 in air, assuming that its total amount in a volume of air is deposited together with the condensed water vapor. For the period 10:30 to 14:30 local time on 8 July 1874, he determined in this way 0.407 µg m^{-3} or 0.268 ppb; values as measured today. Finally, he concludes from his investigation (Schöne 1874, p. 1708)

> [. . .] dass bei der Entstehung des atmosphärischen Wasserstoffhyperoxyds das Sonnenlicht eine hervorragende Rolle spielt [. . . that sunlight plays a singular role in the formation of atmospheric hydrogen peroxide]

Schöne (1878, p. 561) wrote that the best way of determination of H_2O_2 in the air would be using an aspirator method, but he was unsuccessful in doing it; thus, he continued his condensation measurements in 1875 and reports on the results between July 1874 and July 1875 (arithmetic means by the author):

Summer (n = 120)	0.128 ± 0.017 mg L^{-1} in dew, corresponding to 0.37 ± 0.04 ppb in air.
Autumn (n = 18)	0.057 ± 0.011 mg L^{-1} (no reliable air value due to the low humidity)

The diurnal variation of the H_2O_2 values from the condensation experiments shows exactly what is found 120 years later: a maximum around 4 pm (0.6 ppb) and a minimum between 8 pm and 4 am (0.2 ppb) as mean values from all measurements (Schöne 1878, p. 565). The diurnal variation *Schöne* (Schöne 1878, p. 1028) explains by the changing budget: maximum at late afternoon occurs when source = sinks, whereas as source he regarded the sunlight and as sink the chemical decomposition, a correct view (nothing was known at that time about dry deposition and the vertical mixing of the lower atmospheric layer). In summary, *Schöne* (Schöne 1878 p. 1029) found an average of 0.38 ppb with a maximum of 1.4 ppb in the atmosphere over the measurement period. Despite this low concentration, Schöne (1874, p. 1030) argued that its role in natural processes can be huge and challenge other scientists for further systematic and routine measurements at different locations, e. g., in the tropics and at meteorological observatories such as Montsouris in Paris.

Sergius (or *Serge*) *Kern* [Сергей Керн][1224] (unknown living dates) was a Russian chemist working for Obouchoff Steel Works near (12 km north) St. Petersburg who used *Schöne*'s method to determine H_2O_2 in rain between June and September 1878 (Kern 1878); he found values between 0.1 and 1.4 mg L^{-1}, and less with polar wind and more with equatorial wind. He notified that "the occurrence of H_2O_2 in air may be very important for vegetal and animal existence". Schöne (1878, p. 492) comments the paper by *Kern* in a footnote that *Kern* found an average from the four-month observation to be 0.36 mg L^{-1} close to the mean from the same period 1874 at Moscow being 0.33 mg L^{-1} and wondered why *Kern* wrote "[. . .] indeed found a difference between the occurrence of H_2O_2 in rains of St. Petersburg and Moscow".

While studying different combustion processes, Hungarian chemist *Ludwig* (*Lajos*) *Ilosvay von Nagy-Ilosva* (1851–1936) did not find ozone (with the exception while slow combustion of phosphorus) and hydrogen peroxide but always higher oxides of nitrogen, which forms nitrous and nitric acid in the presence of water (Nagy-Ilosva 1889). Thus, he was motivated to prove the presence of O_3 and H_2O_2 in the air of Budapest but believed that the "detection reaction" only shows the presence of nitric acid. *Nagy-Ilosva* argued that atmospheric nitric acid shows all the characters as found from meteorological ozone observations, such as higher levels at daytime than at night, in spring than in summer, and an increase with high above the ground. In response to *Schöne*, who criticized *Nagy-Ilosva*'s view, Nagy-Ilosva (1894) published a long dispute that hydrogen peroxide in rainwater (and ozone in air) cannot be detected with the accepted reagents because they see nitrous acid, and stated that the presence of ozone and hydrogen peroxide in the atmosphere is not proven. Still, only that of higher oxides of nitrogen like nitrogen dioxide (N_2O_2) and nitrogen peroxide or hyperoxide (N_2O_4), acting as oxidizing substances (the formulas are according to *Nagy-Ilosva*). On the other hand, Nagy-Ilosva (1894) correctly found that nitrite in water is completely oxidized to nitrate after a month. Schöne (1893, 1894a) completely rejected *Nagy-Ilosva*'s view but agreed that together with hydrogen peroxide also organic peroxides could be determined. Schöne (1893, p. 3018) cites Chairy (1884), who was unable to detect hydrogen peroxide in rain in Algiers with the comment that (such as for *Houzeau*) the reagent was not sensitive enough. It is worth citing the following phrase from his argumentation with *Nagy-Ilosva* because it is actual until present (Schöne 1893, p. 3019):

> Wir kennen nicht den Ursprung von Tausenden von in der Natur vorkommenden Dingen. Niemand bezweifelt indessen ihre Existenz, sobald die letztere nur durch sichere Mittel nachgewiesen ist. Auch die Frage über die Existenz des atmosphärischen Ozons und Wasserstoffhy-

1224 *Kern* claimed in 1877 to have perceived a new element of the platinum group, called Davydum (in honor to *Davy*) later identified a mixture of iridium, and rhodium.

> peroxyds wird nicht entscheiden durch Betrachtungen über mögliche Quellen desselben, sondern durch exacte Methoden ihres Nachweises in der Luft [We do not know the cause of thousands of things occurring in the nature. Nobody doubt its presence once the latter is detected by reliable means. Also, the question on the presence of ozone and hydrogen peroxide will not be decided by meditations about possible sources but only through exact methods of its determination in air].

The findings by *Emil Schöne* have been included in older textbooks of chemistry (e. g., Smith-D'Ans 1948, p. 210, Jander and Spandau 1952, p. 39): the latter writes (translated from German):

> Hydrogen peroxide can be proven in traces in rain and snow; obviously it is produced in the atmosphere from water and oxygen under the action of ultra-violet light.

After *Schöne*'s investigations, more than a half century later, the first who detected H_2O_2 in atmospheric precipitation was *Helmut Mrose;* Mrose (1949) reported that he detected it at Hoher Sonnblick observatory (3106 m altitude) in snow, fallen daytime but not at night, and concluded that its formation needs solar radiation. Unfortunately, *Mrose* not communicate any detail of this investigation which was very likely done before 1944. Since the late 1930s, *Mrose* was interested in the question of condensation nuclei produced from artificial pollutants (sulfates), so that he was motivated to search after oxidants in the atmosphere.

However, the Israeli chemist *Michael Ardon* (1928–2006; born in Berlin) still doubt in the 1960s on its occurrence due to the instability of H_2O_2 and that the analytical methods used were not specific for peroxides and could easily mistake ozone or nitric oxides for H_2O_2 (Ardon 1965, p. 69).

4.5.3 Understanding hydrogen peroxide chemistry in the atmosphere

After *Thénard* discovered hydrogen peroxide at the action of acids on barium peroxide (BaO_2) in 1818, for about the next half century, it was seen as "oxidized water", namely by *Schönbein*'s view of peroxides, simply due to its formula H_2O_2 to water (H_2O) + oxygen (O); remember that in the early nineteenth century the formula for water was seen as HO, thus HO + O = HOO. *Weltzien* and *Brodie*, and finally *Traube* has shown that hydrogen peroxide is "reduced oxygen". The natural occurrence of H_2O_2 in atmospheric precipitation in the air, in plant and animal organisms, was proved in the second half of the nineteenth century. Its formation mechanism, however, remained for many years in the darkness, although already around 1870 its formation was proved during combustion process in air (*Struve*, *Böttger* and others); with the ending nineteenth century, it became clear that almost all autoxidation processes yield peroxides either as intermediates or even final product (*Traube*, *Engler*, *Bach* and others). The Dutch chemist *Boeke*[1225] stated from own experiments that hydrogen peroxide is gained as a byproduct (with the exception when dry oxygen is used) in all ozone formation processes. The German chemist *Alfred Rieche* (1902–2001) wrote (Rieche 1936, p. 5):

> Die Einwirkung des Luftsauerstoffs auf die verschiedensten Elemente und Verbindungen führt zu Peroxyden [The impact of oxygen onto different elements and compounds leads to peroxides].

Hydrogen peroxide is rather stable in the tropospheric gas phase. It is very water-soluble, hence scavenging (apart from dry deposition) is the ultimate sink of gaseous H_2O_2. In the aqueous phase, H_2O_2 is decomposed mainly by sulfite, but also nitrite; thus, the primarily gaseous pollutants SO_2 and NO_x control H_2O_2 levels. Self-decomposition of H_2O_2 was observed soon after its discovery. An almost complete understanding of the H_2O_2 chemistry has been reached some 170 years after *Thénard*, in following briefly described to get a better view on the history. In the atmospheric gas phase, only

$$HO_2 + H_2O \rightleftarrows [\,H_2O\text{-}HOO] \xrightarrow{HO_2} H_2O_2 + H_2O + O_2$$

is important, thus the HO_2 radical is the essential precursor. But there are competing reactions:

$$\begin{aligned} HO_2 + RO_2 &= O_2 + ROOH \quad \text{(organic peroxides)},\\ HO_2 + NO &= OH + NO_2 \quad \text{(most important), and}\\ HO_2 + O_3 &= OH + 2\,O_2. \end{aligned}$$

There are several sources of HO_2:

$$\begin{aligned} H + O_2 &= HO_2 \quad \text{(ultimate)},\\ HO + O_3 &= HO_2 + O_2\\ &\quad \text{(in competition with numerous reactions OH + X), and}\\ O_2 + RO\ \text{(alkoxy radical)} &= HO_2 + \text{aldehyde}\\ &\quad \text{(special situation: } HCO + O_2 = CO + HO_2). \end{aligned}$$

Alkoxy radicals gained after the attack of OH on hydrocarbons (RH) via the radical chain

$$RH + OH \xrightarrow[-H_2O]{} R \xrightarrow{O_2} RO_2 \xrightarrow{NO} RO + NO_2.$$

1225 Wrong written name as *Boehe* in *Berichte* p. 439 (correction on p. 579 to *Boeke*), and also wrong cited as *Böhl* in Kausch 1938, p. 26). Likely *Jacob Boeke* (1838–1908) who was director of *Rijks Hoogere Burgerschoolte* Alkmaar and brother-in-law of the chemist *Jacob Maarten van Bemmelen* (1830–1911).

H atoms form in the troposphere via $CO + OH = CO_2 + H$ and photolysis of aldehydes ($RC(O)H + h\nu = RCO + H$). Thus, high concentrations of aldehydes from motor exhaust in photochemical smog situation explain the very large H_2O_2 levels measured. Hence in a "pure" atmosphere, H_2O_2 is formed via O_3 photolysis;

$$O_3 \xrightarrow[-O_2]{h\nu} O(^1D) \xrightarrow{H_2O} 2\,OH \xrightarrow[-2\,O_2]{2\,O_3} 2\,HO_2 \xrightarrow[-O_2]{} H_2O_2$$

The presence of carbon monoxide, nitrogen oxides, and hydrocarbons can increase the atmospheric H_2O_2 level but also inhibit its formation. The slow photolysis of H_2O_2 (to OH) plays only in the upper troposphere a role; the fate of H_2O_2 is thus scavenging in clouds and rain and dry deposition. Like for the gas phase, also in the aqueous phase HO_2 and its dissociation product O_2^- (superoxide) is responsible for the formation of H_2O_2, most likely via an electron transfer:

$$HO_2 + O_2^- \xrightarrow[-O_2]{} HO_2^- \xrightarrow{H^+} H_2O_2.$$

Scavenging from the gas phase is the source of aqueous-phase HO_2 but also (less important) the above-mentioned similar OH radical oxidation of dissolved organic compounds. Most important, however, is the photocatalytic reduction of oxygen to O_2^-. Therefore, the cloud water phase could be a source of H_2O_2 after the evaporation of droplets. Precipitation removes H_2O_2 from the atmosphere. But sulfite (dissolved SO_2) and to a much less percentage nitrite (dissolved nitrous acid and aqueous-phase formation from NO_2) decomposes H_2O_2 in the aqueous phase.

4.5.3.1 Gas-phase chemistry

During the 1930s, the radical gas-phase H_2O_2 formation mechanism was described in combustion processes (Hinshelwood and Williamson 1934, Jost 1939/46, Kondratjew 1958)[1226] where peroxide radicals are formed via

$$H + O_2 = HO_2,$$

1226 Sir *Cyril Norman Hinshelwood* (1897–1967) was a British physical chemist and a Nobel Prize laureate. *Friedrich Wilhelm Jost* (1903–1988) was German physical chemist, assistant of *Bodenstein* in Berlin (1926–1929), at MIT (1932/33) in Cambridge, professor in Hannover (1935), Leipzig (1937), Marburg (1943–1953) and Göttingen (since 1953). *Victor Nikolajevich Kondratiev* (German transcription: *Wiktor Nikolajewitsch Kondratjew* [Виктор Николаевич Кондратьев] (1902–1979) was a Soviet physicochemist and well-known for studying combustion chemistry and gas reaction kinetics. Many Soviet scientists studied gas-phase kinetic, almost unknown to the Western world; see Kondratjew (1958) for references.

first proposed by Bates (1933)[1227] and likely independently by Bodenstein and Schenk (1933), who furthermore proposed the reaction

$$HO_2 + HO_2 = H_2O_2 + O_2.$$

Marshall (1926) first studied the formation of H_2O_2 in the photochemical oxygen-hydrogen reaction and its photochemical formation (Marshall 1932). The photodissociation of H_2O_2 into two OH radicals was first shown by Urey et al. (1929) and in detail studied by Volman (1949), who proposed the following reaction scheme:

$$H_2O_2 + h\nu = 2\,OH, \quad OH + H_2O_2 = H_2O + HO_2 \quad \text{and} \quad HO_2 + HO_2 = H_2O_2 + O_2.$$

Haagen-Smit (1952) is the first who reported on the basis of the historic peroxidase-guaiac test on the presence of hydrogen peroxide in the air of the Los Angeles basin. Leighton (1961), who cites *Haagen-Smit*, only describes the photochemistry (photolysis) of peroxides. Junge (1963a) does not mention hydrogen peroxide in his book. Heicklen (1976) mentioned the formation of H_2O_2 as the termination of radical chains (HO_2 removal) and that the photolysis of aldehydes is important for the formation of H and thus HO_2; he writes: "that we (1975) do not have good estimates of the photolysis coefficient".[1228]

The photodissociation of H_2O_2 was first studied by Elbe (1933) and further investigated by Molina et al. (1977) and others. The kinetic of $HO_2 + HO_2$ was first estimated by Foner and Hudson (1962); Hamilton (1975) found that the rate depends form the H_2O vapor, and Hamilton and Naleway (1976) proposed the formation of a complex between HO_2 and H_2O that react fast again with HO_2, later confirmed by Calvert and Stockwell (1983). The self-reaction of the HO_2 radical is the major source of hydrogen peroxide in the atmospheric gas phase and an important radical termination reaction. It was well established in the early 1980s that the rate of this reaction is sensitive to both pressure and water vapor concentrations. Stockwell (1995)[1229] has shown that models which omit these terms will predict incorrect HO_x budgets, underestimate gas

1227 *John R. Bates* (unknown living data) was an American chemist at Johns Hopkins and the Michigan University. *Bates* published 1927 together with *Hugh Stott Taylor* on photosensitization at the Princeton University. *Taylor* begun his academic career in 1914 at Princeton. *Bates* published several articles on photochemistry and kinetics of gas-phase reactions in the 1930s.

1228 *Robert F. Hampson*, Jr. (?–2013) and *David Garvin* (1923–?) from the US National Bureau of Standards in Washington investigated "*Reaction and photolysis rates of interest for atmospheric chemistry*" and published them in several special publication since 1973. Since 1980 the Jet Propulsion Laboratory (U. S.), NASA Panel for Data Evaluation, published "*Chemical kinetics and photochemical data for use in stratospheric modeling*".

1229 *William R. Stockwell* (born 1953), American professor emeritus, well-known atmospheric chemist for his contributions to developed of Eulerian regional air quality models and Atmospheric Chemistry Mechanism.

phase hydrogen peroxide formation rates, and overestimate ozone and organic peroxide formation rates especially in clouds.

In the 1980s, the formation of OH and HO_2 during the ozonolysis of alkenes has been discovered (e. g., Gäb et al. 1985, Becker et al. 1993). In the reaction of the stabilized *Criegee* radical with water vapor, directly H_2O_2 can be formed (Sauer et al. 1999).

4.5.3.2 Aqueous-phase and interfacial chemistry

Already *Thénard* had recognized that one O is only weekly-bonded in H_2O_2 and that the molecule easy decomposes (Thénard 1818).[1230] Remarkably soon after its discovery, it was known that H_2O_2 oxidizes sulfurous acid into sulfuric acid without the formation of free oxygen (Gmelin 1827, p. 240), a mechanism recognized to be important in air chemistry almost 150 years later by Hoffmann and Edwards (1975), Penkett et al. (1979) and Möller (1980) as the most important pathway in the oxidation of dissolved SO_2 in hydrometeors. Mader (1958) was the first to study the kinetics of this process. But one of the most important findings, and finally accepted among the scientists, was the fact that hydrogen peroxide is produced in natural waters under the influence of light.

Schönbein (1861a) first observed the formation of H_2O_2 while slow oxidation of metals in air and oxygen.[1231] *Traube* assumed (in contrast to *Schönbein*, who stated the dissociation of the oxygen molecule) that the O_2 is reduced to H_2O_2 via water degradation (Nasse 1870, p. 206). Schöne (1878, p. 254) writes that the relationship between O_3 and H_2O_2 is expressed by the equation $O_3 + H_2O_2 = 2\,O_2 + H_2O$, soon later confirmed by the British chemist *Herbert McLeod* (1841–1923), who also writes that the decomposition is rapidly in solution, and both reactants loss one atom of oxygen (McLeod 1880). Traube (1893, p. 1479) correctly concluded that – if all three atoms in the ozone molecule would be in the same state – it should destroy three molecules of hydrogen peroxide: $O_3 + 3\,H_2O_2 = 3\,O_2 + 3\,H_2O$.[1232] The formation of H_2O_2 in the process of aqueous-phase ozone decomposition was controversially discussed since *Schönbein*; first Berthelot (1880)[1233] proposed as intermediate H_2O_3 and some years later Mendeleev (1895) H_2O_4 and Gräfenberg (1902, 1903), who named it ozone

---

1230 This (wrong) statement is based on the often generally accepted reaction scheme $A + H_2O_2 \rightarrow AO + H_2O$. However, H_2O_2 cannot transfer O (like O_3); the only way is decomposing according to $H_2O_2 + e^- + H^+ \rightarrow OH + H_2O$; in a subsequent reaction, and hence OH is the oxidizer.

1231 The "hydrogen peroxide theory" while rusting of iron was established by Dunstan et al. (1905).

1232 Today we know that such reaction does not occur and both ozone and hydrogen peroxide decay separately but also react with reaction products like O_2^- and OH.

1233 *Berthelot* studied the reaction between potassium permanganate and H_2O_2 in strong acid solution, writing $Mn^2O^7 + 5\,HO^2 = 2\,MnO + 5\,HO^3$ (Berthelot 1880, p. 659), a very complex reaction, He recognized the odour of ozone and writes $O_2 + O = Oz$ (note Oz was the French formula for O_3), proposing that intermediary O exist. See also p. 468.

acid ($H_2O{\cdot}O_3$) and still speculated by Marshall and Rutledge (1959) and Kobozev et al. (1959); see also p. 638.

There are very few experiments in the nineteenth century which prove the formation of hydrogen peroxide in terms of interfacial chemistry, i. e., in liquids exposed to sunlight and oxygen. *Arthur Richardson* (1858–1912), a British physicochemist, proved the formation of H_2O_2 in urine, exposed to sunlight and oxygen (Richardson 1893); he writes (Richardson 1893, p. 1110):

> While making some experiments to determine the conditions under which oxidation was brought about by light, I noticed that in the case of certain organic compounds, hydrogen peroxide was first formed, and that it acted as a carrier of oxygen from the air to the liquid oxidized.

Adolf Dieudonné (1864–1944), a German physician who studied the influence of light on water sterilization, detected the formation of hydrogen peroxide (with potassium iodide in diluted ferrous sulfate) when exposing agar plates in air, but not in gases free of oxygen (Dieudonné 1894); he concluded from his observations that in each natural water under the influence of light H_2O_2 is gained and named it "self-cleaning of rivers"; he writes (Dieudonné 1894, p. 540):

> Die Bildung von H_2O_2 durch Luft und Licht [. . .] Die Selbstreinigung der Flüsse [. . .] daß auch im Wasser unter dem Einflusse des Lichtes H_2O_2 gebildet wird [. . .] [The formation of H_2O_2 through air and light [. . .] the self-cleaning of rivers [. . .] that also in water under the influence of light H_2O_2 is produced . . .]

Both studies support photocatalytic formation according to the chain $O_2 \rightarrow O_2^- \rightarrow HO_2 \rightarrow H_2O_2$. An indirect evidence Wilson (1899) presents through experiments on fog formation in vessels containing moist pure oxygen under the influence of ultra-violet light, suggesting the correct gross equation O_2 (via O_3) $+ 2\,H_2O \rightarrow 2\,H_2O_2$, which proceeds according to today's knowledge via $O_2 \rightarrow O_3 \rightarrow OH \rightarrow HO_2 \rightarrow H_2O_2$. *Ralph Francis D'Arcy* (1864–1940), a British chemist in Cambridge, who also studied the photochemical decomposition of hydrogen peroxide (D'Arcy 1902), writes[1234] (D'Arcy 1902, p. 324):

> The formation of hydrogen peroxide in nature is a well-recognized fact.
> The main ideas of the present paper are to suggest that hydrogen peroxide when split up yields two parts which are oppositely charged, and that in nature this splitting up may be brought about by the action of light.
> The water formed by the decomposition being positively charged, and the oxygen (whatever its atomicity may be) being negatively charged.

Although *D'Arcy* cites *Downes* and *Blunt* with the pointed remark "the author was unable to find any account of a more varied examination of this effect" (D'Arcy 1902,

1234 *D'Arcy* cites *Wilson* (1899) who indicated the probable formation of hydrogen peroxide by ultraviolet light in moist oxygen (see p. 324).

p. 43), he nothing else contribute to the decomposition of H_2O_2 then it disappears within the range 30–70 % by exposure to intensive sunlight over the day. However, the phrases cited above corresponds to a hypothetical mechanism $H_2O_2 + h\nu \rightarrow H_2O^+$ ($\rightarrow H^+ + OH$) + O^- ($+ H^+ \rightarrow OH$) = 2 OH, or, in another more reliable formulation adopting *Baur*'s model of "molecular electrolysis" (see Chapter 1.3.5.3) $H_2O_2 + h\nu \rightarrow [HO^{\oplus}{-}^{\ominus}OH] \rightarrow HO^+$ ($\rightarrow H^+ + O$) + HO^-, corresponding to the gross equation $H_2O_2 = H_2O + O$. The first mechanism is equivalent to the photolysis and the latter to photosensitization (photocatalytic decomposition).

An almost forgotten investigation by Downes and Blunt (1879b) has shown that hydrogen peroxide in solution is decomposed by sunlight (however, very slowly over a month), but not in the dark. This paper, entitled "The effect of sunlight upon hydrogen peroxide" is only a ¼ page in *Nature* (and shortened reprinted in *Beiblätter Annalen der Physik* **5** (1880) 286), and surely overseen by all further researchers on radicals and H_2O_2 photochemistry as noted by Tidwell (2013).[1235] Thiele (1907) only short communicated that H_2O_2 solution is decomposed under UV light. Benrath (1912, p. 137) makes a remarkable note in relation to the liquid water photolysis, where H_2O_2 is first gained according to $2\,H_2O \rightarrow H_2O_2 + H_2$:[1236]

> Diese Reaktion erklärt das Vorkommen von Wasserstoffsuperoxyd in Schnee- und Regenwasser und macht es wahrscheinlich, daß die desinfizierende Wirkung des ultravioletten Lichtes teilweise der Bildung dieser Substanz zu verdanken ist [This reaction explains the occurrence of hydrogen peroxide in rain and snow water and makes it likely that the disinfecting effect of ultraviolet light is partly due to this substance].

Downes and Blunt (1879b) writes:[1237]

> Here we may regard the hydrogen peroxide as made up of two atomic groupings of the chlorous radicle HO and, if the theory we suggest be correct, the decomposition in this case is brought about by the dissociation of this radicles. We believe that the tendency of sunlight is to dissociate (or "weaken the internal bonds" between) what we have termed "chlorous radicle", whether these be simple, as oxygen or chlorine, or compound of HO, and thus to promote the combining energy, or to bring about a more stable arrangement of their constituent atoms.

Tidwell (2013) correctly writes that "the modernity of this description is breathtaking". The aqueous-phase photodissociation of H_2O_2 into two OH radicals was first shown

1235 However, this is not true; D'Arcy (1902) and Benrath (1912, p. 207) cited it, and also Gmelin (1966, p. 2276).

1236 This is a very remarkable thought. It implicates the primary formation of oxo radicals, being not only precursors of H_2O_2 but also acting as disinfection agents.

1237 The term "radicle" denotes a functional group, not a "free radical". The term "chlorous" (today it denotes chlorine in valency 5 in contrast to "chloric" in valency 3) in historic chemical literature denotes a constituent of a compound, either elementary, or, if compound, performing the function of an element. Thus, OH was termed "monad compound radical" or hydroxyl, SO_2 "dyad radicle" and PO "triad radicle". *Chlorous* was synonym with *basylous*.

by the US chemist *Harold Clayton Urey* (1893–1981)[1238] (Urey et al. 1929). The equation $H_2O_2 + h\nu \rightarrow 2\,OH$ was confirmed later also by Hunt and Taube (1952).[1239]

Between 1876 and 1890, several authors believed they could prove the existence of H_2O_2 in plants.[1240] However, because of insufficient analytical methods, there were doubts about such claims (Machu 1937). After the turn of the nineteenth century, the Russian-Soviet biochemist *Alexei Nikolajevich Bach* [Алексей Николаевич Бах] (1857–1946) and the Swiss botanist *Robert Hippolyte Chodat* (1865–1934), well-known for its oxygenase-peroxidase theory, found evidence for H_2O_2 in living plants (Bach and Chodat 1902). 20 years later it was confirmed by Gallagher (1923), and Tanaka (1925) proved that H_2O_2 was a primary product in respiration.[1241] Because of the significance of biological processes, H_2O_2 formation was also found (in the late 1930s) under visible and UV light influence in different aqueous suspensions containing plant parts (Gmelin 1966). Water containing organic substances produces H_2O_2 under (visible) light influence (Blum and Spealman 1933); cf. Boerhaave (1735) on p. 634 and namely Müller (1873) and Baur and Neuweiler (1927) about H_2O_2 formation in the bleaching process (pp. 102–103).

Mirosław Kernbaum (1882–1911), a Polish physicist in radiation chemistry who worked at the research school of *Marie Curie* (1867–1934) at the Paris Faculty, was one of the pioneers of H_2O_2 formation from water, studying the influence of UV radiation (Kernbaum 1909), electric discharges (Kernbaum 1910) and metals (Kernbaum 1911) on water decomposition, but only in contact with air or free oxygen. For the radiolysis of liquid water, without knowing the transient processes, the generic reaction equation

$$2\,H_2O \xrightarrow{\text{radiation}} H_2O_2 + H_2$$

was established by Kernbaum (1909) during investigating the influence of β radiation (from $RaCl_2$) on pure water. Haïssinsky and Magat (1951) first studied water radiolysis products like H, OH, O^- and suggested the reaction $H_2O + \gamma \rightarrow H_2O^+ + e^-$ ($\lambda_{threshold} = 195\,nm$).[1242] Duane and Scheuer (1913) first studied H_2O_2 formation in water during irradiation with α radiation (from Rn ampulla) and the German radiochemist *Otto Karl Hermann Risse* (1895–1942) the influence of X-rays on water (Risse 1929). In all these studies, H_2O_2 formation was confirmed, but hydrogen often did not escape and was instead dissolved in water with sometimes oxygen even being evolved

1238 US chemist, Nobel Prize in 1934 for the discovery of deuterium and known for the idea of "inorganic life formation" (*Miller-Urey* experiment).

1239 *John P. Hunt* was a chemist at the Washington State University.

1240 Schönbein (1864a) was the first person to detect H_2O_2 in the human body.

1241 See Halliwell and Gutteridge (1984) about literature concerns occurrence of H_2O_2 in vivo. In cells it is both a source of oxidative stress and a second messenger in signal transduction (Georgiou and Masip 2003).

1242 Note that $OH + H^+ = H_2O^+$ and $OH = O^- + H^+$.

(Fricke 1934).[1243] However, X-rays do not produce H_2O_2 in oxygen-free water (Risse 1929, Bonét-Maury and Lefort 1948). All these early observations are consistent with the modern view of H_2O_2 radiolysis through radioactive radiation. Liquid water photolysis under the formation of H_2O_2 was first proven by Tian (1911) for wavelengths < 190 nm. For historical overviews on radiation chemistry and inorganic radicals in solutions, see Uri (1952), Hart (1959), and Allen (1961).

The German photochemist *Waldemar Merckens* (unknown living data) from Mülhausen (Elsass) stated that H_2O_2 is produced according to the reduction potential of metals (Merckens 1906): Mg, Al, Zn, Cd, Ni, Co, Pb with decreasing potential from left to right. *Emil Baur* (Baur and Perret 1924, Baur and Neuweiler 1927) found that H_2O_2 is gained in water under UV radiation (< 400 nm) containing suspended ZnO (see also Fujishima (1972) on p. 493). It was suggested that all metals work in a similar process (The proof that all the oxygen in H_2O_2 comes from molecular oxygen (O_2) was shown by Calvert et al. (1954) using O^{18} as a tracer:

$$\mathrm{Me} + 2\,\mathrm{H_2O} + \mathrm{O_2} \rightarrow \mathrm{Me(OH)_2} + \mathrm{H_2O_2}.$$

The Austrian-British chemist *Joseph Weiss* (1905–1972) published the first general radical theory (Weiss 1944), but the earliest clear idea on the participation of radicals (H and OH) in oxidation mechanisms in aqueous solution was given by Risse (1929). From the kinetic measurements of hydrogen peroxide decomposition, Haber and Willstädter (1931) further concluded that

$$\mathrm{H_2O_2} + \mathrm{OH} \rightarrow \mathrm{H_2O} + \mathrm{HO_2},$$
$$\mathrm{HO_2} + \mathrm{H_2O_2} \rightarrow \mathrm{O_2} + \mathrm{OH} + \mathrm{H_2O}.$$

Haber and Weiss (1932, 1934) added two more reactions: they proposed[1244] the involvement of hydroxyl radicals in the iron(II)/hydrogen peroxide system (termed *Fenton* chemistry):[1245]

$$\mathrm{H_2O_2} + \mathrm{Fe^{2+}} \rightarrow \mathrm{OH} + \mathrm{OH^-} + \mathrm{Fe^{3+}},$$
$$\mathrm{Fe^{2+}} + \mathrm{OH} \rightarrow \mathrm{Fe^{3+}} + \mathrm{OH^-}.$$

These four reactions were termed *Haber-Weiss* cycle, although two reactions were proposed by Haber and Willstädter (1931). Indeed, the only paper by Haber and Weiss

1243 *Hugo Fricke* (1892–1972) was a Danish-American physicist who worked in the field of chemical radiolysis.

1244 They did not mention *Fenton* (Koppenol 2001).

1245 *Henry John Horstman Fenton* (1854–1929) was a British chemist at Cambridge in 1878 and University Lecturer in chemistry from 1904 to 1924. In 1876, *Fenton* described a coloured product obtained by mixing tartaric acid with hydrogen peroxide and a low concentration of a ferrous salt (Fenton 1876) and found that iron acts catalytically (Fenton 1894).

(1934) was written in English and the other papers, published in German by *Haber* et al., were almost ignored (Koppenol 2001). Haber and Willstädter (1931) also proposed

$$O_2^- + H_2O_2 \rightarrow O_2 + OH + OH^-.$$

But this reaction was considered in this cycle to be inefficient (George 1947, Barb et al. 1949) but likely to be important in vivo (Halliwell and Gutteridge 1984). Hence, Weiss and Humphrey (1949) proposed to replace it by the following reaction to maintain the cycling between ferrous and ferric ions:

$$O_2^- + Fe^{3+} \rightarrow O_2 + Fe^{2+}.$$

John Baxendale (1916–1982)[1246] and coworkers (Barb et al. 1949) suggested that superoxide reduces by reaction

$$O_2^- + Fe^{3+}\ (Cu^{2+},\ Mn^{2+}) \rightarrow O_2 + Fe^{2+}\ (Cu^+,\ Mn^+)$$

the concentration of iron(III) formed through

$$H_2O_2 + Fe^{2+} \rightarrow OH + OH^- + Fe^{3+}$$

to explain the catalytic function of the metal. In biological systems, apart from inorganic *Fenton* chemistry, Haber and Willstädter (1931) proposed three basic oxygenic processes:

$$\begin{aligned} O_2 + e_{aq}^- &\rightarrow O_2^- \quad \text{(enzymatic reduction)},\\ 2\,O_2^- + 2\,H^+ &\rightarrow O_2 + H_2O_2 \quad \text{(dismutase)},\\ 2\,H_2O_2 &\rightarrow 2\,H_2O + O_2 \quad \text{(catalase)}. \end{aligned}$$

Hunt and Taube (1952) first proposed the reactions

$$OH + H_2O_2 \rightarrow HO_2 + H_2O \quad \text{and} \quad 2\,HO_2 \rightarrow H_2O_2.$$

In the 1930s, it was observed that paints containing binders and TiO_2 show no durability under the influence of weather and light (Wagner 1939); *Hans Wagner* (unknown living data) was a chemist at I. G. Farben. It was also known that bleaching of light-resistant color paints occurred because of adding TiO_2 (Wagner 1929, Keidel 1929, Wagner 1939), and similar effects were seen when TiO_2 was used as a delustering agent for rayon (Keiner 1934). Keidel (1929) had already proposed a photochemical action where

1246 British chemist at the University of Manchester; pioneer in radiation chemistry.

TiO_2 acts as a sensitizer. Goodeve (1937) suggested that the photosensitization through TiO_2 results in an exciton transport, consisting of a negatively charged electron and a positively charged hole, bound. Schossberger (1942) found that rutile shows no photosensitization but only anatase with disorders in the crystal lattice. It seems to me that this older literature was unknown to many researchers of the last decades (who were unable to read German) to refer that photocatalysis and the formation of ROS were discovered a half century before. It is likely that Schumb et al. (1955), the only monograph on H_2O_2 in English, who cited the other two earlier monographs written in German (Machu 1937, Kausch 1938)[1247] but did not read them and hence not cite one of the abovementioned papers on H_2O_2 formation from “surface-water-sunlight” pathways. It is very remarkable that Leighton (1961) cited the early papers on ZnO and TiO_2 photocatalysis by *Wagner* and *Keidel*. Kormann et al. (1988) reported on H_2O_2 formation in aqueous suspensions containing oxides (TiO_2, ZnO, desert soil) under atmospheric conditions (without citing these older papers).

The photocatalytic mechanism of formation of H_2O_2 in aqueous phase became clear by the work of the Russian chemist G. A. *Korsunovsky* [Корсуновский] (unknown living details) from Leningrad (Korsunovsky 1957, 1960). The German physicochemist *George-Maria Schwab* (1899–1984) from Munich proposed that photosensitizers like ZnO initiate the reaction $O_2 + M + h\nu = O_2^- + M^+$ (Schwab 1957), followed by

$$O_2^- \xrightarrow{H^+} HO_2 \xrightarrow{e^-} HO_2^- \xrightarrow{H^+} H_2O_2.$$

The formation of H_2O_2 is limited due to $M^+ + OH^- = M + OH$, and the subsequent decomposition of H_2O_2 by the radicals OH and HO_2 (proposed mechanisms by *Haber*, *Weiss*, and *Willstätter* in the 1930s). Consequently, in the presence of organic compounds (or OH scavengers in general), the yield of H_2O_2 increases in solution.

The role of organic compounds in photosensitized formation of H_2O_2 in natural water (surface water, river, seawater)[1248] became highlighted with the beginning 1980s by Draper and Grosby (1981, 1983), Cooper and Zika (1983), Zafiriou (1983) and others. The US American chemist *Brian G. Heikes* first suggested the idea of atmospheric aqueous-phase H_2O_2 formation in cloud droplets (Heikes et al. 1982). Schwartz (1984) enlightened scavenging of H_2O_2 and HO_2 from gas phase, and McElroy (1986) stated that superoxide anion O_2^- plays in aqueous phase the same role like HO_2 in gas phase. Weinstein-Lloyd and Schwartz (1991) found that the aqueous-phase production of H_2O_2 rises with light intensity. Years later, it was shown that in all kinds of natural waters chromophoric substances exist being able to produce hydrated electrons after

1247 These books were relatively rare. There are wartime photo-lithoprint reproductions by Edwards Bros, Ann Arbor, Michigan (1943), under the authority of the Alien Property Custodian because H_2O_2 production was important for torpedo engines. *Machu*'s monograph was edited later by Springer-Verlag, Berlin, 1951. All older literature on oxygen species is cited in Gmelin (1966).

1248 The occurrence of H_2O_2 in seawater was first proved by Van Baalen and Marler (1966).

illumination and, consequently, reduce dissolved O_2 in a two-electron step into H_2O_2 (Faust and Allen 1992, Anastasio et al. 1997, Wohlgemuth et al. 2001). Since then, the reduction and oxidation of H_2O_2 was well described in the sense of electron transfers depending on the reduction potential and pH of the aqueous solution (electron donators and/or acceptors).

Aqueous-phase production of H_2O_2 has also been discussed (Staehelin et al. 1984, Martin et al. 1997) via dissolved ozone decay in alkaline medium. Zuo and Hoigné (1992) finally found the formation of hydrogen peroxide (but only in oxygenated solutions) while photodecomposition of iron-III oxalate, proposing an electron transfer from oxalate to the central ferric ion (Zuo and Hoigné 1992, p. 1015), and further reaction $[C_2O_4]^- + O_2 \rightarrow C_2O_4 + O_2^-$, whereas the oxalate radical decays into 2 CO_2.

With the beginning 1980s, the oxidant chemistry has been interlinked with "acid rain" chemistry, i. e., oxidation of man-made emitted NO_x and SO_2 to nitric and sulfuric acid in the gas and aqueous phase (Graedel and Weschler 1981, Chameides and Davis 1982, Graedel and Goldberg 1983, Graedel 1984, Graedel et al. 1986, Mauersberger and Möller 1990). Related to the volume of air, however, the aqueous-phase production of H_2O_2 concerning the radical mechanism (until present, nothing is known on the kinetics of photocatalytic formation of H_2O_2 in atmospheric water) is small, being about 10^{-7} ppb·s^{-1} (Möller and Mauersberger 1992) compared to the gas phase production rate of about 10^{-5} ppb·s^{-1} (Martin et al. 1997).

4.5.4 Studies of hydrogen peroxide after 1945

The first estimate of H_2O_2 after the war was conducted by *Hideo Matsui* (Matsui 1949), who first refers Schöne (1878, 1894a) with his world's first quantitative estimates being 0.04–1.0 mg L^{-1}, and states that there was no other measurement since then. He was motivated by the idea that "water droplets floating in the atmosphere are irradiated with ultra-violet solar light". Because Japanese meteorologists (see Chapter 2.3.4.3) at that time were interested in studying the mechanism of precipitation formation, *Matsui* believed that quantifying H_2O_2 could be helpful. That is surprising from today's perspective, but it shows the first thought on a relation between the oxidation capacity of the atmosphere and formation of condensation nuclei (independently from *Helmut Mrose*, who was the first to detect sulfate from anthropogenic SO_2 sources in nuclei and who also detected H_2O_2). *Matsui* collected rain and snow water in Sake-cho and at Mt. Hakone of the Kanagawa Prefecture (Japan) between April 1948 and March 1949; for the determination of H_2O_2 he used *Schöne*'s colorimetric method slightly improved.[1249] His results from 15 rain and nine snow samples (in mg L^{-1}):

1249 1 mL of 5 % potassium iodide solution and 2 mL of 0.2 % starch solution added to 25 mL of precipitation immediately after sampling well-mixed, then adding 1 drop of 0.5 % ferrous sulfate solution. If

April–September 1848	rain	0.74 ± 0.11 (0.60–0.86)
October 1848–March 1949	rain	0.50 ± 0.09 (0.35–0.61)
November 1858–February 1949	snow	0.12 ± 0.04 (0.08–0.15)

Further, he detected H_2O_2 in traces in dew and frost. Thus, his findings were similar to those *Schöne* obtained: in the annual variation of the H_2O_2 level, a maximum occurs in June and a minimum in January, furthermore exists a parallelism between the yearly variation of the hydrogen peroxide and the ultra-violet ray in the solar radiation; this suggests that the ultra-violet ray produces the hydrogen peroxide in the presence of water particles suspended in the atmosphere he writes. After *Matsui*, advancing investigations of H_2O_2 in rainwater began 1978 by Bufalini et al. (1979) in the USA in 1978; worldwide first direct determination of gas-phase H_2O_2 in the atmosphere also began also in the USA in 1970 by Bufalini et al. (1972).

Hundred years after Schöne (1878), Olszyna et al. (1988) found that the level of H_2O_2 in cloud water at Whitetop Mountain shows a strong seasonal dependence with the highest values in summer; they notified the highest value so far measured to be 247 µM (8.4 mg L^{-1}). Furthermore, they found that gas-phase H_2O_2 concentrations were strongly correlated with ozone. In Europe, the first measurements of H_2O_2 in cloud water using an aircraft (20 flights) have been conducted by a Dutch group of scientists in summer 1981 (Römer et al. 1985); they found at altitudes between 1000 and 3000 m about the same level in rainwater and cloud water. Further, a strong inverse correlation between H_2O_2 and sulfate has been found.

Another half century after *Matsui*, now using modern sampling and measurement technique, the results of measurement H_2O_2 in the air and rainwater between May 2000 and June 2001 at my institute in Berlin-Adlershof were principally nothing else (Möller et al. 2002, Möller 2009); monthly means range 0.005–0.32 mg L^{-1} and 0.02–0.27 ppb (daily maxima 0.05–0.5 ppb). From a historical perspective, and at first glance these three investigations would propose that over 125 years and between very different locations (Russia, Japan, Germany), no significant changes occurred in the chemistry of H_2O_2.

In samples of cloud water and rainwater occurs rapid depletion of aqueous-phase H_2O_2 as it oxidizes S(IV); the instability of H_2O_2 solution over a time scale of hours and days was known since the nineteenth century, but the S(IV)-H_2O_2 reaction has been recognized only end of the 1970s. Thus, only analysis immediately after sampling or inhibition of the S(IV)-H_2O_2 reaction by adding formaldehyde to the sampling device would result in reliable values. Another problem of H_2O_2 observations is the large variations of H_2O_2 levels depending on the time of the day, the season (both ranging by a factor between 10 and 30), from the chemical characteristics of air mass and cloud water, an averaging of values without classification results in questionable, mean levels.

H_2O_2 is present, blue color will appear, and after 10 min the color will be compared with the standard solution prepared by same procedure.

Hence comparison of H_2O_2 levels obtained by different investigators needs caution to draw on general characteristics.

In contrast to investigations on tropospheric ozone (not countable), the number of measurements of H_2O_2 in the atmosphere between 1970 and 2008 (I did not find more studies until 2020) is relatively small (Tables 4.25 to 4.28). However, due to more and more sophisticated measurements concerns time resolution, detection limit, including other substances, simultaneous measurement in gas and aqueous phase, and finally combining the results with modeling simulation, many interesting results in the understanding of atmospheric H_2O_2 chemistry gained.

After *Matsui*'s investigation, the motivation to investigate hydrogen peroxide in the atmosphere arose from the observation of high levels while photochemical smog episodes in the Los Angeles area; however, mainly due to missing adequate sampling and analytical technique,[1250] extensive field measurements (and smog chamber studies) begun only in the 1980s. Another issue to study hydrogen peroxide concerns its role in aqueous-phase chemistry, namely the oxidation of nitrite and sulfite, as well as the formation of radicals. The oxidation of nitrite (or nitrous acid, resp.) by H_2O_2 was known since Schönbein (1847a) and since Carius (1874b) seen as textbook-knowledge according to the gross reaction $H_2O_2 + HNO_2 = H_2O + HNO_3$. Halfpenny and Robinson (1952) found evidence for the intermediate formation of pernitrous acid (HOONO) and radicals but suggested a complicated mechanism.[1251] Lee and Lind (1986) were the first to study this process under concentrations reliable for atmospheric conditions and also proposed the intermediate formation of HOONO (like Anbar and Taube (1954), Benton and Moore (1970), Bhattacharyya and Veeraraghavan (1977)); the rate was determined by all researchers rather agreeing; Lee and Lind (1986) calculated for pH = 4 a lifetime of 84 hours, i. e., nitrite analysis must be done soon after sampling of atmospheric water. Damschen and Martin (1983) additionally studied the kinetics of the nitrite oxidation by O_2 and O_3.

At the same time, the role of H_2O_2 in the formation of sulfate from oxidation of SO_2 dissolved in cloud water has been recognized (Penkett et al. 1979, Möller 1980). In the late 1980s, modeling emphasized the importance of H_2O_2 as radical termination, and together with measurements of ozone, nitrogen oxides, hydrocarbons, and other species large field studies were conducted in the 1990s to understand the oxidizing capacity of the atmosphere.

1250 Kok et al. (1978) developed the first chemiluminescence method that has been later stepwise improved and became in the 1990s commercially available.

1251 Today as initial reaction the oxidative capacity of H_2O_2 in solution is seen as an electron transfer with subsequent decay in OH radical and OH^- ion; the OH radical further acts as oxidant: $H_2O_2 + NO_2^- \rightarrow H_2O_2^-(\rightarrow OH + OH^-) + NO_2(+OH \rightarrow HNO_3)$. Other oxidants such as O_2, OH and O_3 similarly react but also transition metal ions can act as catalyzer with initial electron transfer. Adducts between NO_2 and O_3 and O_2 could lead to intermediary pernitrite.

Table 4.25: Measurements of H_2O_2 in atmospheric gas phase (in ppb).

year	location	levels	reference
1874	Moscow, summer	0.37 ± 0.04	Schöne (1874)
1970	Riverside (CA), summer	40–180	Bufalini et al. (1972)
1978	southern California	10–30	Kok et al. (1978)
1978	southern California	0.3–3	Kelly et al. (1979)
1984	aircraft, fall, east USA	0.4–3.0	Heikes et al. (1987)
1984	USA, aircraft, October	1.3–2.7	Calvert et al. (1985)
1984/85	Dortmund, summer	0.05	Jacob et al. (1986)
1984/85	Dortmund, fall	0.015	Jacob et al. (1986)
1985/86	Dortmund, summer	0.07 (0.01–0.54)	Jacob et al. (1990)
1984/85	Ontario, Canada, summer	<0.3–2.1	Slemr et al. (1986)
1985	Upton, N. Y., August	0.2–2.0	Tanner et al. (1986)
1985	LA Basin, summer and fall	0.02–1.2	Sakugawa and Kaplan (1987)
1986	Whitetop Mt., summer	0.8	Olszyna et al. (1988)
1986	Whitetop Mt., fall	0.15	Olszyna et al. (1988)
1986	west Atlantic, January	0.12 ± 0.07 (<1.2)	Heikes et al. (1988)
1986	aircraft, east USA, February	<(1–2.4)	Barth et al. (1989)
1987	aircraft, central USA, Febr.	<0.1–1.0	Van Valin et al. (1987)
1987	aircraft, Ohio, June	<0.2–7	Daum et al. (1990)
1987	aircraft, NE USA, June	0.6–3.6	Van Valin et al. (1990)
1988	ship, Atlantic, fall	<0.1–3.5	Jacob and Klockow (1992)
1987/90	San Bernardino Mountains	1–3	Sakugawa and Kaplan (1993)
1987/90	Harward, monitoring	0.1–0.3	Dollard and Davies (1993)
1988	Bahia (Brazil)	1.2 (0.2–3.9)	Jacob et al. (1990)
1988	Beijing, May	0.2–1.5	Weihan et al. (1989)
1988	Mount Lushan, China, May	1–2	Weihan et al. (1989)
1988	aircraft, NE US, summer	0.2–5.9	Tremmel et al. (1993)
1989/91	Schauinsland	0.33 (0–0.93)[a]	Gilge (1994)
1990	Mt. Wank (D)	0.3 (0.05–0.85)[a]	Junkermann et al. (1992)
1990/91	Cape Grim (Tasmania)	~0.2–1.4 (wi/su)	Ayers et al. (1996)
1990	Niwort Ridge, CO, summer	0.5–0.8	Watkins et al. (1995)
1990/92	Kinderbish, AL, summer	1.5–2.0	Watkins et al. (1995)
1991	aircraft, global, January	0.42–6.31	Perros (1993)
1991/92	Piedmont, USA, July	0.58 (0–2.2)	Das and Aneja (1994)
1992	Grand Canyon, spring	1.5 (0.7–6)	Tanner and Schorran (1995)
1993	Bretagne (F)	0.2 (<0.1–1.2)	Sauer et al. (1997)
1993	Great Dun Fell (UK), spring	<0.05–1.2	Preiss et al. (1994)
1993	Mt. Norikura (JP), August	0.01–4.5	Watanabe et al. (1995)
1993	Nagoya, Japan, May, day	0.31 ± 0.4 (0–2.4)	Watanabe and Tanaka (1995)
1994	central Portugal, June	0.27 (0–0.63)	Jackson and Hewitt (1996)
1994	Tabu, Portugal, summer	<1.38	Sauer et al. (2001)
1995	Ogasawara Hahajima Island	0.1–1.5	Watanabe et al. (1996)
1996	Mt. Screnica (PL), summer	0.05–2	Acker et al. (1999a)
1997/98	Antarctica	0.15 (0.05–0.4)	Riedel et al. (2000)
1998	Pabtsthum (D), summer	<0.6 (max 1.4)	Grossmann et al. (2003)
1998	near Berlin, summer	<0.015–1.5	Moortgat et al. (2002)

Table 4.25: (continued).

year	location	levels	reference
1998/05	Hohenpeißenberg	0.54 ± 0.62	Gilge (2007)
2000	Beijing, May	2.33	Peng et al. (2003)
2000/01	Berlin, summer	0.16 ± 0.06 (0–3)	Möller et al. (2002)
2000/01	Berlin, winter	0.03 ± 0.02 (0–0.2)	Möller et al. (2002)
2001	Los Angeles, summer	0.5–3.5	Hasson and Paulson (2003)
2001	Mt. Schmücke (D), fall	<0.13	Valverde-Canossa et al. (2005)
2002/03	Taiwan, fall to spring	~1.6 (0.24–3.94)	Peng et al. (2006)
2003	Jungfraujoch (CH)	0.21 (<0.02–1.42)	Walker et al. (2006)
2004	Zagreb, summer	0.3 (<0.05–6.2)	Acker et al. (2008b)
2006	Guangzhou, rural, July	1.26 (0–4.6)	Hua et al. (2008)
2008	Beijing, June-September	<2.34	He et al. (2010)
2013	Beijing, fall	0.55	Zhang et al. (2017)

[a]Monthly means

Pioneering work in the development of H_2O_2 measurement techniques[1252] and ambient measurements was done almost only in the USA between 1970 and the mid-80's by *Bruce W. Gay Jr.*, *Gregory Kok*, *Brian Heikes*, *Joseph Bufalini*, *Rod Zika*, *Eric Saltzman*, *William Chameides*, and others, followed by German scientists (*Dieter Klockow*, *Franz Slemr*, *Wolfgang Junkermann*, and others). It is beyond the scope of this volume to present the individual results from the many investigations of ambient H_2O_2 (see Tables 4.25 to 4.28). A short summary of essential results from early measurements is now followed, often later confirmed by other investigators; see, e. g., Sakugawa et al. (1990) for a review of H_2O_2 measurements.

In the early 1970s, H_2O_2 concentrations between 40 and 180 ppb were observed during photochemical smog episodes in the Los Angeles area (Gay and Bufalini 1971, 1972, Bufalini et al. 1972, Demerjian et al. 1974, Hanst et al. 1975), supporting the formation via the photolysis of aldehydes. Peaking H_2O_2 levels up to 1.4 ppb in the morning rush hour found Das et al. (1983) in Bombay, whereas at daytime levels around 0.4 ppb and in the dark, the levels fall to the detection limit of 0.01 ppb; they correctly concluded on photochemical reactions of pollutants from motor vehicle exhausts.

Zika et al. (1982) concluded for the first from measurements in southern Florida that H_2O_2 is generated to a substantial fraction by aqueous-phase reaction in cloud water (remember that *Heikes* proposed it first in 1982). Zika and Saltzman (1982) also found that H_2O_2 is produced as well as removed while air sampling due to interactions between ozone and water. Heikes (1984) systematically investigated that H_2O_2 is both

1252 It is historically worth to refers the work by Farmer and Dawson (1982) who continued the idea of gas sampling by artificial dew, first applied by Schöne (1874) and then further developed by Quitmann and Cauer (1939), known as "*Cauer*'s condensation dew-cup" by development and theoretical consideration of "sophisticated" condensate collector.

Table 4.26: Measurements of H_2O_2 in rainwater; concentration in µM (1 µM = 34 µg L^{-1}).

year	location	levels	reference
1869	Tbilisi, spring	13.5	Struve (1872)
1874/75	Moscow, summer	9 ± 6	Schöne (1878)
1874/75	Moscow, winter	0.9	Schöne (1878)
1878	St. Petersburg, summer	10.5 (3–41)	Kern (1878)
1949	near Yokohama	18 ± 8.5 (3–29)	Matsui (1949)
1978/79	Research Triangle, summer	6	Bufalini et al. (1979)
1978/79	Research Triangle, winter	0.06–0.2	Bufalini et al. (1979)
1979	Claremont, California,	5–47	Kok (1980)
1981	Tokyo, fall	0.6–0.9	Yoshizumi et al. (1984)
1982	Tokyo, summer	7.5 ± 4.0	Yoshizumi et al. (1984)
1982	Miami, Florida	10–70	Zika et al. (1982)
1985	various locations in Europe	0.01–62	Römer et al. (1985)
1985/86	Dortmund, summer	8.8 (2.9–65)	Jacob et al. (1990)
1987/90	San Bernardino Mountains	22 (3.5–90)	Sakugawa and Kaplan (1993)
1987/88	California, winter, rain	0.6–40.4	Gunz and Hoffmann (1990b)
1987/88	California, winter, snow	2.8 (0.9–7.4)	Gunz and Hoffmann (1990b)
1988	Bahia (Brazil), spring	19–200	Jacob et al. (1990)
1989	Sweden, rural, summer	0.05–2.0	Ross et al. (1992)
1992/94	Wilmington (USA), $n = 61$	12 (0.13–48.3)	Willey et al. (1996)
1993–95	Melpitz, monitoring	~0.2 (0–2.1)	Gnauk et al. (1997)
1993–95	Leipzig, monitoring	~0.4 (0–5.3)	Gnauk et al. (1997)
1995/96	Miami ($n = 34$)	6.9 (0.3–38.6)	Deng and Zuo (1999)
1995	Whiteface Mountains, July	1.61 (0.15–5.15)	Balasubramanian and Husain (1997)
1998	Yokohama	13.1	Takeuchi et al. (2003)
2000/01	Berlin, monitoring, summer	7.2 ± 6.0 (0–28)	Möller et al. (2002)
2000/01	Berlin, monitoring, winter	0.54 ± 0.68 (0–1.8)	Möller et al. (2002)
2002–04	Korogi-machi, Japan	0–67	Watanabe et al. (2005)
2005–09	Imizu City Japan, summer	45 (0.8–190)	Watanabe et al. (2009)
2005–09	Imizu City Japan, winter	1.7 (0–17)	Watanabe et al. (2009)

produced and destroyed in bubblers by atmospheric ozone. Insofar, cloud and rain droplets, when evaporating, transfer H_2O_2 back to the air, and transport it from one site to another and from upper layers to the earth's surface. Additional to the much smaller (or even no) decomposition of dissolved H_2O_2 we can consider its formation in aqueous phase.

Important findings on the latitudinal dependence of the H_2O_2 levels were made by Van Valin et al. (1987) by aircraft measurements and by Jacob and Klockow (1992) on board crossing the Atlantic from Bremerhaven to Rio de Janeiro. Both investigations show an approximately linear increase from north to south by 45 ppt per degree latitude; on the ship the maximum of 3.5 ppb was detected around the equator with decreasing H_2O_2 concentration (but on a significant higher level than northern in October). Ross et al. (1992) concluded on the importance of long-range transport on sur-

Table 4.27: Measurements of H_2O_2 in cloud water. Concentration in μM (1 μM = 34 μg L^{-1}).

year	location	levels	reference
1981	aircraft over Europe, summer	25 (1–88)	Römer et al. (1985)
1982	aircraft, LA basin, May, night	33 (1–88)	Richards (1995)
1984	aircraft, LA basin, June, night	55 (16–127)	Richards (1995)
1985	aircraft, LA basin, June/July	57 (12–163)	Richards (1995)
1982	Riezlern (1200 m), winter	0.15	Römer et al. (1985)
1984	eastern USA by aircraft	0–75	Daum et al. (1984)
1984	Whiteface Mt.	5.4 (0–112)	Lazrus et al. (1985)
1984	Whiteface Mt., summer	~1	Kadlecek et al. (1985)
1984	Whiteface Mt., winter	<0.03	Kadlecek et al. (1985)
1985	cloud and rain eastern USA	10 (0–100)	Kelly et al. (1985)
1986	aircraft, east USA, February	0–112	Barth et al. (1989)
1986	Whitetop Mt. (1689 m), fall	0.15	Olszyna et al. 1988)
1987/90	San Bernardino Mountains	24 (6–62)	Sakugawa and Kaplan (1993)
1993	Mt. Brocken (D), fall	0.06–3.02	Horvath[a]
1993	Mt. Norikura, Japan, summer	3–120	Watanabe et al. (2001)
1995	Kleiner Feldberg (D), fall	<0.01–3.19	Sauer et al. (1996)

[a] *Laszlo Horvath* (unpublished) September–October. Mean values from different air masses and cloud characteristics

Table 4.28: Measurements of H_2O_2 in dew; concentration in μM (1 μM = 34 μg L^{-1}).

year	location	levels	reference
1874	Moscoe	0	Schöne (1878)
1986	Los Angeles	<1	Pierson and Brachaczek (1990)
1987/90	San Bernardino Mountains	0.74 (0.08–1.4)	Sakugawa and Kaplan (1993)
1997	Santiago de Chile	1.6 (0.3–2.6)	Ortiz et al. (2000)
1998	Yokohama	0.4	Takeuchi et al. (2003)
2004	Santiago de Chile, spring	7 ± 5 (1–22)	Rubio et al. (2006)
2008	Imizi City, Japan, fall	0.8–3.6	Watanabe et al. (2009)

face H_2O_2 levels and showed the importance of turbulent mixing processes and dry deposition. Watkins et al. (1995) found while H_2O_2 measurements at two forest sites in Alabama between 1990 and 1992 that the higher the natural isoprene emission (controlled by temperature and vegetation type), the higher the H_2O_2 level.

The French chemist *Pascal Perros* (born 1947)[1253] conducted the most extensive aircraft measurements of hydrogen peroxide in January 1991; the data set covers a latitude range from 60 °N to 60 °S and altitudes up to 11 km. His observations led to fundamental facts for understanding global H_2O_2 chemistry:

1253 He worked at Universités Paris 12 et Paris 7, CNRS "Laboratoire Interuniversitaire des Systèmes Atmosphériques" (LISA).

Hydrogen peroxide concentration was typically low (<2 ppb) in the boundary layer of both hemispheres except for the intertropical belt, even in America, as well as Africa where the mean mixing ratio reached 5 ppb.
In the northern uppermost troposphere, a significant increase in H_2O_2 (>3 ppb) could be noticed. Some episodes of high hydrogen peroxide mixing ratio were encountered in the vicinity of cloud systems; two cloud layers and a significant H_2O_2 increase just above the lowest cloud were noticed.
The SH stratospheric H_2O_2 level is higher (1.17 with a range 0.43–2.31 ppb) then the NH level (0.70 with a range 0.1–1.42 ppb).
Vertical profiles are different for different latitudes as well in NH and SH, reflecting several influencing factors; however, at high altitude, water is the restricting reactive for H_2O_2 production.

From the several experiments and observations (Tables 4.25, 4.26, and 4.27), the following general conclusions can be derived concerning atmospheric hydrogen peroxide:

- distinct seasonal variation with maximum levels in summer (amplitude > 10),
- distinct diurnal cycle of surface-near but not under cloudy conditions,
- no diurnal cycle at mountain tops and in free troposphere,
- a positive correlation between O_3 and H_2O_2 is often observed,
- an anticorrelation between NO_x and H_2O_2 is observed,
- the H_2O_2 concentration in urban areas is mostly lower than in rural areas,
- the H_2O_2 concentration increases from pole to the equator having a maximum around the equator,
- increased H_2O_2 concentration have been often found close to clouds,
- the H_2O_2 concentration is general lower in the boundary layer (<2 ppb), having a pronounced maximum at the top of the PBL,
- in the free troposphere large variations of H_2O_2 concentration occur,
- the gas-phase H_2O_2 concentration decreases during rainfall and within clouds,
- in aqueous phase a strong anticorrelation between H_2O_2 and S(IV) was found,[1254]
- in absence of SO_2/S(IV) the aqueous phase (i. e., in all natural waters) produces H_2O_2 under light,
- H_2O_2 concentrations in cloud water were well below those calculated to be in *Henry*'s law equilibrium with the gas-phase concentrations of H_2O_2 in the cloudy air,

1254 Worth to note here are measurements in cloud water at Mt. Brocken, carried out by *Laszlo Horvath* (Budapest) who spend three month in 1993 as guest scientists in my group. Lowest levels (but high concentration of organic peroxides: 0.46 µM) were detected in clouds from SE in air masses with high SO_2 levels (10–55 ppb) and highest H_2O_2 levels in air from SW with low SO_2 (<2 ppb); however, in raining clouds from SW low H_2O_2 was estimated (0.16 µM). The inverse relations between H_2O_2 (in cloud water, rainwater and gas) and the SO_2 level in air also has been found in our one-year measurement series of the concentrations in air and rainwater 2000–2001 at our institute site in Berlin-Adlershof (Möller et al. 2002, Möller 2009).

- H_2O_2 concentration in cloud and rainwater shows a positive correlation with dissolved organic matter,
- methyl hydroperoxide is the dominant organic hydroperoxide with levels less 1/5 of H_2O_2.

Furthermore, the following factors influence the production and hence the concentration of H_2O_2:
- absolute humidity in gas phase (increase),
- temperature (increase),
- ultra-violet light intensity both in gas and aqueous phase (increase),
- presence of alkenes, namely isoprene (increase due to ozonolysis)
- presence of aldehydes (increase due to photolysis)
- presence of pollutants, namely SO_2 and NO_x (decrease),
- dry deposition at ground (decrease of surface-near H_2O_2)
- presence of dissolved organic matter in aqueous phase (increase)

Remarkable are H_2O_2 measurements in arctic regions, in air, snow, firm, and ice. Hydrogen peroxide is present in polar snow and ice in remarkably high concentrations. With values up to 300 ppb, H_2O_2 is one of the most concentrated impurities in polar ice. In 1984 it was realized that H_2O_2 is present in relatively large amounts in polar snow and ice samples (Neftel et al. 1984). It is quite astonishing that hydrogen peroxide, a rather unstable and reactive species, survives the firnification process and is still detectable in the ice several thousand years old. The observed long-term trend in Greenland deep ice cores points to a very slow disintegration of H_2O_2 with a half-life of the order of several thousand years (Neftel et al. 1986).[1255] Measurements of H_2O_2 in Greenland ice suggested a 50 % increase in the H_2O_2 concentration during the last 200 years, where most of the increase occurred between 1960 and 1988 (Sigg and Neftel 1991). From cores drilled in 1995 at Summit, Greenland, Anklin and Bales (1997) confirmed the H_2O_2 increase found earlier and that show a further increase of the H_2O_2 concentration since 1988, leading to an overall increase of 60 % during the last 150 years; between 1988 and 1994 the increase was 20 %. Besides the mean annual concentration increase, the annual amplitude between winter minima and summer maxima has tripled since 1970. In the record between 1300 and 1850, no trend occurs.

In Antarctica, the annual mean H_2O_2 concentrations of the Siple Station core exhibit large variations but not a significant temporal trend. Sigg and Neftel (1988) found from continuous H_2O_2 firn record, which covers the last 83 years from Siple Station (Antarctica), a very strong seasonality (summer maximum), similar to that also ob-

1255 The process of transfer from the gas phase via snow, firn and ice is very complex; within the first years H_2O_2 is lost and afterwards rather stable (1 % loss per 100 years); see Neftel et al. 1995 for literature).

served in a Greenland ice core. Watanabe et al. (1998) found from an ice core obtained in East Antarctica not only seasonal variation but also about 11-year variations corresponding to the solar cycle.

Atmospheric H_2O_2 concentrations measured at Greenland (summit and at Dye 3) in summer over several years are between 1.5–3 ppb at daytime and 0.2–0.5 ppb with nighttime (Sigg et al. 1992, Neftel et al. 1995, Jacobi et al. 2002, and reference therein), thus higher than at the continental inland; in winter, however, H_2O_2 at Greenland is also very low (<0.01 ppb). Flux measurements have shown that an exchange between the surface snow and gas phase occurs and the snowpack can be considered as a temporary reservoir for H_2O_2 during the night (and also found for HCHO by Jacobi et al. 2002). Modeling, which had rather well reflected surface-near H_2O_2 concentration at several sites in the world, underestimated H_2O_2 levels at Greenland, and Neftel et al. (1995) right concluded on the cause of a missing surface source. Snow scavenging of H_2O_2 from the tropospheric column would result in supersaturation of H_2O_2 in surface snow relative to surface air, and H_2O_2 evaporation could then elevate surface air concentrations substantially above photochemical steady state. Additional, photochemical production of H_2O_2 from the snow surface can occur.

The only long-term measurements of H_2O_2 were conducted by Dollard and Davies (1992, 1993) between late 1987 and summer 1991 and late 1992 and summer 1994 at Harwell (Britain) and indicate an increasing H_2O_2 concentration with time: as annual means from 0.15 to 0.3 ppb in the first and from 0.2 to 0.3 ppb in the second period measurement periods (+0.028 ppb yr^{-1}); but there seem to be discrepancies between period averages, likely due to different measurement techniques. Two other long-term measurement series are available by *Stefan Gilge* (born 1964)[1256] from the stations Schauinsland (1889–1991) and Hohenpeißenberg (1998–2005); Gilge (1994, 2007, Gilge et al. 2000). Considering the H_2O_2 values around 1990 (also, including the one-year measurement from Wank, Junkermann et al. 1992)[1257] being around 0.3–0.33 (0.05–0.93) ppb and the mean (1998–2005) from Hohenpeißenberg to be 0.54 (0.07–3.73) representative for this German region, it would give an increase by 70 % in the atmospheric H_2O_2 content or corresponding about 0.023 ppb yr^{-1} close to the increase at Harwell. Moreover, the Hohenpeißenberg time series shows a similar trend of 0.02 ppb yr^{-1}.

1256 German chemist, who worked 1989–1994 at the ICH-2 in Jülich (at *Dieter Kley*), 1994–2015 at the Meteorological Observatory Hohenpeißenberg and since 2015 in Freiburg at the German Weather Service in Air Hygiene.

1257 *Wolfgang Junkermann* (born 1950) studied physics at Aachen and Heidelberg and worked until his retirement in the institute in Garmisch-Partenkirchen, known for his measurements of atmospheric constituents with a lightweight airplane.

4.5.5 Understanding tropospheric H_2O_2 chemistry and budget

There is little doubt that the atmospheric H_2O_2 content increased similar to O_3 over the past, as shown by modeling, almost due to increasing CH_4 emissions (Thompson et al. 1989, 1991, Thompson 1992, 1995). This is supported by the Central Greenland ice core information that shows the H_2O_2 concentration increasing by 50 % over the past 200 years (Sigg and Neftel 1991). Whereas the long-term increase in ozone is well supported by atmospheric measurements (see Chapter 4.4.8.5), there are only a few measurements that support an increase in atmospheric H_2O_2 between the late 1980s and 2005 by around 70 %. This would be significantly more than expected from production via photochemical O_3 decomposition. Other atmospheric pathways or sources of H_2O_2 are unlikely to cause any increase, even over a period of reducing air pollution. In a clean free troposphere, neglecting dry deposition and wet removal, the relationship among the oxidants (O_3, H_2O_2, OH, and HO_2) is simply given by the photochemical steady state:

$$
\begin{aligned}
O_3 + h\nu &\rightarrow O(^3P) + O_2 \\
O(^3P) + O_2 &\rightarrow O_3 \\
O_3 + h\nu &\rightarrow O(^1D) + O_2 \\
O(^1D) + H_2O &\rightarrow 2\,OH \\
OH + O_3 &\rightarrow HO_2 + O_2 \\
HO_2 + O_3 &\rightarrow OH + 2\,O_2 \\
HO_2 + HO_2 + H_2O &\rightarrow H_2O_2 + H_2O \\
H_2O_2 + h\nu &\rightarrow 2\,OH \\
OH + HO_2 &\rightarrow H_2O + O_2
\end{aligned}
$$

What would be the relation between O_3 and H_2O_2 in a clean atmosphere, was demonstrated by Ayers et al. (1996), who found over one-year measurements of total peroxides (1991/92) at the Cape Grim baseline station that the seasonal and diurnal cycles of ozone were anticorrelated with peroxides concentration, showing the "typical" feature: in the very clean marine boundary layer central processes were daytime photolytic destruction of ozone, transfer of reactive oxygen into the peroxides and continuous heterogeneous removal of peroxides at the ocean surface occurs.

Several model studies were concerning a changing global oxidation power. Modeling by Hough and Derwent (1987) showed that the increase of H_2O_2 due to rising NMVOC is significantly larger than that of O_3. On the other hand, increasing NO emission led to reduced H_2O_2 level. Both relations are facts, also confirmed by several experimental observations. Thompson et al. (1991) and Thompson (1992) calculated using different scenarios an increase in the global H_2O_2 concentration between about 20 % and 50 % between 1980 and 2030. Lelieveld et al. (2002) calculated that global

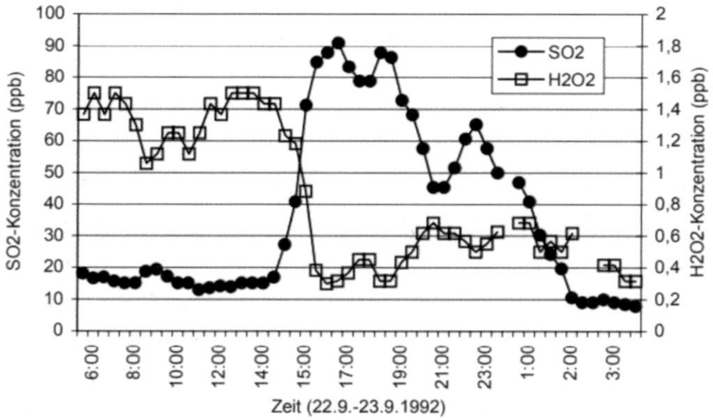

Figure 4.61: Timely variation of concentrations of SO_2 and H_2O_2 at the Mt. Brocken Station from the continuous record in September 1992 (Möller, unpublished); changing air mass around 3 pm with passing high SO_2 polluted air; even the anticorrelation between SO_2 and H_2O_2 is seen in fine structure.

primary (and gross) OH production has increased by about 50 % (and 60 % resp.) since the pre-industrial time. However, the global diurnal mean OH concentration has remained close to about 10^6 molecules cm^{-3}. On the other hand, it decreased merely about five percent. They argue that global OH constancy should not be confused with OH stability. The modeling studies inferred that the relative constancy of mean OH during industrialization is associated with the correlation between NO_x sources and CH_4 and CO. The concurrent growth of CH_4, CO, and NO_x emissions has intensified OH recycling and O_3 formation in the troposphere while increasing CH_4 and CO alone would have reduced global OH by at least a third.

Notwithstanding, since the Industrial Revolution, emissions of SO_2 not only increased exponentially as the main pollutant but also has taken over the role of H_2O_2 consumer via the aqueous phase S(IV)-H_2O_2 reaction, and obviously controlling the H_2O_2 content in the boundary layer (Fig. 4.61).

What now can be seen is that the relation among atmospheric oxidants (O_3, H_2O_2, OH and HO_2) is complex and depends from the evolution of the CH_4, CO, NO_x, and SO_2 emissions. Unarguable the increasing emission of SO_2 would have diminished an increasing production of H_2O_2 as discussed for ozone. On the other site, with decreasing atmospheric SO_2 content due to the beginning emission abatement around 1980, an increase in atmospheric H_2O_2 concentrations, namely in cloud water and precipitation, would be expected. Möller and Mauersberger (1992) have found by modeling that in clouds for 1 ppb gas-phase H_2O_2 an aqueous-phase concentration of H_2O_2 of around 100 μM results, and for gas-phase SO_2 concentrations less than around 0.1 ppb the cloud water act as a source of H_2O_2 (without consideration of photocatalytic processes that would act as further source); on the other hand, for >0.5 ppb SO_2 the aqueous phase act as sink of H_2O_2, and reduce after 10 minutes the primary H_2O_2 concentration to less than 0.5 μM. We have thus concluded (Möller and Mauersberger 1992, p. 162) that in this way, SO_2 limits via the cloud water the gas phase H_2O_2. Because of the practically total reduction of SO_2 emissions in the eastern part of Germany between the years 1990 and 2000, one would expect that the atmospheric H_2O_2 concentration is hypothetically rising. This anticorrelation between SO_2 and H_2O_2 can explain the strong increase in H_2O_2 in Greenland ice between 1975 and 1995 as suggested by Möller (1999a, 2002b).

What now can be further concluded from the nineteenth century atmospheric chemistry, is the assumption that the concentration of H_2O_2 in the troposphere of the mid-nineteenth century was very likely several times higher than today, eventually up to 2 ppb and more (see Möller 2020, p. 546), comparing to a present summer day mean being around 0.5 ppb. Nevertheless, it remains speculative, without modeling, to present a "typical" surface-near H_2O_2 concentration of mid-nineteenth century.

4.6 Nitrogen compounds

Early in the nineteenth century, it has been recognized the importance of bound nitrogen for plant growth. *Cavendish*'s finding that in common atmospheric air under the action of electric sparks nitric acid is gained, and *Schönbein*'s believe that while combustion of hydrocarbons in air ammonium nitrite is produced, and his (and others) idea on the generation of ammonium nitrite from water and atmospheric air under the influence of heat, stimulated much research. Ammonium and nitrate are intriguingly combined in atmospheric air. *Alfred Ditte* writes (Ditte 1904, p. 245):

> A drop of rain allowed to evaporate spontaneously on a bit of glass leaves crystals toward the center. [...] Ammonium nitrate frequently forms remarkable groups of crystals in the shape of crosses and swords, like those obtained by evaporation a drop of snow water. [...] Their presence in the air can not be verified, since, as everyone knows, nitric acid and ammonia unite readily to form ammonium nitrate.

4.6.1 Ammonia (NH_3)

4.6.1.1 Early history and discovery

The ancients not know any compound of ammonia in an isolated state; what later was called "*sal ammoniac*" (NH_4Cl) was obviously rocky salt (NaCl) with traces of ammonium chloride (Kopp 1931, III, p. 237). In the "Historia" written in the fifth century before our era, *Herodotus* reports on a "salt in large lumps on the hill" were the Ammonians also called Ammonites or people of Ammon [Ἀμμώνιοι, in Latin Amonii] lived, a people of Africa, originally emigrants from Egypt and Ethiopia, occupying what is now the Oasis of Siwa of the Libyan Desert in Egypt near the border of modern-day

Libya. These people got their name for the Sanctuary of Ammon; the Egyptians identified the god with their own supreme deity Amun, the god of the Sun, and the first Greeks from Cyrene called the god Zeus Ammon. Siwa was known through the oasis' exportation of salt. *Plinius* referred to a variety of salt called hammoniacum (from ἄμμος = sand) in allusion to it occurring in the sands of the oasis. This shows that a number of salts were confused under one term (Mellor, VIII, p. 144). *Kopp* notes that all these salts were common NaCl and the meaning ἀμμωνιαχόν [ammoniacum] indicates only the origin and that the name sal ammoniacum was used as well for rocky salt and ammonium chloride; the latter was called exclusively sal armoniacum since the 13th century. However, NH_4Cl must be known from ancient times as volcanoes deposit and came from Asia as sal armenicum (Armenian salt) to Europe. Ammonium chloride is first referred by *Geber*, but it is unknown how he designated it; in later prepared translation in Latin, it is named sal ammoniacum or sal armoniacum.

There are different stories on its origin that the *sal ammoniac* was formed by solar distillation in the sands from the urine and dung of camels, and because of the near sea, sea salt (NaCl) deposited. *Kopp* further noted that at this time it was known that putrefied urine is of alkaline nature and thus was used like lye for washing.

Geber described the production of sal ammoniacum from human urine and salt, i. e., the release of NH_3 and HCl that together sublimate into the salt NH_4Cl. The Italian physician and chemist *Angelus Sala* (1576–1637) first noted in his "Aphorismorum Chymiatricorum Synopsis" (1620, pp. 25–26) that the salt gained by mixing of spiritus salis (HCl) with salis volatitis obtained from urine is identical with the sal armoniacum. *Glauber* (in 1648) and *Boyle* (in 1671) knew the analytical composition of this salt. The Mallorquin philosopher *Raymund Lull* (~1232–1316) already noted the great volatility of this salt (Kopp 1931, III, p. 243); he called the evaporate spiritus animalis. It was called by the Flemish alchemist *Isaac Hollandus* (unknown living data, but before 1600) spiritus urinae (*Harngeist* in German).[1258] *Glauber* also said spiritus volatilis salis armoniaci, finally by Bergman (1788) shortened to *Ammoniacum*. Ammonia as gas was first described by *Priestley* in 1774, who named it *alkaline air* and found that due to electric sparks the volume of this gas is enlarged (formation of nitrogen oxides). *Scheele* was the first in 1777 who assumed that it is composed by nitrogen and phlogiston. The composition as NH_3 was established in 1785 by *Berthollet* who found that a volume of ammonia consists from 0.725 hydrogen and 0.275 nitrogen. In air it was found by *Scheele* in 1786 by observing that a precipitate originated on the cork a bottle containing hydrochloric acid, identified as salt ammonia (NH_4Cl) and in the early 1800s confirmed by *Théodore de Saussure* (Saussure 1804, p. 207). Between 1810 and 1820 it became clear that the compound between ammonia and hydrogen (ammonium) a metallic-like compound is.

1258 Tractatus de urina quomodo per spiritum ejus omnes tinctura sint extrahenda. In: Theatrum Chemicum, Volume VI (ed. Lazarus Zetzner), printed 1659–61 in Strasburg, p. 566.

After the works of *Liebig* and *Boussingault* it was generally accepted the ammonia is diffused from the soils and the sea and globally distributed as an important factor in agriculture chemistry. Schloesing (1875)[1259] agreed with *Boussingault*'s finding that the sea is not only a large reservoir of fixed nitrogen but also a source of atmospheric ammonia.[1260] In the early research upon ammonia in the air, no separation with ammonium was made and possible; separate sampling of gaseous ammonia and particulate ammonium became only possible after the introduction of the diffusion denuder by the Swedish *Martin Ferm* in 1979. Smith (1872, p. 277) writes that "the ammonia is one of the sewages of the air: it is the result of decomposition". Still in 1900, it was stated that ammonia never existed freely (i. e., in gaseous form) in air but only in compounds with carbonate and others (Blücher 1900). The history of "ammonia" in rain, fog, dew and cloud water is described in Chapters 2.3 to 2.5. On the occurrence of ammonia in atmospheric air Gmelin (1907, p. 197) writes (translated from German):

> As bicarbonate, nitrite and nitrate, after *Chevallier* in the air of Paris as ammonium sulfide and ammonium acetate. This occurrence is partly attributed to electric impacts in the atmosphere and partly on decomposition of organic substances on the earth's surface.

4.6.1.2 Early measurements of ammonia in the air

Hayes (1851, p. 208) writes:

> Considerable quantities of ammonia were found by several observers at different times; but the most trustworthy results are those of Fresenius, in a series of experiments on the air over a dwelling in Giessen.

In the nineteenth century, two methods for sampling and determination of soluble and insoluble atmospheric bodies were used; first so-termed air-washing, that is, water or a specific solution is shaken in a vessel of the air to be examined (until hundreds of times), and second, passing the air through water or absorbing solution (aspiration method). First quantitative determination of ammonia in air (according to the sampling technique it refers to the sum of gaseous ammonia and particulate ammonium) were carried out by *Kenneth Treasure Kemp* (1805–1842), chemist in Edinburgh, by *Nicolaus Gräger* in Mühlhausen (Thuringia) in 1845, who also translated *Boussingault*'s "*Écomonie rurale*" into German, and *Fresenius* in 1848:

1259 He determined in sea water at Saint-Valery-en-Caux (English Channel) 0.2–0.3 mg L^{-1} nitrate and 0.4–0.5 mg L^{-1} ammonium.

1260 The idea of marine ammonia emission was also supported by the French geologist *Louis Dieulafait* (1829–1886) from Marseille in 1879 (*Ann. Chim. Phys.* **13**, 374–409).

3.68	(ppmm)[1261], before 1842, 300 feet above the Irish Sea (*Kemp*)[1262]
0.323	(ppmm), 4 days in May 1845 in Mühlhausen (Gräger 1845)[1263]
0.133	(ppmm), 40 days Aug.–Sept. 1848 suburb of Wiesbaden (Fresenius 1849)[1264]
1.2–47.6	(ppmm), n = 13, Boston 1849 (Horsford 1850)[1265]
0.017–0.029	(ppmm), 1849–1852, Paris (Ville 1852)
4.5	(mg m^{-3}), Caen, winter 1851–1852 (Pierre 1852)
0.65	(mg m^{-3}), Caen, one year observation 1852–1853 (Pierre 1859, p. 39)
0.02–0.54	(ppmm), n = 9, mean of 0.15 pppm, Lyon (Bineau 1854, 1855a)
0.09–0.19	(mg m^{-3}), n = 6, 1869, Burton-on-Trent, fall (Brown 1870)
0.12–0.14	(mg m^{-3}), n = 5, 1869/70, country, winter (Brown 1870)
0.93–2.32	(mg m^{-3}), n = 8, Clermont-Ferrand (Truchot 1873b)
3.18	(mg m^{-3}), n = 1, Puy de Dôme (Truchot 1873b)
5.55	(mg m^{-3}), n = 2, Pic de Sancy (Truchot 1873b)
0.086	(mg m^{-3}), 1878 Prince's Road in Manchester (Smith 1879)[1266]

1261 Mass ratio, i. e., gram per million grams of air. To recalculate in mg m^{-3}, this value must by multiplicated by factor 1.293 (1 kg air corresponds to 1293 kg under standard conditions).

1262 *Kemp* likely never published a paper; he is first cited by Fresenius (1849) without any source of communication. D. F. Gregory writes (*The London and Edinburgh Philosophical Magazine and Journal of Science* 13 (1838) 434) on *Kemp*: "I do not know whether or not Mr. Kemp communicated a paper [. . .] to any of the journals, but I know that he communicated them freely to other chemists". *Kemp*'s first appointment was that of lecturer on practical chemistry in Surgeon's Square. He proceeded thence to a similar position at the University, which he held until his early death. He seems to have possessed great experimental skill. Among other subjects he investigated the laws of combustion and the liquefaction of gases. He was the first chemist in this country who succeeded in solidifying carbonic acid gas, which he appears to have hoped for equal success in relation to every other gas. He told his students that they might one day see him carrying a stick of solid hydrogen (source: www.kempfamilyhistory.com).

1263 *Gräger* writes 0.6149 parts ammonium carbonate [*kohlensaures Ammoniak*] in 1,000,000 of air (ppmm). Gmelin (1852, p. 837) cites wrongly 0.938 (ppm) "*kohlensaures Ammoniak*" or 0.508 (ppmm) "*Ammoniumoxyd*" or 0.333 (ppmm) ammonia; the latter value is cited by all later scientists. However, there are several "problems" in recalculation; *Gräger* estimated after absorption of 36 cubic feet air into hydrochloric solution and precipitation with platinum chloride the weight of "*Platinsalmiak*" to be 0.006 g = 0.0008466 g "*kohlensaures Ammoniak*". Fresenius (1849) recalculates *Gräger*'s 0.006 g "*Platinsalmiak*" in 0.0007 g Ammon (NH_4O), which gives 0.508 ppmm Ammon = 0.323 ppmm NH_3 = 0.938 ppmm NH_4OCO_3 (thus the values in *Gmelin* are taken from *Fresenius*). The latter value, however, is only correct (0.938/0.323) when the formula $(NH_4)_2CO_3$ for "*kohlensaures Ammoniak*" is used. Assuming "*Platinsalmiak*" to be platinum hexachloroplatinate $(NH_4)_2PtCl_6$ it follows never 0.008466 g for "*kohlensaures Ammoniak*" as given by Gräger (1845) but either 0.0013 g $(NH_4)_2CO_3$ or 0.002 g NH_4OCO_2. The ratios between *Fresenius*' "*Ammon*" and *Gräger*'s "*kohlensaures Ammoniak*" remain obscure too.

1264 Daytime 0.098 ppm and nocturnal 0.257 ppm. *Fresenius* writes that *Gräger* not proved the chemicals to be free of ammonia; the value by *Kemp* he doubts to be too high.

1265 *Eben Norton Horsford* (1818–1892) was a US American chemist and student at *Liebig*. He stated that his value is several times higher than those of *Fresenius* (Horsford 1849).

1266 *Smith* found in his office 0.167 mg m^{-3} ammonia, more than outside.

0.061	($mg\ m^{-3}$), $n = 18$ (0.028–0.20) London 1869 (Smith 1872, p. 433)
0.078	($mg\ m^{-3}$), $n = 4$ (0.037–0.101), Glasgow 1870 (Smith 1872, p. 434)
0.086–0.12	($mg\ m^{-3}$), London (Smith 1879)
0.03–1	($mg\ m^{-3}$), Saint-Valery-en-Caux (Schloesing 1875)
0.012–0.02	($mg\ m^{-3}$), annual means Montsouris (Paris) 1875–1879, Lévy (1880)[1267]
0.032	($mg\ m^{-3}$), Budapest (Fodor 1881)
0.014–0.03	($mg\ m^{-3}$), $n = 8$, August 1882, Pic du Midi (Müntz and Aubin 1882a)
0.011–0.137	($mg\ m^{-3}$), $n = 15$, April–June 1894, Cleveland (Mabery 1895)
0.062–0.072	($mg\ m^{-3}$), Buenos Aires 1905 (Anonymous 1907)

It seems that *Georges Ville* (1825–1897), French agricultural chemist and agronomist at Paris, conducted the first *reliable* and *long-term* measurements of ammonia in air, showing only small variations, corresponding to 22–38 $\mu g\ m^{-3}$ (Ville 1852). His values are close to that gained 25 years later by *Lévy*. The measurements by *Gräger*, *Kemp*, and *Fresenius* are doubtless erroneous. *Horace Brown* (his values are originally calculated as ammonium carbonate) notes that immediately after heavy rain, the quantity falls somewhat down, "but the air is again restored to its normal condition after a lapse of two or three hours". He further notifies "when the air is passed through cotton-wool before entering the absorption-tubes, it is found to be entirely deprived of its ammonia by the filter". *Ippolito Macagno* (no living data known), director of the agricultural station of Palermo, who also determined oxygen, carbon dioxide, and organic matter,[1268] determined unbelievable high ammonia concentrations between February and August 1879: ($n = 10$) of 250 (50–800) $mg\ m^{-3}$ (Macagno 1879).

Carl Alexander Müller (1828–1906), an agricultural chemist in Stockholm, reports in a very short communication (Müller 1865) on a long-term experiment on absorption of ammonia onto shallow glass bowls (cuvette), filled with distilled sulfuric acid and protected with a roof to avoid rainwater deposition, over four month and found a deposition of 0.028 g per square foot or equivalent to about 4 $kg\ ha^{-1}$ (mentioned that 23 kg is the minimum for fertilizing, thus atmospheric deposition cannot replace nitrogen fertilizing, he concluded).

Angus Smith already clearly expressed that "ammonia was everywhere", "and in the air it is always found", and is very soluble in water "it touches all substances and can be found on many" (Smith 1879, pp. 269–270). *Smith* hung clean and empty lab-

1267 Small variations from year to year (1875–1879) and in 1879 from month to month (0.017–0.024) and between 4 sites (0.015–0.019) in Paris. Mean of 1979 from monthly values: 21 (17–24) $\mu g\ m^{-3}$ at Montsouris and at Clichy (June to November 1879): 16 (7–18) $\mu g\ m^{-3}$. Further Lévy (1882) reports on the variation of the monthly means of NH_3 concentration (1877–1881) between 14 and 41 $\mu g\ m^{-3}$ and yearly mean variation 18–32 $\mu g\ m^{-3}$ with a total mean of 22 $\mu g\ m^{-3}$. The concentration is in summer larger than in winter and larger in Paris than at the observatory. The organic N total mean amounts 6 $\mu g\ m^{-3}$ (monthly mean variations 2–10), no seasonal variation observed.

1268 Org ($n = 20$) 1230 (200–3630) $mg\ m^{-3}$ and O_2 19.994–20.977 %.

oratory flasks in various parts of his laboratory and "ammonia could be observed after an hour and a half's exposure at any rate"; he presents several Tables concerns "absorption of ammoniacal substances from air near the ground and other places by pure water placed in a basin under a bell jar" in Manchester 1875. From *Smith's* data, a mean dry deposition value of $27 \cdot 10^{-5}$ mg m^{-2} s^{-1} result; from "typical" today's values (NH_3 concentration 5 µg m^{-3} and dry deposition velocity 0.8 cm s^{-1}) a deposition of $4 \cdot 10^{-5}$ mg m^{-2} s^{-1} result, about 7-fold less – but *Smith* found NH_3 in air to be 86 µg m^{-3}; thus the historical values are reliable, and being much higher than nowadays, namely in towns due to missing or insufficient management of water pollution. *Charles Mabery* who studied air pollution in 1894 in Cleveland denotes (Mabery 1895, p. 119) "the presence of ammonia as a normal constituent of the atmosphere". Comparing his mean values (57 µg m^{-3} NH_3 and 28 µg m^{-3} of albuminoid NH_3) with much smaller values (without giving a reference) from British cities (6–8 µg m^{-3} NH_3).

Heinrich (1881), an unknown German agricultural chemist, conducted over two years measurements of the absorption of ammonia from air onto a glass sampler (78.5 cm^2 square) filled with hydrochloride acid solution (20 %) and determined a total dry deposition of 30.6 kg NH_3-N ha^{-1} yr^{-1}; parallel rainwater sampling resulted in 7 kg NH_3-N ha^{-1} yr^{-1} deposition by rain – this is the first study a simultaneous dry and wet deposition of ammonia. *Max Ludwig Otto Müller*, another unknown German chemist from the Technical College [*Technische Hochschule*] at Braunschweig, was motivated to investigate ammonia in precipitation due to corrosion of roofs made from copper sheets and gutter pipes made from zinc sheets (Müller 1888). He found in freshly fallen snow in 1885 on average 1.5 mg L^{-1}, and this amount increased after a few days upon 5 mg L^{-1}, whereas in the lower snow layers the amount of ammonia did not change. He concluded correctly that the snow absorbs ammonia from the air. As a source of ammonia *Müller* identified the smoke from firing installations in the city. In parts of the city with the highest populations density, he also found the largest amount of ammonia in freshly fallen snow in winter, but in a snow collected at night (no heating and no smoke), he determined only ¼ of the amount of ammonia from daytime. To prove the hypothesis that ammonia is emitted by fuel smoke and diluted while dispersion,[1269] he conducted measurements in Harzburg (Harz Mountains) and found significantly less ammonia in freshly fallen snow increasing with time. Finally – and this is surely the first investigation of ammonia emission from a cow-house – *Müller* determined ammonia near a cow-house lonely situated in the Mountains; near the house, he found more ammonia than in more distant and only at the surface but not in deeper layers of the snow, concluding that the precipitation was free of ammonia.

1269 It is worth to refer the German original phrase here (Müller 1888, p. 244) "[...] *denn das aus den Rauchgasen stammende Ammoniak der Luft wird ja bald in entsprechender Verdünnung überall hingetragen* [... because the ammonia in the air from the smoke will be carried everywhere in the appropriate dilution]."

It was only accumulated from the air by absorption. *Alfred Daniel Hall* (1864–1942), director of the Rothamsted Experimental Station (1902–1915), studied together with *N. H. J. Miller* the absorption of ammonia from the atmosphere; hence, it was one of the first studies on *dry deposition* (Hall and Miller 1911); likely *Miller* was the principal investigator, who studied for many years the rainwater chemical composition (see Chapter 2.3.2.2).

Schloesing (1876) first studied the reservoir distribution of ammonia between natural water and the air, giving values for $r = c(\text{air})/c(\text{aq})$ in dependence on temperature; he reports on different behavior of ammonia dissolved as carbonate or nitrate: from carbonate solution (cloud or rain) ammonia transfers into air also in the case if it already contains relative large amounts of ammonia, but not from nitrate solution (electrolytic dissociation and the pH dependent ammonium/ammonia equilibrium was not yet known in that time).

With the turn of the century, the interest in direct measurements of atmospheric ammonia ceased completely, and only a half century later, new determinations of gaseous ammonia in air were conducted. Stoklasa (1906) detected ammonia during the Vesuvian eruption.

4.6.1.3 Studies of atmospheric ammonia in the twentieth century

The research interest in the 1950s was exclusively focused further on agriculture and the nitrogen cycle, hence, deposition of ammonium-nitrogen. Later, atmospheric ammonia became of interest because of

- formation of particulate matter, namely sulfates,
- as air pollutant (damages to vegetation near large point source),
- eutrophication of lakes and acidification of soils.

First measurements of atmospheric ammonia (and other trace gases such as SO_2 and NO_2) have been conducted with the establishment of the Swedish precipitation network in 1953 by Egnér and Eriksson (1955); small amounts of air were continuously sucked through absorption solutions at these stations to obtain monthly averages. From 4 stations (1954–1955), it was found an average of 3.0 (0.6–9.4) $\mu g\ m^{-3}$, predominantly representing the gas phase the authors notified.[1270]

Junge (1956) conducted measurements of ammonia (note that in that time all "ammonia" measurements in air corresponds to the sum of gaseous ammonia and particulate ammonium) at Hawaii and Florida in 1954, ranging from 8 to 0.7 $\mu g\ m^{-3}$. In his

1270 However, according to our modern experiences in sampling atmospheric ammonia it is important to note that all values obtained by air bubbling through absorption solutions also include partly (or even totally) particulate ammonium. On the other hand, analytical values obtained by sampling particulate matter using filters, include gaseous ammonia which is absorbed, and at higher temperatures a deficit of ammonium due to evaporation of ammonia.

pioneering investigations of atmospheric aerosol in June 1953 at the Round Hill Station, Junge (1954) found that large particles, with $0.08 < r < 0.8\,\mu$, consisted nearly entirely of sulfate and ammonia. Besides these substances (but in significantly smaller concentration), the giant particles, with $0.8 < r < 8\,\mu$, contained variable, though sometimes considerable, amounts of nitrate and sodium chloride. Several measurements in Germany (Georgii 1960, Georgii and Jost 1964) have shown much less ammonia at remote sites (Kleiner Feldberg and Zugspitze with around 7 µg m^{-3}) than at Frankfurt/Main (around 20 µg m^{-3}). Junge (1963a) concluded from the almost similar NH_3 concentrations in summer and winter (in contrast to SO_2 and NO_2) and the fact that the town values (in Frankfurt) are much higher than those from rural sites an anthropogenic production from nearby chemical industries. Further, *Junge* argued that the ocean could be a source of NH_3, but the role of the sea as a source or a sink for NH_3 is not quite clear. Although the ocean was first proposed as a source of ammonia by Boussingault (1856a,c), it was principally assumed until 1980 that the ocean emits no NH_3 (Lenhard and Georgii 1980). Georgii (1963) also concluded from the large ammonia concentrations in air at Frankfurt, namely in winter, that combustion of coal is an important source (cf. the results by Müller (1888) at Braunschweig), whereas, at the Taunus observatory, a more rural site, a spring maximum was observed, indicating another source of ammonia. *Georgii* further stated (Müller 1888, p. 3969) that

> NH_3 in conjunction with SO_2 and cloud droplets, seems to be an important participant in the process of aerosol formation. Both are readily absorbed in cloud droplets were. After catalytic oxidation of SO_2 and subsequent evaporation of the droplets, $(NH_4)_2SO_4$ particles remain.

The process of photochemical sulfuric acid production via homogenous nucleation in the presence of ammonia was not yet known. The question remained open for the following ten years (Georgii and Müller 1974). These authors conducted 75 aircraft-ascents over West Germany to measure profiles of NH_3 and SO_2 between 1969 and 1972. It was learned from these measurements that NH_3 concentration rapidly decreases with altitude and is much larger in summer and on warm days. Through this works, it was recognized that ammonia is the single species to neutralize the acids produced by the oxidation of SO_2 and NO_2 and it enters particulate matter from the gas phase. However, not before the 1980s a complete understanding of ammonia began with the necessary differentiation between gaseous ammonia and particulate ammonium. A remarkable investigation was carried out by *Michael Goethel* in his Diploma-Thesis (Goethel 1980) using the *Ferm* denuder, later meteorologist at the German Weather Service, who likely published the first annual cycle of ammonia and ammonium at a rural site near Frankfurt, having a summer maximum and a midday maximum in the diurnal cycle. Independently *R. Bos* (unknown given name) from Research Institute for Environmental Hygiene in the Netherlands (at TNO) developed the first automatic denuder based on the first denuder by Ferm (1979) for quasi-continuous measurement of ammonia (Bos 1980), writing

Ambient air measurements consistently show a maximum in the NH_3 concentrations at noon. This maximum can be explained as a temperature effect on the dissociation equilibrium of ammonium nitrate aerosol in the atmosphere.

In further ammonia profile aircraft measurement by Lenhard and Gravenhorst (1980)[1271] it was found that atmospheric ammonia is caused mainly by volatilization from domestic animal excrements and that liberation from mineral fertilizer and natural soils contributes only a small amount to the overall production rate. For the conversion of gaseous ammonia into particulate ammonium they deduced from flux density considerations a lifetime of about half-day.

Shizuo Tsunogai (1938–2015)[1272] must be seen as another pioneer in atmospheric ammonia research. Tsunogai and Ikeuchi (1968) published the first diagram of the geochemical cycle of nitrogen compounds. Tsunogai (1971a), who also excluded like Georgii (1963) the ocean as a source of ammonia, found an increasing ratio ammonium/ammonia over the sea with distance from land and that the removal over the sea is exclusively due to precipitation, having a lifetime of about 30 days. The very low ammonia marine background concentration of less than 0.1 ppb was also confirmed by Ayers and Grass (1980). In the 1990s, several researchers concluded that the ocean is a dominant natural source of ammonia (Schlesinger and Hartley 1992, Dentener and Crutzen 1994, Bouwman et al. 1997). Recent work (Paulot et al. 2015) based on a large number of measured concentrations in seawater and air, and supported by different models, identify the ocean as a natural source of ammonia only over the Equatorial Pacific and parts of the Southern Ocean with lower global estimates.

In the 1970s investigation begun to establish global ammonia budgets, i. e., to compile sources and sinks (Søderlund and Svensson 1976, Dawson 1977, Böttger et al. 1978, Möller and Schieferdecker 1982, and others). In the 1980s arose the interest in understanding the acidification of the environment and the role that plays ammonia; Derwent (1987) summarized the possible mechanisms that may involve:
- the participation of ammonia in the cloud droplet phase reaction of sulfur dioxide,
- the influence of ammonia on the dry deposition removal of sulfur dioxide,
- the partial neutralization of strong acids in precipitation,
- the formation of ammonium aerosols and their potential range transport.

Furthermore, agricultural activities like intensive farming with excess fertilizing and large stock-house management led to increasing emission of ammonia and subsequent deposition in many countries, and extensive measurements of atmospheric

1271 *Ulrich Lenhard* is a retired physico-chemist at University Kiel and the meteorologist *Gode Gravenhorst* (born about 1950), a scholar of *Georgii*, and later professor for bioclimatology in Göttingen.
1272 Japanese chemist at the Hokkaido University; title of his PhD: Chemical study of meteoric precipitation: Concentration, depositional rates and sources of chemical constituents in precipitation and air-borne dust (1966).

concentrations, dry deposition fluxes, wet deposition and modeling to obtain NH_x budgets on different scales. Important contributions were made by Dutch scientists, e. g., *Willem A. H. Asman*, *A. F. (Lex) Bouwman* (born 1956), *Ed Buijsman* (born 1948), *Jan Duyzer* (born 1953), *Jan Willem Erisman* (born 1961), *Jack Slanina* (1942–2009), by British scientists, e. g., *Helen ApSimon* (born 1942), *David Fowler* (born 1949), *Mark Sutton* (born about 1963) and many more, e. g., *László Horváth* (Hungary), *Owen Thomas Denmead* (Australia); see also Sutton et al. (2008) and literature therein.

4.6.2 Oxides and oxoacids of nitrogen

Today, we know the presence of N_2O, NO, NO_2, NO_3, N_2O_5, HNO_2 and HNO_3 in the atmospheric air (see Table 1.3 for archaic names); of minor importance are N_2O_3 (in equilibrium with NO + NO_2), N_2O_4 (in equilibrium with 2 NO_2 and peroxonitric acid HNO_4; furthermore, several oxidized organic nitrogen compounds without known natural sources have been detected in the air (not treated in this volume). The oldest known fixed nitrogen compounds are saltpeter (potassium nitrate), since the Middle Age nitric acid (*spiritus nitre*) and in the eighteenth century nitrous gases; however, its constitutions became finally clear after the mid-nineteenth century.

4.6.2.1 Early history and discovery

Saltpeter (German *Salpeter*, French *saltpêtre*), derived from Greek and Latin sal petrae (petra = rock, stone) and likely first named in Latin works of *Geber* in the 8th century and clearly described as KNO_3 because of the production of *aqua regia* (HNO_3 + HCl) and *aqua fortis* (HNO_3); Kopp (1931, III, p. 219). The *nitrum* of the ancients, however, was natron (Na_2CO_3), mentioned in the books of the Old Testament as *neter*, described as *νίτρον* by the Greeks and nitrum by the Romanians (Kopp 1931, IV, p. 23). The name *nitrum* for this alkali (*Geber* called it *Sal Alkali*; today natron or soda) was used for centuries and led to confusion with the term *nitre*, later used for saltpeter. *Marcus Graecus*[1273] named it sal petrosum and described its use in gun powder; it was also known as wall saltpeter. The origin of the term *peter*[1274] is unknown according to *Koop* who also speculates that it could name the city Petra as place of finding (compare the history of *sal ammoniac*). In the thirteenth century (besides sal petrae and sal petrosum) this salt was called sal nitrum, sal nitri or sal-nitro to distinguish it from nitrum (Na_2CO_3). *Basilius Valentinus* lets saltpeter "speak" (Kopp 1931, III, p. 223):

1273 Fictive Byzantine writer: Liber ignium, published in the 12th century in Spain and likely based on Arabic origin.

1274 *Henshaw* (see p. 185) used the term *rockpeeter* and *rockpetre* not for saltpeter but for rocky salt (NaCl), what would mean "rocky rock".

> Zwei Elemente werden in mir am moisten befunden, als Feuer und Luft; Wasser und Erden am wenigsten; drum bin ich feurig und flüchtig [Two elements were mostly found in me, as fire and air; water and earth at least; that's because I am fiery and volatile].

The formation of nitric acid by acid treatment of saltpeter with sulfuric acid (which was seen as the "mother" of acids) and distillation was known since *Geber*, who named it aqua fortis. *Glauber* (17th century) called it spiritus nitri, and about 100 years later the name Acidum nitri was used (in French *acide nitreux* and since 1787 with the new nomenclature *acide nitrique*); only in German the archaic name "*Salpetersäure*" remained. *Helmont* was not only the discoverer of carbon dioxide, he also was acquainted with nitrous gas, produced by the action of aqua fortis when silver is dissolved in it; and he also knew, that when it came in contact with the atmosphere it formed fiery red vapors (cited after Gmelin 1801, p. 194).

In the seventeenth century naturalists (*Evelyn, Digby, Sylvius, Henshaw, Boyle*), based on its finding in dew water, had the view that *nitre* (KNO_3) occurs in the atmosphere (as aerial salt); see pp. 79, 113, 115, 144, 147–148, 287–288. Boyle (1692, pp. 41–42) wrote:

> I know that divers learned Men, some physicians, some Chymists, and some also Philosophers, speak much of *Volatile Nitre* that abounds in the Air, as if that were the only salt wherewith it is impregnated. But though I agree with then, in thinking that the Air is in many Places impregnated with Corpuscles of a Nitrous Nature; yet I confess I have not been hitherto convinc'd of all that is wont to be delivered about the Plenty and Quantity of the Nitre in the Air: For I have not found, that those that build so much upon this volatile Nitre, have made out by and competent Experiment, that there is such a volatile Nitre abounding in the Air. For having often dealt with Salt-peter in the Fire, I do not find it to be easy to be raised by a gentle Heat; and when by stronger Fire, we distil it in close Vessels, 'tis plain that what the Chymists call Spirit of Nitre, has quite different Properties from crude nitre, and from those that air ascribed to the volatile Nitre of the Air; their spirits being so far from being refreshing to the Nature of Animals, that they are exceedingly corrosive.

The soon later following phrase by *Boyle* (Boyle 1692, p. 49) reads like the simultaneous presence of nitrous gases (NO_x), nitric acid (HNO_3), and sulfur dioxide (SO_2):

> We may also hang up in such an Air, Clothes or Silks died with such Colours, that Nitrous (for Instance) or Salino-Sulphureous Spirits (as some Chymists call them) were found peculiarly apt to make to fade, or to discolour them.

Priestley (1772, p. 252) writes on nitrous air:

> Though it was casually observed by Hales, he gave but little attention to it, [...] no name has been given to it [...] I happened to distinguish it by the name nitrous air, because I had procured it by means of spirit of nitre only (p. 66–67).
> This air has a strong disagreeable smell differing but little from that of smoking spirit of nitre when mixing with common air, getting a red or deep orange color, and producing of heat, at the same time the mixture dimishes considerable in it (p. 252).

> [...] nitrous air is exceedingly fatal to vegetables (p. 258).

In 1776, *Lavoisier* has shown that nitric acid consists of nitrous gas and oxygen, but from what the first is composed, he could not decide. After *Cavendish*'s experiments in 1784, *Lavoisier* analyzed nitric acid and found that it consists of 20.5 % nitrogen and 79.5 % oxygen (correctly 22 % and 76 %). Cavendish (1784) and Priestley (1789) described formation of nitrous acid in moist air under the influence of electric discharges. It is a pleasure to read, how *Cavendish* carefully designed his experiments with airs and made his conclusions, showing by the following notes from Cavendish (1785), describing the conversion of phlogisticated air [nitrogen] into nitrous acid (Cavendish 1785, p. 372). By careful experiments, done over mercury and using lime-water as absorbing agent for the acids (afterward he used soap-lees [potash solution] because of better absorption capacity), he excluded that "no fixed air [CO_2] was generated in the operation" (Cavendish 1785, p. 375). *Cavendish* "found, that, when good dephlogisticated air [pure oxygen] was used, the diminution was but small; when perfectly phlogisticated air [pure nitrogen] was used, no sensitive diminution took place" (Cavendish 1785 p. 376). These observations tell us now that nitrogen not dissociates under the influence of electric sparks, but oxygen, whereas the diminution of the gas volume was due to the formation of ozone. In his crucial experiment, he used a mixture of five parts of oxygen and 3 parts of common air, equal to seven parts of oxygen and 3 parts of nitrogen (Cavendish 1785, p. 376), and writes (Cavendish 1785, p. 377)

> [...] and the spark was continued till no more air could be made to disappear. The liquor, when poured out of the tube, smelled evidently of phlogisticated nitrous acid [nitrous gases], and being evaporated to dryness, yielded 1.4 gr. of salt, which is pretty exactly in weight to the nitre.

Thus, *Cavendish* concluded "that phlogisticated air [nitrogen] is nothing else than nitrous acid united to phlogiston" (Cavendish 1785, p. 379). This is hard to understand, but in the view that Cavendish (1784, p. 137) wrote that "[...] inflammable air [hydrogen] is either phlogiston, as Dr. Priestley and Mr. Kirwan suppose, or else water united to phlogiston", and he favored the latter assumption, we see the analogy of both phrases, but we have no idea what he meant with "united to", namely in the background that *Cavendish* clearly stated that the combination of oxygen with nitrogen (due to electric sparks) yield nitrous air (NO) and that of oxygen with hydrogen (by explosion) yield water (H_2O). However, a few pages later (Cavendish 1784, p. 140), *Cavendish* writes that "inflammable air [hydrogen] is phlogisticated water", and, now comparing both phrases' it seems that "united to phlogiston" means "phlogisticated". The next conclusion, "[...] dephlogisticated air [oxygen] is only water deprived of its phlogiston", however, is very clear when adopting that phlogiston = hydrogen. Further, he writes

> [...] phlogisticated air [nitrogen] ought also to be reduced [it is an oxidation] to nitrous acid [...], or forms a chemical combination with dephlogisticated air [oxygen]; only the acid will be more dilute [...].

This phrase could tell us that the first step in the reaction between nitrogen and oxygen is the formation of NO, including further oxidation (in modern terms) to NO_2 and NO_3, followed by the formation of HNO_2 and HNO_3 after dissolution in the potash solution, but not excluding gas-phase formation too due to the presence of humidity.

Van Marum (1786, p. 53) concluded already in 1785 from his experiment with the electrifying machine and different airs, that "*Salpetersäure besteht aus reiner Luft, Mozette und Wasser*" [nitric acid consist from pure air, Mozette and water], surprisingly correct.[1275] It was *Scheele* who first concluded on different kinds of nitrous airs,[1276] first summarizing the observations known since *Glauber* (Scheele 1780, p. 21):

> Chemists have frequently distilled the fuming spirit of nitre, from oil of vitriol spread on saltpetre, observing, that this acid is in the beginning of the observation red; in the segued white and colorless; and at the last again so intensely red, that it is impossible to look across the receiver.[1277]

Scheele's observations clearly indicate the "airs" as nitrogen monoxide (NO), die transformation of NO into nitrogen dioxide (NO_2) and the release of nitrous acid (HNO_2) and formation of its salts (nitrites); (Scheele 1780, p. 23), my comments in squared brackets:

> [...] the acid [HNO_3] is converted into a kind of air [NO], which will neither unite with alkalies, nor with absorbent earth, and with water only in small proportions: if this acid of nitre [NO] analogous to air, meets the air, the latter attracts the phlogiston, and loses its elasticity. The vapours become red [NO_2], the air undergoes likewise this remarkable and natural change, [...] if the acid of nitre [HNO_3] receives a still less proportion of phlogiston, it is likewise changed into a kind of air [HNO_2]; which, like common air is invisible, but capable of uniting with alkalies and terreous substances, and of yielding by their mixtures true neutral salts [nitrites].

Today's nitric oxide [NO] was to *Priestley* "*nitrous air*" and our nitrous oxide [N_2O] he originally named in 1777 as "*nitrous air, diminished*", on account of his preparative method of allowing NO to stand in contact with moist iron filings in 1774 (initially mixed with sulfur, but this was later found not to be essential). Today's equation is $2\,NO + H_2O + Fe \rightarrow N_2O + Fe(OH)_2$. However, the credit to discover N_2O is not to *Priestley* but to *Joseph Black*, who published little of his own work, but accounts of his lectures

1275 Pure air = O_2, Mozette = N_2.

1276 Today, nitrous gas = N_2O. In that time, nitrous air (gases) denoted NO_x (NO and NO_2); German "*nitrose Gase*" and French "*l'air nitreux*" and "*nitreux oxygénés*".

1277 Red fuming nitric acid contains 84 % HNO_3, 13 % N_2O_3 ($NO + NO_2$) and 2 % H_2O. The red color is from NO_2.

and demonstrations survive and one from 1768 records[1278] (cited after Smith 1872, p. 299):

> Ammon Nitros: is the most fusible of the Common Salts; when the heat is increased is copiously converted into Vapour; the degree of heat Sufficient for its fusion is that of boiling water if exposed to a sudden heat undergoes a deflagration although no inflammable matter be added to it.

Today this reaction we write as $NH_4NO_3 \rightarrow N_2O + 2\,H_2O$. *Humphry Davy* (1778–1829) makes his famous statement about the possible use of nitrous oxide in surgery (Davy 1800, p. 556):

> As nitrous oxide in its extensive operation appears capable of destroying physical pain, it may probably be used with advantage during surgical operations in which no great effusion of blood takes place.

4.6.2.2 The compounds of oxidized nitrogen

Gmelin (1827, pp. 423–442) summarizes the knowledge based on researches by *Lavoisier, Cavendish, Dulong, Dalton, Davy, Gay-Lussac, Berzelius*, and others (note that formulas were not yet given, but equimolar mass ratios between nitrogen and oxygen from which the listed formula are derived):

	English	French	German
N_2O	nitrous oxide	oxide d'azote	Stickoxydul
NO	nitric oxide	oxide (gaz) nitreux[a]	Stickoxyd
N_2O_3	hyponitric acid	acide pernitreux (hyponitreux)	untersalpetrige Säure
NO_2	nitrous acid	acide nitreux	salpetrige Säure[b]
N_2O_5	nitric acid	acide nitrique	Salpetersäure[c]

[a] Other names: oxide nitrique, deutoxide dázote
[b] Seen as mixture of *untersalpetrige Säure* and *Salpetersäure* or *unvollkommene* [imperfect] *Salpetersäure*
[c] Was only known as "wässerige Salpetersäure", i. e., in solution with water

Graham (1842, p. 282), referring *Turner* and *Berzelius*, lists the following oxide, and given "modern" formulas: nitrous oxide (NO), nitric oxide (NO_2), nitrous acid (NO_3), peroxide of nitrogen (NO_4), and nitric acid (NO_5). Other names in French for NO: *oxide nitrique, deutoxide dázote*. German chemists have seen "*salpetrige Säure*" as mixture of "*untersalpetrige Säure*" and "*Salpetersäure*" or "*unvollkommene*" [imperfect] "*Salpetersäure*" and "*Salpetersäure*" was only known as "*wässerige Salpetersäure*", i. e., in solution with water. *Dalton, Davy* and *Gay-Lussac* investigated different com-

1278 Manuscript lecture notes made by *Thomas Cochrane* (1775–1860), when he attended *Black*'s lectures in Chemistry in Edinburgh during the session 1767–1768 (edited by McKie 1961).

pounds between nitrogen and oxygen; Gay-Lussac (1816, p. 404) determined the volume ratios:

French name	azote	oxigéne	modern formula	modern name
oxide d'azote	100	50	N_2O	dinitrogen monoxide
gaz nitreux	100	100	NO	nitrogen monoxide
acide pernitreux[1279]	100	150	N_2O_3	dinitrogen trioxide
acide nitreux	100	200	NO_2	nitrogen dioxide
acide nitrique	100	250	N_2O_5	dinitrogen pentoxide

Thus, "*acide pernitreux*" (or *acide hyponitreux*, in German *Untersalpetersäure*), a name which *Gay-Lussac* denoted provisional, was nitrous acid (HNO_2). The anhydrides N_2O_3 and N_2O_5 have been seen as nitrous acid (HNO_2) and nitric acid (HNO_3), respectively. *Gay-Lussac* identified "*acide nitreux*" with "*gaz nitreuse*" or "*vapeur nitreuse*" ($NO_2 = N_2O_4$); but confusing remains the name "*gaz nitreux*" (nitrous gas)[1280] for the monoxide NO. The French chemist *Pierre Louis Dulong* (1785–1838) proposed the names "*l'acide des nitrites*" (= HNO_2) and "*l'acide nitreux anhydre*" (or *sec*) for the gas to distinguish it from the acid and also found that it decomposes to nitrate and nitrite in water. However, it was not understood that the anhydrides are N_2O_3 ($NO + NO_2$) for nitrous acid and nitric acid, not NO_2 but N_2O_5.

In the mid-nineteenth century, Gmelin (1852, pp. 803–829) lists the following compounds; note that the first column denotes the chemical formula used and the second column denotes the chemical composition, based on the N to O ratio (note NO_4 was seen as $N^2O^3 + N^2O^5$):

			English	French	German
NO	or	N_2O	nitrous oxide	oxide d'azote	Stickoxydul
NO^2	or	NO	nitric oxide	oxide nitrique	Stickoxyd
NO^3	or	N_2O_3	nitrous acid	acide hyponitreux	salpetrige Säure
NO_4	or	N_2O_4	hyponitric acid[1281]	acide nitreux	Untersalpetersäure
NO_5	or	N_2O_5	nitric acid	acide nitrique	Salpetersäure

1279 *Acide pernitreux* (pernitric acid) not to confuse with today's pernitric (but mostly termed peroxynitric) acid HNO_4 ($HOONO_2$), produced in the atmosphere in equilibrium with $HO_2 + NO_2$, first evidence obtained spectroscopically by Niki et al. (1977). Pernitrous (or peroxynitrous) acid HNO_3 (ONOOH) plays in biological chemistry an important role $NO + O_2^- \leftrightarrow ONOO^-$ and possibly in atmospheric aqueous chemistry as intermediate.

1280 In German "*nitrose Gases*" denote the sum $NO + NO_2$ as (not equimolar) mixture.

1281 Not to confuse with hyponitrous acid $H_2N_2O_2$, a dimer of HNO that decays to $N_2O + OH^-$. Hydronitrous acid H_2NO_2 ($2\,H_2NO_2 = HNO_2 + NOH + H2O$), also formal seen as anhydride of NO.

Formulas and names are quite confusing to the modern reader; what is seen that only dinitrogen monoxide (N_2O), nitric oxide (NO), and nitric acid (N_2O_5) were clearly recognized from the beginning, whereas nitrogen dioxide (NO_2) and nitrous acid (HNO_2) have been confused. On the other hand, nitrates and nitrites (as salts) were known since the ending eighteenth century. Schönbein (1846, p. 433) supposed two compounds of the formulas $NO_4 + HO_2$ and $NO_2 + HO_2$ (note that he regarded ozone to be HO_2 in that time) formed when the vapors of hyponitric acid (NO_4) are mixed with moist atmospheric air. NO_2 he called "*Untersalpetersäure*" [hyponitric acid] and writes (Schönbein 1852, p. 43): *Stickstoffsuperoxid, wie er diese Verbindung lieber nennen möchte* = $NO^2 + {}_2\overset{\circ}{O}$, *also* [...] *ozonisirtes Stickstoffoxid sein* [nitrogen peroxide, as he prefer to call it = $NO^2 + {}_2\overset{\circ}{O}$, thus ozonized nitrogen oxid]. In German, instead of "*Stick...*" (Table above) also "*Stickstoff...*" was used, and *Schönbein* writes that "*Stickstoffoxid*" NO^2 easily combines in the ordinary cold with oxygen gaining NO^4 (Schönbein 1852, p. 44). The oxygen, however, that reacts with NO^2 is not common [*gewöhnlicher*] but exited or ozonized [*erregt* or *ozonisirt*], denotes with $\overset{\circ}{O}$ in contrast to O. NO^4 *Schönbein* describes as a powerful oxidizer which releases two equivalents of oxygen in contact with water, forming nitric acid. Furthermore, in the presence of sulfuric acid, NO^4 instantly oxidizes it to sulfuric acid. Concerns the "*salpetrichte Säure*" [nitrous acid) (NO^3), *Schönbein* regard it as

> [...] keine primitive Oxidationsstufe des Stickstoffes, sondern nur eine lockere Verbindung oder gar nur ein Gemeng von NO^4 und NO^2 sei ...] [... is not a primitive oxidations state of nitrogen but only lose affiliation or even a mixture of NO^4 and NO^2 ...].

With today's knowledge, we could interpret *Schönbein*'s view as follows. NO_2 reacts with O_3 (ozonized oxygen), forming the nitrate radical NO_3 (and not NO^4), which we know as oxidizer, and that reacts with water under the formation of nitric acid (HNO_3). Moreover, not far from true is his assumption that nitrous acid (now in terms of its anhydride N_2O_3) is a "mixture" of NO and NO_2.

The hyponitric acid (also described as $N_2O_3 + N_2O_5$ by *Gmelin* and as $NO_3 + NO_5$ by *Graham*) should – not be confused with today's hyponitrous acid $H_2N_2O_2$ (HON=NOH or tautomer $H_2N{-}NO_2$) – does not exist, it was in that time often synonymously used for nitrous acid. Weltzien (1860b, p. 226) presents a list of oxygenated nitrogen compound: [...] *Radicale und Verbindungen, wobei ich die gebrauchten Benennungen als nicht maßgebend betrachten bitte* [... radicals and compounds but please do not consider the names as relevant]:

NO^2	Nitroxyd (Stickoxyd)	NO_2
NO^2O^2	Anhydrid der Nitroxydsäure (Untersalpetersäure)	NO_4
NO^2NO^2	Nitroxyl (unknown)	N_2O_4
$NO^2NO^2O^2$	Anhydrid der Nitroxylsäure (salpetrigen Säure)	N_2O_5
NO^4NO^4	Nitrodioxyl (unknown)	NO_8
$NO^4NO^4O^2$	Anhydrid der Nitrodioxylsäure (der Salpetersäure)	N_2O_{10}
$H^2NO^2O^4$	Nitroxydsäure (unknown)	H_2NO_6
HNO^2O^2	Nitroxylsäure (salpetrige Säure) (unknown)	HNO_4
HNO^4O^2	Salpetersäure	HNO_6
$H^2NO^2O^4$	Nitroxydsäure (unknown)	H_2NO_6

Edward Divers (1837–1912) discovered hyponitrite (first called salt of nitrous oxide) by treating a solution of nitrate by sodium (Divers 1871), first done by Schönbein (1861c) obtaining nitrite by reduction of solutions of nitrate by sodium and other metals (in *Divers*'s experiment nitrite is further reduced to HNO – nitroxyl). *Divers* presents a list of hitherto known oxoacids of nitrogen with his formula:

hyponitrous acid[a]	HNO	today known as nitroxyl ($\rightleftarrows NO^- + H^+$)
hyponitroso-nitrous acid	$H_2N_2O_3$	formal $HNO + HONO$ (unknown)
nitrous acid	HNO_2	
nitroso-nitric acid	$H_2N_2O_5$	formal $HNO_2 + HNO_3$ (unknown)
nitric acid	HNO_3	

[a]in German named as "*Hydronitrosylsäure*"

Divers also correctly writes the reaction of nitroxyl: $2\,HNO = N_2O + H_2O$. However, his acids $H_2N_2O_3$ and $H_2N_2O_5$ are speculative derived from N_2O_3 and N_2O_5, respectively, and also led to the name[1282] nitrosyl nitrite (O=N–O–N=O) and nitronium nitrate $O_2N{-}O{-}NO_2$ (N_2O_5). *Odling*, in discussion with Divers (1871) proposed the name nitrite-nitrate or hyponitrate for the salt. The existence of nitrous anhydride N_2O_3 in the gaseous state was experimentally shown by *Georg Lunge* (Lunge 1879). End of the nineteenth century (Gmelin 1907) the compounds have been recognized as we know it today (N_2O, NO, NO_2, N_2O_3, HNO_2, NO_2 or N_2O_4, N_2O_6 or NO_3, N_2O_5 and HNO_3); hyponitrous acid $H_2N_2O_2$ (HON=NOH) finally was also correctly described. NO_2 was also named nitrogen peroxide and nitrogen tetroxide as N_2O_4, respectively until about 1940.

Nevertheless, in *Cavendísh*'s experiments (and similar experiments in the nineteenth century) on the action of electric discharges on air, with the exception of N_2O, all oxidized nitrogen compounds have been gained in different relations depending on

1282 Note that nitrosyl (no to be confused with nitroxyl) is the NO cation NO^+ (also termed nitrosonium) and is the nitryl cation (also termed nitronium).

the experimental conditions, but almost seen as nitrous and nitric acid, and later distinguished by the salt's nitrite and nitrate.[1283] The terms nitrous gas and nitrous acid have thus been used before about 1850 as generic term for that what we know consider as NO_y. Still, Lode (1911, p. 389) writes under the influence of electric discharges and while slow oxidation of many bodies at air, first NO is gained from nitrogen and oxygen and in further oxidation nitrous (N_2O_3) and nitric acid (N_2O_5); however, these compounds do not exist free in air but almost exclusively bond together with ammonium as ammonium nitrite and nitrate he writes.

Yost and Russell (1946) in his "Systematic Inorganic Chemistry" describe the oxides N_2O, NO, N_2O_3, NO_2, N_2O_4, NO_3, and N_2O_5. There was a long dispute whether N_2O_3 exist as gas (molecule) or only as (equimolar) mixture NO + NO_2 (see Raschig and Prahl 1929);[1284] the fast reaction N_2O_3 (or NO + NO_2) + H_2O = 2 HNO_2 was known since end of the ninetheenth century, and extensively investigated in the lead chamber process (e. g., Raschig 1904); see p. 691 upon the reaction of NO with water. *Raschig* writes in 1929 that nitrogen oxides exist only in ratios of nitrogen to oxygen from N_2O_3 and NO_2[1285] (Raschig 1904), p. 254):

> Für die Existenz höherer Stickoxyde sind daher zur Zeit keine Anzeichen vorhanden. Sie sind aus der Literatur zu streichen. [There is no evidence fort the existence of higher nitrogen oxides. They are to delete from literature.]

This phrase must be seen in the background of gas-phase chemistry;[1286] the decomposition of gaseous nitrogen pentoxide is a monomolecular reaction, which takes place at room temperature has been first shown by Daniels and Johnston (1921), who propose $N_2O_5 \rightarrow N_2O_3$ (slow) and $N_2O_3 \rightarrow NO + NO_2$ (very rapid). Busse and Daniels (1927) found that nitric oxide is oxidized immediately by the nitrogen pentoxide and propose $N_2O_5 = NO + NO_2 + O_2$ (slow) and $NO + N_2O_5 = 3\ NO_2$ (fast). Thus, the unimolecular decay of N_2O_5 made some trouble in interpreting the reaction mechanism (there were many more publications on this issue in the 1920s;[1287] to refer here *Bodenstein*, *Christiansen*, *Kramers*, and others).[1288] Schumacher and Sprengler (1928) not only pro-

1283 In French it was named "*azotites*" and "*azotates*" until the late 19th century.

1284 *Friedrich August Raschig* (also called *Fritz Raschig*) (1863–1928) was a German chemist. *Walter Prahl* (1902–?) was a co-worker of *Raschig* and moved 1937 to the USA where he became after War II director of phenol research and development for Hooker Chemical.

1285 NO_2 includes N_2O_4 ($\leftrightarrow$ 2 NO_2); the equilibrium was described by Bodenstein (1922).

1286 Nitrogen pentoxide N_2O_5 was known since the early 19th century and described as a white, volatile, crystalline solid at room temperature (Daniels and Bright 1920).

1287 Today, thermal decomposition of N_2O_5 is described as $N_2O_5 + M = NO_2 + NO_3 + M$ and photolysis as $N_2O_5 + h\nu = NO_2 + NO_3$. Further droplets absorb N_2O_5 under formation of nitrate, and it reacts with particles, namely sea salt. Reaction with water vapor is negligible.

1288 *Jens Anton Christiansen* (1888–1969) was a Danish chemist at Copenhagen University. *Hendrik Anthony "Hans" Kramers* (1894–1952) was a Dutch physicist who worked with *Niels Bohr* in Copenhagen.

posed the "best" mechanism which explains the experimental facts but also suggested "a new nitrogen oxide and thus confirmed the observations made by Warburg and Leithäuser" (see p. 699). *Schumacher* and *Sprengler* proposed the reactions[1289]

$$NO_2 + O_3 \rightarrow NO_3 + O_2,$$
$$NO_3 + NO_2 \rightarrow N_2O_5.$$

4.6.2.3 Dr. *Mitchill*'s septon theory

An interesting historical example of speculative ideas how air pollution causes diseases is the fallacious claim by the American physician and lawyer *Samuel Latham Mitchill* (1763–1835) "to have discovered the demon of all epidemies, particularly that of yellow fever, reigned by virtue of the principle of acidity in the earth, air, and water, causing corruption everywhere" as Kealing (1879, p. 70) writes. *Mitchill* proposed his theory in 1795 (Mitchill 1795, Anonymous 1799) and soon accepted by followers, e. g., the American physician *Samuel Brown* (1769–1830), who describes exhaustive *Mitchill*'s view (Brown 1800). According to "Mitchill, professor of chemistry, natural history and agriculture, in Columbia College", the presence of nitric acid "it is presumed pestilential and malignant diseases depend" (Brown 1800, p. 35), and the following compounds exist (p. 38):

> Septon, in combination with oxygene, gaseous oxyd of septon; (dephlogisticated nitrous air;) 2. septic (nitric) gas; 3. and 4. septous and septic (nitrous and nitric) acid; and 5. septic acid gas.

The proof for the existence of atmospheric acidity and "effects of poisonous atmosphere" was argued by iron rusting, the "leaves of trees, often became spotted, and turned to mortification" (p. 38) and "[...] white cotton [...] spread to dry after washing, [...] when this mist prevailed [...] afterward, by twice boiling in alkaline lie" (Brown 1800, p. 39).[1290] *Brown* further writes (p. 40):

> From chemical combination of these (seption and oxygene) acting upon different parts of the body, seem to spring the common symptoms fevers, dysenteries and plagues. (p. 37)
> [...] that of syphilis from phosphorus, blended with the septon and oxygene; that of measles, from a combination of sulphur, that of pertussis, or croup, from the addition of the unknown radical of the muriatic acid [HCl], forming a nitro-muriatic oxyd".[1291] (p. 38)
> The septic acid, generated by putrefaction, is always on the earth's surface, and it vapours never rise to a great height above it. From these exhalations, the water of dews, mists and fogs, precip-

---

1289 These reactions are considered today as the main pathway in formation of N_2O_5. The nitrate radical NO_3 decomposes via $NO_3 + NO = 2\,NO_2$, photolytic in NO_2 or NO and reacts with hydrocarbons under formation of HNO_3; it is absorbed by droplets under formation of nitrate (und other radicals).
1290 Lye (lie – spelling error?): metal hydroxides obtained by leaching of ashes, containing largely potash (K_2CO_3).
1291 It could be nitrosyl chloride NOCl.

> itated when the atmosphere is cooled, particularly during night, receives a portion of the same acid [. . .] This also accounts for the deleterious effects of fogs and night airs, in warmer latitudes, so often noticed by different medical writers, and which are said sometimes instantly to destroy human life. One night's exposure is often fatal. (p. 40)

As sources of atmospheric trace substances, *Brown* notes marsh exhalations (miasma), animal and vegetable putrefaction and human effluvia. Today, *Mitchill* is forgotten, but at that time, the US Navy was following his idea to protect sailors by impregnating ships with alkaline solutions (basically a nice idea: neutralization after dry deposition that even increases due to reduced surface resistance).

Likely the first statement upon atmospheric acidity was given by the American physician *Lionel Chalmers* (1712–1777), who published "An Account of the Weather and Diseases of South-Carolina" (Chalmers 1776); prefixed to this work is a sketch of the climate of the region (near the Atlantic) which he describes, and giving some extract in an article appeared in 1803 in the *Medical Repository*,[1292] having the title "*Dr. Chalmers on the Acidity prevalent in the Atmosphere of South Carolina*" (p. 93):

> These exhalations do not consist of simply aqueous particles; they partake of the qualities of the substances emitted from the subjacent and surrounding bodies. One of the most predominant of these is an ACID, or some other *saline* principle. The proofs of which acidity in the atmosphere are, the speedy rusting of polishes metals, and the *remarkable fading of such dyed stuffs* as require acids to fix or heighten their colours. For these strongly attracting this salt from the air, &c, etc. (p. 11).

This is a remarkable indication of gaseous hydrochloric acid (HCl) release from sea salt – without any experimental proof.

4.6.2.4 Transformation of nitrogen in water

Concerns aqueous-phase chemistry of nitrogen compounds, the following issues have been investigated:
- combination of free nitrogen with water,
- oxidation of ammonia (ammonium),
- decomposition of ammonium nitrite,
- reactions of NO and NO_2 with water,
- oxidation of nitrous acid (nitrite) to nitrate,
- photolysis of nitrate.

The presence of oxidized nitrogen (in terms of *nitre* and saltpeter), in rainwater and dew has been described in the seventeenth century by *Boyle* (p. 287), *Evelyn* (p. 113, 115) and *Henshaw* (p. 146–147), first detected in the eighteenth century by *Marggraf*

1292 Vol. 6, pp. 92–93 (this is the first scientific journal in the USA).

(p. 154) and systematically determined since the early nineteenth century (see Chapter 2.3.1). After *Cavendish*'s experiments, it was generally accepted that nitric acid (this term was used generically for all oxidized nitrogen) is generated in the atmosphere by lightning. However, some investigations on rainwater have shown that nitric acid is also present in periods of the missing thunderstorm, and ideas arose on further sources:

– the action of ozone on free nitrogen in aqueous solution,
– the combination of free nitrogen with water, and
– the oxidation of ammonia in aqueous solution.

Gay-Lussac (1816, p. 410) writes:

> La vapeur nitreuse, en se combinant avec la potasse, se décompose et produit du nitrate et du pernitrite,

and soon later:

> la vapeur nitreuse elle-même qui, en se décomposant, donne lieu à la formation du l'acide nitrique et de l'acide pernitreux.

This corresponds to the (modern) equation $NO_2 + NO_2 + H_2O = HNO_3 + HNO_2$. Weltzien (1860b, p. 214) writes that "*Untersalpetersäure*" [hyponitric acid] reacts with water to "*Salpetersäure*" [nitric acid] and "*Stickoxyd*" [nitric oxide] according to (note that HO was the formula for water): $3\,NO^2O^2 + 2\,HO = 2\,HNO^4O^2 + NO^2$. Inversely, "*Stickoxyd oxydiert sich zu Untersalpetersäure*" [nitric oxide oxidizes to hyponitric acid], *Weltzien* writes. It is comparable with *Gay-Lussac*'s statement that "*vapeur nitreuse*" (NO and/or NO_2) combines with water to the acids.

Schönbein (1850) assumed that nitric acid (like in nitrification) is gained by oxidation of nitrogen by ozone in the aqueous solution; he always found nitrate when ozone is acting on water that contains carbonates, like lime, much faster than in neutral water.

Baumert (1853b, pp. 42 and 49) observed for the first formation of white fogs on the action of ozonized air on ammonia, consisting of ammonium nitrate. Schönbein (1857a) found that ammonium nitrite is gained in aqueous solution of ammonia, containing oxygen and fine-divided copper. Also, Schönbein (1857b, p. 30) writes that he observed since some years the reaction between ammonia and ozone:[1293]

1293 This phrase reads unusual modern; some 100 years later Cauer (1948a, p. 64) believes that ozone oxidizes ammonia in air to nitrogen oxides and further to nitrogen pentoxide (N_2O_5) in a chain reaction. However, the reaction of ammonia in atmospheric air with OH radicals is too slow (lifetime of about 165 days) to be considered as sink process. Nevertheless, the mechanism of the ozone decomposition in alkaline solution under formation of OH radicals and H_2O_2 have been enlightened only 100 years later; the aqueous-phase oxidation of ammonia by ozone and OH radicals has been studied in

> [...] daß der freie ozonisirte Sauerstoff mit dem Ammoniak salpetersaures Ammoniak erzeugt, $\mathring{\mathrm{O}}$ also mit dem H des freien Ammoniaks zu Wasser und dem N zu Salpetersäure sich vereinigt [...] that free ozonized oxygen with ammonia produces ammonium nitrate, thus $\mathring{\mathrm{O}}$ combines with H of free ammonia to water and with N to nitric acid].

In 1862, *Schönbein* published an extensive series of experiments on the generation of ammonium nitrite from water and atmospheric air under the influence of heat (Schönbein 1862),[1294] supported by Böttger (1862), who claimed the priority of this (wrong) finding. But soon later, the German chemist *Eduard Bohlig* (unknown living data)[1295] rejected *Schönbein*'s view with the argument that in such experiments no precaution apparently was taken to use air which had been purified from its pre-existing ammoniacal and nitrous compounds (Bohlig 1863). On the other hand, Bohlig (1863, p. 32) wrongly assumed that "it is produced by ozone in atmospheric air" from nitrogen. Lösecke (1879) again corroborated *Schönbein*'s statement that the evaporation of water in air produces ammonium nitrites – by the wrong conclusion that he found ammonium nitrite in dew from leaves and in evaporating rainwater.

It is here the point to note that the thermal decomposition (in the range of 40–60 °C) of aqueous solution of ammonium nitrite under formation of nitrogen gas was first shown by Berzelius (1812, p. 206), confirmed by Bohlig (1863) and Berthelot (1874, p. 206) who found that the decomposition accelerates with increasing temperature and slows down with dilution. Without giving experimental details Dhar (1922, p. 153) notified, that solutions of NH_4NO_2 decay under the influence of "tropical sunlight". The atmospheric relevance finally was shown by Takenaka et al. (1999) with the first observation of drying dew water, collected 1996–1997.

It is remarkable that nothing is found in literature on air chemistry on the aqueous phase chemistry of nitrogen monoxide (NO). In textbooks of inorganic chemistry (e. g., Wiberg 2007, p. 710) it is noted that NO does not react with water. However, in older literature (Gmelin 1936) we find that NO slowly reacts with water under the generation of HNO_2, N_2, and N_2O. Russell and Lapraik (1877) first observed that nitric oxide (NO) decomposes slowly in water to nitrous oxide (N_2O) and nitrogen. Cooke (1887) also found

---

waste water treatment (Hoigné and Bader 1978, Kuo et al. 1997), but never regarded in atmospheric chemistry (the oxidants only react with free NH_3 thus in a rather strong alkaline regime). The mechanism of the oxidation of ammonia – in gas and aqueous phase – is until now unknown; very likely is the formation of NH_2 in a first step that can further react in many different ways. To gain HONO, is a long reaction way and several intermediates are imaginably like N_2H_2 (hydrazine), NH_2OH (hydroxyl amine) and HNO (nitrosyl).

1294 *Schönbein* writes that ammonium nitrite is so readily decomposes under the influence of heat into water and nitrogen, that it might be likewise readily regenerates from the same bodies.

1295 Co-owner of the chemical factory Bohlig & Roth in Eisenach, founded 1862. He wrote another interesting article: Bohlig, E. (1863) Ueber ein sehr empfindliches Reagens auf freies oder kohlensaures Ammoniak und auf andere Ammoniaksalze, insbesondere zur Nachweisung dieser Salze im Regenwasser. *Dinglers Polytechnisches Journal* **168**, 131–132.

that NO reacts very slowly with water, and as products, he identified chiefly nitrogen, a small quantity nitrogen and nitrous acids in solution.[1296] Winkler (1901), conducting experiments to determine the solubility of NO, notified that NO also slowly chemical acts with water. Zimmermann (1905), chemist in the laboratory of *Adolf Lieben* (1836–1914) at University Vienna confirmed that with the uptake of NO in the water a chemical process occurs and speculates that NO combines with the OH^- ion and another NO with H^+, finally gaining nitrous acid (HNO_2) and hyponitrous acid ($H_2N_2O_2$). Moser (1911) adopted *Adolf Zimmermann*'s view and argued, citing Hantzsch and Kaufmann (1896), who studied the chemistry of hyponitrous acid, that it decays, namely in acid solution into N_2O (its theoretical anhydride), and to a less percentage in N_2O_3 and NH_3, where the ammonia further reacts with nitrous acid to nitrogen: $NH_3 + HNO_2 = N_2 + H_2O$, explaining the formation of nitrogen. Nichols and Morse (1930) explain the reaction between NO and H_2O in the following way: nitric oxide dissolves in water to form nitrohydroxylaminic acid,[1297] $H_2N_2O_3$, or a compound of the same constitution, which quickly decays in several ways: $H_2N_2O_3 = HNO_2 + NOH$ $(+ NOH = H_2N_2O_2 = N_2O + H_2O)$.

After the works of *Gay-Lussac* and *Raschig*, the reaction of NO_2 with water was studied by Foerster and Koch (1908), proposing two successive equilibria:

$$2\,NO_2 + H_2O \rightleftarrows HNO_2 + HNO_3 \quad \text{and}$$
$$3\,HNO_2 \rightleftarrows HNO_3 + 2\,NO + H_2O.$$

They suggest charge transfer between two NO_2 molecules (verbal in the text, but here expressed in an equation):[1298]

$$NO_2 + NO_2 \rightarrow NO_2^-(+H^+ \rightarrow HNO_2) + NO_2^+(+OH^- \rightarrow HNO_3).$$

Veley (1892) assumes the following equilibrium:

$$2\,NO + HNO_3 + H_2O \rightleftarrows 3\,HNO_2.$$

Veley (an unknown chemist from the Oxford Museum) begins his communication with a remarkable phrase (Veley 1892, p. 1892):

1296 *Cooke* writes (without giving references) that "Humboldt and other declared that nitric oxide is decomposed by contact with water, with the formation of nitrate and ammonium. *Davy*, however, rejects these experiments, and comes to the following conclusions" that NO not decomposes but is dissolved in water.

1297 Today named as hyponitrous acid HO-N(H)NO_2 which does not exist freely, only as salt (termed *Angeli*'s salt).

1298 Thus, from modern view of quickly charge transfer to and from NO and NO_2, in aqueous phase adducts are forms ON–NO, ON–NO_2 and O_2N–NO_2, gaining NO^+, NO^-, NO_2^+, and NO_2^- which react according to $NO^- + H^+ \rightarrow NOH$ $(+ NOH \rightarrow H_2N_2O_2\ (\rightarrow N_2O + H_2O))$, $NO^+ + OH^- \rightarrow HNO_2$, and $NO_2^+ + OH^- \rightarrow HNO_3$.

> Throughout the science of chemistry there is possibly no reagent so frequently represented as taking part in various transformations, but of which so little definite is known, as nitrous acid. In many text-books its properties are curiously discussed in a few lines, while some writers have gone so far as to deny its existence altogether even in the presence of nitric acid.

The photolysis of nitrate to nitrite in water was first mentioned by Thiele (1909). *Otto Warburg* proposed the reaction $KNO_3 \rightarrow KNO_2 + O$, and found gas evolution but that the photochemical equivalent law is not valid (Warburg 1918). Berthelot and Gaudechon (1911) detected oxygen besides nitrite. Villars (1927)[1299] found the photolysis for wavelength larger than 280 nm to be very small. Dhar et al. (1934) propose a photochemical "equilibrium" $2\ HNO_3 \rightleftarrows 2\ HNO_2 + O_2$, whereas the nitrate photolysis proceeds very slow. In the period after the late 1960s, many investigations on the aqueous-phase nitrate photolysis have been made, and the formation of nitrite and oxygen is known as the only stable products. The mechanism, however, is discussed until recent time; see Goldstein and Rabani (2007) with intermediate formation of OH radicals, NO_2, and peroxynitrite.

Honrath et al. (1999) identified the snowpack as an unexpectedly intense source of NO_x. The enhancement in photolysis rate for nitrates adsorbed onto a disordered ice surface is now well accepted and shown to impact the complex coupled interfacial kinetics, yielding a substantial increase in the photochemical NO_2 emissions from ice to the interstitial air (Marcotte et al. 2015).

4.6.2.5 Research in the nineeenth century: oxidation of nitrogen in air

Evidence for the presence of *gaseous* oxidized nitrogen, such as *free* nitric acid, in atmosphere air was already found in the mid-nineteenth century, doubted again and finally confirmed in the 1970s. Like for ammonia and ammonium in air, nitric and nitrous acid in air is always together with nitrate and nitrite, and moreover, nitrite and nitrate are produced from nitrogen oxides while sampling of air for analysis. Furthermore, the sensibility of reagents for both ozone and oxidized nitrogen made any detection questionable (Chapter 4.4.5.2).

After *Schönbein*'s discovery of ozonized air, it remained an open question for about a half century whether in silent electric discharges only ozone or also nitrous gases are produced. *Henry Bence Jones* writes (Bence-Jones 1851, p. 408):

> The experiments on production of nitric acid in all cases of combustion in the air, render it very probable that small quantities of nitric acid exist always in the atmosphere, and that nitric acid will be constantly detected in rain and snow, not only after thunder-storm, but at all other times also.

1299 *Donald Statler Villars* (1900–1988) was a physical chemist and physicist working for the U. S. Government.

This is the first experimental – without understanding the formation mechanism – evidence of the formation of NO_x while combustion processes. *Schönbein*, referring *Cavendish*, rejected the hitherto accepted view that under the influence of the electric spark the ordinary oxygen directly combines with nitrogen into nitric acid; he writes that in an unknown way the electricity generates ozone from ordinary oxygen, and the ozone oxidizes nitrogen to nitric acid (Schönbein 1850, p. 337). Schönbein (1852, p. 15) writes:

> Da in der sauerstoffhaltigen Atmosphäre ununterbrochen electrische Entladungen Platz greifen, so muss in jener hierdurch auch fortwährend Ozon erzeugt werden, zu einer Zeit mehr, zu einer andern weniger, je nach der Stärke und dem Umfange der stattgefundenen Entladungen [Because in the oxygenic atmosphere permanent electric discharges occur, thus constantly ozone must be produced, at a time more, at another time less, depending on the power and extent of occured discharges].

Furthermore, he accepted *Cavendish*'s finding of nitric acid during electric discharges but states (Schönbein 1852, p. 18):

> Es ist aber die Menge von Salpetersäure, welche unter den erwähnten Umständen gebildet wird sehr klein im Vergleich zu der Menge des gleichzeitig auftretenden Ozons [The quantity of nitric acid gained under the mentioned circumstances is very small comparing to the quantity of simultaneously formed ozone].

Whereas in that time most scientists have seen the formation of nitrate (nitric acid) in air together with that of ozone, Boussingault (1856a, p. 40) already expresses subsequent steps: atmospheric oxygen, in *mystérieuse transmutation* to ozone, which combines with nitrogen, in contact *d'un alcali*, and *des plus énergiques*, results in nitrate; whereas the ammonia meets the nitric acid in rain droplets.

Böttger (1858) noted that it is a familiar experience that under repeated electric discharges within atmospheric air, namely within a limited room, after a short time, a distinct smell of "*Untersalpetersäure*" [hyponitric acid, but it is NO_2] occurs. In experiments using the *Ruhmkorff* machine, he obtained in dry air after 15–30 min yellow-colored vapors, and detected "*das Vorhandensein einer höheren Oxydationstufe des Stickstoffs*" [the presence of a higher oxidation level of nitrogen][1300] by detection with an acidic solution of iron vitriol. In the experiment, adding some distilled water absorbs so much nitric acid that with sodium carbonate after evaporation unambiguously crystallized saltpeter forms. *Böttger* concluded that atmospheric air contains free nitric acid, already detected in a rainstorm,

> [. . .] so ist doch nicht zu bezweifeln, dass es über kurz oder lang auch wohl noch gelingen werde, sie *direct* in der *Atmosphäre* nachzuweisen [. . .] so is not to be doubted, that sooner or later it will probably succeed to detect them directly in the atmosphere].

1300 This modern phrase I not found before in other publications.

Andrews and Tait (1860) first recognized that during silent electric discharges besides ozone also nitrous gases gained. However, Cloez (1861, p. 527) writes that the presence of free nitric acid and of nitrogen oxides in the air is not yet clearly shown:

> La présence de l'acide nitrique libre et des composés nitreux oxygénés dans l'air atmosphérique n'a pas encore été démontrée jusqu'ici d'une manière claire et évidente; [...]

Further he states that ("*confusion avec la question de l'ozone*");

> L'air ordinaire, puisé par aspiration à i mètre environ de la surface du sol, fait souvent passer la couleur bleue du papier de tournesol humide à une couleur rouge permanente.

It seems that *Johann Florian Heller* (1813–1871)[1301] was the first who detected nitric acid in air and concluded on its permanent presence (Heller 1851); he exposed filter paper that was impregnated with alkali to the air for some hours or days and identified crystals of nitrate after elution the paper and evaporation of the solution. It is remarkable that he found more free nitric acid suburb (at Mt. Kahlenberg, 484 m altitude) then in Vienna.[1302] After his lecture on May 19, 1851 to the "*Gesellschaft der Aerzte*" [Society of Physicians] a commission[1303] was established to prove *Heller*'s result (Heller 1851, pp. 744–751). This was done by air sampling with an aspirator (about 1200 L over each 2–3 days for a period of 16 days in summer 1851) through a concentrated solution of acidic iron vitriol – and it was not detected any reaction (brown coloration). Furthermore, this detection method was also used while repeating *Heller*'s experiments with filter paper and linen, but without positive signal; however, not looking for nitrate crystals. The commission noted that the detection limit was 1/28.000.00 parts of HNO_3, i. e., about 40 ppb. Today we know that the concentration of free HNO_3 is much less and that in Vienna in 1851 the concentration of total NO_y in air was doubtless much less than 40 ppb: thus, it was the classical case of "no evidence is no evidence of absence". Interestingly, the commission detected ammonia in the solution, which led us to the conclusion that at that time the atmospheric concentration of ammonia was higher than that of NO_y.

1301 Austrian physician and chemist in Vienna; today seen as founder of clinical chemistry.

1302 This should be expected because in the air of Vienna was much more free ammonia which combines with HNO_3 to particulate NH_4NO_3. However, by this passive sampling also NO_2 is absorbed and converted to nitrate. Thus, *Heller* detected the presence of oxidized nitrogen.

1303 Members: *Adolf Martin Pleischl* (1787–1867) was an Austrian physician and chemist in Prague (where he conducted the first chemical analysis of water) and since 1838 in Vienna; *Franz Ragsky* (1814–1875) was an Austrian physician chemist at the geological institute (scholar of Liebig), and *Franz Coelestin Schneider* (1813–1897) was an Austrian chemist and physician in Vienna.

Schönbein (1851b), obviously in response to *Heller*'s publication from May 1851 but before the report of the commissions was published, was in conflict between acceptance of the presence of free nitric acid in air and his own ideas that nitric acid was generated at *Heller*'s filter paper during its exposure:

> Das Ozon als eminent oxidirendes Agens oxidirt bei Anwesenheit kräftiger Salzbasen den freien Stickstoff zu Salpetersäure, welches sich mit den vorhandenen Oxiden zu Nitraten vereinigt [Ozone as important oxidizer oxidizes at presence of strong alkali free nitrogen to nitric acid which combines with oxides present to nitrates]. (Schönbein 1851b, p. 402)

Further *Schönbein* argued that, if the atmospheric air contains free nitric acid, the maximum concentration must be "infinitely less" than the air in a bottle of 10 L volume containing a few grams of pure nitric acid,[1304] and where the smell can be recognized and blue litmus becomes red in short time (Schönbein 1851b, p. 405). Additionally, *Schönbein* (Schönbein 1851b, p. 403) argued (correctly) that

> [...] dieselbe höchst wahrscheinlich völlig neutralisiert wird durch das fortwährend von der Erde in die Atmosphäre aufsteigende Ammoniak [... it will be most likely completely neutralized by the permanent from the earth escaping ammonia].

Cloez believed to have shown the presence of free nitric acid in atmospheric air (Cloez 1861); he communicated that atmospheric air, passed at 1 m above the soil, colors humid litmus paper permanent red, independent from the season. The air, passed through a bottle containing a solution of pure potassium carbonate, produced large quantities of potassium nitrate and traces of alkaline chloride but no sulfate; and air, passed through a long glass tube, filled with pure lead carbonate, forms lead nitrate. The coloration of potassium iodine paper (iodometric papers of *Schönbein* and *Houzeau*) is due to the presence of nitric acid and simultaneous ozone. The same reaction was found in air with artificial nitrous oxides. However, this is no evidence for the presence of HNO_3, but general for the presence of nitrous oxides: NO together with NO_2 in contact with water (namely alkalic) produces nitrite and nitrate. The existence of nitric acid in the air is due to the presence of ammonia (i. e., ammonium) found in rust covers that is formed from the reduction of nitric acid by hydrogen, which is gained while reduction of water, *Cloez* writes.[1305] Moreover, the bronze green, investigated on a bell that was exposed to air since 1793, is also due to the presence of nitric acid in air.[1306] Kolbe (1863) found nitrous acid during the combustion of hydrogen in nitrogen-containing oxygen.

1304 This could be equivalent to 20 mg m^{-3}, a concentration 10.000 times higher than it would be expected in atmospheric air. So far historically, what *Schönbein* understood behind "infinitely".

1305 This phrase is curious. Today we know that ammonia is no gained in atmospheric reactions and nitric acid is no reduced. Absorption of NO_2 on wetted surface forms nitrous acid (HNO_2).

1306 Dunstan et al. (1905) explains rusting of iron that simply presence of H_2O and O_2 led to oxides and hydroxides of iron; the hydroxide further absorbs gaseous acids like nitric acid. Nitrous acid pro-

Iosif Vikentevich Zabelin [Иосиф Викентьевич Забелин] (1834–1875)[1307] wrote a critical review (Zabelin 1864) on previous research by *Schönbein*, *Bohlig*, *Böttger*, and *Meissner*[1308] upon the formation of ammonium nitrite under atmospheric conditions; he conducted a series of careful experiments, and found that when illuminating gas, alcohol and hydrogen burn in the air, nitrous acid and occasionally ammonia formed. The presence of ammonia *Zabelin* precludes by absorption from the atmosphere. *Zabelin* believes that in no case can free nitrogen unite directly with water, but in the conditions of all the forgoing experiments, it enters combination by the action of ozone. Finally, *Samuel William Johnson* (1830–1909)[1309] also critically referring all previous investigators, concluding (Johnson 1869, p. 238):

> The first result of the oxydation of nitrogen is nitrous acid alone when the combustion is complete as in the case of hydrogen, or when organic matters are excluded from the experiment.

The final conclusion by *Johnson* was presented a few years before the discovery that nitrification is a biological process by Schloesing and Müntz (1877); Johnson (1869, p. 241):

> To sum up, the writer believes that in nature, free nitrogen enters into combination, in all cases, by oxidation, that the agent of oxydation is ozone, that in the soil this ozone originates, for the most part, in the slow oxydation of organic matters, and that ammonia and the organic nitrogen of humus, peat and coal are the result of the reduction of oxyds of nitrogen either in the living organism in the acts of nitration, or by the organic matters of the dead plant or animal.

All forgoing researchers had the wrong belief that the union of atmospheric nitrogen and oxygen under the influence of electric tension is preceded by the production of ozone. *Georg Carius* was the first who recognized that ozone is not presupposed for the oxidation of nitrogen in air; he summarized the sources of oxidized nitrogen compounds (Carius 1874b):
- from free nitrogen through electric discharges in air,
- through oxidation of ammonia in electric discharges,

---

duces from copper salts deep green colored complex copper nitrites, hardly soluble (Kurtenacker 1913); copper nitrate is blue and in mixtures with hydroxide green to black. This patina also consists from chlorides, urates, sulfates and organic acids.

1307 Russian chemist and pharmacologist in St. Petersburg who studied 1862–1965 at the physiological institute of *Carl Michael von Voit* (1831–1890) in Munich.

1308 Meissner (1863a, p. 337) imagined the formation of nitric acid, nitrous acid and hyponitrous acid from nitrogen in action with ozone and antozone in presence of water, but not together with ammonia because this is abundant in nature and can combine with the nitrogen acids.

1309 US American agricultural chemist at the Sheffield Laboratory; he visited 1854/55 for one year the station Möckern near Leipzig.

Further, *Carius* also showed with several careful experiments that ozone never reacts with free nitrogen at any temperature and that in the process of condensation or evaporation of water no ammonium nitrite is produced, thus rejected *Schönbein*'s view and collaborating following scientists. He writes (Carius 1874b, p. 39):

> Damit fällt aber auch die bisher vorausgesetzte Bildungsweise von salpetriger Säure und Salpetersäure aus Stickstoff durch Vermittlung des Ozons in der Natur fort, sie *findet nicht statt* [Therewith disappears the previously assumed formation of nitrous and nitric acid from nitrogen under the action of ozone, it *does not take place*].

Carius, also citing Goppelsroeder (1871) who first stated the oxidation of ammonia by ozone in solution under formation of nitrous and nitric acid as well as hydrogen peroxide, conducted appropriate experiments and definitely detected the reaction products. He notified that the oxidation of nitrite to nitrate by ozone and hydrogen peroxide is well established and writes (Carius 1874b, p. 55)

> [...] überaus wichtige Rolle des Ozons bei der Bildung von salpetriger und Salpetersäure in der Natur haben muß [... most important role that ozone must have in formation of nitrous ad nitric acid in nature].

This is the first clear statement that nitric oxide (NO) is formed from atmospheric nitrogen and oxygen (under the action of electric discharges) and ozone is the agent for further oxidation to nitrous and nitric acid; the oxidation of nitrite to nitrate by ozone and hydrogen peroxide in aqueous solution was already known by Schönbein (1847a, p. 227).

To Wurster (1886a), it was already "textbook knowledge" that H_2O_2 fast oxidizes ammonia to nitrite, and O_3, as well as H_2O_2, oxidize nitrite to nitrate. Nagy-Ilosva (1894), who repeated the experiments by *Carius*, did not found H_2O_2 but HNO_2 and HNO_3 (note that H_2O_2 is an intermediate of the O_3 decay and decomposes while oxidation of nitrite); *Nagy-Ilosva* found the reaction to be very slow and (correctly) that dry O_3 and NH_3 do not react and concluded finally in contrast to *Carius* that the role that ozone plays in the formation of nitrite and nitrate is not important. Berthelot (1877c), who also studied the formation of ozone and nitrogen oxides under the action of electric discharges, has definitively shown that a mixture of pure nitrogen and water vapor under electric discharges (two hours with a *Ruhmkorff* induction coil)[1310] does not produce either nitric acid; the existence of nitrites and nitrate in atmospheric air can only be explained by the action of atmospheric electricity on nitrogen and oxygen and afterward dissolution in water, *Berthelot* writes.

1310 *Heinrich Daniel Ruhmkorff* (original *Rühmkorff*) (1803–1877), born in Hannover and since 1839 in Paris, was a German-French instrument maker. His induction coil, patented in 1855, became for many decades a standard laboratory instrument.

Houzeau (1871) and Brodie (1872) first determined nitrous products beside ozone while electric discharges; moreover, Houzeau (1872b) found that electric discharges in air produce only traces of ozone but preferent nitrous acid. In contrast, Berthelot (1877b) did not find determinable amounts of nitrous gases but nitrate after absorption in water. Müntz and Marcano (1889, p. 1062) express very clearly the idea that under the influence of atmospheric electricity nitrogen combines with oxygen gaining nitrates and nitrites that afterwards will by scavenged by rain:

> Sous l'influence de l'électricité atmosphérique, l'azote se combine à l'oxygène pour former des nitrates et des nitrites qui flottent dans l'air et que les eaux pluviales dissolvent sur leur parcours.

Hautefeuille and Chappuis (1881) observed spectroscopically that in a mixture of nitrogen and oxygen a silent discharge produces "*une acide pernitric*" (see also Berthelot 1881). Water or traces of moisture immediately causes the spectrum to disappear, and the substance is, therefore, an anhydrous compound capable of forming an acid, writes Mellor (VIII, p. 383). Italian chemist *Demetrio Helbig* (1873–1954) reported in 1902 that in a gas mixture of N_2O_4 (i. e., NO_2) and ozone in a fast reaction N_2O_5 (nitrogen pentoxide) is obtained (Helbig 1903). Spiegel (1903, p. 367)[1311] represented from an indirect analysis the formula $O_2NO{-}ONO_2$ (N_2O_6) – nitrogen hexoxide. Warburg and Leithäuser (1906) obtained an analogous product by the action on ozone and nitrous gases. It was direct observed that the red color (NO_2) disappears and a colorless higher oxide form, consuming ozone. Ehrlich and Russ (1911)[1312] summarized the previous research of silent electric discharges on nitrogen-oxygen mixtures and studied the oxidation of NO by ozone to N_2O_5 and also found that after consumption of O_3 the pentoxide decays, and finally, a steady state is gained.

The formation of nitrogen oxides in high-temperature processes such as the combustion process was first recognized by Berthelot (1879) and Stohmann (1879). *Berthelot*, studying the combustion of ammonia,[1313] notified a "complication, the formation of nitrogen oxides" (Berthelot 1879, p. 882), but without further comment. The German agricultural chemist *Friedrich Stohmann* (1832–1892) observed during combustion of nitrogen-containing matter (urea and hippuric acid) the formation of nitrogen-oxygen compounds, seeing red vapors and detecting nitric acid. Rubner (1885) was the first who also found oxidation products of nitrogen in the combustion of nitrogen-free bodies and explained it by the combination of atmospheric nitrogen with oxygen, gaining nitrous and nitric acid. Finally, Cramer (1890, pp. 323–326) stud-

1311 *Leopold Spiegel* (1865–1927) was a German chemist at the Berlin University.
1312 *Viktor Ehrlich* (unknown living data) and *Paul Russ* (1876–1949) were chemists in the "*k. k. Staatsgewerbeschule*" [Royal State Trade School] in Vienna; *Russ* was dismissed in 1934 because of its Jewish origin. *Ehrlich* fled with his family 1939 in the USA; his sun *Gert Ehrlich* (1926–2012) became a professor for chemistry at the University of Illinois.
1313 It is worth to cite his formula, i. e., the formation of nitrogen and water: $AzH^3 + O^3 = Az + 3\,HO$.

ied the formation of nitrogen oxides in combustion products of different illumination flames, and, referring Crove (1887), who was the first the measure the temperature of flames spectroscopically to be higher than 1000 °C, concluded (Crove 1887, p. 323)

> Die Entstehung der Untersalpetersäure bei den Beleuchtungsprocessen ist wohl auf eine directe Oxydation des atmosphärischen Stickstoffes in der Flamme zurückzuführen [The formation of hyponitric acid in illumination processes is probably attribute to the direct oxidation of atmospheric nitrogen in the flame].

Eduard Cramer correct writes that "first nitric oxide is formed that instantly gathers O and transforms into hyponitric acid"; however, he refers wrongly *Stohmann* and *Berthelot* concerns combustion of nitrogen-free matter.

4.6.2.6 Research in the twentieth century

The research upon nitrogen oxides and acids in the atmosphere in the twentieth century includes

- development of different measurement techniques,
- detection and systematic measurement in air, precipitation and dry deposition,
- investigation of chemical kinetics and mechanisms among NO_y and with other compounds,
- investigation and quantification of sources,
- establishing (together with ammonia) of regional and global nitrogen budgets,
- understanding the role in tropospheric ozone formation and removal,
- investigation of health effects of NO_2.

Some pioneering research was done before about 1970, but due to still insufficient measurement techniques often only with limited significance or even with failures. Systematic progress was made in the 1970s and 1980s with the development of new techniques and methods. In the mid-80s, the NO_y chemistry was rather well understood, but only the late 1980s and 1990s, due to sophisticated measurement methods, have provided deep insights into the chemistry of the nitrate radical (NO_3) as the night-time oxidant and the day-time chemistry of nitrous acid (HNO_2), namely the photocatalytic reduction of NO_2 to nitrite.

The first determination of nitrous and nitric acid in the atmosphere was conducted by *Charles Mabery* in 1894 in Cleveland, using methods still used nowadays.[1314] A measured volume of air was drawn through a solution of sodium hydroxide prepared from metallic sodium; however, it remains open, which quantities of NO and NO_2 lead to a

1314 Quantitative determination of nitrate is based on phenolsulphonic acid as indicator, and for nitrite, sulphanilic acid and naphthylamine hydrochloride were used as indicator; sulphanilic acid is diazotized by the nitrite to be determined and coupled to 1-naphthylamine-7-sulphonic acid (*Cleve*'s acid 1–7). However, today the red color formed is measured spectrophotometrically.

cross-sensitivity. Note that the values represent the sum of the gaseous acids and particulate nitrite and nitrate, resp., and they are 10 to 20 times higher than maximum values found nowadays:

HNO_2 (n = 15) 46 (0.6–153) µg m^{-3}
HNO_3 (n = 18) 367 (44–1063) µg m^{-3}

Hayhurst and Pring (1910) measured in 1909 (together with O_3) the amount of oxides of nitrogen at various altitudes (2600 to 8500 feet) using absorption vessels, containing pure potassium iodide, attached to a kite at Derbyshire (UK), and found "a rather greater quantity of oxides of nitrogen at higher altitudes than at ground level, although the method used can only serve as a rough approximation". Reynolds (1923) has studied the formation of nitrogen dioxide (and ozone) during a severe thunderstorm over London on 10 July 1923 and reports the following concentrations (in ppb):

	before	during	after
London air	8.3	8.8	7.5
Upminster air (country)	2.8	2.3	2.4

Likely the first reliable method for measurement of NO_2 in ambient air was developed and tested by Francis and Parson (1925) through absorption of oxides of nitrogen by caustic alkali in a closed bottle and by scrubbing and the oxidation of nitrite by means of hydrogen peroxide. In London (St James's Square, Pall Mall), between July and November 1923 (n = 14) they determined 18 ± 15 ppb NO_2 and on two foggy days in November 107 and 168 ppb NO_2, respectively. "Nitrogen peroxide [NO_2] is essentially a constituent of town air" writes *William Colebrook Reynolds*, who continued his measurements from 1923 until 1927 and found in Upminster (23 km distance from London) generally about one sixth of that recorded at Plaistow (8 km distance from London), where it rose to as much as 30 ppb in foggy weather; in summer weather the proportion fell to 8 ppb, somewhat smaller than those recorded by Francis and Parsons, *Reynolds* writes (Reynolds 1930a). Edgar and Paneth (1941) determined 1938 and 1939 (together with ozone) the content of NO_2 in Kew (London suburb) in quantities similar to *Reynold* (1–8 ppb).

Hans Cauer reports (Cauer 1949a, p. 239) on measurements of ammonium, nitrite and pH in "*Nebelkernen*" [fog nuclei], conducted in autumn 1941 at different altitudes (500, 1000, and 1750 m) and every two hours beginning in the early morning until late afternoon in the High Tatra (calm weather, cloudless, weak haze and finally clouds formed). These are the first measurements of N(III), i. e., the sum of nitrite or HNO_2, in air. The results are worth to be cited, and will, together with the following measure-

ments by *Köhler* and *Renger* confirm our view on the heterogeneous formation of nitric acid, more than 60 years later obtained (see Möller 2019, pp. 340–342):
- increase in the content of nitrite with time of the day and increasing altitude,
- decrease of ammonia with increasing altitude,
- decreasing pH with increasing altitude.

Cauer correctly noted the ammonia-rich ground air (its emission from soils was known) and the nitrite-rich mountain and cloudy air; in his further discussion concerns the formation source of nitrite, however, he was curious (similar to that of ground-based ozone) about electrical processes in the air. *Friedrich Renger*, who carried out similar measurements in 1943 (Renger 1944, cited after Cauer (1949a, p. 230)[1315] found the content of nitrite is highest with the occurrence of rain, fog and snow, namely in front of clouds (this observations supports heterogeneous formation). Under the supervision of *Hans Cauer*, the German physician *Erich Köhler* (unknown living dates)[1316] conducted measurements of the diurnal variation of nitrite in fog droplets, recalculated on the volume of air (Köhler 1942). Because he used *Cauer*'s "dew condenser" the analytical result is hard to interpret; *Köhler* found that the concentration of "nitrite" increases with solar radiation and temperature, and decreases with frozen soil, rain and dew. As sources, *Köhler* regards solar radiation and putrefaction in soil. The values vary between 0.5 and 2 μg m^{-3} (given in NO_2) with a few events of concentrations ranging from 4 to 12 μg m^{-3}. Cauer (1949a, p. 230, 1951a, p. 1131) concluded by summarizing that the formation of nitrite depends, similar as to ozone, from the atmospheric electric field and from the activity of soil bacteria.

The presence of N_2O in air was first proposed by *Otto Kestner* (1873–1953),[1317] in down winds from upper layers of the atmosphere (Kestner 1923). First evidence for the presence of nitrous oxide (N_2O) in the atmosphere was obtained by *Arthur Adel* (1908–1994), an American astronomer and astrophysicist (Adel 1938), who also found absorption bands of N_2O_5. Adel (1946) advocates the view that microorganisms (see Hutchinson 1944) are responsible for their presence in the atmosphere. Bates (1952) describes photolysis as the only important decomposition of N_2O in the upper atmosphere.

The Hungarian-born American physicist *Joseph Kaplan* (1902–1991) observed the first nitric oxide (NO) bands in the auroral spectrum or in those of the light of the night sky (Kaplan 1939). The British physicist *William Charles Price* (1909–1993) assumed

1315 *Renger*'s dissertation is lost without trace.

1316 He worked in the sanitorium "*Heilstätte Moltkefels*" in Niederschreiberhau (today Szklarska Poręba in Poland). It is notable that *Friedrich Renger* also studied the variation of nitrite together with those of ozone in air two years later at the same location (Renger 1944).

1317 German physician and lecturer in physiological chemistry in Hamburg; forces retirement in 1934 and emigration 1939–1949 to UK.

that NO may be present as an important atmospheric constituent due to photodissociation of N_2O (Price 1942), and Nicolet (1945) definitely stated NO as an important constituent in the upper atmosphere. The presence of nitrogen oxides in the air of Los Angeles was confirmed by Larson et al. (1953) to be 0.08 ppm in periods of good visibility and to be 0.4 ppm in periods of intensive smog. The US American scientist *Bernard E. Saltzman* (1918–2010) developed a technique[1318] to measure the concentration of nitrogen dioxide (Saltzman 1954) that has been used for decades until a chemiluminescence method has been developed by Fontijn et al. (1970)[1319] and improved by many other in the following years.

In the early 1960s, NO_2 concentrations in the air of towns (Frankfurt/Main) were measured to be 5–8 $\mu g\ m^{-3}$ and in the background to be 2–3 $\mu g\ m^{-3}$ (Georgii 1960). Junge (1963b, p. 84) writes that “no attempt has been made to separate NO and N_2O_3 from NO_2 in the atmosphere”. First attempts in understanding global cycling concerned that of nitrogen. Without having any knowledge, the source of atmospheric ammonia was seen in the putrefaction of organic matter. Early rainwater analyses have further shown that lightning cannot be an exclusive source of oxidized nitrogen; for a long time, the oxidation of ammonia in air was regarded to be an important source of nitric acid. Despite the oxides of nitrogen (N_2O, NO, NO_2, N_2O_3, N_2O_4, N_2O_5, and NO_3) were known, only N_2O and NO_2 have been measured in the air until the early 1960s (Georgii 1963).

Hutchinson (1944) gives the following summary of the sources for combined nitrogen, which can be imagined, namely
- from soils and the ocean,
- from fixation of atmospheric nitrogen (electrically, photochemically, meteoric),
- from industrial contamination;

the latter *Hutchinson* excluded to be “far from sufficient to explain the actual transfer”. Nitrogen fixation in the atmosphere due to lightning have been largely confined to examination of the nitrate content in the accompanying precipitation, but are not regarded as showing a positive result (Viemeister 1960).[1320] The German biometeorologist *Reinhold Reiter* (1920–1998)[1321] get first evidence from field observation

1318 The reagent is a mixture of sulfanilic acid, N-(1-naphthyl)-ethylenediamine dihydrochloride, and acetic acid. A stable direct color is produced with a sensitivity of a few parts per billion for a 10-minute sample at 0.4 liter per minute. Ozone in five-fold excess and other gases in tenfold excess produce only slight interfering effects.

1319 *Arthur Fontijn* (1928–2021) was a Dutch-born chemist in USA.

1320 *Peter Emmons Viemeister* (1929–2011) was an engineer at the *Grumman* Aerospace Corporation, known for “The Lightning Book” (MIT Press, 1972).

1321 He founded 1949 in Munich the “*physikalisch-bioklimatische Forschungsstelle*” [physical bioclimatic institute], which moved 1953 to Farchand (near Garmisch-Partenkirchen) and became in 1962 an institute of the Fraunhofer Society and again moved to Garmisch-Partenkirchen; after his retirement of the ratio between ammonium (assuming only ground-based sources) and nitrate (search for an atmospheric source) that nitrous gases are produced by atmospheric-electrical processes inside clouds (Reiter 1960, 1970). *John F. Noxon* (1928–1985)[1322] first reports on direct observation of NO_2 produced in the troposphere during a lightning storm (Noxon 1976). Finally, the Soviet physicist *Jakov Borisovich Zeldovich* [Яков Борисович Зельдович] (1914–1987) was the first who describes the thermal oxidation of nitrogen and NO_x formation (Zeldovich 1946), known as *Zeldovich* mechanism ($O + N_2 = NO + N$ and $N + O_2 = NO + O$).[1323]

Ångström and Högberg (1952), analyzing the results from the Swedish precipitation chemistry network, first noted: "In our present case the question is open as regards the way in which combined nitrogen is introduced in precipitation". They additional stated "no indication that an electrical production plays an important part". However, due to the strong correlation between NH_4-N and NO_3-N they concluded on production "intimately bond to one another" (Ångström and Högberg 1952, p. 277) and proposed "a photochemical process, occurring for instance in small droplets" according to the equation (*Schönbein*'s idea of generation from water and air) $12\,N + 18\,H_2O = 4\,NH_4NO_3 + 4\,NH_4OH + O_2$.

It is remarkable that Wilson (1959),[1324] who analyzed freshly fallen snow in New Zealand, regarded nitrate and nitrite still as products of electric discharges, but he only found ammonia in the samples, and remarkable is that he found organic nitrogen (he called it *albuminoid*, a term from the nineteenth century) about in the order of ammonium, and suggested its ultimate origin in the ocean. Still, in the early 1960s, speculations on the origin of ammonia and nitrate in tropical rain have been published in renowned journals by the geochemist *Simón Visser* (unknown living data) of Makerere College, Kampala (Uganda), who believed that it originates mainly from local dust (Visser 1961, 1964).

As for ammonia, the first NO_2 measurements (monthly sampling) within the Swedish precipitation network have been conducted still from the agricultural point of view (Egnér and Eriksson 1955), showing values (as NO_x-N) between 0 and 36 µg

---

in 1986, *Wolfgang Seiler* became the new head of the institute which was renamed in "*Institut für atmosphärische Umweltforschung*" [Institute of Atmos. Environ. Res.]- IFU. In 2002 the IFU was integrated in the Institute for Meteorology and Climate Research of the Research Center Karlsruhe (IMK-IFU). The author (DM) worked 1992–1994 in the IFU as head of the department Air Chemistry in the Branch Berlin-Adlershof.

1322 US American physicist, leader of the Optical Aeronomy Program of the Aeronomy Laboratory of NOAA in Boulder, known for spectroscopic studies of stratospheric trace gases (including the first observation of NO_3), and measurements of the global distribution of stratospheric NO_2.

1323 A full description of the chemistry of thermal NO formation including substantial reference is given by Abian et al. (2015).

1324 *Alexander T. Wilson* (unknown living data) was a New Zealand chemist; 1960 professor at the Wellington University.

m^{-3} and a typical value ranging 10–15 µg m^{-3} (corresponding approximately to 20 ppb NO_2 and thus very unreliable).

The focus in research upon nitrogen oxides changed in the late 1950s to air pollution research after recognizing that town concentrations of NO_x are several times larger than suburb; see, e. g., Demerjian et al. (1974), Spicer (1977) and Huebert and Lazrus (1978). PAN has been identified first by Darley et al. (1963), gaseous nitrates which have been identified in the atmosphere include nitric acid by Spicer (1974) and Okita et al. (1976) and the nitrate radical (NO_3) by Platt et al. (1981) and HNO_2 by Perner and Platt (1979).[1325] Further pioneering research was done by *Eric Apel* (at NCAR), *Philip Lincoln Hanst*, *Geoffrey* (*Geoff*) *Harris* (York University, born 1950), *Thomas J. Kelly*, *Dieter Kley*, and others concerns measurements and with respect to chemistry by *Donald L. Baulch* (University of Leeds), *Richard Cadle*, *Jack Calvert*, *Brian Heikes*, *William Stockwell*, and others. Whereas the formation of nitric acid was well understood at the end of the 1970s,[1326] Perner and Platt (1979) write on nitrous acid:

> HNO_2 was investigated by long path optical absorption at three sites in Western Europe. It was definitely identified and measured in moderately polluted air at Jülich. Mixing ratios as high as 0.8 ppb were observed before sunrise. After sunrise the HNO_2 is photolyzed and considerable amounts of OH are produced. The formation of HNO_2 is unclear and possible reactions are discussed.

Akimoto et al. (1987)[1327] proposed the reaction N_2O_4 (NO_2 + NO_2) + $H_2O \rightarrow HNO_2$ + HNO_3 on wetted surface as an artifact OH source via HNO_2 photolysis in smog chambers. The nocturnal increase of HNO_2 measured by Kessler (1984) was later explained by the above-mentioned heterogeneous process and finally regarded as an important source of atmospheric HNO_2. The photochemical formation via NO_2 + OH leads only to very low concentration because of the very fast photolysis of HNO_2. In the years after 2000, numerous field experiments have given evidence for a strong daytime source affecting the OH budget, and a heterogenous photoactinic reduction of NO_2 has been

1325 *Barry J. Huebert* (born 1946) Emeritus professor of Oceanography, University of Hawaii. *Chester William Spicer* (born 1946), American atmospheric chemist at the Battelle Memorial Institute. *Ullrich Platt* (born 1949), German atmospheric physicist, since 1989 at the University Heidelberg (professor emeritus). *Dieter Perner* (1934–2012) was a German physicist at the Institute for Chemistry of the Research Centre Jülich where he was most famous for inventing, supported by *Ulrich Platt*, the Differential Optical Absorption Spectroscopy, DOAS; he also was pioneering in OH detection; he joined the Max Planck Institute Mainz in 1982 to lead the optical spectroscopy group with the Air Chemistry Department, directed by *Paul Crutzen*.

1326 The formation of nitric acid (and nitrate, resp.) was assumed until the beginning 1960s only due to lightning and oxidation of ammonia. Interestingly that Mrose (1961, p. 50) concluded from the absence of the annual variation of nitrate in rainwater that lightning cannot be the main source.

1327 *Hajime Akimoto* (born 1940), Japanese atmospheric chemist, well-known worldwide.

prosed according to[1328]

$$NO_2 \xrightarrow{e^-} NO_2^- (+H^+ \rightleftarrows HNO_2 \xrightarrow{h\nu} NO + OH)$$

4.7 Sulfur compounds

Sulfur[1329] is the oldest known element and has been known from very early times. Burning sulfur was used as fumigant, bleach or incense in the classical world. In *Homer*'s *Odyssey*, *Ulysses* said "Bring my sulfur, which cleanses all pollution, and fetch fire also that I might burn it, and purify the cloisters" (Odyssey Book 22, translated by S. Butcher, Orange Street Press, 1998, p. 278). However, despite sulfur being the main chemical (among mercury) of the alchemists, it was regarded until 1809 by *Humphrey Davy* as a composite body even though *Lavoisier*, in 1777, had already recognized it as an element (simple body). *Gay-Lussac* and *Louis Jacques Thénard* confuted still in 1809 this mistake (Kopp 1931, pp. 310–311) and from this time, sulfur was seen as an element.

This chapter covers in some detail only sulfur dioxide and briefly covers the different sulfur species of interest in the atmosphere, its discovery, and first detection. The first ideas on the sulfur cycle are communicated in Chapter 1.2.3.

4.7.1 Early history and discovery

Today, we know the presence of H_2S, SO_2, CS_2, COS, DMS, and numerous organic sulfur compounds (mainly mercaptans) in the atmospheric air. Sulfite (sulfurous acid, which was not separated from sulfur dioxide in the past; note that H_2SO_3 does not exist free but as $SO_2 \cdot H_2O$ adduct in several molecular structures in water) and sulfate only occur in aqueous solution and in the particulate matter; SO_3 exists in air only as intermediate in the process of SO_2 oxidation (because of $SO_3 + H_2O = H_2SO_4$), and H_2SO_4 only for a short time as nanoparticles due to particle formation. Not in the focus of this volume is the discussion of sources:

- Combustion (SO_2 and H_2SO_4)

1328 The following researchers made significant work: *Karin Acker*, *Christian George*, *Jörg Kleffmann*.
1329 Etymology of sulfur [sulphur]: from Latin sulpur, from Anglo-French *sulfere*, Old French *soufre*; German *Schwefel*, Dutch *zwavel*, Italian *zolfo*. There is no origin from Greek ϑεíον or θεíον [thion]; "thio" is used as prefix in chemical compounds in which oxygen id exchanged by sulfur. *Brimestone* = sulfur in a solidified state (the mineral, now restricted to biblical usage). The meaning *vitriol* for water-containing sulfates of bivalent metals (iron, copper, zinc) comes from Latin vitum [glass] and vitreus [glassy], Medieval Latin vitriolum (the crystals look glass-like). Sulfuric acid was first produced from iron sulfate and thus called oleum vitrioli.

- Industrial (SO_2, H_2S, COS, CS_2, organic sulfur)
- Volcano (H_2S and SO_2)
- Ocean (DMS)
- Wetlands (H_2S)
- Plants and soils (COS, organic sulfur)

Sulfur, or more precisely sulfur dioxide (SO_2), is the oldest known pollutant (cited in the bible but known long before) because it was found residing close to volcanoes and the people quickly recognized its specific properties like burning with a penetrative odor. Without knowing the chemical species, its influence on air quality was described several hundred years ago in European cities where it was prevalent because of coal-burning (e. g., Evelyn 1661). *Johann Sigismund Elsholtz* (1623–1688)[1330] concluded that the air contains sulfuric acid [vitriolis spiriti] and wrote the following exceptional phrase Elsholz (1675, p. 28):

> Continet enim athmosphæra sive circulus aëris circa terræ, qua mari ascendentes: inter qua stamen *nitrosas* ob copiam prævalere vero simile est, utut vitriolacæ, salarmoniacæ, sulphureæ & congeneres non sint excludendæ [The atmosphere or the air circuit around the globe contains what is coming up: among bulk material *saltpeter* because of their abundance, but also vitriolic, sal armoniac, sulphur and not to excluded combinations].

The gas evolved from burning sulfur and thermal treatment of alum was first called (like all "gases") spiritus by *Geber* (and later in the seventeenth century sulphuris spiritus, in German "*Schwefelgeist*") and described as "dissolvent". *Albertus Magnus* (~1193–1280)[1331] said that sulphur philosophorum is not true sulfur, but is rather the spirits of vitriol – scilicet est spiritus vitreoli romani. *Basil Valentine*[1332] said that calcinated vitriol is heated in a retort with a receiver, there first appeared as white spirit, which is mercurius philosophorum, and this is followed by red spirit which is sulphur philosophorum. Spiritus vitrioli romani and sulphur philosophorum could not have been anything other than sulfuric acid. In 1595, *Libavius* noticed that an acid was also formed when the fumes passed into water. Between sulfurous and sulfuric acid was not distinguished; this is not surprising when one considers how easily sulfurous acid is oxidized to sulfuric acid. Sala (1613) describes different sulfur compounds, and it seems that he distinguished already among sulfur dioxide, sulfurous acid and sulfuric acid, and sulfates (of copper and calcium):

1330 Chemist, biologist and Court physician of Elector *Friedrich Wilhelm von Brandenburg* in Berlin (also written *Elßholtz*, *Elßholz*, *Elsholz*, Latinized *Elsholtius*).

1331 German bishop and naturalist. Magnus, A. (1659) Compositum de compositis. In: Theatricum chimicum (ed. Lazarus Zetzner). Argentorati, Straßburg, Vol. 4, pp. 809–862 (first edition 1602).

1332 *Basil Valentine* is the Anglicized version of the name *Basilius Valentinus*, ostensibly a 15th-century alchemist, possibly Canon of the Benedictine Priory of Saint Peter in Erfurt, Germany but more likely a pseudonym used by one or several 16th-century German authors.

- Spiritus vitriolis humidus
- Aqua vitriolis secunda
- Liquor vitrioli acidus primus vulgo spiritus vitrioli
- Spiritus sulphuris vitrioli & liquoris acidi primi rectification
- Liquor vitrioli acidus secundus vulgo oleum vitriolis appelatus
- Sal vitrioli: cuprea vitrioli et terra vitrioli

Further, he notes the observation of nebuli aëris, likely the formation of sulfuric acid particles from fuming sulfuric acid (oleum) and burning sulfur via oxidation of SO_2 to SO_3.

The gas (SO_2), observed while burning of sulfur, was first particularly investigated by *Stahl* in 1702, who also definitively settled the difference between sulfuric acid the stronger (acidum fixum) and sulfurous acid, the weaker acid (acidim volatile). According to *Stahl*'s phlogiston theory, sulfurous acid was called *phlogisticated vitriolic* (or *sulphuric*) *acid*, referred by *Cavendish* in 1765. Further investigated in 1771 by *Scheele* and 1774 by *Priestley*, who prepared the pure gas in 1775 (he called it *vitriolic acid air*); a few years later, *Lavoisier* showed in 1777 (he called it "*gaz acide sulfureux*") that sulfurous acid was an intermediate oxidation compound between sulfur and sulfuric acid. The solution in water *Lavoisier* called "*acide sulfureux*" in contrast to the more highly oxidized "*acide sulfurique*". Further examined by *Claude-Louis Berthollet* in 1782 and 1789, and the salts were studied in 1797 by *Fourcroy* and *Vauquelin* (Gmelin 1852, p. 602).

The Italian mineralogists Monticelli and Covelli (1825, pp. 34 and 58)[1333] discovered SO_2 in fumes of Mount Vesuvius. Since spring 1812, *Humphry Davy* made several observations on Vesuvius (in 1819 together with *Monticelli*); he reports on his expedition in December 1819 (Davy 1828, p. 242):

> On the 5th I ascended the mountain, and examined the crater and the stream of lava. The crater emitted so large a quantity of smoke, with muriatic and sulphurous acid fumes, that it was impossible to approach it except in the direction of the wind.

Davy found in the sublimates of lava chlorides and sulfates of alkali and other metals and recognized in the steam also some sulphuretted hydrogen (H_2S); noticeable is his remark on the effects of SO_2:

> The sulphurous acid gas in the fumes was highly irritating to the organs of respiration, and I suffered so much from the exposure to them that I was obliged to descend; and the effect was not transient, for a violent catarrhal affection ensued, which prevented me for a month from again ascending the mountain (Davy 1828, p. 244).

1333 *Teodoro Monticelli* (1759–1845) was naturalist; *Nicola Covelli* (1790–1829) was professor for chemistry and botany at "*Regia Scuola Veterinaria*" in Napoli.

From his observation, *Davy* concluded on the existence of "great cavities below the Vesuvius filled with aërifom matter" (Davy 1828, p. 249) and supposed the existence of a descending current of air, and thus the excavations become filled with atmospheric air.

Besides sulfur dioxide, hydrogen sulfide (without knowing the compound) was recognized by the ancients because of his smell from some springs and craters; however, according to *Kopp* there was no notice on this particular smell; likely behind the general term fumo sulphuris [sulphureous vapor] all kinds of sulfur gases were included and not distinguished. *Libavius* and *Lémery* first observed H_2S while decomposing of metal sulfides. *Scheele* examines more closely SO_2 and H_2S. Scheele (1777, pp. 102–105) gained liver of sulfur (potassium carbonate heated with sulfur) through mixing powdered with uncalibrated calx (CaO) with sulfur, which he heated without yielding any kind of air but by adding hydrochloric acid, a gas evolved with a hepatic smell [H_2S]. *Scheele* called the gas "*stinkende Schwefelluft*" (it was also called "*Schwefelleberluft*"). Mixing this "air" with common air and inserting a burning light, the inflamed, forming a dense white fog and a smell from volatile spirit of sulfur [SO_2]. This is very likely the first experiment on gas-to particle conversion (from H_2S via SO_2 into sulfuric acid aerosol). The "best way" gaining H_2S, *Scheele* described from adding acids to sulphuretted iron (FeS), (Scheele 1777, p. 105). *Berthollet* showed in 1796 that hydrogen sulfide is an oxygen-free acid (he called it "*gaz hydrogène sulphuré*"); in English *sulphuretted* (and *sulphurated*) hydrogen.

In the early nineteenth century, H_2S was detected in several springs, fumaroles, and volcanic gases, in the air of sewers and cesspools (see Mellor X, p. 115–116 for references). Together with ammonia, hydrogen sulfide (and organic sulfides) caused the *stinking air* in towns before establishing canalization. With the ending nineteenth century, also industrial sources of H_2S and CS_2 have been examined, including methods of determination (Blücher 1900, Rubner 1907), which have been further developed (Flury and Zernik 1931, Quitmann 1937). In the 1950s, Junge (1960) made the first H_2S measurement in the air at Bedford (Massachusetts) and Jacobs et al. (1957)[1334] at New York, proposing almost natural sources. The oxidation of H_2S by O_3 was studied by Cadle and Ledford (1966) and through OH radicals by Stuhl (1974).

Before the discovery of carbon disulfide (CS_2) and carbonyl sulfide (O=C=S, commonly written as COS) in the atmosphere (1976), H_2S was assumed to be the main natural sulfur compound, emitted in large quantities between 100 and 300 Tg S yr^{-1} (Eriksson 1959, 1960, Ericksen 1963, Junge 1963a, Friend 1973). In the mid-1970s, it was recognized that it is almost all oxidized before escaping from anoxic environments into the atmosphere. Lovelock et al. (1972) first assumed that instead of H_2S, dimethyl sulfide (DMS) is the main biogenic sulfur compound in the global budget. Significant contributions to our understanding of emissions and atmospheric distribution of H_2S

1334 *Morris B. Jacobs* (1905–1965), Ukrainian born, spent 30 years in public health work in NY.

was made by *Wolfgang Jaeschke* (Jaeschke et al. 1978, 1980, Jaeschke and Herrmann 1981). The research of *Meinrath Andreae* (born 1949) since the late 1970s contributed mostly to our understanding that DMS is the most important sulfur species because of its large emissions and contribution to climate-relevant atmospheric particulate sulfates (Andreae 1980, 1985, 1990, Andreae and Raemdock 1983). Feedbacks between the climate system and the sulfur cycle was first proposed by Lovelock and Margulis (1974), finally presented as the CLAW hypothesis that highlighted the important potential role in climate regulation of DMS production in the oceans (Charlson et al. 1987).

The Hungarian chemist *Carl* von *Than* discovered carbonyl sulfide (CS_2) in 1867, investigating hot springs at Harkány, Baranya County (Than 1867) and writes:

> Was das Vorkommen des Gases betrifft, so scheint es mir, daß es in der Natur ziemlich verbreitet ist [As for the occurrence of the gas so it seems to me that it is rather abundant in nature].

Crutzen (1976) proposed that COS was a possible source of sulfur for the stratospheric sulfate in the *Junge* layer in quiescent volcanic periods because of its relative abundance and long lifetime compared with other sulfur trace gases. Jaeschke et al. (1976) contributed pioneering to SO_2 measurements in the stratosphere. Crutzen et al. (1979) detected COS in plumes from biomass burning. Turco et al. (1980) proposed COS emission due to coal combustion. Soon later, it was known that COS is the most abundant sulfur compound naturally present in the atmosphere because of its long lifetime. In the 1990s the plant assimilation, as a source as well as a sink for COS, and the importance of antropogenic sources have been emphasized.

4.7.2 Sulfur dioxide (SO_2)

The historic investigations upon atmospheric sulfur dioxide include the following environmental issues:
- impact as air pollutant, namely in town pollution (London smog or winter smog),
- impact as air pollutant on plants (forest damage),
- sulfur cycling (agricultural chemistry),
- contribution to atmospheric acidity (“acid rain”),
- formation of sulfate aerosols (cloud condensation nuclei),
- formation of sulfate aerosols (climate impact).

The research concerns atmospheric measurements, identification of sources and estimating emissions, and understanding atmospheric chemistry.

Air pollution in the industrialized world has undergone drastic changes in the past 50 years. Until World War II, the most important urban atmospheric compound was sulfur dioxide combined with soot from the use of fossil fuels in heat and power production (Möller 1977). At the end of the nineteenth century, almost everything that was known in the 1950s about the causes of smoke and their elimination had already been said, but hardly anything had been done to reduce the smokiness of cities (Halliday 1961).[1335] The reasons for this inactivity were clearly stated by a writer in 1899 in a British building journal; Halliday (1961, pp. 15–16):

- smoke was a by-product of activities producing goods and profits, and therefore any effort to avoid smoke was very limited,
- since the damage done by smoke was to a very large number of small producers of smoke, a clear relationship between causes and effects was difficult to establish so that interest in the reduction of smoke was lacking.

It is amazing, what was known in the middle of the nineteenth century on the impact of SO_2 onto vegetation and the emission as well as technical problems from different technological processes like brick fabrics, sulfuric acid manufacture, soda factory, copper smelter, etc.; the emission of SO_2 for the year 1872 in Germany is given to be 1 Mt from combustion of 33.7 Mt hard coal (Schubert 1857, Anonymous 1876). *Clemens Winkler* denoted[1336] this era as "the era of combustion" in a lecture at Dresden in 1883 (Winkler 1883), and in a later lecture, he said that this era will still last several hundreds of years (Winkler 1900). Liefmann (1908) wrote on fume and soot concerns hygienic and measurement aspects;[1337] remarkable is a large number of references. He cites *Ramsay*[1338] that an important effect of town pollution is due to reducing sunshine. Besson (1928) studied the visibility in Paris between 1893–1903 and 1919–1926 and found a decrease by 50 % based on the same season, daytime and humidity between both periods. He further stated the visibility increases with decreasing humidity and is higher in evening and summer. Clarke (1920, p. 46) expresses the existing knowledge on SO_2 at the turn of the century in the following words:

> It undergoes rapid oxidation in the presence of moisture, being converted into sulphuric acid, and that compound, either free or represented by ammonium sulphate, is brought back to the surface of the earth by rain.

1335 *Eric C. Halliday* (1904–1996) was a South African physicist at the University of the Witwatersrand in Johannesburg and in 1945 he was appointed to the newly founded Council of Scientific and Industrial Research (CSIR), and a later international authority on atmospheric pollution. *Halliday* refers the journal Builder (London); it is *The Building News*, Vol 77 (1999); unfortunately, I did not find an article concerns this notice. However, in the edition from March 3, 1899 (p. 300 in Vol. 77) at a meeting in Glascow Mr. *Oscar Paterson* referred about "atmospheric influence on architecture and decorations" and "pointed out that the very offensive and noxious condition of the atmosphere of Glasgow was the resultant of the free escape of factory smoke and the subsequent oxidation od the sulphur in the coal into sulphuric acid." *Paterson* (1863–1936) was a famous glass stainer.

1336 Original German: *Unser Zeitalter ist im wahrsten Sinne des Wortes das Zeitalter der Verbrennung.*

1337 See footnote 588 on p. 297.

1338 Ramsay, W. (1897) Démonstration de la suie dans l'air. *Revue d'hygiene et de prophylaxie sociales* **18**, 857.

4.7.2.1 Early research: town pollution and smelter smoke

Sulfur dioxide (termed sulfurous acid in the nineteenth century, and not always clearly separated from sulfuric acid or SO_3) became in the interest of research first in England due to coal combustion (Smith 1845) and soon later in Germany, namely Saxony due to observation of forest damages in relation with mining and ore smelting (Stöckhardt 1850). *Chevallier*'s remark from 1834 (p. 13) is likely the first written source on sulfuric acid in the air (of London). Atmospheric concentrations of SO_2 at remote sites in the mid-ninetheenth century were fully negligible; however, in cities like London and Amsterdam, and everywhere with increasing coal combustion in the late nineteenth century, orders of magnitude higher SO_2 levels than nowadays were common.

The French chemist *Jean Pierre Joseph d'Arcet*, in 1831 travelling to London (p. 166), found that litmus paper, exposed to air (also in rooms), turns red; in rainwater he detected sulfurous and sulfuric acid – without any doubt caused from sulfur dioxide in the air. The first direct evidence for SO_2 in atmospheric air must be credited to *Richard Dundas Thomson* (p. 440) who passed air of London 1854 using an aspirator through distilled water which showed a "strongly acid reaction with litmus paper" (Thomson 1855, p. 123); after addition of nitrate of baryta a copious precipitate fells being insoluble in hydrochloric acid. He writes: "the source of this acid was no doubt the product of the combustion of the sulphur of coal used by heating"; no chlorine could be detected. About at the same time, the German chemist *Ferdinand Reich* (1799–1882) at Freiberg studied different methods to reduce SO_2 from furnace gases, and constructed likely the first apparatus for "quick and simple" determination of SO_2 in the air (Reich 1858); finally, he recommended the use of tall chimneys (Reich 1858, p. 176) to dilute the exhaust gases. The idea of a tall chimney was a result of the attempts in 1822 by the British copper smelter *John Henry Vivian* (1785–1855) to find a solution concerns the smoke problem. *Reich* refers *Carl Friedrich Plattner* (1800–1858), a German metallurgist and chemist at Freiberg, who carried out the first SO_2 measurements in flue gases (in percent by weight): 0.003–0.4 (H_2SO_4) and 0.015–1.7 (H_2SO_3). *Reich* likely conducted the first reliable determinations of SO_2 in the air of Freiberg in 1858: 0.0056 % (brisk wind) and 0.01 % (calm air), continued in 1863, 60 steps from a furnace to be 0.00078 %; unpublished documents, quoted by Voland (1987, p. 50). It is worth presenting deposition measurements by *Reich* carried out February 6–8, 1864, at 100 m and 600 m distance from the smelter Muldenhütte near Freiberg (in ppm in snow):[1339]

H_2SO_4	3200	150
ZnO	1900	100
PbO	3400	150
As_2O_3	800	100

1339 Ueber die Einwirkung des Hüttenrauches. *Reich* estate, folio 13 and 14, old stock mining academy, call number XVII/636 (cited by Voland 1987, p. 50).

Robert Angus Smith is the first who systematically measured SO_2 and HCl in the air of Manchester and London, published in the "Annual Report by the Inspector of the Alkali Act" (first report for the year 1863 published in 1864, London; 18 reports until 1883); Smith (1872, pp. 427–428); in μg m^{-3}:

SO_2 2000 (127–5247), $n = 29$ (Manchester 1868)
SO_2 1669 (884–2257), $n = 5$ (London 1869)
HCl 250 (46–715), $n = 29$ (Manchester 1868)
HCl 211 (119–265), $n = 3$ (London 1869)

Albert Ladureau,[1340] director of the agricultural experiment station at Lille, estimated the atmospheric SO_2 concentration to be 1.8 ppm(v) (that corresponds to 4.7 mg m^{-3}) and found in rainwater 22 μg L^{-1} sulfuric acid (Ladureau 1883). Mabery (1895), who determinated ammonia and nitrous gases in the air of Cleveland in 1894, also obtained values upon SO_2 by passing the air through a solution of sodium hydroxide, oxidation with bromine, and precipitation with barium chloride; he found between 7 and 64 mg m^{-3} (given as H_2SO_4). It is remarkable that *Charles Mabery* did try to separate sulfuric from sulfurous acid by precipitation sulfuric acid in the solution before and after oxidation by bromine; the difference in weight corresponded to 33 mg m^{-3} H_2SO_3, whereas the amount of H_2SO_4 was determined to 20 mg m^{-3} H_2SO_4.

The forestry commissioner ["*Forstrat*"] *Heinrich Constantin Gerlach* (1850–1941) in Saxony experimented since 1891 with different techniques to sample polluted air (bell-jar and frame absorber) and finally constructed in 1896 an apparatus called "*Rauchluftanalysator*" [smoke air analyzer], stepwise improved and 1908 patented (Fig. 4.62); Gerlach (1907, 1908). This rather large instrument is simply an aspirator, whereas in one of the cylinders a vacuum is gained through flowing water from the other one; the air passes in a *Woulf* bottle filled with alkaline bromine solution; the amount of passing air corresponds to the level of water in the second cylinder. *Emil Ramann* (1851–1926), professor for agricultural chemistry and soil scientist at University Munich, writes (Ramann 1908, p. 233):

> Gerlach verdanken wir den unmittelbaren Nachweis der schwefligen Säure in der atmosphärischen Luft [We owe Gerlach the immediate evidence of sulfurous acid in atmospheric air].

Between 1896 and 1907, *Gerlach* determinates at different forest locations in Saxony near manufactories 03–0.8 mg m^{-3} sulfurous acid gas; in a distance of 400–1000 m of the "*Remser Cellusen- und Papierfabrik*" [papermill] he found 55–331 mg m^{-3}

1340 No living data known; he graduated in 1867 as "bachelier ès lettres et ès sciences, est nommé préparateur de chimie à la faculté des sciences de Lille".

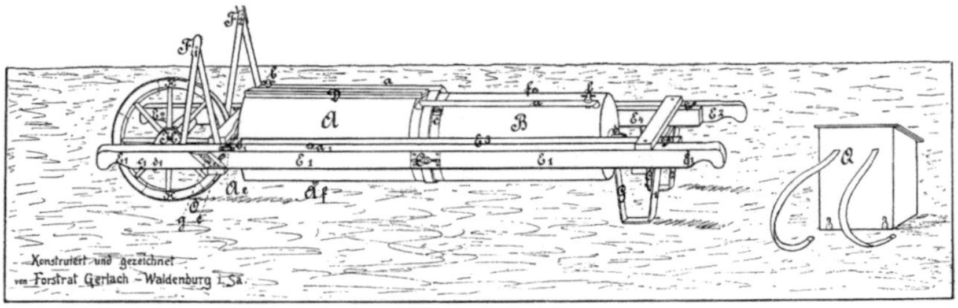

Abb. 3. Der Gerlach'sche Rauchanalysenapparat, Modell 1908 und 1909. D. R. G. M. Nr. 347 384 und 373 496. Zusammengeschlagen und zum Transport bereit. Maßstab 1 : 20.

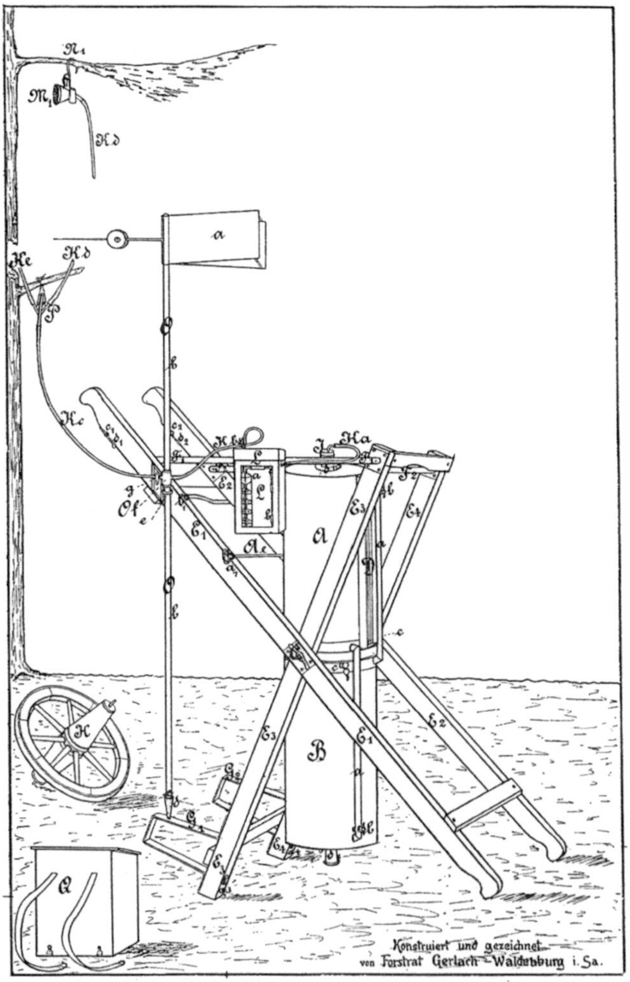

Abb. 4. Der Gerlach'sche Rauchanalysenapparat, Modell 1908 u. 1909 D. R. G. M. Nr. 347 784 und 373 496. Aufgestellt und in Tätigkeit befindlich. Maßstab 1 : 20.

Figure 4.62: Apparatus for sampling and determination of SO_2 in the air ("*Gerlach*'s *Rauchluftanalysator*"). Source: Gerlach (1908, pp. 10 and 11).

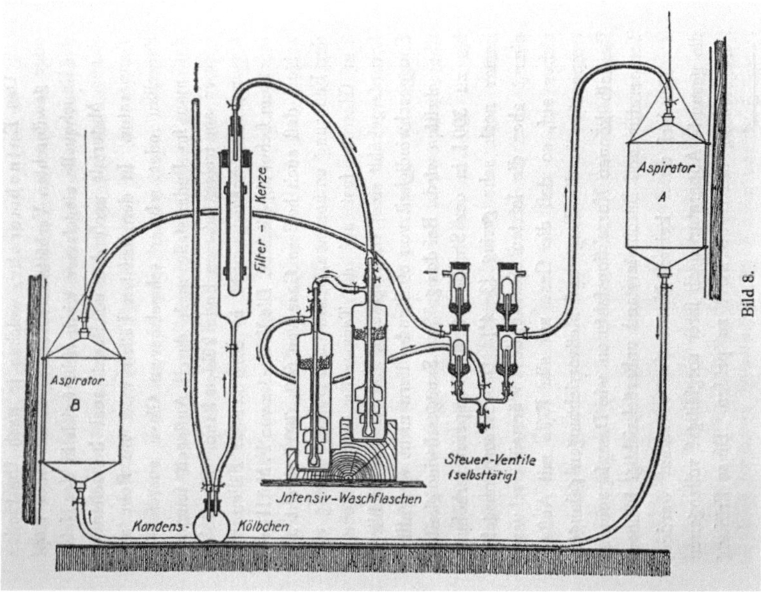

Figure 4.63: Apparatus for SO_2 sampling (Jentsch 1917, p. 36).

(Gerlach 1908);[1341] Some years later, the chemist *Siegfried Jentsch* (1884–?) in his dissertation (1916) at the University Dresden (at *Walter Hempel*)[1342] constructed a sampling device (Fig. 4.63) which had already a "modern look".

Rubner (1906, pp. 114–128) notified that the data on the presence of SO_2 in towns air are very uncertain. He refers[1343] *Friedrich Knapp* (1814–1904), who communicated that 1 kg coal yield about 15 g SO_2 and *Wilhelm Stein* (1811–1904), who found that 24–71 % of fuel sulfur transfers into SO_2. *Rubner* who tested different methods for di-

1341 *Gerlach*, a pioneer of research on forest damage, caused by air pollution, in Saxony, remained livelong interested in this research. After his retirement in 1913 he published "*Waldrauchschäden und ihre Folgen, insbesondere an Fichte und Kiefer*" [forest smoke damages and consequences for spruces and pines]. Verlag J. Neumann,1925, 45 pp.); "*Der Nachweis der schädlichen Rauchsäuren im Walde mittels dreier hierfür gefertigter Apparate*" [Detection of harmful smoke acids by three instruments manufactured for it]. *Silva* **17** (1929) 145–148; "*Dauerwald und Rauchschäden* [Permanent forest cover and smoke damages]. *Der deutsche Forstwirt* **20** (1938) 967–969; "*Ein selbstregulierender transportabler Rauchluft- und Rauchniederschlagswasser-Separator zur Feststellung der innerhalb eines Rauchgebietes befindlichen Rauchquellen und deren Abgase*" [A self-controlling transportable separator of smoke and smoke precipitation water to detect smoke sources and its exhaust gases within a smoke area]. *Der deutsche Forstwirt* **21** (1939) 1045–1046); further reading on *Gerlach* see Seibt (1991).

1342 The work he conducted since 1913 in Tharandt, supported by *Wislicenus*.

1343 Knapp, F. (1865) Lehrbuch der chemischen Technologie. Vol. 1. Vieweg, Braunschweig 655 pp. Stein, W. (1853) Chemische und technische Untersuchungen der Steinkohlen Sachsens. Engelmann, Leipzig, 98 pp.

rect determination of SO_2 in air, concluded that all remain insufficient; he finally used absorption tubes filled with coconut coal, followed by thermal desorption, afterward absorption in bromine water and precipitation as barium sulfate. He found in Berlin "typical" SO_2 values of 1.5–2 mg m^{-3}. Similar SO_2 values *Julius Stoklasa* measured in Prague (Stoklasa 1923), in mg m^{-3}:

1.2–2.5	1895
4.1–13.1	1913
5.9–58.5	1913, industrial area

The Slovakian chemist *Stanislaw Landa* (1898–1981) determined between February 1929 and January 1930 in Prague SO_2 by passing the air through 3 % H_2O_2 solution and measuring SO_2 gravimetrically and by titration; he found in summer around 0.05 mg m^{-3} (but also less than 0.001) and in winter 0.5 (0.16–0.99) mg m^{-3} (Landa 1931).

Reynolds (1930a), who also measured ozone (p. 604) and nitrogen oxides (p. 701) in Plaistow near London, found in winter that included fog, over 400 ppb(v) SO_2, and temporarily it doubtless considerably exceeded this figure; in summer it fells to 80 ppb(v). During Christmas in 1923–1927, *Reynolds* observed a considerable drop of the SO_2 concentration, and explained it by the fact that "factory chimneys are least active during the Christmas vacation". He argued that domestic consumers are the principal source of soot that "travels somewhat slowly up narrow soot-covered passages, and the absorption of SO_2 by the latter may remove an important proportion". On the other hand, factory boilers "are almost free of soot, and no appreciable absorption of SO_2 from the wide rapidly-travelling column of the combustion products occurs" (cf. Fig. 3.1, on p. 293). Concerns the seasonal variation, he further writes (Reynolds 1930a, p. 170):[1344]

> Ordinary diffusion is hindered during the shorter days by frequent mists and fogs, which exercise strong solution and precipitating influences, whereas both diffusion and hot air convection currents, from the sun-warmed ground and buildings, tend to lower the concentrations in summer, and these meteorological influences may be the dominating factors in the case.

Fig. 4.64 is a rare example of historical seasonal SO_2 cycles. Manufacturing (attributing them to the summer value) led to 2–4 mg m^{-2} and household fires in winter to an additional 4–8 mg^{-3} in London and Manchester by the 1890s. City chemist [*Stadt-Chemiker*] *Franz Hurdelbrink* (unknown living data) measured SO_2 at four different

1344 *Reynolds* also short discusses the removal of SO_2 from its products of combustion which was considered by the London Power Company's chief chemist Mr. *Hewson*, and although theoretically simple, practically and economically, obviously difficult problems are presented. The flue gas can be sprayed with water, and supplementary treatment with alkali could be necessary. Further he asked, considering the flue dust involved, "can anything done with the crude sulfuric and sulfites, or will the whole of the costs of plant labour, and materials have to be added to the electricity bill?"

Monat	Jan.	Febr.	März	Apr.	Mai	Juni	Juli	Aug.	Sept.	Okt.	Nov.	Dez.
Manchester	2,9	2,4	3,3	3,4	1,4	1,2	1,3	1,3	0,6	0,9	1,7	3,4
Königsberg	0,19	0,36	0,28	0,07	0,05	0,02	0,04	0,03	0,04	0,08	0,4	0,28

Betreffs des niedrigeren Gehaltes an schwefliger Säure in Königsberg ist zu bemerken, daß in Manchester auch der Gehalt des Rußes an schwefliger Säure mitbestimmt wurde, während in Königsberg der Ruß durch ein Rubnersches Filter abgefangen wurde.

Von dieser Differenz abgesehen, zeigte die Luft beider Städte — gleich der Londoner Luft — im Sommer das Minimum an SO_2-Gehalt, wie aus der graphischen Darstellung, Fig. 6, ersehen werden kann.

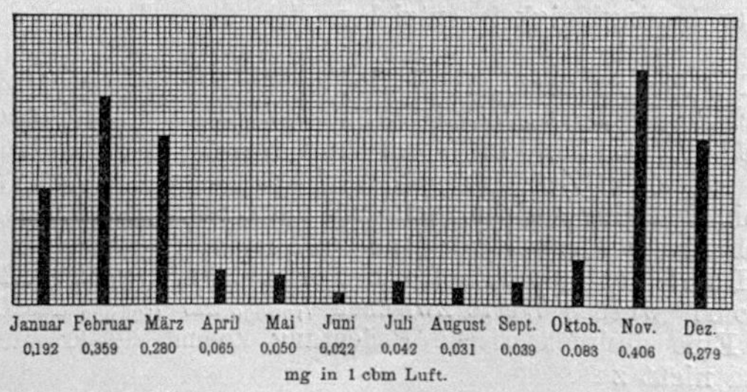

Fig. 6. Durchschnittlicher Gehalt der Königsberger Luft an schwefliger Säure in den verschiedenen Monaten. 1 mm = 0,01 mgr in 1 cbm.

Figure 4.64: Historical seasonal cycle of SO_2 concentration in Königsberg (1907). Source: Rubner et al. (1911, p. 391). The Königsberg values are originally from Hurdelbring (1907, 1909).

locations in Königsberg (1907), but only at "*Wasseramt*" [water authority] over the whole year, using *Rubner*'s aspirator method. It is notable that the summer values from Königsberg (1907) are comparable (mean 47 and 62 µg m^{-3}, resp.) with that from Dresden (1965/1968), probably because in Königsberg was no important industrial sector and quick mixing town air with surrounding pristine air. However, in winter in Königsberg much higher values (mean 302 µg m^{-3}) were found, likely due to much lower temperatures compared to Dresden (mean 158 µg m^{-3}). *Charles Baskerville* (1870–1922), professor of chemistry at the College of New York since 1904, who communicates that in New York daily 1300 tons of SO_2 is discharged into the atmosphere, writes (Baskerville 1910):

> Its presence in atmospheric air is a menace to hygienic welfare [. . .] From an economic standpoint, this is an enormous, partly avoidable, waste, [. . .]

Baskerville made in 1907 a number of determinations of the SO_2 content of the air of New York City (in ppm(v)):

Downtown region	1.05–5.60
Railroad tunnels	8.54–31.50
Subway	none
Various parks	0–0.84

Such as in England and Germany, the smoke of smelters, recognized by the public as "smelter-smoke smell", led to intensive research in the USA (Holmes et al. 1915).[1345] *Arthur Edward Wells* (1884–1939), metallurgist and chief chemist in charge of smelter investigations and field work conducted an extensive investigation on the SO_2 content in air around the Selby Smelting Works in Contra Costa County, near San Francisco, one of America's busiest smelters (Wells 1915); between September 1913 and May 2014 he and his co-workers found concentrations (including the air of San Francisco) ranging from 0.00 to 3.6 ppm, with a mean of 0.3 ppm ($n = 173$). Notable is further the work of Monnett et al. (1926),[1346] who reported on SO_2 concentrations in Salt Lake City 1919–1921 to be on average 0.15 ppm in winter and 0,01 in summer (maximum 0.8 ppm); noticeable are airplane measurements between 1000- and 3000-feet altitudes (0.0–0.2 ppm SO_2). Burdick and Barkley (1939) compiled data on SO_2 measurements made in the USA, Great Britain, and other European towns; in following some examples (in ppm(V) – note that 1 ppm(v) = 2.63 mg m^{-3} SO_2):

Portsmouth (1927–1927)	0.97 (0.2–1.3)
Washington D. C. (1926–1928)	0.18 (0.004–0.54)
Pittsburgh (1936–1937)	0.078 (summer 0.06, winter 0.09, maximum 0.9)
Chicago (1937)	0.3 (maximum 3.16)

From Great Britain, it is reported that in 14 cities 24-hour continuous samples were taken with a yearly average ranging between 0.6 and 0.15 ppm (maximum daily value 1.6 ppm). Wilkins (1954) reports on the highest level over 48 hours during the London fog in 1952 of 1.34 ppm (and 4.48 mg m^{-3} smoke).

In the air of Berlin (Potsdamer Platz at 20 m heigh in spring), Haehnel (1922)[1347] determined the amount of SO_2 with a surprising quantity to be 33 mg m^{-3} (it is notable that CO_2 he determined to be 2000 ppm) – that's on Berlin air[1348] in the 1920s; how-

1345 This "Report of the Selby Smelter Commission" contains several further interesting studies: fumigation experiments to determines the effect of highly diluted sulphur dioxide on a growing grain crop by *A. E. Wells* (pp. 213–303); disappearance of sulphur dioxide from diluted mixtures of sulphur dioxide with air by *G. C. Bartels* (pp. 308–323); bibliography on effects of sulphur dioxide in vegetation and animals (pp. 503–528). *Bartels* found that SO_2 disappears almost simply due to absorption and to a small degree due to oxidation to sulphur trioxide in an excess of moist air.

1346 *Osborn Monnett* (unknown living data) was a fuel engineer for the Bureau of Mines; 1914 chief smoke inspector for Chicago, charged for the Salt Lake City investigation (1920–1926), and later consultant engineer for the Citizens Smoke Abatement League of St. Louis. *George St. John Perrott* (1893–1980) was another smoke engineer and co-worker of *Monnett* in Salt Lake City.

1347 *Friedrich August Otto Haehnel* (1884–?) was a German chemist and since 1935 professor at Technical University Berlin.

1348 "*Das ist die Berliner Luft, Luft, Luft...*" is an operetta song written by *Heinrich Bolten-Beackers* (1871–1838) and set to music in 1904 by *Paul Lincke* (1866–1946), first performed with the operetta *Frau Luna* in Berlin 1922, to be regarded as unofficial anthem.

ever, these high values are unreliable! Liesegang (1931, 1933) discussed the problems of absolute determination of SO_2 in air and suggested that relative values obtained by passive sampling[1349] are sufficient for an assessment of the air pollution in the distance from sources. For remote areas, *Liesegang* found (the numbers correspond to 100 hours sampling with apparatus) the 1–3 mg S, in Berlin-Dahlem 10–20 mg S, and close to manufactures typically 50 mg S and more. *Olga Waldbauer* (unknown living data), a chemist at the Metropolitan Institute for Hygiene and Bacteriology in Budapest, determined using *Liesegang*'s method in Budapest from October 1932 to May 1934 values between 3 and 45 mg (Waldbauer 1935). Katz (1961) and Junge (1963a) report on the following SO_2 values (in µg m^{-3})

London (1940–1950)	600
London (1949–1954)	350 (210–500; maximum 3520)
Los Angeles (1950–1955)	60–600
Frankfurt/Main (~1955)	87 (summer), 443 (winter)
New York (1957)	280–1500
USA, rural (~1955)	23
Florida and Hawaii (~1955)	1–3
Sweden, background (1954/55)	12 (2–19)

Although SO_2 concentrations in cities were in the range of 1–10 mg m^{-3} around 1900, they dropped to the upper range of µg m^{-3} (some hundreds) by the 1950s and at present they are in the lowest range of µg m^{-3} (5–10), close to the remote background values of the order of 0.5 to 4 µg m^{-3} (Georgii 1970). In West Berlin, SO_2 is continuously measured since 1968, showing a strong diurnal as well as seasonal variation with a similar amplitude (1969–1973): maximum (winter and noon, resp.) 250 µg m^{-3} and minimum (summer and early morning, resp.) 50 µg m^{-3} (Lahmann and Prescher 1974).[1350] Nowadays, SO_2 in the atmospheric air plays no role with respect to the sanitary question and the concentration is around 1 µg m^{-3}, which is a thousandfold less than 100 years ago.

1349 He introduced the so-called "*Glockenmethode*" [absorption cylinder], called later *Liesegang*'s method (bell-jar process of *Liesegang*): a filter-paper shell soaked with a solution of potassium carbonate in water and glycerin is exposed for 100 hours in a cover made from porcelain, and after elution the amount of sulfate is determined (in mg sulfur). *Liesegang* found that at values less than 30 mg S, no plant damages have been observed.

1350 *Erwin Lahmann* (1925–2001) was a German chemist, since 1971 professor at the Technical University Berlin and since 1976 head of the department Air Hygiene of the Institute for Water, Soil and Air Hygiene in Berlin. For SO_2 measurements in the former East Germany see Möller (2020, pp. 422–436).

4.7.2.2 Early research: forest damage and SO_2

Forest damages have been observed without any doubt soon after the establishment of smelters around Freiberg (Saxony) in the "*Erzgebirge*" [Ore Mountains] in 1530. *Christian Lehmann* (1611–1688), state historian and pastor, writes "*Gifftiger Thau fällt auf die Wälder und verursacht eine große Fäulnis*" [Poisonous dew falls on the woods and causes a large corruption], see Fig. 4.65 (Lehmann 1699; note that all his works have been published posthumous). This statement is close to *John Evelyn*'s remark on dew deposition (p. 113). A next note upon the cause of forest damage was given by *Hermann Friedrich von Göchhausen* (1663–1734), "*Oberlandjägermeister*" [Supreme District Huntsman] in Saxony, writing that pinewood is damaged "*durch das wiedrige schweflichte Erdreich entkräftet*" [through adverse sulfureous soil devitalized], see Fig. 4.65 (Göchhausen 1710).

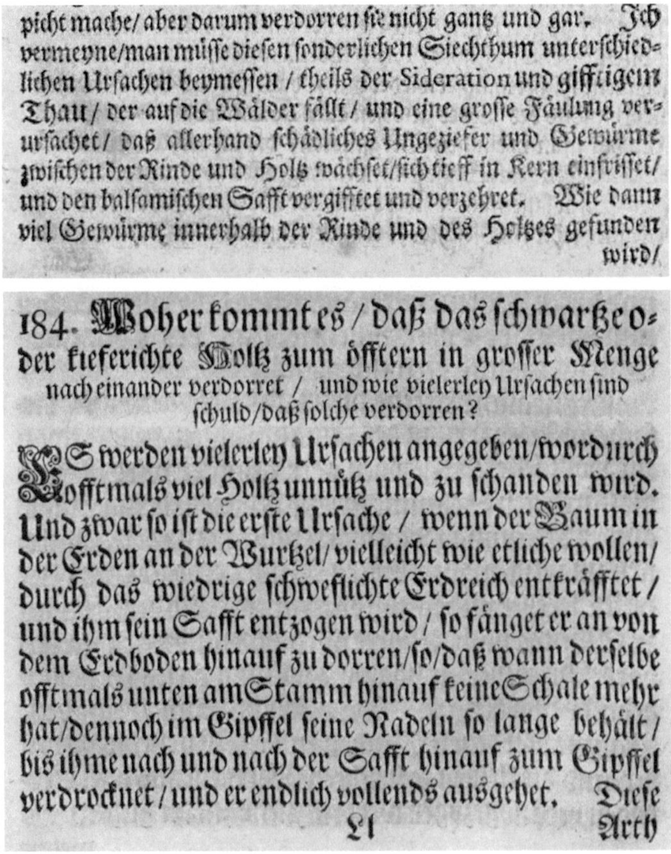

picht mache/ aber darum verdorren sie nicht gantz und gar. Ich vermeyne/man müsse diesen sonderlichen Siechthum unterschiedlichen Ursachen beymessen / theils der Sideration und gifftigem Thau/ der auf die Wälder fällt / und eine grosse Fäulung verursachet/ daß allerhand schädliches Ungezieffer und Gewürme zwischen der Rinde und Holtz wächset/sich tieff in Kern einfrisset/ und den balsamischen Safft vergifftet und verzehret. Wie dann viel Gewürme innerhalb der Rinde und des Holtzes gefunden wird/

184. Woher kommt es / daß das schwartze oder kieferichte Holtz zum öfftern in grosser Menge nach einander verdorret / und wie vielerley Ursachen sind schuld/daß solche verdorren?

ES werden vielerley Ursachen angegeben/wordurch offtmals viel Holtz unnütz und zu schanden wird. Und zwar so ist die erste Ursache / wenn der Baum in der Erden an der Wurtzel/ vielleicht wie etliche wollen/ durch das wiedrige schweflichte Erdreich entkräfftet / und ihm sein Safft entzogen wird / so fänget er an von dem Erdboden hinauf zu dorren/so/daß wann derselbe offtmals unten am Stamm hinauf keine Schale mehr hat/dennoch im Gipffel seine Nadeln so lange behält / bis ihme nach und nach der Safft hinauf zum Gipffel verdrocknet / und er endlich vollends ausgehet. Diese

Ll Arth

Figure 4.65: Copy of p. 175 (Lehmann 1699) concerns causes of forest damages (poisonous dew), (upper figure) and (lower figure) of p. 133 (Göchhausen 1710) concerns "sulfureous soil".

Edward Turner (p. 72) and the Scottish toxicologist and physician *Robert Christison* (1797–1882) were the first who conducted laboratory observation on the effect of SO_2 on plants (Turner and Christison 1827); they found in a bottle filed with 2/470 parts of SO_2 (0.4 %) that "the leaves of the plant became greenish-gray, and dropped much in less than two and a half hours". Even at a "small" concentration of 1/10000, after 2 days the effects were seen; however, *Turner* and *Christison* further comment:

> We do not pretend to infer, as an absolute conclusion from these experiments, that manufactories from which sulphurous acid gas is disengaged even in large quantities must injurious to surrounding vegetation. It is probable, nay certain, that, small as the proportion was which we have tried, the proportion contained in the air in the vicinity of the largest manufactories is a great deal less.

A few years later in 1849, relationships between SO_2 polluted air and smoke damages on plants and trees in Saxony were described by *Julius Adolph Stöckhardt* (1809–1886), a German pioneer in smoke damage research at the "*Tharandter Forstliche Akademie*" [forestry academy] in Tharandt near Dresden[1351] (Stöckhardt 1850, Schroeder and Reuß 1883, Wislicenus et al. 1914, Wienhaus and Däßler 1991, Wienhaus 1999). Stöckhardt (1850) writes:[1352]

> Die Bestandtheile des Halsbrücker Hüttenrauchs zerfallen in zwei Klassen: a) in unschädliche (Kohlensäure, Kohlenoxydgas, Ruß, Kohlenwasserstoff, Wasserdampf, Stickstoff und Schwefeldampf), b) in schädliche (schweflige Säure, Schwefelsäure, Chlorgas, arsenige Säure, Arsensuboxyd, Bleioxyd, Zinkoxyd und Antimonoxyd). Rücksichtlich der Wirkungsweise der zur letzten Classe gehörenden Substanzen auf das Pflanzenleben sind die angestellten Erörterungen zufolge zwei specifisch verschiedene Fälle zu unterscheiden, nämlich eine *directe* oder *acute Vergiftung* der Pflanzen durch die mit ihnen in Berührung kommenden gas- oder dampfförmigen Säuren, und eine *indirecte* oder *chronische Vergiftung*, welche der Erdboden, auf dem die Pflanzen wachsen, durch lang fortgesetzte Zuführung von metallischen Dämpfen erfährt. Acute Vergiftung der Pflanzen durch saure Dämpfe oder Gase. Diese wird hauptsächlich durch die in dem Rauche enthaltene schweflige Säure (schwefligsaures Gas) bewirkt [The constituents of smelter smoke from Halsbrück fall into two classes. a) in harmless (carbonic acid, carbon monoxide, soot, hydrocar-

---

1351 *Heinrich von Cotta* (1763–1844) established 1811 a private "*Forstlehranstalt*" [forestry school] at Tharandt, which was 1816 named "*Königlich-Sächsische Forstakademie*" [Royal Saxon Academy of Forestry], renamed 1830 in "*Akademie für Forst- und Landwirthe zu Tharandt*" [academy for forester and farmer] and 1871 in "*Akademie für Forstwirthe*" (the department for agriculture that was integrated 1830, was relocated in 1869 to the University Leipzig) and raise in 1904 the rang of college, in 1918 renamed in "*Forstliche Hochschule Tharandt*" [forest college] and finally 1929 integrated as department to the Technical University Dresden (renamed 1961 in Technical University) that raised to a faculty in 1941. *Cotta* was the director until 1844. The institute was renamed in "*Wissenschaftsbereich Pflanzenchemie an der Sektion Forstwirtschaft*" [scientific department plant chemistry at the section forest management] and 1990 finally renamed in "*Institut für Planzen- und Holzchemie*" belong the discipline ["*Fachrichtung*"] forest sciences within the faculty environmental sciences at TU Dresden.
1352 The exhausted SO_2 from the smelters smoke was reused with the first sulfuric acid factory built in 1857 at Muldenhütten near Freiberg (Thiele 2000).

bon gas, steam, nitrogen and sulfur vapor), b) in harmful (sulfureous acid, sulfuric acid, chlorine gas, arsenic acid, arsenic suboxide, lead oxide, zinc oxide and antimony monoxide). Respectively the effects of the substances of the latter class, two different cases must be distinguished, namely a *direct* or *acute poisoning* of plants through contact with the gaseous or vaporous acids, and an *indirect* or *chronic poisoning*, which the soil suffers where the plants grow through continues feed by metallic vapors. Acute poisoning of plants by acid vapors and gases. This is mainly affected by the sulfureous acid (sulfureous gas) containing in the smoke.

The first chair for agricultural chemistry and agricultural technology in Tharandt was established by *Stöckhardt* in 1847, which he held until 1883, succeeded 1883–1895 by *Julius von Schroeder* (1843–1895), professor for agricultural chemistry. 1896 foundation of the "*Institut für Pflanzenchemie und Holzforschung*" [institute for plant chemistry and wood research], the first director was 1896–1934 *Hans Adolf Wislicenus* (1867–1951). The historical smoke damage research at Tharandt, carried out 1847–1935 by *Stöckhardt*, *Schroeder* and *Wislicenus*, was restored in 1961 after a suggestion by *Heinrich Wienhaus* (1882–1959), in 1958. *Hans-Günther Däßler* (born 1925) became a docent (1961) and professor (1963) for "chemistry of smoke damages" and later (1968) professor for plant chemistry at the institute.[1353] He was succeeded 1990–2002 by *Otto Wienhaus* (born 1936), son of *Heinrich Wienhaus*, as a professor for plant chemistry and ecotoxicology. *Däßler* also coordinated the smoke damage research ["*Rauchschadensforschung*"] in East Germany 1961–1990.[1354]

4.7.2.3 Atmospheric chemistry: oxidation of SO_2

Today we know that SO_2 is oxidized in the gas phase only by OH radicals with a moderate rate (lifetime about a week), but in the aqueous phase much faster by different oxidants. Since *Thénard* in 1818 mentioned the oxidation of sulfite by hydrogen peroxide, a large number of investigations were made on the oxidation of sulfuric acid, sulfites, and dissolved sulfur dioxide in aqueous solutions as well as gaseous SO_2 (see *Gmelin*, *Mellor*). The oxidation of SO_2 in aqueous solutions by O_3 was first noticed by Houzeau (1861, p. 151). Witz (1885)[1355] argued from the observations at Paris and Rouen that the minimum of ozone occurs in December when the presence of SO_2 in air should be highest (and detected as sulfate in precipitation), and thus concluded that SO_2 is

1353 The institute was renamed in "*Wissenschaftsbereich Pflanzenchemie an der Sektion Forstwirtschaft*" [scientific department plant chemistry at the section forest management] and 1990 finally renamed in "*Institut für Planzen- und Holzchemie*" belong the discipline ["*Fachrichtung*"] forest sciences within the faculty environmental sciences at TU Dresden.

1354 The author (DM) took part at several meetings between 1975 and 1989 on this subject. *Däßler* was referee and "doctor father" (beside *Karlheinz Lohs*) of *Möller*'s habilitation (*Zur Untersuchung des atmosphärischen Schwefel-Zyklus unter dem Einfluss anthropogener Aktivitäten*. AdW, Berlin, 1982, 136 pp.).

1355 Unknown German chemist in Rouen.

effectively oxidized to sulfate in mists and fogs by O_3. *Witz* also found that the atmospheric concentration of SO_2 decreases with the distance from the town (Rouen) and is at 3 km no longer detectable.

Only a few laboratory studies I refer in the following before we are going to historic investigations upon the oxidation of SO_2 under atmospheric conditions. *Oscar Loew* (footnote 937 on p. 473) exposed in 1869 diluted sulfurous acid in sealed tubes to the sunlight for two month and found that it was gradually reduced to sulfur whereas "another part of the acid having been oxidized by it to sulphuric acid" (Loew 1870a). *Charles Lee Reese* (1862–1940)[1356] writes that "it is well known that solutions of the sulphites are oxidised on exposure to the air" and found that the weaker solution oxidize more rapidly: a solution of 0.07 % SO_2 in water is completely oxidized after 3 weeks (Reese 1884). Berthelot (1878b) reported on a reaction O_3 + NO_x in aqueous solution and Traube (1893, p. 1479) communicates $O_3 + SO_2 = O_2 + SO_3$. *Frederick Ferraboschi* (?–1969)[1357] found when SO_2 is rapidly led into a solution of H_2O_2, it is entirely destroyed, and ozone is produced (Ferraboschi 1909). The French chemist *Henry Baubigny* (1842–1913) found (Baubigny 1912) that sulfite transforms to dithionite by Cu(II) which is reduced to Cu(I). This was the first step in understanding the role of transition metal ions (now called TMI) in the autoxidation of S(IV), now called catalytic oxidation. The Swedish chemist *Hans Bäckström* (1896–1977) first concluded that the sulfite oxidation in aqueous solution proceeds as chain reactions and that the mechanism of the thermal autoxidation is the same as that of the photochemical reactions (Bäckström (1927). *James Franck* and *Fritz Haber* combined the findings by *Baubigny* and *Bäckström* and first proposed a radical chain mechanism of S(IV) oxidation; the sulfite radical SO_3^- they regarded as the anion of the weak radical acid HSO_3, and named it monothionic acid; it forms as dimer dithionite, but it also adds oxygen, gaining sulfate and generating the OH radical (Franck and Haber 1931):

$$Cu^{2+} + SO_3^{2-} = Cu^+ + SO_3^- \quad \text{and} \quad 2SO_3^- = S_2O_6^{2-}$$

$$HSO_3 + O_2 + SO_3^{2-} + H_2O = 2SO_4^{2-} + 2H^+ + OH.$$

The latter net reaction *Franck* and *Haber* segment in the addition of O_2 according to the view of *Bach* and others, a so-called moloxide (HSO_5) that further reacts with sulfite; the chain is provided by the reaction

$$SO_3^{2-} + OH = SO_3^- + OH^-.$$

1356 American chemist; undergraduate at University of Virginia in 1884, he moved to Germany for his graduate studies at *Bunsen* in Heidelberg and *Victor Meyer* (1848–1897) in Göttingen, completed his PhD in 1886; chemical director of DuPont serving from 1911 to 1924.

1357 He entered the Clare College Cambridge in 1906, changes name to *Ferrar* in 1940.

In discussing another interpretation than given by *Bäckström* concerns the light-induced oxidation, *Franck* and *Haber* propose the release of an electron from the anion, which transfers to the added water molecule that further "decays" according to

$$SO_3^{2-} + H_2O + h\nu = SO_3^- + H + OH^-.$$

Today we would write this process as an electron transfer onto water (that also represent the aquated electron e_{aq}^- which is equivalent to $H + OH^-$):

$$SO_3^{2-} + H_2O + h\nu = SO_3^- + H_2O^-.$$

Albu and Schweinitz (1932) adapted these mechanisms for the metal-catalyzed autoxidation by Cu(II) and Fe(II). Bäckström (1934) further exanimated the radical mechanisms, explaining the pH dependency of the rate, and proposed as a first subsequent reaction

$$SO_3^- + O_2 = SO_5^-(+HSO_3^- = HSO_5^- + SO_3^-),$$

followed by the fast reaction

$$HSO_5^- + SO_3^{2-} = HSO_4^- + SO_4^{2-}.$$

Additional *Bäckström* suggested the participation of the sulfate radical SO_4^- with the reactions

$$SO_4^- + SO_3^{2-}(HSO_3^-) = SO_4^{2-}(HSO_4^-) + SO_3^-.$$

All these reactions are today included in any aqueous-phase model; the reaction rate constants have been estimated in the 1980s.

The German meteorologist *Wilhelm Krebs* (1858–1924) at Berlin first asked the question "sulfuric acid or sulfurous acid in town air" (Krebs 1893); he first states from "chemical experience" that coal burned only into sulfurous acid (SO_2) which in the atmosphere oxidizes into sulfuric acid. Further he refers some previous studies upon the oxidation of sulfurous acid (*Reese, Loew, Bachmann*) and concluded that the oxidation of sulfurous acid in sunlight needs some time. Somewhat circumstantial *Krebs* emphasized that the presence of fog and cloud droplets could be essential for the transformation of sulfurous acid into sulfuric acid, and, with respect to the health effects of the London fog he emphasizes the need to investigate the transformation process "in all directions" (Krebs 1893, p. 364):

> [. . .] in den angegebenen Richtungen, einerseits am trockenem Gas, andererseits in seiner wässrigen Lösung, vor allem in Nebelform, nicht zu verkennen [. . . in the indicated directions, on the one hand at dry gas, on the other hand in aqueous solution, namely as fog, not to underate].

The rather fast oxidation of dissolved SO_2 in snow and rain samples has been recognized by Küppers (1918) and Stoklasa (1923), see p. 215.

Since Helmholtz (1887) it was known that sulfuric acid, as formed, will rapidly pass almost entirely into a liquid phase. Fuchs and Oschman (1935)[1358] first produced sulfuric acid aerosol and found them in the size of *Aitken* nuclei, but having the tendency to combine to form larger droplets. Coste and Courtier (1936)[1359] developed a method to sample these sulfuric acid particles by inducing condensation of water vapor through cooling, so as to enable them to fall to the bottom of a vessel; they conclude from the presence of such very small proportion of sulfuric acid in air (1–20 µg m^{-3} in London) containing relatively large proportions of SO_2 on its slow.

The oldest publication concerns action of solar light on SO_2 is known by French *Jean François Auguste Morren* (p. 321), who first studied the photochemical oxidation of SO_2 and discusses for the first that under the influence of light the vibrations between the atoms of the molecule SO_2 enlarge which finally broke down and combine newly (Morren 1870, pp. 346–347). *Alfred Coehn* (1863–1938), German electro- and photochemist in Göttingen, first studied the photochemical oxidation of SO_2 using a mercury halogen lamp and found SO_3 as product but only in the presence of humidity (intensive drying of reactant gases excluded any oxidation (Coehn 1907); see also Coehn and Becker (1910) for photochemistry of sulfuric acid.

The English chemist *Ashton Hill* (unknown living data) at the University of Liverpool studied the photodecomposition of SO_2 at 313 nm according to $SO_3 + S$ (Hill 1924). These experiments were continued in dry atmosphere for different wavelengths in the ultraviolet range by Kornfeld and Weegmann (1930).[1360] First, Blacet (1952) proposed the atmospheric photolysis of SO_2 to be likely important in photochemical smog. Gerhard and Johnstone (1955) found the photochemical oxidation of SO_2 to be very slow. Katz and Gale (1971) summarize the mechanism of photooxidation of SO_2 in atmosphere. Until the late 1970s, a large number of publications upon the SO_2 photochemistry appeared, but after about 1975, it was accepted that only the (relatively slow) homogeneous reaction SO_2 + OH is of significance in the gas phase (see papers by *Cox*, *Calvert*, *Stockwell* and others).[1361]

1358 *Nikolaj Albertovich Fuchs* [Никола́й Альбе́ртович Фукс] (1895–1982) was a Sowjet physicochemist and the founder of aerosol research in the USSR. See Spurny (1996).

1359 *John Coste*, see footnote 651 on p. 228.

1360 *Gertrud Kornfeld* (1891–1955) was a German chemist, the first and only woman who made habilitation in chemistry in the Weimar Republic, born in Prague and worked together with *Max Bodenstein* at Hannover and Berlin; immigrated 1933 via England, Vienna to USA, but was unable to continue his academic carrier. *Eugen Weegmann* (1897–?) was a German chemist who made his dissertation 1930 on this project and further worked at the *Beilstein* editing administration.

1361 Cox and Penkett (1970) Calvert et al. (1978), Calvert and Stockwell (1983), Stockwell and Calvert (1983).

As mentioned above, the oxidation of SO, in aqueous solutions (uncatalyzed and catalyzed) is a classic process, which has been studied extensively over the past, but its atmospheric importance was first emphasized by Junge and Ryan (1958), after Junge (1957) showed that *Aitken* particles mainly consist from ammonium sulfate. This was later confirmed also for Europa by Mészaros (1966), who showed that sub-micron particles consist of about 30 % sulfate. This finding obviously stimulated *Basil John Mason* (1923–2015)[1362] to laboratory studies upon the rate of formation of ammonium sulfate in water drops exposed to air containing known concentrations of gaseous sulfur dioxide and ammonia (Heuvel and Mason 1963). Since then, for many years, the role of NH_3 in aqueous-phase SO_2 oxidation was seen as a catalyst (Georgii 1969), and the rainout and washout of SO_2 was studied (Beilke and Georgii 1968).[1363]

Uptake and oxidation of SO_2 was further in detail reviewed and studied by Scott and Hobbs (1967), Tsunogai (1971b), Larson and Harrison (1977), Larson et al. (1978), Penkett (1972), Penkett et al. (1979), Beilke et al. (1975), Hegg and Hobbs (1978). Before the works of *Stuart Penkett*, it was assumed that oxygen is the oxidizing agent, and since then O_3 and H_2O_2 have been considered as the key oxidizing agents in aqueous solution. Möller (1980) first calculated the relative importance of different atmospheric SO_2 oxidation pathways. Reaction rate constants for S(IV) + H_2O_2 were given by Hoffmann and Edwards (1975), Penkett et al. (1979), Martin and Damschen (1981), McArdle and Hoffmann (1983), Kunen et al. (1983), Lee et al. (1986), Lagrange et al. (1993, 1996), and Maaß et al. (1999). Reaction rate constants for S(IV) + O_3 for the nucleophilic mechanism were given by Erickson et al. (1977), Hoigné et al. (1985), and Hoffmann (1986). Reaction rate constants for S(IV) + O_3 for the radical mechanism were given by Penkett et al. (1979), Maahs (1983), Martin (1984), Nahir and Dawson (1987), Lagrange et al. (1993), and Botha et al. (1994). Freiberg (1974, 1975, 1976) studied the iron-catalyzed SO_2 oxidation. Hoffmann (1977) studied the reaction $H_2S + H_2O_2$.

At the end of the 1990s, researchers in the field of atmospheric sulfur chemistry agreed that air pollution and climate models simulate the formation of sulfate using acid rain chemistry with sufficient accuracy, and no further research is needed. Model studies suggest that aqueous oxidation comprises a large majority of the global sulfate production. Recent arguments, however, concern that the range of model estimates is rather large, reflecting the difficulty in reproducing cloud processes in large-scale models. This "problem" of cloud modeling is well known for decades because of the stochastic properties[1364] of cloud microphysics and dynamic; there is little hope in fu-

1362 Sir *Basil John Mason* (1923–2015) was an expert on cloud physics; from 1965 to 1983 he was Director of the UK Meteorological Office at Bracknell.

1363 Later it became clear that NH_3 does not act as catalyst but buffers the formation of acid due to S(IV) oxidation and thus allowed further absorption of SO_2.

1364 In 1932, the British physicist and expert on fluid dynamics, *Horace Lamb* (1849–1934), told at a meeting of the British Association for the Advancement of Science: "I am an old man now, and when I die and go to Heaven there are two matters on which I hope for enlightenment. One is quantum ture to model cloud chemistry (whereas the aqueous-phase chemistry is rather good described) more "reliable" (and the question, why we do need it, remains open). Nevertheless, research is going forward, and recently laboratory stadies found that the H_2O_2-driven oxidation of SO_2 in aqueous aerosol particles can contribute to the missing sulfate source during severe haze pollution events (Liu et al. 2020).

4.7.3 A very short sketch on atmospheric sulfur research 1975–1995

There are no known biogenic sources of sulfur dioxide, and the only known natural sources are volcanic activities. It is the oldest known air pollutant and became the icon for the adverse effects of the 19th-century industrialization (Robinson and Robbins 1970). To the post-war German pioneers of atmospheric sulfur dioxide and particulate sulfate research must be viewed a generational sequence of meteorologists; *Christian Junge*, (1912–1996) his scholar *Hans-Walter Georgii*, (1924–2018), his scholars *Siegfried Beilke* (born 1938), *Dieter Jost* (1942–2011), *Harald Berresheim* (born 1955), and the chemist *Wolfgang Jaeschke* (born 1942).

In the 1970s, it was recognized that sulfur dioxide enlarged the air pollution problem from a local (urban pollution), to regional (Europa and North America) and finally to the global scale – not because of a long residence time such as for carbon dioxide and other "greenhouse" gases but due to exponential use of coal at almost all places of the world. "Acid rain" was the trigger not only of much of the research but also led to the first international sulfur abatement agreement in Sofia in 1980. Despite some early research and notable results (see previous chapters), the 1980s were the "high noon" of atmospheric sulfur research, gaining a sufficient understanding upon

- the sulfur cycling (emission, distribution, deposition),
- the atmospheric oxidation of SO_2,
- the formation of sulfate aerosol and nanosized sulfuric acid (homogeneous nucleation process),
- the stratospheric sulfate aerosol (*Junge* layer),
- the adverse effects caused by SO_2 (acidification of soils and lakes, plant damages, material corrosion),
- the marine emission of dimethyl sulfide (DMS), and its role in climate (CLAW hypothesis),
- organic sulfur compounds: sources and atmospheric chemistry

electrodynamics, and the other is the turbulent motion of fluids. And about the former I am really rather optimistic" (cited after Pletcher et al. 1997, p. 272). Mullin (1989) writes: "nearly 60 years later, the fundamental nature of turbulence remains a mystery although new and exciting developments continue to emerge. One of these developments is the application of some of the ideas of chaos".

- modeling of "acid rain",
- modeling of climate, etc.

The literature on this subject is uncountable, and research is not ending.

5 An outlook: from the past into future

> The longer you can look back, the farther you can look forward (speech of *Winston Churchill*, March 1944, cited after Ratcliffe, S., ed. (2016) Oxford Essential Quotations)

At the very beginning of this volume, I wrote that I am not a historian but a person who spent almost forty years studying atmospheric chemistry and air pollution. Now, after finishing this book, I am still not a historian, but I have learned to understand that a *voyage through the history* needs two different views, first, to empathize with the past and our scientific predecessors, and second, to look at the past with today's knowledge. Moreover, I got doubts about whether we should name this discipline *atmospheric chemistry*. It sounds perhaps arrogant (to the younger generation), but atmospheric chemistry was an intermittent discipline over the past fifty years. What will be history or will be fully forgotten in another fifty years, we do not know. Before about 1970, it was chemical meteorology, and in the future, it returns back again as *chemical meteorology*. The American atmospheric chemist *Gregory Kok* nicely describes the five decades of atmospheric chemistry since 1950 (Kok 1992, p. 55–56), better than I could have expressed:

> The 50s can be called "Recognition of Atmospheric chemistry" years; Los Angeles gained its famous nature of smog. Ozone as a product of atmospheric chemistry became better known and atmospheric photochemistry became recognized.
> The 60s can one call the "Adolescence" years; books on "Photochemistry of Air Pollution" and "Air Chemistry and Radioactivity" were published and set the stage. Measurement techniques were developed; the chemistry of the ozone/NO reaction was explored, as well as recognition of the OH radical.
> The 70s were the "Focus on Measurement" years; smog chambers were used to measure every possible reaction. The products were looked at and some kinetics were developed.
> The 80s can be called "Unravelling the Chemistry" years; acid precipitation studies addressed questions like whether sulfur conversion was oxidant limited, or whether hydrocarbon or NO_x were limiting in ozone formation. Modelling brought all of the chemistry together. Finally, the ozone hole came along, and there was a new focus on the stratosphere.
> The decade of nineties is "Putting it All Together"; chemists were now involved in multidisciplinary activities.

What *Greg Kok* in the early 1990s expected in atmospheric chemistry research, after the 90s, was done: the beginning to link with biologists, meteorologists, and oceanographers, and balancing multidisciplinary studies including air/sea exchange, flux studies, and fast-response instruments. Over the five decades of atmospheric chemistry since 1950, we found out what to measure; we measured it; we found out where it is going, where it is coming from, and how it is getting there. Now,

> the two first decades of the 21th century can be called "Chemistry of the Climate System"; chemistry´s influence is moving into a new frontier. From an understanding how anthropogenic emission may influence clouds, precipitation and potentially radiation we now study the feedback of climate change on atmospheric chemistry and natural emissions.

https://doi.org/10.1515/9783110732467-005

For about thirty years, *human-induced* climate change has been recognized and investigated – and the just-published newest IPCC report brings nothing new but emphasizes the *acceleration* – and a key question that should be answered is what role does chemistry play. I would answer, not a large one, *only* a slight change in the chemical composition of the atmosphere. The players in weather and climate, however, are radiation, temperature, and water (vaporous and condensed), a triunity, so complex in time and space that any description is imperfect so far and likely will also remain in the future. This is the physical part of meteorology, further to be investigated. Concerning the chemical part, *atmospheric chemistry* generated the insights we needed for an understanding of *changes*:
- understanding atmospheric acidity,
- understanding ozone chemistry in the stratosphere and troposphere,
- understanding the chemistry of other reactive oxygen species (ROS),
- understanding the chemistry of nitrogen oxides, nitrous, and nitric acid,
- understanding the oxidation of sulfur dioxide and the formation of sulfuric acid.

The list can be extended concerns the chemistry of halogen compounds and the oxidation of organic compounds, mercury, etc. Notwithstanding, we do not know and understand everything completely. New insights like marine halogen chemistry and its relation to atmospheric chemistry recently came. Still unknown insights will come.

But our knowledge was sufficient enough to implement in the past (successful) measures to abate air pollution (in other terms, to control the chemical composition of our air), and to *know* what must be done now and in the future to control our climate. To control the measures and their effects as well as to control the atmospheric chemical composition remains the task of *chemical meteorology*, but not atmospheric chemistry. This also includes in future two of the previous legs of atmospheric chemistry:
- measurement of substances in atmospheric air, and
- modeling of distribution and transformation of substances in air.

Atmospheric chemistry *per se* includes (the first leg):
- studying kinetics and mechanisms of chemical reactions being relevant in air.

However, this latter issue is (pure) *chemistry*, to describe it as academic or applied, plays no role; and because the number of chemical compounds and reactions are (theoretically) endless, this kind of laboratory research will remain in the future. The big time of large and international *field* experiments/campaigns to understand tropospheric ozone formation, cloud chemistry, and particle formation, however, is over (about 1980–2010). What, however, remains and becomes more important are geophysical, geochemical, biochemical, and biological studies of our climate system, likely best described as the discipline of *biogeochemistry*.

Abbreviations used in this volume

AdW	*Akademie der Wissenschaften der DDR*
a. s. l.	above sea level
BAPMoN	Background Air Pollution Monitoring Network (WMO)
CAO	Central Aerological Observatory (at Dolgoprudny, Russia)
CCN	cloud condensation nuclei
CET	central European time
CLAW	hypothesis called after *Charlson*, *Lovelock*, *Andreae* and *Warren*
CNRS	*Centre national de la recherche scientifique* (French research organization)
DDR	*Deutsche Demokratische Republik* (former East Germany)
DIC	dissolved inorganic carbon
DMS	dimethyl sulfide
DU	Dobson unit
EMEP	Co-operative Programme for Monitoring and Evaluation of the Long-range Transmission of Air Pollutants in Europe
ETH	Eidgenössische technische Hochschule (Zürich)
EUROTRAC	European Experiment on Transport and Transformation of Environmentally Relevant Trace Constituents in the Troposphere over Europe
GAW	Global Atmosphere Watch (WMO)
GDR	German Democratic Republic (former East Germany)
GGO	transcription of Russian ГГО (see MGO)
FCT	*Forschungsstelle für chemische Toxikologie* (Leipzig)
IGY	International Geophysical Year (1957/1958)
IPCC	Intergovernmental Panel on Climate Change
IR	infrared
IUPAC	International Union of Pure and Applied Chemistry
LWC	liquid water content
MGO	Main Geophysical Observatory (Leningrad, resp., St. Petersburg)
MISU	Meteorological Institute of the Stockholm University
MIT	Massachusetts Institute of Technology
MPI	Max-Planck Institute
NCAR	National Center for Atmospheric research (Boulder)
NMHC	non-methane hydrocarbons (= NMVOC)
NMVOC	non-methane volatile organic compounds
PAN	peroxyacetyl nitrate
POP	persistent organic pollutants
ROS	reactive oxygen species
SCOPE	Scientific Committee of Problems of the Environment
SOA	secondary organic aerosol
STP	standard temperature and pressure
TNO	*Nederlandse Organisatie voor toegepast-natuurwetenschappelijk onderzoek* (Netherlands Organization for Applied Scientific Research)
TROPOS	*Institut für Troposphärenforschung* (Leipzig); former IfT
UBA	*Umweltbundesamt* (Federal Environmental Agency in Germany)
US	United States (of Amerika)
USSR	Union of Soviet Socialist Republics
UV	ultraviolet

https://doi.org/10.1515/9783110732467-006

VOC	volatile organic compounds
WMO	World Meteorological Organization

Name index

Abbot, Charles Greeley 581
Abbott, Francis 503, 508, 567
Achard, Franz Carl 383
Acker, Karin 706
Aderkas, Olga Jurevna 592
Addison, William 543
Adel, Arthur 702
Adelung, Johann Christoph 76
Adet, Pierre August 71
Agricola, Georg 87
Airy, George Biddell 2
Aitken, John 129, 322, 347
Akimoto, Hajime 705
Akimoto, Minoru 222, 236
Albinus, Bernhardt Siegfried 460
al-Khazini, Abd al-Rahman 146
Allaire, Paul Léon Félix 555
Alt, Eugen Johann 432
Altshuller, Aubrey P. 42
Ambarzumjan, Victor Amazaspovich 593
Ambrook, Charles 564f
Amelung, Walther 10
Anastasio, Cort 274
Anaxagoras 120
Anaximander 119
Anaximenes 119, 347
Ancelle, Sébastien Wisse 390
Anderson, Valentine George 219f
Andreae, Meinrath O. 16, 709
Andrée, Salomon August 416
Andrews, Thomas 347, 452, 467f, 509, 516
Aneja, Viney 274
Ångström, Anders Knutsson 238, 574
Ångström, Knut Johan 574
Anschütz, Richard 451
Apel, Eric 705
ApSimon, Helen 270, 679
Arago, Dominique François Jean 301, 460
Archenhold, Friedrich Simon 141
Ardon, Michael 647
Aristoteles (in English: Aristotle) XIII, 119f, 347
Armstrong, George Frederick 424
Arndt, Johann Albert 525
Arnold, Frank 591
Arny, Henry Vinecome 594
Arrhenius, Svante August 93, 430
Ascher, Ludwig 298
Ascherson, Paul 505, 531f
Asman, Willem 679
Aßmann, Richard 122, 132, 271
Atkinson, Roger 42f
Attmannspacher, Walter 620
Attems, Joseph Graf von 138
Aubin, E. 194, 234, 255, 392, 404, 409, 415, 419, 422ff, 433, 674
Audubert, René 19
Auel, Renate 249
Auer, R. 602
Aufray, Maurice 192
Avery, David 220
Avogadro, Amedeo 93
Ayers, Greg 16

Babo, Lambert Heinrich von 467, 581
Bach, Alexei Nikolajewich 654
Back, William IX
Bäckström, Hans 723f
Baeyer, Adolf von 122, 638
Bailey, George Herbert 122
Balard, Antoine Jérôme 70
Baly, Edward Charles Cyril 82
Bamber, Montague Kelway 189, 234
Bandelier, Adolph Francis Alphonse 503, 564
Banks, Joseph 331
Barbier, Daniel 598
Barker, Tomas Herbert 502, 545, 561
Barkow, Erich 325
Barnes, Ian 43
Barral, Jean-Augustin 171f, 179f, 201, 233
Bartels, Julius 585
Bartels, G. C. 718
Basilius Valentinus 112
Baskerville, Charles 717
Bastelaer, Désiré Alexandre Van 505, 508, 569
Bates, David Robert 589
Bates, John R. 486, 650
Baubigny, Henry 723
Baudrimont, Alexandre Edouard 341
Bauer, Alexander 374
Bauer, Karl Hugo 62
Bauer, Josef 594
Baulch, Donald L. 44
Baumann, Anton 3
Baumé, Antoine 353

https://doi.org/10.1515/9783110732467-007

Baumert, Friedrich Moritz 452, 462ff, 483, 490
Baumgartner, Andreas Freiherr von 388
Baumhauer, Eduard Henrich von 172
Baur, Emil 102f, 226, 494, 653
Baur, Ferdinand Christian 521
Baxendale, John 656
Baxendell, Joseph 476f, 504, 546ff, 571
Beard, George Miller 458
Bebber, Wilhelm Jacob van 447
Becher, Johann Joachim 90f, 354, 443
Bechi, Emilio 196, 234
Beck, Johann Gottlieb 532
Beck, John Brodhead 331f
Becker, Karl-Heinz 43, 628
Becker, Richard 131
Becquerel, Alexandre Edmond 453, 462
Beddoes, Thomas 158
Beddow, O. K. 197, 201, 206, 235
Beer, August 93, 376
Begemann, Carl 504, 528f
Beguin, Jean 83
Behrens, Theodor Heinrich 499
Beijerinck, Martinus Willem 59
Beilke, Siegfried 727
Bellecroix, James de 390
Bellucci, Giuseppe 186, 234
Bemmelen, Jacob Maarten van 648
Bence-Jones, Henry 182, 693
Benedict, Francis 387, 393, 425ff
Beneke, Friedrich Wilhelm 522f
Benoist, Louis 614
Benrath, Alfred Hermann IX, 97, 194, 653
Bentley, Richard 461
Berg, Helmut 231
Berger 129f
Bergeron, Tor 140
Bergey, David Hendricks 445f
Bergman, Torbern Olof 83, 91, 305, 316, 350, 355ff, 358, 395, 406
Bergson, Joseph 140
Bérigny, Adolphe 83, 91, 452, 499f, 503f, 507, 541, 551f
Bernheim, Geneviève 596
Bernoulli, Daniel 129
Bernstein, Basil X
Berresheim, Harald 42, 727
Bertalisse, A. 188
Bertels, C. 145, 169f, 172
Berthelot, Daniel 104f
Berthelot, Marcellin Pierre Eugène 104, 170, 468, 637, 651, 698f, 723
Bertherand, Alphonse François 502, 568
Bertherand, Èmile Louis 568
Berthollet, Claude-Louis 352, 384, 435, 437, 637, 671, 708f
Bertrand, Gabriel Emile 221, 235
Berzelius, Jöns Jakob 65, 72, 159f, 183
Bessels, Emil 565
Bestushew-Rjumin, Alexei Petrowitsch 97
Betz, Philipp Friedrich 502, 520f
Beyer, A. 185, 233
Beysens, Daniel 282
Bider, Max 29
Bieber, Willie 484
Bieling, Kurt 598
Billings, John Shaw 399, 407, 409, 414
Billiard 551, 563
Bineau, Armand 172, 178, 180, 193, 233
Bingemer, Heinz G. 16, 42
Biquard, Robert 375
Blacet, Francis Edward 485f, 490, 724
Black, Joseph 83, 91, 347, 349f, 355, 372, 394f, 682
Black, W. J. 550
Blackwell, John 162
Blagden, Charles Brian 352, 372
Blake, Robert Frederick 401, 409, 421, 424f
Blake, William Phipps 301
Blanchard, Duncan C. 333f
Blasius, Ewald 62
Bleibaum, Irma 598
Bliss, John 504, 567
Blochmann, Georg Rudolf Reinhart 398, 407, 409, 415
Bloxam, John Charlton 542
Blücher, Hans 133, 447
Blunt, Thomas Porter 103, 652f
Bobierre, Pierre Adolphe 182, 193, 202, 233
Bodenstein, Max Ernst August 94, 101f
Boë, Franz de le (see Sylvius) 144
Boeckel, Eugène 453, 551, 563
Boeckel, Thèodor 502, 537, 540f, 551f
Boehm, Joseph Georg 454, 502, 506, 539, 541, 553, 563
Boeke, Jacob 96, 647
Boerhave, Herman 54, 73, 88f, 142ff, 148ff, 168, 233, 237, 283, 285, 336f, 441
Bohlig, Eduard 691, 697

Bohlig, Franz Josef Dominicus 166, 208, 233
Bohr, Niels 93, 687
Bojkov, Rumen D. 557, 586, 621
Bolin, Bert 5f, 16, 46f, 244
Boltzmann, Ludwig 93f
Bona, Henri 455
Bonâme, Philipp 234, 236f
Bondt, Nicolas 443
Bonhoeffer, Karl-Friedrich 226, 487, 591
Boothby, Walter M. 392
Borch, Ole (see also Borrichius) 64, 147, 357
Boresch, Karl 59
Borius, Alfred 500, 505, 567f
Born, Max 93
Borovikov, Aleksandr Moiseevich 259
Borrichius, Olaus 64, 147
Bos, R. 677
Böttger, Rudolf Christian 452, 472f, 498f, 647, 691, 694, 697
Bottini, Ottaviano 146, 221, 235
Bouguer, Pierre 130
Boussingault, Jean-Baptiste 3, 53, 57f, 121, 136, 145, 168, 172, 174, 178f, 189f, 186, 193, 203, 233, 255, 280, 307, 378, 388f, 407f, 415, 433, 438
Bouwman, A. F. 679
Bowen, Irby Gerald 610
Boylan, R. F. 327
Boyle, Robert 82ff, 197, 142, 254, 277ff, 286f, 320f, 347, 355, 671, 680
Brand, Hennig 460
Brandes, Heinrich Wilhelm 125, 145f, 163f, 202, 208, 233
Brandes, Simon Rudolph 163
Brenner, August Rudolf 502, 519f
Bretschneider, Paul 185, 193, 233
Brewer, Alan West 588, 611
Bridge, James Howard 458
Brimblecombe, Peter 245, 289f
Briner, Emile 599
Brodie, Benjamin Collins 482, 647, 699
Brodie, Frederick J. 135
Brodin, Gunnar 239
Brøgger, Waldemar Christofer 214, 305
Broglie, Maurice de 599
Brorsen, Theodor 538
Brown, Alexander Crum 471, 474, 512
Brown, Horace Tabberer 409, 421f, 424f, 673f
Brown, Samuel 688
Browne, Sir Thomas 144
Bruckmann, Peter 625
Brüggemann, Erika 249
Brugghe, Simon Steven van 82
Brugmans, Sebald Justinus 254f
Brühl, Julius Wilhelm 632
Bruhns, Karl Christian 504, 530ff
Brunner, Carl Emanuel 388, 397f, 408f
Brünnich, Johannes Christian 195, 200, 234
Brusilovsky, Efrem Moisseevich 59
Buch, Kurt 426ff
Büchner, Ludwig 55
Buchner, Johann Andreas 62
Bucquet, Jean-Baptiste-Michel 395
Budde, Emil Arnold 99
Bufalini, Joseph James 42, 491, 659, 661ff
Buijsman, Ed 679
Buisson, Henri 480, 580f, 598
Bujwid, Odo 345
Bunsen, Robert Wilhelm 93, 99, 171f, 176, 184, 292, 376f, 387, 392, 407, 443, 462, 723
Burgaud, Maurice 581
Burgstaller, Alexander 597
Burckhardt, Hans 10, 39, 320, 329
Burkser, Evgenij Samojlovitsch 4, 46, 205, 224, 236
Butenandt, Adolf 19
Butterfield, Herbert XI
Büttner, Konrad Johannes Karl 10
Bykov, A. 190, 195, 199, 235

Cadle, Richard D. 18, 42, 447f, 491, 592, 625, 705, 708
Callendar, Guy Stewart 422, 425ff, 432
Calvert, Jack G. 18, 42, 492, 650, 655, 661, 705, 725
Cann-Lippincott *see* Lippincott
Cantoni, Giovanni 511
Capelo, João Carlos de Brito 506
Carius, Georg Ludwig 499, 697
Carmody, Patrick 198, 234, 237
Carnelley, Thomas 409f
Carus, Titus Lucretius 285
Casali, Adolfo 192, 234
Castelli, Benedetto 113
Castillo, Raymond A. 265f, 274
Cauer, Grete 20ff, 27

Cauer, Hans 4, 6, 10f, 17ff, 146, 206, 228ff, 256ff, 330, 338ff, 480, 517, 600f, 603, 607ff, 613f, 624, 644, 662, 690, 701f
Cavallo, Tiberius 384f, 458
Cavendish, Henry 91, 347, 349f, 352f, 355f, 370ff, 374, 382, 386f, 681f, 708
Chachalin, V. S. 259
Chadwick, Edwin 379
Chairy 146, 190f, 201, 204, 237
Chalmers, Lionel 187
Chalonge, Daniel 595, 598
Chaluppa, Kurt 231
Chamberlin, Thomas Chrowder 429f
Chameides, William 517, 625, 662
Chapman, Ernest Theophron 445f
Chapman, David Leonard 468, 492
Chapman, Sydney 7, 480, 577, 595, 589f
Chappuis, Joseph 468, 480, 572ff, 699
Chaptal, Jean-Antoine 356, 385
Chardin, Pierre Teilhard de 51
Charles, Jacques 93
Charlson, Robert 16, 47, 49
Chatin, Gaspard Adolphe 145, 170f, 233
Chernikov, Albert Alekseevich 141
Chevallier, Jean-Baptiste-Alphonse 12, 672
Chiplonkar, M. W. 588, 598
Chladni, Ernst Florens Friedrich 161, 308f, 311
Chlopin, Witali Grigorjewitsch 575, 641
Chodat, Robert Hippolyte 654
Chomet, Joseph Antoine Hector 503, 568
Choularton, Tom 274
Christiansen, Jens Anton 687
Christison, Robert 721
Church, Sir Arthur Herbert 204, 233
Churchill, Winston 729
Ciamician, Giacomo Luigi 100f
Cicerone, Ralph 16, 489, 589f
Civilale, Aimé 182
Claesson, Johan Peter 408, 415, 423, 425
Clapham, Arthur Roy 51
Clarke, Frank Wigglesworth 44f, 189, 205, 392, 711
Clark, James (Sir) 500
Claubry, Henri-François Gaultier de 313
Clayden, Arthur William 135
Cloud, Preston Ercelle 60
Clausius, Rudolf Julius Emanuel 94, 129f, 452, 463, 466f, 489f
Clemens, Theodor 502, 518f, 541
Cloez, François Stanislas 452, 475, 503, 563, 695f
Clyde, James 204
Cnöffel, Andreas 277
Coblentz, William Weber 585f
Cochrane, Thomas 683
Coehn, Alfred 725
Cohen, Julius Berend 17, 122, 195, 235, 254, 290, 292, 294f
Collett, Jeff 266, 269, 274
Comel, Alvise 221, 235
Comes, Franz-Josef 43
Condon, Edward Uhler 94
Configliachi, Pietro 406
Conrad, Viktor 136, 141
Conraux, François Joseph 502, 548, 551
Conring, Hermann 147
Cook, Henry 504, 568
Cordeiro, Frederick Joaquium Barbosa 400
Cornu, Alfred 5, 572
Cossa, Alfonso 177, 185, 233, 492
Coste, John Henry 328, 724
Cotta, Heinrich von 721
Cotte, Louis 317
Coulier, Paul-Jean 129, 329
Covelli, Nicola 708
Cox, Richard Anthony 16, 43f, 725
Coxe, John Redman 66, 70
Coxwell, Henry 544
Craig, Richard Ansel 578
Cramer, Carl Eduard 444f, 699f
Criegee, Rudolf 482
Crowther, Charles 194, 214f
Cruickshank, William 432, 458, 480
Crusius, Heinrich Wilhelm Leberecht 175
Crutzen, Paul 5, 16, 41, 46, 48f, 470, 480, 579, 590f, 610, 625, 705, 710
Cullen, William 83
Cunningham, Robert M. 257
Curie, Marie Skłodowska 654
Curry, Manfred 21f, 606f, 609, 614
Czerny, Marianus 93

Dachauer, Gustav 452f
Dafert, Franz Wilhelm 177
Dallowe, Timothy 89
Dalton, John 72, 92, 94, 125, 145, 162, 172, 203, 233, 346, 376, 386, 388, 395ff, 406f, 433, 440, 683

Damköhler, Gerhard 375
Dancer, John Benjamin 458, 475ff
D'Ans, Jean IX
d'Arcet, Jean Pierre Joseph 145, 166, 233
D'Arcy, Ralph Francis 652
Daremberg, Georges 505, 559
Darwin, Charles 50, 301f, 321, 515
Däßler, Hans-Günther 722
Daubeny, Charles 110, 161, 479, 504, 548
Daubrée, Gabriel Auguste 305
Daum, Peter H. 266f, 274, 661, 664
Daumas, Maurice 368
Dauvillier, Alexandre 599
Davies, Herbert 515
Davy, Edmund 340
Davy, John 458f
Davy, Sir Humphry 70, 156, 331, 336, 340, 382, 384, 440, 459, 683, 692, 706, 708f
Day, Henry 452, 501, 572
Day, John 504, 567
de la Rive, Auguste Arthur 453, 460, 462, 464, 480, 541
Delwiche, Constantin Collin 59
Denmead, Owen Thomas 679
Decharmes, Joseph Constantin 502, 506, 552, 563
Deimann, Johann Rudolph, 443
Dellmann, Johann Friedrich 131, 319
Deluc, Jean-André (*also* De Lüc *or* De Luc) 121, 125f, 128, 271, 371f, 457
Demerjian, Kenneth L. 491, 625, 662, 705
Denison, Sir William 187
Denza, Francesco 504, 506, 542
Derwent, Richard G. 43, 625, 668, 678
Desplats, Victor 455
Des Voeux, Harold Antoine 296f
Descartes, René 73, 121ff
Dewar, Michael James Stuart 467, 487
Dhar, Nil Ratan 103f, 146, 195, 226f, 235, 280, 691, 693
Dickinson, Edmund 147
Dieudonné, Adolf 652
Dieulafait, Louis 672
Dietrich, Theodor 3, 504, 529, 548
Dietze, Gerhard 34
Digby, Kenelm 112, 144, 680
Dines, William Henry 136
Diness, George 132, 136
Dirnagl, Karl 609
Ditte, Alfred 301, 305, 307, 670
Dittmar, Wilhelm (William) 62
Divers, Edward 686
Di Vestea, Alfonso 221, 234
Döbereiner, Johann Wolfgang 384, 388, 435
Dobson, Gordon 13, 290, 480, 548, 578, 580ff, 588, 596f, 600, 611
Dohrandt, Ferdinand 504, 512ff, 554, 564, 570
Dorfwirth, Maria 586f
Döring, Werner 131
Dorno, Carl 8, 327
Dove, Heinrich-Wilhelm 523f
Downes, Arthur Henry 103, 652f
Downs, Thomas 544
Doyle, G. J. 42, 132
Dragendorff, Georg 416
Draper, John William 98
Drasche, Anton 522
Drescher, Wilhelm 601
Drew, John 469
Driessen, Jan Constantin 157
Driessen, Peter van 157
Drischel, Hans 124, 195f, 200, 207, 221, 228f, 235, 602
Drozdova, Valentina Ivanovna 146, 225, 260f, 275
Du Clos, Samuel Cottereau 75
Duce, Robert A. 16, 40, 333, 340
Ducros, Hippolyte 145, 168f, 228, 233
Duerst, Johann Ulrich 427
Düggeli, Max 59
Dufour, Louis 432
Dulong, Pierre Louis 683f
Dumas, Jean-Baptiste André 357, 369, 373, 378, 388f, 408, 420, 423, 437, 440, 463, 638
Dunstan, Wyndham 220, 651
Dütsch, Hans Ulrich 579, 588
Duyzer, Jan 679
Dyrenforth, Robert Saint George 140

Eaton, Henry Storks 542
Eaton, S. V. and J. H. 218, 235
Ebel, Adolf 586
Ebermayer, Ernst 391f, 409, 475, 504, 507, 530, 536, 566
Ebert, Hermann 326
Ecker, Johann Alexander 498
Eckardt, Wilhelm Richard 9
Edelstein, Sidney 374

Eder, Josef Maria 100
Edrisi 304
Effenberger, Ernst Friedrich 607f, 613, 615
Egnér, Hans Gabriel 230, 238f
Ehhalt, Dieter 16, 42f, 439f, 480, 592, 605
Ehmert, Albert 30, 585, 603, 608, 611f, 616, 618
Ehrenberg, Christian Gottlieb 301ff, 309
Ehrenberg, Paul 58
Ehrlich, Viktor 699
Eichborn, Johann-Ludwig von 28
Eichhorn, Carl Herrmann Alexander 185
Einstein, Albert 93, 100
Elbe, Guenther von 486f, 650
Ellery, Robert Lewis John 504, 567
Elster, Julius 320
Elsholtz (Elsholz), Johann Sigismund 707
Emanuelsson, Arne 238
Emich, Friedrich 493
Emmett, Anthony 408, 410
Empedocles 119
Engler, Carl Oswald Viktor 53, 96, 451, 453, 470, 481, 484, 555, 638, 647
Erdman, Lewis Wilson 218, 235
Erdmann, Otto Linné 56, 758
Ericksen, George Edward 343
Eriksson, Erik 146, 205, 236, 238f, 244, 339f
Erismann, Friedrich Huldreich 444
Erisman, Jan-Willem 679
Erk, Fritz 3
Erndtel, Christian Heinrich 111
Ertel, Hans 29
Escombe, Fergusson 396, 421f, 424f
Esser, Hermann 22
Etheridge, David M. 424, 439
Eugling, Carl Friedrich Wilhelm 193, 196, 234
Eugster, Werner 135, 274
Eucken, Arnold Thomas 489
Eulenberg, Hermann 12
Evans, Conway 503, 545
Evelyn, John 113, 115, 146, 286, 290, 296, 680, 689, 707, 720
Everdingen, Ewoud van 596
Eversmann, Alexander Eduard Friedrich 167
Eyring, Henry 94

Faber, Wilhelm Eberhard 502, 519
Fabian, Peter 439f, 450
Fabian, Rolf 583
Fabri, Honoré 308
Fabricius, Thomas Balthasar 61
Fabry, Charles 480, 578f, 580f, 595, 598
Fachini, Maria Christina 246
Fahrenheit, Daniel Gabriel XIII
Failyer, George Henry 146, 190, 194, 234, 237
Falkner, Johann Ludwig 388
Fallati, Karl Nicolaus 505, 520f
Faraday, Michael 93, 469, 541
Farkas, László 131
Farman, Joseph Charles [Joe] 13, 480, 589
Farsky, Franz (František) 415, 423, 498
Fay, Charles François de Cisternay du 126
Feilitzen, Carl Axel Hjalmar von 195, 198, 235, 237
Feilitzen, Carl Henrik Jobst von 198
Feister, Uwe 35, 271, 586, 588, 620, 627
Feldt, Victor Richard von 409, 416f, 423, 425
Fellenberg, Ludwig Rudolf von 460
Ferm, Martin 239f, 385, 672, 677
Fermi, Enrico 93
Ferraboschi, Frederick 723
Fersman, Alexander Jewgenjewich 44
Festing, Augustus Morton 500
Ficinus, Heinrich David August 311
Figuier, Louis 554
Filhol, Jean Pierre Bernard Édouard 172, 182, 193, 201, 208, 233
Findeisen, Walter 140f, 329, 333
Finke, Leonhard Ludwig 317
Finlayson-Pitts, Barbara 5, 340, 624, 630
Finnell, Henry Howard 195, 220, 235
Firmicus, Julius Maternicus 64
Fišák, Jaroslav 274
Fischer, Emil 450
Fischer, Ferdinand 12, 97, 433
Fischer, Johann Carl 79
Fischer, Nathaniel Wolfgang 152, 460
Fittbogen, J. 405, 408
FitzRoy, Robert 398, 514
Flach, Emil Friedrich 10f
Flammarion, Camille 133f, 136, 292, 300, 308, 315, 544
Flechner, Anton 502, 537f, 563
Fleming, James Rodger XII
Flemming, Ferdinand 503, 523
Floderus, Manfred Mustafa 453
Flohn, Hermann 10, 320, 606
Flood, Håkon 132
Flügge, Carl 445, 507

Flury, Ferdinand 12, 709
Fodor, Jozef von 405, 408, 445, 674
Fonrobert, Ewald 451, 493
Fonselius, Stig 427f
Fontana, Felice 380, 382f, 384, 395, 400
Fontenelle, Jean-Sébastien-Eugène Julia de 278f, 388
Fontijn, Arthur 703
Formey, Ludwig 287
Forster, Hans 320
Fort, Charles 309
Fourcroy, Antoine François Comte de 70f, 87, 92, 107, 352, 373, 406, 433, 708
Fournet, Joseph 551
Fowle, Frederick Eugene 581
Fowler, Alfred 574, 580
Fowler, David 270, 679
Fox, Cornelius Benjamin 5, 12, 96, 447, 450, 456, 459, 470, 500f, 504, 511f, 515, 545, 548ff, 554f, 571
Franck, James 94, 723f
Franke, Wilhelm 638
Frankland, Sir Edward 145, 187, 193, 234, 391, 423, 437, 511
Fraps, George Stronach 218, 235f
Fraser, Paul 439f
Free, Edward Elway 289, 414, 319, 323
Frémy, Edmond 452f, 462, 464, 511f, 638, 555, 638
Freney, John Raymond 47, 60
Frenkel, Jakov Iljich 132, 329
Frenreisz, Franz von 503, 538
Fresenius, Carl Remigius 61f, 117, 178, 404, 483, 672ff
Frey, Eugen von 409, 416f, 423, 425
Freytag, Karl 214
Fricke, Hugo 655
Fricke, Wolfgang 626
Fridlander, Ernest David 326
Friend, James P. 59
Friese, Walther 294
Friesenhof, Gregor 505, 507, 535, 641
Fritsch, Carl 539
Fuchs, Gottlob 401
Fuchs, Nikolaj Albertovich 725
Fugger, Eberhard 136
Fujishima, Akira 493
Fuller, John 203
Fuzzi, Sandro 246, 262, 267, 274
Gaffron, Hans 104
Gairdner, Meredith 156, 202, 279
Gaillard, Samuel Edwin 505, 551, 563f
Gaivoronsky, Ivan Ivanovich 224
Galbally, Ian Edward 619
Galileo Galilei 113, 347
Garnerin, André-Jacques 406
Garvin, David 650
Gasparin, Adrien Étienne Pierre de 178
Gaudechon, Henry 490, 575, 693
Gautier, Armand 201, 349, 355, 438, 443ff, 448, 453, 559
Gay, Bruce 662
Gay-Lussac, Joseph Louis 73, 81, 93f, 98, 313, 335, 341ff, 347, 373, 386f, 435, 511, 683f, 690, 692, 706
Geber (*see also* Ibn Hayyān) 85f, 335, 707
Gehlen, Adolph Ferdinand 758
Gehler, Johann Samuel Traugott 66, 77, 125f, 290, 406, 456
Geißler, Heinrich 575
Geitel, Hans 320, 326
Gellert, Christlieb Ehregott 84, 287ff, 336f
Gellicus, Aulus 111
Geoffroy, Etienne 83
George, Christian 706
Georgii, Hans-Walter 16, 19, 28, 33, 36, 40ff, 190, 241, 267, 603, 626, 677ff, 703, 719, 726f
Georgii, Walter 40, 603
Gerhardt, Charles Frédéric 436, 463
Gerlach, Heinrich Constantin 713ff
Gersten, Christian Ludwig 126
Giacosa, Piero 186, 234
Gibor, Aharon 59
Gibbs, Josiah Willard 94
Gibbs, William Edward 131, 327
Giberne, Agnes 109, 133, 292
Gilbert, Joseph Henry 58, 174, 186
Gilbert, Ludwig Wilhelm 758
Gilge, Stefan 620, 661f, 667
Gilgner, Juri A. 259
Gilm, Hugo von 398, 407f, 415, 423
Giobert, Giovanni Antonio 383f
Girardin, Jean Pierre Louis 145, 167, 233, 237
Girtanner, Christoph 65, 71, 356, 373, 406
Glaisher, James 2, 322, 327, 476, 502ff, 543ff, 561
Glauber, Johann Rudolf 87, 671, 680, 682

Glueckauf, Eugen (Glückauf before 1947) 375f, 379, 393, 439, 584, 605f, 615f, 621f
Gmelin, Johann Friedrich 2, 61, 385, 395
Gmelin, Karl 10
Gmelin, Leopold 2, 56, 61, 92, 435ff, 672f, 683f, 685
Gnauck, Thomas 37
Göchhausen, Hermann Friedrich von 720f
Goethel, Michael 677
Goldschmidt, Johannes 33f
Goldschmidt, Johannes Wilhelm 341
Goldschmidt, Victor Moritz 44f
Goldstein, Gotthilf-Eugen 575f
Goppelsroeder, Christoph Friedrich 3, 191
Gorham, Eville 29, 47, 52, 231f, 245
Gorup-Besanez, Eugen Freiherr von 452, 476, 493, 635
Göttling, Johann Friedrich August 61
Götz, Paul 29, 480, 517, 579ff, 584f, 587, 598, 600, 611f, 625
Grabowski, Rostislav Ivanowitsch 259, 333
Graedel, Thomas E. 16, 439, 449, 658
Gräfenberg, Leopold 637, 651
Graftiau, Jean 192, 194, 234, 237, 396, 415, 420, 423
Graecus, Marcus 679
Gräger, Nicolaus 178, 502, 521, 538, 672ff
Graham, Richard Allan 490
Graham, Thomas 124, 284, 683, 685
Grandeau, Louis Nicolas 191
Grasnick, Karl-Heinz 35, 587f
Gratacap, Louis Pope 505, 566
Gravenhorst, Gode 41, 678
Gray, George 186, 194f, 214, 234, 237
Grellois, Eugène 503, 508, 552f, 563
Gren, Friedrich Albrecht Carl 63f, 68, 385
Gresset 386
Griesinger, Wilhelm 505, 520f
Griffith, Francis Llewellyn 63
Grindel, David Hieronymus 406
Gröger, Max 465
Groth, Wilhelm 42f, 488
Grotthuß, Theodor von 98, 311
Gruber, Max von 385, 433f
Grüneisen, Eduard 755
Grunow, Johannes 261
Guderian, Robert 29
Guericke, Otto von 75, 121, 123f, 271
Guéron, Jules 594f
Guldberg, Cato Maximilian 94
Gushchin, Gennadij Petrovich 580, 592f, 596
Guicherit, Robert 625
Güsten, Hans 43

Haagen-Smit, Arie Jan 13, 17, 448, 480, 489, 491, 606, 650
Haber, Fritz 492, 487, 489, 655ff, 723
Häberlin, Carl 10
Hadfield, William 396, 407
Haehnel, Friedrich August Otto 718
Hagen, Ernst Bessel 392
Hagen, Wilhelm 26
Hagenbach-Bischoff, Eduard 639
Haggard, Howard Wilcox 12
Hahn, Amandus 58
Hahn, Jürgen 16
Haldane, John Burdon Sanderson 53
Haldane, John Scott 392
Halem, Friedrich Wilhelm von 522
Hales, Stephen 67, 347ff, 366, 394f, 680
Hall, Alfred Daniel 59, 201, 676
Hall, Alfred Robert 368
Hall, David 60
Hall, Edward Swarbreck 503, 567
Haller, Carl 502, 505f, 522, 540
Halliday, Eric C. 711
Hamlet, William Mogford 454
Hammerschmied, Johann 477f
Hampson, John 480, 590
Hampson, Robert F., Jr. 650
Hann, Julius von XII, 6, 9, 136, 233
Hänschen, Hans-Adolf 21
Hanst, Philip Lincoln 42, 662, 705
Hantzsch, Arthur Rudolf 437, 692
Hara, Hiroshi 223f
Harper, Horace J. 219, 235
Harrassowitz, Hermann 230
Harries, Carl Dietrich 450f, 482, 574
Harris, Geoffrey 705
Harris, William 549
Harrison, Douglas Neill 580
Harrison, John Burchmore 189, 194, 234, 237
Harteck, Paul 42f, 486, 489, 576, 581
Harting, Pieter 166
Hartley, Walter Noel 480, 573f
Hässelbarth, Paul 405, 408
Hassenfratz, Jean Henri 71
Hatcher, Robert Anthony 564, 593f

Haurwitz, Bernhard 594
Hautefeuille, Paul 468, 517, 572, 699
Haviland, Alfred 502, 545
Hawksbee, Francis 355
Hayes, August A. 2, 672
Hayhurst, Walter 596f, 701
Hayyan, Jabir Ibn 70, 85
Heelis, Edward 505, 568
Hegg, Dean A. 266, 274
Heicklen, Julian 5, 42, 434, 450, 491, 650
Heikes, Brian G. 657, 661f, 705
Heiden, Eduard 183
Heidenschreider, Anton 504, 527
Heimann, Jacob 409, 416f, 423, 425
Heine, Heinrich 401, 403, 408, 416f, 423, 425, 430
Heinrich, Heine (poet) 271
Heinrich, Reinhold 185, 194, 234
Heintzenberg, Jost 272, 587
Heisenberg, Werner Karl 93f
Heitler, Walter 93
Helbig, Demetrio 699
Heldt, Wilhelm 464, 481, 503, 510f, 636
Heller, Arnold 11, 115, 216, 298
Heller, Johann Florian 695f
Heller, Victor G. 221
Heller, Wilfried 614
Hellmann, Johann Georg Gustav XII, 31, 128, 130, 135, 271, 304, 314
Hellriegel, Herman 58, 191
Helmholtz, Hermann von 93, 130, 479, 481, 524, 725
Helmholtz, Robert von 130f
Helmont, Johann Baptist van 67, 73, 75f, 77, 88, 91, 107, 112, 347f, 355, 394, 680
Helmts 162, 168, 233
Hempel, Walter 294, 377, 391f, 464, 409, 423, 715
Henderson, Yandell 12
Henneberg, Johann Wilhelm Julius 408f, 423
Henriet, Henri 4, 445, 575, 596
Henry, Joseph 515
Henry, Thomas 368
Henry, William 94, 266, 440
Henshaw, Thomas 145, 275f, 680f, 689
Hentschel, Gerhard 10f, 30, 33
Heraclites 119
Hermbstädt, Sigismund Friedrich 337
Herodotus of Halicarnassus 118
Heron 347
Herrmann, Günter 36
Herrmann, Hartmut 43, 274
Hertz, Heinrich 93
Hess, Hermann Heinrich 94
Hesse, Fritz 28
Hesse, Ludwig Otto 408f
Hesse, Walther 400
Hessen-Rotenburg, Hermann von 135
Heydloff, Reinhold 502, 522, 571
Higgin, Robert 158
Hill, Ashton 724
Hill, Henry Barker 408f
Hinshelwood, Cyril Norman 93, 649
Hinzpeter, Hans 35, 586f
Hippke, Erich 22
Hippocrates 111, 118
Hirasawa, Kenzō 223, 236
Hirayama 222
Hirt, Ludwig 12
His, Wilhelm 452
Hjärne, Urban 147
Hlasiwetz, Heinrich Hermann 398, 408, 410
Hobbs, Peter Victor 266, 274, 726
Hobbs (Sergeant) 564
Hoelper, Otto 585
Hoff, Karl Ernst Adolf von 318
Hoffmann, Friedrich 433
Hoffmann, Michael 257, 263, 265, 274, 494, 651, 663, 726
Hofmann, August Wilhelm von 436
Hofmeister, Jürgen 37
Högberg, Linus Vilhelm 238
Högbom, Arvid Gustaf 430
Hohenheim *see* Paracelsus
Hollandus, Isaac 671
Holmén, Kim 60
Holmes, Frederic Lawrence 368
Holzapfel, Luise 28
Homberg, Wilhelm 90
Home, Francis 113
Homer 119, 307, 456
Hooke, Robert 357
Hooper, Robert 335f
Hope, Thomas Charles 382, 384
Höring, Franz Jacob 520
Horn, Franz Xaver Herman 481, 498
Hornberger, Richard 3
Horsfall, Thomas Coglan 291

Horváth, László 192, 197, 240, 243, 664, 665, 679
Houghton, Henry Garrett 140f, 195, 200, 235
Houzeau, Jean Auguste 96, 133, 181, 189, 196, 347, 456f, 464, 476, 480, 503f, 516f, 539, 553ff, 559, 570f, 594f, 638f, 642, 696, 699, 722
Houzeau, Jean-Charles 133
Hov, Øystein 621
Howard, George Shelby 275
Howard, Luke 121
Hoyer, Carl 316f
Hube, Jan Michał 126
Huber, Bruno 427
Hückel, Erich 94
Hudig, Joost 195, 199, 235
Huebert, Barry J. 705
Huggins, Margaret Lindsay 574
Huggins, William 574
Huizinga, Dirk 500, 504, 508, 511f, 570
Hüller, Oswald 504, 532
Humboldt, Alexander von 8, 73, 77, 89f, 117, 126, 301, 304, 321, 347, 352, 373, 386f, 390, 395f, 406, 525, 692
Humphreys, William Jackson 580
Hunt, John P. 590, 654, 656
Hunt, Robert 453
Hunt, Thomas Sterry 462, 480
Hunter, John 1
Hurdelbrink, Franz 298, 716
Hutchinson, George Evelyn 50
Hutchinson, W. L. 197, 234, 236
Hutzinger, Otto 47

Ibn Hayyān, Jābir (Dschabir) 85
Ideler, Julius Ludwig 167
Igawa, Manabu 267, 270f, 274
Ihle, Peter 36
Iljenkov, Pavel Antonovich 515
Ingenhousz, Jan 351, 380, 385
Ireland, William Wothesporn 503f, 508, 568
Isaksen, Ivar 623
Isidorov, Valerij Alekseevitch 449f
Israel, Hans 39, 326, 328
Israel, Juri Antonovich 225
Izarn, Joseph 340
Ivanov, Mikhail Vladimirovich 46

Jacob, Daniel 257, 262f, 265, 274, 661, 663
Jacobs, Morris B. 709
Jacobs, Woodrow Cooper 333
Jacobsen, Oscar Georg Friedrich 499
Jacolot, Alphonse Aristide Marie 503, 508, 568
Jaenicke, Ruprecht 16, 38f, 41f, 334
Jaeschke, Wolfgang 42, 267, 269, 274, 710, 727
Jame, Édouard 500, 552, 558, 568
Jander, Gerhard 62
Janssen, Pierre Jules César 560
Jelinek, Karl 535, 539
Jenkin, Michael E. 44
Jennings, Stephan G. 335
Jentsch, Siegfried 715
Jenrich, Georg 36
Jessel, Uwe Karl Albert Werner 608f
Jevons, William Stanley 476
John, Johann Friedrich 61
Johnson, Jacob 176
Johnson, Samuel William 697
Johnston, Harold Sledge 384, 448, 490f, 582, 589, 624
Johnston, James Finlay Weir 56, 72
Jolly, Johann Philipp Gustav von 391f
Jorissen, Paulinus 204, 234
Jost, Dieter 727
Jost, Friedrich Wilhelm 649
Junge, Christian 5, 7, 10, 13, 16, 18, 28, 31, 33, 36ff, 231f, 239f, 256, 280, 329f, 334, 339f, 486, 607, 614ff, 650, 676f, 703, 709, 719, 726f
Junkermann, Wolfgang 661f, 667
Juritz, Charles Frederick 195, 198, 235, 237
Jursch, Heinz 250

Kähler, Karl Wilhelm 328
Kalaß, Dieter 37
Kalle, Kurt Albert Edmund Theodor 29
Kampmann, Friedrich Eduard 504, 559
Kämtz, Ludwig Friedrich 128f, 406
Kaplan, Joseph 661, 663f, 702
Karandikar, V. R. 588
Kargin, Valentin Alekseevich 638
Karliński, Michał Franciszek Ignacy 502, 507f, 526, 536, 538f, 541
Karpov, K. 190, 195, 199, 235
Karsten, Herrmann 413
Kastner, Karl Wilhelm Gottlob 161, 757
Katō, Takeo 222
Katz, Walter 595, 632, 719, 725

Kawamura, Kiyoshi 588, 613
Kay, Robert Henry 611
Kazarnovski, Isaak Abramovich 482
Kazay, Endre 192, 194, 237, 240
Kedzie, Robert Clark 505, 564ff
Keeling, Charles David 391, 394, 422, 427, 429
Keeling, Ralph F. 427, 429
Keene, William C. 59
Keir, James 395
Kekulé, August 176, 436
Kellner, Oskar 177, 175, 191, 194, 222, 234, 237
Kellog, William Welch 59
Kelly, Thomas 661, 664, 705
Kelvin *see* Thomson, William
Kemp, Kenneth Treasure 178, 672ff
Kern, Serge 646, 663
Kernbaum, Mirosław 641, 654
Kerr, J. Alistair 491, 625
Kerr, Robert 352
Kestner, Otto 702
Khalil, Mohammad Aslam Khan 439
Khrgian, Aleksandr Khristoforovich XII, 258f, 592
Kiepenheuer, Karl-Otto 582, 584f
Kiessling, Johann 325
Kinch, Eduard 204, 234
Kingston, George Templeman 563
Kingzett, Charles Thomas XI, 471f, 533
Kirchhoff, Gustav Robert 94, 376
Kirwan, Richard 67, 70, 355, 681
Kistiakowsky, George Bogdan 577
Kistiakowsky, Vladimir Alexandrovitch 99, 577
Klaproth, Martin Heinrich 92
Kleffmann, Jörg 706
Kleist, Ewald G. 575
Klemm, Otto 267, 271, 274
Kley, Dieter 42, 605, 621f, 626f, 667, 705
Klockow, Dieter 16, 661ff
Klossowski, Alexander Vikentievich 199
Klumb, Hans 30, 33
Knapp, Friedrich 715
Knieriem, Johann Karl Woldemar von 176
Knoch, Karl Heinrich 26, 29, 606, 610
Knoche, Walter 343
Knop, Johann August Ludwig Wilhelm 53, 56, 175, 183, 193, 233, 280, 346, 539
Knott, Cargill Gilston 322
Koch, Robert 345, 400, 445
Koene, Corneille Jean 341
Köhler, Erich 702
Köhler, Hilding 136, 141, 256, 260, 328, 334
Kok, Gregory 632, 660ff, 729
Kolbe, Hermann 436, 443, 452, 473, 696
Kolbig, Joachim 617
Kondratiev, Viktor Nikolajevich 649
König, Joseph 50, 136, 187
König, Willi 31
Konstantinova-Schlesinger, Maria 599, 614
Kopp, Hermann Franz Moritz 61ff, 76, 91, 157, 353, 358, 394, 442f, 670f, 679, 706, 709
Kopp, Johann Heinrich 61
Köppen, Wladimir Peter 9, 128
Kordes, Ernst Karl Ferdinand 45
Kornfeld, Gertrud 725
Koroleff, Folke 428
Koschmieder, Harald 585
Kosmann, Constant Philippe 475, 479
Kossovich, Peter Samjonowich 60, 205f, 235
Kotze, Robert Nelson 39
Kovalskij, Viktor Vladislavovich 46
Kovda, Viktor Abramovich 47, 60
Kowalczyk, Jan 539
Kramers, Hendrik Anthony "Hans" 687
Kratzenstein, Christian Gottlieb 124, 128
Krauß, Friedrich 505, 521
Krebs, Wilhelm 724
Kreil, Karl 535f
Kremer, Albert 498f
Kreusler, Hans 576
Kreusler, Ulrich 392f
Kreutz, Wilhelm 427
Krogh, August 387, 391, 393, 425ff
Krügel, Curt Gustav Adolf 465
Krüger, Friedrich Wilhelm 337
Krüger, Johann Friedrich 312
Krummel, Paul B. 439
Krünitz, Johann Georg 107
Kühn, Gustav 175
Kunckel, Johann 90
Kunze, Max Friedrich 505, 532
Kupffer, Adolph Theodor 224
Küppers, E. 215, 572, 725
Korsunovsky, G. A. 426
Kurzmann, Johann Philipp 304
Küttlinger, Carl Julius Adelberg 502, 524

Ladenburg, Albert 379, 465, 467, 574
Ladenburg, Erich Robert 574

Ladenburg, Walter Rudolf 574, 584, 598
Ladureau, Albert 713
Lahmann, Erwin 217
Laj, Paolo 270, 274
Lakatos, Imre X
Lamanon, Robert de 316
Lamarck, Jean-Baptiste 121
Lamb, Horace 726
Lambert, Charles 551
Lambert, Johann Heinrich 94
Lamont, Johann von 537
Lampadius, Wilhelm August 126, 107, 126, 145, 160, 164f, 168, 17, 202, 233, 253, 278
Lamy, Claude Auguste 504, 558
Landa, Stanislaw 716
Landau, Lew Dawidowitsch 38f, 94
Landriano, Marsilio 380, 384
Landriset, Auguste 409, 421
Landsberg, Helmut Erich 38f, 40, 329
Langenfelds, Ray L. 439
Langevin, Paul 326, 328
Langlo, Kaare 586
Laplace, Pierre-Simon 131
Laskowski, Nikolai Erastowitsch 388
Lassone, Joseph-Marie-François de 433
Latour, Bruno 49
Laurent, Auguste 357, 463
Lauscher, Friedrich 256, 260, 557, 594, 622
Lauter, Ernst August 587
Lauwerenburgh, Anthoni 443
Lavoisier, Antoine-Laurent de 61f, 65, 70, 77, 83, 90, 92, 108, 14, 156, 233, 237, 347, 351ff, 354, 356f, 362, 364ff, 372, 386, 395, 433, 435, 460, 681, 683, 706, 708
Lavrinenko, Rimma Fedorovna 225
Lawes, John Bennet 174, 186, 193, 233, 280
Lazrus, Allen L. 59, 262, 664, 705
Leather, John Walter 192, 194, 234, 280
Lebedinzeff, Arsenius 400, 409, 423
Le Chatelier, M. Henry 453
Lechler, Paul 298
Leck, Caroline 5, 248
Ledokhovich, Aleksej Aleksandrovich 259
Leeds, Albert Ripley 345, 408, 421, 423, 451, 507, 517, 636f
Leeuwenhoek, Antoni van 145, 148, 277
Lefohn, Allen S. 620f
Lehmann, Christian 720
Lehmann, Erich 574
Lehmann, Hans 11, 115, 215f, 290
Lehmann, Karl Bernhard 210, 445, 447
Leibniz, Gottfried Wilhelm XI, 49, 82
Leighton, Philipp Albert 5, 13, 18, 42, 486, 490f, 494, 577, 650, 657
Lejay, Pierre 581
Lejeune, Phocas 176
Lelieveld, Jos 610, 631, 668
Lémery, Nicolas 90, 435, 709
Lenard, Philipp 323f, 326f, 480, 533, 575f
Lender, Carl Friedrich Constantin 453f, 476, 479, 498f, 504f, 507, 526, 529f, 544, 610, 635
Lenhard, Ulrich 677f
Lepape, Adolphe 375, 598
Lerebours, Noël Marie Paymal 552
Le Roy, Charles 121, 124, 126f
Le Roy, Édouard Louis Emmanuel Julien 51
Lespieau, Robert 596
Lessing, Rudolf 29
Lettau, Heinz 329
Letts, Edmund Albert 396, 401, 403, 405, 421f, 424f
Leuchs, Johann Carl 471
Levsen, Karsten 43
Lévy, Albert 145, 188, 193f, 234, 280, 409, 413, 423, 555ff, 571, 574, 593, 595, 674
Levy, Hiram 440, 491
Lewis, Gilbert Newton 93, 467
Lewy, Bernhard Carl 388f, 391, 408, 411, 415, 423
Libavius, Andreas 85, 525f
Lichtenstein, Eduard 464, 478f, 524f
Lieben, Adolf 491
Liebhafsky, Hermann Alfred 512
Liebig, Justus von 6, 56f, 58, 72, 96, 117, 145, 157, 160, 164, 171, 173f, 175, 177f, 182ff, 196, 208, 230, 233, 347, 377, 388, 435f, 467, 473, 501, 510, 516, 672f, 695
Liebknecht, Otto 595, 632
Liebreich, Matthias Eugen Oskar 514
Liefmann, Harry 294, 297, 711
Liesegang, Wilhelm 115ff, 235
Likens, Gene 243ff
Liljequist, Gösta Hjalmar 7
Lindgren, Waldemar 59
Lindner 503, 523
Lindström, Gustav 305
Link, Heinrich Friedrich 202, 233

Linke, Franz 38f
Linow, Carl 345
Lippincott 500, 504, 506, 544, 548f
Liss, Peter 16
Lissauer, Abraham 452
Liveing, George Downing 487
Locke, John 79
Lode, Alois 396, 404, 686
Loew, Oskar 473, 723f
Loewy, Adolf 8, 396
Löffler, B. 298
Logan, Jennifer 620f, 630
Lohs, Karlheinz 37, 721
Lomonossov, Michail Wassiljewitsch 89, 150
London, Fritz Wolfgang 94
Loschmidt, Josef 94
Lösecke, August Georg von 505, 532, 691
Lossnitzer, Heinz 10
Louvet, Albéric 193, 196, 234, 505, 508, 568
Low, David 367
Lowe, Edward Joseph 545
Löwe, Fritz 39
Löwig, Carl Jacob 443
Luca, Sebastiano de 170, 233, 237
Lucke, Otto 601, 606
Ludolf, Hieronymus von 336
Ludwig, Friedrich Hermann 117, 157, 166
Luedicke, Otto Carl Friedrich 504, 506, 530, 532
Lugin, N. P. 592f
Lugner, Ivar Henrik 195, 198, 235, 237, 239
Lull, Raymund 335, 671
Luna, Ramon Torres Muñoz de 3, 408, 410
Lundegårdh, Henrik Gunnar 427
Lunge, Georg 200, 332, 342, 400f, 686
Luther 76
Lux, Herbert 208, 250
Luz, Johann Friedrich 382f, 384
Lymann, Theodore 572

Maack, Petrus Heinrich Karl von 502, 509f, 521, 570
Mabery, Charles 214, 234, 296, 674f, 700, 713
Macadam, Stevenson 171
Macagno, Ippolito 409f, 674
Macartney, George Lord 387
MacDonnell, Edmund 504, 567
Mach, Edmund 186, 194, 234
Mackereth, Thomas 450, 476f, 480, 503, 505, 507, 546f
Mackie, William 410
Macquer, Pierre-Joseph 66, 372, 433
Magellan, Jean Hyacinthe de 384f
Magill, Paul L. 447f
Magnus, Albertus 707
Magnus, Gustav 390
Maier-Leibnitz, Heinz 598
Maimonides, Moses 285
Main, James 292
Main, Robert 548
Majo, Ester 221, 235
Mangin, Arthur 133
Mani, Anna Modayil 588
Mantegazza, Paolo 475, 541
Manzevich, V. V. 259
Marcano, Vicente 189, 194, 197, 234, 237, 699
Marcet, William 409, 421
Marchand, Eugène 145, 169f, 172, 233, 452
Marchand, Richard Felix 408, 410f, 414, 421, 423
Marchi, Luigi De 429
Marchlewski, Leon Paweł Teodor 401
Marcuse, Adolf 133
Marenco, Alain 579, 621ff
Marggraf, Andreas Sigismund 91, 117, 143, 145, 151ff, 159, 178, 182, 233, 237f
Marié-Davy, Edme Hippolyte 188, 390, 405, 408, 413, 556
Marignac, Jean-Charles de 388, 452f, 462ff, 541
Mariotte, Edme 54
Mark, Hermann Franz 599
Märklin, Georg Friedrich 312
Marquardt, Wolfgang 190, 248ff
Marsh, Arnold 290, 293, 298
Marshall, Abraham Lincoln 487, 650, 652
Marshall, Samuel 506, 546
Martell, Edward Ambrose 16, 59
Marti y Franquès, Antonio de 387
Martin 193, 208
Martin, Charles-Noël 230
Martius, Theodor 528f
Marum, Martinus van 317, 457f, 480, 555, 682
Mason, Basil John 724
Mason, Charlotte 286
Matsudaira, Yasuo 222, 236
Matsui, Hideo 146, 223, 236, 658f
Mauersberger, Günther XI, 36, 610, 631, 658, 669f
Mayer, Adolf Eduard 176

Mayow, John 4, 66, 89, 347, 349, 355, 357f, 394
Mazin, Ilja Pavlovich 259
McConnell, John C. 440, 491
McLeod, Herbert 651
Mecke, Reinhard 577f, 585
Meetham, Alfred Roger 582, 600
Metnieks, Arvids Leons 334
Meier 504, 532
Meigen, Wilhelm 20
Meissner, Georg 145, 184, 233, 355, 453, 472
Meixner, Franz 16, 42
Melander, Gustaf 333f
Meldrum, Andrew Norman 368
Menchikowsky, Felix 237
Mendeleev, Dmitrii Ivanovich 637, 651
Mène, Ch. 167f
Mensbrugghe, Gustave Léonard van der 131
Mensching, J. E. 195, 235, 247
Menzl, Oswald 207
Merckens, Waldemar 655
Mészáros, Ernö 5, 192, 243
Mettenheimer, Carl von 502, 510, 521
Meyer, Edgar 480, 573f
Meyer, Friedrich (Frederick) 504, 508, 565
Meyer, Victor 200, 723
Meyer *or* Meijer 160
Meyrac, Victor 172, 180, 208, 233, 255, 280
Middleton, J. T. 42
Middleton, William Edgar Knowles XIII
Migeotte, Marcel Victor 439
Miller, N. H. J. 115, 145, 177, 187, 189, 194ff, 198, 200f, 205f, 220, 234, 237
Minervin, Vladislav Evgenjevich 259
Miquel, Pierre 313, 409, 413, 423, 425, 593
Mitchell, Arthur Crichton 451, 503f, 548
Mitchill, Samuel Latham 65, 70, 330, 688f
Mitscherlich, Eilhard 513
Miyake, Yasuo 145, 190, 195, 222ff, 228, 236, 258f, 260, 274, 588, 613
Moffat, Thomas Barbour 452, 454, 469, 500, 502, 508, 542ff, 550, 552, 561, 568
Möhl, Heinrich 451, 504, 529, 548
Mohnen, Volker Arnim 132, 265f, 269, 272f, 583, 605
Mohr, Karl Friedrich 410, 456, 459
Mohry, Herbert 37
Moissan, Henri 376
Moldan, Bedřich 242f
Moleschott, Jacob 117, 157, 171f
Molina, Mario José 16, 40, 480, 590f
Monceau, Henri-Louis Duhamel du 75
Monge, Gaspard 370
Monks, Paul S. 623, 627
Monnett, Osborn 718
Mons, Jean-Baptiste van 162, 254f
Montani, Francesco Fabi 452
Monticelli, Teodoro 708
Moortgat, Geert 43, 661
Morgan, James J. 16
Mörikofer, Walter 29
Morley, Edward Williams 373f, 391f
Morren, Jean François Auguste 321f, 327, 725
Morse, Philip McCord 93, 691
Morton, John Chalmers 187
Morton, William James 314, 454
Morveau, Louis Bernard Gyuton de 352, 433
Moser, Helmut 585, 692
Mosheim, Johann Lorenz von 276
Moss, William Amphlett 445f
Moureu, Charles 375f
Mrose, Helmut 7, 10f, 18f, 28, 30ff, 40, 190, 231, 241, 256f, 259ff, 267ff, 274, 329f, 607, 617, 647, 658, 705
Mügge, Ratje 40
Mühry, Adolf Albrecht 9
Mulder, Eduard 468
Mulder, Gerardi Johannis 160, 168, 172, 233
Mulder, Louis 176
Müller, Armand 5, 101
Müller, Carl Alexander 176, 493, 674
Müller, Karl 505, 519, 521
Müller, Max Ludwig Otto 215, 675, 677
Müller, Richard 342
Müller, Wilhelm Josef 31
Mulliken, Robert 94
Mulvany, John 504, 550
Muncke, Georg Wilhelm 166f, 290, 314
Munger, William 263, 265, 274
Müntz, Charles Achille 57f, 145, 183, 189, 194, 197, 234, 237, 255, 392, 404, 408, 415, 419f, 423ff, 433, 445, 674, 697, 699
Muspratt, James Sheridan 62
Musschenbroek, Peter van 126
Myers, Carl 140

Nägeli, Carl Wilhelm von 473
Nagy-Ilosva, Lajos Ilosvay von 473, 646, 698
Nakazawa, Zenichi 223

Nash, William Carpentier 506, 544
Nasse, Otto Johann Friedrich 96, 464, 481, 484f, 631, 651
Nathanson, Alexander 59
Neftel, Albrecht XI, 424, 666f
Negretti, Henry 500
Nehls, H. 217
Nehse, Claus Eduard 271
Nékám, Lajos Alexander 345, 445
Neljubin, Alexander Petrowitsch 145, 167, 199, 237
Neng-Huei (George) 274
Nernst, Walther 94, 101f, 578, 637
Neßler, Julius 175f
Neumann, Franz 452, 537
Neumayer, Georg von 9, 567
Neuwirth, Robert 10
Neven, Lucien 439
Newton, Isaac 49, 124, 147
Nicholson, A. W. 505, 564
Nicolet, Marcel 480, 573, 589f, 703
Niebergall, Carl 502, 522
Nielsen, Harald Herborg 439
Niki, Hiromi 16, 440, 491, 684
Nippoldt, Johann 31
Noack, Rolf 10
Noble, Robert Hamilton 439
Nolan, John James 327f
Nollet, Jean 574
Nordenskiöld, Adolf Erik 303, 305, 351, 358, 361
Norrish, Ronald George Wreyford 489f, 590f
Noxon, John F. 704

Obenland, Eugen 609f
Obolensky, Vladimir Nikolaevich 141
Odén, Svante 244, 246
Odling, William 465, 480, 686
O'Dowd, Colin 335
Oeffinger, Hermann 532
Oettel, Felix 393, 409, 424
Ogburn, William Fielding X
Ogiwaraa, Sekiji 334
Ogren, John 267, 269, 274
Ogura, Yutaka 146, 223, 236
Ohtake, Takeshi 259, 274
Okita, Toshiichi 223f, 260f, 334
Okochi, Hiroshi 274
Oltmans, Samuel J. 620f
Oparin, Alexander Iwanowitsch 53
Oppenheimer, Julius Robert 94
Osann, Emil 156, 161, 168
Osann, Gottfried Wilhelm 452, 464, 469, 474, 480, 482f, 499f, 502, 549, 631, 636, 638
Ostwald, Friedrich Wilhelm 60, 94, 100, 124, 131, 226, 353, 358f
Ostwald, Wolfgang 124, 284
Otto, Marius Paul Antoine Gabriel 482
Owens, John Switzer 297f

Paetzold, Hans-Karl 589, 612f
Palissy, Bernard 87
Palmieri, Luigi 503, 542
Palmqvist, Laura Augusta 401, 409, 416, 421, 423
Paneth, Friedrich Adolf 375f, 379, 393, 595, 598, 603ff, 614, 702
Panopolis, Zosimos of 64
Papée, Henry M. 344
Paping, Bernard Joseph 145, 157, 228
Paracelsus 74f, 86, 88, 90, 144, 147, 173, 277, 335, 347
Parkes, Edmund Alexander 515
Parrot, Georg Friedrich 384, 406
Partington, James Riddick 75f, 80, 89, 349, 364, 372, 374, 382
Pascal, Blaise 347
Passerini, Napolenoe Pio 146, 190f, 221, 234
Pasteur, Louis 313, 347
Patat, Franz Xaver Maria Theresia 578
Paterson, Oscar 711
Pauli, Simon 308
Pauli, Wolfgang Ernst 93
Pauling, Linus 60, 93f
Payen, Anselme 179
Peiresc, Nicolas Claude Fabri de 307
Péligot, Eugene-Melchior 172, 178
Pendleton, Edmund Monroe 3
Penkett, Stuart 16, 651, 660, 726
Penndorf, Rudolf 585
Perner, Dieter 705
Pernter, Josef Maria 138, 430, 514f, 570
Perros, Pascal 661, 664
Perrott, George St. John 718
Perrottet, Ernest 600
Peter, Robert 560
Petermann, Arthur 176, 191, 194, 234, 237, 396, 405, 409, 415, 420, 423

Petrenchuk, Olga Petrovna 225, 246, 259ff, 269, 272
Pettenkofer, Max 398ff, 404, 408f, 433, 444f, 446, 501, 507, 530, 551
Pettersson, Sven Otto 397, 401, 421
Petzholdt, Alexander 204
Pfaff, Christoph Heinrich 337
Pfaff, Emil Richard 504, 527
Pfeffer, Wilhelm 431
Pfeiffer, Arthur 600
Pfleiderer, Alfred Otto Heinrich 10, 23
Pictet, Marc-Auguste 167
Pieraggi, Leonard Endymion 504, 552
Pierre, Joachim Isidore 145, 157, 179, 302, 208, 279, 673
Pierre, Victor 480, 502, 508, 514, 538f
Pierrou 60
Pietra Santa, Prosper de 558f
Pilling, Michael J. 43
Pincus, Salomon 184, 233, 472f
Pink, William E. 62
Pissis, Pierre Joseph Aimé 343
Pitts Jr., James N. 5, 18, 42, 340, 491, 624, 630
Planck, Max 93
Plantamour, Emilie 390
Plass, Gilbert Norman 432
Platon 120
Plattner, Carl Friedrich 712
Platt, Ulrich 705
Playfair, Lyon 208f
Pleischl, Adolf Martin 695
Plieninger, Heinrich Theodor 502, 521
Plinius (in English: Pliny) 284, 335, 394, 442f, 521, 671
Plotnikov, Ivan Stepanovich 100
Plutarch 63
Poey, Felipe 500
Poey, y Aguirre, Andrés 475, 500, 504, 508, 568
Poggendorff, Johann Christian 756
Polanyi, Michael 94
Pollak, Leo Wenzel 29, 334
Polli, Giovanni 500, 542
Pollock, James Arthur 326
Pope, John 503, 505, 563
Popp, Christian 587
Popper, Karl X
Porta, Johann Baptista 124
Pošepný, František (Franz) 57, 201f, 322, 333
Powers, Edward 140
Prahl, Walter 687
Prather, Michael J. 624
Pregl, Fritz 435
Preining, Othmar 29
Prestel, Michael August Friedrich 301, 318, 451, 484, 503, 507, 524, 526ff
Prettner, Johann 502, 537ff
Prettre, Marcel 595f
Prévost, Pierre 127
Prévot, Edward William 204, 234
Preyer, William Thierry 498
Price, William Charles 702
Priestley, Joseph 5, 80, 91, 128, 158, 164, 203, 346f, 349ff, 354ff, 356, 258, 362ff, 368, 370ff, 380, 395, 406, 433, 671, 680ff, 708
Prince, Charles Leeson 543
Pring, John Norman 480, 595f, 701
Pringal, Erich Julius Otto 325, 327
Pringsheim, Ernst 99
Proeschel, Frederick 567
Proust, Joseph Louis 92
Prout, William 109, 128, 254, 309, 388, 435, 633
Pruppacher, Hans R. 16
Puchner, Heinrich Ruprecht 409f
Pusinelli, Fritz 393
Puxbaum, Hans 267

Quetelet, Adolphe Jacques 502, 506, 537
Quitmann, Eugen 21ff, 229, 256, 327, 709

Radau, Jean Charles Rodolphe 555
Radford, William H. 140f, 257f, 274
Ragona, Domenico 319, 504, 542
Ragsky, Franz 695
Rafinesque, Constantine Samuel 284
Ram, Atma 195, 226f, 235, 280, 445, 448
Ramanathan, Kalpathi Ramakrishnan 588f
Ramann, Emil 713
Ramazzini, Bernardino 54, 145, 148, 276
Rammelsberg, Carl Friedrich 499, 515
Ramsay, Sir William 35, 79, 346f, 354, 356, 470, 374f, 387, 711
Randall, John 113
Rao, Basrur Sanjiva 597f, 604
Raoult, François Marie 94
Raschig, Fritz 687, 692
Rasmussen, Reinhold Albert 439, 449
Rasumowski, Kirill Grigorjewitsch 515

Rayleigh, Lord 327, 346f, 349, 355, 357, 374f, 573, 580
Réaumur, René-Antoine Ferchault de 307
Reboul, Henri Paul Irénée 383f
Reclus, Jacques Élisée 133
Redi, Francesco 148
Redtenbacher, Josef 177, 310
Reese, Charles Lee 723f
Regener, Erich 480, 576, 581, 585, 594, 600, 603, 607ff, 612ff, 618
Regener, Victor 480, 584, 600, 602, 611, 613
Regnault, Henri Victor 388f, 392, 408
Rehberg, Poul Brandt 428
Reich, Albert 297
Reich, Ferdinand 712
Reichardt, Eduard 193, 233, 408, 410, 425
Reichenbach, Karl von 306
Reid, Boswell 399
Reimann, Ernst Julius 133
Reimann, Karl Ludwig 338
Reimer, Eberhard 628
Reinau, Erich Hellmut 427
Reinsch, Edgar Hugo Emil 377f
Reiset, Jules 408, 413, 418f, 422f, 425
Reiss, Howard 132
Reiter, Reinhold 10, 704
Remsen, Ira 445f, 468
Renatus, Flavius Vegetius 111
Renger, Friedrich 30, 602, 607, 702
Renk, Friedrich Georg 396, 404, 447
Renner, Eberhard 36, 623
Reslhuber, Augustin 451, 501f, 508, 523f, 534ff, 538f, 555
Revelle, Roger Randall Dougan 429
Rey, Jean 75
Reynolds, William Colebrook 517, 604, 701, 716
Richards, Eric Hannaford 194, 201, 234f
Richardson, Arthur 515, 652
Richter, Jeremias Benjamin 116
Rideal, Eric Keightley 451, 453, 572
Rieche, Alfred 648
Riesenfeld, Ernst Herrmann 468f, 473, 574
Risler, Charles Eugene 408f
Risse, Otto Karl Hermann 654f
Ritschl, Rudolf Karl Ludwig 585
Ritthausen, Heinrich 175
Rive *see* de la Rive
Robert, Aimè-Augustin Joseph 503, 552, 563
Robinet, Stéphane 171
Rocha, Manuel 505, 508, 568f
Rodhe, Henning 5, 16, 47, 239, 244f, 248, 267, 269
Rodionov, Sergey Fedorovich 592f
Rogers, Lewis H. 489
Rogers, William Barton 475, 503, 548, 563
Rolle, Wolfgang 36f, 43, 248, 587, 623, 625
Röntgen, Wilhelm Conrad 401, 430
Roscoe, Henry Enfield 99, 122, 292, 411, 437, 443, 475
Rose, Heinrich 62
Rosing, Hans Anton 186, 233, 237
Rossby, Carl-Gustaf Arvid 5, 7, 29, 239, 339
Roster, Giorgio 409f
Rotch, Albert Lawrence 505, 566
Rotheram, Caleb 364
Rothmund, Viktor 327, 484, 597, 634
Roussin, Albin Reine 304
Rowland, Sherwood 16, 40, 590f
Rubens, Heinrich 574
Rubin, Morcedal 452, 459, 462, 477, 576, 633
Rubner, Max 9, 400, 404, 445, 699, 709, 715, 717
Rudeck, Eugen 505, 533f
Rudolph, Jochen 16
Ruhmkorff, Heinrich Daniel 694, 698
Runge, Carl 573
Russ, Paul 699
Russell, Colin Archibald X
Russell, Edward John 186f, 194, 200f, 220, 234
Russell, Francis Albert Rollo 135, 291ff
Russell, William James 325, 341, 421, 691
Ruston, Arthur Gough 194f, 214f, 235, 294f
Rutherford, Daniel 347, 350ff, 355ff
Rutherford, Ernest 489
Ryaboshapko, Alexey 16, 60, 355f

Sachse, Karl Traugott 503, 523
Sainte-Claire Deville, Charles Joseph 504, 558
Sakurai, Sumiko 613
Sala, Angelus 671, 707
Salisbury, Richard Anthony 331
Saltzman, Bernard E. 703
Saltzman, Eric 662
Sand, Johann David 406
Sandiford, Irene 392
Santorio, Santorio XIII
Saruhashi, Katsuko 588
Saunders, Sandra 44

Saussure, Horace-Bénédict de 121, 124f, 128
Saussure, Nicolas Théodore de 171, 173, 384f, 387, 395, 397, 403, 405ff, 414, 423, 433, 472, 671
Sawjalow, Wassili Wassiljewitsch 59
Sawyer, Frederick Ernest 549
Schaefer, Vincent Joseph 140f
Schaller, Eberhard 248
Schaper, Carl Wilhelm Ludwig 502f, 523
Scharnow, Ulrich 7
Scheby-Buch, Oskar Louis 332
Scheele, Carl Wilhelm 5, 66, 70, 91, 95, 98, 144, 155, 345, 347, 350ff, 355ff, 258ff, 264, 368, 370, 382, 384ff, 406, 460, 516, 671, 682, 708f
Schelenz, Herrmann Emil 332, 505, 533
Schemenauer, Bob 266, 271, 274
Schenzl, Guido 505, 538
Scherer, Alexander Nicolas 758
Scherer, Johann Baptist Andreas Ritter von 65, 380, 385f
Scherf, Johann Christian 385
Schidlowski, Manfred 41
Schiedermayr, Karl 528
Schiefferdecker, Wolfgang 451, 502, 523f, 537, 555, 570
Schlagintweit, Adolph 122, 136
Schlagintweit, Hermann 122, 136
Schlipköter, Hans-Werner 29
Schloemer, Wolfgang 587
Schloesing, Jean-Jacques Théophile 58, 186, 200, 396, 419, 672, 674, 696f
Schlossberger, Julius Eugen 505, 520f
Schlotfeldt 451, 502, 520
Schmauß, August 284, 320
Schmid, Werner 483f, 641f
Schmidt, Ulrich 16
Schmirjan, Hans 29
Schneider, Franz Coelestin 695
Schneider, Jacob Sparre 393
Schneider-Carius, Karl 9
Schofield, Robert 372, 374
Scholz, Joachim 328
Schön, Johann 318
Schönbein, Christian Friedrich 44, 96f, 184, 322, 327, 341, 347, 355, 450ff, 455ff, 460ff, 468ff, 474ff, 479ff, 492, 494ff, 499ff, 508ff, 516, 518ff, 524, 534, 541, 543, 545, 550, 552, 572, 591, 633ff, 638f, 651, 654, 670, 685f, 690, 694, 696ff, 704
Schöne, Herman Emil 234, 256, 355, 478, 480, 499, 505, 511f, 514ff, 570f, 573, 594f, 638, 640ff, 651, 658f, 661ff
Schoof, Christian Ludwig 503, 524
Schreiber, Joseph 540
Schröder, Johannes 170, 195, 200
Schrödinger, Erwin 94
Schroeder, Julius von 207f, 214, 233, 235, 237
Schröer, Erich 578, 612
Schrötter, Anton 452, 474
Schuck, E. A. 42
Schultz, August Wilhelm Ferdinand 502, 522
Schultz, Hermann 172, 400
Schulz, Leo 36, 39, 328f, 333
Schulze, Franz Ferdinand 408, 413f, 416, 421, 423, 425
Schulze, Johann Heinrich 98, 355
Schulze, Max 529
Schumacher, Hans-Joachim 577f, 687f
Schumann, Victor 573, 575
Schurath, Ullrich 43
Schürmann 452
Schwab, George-Maria 657
Schwabe, Samuel Heinrich 477
Schwartz, Stephen E. 252, 657
Schwarzenbach, Valentin 498, 541
Schweigger, Johann Salomo Christoph 758
Scoresby-Jackson, Robert Edmund 514
Scott, Robert Henry 135
Scott, Alexander 373
Scotti, Giberto 542
Scoutetten, Robert Joseph Henri 453, 474, 479, 497, 502, 506, 526, 554
Secci, Angelo 319
Séguin, Armand Jean François 370, 373, 386
Seiler, Wolfgang 16, 37, 40f, 630, 704
Seilern, Carl Max Graf 191
Seinfeld, John Hersh 5, 42, 485
Seip, Hans Martin 245
Seitz, Franz 502, 523, 563
Sekihara, Kyo 613
Selander, Nils Edvard 421
Selezneva, Evgenia Semenovna 224f
Selinski, Nikolai Dmitrijewitsch 59
Senebier, Jean 347, 382, 385
Sendtner, Rudolf 215
Serpieri, Alessandro 502, 541

Sertürner, Friedrich Wilhelm Adam 463
Shalamyansky, Arkady Mordukhovich 593
Shapter, Thomas 469f, 503, 545, 548
Shaw, Peter 89
Shaw, Sir William Napier 138, 220, 297f
Shimizu, Mitsuo 222
Shmeter, Solomon Moiseevich 258f, 274
Shutt, Frank Thomas 195, 220, 235, 237
Silbermann, Johann Theobald 502, 541, 552, 563
Silberschlag, Johann 130
Sillman, Sanford 627
Silvestri, Orazio 303, 503, 542
Simon, Karl-Hermann 250
Simonin, Jean Baptiste 502, 541, 551, 553, 563, 571
Simpson, Georg Clarke 329, 333
Singh, Hanwant B. 439, 450
Six, James 127
Skeib, Günther 587
Skryabin, Georgij Konstantinovitsch 47
Slanina, Jacob (Sjaak and Jack) 251, 679
Slemr, Franz 661f
Slinn, W. George N. 16
Slocum, Giles 428
Smallwood, Charles 451, 453, 502, 506, 508, 542, 561ff
Smith, Robert Angus 6f, 14, 17ff, 63, 118, 145, 155, 157, 168, 193, 201, 204, 207ff, 213ff, 227ff, 233, 238, 243, 245, 247, 279, 307, 312f, 346f, 391f, 398ff, 407f, 416, 441f, 461, 546, 548, 613, 674ff, 683, 712f
Smyth, John 455, 500, 504, 545f
Smyth, Robert Brough 503, 567
Snell, Willebrord van Roijen 82
Sobik, Mieczysław 274
Solomon, Susan 480, 591f
Sondén, Klas 401f, 421
Soret, Jacques-Louis 390, 452, 465f, 467f, 477, 480f, 492
Spengler, Ludwig 502, 508, 518, 520, 541
Spicer, Chester William 705
Spiegel, Leopold 699
Spindler, Gerald 37
Spurný, Květoslav Rudolph 29
Staedel, Wilhelm 465
Staehelin, Johannes 589, 592, 630, 622f, 658
Stahl, Georg Ernst 77, 90f, 354f, 708
Stair, Ralph 586
Stamp, Josiah Charles XIV
Stark, Johannes Nikolaus 100
Stark, William 145f, 158f, 168, 228, 233, 237
Stedman, Donald 591
Steele, Paul A. 439f
Stein, Gabriel 494
Stein, Wilhelm 715
Steinhauser, Ferdinand 10, 231, 613f, 616
Steinmann, Josef Johann 311
Stella, Tilemann 271
Stephans, E. R. 42
Stewart, Richard W. 625
Stewart, Robert 218, 235
Stöckhardt, Ernst Theodor 174
Stöckhardt, Julius Adolph 1, 3, 58, 174f, 186, 214f, 267, 347, 442, 712, 721f
Stockwell, William R. 650, 705, 725
Stohmann, Friedrich 699f
Stoklasa, Julius 146, 191, 194, 215f, 234, 237, 255, 676, 716, 725
Stoney, George Johnstone 320
Stolarski, Richard 480, 590f
Storer, Francis Humphreys 408f
Stow, Fenwick William 550
Strambio, Gaetano 502, 508, 541, 551
Stranz, Hubert Dietrich 583
Stromeyer, Friedrich 66, 70
Strumpf, Ferdinand Ludwig 80
Strutt, John William (3rd Rayleigh) 346, 580
Strutt, Robert John (4th Rayleigh) 574, 580
Struve, Heinrich Wilhelm von 96, 145, 190f, 199, 233, 473, 514, 634, 638ff, 647, 663
Strzyżowski, Anton Stanislavovich 455
Stuhl, Friedrich 43, 710
Stumm, Werner 246, 493
Suess, Eduard 50
Suess, Hans Eduard 42, 429, 488
Sugawara, Ken 146, 223, 236
Sundt, Lorenzo 343
Süpfle, Karl 434
Sutton, Mark 242, 679
Svistov, Peter Filippovitsch 225f
Swederus, Magnus 358
Sylvius, Francis 144
Symons, George James 550

Tacchini, Pietro 305
Tait, Peter Guthrie 452, 465f, 471, 695
Takematsu, Okada 222
Takeuchi, Ushio 223

Tamm, Carl Olof 240, 244
Tans, Pieter 427
Tansley, Arthur George 51
Taube, Henry 338, 488, 493, 654, 656, 660
Taylor, Eva Germaine Rimington 298
Taylor, Geoffrey Ingram 254
Taylor, Hugh Stott 102, 488, 650
Taylor, William Henry 603
Teichert, Friedrich 11, 19, 28, 33f, 35f, 241, 249, 614f, 617
Teilhard de Chardin, Pierre 51
Thales of Miletus 119
Than, Carl von 473, 490
Thénard, Arnould Paul Edmond 555
Thénard, Louis Jacques 61f, 98, 386, 406ff, 435, 481, 555, 632, 651, 706, 722
Thiel, Carl Emil Hugo 759
Thiele, Hermann 434, 641, 653, 693
Thierry, Maurice de 409, 425, 559f
Thoms, George 176
Thomson, Joseph John 131
Thomson, Richard Dundan 355, 440, 712
Thomson, Thomas 64, 81, 174, 381, 388, 632
Thomson, William, Lord 131
Thorpe, Sir Thomas Edward 408, 414f, 421, 423, 425
Thurneisser, Leonhard 144
Tichy, Hans 23, 34, 601ff, 614, 616
Tiggelen, Adolphe Van 440
Tilden, Sir William Augustus 341
Timirjasew, Kliment Arkadjewitsch 515
Tissandier, Albert 416f
Tissandier, Gaston 203, 404ff, 408, 416f, 418, 425
Tizenhauz, Konstantin 310
Toel 503, 522
Tomaschek, Rudolf 31
Tomlinson, Charles 127, 279
Torrey, John 332
Torricelli, Evangelista 113
Torstensson, Gunnar 195, 238
Trabezkoj, Pawel Peter 199
Tracy, Samuel Mills 197, 234, 236
Traube, Moritz 437, 567, 474, 485, 635f, 647, 651, 723
Traube, Wilhelm 103, 437
Travers, Morris William 355, 375
Trimble, Henry Macagno 62
Tripe, John William 549, 550
Trommsdorff, Christian Wilhelm Hermann 191
Trommsdorff, Hugo 191
Trommsdorff, Johann Bartholomäus 62, 191
Trommsdorff, Johann Samuel 277
Trommsdorff, Wilhelm Bernhard 191
Troostwyck, Adrien Paets van 443
Truchot, Paul 408f, 417, 425, 673
Tschudi, Johann Jakob von 503, 508, 568
Tsunogai, Shizuo 678, 726
Turner, Edward 72, 719f
Tuxen, Christian Frederik August 186, 194, 234, 237
Twomey, Sean 49, 335
Tyndall, John 127, 321ff, 429ff, 448, 477, 573

Uffelmann, Julius August Christian 345, 347
Ule, Otto Eduard Vincenz 452
Ulloa, Antonio de 130
Umlauft, Friedrich 82, 133
Ungeheuer, Hans 609, 611, 614
Urey, Harold Clayton 488, 491, 650, 654
Usher, Francis Lawry 596f, 603

Valentine, Basil 706
Valle, Manuel Remón Zarco de 390
van Stoop 160
van't Hoff, Jacobus 84, 94
Varigny, Henry de 313, 594f
Vassy, Étienne 595
Vauquelin, Louis-Nicolas 12, 61, 279, 373, 708
Vavilov, Sergei Ivanovich 599
Veltmann, Christian Heinrich 318
Veraart, August Willem 140
Verdét, Marcel Émile 464
Vernadsky, Vladimir Ivanovich 4, 19, 40, 44ff, 50ff, 224
Vernon, George Venables 504, 546
Verrier, Urbain Jean Joseph Le 553
Verver, Bernardus 388, 423f
Viemeister, Peter Emmons 704
Vigroux, Ernest 595
Villars, Donald Statler 693
Ville, Georges 673f
Villiger, Victor 482, 637
Vinogradov, Alexander Pavlowitsch 45
Visser, Simón 705
Vitkevich, Vitold Ignatevich 140
Vitruvius, Marcus 111
Vityn, Jan 205, 224, 235

Vivian, John Henry 712
Voeikov, Alexander Iwanowitsch 224
Vogel, August 173, 181f, 234, 377, 388, 408, 423
Vogel, Heinrich August von 146, 166
Vogel, Hermann Wilhelm 100, 337
Vogel, Klaus 40
Vogler, Christian August 391
Vogt, Heinrich 21, 228f
Voit, Carl Michael von 697
Vollmer, Carl Gottfried Wilhelm 133
Volman, David H. 491, 649
Volmer, Max 131f
Volta, Alessandro 372, 381f, 384, 386f, 437
Voltolini, Friedrich Eduard Rudolf 521
Volz, Friedrich E. 609ff
Volz-Thomas, Andreas XIV, 556f, 621f, 627f
Vonnegut, Bernard 140
Vosmaer, Alexander 451, 453

Waage, Peter 94
Wackenroder, Ferdinand 157
Wafer, Lionel 456
Wagner, Hans 656f
Wahner, Andreas 43, 605
Waldbauer, Olga 718
Waldman, Jed 263f, 269, 274
Waldmann, Wilhelm 498
Walker, James 401
Walker, James C. G. 625, 662
Waller, Augustus Volney 121, 132, 167
Wallerius, Johann Gottschalk 147
Walther, Reinhold von 144
Wang, Tao 274
Wankel, Georg Reinholdt 186, 237
Wanklyn, James Alfred 377, 445
Warburg, Emil Gabriel 100, 104, 414, 572, 574f, 687
Warburg, Otto Heinrich 104, 693
Warington, Robert 186f, 193f, 233
Warmbt, Wolfgang 7, 10f, 30, 33ff, 606, 614ff, 627
Wärme, Karl Erik 428
Warneck, Peter 5, 40ff, 490, 587
Watanabe, Koichi 270, 274, 661, 663f, 667
Watson, Henry Hough 407
Watson, Herbert Edmeston 375, 398, 408, 423
Watson, William Weldon 487
Watt, James 353, 370ff
Way, John Thomas 186f, 193, 233
Wayne, Richard 590
Weaver, Richard 408, 410
Weber, Eduard 503, 525
Webster 503, 563
Webster, George E. 62
Weedon, Thornhill 200, 237
Weegmann, Eugen 725
Weeks, Mary Elvira 356ff
Wegener, Alfred 140
Weickmann, Ludwig 329
Weigert, Fritz 99, 103, 226, 576, 591
Weinstock, Bernard 434, 440, 491
Weiss, Joseph 492, 655ff
Welbel, Benzion Movscha-Morduchovitsch 190, 194, 198, 205, 224, 234, 280
Wells, Arthur Edward 718
Wells, Thomas Spencer 502, 550
Wells, William Charles 110, 114, 126f, 279, 282
Welsh, Harry Lambert 576
Welt, H. 195, 199
Weltzien, Carl 452f, 466, 470, 481f, 484, 647, 685, 690
Went, Frits Warmolt 321, 448f
Westendorff, Klaus 250
Wetherill, Charles Mayer 346, 440f, 504, 515, 564
Wette, Ludwig de 522
Wetzel, Walter 343f
Wexler, Harry 29, 617
Whipple, George Matthew 550
Whitehead, Walter Lucius 343
Whytlaw-Gray, Robert 327f
Wiedeburg, Johann Ernst Basilius 316
Wiedemann, Gustav Heinrich 758
Wiegmann, Arend Friedrich August 166
Wiegmann, Arend Joachim Friedrich 161ff, 166, 312, 318
Wieland, Heinrich Otto 638
Wien, Wilhelm 758
Wienhaus, Heinrich 722
Wienhaus, Otto XV, 721f
Wieprecht, Wolfgang 37, 267
Wierzbicki, Daniel 539
Wiesen, Peter 43
Wigand, Albert 18, 36, 284, 320, 327
Wildt, Rupert 453
Wilfarth, Hermann 58
Willard, Julius Terrass 146, 190, 194, 234, 237
Willett, Hurd Curtis 141

Williams, Charles Theodore 550
Williams, John William 189, 234, 237
Williams, S. Francis 197, 201, 206, 235
Williams, William Carleton 409, 422, 425
Williamson, Alexander William 436, 452, 464, 649
Willstätter, Richard Martin 29, 492
Wilson, Alexander 127
Wilson, Alexander T. 704
Wilson, Benjamin Dunbar 195, 218, 220f, 235
Wilson, Charles Thomson 324f, 327, 652
Wilson, George 372
Wilson, Patrick 127
Wilson-Barker, David 220
Winkler, Clemens Alexander 54, 711
Winkler, Lajos (Ludwig Wilhelm) 21, 692
Winkler, Peter 247, 267, 330
Winogradsky, Sergej Nikolajewitsch 59
Wislicenus, Hans Adolf 437, 715, 721f
Wislicenus, Johannes Adolf 437
Wislizenus, Friedrich Adolph 504, 564
Witt, Otto Nikolaus 493
Witting, Ernst Julius Heinrich 145, 161f, 168, 202, 233, 254, 279, 317f, 337
Wittwer, Wilhelm Constantin 98f
Wofsy, Steven C. 491
Wöhler, Friedrich 56, 436f, 457
Woldřich, Jan Nepomuk 503, 638f
Wolf, Johann Rudolf 453, 502, 537, 541, 563
Wolf, Max (Maximilian Franz Joseph Cornelius) 324
Wolf, W. 183, 193, 196, 233, 280
Wolff, Christian 113
Wolff, Emil Theodor von 174f, 183, 191, 378
Wolffenstein, Otto 176, 472, 533
Wolffenstein, Richard 176
Wolffhügel, Gustav 408f, 451, 505, 530f
Wolkowicz, Aleksander 467
Wollny, Ewald 3, 184, 193, 233
Wolpert, Heinrich 400, 445, 448
Woodcock, Alfred H. 333f
Woodhouse, James 433
Worm, Ole 307
Worpert, Adolf 400
Wourtzel, Eugène 342
Wragge, Clement Lindley 505, 568
Wrede, Josef Johann Nepomuk 520
Wright, Henry Lionel 327f, 333
Wulf, Oliver Reynold 576, 584
Wüllerstorf, Bernhard Freiherr von 503, 540
Würfel, Heinz 10
Wurster, Casimir 478f, 480, 492, 698
Wurtz, Charles Adolphe 379, 436, 443, 482
Wurzer, Ferdinand 160

Young, Thomas 127

Zabelin, Iosif Vikentevitch 697
Zaizev, Vasili Aleksandrovitsch 224, 259
Zambra, Joseph 500
Zedler, Johann Heinrich 276
Zeldovich, Yakov Borisovich 704
Zellner, Reinhard 16, 43
Zenger, Karl Wenzel 503, 540
Zenker, Helmut 10f
Zernik, Franz 12
Zetzsch, Cornelius 43, 340
Zhao, Dianwu 243, 245
Zier, Manfred 262, 267f
Zierath, Reinhard 232, 241, 245, 249f
Zika, Rod 657, 662f
Zimmermann, Adolf 692
Zimmermann, Eberhard August Wilhelm 55
Zimmermann, Frank 268
Zimmermann, Friedrich 505, 532
Zimmermann, Gerhard 610f
Zimmermann, H. 503f, 523
Zimmermann, Siegfried 491
Zimmermann, Wilhelm Ludwig 145f, 169f, 164, 168, 177, 202, 208, 230, 233, 255, 312
Zimmermann, W. F. A. 116, 133, 137
Zittel, Karl Alfred Ritter von 408, 410, 479, 505, 507f, 531
Zobrist, Jürg 246, 267
Zosimus 64
Zschörner, Helmut 586, 588

Bibliography

Comments to historical journal titles

In the following list of references, the complete title of the journal is given but the following listed journals or periodicals are abbreviated:

ACP	Atmospheric Chemistry and Physics
Ann.	Annalen der Chemie und Pharmacie (1840–1872)
Ann. chim. phys.	Annales de chimie et de physique
Ann. Phys.	Annalen der Physik (1900–1943)
Annu. soc. mét. de France	Annuaire de la Société Météorologique de France
Arch. Pharm.	Archiv der Pharmacie
Atmos. Env.	Atmospheric Environment
Atmos. Res.	Atmospheric Research
Ber.	Berichte der Deutschen Chemischen Gesellschaft
C. r.	Comptes rendus hebdomadaires des séances de l'Académie des sciences
Environ. Sci. Technol.	Environmental Science and Technology
Geophys. Res. Lett.	Geophysical Research Letters
Gilb. Ann.	Annalen der Physik (1799–1824)
J. Aerosol. Sci.	Journal of Aerosol Sciences
J. Am. Chem. Soc.	Journal of the American Chemical Society
J. Atmos. Chem.	Journal of Atmospheric Chemistry
J. Chem. Soc.[a]	Journal of the Chemical Society (London)
JGR	Journal of Geophysical Research
J. Chem. Phys.	Journal of Chemical Physics
J. phys. Chem.	Journal of Physical Chemistry
J. prakt. Chem.	Journal für praktische Chemie
Met. Z.	Meteorologische Zeitschrift
Phil. Mag.	Philosophical Magazine
Phil. Trans.	Philosophical Transactions of the Royal Society, London
Pogg. Ann.	Annalen der Physik und Chemie (1824–1877)
Proc. R. Soc. Edinb.	Proceedings of the Royal Society of Edinburgh
Proc. R. Soc. Lond.	Proceedings of the Royal Society, London
Sitzungsber. Berlin[b]	Sitzungsberichte der Königlich Preußischen Akademie der Wissenschaften zu Berlin
Sitzungsber. München[c]	Sitzungsberichte der mathematisch-physikalischen Classe der k. b. Akademie der Wissenschaften zu München
Sitzungsber. Wien[d]	Sitzungsberichte der Kaiserlichen Akademie der Wissenschaften in Wien – mathematisch-naturwissenschaftliche Classe
Q. J. Roy. Meteor. Soc.	Quarterly Journal of the Royal Meteorological Society
Trans. Faraday Soc.	Transaction of the Faraday Society
Trans. R. Soc. Edinburgh	Transactions of the Royal Society of Edinburgh
Wied. Ann.	Annalen der Physik und Chemie (1877–1899)
Z. anal. Chem.	Fresenius, Zeitschrift für analytische Chemie
Z. angew. Chem.	Zeitschrift für angewandte Chemie
Z. angew. Met.	Zeitschrift für angewandte Meteorologie

https://doi.org/10.1515/9783110732467-008

ACP	Atmospheric Chemistry and Physics
Z. anorg. Chem.	Zeitschrift für anorganische Chemie
Z. Elektrochem.	Zeitschrift für Elektrochemie und angewandte physikalische Chemie
Z. Met.	Zeitschrift für Meteorologie
Z. Öster. Ges. Met.	Zeitschrift der Österreichischen Gesellschaft für Meteorologie
Z. phys. Chem.	Zeitschrift für physikalische Chemie

[a]Different titles and numerations:
Quarterly Journal of the chemical Society of London (1849–1862),
Journal of the Chemical Society of London (1862–1877),
Journal of the Chemical Society of London, Transactions (1878–1925),
Journal of the Chemical Society of London, Abstracts (1878–1925),
Journal of the Chemical Society of London (1926–1965),
1965–1971 the journal was split into Section A (Inorganic, Physical, Theoretical), B (Physical Organic), C (Organic).

[b]Proceedings of the Prussian academy of sciences (Berlin): "Monatsberichte der Königlichen Preußischen Akademie der Wissenschaften zu Berlin" (1856–1881); succeeded by "Sitzungsberichte der Königlich Preußischen Akademie der Wissenschaften zu Berlin" (1882–1918); succeeded 1920 by "Sitzungsberichte der Preussischen Akademie der Wissenschaften" (*Other:* "Jahrbuch der Preussischen Akademie der Wissenschaften zu Berlin"); succeeded by "Deutsche Akademie der Wissenschaften zu Berlin. Mathematisch-Naturwissenschaftliche Klasse. Sitzungsberichte der Deutschen Akademie der Wissenschaften zu Berlin".

[c]Proceeding of the Bavarian academy of sciences (Munich): "Gelehrte Anzeigen" Vol. 1–50 (1835–860); 1860–1870 (no volume numeration) "Sitzungsberichte der Königl. Bayerischen Akademie der Wissenschaften zu München". In 1871 split into "Sitzungsberichte der Mathematisch-Physikalischen Classe der k. b. Akademie der Wissenschaften zu München" Vol. 1–38 (1871–1909). Succeeded by "Sitzungsberichte der Mathematisch-Physikalischen Klasse der Bayerischen Akademie der Wissenschaften zu München".

[d]The proceeding of the academy of sciences at Vienna (note "*Classe*" is written as "*Klasse*" since 1903): "Sitzungsberichte der Mathematisch-Naturwissenschaftliche Classe der Kaiserlichen Akademie der Wissenschaften" Vol. 1–42 (1848–1860), title changed in 1861 in "Sitzungsberichte der Kaiserlichen Akademie der Wissenschaften. Mathematisch-Naturwissenschaftliche Classe'. In 1861 split into "Sitzungsberichte der Kaiserlichen Akademie der Wissenschaften. Mathematisch-Naturwissenschaftliche Classe" Vol. 43–126 (1861–1917); "Abt. 1, Mineralogie, Botanik, Zoologie, Anatomie, Geologie und Paläontologie" (Vol. 97 in 1889: "Abt. 1 Mineralogie, Krystallographie, Botanik, Physiologie der Pflanzen, Zoologie, Paläontologie, Geologie, physische Geographie und Reisen"; and "Sitzungsberichte der Kaiserlichen Akademie der Wissenschaften. Mathematisch-Naturwissenschaftliche Classe. Abt. 2, Mathematik, Physik, Chemie, Physiologie, Meteorologie, physische Geographie und Astronomie". Succeded 1918 by "Sitzungsberichte / Akademie der Wissenschaften Wien" Vol. 127–156 (1918–1946). Succeeded in 1947 by "Sitzungsberichte / Österreichische Akademie der Wissenschaften in Wien Mathematisch-Naturwissenschaftliche Klasse, Abteilung 1 – Biologie, Mineralogie, Erdkunde und verwandte Wissenschaften" Vol. 157–200 (1947–1993).

In historical publications before 1900, citations are almost incomplete or even missing (often only the name of the person is mentioned). It is interesting that often no given name is printed but M. (*monsieur*) in French and H. or Hr. (*Herr*) in German is set before the family name.

Russia:
For transcription of Cyrillic to Latin different rules exist (e. g., ISO, DIN, GOST, BSI, ALA, Duden), which are partly different; only the Latin letters a, b, g, d, e, i, k, l, m, n, o, p, r, s, t, u, f, have an identical synonym with Cyrillic letters (а, б, г, д, е и, к, л, м, н, о, п, р, с, т, у, ф). International recommended is now the so-called scientific transcription. In popular literature (e. g., Wikipedia) is almost the English standard (BSI) or the German standard (Duden) used. The lately proposed scientific transcription is close to ISO, but not fully identical. Moreover, in literature, namely the older one, often "individual" transcription is used. Russian/Soviet scientists who published in English (or German in past) often used an "individual" transcription of his name that is adopted here (only *this* spelling should be accepted).[1365] However, there is also confusion for some names (notified in footnotes in this text). Some western scientists (namely in past), quoting Russian authors, used an "exotic" transcription, which makes it difficult and even impossible, namely for less known persons, to identify the name originally written in Cyrillic. Hence, in this volume the name of Russian/Soviet persons and journals are additional given in original Cyrillic. The following Cyrillic letters have different scientific transcriptions (in parenthesis German transcription and English if different): в = v (w), ё = ë (je), ж = ž (sh), з = z (s), х = ch (ch, but in English kh), ц = c (z), ч = č (tsch but in English ch), ш = š (sch, but in English sh), щ = šč (schtsch, but in English shch), э = è (e), whereas ы = y, й = j, ю = ju and я = ja only.

In the Soviet Union, scientific results mostly were published by institutions in "*Труды*" (+ name of the institute) [*Trudy* = *Transactions*]. Universities edited "*Вестник*" (+ name of the University) [*Vestnik* = *News*], Academies "*Доклады* or *Известия*" (+ name of the academy) [*Doklady* or *Isvestija* = *Annals*, *Reports* or *Communications*]; the name of the institution in reference is only given as acronym, for example "ЛГУ" or "МГУ" = University of Leningrad or Moscow (more exactly: State University of Leningrad), "АН" = academy of science (and added either СССР for USSR or the union republic "X" such as XССР). In historic literature, *Доклады АН СССР* is often referred as *C. r. Acad. URSS* and *Известия АН СССР* as *Bl. Acad. URSS*.

France:
The oldest chemical journal in France is "*Annales de Chimie*", founded by *Morveau*, *Lavoisier* and others in 1789; the title of this journal changed: "*Annales de chimie ou recueil de mémoires concernant la chimie et les arts qui en dependent*" (1789–1814), "*Annales de chimie et de physique*" (1815–1914), abbreviated as *Ann. chim. phys.*

Almost all publications in the field of atmospheric research in the nineteenth century were published in *C. r.*, *Compt. Rend.* or *Compte Rendu* (*Comptes rendus de l'Acadêmie des sciences*), complete title: "*Comptes rendus hebdomadaires des séances de l'Acadêmie des sciences*" (1835–1965), edited from the French Academy of Sciences.

A third French journal is important in our field of research: "*Journal de Pharmacie et de Chimie*" (1842–1942), abbreviated as *J. Pharm. Chim.* that appeared in nine series: sér. 1 Vol. 1–6 (1809–1814), sér. 2 Vol. 1–27 (1814–1841), sér. 3 Vol. 1–46 (1842–1864), sér. 4 Vol. 1–30 (1865–1879), and further in each 30 volumes over each 10 years from series 5–8 (1880–1939).

1365 For example, my Russian friend (*Alexey*) Рябошапко is internationally known with the name *Ryaboshapko* (BSI standard) but *Rjaboschapko* would be the German standard (Duden) and *Rjabošapko* the scientific standard (in a French publication I found *Ryaboschapko*, an unusual mixture of rules). The given name Александр is often in German transcript to *Alexander*, but also *Alexandr* (the British rule would be "Aleksandr", but never used for original Russian names); and finally, Петрович (*Petrovich* according to the British rules)] is only found in old literature as *Petrowitsch* (German transcription) which is generally valid for the ending "вич" ("witsch") instead of modern "vich" or "vič".

Germany:
The following journals were the most important in the field of atmospheric research in the nineteenth century.

One of the most prestigious journals (worldwide) was "*Annalen der Physik*," founded 1795 by *Friedrich Albrecht Carl Gren* (1760–1798), continued 1799–1824 by *Ludwig Wilhelm Gilbert* (1769–1824), 1824–1877 by *Johann Christian Poggendorff* (1796–1877), 1877–1899 by *Gustav Heinrich Wiedemann* (1826–1899) and 1899–1943 by *Wilhelm Wien* (1864–1928), *Max Planck* (1858–1947) and *Eduard Grüneisen* (1877–1949), all together in 435 volumes. The title of this journal changed over time and different citations are in use (note that citations *Ann. Phys.* use counting of the complete series but *Gilb. Ann.* and *Pogg. Ann.* counting of the subseries; e. g., *Pogg. Ann.* Vol. 24 = *Ann. Phys.* Vol. 100, and *Pogg. Ann.* Vol. 100 = *Ann. Phys.* Vol. 176; however, it becomes even more confusing due to subcounting in some citations, for example: *Ann. Phys.* (1853) = Vol. 88 = Vol. 164 *der ganzen Folge* [of the complete series] = Vol. 28 *der dritten Reihe* [of the 3rd sequence]):
- *Journal der Physik*, Vol. 1–8 (1795–1794), (*Grens Journal*) *Neues Journal der Physik*, Vol. 1–4 (1795–1797),
- *Annalen der Physik*, Vol. 5–30 (1799–1809),
- *Annalen der Physik, Neue Folge* (Vol. 1–30), Vol. 31–60 *der ganzen Folge* [of the complete series] (1809–1818),
- *Annalen der Physik, Neueste Folge* (Vol. 1–16) = *Annalen der Physik und physikalischen Chemie* (Vol. 61–76 *der ganzen Folge* [of the complete series] (1819–1824), 1799–1824 also termed *Gilberts Annalen* (*Gilb. Ann.*),
- *Annalen der Physik und Chemie* (Vol. 1–60), Vol. 77–236 *der ganzen Folge* [of the complete series] (1824–1877), 1824–1877 also termed *Poggendorffs Annalen* (*Pogg. Ann.*),
- *Annalen der Physik und Chemie* = *Annalen der Physik, Vierte Folge* (Vol. 1–69), Vol. 237–305 *der ganzen Folge* [of the complete series] (1877–1899), also termed *Wiedemanns Annalen* (*Wied. Ann.*),
- *Annalen der Physik, Vierte Folge* (Vol. 1–87), Vol. 306–392 *der ganzen Folge* [of the complete series] (1900–1928),
- *Annalen der Physik, Fünfte Folge* (Vol. 1–42), Vol. 393–435 *der ganzen Folge* [of the complete series] (1929–1943).

Another important German Journal is "*Journal für Chemie und Physik*" (1798–1833) that appeared with different titles, founded in 1798 by *Alexander Nicolas Scherer* (1777–1824), 1803–1810 continued by *Adolph Ferdinand Gehlen* (1775–1815) and 1811–1833 by *Johann Salomo Christoph Schweigger* (1779–1857); succeeded in 1834 by "*Journal für praktische Chemie*":
- *Allgemeines Journal für Chemie* (*Scherers Journal*) in 10 Vol. (1798–1803),
- *Neues allgemeines Journal der Chemie*, Vol. 1–6 (1803–1806),
- *Journal für die Chemie und Physik*, Vol. 1–3 (1806–1807),
- *Journal für die Chemie, Physik und Mineralogie*, Vol. 4–9 (1907–1810), also called *Gehlens Journal*,
- *Beiträge zur Chemie und Physik*, Vol. 1–9 (1811–1813),
- *Journal für Chemie und Physik*, Vol. 10–30 (1814–1820),
- *Journal für Chemie und Physik*, Vol. 31–60 = *Jahrbuch der Chemie und Physik*, Vol. 1–30 (1821–1830), and *Neues Journal für Chemie und Physik*, Vol. 61–69 – *Neues Jahrbuch der Chemie und Physik*, Vol. 1–8 (1830–1833), all also called *Schweiggers Journal*.
- *Journal für praktische Chemie* (*J. prakt. Chem.*) was 1828 founded by *Otto Linné Erdmann* (1804–1869) as *Journal für technische und ökonomische Chemie* and 1834 renamed in *Journal für praktische Chemie* (1834–1869 in 108 Vol.), and as *Neue Folge* continued from 1870–1943 (163 Vol.)

The "*Archiv der Pharmacie*" was founded in 1822 (until present) by *Rudolph Brandes* (1795–1842) and appeared also under different names to the present:

- *Archiv des Apothekervereins im nördlichen Teutschland* (*Br. Archiv*), Vol. 1–10 (1822–1824),
- *Archiv des Apotheker-Vereins im nördlichen Teutschland für Pharmacie und ihre Hülfswissenschaften* (*Br. Archiv*), Vol. 11–39 (1825–1834),
- *Archiv des Apotheker-Vereins im nördlichen Teutschland* (*als Beilage zu Annalen der Pharmacie des Justus Liebig*), Vol. 40–51 (1835–1838), Zweite Reihe Vol. 1
- *Archiv der Pharmacie, eine Zeitschrift des Apotheker-Vereins in Norddeutschland*, Vol. 68–114 (1839–1850),
- *Archiv der Pharmacie. Eine Zeitschrift des allgemeinen deutschen Apotheker-Vereins, Abtheilung Norddeutschland*, Vol. 115–200 (1851–1872), *Zweite Reihe* Vol. 65,
- *Archiv der Pharmacie*, Vol. 201–261 (1872–1923), *Dritte Reihe* Vol. 1.

The "*Archiv für die gesammte Naturlehre*" (*Kastner's Archiv*) was founded by the German chemist *Karl Wilhelm Gottlob Kastner* (1783–1857) and appeared in Vol. 1–18 (1824–1829) and as "*Archiv für Chemie und Meteorologie*" in Vol. 1–9 (1830–1835) as Vol. 19–27 of *Kastner's Archiv*.

Another important journal concerns contributions on atmospheric composition is "*Annalen der Chemie*," founded as *Annalen der Pharmacie* in 1832 by *Rudolph Brandes* and continued by *Justus von Liebig;* from Vol. 11 (1834) combined with "*Trommsdorffs Neuem Journal der Pharmacie*" (in old references called *Ann.* or *Liebigs Ann.*):

- *Annalen der Pharmacie* (1832–1839) Vol. 1–32,
- *Annalen der Chemie und Pharmacie* (1840–1872) Vol. 33–164,
- *Justus Liebig's Annalen der Chemie und Pharmacie* (1873–1874) Vol. 165–174,
- *Justus Liebig's Annalen der Chemie* (1875–1944) Vol. 175–556.

Further important sources of historical works on chemical constituents in air are German agricultural journals (see also Wulff (1920, pp. 90–92)).

- *Annalen der Landwirthschaft in den Königlich Preußischen Staaten* (in 58 Volumes 1843–1871). Wiegandt & Hempel, Berlin. Succeeded by *Landwirtschaftliche Jahrbücher.*
- *Landwirtschaftliche Jahrbücher. Zeitschrift für wissenschaftliche Landwirtschaft und Archiv des Königlich Preussischen Landes-Ökonomie-Kollegiums*; also cites as *Thiel's Landwirthschaftliche Jahresbücher*[1366] (in 93 Volumes 1872–1943/1944). Paul Parey, Berlin.
- *Jahresbericht über die Fortschritte der Agricultur-Chemie* (Vol. 1–6, 1860–1865). Springer, Berlin. Succeeded by *Jahresbericht über die Fortschritte auf dem Gesammtgebiete der Agricultur-Chemie* (1866–1919): Vol. 7 (1866) – 20 (1877); *Neue Folge* (N. F.) Vol. 1 (1878) – 21 (1897), and 3. *Folge* Vol. 1 (1898) – 41 (1919).[1367] Succeeded by *Jahresbericht für Agrikulturchemie* (1920–1936). Paul Parey, Berlin.
- *Forschungen auf dem Gebiet der Agricultur-Physik*; also cited in literature as "*Wollny's Forschungen*" (in 20 Volumes 1878–1898). Carl Winter's Universitätsbuchhandlung, Heidelberg.
- *Landwirthschaftliches Centralblatt für Deutschland. Repertorium der wissenschaftlichen Forschungen und praktischen Erfahrungen im Gebiete der Landwirthschaft* (in 24 Volumes 1853–1876). Wiegandt, Hempel & Parey, Berlin.

1366 *Carl Emil Hugo Thiel* (1839–1918), Ministerial Director in the Prussian Ministry for Agriculture.

1367 The 3rd series contains: Jahresbericht über die Fortschritte der Chemie des Bodens und der Luft (Meteorologie, Gewässer) und der Pflanze; Vol. 2: Jahresbericht über die Fortschritte der Chemie der Thierernährung und der landwirtschaftlichen Nebengewerbe; Vol. 3: Jahresbericht über die Fortschritte der Chemie des Düngers.

And finally, the most important source of chemical knowledge and likely the world's most important journal in chemistry in the nineteenth century is "*Berichte der Deutschen Chemischen Gesellschaft*" (1868–1919), abbreviated often as *B.* or *Ber.*

Other German abbreviations:

Abt.	*Abteilung* [Section]
Ann.	*Annalen*
Abh.	*Abhandlungen* [Transactions]
Ber.	*Berichte* [Proceedings]
Ges.	*gesamte* [total]
Nachr.	*Nachrichten* [News]
Z.	*Zeitschrift* [Journal]

Abbot, C. G. and F. E. Fowle (1908) Recent determination of the solar constant of radiation. *JGR* **13**, 79–82.

Abeledo, C. A. and I. M. Kolthoff (1932) The reaction between nitrite and iodide and its application to the iodometric titrations of these anion. *J. Am. Chem. Soc.* **53**, 2893–2897.

Abian, M., M. U. Alzueta and P. Glarborg (2015) Formation of NO from N_2/O_2 mixtures in a flow reactor: Toward an accurate prediction of thermal NO. *Journal of Chemical Kinetics* **47**, 518–532.

Acker, K. and D. Möller (2007) Atmospheric variation of nitrous acid at different sites in Europe. *Environmental Chemistry* **4**, 242–255.

Acker, K., W. Wieprecht, D. Möller, G. Mauersberger, S. Naumann and A. Oestreich (1995) Evidence of ozone destruction in clouds. *Naturwissenschaften* **82**, 86–89.

Acker, K., D. Möller, W. Wieprecht and S. Naumann (1996) Mt. Brocken, a site for a cloud chemistry programme in Central Europe. *Water, Air and Soil Pollution* **85**, 1979–1984.

Acker, K., D. Möller, W. Marquardt, E. Brüggemann, W. Wieprecht, R. Auel and D. Kalaß (1998a) Atmospheric research program for studying changing emission pattern after German Unification. *Atmos. Env.* **32**, 3435–3443.

Acker, K., D. Möller, W. Wieprecht, D. Kalaß and R. Auel (1998b) Investigation of ground-based clouds at Mt. *Brocken. Z. anal. Chem.* **361**, 59–64.

Acker, K., W. Wieprecht D, R. Möller, D. Auel, D. Kalaß, A. Zwozdziak, J. Zwozdziak and G. Kmiec (1999a) Results of cloud water and air chemistry measurements at Mt. Szrenica in Poland. In: Proceedings of EUROTRAC-2 Symposium '98, vol. 1. Transport and Chemical Transformation in the Troposphere (eds. P. M. Borrell and P. Borrell). WIT press Computational Mechanics Publications, Southampton, pp. 448–452 (*see also*: *Transactions on Ecology and the Environment* **28**, 448–452). *See also*: Acker, K., W. Wieprecht, D. Möller, R. Auel, D. Kalaß, A. Zwozdziak, J. Zwozdziak and G. Kmiec (1999b) Results of cloud water and air chemistry measurements at three different sites in Europe. *Biology Bulletin*, 26, No 6, pp. 606–616, Official English Translation of Izvestiya Rossiskoi Akademii Nauk – Seriya Biologicheskaya.No 6 (Nov.-Dec. 1999), 736–747.

Acker, K., D. Möller, W. Wieprecht, R. Auel and D. Kalaß (1999b) Clouds – a source of HNO_2? *Geophysical Research Abstracts* Vol. 1, No. 2, p. 511.

Acker, K., W. Wieprecht, R. Auel, D. Kalaß and D. Möller (2000) Evidence for heterogeneous formation of nitrous acid on cloud droplets. *J. Aerosol. Sci.* **31**, S1, 352–353.

Acker, K., D. Möller, W. Wieprecht, R. Auel, D. Kalaß and W. Tscherwenka (2001) Nitrous and nitric acid measurements inside and outside of clouds at Mt. Brocken. *Water, Air and Soil Pollution* **130**, 331–336.

Acker, K., S. Mertes, D. Möller, W. Wieprecht, R. Auel and D. Kalaß (2002) Case study of cloud physical and chemical processes in low clouds at Mt. Brocken, Germany. *Atmos. Res.* **64**, 41–51.

Acker, K., W. Wieprecht and D. Möller (2003) Chemical and physical characterisation of low clouds: results from the ground-based cloud experiment FEBUKO. *Archives of Industrial Hygiene and Toxicology* **54**, 231–238.

Acker, K., D. Möller, W. Wieprecht, F. X. Meixner, B. Bohn, S. Gilge, C. Plass-Dülmer and H. Berresheim (2005) Strong daytime production of OH from HNO_2 at a rural mountain site. *Geophys. Res. Lett.* **33**, L02809, doi:10.1029/2005GL024643.

Acker, K., D. Beysens and D. Möller (2008a) Nitrite in dew, fog, cloud and rainwater – indicator for heterogeneous processes on surfaces. *Atmos. Res.* **87**, 200–212.

Acker, K., N. Kezele, L. Klasinc, D. Möller, G. Pehnec, G. Šorgo, W. Wieprecht and S. Žužul (2008b) Atmospheric H_2O_2 measurements and modeling campaign during summer 2004 in Zagreb, Croatia. *Atmos. Env.* **42**, 2530–2542.

Addison, W. (1853) Requirements in meteorological tables. *Association Medical Journal* **1**, 80.

Adel, A. (1938) Further detail in the rock-salt prismatic solar spectrum. *Astrophysical Journal* **88**, 186–188; *See also*: Note on the atmospheric oxides of nitrogen. Ibid. **90** (1939) 627 and **93** (1941) 509.

Adel, A. (1946) A possible source of atmospheric N_2O. *Science* **103**, 280; see also: Atmospheric nitrous oxide and the nitrogen cycle. *Astrophysical Journal* **113** (1950) 624.

Adel, A. (1952) Identification of carbon monoxide in the atmosphere above Flagstaff, Arizona. *Astrophysical Journal* **116**, 442–443.

Adelung, J. Chr. (1796) Grammatisch-kritisches Wörterbuch der Hochdeutschen Mundart (in 4 Vol., 7690 pp.), Band 2. Breitkopf und Härtel, Leipzig, p. 425.

Aderkas, O. [Адеркас] (1928) О содержании озона в атмосфере [On the amount of ozone in the atmosphere]. *Метеорологический вестник* [Meteorological Journal] **37**, 103–105.

Aikawa, M., T. Hiraki, M. Shoga and M. Tamaki (2001) Fog and precipitation chemistry at Mt. Rokko in Kobe, April 1997–March 1998. *Water, Air and Soil Pollution* **130**, 1517–1522.

Aitken, J. (1880) On dust, fogs and clouds. *Nature* **23**, 195–197.

Aitken, J. (1881) Dust and fogs. *Nature* **23**, 311–312; Dust fogs and clouds. *Nature* **23**, 384–385.

Aitken, J. (1883) On dust, fogs and clouds. *Trans. R. Soc. Edinburgh* **30**, 337–368.

Aitken, J. (1886) Mr. Aitken on dew. *Nature* **33**, 256–258.

Aitken, J. (1887) On dew. *Trans. R. Soc. Edinburgh* **33**, 9–64.

Aitken, J. (1888) On the number of dust particles in the atmosphere. *Nature* **37**, 428–430.

Aitken, J. (1889a) I.- On the number of dust particles in the atmosphere. *Proc. R. Soc. Edinb.* **35**, 1–19.

Aitken, J. (1889b) On improvements in the apparatus for counting the dust particles in the atmosphere. *Proc. R. Soc. Edinb.* **16**, 135–172.

Aitken, J. (1890) On the number of dust particles in the atmosphere of certain places in Great Britain and on the continent, with remarks on the relation between the amount of dust and meteorological phenomena. *Nature* **41**, 394–396.

Aitken, J. (1892) On a simple pocket dust-counter. *Proc. R. Soc. Edinb.* **18**, 39–52.

Aitken, J. (1894) III.- On the number of dust particles in the atmosphere. *Proc. R. Soc. Edinb.* **37**, 675–672.

Aitken, J. (1911) On some nuclei of cloudy condensation. Part II. *Proc. R. Soc. Edinb.* **31**, 478–498. *And*: Collected scientific papers of John Aitken (ed. G. Knott), Cambridge University Press, 1923, pp. 495–511.

Aitken, J. (1912) The sun as a fog producer. *Proc. R. Soc. Edinb.* **32**, 103–215, *and* Collected Papers, pp. 512–536.

Aitken, J. (1917) On some nuclei of cloudy condensation. Part III. *Proc. R. Soc. Edinb.* **37**, 215–245, *and* Collected Papers, pp. 545–572.

Akimoto, M. (1929) On the nature of rainfall in Osaka (in Japanese). *Journal of the Meteorological Society of Japan* **6**, 463–470.

Akimoto, H., K. Takagi and F. Sakamaki (1987) Photoenhancement of the nitrous acid formation in the surface reaction of nitrogen dioxide and water vapor: Extra radical source in smog chamber experiments. *International Journal of Chemical Kinetics* **19**, 539–551.

Albini, A. and V. Dichiarank (2009) The "belle èpoque" of photochemistry. *Photochemical and photobiological sciences* **2**, 248–254.

Albrecht, F. (1925) Wolkenuntersuchungen auf dem Hohen Sonnblick im Sommer 1924. 33. *Jahresbericht des Sonnblick-Vereins für das Jahr 1931.* J. Springer, Wien, 14–20.

Albu, H. W. and H. D. Graf von Schweinitz (1932) Über die Bildung von Dithionat bei der Oxydation wäßriger Sulfit-Lösungen (V. Mitteil. über die Autoxydation). *Ber.* **65**, 729–737.

Aleksandrov, N. N. and O. P. Petrenchuk [Александров, Петренчук] (1959) Метод измерения ядер конденсации в свободной атмосфере при самолетных зондированиях [Method to measure condensation nuclei in the free atmosphere using aircrafts]. *Труды ГГО* **93**, 81–87.

Aleksandrov, N. N. and O. P. Petrenchuk [Александров, Петренчук] (1962) Методика взятия проб облачной воды с самолета [Method for sampling cloud water by an aircraft]. *Труды ГГО* **134**, 126–130.

Allen, A. O. (1961) The radiation chemistry of water and aqueous solutions. Van Nostrand, New York, 204 pp.

Allmand, A. and L. Reeve (1926) The photochemical decomposition of aqueous oxalic acid solutions. *J. Chem. Soc.* **129**, 2834–2851.

Allmand, A. and J. Spinks (1929) Photosensitised decomposition of ozone. *Nature* **124**, 651.

Alt, E. (1916) Die Physik des Klimas. In: Handbuch der Balneologie, medizinischen Klimatologie und Balneographie, Band 1 (eds. E. Dietrich and S. Kaminer), Georg Thieme, Leipzig, pp. 423–503.

Altshuller, A. P. (1983) Review: natural volatile organic substances and their effects on the air quality in the eastern United States. *Environ. Sci. Techn.* **14**, 1337–1349.

Altshuller, A. P. and J. J. Bufalini (1965) Photochemical aspects of air pollution: a review. *Photochemistry and Photobiology* **4**, 97–146.

Altshuller, A. P. and J. J. Bufalini (1971) Photochemical aspects of air pollution: a review. *Environ. Sci. Techn.* **5**, 39–60.

Alway, F. J., A. W. Marsch and W. J. Methley (1937) Sufficiency of atmospheric sulfur for maximum crop yield. *Soil Science Society of America Proceedings* **2**, 229–238.

Ambarzumjan, V. A. [Амбарцумян] (1934) К вопросу о распределении озона в атмосфере [On the question of distribution of ozone in the atmosphere]. *Бюллетень Комиссии по исследованию Солнца* [Bulletin of the Commission on investigation of the Sun] **5–6**, 30–32.

Ambrook, Ch. (1877) Ozone. *Report of Colorado State Board of Health* **2**, 107–115. See also: Tables of ozone observations in Colorado, showing its relation with meteorology and altitude. Ibid., pp. 100–105.

Amelung, W. and H. Landsberg (1934) Kernzählungen in Freiluft und Zimmerluft. *Met. Z. Bioklimatische Beiblätter* **1**, 49–53.

Anastasio, C., B. C. Faust and C. J. Rao (1997) Aromatic carbonyl compounds as aqueous-phase photochemical sources of hydrogen peroxide in acidic sulfate aerosols, fogs and clouds. 1. Non-phenolic methoxybenzaldehydes and methoxyacetophenones with reductants (phenols). *Environ. Sci. Techn.* **31**, 218–232.

Anbar, M. and H. Taube (1954) Interaction of nitrous acid with hydrogen peroxide and with water. *J. Am. Chem. Soc.* **76**, 6243–6247.

Anderson, R. P. (1913) Ein tragbarer Petterson-Palmqvist-Apparat. *Zeitschrift Für Hygiene und Infektionskrankheiten* **53**, 549–555.

Anderson, V. G. (1915) The influence of weather conditions upon the amount of nitric acid and nitrous acid in rainfall near Melbourne. *Q. J. Roy. Meteor. Soc.* **41**, 99–122; *see also*: *Chemical News* **110**, No. 2849.

Andreae, M. O. (1980) Dimethylsulfoxide in marine and freshwaters. *Limnology and Oceanography* **25**, 1054–1063.

Andreae, M. O. (1985) The emission of sulfur to the remote atmosphere: Background paper. In: The biogeochemical cycling of sulfur and nitrogen in the remote atmosphere (eds. J. N. Galloway, R. J. Charlson, M. O. Andreae and H. Rodhe), Reidel, Dordrecht, pp. 5–25.

Andreae, M. O. (1990) Ocean-atmosphere interactions in the global biogeochemical sulfur cycle. *Marine Chemistry* **30**, 1–29.

Andreae, M. O. and H. Raemdock (1983) Dimethyl sulfide in the surface ocean and the marine atmosphere: a global view. *Science* **221**, 744–747.

Andreae, M. O. (2012) Biogeochemische Forschung am Kaiser-Wilhelm-/Max-Planck-Institut für Chemie. In: 100 Jahre Kaiser-Wilhelm/Max-Planck-Institut für Chemie (Otto-Hahn-Institut). Facetten seiner Geschichte (eds. H. Kant and C. Reinhardt), Archiv der Max-Planck-Gesellschaft, Berlin, pp. 133–185.

Andrée, S. A. (1894) Über die Kohlensäure in der Atmosphäre. Öfersigt af Kongl. *Vetenskaps-akademiens förhandlingar* **51**, 355–371. *See also*: *Forschungen auf dem Gebiet der Agricultur-Physik* (*Wollny's Forschungen*) **18**, 409–411.

Andrews, Th. (1855) On the constitution and properties of ozone. *The Chemical Gazette* **13**, 339–340. See also: Ueber Ozon. *Ann.* **97** (1855) 371–373.

Andrews, Th. (1856) On the constitution and properties of ozone. *Phil. Trans.* **146**, 1–13. See also: Ueber die Beschaffenheit und Eigenschaften des Ozons. *Pogg. Ann.* **98**, 435–454.

Andrews, Th. (1867) On the identity of the body in the atmosphere which decomposes iodide of potassium with ozone. *Proc. R. Soc. Lond.* **16**, 63–64.

Andrews, Th. (1874) Ozone. *Nature* **9**, 349–349 and 364–366. See also: Ueber das Ozon. *Pogg. Ann.* **152**, 311–331.

Andrews, Th. and P. G. Tait (1858) Sur la densité de ozone. *Ann. chim. phys.* **52**, 333–334.

Andrews, Th. and P. G. Tait (1860) On the volumetric relations of ozone and the action of the electrical discharge on oxygen and other gases. *Phil. Trans.* **150**, 113–131.

Anfossi, D. and S. Sandroni (1997) Ozone levels in Paris one century ago. *Atmos. Env.* **20**, 3481–3482.

Anfossi, D., S. Sandroni and S. Viarengo (1991) Tropospheric ozone in the nineteenth century: the Moncalieri series. *JGR* **96**, 17349–17352.

Ångström, A. and L. Högberg (1952) On the content of nitrogen (NH_4-N and NO_3-N) in atmospheric precipitation. *Tellus* **4**, 31–42.

Ångström, K. (1904) Die Ozonbänder des Sonnenspektrums und die Bedeutung derselben für die Ausstrahlung. *Arkiv för matematik, astronomi och fysik / utgifvet af K. Svenska Vetenskaps-Akademien* **1**, 395–400.

Anklin, M. and R. C. Bales (1997) Recent increase in H_2O_2 concentration at Summit, Greenland. *JGR* **102**, 19099–19104.

Anonymous (1693) The Extract of a Letter from Lisle in Flanders, May 25. N. S. 1686. Giving an Account of an Unusual Storm of Hail which Fell There. *Phil. Trans.* **17**, 858.

Anonymous (1755) Eines Engländers Auszug aus Hermann Boerhaave Anfangs-Gründen der Chimie, übersetzt von *** Medic. Doctor. J. Chr. Richter, Hannover, 588 pp.

Anonymous (1799) IV. Progress of Dr. Mitchill's mind in investigating the cause of the pestilential distempers which visit the cities of America in Summer and Autumn. Being a development of his theory of pestilential fluids, as published to the world in 1795 and the succeeding years. *Phil. Mag.* **4**, 35–43 and 132–139.

Anonymous (1821) Neues merkwürdiges Beyspiel eines Raupenregens. *Allgemeiner Anzeiger der Deutschen* **62**(312), 3373–3377.

Anonymous (1822) Verdunklung der Luft und schwarzer Regen in Kanada, und Seiden-Regen in Brasilien. *Gilb. Ann.* **67**, 218–219.

Anonymous (1823) On the muriatic acid in the air of the atmosphere. *Annals of Philosophy, New Series* **6**, 25–29.
Anonymous (1835a) Versuche über die atmosphärische Luft zu London. *Arch. Pharm.* **51**, 132.
Anonymous (1835b) Ueber das Bleichen an der Sonne. *Allgemeine Polytechnische Zeitung* p. 156.
Anonymous (1845) Neue Bleichart ohne Lauge, Seife, Licht, Chlor und Säuren. *J. prakt. Chem.* **35**, 191–192.
Anonymous (1847) The dew-drop and the mist, or, an account of the nature, properties, dangers and uses of dew and mist, in various parts of the world. London, Soc. for Promot. Christian Knowledge. 115 pp.
Anonymous (1853) Ozone: Dr. Schönbein's ozonometer. *Association Medical Journal* **1**, 763.
Anonymous (1855) Local reports of epidemic and endemic diseases during the months of June, July and August 1855. *Journal of Public Health and Sanitary Review* **1**, 299–308. During the Months of Dec. 1854 and Jan. and Feb. 1855, pp. 55–61.
Anonymous (1856) Übersicht der Beobachtungen des Ozongehalts der Luft im Jahr 1855; zu Krakau im Jahre 1855; Nachträge und Verbesserungen. *Sitzungsber. Wien* **20**, 298–300.
Anonymous (1871a) Air and some of its impurities. In: Public Documents of Massachusetts: Being the Annual Reports of various Public Officers and Institution for the year 1870. Vol. IV, Nos. 27 to 37, Wright & Potter, Boston, pp. 296–408.
Anonymous (1871b) Ueber den Gehalt des Regenwassers an Ammoniak und Salpetersäure. In: *Jahresbericht über die Fortschritte auf dem Gesamtgebiete der Agrikultur-Chemie*. Die Jahre 1868 und 1869. Elfter und Zwölfter Jahrgang: Die Jahre 1868 und 1869. Mit einem Vollständigen Sach- und Namen-Register. Verlag von Julius Springer, Berlin, pp. 150–154. *See also*: 1866, 67–71 and 1867, 58–62.
Anonymous (1872) The formation of ozone by flowers. *The American Naturalist* **6**, 428.
Anonymous (1873) Report on ozone observations. Conducted under the supervision of a Committee appointed by the Council on 14th January 1869. *Journal of the Scottish Meteorological Society, new series* **3**, 240–244.
Anonymous (1876) Verunreinigung der Atmosphäre durch Fabriken und Gewerbe. *Polytechnisches Journal (Dingler's J.)* **220**, 87–89.
Anonymous (1881a) Die chemische Beschaffenheit der Luft in Brandenburg a. H.; ein Beitrag zur Kenntnis der quantitativen Zusammensetzung der atmosphärischen Luft. *Deutsche Medicinal-Zeitung* **7**, 495–461.
Anonymous (1881b) Organization of the meteorological service in some of the principal countries of Europa. *Symons's Monthly Meteorological Magazine* **182**, 33–39.
Anonymous (1889) Societies and Academies. *Nature* **39**, 310–311.
Anonymous (1890) Apparate zur Erkennung des Kohlensäuregehaltes der Luft. *Polytechnisches Journal (Dingler's J.)* **276**, 302–303.
Anonymous (1892) The Proposed Observatory on Mt. Blanc. *Publications of the Astronomical Society of the Pacific* **4**, 181–183.
Anonymous (1895) Gegen die Rußplage. *Die Gartenlaube* **11**, Heft 23, 258.
Anonymous (1898) Vingt annés d'observations ozonométriqes, Ciel et terre: bulletin de la Société Belge d'Astronomie. *de Météorologie et de Physique du Globe* **19**, 191–196.
Anonymous (1902a) Bericht über die internationale Expertenkonferenz für Wetterschießen in Graz. In: Jahrbuch der k. k. Zentralanstalt für Meteorologie und Erdmagnetismus. Bd. 39, Anhang, 154 pp.
Anonymous (1902b) Pamphlets relative to Wetterschiessen. Monthly Weather Review, January, p. 33.
Anonymous (1903) International Conference on Weather-Shooting. *Nature* **67**, 213.
Anonymous (1907) Analysis of rain water. *Statistical Year-Book of the City of Buenos Aires* **17**, 50–52 and **16** (1906) 3–8.

Anonymous (1913) Les Classiques de la Science. III. Eau Oxygénée et Ozone (Memoirs de Thénard, Schoenbein, De Marignac, Soret, Troost, Hautefeulle, Chappius). Libr. Armand Colin, Paris, 111 pp.

Anonymous (1931) Our Bookshelf.: Geology. *Nature* **127**, 368–369.

Anthes, R. A., H. A. Panofsky, J. J. Cahir and A. Rango (1975) The atmosphere. Charles E. Merril Publ. Comp., Columbus (USA), 339 pp.

Anthon, E. F. (1833) Handwörterbuch der chemisch-pharmazeutischen, technisch-chemischen und pharmakognostischen Nomenklaturen: oder, Uebersicht aller lateinischen, deutschen und französischen Benennungen sämmtlicher chemischen Präparate des Handels und sämmtlicher rohen Arzneistoffe für Aerzte, Apotheker und Droguisten. J. L. Schrag, Nürnberg, 724 pp.

ApSimon, H., ed. (1997) Emission Abatement Strategies and the Environment (EASE), final EU project report, Imperial College, London.

Arago, F. (1857) Populäre Astronomie. Band 4, Wigand, Leipzig, 693 pp.; Astronomie populaire. Gide et J. Bandes, Paris, 1854–1857 in 4 Vol.

Archenhold, F. S. (1913) Vorrichtung zum Niederschlagen von atmosphärischem Nebel. Patentschrift Nr. 306293 vom 25.4.1913.

Ardon, M. (1965) Oxygen: elementary forms and hydrogen peroxide. W. A. Benjamin, New York, 106 pp.

Aristoteles (1923) The works of Aristotle. Translated into English. Meteorologica by E. W. Webster, Clarendon Press, Oxford. *See also*: Aristoteles Werke in deutscher Übersetzung. Teil 1 Meteorologie, Teil II Über die Welt. Übersetzt von Hans Strom. Wissenschaftliche Buchgesellschaft, Darmstadt, 1970, 352 pp. Aristotelis Meteorologica. Ex recensione Immanuelis Bekkeri. Reimeri, Berolini, 1829, 123 pp. Aristotelis Meteoroloicorum Libri Quattuor. Recensuit indicem verborum addidit F. H. Fobes. Georg Olms Verlagsbuchhandlung, Hildesheim, 1967, 235 pp.

Armstrong, G. F. (1879) On the variations in the amount of carbon dioxide in the air. *Proc. R. Soc. Lond.* **30**, 343–355.

Arnts, R. R., W. B. Petersen, R. L. Seila and B. W. Jr. Gay (1982) Estimates of alpha-pinene emissions from a loblolly pine forest using an atmospheric diffusion model. *Atmos. Env.* **16**, 2127–2137.

Arrhenius, S. (1896) On the influence of carbonic acid in the air upon the temperature of the ground. *Phil. Mag.* **41**, 237–276.

Arrhenius, S. (1903) Lehrbuch der kosmischen Physik. In zwei Teilen. S. Hirzel, Leipzig, 1026 pp.

Artz, R. S. and R. F. Lavrinenko (1993) Background precipitation monitoring in the Soviet Union. *Tellus* **26**, 1–25.

Ascher, L. (1907) Die Bekämpfung des Rauches in Königsberg vom gesundheitlichen Standpunkte. *Schriften der physikalisch-ökonomischen Gesellschaft in Königsberg in Preussen* **48**, 124–137.

Ascherson, P. (1877) Neue Beobachtungen über Ozon in der Luft der Libyschen Wüste. *Sitzungsber. München* **7**, 77–89.

Ashworth, J. R. (1933) Smoke and the atmosphere. Studies from a factory town. Manchester Univ. Press, 131 pp.

Asman, W. A. H. and L. A. Conrads (1975) Bibliography on precipitation chemistry: subject index. Rapport, Instituut voor Meteorologie en Oceanografie van de Rijksuniversiteit Utrecht.

Aßmann, R. (1885) Mikroskopische Beobachtung der Wolken-Elemente auf dem Brocken. *Met. Z.* **2**, 41–47.

Atkinson, R. (1986) kinetics and mechanisms of gas phase reactions of the hydroxyl radical with organic compoubnds under atmospheric conditions. *Chemical Review* **86**, 69–201.

Atkinson, R. and W. P. L. Carter (1984) Kinetics and mechanisms of the gas-phase reactions of ozone with organic compounds under atmospheric conditions. *Chemical Review* **84**, 37–470.

Atmannspacher, W. (1976) Über Extremwerte des natürlichen und anthropogenen bodennahen Ozons. *Meteorologische Rundschau* **29**, 33–38.

Atmannspacher, W., R. Hartmannsgruber and P. Lang (1984) Langzeittendenzen des Ozons der Atmosphäre aufgrund der 1967 begonnenen Ozonmessreihen am Meteorologischen Observatorium Hohenpeißenberg. *Meteorologische Rundschau* **37**, 193–199.
Attems, J. (1787) Verordnungen. *Salzburger Intelligenzblatt* vom 17. Februar (3. Stück), pp. 49–51.
Aubel, Edm. van (1909) Sur la production d'ozone sous l'influence de la lumière ultra-violette. *C. r.* **149**, 983–985.
Audubert, R. (1938) Die Emission von Strahlung bei chemischen Reaktionen. *Z. angew. Chem.* **51**, 153–163.
Auer, R. (1939) Ueber den täglichen Gang des Ozongehalts der bodennahen Luft. *Gerlands Beiträge zur Geophysik* **54**, 137–145.
Aufray, M. (1909) Richesse des eaux de pluie en acide azotique et en ammoniaque au Tonkin. *Bulletin Économique de l'Indochine* **12**, 595–616.
Aulie, R. P. (1970) Boussingault and the nitrogen cycle. *Proceedings of the American Philosophical Society* **114**, 435–479.
Ayers, G. P. and J. L. Grass (1980) Ammonia gas concentration over the Southern Ocean. *Nature* **284**, 539–540.
Ayers, G. P. and J. P. Ivey (1988) Precipitation composition at Cape Grim, 1977–1985. *Tellus* **40B**, 297–307.
Ayers, G. P., S. A. Penkett, R. W. Gillett, B. Bandy, I. E. Galbally, C. P. Meyer, C. M. Eisworth, S. T. Bentley and B. W. Forgan (1996) The annual cycle of peroxides and ozone in marine air at Cape Grim. *J. Atmos. Chem.* **23**, 221–252.
Babo, L. von (1861) Apparat zur Darstellung von Ozone. *Berichte der Naturforschenden Gesellschaft zu Freiburg i. Br.* **2, 3**, 331–334.
Babo, L. von (1863) Beiträge zur Kenntnis des Ozons. *Annalen der Chemie und Pharmacie, II.* Supplementband, 3. Heft, pp. 265–296.
Babo, L. von (1865) Beiträge zur Kenntnis des Ozons. *Annalen der Chemie und Pharmacie*, Supplementband II, 265–296.
Bach, A. (1893) Ueber die Herstammung des Wasserstoffhyperoxyds der atmosphärischen Luft und der atmosphärischen Niederschläge. *Ber.* **27**, 340–344.
Bach, A. (1897) Du rôle des peroxydes dans les phénomènes d'oxydation lente. *C. r.* **124**, 951–954.
Bach, A. (1902) Hydrotetroxyd und Ozonsäure. *Ber.* **35**, 3424–3425.
Bach, A. and R. Chodat (1902) Untersuchungen über die Rolle der Peroxide in der Chemie der lebenden Zelle. Erste Mitteilung: Ueber das Verhalten der lebenden Zelle gegen Hydroperoxyd. *Ber.* **35**, 240–262.
Bäckström, H. L. J. (1927) The chain-reaction theory of negative catalysis. *J. Am. Chem. Soc.* **49**, 1460–1472.
Bäckström, H. L. J. (1934) Der Kettenmechanismus bei der Autoxydation von Natriumsulfitlösungen. *Z. phys. Chem.* **25B**, 122–138.
Baeyer, A. and V. Villiger (1900) Ueber die Nomenclatur der Superoxyde und die Superoxyde der Aldehyde. *Ber.* **33**, 2479–2487.
Baeyer, A. and V. Villiger (1902) Ueber Ozonsäure. *Ber.* **35**, 3038–3039.
Bailey, G. H. (1893) The atmosphere of Manchester and Salford. Second Report of the Air Analysis Committee. *Report and proceedings of The Manchester Field Naturalists' and Archaeologists' Society* pp. 73–98. See also: Some aspects of town air as contrasted with that of the country. *Literary and Philosophical Society of Manchester, Memoirs* 4th series **8** (1893) 11–17. *Furthermore*: The air of large towns. *Science* **22** (1893) 197–198.
Baillaud, J. (1903) Rapport sur la conférence internationale des experts sur le tir contre la grêle à Graz. *Journal de Physique Théorique et Appliquée* **2**, 915–916. *See also*: Bericht über die internationale Expertenconferenz für Wetterschiessen in Graz. Jahrbücher der K. K. Central-Anstalt für Meteorologie und Erdmagnetismus, 1902.

Bainbridge, A. E. and L. E. Heidt (1966) Measurement of methane in the troposphere and lower stratosphere. *Tellus* **18**, 221–224.

Balakov, V. V., V. G. Vafidi and S. G. Krivich [Балаков, Вафиди, Кривич] (1936) Измерение концентрации атмосферного озона методом поглощения ультрафиолетовых лучей [Measurement of the concentration of atmospheric ozone based on ultra-violet absorption]. In: Труды Эльбрусской экспедиции АН СССР и ВИЭМ, 1934 и 1935 г. [Transactions of the Elbrus Expedition of the Academy of Sciences of the USSR and All-Union Institute of Experimental Medicine, 1934 and 1935], Изд-во АН СССР, Moscow and Leningrad, pp. 109–116.

Balasubramanian, R. and L. Husain (1997) Observations of gas-phase hydrogen peroxide at an elevated rural site in New York. *JGR* **102**, 21,209–21,220.

Baly, E. C. C., I. M. Heilbron and W. F. Barker (1921) Photocatalysis. Part I. The synthesis of formaldehyde and carbohydrates from carbon dioxide and water. *J. Chem. Soc. Transactions* **119**, 1025–1035.

Bamber, M. K. (1900) Report on Ceylon tea soils and their effects on the quality of tea. Tea Research Institute, Ceylon, 7 pp. (cited after Miller 1905a).

Bandelier, Ad. F. (1868) Observations of ozone made in Hughland, Madison Co., Ills. *The Transactions of the Academy of St. Louis* 1861–1868, **2**, 417–418.

Barb, W. G., J. H. Baxendale, P. George and K. R. Hargrave (1949) Reactions of ferrous and ferric ions with hydrogen peroxide. *Nature* **163**, 692–694.

Barbier, D., D. Chalonge and E. Vassy (1935) Mesure de l'épaisseur réduite de l'ozone atmosphérique pendant l'hiver polaire. *C. r.* **201**, 787–789.

Barbier, D. and D. Chalonge (1939) Recherches sur l'ozone atmospherique. *Journal de Physique et le Radium* **10**, 113–123.

Barker, T. H. (1853) Medical meteorology. *Association Medical Journal* **1**, 1152–1153.

Barker, T. H. (1856) Remarks on the relative value of the ozonometer of Drs. Schönbein and Moffat, based upon daily observations at Bedford from December 1, 1854 to September 1. *Report of the Council of the British Meteorological Society* **6**, 29–46. 1856.

Barkow, E. (1907) Versuche über Entstehung von Nebel bei Wasserdampf und einigen anderen Dämpfen. *Ann. Phys.* **328**, 317–344 (Dissertation Marburg 1906).

Barral, J.-A. (1852) Premier Mémoire sur les eaux de pluie recueillies à l'Observatoire de Paris. *C. r.* **34**, 283–284; *and*: Deuxième Mémoire sur les eaux de pluie recueillies à l'Observatoire de Paris (i^er semestre 1852). **35**, 427–431. *See also*: Gehalt des zu Paris und Lyon gesammelten Regenwassers an fremden Substanzen. *Pogg. Ann.* **862** (1852) 332–334. Ueber das auf dem Observatorium zu Paris gesammelte Regenwaser. *Archiv der Pharmacie* **123** (1852) 192–194, *and*: Ueber das Regenwasser zu Paris. Ibid., p. 305.

Barral, J.-A. (1860) mémoire sur la présence des matières phosphorées dans l'atmosphère. *C. r.* **51**, 769–773.

Barrett, E. and G. Brodin (1955) The acidity in Scandinavian precipitation. *Tellus* **7**, 251–257.

Barth, M. C., D. A. Hegg, P. V. Hobbs, J. G. Walega, G. L. Kok, B. G. Heikes and A. L. Lazrus (1989) Measurement of atmospheric gas-phase and aqueous-phase hydrogen peroxide concentration in winter on the east coast of the United States. *Tellus* **41B**, 61–69.

Bartlett, F. W. (1883) The practical utilization of ozone in the treatment of diseases. *Buffalo Medical and Surgical Journal* **23**, 1–14.

Basile, G. (1895) Analisi delle acque meteoriche cadute a Catania da tuto giugno 1888 a tutto settembre 1889. *Staz. Sper. Agrar. Ital. (Le Stazioni Sperimentali Agrarie Italiane)* **28**, 545–574.

Baskerville, C. (1910) The smoke problem and the community. *Journal of Industrial and Engineering Chemistry* **2**, 355–359.

Bastelaer, D. A. Van (1893) L'ozone atmosphérique et l'ozonométrie en Belgique: résume de six années d'observations faites en 1886, 1887, 1888, 1889, 1890, et 1891 dans toutes les diverses stations. *Ciel et terre: bulletin de la Société Belge d'Astronomie, de Météorologie et de Physique du Globe* **13**, 509–519. *See also*: *Bulletin de l'Académie royale de médecine de Belgique* **4** (1892) 671–708. Ibid. (1894) 617–641.

Bates, D. R. (1952) Some reactions occurring in the earth's upper atmosphere. *Annals of Geophysics* **8**, 194–202.

Bates, D. R. and M. Nicolet (1950) The photochemistry of atmospheric water vapor. *JGR* **55**, 301–327.

Bates, D. R. and A. E. Witherspoon (1952) The photochemistry of some minor constituents of the Earth's atmosphere. *Monthly Notices of the Royal Astronomical Society* **112**, 101–112.

Bates, J. R. (1933) The Mechanism of the reaction of hydrogen atoms with oxygen. *J. Chem. Phys.* **1**, 456–465.

Battle, M., S. M. Fletcher, et al.(2006) Atmospheric potential oxygen: new observations and their implications for some atmospheric and oceanic models. *Global Biogeochemical Cycles* **20**, GB1010.

Baubigny, H. (1912) Recherches sur le process de formation de l'acide dithionique dans l'action sulfites sur les sels de cuivre. *C. r.* **155**, 701–703.

Baudrimont, A. (1846) Recherches sur l'eau regale et sur un produit auquel elle doit ses principles propértiés. *Ann. chim. phys.* **17**, 24–42.

Bauer, A. (1895) Stickstoff und Argon. Vortrag, gehalten den 27. November 1895. *Schriften des Vereines zur Verbreitung naturwissenschaftlicher Kenntnisse in Wien*, XXXVI. Band, pp. 178–221.

Bauer, H. (1914) Geschichte der Chemie. I. Von der ältesten Zeit bis Lavoisier. Sammlung Göschen, Berlin und Leipzig, 96 pp.

Bauer, H. (1921) Geschichte der Chemie. II. Von Lavoisier bis zur Gegenwart. Sammlung Göschen, Berlin und Leipzig, 1446 pp.

Bauer, J. (1912) Über das atmosphärische Ozon. *Met. Z.* **47**, 372–378.

Baumann, A. (1893) Chemie der Atmosphäre. In: *Jahresbericht über die Fortschritte auf dem Gesamtgebiete der Agricultur-Chemie. Neue Folge*, XV. 1892, Verlag Paul Parey, Berlin, pp. 3–18.

Baumert, M. (1853a) Chemische Untersuchung über die Respiration des Schlammpeizgers (Cobiti fossilis). *Ann.* **88**, 1–56.

Baumert, M. (1853b) Ueber eine neue Oxydationsstufe des Wasserstoffs und ihr Verhältniss zum Ozon. *Pogg. Ann.* **89**, 38–55. *See also*: *Ann.* **88** (1853) 221–224.

Baumert, M. (1856) Zur Ozonfrage. *Pogg. Ann.* **99**, 88–94.

Baumert, M. (1857) Ueber das electrolytisch entwickelte Ozon. *Ann.* **101**, 88–90.

Baumgardner, Jr., R. E., S. S. Isil, T. F. Lavery, C. M. Rogers and V. A. Mohnen (2003) Estimates of cloud water deposition at Mountain Acid Deposition Program Sites in the Appalachian Mountains. *Journal of the Air & Waste Management Association* **53**, 291–308.

Baumgartner, A. (1832) Untersuchung der atmosphärischen Luft und der Luft-Electricität in Wien während des epidemischen Brechdurchfalls. *Medicinische Jahrbücher des k. k. österreichen Staates* **12**, 83–88.

Baur, E. (1918) Photolyse und Elektrolyse. *Helvetica Chimica Acta* **1**, 186–201.

Baur, E. and A. Perret (1924) Über die Einwirkung von Licht auf gelöste Silbersalze in Gegenwart von Zinkoxyd. *Helvetica Chimica Acta* **7**, 910–915.

Baur, E. and C. Neuweiler (1927) Über photolytische Bildung von Hydroperoxyd. *Helvetica Chimica Acta* **10**, 901–907.

Baxendell, J. (1869) On observations of atmospheric ozone. *Proceedings of the Literary and Philosophical Society of Manchester* **8**, 21–32. Note the same paper appeared in: *Proceedings of the Literary and Philosophical Society of Manchester* **4** (3 Ser.), 191–202.

Baxendell, J. (1876) On a source of ozone. *Z. Öster. Ges. Met.* **15**, 113–121. *See also*: *Z. Öster. Ges. Met.* **13** (1878) 208.

Beard, G. M. (1874) Atmospheric electricity and ozone: their relation to health and diseases. *The Popular Science Monthly* **4**, 456–469.

Bebber, W. J. von (1895) Hygienische Meteorologie für Ärzte und Naturforscher. Enke, Stuttgart, 330 pp.

Bechi, E. (1873) Saggi di esperienze agrarie [Essays of agricultural experiences]. Florence. (1870–1876). *See also Abstracts*: *Ber.* (1873) 5, 292; **6**, 1203 and (1876) **9**, 346.

Beck, J. B. (1819) Observations on salt storms and the influence of salt and saline air upon animal and vegetable life. *American Journal of Science* **1**, 388–397.

Beck, J. P. and P. Grennfelt (1994) Estimate of ozone production and destruction over Northwestern Europe. *Atmos. Env.* **28**, 129–140.

Becker, K.-H. (1977) Einführung. In: Ozon und Begleitsubstanzen in photochemischen Smog. VDI-Berichte 270, VDI-Verlag GmbH, Düsseldorf, pp. 7–12.

Becker, K.-H. (1991) Bildung von Photooxidantien (Laboruntersuchungen). In: Proceedings Ozon-Symposium in München 1.-4. Juli 1991, TÜV Akademie Bayern/Hessen, 13 pp.

Becker, K.-H., J. Bechara and K. J. Brockmann (1993) Studies on the formation of H_2O_2 in the ozonolysis of alkenes. *Atmos. Env.* **27A**, 57–61.

Becker, K. H., B. Donner and S. Gäb (1999) BERLIOZ: A field experiment within the German Tropospheric Research Programs (TFS). In: Proceedings of EUROTRAC Symposium 98, vol. 2 (eds. P. M. Borrell and P. Borrell), WIT, Southhampton, UK, pp. 669–672.

Becker, K. H., B. Scherer, M. Memmesheimer and H. Geiss (2002) Studying the city plume of Berlin on 20 July 1998 with three different modelling approaches. *J. Atmos. Chem.* **42**, 41–70.

Becker, R. and W. Döring (1935) Kinetische Behandlung der Keimbildung in übersättigten Dämpfen. *Ann. Phys.* **416**, 719–752. Fünfte Folge, 24.

Beckham, L. J., W. A. Fessler and M. A. Kise (1951) Nitrosyl chloride. *Chemical Reviews* **48**, 319–396.

Becquerel, E. (1850) Relative aux expériences de M. Schönbein sur l'ozone. *C. r.* **30**, 13–16.

Begemann, C. (1864) Ueber den Ozongehalt der atmosphärischen Luft. *Arch. Pharm.* **158**, 1–9.

Behrens, H. (1873) Ueber Ozonwasser. Vierteljahresschrift für praktische. *Pharmacie* **22**, 230–231. *See also*: *Polytechnisches Journal* (*Dingler's J.*) **208** (1873) 78.

Beilke, S. and H.-W. Georgii (1968) Investigation on the incorporation of sulfur-dioxide into fog- and rain-droplets. *Tellus* **20**, 435–442.

Beilke, S., D. Lamb and J. Müller (1975) on the oxidation of SO_2 in aqueous solution. *Atmos. Env.* **9**, 1083–1090.

Bellucci, G. (1869) Sull' ozono: Note e riflessioni. Tipogr. Giachetti, Prato, 456 pp.

Bellucci, G. (1876) Sulla produzione dell'ozono durante la nebulizzazione dell'acqua. *Gazzetta Chimica Italiana* **6**, 88–97; *see also*: *Ber.* **9** (1876) 581.

Bellucci, G. (1888) Contribuzione allo studio delle acque meteoriche. *Staz. Sper. Agrar. Ital. (Le Stazioni Sperimentali Agrarie Italiane)* **14**, 255–258.

Bellucci, J. (1874) Sur la prétendu dégagement de l'ozone des plantes. *C. r.* **78**, 362–365. (note that the given name "J" likely is a misprint and "G" (Giuseppe) would be correct). See also: On the supposed disengagement of ozone from plants. *J. Chem. Soc.* **27** (1874) 596 (abstract).

Bence-Jones, H. (1851) Contributions to animal chemistry. Paper V. On the oxidation of ammonia in the human body, with some remarks on nitrification. *Phil. Trans.* **141**, 399–409.

Benedict, F. G. (1912) The composition of the atmosphere with special reference to its oxygen content. Publication by the Carnegie Institution, Washington, USA, 115 pp.

Beneke, F. W. (1855) Ueber die Wirkungen des Nordsee-Bades. *Allgemeine Medicinische Central-Zeitung* **24**, 489–490.

Beneke, F. W. (1858) Morbilitäts-Nachrichten aus dem Jahre 1855. *Archiv des Vereins für Gemeinschaftliche Arbeiten zur Förderung der Wissenschaftlichen Heilkunde (Marburg)* **3**, 173–256. And: Anhang zu den Morbititäs- und Mortalitäts-Nachrichten vom Jahre 1855, ibid. pp. 257–268.

Benoist, L. (1919) Réactif et méthode de dosage de l'ozone. *C. r.* **168**, 612–615.

Benrath, A. (1912) Lehrbuch der Photochemie. C. Winkler's Universitätsbuchhandlung, Heidelberg, 287 pp.

Bentley, R. (1906) The meteorology of daily life. *Q. J. Roy. Meteor. Soc.* **32**, 81–112.

Benton, D. J. and P. J. Moore (1970) Kinetics and mechanism of the formation and decay of peroxynitrous acid in perchloric acid solutions. *J. Chem. Soc. A* 3179–3182.

Beretta, U. and H.-J. Schumacher (1932) Photokinetik des Ozons. II. Der Zerfall des Ozons im ultravioletten Licht. *Z. phys. Chem., Abteilung B: Chemie der Elementarprozesse, Aufbau der Materie B* **17**, 417–428.

Berg, H. (1959) pH-Werte des Niederschlags in Köln. *Meteorologische Rundschau* **12**, 47–52.

Berger (1863) Ueber Nebel. *Pogg. Ann.* **118**, 456–471.

Bergerhoff, H. (1928) Untersuchung über die Berg- und Rauchschädenfrage mit besonderer Berücksichtigung des Ruhrgebietes. Dissertation Landwirtschaftliche Hochschule Bonn-Poppelsdorf. Bonner Universitätsbuchdruckerei Gebr. Scheur.

Bergeron, T. (1935) On the physics of cloud and precipitation. *Proceeding's 5th Assembly U. G. G. I.*, Lisbon, Vol. 2, pp. 156–175.

Bergey, D. H. (1896) Methods for the determination of organic matter in air. Volume 39. Smithsonian Institution, Washington, 28 pp.

Bergman, T. (1788) Physical and chemical essays. Translated from original Latin by by E. Cullen, Vol. 1, London (J. Murray), 467 pp.; Original: Opuscula Physica et Chemica. Upsala (1779), 411 pp. (in 5 Vol. 1779–1788).

Bérigny, A. (1855) Observations faites à l'observatoire météorologique de Versailles avec le papier dit ozonometrique de M. Schonbein's (de Bâle), pendant le mois d'aout 1855, à 6 heuires du matin, midi, 6 heures du soir et minuit. *C. r.* **41**, 426–428. *See also*: *Gazette Médicale de Paris* **25**, 3rd series, Vol 10, 600–601; *and*: *Allgemeine Medicinische Central-Zeitung* **25** (1856) 451.

Bérigny, A. (1856) Sur les observations ozonométriques faites avec le papier Schönbein, autour de la caserne Saint-Cloud et au milieu de la cour, à deux altitudes differentes, à sept heures et demie du matin et à sept heures et demie du soir, pendant trente et un jours (du 6 octobre au 5 novembre 1855 inclusivement), et les relations qui existent entre l'ozone et les phénoménes physiologique. *Annu. soc. mét. de France* **4**, 81–98.

Bérigny, A. (1857) Recherches et observations pratique sur la papier ozonométriques. *Annu. soc. mét. de France* **5**, 149–156. Same title (troisème Mémoire). *C. r.* 44 (1857) 1104–1108. Also published separately: H. et C. Noblet, Paris, 15 pp.

Bérigny, A. (1858a) Quatrième mémoire sur l'ozonometrie. *C. r.* **46**, 237–239.

Bérigny, A. (1858b) Sur les observations ozonométriques et météorologiques faites en Crimée au milieu dès campements et des ambulances. *C. r.* **47**, 947–949.

Bérigny, A. (1858c) Mémoire sur les observations ozonométrique rt météorologique faites en Crimèe, au milieu des campements at de ambulances, d'après les orders de M.le marèchal Vaillant, ministre de la Guerre, member de l'institut, du 7 mai au 4 juin 1856 inclusivement. *Annu. soc. mét. de France* **6**, 239–246.

Bérigny, A. (1859) Proportion de l'ozone avant, pendant et après la période de l'influence de l'aurore boréale du 28 au 29 août. *C. r.* **49**, 391–392.

Bérigny, A. (1860) Observations ozonométriques. *C. r.* **51**, 947.

Bérigny, A. (1867) Remarques sur les colorations ozonoscopiques obtenues à l'aide d'un réactif de Jame (de Sedan) et sur l'échelle ozonométriques. *C. r.* **65**, 708–711.

Bérigny, A. (1870) Ozone. Observations faites à Versailles, de 1855 à 1858. *Annu. soc. mét. de France* **18**, 25–28.

Bernstein, B. (2000) Pedagogy, Symbolic Control and Identity. Revised Edition. Rowman & Littlefield Publishers, 256 pp.

Bertels, C. (1842) Das Regen- und Schneewasser in Hinterpommern, chemisch untersucht. *J. prakt. Chem.* **26**, 89–96.

Berthelot, D. and H. Gaudechon (1910a) Effects chimiques des rayons ultraviolets sur les corps gazeux. Actions de polymerization. *C. r.* **150**, 1169–1672.
Berthelot, D. and H. Gaudechon (1910b) Décomposition photochimique des alcools, des aldehydes, des acides at des cétones. *C. r.* **151**, 478–481.
Berthelot, D. and H. Gaudechon (1911) La nitrification par les rayons ultraviolet. *C. r.* **152**, 522–524.
Berthelot, D. (1914) Sur les divers modes de photolyse de l'acide oxalique par les rayons ultraviolets de différentes longueurs d'onde. *C. r.* **158**, 1793–1795.
Berthelot, M. and G. André (1886) Sur la matiéres azotées contenue dans l'eau de pluie. *C. r.* **102**, 957. *See also*: Ueber die stickstoffhaltigen Substanzen im Regenwasser. *Forschungen auf dem Gebiet der Agricultur-Physik* (*Wollny's Forschungen*) 1886, **15**, 342–343.
Berthelot, M. (1874) Sur la chaleur dégagée dans les combinaisons de l'azote aved l'oxygène. *C. r.* **78**, 99–106.
Berthelot, M. (1876) Sur la formation thermique de l'ozone. *C. r.* **82**, 1281–1283.
Berthelot, M. (1877a) L'ozone se combine-t-il-avec l'azote libre en présence des alcalis, pour former des composés nitreux et des nitrates? *C. r.* **84**, 61–62. See also: Ann. chim. Phys. 12 (1877) 440–445.
Berthelot, M. (1877b) Sur la reaction entre l'azote et l'eau. *Ann. chim. Phys.* **12**, 445.
Berthelot, M. (1877c) Nouvelles recherches sur le phénomènes chimiques produits per l'électricité de tension. *Ann. chim. Phys.* **12**, 446–452.
Berthelot, M. (1878a) Sur la stabilité de l'ozone. *C. r.* **86**, 76–77. *See also*: *Ann. chim. phys.* **14** (1878) 361–363.
Berthelot, M. (1878b) Sur l'incompatibilité entre l'acid azoteus, l'ozone et l'oxygène humidide. *Ann. chim. phys.* **14**, 367–368.
Berthelot, M. (1879) Sur la chaleur de formation de l'ammonique. *C. r.* **89**, 877–883.
Berthelot, M. (1880) Observations sur la décomposition du permanganate de potasse par l'eau oxygénée. *C. r.* **90**, 656–660.
Berthelot, M. (1881) Observations sur l'acide perazotique. *C. r.* **92**, 80–82; *see also* **94** (1882), pp. 111 and 1114.
Berthollet, C. L. (1805) Einige Bemerkungen zu dem vorstehenden Auffsatze [see Humboldt and Gay-Lussac 1805]. *Gilb. Ann.* **20**, 93–98.
Berthollet, C.-L. (1788) Suite des recherches sur la nature des substances animales, et sur leurs rapports avec les substances végétables. *Histoire de l'Académie royale des sciences* [Mémoires] Année 1784, Paris, pp. 331–349.
Bertramson, G. R., M. Fried and S. L. Tisdale (1950) Sulfur studies of Indiana soils and crops. *Soil Science* **70**, 27–41.
Bertrand, G. (1935) A propos des apports atmosphérique de soufre combiné aux terres arables. *Comptes Rendus Acad. Agr. France* **21**, 1015–1018; *Congr. Chim. Ind. Bruxelles* **15II**, 823; Observations à propos des apports atmosphérique de soufre combiné aux terres arables. *C. r.* **201**, 309–312.
Berzelius, J. (1812) Die Salpetersäure und die salpetrige Säure, als Beweise, daß der Stickstoff nicht chemisch einfach ist. *Gilb. Ann.* **40**, 162–209.
Berzelius, J. (1813a) Experiments on the nature of azote, of hydrogen and of ammonia and upon the degrees of oxidation of which azote is susceptible. *Annals of Philosophy* **2**, 276–284, 357–368.
Berzelius, J. (1813b) On the nature of muriatic acid. *Annals of Philosophy* **2**, 254–259.
Berzelius, J. (1814) Essay on the cause of chemical proportions and on some circumstances relating to them: together with a short and easy method of expressing them. *Annals of Philosophy* **3**, 51–52, 93–106, 244–255, 353–364; and (1813) **2**, 443–454.
Berzelius, J. (1820) Lehrbuch der Chemie nach der zweyten schwedischen Originalausgabe und den eigenhändigen Zusätzen und Berichtigungen des Verfassers übersetzt und bearbeitet von K. A. Böde. Verlag Arnold, Dresden, 502 pp.

Berzelius, J. (1847) Ueber das Ozon. In: Auszüge aus Berzelius's Jahresbericht (26. Jahresgang). *J. Prakt. Chem.* **40**, 242–246.

Bessels, E. (1879) Die amerikanische Nordpol-Expedition. Engelmann, Leipzig, 648 pp.

Besson, M. (1928) La visibilité et la teneur de l'air en poussierére à Paris. *Le Géne Civil* Tom 92, No 14 (7.4.28), p. 344.

Best, N. W. (2015) Lavoisier's "Reflections on phlogiston" I: Against phlogiston theory". *Foundations of Chemistry* **17**, 137–151.

Best, N. W. (2016) Lavoisier's "Reflections on phlogiston" II: On the nature of heat. *Foundations of Chemistry* **18**, 3–13.

Betz, F. (1848) Bericht über die Beobachtungen auf Ozon im Monate Januar. *Medicinisches Correspondenzblatt des Württembergischen Ärztlichen Vereins* **18**, 29–31.

Betz, F. (1849) Bericht über die Beobachtungen auf Ozon im Monate Januar. *Medicinisches Correspondenzblatt des Württembergischen Ärztlichen Vereins* **19**, 29–31.

Beyreis, O., A. Heller and E. M. Bursche (1955) Außenlufthygiene. G. Fischer Verlag, Stuttgart, 83 pp.

Beysens, D. (1995) The formation of dew. *Atmos. Res.* **39**, 215–237.

Beysens, D. (2018) Dew water. River Publishers, Denmark, 305 pp.

Beysens, D., C. Ohayon, M. Muselli and O. Clus (2006) Chemical and biological characteristics of dew and rain water in an urban coastal area (Bordeaux, France). *Atmos. Env.* **40**, 3710–3723.

Bhattacharyya, P. K. and R. Veeraraghavan (1977) Reaction between nitrous acid and hydrogen peroxide in perchloric acid medium. *International Journal of Chemical Kinetics* **9**, 629–640.

Bieber, W. (1911) Untersuchungen über die Kondensation von Wasserdampf in Gegenwart von Ozon, Stickstoffoxiden und Wasserstoffsuperoxyd. Insbesondere über die Kerne des blauen Nebels. Dissertation, University Marburg, Enz. & Rudolph, Frankfurt a. M., 46 pp.

Bieber, W. (1912) Untersuchungen über die Kondensation von Wasserdampf in Gegenwart von Stickstoffoxyden und Wasserstoffperoxyd. Nachweis der Bildung von H_2O_2 durch Oxydation von Wasserdampf. Die Wirkung des ultravioletten Sonnenlichtes auf die Erdatmosphäre. *Ann. Phys.* **344** (*Vierte Folge* **39**), 1313–1337.

Biechle, M. (1903) Pharmazeutische Übungspräparate. Springer, Berlin, 310 pp.

Bieling, K. and J. Bleibaum (1934) Untersuchung über bodennahes Ozon nach der spektrographisch-photometrischen Methode im Thüringer Wald. *Der Balneologe* **1**, 283–284.

Billiard (1854) Cause secondaire du cholera-morbus. *C. r.* **39**, 1097.

Billings, J. S. (1893) Ventilation and heating. chapter IV: Carbonic Acid. The Engineering Record, New York, pp. 61–84.

Bineau, A. (1852) Recherches sur la composition chimique des eaux de pluie recueillies dans l'hiver de 1851–1852 à l'Observatoire de Lyon. *C. r.* **34**, 357–359. *See also*: Ueber die chemische Zusammensetzung der Regenwässer, welche im Winter 1851–1852 auf dem Observatorium von Lyon aufgefangen wurden. *J. prakt. Chem.* **55**, 476–478.

Bineau, A. (1854) Ètudes chimiques sur les eaux pluviales et sur l'atmosphère de Lyon et de quelques points des environs, pendant les années 1852 et 1853. *Annales de Chimie* **42**, 428–484.

Bineau, A. (1855a) Sur les eaux pluviales et sur l'atmosphère de Lyon, et quelques points des environs, pendant les années 1852 et 1853. *C. r.* **38**, 272–274; *and*: *Ann. Chim. Phys.* [iii] **42**, 428–484.

Bineau, A. (1855b) Note sur l'ozone atmosphérique. *Annales de sciences physique et naturelles, d'agriculture at d'industrie* **7**, 250–255.

Birckenbach, L. (1909) Die Untersuchungsmethoden des Wasserstoffperoxyds. VII. Band. Die chemische Analyse (ed. B. M. Margosches), F. Enke, Stuttgart, 142 pp.

Biswas, A. K. (1970) History of hydrology. North Holland, Amsterdam and London, 336 pp.

Bitter, H. (1890) Über Methoden zur Bestimmung des Kohlensäuregehaltes der Luft. *Zeitschrift Für Hygiene und Infektionskrankheiten* **9**, 1–40.

Blacet, F. E. (1952) Photochemistry in the lower atmosphere. *Industrial & Engineering Chemistry* **44**, 1339–1342.
Black, J. (1754) De humore acido a cibis orto, et magnesia alba. Doctoral Thesis; published a fuller account of them in 1756 as "Experiments upon magnesia, quicklime and some other alkaline substances". On acid humor arising from foods and on white magnesia: A translation of the Latin thesis De humore acido a cibis orto, et magnesia alba (1973) Bell Museum of Pathobiology (University of Minnesota Medical School), 40 pp.
Black, J. (1756) Experiments upon magnesia alba, quicklime and some other alkaline substances. Reprint by The Alembic Club, Edinburgh (1944), 46 pp.
Black, J. (1803) Lectures on elements of chemistry (Ed. by J. Robinson). Mundell & Son, Edinburgh, 762 pp.
Black, W. (1886) Ozone papers in towns. *Nature* **35**, 76.
Black, W. J. (1885) Ozone at sea. *Nature* **32**, 416.
Blake, W. P. (1855) On the grooving and polishing of hard rocks and minerals by dry sand. *The American Journal of Science and Arts (2nd ser.)* **70**, 178–181.
Blanchard, M. S. and S. F. Pickering (1927) A review of the literature relating to the normal densities. In: Scientific papers of the Bureau of Standard. Vol. 21. 1926–1927, US Govn. Printing Office, pp. 141–177.
Blanchard, D. C. and A. H. Woodcock (1957) Bubble formation and modification in the sea and its meteorological significance. *Tellus* **9**, 145–158.
Blanchard, D. C. (1984) The life and science of Alfred H. Woodcock. *Bulletin of the American Meteorological Society* **65**, 457–463. *See also*: Alfred H. Woodcock (1905–2005). *Eos* (Transactions of the American Geophysical Union) **86** (2011) 542.
Bleier, O. (1898) Ein tragbarer Apparat für hygienische Luftanalysen (Kohlensäure-bestimmung). *Zeitschrift Für Hygiene und Infektionskrankheiten* **27**, 111–115.
Blinov, L. K. [Блинов] (1950) О поступлении морских солей в атмосферу и о значении ветра в солевом балансе Каспийского моря [On emissions of sea salt and the importance of wind in the salt budget of the Caspian Sea]. *Труды Гос. Океаногр. Университета* [Transaction of the State Oceanographic University] **15**, 67–112.
Blochmann, R. (1884) Ueber ein einfaches Verfahren zur annähernden Bestimmung der Kohlensäure in der Luft bewohnter Räume und in anderen Gasgemischen. *Z. anal. Chem.* **23**, 333–345.
Blochmann, R. (1886) Ueber den Kohlensäuregehalt der atmosphärischen Luft. *Liebigs Ann.* **237**, 39–90.
Blücher, H. (1900) Die Luft. Ihre Zusammensetzung und Untersuchung, ihr Einfluss und ihre Wirkung sowie ihre technische Ausnutzung. Verlag Otto Wigand, Leipzig, 322 pp.
Blum, H. F. and C. R. Spealman (1933) Photochemistry of fluorescein dyes. *J. phys. Chem.* **37**, 1123–1133.
Blume, H.-P. and M. Bölter (2011) Christian G. Ehrenberg und die Geburt der Bodenmikrobiologie Mitte des 19. Jh. In: Persönlichkeiten der Bodenkunde II, Inst. für Pflanzenernährung und Bodenkunde, Universität Kiel, pp. 1–30.
Bobierre, A. (1864) Recherches sur la composition chimique de l'eau pluviale recueillie dans les villes à diverses altitudes. *C. r.* **58**, 755–757.
Bodenstein, M. (1897) Zersetzung und Bildung von Jodwasserstoff. *Z. phys. Chem.* **22U**, 1–22.
Bodenstein, M. (1913) Photochemische Kinetik des Chlorknallgases. *Z. Elektrochem.* **19**, 836–856.
Bodenstein, M. (1922) Bildung und Zersetzung der höheren Stickoxyde. *Z. phys. Chem.* **100**(U1), 68–123.
Bodenstein, M. and H. Lütkemeier (1924) Die photochemische Bildung von Brom-Wasserstoff und die Bildungsgeschwindigkeit der Brommolekel aus den Atomen. *Z. phys. Chem.* **114**, 208–236.
Bodenstein, M., E. Padelt and H.-K. Schumacher (1929) Die thermische Reaktion zwischen Chlor und Ozone. *Z. phys. Chem.* **5BB**, 209–232.

Bodenstein, M. and P. W. Schenk (1933) Die photochemische Kinetik der Reaktion zwischen Chlor, Wasserstoff und Sauerstoff. *Z. phys. Chem.* **20B**, 420–450.
Boeckel, E. (1856) De l'ozone: thèse présentée à la faculté de médecine de Strasbourg, G. Silbermann, Strassbourg, 70 pp.
Boeckel, Th. (1865) De l'ozone comme element météorologique. *Ann. chim. phys.* **6**, 235–248. *See also*: *Die Fortschritte de Physik* **21** (1865) 628–629. De l'ozone. *Mém. Soc. méd. Strasb.* II, 1854. 3. Météorologie et constit. médic. du dép. du Bas-Rhin. Ib. 1859, p. 1 u. *Mém. Soc. méd. Strasb.* III, p. 297. 4. Obss. ozonométriques et critique de ques public. sur l'ozone. *Gaz. médic. Strasb.* 1860, p. 67. 5. De l'ozone. Résultat des obss. de 7 années, de 1854 à 60. *Gaz. médic. Strasb.* 1862, p. 1. 6. Obss. sur l'ozone, de 1861 à 1864. Ib. 1865, p. 133. 7. De l'ozone comme élément météorol. *Mém. Soc. méd. Strasb.* III, 1861, p. 390 u. V, 1866, p. 53, sowie *Ann. de chimie et phys.* 4me Sér. VI, 1866.
Boehm, J. (1858) Untersuchungen über das atmosphärische Ozon. In: *Sitzungsber. Wien* **29**, 409–440.
Boeke, D. J., cited by J. Myers (1873) Correspondenz aus Amsterdam. *Ber.* **6**, 439–440 [note it is wrong quoted as *Boehe*].
Boerhaave, H. (1735) Elements of Chemistry: being the Annual Lectures of Herman Boerhaave, formerly professor of Chemistry and Botany, And at present professor of Physick in the University Leyden. Translated from the Original Latin by Timothy Dallowe, London, for J. and J. Pemberton, J. Clarke, A. Millar and J. Gray, in two Volumes, 528+11 pp. and 376+13 pp.
Boerhaave, H. (1762) Hermann Boerhaavs weil. berühmten Professors der Artzneygelehrheit zu Leiden. Anfangsgründe der Chymie: oder gründliche Anweisung auf was Art die natürlichen Cörper können chymisch aufgeschlossen, und daraus heilsame Artzeneyen bereitet werden. aus dem Lateinischen ins Deutsche übersetzt. Nebst einem nützlichen Angange von chymischen Geräthschaften, von Anwendung sowohl des Florentischen als Fahrenheitischen Thermometers und von den Graden des Feuers bey Bearbeitung chymischer Processe. Mit nöthigen Kupfern und allergnädigster Freyheit. Friedrich Nicolai, Berlin, 712 pp.
Bohlig, J. D. (1834) Zur Kenntnis der Meteorwässer und des gewöhnlichen destillierten Wassers. *Archiv für die gesammte Naturlehre (Kastn. Arch.)* **26**, 419–424.
Bohlig, E. (1863) Ueber das salpetrigsaure Ammoniak der Atmosphäre und dessen Entstehung. *Ann.* **125**, 21–33.
Bohn, M. (2011) Concentrating on CO_2: the Scandinavian and Arctic measurements. *Osiris* **26**, 165–179.
Boillot, A. (1873) Action des effluves électriques sur l'air atmosphérique. *C. r.* **76**, 869–871.
Bojkov, R. D. (1986) Surface ozone during the second half of the nineteenth century. *Journal of Climate and Applied Meteorology* **25**, 343–352.
Bojkov, R. D. (1995) The changing ozone layer. Special Publication WMO, Geneve, 26 pp.
Bojkov, R. D. (2012) International ozone commission: history and commission. IAMAS Publications Series No. 2, Oberpfaffenhofen, 74 pp.
Bolin, B. (1959) Atmospheric chemistry and broad geophysical relationships. *Geophysics* **45**, 1663–1672.
Bolin, B. (1970) The carbon cycle. *Scientific American* **223**, 124–135.
Bolin, B. (1983) Changing global biogeochemistry. In: Oceanography (ed. P. G. Brewer), Springer, New York, pp. 305–326.
Bolin, B., E. T. Degens, P. Duvigneaud and S. Kempe (1979) The global biogeochemical carbon cycle. In: The global carbon cycle, SCOPE Report No. 13 (Eds B. Bolin, E. T. Degens, S. Kempe and P. Ketner), John Wiley & Sons, Chichester, pp. 1–56.
Bonâme, P. (1896) Rapport annuel de la Station Agronomique pour 1895, Mauritius. 81 pp.
Bonét-Maury, P. and M. Lefort (1948) Formation of hydrogen peroxide in water irradiated with X- and alpha-rays. *Nature* **162**, 381–382.

Bonhoeffer, K.-F. (1928) Über die Existenz von gasförmigem Siliciummonoxyd. *Z. phys. Chem.* **131U**, 163–165.

Bonhoeffer, K.-F. and H. Reichardt (1928) Thermische Dissoziation von Wasserdampf in Wasserstoff und freies Hydroxyl. *Z. Elektrochem.* **34**, 652–654.

Bonhoeffer, K.-F. and F. Haber (1928) Bandenspektroskopie und Flammenvorgänge. *Z. phys. Chem.* **137A**, 263–288.

Bordas, M. M. and A. Desfemmes (1928) La répartition et le transport des chlorures dans l'atmosphère. *C. r.* **185**, 603–695.

Boresch, K. (1931) Kreislauf der Stoffe in der Natur. In: Pflanzenernährung (eds. K. Boresch et al.), Springer, Berlin, Heidelberg.

Borius, A. (1874) Note sur le climate du Sénégal. *Annu. soc. mét. de France* **22**, 175–185.

Borius, A. (1875) Recherches sur le climat du Sénégal. Gauthier-Villars, Paris, 327 pp.

Borius, A. (1879) De l'identité des résultats fournis au Sénégal par l'observation de l'évaporomètre de Piche et du papier ozonométrique de Jame (de Sedan). Congrès international de météorologie tenu à Paris du 24 au 28 août 1878. *Comptes rendus sténographique* No. 20. Imprimerie Nationale, Paris, pp, 187–188.

Borrichius, O. (1674) Hermetis, Aegyptiorum et Chemicorum sapienta ab Hermanni Conringii anidmadversionibus vindicate. Sumptibus Petri Hauboldi, Reg. Acad. Bibl., 448 pp.

Bos, R. (1980) Automated measurement of atmospheric ammonia. *Journal of the Air Pollution Association* **30**, 1222–1224.

Botha, C. F., J. Hahn, J. J. Pienaar and R. van Eldik (1994) Kinetics and mechanism of the oxidation of sulfur(IV) by ozone in aqueous solution. *Atmos. Env.* **28**, 3207–3212.

Böttger, R. (1858) Ueber die Entstehung der Untersalpetersäure und Salpetersäure durch Decomposition der atmosphärischen Luft mittels Inductionselectrizität. *J. prakt. Chem.* **73**, 494–496.

Böttger, R. (1862) Notizen. 2. Ueber die Bildung von salpetrigsaurem Ammoniak bei Verbrennungsprozesses. *Zeitschrift für praktische Chemie* **85**, 396–397.

Böttger, R. (1871) Ueber Ozon und das sogenannte Ozonwasser. *Pharmaceutische Centralhalle für Deutschland* **12**, 489–490. *See also*: Ueber die Bestandteile des käuflichen sogenannten Ozonwassers. *Jahresberichte des physikalischen Vereins zu Frankfurt a. M. für 1871–1872* (1873) 12–14.

Böttger, R. (1873a) Ueber einen experimentellen Beweis, dass nicht bloss bei der Zerlegung des Wassers auf elektrolytischem Wege, sondern auch bei dessen Bildung, d. h. bei der Vereinigung reinsten Sauerstoff- und Wasserstoffgases, nachweisbare Mengen an Ozon auftreten. *Jahresberichte des physikalischen Vereins zu Frankfurt a. M. für 1871–1872* (1873) 14.

Böttger, R. (1873b) Ueber die Einwirkung der Inductionselektricität auf atmosphärischer Luft und die hierbei auftretenden Produkte. Jahresberichte des physikalischen Vereins zu Frankfurt a. 1871–1872. M. für. 1873, 12.

Böttger, A., D. H. Ehhalt and G. Gravenhorst (1978) Atmosphärische Kreisläufe von Stickoxiden und Ammoniak. Berichte Kernforschungsanlage Jülich Nr. 1558. 243 pp.

Bottini, O. (1935) Fattori pedogenetici particolari della Regione Vesuviana: gas e sublimazioni vulcaniche. *Annali del Reale Osservatorio Vesuviano, ser. IV* **3**, 141.

Bottini, O. (1939) Le piogge caustiche della Regione Vesuviana. *Annali di chimica applicata* **29**, 425.

Bouguer, P. (1749) La figure de la terre, déterminée par les observations de Messieurs Bouguer, & de la Condamine, de l' Académie Royale des Sciences, envoyes par ordre du Roy au Pérou, pour observer aux environs de l' Equateur: avec une relation abregée de ce voyage, qui contient la description du pays dans lequel le opérations ont été faites. Charles-Antoine Jombert, Paris, 396 pp.

Boussingault, J. B. (1834) Recherches sur la composition de l'atmosphère. Premier mémoire, sur la possibilité de constater l'existence des miasmes. – Sur la présence d'un principe hydrogéné dans l'air. *Ann. chim. phys.* **57**, 148–182. *See also*: Boussingault, J. B. (1834) Ueber die Zusammensetzung der Atmosphäre, die Möglichkeit die Existenz von Miasmen darin nachzuweien, und über das Vorhandensein des wasserstoffhaltigen Prinzips in der Luft. *J. prakt. Chem.* **3**, 151–161. *And*: Boussingault, J. B. (1835) Untersuchung über die Zusammensetzung der Atmosphäre. Erste Abhandlung. Ueber die Möglichkeit, das Daseyn von Miasmen zu erweisen, und über die Gegenwart einer wasserstoffhaltigen Substanz in der Luft. *Pogg. Ann.* **36**, 436–456.

Boussingault, J. B. (1838) Recherches chimiques sur la végétation entreprises dans le but d'examiner si les plantes prennent de de l'azote de l'atmosphere. *Annales de Chemie* **69**, 353–367.

Boussingault, J. B. (1845) Rural Economy, in its relation with Chemistry, Physics and Meteorology; or, Chemistry Applied to Agriculture. Transl. by G. Law, D. Appleton & Co., New York, 507 pp. Original: *Economie rurale considérée dans ses rapports avec la chimie, la physique et la météorologie. 2 volumes*. Paris 1841–1842 (1st ed.: 648 + 742 pp.); 1851 (2nd ed.: 819 + 767 pp.

Boussingault, J. P. (1853) Mémoire sur le dosage de l'ammoniaque contenue dans les eaux. *Ann. chim. phys. 3th ser.* **39**, 257–293. *And*: Sur la quantité d'ammoniaque contenue dans l'eau de pluie recueillis loin des ville. *C. r.* **37**, 207–208, 798–806. *See also*: Ueber den Ammoniakgehalt des fern von Städten gefallenen Regenwassers. *J. prakt. Chem.* **61** (1854) 113–122.

Boussingault, J. P. (1854) Mémoiré sur la quantité d'ammoniaque contenue dans la pluie, la rosée et la brouillárd recueillis loin des villes. *Ann. chim. phys. 3th ser.* **40**, 129–155. Sur la quantité d'ammonique contenue dans la pluie et dans l'eau déposée par le brouillard. *C. r.* **38** (1854) 249–251. *See also*: Boussingault (1856c).

Boussingault, J. B. (1856a) Recherches sur la vegetation. Troisiéme mémoire. De l'action du salpêtre sur le développement des plants. *Ann. chim. phys. 3th ser.* **45**, 5–41.

Boussingault, J. B. (1856b) Ueber die Bestimmung des im Wasser verschiedenen Ursprungs enthaltenen Ammoniaks. Vorgetragen in der Akademie der Wissenschaften in der Sitzung vom 9. Mai 1853. In: Beiträge zur Agricultur-Chemie und Physiologie. Deutsch bearbeitet von R. Graeger, Ch. Grager, Halle, pp. 217–245.

Boussingault, J. B. (1856c) Ueber die Menge des im Regen, Thau und Nebel, entfernt von Städten genommen, enthaltenen Ammoniaks. In: Beiträge zur Agricultur-Chemie und Physiologie. Deutsch bearbeitet von R. Graeger, Ch. Graeger, Halle, pp. 245–267.

Boussingault, J. B. (1863) Sur l'apparition du gaz oxygen de carbone pendant l'absorption de l'oxygene par certaines substances végétales. *C. r.* **57**, 885–893.

Boussingault, J. B. (1882) Recherches sur la présence de l'acide nitrigue et de l'ammoniadue dans les eaux et la neige, recueillies dans les glaciers des Alpes par M. Civiale. *C. r.* **95**, 1121–1123.

Boussingault, J. B. and A. Lewy (1844) Observations faites à Paris et à Andilly, près Montmorency, pour rechercher la proportion d'acide carbonique contenu dans l'air atmosphèrique. *Annuire de Chimie* **10**, 456–469.

Bouwman, A. F., D. S. Lee, W. A. H. Asman, F. J. Dentener, K. W. van der Hoek and J. G. J. Olivier (1997) A global high-resolution emission inventory for ammonia. *Global Biogeochemical Cycles* **11**, 561–588.

Bowen, I. G. (1949) A study of atmospheric ozone near the surface of the Earth. Master Thesis, University of New Mexico, 17 pp.

Bowen, I. G. and V. H. Regener (1951) On the automatic chemical determination of atmospheric ozone. *JGR* **65**, 307–324.

Bower, K. N., T. W. Choularton, M. W. Gallagher, R. N. Colvile, M. Wells, K. M. Beswick, A. Wiedensohler, H.-C. Hansson, B. Svenningsson, E. Swietlicki, M. Wendisch, A. Berner, C. Kruisz, P. Laj, M. C. Facchini, S. Fuzzi, M. Bizjak, G. Dollard, B. Jones, K. Acker, W. Wieprecht, M. Preiss, M. A. Sutton, K. J. Hargreaves, R. L. Storeton-West, J. N. Cape and B. G. Arends (1997) Observations and modelling of the processing of aerosol by a hill cap cloud. *Atmos. Env.* **31**, 2527–2543.

Boylan, R. F. (1927) Atmospheric dust and condensation nuclei. *Proceedings of the Royal Irish Academy. Section A: Mathematical and Physical Sciences* **37** (1924–1927), 58–70.

Boyle, R. (1662) New experiments physico-mechanical touching the springs of air and its effects. H. Hall, Oxford. Reprint (2008) UMI, Ann Arbor, Michigan, USA, 207 pp.

Boyle, R. (1680) Chymista Scepticus uel Dubia et Paradoxa Chymico-Physica, circa Spagyricorum Principia, vulgo dicta Hypostatica, Prout proproni & propugnari solent a Turba Alchymistarum... 5th Ed. (in Latin), Samuel de Tournes, Geneve, 148 pp.

Boyle, R. (1690) An experimental discource of some unheeded causes of the insalubrity and salubrity of the air being a part of an intended natural history of air. Sam. Smith, London 113 pp. (imprint EEBO editions).

Boyle, R. (1692) The general history of the air designed and begun by the Honble. Printed for Awnsham and John Churchill, London, 259 pp.

Braadlie, O. (1930) Inneholdet av ammoniak og nitratkvelstoff I nedbøren ved Trondhjem. *Kgl. Norske Videnskab. Selskab. Forhandl.* **3**, 78 (quoted after Eriksson 1952).

Brandes, R. (1826a) Beiträge zur Kenntnis der Meteorwässer. *Journal für Chemie und Physik* **48**, 153–177 (identical with *Jahrbuch der Chemie und Physik*, Vol. 18).

Brandes, R. (1826b) Nachschrift zu vorstehendem Aufsatz. *Arch. Pharm.* **19**, 189–193.

Brandes, R. (1826c) Pharmaceutischer Jahresbericht. *Arch. Pharm.* **18**, 1–368.

Braslavsky, S. E. and M. B. Rubin (2011) The history of ozone. Part VIII. Photochemical formation of ozone. *Photochemistry and Photobiological Sciences* **10**, 1515–1520.

Bray, J. R. (1959) An analysis of the possible recent change in atmospheric carbon dioxide concentration. *Tellus* **11**, 220–230.

Bray, W. (1827) Memoirs of John Evelyn ESQ, F. R. S. Author of the "sylva" &c, &c, Comprising His Diary, from 1641–1705-6 and a Selection of His Familiar Letters, to which is Subjoined, the Private Correspondence Between King Charles I. and Sir Edward Nicholas; Also Between Sir Edward Hyde, Afterwards Earl of Clarendon and Sir Richard Browne, Ambassador to the Court of France, in the Time of King Charles I. and the Usurpation. Vol. 3. H. Colburn, London, 441 pp.

Brenner, A. R. (1849) Ueber den Gehalt der Atmosphäre an Ozon und dessen Eigenschaften als krankmachendes Agens. *Berichte des Naturwissenschaftlichen Vereins des Harzes zu Blankenburg für die Jahre 1848 und 1849*. 5–7.

Brenner, A. R. (1851) Ueber das sogenannte Ozon. *Berichte des Naturwissenschaftlichen Vereins des Harzes zu Blankenburg für die Jahre 1849 und 1850*. 14–17.

Bretschneider, P. (1872) Ueber den Regenfall zu Ida-Marienhütte in den Jahren 1865–1872 und den Gehalt des meteorischen Wassers an Stickstoff. Breslau. *See also*: *Landwirthschaftliches Centralblatt für Deutschland. Repertorium der wissenschaftlichen Forschungen und praktischen Erfahrungen im Gebiete der Landwirthschaft* 20 (1872) 291–295. And: *Jahresbericht ueber die gesammten Fortschritte auf dem Gebiet der Agricultur-Chemie* 13 (1870) 142–147.

Brevik, E. C. (2004) Contributions of Edward Elway Free to American soil science in the early 1900s. *Soil Science Society of America Journal* **68**. doi:10.2136/sssaj2004.0904.

Brewer, A. W. (1955) Ozone concentration measurements from an aircraft in N. Norway. *Meteorol. Research Papers* No. 946, Meteorol. Res. Comm., London, 6 pp.

Brewer, R. L., R. J. Gordon and L. S. Shephard (1983) Chemistry of mist and fog from the Los Angeles urban area. *Atmos. Env.* **17**, 2267–2270.

Bridge, J. H. (1907) Ozone: its nature, production and uses. *Journal of the Franklin Institute* **163**, 355–381.
Bridges, K. S., T. D. Jickells, T. D. Davies, Z. Zeman and I. Hůnová (2002) Aerosol, precipitation and cloud water observations on the Czech Krusne Hory plateau adjacent to a heavily industrialised valley. *Atmos. Env.* **36**, 353–360.
Brierly, W. E. (1970) Bibliography of atmospheric (cyclic) sea-salt. US Earth Science Laboratory, Technical Report 70–63-ES. In: Marine aerosols: a selective bibliography (Ed M. F. Haymes), NCAE Techn. Note IA-96, Boulder (1974), pp. 55–92.
Brierly, W. E. (1979) Bibliography of atmospheric (cyclic) sea-salt. US Earth Science Laboratory, Natick (Mass.), 76 pp.
Brimblecombe, P. (1977) London air pollution, 1500–1900. *Atmos. Env.* **11**, 1157–1162.
Brimblecombe, P. (1978) "Dew" as a sink for sulphur dioxide. *Tellus* **30**, 151–157.
Brimblecombe, P. (1987) The big smoke: a history of air pollution in london since medieval times. Methuen & Co., New York, 185 pp.
Brimblecombe, P. (1991) History of atmospheric acidity. In: Atmospheric acidity (eds. M. Radojevic and R. H. Harrison), Elsevier, pp. 267–305.
Brimblecombe, P. (1996) Air composition and chemistry. Cambridge University Press, 253 pp.
Brimblecombe, P. (2003) The global sulfur cycle. In: Treatise on geochemistry, Volume 8 (ed. W. H. Schlesinger), Elsevier, pp. 645–682.
Brimblecombe, P. (2004) Perceptions and effects of late Victorian air pollution. In: Smoke and mirrors. The politics and culture of air pollution (ed. E. M.- DuPois), New York University Press, pp. 15–26.
Brimblecombe, P. and I. Todd (1977) Sodium and potassium in dew. *Atmos. Env.* **11**, 649–650.
Brimblecombe, P. and J. Pitman (1980) Long-term deposit at Rothamsted, southern England. *Tellus* **32**, 261–267.
Brimblecombe, P. and S. L. Clegg (1988) The solubility and behaviour of acid gases in the marine aerosol. *J. Atmos. Chem.* **7**, 1–18.
Briner, E. and R. Perrottet (1937) Méthode d'analyse de l'ozone, à l'état très dilué, fondée sur l'action catalytique exercée par ce gaz dans l'oxydation des aldéhydes. *Helvetica Chimica Acta* **20**, 293–298. Méthode d'analyse de l'ozone très dilué. II. Détermination de la concentration de l'ozone dans l'air à Genève. *Ibid.*, pp. 458–461. Méthode d'analyse de l'ozone très dilué. III. Action sensibilisatrice du peracide présent dans l'aldéhyde. Mesure de la teneur de l'air en ozone à différentes altitudes. *Ibid.*, pp. 1200–1207.
Brodie, B. C. (1862) Note on the oxidation and disoxidation effected by the peroxide of hydrogen. *Proc. R. Soc. Lond.* **11**, 442–444.
Brodie, B. C. (1872) An experimental inquiry on the action of electricity on gases. – I. On the action of electricity on oxygen. *Phil. Trans.* **162**, 435–484.
Brodie, F. J. (1892) On the prevalence of fog in London: During the 20 years 1871 to 1890. *Q. J. Roy. Meteor Soc.* **18**, 40–45.
Brøgger, W. C. (1881) Om smudsig nyfallen Sne paa Selø. *Naturen* **5**, 47.
Brönnimann, S., J. Staehelin, S. F. G. Farmer, J. C. Cain, T. Svendby and T. Svenøe (2003) Total ozone observations prior to the IGY. I: A history. *Q. J. Roy. Meteor. Soc.* **129**, 2797–2817.
Brosset, C. (1976) Air-borne particles: Black and white episodes. *Ambio* **5**, 157–163.
Brown, A. C. (1869) Ozone. *Journal of the Scottish Meteorological Society* **2**, 253–256.
Brown, H. T. (1870) On the estimation of ammonia in atmospheric air. *Proc. R. Soc. Lond.* **23**, 286–288.
Brown, H. T. and F. Escombe (1905a) Researches on some of the physiological processes of green leaves, with special reference to the interchange of energy between the leaf and its surroundings. *Proc. R. Soc. Lond. B.* **76**, 29–111.

Brown, H. T. and F. Escombe (1905b) On the variations in the amount of carbon dioxide in the air of Kew during the years 1898–1901. *Proc. R. Soc. Lond. B.* **76**, 118–121.

Brown, S. (1800) A treatise on the nature, origin and progress of the yellow fever: with observations on its treatment, comprising an account of the disease in several of the capitals of the United States, but more particularly as it has prevailed in Boston. Manning and Loring. Boston. 112 pp.

Brücke, E. (1853) Ueber die Farben, welche trübe Medien im auffallenden und durch-fallenden Lichte zeigen. *Pogg. Ann.* **88**, 363–385.

Bruckmann, P. (1991) Synoptische Darstellung der Ozonbelastung des Jahres 1990 in Deutschland. In: Proceedings Ozon-Symposium in München 1.-4. Juli 1991, TÜV Akademie Bayern/Hessen, 17 pp.

Brüggemeier, F.-J. (1996) Das unendliche Meer der Lüfte. Luftverschmutzung, Industrialisierung und Risikodebatten im 19. Jahrhundert. Klartext Verlag, Essen, 344 pp.

Brugmans, S. J. (1783) Natuurkundige verhandeling over een zwavelagtigen nevel den 24 Juni 1783 in de provintie van Stad en Lande en naburige landen waargenomen. Petrus Doekman, Groningen, 58 pp.

Brühl, J. W., ed. (1897) Schorlemmers Lehrbuch der Kohlenwasserstoffverbindungen oder der organischen Chemie. 3. Auflage. Vieweg, Braunschweig.

Bruhns, C. (1867) Resultate aus den meteorologischen Beobachtungen angestellt an mehreren Orten im Königreich Sachsen in den Jahren 1848–1863 und an den 22 Kgl. Saechsischen Stationen im Jahre 1864: Nach den monatlichen Zusammenstellungen im statistischen Bureau des Kgl. Ministeriums des Innern bearbeitet von C. Bruhns, Band 2. E. J. Günther, Leipzig, 147 pp.

Bruhns, C. (1870) Resultate aus den meteorologischen Beobachtungen angestellt an den 25 Kgl. Sächsischen Stationen im Jahre 1868. Bearbeitet v. C. Bruhns. Jg. 5, Teubner, Dresden, 65 pp.

Bruhns, C. (1877) Resultate aus den meteorologischen Beobachtungen angestellt an 24 kgl. sächsischen Stationen in den Jahren 1872 und 1873. IX. und X. Jahrgang. Dresden und Leipzig.

Bruhns, C. (1880) Resultate aus den meteorologischen Beobachtungen angestellt an 25 kgl. sächsischen Stationen in den Jahren. 1874 und 1875. XI. und XII. Jahrgang. Dresden und Leipzig.

Brunner, C. (1832) Beschreibung einer Methode, die Menge der in der Atmosphäre enthaltenen Kohlensäure zu bestimmen. *Pogg. Ann.* **24**, 569–575.

Brunner, C. (1853) Ueber die Analyse der atmosphärischen Luft. *Mittheilungen der naturforschenden Gesellschaft in Bern aus dem Jahre 1853*, No. 265–309, 249–262.

Brünnich, J. C. (1909) Fertilizing value of rain water. *Annual Report of the Department of Agriculture and Stock*, Queensland (1908/09), pp. 54–78.

Brush, S. and H. E. Landsberg (1985) The History of Geophysics and Meteorology: An Annotated Bibliography. Garland Pub., 450 pp.

Buch, K. (1939) Beobachtungen über das Kohlensäuregleichgewicht und über den Kohlensäureaustausch zwischen Atmosphäre und Meer im Nordatlantischen Ozean. *Acta Academiae Aboensis. Mathematica et Physica* **11**, 28–31.

Buch, K. (1945) Kolsyrejämvikten i Baltiska Havet. *Fennia* **68**, 1–208. (on pp. 151–153); quoted after Bohn (2011, p. 170).

Buch, K. (1948) Der Kohlendioxydgehalt der Luft als Indikator der Meteorologischen Luftqualität. *Geophysica* **3**, 63–79.

Büchner, J. (1855) Kraft und Licht. Empririsch-naturwissenschaftliche Studien. In allgemeinverständlicher Darstellung. Verlag von Meidinger, Frankfurt am Main, 269 pp. (several editions and translated into many languages). First edition in English 1862: Force and matter: Empirico-philosophical studies, intelligibly rendered. With an additional introduction expressly written for this edition. Edited, from the last edition of "Kraft und Stoff", by J. Frederick Collingwood, London, Trubner & Co.

Büchner, J. (1884) Force and matter: empirico-philosophical studies. Translated from the 15th German edition (fours English edition), Asher and Co., London, 512 pp.

Buchner, J. A. (1826) Vollständiger Inbegriff der Pharmacie in ihren Grundlehren und praktischen Theilen. Ein Handbuch für Ärzte und Apotheker. Drittes Theil. Erster Band. Grundriß der Chemie. J. L. Schrag, Nürnberg, 686 pp.

Bucquet, J. B. M. (1778) Mémoire sur la manière dont les animaux sont affectés par différens fluides aériformes, méphitiques; & sur les moyens de remédier aux effets de ces fluides. Précédé d'une histoire abrégée des différens fluides aériformes ou gas. De l'Impimerie Royale, Paris, 86 pp.

Budde, E. (1871) Ueber die Einwirkung des Lichtes auf Chlor und Brom. *Pogg. Ann.* **144**, 213–219.

Budde, E. (1874) Ueber die Einwirkung des Lichtes auf freies Chlor. *Pogg. Ann.* **153**, 477–498.

Bufalini, J. J., B. W. Jr. Gay and K. L. Braubaker (1972) Hydrogen peroxide formation from formaldehyde photooxidation and its presence in urban atmosphere. *Environ. Sci. Technol.* **6**, 816–821.

Bufalini, J. J., H. T. Lancaster, G. R. Namie and B. W. Jr. Gay (1979) Hydrogen peroxide formation from the photooxidation of formaldehyde and its presence in rainwater. *Journal of environmental Science and Health A* **14**, 135–141.

Bugge, G., Ed. (1929) Das Buch der großen Chemiker. Vol. I: Von Zosimos bis Schönbein, Verlag Chemie Berlin, 496 pp.

Buijsman, E. (2010) Een kleine geschiedenis van het chemische neerslagonderzoek in Nederland. Uitgeverij Tinsentiep, Houten, 50 pp.

Buijsman, E. (2011) Een geannoteerd overzicht van publicaties over de chemische samenstelling van lucht en neerslag in Nederland. Uitgeverij Tinsentiep, Houten, 45 pp.

Bujwid, O. (1881) Poszukiwania jodu, bromi I ozonu w powietrzy težni Ciechocińskiej [Analysis of atmospheric air at Ciechocinek]. *Polska gazeta lekarska* **2**, 878–883.

Bunsen, R. (1855) Ueber das Gesetz der Gasabsorption. *Ann.* **93**, 1–51.

Bunsen, R. (1857) Gasometrische Methoden. Zweite Auflage. Vieweg und Sohn, Braunschweig, 387 pp.

Bunsen, R. and H. E. Roscoe (1855, 1857) Photochemische Untersuchungen. *Pogg. Ann.* **96**, 373–394, *Pogg. Ann.* **100** (1857) 43–88, 481–515; *Pogg. Ann.* **101** (1857) 234–263 (note that under the same title 4 papers appeared).

Burckhardt, H. and H. Flohn (1939) Die atmosphärischen Kondensationskerne in ihrer physikalischen, meteorologischen und bioklimatischen Bedeutung. Mit Berichten über Meßergebnisse von E. Flach und L. Schulz. In: Abhandlungen aus dem Gebiet der Bäder- und Klimaheilkunde, Heft 3 (eds. H. Vogt and K. Knoch), Springer, Berlin, 119 pp.

Burdick, L. R. and J. F. Barkley (1939) Concentration of sulfur compounds in city air. United States, Bureau of Mines, 14 pp.

Burkard, R., P. Bützberger and W. Eugster (2003) Vertical fogwater flux measurements above an elevated forest canopy at the Lägeren research site. *Switzerland. Atmos. Env.* **37**, 2979–2990.

Burkser, E. S. [Бурксер] (1924) Исследование радиоактивности воздуха в лечебных местностях побережья Черного и Азовского морей [Investigation of radioactivity at therapeutic sites near the coast of the Black and Azov Sea]. *Журн. науч.-исслед. кафедр в Одессе [Journal of the scientific Department in Odessa]* **1**, 1–7.

Burkser, E. S. [Бурксер] (1937) Определение йода и брома в атмосферном воздухе при их совместном присутствии [Determination of iodine and bromine in atmospheric air with simultaneous presence]. *Журнал прикладный Химии* [Journal of Applied Chemistry] **12**, 2152–2161.

Burkser, E. S. [Бурксер] (1940) Опыт определения содержания скипидара и озона в воздухе соснового леса [Experience in determination of the content of and ozone in the air of a pine forest]. *Вопр. Курортологии* [Problems of Balneology] **4**, 36–43.

Burkser, E. S. and E. V. Gernet [Бурксер, Гернет] (1939) Спроба аерохімічного здіймання Євпаторійського і Мойнакського курортів [Attempt of an air chemical survey at Eupatorian and Moynakian health resorts] *Геологичний журнал* [Geological Journal] **6**, No. 3.
Burkser, E. C. [Бурксер] and N. E. Fedorova [Федорова] (1949) Опыт исследования химического состава атмосферных вод [Investigation of the chemical composition of atmospheric water]. *Гидрохимические материалы* [Hydrochemical Materials] **16**, 107–112.
Burkser, E. S. and V. V. Burkser [Бурксер, Бурксер] (1951) Аэрохимические исследования на Украине [Air chemical research in the Ukraine]. *Тр. Института Геологических наук ССР петрографии, минералогии и геохимии* [Transactions of the Institute of Geological Sciences SSR for Petrography, Mineralogy and Geochemistry] **1**, 7–127.
Burns, D. T. (1997) Early problems in the analysis and the determination of ozone. *J. anal. Chem.* **357**, 178–183.
Burns, D. T. and M. J. Walker (2015) Edmund Alberts Letts (1852–1918) – pioneer environmental analytical chemist and his association with official analytical post in Ulster. *Journal of the Association of Public Analyst* (online) **43**, 013–026, pp. 13–25.
Burt, F. and E. Edgar (1916) The combining volumes of hydrogen and oxygen. *Phil. Trans. Series A* **216**, 393–427.
Busch, G. C. B. (1807) Handbuch der Erfindungen. Vierten Theils erste Abtheilung, den Buchstaben E enthaltend. J. G. E. Wittekindt, Eisenach, 292 pp.
Busse, W. F. and F. Daniels (1927) The decomposition of nitrogen pentoxide in the presence of foreign gases. *J. Am. Chem. Soc.* **49**, 1257–1269.
Butcher, S. S. and R. J. Charlson (1972) An introduction to air chemistry. Academic Press, New York and London, 241 pp.
Butler, J. D. (1979) Air pollution chemistry. Academic Press, New York, 408 pp.
Butterfield, H. (1931) The Whig interpretation of history. G. Bell and sons, London, 132 pp.
Buxton, G. V. (1982) Basic radiation chemistry of liquid water. In: The study of fast processes and transient species by electron pulse radiolysis (eds. J. H. Baxendale and F. Busi), D. Reidel Publ. Comp., pp. 241–266. *See also*: Buxton, G. V., G. A. Salmon, S. Barlow, T. N. Malone, S. McGowan, S. A. Murray, J. E. Williams and N. D. Wood (1996) Photochemical sources and reactions of free radicals in the aqueous phase. In: Heterogeneous and liquid-phase processes. Vol. 2: Transport and chemical transformation of pollutants in the troposphere (ed. P. Warneck), Springer-Verlag, Berlin, pp. 133–145.
Bykov, A. and K. Karpov [Быков, Карпов] (1911) Исследование химического состава атмосферных осадков в Томске зимой 1909–10 г. въ зависимости отъ метеорологичекихъ факторовъ [Investigation of the chemical composition of atmospheric precipitation in dependence from meteorological factors]. *Известия томского политехнического университета* [Bulletin Tomsk Polytechnical University] **24**, No. 4, 1–13.
Cacace, F., G. de Petris and A. Troiani (2001) Experimental detection of tetraoxygen. *Z. angew. Chem. International Edition* **40**, 4062–4065.
Cadle, R. D. and H. S. Johnston (1952) Chemical reactions in Los Angeles smog. In: Proceeding's 2nd National Air Pollution Symposium, Stanford Research Institute, Menlo Park, CA, pp. 28–34.
Cadle, R. D. and P. L. Magill (1951) Study of eye irritation caused by Los Angeles smog. *Archiv of Industrial and Occupational Medicine* **4**, 74–84.
Cadle, R. D. and P. L. Magill (1956) Chemistry of contaminated atmosphere. In: Air pollution handbook (eds. P. L. Magill, F. R. Holden and C. Ackley), McGraw-Hill, New York, Chapter 3 (27 pp.).
Cadle, R. D. and M. Ledford (1966) The reaction of ozone with hydrogen sulfide. *Air and Water Pollution International Journal* **10**, 25–30.
Cadle, R. D. and E. R. Allen (1970) Atmospheric photochemistry. *Science* **167**, 243–249.

Cadle, R. D., P. J. Crutzen and D. Ehhalt (1975) Heterogeneous reactions in the stratosphere. *JGR* **80**, 3381–3385.

Cadle, S. H. and P. J. Groblicki (1983) Composition of dew in an urban area. In: Transactions of the APCA Specialty Conference: The Meteorology of Acid Deposition (ed. P. J. Samson), pp. 17–29.

Cady, G. H. (1935) Reaction of fluorine with water and with hydroxides. *J. Am. Chem. Soc.* **57**, 246–249.

Cain, W. S., R. Schmidt and P. Wolkoff (2007) Olfactory detection of ozone and d-limonene: reactants in indoor spaces. *Indoor Air* **17**, 337.347.

Callendar, G. S. (1938) The artificial production of carbon dioxide and its influence on temperature. *Q. J. Roy. Meteor. Soc.* **64**, 223–240.

Callendar, G. S. (1940) Variations of the amount of carbon dioxide in different air currents. *Q. J. Roy. Meteor. Soc.* **66**, 395–400.

Callendar, G. S. (1949) Can carbon dioxide influence climate? *Weather* **4**, 310–314.

Callendar, G. S. (1958) On the amount of carbon dioxide in the atmosphere. *Tellus* **10**, 243–248.

Calvert, J. G., K. Theurer, G. T. Rankin and W. M. MacNevin (1954) A study of the mechanisms of the photochemical synthesis of hydrogen peroxide at zinc oxide surfaces. *J. Am. Chem. Soc.* **76**, 2575–2578.

Calvert, J. G., J. A. Kerr, K. L. Demerjian and R. D. McQuigg (1972) Photolysis of formaldehyde as a hydrogen atom source in the lower atmosphere. *Science* **175**, 751–752.

Calvert, J. G., F. Su F, J. W. Bottenheim and O. P. Strausz (1978) Mechanism of the homogeneous oxidation of sulfur dioxide in the troposphere. *Atmos. Env.* **12**, 197–226.

Calvert, J. G. and W. R. Stockwell (1983) Acid generation in the troposphere by gas-phase chemistry. *Environmental Science & Technology* **17**, 428–443.

Calvert, J. G., A. Lazrus, G. L. Kok, B. G. Heikes, J. G. Walega, J. Lind and C. A. Cantrell (1985) Chemical mechanisms of acid generation in the troposphere. *Nature* **317**, 27–35.

Calvert, J. G., R. G. Derwent, J. J. Orlando, G. S. Tyndall and T. J. Wallington (2008) Mechanisms of atmospheric oxidation of the alkanes. Oxford University Press, 1008.

Camuffo, D. (1992) Acid rain and deterioration of monuments: How old is the phenomenon? *Atmos. Env.* **26B**, 241–247.

Canseliet, E., Ed. (1991) Die Alchemie und ihr Stummes Buch (Mutus Liber), Edition Weber, Amsterdam, 149 pp.

Cantoni, G. (1866) *Meteorologica Italiana* **16**, 15. *See also*: Ueber die Sicherheit ozonometrischer Bestimmungen nach Fremy und Cantoni. *Z. Öster. Ges. Met.* **1** (1866) 305–312.

Cape, J. N., K. J. Hargreaves, R. L. Storeton-West, B. Jones, T. Davies, R. N. Colvile, M. W. Gallagher, T. W. Choularton, S. Pahl, A. Berner, C. Kruisz, M. Bizjak, P. Laj, M. C. Facchini, S. Fuzzi, B. G. Arends, K. Acker, W. Wieprecht, R. M. Harrison and J. D. Peak (1997) The budget of oxidised nitrogen species in orographic clouds. *Atmos. Env.* **31**, 2625–2636.

Carius, L. (1872) Absorption von Ozon in Wasser. *Ber.* **5**, 520–526.

Carius, L. (1873a) Ueber Ozonwasser. *Vierteljahresschrift für praktische Pharmacie* **22**, 77–79.

Carius, L. (1873b) Absorption von Ozon in Wasser. *Ber.* **6**, 806–809.

Carius, L. (1874a) Verhalten des Ozons gegen Wasser und Stickstoff. *Liebigs Ann.* **174**, 1–30.

Carius, L. (1874b) Ueber Bildung von salpetriger Säure, Salpetersäure und Wasserstoffsuperoxyd in der Natur. *Liebigs Ann.* **174**, 31–44.

Carlson, P. and A. A. Watson (2014) Erich Regener and the maximum in ionisation of the atmosphere. *History of Geophysics and Space Science* **5**, 163–174.

Carmody, P. (1902) Annual Report of the Government Analyst. In: Colonial Reports – Annual. No. 382, Trinidad and Tobago 1901–1902 (quoted after Drischel 1940).

Carnelley, T. and W. Mackie (1886) The determination of organic matter in air. *Proc. R. Soc. London* **41**, 238–247.

Carnelley, T. and A. Wilson (1888) A new method of determining the number of microorganisms in air. *Proc. R. Soc. London* **44**, 455–464.
Carpenter, Th. M. (1937) The constancy of the atmosphere with respect to carbon dioxide and oxygen content. *J. Am. Chem. Soc.* **59**, 358–360.
Carrol, D. (1962) Rainwater as a chemical agent of geological processes – a review. Geological Survey Water-Supply Paper 1535-G, US Gov. Print. Office, Washington, 23 pp.
Cartalis, C. and C. Varotsos (1994) Surface ozone in Athens, Greece, at the beginning and at the end of the twentieth century. *Atmos. Env.* **28**, 3–8.
Casali, A. (1901) L'ammoniaca nelle acque meteoriche e la pioggia sanguinante. *Staz. Sper. Agrar. Ital. (Le Stazioni Sperimentali Agrarie Italiane)* **34**, 833–848.
Castillo, R. A. (1979) An investigation of the acidity of stratus cloud water and its relationship to droplet distribution, pH of rain and weather parameters. Ph. D. thesis, Dep. of Atmos. Sci., State Univ. of N. Y., Albany.
Castillo, R. A. and J. E. Jiusto (1983) The pH and ionic composition of stratiform cloud water. *Atmos. Env.* **17**, 1497–1505.
Castillo, R. A., J. Kadlecek and S. McLaren (1985) Selected Whiteface Mountain cloud water concentrations summer 1981 and 1982. *Water, Air and Soil Pollution* **24**, 323–328.
Catling, D. C., M. W. Claire, K. J. Zahnle, R. C. Quinn, B. C. Clark, M. H. Hecht and S. Kounaves (2010) Atmospheric origins of perchlorate on Mars and in the Atacama. *Journal of Geophysical Research. Planets Banner* **115**, doi:10.1029/2009JE003425.
Cauer, H. (1929) Über das Vorkommen von Jod in Gesteinen, Erden und Wässern und seine Beziehung zum Kropf. *Journal für Landwirtschaft* **77**, 251–274.
Cauer, H. (1932) Über das Jod der Luft, Chemismus, Transport und bioklimatische Bedeutung. *Zeitschrift für die gesamte physikalische Therapie* **43**, 135–147.
Cauer, H. (1934) Chemische Klimatologie. *Der Balneologe* **1**, 242–245.
Cauer, H. (1935a) Entnahmeapparatur für chemisch-klimatologische und technische Untersuchungen. *Z. anal. Chem.* **103**, 166–180.
Cauer, H. (1935b) Bestimmung des Gesamtoxydationswertes, des Nitrits, des Ozons und des Gesamtchlorgehaltes roher und vergifteter Luft. *Z. anal. Chem.* **103**, 321–334, 385–416.
Cauer, H. (1936) Chemisch-bioklimatische Studien in der Hohen Tatra und ihrem Vorland. *Der Balneologe* **2**, 7–23.
Cauer, H. (1937a) Möglichkeiten und Wege zum Studium chemisch-bioklimatischer Fragen. *Der Balneologe* **4**, 286–292.
Cauer, H. (1937b) Chemisch-bioklimatische Studien im Glatzer Bergland. *Der Balneologe* **4**, 545–565.
Cauer, H. (1938) Einiges über den Einfluss des Meeres auf den Chemismus der Luft. *Der Balneologe* **5**, 409–415.
Cauer, H. (1948a) Der pH-Wert der Nebelkerne und das Ozon bodennaher Luft. *Schriftenreihe des Deutschen Bäderverbandes* **2**, 61–69.
Cauer, H. (1948b) Chemie der Atmosphäre. In: Meteorologie und Physik der Atmosphäre. Vol. 19: Naturforschung und Medizin in Deutschland 1936–1946 (ed. R. Mügge), Dieterich'sche Verlagsbuchhandlung W. Klemm, Wiesbaden, pp. 277–291.
Cauer, H. (1949a) Ergebnisse chemisch-meteorologischer Forschung. *Archiv für Meteorologie, Geophysik und Bioklimatologie. Serie B* **1**, 221–256.
Cauer, H. (1949b) Chemisch-bioklimatische Studien in Königsstein im Taunus. *Archiv für physikalische Therapie, Balneologie und Klimatologie* **1**, 87–102.
Cauer, H. (1950) Chemisch-bioklimatische Studien in Wyk auf Föhr. Unpublished research report. Institut für Chemische Klimatologie (Hohenberg) and Bioklimatische Forschungsstelle des NWD-Wetterdienstes (Wyk), 37 pp.

Cauer, H. (1951a) Some problems of atmospheric chemistry. In: Compendium of Meteorology (ed. T. F. Malone), American Meteorological Society, Boston, pp. 1126–1138; note that is the translation of Cauer (1949b) but shortened (more interesting is the German version, but noted and cited was only the English version).
Cauer, H. (1951b) Chemische Spurenstoffe in der Hochgebirgsluft und an der Nordseeküste. *Archiv für physikalische Therapie, Balneologie und Klimatologie* **3**, 42–45.
Cauer, H. (1951c) Ergebnisse der Ozonuntersuchungen in den Jahren 1949/50. *Archiv für physikalische Therapie, Balneologie und Klimatologie* **3**, 200–210.
Cauer, H. (1954) Physikochemische Vorgänge bei der Kondensation yon Wasserdampf in der freien Atmosphäre. *Geofisica pura e applicata (Milano)* **28**, 199–207.
Cauer, H. (1956) Die pH-Werte on Aerosolen und Niederschlägen und ihre lufthygienische und bioklimatische Indikatorbedeutung. *Zeitschrift für Aerosol-Forschung und -Therapie* **5**, 459–508.
Cauer, H. (1957) Bioklimatische Bedeutung potentieller Aerosole. *International Journal of Bioclimatology Biometeorology* **1**, 31–33.
Cauer, H. and G. Cauer (1941) Studium über den Chemismus der Nebelkerne in Oberschreiberhau. *Der Balneologe* **8**, 345–353.
Cauer, H. and G. Cauer (1942) Das Magnesiumchlorid der Nebelkerne. *Der Balneologe* **9**, 301–309.
Cauer, H. and G. Cauer (1943) Chemisch-bioklimatische Studie des Jodmilieus von Bad Darkau (Oberschlesien). *Der Balneologe* **10**, 285–292.
Cavallo, T. (1782) A complete treatise on electricity in theory and practice with original experiments. 2nd ed. with considerable additions and alterations. London, 495 pp.
Cavallo, T. (1803) The elements of natural or experimental philosophy, in four volumes, Vol. 3. T. Cadell and W. Davies, London, 572 pp.
Cavanaugh, L. A., C. F. Scbadt and E. Robinson (1969) Atmospheric hydrocarbon and carbon monoxide measurements at Point Barrow, Alaska. *Environ. Sci. Technol.* **8**, 251–257.
Cavendish, H. (1766) Three papers containing experiments on factitious airs. *Phil. Trans.* **56**, 141–184.
Cavendish, H. (1783) An account of a new eudiometer. *Phil. Trans.* **73**, 106–135.
Cavendish, H. (1784) Experiments on air. *Phil. Trans.* **74**, 119–153 and **75** (1785) 372–384. Alembic Club Reprint (1893) No. 3, W. F. Clay, Edinburgh, 52 pp.
Cavendish, H. (1785) Experiments on air. *Phil. Trans.* **75**, 352–384.
Chabrier (1871) Dosages des acides nitreux et nitrique dans les eaux de pluie. *C. r.* **73**, 485–488. *See also*: Observations nouvelles sur la prédominance alternative de l'acide nitrique et de l'acide nitreux dans les eaux de pluie. Ibid., 1273–1274.
Chairy (1884) Sur les eaux de pluie de la ville d'Algier. *C. r.* 99, 869–871. *See also*: Chairy (1884) Étude de l'air de la ville d'Alger. *C. r.* **99**, 798–800.
Chakravarti, D. N. and N. R. Dhar (1925) Fluorescenz und chemische Umwandlungen. *Zeitschrift für anorganische und allgemeine Chemie* **142**, 299–328.
Chalmers, L. (1776) An account of the weather and diseases of South-Carolina. In two volumes. Edward and Charles Dilly, London, 222 pp.
Chalonge, D., F. W. P. Götz and E. Vassy (1934) Simultanmessungen des bodennahen Ozons auf Jungfraujoch und in Lauterbrunnen. *Naturwissenschaften* **22**, 297.
Chamberlin, T. C. (1899) An attempt to frame a working hypothesis of the cause of glacial periods on an atmospheric basis. *The Journal of Geology* **7**, 545–584.
Chameides, W. and J. C. G. Walker (1977) A time dependent photochemical model for ozone near the ground. *JGR* **81**, 413–420.
Chameides, W., D. H. Stedman, R. R. Dickerson, D. W. Rasch and R. J. Cicerone (1977) NO_x production in O(1D). *Journal of Atmospheric Sciences* **34**, 143–149.

Chameides, W. (1978) The photochemical role of tropospheric nitrogen oxides. *Geophys. Res. Lett.* **5**, 17–20.
Chameides, W. and D. D. Davis (1982) The free radical chemistry of cloud droplets and its impact upon the composition of rain. *JGR* **87**, 4863–4877.
Chapman, E. T. (1870) A note on the organic matter contained in the air. *Chemical News* **21**, 65.
Chapman, D. H. and H. E. Jones (1910) The homogeneous decomposition of ozone in the presence of oxygen and other gases. *J. Chem. Soc., Transactions* **97**, 2463–2477.
Chapman, D. H. and H. E. Jones (1911) Decomposition of dry ozone. *J. Chem. Soc., Transaction* **99**, 1811–1819.
Chapman, S. (1930) A theory of upper-atmospheric ozone. Memoirs of the Royal Meteorological Society 3, 103–125. See also: Discussion of memoirs. On a theory of upper-atmospheric ozone. *Q. J. Roy. Meteor. Soc.* **58** (1932), 11–13.
Chapman, S. (1951) Photochemical processes in the upper atmosphere and resultant composition. In: Compendium of Meteorology (ed. T. F. Malone). American Meteorological Society, Boston, pp. 262–274.
Chappuis, J. (1880) Sur le spectra d'absorption de l'ozone. *C. r.* **91**, 985–986.
Chappuis, J. (1882a) Sur le spectra d'absorption de l'ozone. *C. r.* **94**, 858–860.
Chappuis, J. (1882b) Étude spectroscopique sur l'ozone. *Annales scientifiques de l'École Normale Supérieure, Série 2* **11**, 137–186.
Charlson, R. and H. Rodhe (1982) Factors controlling the acidity of natural rainwater. *Nature* **295**, 683–685.
Charlson, R. J., J. E. Lovelock, M. O. Andreae and S. G. Warren (1987) Oceanic phytoplankton, atmospheric sulphur, cloud albedo and climate. *Nature* **326**, 655–661.
Chatin, A. (1851) Présence de l'iode dans l'air, et absorption de ce corps dans l'acte de la respiration animale. *C. r.* **32**, 668–672.
Chatin, A. (1854) Rapport sur les travaux de M. Chatin, relatifs à la recherche de l'iode, et sur différentes Notes ou Mémoires présentés sur le même sujet, par MM. Marchand, Niepce, Meyrac. *C. r.* **34**, 503–508.
Chatin, A. (1855) Présence de l'iode dans les eaux pluviales, les eaux courantes, et les plants des Antilles at les côtes de la Méditeranée. *C. r.* **37**, 958–959.
Chatin, A. (1859) Existence de l'iode dans l'air, les eaux, le sol at les produits alimentaire. *Annu. soc. mét. de France* **7**, 50–112.
Chen, C. T. and A. Poisson (1984) Excess carbon dioxide in the Weddell Sea. *Antarctic Journal* 1984 Rev., pp. 74–75.
Chernikov, A. A. (2009) Artificial rainfall. In: Hydrological cycle. Vol. II. Encyclopedia of Life Support Systems, pp. 272–294.
Chesselet, R., J. Morelli and P. Buat-Menard (1972) Variations in ionic ratios between reference sea water and marine aerosol. *JGR* **77**, 5116–5131.
Chevallier, A. (1834) Composition de l'atmospère. *Journal de Pharmacie et des Sciences Accessoires* **20**, 655–656.
Chiplonkar, M. W. (1939) Measurements of atmospheric ozone at Bombay. *Proceedings of the Indian Academy of Sciences, Sect A* **10**, 105–120.
Chladni, E. F. F. (1821) Neue Beiträge zur Kenntniß der Feuermeteore und der herabgefallenen Massen. *Gilb. Ann.* **68**, 329–367.
Chladni, E. F. F. (1826) Neue Beiträge zur Kenntniß der Feuermeteore und der herabgefallenen Massen. *Pogg. Ann.* **82**, 21–35, 161–183; **84**, 45–60.
Chlopin, W. [Хлопин] (1911) Об образовании окислителей в воздухе под влиянием действия ультра-фиолетовых лучей [Formation of oxidizing agents in air under the influence of ultra-violet light]. *Журнал руского физико-химического общества* [Journal of the Russian Physico-Chemical Society] **43**, 554–561. *See also*: Über die Entstehung von Oxydationsmittel in der atmosphärischen Luft bei Einwirkung von ultravioletten Strahlen. *Z. anorg. Chem.* **71**, 198–205.

Choularton, T. W., R. N. Colvile, K. N. Bower, M. W. Gallagher, M. Wells, K. M. Beswick, B. G. Arends, J. J. Möls, G. P. A. Kos, S. Fuzzi, J. A. Lind, G. Orsi, M. C. Facchini, P. Laj, R. Gieray, P. Wiesner, T. Engelhardt, A. Berner, C. Kruisz, D. Möller, K. Acker, W. Wieprecht, J. Lüttke, K. Levsen, B. Bizjak, H.-C. Hansson, S.-I. Cederfelt, G. Frank, B. Mentes, B. Martinsson, D. Orsini, B. Svenningsson, E. Swietlicki, A. Wiedensohler, K. J. Noone, S. Pahl, P. Winkler, E. Seyffer, G. Helas, W. Jaeschke, H. W. Georgii, W. Wobrock, M. Preiss, R. Maser, D. Schell, G. Dollard, B. Jones, T. Davies, D. L. Sedlak, M. M. David, M. Wendisch, J. N. Cape, K. J. Hargreaves, M. A. Sutton, R. L. Storeton-West, D. Fowler, A. Hallberg, R. M. Harrison and J. D. Peak (1997) The Great Dun Fell Cloud Experiment 1993: an overview. *Atmos. Env.* **31**, 2393–2405.

Christie, J. R. (1994) Historiography of chemistry in the eighteenth century: Hermann Boerhaave and William Cullen. *Ambix* **41**, 4–19.

Christensen, P. A., T. Yonar and K. Zakaria (2013) The electrochemical generation of ozone. *Ozone: Science & Engineering* **35**, 149–167.

Chromov, S. P. and L. I. Mamontova [Хромов, Мамонтова] (1974) Метеорологический словарь [Meteorological dictionary]. Гидрометеоиздат, Ленинград. 568 pp.

Ciamician, G. and P. Silver (1908) Chemische Lichtwirkungen. *Ber.* **41**, 1928–1935.

Ciamician, G. (1912) The photochemistry of the future. *Science* **36**, 385–394.

Claesson, P. (1876) Barythydrat, in geeigneter Weise präparirt, zu allen Zwecken der Kohlensäureabsorption anwendbar. *Ber.* **9**, 174–176.

Clarke, F. W. (1908) The data of geochemistry. Bulletin No. 330. US Govn. Printing Office, Washington, 716 pp.

Clarke, F. W. (1911) The data of geochemistry. 2nd ed. Government Printing Office, Washington D. C. (USA), 782 pp.

Clarke, F. W. (1920) The data of geochemistry. Government Printing Office, Washington D. C. (USA), 832 pp.

Clarke, A., S. R. Owens and J. Zhou (2006) An ultrafine sea-salt flux from breaking waves: Implications for cloud condensation nuclei in the remote marine atmosphere. *JGR* **111**, D06202, doi:10.1029/2005JD006565.

Claubry, H. F. Gaultier de (1827) XCVI. Auszug aus einem Schreiben des Hrn. Gaultier de Glaubry an Hrn. Gay-Lussac, über die Art, wie die alkalischen Chlorüre als Luft verbessernde, und Fäulniß zerstörende Körper wirken. *Polytechnisches Journal* (*Dingler's J.*) **23**, 447–450; *see also*: Lettre a M. Gay-Lussac sur la Manière d'Agir des Chlorures Alcalins Comme Corps Désinfectants. *Ann. chim. phys.* **33** (1826) 271–272.

Claude, H.-J. (1996) Das Ozonmeßprogram am Hohenpeißenberg. *Meteorologische Fortbildung* **25**, 104–108.

Clausius, R. (1849a) Ueber die Natur der geringen Bestandteile der Erdatmosphäre, durch welche die Lichtreflexion in derselben bewirkt wird. *Pogg. Ann.* **76**, 161–188.

Clausius, R. (1849b) Ueber die blaue Farbe des Himmels und die Morgen- und Abendröthe. Pogg. Ann. 76, 188–195.

Clausius, R. (1853) Ueber das Vorhandenseyn von Dampfbläschen in der Atmosphäre und ihren Einfluss auf die Lichtreflexion und die Farben derselben. *Pogg. Ann.* **88**, 543–556.

Clausius, R. (1857) Ueber die Art der Bewegung welche wir Wärme nennen. *Pogg. Ann.* **100**, 353–380.

Clausius, R. (1858) Ueber die Natur des Ozons. *Pogg. Ann.* **103**, 644–652. *See also*: On Ozone. *Philosophical Magazine*, Ser. 4. **16**, 45–51.

Clausius, R. (1864a) Ueber den Unterschied zwischen activem und gewöhnlichen Sauerstoffe. *Pogg. Ann.* **121**, 250–268.

Clausius, R. (1864b) Abhandlungen über die mechanische Wärmetheorie. Vieweg, Braunschweig, 351 pp.

Clausius, R. (1869) Zur Geschichte des Ozons. *Pogg. Ann.* **136**, 102–105.

Clay, K. and W. Troesken (2010) Did Frederick Brodie discover the world's first environmental Kuznev curve? Coal smoke and the rise and fall of the London fog. NBER Working Paper Ser., Nat. Bureau of Econ. Res., Cambrigde, 77 pp.

Clayden, A. W. (1891) On brocken spectres in a London fog. *Q. J. Roy. Meteor. Soc.* **17**, 209–216.

Clemens, Th. (1849) Wirkungen Ozon-zerstörender Gase auf den menschlichen Organismus, nebst einigen Beobachtungen über Ozon als Krankheitsursache. *Zeitschrift für rationelle Medicin* **7**, 237–244.

Clemens, Th. (1853) Malaria und Ozon, oder Untersuchung der Frage, inwiefern stehende Wässer durch Gasexhalationen oder Miasmen der menschlichen Gesundheit nachtheilig seien, mit besonderem Hinblick auf den, neben dem Hospitale zum heiligen Geist in Frankfurt a. M. belegenen Weiher. *Adolph Henke's Zeitschrift für die Staatsarzneikunde* **65**, 1–98. *See also* extended review by Moore, W. D. (1855) Malaria and Ozone; or, Examination of the Question, How Far Stagnant Waters are, through Gaseous Exhalations or Miasmata, Injurious to Human Health. With Especial Reference to the Fish-Ponds Situated near the Hospital of the Holy Ghost, at Frankfort on the Maine. *British Foreign Medical Chirurgical Review* **16** (31),123–136.

Cloez, F. S. (1856) Observations et expériences sur remploi de l'iodure de potassium comme réactif de l'ozone. *C. r.* **43**, 38–41.

Cloez, F. S. (1861) Sur la présence de l'acide nitrique libre et des composés nitreux oxygénés dans l'air atmosphérique. *C. r.* **52**, 527–530.

Cloud, P. and A. Gibor (1970) The oxygen cycle. *Scientific American* **223**, 110–123.

Cnöffel, A. (1676) Observatio Dn, Andreæ Cnoeffeli. In: Miscellanea curiosa medico-physica academiae naturae curiosorum sive ephemeridum medico physicarum germanicarum annus quartus et quintus Anni MDCLXXIII & MDCLXXIV. Continens celeberrimorum cirorum tum medicorum tum aliorum eruditorum in Germania & extra eam observatziones medicas physicas chymicas, accessit appendix, in qua non nulla lectu haut indigna aut ingrata occurrent. Francofurti & Lipsiæ, Johannis Fritsch, pp. 44–55.

Coblentz, W. W. and R. Stair (1939) Distribution of ozone in the stratosphere. *Journal of the Research of the National Bureau of Standards* **22**, 573–606.

Coblentz, W. W. and R. Stair (1941) Distribution of ozone in the stratosphere: measurements of 1939 and 1940. *Journal of the Research of the National Bureau of Standards* **26**, 161–174.

Coehn, A. (1907) Über die Einwirkung des Lichtes auf die Bildung der Schwefelsäure. *Z. Elektrochem.* **13**, 545–572.

Coehn, A. and H. Becker (1910) Zur Photochemie der Schwefelsäure. *Z. phys. Chem.* **70**, 88–115.

Cofer, W. R., V. G. Collins and R. W. Talbot (1985) Improved aqueous scrubber for collection of soluble atmospheric trace gases. *Environ. Sci. Technol.* **19**, 557–560.

Cogbill, C. (1976) The history and character of acid precipitation in eastern North America. In: Proceedings of the first international symposium on acid precipitation and the forest ecosystem (eds. L. S. Dochinger and T. A. Seliga). Gen. Tech. Rep. NE-23. Upper Darby (PA), U. S. Department of Agriculture, Forest Service, Northeastern Forest Experiment Station, pp. 363–370.

Cohen, J. B. (1896) The air of towns. In: Smithsonian Miscellaneous Collections v. 39, No. 1073, 42 pp. Published also in: Annual Report of the Board of Regents of the Smithsonian Institution for 1895. Washington (1896), pp. 349–387.

Cohen, J. B. and A. G. Ruston (1911) Soot: its character and composition. *Journal of the Society of the Chemical Industry* **30**, 1360–1364.

Cohen, J. B. and A. G. Ruston (1925) Smoke. A study of town air. 2nd edition (first: 1912), Arnold & Co., London, 108 pp.

Collett, Jr., J., B. Oberholzer and J. Staehelin (1993) Cloud chemistry at Mt. Rigi, Switzerland: dependence on drop size and relationship to precipitation chemistry. *Atmos. Env.* **27A**, 33–42.

Collins, W. D. and K. T. Williams (1933) Chloride and sulfate in rain water. *Industrial and Engineering Chemistry* **25**, 944–945.
Collison, R. C. and J. E. Mensching (1932) Lysimeter investigations. II. Composition of rain water at Geneva, N. Y., for a 10-year period. *New York State Agricultural Experiment Station, Technical Bulletin* **193**, 19 pp.
Colvile, R. N., K. N. Bower, T. W. Choularton, M. W. Gallagher, K. M. Beswick, B. G. Arends, G. P. A. Kos, W. Wobrock, D. Schell, K. J. Hargreaves, R. L. Storeton-West, J. N. Cape, B. M. R. Jones, A. Wiedensohler, H.-C. Hansson, M. Wendisch, K. Acker, W. Wieprecht, S. Pahl, P. Winkler, A. Berner, C. Kruisza and R. Gieray (1997) Meteorology of the Great Dun Fell Cloud Experiment 1993. *Atmos. Env.* **31**, 2407–2420.
Comel, A. (1933) Un secondo biennio do oscervazioni sull comossizine chimica delle acque piovane della medie pianura frincana (1930–1931) a resultatu riassunfivi del quadrienno 1918–31. *Annali della Stazione chimico-agraria sperimentale di Undine, Serie III* **2**, 12–179.
Configliachi, P. (1811) Ueber die Analyse der Luft aus der Schwimmblase der Fische. *Journal für Chemie und Physik* **1** (1811), 137–163.
Connick, R. E. (1947) The interaction of hydrogen peroxide and hypochlorous acid in acidic solutions containing chloride. *J. Amer. Chem. Soc.* **69**, 1509–1514.
Connolly, J. I., M. J. Martinek and J. J. Aeberley (1928) The carbon monoxide hazard in city streets. *American Journal of public Health* **18**, 1375–1383.
Conrad, V. (1901) Über den Wassergehalt der Wolken. Denkschrift der mathematisch-naturwissenschaftlichen Classe der Kaiserlichen Akad. der Wiss., Wien, Band **73**, 2, 1–7.
Cook, H. (1865) Reports on the registration of ozone in the Bombay Presidency for the years 1863–1864; 1864–1865, Bombay, 34 pp and 32 pp.
Cook, H. (1870) On the registration of ozone in the Bombay Presidency and the chief causes which influence its appreciable amount in the atmosphere. *Report of the British Association for the Advancement of Science*. Thirty-ninth Meeting held at Exeter in August 1869. pp. 64–67.
Cooke, S. (1887) Note on the decomposition of nitric oxide in contact with water and alkaline solutions. *Proceedings of the Philosophical Society of Glasgow* **18**, 299–302. *See also*: *Chemical News* **28** (1888) 155–156.
Cooper, W. S. (1957) Sir Arthur Tansley and the science of ecology. *Ecology* **38**, 658–659.
Cooper, W. J. and R. G. Zika (1983) photochemical formation of hydrogen peroxide in surface and ground water exposed to sunlight. *Science* **220**, 711–712.
Cooper, O. R., R.-S. Gao, D. Tarasick, T. Leblanc and C. Sweeney (2012) Long-term ozone trends at rural ozone monitoring sites across the United States, 1990–2010. *JGR* **117**, D22307.
Cooper, O. R., D. D. Parrish, J. Ziemke, N. V. Balashov, M. Cupeiro, I. E. Galbally, S. Gilge, L. Horowitz, N. R. Jensen, J.-F. Lamarque, V. Naik, S. J. Oltmans, J. Schwab, D. T. Shindell, A. M. Thompson, V. Thouret, Y. Wang and R. M. Zbinden (2014) Global distribution and trends of tropospheric ozone: An observation-based review. *Elementa: Science of the Anthropocene* **2**, 000029.
Cordeiro, F. J. B. (1885) A ready method of determination the amount of carbonic dioxide in the atmosphere. *The Sanitarian (New York)* **14**, 299–302.
Cornu, A. (1879a) Sur la limite ultra-violettes du spectra solaire. *C. r.* **88**, 1101–1108.
Cornu, A. (1879b) Sur l'absorption par l'atmosphère des radiations ultra-violettes. *C. r.* **88**, 1285–1290.
Cossa, A. (1865) Residuo delle acque piovane cadute a Pavia nel 1865. *Bollettino della Società Meteorologica Italiana* (cited after Mosello et al. (1992)).
Cossa, A. (1867) Ueber die Ozonometrie. *Z. anal. Chem.* **6**, 24–28.
Coste, J. H. and G. B. Courtier (1936) Sulphuric acid as disperse phase in town air. *Trans. Faraday Soc.* **32**, 1198–1201.
Coulier, P.-J. (1875) Note sur une nouvelle propriété de l'air. *J. de Pharmacie et de Chimie* **22**, 165–173.

Cowling, E. B. (1982) Acid precipitation in historical perspective. *Environ. Sci. Technol.* **16**, 110A–123A.
Cox, R. A. and S. A. Penkett (1970) The photooxidation of SO_2 in sunlight. *Atmos. Env.* **4**, 425–433.
Coxe, J. R. (1808) The Philadelphia Medical Dictionary: Containing a Concise Explanation of All the Terms Used in Medicine, Surgery, Pharmacy, Botany, Natural History, Chymistry and Materia Medica, Publ. Th. Dobson, Philadelphia, 433 pp.
Craig, R. A. (1950) The observations and photochemistry of atmospheric ozone and their meteorological significance. In: Meteorological Monographs, Vol. 1, American Meteorological Society, Boston, pp. 1–50.
Cramer, C. E. (1890) Die Luftverunreinigung. *Archiv für Hygiene* **10**, 313–330.
Crawford, M. F., H. L. Welsh and J. L. Locke (1949) Infra-red absorption of oxygen and nitrogen induced by intermolecular forces. *Physical Review* **75**, 1607.
Crosland, M. P. (1962) Historical studies in the language of chemistry. Heinemann, London, 406 pp.
Crove, A. (1887) Sur la mesure spectrométrique des hautes températures. *C. r.* **87**, 979–981.
Crowther, C. and A. G. Ruston (1911) The nature, distribution and effects upon vegetation of atmospheric impurities in and near an industrial town. *Journal of Agricultural Science* **4**, 25–55.
Crowther, C. and D. A. Steuart (1913) The distribution of atmospheric impurities in the neighbourhood of an industrial city. *Journal of Agricultural Science* **5**, 391–408.
Cruickshank, W. (1801) Additional remarks on galvanic electricity. *Journal of Natural Philosophy, Chemistry and the Arts* **4**, 254–264. *See also*: Fortgesetzte Beobachtungen über chemische Wirkungen der galvanischen Electricität. *Gilb. Ann.* **7**, 88–113 (on p. 107).
Crutzen, P. J. (1969) Determination of parameters appearing in dry and wet photochemical theories for ozone in the stratosphere. *Tellus* **21**, 368–388.
Crutzen, P. J. (1970) The influence of nitrogen oxides on atmosphere ozone content. *Q. J. Roy. Meteor. Soc.* **96**, 320–325.
Crutzen, P. J. (1971) Ozone production rates in an oxygen-hydrogen-nitrogen oxide atmosphere. *JGR* **76**, 7311–7327.
Crutzen, P. J. (1972) Gas phase nitrogen and methane chemistry in the atmosphere. Rept. AP-10, Institute of Meteorology, Univ. Stockholm. 20 pp.
Crutzen, P. J. (1973) Gas-phase nitrogen and methane chemistry in the atmosphere. In: Physics and Chemistry of the Upper Atmosphere (ed. B. McCormac), D. Reidel, Dordrecht, pp. 110–124.
Crutzen, P. J. (1974a) Photochemical reactions initiated by and influencing ozone in unpolluted tropospheric air. *Tellus* **26**, 47–57.
Crutzen, P. J. (1974b) Estimates of possible future ozone reductions from continued use of fluoro-chloro-methanes (CF_2Cl_2, $CFCl_3$). *Geophys. Res. Lett.* **1**, 205–208.
Crutzen, P. J. (1976) The possible importance of COS for the sulphur layer of the stratosphere. *Geophys. Res. Lett.* **3**, 73–76.
Crutzen, P. J. (1986) Is the detergent of the atmosphere decreasing: importance of methane for the OH radical concentration and atmospheric photochemistry. Luxembourg, Commission of the European Communities, 28 pp.
Crutzen, P. J. (1988) Tropospheric ozone: an overview. In: Tropospheric Ozone (ed. I. S. A. Isaksen), Reidel, Dordrecht, pp. 3–32.
Crutzen, P. J. (2002) A critical analysis of the Gaia hypothesis as a model for climate/biosphere interactions. *GAIA* **11**, 96–103.
Crutzen, P. J. and F. Arnold (1986) Nitric acid cloud formation in the cold Antarctic stratosphere. A major cause for the springtime "ozone hole". *Nature* **342**, 651–655.
Crutzen, P. J. and J. Birks (1982) The atmosphere after a nuclear war: twilight at noon. *Ambio* **11**, 114–125.
Crutzen, P. J. and D. Ehhalt (1977) Effects of nitrogen fertilizers and combustion on the stratospheric ozone layer. *Ambio* **6**, 112–117.

Crutzen, P. J., L. E. Heidt, J. P. Krasnec, W. H. Pollock and W. Seiler (1979) Biomass burning as a source of atmospheric gases CO, H_2, N_2O, NO, CH_3Cl. *Nature* **282**, 253–256.

Crutzen, P. J. and P. H. Zimmermann (1991) The changing photochemistry of the troposphere. *Tellus* **43B**, 136–151.

Crutzen, P. J., M. L. Lawrence and U. Pöschl (1999) On the background photochemistry of tropospheric ozone. *Tellus* **51A–B**, 123–146.

Crutzen, P. J. and E. F. Stoermer (2000) The “Anthropocene”. *Global Change Newsletters* **41**, 17–18.

Cuevas, C. A., N. Maffezzoli, J. P. Corella, A. Spolaor, P. Vallelonga, H. A. Kjær, M. Simonsen, M. Winstrup, B. Vinther, C. Horvat, R. P. Fernandez, D. Kinnison, J. F. Lamarque, C. Barbante and A. Saiz-Lopez (2018) Rapid increase in atmospheric iodine levels in the North Atlantic since the mid-20th century. *Nature communications* **9**, 1452.

Cunliffe, R. J. (1963) A lexicon of the homeric dialect. University of Oklahoma Press, Norman, 445 pp.

Cunningham, R. M. (1941) Chloride content of fog water in relation to air trajectory. *Bulletin of the American Meteorological Society* **22**, 17–21.

Curry, M. (1946) Bioklimatik. Die Steuerung des gesunden und kranken Organismus durch die Atmosphäre. Selbstverlag, Riederau (Ammersee), 2. Vol., 1434 pp. *See also*: Curry, M. (1950) Abhängigkeit klinischer Beobachtungen vom Arangehalt der Luft. *Acta Neurovegetativa* **1**, 408–426.

Dachauer, G. (1864) Ozon. Eine gedrängte Zusammenstellung bisher gewonnener Resultate. E. H. Gummi, München, 204 pp.

Dainton, F. S. and H. M. Kimberley (1950) The reaction between phosphorus vapour and oxygen. *Trans. Faraday Soc.* **46**, 629–641.

Dalton, J. (1802) Experimental essays, on the constitution of mixed gases; on the force of steams or vapours from water and other liquids in different temperatures, both in a torricellian vacuum and in air; on evaporation; and on the expansion of gases by heat. *Memoirs of the Literary and Philosophical Society of Manchester* **5** (Part 2), 535–603.

Dalton, J. (1805) Experimental enquiry into the proportions of the several gases or elastic fluids constituting the atmosphere. *Memoirs of the Literary and Philosophical Society of Manchester, 2nd series* **1**, 244–258. *See also*: *Philosophical Magazine* Series 1, **23** (1806) 349–356.

Dalton, J. (1813) On respiration and animal heat. *Memoirs of the Literary and Philosophical Society of Manchester, 2nd series* **2**, 15–44.

Dalton, J. (1824) Saline impregnation of rain which fell during the late storm, December 5th 1822. *Memoirs of the Literary and Philosophical Society of Manchester, 2nd series* **4**, 324–331 and (ibid.) Appendix to the essay on salt rain, page 324, with additional observations on the succeeding storms of wind and rain. 363–372 and *The Edinburgh Journal of Sciences* (1825) **2**, 176.

Damköhler, G. (1935) Neubestimmung des Krypton- und Xenon-Gehaltes der atmosphärischen Luft. *Z. Elektrochem.* **41**, 74–80.

Damschen, D. E. and L. R. Martin (1983) Aqueous aerosol oxidation of nitrous acid by O_2, O_3 and H_2O_2. *Atmos. Env.* **17**, 2005–2011.

Dancer, J. B. (1876) An account of some early experiments with ozone and remarks, upon its electric origin. *Proceedings of the Literary and Philosophical Society of Manchester* **15**, 121–127.

Daniels, F. and A. C. Bright (1920) Pressure measurements of corrosive gases. The vapor pressure of nitrogen pentoxide. *J. Am. Chem. Soc.* **42**, 1131–1141.

Daniels, F. and E. H. Johnston (1921) The thermal decomposition of gaseous nitrogen pentoxide. A monomolecular reaction. *J. Am. Chem. Soc.* **43**, 53–71.

Danielson, E. E. (1968) Stratospheric-tropospheric exchange based on radioactivity, ozone and potential vorticity. *Journal of Atmospheric Sciences* **25**, 502–518.

D’Ans, J. and E. Lax, eds. (1943) Taschenbuch für Chemiker und Physiker. Springer-Verlag, Berlin, 1896 pp.

d'Arcet, J. P. J. (1834) Essais sur l'air atmospherique de Londres. *Journal de Chimie Médicale de Pharmacie at de Toxicologie* **10**, 292–295.

D'Arcy, R. F. (1902) The decomposition of hydrogen peroxide by light and the electrical discharging action of this decomposition. *The London, Edinburgh and Dublin Philosophical Magazine and Journal of Science* **3**, 42–52.

Daremberg, G. (1878) Sur la recherche de l'ozone dans l'air atmosphérique. *C. r.* **86**, 1203–1205. *See also*: Daremberg: Communication. *Annu. soc. mét. de France* **26** (1878) 161.

Daremberg, G. (1879) Sur la valeur des papiers ozonoscopiques iodure-amidonniés. Congrès international de météorologie tenu à Paris du 24 au 28 août 1878. *Comptes rendus sténographique* No. 20. Imprimerie Nationale, Paris, pp. 204–206.

Darley, E. F., K. A. Kettner and E. R. Stephens (1963) Analysis of peroxyaceyl nitrates by gas chromatography with electron capture detection. *Analytical Chemistry* **35**, 589–591.

Darmstaedter, L. (1908) Ludwig Darmstaedters Handbuch zur Geschichte der Naturwissenschaften und der Technik. In chronologischer Darstellung. 2., umgearb. und verm. Aufl. Unter Mitwirkung von Professor Dr. R. du Bois-Reymond und Oberst z. d. C. Schaefer. Springer, Berlin, 1262 pp.

Darwin, C. (1839) The voyage of the Beagle. Wordsworth Classics. (1997), 480 pp.

Das, T. N., P. N. Moorthy and K. N. Rao (1983) Chemiluminescent measurement of atmospheric peroxide in the Bombay area. *Atmos. Env.* **17**, 79–82.

Das, M. and V. P. Aneja (1994) Measurements and analysis of mixing ratios of gaseous hydrogen peroxide and related species in the rural central Piedmont region of North Carolina. *Atmos. Env.* **28**, 2473–2483.

Dasch, J. M. (1988) Hydrological and chemical inputs to fir trees from rain and clouds during a 1-month study at Clingmans Peak. *N. C. Atmos. Env.* **22**, 2255–2262.

Da Silva, L. M., L. A. De Faria and J. F. C. Boodts (2003) Electrochemical ozone production: influence of the supporting electrolyte on kinetics and current efficiency. *Electrochimica Acta* **6**, 699–709.

Däßler, H.-G. and G. Stein (1968) Luftanalytische Untersuchungen im Erz- und Elbsandsteingebirge mit ständig betriebenen SO_2- und Staubmessstellen. *Luft- und Kältetechnik* **7**, 315–318.

Daube, B. C., K. D. Kenneth, P. A. Lamar and K. C. Weathers (1987) Two ground-level cloud water sampler designs which reduce rain contamination. *Atmos. Env.* **21**, 893–900.

Daubeny, C. (1836) Report on the present state of knowledge with respect to mineral and thermal waters. In: Report of the Sixth Meeting of the British Association for the Advanced Science. Vol. 7, J. Murray, London, pp. 1–97.

Daubeny, C. (1867) On ozone. *J. Chem. Soc.* **20**, 1–28.

Daum, P. H., T. J. Kelly, S. E. Schwartz and L. Newman (1984) Measurement of the chemical composition of stratiform clouds. *Atmos. Env.* **18**, 2671–2684.

Daum, P. H., L. I. Kleinman, A. J. Hills, A. L. Lazrus, A. C. D. Leslie, K. Busness and J. Boatman (1990) Measurement and interpretation of concentrations of H_2O_2 and related species in the Upper Midwest during summer. *JGR* **95**, 9857–9871.

Daumas, M. (1941) Lavoisier. Gallimard, Paris, 161 pp.

Dauvillier, A. (1933) Sur l'originie de l'ozone atmosphérique. Recherches faites au Scoresby Sund durant l'année polaires. *C. r.* **197**, 1339–1341.

Davenport, S. I. and G. G. Morgis (1954) Air pollution: a bibliography. Bulletin 537. Bureau of Mines, U. S. Governmental Printing office, Washington, 448 pp.

Davies, P. B. and B. F. Thrush (1968) Reactions of oxygen atoms with hydrogen cyanide, cyanogen chloride and cyanogen bromide. *Trans. Faraday Soc.* **64**, 1836–1843.

Davy, E. (1831) On a new combination of chlorine and nitrous gas. *Philosophical Magazine* **9**, 355–356. *See also*: *Proc. R. Soc.* **3** (1837) 27–29.

Davy, H. (1800) Researches chemical and philosophical; chiefly concerning nitrous oxide, or dephlogisticated nitrous air and its respiration. J. Johnson, London, 580 pp.

Davy, H. (1801) An account of a new Eudiometer. *Journals of the Royal Institution* **1**, 45–48; republished in *Philosophical Magazine*, 1801, **10**, 56–58 and in *A Journal of Natural Philosophy, Chemistry and the Arts*, 1801, **5**, 175–177.

Davy, H. (1805) Davy's neues Eudiometer, und Versuche damit. *Gilb. Ann.* **19**, 394–399.

Davy, H. (1828) On the phænomena of volcanoes. *Phil. Trans.* **118**, 241–250.

Dawson, G. A. (1977) Atmospheric ammonia from undisturbed land. *JGR* **82**, 3125–3133.

Day, H. (1868) On ozone. *The Lancet* **91**, 79–81 and 124–126.

Day, J. (1866) Some observations of ozone, as recorded in Greelong during the month of August and September 1866. *Australian Medical Journal* **10**, 397–402.

Decharmes, C. (1856) Météorologie. De l'ozone et de l'importance des observations ozonométriques, relativement au degré de salubrité de l'air, et de la marche des épidémies. *Memoire lu a l'Academie d'Amiens*. 12 pp.

DeFelice, T. P. and V. K. Saxena (1991) The characterisation of extreme episodes of wet and dry deposition of pollutants on an above cloud-base forest during its growing season. *Journal of Applied Meteorology* **30**, 1548–1561.

de la Rive, A. A. (1841) Neue Untersuchungen über die Eigenschaften der discontinuierlichen electrischen Ströme von abwechselnden entgegengesetzten Richtungen. *Pogg. Ann.* **54**, 378–408.

de la Rive, A. A. (1845) Sur les mouvements vibratoires que déterminent dans les corps, soit la transmission des courants électriques, soit leur influence extérieure. *C. r.* **20**, 1287–1291.

Dellmann, J. F. (1869) Ueber atmosphärische Electricität. III. Der Höhenrauch. *Z. Öster. Ges. Met.* **4**, 513–529.

Deluc, J.-A. (1777) Barometric observations on the depth of the mines in the Hartz. Phil. Trans. Vol. LXVII, Part II, pp. 401–449.

Deluc, J.-A. (1787) Idées sur la météorologie. Duchesne, Paris. 543 pp.; In German: Lüc, J. A. (1787) Neue Ideen über die Meteorologie. Aus dem Französischen übersetzt. F. Nicolai, Berlin und Stettin, 459 pp.

de Luca, S. (1854) Recherches d'iode dans l'air, dan l'eau de pluie er dans la neige. *Journal de Parmacie et de Chimie* **16**, 250–263.

de Luca, S. (1888) Recherches sur les matiéres organique et minerales des eaux de pluie. *C. r.* **53**, 153–156.

Delwiche, C. C. (1970) The nitrogen cycle. *Scientific American* **223**, 136–147.

Demarée, G. R. and A. E. J. Ogilvie (2017) L'eruption du Lakagígar en Islande ou "Annus Mirabilis 1783". Chronique d'une année extraordinaire. *SÉMATA, Ciencias Sociais e Humanidades* **29**, 239–260.

Demerjian, K., J. A. Kerr and J. Calvert (1974) Mechanism of photochemical smog formation. In: Advances in Environmental Science and Technology. Vol. 4 (eds. J. N. Pitts, R. L. Metcalf and A. C. Lloyd), John Wiley, New York, pp. 1–262.

Demuth, N. and K. G. Heumann (1999) Determination of trace amounts of boron in rainwater by ICP-IDMS and NTI-IDMS and the dependence on meteorological and anthropogenic influences. *Journal of Analytical Atomic Spectrometry* **14**, 1449–1453.

Deng, Y. and Y. Zuo (1999) Factors affecting the levels of hydrogen peroxide in rainwater. *Atmos. Env.* **33**, 1469–1478.

Denisov, P. V. [Денисов] (1956) Химический состав атмосферных осадков Северного Тянь-Шаня [Chemical composition of atmospheric precipitation in northern Tien-Shan]. *Доклады Академий Наук СССР* **110**, No 5.

Denisov, P. V. and A. L. Bugaev [Денисов, Бугаев] (1956) Химический состав атмосферных осадков северовосточной части Украины [Chemical composition of atmospheric precipitation in northeastern Ukraine]. *Доклады Академий Наук СССР* **108**, 879–881.

Dentener, F. J. and P. J. Crutzen (1994) A three-dimensional model of the global ammonia cycle. *J. Atmos. Chem.* **19**, 331–369.
Denza, F. (1868) Valeurs de l'électricité et l'ozono à Moncalieri à l'époque du choléra in. *C. r.* **62**, 105–108. *See also*: Sui valori delle'electricità e dell'ozono ostervati a Moncalieri nel tempo di cholera. *Bullettino meteorologico dell' Osservatorio del R. Collegio Carlo Alberto* **2** (1868) 399.
Derwent, R. G. (1987) Modelling the long-range transport of ammonium and ammonia compounds. In: Proceedings of the Symp. Ammonia and Acidification, Bilthoven, The Netherlands, pp. 223–238.
Derwent, R. G. (1990) Evaluation of a number of chemical mechanisms for their application in models describing the formation of photochemical ozone in Europe. *Atmos. Env.* **24**, 2615–2624.
Derwent, R. G., M. E. Jenkin and T. J. Wallington (2017) Photochemical ozone creation potentials for volatile organic compounds: Rationalization and estimation. *Atmos. Env.* **163**, 128–137.
Descartes, R. (1637) Les Météores (reprint). Die Meteore (German translation), Ed. C. Zittel. In: Zeitsprünge – Forschungen zur Frühen Neuzeit, Band 10 (2006), 339 pp.
Dewar, J. M. S. (1948) The structure of ozone. *J. Chem. Soc.* 299–1305.
Dhar, N. R. (1922) Katalyse XVI. Strahlung als Faktor bei thermischen und photochemischen Reaktionen. *Zeitschrift für anorganische und allgemeine Chemie* **122**, 151–158.
Dhar, N. H. (1932) Formaldehyde in rain water. *Nature* **130**, 313–314.
Dhar, N. H. (1933) Formaldehyde in dew. *Nature* **131**, 800.
Dhar, N. H. (1934) Formation of hydroxyl radicals from photolysis of water and the generation of formaldehyde. *Trans. Faraday Soc.* **30**, 142–148.
Dhar, N. H. and A. Ram (1932a) Photoreduction of carbonic acid, hydrogen carbonates and carbonates to formaldehyde. *Nature* **129**, 205.
Dhar, N. H. and A. Ram (1932b) Radiation hypothesis of chemical reactions and concepts of threshold wave length. *J. phys. Chem.* **36**, 646–657.
Dhar, N. H. and A. Ram (1933a) Variation in the amounts of ammoniacal and nitric nitrogen in rain water of different countries and the origin of nitric acid in the atmosphere. *Journal Indian Chemical Society* **10**, 125–133.
Dhar, N. H. and A. Ram (1933b) Presence of formaldehyde in the terrestrial and solar atmosphere. *Journal Indian Chemical Society* **10**, 161–166.
Dhar, N. H. and A. Ram (1933c) Formaldehyde in rain and dew and its formation by photo-oxidation of organic compounds and the problem of carbon assimilation. *Journal Indian Chemical Society* **10**, 287–298.
Dhar, N. H. and A. Ram (1933d) Photosynthesis in tropical sunlight. VI. Presence of formaldehyde in rain water. *J. phys. Chem.* **37**, 525–532.
Dhar, N. H. and A. Ram (1933e) Formaldehyde in the upper atmosphere. *Nature* **132**, 819–820.
Dhar, N. R., S. P. Tandon, N. N. Biswas and A. K. Bhateacharya (1934) Photo-oxidation of nitrite to nitrate. *Nature* **133**, 213–214.
Diaz, G. C. (1994) The Nitrate Deposits of Chile. In: Tectonics of the Southern Central Andes (eds. K. J. Reutter, E. Scheuber and P. J. Wigger), Springer, Berlin, pp. 303–316.
Dieudonné, A. (1894) Über die Bedeutung des Wasserstoffperoxyds für die bakterientötende Kraft des Lichtes. *Arbeiten aus dem Kaiserlichen Gesundheitsamte* **9**, 537–540.
Dines, W. H. (1894) On the relation between the mean quarterly temperature and the death rate. *Q. J. Roy. Meteor. Soc.* **20**, 173–179.
Diness, G. (1879) Dew, Mist and Fog. *Q. J. Roy. Meteor. Soc.* **5**, 157–166.
Diogenes (1921) Diogenes Laertius. Leben und Meinungen berühmter Philosophen. Übers. und erl. von Otto Apelt. In 6 Vol. Meiner, Leipzig.
Dirnagl, K. (1949) An instrument that records the oxidizing effect of the air. *Bulletin of the American meteorological Society* **30**, 214–217.

Dirnagl, K. (2002) Erlebte Bäderheilkunde. Ein Physiker auf der Suche nach den Quellen. Selbstverlag, Riederau, 172 pp.

Ditte, A. (1904) Metals in the atmosphere. *Annual Report of the Smithsonian Institution* for 1904, pp. 235–247.

Divers, E. (1871) On the existence and formation of salts of nitrous oxide. *Chemical News* **23**, 206–207.

Di Vestea, A. (1916) La pioggia considerata sopra piani verticali orientati: pioggia obliqua. *Annali delle Università Toscane, Nuova serie*, vol. 1, fasc. 1, 26 pp.

Dobell, C. (1932) Antony van Leuvenhoeck and his "little animals". Being some account of the father of protozoology and bacteriology and his multifarious discoveries in these disciplines. Collected, translated and edited, from his printed works, unpubl. manuscripts and contemporary records. Harcourt, Brace and Company, New York, 516 pp.

Dobson, G. M. B. (1930) The ozone in the earth's atmosphere. *Beiträge zur Physik der freien Atmosphäre* **16**, 76–85.

Dobson, G. M. B. (1931) A photoelectric spectrophotometer for measuring the amount of atmospheric ozone. *Proceedings of the Physical Society, London* **43**, 324–339.

Dobson, G. M. B. and A. R. Meetham (1934) Some problems of modern meteorology, No. 15 atmospheric ozone and meteorology. *Q. J. Roy. Meteor. Soc.* **60**, 265–278.

Dobson, G. M. B. (1968) Forty years' research on atmospheric ozone at Oxford: a history. *Applied Optics* **7**, 387–405.

Dobson, G. M. B. and D. N. Harrison (1926) Measurements of the amount of ozone in the earth's atmosphere and its relation to other geophysical conditions. *Proc. R. Soc. Lond.* **110**, 660–693.

Dohrandt, F. (1874) Zur Kritik der Ozonbeobachtungen. *Repertorium für Meteorologie* (ed. Kaiserliche Akademie der Wissenschaften St. Petersburg) **3**, no. 4, 1–16.

Dollard, G. J. and T. J. Davies (1992) Observations of H_2O_2 and PAN in a rural atmosphere. *Environmental Pollution* **75**, 45–52.

Dollard, G. J. and T. J. Davies (1993) Measurements of rural photochemical oxidants. In: The Chemistry and Deposition of Nitrogen Species in the Troposphere (ed. A. T. Cox), Royal Society of Chemistry, Cambridge, pp. 46–77.

Downes, A. and T. P. Blunt (1877) Researches on the effect of light upon bacteria and other organisms. *Proc. R. Soc. Lond.* **26**, 488–500.

Downes, A. and T. P. Blunt (1878) On the influence of light upon protoplasm. *Proc. R. Soc. Lond.* **28**, 199–212.

Downes, A. and T. P. Blunt (1879a) Note upon the behaviour of oxalic acid and the oxalates of the alkalies and of potassium iodide, in sunlight. *Proc. R. Soc. Lond.* **29**, 219–221.

Downes, A. and T. P. Blunt (1879b) The effect of sunlight upon hydrogen peroxide. *Nature* **20**, 521.

Doyle, G. J. (1961) Self-nucleation in the sulfuric acid-water system. *J. phys. Chem.* **35**, 795–799.

Dragendorff, G. (1892) Kohlensäuregehalt der Luft in und bei Dorpat. *Sitzungsberichte der Naturforscher-Gesellschaft bei der Universität Dorpat* **9**, 158–160. *See also*: Über den Kohlensäuregehalt der Luft. *Chemisches Central-Blatt* 3. *Folge* (1887) **18**, 1367.

Draper, H. N. (1883) Salt and dew. *Nature* **22**, 77.

Draper, W. M. and D. G. Grosby (1981) Hydrogen peroxide and hydroxyl radical: Intermediates in indirect photolysis reactions in water. *Journal of Agricultural and Food Chemistry* **29**, 699–702.

Draper, W. M. and D. G. Grosby (1983) Photochemical generation of hydrogen peroxide in natural waters. *Archive of Environmental Contamination and Toxicology* **12**, 121–126.

Drasche, A. (1860) Die epidemische Cholera: Eine monographische Arbeit. C. Gerold, Wien, 387 pp.

Driessen, J. C. (1822) De Acidi Muriatici praesentia in aëre atmosphaerico. *Schweigger's Journal für Chemie und Physik* **36**, 139–156.

Drischel, H. (1940) Chlorid-, Sulfat- und Nitratgehalt der atmosphärischen Niederschläge in Bad Reinerz und Oberschreiberhau im Vergleich zu bisher bekannten Werten anderer Orte. *Der Balneologe* **7**, 321–334.

Drozdova, V. M. [Дроздова] (1962) Характеристика минерализации и химического состава воды атмосферных осадков, собранных в различных пунктах СССР за период МГГ и МГС [Characterization of the salinity and chemical composition of atmospheric precipitation collected at different sites of the USSR during IGY and IGS]. *Труды ГГО* [Trudy GGO] **134**, 26–32.

Drozdova, V. M., O. P. Petrenchuk and P. F. Svistov [Дроздова, Петренчук, Свистов] (1962) Некоторые данные о составе облачной воды [Some data on the composition of clouds]. *Труды ГГО* **134**, 131–134.

Drozdova, V. M., O. P. Petrenchuk, E. S. Selezneva and P. F. Svistov [Дроздова, тренчук, Селезнева, Свистов] (1964) Химический состав атмосферных осадков на Европейской территории СССР [Chemical composition of atmospheric precipitation over the European territory of the USSR]. Leningrad, Gidrometeoizdat, 212 pp.

Drozdova, V. M. and E. P. Makhonko (1970) Content of trace elements in precipitation. *JGR* **75**, 3600–3612.

Duane, W. and O. U. Scheuer (1913) Recherches sur la décomposition de l'eau par les rayons α. Le Radium 10, 33–46;Décomposition de l'eau par les rayons α. *C. r.* **156**, 466–467.

Duce, R. A. (1969) On the source of gaseous chloride in the marine atmosphere. *JGR* **74**, 4597–4599.

Duce, R. A. (1979) Marine aerosols. In: Preprints National 177th Meeting of the American Chemical Society, Division of Environmental Chemistry, Honolulu, Hawaii, Vol. 19, No. 1, pp. 385–388.

Duce, R. E. (1997) Christian Junge (1912–1997). *Eos* (Transactions of the American Geophysical Union) **78**, 39–50.

Ducros, H. (1845) Observations d'une pluie acide. *Journal de Pharmacie et de Chimie* **7**, 273–277.

Due, S. (1995) John Day, a colonial medical life. *Journal of Medical Biography* **3**, 99–104.

Duerst, U. (1939) Neue Forschungen über Verteilung und analytische Bestimmung der wichtigsten Luftgase als Grundlage for deren hygienische und tierzüchterische Wertung. *Schweizer Archiv für Tierheilkunde* **81**, 305–317.

Dufour, L. (1870) Notes sur le problème de la variation du climat. *Bulletin de la Société Vaudoise des Sciences Naturelles* **10**, 359–436.

Düggeli, M. (1919) Die Schwefelbakterien. Neujahrsblatt der Naturforschenden Gesellschaft zu Zürich. Beer & Cie, Zürich, 46 pp.

Dumas, J.-B. and E. M. Peligot (1835) Mémoire sur l'esprit de bois et sur le divers composes ehérés qui en proviennent. *Ann. chim. phys.* **58**, 5–74.

Dumas, J.-B. (1882) Sur l'acide carbonique normal de l'air atmosphérique. *Ann. chim. phys.* **26**, Ser. 5, 254–261.

Dumas, J.-B. and J. P. Boussingault (1841) Untersuchungen über die wahre Zusammensetzung der atmosphärischen Luft. *J. prakt. Chem.* **24**, 65–89. And: Untersuchungen über die wahrhafte Zusammensetzung der Atmosphäre. *Pogg. Ann.* **53**, 391–408. *See also*: Recherches sur la véritable constitution de l'air atmosphérique. Bibliothèque Universelle de Genève, Nouvelle Sèrie **33**, 362–388. Furthermore: *C. r.* **23**, 1005–1025.

Dunstan, W. R., H. A. D. Jowett and E. Goulding (1905) CLIII. – The rusting of iron. *Transaction of the Chemical Society* **87**, 1548–1574.

Dütsch, H. U. (1980) Ozon in der Atmosphäre. Gefährdet die Stratosphärenverschmnutzung die Ozonschicht? *Vierteljahreschrift der Naturforschenden Gesellschaft zu Zürich* **124**, Nr. 5, 48 pp.

Duveen, D. and H. Klickstein (1954) The introduction of Lavoisier's chemical nomenclature into America. *Isis* **45**, 278–292.

Eaton, S. V. and J. H. Eaton (1926) Sulfur in rainwater. *Plant Physiology* **1**, 77–87.

Ebermayer, E. (1873a) Das Ozon. *Z. Öster. Ges. Met.* **8**, 353–359, 369–372.

Ebermayer, E. (1873b) Ozongehalt der Luft im Walde und freiem Felde, oder hygienische Bedeutung des Waldes. In: Die physikalischen Einwirkungen des Waldes auf Luft und Boden und seine klimatologische und hygienische Bedeutung: begründet durch die Beobachtungen der forstl.-meteorolog. Stationen im Königreich Bayern, C. Krebs, Aschaffenburg, pp. 237–246.

Ebermayer, E. (1886) Untersuchungen über der Sauerstoffgehalt der Waldluft. *Forschungen auf dem Gebiet der Agricultur-Physik (Wollny's Forschungen)* **9**, 229–243.
Ebers, J. (1799) The new and complete dictionary of the German and English languages: composed chiefly after the German dictionaries of mr. Adelung and of mr. Schwan: every German word being rendered into proper English and thoroughly enriched with phrases and terms of arts and sciences: a work, which will be useful and even indispensible and therefore welcome to all such as have a mind to translate or read the works of either of the two languages, Band 3. Breitkopf and Haertel, Leipzig.
Ebert, H. (1901) Verteilung der electrischen Ionen in den höheren Schichten der Atmosphäre. *Terrestrial Magnetism and Atmospheric Electricity* **6**, 97–118. *See also*: Aspirationsapparat zur Bestimmung des Ionengehalts der Atmosphäre. *Physikalische Zeitschrift* **2** (1901) 662–664.
Eckardt, W. R. (1912) Klima und Leben (Bioklimatologie). Göschen, Berlin und Leipzig, 83 pp.
Ecker, A. (1847) Das Ozon vielleicht Ursache von Krankheiten. Beobachtungen von C. J. Schönbein. *Zeitschrift für rationelle Medicin* **6**, 178–184.
Edelstein, S. (1948) Priestley settles the water controversy. *Chymia* **1**, 123–137.
Eder, J. M. (1906) Photochemie (die chemischen Wirkungen des Lichtes). Wilhelm Knapp, Halle, 533 pp.
Edgar, J. L. and F. A. Paneth (1941) The separation of ozone from other gases. J. Chem. Soc.0, 511–519. The determination of ozone and nitrogen dioxide in the atmosphere. Ibid., pp. 519–527.
Effenberger, E. (1940) Kern- und Staubuntersuchungen am Collmberg. *Veröffentlichungen des Geophysikalischen Institutes Universität Leipzig* **12**, 305–359. (Dissertation).
Effenberger, E. (1948) Kritische und vergleichende Untersuchungen der luftchemischen Meßmethoden des Ozongehaltes und Gesamtoxydationswertes. *Meteorologische Rundschau* **1**, 488–491.
Effenberger, E. (1949) Messungen des Ozongehalts und des Gesamtoxydationswertes der bodennahen Luftschicht auf Sylt. *Meteorologische Rundschau* **2**, 94–97.
Effenberger, E. (1951) Messmethoden zur Bestimmung des CO_2-Gehaltes der Atmosphäre und die Bedeutung derartiger Messungen für die Biometeorologie und Meteorologie. Zweiter Teil: Ergebnisse der bisherigen CO_2-Messungen. *Annalen der Meteorologie* **4**, 417–427.
Egli, J. (1947) Geschichte des Wortes Gas. *Das Mosaik. Kunst Kultur Natur* **2/4**, 125 ff.
Egnér, H. and E. Eriksson (1955) Current data on the chemical composition of air and precipitation. *Tellus* **7**, 134–139.
Egnér, H., E. Eriksson and A. Emanuelsson (1947) Composition of atmospheric precipitation, I. Sampling Technique. *Kungl. Lantbukshögskolans Annaler – Annals of the Royal Agricultural College of Sweden* **16**, 593–602.
Ehhalt, D. H. (1967) Methane in the atmosphere. *Journal of the Air Pollution Control Association* **17**, 518–519. *See also*: Ehhalt, D. H. (1974) The atmospheric cycle of methane. *Tellus* **26**, 58–70.
Ehmert, A. (1941a) Ein einfaches Verfahren zur Messung kleinster Jodkonzentrationen, Jod- uns Natriumthiosulfatmengen in Lösung. *Zeitschrift für Naturforschung* **46**, 321–327.
Ehmert, A. (1941b) Über den Ozongehalt der unteren Atmosphäre bei winterlichem Hochdruckwetter nach Messungen. *Forschungs- und Erfahrungsberichte des Reichswetterdienstes* **13**. And: Über den Tagesgang des bodennahen Ozons. *Berichte des Deutschen Wetterdienstes US-Zone* **11** (1949) 63–66.
Ehmert, A. and H. Ehmert (1941) Über die chemische Bestimmung des Ozongehalts der Luft. *Forschungs- und Erfahrungsberichte des Reichswetterdienstes* **13**. And: *Berichte des Deutschen Wetterdienstes US-Zone* **11** (1949) 67–71. *See also*: Über den Tagesgang des bodennahen Ozons. *Ibid.* **11** (1949) 53–62.
Ehmert, A. (1952) Gleichzeitige Messungen des Ozongehalts bodennaher Luft an mehreren Stationen mit einem einfachen absoluten Verfahren. *Journal of Atmospheric and Terrestrial Physics* **2**, 189–195.

Ehrenberg, C. G. (1830) Neue Beobachtungen über blutartige Erscheinungen in Aegypten, Arabien und Sibirien, nebst einer Uebersicht und Kritik der früher bekannten. *Pogg. Ann.* **17**, 477–514.

Ehrenberg, C. G. (1849) Passatstaub und Blutregen: Ein grosses organisches unsichtbares Wirken und Leben in der Atmosphäre. Königliche Akademie der Wissenschaften, Berlin, 191 pp.

Ehrenberg, C. G. (1872) Übersicht der seit 1847 fortgesetzten Untersuchungen über das von der Atmosphäre unsichtbar getragene reiche organische Leben. Nachtrag zur Übersicht der organischen Atmosphärilien. *Abhandlungen der Königlichen Akademie der Wissenschaften zu Berlin aus dem Jahr 1971.* pp. 1–150 and 233–275.

Ehrenberg, P. (1907) Die Bewegung des Ammoniakstickstoffs in der Natur. Kritische Monographie aus dem Kreislauf des Stickstoffs. Parey, Berlin, 254 pp.

Ehrlich, V. and F. Russ (1911) Über den Verlauf der Stickstoffoxydation bei elektrischen Entladungen in Gegenwart von Ozon. *Sitzungsber. Wien. (Abt. 2)* **120**, 713–793. *See also*: *Monatshefte für Chemie* **32** (1911) 917–996.

Eichhorn (1868) Bericht der Central-Kommission für das agrikulturchemische Versuchswesen über die im Jahre 1866/67 ausgeführten Messungen der mit dem regen niedergefallenen Ammoniak- und Salpetersäuremengen. *Annalen der Landwirthschaft* **51**, 223–233.

Elbe, G. von (1933) The photodissociation of hydrogen peroxide in the presence of hydrogen and carbon monoxide and the recombination of OH Radicals. *J. Am. Chem. Soc.* **55**, 62–69.

Elbe, G. von and B. Lewis (1942) Mechanism of the thermal reaction between hydrogen and oxygen. *J. Chem. Phys.* **10**, 366–399.

Ellery, R. L. J. (1866) On Ozone. *Transactions and Proceedings of the Royal Society of Victoria* **7**, 131–135.

Ellett, W. B. and H. H. Hill (1929) effect of lime materials on outgo of sulphur from Hagerstown silt loam soil. *Journal of Agricultural Research* **38**, 697–711.

Ellis, R. (1850) The chemistry of creation: being a sketch of the chemical phenomena of the earth, the air, the ocean. Society for promoting Christian knowledge, London, 512 pp.

Elsholz, J. S. (1675) Sale ex aerie, et sulphure ex sole colligendo. *Miscellanea Curiosa Medico-Physica Academiæ Naturæ Curiosorum, sive Ephemerididium Medico-Physicarum Germanicarum* (Ephemeriden deutscher Naturforscher). Annus sextus at septus (1675 & 1676). J. Fritschii, Francforti & Lipsiæ, 1677, 28–29.

Elster, J. and H. Geitel (1899) Über die Existenz electrischer Ionen in der Atmosphäre. *Terrestrischer Magnetismus and Atmospheric Electricity* **4**, 213–234.

Emanuelsson, A., E. Eriksson and H. Egnér (1954) Composition of atmospheric precipitation in Sweden. *Tellus* **6**, 261–267.

Emeis, S. (2008) History of the Meteorologische Zeitschrift. *Met. Z.* **17**, 685–693.

Emmett, A. (1837) Experiments made during a Voyage and at Bermuda on the carbonic acid in the air. *Phil. Mag.* **11**, 225–227.

Encyclopædiaa Britannica (1771) in 3 Vol., Bell and Macfarquhar, Edinburgh.

Engler, C. (1879) Historisch-kritische Studien über das Ozon. *Leopoldina* Heft 15, Halle, 67 pp.

Engler, C. (1911) Über Zerfallsprozesse in der Natur. S. Hirzel, Leipzig, 33 pp.

Engler, C. and O. Nasse (1870) Ozone und Antozone. *Liebigs Ann.* **154**, 215–237.

Engler, C. and J. Weissberg (1904) Kritische Studien über die Vorgänge der Autoxydation. Braunschweig. F. Vieweg und Sohn, Braunschweig, 204 pp.

Erdman, L. W. (1923) The effect of gypsum on Iowa soils. *Soil Science* **15**, 137–155.

Ericksen, G. E. (1963) Geology of the Salt Deposits and the Salt Industry of Northern U. S. Geological Survey, Open-File Report 63–31, 164 pp.

Ericksen, G. E. (1981) Geology and origin of the Chilean nitrate deposits. U. S. Geological Survey Professional Paper 1188, 37 pp.

Ericksen, G. E. (1983) The Chilean nitrate deposits. *American Scientist* **71**, 366–374.

Erickson, R. E., L. M. Yates, R. L. Clarke and D. McEwen (1977) The reaction of sulfur dioxide with ozone in water and is possible atmospheric significance. *Atmos. Env.* **11**, 813–817.
Eriksson, E. (1952) Composition of atmospheric precipitation. I. Nitrogen compounds. *Tellus* **4**, 215–232 and II. Sulfur, chloride, iodine compounds. Bibliography. pp. 280–303.
Eriksson, E. (1954) Report on an informal conference in atmospheric chemistry held at the Meteorological Institute, University of Stockholm, May 24–26, 1954. *Tellus* **6**, 302–307.
Eriksson, E. (1957) The chemical composition of Hawaiian rainfall. *Tellus* **9**, 509–520.
Eriksson, E. (1959, 1960) The yearly circulation of chloride and sulfur in nature: meteorological, geochemical and pedological implications. Part 1. *Tellus* **11**, 375–403; Part 2. **12**, 63–109.
Eriksson, E. (1963) The yearly circulation of sulfur in nature. *JGR* **93**, 14079–14088.
Erismann, F. (1876) Kohlensäure in geschlossenen Räumen. *Zeitschrift für Biologie* **12**, 315–365.
Erndtel, Ch. H. (1730) Warsavia physice illustrata, sive de aere, aquis, locis et incolis Warsaviae, eorundemque moribus et morbis tractatus (1730) Zimmermann et Gerlach, Dresden, 132 pp.
Ernst, W. (1938) Über pH-Wert-Messungen von Niederschlägen. *Der Balneologe* **5**, 545–549.
Erwin, R. W. (1893) Nitrogen in rainwater. *Utah State Report*, pp. 252–153.
Etheridge, D. M., L. P. Steele, R. L. Langenfelds, R. J. Francey, J.-M. Barnola and V. I. Morgan (1996) Natural and anthropogenic changes in atmospheric CO_2 over the last 1000 years from air in Antarctic ice and firn. *JGR* **101**, 4115–4128.
Eugling, W. (1878) Ueber den Ammoniak- und Salpetersäuregehalt der atmosphärischen Niederschläge und den Ozongehalt der Luft. Bericht über die Thätigkeit der landwirtschaftlich-chemischen Versuchsstation des Landes Vorarlberg 1876–1877, Bregenz. *See also*: *Jahresbericht ueber die gesammten Fortschritte auf dem Gebiet der Agricultur-Chemie* **21**, N. F. (1. Jahrgang) für das Jahr 1878, pp. 73–74.
Eucken, A. and F. Patat (1936) Die Temperaturabhängigkeit der photochemischen Ozonbildung. *Z. phys. Chem.* **33B**, 459–474.
Eugster, W. (2008) Fog research. *Die Erde* **139**, 1–13.
Eulenberg, H. (1865) Die Lehre von den schädlichen und giftigen Gasen: toxikologisch, physiologisch, pathologisch, therapeutisch, mit besonderer Berücksichtigung der öffentlichen Gesundheitspflege und gerichtlichen Medicin systematisch und nach eigenen Versuchen bearbeitet. Vieweg, Braunschweig, 550 pp.
Evans, C. (1857) Strand district. Special report of the medical officier of health to the Strand district, on slaughter-houses. London.
Evans, M. G. and N. Uri (1949) Dissociation constant of hydrogen peroxide and the electron affinity of the HO_2 radical. *Trans. Faraday Soc.* **45**, 224–230.
Evans, S. and S. R. Pittard (1858) Of the district meteorology of London, according to daily observations, from information contributed by the medical officers of health, with assistance from other officers of public institutions. Vol. 1, Nos. 1–36 and Vol. 2, Nos. 1–42, London.
Evelyn, J. (1661) Fumifigium: or, The Inconveniencies of the AER and Smoake of London Dissipated. First published by W. Godbid for G. Bedel and Th. Collins, London. Reprinted in 1772 "for B. White with preface by S. Pegge," and in 1825 in the edition of "Evelyn's Miscellaneous Writings ed. By W. Upcott." Here, I used a reprint in original form in 100 numbered Copies (Swan Press in Baskerville), no year but before 1955, No. 10, 49 pp.
Evelyn, J. (1676) A philosophical discourse of earth relating to the culture and improvement of it for vegetation and the propagation of plants, etc. as it was presented to the Royal Society. London, J. Martyn, 182 pp.
Eversmann, E. von (1824) Hagel mit metallischem Kern. *Pogg. Ann.* **76**, 340–343. *See also*: Aerolithen-Hagel, beobachtet zu Sterlitamak, 200 Werst von Orenburg in Sibirien. *Archiv der gesammten Naturlehre* **4** (1825) 196–198.
Eyler, J. M. (1980) The conversion of Angus Smith: the changing role of chemistry and biology in sanitary science, 1850–1880. *Bulletin of the History of Medicine* **54**, 216–234.

Faber, W. E. (1849) Ueber das in der Atmosphäre enthaltene Ozon. *Medicinisches Correspondenzblatt* des *Würtembergischen Ärztlichen Vereins*, Stuttgart **19**, 1–3.

Fabian, P., R. Borchers, K. H. Weiler, U. Schmidt, A. Volz, D. H. Ehhalt, W. Seiler and F. Müller (1979) Simultaneously measured vertical profiles of H_2, CH_4, CO, N_2O, $CFCl_3$ and CF_2C1_2 in the mid-latitude stratosphere and troposphere. *JGR* **84**, 3149–3154.

Fabian, P., R. Brochers, G. Flentje, W. A. Matthews, W. Seiler, H. Giehl, K. Bunse, F. Muller, U. Schmidt, A. Volz, A. Khedim and F. J. Johnen (1983) The vertical distributions of stable trace gases at mid-latitudes. *JGR* **86**, 5179–5184.

Fabri, H. (1670) Physica, id est Scientia Rerum Corporearum. Vol. 2, Anisson. Lyon.

Fabry, Ch. and H. Buisson (1913) Sur l'absorption de l'ultraviolet pur l'ozone et l'extrémité du spectra soilaire. *C. r.* **156**, 782–785.

Fabry, Ch. and H. Buisson (1931) Sur l'absorption de radiations dans le basse atmosphere et de dosage de l'ozone. *C. r.* **192**, 457–461.

Fabry, Ch. (1950) L'ozone atmospherique. Centre National de la Recherche Scientifique, Paris, 278 pp.

Facchini, M.-C., S. Fuzzi, M. Kessel, W. Wobrock, W. Jaeschke, B. G. Arends, et al.(1992) The chemistry of sulphur and nitrogen species in a fog system – A multiphase approach. *Tellus* **44B**, 505–521.

Facchini, M. C. and S. Fuzzi (1993) Atmospheric acidity: a useful tool to describe the distribution of chemical species among the different phases in fog. In: Photo-oxidants: Precursors and products, Proceedings. EUROTRAC Symp. 1992 (eds. P. M. Borrell, P. Borrell, T. Cvitas, W. Seiler), SPB Academic Publ., Den Haag, pp. 505–509.

Failyer, G. H. and J. T. Willard (1889) Ammonia and nitric acid in atmospheric waters. Second *Ann. Rep. Kansas Agric. Exper. Stat.* pp. 123–132 (cited after Miller 1905).

Failyer, G. H. (1889–1890) Ammonia and nitric acid in rain-waters collected at the Agricultural College. *Transactions of the Annual Meetings of the Kansas Academy of Sciences* **12**, 21–24.

Fairley, R. (1907) The early history of distillation. *Journal of the Institute of Brewing* **13**, 559–582.

Falconer, R. E. and P. D. Falconer (1980) Determination of cloud water acidity at a mountain observatory in the Adironback Mountains of New York State. *JGR* **85**, 7465–7470.

Farcy, L. (1931) La composition des eaux de pluie de la région de Brest et du Nord finistère. *Annales d'hygiène publique, industrielle et sociale* **9**, 504–515 (cited after Cauer 1938 and Eriksson 1952).

Farkas, L. (1927) Keimbildungsgeschwindigkeit in übersättigten Dämpfen. *Z. phys. Chem.* **125**, 236–242.

Farman, J. C., B. G. Gardiner and J. D. Shanklin (1985) Large losses of total ozone in Antarctica reveal seasonal ClO_x/NO_x interaction. *Nature* **315**, 207–210.

Farmer, J. C. and G. A. Dawson (1982) Condensation sampling of soluble atmospheric trace gases. *JGR* **87**, 8931–8942.

Farrar, K. R. (1963) A note on a eudiometer supposed to have belonged to Henry Cavendish. *The British Journal for the History of Science* **1**, 375–380.

Farsky, F. (1877) Bestimmungen der atmosphärischen Kohlensäure in den Jahren 1874–1875 zu Tabor in Böhmen. *Sitzungsber. Wien* **74**, 67–77.

Faust, B. C. and J. M. Allen (1992) Aqueous-phase photochemical sources of peroxyl radicals and singlet molecular oxygen in clouds and fog. *JGR* **97**, 12913–12926.

Faust, B. V., C. Anastasio, J. M. Allen and T. Arakaki (1993) Aqueous-Phase Photochemical Formation of Peroxides in Authentic Cloud and Fog Waters. *Science* **260**, 73–75.

Feilitzen, H. von and I. Lugner (1910) On the quantity of ammonia and nitric acid in the rain-water collected at Flahult in Sweden. *Journal of agricultural science* **3**, 311–313; *see also*: Undersökningar över de mängder bundet kväve, som tillföres jorden med nederbörden. *Kgl. Lantbr. Akad. handl. o. tidskrifter* (1910) 151.

Feister, U. (1985) Zum Stand der Erforschung des atmosphärischen Ozons. *Veröffentlichungen des meteorologischen Dienstes der Deutschen Demokratischen Republik* Nr. 26, Akademie-Verlag, Berlin, 54 pp.
Feister, U. (1994) Comparison between Brewer spectrometer, M 124 filter ozonometer and Dobson spectrophotometer. In: NASA. Goddard Space Flight Center, Ozone in the Troposphere and Stratosphere, Part 2, pp. 770–773.
Feister, U. (2009) Long-term measurements of atmospheric ozone. In: Berichte zur Polar- und Meeresforschung [Reports on polar and marine research]. Selected Contributions on Results of Climate Research in East Germany (the former GDR), No. 588 (eds. P. Hupfer and K. Dethloff), Alfred Wegener Institute for Polar and Marine Research in Bremerhaven, pp. 22–35.
Feister, U. and W. Warmbt (1985) Long-term surface ozone increase at Arkona (54.68°N, 13.43°E). In: Atmospheric Ozone, Proc. Quadrenn. Ozone Symposium, Halkidiki, Greece, 3–7 September 1984, D. Reidel Publ., Holland, pp. 782–787.
Feister, U. and W. Warmbt (1987) Long term measurements of surface ozone in the German Democratic Republic. *J. Atmos. Chem.* **5**, 1–21.
Feister, U., K.-H. Grasnick and G. Jakobi (1989) Surface ozone and solar radiation. In: Ozone in the Atmosphere (eds. R. D. Bojkov and P. Fabian), A. Deepak Publishing, pp. 37–40.
Feldt, V. (1887) Der Kohlensäuregehalt der Luft in Dorpat bestimmt in den Monaten Februar bis Mai 1887: Inaugural-Dissertation zur Erlangung des Grades eines Doctors der Medicin. Schnakenburg's Buchdruckerei, Dorpat, 49 pp.
Fellenberg, L. R. and L.-Th. Rivier (1845) Second mémoir sur l'ozone. *Archives de l'Electricitie*. 20, No. 5. *See also*. *Verhandlungen der Schweizerischen Naturforschenden Gesellschaft* **29** (1845) 240–151. And: Anonymous (1846) Ozone. *The American Journal of Science and Arts*, 2nd ser. **2**, 106–110.
Feng, J. and D. Möller (2004) Characterization of water-soluble macromolecular substances in cloud water. *J. Atmos. Chem.* **48**, 217–233.
Ferm, M. (1979) Method for determination of atmospheric ammonia. *Atmos. Env.* **13**, 1385–1393.
Ferm, M., L. Granat, M. Engardt, G. Pihl Karlsson, H. Danielsson, P. E. Karlsson and K. Hansen (2019) Wet deposition of ammonium, nitrate and non-sea-salt sulphate in Sweden 1955 through 2017. *Atmos. Env.* **X**, 100015. doi:10.1016/j.aeaoa.2019.
Ferraboschi, F. (1909) the production of ozone in the interaction between hydrogen dioxide and sulphur dioxide. *Proceedings of the Chemical Society, London* **25**, 179.
Fersman, A. E. [Ферсман] (1923) Химические элементы Земли и космоса [Chemical elements of the earth and the cosmos], Petersburg [Петербург, Научное Химико-Техническое Издательство Научно-Технического Отдела В. С. Н. Х.], 180 pp.
Festing, M. (1869) Description of a new ozone-cage. *Proceedings of the Meteorological Society* **4**, 198–199.
Feth, J. H. (1981) Chloride in natural continental water – a review. US State Geological Survey Paper 2176, Washington, 36 pp.
Figuier, L. (1866) L'ozon – changement curieux observés dans les propriétés de l'air atmosphérique. *L'Annee Scientifique et Industrielle* **10**, 120–128.
Filhol, E. (1855) Note sur la composition chimique de l'eau de pluie tombée aux environs de Toulouse pendant le premier semestre de l'année 1855. *C. r.* **41**, 838–839.
Findeisen, W. (1932) Messungen der Größe und Anzahl der Nebeltropfen zum Studium der Koagulation inhomogenen Nebels. *Gerlands Beiträge zur Geophysik* **35**, 295–340.
Findeisen, W. (1937) Entstehen die Kondensationskerne an der Meeresoberfläche? *Met. Z.* **54**, 377–379.
Findeisen, W. (1938) Die kolloidmeteorologischen Vorgänge bei der Niederschlagsbildung. *Met. Z.* **55**, 121–133.

Findeisen, W. (1939) Die Kondensationskerne. *Beiträge zur Physik der Atmosphäre* **25**, 377–379. *See also*: Die Kondensationskerne. Entstehung, chemischer Natur, Grösse und Anzahl. *Beiträge zur Physik der freien Atmosphäre* **25**, 220–232.

Finer, S. E. (1952) The life and times of Sir Edwin Chadwick. Methuen & Co., London, 555 pp.

Finke, L. L. (1820) Naturkundliche Bemerkungen, betreffend eine auf vieljährige Beobachtungen sich stützende Beschreibung des Moordampfes zu Westphalen und seine nachtheiligen Einflüsse auf die hiesige Witterung, nebst Beurtheilung des großen Unterschiedes, der zwischen Moordampf und Höhenrauch stattfindet, und der oft irrigen Verwechselung des letzteren mit dem ersteren. Hannover, 96 pp.

Finlay, M. R. (1988) The German agricultural experiment stations and the beginnings of American agricultural research. *Agricultural History* **62**, 41–50; *see also*: Finlay (1992) Science, practice and politics: German agricultural experiment stations in the nineteenth century (1992) retrospective thesis and dissertations, Iowa University, 447 pp.

Finlay, M. R. (1992) Science practice and politics: German agricultural experimental stations in the nineteen's century. Retrospektive Thesis and Dissertation 9830, Iowa State University, 447 pp.

Finlayson-Pitts, B. J. (1983) Reaction of NO_2 with NaCl and atmospheric implications of NOCl formation. *Nature* **306**, 676–677.

Finlayson-Pitts, B. J. and J. N. Pitts, Jr. (1986) Atmospheric chemistry: fundamentals and experimental techniques. John Wiley & Sons, New York, 1098 pp.

Finlayson-Pitts, B. J., M. J. Ezell and J. N. Pitts (1989) Formation of chemically active chlorine compounds by reactions of atmospheric NaCl particles with gaseous N_2O_5 and $ClONO_2$. *Nature* **337**, 241–244.

Finlayson-Pitts, B. J. and J. N. Pitts (2000) Chemistry of the upper and lower atmosphere. Academic Press, San Diego, 969 pp.

Finlayson-Pitts, B. J., L. M. Wingen, A. L. Sumner, D. Syomin and K. A. Ramazan (2003) The heterogeneous hydrolysis of NO_2 in laboratory systems and in outdoor and indoor atmospheres: An integrated mechanism. *Physical Chemistry and Chemical Physics* **5**, 223–242.

Finnell, H. H. and H. G. Houghton (1931) Soil nitrogen income from rainwater. *The Panhandle Bulletin. Panhandle Agricultural Experimental Station* **23**, 7–24. *See also*: Nitrogen in rainwater. Ibid. **34** (1932) 3–8.

Fischer, F. (1878) Ueber die Untersuchung der Rauchgase. *Polytechnisches Journal (Dingler's J.)* **227**, 171–177.

Fischer, F. (1880) Können eiserne Oefen durch Entwicklung von Kohlenoxyd gesundheitsschädlich werden? *Polytechnisches Journal (Dingler's J.)* **235**, 438–443.

Fischer, F. and O. Ringe (1908) Die Darstellung von Argon aus Luft mit Calciumcarbid. *Ber.* **41**, 2017–2030.

Fischer, J. C. (1802) Geschichte der Physik seit der Wiederherstellung der Künste und Wissenschaften. Zweyter Band. J. F. Röwer, Göttingen, 617 pp.

Fischer, N. W. (1845) Bemerkungen über das sogenannte Ozon. *Pogg. Ann.* **66**, 163–167; Bemerkungen zu Hrn. Schönbein's Beleuchtung meiner Meinung, betreffend das Ozon. Ibid., pp. 168–173; Ueber das Vermögen mehrerer gas- und dunstförmiger Körper, Metalle zu polarisieren und auf Jodkalium, Cyaneisenkalium etc. zersetzend einzuwirken. *J. prakt. Chem.* **34** (1845) 186–191; Ueber die Bemerkungen des Hrn. Prof. Schönbein zu meiner Notiz, das Verhalten des Jodkaliums zu verschiedenen Gasarten etc. betreffend. Ibid., **35** (1845) 180–181; Ueber das Leuchten des Phosphors. Ibid., pp. 342–356; Bemerkungen zu dem Aufsatz des Herrn Schoenbein "Über die Erzeugung des Ozons durch Phosphor". *Pogg. Ann.* **76** (1849) 158–160.

Fishman, J. and P. J. Crutzen (1978) The origin of ozone in the troposphere. *Nature* **274**, 855–858.

Fishman, J., V. Ramanathan, P. J. Crutzen and S. C. Liu (1979) Tropospheric ozone and climate. *Nature* **282**, 818–820.

Fittbogen, J. and P. Hässelbarth (1879) Ueber locale Schwankungen im Kohlensäuregehalt der atmosphärischen Luft. Jahresbericht ueber die gesammten Fortschritte auf dem Gebiet der. *Agricultur-Chemie* **22**, 67–71. *See also*: *Biedermann's Centralblatt für Agrikulturchemie und Rationellen Landwirtschafts-Betrieb* (1880) **9**, 1616–164.

FitzRoy, R. (1863) The weather book: A manual of practical meteorology. Vol. 2. Longman & Green, London. 464 pp.

Flagan, R. C. (1998) History of electrical aerosol measurements. *Aerosol Science and Technology* **28**, 301–380.

Flammarion, C. (1874) The atmosphere (translated from French). Harper & Brothers, New York, 451 pp.

Flammarion, C. (1888) L'Atmosphére. Météorologie populaire. Paris, Hachette, 808 pp.

Flechner, A. E. (1855–1860) Meteorologische Beobachtungen und Krankheitscharakter in Wien vom 29. Jänner bis Ende Februar 1855. *Östereichische Zeitschrift für praktische Heilkunde* **1** (1855) 84–86. Continued: 327–329, 2 (1856) 62–64, 574–576, 816–818, 3 (1857) 64–65, 750–752, 4 (1858) 76–78, 326–328, 5 (1859) 353–356, 532–533, 756–761, 6 (1860) 115–116. Note that surname was different given as E. A., A. E., E and A.

Fleming, J. R. (1990) Meteorology in America 1800–1870. The Johns Hopkins University Press, Baltimore, 264 pp.

Fleming, J. R. (2010) Gilbert N. Plass: Climate science in perspective. *American Science* **98**, 58, doi:10.1511/2010.82.58.

Fleming, J. R. (2012) Fixing the sky – the checkered history of weather and climate control. Columbia University Press, 344 pp.

Fleming, J. R. (2016) Inventing atmospheric science. Bjerknes, Rossby, Wexler and the foundations of modern meteorology. The MIT Press, Cambridge, 296 pp.

Flemming, F. (1855) Cholera und Ozon. *Allgemeine Medicinische Central-Zeitung* **24**, 785–791.

Floderus, M. M. (1859) De vigtigaste åsigterna om ozon, historisk-kritisk afhandling. [Konsistorieavh.] Upsala, 42 pp.

Flohn, H. (1949) Bioklimatik. Von Manfred Curry (Book review). *Zeitschrift für Naturforschung B* **4**, 62–63.

Flood, H. (1934) Tröpfchenbildung in übersättigter Äthylalkohol-Wasserdampfgemischen. *Z. phys. Chem. A* **170**, 286–294.

Flügge, C. (1881) VI. Bestimmung des Ozons (pp. 97–109). In: Lehrbuch der hygienischen Untersuchungsmethodn. Eine Anleitung zur Anstellung hygienischer Untersuchungen und zur Begutachtung hygienischer Fragen für Ärzte und Chemiker, Sanitäts- und Verwaltungsbeamte, sowie Studirende. Veit, Leipzig, 602 pp.

Flury, F. and F. Zernik (1931) Schädliche Gase. Dämpfe, Nebel, Rauch- und Staubarten. Springer, Berlin, 631 pp.

Fodor, J. (1879) Experimentelle Untersuchungen über Boden und Bodengase. *Vierteljahresschrift für öffentliche Gesundheitspflege* **7**, 205–237.

Fodor, J. (1881) Hygienische Untersuchungen über Luft, Boden und Wasser. Erste Abteilung: Die Luft. F. Vieweg, Braunschweig, 184 pp.

Foerster, F. and M. Koch (1908) Über die Einwirkung von nitrosen Gasen und Sauerstoff auf Wasser. *Z. angew. Chem.* **21**, 2161–2172.

Fogg, T. R. and R. A. Duce (1985) Boron in the troposphere: Distribution and fluxes. *JGR* **90**, 3781–3796.

Foner, S. N. and R. L. Hudson (1953) Detection of the HO_2 radical by mass spectrometry. *J. Chem. Phys.* **21**, 1608–1609.

Foner, S. N. and R. L. Hudson (1962) Mass spectrometry of inorganic free radicals. *Advances in Chemistry Series* **36**, 34–49.

Fonrobert, E. (1916) Das Ozon. Verlag von F. Enke, Stuttgart, 282 pp.

Fonselius, S., F. Koroleff and K.-E. Wärme (1956) Carbon dioxide variations in the atmosphere. *Tellus* **8**, 176–183.
Fontenelle, S.-E. J. de (1823) Dissertation sur les eaux minérales connues sous le nom de Bains. De Renses. *Also*: Recherches historiques, chimiques et médicales sur l'air marécageux. Ouvrage couronné par l'Académie des sciences de Lyon. Gabon, Paris, 155 pp.; *see also*: Recension by unknown author: *Kritisches Repertorium für die gesamte Heilkunde*. Ed. by J. N. Rust, Vol. 3, Berlin (1824), pp. 138–144.
Fontenelle, S.-E. J. de (1832) Analyse der atmosphärischen Luft in Paris. *Notizen aus dem Gebiete der Natur- und Heilkunde* **23**, 229–230.
Fontijn, A., A. J. Sabadell and R. J. Ronco (1970) Homogeneous chemiluminescence measurement of nitric oxide with ozone. *Analytical Chemistry* **42**, 575–579.
Fonvielle de, W. (1876) New meteorological observatory at Montsourios. *Nature* **14**, 156–157.
Formey, L. (1796) Versuch einer medicinischen Topographie von Berlin. E. Felisch, Berlin, 382 pp.
Forster, H. (1940) Studien über Kondensationskerne. Ihre physikalische und biologische Bedeutung im Außen- und Innen-klima. Dissertation, ETH Zürich, 180 pp.
Fourcroy, M. (1790) Elements of natural history and chemistry. Translated from the last Paris edition, 1789, in three Volumes. Elliot and Kay, London. Vol. 1, 526 pp.
Fourcroy, M. (1800) Système des connaissance chimiques et leurs applocations aux phènomènes de la nature at de l'art. Tome II. Brumaier, Paris, 352 pp.
Fowle, F. E. (1929) Atmospheric ozone: its relation to some solar and terrestrial phenomena. *Smithsonian Miscellaneous Collection* **81**, No. 11, 27 pp.
Fowler, A. and R. J. Strutt (1917) Absorption bands of atmospheric ozone in the spectra of sun. *Proc. R. Soc. Lond. A* **93**, 577–587.
Fowler, D., I. D. Leith, J. Binnie, A. Crossley, D. W. F. Inglis, T. W. Choularton, M. Gay, J. W. S. Longhurst and D. E. Conland (1995) Orographic enhancement of wet deposition in the United Kingdom: continuous monitoring. *Water, Air and Soil Pollution* **85**, 2107–2112.
Fox, C. B. (1873) Ozone and antozone. Their history and nature. When, where, why, how ozone is observed in the atmosphere? J. & J. Churchill, London, 329 pp.
Fox, C. B. (1876) On the employment of aspirators in atmospheric ozonometry. In: Conference held in connection with the special loan collection of scientific apparatus. Vol. 2. Chapman and Hill, London, pp. 272–276.
Fox, C. B. (1886) Sanitary examination of water, air and food. J. & A. Churchill, London, 563 pp.
Francis, S. G. and A. T. Parson (1925) The determination of oxides of nitrogen (except nitrous oxide) in small concentration in the products of combustion of coal gas and in air. *The Analyst* **50**, 262–272.
Franck, J. and F. Haber (1931) Zur Theorie der Katalyse durch Schwermetallionen in wäßriger Lösung und insbesondere zur Autoxidation der Sulfitlösungen. *Sitzungsber. Berlin* **13**, 250–256.
Franckowiak, R. (2011) Mechanical and chemical explanations in Du Clos' chemistry. *Ambix* **58**, 13–28.
Frankland, E. (1861) On the composition of air at Mont Blanc. *Q. J. Chem. Soc.* **13**, 22–30.
Frankland, E. (1874) Sixth report of the River's Pollution Commission. London.
Fraps, G. S. (1930) Sulphur in rain water. In: Possibilities of sulfur as a soil amendment. *Texas Agronomic Experimental Station Bulletin* **414** (56 pp.), pp. 9–15.
Fraser, P. J., M. A. K. Khalil, R. A. Rasmussen and A. J. Crawford (1981) Trends of atmospheric methane in the Southern Hemisphere. *Geophys. Res. Lett.* **8**, 1063–1066.
Fraser, K. A. (2004) Zosimos of Panopolis and the Book of Enoch: Alchemy as forbidden knowledge. *Aries* **4**, 125–147.
Frear, G. L. (1934) Kinetics of the methane-oxygen reaction. *J. Am. Chem. Soc.* **56**, 305–307.
Free, E. E. (1911) The movement of soil material by the wind. With a bibliography of eolian geology by S. C. Stuntz and E. E. Free. U. S. Dept. of Agriculture, Bureau of Soils, Bulletin No. 68, Washington, 292 pp.

Freeman, Y. F. (1924) Nitrogen in the rainwater of different points in Kentucky. *Journal of the American Society for Agronomy* **16**, 356–358 (cited after Eriksson 1952).

Freiberg, J. (1974) Effects of relative humidity and temperature on iron-catalyzed oxidation in atmospheric aerosols. *Env. Sci. Technol.* **8**, 731–734.

Freiberg, J. (1975) The mechanism of iron catalyzed oxidation of SO_2 in oxygenated solution. *Atmos. Env.* **9**, 661–672.

Freiberg, J. (1976) The iron catalyzed of SO_2 to acid sulfate mist in dispersing plumes. *Atmos. Env.* **10**, 121–130.

Frémy, E. (1865) Observations relatives aux incertitudes de l'ozonométrie atmosphérique. *C. r.* **61**, 939–941. *See also*: Ueber die Sicherheit ozonometrischer Bestimmungen nach Fremy und Cantoni. *Z. Öster. Ges. Met.* **1** (1866) 305–312.

Frémy, E. (1866) Sur l'oxygène et l'ozone. Deuxième Conférence du 10 April, 1866, sous le patronage de l'Impératrice. Au bénéfit de la Sociéte de Secours des Amis des Sciences. Gauthier-Villars, Paris, 20 pp.

Frémy, E. and E. Becquerel (1852) Recherches èlectrochimiques sur les propriétés des corps électrisés (premier Mémoire). *Ann. Chim. Phys* **25**, 62–105. *See also*: *Quart. J. Chem Soc.* **5** (1852) 272.

Frenkel, J. I. [Френкель Яков Ильич] (1939) Общая теория гетерофазных флуктуаций и предпереходных явлений [A general theory of heterophase fluctuations and pretransition phenomen]. *Журнал Экспериментальной и Теоретической Физики* (ЖЭТФ) [Journal of Experimental and Theoretical Physics] **12**, 952–961. *See also*: *J. Chem. Phys.* **7** (1939) 538–547.

Frenkel, J. I. [Френкель, Я. И.] (1953) О фотохимическом механизме возникновения ядер конденсации в земной атмосфере [On the photochemical mechanism of the emergence of condensation nuclei in the Earth's atmosphere]. *Известия Академии наук СССР. Серия геофизическая* [Izvestiya Akademii Nauk SSSR. Seriya Geofizicheskaya] **2**, 191–192.

Fresenius, R. (1841) Anleitung zur qualitativen chemischen Analyse oder systematisches Verfahren zur Auffindung der in der Pharmacie, den Künsten und Gewerben häufiger vorkommenden Körpern. Bonn, 82 pp. (2nd – seventeenth edns. 1842–1896).

Fresenius, R. (1847) Anleitung zur quantitativen chemischen Analyse. 2. Auflage, Vieweg, Braunschweig, 518 pp. (1st – 6th edns. 1842–1875).

Fresenius, F. (1849) Ueber den Ammongehalt der atmosphärischen Luft. *J. prakt. Chem.* **46**, 100–106 and *Arch. Pharm.* **109**, 52–55 and: Sur la quantité d'ammoniaque contenue dans l'air atmosphérique. *Ann. chim. phys.* 3th ser. **26**, 208–214.

Fresenius, R. (1875) Anleitung zur quantitativen chemischen Analyse. 6. Auflage in zwei Bänden. Vieweg, Braunschweig, 668+871 pp. (newprints until 1903).

Frey, E. von (1889) Der Kohlensäuregehalt der Luft in und bei Dorpat bestimmt in den Monaten September 1888 bis Januar 1889. Doctor Thesis, Medizinische Fakultät der Kaiserlichen Universität Dorpat, 49 pp.

Fricke, H. (1934) Reduction of oxygen to hydrogen peroxide by the irradiation of its aqueous solution with X-rays. *J. Chem. Phys.* **2**, 556–557.

Fricke, W. (1983) Großräumige Verteilung und Transport von Ozon und Vorläufern. VDI Berichte 500, VDI-Verlag GmbH, Düsseldorf, pp. 55–62.

Fridlander, E. D. (1896) Atmospheric dust observations from various parts of the world. *Q. J. Roy. Met. Soc.* **22**, 184–203.

Friedheim, C., Ed. (1907) Gmelin-Kraut's Handbuch der anorganischen Chemie. Band 1, Abteilung 1. Carl Winters Universitätsbuchhandlung, Heidelberg, 888 pp.

Friedli, H., H. Lötscher, H. Oeschger, U. Siegenthaler and B. Stauffe (1986) Ice core record of 13C/12C ratio of atmospheric CO_2 in the past two centuries. *Nature* **324**, 37–38.

Friend, J. P. (1973) The global sulfur cycle. In: Chemistry of the lower atmosphere (Ed. I. E. Rasool), Plenum Press, New York, pp. 177–201.

Friese, W. (1909) Beiträge zur Kenntnis des Staubs in der Stadtluft. Dissertation. Kgl. Sächs. Technische Hochschule Dresden, 53 pp.
Friesenhof, G. (1904) Einiges über Ozonbeobachtung. *Met. Z.* **21**, 380–382.
Frisinger, H. H. (1977) The history of meteorology: to 1800. Sci. Hist. Publ., New York, 147 pp.
Fritsch, C. (1866) Weitere Bemerkungen zu den Ozonometer-Beobachtungen. *Z. Öster. Ges. Met.* **1**, 312–314.
From, E. and C. D. Keeling (1986) Reassessment of late nineteenth century atmospheric carbon dioxide variations in the air of western Europe and the British Isles based on an unpublished analysis of contemporary air masses by G. S. Callendar. *Tellus* **38B**, 87–105.
Fuchs, N. and N. Oschman (1935) Über die Bildung von Aerosolen. 1. Mitt. Methodik. Schwefelsäurenebel. *Acta Physicochimica U. R. S. S.* **3**, 61–78.
Fuchs, G. (1889) Beiträge zur Untersuchung der Luft auf ihren Kohlensäuregehalt. Inaugural-Dissertation, Universität Würzburg. Becker's Universitäts-Buchdruckerei, 17 pp.
Fuhrer, J. (1986) Chemistry of fogwater and estimated rates of occult deposition in an agricultural area of central Switzerland. *Agriculture and Ecosystems Environment* **17**, 153–164.
Fujishima, A. (1972) Electrochemical photolysis of water at a semiconductor electrode. *Nature* **238**, 37–38. *See also*: Fujishima, A., K. Hashimoto and T. Watanabe (1999) TiO_2 photocatalysis – fundamentals and applications. BKC Inc. Tokyo, 176 pp.
Futter, M. N., S. Valinia, S. Löfgren, S. J. Köhler and J. Fölster (2014) Long-term trends in water chemistry of acid-sensitive Swedish lakes show slow recovery from historic acidification. *Ambio* **43**, 157–163.
Fuzzi, S. (1988) Fog chemistry and deposition in the Po Valley, Italy. In: Acid deposition at high elevation sites (eds. M. H. Unsworth and D. Fowler), Kluwer Acad. Publ., pp. 443–452.
Fuzzi, S., Ed. (1995) The Kleiner Feldberg Cloud Experiment 1990. Kluwer Academic Publ., Dordrecht, The Netherlands, 264 pp.
Fuzzi, S., G. Orsi and M. Mariotti (1983) Radiation fog liquid water acidity at a field station in the Po valley. *J. Aerosol. Sci.* **14**, 135–138.
Fuzzi, S., R. A. Castillo, J. E. Juisto and G. G. Lala (1984) Chemical composition of radiation fog water at Albany, New York and its relationship to fog microphysics. *JGR* **89**, D5, 7159–7164.
Fuzzi, S., M. C. Fachini, G. Orsi, J. A. Lind, W. Wobrock, M. Kessel, R. Maser, W. Jaeschke, K. H. Enderle, B. G. Arends, A. Berner, I. Solly, C. Kruisz, G. Reischl, S. Pahl, U. Kaminski, P. Winkler, J. A. Ogren, K. J. Noone, A. Hallberg, H. Fierlinger-Oberlinninger, H. Puxbaum, A. Marzorati, H.-C. Hansson, A. Wiedensohler, B. Svenningsson, B. G. Martinsson, D. Schell and H.-W. Georgii (1992) The Po valley experiment 1989: an overview. *Tellus* **44B**, 448–468.
Fuzzi, S., M. C. Facchini, G. Orsi, G. Bonforte, W. Martinotti, G. Ziliani, P. Mazzali, P. Rossi, P. Natale, M. M. Grosa, E. Rampado, P. Vitali, R. Raffaelli, G. Azzini and S. Grotti (1996) The NEVALPA project: a regional network for fog chemical climatology over the Po Valley basin. *Atmos. Env.* **30A**, 201–213.
Fuzzi, S., M.-C. Facchini, S. Decesari, E. Matta and M. Mircea (2002) Soluble organic compounds in fog and cloud droplets: what have we learned over the past few years? *Atmos. Res.* **64**, 89–98.
Gäb, S., E. Hellpointner, W. V. Turner and F. Korte (1985) Hydroxymethyl hydroperoxide and bis-(hydroxymethyl) peroxide from gas-phase ozonolysis of naturally occurring alkenes. *Nature* **316**, 535–536.
Gaffron, H. (1927) Die photochemische Bildung von Peroxyd bei der Sauerstoff-Übertragung durch Chlorophyll. *Ber.* **60**, 2229–2238.
Gaillard, E. S. (1864) Dr. Gaillard's prize essay on ozone;its relation to health and diseases. *Boston Medical and Surgical Journal* **71**, 129–136, 152–162, 171–181, 218–222, 232–237, 252–256, 272–276, 294–297. Also published as book (dissertation) in Boston, Medical Society 1864 (53 pp.), reprinted from Journal (renamed in 1929 to *The New England Journal of Medicine*).

Gairdner, M. (1832) Essay on the natural history, origin, composition and medicinal effects, of mineral and thermal springs. W. Blackwood, Edinburgh and T. Cadell, Strand, London, 420 pp.
Galbally, I. A. (1972) Ozone and oxidants in the surface air near Melbourne/Victoria. In: Proceedings of the International Clean Air Congress, Melbourne, pp. 192–201.
Gallagher, P. H. (1923) Mechanism of oxidation in the plant: Part I. The oxygenase of Bach and Chodat: function of lecithins in respiration. *Biochemical Journal* **17**, 515–529.
Galloway, J. N., E. B. Cowling, E. Gorham and W. W. McFee (1978) A national program for assessing the problem of atmospheric deposition (acid rain). In: Effects of acid rain. Publ. No. 96–126, Govern. Print. Office, Washington (1980), pp. 449–550.
Galloway, J. N., G. E. Likens, W. C. Keene and J. M. Miller (1982) The composition of precipitation in remote areas of the world. *Journal of Geophysical Research* **87**, 8771–8786.
Galloway, J. N., F. J. Dentener, D. G. Capone, E. W. Boyer, R. W. Howarth, S. P. Seitzinger, G. P. Asner, C. C. Cleveland, P. A. Green, E. A. Holland, D. M. Karl, A. F. Michaels, J. H. Porter, A. R. Townsend and C. J. Vosmarty (2004) Nitrogen cycles past, present and future. *Biogeochemistry* **70**, 153–226.
Galloway, J. N., A. M. Leach, A. Bleeker and J. W. Erisman (2013) A chronology of human understanding of the nitrogen cycle. *Phil. Trans.* **368** (1621), 20130120.
Gambell, A. W. and D. W. Fisher (1964) Occurrence of sulphate and nitrate in rainfall. *JGR* **69**, 4203–4210.
Gasparin, de, A. (1844) Cours d'agriculture. Tome second. Paris, 562 pp. (in 6 Vol. 1843–1848).
Gast, J. A. and T. G. Thompson (1959) Evaporation of boric acid from sea water. *Tellus* **11**, 344–347.
Gautier, A. (1898) Note préliminaire sur la presence de l'hydrogène libre dans l'air atmosphérique. *C. r.* **127**, 693–694.
Gautier, A. (1899) Quantité maximum de chlorures contenus dans l'air de la mer. *C. r.* **128**, 715–716.
Gautier, A. (1900) Gaz combustibles de l'atmosphere: Air de villes. *C. r.* **130**, 1677–1684.
Gautier, A. (1901a) Gaz combustibles de l'air: air des bois;air des hautes montagnes. *C. r.* **131**, 13–18.
Gautier, A. (1901b) Gaz combustibles de l'air: air de la mer. Existence de l'hydrogène libre dans l'atmosphère terrestre. *C. r.* **131**, 86–90. *See also*: Gautier, A. (1901) Hydrogen in air. *Nature* **63**, 478–479.
Gautier, A. (1901c) Origines de l'hydrogène atmosphérique. *C. r.* **131**, 647–652.
Gautier, A. (1901d) Les fumeés de Paris. Inference excerceé par les produits de combustion sur l'atmosphere de la ville. *Revue d'hygiène et de police sanitaire* **23**, 99–101.
Gautier, A. (1901e) Les gaz combustible de l'air: l'hydrogéne atmosphérique. *Ann. chim. phys.* **22**, 5–110.
Gay, B. W., Jr. and J. J. Bufalini (1971) Hydrogen peroxide in the urban atmosphere. In: Photochemical smog and ozone reactions. Advances in Chemistry Series No. 113, American Chemical Society, Washington D. C., pp. 255–263.
Gay, Jr., B. W. and J. J. Bufalini (1972) Hydrogen peroxide in the urban atmosphere. *Environmental Letters* **3**, 21–24.
Gay-Lussac, J. (1804) Relation d'un voyage aérostatique fait par M. Gay-Lussac, le 29 fructitor an 12. *Annales de Chimie* **52**, 75–94. *See also*: *Journal de Physique* **59**, 454–161.
Gay-Lussac, J. (1805a) Account of an aerostatic voyage performed by M. Guy-Lussac on the 29th of Fructidor, Year 12 and read in the National Institute, Vendemiaire 9th Year 13. *Phil. Mag.* (1805) **21**, 222–228.
Gay-Lussac, J. (1805b) Bericht Gay-Lussac's von seiner aerostatischen Reise, am 16ten Sept. 1804. *Gilb. Ann.* **20**, 19–37. *See also*: Ueber die Mischung der atmosphärischen Luft in großer Höhe. *Neues Allgemeines Journal der Chemie* **4** (1805) 446–448.
Gay-Lussac, J. (1805c) Experiences of the eudiometer and the proportion of the principal constituents of the atmosphere. *Journal de Physique* **60**, 129–168.

Gay-Lussac, J. (1816) Sur la combinaisons de l'azote avec l'oxigène. *Ann. chim. phys.* **1**, 394–410.
Gay-Lussac, J. (1848) Mémoire sur l'eau regale. *Ann. chim. phys.* **23**, 203–229. *See also*: Extrait d'un Mémoire sur l'eau régale. *C. r.* **26**, 619–625 (and identical text): *Journal de Pharmacie et de Chimie* **14** (1848) 92–98.
Gay-Lussac, J. and L. Thénard (1809) De la nature et des propriétés de l'acide muriatique et de l'acide muriatique oxygéné. *Mémoires de Physique at de Chimie de la Société d'Arcueil* **2**, 339–358.
Geber (1530) Das Buoch geberi: Des hoch berümpten Phylosophy vonn derverborgenheyt der Alchimia, kützlich in dreyer Bücher getheylt, undgeschiklicher weisz eroffnet, dise Kunst wie sye zu er ergründenn oder zufynden sy. Auch dreyen hochverstenlichen Wörterenn dise loblich Kunstausz trukende. Auch ein Epistel des Keysers Alexandri. Amandus Farckal für Johannes Grüninger, Straßburg, 140 pp.
Geever, M., C. D. O'Dowd, S. van Ekeren, R. Flanagan, E. D. Nilsson, G. de Leuw and U. Rannik (2005) Submicron sea spray fluxes. *Geophys. Res. Lett.* **32**, 15 Art. No. L15810.
Gehler, J. S. T. (1789) Physikalisches Wörterbuch oder Versuch einer Erklärung der vornehmsten Begriffe und Kunstwörter der Naturlehre mit kurzen Nachrichten von der Geschichte der Erfindungen und Beschreibungen der Werkzeuge begleitet. Vol. 2. Ed. B. Schwickert, Leipzig, pp. 762–770. Also online (Archimedes Projekt).
Gehler, J. S. T. (1825) Physikalisches Wörterbuch. I. Band. Zweite Abtheilung. neu bearbeitet von Brandes. Gmelin, Horner, Muncke, Pfaff. Schwickert, Leipzig.
Gehler, J. S. T. (1826) Physikalisches Wörterbuch, neu bearbeitet von Brandes, Gmelin, Horner, Muncke, Pfaff. 2. Band. Schwickert, Leipzig.
Gehler, J. S. T. (1833) Physikalisches Wörterbuch, neu bearbeitet von Brandes, Gmelin, Horner, Muncke, Pfaff. 7. Band. Schwickert, Leipzig.
Gellert, C. E. (1750) Anfangsgründe zur Metallurgischen Chimie, In einem theoretischen und practischen Theile nach einer in der Natur gegründeten Ordnung. Johann Wendler, Leipzig, 339 pp. Zweyte, vermehrte und verbesserte Auflage [2nd ed.] bey Caspar Friedrich, Leipzig, 1776, 498 pp. (as Vol. 1 together with Vol. 2: Anfangsgründe zur Probierkunst, als der Zweyte Theil der practischen Metallurgischen Chimie, worinnen verschiedene neue Arten zuverlässig zu probieren gezeiget werden. Wendler, Leipzig, 168 pp.)
Gellert, C. E. (1776) Metallurgic chymistry. Being a system of mineralogy in general and of all the arts arising from this science. To the great improvement of manufactures, and the most capital branches of trade and commerce. Theoretical and practical. In two parts. Translated from the original German of C. E. Gellert by John Seiferth, London, 1776. 410 pp.
Genz, H. (1994) Die Entdeckung des Nichts. Leere und Fülle im Universum. Rororo, München, 416 pp.
George, P. (1947) Some experiments on the reactions of potassium superoxide in aqueous solution. *Discussions of the Faraday Society* **2**, 196–205.
Georgii, H. (1960) Untersuchungen über atmosphärische Spurenstoffe und ihre Bedeutung für die Chemie der Niederschläge. *Geofisica pura e applicata (Milano)* **47**, 155–171.
Georgii, H. W. (1963) Oxides of nitrogen and ammonia in the atmosphere. *JGR* **68**, 3963–3970.
Georgii, H. W. (1965) Untersuchungen über das Ausregnen und Auswaschen atmosphärischer Niederschläge durch Wolken und Niederschläge. Berichte des Deutschen Wetterdienstes Nr. 100 (Band 14), Offenbach, 23 pp.; *see also*: Georgii, H. W. and E. Weber (1960) The chemical composition of individual rainfalls. Technical Note. Contract AF 61(052)-249, pp. 1–28. Air Force Cambridge Research Centre, Bedford (Mass); cited after Junge 1963.
Georgii, H. W. (1969) Beitrag zum Schwefelhaushalt aufgrund von SO_2- und Sulfatmessungen in der freien Atmosphäre. *Annalen der Meteorologie* **4**, 117–121.
Georgii, H. W. (1970) Contribution to the atmosphere sulfur budget. *JGR* **75**, 2365–2371.
Georgii, H. W. and D. Jost (1964) Untersuchung über die Verteilung von Spurengasen in der freien Atmosphäre. *Pure and applied Geophysics* **9**, 217–224.

Georgii, H. W. and W. J. Müller (1974) On the of ammonia in the middle and lower troposphere. *Tellus* **26**, 180–184.

Georgii, H.-W., W. Fricke, W. Rudolf, M. Deimel, K.-H. Becker and U. Schurath (1977) Bildung und Transport von Photooxidantien im Raum Bonn-Köln und Frankfurt/M. VDI-Berichte 270, VDI-Verlag GmbH, Düsseldorf, pp. 19–24.

Georgiou, G. and L. Masip (2003) An overoxidation journey with a return ticket. *Science* **300**, 592–594.

Geralevich, K. G. [К Г Гералевич] (1891) К вопрос об определения углекислоты в воздухе [On the question of determination of carbonic acid in air]. St. Petersburg. *Врач* [Vrach – The Physician] **12**, 197 (a weekly paper devoted to all branches of clinical medicine, public and private hygiene).

Gerber, H. (1984) Liquid water content of fogs and hazes from visible light scattering. *Journal of Climate and Applied Meteorology* **23**, 1247–1252.

Gerhard, E. R. and H. F. Johnstone (1955) Air pollution studies-photochemical oxidation of sulfur dioxide in air. *Industrial & Engineering Chemistry* **47**, 972–976.

Gerlach, C. (1907) Ein Apparat zur qualitativen und quantitativen Ermittlung der aus industriellen Etablissements etc. entweichenden sauren Rauch- und Abgase. *Allgemeine Forst- und Jagdzeitung* **83**, 150–155.

Gerlach, C. (1908) Die Ermittlung des Säuregehaltes der Luft in der Umgebung von Rauchquellen und der Nachweis seines Ursprungs. In: Sammlung von Abhandlungen über Abgase und Rauchschäden (ed. H. Wislicenus), Paul Paray, Berlin, Heft 3, 29 pp. Reprint: Waldsterben im 19. Jahrhundert. VDI-Verlag Düsseldorf 1985.

Gersten, C. L. (1733) Tentamina systematis novi ad mutationes barometri ex natura elateris aerei demonstrandas, cui adjecta sub finem Dissertatio roris decidui errorem antiquum et vulgarem per observationes et experimenta nova executiens. Franciscum Varentrapp, Frankfurt. 222 pp.

Ghilarov, A. M. (1995) Vernadsky's biosphere concept: a historical perspective. *The Quarterly Review of Biology* **70**, 193–203.

Giacosa, P. (1895) Indagini sulle acque e sulle nevi nelli alle regioni. *Bolletino della Società Meteorologica Italiana* **20**, 45–72.

Giberne, A. (1890) The Ocean of Air: meteorology for beginners. London, Seeley & Co, 340 pp.

Gibbs, W. E. (1924) Clouds and smoke. The properties of disperse systems in gases and their practical application. Churchill, London, 240 pp.

Gibson, A. and W. V. Farrar (1974) Robert Angus Smith, F. R. S. and "sanitary science". *Notes and Records of the Royal Society* **28**, 241–262.

Gilbert, O. (1907) Die meteorologischen Theorien des griechischen Altertums. Teubner, Leipzig, 746 pp.

Gilge, S. (1994) Messung von Wasserstoffperoxid und organischen Hydroperoxiden am Schauinsland im Schwarzwald. Dissertation, TH Aachen, Germany, 166 pp.

Gilge, S. (2007) Measurements of H_2O_2 and ROOH at the meteorological observatory Hohenpeißenberg (Germany) within the Global Atmospheric Watch Program www.dwd.de/gaw.

Gilge, St., D. Kley and A. Voltz-Thomas (2000) Messungen von Wasserstoffperoxid – ein Beitrag zur Charakterisierung der limitierenden Faktoren bei der Ozonproduktion. In: *Troposphärisches Ozon. Proceedings VDI Symposium*, Band 32, Schriftenreihe Kommission Reinhaltung der Luft, Düsseldorf, pp. 379–382.

Gilge, S., C. Plass-Duelmer, W. Fricke, A. Kaiser, L. Ries, et al.(2010) Ozone, carbon monoxide and nitrogen oxides time series at four alpine GAW mountain stations in central Europe. *ACP* **10**, 12295–12316.

Gillman, F. (1883) Salt and dew. *Nature* **22**, 172.

Gilm, V. H. (1857) Über die Kohlensäure-Bestimmung der atmosphärischen Luft. *Sitzungsber. Wien* **24**, 279–284.

Girardin, J. (1839) Analyse des grêlons. *Journal de Pharmacie et des Siences Asseccoires* **25**, 390–392.
Girsberger, J. and J. Stahel (1901) Die Bekämpfung der Hagelwetter. Studie über das Wetterschiessen in Oesterreich, Italien und im Kanton Tessin und Vorschläge zur Einführung desselben im Kanton Zürich. Reisebericht. Müller, Werder & Cie, Zürich, 165 pp.
Girtanner, C. (1792) Anfangsgründe der antiphlogistischen Chemie. J. Unger, Berlin, 494 pp.
Glaisher, J. (1845) Magnetical and meteorological observations at the Royal Observatory, Greenwich, in the year 1843, London. pp. 43, 89, 137, 139, 145, 151, 173, 201, 207, 215.
Glaisher, J. (1855) Report upon the meteorology of London in relation to the chorela-epidemic of 1853–4. Ozone (pp. 71–73). In: Appendix to report on the Committee for Scientific Inquiries in relation to the chorela-epidemic of 1854., London, pp. 1–118.
Glaisher, J. (1866) There are several published sources on *Glaisher*'s blue mist observations: Remarks on the weather during the quarter ending 30th September. In: *Journal of the Statistical Society of London*, Vol. 29, 1866, pp. 641–642 and p. 637; Accounts of the meteorological and physical observations made in balloon ascents, 1862–66. *Reports to the British Association for the Advancement of Science*, London 1866. *The Illustrated London News*. Vol. 49, p. 191 (August 25, 1866).
Glaser, E. M. (1957) Bioclimatology and biometeorology. *Nature* **180**, 1328–1329.
Glindemann, D., M. Edwards and P. Kuschk (2003) Phosphine gas in the upper troposphere. *Atmos. Env.* **37**, 2429–2433.
Glindemann, D., M. Edwards and O. Schrems (2004) Phosphine and methyl phosphine production by simulated lightning – a study for the volatile phosphorus cycle and cloud formation in the earth atmosphere. *Atmos. Env.* **38**, 6867–6874.
Glindemann, D., M. Edwards, J. Liu and P. Kuschk (2005) Phosphine in soils, sludges, biogases and atmospheric implications – a review. *Ecological Engineering* **24**, 457–463.
Glotfelty, D. E., J. N. Seiber and L. A. Liljedahl (1987) Pesticides in fog. *Nature* **325**, 602–605.
Glückauf, E. (1944) The ozone content of the surface air and its relation to some meteorological conditions. *Q. J. Roy. Meteor. Soc.* **70**, 13–21.
Glueckauf, E. (1946) A micro-analysis of the helium and neon contents of air. *Proceeding of the Royal Society (A)* **70**, 13–19.
Glueckauf, E. (1951) The composition of atmospheric air. In: Compendium of Meteorology (ed. T. F. Malone), American Meteorological Society, Boston, pp. 3–10.
Glueckauf, E. and G. P. Kitt (1956) The krypton and xenon contents of atmospheric air. *Proc. R. Soc. Lond. (A)* **234**, 557–565.
Gmelin, J. F. (1798) Beitrag zur Geschichte der chemischen Kenntnisse der sogenannten Gasarten aus früheren Zeiten. *Göttingisches Journal der Naturwissenschaften* **1** (1798) Heft 4, 1–22.
Gmelin, J. F. (1799) Geschichte der Künste und Wissenschaften seit der Wiederherstellung bis an das Ende des achtzehnten Jahrhunderts. Achte Abtheilung. Geschichte der Naturwissenschaften. II. Geschichte der Chemie. Dritter und letzter Band. J. G. Rosenbusch, Göttingen, 1288 pp.
Gmelin, J. F. (1801) On the chemical knowledge which the philosophers of the 16th and 17th century had of different gases. *Phil. Mag. Series 1* **11**, 193–292. From Gmelin (1798) – but not 1:1 translation.
Gmelin, L. (1817) Handbuch der theoretischen Chemie. Erster Band. 1st ed. F. Varrentrapp, Frankfurt am Main. pp. 1–354.
Gmelin, L. (1819) Handbuch der theoretischen Chemie. Dritter Band. 1st ed. F. Varrentrapp, Frankfurt am Main. pp. 935–1588.
Gmelin, L. (1827) Handbuch der theoretischen Chemie. Erster Band. 3th ed. by F. Varrentrapp, Frankfurt am Main.

Gmelin, L. (1848) Handbuch der Chemie. Vierter Band. Organische Chemie im Allgemeinen, organische Verbindungen mit 2 und 4 Atomen Kohlenstoff. Handbuch der organischen Chemie. Erster Band. 4. Umgearbeitete und vermehrte Auflage. Univ.-Buchhandl. K. Winter, Heidelberg, 933 pp.

Gmelin, L. (1852) Handbuch der Chemie. 1. Band. Handbuch der anorganischen Chemie, 1. Band (Cohäsion, Adhäsion, Affinität, unwägbare Stoffe und unorganische Verbindungen der nichtmetallischen wägbaren Stoffe). 5. Auflage Univ.-Buchhandl. K. Winter, Heidelberg, 916 pp.

Gmelin, L. (1867) Handbuch der Chemie. Supplementsband. Erste Abteilung. Univ.-Buchhandl. K. Winter, Heidelberg, 740 pp.

Gmelin, L. (1907) Gmelin-Kraut's Handbuch der anorganischen Chemie. 7. Auflage, Band 1, Abteilung 1. C. Winter, Heidelberg, 888 pp.

Gmelin, L. G. (1927) Gmelins Handbuch der anorganischen Chemie. 8. Auflage, System-Nr. 6: Chlor. Verlag Chemie, Berlin, 442 pp.

Gmelin, L. G. (1936) Handbuch der anorganischen Chemie. 8. Aufl., System-Nr. 4: Stickstoff, Lieferung 3, Verbindungen des Stickstoffs mit Sauerstoff, Verlag Chemie Weinheim, pp. 507–854.

Gmelin, L. G. (1943) Gmelins Handbuch der anorganischen Chemie. 8. Auflage, System-Nr. 3: Sauerstoff, Lieferung 1–2, Verlag Chemie, Weinheim, pp. 298–300.

Gmelin, L. G. (1952) Handbuch der anorganischen Chemie. 8. Auflage, System-Nr. 3, Sauerstoff, Lieferung 2. Verlag Chemie Weinheim, pp. 101–111.

Gmelin, L. G. (1958) Handbuch der anorganischen Chemie (1952), 8. Aufl., System-Nr. 3, Sauerstoff, Lieferung 2 (Vorkommen – Technologie). Verlag Chemie Weinheim, 300 pp.

Gmelin, L. G. (1960) Handbuch der anorganischen Chemie (1960) 8. Aufl., System-Nr. 3, Sauerstoff, Lieferung 4, Verlag Chemie Weinheim, pp. 819–1184.

Gmelin, L. G. (1966) Handbuch der anorganischen Chemie. 8. Auflage, System-Nr. 3, Sauerstoff, Lieferung 7 (Wasserstoffperoxid). Verlag Chemie Weinheim, pp. 2097–2526.

Gmelin, L. G. (1969) Handbuch der anorganischen Chemie. 8. Auflage, System-Nr. 3, Sauerstoff, Lieferung 8 (Sauerstoff). Verlag Chemie Weinheim, pp. 2527–2947.

Gnauk, T., W. Rolle and G. Spindler (1997) Diurnal variations of atmospheric hydrogen peroxide concentrations in Saxony (Germany). *J. Atmos. Chem.* **27**, 79–103.

Göchhausen, H. F. (1710) Notabilia Venatoris, oder Jagd- und Weidwercks-Anmerkungen. Bey Johann Leonhard Mumbachen, Fürstl. Sächs. Hof-Buchdruckern. 188+16 pp. (note there are several later editions).

Goethel, M. (1980) Untersuchung ausgewählter Quellen und senken des atmosphärischen Ammoniaks mit einem neuen Probenahmeverfahren. Diploma-Thesis, Institut für Meteorologie und Geophysik der Universität Frankfurt/M.

Goldberg, E. D., ed. (1982) Atmospheric chemistry. Report of the Dahlem Workshop, Springer, Berlin Heidelberg New York, 385 pp.

Goldschmidt, H. (1880) Ueber die Unterchlorsalpetersäure von Gay-Lussac. *Sitzber. Akad. Wiss. Wien* **80**, 242–250. *See also*: *Ber.* **13** (1880) 925; *Liebig's Ann.* **205** (1880) 372–380.

Goldstein, E. (1903) Über Ozonbildung. *Ber.* **36**, 304–309.

Goldstein, S. and J. Rabani (2007) Mechanism of nitrite formation by nitrate photolysis in aqueous solutions: The role of peroxynitrite, nitrogen dioxide and hydroxyl radical. *J. Am. Chem. Soc.* **129**, 10597–10601.

Goodeve, C. F. (1937) The absorption spectra and photo-sensitising activity of white pigments. *Trans. Faraday Soc.* **33**, 340–347.

Goppelsroeder, F. (1871) Mitteilungen. 1. Beitrag zur Chemie der atmosphärischen Niederschläge mit besonderer Berücksichtigung ihres Gehalts an Salpetersäure. *Zeitschrift für praktische Chemie* **122** (NF 4), 139–154. *See also*: Ueber Schwankungen im Gehalt der Trinkwasser an Salpetersäure und deren Menge in den atmosphärischen Niederschlägen. *Z. anal. Chem.* **9**, 177–178; Beitrag zur Kenntnis der Chemie der atmosphärischen Niederschläge und des Salpetersäure- resp. Nitratgehaltes verschiedener Quell-, Bach-, Fluss- und Seewasser. Ibid. **10** (1871) 259–276 and (Fortsetzung) Ibid. **11** (1872) 16–22.

Gorbushina, A. A., R. Kort, A. Schulte, D. Lazarus, B. Schnetger, H.-J. Brumsack, W. J. Broughton and J. Favet (2007) Life in Darwin's dust: intercontinental transport and survival of microbes in the nineteenth century. *Environmental Microbiology* **9**, 2911–2922.

Gorham, E. (1955) On the acidity and salinity of rain. *Geochimica et Cosmochimica Acta* **7**, 231–239.

Gorham, E. (1958a) The influence and importance of daily weather conditions in the supply of chloride, sulphate and other ions to fresh waters from atmospheric precipitation. *Phil. Trans., Series B* **241**, 147–178.

Gorham, E. (1958b) Atmospheric pollution by hydrochloric acid. *Q. J. Roy. Meteor. Soc.* **84**, 274–276.

Gorham, E. (1981) Scientific understanding of atmosphere-biosphere interactions: a historical overview. In: Atmosphere-biosphere interactions: toward a better understanding of the ecological consequences of fossil fuel combustion, National Academy Press, Washington, pp. 9–13.

Gorham, E. (1989) Scientific understanding of ecosystem acidification: a historical review. *Ambio* **18**, 150–154.

Gorham, E. (1991) Biogeochemistry: its origins and development. *Biogeochemistry* **13**, 199–239.

Gorham, E. (1994) Neutralizing acid rain. *Nature* **367**, 321.

Gorham, E. (1998) Acid deposition and its ecological effects: a brief history of research. *Environmental Science & Policy* **1**, 153–166.

Gorham, E., F. B. Martin and J. T. Litzau (1984) Acid rain: ionic correlations in the Eastern United States, 1980–1981. *Science* **225**, 407–409.

Gorup-Besanez, E. von (1859) Ueber die Einwirkung des Ozons auf organische Verbindungen. *Ann.* **110**, 86–107.

Gorup-Besanez, E. von (1872) Ueber die Ozonreactionen der Luft in der Nähe von Gradirhäusern. *Ann.* **161**, 232–251.

Götz, F. W. P. (1931) Das atmospharische Ozon. *Beiträge zur Geophysik* (Supp.) **1**, 180–235.

Götz, F. W. P. (1934) Die heilklimatische Bedeutung des Ozons. *Der Balneologe* **1**, 23–26.

Götz, F. W. P. (1951) Ozone in the atmosphere. In: Compendium of Meteorology (ed T. F. Malone), American Meteorological Society, Boston, MA, pp. 275–291.

Götz, F. W. P. and G. M. B. Dobson (1928) Observations of the height of the ozone in the upper atmosphere. *Proc. R. Soc. Lond. A* **120**, 252–259.

Götz, F. W. P., G. M. B. Dobson and A. R. Meetham (1934) Vertical distribution of ozone in the atmosphere. *Nature* **132**, 281. *See also*: Götz, F. W. P., A. R. Meetham and G. M. B. Dobson (1934) The vertical distribution of ozone in the atmosphere. *Proc. R. Soc. Lond. A* **145**, 416–446.

Götz, F. W. P. and R. Ladenburg (1931) Ozongehalt der unteren Atmosphärenschichten. *Naturwissenschaften* **19**, 373–374.

Götz, F. W. P. and H. Maier-Leibnitz (1933) Zur Ultraviolettabsorption bodennaher Luftschichten. *Zeitschrift für Geophysik* **9**, 253–260.

Götz, F. W. P., M. Schein and R. Stoll (1935) Messungen des bodennahen Ozons in Zürich. *Gerlands Beiträge zur Geophysik* **45**, 237–242.

Götz, F. W. and F. Volz (1951) Aroser Messungen des Ozongehalts der unteren Troposphäre und sein Jahresgang. *Zeitschrift für Naturforschung* **6a**, 634–639.

Grabowski, R. I. [Грабовский] (1951a) О концентрации хлорида в осадках и облачных элементах [On the concentration of chloride in rainfall and clouds]. *Вестник Ленинград университета* [Bulletin of the Leningrad University] **10**, 36–49.

Grabowski, R. I. [Грабовский] (1951b) О природе атмосферных ядер конденсации. *Метеорол. и гидрол.* [On the nature of atmospheric condensation nuclei]. [Meteorology and Hydrology] No 4.
Grabowski, R. I. [Грабовский] (1952) Мировой океан как источник атмосферных ядер конденсации [The ocean as source of atmospheric condensation nuclei] *Известия АН СССР, серия геофизическая* [Communicfations of the Academy of Sciences USSR, ser. Geophysics], No 2.
Grabowski, R. I. [Грабовский] (1953a) О происхождении атмосферных ядер конденсации [On the occurrence of atmospheric condensation nuclei]. *Природа* [Prioroda], No 1.
Grabowski, R. I. Грабовский (1953b) К вопросу о «вымывании»хлоридов из атмосферы осадками [On the question of wash-out of chlorides by atmospheric precipitation]. *Вестник Ленинградского государственного университета* **12**, 87–91.
Grabowski, R. I. [Грабовский] (1956) Атмосферные ядра конденсации [Atmospheric condensation nuclei]. Leningrad, Gidrometeoizdat, 240 pp.
Graedel, T. E. (1978) Chemical compounds in the atmosphere. Academic Press, New York et al., 440 pp.
Graedel, T. E. (1984) Effects of below-cloud gas scavenging on raindrop chemistry over remote ocean regions. *Atmos. Env.* **18**, 1835–1842.
Graedel, T. E. and C. J. Weschler (1981) Chemistry within aqueous atmospheric aerosol and raindrop. *Reviews of Geophysics and Space Physics* **19**, 505–539.
Graedel, T. E. and K. T. Goldberg (1983) Kinetic studies of raindrop chemistry. I. Inorganic and organic processes. *JGR* **88**, 865–882.
Graedel, T. E., M. L. Mandich and C. J. Weschler (1986) Kinetic model studies of atmospheric droplet chemistry. 2. Homogeneous transition metal chemistry in raindrops. *JGR* **91**, 5205–5221.
Graedel, T. E. and W. C. Keene (1990) The budget and cycle of Earth's natural chlorine. *Pure & Applied Chemistry* **68**, 1689–1697.
Graedel, T. E. and P. J. Crutzen (1993) Atmospheric change: An earth system perspective. W. H. Freeman, New York, 446 pp.
Graedel, T. E. and P. J. Crutzen (1995) Atmosphere, climate and change. W. H. Freeman, New York, 208 pp.
Gräfenberg, L. (1902) Das Potential des Ozons. *Z. Elektrochem.* **8**, 297–301.
Gräfenberg, L. (1903) Beiträge zur Kenntnis des Ozons. *Z. anorg. Chem.* **36**, 355–379.
Gräger, N. (1845) Ammoniakgehalt der Atmosphäre. *Ann.* **56**, 208–209; and: *Arch. Pharm.* **94**, 35–36.
Gräger, N. (1852) Ueber den Ozongehalt der Atmosphäre. *Arch. Pharm.* **119**, 278–286.
Graham, Th. (1842) Elements of chemistry, including the applications od science in the arts. H. Balliere, London, 1088 pp.
Graham, Th. (1861) Liquid diffusion applied to analysis. *Phil. Trans.* **151**, 183–224.
Graham, R. A. and H. D. Johnston (1978) The photochemistry of NO_3 and the kinetics of the N_2O_5-O_3 system. *J. phys. Chem.* **82**, 254–268.
Granat, L. (1972) On the relation between pH and the chemical composition in atmospheric precipitation. *Tellus* **24**, 550–560.
Granat, L. (1978) Sulphate in precipitation as observed by the European Atmospheric Chemistry Network. *Atmos. Env.* **12**, 413–424.
Granat, L. (1988) Ur Arsrapport till SNV. Institute for Meteorology, Stockholm University, Report 3475 (in Swedish).
Granat, L., H. Rodhe and R. O. Hallberg (1976) The global sulfur cycle. In: Nitrogen, phosphorus and sulphur – global cycles (eds. B. H. Svensson and R. Søderlund), SCOPE Report 7, Ecol. Bull. (Stockholm) 22, pp. 89–134.
Grapi, P. (2019) Inspiring air. A history of air-related science. Vernon Press, 380 pp.

Grasnick, K.-H. (1963) Ergebnisse der Ozonmessungen in Potsdam während des IGJ und der IGC 1957–1959. Veröffentlichungen des Meteorologischen und Hydrologischen Dienstes der DDR Nr. 19. Akademie-Verlag Berlin. pp. 1–35.

Grasnick, K.-H. and W. Warmbt (1973) Atmospheric ozone – surface ozone. Meteorology in the GDR. News and activities. *Meteorological Service of the German Democratic Republic* **2**, 23–29.

Gratacap, L. P. (1881) Atmospheric ozone for January, 1881. *Science* **2**, 104–105.

Gray, G. (1888) On the dissolved matter contained in the Rain-water collected at Lincoln, Canterbury, New Zealand. *Proceedings of the Australasian Association for the Advancement of Science* **1**, 138–152.

Gray, G. (1910) On the dissolved matter contained in the rain water collected at Lincoln. *N. Z. The Canterbury Agricultural College Magazine* **25**, 183–195.

Green, H. L., W. R. Lane and H. Hartley (1964) Particulate clouds: dusts, smokes and mists. Van Nostrand, Toronto, 471 pp.

Greiner, N. R. (1970) Hydroxyl radical kinetics by kinetic spectroscopy, VI. Reactions with alkanes in the range 300–500°K. *J. Chem. Phys.* **53**, 1070–1076.

Grellois, E. (1857a) Notice sur les observations ozonométriques faites pendant neuf mois, en 1855 et 1856, a la pointe du Sérail, a Constantinople. *Annu. soc. mét. de France* **5**, 58–70.

Grellois, E. (1857b) Ozonometre. Résultats de quelques experiences faites a Thionville. *Annu. soc. mét. de France* **5**, 185–191.

Gren, F. A. C. (1800) Principles of modern chemistry: systematically arranged. Vol. 1, translated from German edition (1796), T. Cadell, Jun. & W. Davies, London, 488 pp.

Grimm: Deutsches Wörterbuch von Jacob Grimm und Wilhelm Grimm (http://dwb.uni-trier.de).

Gröger, M. (1898) Ueber Dichte und Molekulargewicht des Ozons. *Ber.* **34**, 3174–3176.

Grosjean, D. (1982) Formaldehyde and other carbonyls in Los Angeles ambient air. *Environ. Sci. Technol.* **16**, 254–262.

Grosjean, D. and B. Whright (1983) Carbonyls in urban fog, ice fog, cloud water and rainwater. *Atmos. Env.* **17**, 2093–2096.

Grossmann, D., G. K. Moortgat, M. Kibler, S. Schlomski, K. Bächmann, B. Alicke, A. Geyer, U. Platt, M.-U. Hammer, B. Vogel, D. Mihelcic, A. Hofzumahaus, F. Holland and A. Volz-Thomas (2003) Hydrogen peroxide, organic peroxides, carbonyl compounds and organic acids measured at Pabstthum during BERLIOZ. *JGR* **108**, D4, 8250.

Groth, W. and H. Suess (1938) Bemerkungen zur Photochemie der Erdatmosphäre. *Naturwissenschaften* **26**, 77.

Grotthuß, T. von (1820) Physisch-chemische Forschungen. Schragsche Verlagsbuchhandlung, Nürnberg, 158 pp.

Grotthuß, T. von (1821) Untersuchung eines in Kurland, im Dünaburg'schen Kreise, am 30 Juni (12 Juli) herabgefallenen Meteorsteines. *Gilb. Ann.* **67**, 337–370.

Gruber, M. (1881) Ueber den Nachweis und die Giftigkeit des Kohlenoxydes und sein Vorkommen in Wohnräumen. *Polytechnisches Journal (Dingler's J.)* **241**, 219–224.

Grundmann, W. (1942) Verfahren und Geräte zur Bestimmung der Staub- und Kernbeimengungen der Luft. Glasinstrumentenkunde 5. Band. R. Wagner Sohn, Weimar, 75 pp.

Grunow, J. (1955) Der Niederschlag im Bergwald – Niederschlagszurückhaltung und Nebelniederschlag. *Forstwissenschaftliches Zentralblatt* **74**, 21–36.

Guericke, O. von (1672a) Experimenta nova (ut vocantur) de vacuo spatio. Reprint from the first edition in Latin. Verlag Stekovics, Halle (2002), 278 pp.

Guericke, O. von (1672b) Neue "Magdeburgische" Versuche über den leeren Raum. Drittes Buch: Ueber eigene Versuche (Ed. F. Dannemann), Ostwald's Klassiker der exakten Naturwiss. Nr. 59. Akad. Verlagsgesell., Leipzig (1894), 116 pp.

Guericke, O. von (1672c) Neue (sogenannte) Magdeburger Versuche über den leeren Raum. Drittes Buch: Die eigenen Experimente. Dt. Übersetzung aus dem Lat. von A. Kauffeldt, VEB Deutscher Verlag für Grundstoffindustrie, Leipzig (1986), 256 pp.

Guéron, J. and M. Prettre (1935) Complexité de la reaction entre l'ozone de l'iodine. *C. r.* **200**, 2084–2086.
Guéron, G., J. Guéron and M. Prettre (1935) Oxydation inquite de l'iodine de potassium per l'ozone. *C. r.* **201**, 1376–1378.
Guicherit, R. (1988) Ozone on an urban and regional scale with special reference to the situation in the Netherlands. In: Tropospheric Ozone, Regional and Global Scale Interactions. NATO ASI series, Vol. 227 (ed. I. A. A. Isakssen), D. Reidel, Dordrecht, pp. 49–62.
Gunz, D. W. and M. R. Hoffmann (1990a) Atmospheric chemistry of peroxides: A review. *Atmos. Env.* **24**, 1601–1633.
Gunz, D. W. and M. R. Hoffmann (1990b) Field investigations on the snow chemistry in central and southern California. I. Inorganic ions and hydrogen peroxide. *Atmos. Env.* **24**, 1661–1671.
Guo, J., Y. Wang, X. H. Shen, Z. Wang, T. Lee, X. F. Wang, P. H. Li, M. H. Sun, J. L. Collett and W. X. Wang (2012) Characterization of cloud water chemistry at Mount Tai, China: seasonal variation, anthropogenic impact and cloud processing. *Atmos. Env.* **60**, 67–476.
Gushchin, G. P. [Гущин] (1957) К вопросу об измерении общего содержания атмосферного озона и его вертикального распределения [On the question of measurements of total ozone and its vertical distribution]. *Метеорология и Гидрология* [Meteorology and Hydrology] **6**, 26–32.
Gushchin, G. P. [Гущин] (1960) Преварительные резултаты измерение общево содержания атмосферного озона во время МГГ в СССР [Preliminary results from measurements of the total atmospheric ozone amount during IGY in the USSR]. *Труды ГГО* **105**, 3–16.
Gushchin, G. P. [Гущин] (1963) Исследование атмосферного озон [Investigation of atmospheric ozone]. Гидрометеорологическое изд-во, Ленинград, 266 pp.
Gushchin, G. P. [Гущин] (1964) Озон и аэросиноптические условия в атмосфере [Ozone and atmospheric synoptic conditions in the atmosphere]. Гидрометеоиздат, Ленинград, 341 pp.
Gushchin, G. P. [Гущин] (1995) К вопросу об истории создания озонометрической сети в Россия [On the history of the ozonometric network in Russia]. *Известия АН. Физика Атмосфера и Океана* **31**, 6–9.
Gushchin, G. P. and E. S. Selezneva, eds. [Гущин, Селезнева] (1961) Данные по химическому составу атмосферных осадков и общему содержанию озона в атмосфере в различных пунктах СССР (материалы МГГ и МГС за 1957–1959 гг. [Data on chemical composition of atmospheric precipitation and total ozone content in the air at different sites of the USSR in the International Geophysical Year and International Sun Year, 1957–1959). Gidrometeoizdat, Leningrad, 124 pp.
Gushchin, G. P. and N. N. Vinogradova [Гущин, Виноградова] (1983) Суммарный озон в атмосфере [Total ozone in the atmosphere]. Гидрометеоиздат, Ленинград, 241 pp.
Guth, F. (1919) Die Bekämpfung der Rauchplage mit besonderer Berücksichtigung der Stadt Saarbrücken und die zukünftige Entwicklung der Brennstoffwirtschaft. *Gesundheits-Ingenieur* **42**, 457–468, 472–477 and 497–505.
Haagen-Smit, A. J. (1952) Chemistry and physiology of Los Angeles smog. *Industrial and Engineering Chemistry* **44**, 1342–1346.
Haagen-Smit, A. J. and M. M. Fox (1956) Ozone formation in photochemical oxidation of organic substances. *Industrial and Engineering Chemistry* **48**, 1484–1487.
Haber, F. and J. Weiss (1932) Über die Katalyse des Hydroperoxides. *Naturwissenschaften* **51**, 948–950.
Haber, F. and J. Weiss (1934) The catalytic decomposition of hydrogen peroxide ba iron salts. *Proc. R. Soc. Lond.* **147**, 332–352.
Haber, F. and R. Willstädter (1931) Unpaarigkeit und Radikalketten im Reaktionsmechanismus organischer und enzymatischer Vorgänge. *Ber.* **64**, 2844–2856.
Hadfield, W. (1842) An account of some experiments to determine the quantity of carbonic acid in the air. *Memoirs of the Literary and Philosophical Society of Manchester, 2nd series* **6**, 10–18.

Haehnel, O. (1922) Kohlendioxyd- und Schwefeldioxydgehalt der Berliner Luft. *Z. angew. Chem.* **35**, 618–620.
Hahn, A. (1941) Der Kreislauf der Stoffe in der Natur. J. F. Lehmanns Verlag, München und Berlin, 104 pp.
Haïssinsky, M. and M. Magat (1951) Sur les réactions primaires produites par les radiations ionisantes dans l'eau. *C. r.* **233**, 954–956.
Haldane, J. S. (1895) A method of detecting and estimating carbonic dioxide in air. *Journal of Physiology* **18**, 462–469.
Haldane, J. S. (1918) Methods of air analysis. Ch. Griffin, London, 137 pp.
Haldane, J. S. and J. L. Smith (1892) The physiological effects of air vitiated by respiration. *The Journal of Pathology and Bacteriology* **1**, 168–186.
Haldane, J. S. B. (1936) Carbon dioxide content of atmospheric air. *Nature* **137**, 575. [Note: *John Burdon Sanderson Haldane* (1892–1964), British theoretical geneticist, is the son of J. S. Haldane who died 1936 but conducted the measurements together with *Makgill*].
Hales, S. (1727) Vegetable staticks: or an account of some statical experiments on the sap in vegetables: being an essay towards a natural history of vegetation. Also, a specimen of air, by a great varity of chamio-statical experiments, which were read at several meetings before to the Royal Society. W. and J. Innys, London, 376 pp. *See also*: The Scientific Book Guild. Reprinted with forward by M. A. Hoskin, London 1961.
Halfpenny, E. and P. L. Robinson (1952) Pernitrous acid. The reaction between H_2O_2 and nitrous acid and the properties of the intermediate products. *J. chem. Soc.* **168**, 928–938.
Hall, A. D. and N. H. J. Miller (1911) On the absorption of ammonia from the atmosphere. *Journal of Agricultural Science* **4**, 56–68.
Hall, E. S. (1860) On the influenza epidemics in Hobart (Tasmania) in July, 1860. *The Australian Medical Journal* October 1860, 252–269.
Hall, R. A. (1954) The scientific revolution, 1500–1800: the formation of the modern scientific attitude. Longmans, London, pp. 394.
Hall, R. E. (1973) Al-Khazini. In: Dictionary of scientific biography (ed. C. C. Gillispie), Vol. 7, New York, pp. 335–351.
Haller, C. (1874) Das Ozon der Gebirgs-Atmosphäre. *Z. Öster. Ges. Met.* **9**, 81–84.
Halliday, E. C. (1961) A historical review of atmospheric pollution. In: Air pollution (ed. WHO), Columbia Univ. Press, New York, pp. 9–35.
Halliwell, B. and J. M. Gutteridge (1984) Oxygen toxicity, oxygen radicals, transition metals and disease. *Biochemical Journal* **219**, 1–14.
Hamilton, Jr, E. J. (1975) Water vapor dependence of the kinetics of the self-reaction of HO_2 in the gas phase. *J. Chem. Phys.* **63**, 3682–3683.
Hamilton, Jr., E. J. and C. A. Naleway (1976) Theoretical calculation of strong complex formation by the HO_2 radical: $HO_2{\cdot}H_2O$ and $HO_2{\cdot}NH_3$. *J. Chem. Phys.* **80**, 2037–2040.
Hamlet, W. M. (1881) On the action of compounds inimical to bacterial life. *J. Chem. Soc. Transactions* **39**, 326–331.
Hammerschmied, J. (1873) Das Ozon vom chemischen, physiologischen und sanitären Standpunkte. *Schriften des Vereines zur Verbreitung Naturwissenschaftlicher Kenntnisse in Wien* **13**, 393–525. *See also*: Das Ozon und seine Wichtigkeit im Haushalte der Natur und des menschlichen Körpers. Mit einem Anhange über allgemeine chemisch-physikalische Fragen. K. Gerold's Sohn, Wien, 1873, 186 pp.
Hampson, J. (1964) Photochemical behaviour of the ozone layer. CARDE techn. Rep. note 1627, Canadian Armament Res. and Development Establishment. *See also*. Photolysis of wet ozone and its significance to atmospheric heating of the ozone layer. In: Procceedings International Council of the Aeronautical Sciences, Third Congress, Stockholm 1962. Macmillan, London, pp. 215–234.

Hampson, J. (1974) Photochemical war on the atmosphere. *Nature* **250**, 189–191.
Hann, J. von (1883) Handbuch der Klimatologie. Bibliothek geographischer Handbücher (Ed. F. Ratzel), J. Engelhorn, Stuttgart, 764 pp.; 2nd ed. published in 1897 in 3 vols; 3rd edn in 1908–1911 in 3 vols.; translated into English: Handbook of Climatology: Part I. General Climatology, Macmillan Company, 1903); 3rd ed.: Handbuch der Klimatologie. I. Band: Allgemeine Klimalehre (Dritte Auflage). Engelhorn, Stuttgart, 394 pp.
Hann, J. von (1889) Wassergehalt der Wolken- und Nebel-Luft. *Met. Z.* **6**, 304–306.
Hansen, F. (1926) Om bestemmelse af nitratkvælstoff i regnvand, dranvand og jorg. *Tidsskrift for Planteavl* **32**, 69–120 (cited after Eriksson 1952).
Hansen, F. (1931) Undersøgelser af regnvand. *Tidsskrift for Planteavl* **37**, 123–150 (cited after Eriksson 1952).
Hanst, P. L., W. E. Wilson, R. K. Patterson, B. W. Gay Jr., L. W. Chancy and C. S. Burton (1975) A spectroscopic study of California smog. EPA publication No. EPA-650/4 75-OM. EPA, U. S. A.
Hantzsch, A. and L. Kaufmann (1896) Zur Kenntniss der untersalpetrigen Säure. *Liebigs Ann.* **292**, 317–340.
Harper, H. J. (1942) Sulfur content of Oklahoma rainfall. *Proceedings of the Oklahoma Academy of Sciences for 1942*, 73–82.
Harper, K. C. (2008) Weather by the Numbers: The Genesis of Modern Meteorology (Transformations: Studies in the History of Science and Technology. Mit Press, 308 pp.
Harrassowitz, H. (1956) Atombombenexplosionen und Regen-p_H. *Naturwissenschaften* **43**, 11–12.
Harries, C. (1911) Über Bildung des Ozons: Sitzung im großen Hörsaale des chemischen Instituts der Universität. *Z. Elektrochem.* **45**, 629–633.
Harries, C. (1912) Zur Kenntnis der Bestandteile des Ozons. *Ber.* **45**, 936–944.
Harries, C. (1916) Untersuchungen über das Ozon und seine Einwirkung auf organische Verbindungen (1903–1916). Springer, Berlin, 720 pp.
Harrison, J. B. (1911) Composition of rain-water: British-Guiana. *Journal of the Chemisl Society. Abstract of Papers*. Vol. C. Part II, pp. 530–531; original paper: *Brit. Guiana Dept. Sci. Agric. Rep for 1909–10*, 15–18.
Harrison, J. B. and J. Williams (1897) The proportions of chlorine and nitrogen as nitric acid and as ammonia in certain tropical rain waters. *J. Am. Chem. Soc.* **19**, 1–9.
Harrison, R. M. and C. A. Pio (1983) A comparative study of the ionic composition of rainwater and atmospheric aerosols: implications for the mechanism of acidification of rainwater. *Atmos. Env.* **17**, 2539–2543.
Hart, E. B. and W. H. Peterson (1911) The sulfur requirements of farm crops in relation to the soil and air supply. *J. Am. Chem. Soc.* **33**, 549–564.
Hart, E. J. (1959) Development of the radiation chemistry of aqueous solutions. *Journal of Chemical Education* **36**, 266–272.
Hart, E. J. and M. Anbar (1970) The hydrated electron. John Wiley & Sons, New York, 267 pp.
Harteck, P. (1931) Die Schwankungen des Ozongehaltes der Atmosphäre. *Naturwissenschaften* **19**, 858–860.
Harteck, P. and U. Kopsch (1931) Gasreaktionen mit atomarem Sauerstoff. *Z. phys. Chem. B* **12**, 327–347.
Hartley, H. (1971) Studies in the history of chemistry. Clarendon Press, 256 pp.
Hartley, W. N. (1880) On the probable absorption of the solar ray by atmospheric ozone. *Chemical News* **42**, 268–274.
Hartley, W. N. (1881) On the absorption of the solar ray by atmospheric ozone. *J. Chem. Soc. Transactions* **39**, 111–128.
Hasenbach, C. W. (1871) Beitrag zur Kenntnis der Untersalpetersäure und der salpetrigen Säure. *J. prakt. Chem.* **112** (NF 4), 1–19.

Hasson, A. S. and S. E. Paulson (2003) An investigation of the relationship between gas-phase and aerosol-borne hydroperoxides in urban air. *J. Aerosol. Sci.* **34**, 459–468.
Hatcher, R. A. and H. V. Arny (1900) Atmospheric ozone. *American Journal of Pharmacy* **72**, 423–429.
Haurwitz, B. (1938) Atmospheric ozone as a constituent of the atmosphere. *Bulletin of the American Meteorological Society* **19**, 417–424.
Hautefeuille, P. and J. Chappuis (1881) Quelques faits pour server à l'histoire de la nitrification. *C. r.* **92**, 134–137.
Hautefeuille, P. and J. Chappuis (1884) Recherches sur l'ozone. Première partie: étude de la préparation de l'ozone au moyen de l'effluve électrique. Deuxième partie. propriétés de l'ozone. *Annales scientifiques de l'École Normale Supérieure, Série 3* **1**, 55–84.
Haviland, A. (1855) Climate, weather and disease; a sketch of the opinions of antient and modern writers with regard to the influence of climate and weather in producing disease. Churchill, London 144 pp.
Hayes, A. A. (1851) On the existence of ammonia in the general atmosphere. In: Proceedings of the American Association for the Advancement of Science, fourth meeting, S. F. Baird, Washington, pp. 207–213.
Hayhurst, W. and J. N. Pring (1910) The examination of the atmosphere at various altitudes for oxides of nitrogen and ozone. *J. Chem. Soc., Transactions* **97**, 868–877.
He, Z., Z. M. Chen, X. Zhang, Y. Zhao, D. M. Huang, J. N. Zhao, T. Zhu, M. Hu and L. M. Zeng (2010) Measurement of atmospheric hydrogen peroxide and organic peroxides in Beijing before and during the 2008 Olympic Games: Chemical and physical factors influencing their concentrations. *JGR* **115**, D17307.
Heelis, E. (1873–1876) Results of meteorological observations, taken at Langdale, Dimbula Ceylon, in the year 1873. *Proceedings of the Literary and Philosophical Society of Manchester* **13–15**, 93–96.
Hegg, D. A. and P. V. Hobbs (1978) Oxidation of sulfur dioxide in aqueous systems with particular reference to the atmosphere. *Atmos. Env.* **12**, 241–253.
Hegg, D. A. and P. V. Hobbs (1981) Cloud water chemistry and the production of sulfates in clouds. *Atmos. Env.* **15**, 1597–1604.
Hegg, D. A. and P. V. Hobbs (1982) Measurements of sulfate production in natural clouds. *Atmos. Env.* **16**, 2663–2668.
Hegg, D. A. and P. V. Hobbs (1986) Sulfate and nitrate chemistry in cumuliform clouds. *Atmos. Environ.* **20**, 901–909.
Heicklen, J. (1976) Atmospheric chemistry. Academic Press, New York, 406 pp.
Heicklen, J., K. Westberg and N. Cohen (1971) Discussion remark. In: Chemical Reactions in Urban Atmospheres (ed. C. S. Tuesday). Elsevier, Amsterdam (287 pp.), pp. 55–59.
Heidenschreider, A. (1868) Meteorologische Beobachtungen in Herrieden, in Verbindung mit den herrschenden Krankheiten im Etatsjahre 1866–1867, resp., Kalenderjahr 1867: *Abhandlungen der Naturhistorischen Gesellschaft zu Nürnberg* **4**, 62–85.
Heikes, B. G. (1984) Aqueous H_2O_2 production from O_3 in glass impingers. *Atmos. Env.* **18**, 1433–1445.
Heikes, B. G., A. L. Lazrus, G. L. Kok, S. M. Kunen, B. W. Gandrud, S. N. Gitlin and P. D. Sperry (1982) Evidence for aqueous phase hydrogen peroxide synthesis in the troposphere. *JGR* **87**, 3045–3051.
Heikes, B. G., G. L. Kok, J. G. Walega and A. L. Lazrus (1987) H_2O_2, O_3 and SO_2 measurements over the eastern United States during fall. *JGR* **92**, 915–931.
Heikes, B. G., J. G. Walega, G. L. Kok, J. A. Lind and A. L. Lazrus (1988) Measurement of H_2O_2 during WATOX-86. *Global Biogeochemical Cycles* **2**, 57–611.
Heimann, J. (1888) Der Kohlensäuregehalt der Luft in Dorpat bestimmt in den Monaten Juni bis September 1888. Dissertation, Medizinische Fakultät der Kaiserlichen Universität Dorpat, 53 pp.

Heine, H. (1882) Ueber die Absorption der Wärme durch Gase und eine darauf beruhende Methode zur Bestimmung des Kohlensäuregehalts der atmosphärischen Luft. In: Einundzwanzigster Bericht der Oberhessischen Gesellschaft für Natur- und Heilkunde, Giessen, pp. 17–60. *See also*: *Wied. Ann.* **16** (1882) 441–481.

Heinrich, R. (1881) Ueber die Ammoniakmengen welche der Atmosphäre im Laufe eines Jahres durch Salzsäure entzogen werden. *Forschungen auf dem Gebiet der Agricultur-Physik (Wollny's Forschungen)* **4**, 446–452.

Helbig, D. (1903) Nuova sintesi della anidridie nitrica. *Atti della Accademia nazionale dei Lincei; ser. 5* **12**, 166–170. *See also*: Über die synthetische Darstellung des Stickstoff-trioxyds und die Eigenschaften dieses Körpers. *Z. Elektrochem.* **16** (1910) 205–206.

Heldt, W. (1861) Die Fundamental-Eigenschaften des Sauerstoffs und Wasserstoffs. Hickethier, Berlin, 69 pp.

Heller, W. (1935) Le dosage de l'ozone atmosphérique par la fluorescéine. *C. r.* **200**, 1936–1938.

Heller, J. F. (1851) Salpetersäure als constanter Bestandteil der athmosphärischen Luft, und die Verhältnisse jener zum Ozon. *Zeitschrift der k. k. Gesellschaft der Ärzte zu Wien* **7**, 738–742. *See also*: Bericht der von der k. k Gesellschaft der Aerzte zu Wien bestimmten Kommission zur Ausmittelung des angeblichen Salpetersäure-Gehalts der atmosphärischen Luft. Ibid. pp. 744–751.

Heller, V. G. (1938) The chemical content of Oklahoma rainfall. Techn. Bull. No. 1, Oklahoma Agricultural and Mechanical College, Agricultural Experimental Station, Stillwater, 23 pp.

Heller, A. (1974) 50 Jahre Arbeitsgebiet "Lufthygiene" beim Institut für Wasser-, Boden- und Lufthygiene des Bundesgesundheitsamtes. In: Lufthygiene 1974. *Schriftenreihe des Vereins für Wasser-, Boden- und Lufthygiene* **41**, 35–82.

Hellmann, G. (1879) Über die auf dem Atlantischen Ocean in der Höhe der Capverdischen Inseln häufig vorkommenden Staubfälle. *Monatsberichte der Königlichen Preussischen Akademie der Wissenschaften für das Jahr 1878*, 365–403.

Hellmann, G. (1883) Repertorium der Deutschen Meteorologie. Leistungen der deutschen Schriften, Erfindungen und Beobachtungen auf dem Gebiete der Meteorologie und des Erdmagnetismus von den ältesten Zeiten bis zum Schluss des Jahres 1881. W. Engelmann, Leipzig, 995 pp.

Hellmann, G. (1904) Denkmäler Mittelalterlicher Meteorologie. Neudrucke von Schriften und Karten über Meteorologie und Erdmagnetismus. Asher & Co., Berlin, 282 pp.

Hellmann, G. (1908) The dawn of meteorology. *Q. J. Roy. Meteor. Soc.* **34**, 221–232.

Hellmann, G. (1914) Beiträge zur Geschichte der Meteorologie. Erster Band (Nr. 1–5). Anhang. Behrend, Berlin, 147 pp.

Hellmann, G. (1917) Beiträge zur Geschichte der Meteorologie. Zweiter Band (Nr. 6–10). Anhang. Behrend, Berlin, 340 pp.

Hellmann, G. (1920) Beiträge zur Erfindungsgeschichte meteorologischer Instrumente. Abhandlungen der Preussischen. Akademie der Wissenschaften, Berlin, 60 pp.

Hellmann, G. (1921) Die Meteorologie in den deutschen Flugschriften und Flugblättern des XVI. Jahrhunderts. Ein Beitrag zur Geschichte der Meteorologie. Abhandlungen der Preussischen. Akademie der Wissenschaften, Berlin, 96 pp.

Hellmann, G. (1922a) Beiträge zur Geschichte der Meteorologie. Dritter Band (Nr. 11–15). Anhang. Behrend, Berlin, 102 pp.

Hellmann, G. (1922b) Entwicklungsgeschichte des klimatologischen Lehrbuches. In: *Beiträge Geschichte Meteorologie* **11**, 1–14.

Hellmann, G. (1924) Versuch einer Geschichte der Wettervorhersage. Aus d. Abhandl. der Preuss. Akad. d. Wiss., Phys.-Math. Klasse Nr. 1, Verlag der Akad. d. Wiss., Berlin, 54 pp.

Hellmann, G. (1927) Die Entwicklung der meteorologischen Beobachtungen bis zum Ende des XVIII. Jahrhunderts. *Abhandlungen der Preussischen. Akademie der Wissenschaften*, Phys.-Math. Klasse Nr. 1, Verlag der Akademie der Wissenschaften, Berlin, 48 pp.

Helmholtz, R. von (1886) Ueber Nebelbildung. *Naturwissenschaftliche Rundschau* **1**, 69–71.
Helmholtz, R. von (1887) Versuche mit einem Dampfstrahl. *Wied. Ann.* **32**, 1–18.
Helmont, J. B. (1648) Ortus Medicinae. Id Est, Initia Physicae Inaudita: Progressus mediccinae novus, in Morborum Ultionem, Ad Vitam Longam. Elzevirius, Amsterdami, 800 pp. (reprinted Brussels: Culture et Civilisation, 1966); translated into English by John Chandler (1662): Oriatrike or, Physicks Refined...
Helmont, J. B. (1662) Oriatrike or, Physick Refined. The common Errors therein refuted and the whole art reformed and rectified: being a new Rise and Progress of Phylosophy and Medicine, for the Destruction of Diseases and Prolongation of Life. Written by that most Learned, Famous, Profound and Acute Phylosopher and chemical Physitian, John Baptista Van Helmont and now faithfully rebndered into English, in tendency to a common good and the increase of true Science; by JC, sometimes of MH Oxon. Lodowick Loyd, London, 814 pp.
Helmts (1826) Fremde Beimischungen im Regenwasser. *Arch. Pharm.* **17**, 77.
Hempel, W. (1885) Die Sauerstoffbestimmung in der atmosphärischen Luft. *Ber.* **18**, 267–282.
Hempel, W. (1890a) Neue Methoden zur Analyse der Gase. Fr. Vieweg & Sohn, Braunschweig, 129 pp.
Hempel, W. (1890b) Gasanalytische Methoden, first ed. Fr. Vieweg & Sohn, Braunschweig, 367 pp.
Hempel, W. (1913) Gasanalytische Methoden. 4th ed. (first 1889). Fr. Vieweg & Sohn, Braunschweig, 427 pp.
Henderson, Y. and H. W. Haggard (1927) Noxious gases and the principles of respiration influencing their action. 200 pp. (2nd ed. 1943, Reinhold, New York).
Henneberg, W. (1873) Der Kohlensäuregehalt der atmosphärischen Luft. *Die landwirthschaftlichen Versuchs-Stationen* **16**, 70–71. *See also*: Anonymous (1872) Kohlensäuregehalt der atmosphärischen Luft. *Arch. Pharm.* **202**, 463–463.
Henriet, H. (1897) Les gaz de l'atmosphére. Gauthiers-Vilars et Masson, Paris, 192 pp.
Henriet, H. (1904) Sur la présence de l'aldèhyde formiquie dans l'air atmosphèrique. *C. r.* **138**, 203–205; and: Dosage de la formaldéhyde atmosphérique, pp. 1272–1274. *See also*: *Naturwissenchaftliche Rundschau* **19** (1904) 167.
Henriet, H. and M. Bonyssy (1908) Sur l'origine de l'ozone atmosphérique er ies causes de variations de l'acide carbonique de l'air. *C. r.* **146**, 977–978.
Henson, R. (2010) Weather on the air: a history of broadcast meteorology. Springer, 264 pp.
Henshaw, Th. (1665) Some observations and experiments upon May-dew. *Phil. Trans.* (for 1665 and 1666) **1** (1867) 33–36. *See also in a shortened and changed version in*: The *Philosophical Transactions of the Royal Society of London from their commencement, in 1665, to 1800*. Vol. 1 (from 1665–1672), London 1809, pp. 13–14.
Henshaw, Th. (1667) The history of the making salt peter. In: The history of the Royal Society of London for the improving of natural knowledge by Thomas Sprat, London, pp. 260–268.
Herman, F. A. and E. Gorham (1957) Total mineral material, acidity, sulphur and nitrogen in rain and snow at Kentville, Nova Scotia. *Tellus* **9**, 180–183.
Hermann, A. (1954) Das Buch "Kmj.t" und die Chemie. *Zeitschrift für Ägyptische Sprache und Altertumskunde* **79**, 99–105.
Hermann, E. (1868) Die Wunder der Schöpfung. Zakarija Ben Muhammed Ben Mahmûd El-Kazwîni's Kosmographie: nach der Wüstenfeldschen Textausgabe, mit Benutzung und Beifügung der reichhaltigen Anmerkungen und Verbesserungen des Herrn Prof. Dr. Fleischer in Leipzig, aus dem Arabischen zum ersten Male vollständig übersetzt. Fues's Verlag, Leipzig, 532 pp.
Hermbstädt (1821) Beobachtungen über die Atmosphäre und das Wasser der Ostsee. *Neues Journal für Chemie und Physik* **32**, 281–291.
Herrmann, G. (1962) Bestimmung und Herstellung kleiner SO_2-Konzentrationen in Luft. *Zeitschrift für Chemie* **2**, 42–50.

Herrmann, G. (1963) Ein leicht transportables Gerät zur Schnellbestimmung von SO_2-Spuren in der Luft auf der Grundlage der Pararosanilin-Methode. *Chemische Technik* **15**, 342–349.
Herrmann, G. (1972) Beiträge zur Bestimmung von Schwefeldioxid in der Atmosphäre und Herstellung von Schwefeldioxid-Luftgemischen. *Abhandlungen der Sächsischen Akademien der Wissenschaften zu Leipzig. Mathematisch-naturwissenschaftliche Klasse*, Band 51, Heft 2, 179 pp.
Herrmann, G., N. Hesse, G. Scheibe and M. Zier (1991) Von der Wetterwarte zum Landesamt für Geologie. 75 Jahre Meteorologisches Observatorium Wahnsdorf. Radebeul, 40 pp.
Herrmann, H., ed. (2005) FEBUKO and MODMEP: A combined study of aerosol-cloud interaction by field experiments and model development. *Atmos. Env.* **39**, 4167–4418.
Hesse, W. (1879) Anleitung zur Bestimmung der Kohlensäure in der Luft, nebst einer Beschreibung des hierzu nöthigen Apparates. *Vierteljahreszeitschrift für gerichtliche Medizin und öffentliches Sanitätswesen* **31**, 357–370.
Hettner, G., R. Pohlmann and H. J. Schumacher (1935) Die Struktur des Ozon-Moleküls und seine Banden im Ultrarot. *Z. Elektrochem.* **41**, 372–387.
Heuvel, A. P. van den and B. J. Mason (1963) The formation of ammonium sulfate in water droplets exposed to gaseous sulfur dioxide and ammonia. *Q. J. Roy. Meteor. Soc.* **89**, 271–275.
Heydloff, R. (1853–54) Beobachtungen des Herrn Kreisphysicus Dr. Heydloff zu Erfurt über den Gehalt der atmosphärischen Luft an Ozon. *Wissenschaftliche Berichte* [Akademie gemeinnütziger Wissenschaften zu Erfurt] 1, 149–156. Fortgesetzte Beobachtungen des Kreis-Physikus Dr. Heydloff zu Erfurt über den Gehalt der atmosphärischen Luft an Ozon vom 1. März 1853 – 1. März 1854. **2** (1854) 293–306.
Hileman, B. (1983) Acid fog. *Environ. Sci. Technol.* **17**, 117A–120A.
Hill, R. A. (1924) The photochemical decomposition of gaseous sulphur dioxide. *Trans. Faraday Soc.* **20**, 107–112.
Hill, J. F. (2013) Chemical research on plant growth. A translation of Théodore de Saussure's Recherches chimique sur la Végétation. Springer, New York, 192 pp.
Hinshelwood, C. N. and A. T. Williamson (1934) The reaction between hydrogen and oxygen. Clarendon Press, Oxford, 108 pp.
Hinzpeter, H. (1952) Ergebnisse der in den Jahren 1941–1945 in Potsdam durchgeführten Ozonmessungen. Veröffentlichungen des Meteorologischen und Hydrologischen Dienstes der Deutschen Demokratischen Republik Nr. 9. Akademie-Verlag Berlin, 22 pp.
Hirasawa, K. (1941) On the results of chemical analysis of rainwater in Miyako (in Japanese). *Journal of the Meteorological Society of Japan* **19**, 395–400.
Hirt, L. (1871) Die Krankheiten der Arbeiter: Beiträge zur Förderung der öffentlichen Gesundheitspflege. Erste Abtheilung: Die inneren Krankheiten der Arbeiter. Erster Theil: Die Staubinhalations-Krankheiten und die von ihnen besonders heimgesuchten Gewerbe und Fabrikbetriebe, Band 1, Ausgabe 1. F. Hirst, Breslau. 304 pp.
Hitchcock, D. R., L. L. Spiller and W. E. Wilson (1980) Sulfuric acid aerosols and HCl release in coastal atmosphere: evidence of rapid of sulfuric acid particulates. *Atmos. Env.* **14**, 165–182.
Hlasiwetz, H. H. (1856) Über Kohlensäure-Bestimmungen der atmosphärischen Luft. *Sitzungsber. Wien* **20**, 189–200.
Hobbs, P. V. (1995) Basic physical chemistry for the atmospheric sciences. Cambridge University Press, 206 pp.
Hobbs, P. V. (2000) Introduction to atmospheric chemistry. A companion text to basic physical chemistry for the atmospheric sciences. Cambridge University Press, 206 pp.
Hoff, K. E. A. von (1827) Ueber die Natur des Höhenrauchs, fortgesetzte Bemerkung, vom Jahr 1827. *Archiv für die gesammte Naturlehre* **11**, 438–459. 1827.

Hoffmann, F. (1719) Herrn Friedrich Hoffmanns weitberühmten Medici, Gründliche Anweisung Wie ein Mensch Durch vernünfftigen Gebrauch der Leibes=Bewegungen und Mäßigkeit Wie auch Vermeydung ungesunder und mit schädlichen Dämpffen angefülleter Lufft Seine Gesundheit erhalten / und sich von schweren Kranckheiten befreyen könne. Fünffter Theil. Renger, Halle.
Hoffmann, M. R. (1977) Kinetics and mechanism of oxidation of hydrogen sulphide by hydrogen peroxide in acidic solution. *Environ. Sci. Technol.* **11**, 61–66.
Hoffmann, M. R. (1984) Acid fog. *Engineering and Science* **48**, 1, 5–11.
Hoffmann, M. R. (1986) On the kinetics and mechanism of the oxidation of aquated sulfur dioxide by ozone. *Atmos. Env.* **20**, 1145–1154.
Hoffmann, M. R. (1990) Catalysis in aquatic environment. In: Aquatic chemical kinetics (ed. W. Stumm), John Wiley & Sons, New York, pp. 71–112.
Hoffmann, M. R. and J. O. Edwards (1975) Kinetics of the oxidation of sulphite by hydrogen peroxide in acidic solution. *J. phys. Chem.* **79**, 2096–2098.
Hoffmann, M. R., S. T. Martin, W. Choi and D. Bahnemann (1995) Environmental applications of semiconductor photocatalysis. *Chemical Reviews* **95**, 69–75.
Högbom, A. (1894) Om Sannolikheten För Sekuläara Fööräandringar I Atmosfärens Kolsyrehalt [On probabilities of changing atmosphäric carbonic acid content]. *Svensk kemisk Tidskrift* **6**, 169–177.
Hoigné, J. and H. Bader (1978) Ozonation of water: Kinetics of oxidation of ammonia by ozone and hydroxyl radicals. *Environ. Sci. Technol.* **12**, 79–84.
Hoigné, J., H. Bader, W. R. Haag and J. Staehelin (1985) Rate constants of reactions of ozone with organic and inorganic compounds in water-III. *Water Research* **19**, 993–1004.
Holmén, K. (1992) The global carbon cycle. In: Global biogeochemical cycles (eds. S. S. Butcher, R. J. Charlson, G. H. Orian and G. V. Wolfe), Acad. Press London, pp. 239–262.
Holmes, J. A., E. C. Franklin and R. A. Gould (1915) Report of the Selby Smelter Commission. Bull. 98, Bureau of Mines, Washington, DC (USA), 528 pp.
Holmes, H. H. and F. Daniels (1934) The photolysis of nitrogen oxides: N_2O_5, N_2O_4 and NO_2. *J. Am. Chem. Soc.* **56**, 630–637.
Holmes, F. L. (1998) Antoine Lavoisier – the next crucial year or the source of his quantitative method in chemistry. Princeton University Press, 184 pp.
Honrath, R. E., M. C. Peterson, S. Guo, P. B. Shepson and B. Campbell (1999) Evidence on NOx production within or upon ice particles in the Greenland Snowpack. *Geophys. Res. Lett.* **26**, 695–698.
Hooper, R. (1845) Lexicon Medicum, or, medical dictionary: containing an explanation of the terms in anatomy, botany, chemistry, materia medica, midwifery, mineralogy, pharmacy, physiology, practice of physic, surgery and the various branches of natural philosophy connected with medicine. Harper & Brothers, New York (first published in 1798 with many editions), about 1400 pp.
Hope, T. C. (1805) Beschreibung des eudiometrischen Apparats. *Gilb. Ann.* **19**, 385–388.
Horn, H. (1856) Durch Reibungs-Elektricität alle Wohlgerüche in widerliche Gerüche und alle widerlichen in Wohlgerüche zu verändern, sowie über das Wesen des Ozon's und Jodosmon's. 9. Heft. Dr. Wild'sche Buchdruckerei, München, 30 pp.
Horn, H. (1857) Das Wirken der Elektricität in den Organismen. 11. Heft. Das Ozon als Heilmittel. J. J. Lentner'sche Buchhandlung, München, 31 pp.
Horsford, E. N. (1849) Moisture, ammonia and the organic matter in the atmosphere. Paper, presented at the Convention of the Amer. Assoc. for the Advancement of Sciences, held at Cambridge (Mass), August 1849. In: Ammonia in the atmosphere. *The Cultivator, new Ser.* **5**, 315.
Horsford, E. N. (1850) Notiz über den Ammoniakgehalt der Atmosphäre. *Ann.* **74**, 243–244.

Horváth, L. (1978) A csapadékvíz kémiai összetétele és a légköri nyomanya-gok depozíciója Budapesten. *Időjárás* **82**, 211–222.
Horváth, L. (1983) Trend of the nitrate and ammonium content of precipitation water in Hungary for the last 80 years. *Tellus* **35B**, 304–308.
Horváth, L. and E. Mészáros (1984) The composition and acidity of precipitation in Hungary. *Atmos. Env.* **18**, 1843–1847.
Hough, A. M. and R. G. Derwent (1987) Computer modelling studies of the distribution of photochemical ozone production between different hydrocarbons. *Atmos. Env.* **21**, 2015–2034.
Houghton, H. G. (1931) The transmission of visible light through fog. *Physical Review* **38**, 152–158.
Houghton, H. G. (1932) The size and size distribution of fog particles. *Journal of Applied Physics* **2**, 467–475.
Houghton, H. G. (1955) On the chemical composition of fog and cloud water. *Journal of Meteorology* **12**, 355–357.
Houghton, H. G. and A. C. Bemis (1939) Cloud particle studies on Mount Washington. *Bulletin of the meteorological Society of America* **20**, 400–401.
Houghton, H. G. and W. H. Radford (1938a) I. Microscopic measurement of the size of natural fog particles. In: Houghton, H. G. und W. H. Radford (1938) On the measurement of drop size and liquid water content in fogs and clouds. Paper in Physical Oceanography and Meteorology, Mass. Inst. of Technol. and Woods Hole Oceanographic Institution, Vol. VI, No. 4, pp. 5–18.
Houghton, H. G. and W. H. Radford (1938b) On the local dissipation of natural fog. Paper in Physical Oceanography and Meteorology, Mass. Inst. of Technol. and Woods Hole Oceanographic Institution, Vol. VI, No. 3, 63 pp.
Houzeau, A. (1855) Ueber den Sauerstoff im activen Zustande (état naissant). *Pogg. Ann.* **95**, 484–489.
Houzeau, A. (1856) Recherches sur l'oxygène à L'état naissant. *C. r.* **43**, 34–38.
Houzeau, A. (1857) Observations sur la valeur du papier dit ozonométrique et exposition d'une nouvelle méthode analytique pourrie connaitre et doser l'ozone. *Annu. soc. mét. de France* **5**, 43–53. Preuve de la présence dans l'atmosphérique d'un nouveau principe gazeux l'oxygène naissant ou ozone. Ibid. 288–292; *See also*: Houzeau, A. (1857) Ueber den activen Sauerstoff. *J. prakt. Chem.* **70**, 340–344.
Houzeau, A. (1858) Preuve de la présence dans l'atmosphère d'un nouveau principe gazeux-l'oxygen naissant. *C. r.* **46**, 89–91.
Houzeau, A. (1860) Neue Methode, das Ozon zu erkennen und quantitativ zu bestimmen. *Pogg. Ann.* **109**, 180–181.
Houzeau, A. (1861) Recherches sur l'oxygène a l'état naissant (oxygène odorant, oxygène naissant). *Ann. chim. phys. 3. ser.* **62**, 129–159.
Houzeau, A. (1863) Nouvelle méthode pour reconnaitre et doser l'ozone (oxygène odorant, oxygène naissant). *Ann. chim. phys. 3. ser.* **67**, 466–484.
Houzeau, A. (1865) Remarques sur l'ozone atmosphérique. *C. r.* **61**, 1113–1116. *See also*: *Die Fortschritte der Physik* **21** (1865) 630–631. Additional: De l'influence des saisons sur les propriétés de l'air atmosphérique. *C. r.* **60**, 788–793, *see also*: *Die Fortschritte der Physik* **21** (1865) 631–632.
Houzeau, A. (1868) Sur l'eau oxygénée, considérée comme ri étant pas la cause des altérations que l'air atmosphérique fait subir aux papiers de tournesol mi-ioduré, employés comme réactifs de l'ozone. *C. r.* **66**, 314–317 and: Observations sur la présence dans l'atmosphère de l'oxygéne actif ou ozone (deuxième partie), 491–494.
Houzeau, A. (1871) Expériences sur l'èlectrisation de l'air ou de l'oxygène comme moyen de production de l'ozone. *Ann. chim. phys. 4. ser* **22**, 150–171.
Houzeau, A. (1872a) Sur la proportion d'ozone contenue dans l'air de la campagne et sur son origine. *C. r.* **74**, 712–715.

Houzeau, A. (1872b) Sur l'ozone atmosphérique. *Ann. chim phys. 4. ser* **27**, 5–68.
Houzeau, A. (1883) Sur les causes capables d'influence sur la teneur en ammoniaque des eaux pluviales. *C. r.* **96**, 259–260.
Houzeau, J. Ch. (1851) Physique du globe et météorologie, Bruxelles, Jamar, 119 pp.
Hov, Ö., E. Hesstvedt and I. Isaksen (1978) Long-range transport of tropospheric ozone. *Nature* **273**, 341–344.
Howard, L. (1803) Essay on the modification of clouds. *Tilloch's Philosophical Magazine*, Vol. XVII, No. 65, London.
Hoyer, C. (1846) Physikalisch-chemische Abhandlung über den Höherauch. *Arch. Pharm.* **97**, 299–310.
Hu, L., D. J. Jacob, X. Liu, Y. Zhang, L. Zhang, P. S. Kim, M. P. Sulprizio and R. M. Yantosca (2017) Global budget of tropospheric ozone: Evaluating recent model advances with satellite (OMI), aircraft (IAGOS) and ozone sonde observations. *Atmos. Env.* **167**, 323–334.
Hua, W., Z. M. Chen, C. Y. Jie, Y. Kondo, A. Hofzumahaus, N. Takegawa, K. D. Lu, Y. Miyazaki, K. Kita, H. L. Wang, Y. H. Zhang and M. Hu (2008) Atmospheric hydrogen peroxide and organic hydroperoxides during PRIDE-PRD'06, China: their concentration, formation mechanism and contribution to secondary aerosols. *ACP Discussion* **8**, 10481–10530.
Hubbs, C. L. (1930) The high toxicity of nascent oxygen. *Physiological Zoology* **3**, 441–460.
Hube, J. M. (1790) Ueber die Ausdünstung und ihre Wirkungen auf die Atmosphäre. Göschen, Leipzig, 440 pp.
Huber, B. (1952) Der Einfluss der Vegetation auf die CO_2-Schwankungen der Atmosphäre. *Archiv für Meteorologie, Geophysik und Bioklimatologie B* **3**, 154–167.
Hudig, J. (1910) Het stikstofgehalte von regenwater. *Verslagen van landbouwkundje onderzoekingen der Rijkslandbouwproefstations* **9**, 78–82.
Hudig, J. (1912) The amounts of nitrogen as ammonia and as nitric acid (and nitrous) in the rainwater collected at Uithuizermeeden, Groningen. *Journal of Agricultural Science* **4**, 160–269.
Hudig, J. and H. Welt (1911) Het drainage proefveld te Uithuizermeeden in de jaren 1900–1910. *Verslagen van landbouwkundje onderzoekingen der Rijkslandbouwproefstations* **10**, 123–244.
Huebert, B. J. and A. L. Lazrus (1978) Global tropospheric measurements of nitric acid vapor and particulate nitrate. *Geophys. Res. Lett.* **5**, 577–580.
Huggins, W. and M. L. Huggins (1890) On a new group of lines in the photographic spectrum of Sirius. *Proc. R. Soc. Lond.* **48**, 216–217.
Hughes, G. and C. R. Lobb (1976) Reactions of solvated electrons. In: Comprehensive chemical kinetics, Vol. 18 (eds. C. H. Bamford and C. F. H. Tipper), Elsevier Sci. Publ., Amsterdam, pp. 429–461.
Huizinga, D. (1867) Ueber den Nachweis des Ozons und die Anwesenheit desselben in der Atmosphäre. *J. prakt. Chem.* **102**, 193–204.
Hüller, O. (1872) Ozonometrische Beobachtungen. *Deutsche Klinik* **24**, 417–418.
Humboldt, A. von (1798) Lettre de Fréderic von Humboldt à Garnerin l'ainé sur l'analyse de l'air atmosphérique, pris à l'hauter de 669 toises avec aérostat. *Journal de Physique, de Chimie, d'Histoire Naturelle*, Fructidor an VI. **4**, 202–203.
Humboldt, A. von (1799) Versuche über die chemische Zerlegung des Luftkreises und über einige andere Gegenstände der Naturlehre. Vieweg, Braunschweig, 258 pp.
Humboldt, A. von (1800) Versuche über die chemische Zerlegung des Luftkreises. *Gilb. Ann.* **3**, 77–83.
Humboldt, A. von (1836) Kritische Untersuchungen über die historische Entwicklung der geografischen Kenntnisse von der neuen Welt und die Fortschritte der nautischen Astronomie in dem 15ten und 16ten Jahrhundert (aus dem Französischen übersetzt von J. L. Ideler). Nicolai, Berlin. 562 pp.

Humboldt, A. von (1849) Ansichten der Natur. 3. Auflage, Band 2, Cotta, Stuttgart and Tübingen, 407 pp.
Humboldt, A. von (1850–52) Cosmos: a sketch of a physical description of the universe. Translated from German by E. C. Otté. Harper & Brothers, New York, Vol. 1 (1850) 275 pp., Vol. 2 (1850) 367 pp., Vol. IV (1851) 219 pp., Vol. IV (1852) 234 pp.
Humboldt, A. von and J. F. Gay-Lussac (1805) Versuche über die eudiometrischen Mittel, und über das Verhältniss der Bestandtheile der Atmosphäre. *Gilb. Ann.* **20**, 38–92.
Humphreys, W. J. (1910) Solar disturbances and terrestrial temperatures. *Astrophysical Journal* **32**, 97–111.
Hůnová, I., J. Šantroch and J. Ostatnická (2004) Ambient air quality and deposition trends at rural stations in the Czech Republic during 1993–2001. *Atmos. Env.* **38**, 887–898.
Hůnová, I., P. Kurfürst, J. Maznová and M. Coňková (2011) The contribution of occult precipitation to sulphur deposition in the Czech Republic. *Erdkunde* **65**, 247–259.
Hunt, R. (1854) The poetry of science or, studies of the physical phenomena of nature. Third edition. H. G. Gohn, London: 419 pp.
Hunt, B. G. (1966) Photochemistry of ozone in moist atmosphere. *JGR* **71**, 1385–1398.
Hunt, J. P. and H. Taube (1952) The photochemical decomposition of hydrogen peroxide. Quantum yields, tracer and fractionation effects. *J. Am. Chem. Soc.* **74**, 5999–6002.
Hurdelbring, F. (1907) Angewandte Methoden zur Untersuchung der Luft auf schweflige Säure und Russ. *Schriften der physikalisch-ökonomischen Gesellschaft in Königsberg in Preussen* **48**, 145–151.
Hurdelbring, F. (1909) Untersuchung der Königsberger Luft auf Russ und schweflige Säure. *Deutsche Vierteljahresschrift für öffentliche Gesundheitspflege* **41**, 369–384.
Hutchinson, G. E. (1944) Nitrogen in biochemistry of the atmosphere. *American Scientist*, pp. 178–195.
Hutchinson, G. E. (1954) The biochemistry of the terrestrial atmosphere. In: The Earth as a planet (ed. G. P. Kuiper), Chicago University Press, pp. 375–392.
Hutchinson, G. E. (1970) The biosphere. *Scientific American* **223**, 44–53.
Hutzinger, O., Ed. (1980) The natural environment and the biogeochemical cycles (part of the the handbook of environmental chemistry, book series, volume 1/1A). Springer, Berlin Heidelberg, 258 pp.; note that under this title, 6 volumes did appear between 1980 and 1992.
Ideler, J. L. (1829) Ueber den Hagel und die elektrischen Erscheinungen unserer Atmosphäre. *Pogg. Ann.* **27**, 435–472.
Igawa, M. (2016) personal information.
Igawa, M., Y. Tsutsumi, T. Mori and H. Okochi (1998) Fog water chemistry at a nountainside forest and the estimation of the air pollutant deposition via fog droplets based on the atmospheric quality at the mountain base. *Environ. Sci. Technol.* **32**, 1566–1572.
Igawa, M., K. Matsumura and H. Okochi (2001) Fog water chemistry at Mt. Oyama and its dominant factors. *Water, Air and Soil Pollution* **130**, 607–612.
Ilkov, V. [Илков, В.] (1925) Дъждовната вода е азотен тор? [Is rain water a nitrogen fertilizer]. *Spis. Zeml. Isp. Inst. B'lgarii* (*Списание на земеделските опитни институти в България* [Journal of the Institute of agricultural Experiments in Bulgaria (Sofia)]) **3**, 325–338; cited after *Experimental Station Record* **54** (1926) 615.
Ingenhousz, J. (1779) Experiments upon vegetables, discovering their great power of purifying the common air in the sunshine, and of injuring it in the shade and at night: to which is joined, a new method of examining the accurate degree of salubrity of the atmosphere. P. Elmsly, London, 68+302+15 pp.
Ireland, W. W. (1862) Notes on the medical topography of Kussouli with especial reference to the quantity of ozone at different elevations and the effect of that agent on malaria. *Edinburgh Medical Journal* **8**, 12–22.

Isaac, G. A. and P. H. Daum (1987) A winter study of air, cloud and precipitation chemistry in Ontario, Canada. *Atmos. Env.* **21**, 1587–1600.

Isaksen, I. S. A. and O. Hov (1987) Calculation of trends in the tropospheric concentration of O_3, OH, CH_4 and NOX. *Tellus* **39B**, 271–285.

Ishikawa, Y. and H. Hara (1997) Historical change in precipitation pH at Kobe, Japan: 1936–1961. *Atmos. Env.* **31**, 2367–2369.

Isidorov, V. A. [Исидоров] (2001) Органическая химия атмосфера. Химия, Ленингад, 351 pp. (third edition in Russian, first ed. in 1985, 264 pp.).

Isidorov, V. A. (1990) Organic chemistry of the Earth's atmosphere. Springer-Verlag, Berlin, 215 pp. (translation of the Russian edition from 1985).

Isidorov, V. A., I. G. Zenkevich and B. V. Ioffe (1985) Volatile organic compounds in the atmosphere of forests. *Atmos. Env.* **19**, 1–8.

Isono, K. (1957) On sea-salt nuclei in the atmosphere. *Geofisica pura e applicata (Milano)* **36**, 156–164.

Israel, H. and L. Schulz (1932) Über die Größenverteilung der atmosphärischen Ionen. *Met. Z.* **49**, 226–233.

Israel, H. (1970) Atmospheric Electricity. Israel Program for Scientific Translations, Jerusalem, 317 pp.

Ivanov, M. V. and J. R. Freney, eds. (1983) The global biogeochemical sulphur cycle. Wiley, New York, 470 pp.

Ives, J. E., R. H. Britten, D. W. Armstrong, W. A. Gill and F. H. Goldman (1936) Atmospheric Pollution of American Cities for the Years 1931 to 1933 with special reference to the solid constituents of the pollution. Treasury Department, Public Health Bulletin No. 224, 75 pp.

Jackson, W. F. (1934) A study of the photochemical carbon monoxide oxidation. *J. Am. Chem. Soc.* **56**, 2631–2635.

Jackson, A. V. and C. N. Hewitt (1996) Hydrogen peroxide and organic hydro-peroxide mixing ratios in air in a eucalyptus forest in central. *Portugal. Atmos. Env.* **30**, 819–830.

Jackson, A. V. and C. N. Hewitt (1999) Atmosphere hydrogen peroxide and organic hydroperoxides: A review. *Critical Reviews in Environ. Sci. Technol.* **29**, 175–228.

Jacob, D. J., R. Fang, T. Wang and R. C. Flagan (1984) Fogwater collector design and characterization. *Environ. Sci. Technol.* **18**, 827–833.

Jacob, D. J., J. M. Waldman, J. W. Munger and M. R. Hoffmann (1985) Chemical composition of fog water collected along the California coast. *Environ. Sci. Technol.* **19**, 730–736.

Jacob, P., T. M. Tavares and D. Klockow (1986) Methodology for the determination of gaseous hydrogen peroxide in ambient air. *Z. anal. Chem.* **325**, 359–364.

Jacob, P., T. M. Tavares, V. C. Rocha and D. Klockow (1990) Atmospheric H_2O_2 field measurements in a tropical environment: Bahia, Brazil. *Atmos. Env.* **24**, 377–382.

Jacob, P. and D. Klockow (1992) Hydrogen peroxide measurements in the marine atmosphere. *J. Atmos. Chem.* **15**, 353–360.

Jacobi, H.-W., M. M. Frey, M. A. Hutterli, R. C. Bales, O. Schrems, N. J. Cullen, K. Steffen and C. Koehler (2002) Measurements of hydrogen peroxide and formaldehyde exchange between the atmosphere and surface snow at Summit, Greenland. *Atmos. Env.* **36**, 2619–2628.

Jacobs, M. B., M. M. Bravermann and S. Hochheiser (1957) Ultramicro determination of sulfides in air. *Analytical Chemistry* **29**, 1349–1351.

Jacobs, A. F. G., B. G. Heusinkveldand and S. M. Berkowicz (2000) Dew measurements along a longitudinal sand dune transect, Negev Desert, Israel. *International Journal of Biometeorology* **43**, 184–190.

Jacobs, W. C. (1937) Preliminary report on a study of atmospheric chloride. *Monthly Weather Review* **65**, 147–151.

Jacobson, M. Z. (2002) Atmospheric pollution. History, science and regulation. Cambridge Univ. Press, 399 pp.

Jacolot, A. A. M. (1865) Recherches ozonométrique faites pendant la campagne des la frégate la Danaé en Islande (1864). In: Archives de médecine navale, Vol. 3, J.-B. Baillière, Paris, pp. 115–131.

Jaenicke, R. (1982) Physical aspects of the atmospheric aerosol. In: Chemistry of the unpolluted and polluted troposphere (eds. H.-W. Georgii and W. Jaeschke), D. Reidel Publ., Dordrecht, pp. 341–373.

Jaenicke, R. (2012) Die Erfindung der Luftchemie – Christian Junge. In: 100 Jahre Kaiser-Wilhelm-/Max-Planck-Institut für Chemie (Otto-Hahn-Institut) (eds. H. Kant and C. Reinhardt), MPG, Berlin, pp. 187–202. The English version is found under www.blogs.uni-mainz.de. *Jaenicke* further wrote an extended biography concerns *Junge*'s aerosol research: http://www.iara.org: aerosol pioneers.

Jaenicke, R. and C. N. Davis (1976) The mathematical expression of the size distribution of atmospheric aerosol. *J. Aerosol. Sci.* **7**, 255–259.

Jaeschke, W., R. Schmitt and H.-W. Georgii (1976) Preliminary results of stratospheric SO_2-measurements. *Geophys. Res. Lett.* **3**, 517–519.

Jaeschke, W. and J. Stauff (1978) Die Chemilumineszenz der SO_2-Oxidation und ihre Anwendung in der Chemie der Atmosphäre. *Ber.* **82**, 1180–1184.

Jaeschke, W., H.-W. Georgii, H. Claude and H. Malewski (1978) Contributions of H_2S to the atmospheric sulfur cycle. *Pure Applied Geophysics* **116**, 465–475.

Jaeschke, W., H. Claude and J. Herrmann (1980) Sources and sinks of atmospheric of H_2S. *JGR* **85**, 5639–5644.

Jaeschke, W. and J. Herrmann (1981) Measurements of H_2S in the atmosphere. *Journal of Environmental Analytical Chemistry* **10**, 107–120.

Jaffe, D. A. (1992) The nitrogen cycle. In: Global biogeochemical cycles (eds. S. S. Butcher, R. J. Charlson, G. H. Orian and G. V. Wolfe), Acad. Press, London, pp. 263–284.

Janach, W. E. (1989) Surface ozone: trend details, seasonal variations and interpretation. *JGR* **94**, 18289–18295.

Jander, G. and H. Spandau (1952) Kurzes Lehrbuch der anorganischen und allgemeinen Chemie. Fünfte Auflage. Springer-Verlag, Berlin, 563 pp.

Janssen, J. (1895) Sur une troisième ascension au sommet u Mont-Blanc et les travaux exécutés pendant l'été de 1895 dans le massif de cette montagne. *La Nature* **23**, 326–330.

Jelinek, C. (1866) Ueber ozonometrische Bestimmungen in Österreich. *Z. Öster. Ges. Met.* **1**, 292–298, 445–452.

Jenner, M. (1995) The politics of London air John Evelyn's Fumifugium and the restoration. *The Historical Journal* **38**, 535–551.

Jennings, S. G. (2000) Atmospheric aerosol measurements at Mace Head, Ireland. *J. Aerosol. Sci.* **31**, S580–S583.

Jenrich, G. (1914) Gleichzeitige Beobachtungen von Empfangsstörungen elektrischer Wellen und der meteorologischen Elemente, besonders der Kondensationskernzahl. Dissertation, Halle. Mitteilungen der physikalischen Versuchsstation Halle-Cröllwitz, No. 37. C. A. Kaemmerer, Halle. 29 pp.

Jensen, W. B. (1990) What ever happened to the nascent state? *Bulletin for the History of Chemistry* **6**, 26–36.

Jensen, W. B. (2011) Plagiarizing the professor. Notes from the Oesper Collection, Museum Notes, May 2011, 3 pp.

Jentsch, S. (1917) Über die Erfahrungen bei Abgasanalysen und die Bestimmung geringer Säuremengen in den Gasen industrieller Rauchquellen. Dissertation (Univ. Dresden). Noske Borna-Leipzig, 64 pp.

Jevons, W. S. (1869) Remarks on Mr. Baxendell's laws of atmospheric ozone. *Proceedings of the Literary and Philosophical Society of Manchester* **8**, 33–34.

Jessel, U. (1952) Über die Eignung von Frittwaschflaschen für Ozonanalysen atmosphärischer Luft. *Berichte des Deutschen Wetterdienstes* **42**, 344–349.
Jessel, U. (1955) Beiträge zur Spurenstoffchemie der Meeres- und Brandungsluft. *Archiv für physikalische Therapie, Balneologie und Klimatologie* **7**, 230–234.
Johnson, E. M. (1924) Sulfur in rainfall in Kentucky. *Journal of the American Society of Agronomy* **16**, 363–366. *See also*: Analyses of rainfall from a protected and an exposed gauge for sulfur, nitrate nitrogen and ammonia. *Ibid.* 17 (1926) 589–591.
Johnson, S. W. (1869) On nitrification. *American Journal of Science 2nd series* **47**, 234–242.
Johnson, F. M. G. and D. McIntosh (1909) On the formation of ozone by the ultra-violet rays. *J. Am. Chem. Soc.* **31**, 1146–1147.
Johnston, J. F. W. (1853) The circulation of matter: A review with excerpts from Johnston's letters on agricultural chemistry and geology. *Blackwood's Edinburgh Magazine* **73**, 550–562.
Johnston, H. S. and H. J. Crosby (1951) Rapid gas phase reaction between nitric oxide and ozone. *J. Chem. Phys.* **19**, 799.
Johnston, H. S. (1971) Reductions of stratospheric ozone by nitrogen oxide catalysts from supersonic transport exhaust. *Science* **173**, 517–522.
Johnston, F. S., J. D. Purchell and R. Tousey (1951) Measurements of the vertical distribution of atmospheric ozone from rockets. *JGR* **56**, 583–594.
Jolly, P. (1879) Die Veränderlichkeit in der Zusammensetzung der atmosphärischen Luft. *Abhandlungen der Königlichen Bayerischen Akademie der Wissenschaften* **13**, 52–74; *see also*: *Z. Öster. Ges. Met.* **14** (1879) 227–229.
Jorissen, W. P. (1906) Het chloorgehalte von regenwater. *Chemisch Weekblad* **3**, 647–649.
Jost, W. (1939/46) Explosions- und Verbrennungsvorgänge in Gasen. Springer, Berlin, 608 pp. English translation: Explosion and combustion processes in gases. McGraw-Hill Book Company, London (1946), 621 pp.
Junge, C. E. (1935) Neuere Untersuchungen an den großen atmosphärischen Kondensationskernen. *Met. Z.* **52**, 467–470. *See also*: Übersättigungsmessungen an atmosphärischen Kondensationskernen. *Gerlands Beiträge zur Geophysik* **46** (1935) 108–129.
Junge, C. (1936) Zur Frage der Kernwirksamkeit des Staubes. *Met. Z.* **53**, 186–188.
Junge, C. (1951) Nuclei of atmospheric condensation. In: Compendium of Meteorology (ed. T. E. Malone), American Meteorological Society, Boston, pp. 182–191. *See also*: Das atmosphärische Aerosol. *Umschau* **51**, 64–649.
Junge, C. (1952) Gesetzmäßigkeiten in der Größenverteilung der atmosphärischen Aerosole über dem Kontinent. *Berichte des Deutschen Wetterdienstes in der US-Zone* Nr. 35, 261–277. *See also*: Das Größenwachstum der Aitkenkerne. Ibid., pp. 264–267.
Junge, C. (1953) Die Rolle der Aerosole und der gasförmigen Beimengungen der Luft im Spurenstoffhaushalt der Troposphäre. *Tellus* **5**, 1–26.
Junge, C. (1954) The chemical composition of atmospheric aerosols. I. Measurements at Round Hill Field Station, June-July 1953. *Journal of Meteorology* **11**, 323–333.
Junge, C. (1955) The size distribution and aging of natural aerosols as determined from electrical and optical data on the atmosphere. *Journal of Meteorology* **12**, 13–25.
Junge, C. (1956) Recent investigations in air chemistry. *Tellus* **8**, 127–139.
Junge, C. (1957) Chemical analysis of aerosol particles and gas traces on the island of Hawaii. *Tellus* **9**, 528–537.
Junge, C. (1958a) The distribution of ammonia and nitrate in rain water over the United States. *Transactions of the American Geophysical Union* **39**, 241–248.
Junge, C. (1958b) Atmospheric chemistry. In: Advances in geophysics. Vol. IV, Acad. Press, New York, (separate volume), 108 pp.
Junge, C. (1960) Sulfur in the atmosphere. *JGR* **65**, 227–237.

Junge, C. (1962) Global ozone budget and exchange between stratosphere and troposphere. *Tellus* **14**, 363–377.

Junge, C. (1963a) Air chemistry and radioactivity. Vol. 4 Intern. Geophys. Ser. (Ed. J. van Miegham). Academic Press, New York and London, 382 pp. Translation into Russian: Юнге, Х. (1965) Химический состав и радиоактивность атмосферы (пер. с англ. В. Н. Петрова и А. Я. Прессмана, под ред. Ю. А. Израэля), Москва, Мир, 424 pp.

Junge, C. (1963b) Neuere Ergebnisse der Luftchemie und ihre Bedeutung für die Meteorologie. *Berichte des Deutschen Wetterdienstes* **91**, 80–88.

Junge, C. (1972) Cycle of atmospheric gases, natural and man-made. *Q. J. Roy. Meteor. Soc.* **98**, 711–729.

Junge, C. and P. E. Gustafson (1956) Precipitation sampling for chemical analysis. *Bulletin of the American Meteorological Society* **37**, 244–245.

Junge, C. and P. E. Gustafson (1957) On the distribution of sea salt over the United States and its removal by precipitation. *Tellus* **9**, 164–173.

Junge, C. and T. G. Ryan (1958) Study of the SO_2 oxidation in solution and its role in atmospheric chemistry. *Q. J. Roy. Meteor. Soc.* **84**, 46–55.

Junge, C. and R. T. Werby (1958) The concentrations of chlorite, sodium, potassium, calcium and sulphate in rainwater over the United States. *Journal of Meteorology* **3**, 417–425.

Junge, C., C. W. Chagnon and J. E. Manson (1961) Stratospheric aerosols. *Journal of Meteorology* **18**, 81–108. *See also*: Junge, C. and J. E. Manson (1961) Stratospheric aerosol studies. *JGR* **66**, 2163–2182.

Junkermann, W., M. Fels, P. Petruk, F. Slemr and J. Hahn (1992) Peroxide measurements at remote mountain field sites (Wank 1780 m asl and 1175 m asl): seasonal and diurnal variations of hydrogen peroxide and organic peroxides. In: Proceedings of the EUROTRAC Symposium '92 (eds. P. M. Borrel, P. Borrel, T. Cvitas and W. Seiler). SPB Acad. Publ, The Hague (Netherlands), pp. 180–184.

Juritz, C. F. (1910) Chemical composition of rain in the Union of South Africa. *South African Journal of Science* **10**, 170–193.

Kadlecek, J., S. McLaren, V. Mohnen, B. Mossel, A. Kadlecek and N. Camarota (1985) Wintertime cloud water chemistry studies, Whiteface Mountain 1982–1984. Atmospheric Sciences Research Center/State University of New York Publ. No. 1008. *See also*: Kadlecek, J., S. McLaren, N. Camarota, V. Mohnen and J. Wilson (1983) Cloud water chemistry at Whiteface Mountain. In: Precipitation scavenging, dry deposition and resuspension. Vol. 1 (eds. H. R. Pruppacher, G. R. Semonin and W. G. N. Slinn). Elsevier, Amsterdam, pp. 103–114.

Kahlbaum, G. W. A. (1899) Christian Friedrich Schönbein 1799–1868 – ein Blatt zur Geschichte des 19. Jahrhunderts, 1. Teil. Monographien aus der Geschichte der Chemie (ed. G. W. A. Kahlbaum) 4. Heft, J. A. Barth Leipzig, Reprint Zentralantiquariat der Deutschen Demokratischen Republik Leipzig 1970, 230 pp.

Kahlbaum, G. W. and E. Thon, eds. (1900) Justus von Liebig und Christian Friedrich Schönbein. Briefwechsel 1853–1868. Johann Ambrosius Barth, Leipzig, 278 pp.

Kahlbaum, G. W. A., E. Schaer and E. Thon (1901) Christian Friedrich Schönbein 1799–1868 – ein Blatt zur Geschichte des 19. Jahrhunderts, 2. Teil. Monographien aus der Geschichte der Chemie (ed. G. W. A. Kahlbaum) 6. Heft, J. A. Barth Leipzig, Reprint Zentralantiquariat der Deutschen Demokratischen Republik Leipzig 1970, 230 pp.

Kähler, K. (1932) Über das Ionenspektrum der Atmosphäre. *Naturwissenschaften* **20**, 783–786.

Kallend, A. S., A. R. W. Marsh, J. H. Pickles and M. V. Proctor (1983) Acidity of rain in Europe. *Atmos. Env.* **17**, 127–137.

Kaminsky, U. and P. Winkler (1994) Die luftchemischen Meßreihen des Meteorologischen Obervatoriums Hamburg. Berichte des Deutschen Wetterdienstes Nr. 192. Offenbach, 124 pp.

Kampmann, F.-E. (1868) L'ozone atmosphérique. Résume des observations ozononométric faites à Colmar du 1 mars 1866 au 29 fév. 1868. *Bulletin de la Société d'histoire naturelle de Colmar* **8/9**, 59–80. And: Observations ozonometriques du 1er mars 1868 au 23 fév. Ibid. **10** (1869) 315–318.

Kämtz, L. F. (1836) Lehrbuch der Meteorologie. 1. Band. Gebauersche Buchhandlung, Leipzig, 510 pp.

Kämtz, L. F. (1840) Vorlesungen über die Meteorologie. Gebauer, Halle, 591 pp.

Kaplan, J. (1939) Nitric oxide in the Earth's upper atmosphere. *Nature* **144**, 152.

Kapustin, M. J. [Капустин] (1879) Определение углекислоты в воздухе посредством спиртовой раствора уегкаво натрий и титрования [Quantitative estimation of carbonic acid by means of an alcoholic solution of caustic soda and titration with water]. St. Petersburg.

Karandikar, R. V. (1948) Studies in atmospheric ozone. *Proceedings of the Indian Academy of Sciences (Math. Sci.)* **28**, 63–85.

Kargin, V. A. (1929) Über die elektrolytische dissoziationskonstante von Wasserstoff-superoxyd. *Zeitschrift für anorganisch und allgemeine Chemie* **183**, 77–81.

Karliński, F. M. (1874) Zwanzigjährige Ozon-Beobachtung zu Krakau. *Z. Öster. Ges. Met.* **9**, 94.

Katz, M. (1961) Some aspects of the physical and chemical nature of air pollutants. In: Air pollution. World Health Organization Monograph Series No. 46, Columbia Press, New York, pp. 97–158.

Katz, M. and S. E. Gale (1971) Mechanism of photooxidation of SO_2 in atmosphere. In: Procceedings of the 2nd Intern. Clean Air Congress (eds. H. M. Englung and W. T. Berry), Academic Press, New York, pp. 336–343.

Kausch, O. (1938) Das Wasserstoffsuperoxyd. Eigenschaften, Herstellung und Verwendung. Verlag W. Knapp, Halle, 254 pp.

Kavarnos, G. J. and N. J. Turro (1987) Photosensitization by reversible electron transfer: Theories, experiment al evidence and examples. *Chemical Reviews* **86**, 401–449.

Kawamura, K. and I. R. Kaplan (1984) Capillary gas chromatography determination of volatile organic acids in rain and fog samples. *Analytical Chemistry* **56**, 1616–1620.

Kay, R. H. (1953) An interim report on the measurement of the vertical distribution of atmospheric ozone by a chemical method, to heights of 12 km, from aircraft. Meteorol. Res. Pap. No. 817: 1–15. Meteorol. Res. Comm. (London).

Kazarnovski, I. A., G. P. Nikolski and T. A. Abletzov [Казарнóвский, Никольский, Аблецова] (1949) Новый окисел калия [A new oxide of potassium]. *Доклады АН СССР. Новая серия* **64**, 69–72.

Kazarnovski, I. A., D. S. Gorbenko-Germanov and I. V. Koslova [Казарновский, Горбенко-Германов, Козлова] (1969) Спектроскопическое исследование образования и распада озонид-иона O_3^- в водных средах [Spectroscopic investigation of formation and decomposition of ozonide O_3^- in aqueous solution]. *Изв. АН СССР. сер. хим.* 1 том 1969. No. 1. C. 198–200.

Kazay, E. (1904) A légköri csapadékok chemiaia analysise. Idöjaras 8, 301–396 and Az ó-gyallai observatoriumon végzett esővíz analisisek 1902–1903: Regenwasser Analysen im Jahre 1902 am Observatorium zu Ó-Gyalla. *Magyar Királyi Orszagos Meteorológiai és Földmágnességi Intézet Kiadványai* (Annals of the Hungarian Royal Institut for Meteorology and Magnetism), Budapest **32**, 114–117.

Kealing, J. M. (1879) The history of the yellow fever epidemy of 1878 in Memphis, Tennessee. Howard Association, 442 pp. (reprint Heritage Books, 2009, 560 pp.).

Kedzie, R. C. (1876) Ozone. *Annual report of the State Board of Health of Michigan 1875* **3**, 135–146.

Keeling, C. D., R. B. Bacastow, A. E. Bainbridge, C. A. Ekdahl, P. R. Guenther, L. S. Waterman and J. F. S. Chin (1976) Atmospheric carbon dioxide variations at Mauna Loa Observatory, Hawaii. *Tellus* **28**, 538–551.

Keeling, C. D. and T. P. Whorf (2004) Atmospheric carbon dioxide concentrations at 10 locations spanning latitudes 82°N to 90°S (Eds. T. J. Blasing and S. Jones), ORNL/CDIAC-147, NDP-001a. Carbon Dioxide Information Analysis Center, Oak Ridge National Laboratory, U. S. Department of Energy, Oak Ridge, Tennessee, 30 pp. doi:10.3334/CDIAC/atg.ndp001.2004.

Keeling, R. A. and S. R. Shertz (1992) Seasonal and interannual variation in atmospheric oxygen and implications for the global carbon cycle. *Nature* **358**, 723–727.

Keeling, R. F. (1988a) Measuring correlations between atmospheric oxygen and carbon dioxide mole fractions: a preliminary study in urban air. *J. Atmos. Chem.* **7**, 153–176.

Keeling, R. F. (1988b) Development of an interferometer oxygen analyser for precise measurement of the atmospheric O_2 mole fraction. Doctoral Thesis, Harvard University, Cambridge, Mass. (USA).

Keeling, R. F., B. B. Stephens, R. G. Najjar, S. C. Doney, D. Archer and M. Heimann (1998) Seasonal variations in the atmospheric O:/N: ratio in relation to the kinetics of air-sea gas exchange. *Global Biogeochemical Cycles* **12**, 141–163.

Keidel, E. (1929) Die Beeinflussung der Lichtechtheit von Teerfarblacken durch Titanweiss. *Farben-Zeitung* **34**, 242–243.

Keiner, L. (1934) Die Mattkunstseide. *Mellior Textilberichte* **15**, 118–120.

Keir, J. (1777) A treatise on the various kinds of permanently elastic fluids, or gases. London, printed for T. Cadell and P. Elmsley, 120 pp.

Kellner, O., J. Sawano, T. Yoskii and R. Makino (1886) Untersuchungen über den Gehalt der atmosphärischen Niederschläge an Stickstoffverbindungen. *Landwirtschaftliche Jahresberichte* **15**, 701–711. *See also*: *Forschungen auf dem Gebiet der Agricultur-Physik* (*Wollny's Forschungen*) 1886, **15**, 701–708.

Kellog, W. W., R. D. Cadle, E. R. Allen, A. L. Lazrus and E. A. Martell (1972) The sulfur cycle. *Science* **175**, 587–595.

Kelly, T. J., F. H. Stedman and G. L. Kok (1979) Measurements of H_2O_2 and HNO_3 in rural air. *Geophys. Res. Lett.* **6**, 375–378.

Kelly, T. J., P. H. Daum and S. E. Schwartz (1985) Measurements of peroxides in cloud water and rain. *JGR* **90**, 7861–7871.

Kern, S. (1878) On the presence of hydrogen peroxide in the atmosphere. *The Chemical News and Journal of Physical Sciences* **37**, 35.

Kernbaum, M. (1909) Action chimique sur l'eau des rayons pénétrants de radium. *C. r.* **148**, 705–707. *See also*: Décomposition de l'eau par les rayons ultra-violets. *C. r.* **149**, 273–275.

Kernbaum, M. (1910) Décomposition de l'eau par l'aigrette. *C. r.* **151**, 275–278.

Kernbaum, M. (1911) Sur la décomposition de l'eau par les métaux. *C. r.* **152**, 1668–1670.

Kerr, J. A., J. G. Calvert and K. L. Demerjian (1972) The mechanism of photochemical smog formation. *Chemistry in Britain* **8**, 252–257.

Kessler, C. (1984) Gasförmige salpetrige Säure (HNO_2) in der belasteten Atmosphäre. Doctoral Thesis, University of Cologne, 97 pp.

Kestner, O. (1923) Die Ursache der Schwüle. *Klinische Wochenschrift* **2**, 1874–1875.

Khrgian, A. Kh. [Хргиан] (1952) Некоторые данные о микроструктуре облаков [Some data on the microstructure of clouds]. *Trudy Centralnaja Aerologičeskaja Observatoria* (*TsAO, CAO*), No. 7.

Khrgian, A. Kh. (1970) Meteorology. A historical survey. Vol. 1 (translation from Russian). Jerusalem, 387 pp.

Khrgian, A. Kh. [Хргиан] (1973) Физика атмосферного озона. Гидрометеоиздат, Ленинград, 201 pp. English edition: The physics of atmospheric ozone. Keter Publ. House, Jerusalem, 1975, 262 pp.

Khrgian, A. Kh. and I. P. Mazin [Хргиан, Мазин] (1952) О разпределение капель по размерам воблаках [The size distribution of droplets in clouds]. *Trudy Centralnaja Aerologičeskaja Observatoria* (*TsAO, CAO*), No. 7, 56.

Khrgian, A. Kh., A. M. Borovikov, I. I. Gaivoronskij, E. G. Zak, V. V. Kostarev, I. P. Mazin, V. E. Minervin und S. M. Shmeter [Хргиан, Боровиков, Гайворонский, Мазин, Минервин, Шметер] (1961) Физика облаков [cloud physics]. Gidrometeoroizdat, Leningrad, 558 pp.; English version: (1963) Cloud Physics, Israel Program for Scientific Translations, Jerusalem, 392 pp.

Kiepenheuer, K. O. (1938) Über die Sonnenstrahlung zwischen 2000 and 3000 A. *Veröffentlichungen der Universitätssternwarte Göttingen* **57**, 348–356.
Kiessling, J. (1885) Nebelglühapparat. *Z. Öster. Ges. Met.* **20**, 57–62.
Kim, D.-S. and V. P. Aneja (1992) Chemical composition of clouds at Mt. Mitchell, North Carolina. *Tellus* **44B**, 41–53.
Kinch, E. (1887) The amount of chlorine in rain-water collected at Cirencester. *Journal of the Chemical Society London, Transactions* **51**, 92–94 and **77** (1900) 1271–1274.
Kingzett, C. T. (1874) XXVII. On ozone as a concomitant of the oxidation of the essential oils. Part I. *J. Chem. Soc.* **27**, 511–524.
Kingzett, C. T. (1880) Report on the atmospheric oxidation of phosphorus and some reactions of ozone and hydric peroxide. *J. Chem. Soc., Trans.* **37**, 792–807.
Kirchhoff, G. and R. Bunsen (1860) Chemische Analyse durch Spectralbeobachtungen. *Pogg. Ann.* **110**, 161–189.
Kirwan, R. (1787) An essay on phlogiston and the constitution of acids. J. Davis, London, 146 pp.
Kistiakowsky, Wl. (1900) Versuche über die Lichtempfindlichkeit des Wasserstoffsuperoxyds in wässerigen Lösungen beim Zusatz von Blutlaugensalzen. *Z. phys. Chem.* **35U**, 431–439.
Kistiakowsky, G. B. (1925) Ozonzerfall im roten Lichte. *Z. phys. Chem.* **117U**, 337–360.
Kistiakowsky, G. B. (1928) Photochemical processes. McGraw-Hill, New York, 270 pp.
Klein, R. M., S. Adamowicz, T. D. Perkins and H. Liedeker (1988) Precipitation as a source of assimilable nitrogen: a historical survey. *American Journal of Botany* **75**, 928–937.
Klemm, O. and T. Wrzesinsky (2007) Fog deposition fluxes of water and ions to a mountainous site in Central Europe. *Tellus B* **59**, 705–714.
Kley, D., A. Volz and F. Mülheims (1988) Ozone measurements in historic perspective. In: Tropospheric ozone. NATO ASI Series, Series C: Mathematical and Physical Sciences, Vol. 227 (ed. I. S. A. Isaksen), Springer, Dordrecht, pp. 63–72.
Kley, D., H. Geiss and V. A. Mohnen (1994) Tropospheric ozone at elevated sites and precursor emissions in the United States and Europe. *Atmos. Env.* **28**, 149–158.
Kmiec, G., A. Zwozdziak, K. Acker and W. Wieprecht (1998) Cloud/fog water chemistry at high elevation in the Sudeten Mountains, south-western Poland. In: Proceedings of the First International Conference on Fog and Fog Collection, Vancouver, Canada, July 1998, pp. 69–72.
Knoche, W. (1930) Zur Entstehung des Chile-Saltpeters. *Forschungen und Fortschritte* **6**, 196–197.
Knoeff, R. (2002) Hermann Boerhaave (1668–1738). Calvinist chemist and physician. Kon. Ned. Akad. Van Wettenschappen, Amsterdam, 253 pp.
Knollenberg, R. G. (1981) Techniques for probing cloud microstructure. In: Clouds. Their formation, optical properties and effects (eds. P. V. Hobbs and A. Deepak), Academic Press, 495 pp.
Knop, W. (1868) Der Kreislauf des Stoffs. Lehrbuch der Agricultur-Chemie. H. Haessel, Leipzig, 925+309 pp.
Knop, W. and W. Wolf (1861) Ueber das Vorkommen des Ammoniaks in den Regen-, Teich- Fluß-, Brunnenwässern, in den Ackererden und sein Verhalten in derselben. Die landwirthschaftlichen Versuchs-Stationen. *Organ für wissenschaftliche Forschungen auf dem Gebiete der Landwirthschaft* **3**, 120.
Knott, C. C., ed. (1923) Collected scientific papers of John Aitken, LL. D., F. R. S. Univ. Press, Edinburgh, 590 pp.
Kober, J. (1872) Ueber die angebliche Bläschenform des Wassers bei seiner Condensation, Dissertation. W. Ratz, Jena, 38 pp. *See also*: *Pogg. Ann.* **144** (1871) 395–427.
Kobozev, N. I., L. I. Nekrasov and I. I. Skorokhodov (1959) The physical chemistry of concentrated ozone: II. A study of the synthesis of the higher peroxides of hydrogen H_2O_4 by interaction of concentrated ozone with atomic hydrogen. National Research Council of Canada, 36 pp.
Koch, W. F., G. Marinenko and R. C. Pauls (1986) Development of a standard reference method for rainwater analysis. *Journal of Research of the National Bureau of Standard* **91**, 33–41.

Kockmann, N. (2014) History of distillation. In: Distillation. Fundamentals and principles (eds. A. Górak and E. Sorensen), Elsevier, pp. 1–43.
Koene, C. J. (1844) Om Undersalpetersyra och om Kungsvatten. *Öfversigt af Kongl. Vetenskaps-akademiens forhandlingar (Stockholm)* **1**, 183–187.
Köhler, E. (1942) Tagesgang des Nitrits der Luft in Ober-Schreiberhau direkt über dem Boden (0,10 m) und in höheren und bodennahen Schichten (15 m). *Der Balneologe* **9**, 365–369.
Köhler, H. (1923) Zur Kondensation des Wasserdampfes in der Atmosphäre. *Geofysiske publikasjoner* Vol. II, Oslo, 53 pp.
Köhler, H. (1925) Untersuchungen über die Elemente des Nebels und der Wolken. *Meddelanden från statens meteorologisk-hydrografiska Anstalt* **2**, No. 5 (Stockholm).
Köhler, H. (1926) On water in the clouds. Geofysiske publikasjoner Vol. V, Oslo, 16 pp.
Köhler, H. (1929) Wolkenuntersuchungen auf dem Sonnblick. 37. *Jahresbericht des Sonnblick-Vereins für das Jahr 1931*. J. Springer, Wien, 14–20.
Köhler, H. (1936) The nucleus in and the growth of hygroscopic droplets. *Trans. Faraday Soc.* **32**, 1152–1161.
Köhler, H. (1937) Studien über Nebelfrost und Schneebildung und über den Chlorgehalt des Nebelfrostes, des Schnees und des Seewassers im Halddegebiet [Studies on fog frost and snow formation and on the chlorine content of fog frost, snow and lake water in the Haldde area]. *Bulletin of the Geological Institution of the University of Upsala* **26**, 279–308.
Kok, G. L. (1980) Measurements of hydrogen peroxide in rainwater. *Atmos. Env.* **14**, 653–656.
Kok, G. L. (1992) Five decades of atmospheric Chemistry. In: Meeting Review: Third NCAR Research Aircraft Fleet Workshop (eds. L. F. Radke and P. Spyers-Duran). NCA Technical Note 364, 79 pp.
Kok, G. L., K. R. Darnall, A. M. Winer, J. Pitts and B. Gay (1978) Ambient air measurements of hydrogen peroxide in the California south coast air basin. *Environ. Sci. Technol.* **12**, 1077–1080.
Kolbe, H. (1854) Graham-Otto's ausführliches Lehrbuch der Chemie. Dritte umgearbeitete Auflage. Dritter Band: Organische Chemie (Subtitle: Kolbe's Organische Chemie). Vieweg & Son, Braunschweig, 1056 pp.
Kolbe, H. (1863) Ueber die Erzeugung von salpetriger Säure beim Verbrennen von Wasserstoff im stickstoffhaltigen Sauerstoff. *Neues Repertorium für Pharmacie* **12**, 308–310.
Kolbig, J. and W. Warmbt (1978) Messungen des bodennahen Ozons in Mirny, Antarktika. *Z. Met.* **28**, 270–273.
Komar, N. (1927) Zur Bestimmung geringer Mengen von Kohlenoxyd in der Luft. *Z. anal. Chem.* **72**, 476–477.
Kondratjew, W. N. [Кондратьев] (1958) Кинетика химических газовых реакций. Изд-во АН СССР, Москва. 688 pp. English edition: Kondratiev, V. N. (1964) Chemical kinetics of gas reaction. Pergamon Press, Oxford, 812 pp.
König, J. (1878) Der Kreislauf des Stickstoffs und seine Bedeutung für die Landwirthschaft. Theissing'sche Buchhandlung, Münster, 20 pp.
König, J. (1893) Das Meteorwasser. In: Die menschlichen Nahrungs- und Genussmittel. 3. Auflage, Springer, Berlin, pp. 1141–1143.
König, J. (1904) Die menschlichen Nahrungs- und Genussmittel. Springer, Berlin, 1563 pp.
Konstantinova-Schlesinger, M. A. [Константинова-Шлезингер] (1936a) Новый флюоресцентный метод определения малых концентраций озона [A new fluorescence method for determination of small concentrations of ozone]. *Труды ФИАН* [Transactions of the Physical Institute of the Academy of Sciences] **1**, 119–130.
Konstantinova-Schlesinger, M. A. [Константинова-Шлезингер] (1936b) Флюоресцентный метод определения содержания озона в воздухе [Fluorescence method for determination of ozone in air]. In: Труды Эльбрусской экспедиции АН СССР и ВИЭМ, 1934 и 1935 г. [Transactions of the Elbrus Expedition of the Academy of Sciences of the USSR and All-Union Institute of Experimental Medicine, 1934 and 1935], Изд-во АН СССР, Moscow and Leningrad, pp. 49–60.

Konstantinova-Schlesinger, M. A. [Константинова-Шлезингер] (1937a) Определение флуоресцентным методам содержания озона в воздухе на высоте 9620 м. [Estimation of ozone in air at an altitude of 9620 m by a fluorescence method]. *Доклады АН СССР* [Annals of the Academy of Sciences of the USSR] **14**, 187–188.

Konstantinova-Schlesinger, M. A. [Константинова-Шлезингер] (1937b) Результаты определения содержания озона в воздухе флуоресцентным методам [Results of ozone measurements in air with a fluorescence method]. *Известия АН СССР, Серия геофизическая* pp. 213–222.

Konstantinova-Schlesinger, M. A. [Константинова-Шлезингер] (1938) Определения содержания озона в пробах воздухе с высот 13 и 14 км на уровнем моря [Estimation of ozone in air samples from altitudes 13 and 14 km above sea level]. *Доклады АН СССР* [Annals of the Academy of Sciences of the USSR] **18**, 337–338.

Kopp, H. (1869) Die Entdeckung der Zusammensetzung des Wassers. In: Beiträge zur Geschichte der Chemie (Hermann Kopp), Fr. Vieweg und Sohn, Braunschweig, pp. 235–310.

Kopp, H. (1886) Die Alchemie in Älterer und Neuerer Zeit. Ein Beitrag zur Kulturgeschichte (reprint 1971). Two Vol. in one. G. Olms Verlag, Hildesheim/New York, 260+425 pp.

Kopp, H. (1931) Geschichte der Chemie (reprint of the first edition 1843–1847). Alfred Lorentz Buchhandlung, Leipzig, 4 Vol. in 2 books, 455+426 pp., 372+448 pp.

Köppen, W. (1899) Klimalehre. Sammlung Göschen, Leipzig, 122 pp. and (1906) Klimakunde I: Allgemeine Klimakunde. Göschen, Leipzig, 132 pp.

Köppen, W. (1906) Klimakunde I: Allgemeine Klimakunde. Göschen, Leipzig, 132 pp.

Köppen, W. (1923) Die Klimate der Erde. Grundriß der Klimakunde. 1. Auflage. de Gruyter, Berlin, 369 pp.

Köppen, W. (1931) Grundriß der Klimakunde. 2. verbesserte Auflage der "Klimate der Erde.". De Gruyter, Berlin, 388 pp.

Koppenol, W. H. (2001) The Haber-Weiss cycle – 70 years later. *Redox Report* **6**, 229–234.

Körber, H.-G. (1987) Vom Wetteraberglauben zur Wetterforschung. Edition Leipzig, 231 pp.

Kormann, C., D. W. Bahnemann and M. R. Hoffmann (1988) Photocatalytic production of H_2O_2 and organic peroxides in aqueous suspensions of TiO_2, ZnO and desert sand. *Environ. Sci. Technol.* **22**, 798–806.

Kornfeld, G. and E. Weegmann (1930) Die Oxydation von Schwefeldioxyd im ultravioletten licht. *Z. Elektrochem.* **36**, 789–794.

Korsunovsky, G. A. [Корсуновский] (1957) Фотоокисление воды красителями на поверх-ности полупроводников [Photooxidation of water by dyes on the surface of semicunductors]. *Доклады Академий Наук СССР* **113**, 853–885. *See also*: The photo-oxidation of waters on heterogeneous sensitizers. In: Problems of Photosynthesis: Reports of the Second All-Union Conference on Photosynthesis, Moscow, January 21–6, 1957 (ed. A. L. Kursanov). Oak Ridge, Tenn. U. S. Atomic Energy Commission Division of Technical Information, 1962, pp. 115–124.

Korsunovsky, G. A. [Корсуновский] (1960) Влияние кислорода и воды на электро-проводность окиси цинка, окрашенной эритрозином [Influence of oxygen and water on the electric conductivity of zinc oxide, painted with erythrosine]. *Доклады Академий Наук СССР* **134**, 1394–1396.

Kosmann, C. (1862) Études sur l'ozone exhale par les plants. *C. r.* **55**, 731–736. *See also*: *Annales des Sciences Naturelles*, quatrieme sèrie, Botanique **8**, 111–117.

Kossovitch, P. S. [Коссович] (1913) О круговороте серы и хлора на земном шаре и о значении этого процесса в природе, почве и в культуре сельскохозяйственных растений [On the cycle of sulfur and chlorine on Earth and the importance of this process in nature, for soils and vegetation]. СПб (Сообщения из Бюро по земледелию и почвоведению Ученого комитета Главного управления землеустройства и земледелия) [Mitteilungen aus dem Bureau für Ackerbau und Bodenkunde am Gelehrtenkomite der Hauptverwaltung für Landorganisation und Ackerbau – Transactions of the office for agriculture and pedology of the scientific committee Scholarly committee for country organization and agriculture]. Vol. 12, 86 pp. (German summary: pp. 76–86).

Kostin, S. (1901) Ueber den Nachweis minimaler Mengen Kohlenoxyd in Blut und Luft. *Archiv für die gesammte Physiologie des Menschen und der Thiere* **83**, 572–608.

Kovda, V. A. [Ковда], ed. (1976) Биогеохимические циклы в биосфере [Biogeochemical Cycles in the Biosphere]. Наука, Москва, 355 pp.

Kraft, F., ed. (1996) Otto von Guerickes neue (sogenannte) Magdeburger Versuche über den leeren Raum. VDI Verlag, Düsseldorf, 306 pp.

Kratzenstein, Chr. G. (1744) Abhandlungen von dem Aufsteigen der Dünste und Dämpfe, welche von der Akademie zu Bordeaux den Preis erhalten. Halle, 80 pp. (reprint Antiqua-Verlag, Lindau, 1977).

Kraus, C. A. (1908) Solutions of metals in non-metallic solvents. IV. Material effects accompanying the passage of an electrical current through solutions of metals in liquid ammonia. Migration experiments. *J. Am. Chem. Soc.* **30**, 1323–1344.

Krebs, W. (1893) Schwefelsäure oder Schweflige Säure in der Stadtluft? *Centralblatt der Bauverwaltung* **13**, 363–364.

Kremer, A. (1872) Untersuchung von Sauerstoffwasser und Ozonwasser der Firma Krebs, Kroll und Comp. *Pharmaceutische Centralhalle für Deutschland* **13**, 2–3.

Kreusler, U. (1887) Ueber den Sauerstoffgehalt der atmosphärischen Luft. *Ber.* **20**, 991–999.

Kreusler, H. (1901) Anwendung des photoelektrischen Stromes zur Photometrie der ultravioletten Strahlen. *Ann. Phys.* **311**, 412–423.

Kreutz, W. (1941) Kohlensäuregehalt der unteren Luftschichten in Abhängigkeit von Witterungsfaktoren. *Angewandte Botanik* **23**, 89–117.

Krogh, A. (1919) The composition of the atmosphere. Det Kgl. Danske Videnskabernes Selskab. Math.-fys. Meddelser. I, 12, København, 19 pp.

Krüger (1822) Ueber das färbende Princip in der Atmosphäre der Ostsee. *Journal für Chemie und Physik* (*Schweigg J.*) **35**, 179–195. *See also*: Ueber die eigenthümliche Färbung des salpetersauren Silbers, durch das Wassergas der Atmosphäre. Ibid., 440–451.

Krüger, P., B. Voland and D. Wolf (1990) W. I. Wernadskij und die globalen Probleme der Menschheit. *Schriftenreihe für den Referenten der Urania* **1**, 39–53.

Krünitz, J. G. (1779) Vol. 16: Oekonomische Encyklopädie oder allgemeines System der Staats- Stadt- Haus- und Landwirthschaft: in alphabetischer Ordnung. Bd. [Vol.] 1–242. Pauli, Berlin (1773–1858). online: http://www.kruenitz1.uni-trier.de/.

Kudrin, S. A. [Кудрин, С. А.] (1948) О поступлении азота с атмосферными осадками в почвы сероземной зоны [On the nitrogen deposition with rainfall to soils in the Black Earth area]. *Почвоведение* [Pedology] **10**, 608–611.

Kumar, R., A. Rani, S. P. Singh, K. M. Kumari and S. S. Srivastava (2002) A long term study on chemical composition of rainwater at Dayalbagh, a suburban site of semiarid region. *J. Atmos. Chem.* **41**, 265–279.

Kunen, S. M., A. L. Lazrus, G. L. Kopk and B. H. Heikes (1983) Aqueous oxidation of SO_2 by H_2O_2. *JGR* **88**, 3671–3674.

Kunze, M. (1877) Ueber den Einfluß der Laub- und Nadelholzwaldungen auf die Temperatur und den Ozongehalt der Luft. *Tharander forstliches Jahrbuch* **27**, 159–176.

Kuo, C.-H., F. Yuan and D. O. Hill (1997) Kinetics of oxidation of ammonia in solutions containing ozone with or without hydrogen peroxide. *Industrial & Engineering Chemical Research* **36**, 4108–4113.

Küppers, E. (1918) Über den Schwefelsäuregehalt von Schnee und Regenwasser. *Z. angew. Chem.* **29**, 74–76.

Kurashige, E. (1934) Composition of rain. *Tenki to Kikō* [Weather and Climate] **1**, 390–393.

Kurashige, E. and G. Kagei (1935) Chemical investigation on the precipitation at Tokyo and its suburbs. *Journal of the Meteorological Society of Japan* **13**, 211–216.

Kurtenacker, A. (1913) Über komplexe Kupfernitrite. *Z. anorg. Chem.* **82**, 204–215.

Küttlinger, A. (1859) Ueber ozonometrische Beobachtungen. *Wissenschaftliche Mittheilungen der physicalisch-medicinischen Societät zu Erlangen* **1**, 39–51. *And*: Einfluss des Ozongehalts der Luft auf den Gesundheitszustand. *Froriep's Notizen aus dem Gebiete der Natur- und Heilkunde* (Jena), Jahrgang 1859, Band 4, 360–365.

Kuum, J. (2018) Eesti põllumajandusteaduse hällist ja selle esiklapsest Jakob Johnsonist [The formation of Estonian agricultural science and the first Estonian agriculturist Jakob Johnson]. *AGRAARTEADUS: Journal of Agricultural Science* **29**, 102–104.

Ladenburg, A. (1898a) Ueber Ozon. *Ber.* **31**, 2508–2513.

Ladenburg, A. (1898b) Ueber Dichte und Molekulargewicht des Ozons. *Ber.* **31**, 2830–2831.

Ladenburg, A. (1899) Ueber Dichte und Molekulargewicht des Ozons. *Ber.* **31**, 221–222.

Ladenburg, A. (1900) Ueber das Ozon. IV. *Ber.* **32**, 2282–2284.

Ladenburg, A. (1901) Eine neue Methode zur Molekulargewichtsbestimmung des Ozons. *Ber.* **34**, 631–635.

Ladenburg, E. and E. Lehmann (1906) Über Versuche mit hochprozentigem Ozon. *Ann. Phys.* **326**, 305–318. *See also*: Über das Absorptionsspektrum des Ozons. *Verhandlungen der physikalischen Gesellschaft* **8** (1906) 124–135.

Ladenburg, R. W. (1935) Light absorption and distribution of atmospheric ozone. *Journal of the Optical Society of America* **25**, 259–269.

Ladureau, A. (1883) L'acide sulfureaux dans l'atmospère de Lille. *Ann. chim. phys.* **29**, 427–433.

Lagrange, J., C. Pallares, G. Wenger and P. Lagrange (1993) Electrolyte effects on aqueous atmospheric oxidation of sulphur dioxide by hydrogen peroxide. *Atmos. Env.* **27**, 129–137.

Lagrange, J., C. Pallares, G. Wenger and P. Lagrange (1996) Kinetics of sulphur(IV) oxidation by hydrogen peroxide in basic aqueous solution. *Atmos. Env.* **30**, 1013–1018.

Lahmann, E. and K.-E. Prescher (1974) Schwefeldioxid-Immissionen in Berlin. In: Lufthygiene 1974. *Schriftenreihe des Vereins für Wasser-, Boden- und Lufthygiene* **41**, 83–114.

Laing, M. (1995) The bonding in the ozone molecule: from a different perspective. *Structural Chemistry* **6**, 397–402.

Laj, P., S. Fuzzi, M. C. Facchini, J. A. Lind, G. Orsi, M. Preiss, R. Maser, W. Jaeschke, E. Seyffer, G. Helas, K. Acker, W. Wieprecht, D. Möller, B. G. Arends, J. J. Möls, R. N. Colvile, M. W. Gallagher, K. M. Beswick, K. J. Hargreaves, R. L. Storeton-West and M. A. Sutton (1997) Cloud processing of soluble gases. *Atmos. Env.* **31**, 2589–2598.

Lajtha, K. and J. Jones (2013) Trends in cation, nitrogen, sulfate and hydrogen ion concentrations in precipitation in the United States and Europe from 1978 to 2010: a new look at an old problem. *Biogeochemistry* **116**, 303–334.

Lakatos, I. (1981) History of science and its rational reconstructions. In: Scientific revolutions (ed. J. Hacking), Oxford University Press, pp. 107–127.

Lal, B. (1952) Photochemical after-effect in H2O2-$K_4Fe(CN)_6$ reaction. *Proceedings of the Indian Academy of Sciences – chapter A* **36**, 244–257.

Lamarck, J.-B. (1802) De nouvelles observations sur les vents, sur les circonstances propres à la formation des nuages, sur l'ordre de température des différentes couches de l'atmosphère, et sur les faits météorologiques que l'on recueille dans divers points de la République et qui sont réunis, comparés, conservés à Paris dans les Bureaux de la statistique de France. In: *Annuaire météorologique pour l'année XI de l'ère de la République Françoise, à l'usage des Agriculteurs, des Médecins, des Marins, etc.* N° 4, l'Auteur et Maillard, Paris. 163 pp.

Lamarque, J.-F., P. Hess, L. Emmons, L. Buja, W. Washington and C. Granier (2005) Tropospheric ozone evolution between 1890 and 1990. JGR **110**, D08304, 15b pp.

Lampadius, W. A. (1806) Systematischer Grundriß der Atmosphärologie. Craz and Gerlach, Freyberg (Saxony), 392 pp.

Lampadius, W. A. (1820) Bericht über einige neue chemische und hüttenmännische Erfahrungen in den Jahren 1819–1820. *Journal für Chemie und Physik (Schweigg. J.)* **30**, 253–258.

Lampadius, W. A. (1834) Ueber die Quellwässer des sächsischen Erzgebirges, so wie über die atmosphärischen Wässer. *J. prakt. Chem.* **1**, 100–111 and 281–290.
Lampadius, W. A. (1835) Fortgesetzte Beiträge zur Kenntnis verschiedener Wasser. *J. prakt. Chem.* **6**, 365–382.
Lampadius, W. A. (1837) Fortgesetzte Beiträge zur Kenntnis verschiedener Wasser. *J. prakt. Chem.* **10**, 78–88.
Lampadius, W. A. (1838a) Mittheilungen vermischten Inhalts. *J. prakt. Chem.* **13**, 237–244.
Lampadius, W. A. (1838b) Ueber den Gehalt an Salpetersäure in dem Atmosphärwasser, welches am 15. Mai 1838 in Freiberg aus einem Gewitter niederfiel. *J. prakt. Chem.* **14**, 5–58.
Lamy, C. (1869) Observations relative à l'emploi du protoxyde de thallium substance comme oconoscopique. *Bulletin mensuel de la Société Chimique de France* **11**, 210–211. *See also*: Ueber Thalliumoxydul als ozonoskopische Substanz. *Z. anal. Chem.* **9** (1870) 74.
Landa, S. (1931) O cistote prazského ovzdusi [The purity of the atmosphere in Prague]. *Chemické listy* **25**, 448–450.
Landa, E. R. and S. Ince, eds. (1987) The history of hydrology. American Geophysical Union, Washington, DC, 122 pp.
Landener, X. (1856) Ueber verschiedene Dunsthöhlen im Oriental. *Archiv für Pharmazie* **25**, 123–125.
Landsberg, H. (1934) Zählungen von Kondensationskernen auf dem Taunus-Observatorium. Bioklimatische Beiblätter, Meteorologische Zeitschrift 1, 125–128. See also: Observations of condensations nuclei in the atmosphere. *Monthly Weather Review* **62** (1934), 442–445.
Landsberg, H. (1937) The ultra-violet dosimeter. A review on principles, use and results of a simple instrument. *Bulletin of the American Meteorological Society* **18**, 161–167.
Landsberg, H. (1938) Atmospheric condensation nuclei. *Gerlands Beiträge zur Geophysik* (Supp.) **3**, 155–252.
Landsberg, H. (1954) Some observations of the pH of precipitation elements. *Archiv für Meteorologie, Geophysik und Bioklimatologie, Serie A* **7**, 219–226.
Lange, C. A., J. Matschullat, F. Zimmermann, G. Sterzik and O. Wienhaus (2003) Fog frequency and chemical composition of fog water – a relevant contribution to atmospheric deposition in the eastern Erzgebirge, Germany. *Atmos. Env.* **37**, 3731–3739.
Langevin, P. (1905) Les ions dans l'atmosphère. *C. r.* **140**, 232–234.
Langlo, K. (1952) On the amount of atmospheric ozone and its relation to meteorological conditions. *Geofysiske Publikasjoner* Vol. XVIII, No. 6 (utgitt av det Norske Videnskaps-Akademi) Grøndal & Søns, Oslo), 42 pp.
Larson, G. P., G. I. Fischer and W. J. Hamming (1953) Evaluating sources of air pollution. *Industrial & Engineering Chemistry* **45**, 1070–1074.
Larson, T. V. and H. Harrison (1977) Acidic sulfate aerosols: formation from heterogeneous oxidation by O_3 in clouds. *Atmos. Env.* **11**, 1133–1141.
Larson, T. V., N. R. Horike and H. Harrison (1978) Oxidation of SO_2 by O_2 and O_3 in aqueous solution: a kinetic study with significance to atmospheric rate processes. *Atmos. Env.* **12**, 1597–1611.
Laskowsky, N. (1850) Ueber die Zusammensetzung der atmosphärischen Luft zu Moskau während der Cholerazeit. *Ann.* **75**, 176–190.
Latour, B. (2015) Face à Gaïa: Huit conférences sur le nouveau régime climatique. La Découverte, Paris, 398 pp. In German: Kampf um Gaia. Acht Vorträge über das neue Klimaregime. Aus dem Französischen von Achim Russer und Bernd Schwibs. Suhrkamp, Berlin 2017, 523 pp.
Laurent, A. (1846) Recherches sur les combinaisons azotée. *Ann. chim. phys.* **18**, 266–298.
Lauscher, F. (1983) Aus der Frühzeit atmosphärischer Ozonforschung. Wetter und Leben. *Z. angew. Met.* **35**, 69–80.
Lauscher, F. (1984) Ozonbeobachtungen in Wien von 1852 bis 1981. Zusammenhänge zwischen Ozon und Wetterlagen. Publikation Nr. 284, Zentralanstalt für Meteorologie und Geodynamik in Wien, 29 pp.

Lauscher, F. and K. Schwarz (1932) Über den Chlorgehalt des Nebelfrosts auf dem Sonnblick. *Jahresberichte des Sonnblick-Vereins für das Jahr 1931*. J. Springer, Wien, pp. 32–34.

Lavin, G. I., A. D. Coates and J. A. Rakaczky (1961) Free radical bibliography and survey of publications (up to 1959). Ballistic Res. Lab., Maryland, Report No. 1142, 400 pp.

Lavoisier, A.-L. (1770) Prémier mémoire sur la nature de l'eau et sur les experiences, parlesquelles on a preténdu prouver la possibilité de son changement en terre, pp. 73–82; Second mémoire sur la nature de l'eau et sur les experiences, parlesquelles on a preténdu prouver la possibilité de son changement en terre, par le même, pp. 90–107. *Memoires de l'Academie Royale des Sciences de Paris*.

Lavoisier, A.-L. (1774) Opuscules physiques et chimiques. Tom Prémiere. Paris, 436 pp. (only the first Volume of several planned did appear; English translation in 1776 and German by C. E. Weigel: *Lavoisier physikalisch-chemische Schriften*. Greifswald, 1783–1794).

Lavoisier, A. L. (1776) Essays physical and chemical. Translated by Thomas Henry. J. Johnson, London, 474 pp. (from the French original *Opuscules physiques et chimiques*).

Lavoisier, A. L. (1780) Mémoire sur la combustion en général. *Mémoires de l'Académie Royale des Sciences Année 1777*, 592–600.

Lavoisier, A. (1781) Considérations générales sur la nature des acides. *Mémoires de l'Académie Royale des Sciences Année 1778*, 535–547.

Lavoisier, A. L. (1783) Extrait d'un mémoire lu par M. Lavoisier, à la séance publique de i'Académie royale des sciences, du 12 novembre, sur la nature de l'eau [Report of a memoir read by M. Lavoisier at the public session of the Royal Academy of Sciences of November 12, on the nature of water]. *Observations sur la Physique* **23**, 452–455.

Lavoisier, A.-L. (1786) Réflexions sur le phlogistique, pour servir de développement à la théorie de la combustion & de la calcination, publiée en 1777. *Mémoires de l'Académie Royale des Sciences Année 1783*, 505–538.

Lavoisier, A. L. (1789) Elements of Chemistry in a new systematic order, containing all the modern discoveries. Transl. by R. Kerr, reprint 1965, Dover Publ., New York, 511 pp.

Lavrinenko, R. F. [Лавриненко] (1968) О содержании серы в атмосферных осадках [On the amount of sulfur in atmospheric precipitation]. *Труды ГГО* **207**, 87–91.

Lavrinenko, R. F. [Лавриненко] (1979) Баланс сульфатов в атмосферных осадках [The balance of sulfate in atmospheric precipitation]. *Труды ГГО* **418**, 34–42.

Lawes, J. B. and J. H. Gilbert (1854) On the amount of and methods of estimating, ammonia and nitric acid in rain-water. *Reports British Association, Rothamsted Memoirs* **1**, No. 6 (cited after Miller 1905).

Lawes, J. B., J. H. Gilbert and E. Pugh (1863) On the source of the nitrogen of vegetation; with special reference to the question whether plants assimilate free or uncombined nitrogen. *Journal of the Chem. Society* **16**, 100–186.

Lawes, J. B., J. H. Gilbert and R. Warington (1881) The amount and composition of the rain and drainage waters collected at Rothamsted. Part I, II, III. *Journal of the Royal Agricultural Society of England* **17**, 241–279 and 311–359; **18** (1882) 1–71.

Lawes, J. B., J. H. Gilbert and R. Warington (1882) Amount and composition of rain and drainage water collected at Rothamsted. W. Clones and Sons, London, 167 pp.

Lawther, R. J. (1959) Some analytical and clinical aspects of British urban air pollution. In: Atmospheric Chemistry of Chlorine and Sulfur Compounds: Proceedings of a Symposium Held at Taft Sanitary Engineering Center, Geophysical Monograph 3 (ed. J. J. Lodge, Jr.), American Physical Union, Washington, pp. 88–97.

Lax, G. (2016) Zum Aufbau der Atmosphärenwissenschaften in der BRD seit 1968. *NTM Zeitschrift für Geschichte der Wissenschaften, Technik und Medizin* **24**, 81–107.

Lax, G. (2018) Von der Atmosphärenchemie zur Erforschung des Erdsystems. Beiträge zur jüngeren Geschichte des Max-Planck-Instituts für Chemie (Otto-Hahn-Institut), 1959–2000. MPG, Berlin, 128 pp. Also in English: From atmospheric chemistry to earth system science. Contributions to the recent history of the Max Planck Institute for Chemistry, 1959–2000. 164 pp.

Lazrus, A. L., H. W. Blodgett and J. P. Lodhe, Jr. (1970) Trace constituents in oceanic cloud water and their origin. *Tellus* **22**, 106–114.

Lazrus, A. L., G. L. Kok, S. N. Gitlin, J. A. Lind and S. E. McLaren (1985) Automated fluorimetric method for hydrogen peroxide in atmospheric precipitation. *Analytical Chemistry* **57**, 917–922.

Leaf, W. (2010) Homer, the Illiad. Cambridge University Press, 692 pp.

Leather, W. (1906) Composition of Indian rain and dew. *Memoirs of the Department of Agriculture in India*. Chemical Series **1**, No. 1, Printed by Thacker, Spink & Co. Calcutta and London, pp. 1–12.

Lebedinzeff, A. (1891) Neue Modification der Dalton-Pettenkofer'schen Methode zur Bestimmung der Kohlensäure in der Luft. *Z. anal. Chem.* **30**, 267–279.

Le Chatelier, H. (1913) Les Classiques de la Science. III. Eau Oxygénée et Ozone (Memoirs de Thénard, Schoenbein, De Marignac, Soret, Troost, Hautefeuille, Chappius). Libr. Armand Colin, Paris, 111 pp.

Leck, C. and H. Rodhe (1989) On the relation between anthropogenic SO_2 emissions and concentration of sulfate in air and precipitation. *Atmos. Env.* **23**, 959–966.

Lee, Y.-N., J. Shen, P. J. Klotz, S. E. Schwartz and L. Newman (1986) Kinetics of hydrogen peroxide – sulfur(IV) to sulfur(VI) reaction in rainwater collected at a northwestern U. S. site. *J. Geophy. Res.* **91**, 13264–13274.

Lee, M., B. G. Heikes and D. W. O'Sullivan (2000) Hydrogen peroxide and organic hydroperoxide in the troposphere: a review. *Atmos. Env.* **34**, 3475–3494.

Lee, Y.-N. and J. A. Lind (1986) Kinetics of aqueous-phase oxidation of nitrogen(III) by hydrogen peroxide. *JGR* **91**, 2793–2800.

Leeds, A. R. (1878) Contributions towards a knowledge of the chemical constituents of the atmosphere. *Annals of the New York Academy of Sciences* **1**, 193–219.

Leeds, A. R. (1879a) Upon ammonium nitrite and upon the by-products obtained in the ozonisation of air by moist phosphorus. *J. Amer. Chem. Soc.* **1**, 145–157. *See also*: Ueber Ammoniumnitrit und die bei der Ozonisation der Luft durch feuchten Phosphor erhaltenen Nebenproducte. *Liebigs Ann.* **200** (1879) 286–301.

Leeds, A. R. (1879b) Lines of discovery in the history of ozone, with an index of its literature and an appendix upon the literature of peroxide of hydrogen. *Annales of the New York Academy of Sciences* **1**, 363–426.

Leeds, A. R. (1881) Upon the invariable production, not only of ozone and hydrogen peroxide, but also of ammonium nitrate, in the ozonation of purified air by moist phosphorus. *J. Am. Chem. Soc.* **3**, 5–15.

Leeds, A. R. (1883) The literature of ozone and peroxide of hydrogen;second memoir, including: I. Historical-critical resumë of the progress of discovery since 1879. II. Index to the literature of ozone (1879–1883) III. Index to the literature of peroxide of hydrogen (1879–1883). *Annals of the New York Academy of Sciences* **3**, 137–155.

Leeflang, K. W. H. (1938) De chemische samenstelling van den neerslag in Nederland. *Chemisch Weekblad* **35**, 658–664.

Leeuwenhoek, A. van (1677) Observations, communicated to the publisher by Mr. Antony van Leewenhoeck, in a dutch letter of the 9th Octob. 1676. here English'd: concerning little animals by him observed in rain-well-sea- and snow water; as also in water wherein pepper had lain infused. *Phil. Trans.* **13**, 821–831.

Leeuwenhoek, A. van (1697) Continuatio Arcanorum Naturae derectorum. Delphis Batavorum, Koonevelt, 199 pp.

Lefohn, A. S., D. S. Shadwick, U. Feister and V. A. Mohnen (1992) Surface-level ozone: Climate change and evidence for trends. *Journal of the Air Waste Management Association* **42**, 136–144.
Lehmann, Ch. (1699) Historischer Schauplatz derer natürlichen Merckwürdigkeiten im Meißenischen Obererzgebirge. Lanckisch. Erben, Leipzig, 1005 pp.
Lehmann, G. (1884) Meteorologische Literatur Thüringens. In: Mitteilungen der Geographischen Gesellschaft für Thüringen zu Jena. Band II, Jena, pp. 152–178.
Lehmann, K. B. (1890) Die Methoden der praktischen Hygiene. Anleitung zur Untersuchung und Beurteilung der Aufgaben des täglichen Lebens fur Aerzte, Chemiker und Juristen. J. F. Bergmann, Würzburg, 612 pp.
Lehmann, K. B. (1921) Die Luft. In: Lunge-Berl: Chemisch-technische Untersuchungs-methoden (ed. E. Berl), Vol. 1, Springer, Berlin, pp. 635–683.
Lehmann, H. and A. Heller (1940) Luft. In: Handbuch der Lebensmittelchemie. Achter Band: Wasser und Luft. Zweiter Teil: Untersuchung und Beurteilung des Wassers – Luft, Springer, Berlin, pp. 487–599.
Leicester, H. M. and H. S. Klickstein (1952) A source book in chemistry, 1400–1900. Harvard Univ. Press, Cambridge, 550 pp.
Leighton, P. A. (1961) Photochemistry of air pollution. Vol. 9 in "Physical chemistry" series of monographs (Ed. E. M. Loebl), Academic Press, New York and London, 300 pp.
Leighton, P. A. and F. E. Blacet (1932) The photolysis of the aliphatic aldehydes. I. Propionaldehyde. *J. Am. Chem. Soc.* **54**, 3165–3178.
Lejay, P. (1937) Mesures de la quantité d'ozone contenue dans l'atmosphère à l'observatoire de Zô-sè, 1934–1935–1936; les variations de l'ozone et les situations météorologiques. *Notes de météorologie physique* [Observatoire magnétique et météorologique de Zi-ka-wei (Chine)] fascicule 7 (Shanghai: Imprimerie "Union commerciale"). *See also*: Lejay, P. (1934) Mesures de la quantité d'ozone contenue dans l'atmosphère, janvier 1932-décembre 1933. Ibid., fascicule 3, 15 pp. Furthermore: Burgaud, Maurice and Kion Wei-zen (1948) Vapeur d'eau, trouible atmosphérique, ozone, radiation solaire 1937–1944. Ibid., fascicule 10, 30 pp.
Lelieveld, J. and P. J. Crutzen (1990) Influence of cloud and photochemical processes on tropospheric ozone. *Nature* **343**, 227–233.
Lelieveld, J. and P. J. Crutzen (1991) The role of clouds in tropospheric photochemistry. *J. Atmos. Chem.* **12**, 229–267.
Lelieveld, J., W. Peters, E. J. Dentener and M. C. Krol (2002) Stability of tropospheric hydroxyl chemistry. *JGR* **107**, pp. ACH 17-1–ACH 17-11.
Lemmerich, J. (1990) Die Entdeckung des Ozons und die ersten 100 Jahre der Ozonforschung. SIGMA, Berlin, 125 pp.
Lenard, P. and M. Wolf (1889) Zerstäuben der Körper durch ultraviolettes Licht. *Wied. Ann.* **37**, 443–456.
Lenard, P. (1900) Ueber Wirkungen des ultravioletten Lichtes auf gasförmige Körper. *Ann. Phys.* **306** (*Vierte Folge* **1**), 486–507.
Lenard, P. (1936) Deutsche Physik, Vol. 1: Einleitung und Mechanik. Lehmanns, München, 246 pp.
Lender, C. (1870) Zur Anwendung von Sauerstoff und Ozon-Sauerstoff. *Deutsche Klinik* **22**, 435–436.
Lender, C. (1872a) Das atmosphärische Ozon, nach Messungen in Marienbad, Kissingen und Mentone. *Deutsche Klinik* **24**, 173–174, 181, 221–225, 233–234, 241–243, 249–252, 269–270, 282–284, 298–302, 305–307. *See also*: Separat-Abdruck aus Göschen's "Deutscher Klinik" No. 19. G. Reimer, Berlin, 1873, 84 pp. and 1873, 70 pp.
Lender, C. (1872b) Das Ozonwasser. *Deutsche Klinik* **24**, 109–110. *See also*: *Allgemeine Medicinische Central-Zeitung* **41** (1872) 322–323.
Lender, C. (1872c) Der Giftstoff und der Arzneikörper der Luft. Vortrag gehalten zum Besten der Verwundeten am 9. August 1871 im Conversations-Saale zu Kissingen. T. A. Schachenmayer, Kissingen, 32 pp. *See also*: Separat-Abdruck aus Göschen's "Deutscher Klinik" No. 19. G. Reimer, Berlin, 1872, 84 pp. and: Das atmosphärische Ozon. II. Theil. Separat-Abdruck aus Göschen's "Deutscher Klinik". G. Reimer, Berlin, 1873, 70 pp.

Lender, C. (1874) Das atmosphärische Ozon. II. Theil. Separat-Abdruck aus Göschen's "Deutscher Klinik". G. Reimer, Berlin, 70 pp.
Lender, C. (1878) Ueber den Ursprung des atmosphärischen Ozons. *Z. Öster. Ges. Met.* **13**, 301–302.
Lenhard, U. and H.-W. Georgii (1980) Der Ozean als Quelle reaktiver Stickstoffverbindungen. *Pageoph* **118**, 1145–1154.
Lenhard, U. and G. Gravenhorst (1980) Evaluation of ammonia fluxes into the free atmosphere over Western Germany. *Tellus* **32**, 48–55.
Lepape, A. and G. Colange (1929) Relation entre les titre en ozone de l'air du sol at de l'air de la haute atmosphere. *C. r.* **189**, 53–54.
Lersch, M. B. (1863) Geschichte der Balneologie und Pedologie oder des Gebrauches des Wassers. Stahelßsche Buchhandlung, Würzburg, 242 pp. (reprint Zentralantiquariat der DDR, Leipzig 1987).
Le Roy, Ch. (1751) Mémoire sur élévation & suspension de l'eau dans l'air et la rosée. *Mém. Acad. Roy. Sci.* Paris, 481–518. See also: In: Mélanges de physique et de médecine (398 pp.). P. G. Cavelier, Paris, pp. 1–60.
Le Roy, E. (1927) L'exigence idéaliste et le fait de l'évolution. Boivin & Cie, Paris, 270 pp.
Le Roy, E. (1928) Les Origines humaines et l'évolution de l'intelligence. Boivin & Cie, Paris, 375 pp.
Lespieau, R. (1906) Ètude du pouvoir oxidant de l'air sur un mélange d'ioidine et d'arsenite de potassium en divers points du Mont-Blanc. *Bulletin de la Société Chimique de France* **35**, 616–619.
Lettau, H. (1939) Versuch einer Bilanz im Kondensationskern-Haushalt der Troposphäre im Durchschnitt für die ganze Erdoberfläche. *Annalen der Hydrographie und maritimen Meteorologie* **67**, 551–559.
Letts, E. A. and R. F. Blake (1900) The carbonic anhydride of the atmosphere. *Proceedings of the Royal Dublin Society* **9**, 107–270.
Letts, E. A. and R. F. Blake (1901) On some problems connected with atmospheric carbonic anhydride and on a new and accurate method for determining its amount suitable for scientific expeditions. *Proceedings of the Royal Dublin Society* **9**, 436–453.
Levine, B. S. (1966) Survey of U. S. S. R. Literature on Air Pollution and Related Occupational Diseases. Part I. Atmospheric ozone. Results of U. S. S. R. International Geophysical Year studies presented at the 28–31 October 1959 Conference. Part II. Atmospheric ozone. Data presented at the 21–23 August 1963 Conference on Atmospheric Ozone. U. S. Department of Commerce, Office of Technical Services, 275 pp.
Levit, G. S. (2001) Biogeochemistry – Biosphere – Noosphere. The Growth of the Theoretical System of Vladimir Ivanovich Vernadsky. Verlag für Wissenschaft und Bildung, Berlin (Studien zur Theorie der Biologie, Vol. 4), 116 pp.
Lévy, A. (1878a) Analyse de l'air. Ozone. *Annuaire de l'Observatoire de Montsouris*. 495–505.
Lévy, A. (1878b) Sur la recherche de l'ozone dans l'air atmosphérique. *C. r.* **86**, 1263–1265.
Lévy, A. (1878c) Zur Ozonmessung. *Z. Öster. Ges. Met.* **13**, 300–301 (note that his name is wrong written as Lewy).
Lévy, A. (1879) Histoire de l'air. G. Baillière, Paris, 184 pp.
Lévy, A. (1880) Ammoniaque de l'air et des eaux. *C. r.* **91**, 94–97.
Lévy, A. (1882) *Annuaire de l'Observatoire de Montsouris pour l'année 1882*. Paris, pp. 372–405. *See also*: *Jahresbericht über die Fortschritte auf dem Gesammtgebiete der Agricultur-Chemie. Neue Folge* VI. (Das Jahr 1883), Ed. A. Hilger, Berlin, Verlag von Paul Parey. 1884. p. 84.
Lévy, A. (1891) L'ammoniaque dans les eaux météorique. *C. r.* **113**, 804–805.
Lévy, A. (1899) Vingt annees d'observations ozonometriques. *Ciel et terre* **19**, 291–296.
Lévy, A. (1903a) Analyse de l'air. *Annales de l'Observatoire Municipal (Observatoire de Montsouris)* **4**, 128–134, 179–183, 259–268 and 1904, **5** 22–24, 156–160.

Lévy, A. (1903b) Analyse des eaux météoriques. *Annales de l'Observatoire Municipal (Observatoire de Montsouris)* **4**, 105–106, 153–154, 235–239 and 1904, **5**, 5–8, 141–143.
Lévy, A. (1907) Analyse chimique de l'air – ozone. *Annales de l'Observatoire Municipal (Observatoire de Montsouris)* **8**, 495–505.
Lévy, A. and P. Miquel (1891) *Annuaire de l'Observatoire Municipal de Montsouris pour l'année 1891*; cited from Baumann (1893).
Lévy, A., H. Henriet and M. Bouyssy (1905) L'ozone atmospherique. *Annales de l'Observatoire Municipal (Observatoire de Montsouris)* **6**, 18–22, 315–321.
Levy, H. (1971) Normal atmosphere: large radical and formaldehyde concentrations predicted. *Science* **173**, 141–143.
Levy, H. (1973) Photochemistry of minor constituents in the troposphere. *Planetary and Space Science* **21**, 575–591.
Lewis, G. N. (1923) Valence and structure of molecules. American Chemical Monograph Series. New York, 172 pp.
Lewis, G. N. (1924) The magnetism of oxygen and the molecule O_4. *J. Am. Chem. Soc.* **46**, 2027–2032.
Lewis, W. M., M. C. Grant and S. K. Hamilton (1985) Evidence that filterable phosphorus is a significant atmospheric link in the phosphorus cycle. *Oikus* **45**, 428–432.
Lewis, E. R. and S. E. Schwartz (2004) Sea salt aerosol production. Mechanisms, methods, measurements and models – a critical review. Geophysical Monograph 152, Amer. Geophys. Union, Washington, 413 pp.
Lewy, B. (1843) XXVI. Untersuchungen über die Zusammensetzung der atmosphärischen Luft. *J. prakt. Chem.* **30**, 207–224. *See also*: Recherches sur la composition de l'air atmosphérique. *C. r.* **17**, 235–248.
Lewy, B. (1851) Ueber die Zusammensetzung der Atmosphäre. *J. prakt. Chem.* **54**, 249–253. *See also*: Recherches sur la composition de l'atmosphère. *C. r.* **33**, 345–349.
Li, J., X. Wang, J. Chen, C. Zhu, W. Li, C. Li, L. Liu, C. Xu, L. Wen, L. Xue, W. Wang, A. Ding and H. Herrmann (2017) Chemical composition and droplet size distribution of cloud at the summit of Mount Tai, China. *ACP* **17**, 9885–9896.
Li, W., G. V. Gibbs and S. T. Oyama (1998) Mechanism of Ozone Decomposition on a Manganese Oxide Catalyst. 1. In Situ Raman Spectroscopy and Ab Initio Molecular Orbital Calculations. *J. Am. Chem. Soc.* **120**, 9041–9046.
Li, Z. and V. P. Aneja (1992) Regional analysis of cloud chemistry at high elevations in the Eastern United States. *Atmos. Env.* **26A**, 2001–2017.
Lichtenstein, E. (1860a) No title. *Allgemeine medicinische Central-Zeitung* **29**, 655–656.
Lichtenstein, E. (1860b) Ozon und Polarlicht. *Archiv für gemeinschaftliche Arbeiten zur Förderung der wissenschaftlichen Heilkunde* **4**, 594–598.
Lichtenstein, E. (1862) Direkte Einführung der Ozonometrie in die Heilkunde, sowohl in physiologischer als klinischer Bedeutung. *Allgemeine medicinische Central-Zeitung* **31**, 258–262.
Lichtenstein, E. (1876) Beitrag zur Polarforschung. *Leopoldina* **12**, 122–128.
Lichtheim, M. (1980) Ancient Egyptian literature. Vol. III: the late period. Berkeley 1980, 227 pp.
Liebhafsky, H. (1932) The catalytic decomposition of hydrogen peroxide by the iodine-iodide couple at 25 °C. *J. Am. Chem. Soc.* **54**, 1792–1806; The catalytic decomposition of hydrogen peroxide by the iodine-iodide couple. II and III. The rate of oxidation in neutral and in acid, solution of hydrogen peroxide by iodine, pp. 3499–3508.
Liebig, J. von (1827) Sur la Nitrification. *Ann. chim. phys.* **35**, 329–333.
Liebig, J. von (1840) Die organische Chemie in ihrer Anwendung auf Agricultur und Physiologie [Organic chemistry in its applications to agriculture and physiology]). Friedrich Vieweg und Sohn, Braunschweig, 344 pp.

Liebig, J. von (1843) Handbuch der Chemie mit Rücksicht auf Pharmacie. Erste Abtheilung. Anorganische Chemie. C. F. Winter, Heidelberg, 629 pp.

Liebig, J. von (1865) Die Chemie in ihrer Anwendung auf Agricultur und Physiologie. 1. Theil. Vieweg, Braunschweig, p. 317 (*see also*: *Ann. chim. phys.* **35**, 1827, 329).

Liebig, J. von and H. Kopp, eds. (1853) Jahresbericht über die Fortschritte der reinen, pharmazeutischen und technischen Chemie, Physik, Mineralogie und Geologie. Ricker, Giessen. pp. 705–709.

Liebknecht, O. and W. Katz (1953) Ozon. In: Handbuch der analytischen Chemie (eds. W. Fresenius and G. Jander), Springer-Verlag, Berlin, pp. 100–182. Wasserstoffperoxyd. Ibid., pp. 183–342.

Liebreich, M. E. O. (1880) Ueber Ozon. *Deutsche Medizinische Wochenschrift* **6**, 317–319.

Liefmann, H. (1908) Über die Rauch- und Russfrage, insb. vom gesundheitlichen Standpunkt und eine Methode des Russnachweises in der Luft (Sonderabdruck aus der "Dt. Vierteljahreszeitschrift für öffentliche Gesundheitspflege", Vol. XL, 2. Heft). Fr. Vieweg, Braunschweig, 90 pp.

Liesegang, W. (1927) Chemische Ermittlungen auf dem Gebiete der Lufthygiene. *Kleine Mitteilungen für die Mitglieder des Vereins für Wasser-, Boden und Lufthygiene* **3**, 4–10. *See also*: Chemische Fragen der Lufthygiene. *Ibid.* 329–344.

Liesegang, W. (1927/29) Die Untersuchung atmosphärischer Niederschläge. *Kleine Mitteilungen für die Mitglieder des Vereins für Wasser-, Boden und Lufthygiene* **3**, 317–327, **4**, 178–182, 261–265, 329–330, **5**, 84–86.

Liesegang, W. (1931) Die Verteilung schwefelhaltiger Abgase in freier Luft. *Der Gesundheitsingenieur* **54**, 705–709.

Liesegang, W. (1932) Über den Nachweis von Verunreinigungen durch Industrieabgase in der freien Atmosphäre. *Kleine Mitteilungen für die Mitglieder des Vereins für Wasser-, Boden und Lufthygiene* **8**, 174–181.

Liesegang, W. (1933) Über einige gesundheitlich wichtige gas- und dampfförmige Verunreinigungen der atmosphärischen Luft. III. Schwefeldioxid. *Kleine Mitteilungen für die Mitglieder des Vereins für Wasser-, Boden und Lufthygiene* **9**, 286–296. *See also*: I. Schwefelwasserstoff. *Ibid.* **6** (1930) 216–222, II. Kohlenoxyd. *Ibid.* **9** (1933) 45–55.

Liesegang, W. (1934/36) Untersuchungen über die Mengen der in Niederschlagswässern enthaltenen Verunreinigungen. *Kleine Mitteilungen für die Mitglieder des Vereins für Wasser-, Boden und Lufthygiene* **9**, 306–312, 350–355, **10** (1935), 248–249 and together with H. Nehls: **11** (1936) 314–318.

Liesegang, W. (1935) Reinhaltung der Luft. In: Ergebnisse der angewandten physikalischen Chemie, Band III, Teil 1 (ed. M. le Blanc). Akademische Verlagsgesellschaft, Leipzig, pp. 1–109.

Likens, G. E., F. H. Bormann and N. M. Johnson (1972) Acid rain. *Environment* **3**, 33–40.

Likens, G. E. and F. H. Borman (1974) Acid rain: A serous regional environmental problem. *Science* **184**, 1176–1179.

Likens, G. E., W. C. Keene, J. M. Miller and J. N. Galloway (1987) Chemistry of precipitation from a remote, terrestrial site in Australia. *JGR* **92**, 13299–13314.

Likens, G. E., R. F. Wright, J. N. Galloway and T. J. Butler (1979) Acid rain. *Scientific American* **241**, 43–51.

Liljequist, G. H. and K. Cehak (1984) Allgemeine Meteorologie. Fr. Vieweg und Sohn, Braunschweig, 368 pp.

Lin, N.-H. (2015) Cloud Water Chemistry Measured at Mt. Bamboo in East Asia since 1996. 95th American Meteorological Society Annual Meeting, January 04–08, Phoenix, AZ.

Lindgren, W. (1923) Concentration and circulation of the elements from the standpoint of economic geology. *Economic Geologist* **18**, 419–442.

Lindner (1855) Beobachtungen über die Wirkung des Ferrum hydricum gegen Cholera. *Allgemeine medicinische Central-Zeitung* **24**, 705–708.

Link, H. F. (1826) Handbuch der physikalischen Erdbeschreibung. Theil 1. F. Dümmler, Berlin, 404 pp.
Linke, F. (1942) Das atmosphärische Aerosol. In: Handbuch der Geophysik, Vol. 8. Physik der Atmosphäre 1 (ed. B. Gutenberg), Bornträger, Berlin, pp. 14–27.
Linow, C. (1892) Die Untersuchung der Luft auf organische Substanzen. C. Boldt, Rostock, 36 pp.
Linvill, D. E., W. J. Hooker and B. Olson (1980) Ozone in Michigan's environment 1876–1880. *Journal of the American Meteorological Society* **108**, 1883–1891.
Lipp, H. (1932) Über die chemische Zusammensetzung des Nebelfrostes auf der Zugspitze. *40. Jahresbericht des Sonnblick-Vereins für das Jahr 1931*. J. Springer, Wien, 27–32.
Lippincott, R. C. C. (1869) Influence of the sun's hour-angle on the development of ozone. *Proceedings of the Meteorological Society* **4**, 155–160.
Lippincott, R. C. C. (1871) Ozone-observations. *Proceedings of the Meteorological Society* **5**, 51.
Lissauer, A. (1864) Ueber das Ozon und Antozon, Vortrag gehalten im Gewerbeverein zu Danzig. C. Ziemssen, Danzig, 16 pp. *See also*: *Allgemeine Medicinische Central-Zeitung* **33** (1864) 822–823, 837–839.
Liu, C., C. Chung, F. Zhang and Y. Yin (2016) The colors of biomass burning aerosols in the atmosphere. *Scientific Reports* **6**, 28267.
Liu, T., S. L. Clegg and J. P. D. Abbatt (2020) Fast oxidation of sulfur dioxide by hydrogen peroxide in deliquesced aerosol particles. *Proceedings of the National Academy of Sciences* **117**, 1354–1359.
Liveing, G. D. and J. Dewar (1880) XV. On the spectrum of water. *Proc. R. Soc. Lond.* **30**, 580–582.
Lloyd, H. E. and G. H. Noehden (1836) A new dictionary of English and German languages. Printed for A. Campe, Hamburg, 519 pp.
Lockemann, G. (1950) Geschichte der Chemie. II. Von der Entdeckung des Sauerstoffs bis zur Gegenwart. Sammlung Göschen, Bd. 265/165a. de Gruyter, Berlin, 149 pp.
Lode, A. (1911) Atmosphäre. In: Handbuch der Hygiene. I. Band (eds. M. Rubner, M. v. Gruber and M. Ficker), Verlag Hirzel, Leipzig, pp. 367–518.
Lodge, Jr, J. P., J. E. McDonald and F. Baer (1954) An investigation of the Melander effect. *Journal of Meteorology* **11**, 318–322.
Loew, O. (1870a) On the action of sunlight on sulphurous acid. *The American Journal of Science and Arts 2nd series* **49**, 368.
Loew, O. (1870b) On the formation of ozone by rapid combustion. *American Journal of Science 2nd series* **49**, 369.
Loewe, R. (1936) Gas. *Zeitschrift für vergleichende Sprachforschung* **63**, 118–122.
Loewy, A. (1924) Allgemeine Klimatophysiologie. In: Klimaphysiologie und Strahlen-physiologie, Vol. 3 Handbuch der Balneologie, medizinischen Klimatologie und Balneo-graphie (eds. E. Dietrich and S. Kaminer), Georg Thieme, Leipzig, pp. 3–166.
Logan, J. A., M. J. Prather, S. C. Wopsy and M. B. McElroy (1981) Tropospheric chemistry: a global perspective. *JGR* **86**, 7210–7254.
Logan, J. A. (1985) Tropospheric ozone, seasonal behavior, trends and anthropogenic influence. *JGR* **90D**, 10,463–10,482.
Logan, S. R. (1967) The solvated electron – The simplest ion and reagent. *Journal of Chemical Education* **44**, 344–349.
Lomonossov, M. V. (1752) Tentament chymiæ physicæ in usus studiosus juventutia adornatum. Dromus ad veram chymiam physicam (introduction into the true physical chemistry). In: Полное собрание сочинений. Том второй. Труды по физике и химии 1747–1752 гг. Издательство Академии наук СССР (1951). 732 pp., (pp. 481–576); *see also*: Классики естествознания. М. В. Ломоносов: Физико-химические работы (Ed. Б. Н. Меншуткин), государственное издательство, Moscow and Petrograd, 1923, 124 pp.; and: Physikalisch-chemische Abhandlungen M. W. Lomonossows 1741–1752. Herausgegeben von B. N. Menschutkin und Max Speter. Ostwald's Klassiker No. 178, Leipzig: Engelmann 1910, 60 pp.

Long, J. H. (1881–1882) On the oxidizing gases of the atmosphere. *National Board Health Bulletin*, Washington **3**, 367.

Lorenz, N. von (1887) Kohlensäuregehalt der Luft auf dem Sonnblick (3100). *Met. Z.* **4**, 464–466. *See also*: *Naturwiss. Rundschau* (1887) Heft 10, p. 129.

Lösecke, A. von (1874) Ozonbeobachtungen. *Arch. Pharm.* **205**, 427–432.

Lösecke, A. von (1879) Ueber die Bildung des salpetrigsauren Ammoniaks. *Arch. Pharm.* **214**, 54–58.

Louvet, A. (1876) Observations recuellies sur les eaux pluviales à Saint-Louis (Sénégal). *Annu. soc. mét. de France* **24**, 178–188.

Lovelock, J. E. (1979) Gaia: a new look at life on Earth. Oxford University Press, Oxford, 157 pp.

Lovelock, J. E., R. J. Maggs and R. A. Rasmussen (1972) Atmospheric dimethyl sulphide and the natural sulphur cycle. *Nature* **237**, 462–463.

Lovelock, J. E. and L. Margulis (1974) Atmospheric homeostasis by and for the biosphere: the Gaia hypothesis. *Tellus* **26**, 2–10.

Low, D. (1848) The simple bodies of chemistry. 2nd ed. Longman, London, 85 pp.

Low, D. (1856) An inquiry into the nature of the simple bodies of chemistry. A. and C. Black, 386 pp.

Lowe, E. J. (1863) On ozone. *Proceedings of the Royal Society, London* **12**, 518–523.

Löwig, C. (1832) Lehrbuch der Chemie mit besonderer Berücksichtigung des technischen und medicinischen Theils. Engelmann, Heidelberg, 481 pp.

Lu, C., S. Niu, L. Tang, J. Lv, L. Zhao and B. Zhu (2010) Chemical composition of fog water in Nanjing area of China and its related fog microphysics. *Atmos. Res.* **97**, 47–69.

Ludolf, H. (1752) D. Hieron. Ludolfs, öffentlichen Lehrers der Chymie und Mathematik, der philos. und medicinis. Facultät Decanus und ordentl. Beysitzers, wie auch Stadtphysicus allhier, Vollständige und gründliche Einleitung in die Chymie: darin nicht allein alle Chymische Arbeiten deutlich gezeiget und gründlich erkläret, sondern auch zu derselben Erläuterung die wichtigsten Versuche aus der Pharmacie, Metallurgie und Alchymie nebst allen Vortheilen treulich ausgeführet worden: mit nöthigen Kupfern versehen. Joh. Heinr. Nonnens, Erfurt, 1104 pp.

Ludwig, H. (1856) Ueber den Hagel. *Arch. Pharm.* **135**, 30–41.

Ludwig, H. (1862) Die natürlichen Wässer in ihren chemischen Beziehungen zu Luft und Gesteinen. F. Enke, Erlangen. 336 pp. ("on atmospheric water": pp. 1–29).

Ludwig, H. (1871) Classification der Gerüche. *Arch. Pharm.* **197**, 225–245.

Ludwig, H. (1882) Ueber die Methoden der Kohlensäurebestimmung in der atmosphärischen Luft. *Mitteilungen des Vereins der Aerzte in Nieder-Öserreich, Wien* **8**, 73–75.

Luedicke, O. (1871) Ueber das Auftreten des Ozons in der Atmosphäre. *Z. Öster. Ges. Met.* **6**, 273–277.

Luedicke, O. (1875) Ozonoscopische Untersuchungen an Localitäten, welche die Entwicklung oxydabler Körper (Ozonräuber) begünstigen. *Correspondenz-Blätter des allgemeinen ärztlichen Vereins von Thüringen* **4**, 57–59.

Lugin, N. P. [Лугин] (1936) Результаты измерения содержания озона в 1935 г. по наблюдениям в Кучино. *Астрономический журнал* [The Astronomical Journal] **13**, 348.

Luna, T. M. de (1860) Estudios quimicos sobre el Aire atmosférico de Madrid. Imprementa Alvarez, Madrid, 54 pp.; *see also*: Étude chimique sur l'air atmosphérique de Madrid, par Ramón Torrez Muños de Luna, ... Traduit de l'espagnol, avec des notes, par M. H. Gaultier de Claubry. J.-B. Baillière, Paris (1861), 55 pp.

Lundegårdh, H. (1924) Der Kreislauf der Kohlensäure in der Natur. Ein Beitrag zur Pflanzenökologie und zur landwirtschaftlichen Düngungslehre. Gustav Fischer, Jena, 308 pp.

Lunge, G. (1877) Zur Frage der Ventilation mit Beschreibung des 'minimetrischen' Apparates zur Bestimmung der Luftverunreinigung. Caesar Schmidt, Zürich, 47 pp. (lecture hold on 18th January 1877 in Zürich).

Lunge, G. (1879) Ueber die Existenz des Salpetrigsäureanhydrids im Dampfzustande. *Ber.* **12**, 357–359.
Lunge, G. (1890) Das Gasvolumeter, ein Apparat zur völligen Ersparung aller Reductionserscheinungen bei Ablesungen von Gasvolumen. *Z. angew. Chem.* **3**, 139–144.
Lunge, G. and A. Zeckendorf (1888) Neue Methode zur Bestimmung des Kohlensäuregehaltes der Luft für hygienische Zwecke. *Z. anal. Chem.* **27**, 395–399.
Lunge, G. and L. Marchlewski (1891) Neuer Apparat zur Bestimmung der Kohlensäure (insbesondere der gebundenen) auf gasvolumetrischen Wege. *Z. angew. Chem.* **4**, 229–235.
Lunge, G. and L. Pelet (1895) Über die Darstellung von Chlor aus Salzsäure mittels Salpetersäure. *Z. angew. Chem.* **8**, 3–13.
Lüttke, J., K. Levsen, K. Acker, W. Wieprecht and D. Möller (1999) Phenols and nitrated phenols in clouds at Mt. Brocken. *International Journal of Environmental Analytical Chemistry* **74**, 69–99.
Lüttke, J., V. Scheer, K. Levsen, G. Wünsch, J. N. Cape, K. J. Hargreaves, R. L. Storeton-West, K. Acker, W. Wieprecht and B. Jones (1997) Occurrence and formation of nitrated phenols in and out of clouds. *Atmos. Env.* **31**, 2637–2648.
Luz, J. F. (1807) Anweisung, das Eudiometer des Herrn Abt von Fontana zu verfertigen und zum Gebrauch bequemer zu machen, ingleichen durch eine sehr einfache Einrichtung in kurzer Zeit Mineralwasser zu verfertigen. E. Weigel und A. G. Schneider, Nürnberg und Leipzig, 78 pp.
Maack, von (1856) Beiträge zur Ozonometrie. Erster Artikel. Kritik des Schönbeinschen Ozonometers. *Archiv des Vereins für gemeinschaftliche Arbeiten zur Förderung der wissenschaftlichen Heilkunde (Marburg)* **2**, 24–30.
Maahs, H. G. (1983) Measurement of the oxidation rate of sulphur(IV) by ozone in aqueous solution and their relevance to SO_2 conversion in nonurban tropospheric clouds. *Atmos. Env.* **17**, 341–345.
Maaß, F., H. Elias and K. J. Wannowius (1999) Kinetics of the oxidation of hydrogen sulfite by hydrogen peroxide in aqueous solution: ionic strength effects and temperature dependence. *Atmos. Env.* **33**, 4413–4419.
Mabery, C. F. (1895) An examination of the atmosphere of a large manufacturing city. *J. Am. Chem. Soc.* **17**, 105–122.
Macadam, S. (1858) On M. S. de Luca's claim to be the discoverer of the non-presence of iodine in the atmosphere, in rain-water and in snow. *The Edinburgh New Philosophical Journal* **7**, 106–110.
Macagno, I. (1879) Ricerche chimico-idrolgiche suttee acque potabili e d'irrigazione die Palermo e suoi dintorni. *Bollettino della Stazione Agraria di Palermo, "A. A. S."* **9**, 24. *See also*: Jahresbericht über die Fortschritte auf dem Gesammtgebiet der Agricultur-Chemie N. F. 3 für 1880, Vol. 23 (1881) pp. 80–81).
Mach, E. (1888) Ueber den Gehalt des Regenwassers in S. Michele an Ammoniak und Salpetersäure. *Tiroler Landwirtschaftliche Blätter* **7**, 16–17.
Machta, L. and E. Hughes (1970) Atmospheric oxygen in 1967 to 1970. *Science* **168**, 1582–1584.
Machu, W. (1937) Das Wasserstoffperoxyd und die Perverbindungen. Springer-Verlag, Wien, 408 pp. (2nd edition 1951).
MacIntire, W. H. and J. B. Young (1923) Sulfur, calcium, magnesium and potassium content and reaction of rainfall at different points in Tennessee. *Soil Science* **15**, 205–227.
Mack, E. J., U. Katz, C. W. Rogers and R. J. Pilie (1974) The microstructure of California coastal stratus and fog at sea. Calspan Corp., Buffalo, NY, 67 pp.
Mack, E. J., U. Katz, C. W. Rogers, D. W. Gaucher, K. R. Piech, C. K. Akers and R. J. Pilie (1977) An investigation of the meteorology, physics and chemistry of marine boundary layer processes. Report No. CJ-6017-M-1, Calspan Corp., Buffalo, NY (unpublished).
Mack, K. (1900) Die Bekämpfung des Hagels durch das sogenannte Wetterschiessen. *Band 56 von Jahreshefte des Vereins für vaterländische Naturkunde in Württemberg*. 14 pp.

Mackenzie, F. T. and J. A. Mackenzie (1995) Our changing planet. Prentice-Hall, Upper Saddle River, NJ, pp 288–307.

Mackereth, Th. (1867) Results of raingauge, anemometer and ozone observations, made in Eccles, near Manchester, during the year 1866. *Proceedings of the Literary and Philosophical Society of Manchester* **6**, 94–98.

Mackereth, Th. (1868) On ozone and its probable connection with solar radiation. ozone. *Proceedings of the Literary and Philosophical Society of Manchester* **7**, 213–219.

Macků, M., J. Podzimek and L. Šrámek (1959) Results of chemical analysis of precipitation collected on territory of Czechosloval Republic in IGY. *Geofyz. Sbornik* **124**, 441–519 and: Distribution of some chemical compositions in atmospheric precipitation over Czechoslavakia. *Geofisica pura e applicata* (Milano) **42**, 24–31.

McMath, R. R., O. C. Mohler and L. Goldberg (1948) The abundance and temperature of methane in the earth's atmosphere. *Physical Review* **74**, 629–624.

Macquer, P. J. (1788) Chymisches Wörterbuch oder allgemeine Begriffe der Chymie nach alphabetischer Ordnung. Aus dem Französischen nach der zweyten Ausgabe übersetzt und mit Anmerkungen und Zusätzen vermehrt von Johann Gottfried Leonhardi. 2. verbesserte und vermehrte Auflage. 7 Bände. Verlag: Leipzig Weidmann (Zweyter Theil von D bis Gas, entzündbares, 853 pp.).

Mader, P. M. (1958) Kinetics of the hydrogen peroxide-sulfite reaction in alkaline solution. *J. Am. Chem. Soc.* **80**, 2634–2639.

Magill, P. L. (1949) The Los Angeles smog problem. *Industrial & Engineering Chemistry* **41**, 2476–2486.

Magill, P. L., F. R. Holden and C. Ackley, eds. (1956) Air pollution handbook. McGraw-Hill, New York, pp. 2–45.

Main, J. (1832) London fogs. *Magazine of Natural History* **5**, 303.

Majo, E. (1927) [unknown title] *Il coltivatore giornale di agricoltura pratica* **28**, 327 (quoted after Mosello et al. 1992).

Maksimovich, G. A. [Максимович] (1955) Химическая география вод суши [Chemical geography of continental waters). Государственное издательство географической литературы, Москва, 329 pp.; chapter 1: Химический состав атмосферных осадков [Chemical composition of atmospheric precipitation], pp. 23–37.

Malley, C. S., M. R. Heal and C. F. Braban (2016) Insights from chronology of the development of atmospheric composition monitoring networks since the 1800s. *Atmosphere* **7**, 160. 22 pp.

Mangin, A. (1866) Das Reich der Luft. Das Weltall, Teil 4. R. Schlingmann-Verlag, Berlin, 432 pp.

Mani, A. (1990) Ozone studies in India. *Indian Journal of Radio and Space Physics* **19**, 542–549.

Mantegazza, P. (1870) Dell'azione delle essenze e dei fiori sulla produzione dell'ozono atmosferico e della loro utilità igienica: ricerche esperimentali. *Rendiconti del Reale Istituto Lombardo di Scienze e Lettere*. Serie 2, Volume 3. *See also*: Ueber den Einfluss der aromatischen Substanzen (Essenza) und der Blumen auf die Production des atmosphärischen Ozons und über deren hygienischen Nutzen. *Allgemeine medicinische Central-Zeitung* 39 (1879) 715. Additional: Anonymous (1872) The formation of ozone by flowers. *The American Naturalist* **6**, 428.

Marcano, V., A. Müntz and A. Lévy (1891) Ammoniaque dans látmosphere et dans les pluies d'une region tropicale. *C. r.* **113**, 779–781.

Marcet, W. (1889) On fogs. An address delivered at the Annual General Meeting of the Royal Meteorological Society on January 16th. *Q. J. Roy. Meteor. Soc.* **15**, 59–72.

Marcet, W. (1890) On atmospheric dust. *Q. J. Roy. Meteor. Soc.* **16**, 73–85.

Marcet, W. and A. Landriset (1887) The influence of weather on the proportion of carbonic acid in the air of plains and mountains. *Q. J. Roy. Meteor. Soc.* **13**, 166–173.

Marchand, R. F. (1848) Über das Verhalten des Stickstoffs bei dem Respirationsprocesse. *J. prakt. Chem.* **48**, 1–37; Lecture (offprint, Halle, 1845, 64 pp): vorgetragen in der naturforschenden Gesellschaft zu Halle, im Juni 1845 mit einigen späteren Zusätzen.

Marchand, R. F. (1850) Ueber die Eudiometrie. *J. prakt. Chem.* **49**, 449–468.
Marchand, E. (1852) Sur la constitution physique et chimique des eaux naturelles. *C. r.* **34**, 54–56.
Marchi, L. de (1895) Le cause dell'era glaciale. Bernardoni e Rebeschini, Milano, 231 pp.
Marcotte, G., P. Marchand, S. Pronovost, P. Ayotte, C. Laffon and P. Parent (2015) Surface-enhanced nitrate photolysis on ice. *J. phys. Chem. A* **119**, 1996–2005.
Marcuse, A. (1896) Atmosphärische Luft. Verlag Friedländer und Sohn, Berlin, 77 pp.
Marenco, A., H. Gouget, P. Nédélec and J.-P. Pagés (1994) Evidence of a long-term increase in tropospheric ozone from Pic du Midi series: consequences: positive radiative forcing. *JGR* **99** (1994) 16,617–16,632. *See also*: Marenco, A., N. Philippe and G. Hérve (1994) Ozone measurements at Pic du Midi observatory. *EUROTRAC Annual report part 9: TOR*, EUROTRAC ISS, Garmisch-Partenkirchen, pp. 121–130.
Marggraf, A. S. (1753) Examen chymique de l'eau. *Histoire de l'Academie Royale des Sciences et Belles Lettres*, Berlin, pp. 131–157.
Marggraf, A. S. (1786) Chemische Untersuchung des gemeinen Wassers. *Physikalische und medicinische Abhandlungen der Königlichen Academie der Wissenschaften zu Berlin* **4**, 69–95.
Marié-Davy, M. (1876) Note sur l'ozone de l'air atmosphérique. *C. r.* **82**, 900–902.
Marié-Davy, M. (1880) L'acide carbonique de l'air, dans ses rapports avec les grands mouvements de l'atmosphére. *C. r.* **90**, 32–36.
Marimpietri, L. and F. Simoncelli (1932) Il pH e il contenuto in azote delle precipitazioni meteoriche vute in Roma nel triennio 1929–1931. Pubblicazioni della R. Stazione Chimico-Agraria di Roma. Pubbl. No. 299.
Marquardt, W., D. Möller and W. Rolle (1984) Saure Niederschläge – Bildung und Einschätzung. *Nachrichten Mensch-Umwelt, Berlin* **12**, 37–69.
Marquardt, W. and P. Ihle (1988) Acidic and alkaline precipitation components in the mesoscale range under the aspect of meteorological factors and the emissions. *Atmos. Env.* **22**, 2707–2716.
Marquardt, W., E. Brüggemann and P. Ihle (1996) Trends in the composition of wet deposition: effects of the atmospheric rehabilitation in East-Germany. *Tellus* **48B**, 361–371.
Marquardt, W., E. Brüggemann, R. Auel, H. Herrmann and D. Möller (2001) Trends of pollution in rain over East Germany caused by changing emissions. *Tellus* **53B**, 529–545.
Marr, Th. (1891, 1893, 1898) Over nitreuse stikstof in het regenwater. Meded. Proefstation. Oost Java, 1. ser., No. 38, 1891, pp. 5–8; *Bijlage Arch. Java Suikerindus.*, 1898, pp. 151–154; Over den regen val en het stikstofgehalte van het regenwater. Meded. Proefstation Oost Java, 1. ser., No. 47, 1893, pp. 63–67; *Arch. Java Suikerindus.* **1**, 1893, pp. 96–101.
Marriott, W. (1904) Some account of the meteorological work of the late James Glaisher. *Q. J. Roy. Meteor. Soc.* **30**, 1–28.
Marsh, A. (1947) Smoke. The problem of coal and the atmosphere. Faber and Faber, London, 306 pp.
Marshall, A. L. (1926) Photosensitization by optically excited mercury atoms. The hydrogen oxygen reaction. *J. phys. Chem.* **30**, 34–46.
Marshall, A. L. (1932) The photochemical formation of hydrogen peroxide. *J. Am. Chem. Soc.* **54**, 4460–4461.
Marshall, J. G. and P. V. Rutledge (1959) A higher hydrogen peroxide, H_2O_4?. *Nature* **184**, 2013–2014.
Martens, C. S., J. J. Wesolowski, R. C. Harriss and R. Kaifer (1973) Chlorine loss from Puerto Rican and San Francisco Bay area marine aerosols. *JGR* **78**, 8778–8792.
Marti, A. de (1805) Antonio de Marti's eudiometrische Untersuchungen. *Gilb. Ann.* **19**, 389–393.
Martin (1853) Mémoire sur l'analyse chimique de l'eau de pluie. *C. r.* **37**, 487–488.
Martin, Ch-N. (1954) Sur les effets accumulatifs provoqués par les explosions thermonucléaires à la surface du globe. *C. r.* **239**, 1287–1289.

Martin, D., M. Tsivou, B. Bonsang, C. Abonnel, T. Carsey, M. Springer-Young, A. Pszenny and K. Suhre (1997) Hydrogen peroxide in the marine atmospheric boundary layer during the Atlantic Stratocumulus Transition Experiment/Marine Aerosol and Gas Exchange Experiment in the eastern subtropical North Atlantic. *JGR* **102**, 6003–6016.
Martin, L. R. and D. E. Damschen (1981) Aqueous oxidation of sulfur dioxide by hydrogen peroxide at low pH. *Atmos. Env.* **15**, 1615–1621.
Martin, L. R. (1984) Kinetic studies of sulfite oxidation in aqueous solution. In: SO_2, NO and NO_2 oxidation mechanisms: atmospheric considerations (ed. J. G. Calvert), Butterworth Publishers, Boston, MA, USA, pp. 63–100.
Martius, Th. (1862) Ozon, der eigenthümliche Geruch bei einem Blitzschlag. *Arch. Pharm.* **161**, 211–212.
Mason, C. (1780) The lady's assistant. A new edition – to which is now added, an Appendix. J. Walker, London, 478 pp.
Matsui, H. (1939) Atmospheric impurities in the central part of Tokyo (in Japanese). *Journal of the meteorological Society of Japan* **17**, 367–372.
Matsui, H. (1942) The Chemistry of rain water. Part II (in Japanese). *Journal of the meteorological Society of Japan* **20**, 291–296.
Matsui, H. (1949) On the content of hydrogen peroxide of the atmospheric precipitates (in Japanese). *Journal of the meteorological Society of Japan* **27**, 380–381.
Mauersberger, G. and D. Möller (1990) Auswaschen von Gasen und Aerosolen durch Niederschläge unter Berücksichtigung einer komplexen Flüssigphasenchemie. 1. Modellbeschreibung. *Z. Met.* **40**, 322–329.
Mayow, J. (1674) Tractateus quinque medico-physici. Quorum primus agit de sal-nitro, et spiriti nitro-aereo. Secundus de respiratione. Tertius de respiratione foetus in utero, et ovo. Quartus de motu musculari, et spiritius animalibus. Ultimus de rhachixite. Oxonii [Oxford], E Theatro Sheldoniano, 487 pp.
McArdle, J. V. and M. R. Hoffmann (1983) Kinetics and mechanism of the oxidation of aquated sulfur dioxide by hydrogen peroxide. *J. Phys. Chem.* **87**, 5425–5429.
McCabe, I. M. (2012) Second best as a researcher, second to none as a populariser? The atmospheric science of John Tyndall FRS (1820–1893). Dissertation, University College London, 318 pp.
McConnell, J. L. C., M. B. McElroy and S. C. Wofsy (1971) Natural sources of atmospheric CO. *Nature* **233**, 187–188.
McElroy, W. J. (1986) Sources of hydrogen peroxide in cloud water. *Atmos. Env.* **20**, 427–438.
McElroy, M. B., R. J. Salawitch, S. C. Wofsy and J. A. Logan (1986) Reductions of Antarctic ozone due to synergistic interactions of chlorine and bromine. *Nature* **321**, 759–762.
McEvoy, J. G. (2014) Gases, God and the balance of nature: a commentary on Priestley (1772) Observations on different kinds of air. *Phil. Trans. A* **373**, 20140229.
McGrath, W. D. and R. G. W. Norrish (1957) The flash photolysis of ozone. *Proc. R. Soc. Lond. A* **242**, 265–276.
McKie, D. (1961) Joseph Priestley and the Copley Medal. *Ambix* **9**, 1–22.
McLeod, H. (1880) Note on formation of ozone during slow oxidation of phosphorus. *J. Chem. Soc.* **37**, 118–120.
Mecke, R. (1931) Zur Deutung des Ozongehalts der Atmosphäre. *Z. phys. Chem.* **1931A** (Issue Supplement: *Bodenstein-Festband*) 392–404. *See also*: The photochemical ozone equilibrium in the atmosphere. *Trans. Faraday Soc.* **27** (1931) 375–377.
Meetham, A. R. and G. M. B. Dobson (1935) The vertical distribution of atmospheric ozone in high latitudes. *Proc. R. Soc. Lond.* **148A**, 598–603.
Mégie, G. (1991) Ozon – Atmosphäre aus dem Gleichgewicht. Springer-Verlag, Berlin u. a., 177 pp.
Meissner, G. (1863a) Untersuchungen über den Sauerstoff. Hahn'sche Hofbuchhandlung, Hannover, 370 pp.

Meissner, G. (1863b) Ueber die Bestandtheile des Regenwassers. Nachrichten von der Georg-August-Universität und der königl. 264–268. Gesellschaft der Wissenschaften zu Göttingen. Vom Jahre. 1863.
Meissner, G. (1869) Neue Untersuchungen über den elektrisierten Sauerstoff. Bd. 14 Abhandlungen der Kgl. Gesellschaft der Wissenschaften zu Göttingen. Dieterische Buchhandlung, Göttingen, 110 pp.
Melander, G. (1897) Sur la condensation de la vapeur d'eau dans l'atmosphère. Imprimerie des Heritiers de Simelius, Helsinki, 141 pp.
Meldrum, A. N. (1930) The eighteenth-century revolution in science – the first phase. Longmans, London, 60 pp.
Meldrum, A. N. (1933) Lavoisier's early work in science 1763–1771. *Isis* **19**, 338–363.
Mellor, J. W. (1922–1940) A comprehensive treatise of inorganic and theoretical chemistry. In XVI Vol., Longmans, London.
Menchikowsky, F. (1924) Composition of rain falling at Tel-Aviv. *Institute for Agronomy and National History Bulletin* **2**, IV, 25–38 (cited after Eriksson 1960) and: Agricultural Experimental Station Tel-Aviv (1925), pp. 26–36 (in Hebrew).
Mendeleev, D. I. [Менделеев] (1895) Основые химий. Типография В. Демакова, С.-Петербургъ; The principles of chemistry. London 1897, 2nd ed., in 2 Vol.
Mène, Ch. (1851a) Présence de l'ammoniaque dans des grêlons recueillis près de Paris, le 5 mai 185l. *C. r.* **32**, 770–771.
Mène, Ch. (1851b) Note sur la quantité d'acide carbonique dans l'air à différentes hauteurs. Première série d'expériences faites au Panthéon, les 8, 10 et 11 juillet 1851. *C. r.* **33**, 39–40. And: Nouvelle manière de doser l'acide carbonique; mesures et quantités de ce gaz dans l'atmosphère; remarque sur les analyses organiques, ibid. 222–224.
Mensbrugghe, G. L. (1892) The formation of fog and of clouds. *Symons's Monthly Meteorological Magazine* **27**, 40–41 (translated from *Ciel et Terre*).
Menz, F. C. and H. M. Seip (2004) Acid rain in Europe and the United States: an update. *Environmental Sciences and Policy* **7**, 253–265.
Menzl, O. (1948) Über den Chlorgehalt der Niederschläge. *Z. Met.* **2**, 289–295.
Merckens, W. (1906) Über strahlenartige Einwirkungen auf die photographische Bromsilbergelatine. *Gilb. Ann.* **16**, 667–683.
Mészáros, E. (1965) On the origin and the composition of atmospheric calcium. *Tellus* **18**, 262–265.
Mészáros, E. (1966) Some data on the chemical composition of atmospheric particles in the submicron range of sizes. *Idöjaras* **70**, 257–261.
Mészáros, E. (1973) A csapadékviz nyom-anyag koncentrációjának évi menete Magyarországom. *Idöjaras* **77**, 304–310.
Mészáros, E. (1974) On the spring maximum of the concentration of trace constituents in atmospheric precipitation. *Tellus* **26**, 402–407.
Mészáros, E. (1981) Atmospheric chemistry. Fundamental Aspects. Akadémiai Kiadó, Budapest, 201 pp.
Metnieks, A. L. (1958) the size spectrum of large and giant sea-salt nuclei under maritime conditions. *Geophysical Bulletin, School of Cosmic Physics, Dublin* **15**, 1–50.
Mettenheimer, C. (1863) Beobachtungen über den Ozongehalt der Atmosphäre in Frankfurt a. M. in dem Zeitraum vom 18. April 1852 bis 30. April 1853. *Archiv des Vereins für gemeinschaftliche Arbeiten zur Förderung der wissenschaftlichen Heilkunde* **6**, 492–559.
Meyer, E. (1903) Über die Absorption der ultravioletten Strahlung in Ozon. *Ann. Phys.* **317** (Vierte Folge 12), 849–859.
Meyer, E. von (1914) Geschichte der Chemie von den ältesten Zeiten bis zur Gegenwart. 4. ed. Veit & Comp., Leipzig, 616 pp.

Meyer and Stoop (1819) Pluie rouge, tombée a Blankenburg; analyse de cette eau. *Annales Générales des Sciences Physique* **2**, 269–271. *See also:* Analysis of red rain which fell at Blankenburg November 2, 1819 by Messrs. Meyer and Stoop. Chemists at Bruges. *Phil. Mag. and Journal* **55** (1820) 231–232. Anonymous (1820) Lets over den gekleurden regen en sneeuw. *Letterkundig Magazijn Van Wetenschap, Kunst En Smaak* **2** (1822) 672 (note, it is *Meijer* given as name). Meyer and van Stoop (1820) Untersuchung des zu Blankenberge in Flandern am 2. November 1819 herabgefallenen rothen Regen. *Gilb. Ann.* **64** (1821) 335–337.

Meyrac, V. (1852) Observations sur les eaux de pluie, des neiges et des rosées. *C. r.* **34**, 714–717.

Michaud, A. B., J. E. Dore, D. Leslie, W. B. Lyons, D. C. Sands and J. C. Priscu (2014) Biological ice nucleation initiates hailstone formation. *JGR* **119**, 12,186–12,197.

Middleton, W. E. K. (1966) A history of the theories of rain and other forms of precipitation. Franklin Watts, New York, 223 pp.

Migeotte, M. V. (1948) Spectroscopic evidence of methane in the earth's atmosphere. *Physical Review* **73**, 519–520. *See also*: Migeotte, M. V. (1948) Methane in the Earth's atmosphere. *Astrophysical Journal* **107**, 400–403.

Migeotte, M. V. and L. Neven (1950) Détection du monoxyde de carbone dans l'atmosphère terrestre à 3580 mètres d'altitude. *Physica* **16**, 423–424.

Miller, D. P. (2004) Discovering water: James Watt, Henry Cavendish and the nineteenth-century "water controversy." Aldershot: Ashgate, 316 pp.

Miller, N. H. J. (1905a) The amount of nitrogen as ammonia and as nitric acid and of chlorine in the rain-water collected at Rothamsted. *Journal of Agricultural Science* **1**, 280–303.

Miller, N. H. J. (1905b) The amount of nitrogen as ammonia and nitrates, chlorine and sulphuric acid in the Rain-water at Rothamsted. In: The book of the Rothamsted Experiments (Ed. A. D. Hall), J. Murray, London, pp. 18–23.

Miller, N. H. B. (1913) The compositi-on of rainwater collected in the Hebrides and in Iceland. *Journal of the Scottish Meteorological Society* **3**, 141–158.

Minami, Y. and Y. Ishizaka (1996) Evaluation of chemical composition in fog water near the summit of a high mountain in Japan. *Atmos. Env.* **30**, 3363–3376.

Miquel, P. (1878) Des poussières organisées tenues en suspension dans l'atmospherique. *C. r.* **86**, 1552–1554.

Miquel, P. (1883) Les Progrès de la Micrographie atmosphérique. Les Organismes vivans de l'atmosphère. *Revue des Deux Mondes, 3e période* **57**, 442–454.

Mitchell, A. (1869) Report of Committee on Ozone Observations and methods of observing ozone. *Journal of the Scottish Meteorological Society* **2**, 252–253. Note on ozonometry. Ibid., pp. 256–258.

Mitchell, A. (1880) Report of the Ozone Committee. *Journal of the Scottish Meteorological Society, new series* **5**, 3–8.

Mitchill, S. M. (1795) Remarks on the gaseous oxide of azote or nitrigene and on the effects it produces when generated in the stomach, inhaled into the lungs and applied to the skin: being an attempt to ascertain the true thereupon of contagion the phenomena of fever. Printed by T. and J. Swords, New York, 43 pp.

Miyake, Y. (1939) The chemistry of rainwater. *Journal of the Meteorological Society of Japan* **17**, 20–37 (in Japanese, English abstract pp. 45–46).

Miyake, Y. (1948) The chemical nature of the saline matter in the atmosphere. *Geophysical Magazine* **16**, 64–65.

Miyake, Y. (1965) Elements of geochemistry. Maruzen Comp., Tokyo, 475 pp.

Miyake, Y. (1978) Geochemical study of the ocean and the atmosphere. Yasuo Miyake seventieth anniversary. Publ. by the Geochem. Lab., Met. Inst. Tokyo, Japan, 850 pp.

Miyake, Y. and K. Kawamura (1954) Studies on the atmospheric ozone at Tokyo. *Papers in Meteorology and Geophysics* **5**, 178–181.

Miyake, Y., K. Kawamura and S. Sakurai (1961) Ozone and nitrogen dioxide in an urban atmosphere. *Papers in Meteorology and Geophysics* **12**, 75–85.
Miyake, Y., K. Kawamura and Y. Sugiura (1962) Atmospheric ozone and nitrogen dioxide observed at Mt. Norikura. *Papers in Meteorology and Geophysics* **12**, 310–317.
Moeller, M. (1921) Das Ozon. Sammlung Vieweg Tagesfragen aus den Gebieten der Naturwiss. und d. Techn., Heft 52. Fr. Vieweg & Sohn, Braunschweig, 155 pp.
Moffat, T. (1853) On medical meteorology. *Association Medical Journal* **1**, 129–130.
Moffat, T. (1854) Medical meteorology and atmospheric ozone. *Association Medical Journal* **2**, 359.
Moffat, T. (1861) On atmospheric ozone. In: Report British Association for the Advancement of Science, J. Murray, London, pp. 88–89.
Moffat, T. (1862a) Medical meteorology and atmospheric ozone, as based on observations taken at Harwarden during a period of six years. *Report of the Council of the British Meteorological Society* **11**, 5–34.
Moffat, T. (1862b) On the cause of the loss of colour of ozone test-paper. *Report of the Council of the British Meteorological Society* **11**, 59–62.
Moffat, T. (1873) On atmospheric ozone and its sources. *Journal of the Scottish Meteorological Society* **4** (1873–1876), 99–107.
Möhl, H. and Th. Dietrich (1868) Beobachtungen über den relativen Ozongehalt. In: *Jahresbericht über die Fortschritte auf dem Gesammtgebiet der Agricultur-Chemie*. Die Luft. Band 10 für 1867, pp. 51–52.
Mohnen, V. A. and J. P. Lodge, Jr. (1969) Review and general survey of gas-to-particle conversion. In: Proceedings of 7th ICCN Conference, Prague/Vienna, September 1969, pp. 69–91.
Mohnen, V. A. (1980) Cloud water collection from aircraft. *Atmospheric Technology* **12**, 20–25.
Mohnen, V. A. (1989) Cloud chemistry research at Whiteface Mountain. *Tellus* **41B**, 79–91.
Mohnen, V. A. and J. A. Kadlecek (1989) Cloud chemistry research at Whiteface Mountain. *Tellus* **41B**, 79–91.
Mohnen, V. A. and R. J. Vong (1993) A climatology of cloud chemistry for the eastern United States derived from the cloud chemistry project. *Environmental Review* **1**, 38–54.
Mohr, F. (1854) Aelteste Nachricht über Ozon und seine Benennung. *Pogg. Ann.* **91**, 625–627.
Moissan, H. (1903) Sur le dosage de l'argon dans l'air atmosphérique. *C. r.* **137**, 600–604.
Moldan, B. and M. Veselý (1987) Chemical composition of atmospheric precipitation in Czechoslovakia, 1976–1984. I. Monthly samples. *Atmos. Env.* **21**, 2383–2395.
Moldan, B., J. Kopaček and J. Kopaček (1988) Chemical composition of atmospheric precipitation in Czechoslovakia, 1976–1984. II. Event samples. *Atmos. Env.* **22**, 1901–1908.
Moleschott, J. (1852) Kreislauf des Lebens. Physiologische Antworten auf Liebig's Chemische Briefe. Victor von Zahen, Mainz, 485 pp.
Moleschott, J. (1859) Physiologie der Nahrungsmittel. Ein Handbuch für die Diätetik. Ferber, Giessen. 572+254 pp. (on rainwater and snow water: pp. 387–390).
Molina, M. J. and F. S. Rowland (1974) Stratospheric sink for chlorofluoromethanes. Chlorine atom catalyzed destruction of ozone. *Nature* **249**, 810–812.
Molina, L. T., S. D. Schinke and M. J. Molina (1977) Ultraviolet absorption spectrum of hydrogen peroxide vapor. *Geophys. Res. Lett.* **4**, 580–582.
Möller, D. (1977) Der Schadstoff Schwefeldioxid. *Zeitschrift für die gesamte Hygiene* **23**, 310–316.
Möller, D. (1980) Kinetic model of atmospheric SO_2 oxidation based on published data. *Atmos. Env.* **14**, 1067–1076.
Möller, D. (1982) The global sulphur cycle. *Idöjaras* **87**, 121–143. *See also*: Möller, D. (1884) On the global natural sulphur emission. *Atmos. Env.* **18**, 29–39. Estimation of the global man-made sulphur emission. *Ibid.* **18**, 19–27.
Möller, D. (1985) Der globale biogeochemische Schwefelzyklus. In: Technik und Umweltschutz, Bd. 31 Umweltschutz in der Land- und Fortwirtschaft, VEB Dt. Verlag für Grundstoffindustrie, Leipzig, pp. 35–65.

Möller, D. (1988) Chemie der Atmosphäre und ihre Bedeutung im biogeochemischen Stoffkreislauf. In: Belastung von Ökosystemen und Organismen durch abiotische Umweltfaktoren (eds. E.-G. Mahn, Chr. Hänsel und K. Ungar), Kongr. und Tagungsber. MLU Halle, pp. 14–20.
Möller, D. (1989) The possible role of H_2O_2 in new-type forest decline. *Atmos. Env.* **23**, 1187–1193.
Möller, D. (1990) The Na/Cl ratio in rainwater and the sea salt chloride cycle. *Tellus* **42B**, 254–262.
Möller, D. (1996) Global sulfur and nitrogen biogeochemical cycles. In: Physics and chemistry of the atmosphere of the Earth's and other objects of the solar system, ERCA Vol. 2, les éditions de physique Les Ulis (Frankreich) (Ed. C. Boutron), pp. 125–156.
Möller, D. (1998) Rethinking the acid rain problem (Lecture hold at Int. Symp. on Env. Sci. and Sustainable Develop., Peking Univ., China, May 2–5, 1998). In: Environmental Sciences and Sustainable Development towards 21st Century (ed. X. Tang), China Science Press, Peking (China), 2000, pp. 284–296.
Möller, D. (1999a) Explanation for the recent dramatic increase of H_2O_2 concentrations found in Greenland ice cores. *Atmos. Env.* **33**, 2435–2437.
Möller, D. (1999b) Acid rain – gone? In: Prediction of atmospheric environmental problems between technology and nature (ed. D. Möller), Springer-Verlag, Berlin, pp. 141–178.
Möller, D. (2000) Troposphärisches Ozon: Entstehung, Konzentrationsvariabilität und Wirkung unter dem Gesichtspunkt der Ozonminderung. *Z. Umweltchem. Ökotox.* **12**, 201–208.
Möller, D. (2002a) Rethinking the tropospheric ozone problem. In: Problems of atmospheric boundary-layer physics and air pollution – to the 80th birthday of Professor M. E. Berlyand (ed. S. S. Chicherin), Hydrometeoizdat, St. Petersburg, pp. 252–269.
Möller, D. (2002b) Hydrogen peroxide trends in Greenland glaciers. In: Encyclopedia of global environmental change (ed. T. Munn) Vol. 1: The Earth's system: physical and chemical dimensions of global environmental change (eds. M. C. MacCracken and J. S. Perry), John Wiley & Sons, Chichester, pp. 447–450.
Möller, D. (2003) Luft. De Gruyter, Berlin, 750 pp.
Möller, D. (2004) The tropospheric ozone problem. *Archives of industrial hygiene and toxicology* **55**, 11–23.
Möller, D. (2006) Wieviel Chemie ist im Klima? Eine chemische Klimatologie. In: Klimawandel – vom Menschen verursacht? 8. Symposium Mensch – Umwelt (Ed. D. Möller). *Acta Academiae Scientiarum* **10**, 157–217.
Möller, D. (2008) On the history of the scientific exploration of fog, dew, rain and other atmospheric water. *Die Erde* **139**, 11–44.
Möller, D. (2009) Atmospheric hydrogen peroxide: Evidence for aqueous-phase formation from a historic perspective and a one-year measurement campaign. *Atmos. Env.* **43**, 5923–5936.
Möller, D. (2010) Chemistry of the climate system. De Gruyter, Berlin und New York, 722 pp.
Möller, D. (2011) Atmospheric chemistry – bridging the chemical air composition with the climate. *Idöjaras* **115**, 123–145.
Möller, D. (2014a) On the origin and meaning of the German word Luft and some meteorological terms concerning atmospheric water, especially fog. *Die Erde* **145**, 212–227.
Möller, D. (2014b) Zur Geschichte der Nebelforschung. In: Von Kometen, Windhosen, Hagelschlag und Wetterballons – Beiträge zur Geschichte der Meteorologie (eds. I. Kästner und J. Kiefer), Shaker Verlag, pp. 169–198.
Möller, D. (2014c) Le brouillard dans la "science des météores", depuis l'Antiquité jusqu'à l'époque moderne. In: La brume et le brouillard dans la science, la littérature et leart (eds. K. Becker and O. Leplatre), Hermann, Paris, pp. 65–102.
Möller, D. (2019) Chemistry of the climate system, 3rd ed. Vol. 1: Fundamentals and processes. De Gruyter, Berlin, 619 pp.
Möller, D. (2020) Chemistry of the climate system, 3rd ed. Vol. 2: History, change and sustainability. De Gruyter, Berlin, 776 pp.

Möller, D. and L. Horváth (1989) Estimation of natural acidity of precipitation water on global scale. *Idöjaras* **93**, 324–335.
Möller, D. and H. Lux, eds. (1992) Deposition atmosphärischer Spurenstoffe in der ehemaligen DDR. Kommission Reinhaltung der Luft, Schriftenreihe Band 18, Düsseldorf, 308 pp.
Möller, D. and G. Mauersberger (1990) Auswaschen von Gasen und Aerosolen durch Niederschläge unter Berücksichtigung einer komplexen Flüssigphasenchemie. 2. Modellergebnisse. *Z. Met.* **40**, 330–339.
Möller, D. and G. Mauersberger (1992) Cloud chemistry effects on tropospheric photooxidants in polluted atmosphere. *J. Atmos. Chem.* **14**, 153–165.
Möller, D. and H. Schieferdecker (1982) Zur Rolle des atmosphärischen Ammoniaks im biogeochemischen Stickstoff-Zyklus. *Zeitschrift für die gesamte Hygiene* **28**, 797–802.
Möller, D. and H. Schieferdecker (1989) Ammonia emission and deposition of NH_x in the GDR. *Atmos. Env.* **23**, 1187–1193.
Möller, D. and R. Zierath (1986) On the composition of precipitation water and its acidity. *Tellus* **38B**, 44–50.
Möller, D., K. Acker and W. Wieprecht (1993) Cloud chemistry at the Brocken in the Harz Mountains, Germany. *EUROTRAC Newsletters* **12**, 24–29.
Möller, D., K. Acker, W. Marquardt and E. Brüggemann (1996) Precipitation and cloud chemistry in the Neue Bundesländer of Germany in the background of changing emissions. *Idöjaras* **100**, 117–133.
Möller, D., K. Acker, W. Wieprecht and R. Auel (1997a) Study of the interaction of photo-oxidants and acidic components between gas and liquid phase. In: Cloud multi-phase processes and high Alpine air and snow chemistry (eds. S. Fuzzi and D. Wagenbach). Vol 5: Transport and Chemical Transformation of Pollutants in the Troposphere, Springer, pp. 138–145.
Möller, D., K. Acker and W. Wieprecht (1997b) Influence of cloud chemical processes on the mesoscale ozone budget. In: 7th Europ. Symp. on Physico-Chem. Behaviour of Atmos. Poll. "The Oxidizing Capacity of the Troposphere" (eds. B. Larsen und B. Versino), Report EUR 17482 EN, pp. 604–608.
Möller, D., K. Acker, D. Kalaß and W. Wieprecht (1999) Five-Year Record of Ozone at Mt. Brocken (Germany) – Implications for Changing Heterogeneous Chemistry. In Atmos. Env.al Research Critical Decisions between Technological Progress and Preservation of Nature (ed. D. Möller), Springer, Berlin, pp. 133–139.
Möller, D., W. Wieprecht, J. Hofmeister, D. Kalass, F. Elbing and M. Ulbricht (2001) Fog dissipation by dry ice blasting: process mechanism. In: Proceeding's 2nd Int. Conf. on Fog and Fog Collection, St. John's (Canada), pp. 485–491.
Möller, D., W. Wieprecht, J. Hofmeister, D. Kalass, F. Elbing und M. Ulbricht (2003) Fog dissipation by nucleation scavenging using ice particle blasting. Eight WMO Scientific Conference on Weather Modification. WMP Report No 39 (Technical Document WMO/TD – No. 1146), Geneve, pp. 389–392.
Möller, D., U. Biermann and X. Tian-Kunze (2002) Anstieg der atmosphärischen Konzentration von Wasserstoffperoxid als Konsequenz der Rauchgasentschwefelung? Research Report DFG-MO 540/3-3.
Monks, P. S. (2000) A review of the observations and origins of the spring ozone maximum. *Atmos. Env.* **34**, 3545–3561.
Monnett, O., G. St. J. Perrott and H. W. Clark (1926) Smoke-abatement investigations at Salt-Lake City, Utah. Bull. 254, Bureau of Mines, Washington, DC (USA). pp. 18–22, 33–37, 72–76, 85–86, 90–91, 94–95. *See also*: Monnett, O. (1920) The Salt Lake City Smoke Investigation. *American City* **23**, 14–16. Monnett, O. (1923) Smoke Abatement. U. S. Bureau of Mines, Technical Paper No. 273, pp. 9–13.

Mons, I. B. van (1817) Über einige Eigenheiten der verschiedenen Nebel. *Journal für Technische und Ökonomische Chemie* **2**, 41–48.
Mons, I. B. van (1827) Einige Eigenheiten der verschiedenen Nebel. *Archiv für die gesammte Naturlehre* **12**, 477–496.
Mons, I. B. van (1828) Einige Eigenheiten der verschiedenen Nebel. *Archiv für die gesammte Naturlehre* **13**, 55–70.
Montani (1865) [no title]. *C. r.* **67**, 193.
Monticelli, T. and N. Covelli (1825) Prodoma della mineralogia Vesuvian. Volume 1: Orittognosia. Da'torchi del Tramater, Napoli, 428 pp.
Moortgat, G. K., D. Grossmann, A. Boddenberg, G. Dallmann, A. P. Ligon, W. V. Turner, S. Gäb, F. Slemr, W. Wieprecht, K. Acker, M. Kibler, S. Schlomski and K. Bächmann (2002) Hydrogen peroxide, organic peroxides and higher carbonyl compounds determined during the BERLIOZ campaign. *J. Atmos. Chem.* **42**, 443–463.
Mordy, W. A. (1953) A note on the chemical composition of rainwater. *Tellus* **5**, 470–474.
More, L. T. (1941) Boyle as Alchemist. *Journal of the History of Ideas* **2**, 61–76.
Móring, A. and L. Horváth (2014) Long-term trend of deposition of atmospheric sulfur and nitrogen compounds in Hungary. *Idöjaras* **118**, 168–191.
Morley, E. W. (1895) On the densities of hydrogen and oxygen and on the ratio of their atomic weights. Smithsonian Institution Contributions to Knowledge, No. 980. Smithsonian Institution, Washington, D. C., 117 pp.
Morren, A. (1869) Sur quelques phénomènesde décompoisition produits per la lumiére. *C. r.* **69**, 397–400.
Morren, A. (1870) Sur les reactions chimique produits par la lumiére solaire. *Ann. chim. phys.* **21**, 323–349.
Morton, W. J. (1894) Ozone and its use in medicine. *Transactions of the New York Academy of Medicine, second series* **10**, 329–381.
Mosello, R., A. Marchetto and F. Decet (1992) Chemistry of atmospheric deposition and freshwater acidification: Research in Italy. In: Limnology in Italy (eds. P. Guilizzoni, G. Tartari and G. Giussani), Memorie dell'Istituto Italiano di Idrobiologia 50, pp. 417–455.
Moser, L. (1911) Die Darstellung und Bestimmung von Stickoxyd und sein Verhalten zu Wasser. *Z. anal. Chem.* **50**, 401–433.
Mosley, S. (2009) A network of trust': measuring and monitoring air pollution in British cities, 1912–1960. *Environment and History* **15**, 273–302.
Moss, W. A. (1871) On the estimation of nitrogeneous organic matter in air. *The Lancet* **100**, 627–630.
Moureu, C. (1923) Les gaz rares des gaz naturels. *J. Chem. Soc., Transactions* **123**, 1905–1947.
Moureu, C. and A. Lepape (1926) Titre de l'air atmosphérique en krypton et en xenon. *C. r.* **183**, 171–175.
Mrose, H. (1940) Chemische Untersuchung von Raureifschmelzwasser in Thüringen. *Zeitschrift für angewandte Meteorologie. Das Wetter* **57**, 241–254.
Mrose, H. (1941) Die Verbreitung baumbewohnender Flechten in Abhängigkeit vom Sulfatgehalt der Niederschlagwässer. *Bioklimatische Beiblätter, Meteorologische Zeitschrift* **8**, 58–60.
Mrose, H. (1949) Chemische Aufgaben der Meteorologie. *Z. Met.* **3**, 110–113.
Mrose, H. (1952) Über die Entstehung nicht-anthropogener Kondensationskerne. *Berichte des Deutschen Wetterdienstes (US-Zone)* **42**, 112–125.
Mrose, H. (1954) Bildung von Kondensationskernen im Sonnenlicht. *Z. angew. Met.* **2**, 82–86.
Mrose, H. (1955) Klima und Wetter in ihrer Wirkung auf den Menschen. Ergebnisse bioklimatischer Forschung. Ziemsen, Wittenberg, 106 pp. (2nd ed. 1956, 112 pp.).
Mrose, H. (1958) Die Kontrolle der Luftreinheit in der freien Atmosphäre. *Zeitschrift für Aerosol-Forschung und -Therapie* **7**, 205–217.

Mrose, H. (1959a) Ein chemisches Verfahren zur Rußbestimmung in atmosphärischer Luft. *Z. Met.* **11**, 139–142.
Mrose, H. (1959b) Reduzierende Spurenstoffe in der Atmosphäre. *Z. Met.* **13**, 51–58.
Mrose, H. (1961) Ergebnisse von Spurenstoff-Bestimmungen im Niederschlag. *Z. Met.* **15**, 46–54.
Mrose, H. (1962) Erfahrungen mit dem SO_2-Bestimmungsverfahren von Zepf und Vetter in Wahnsdorf bei Dresden. *Zeitschrift für die gesamte Hygiene und ihr Grenzgebiete* **8**, 767–773.
Mrose, H. (1966) Measurements of pH and chemical analyses of rain-, snow and fog water. *Tellus* **18**, 266–270.
Mrose, H. and W. Warmbt (1964) Bestimmung des SO_2-Gehaltes der Atmosphäre durch amperometrische Titration. *Z. angew. Met.* **4**, 345–355.
Mueller, S. P. and F. Weatherford (1988) Chemical deposition to a high elevation red spruce forest. *Water, Air and Soil Pollution* **38**, 345–363.
Mukharji, P. B. (2012) The "cholera cloud" in the nineteenth-century "British World": history of an object-without-an-essence. *Bulletin of the History of Medicine* **86**, 303–332.
Mulawa, P. A., S. H. Cadle, F. Lipari, C. C. Ang and R. T. Vandervennet (1986) Urban dew: Its composition and influence on dry deposition rates. *Atmos. Env.* **20**, 1389–1396.
Mulder, G. J. (1825) Rheno-Trajectini: Responsio ad quaestionem ab ordine matheseos et philosophiae naturalis, propositam: Quaeritur accurata analysis chemica aquarum pluviae, fontium et fluminum urbis Rheno-Trajectinae et agri suburbani, ut inde efficiatur, quales ad potum quotidianum sint saluberrimae, quales in artibus quibusdam usurpari possint, quales evitandae sint. Vol. 1. O. J. van Paddenburg and J. van Schoonhoven, Amsterdam, 158 pp.
Mulder, G. J. (1844) Versuch einer allgemeinen physiologischen Chemie. Aus dem Holländischen übersetzt von J. Moleschott. Heidelberg, C. F. Winter, 864 pp.
Mulder, E. (1885) *Over een effluve-ozonometer* en ontledingssnelheid van ozon. *Verslagen der Koninklijke Akademie van Wetenschappen Afdeling Natuurkunde*, Derde reeks, Eerste deel, pp. 400–408. *See also*: Sur un ozonomètre à effluve et sur la vitesse de dècomposiition de l'ozone. *Recueil des Travaux Chimiques des Pays-Bas* **4**, 135–146.
Müller, A. (1865) Ueber den Ammoniakgehalt der atmosphärischen Luft. *J. prakt. Chem.* **96**, 339.
Müller, A. (1873) Ueber die Einwirkung des Lichtes auf Wasser: Ein Versuch zur Erklärung der chemischen Lichtreactionen, des Gewitters und der Production organischer Substanz. Schabelitz'sche Buchhandlung, Zürich, 124 pp.
Müller, M. (1888) Über eine auffällige Zerstörung von aus Zinkblech gefertigten Fallröhren; über den Ammoniakgehalt des Meteorwassers in der kälteren Jahreszeit. *Z. angew. Chem.* **1**, 240–244.
Müller, R. (1862) Beitrag zur Kenntnis der Untersalpetersäure. *Ann.* **122**, 1–22.
Müller, R. (2009) A brief history of stratospheric ozone research. *Met. Z.* **18**, 3–24.
Mullin, T. (1989) Turbulent times for fluids. *New Science* **1690**, 52–55.
Mulvany, J. (1880) Ozon in nature. Its relations, sources and influences etc. from fifteen years' observations afloat and ashore, under all conditions of climate. *Q. J. Roy. Meteor. Soc.* **6**, 184–190.
Muncke, G. W. (1829) Hagel. In: Gehler's Physikalisches Wörterbuch, Band 5, pp. 30–82.
Munger, J. W., D. J. Jacob, J. M. Waldman and M. R. Hoffmann (1983) Fog water chemistry in an urban atmosphere. *JGR* **88**, 5109–5121.
Munger, J. W., D. J. Jacob and M. R. Hoffmann (1984) The occurrence of bisulfite-aldehyde addition. *J. Atmos. Chem.* **1**, 335–350.
Munger, J. W., C. Tiller and M. R. Hoffmann (1986) Identification of hydroxymethane-sulfonate in fog water. *Science* **231**, 247–249.
Munn, R. E. and H. Rodhe (1971) On the meteorological interpretation of the chemical composition on monthly precipitation samples. *Tellus* **23**, l–13.
Münnich, J. (1880) Ueber die Bestimmung der Kohlensäure in der Luft und dem von Dr. Hesse angegebene transportablen Apparat. *Deutsche militär-ärztliche Zeitschrift*, Berlin **9**, 97–117.

Müntz, A. (1881) Sur la présence de l'alcool dans le sol, dans les eaux, dans l'atmosphère. *C. r.* **92**, 499–502.

Müntz, A. (1891) Sur la répartition du sel marin suivant les altitude. *C. r.* **112**, 447–479.

Müntz, A. (1892) Ammoniaque dans l'atmosphere et dans les pluies. *C. r.* **114**, 184–186.

Müntz, A. and E. Aubin (1881a) Sur la proportion d'acide carbonique contenu dans l'air. *C. r.* **92**, 1229–1230.

Müntz, A. and E. Aubin (1881b) Sur les proportions d'acide carbonique dans les hautes régions de l'atmosphère. *C. r.* **93**, 797–799.

Müntz, A. and E. Aubin (1882a) De la distribution de l'ammoniaque dans l'air et les météores aqueux aux grandes altitudes. *C. r.* **95**, 788–790.

Müntz, A. and E. Aubin (1882b) Sur la proportion d'acid carbonique dans les hautes regions de l'atmosphére. *Ann. chim. phys., Ser. 5* **26**, 222–254.

Müntz, A. and E. Aubin (1884a) Determination d'acide carbonique de l'air effecttuée par la Mission dui Cap Horn. *C. r.* **98**, 487–494.

Müntz, A. and E. Aubin (1884b) Sur les composés carbonés combustibles existant dans l'air atmosphérique. *C. r.* **99**, 871–874.

Müntz, A. and E. Aubin (1886) Analyse de l'air pris au cap Horn. *C. r.* **102**, 421–423. *See also*: Recherches sur la constitution chemique de l'atmosphere, d'après les expériences de M. le Dr. Hyades'. In: Mission Scientifique de Cap Horn 1882–1883, Tome III, Gauthiers-Villars, Paris, pp. A2–A78. *See also*: Détermination des quantités d'oxygene contenue dans l'air. Ibid., pp. A83–A87.

Müntz, A. and E. Laine (1911) Les nitrates dans l'atmosphere des regions australes. *C. r.* **152**, 166–169.

Müntz, A. and V. Marcano (1889) Sur la proportion de nitrates contenus dans les pluies des regions tropical. *C. r.* **108**, 1062–1064.

Muselli, M., D. Beysens and E. Soyeux (2006) Is dew water potable? Chemical and biological analyses of dew water in Ajaccio (Corsica Island, France). *Journal of Environmental Quality* **35**, 1812–1817.

Nagy-Ilosva, L. I. de (1889) Se forme-t-il de l'ozone ou du peroxyde d'hydrogéne pendant la combustion vive, ou bien sout-ee les oxydes supérieur de l'azote qui se forment alors qu'on peut constater avee les reactions de l'acide azoteux etazotique. *Bulletin de la Societé chimique de Paris, Ser. 3*, **2**, 360–377. Ya-t-il de l'ozone et du peroxydes d'hydrogéne dans l'air. Ibid. 377–391. *See also*: Bildet sich Ozon oder Wasserstoffperoxyd bei der lebhaften Verbrennung? Kommen Ozon und Wasserstoffperoxyd in der Luft vor? *Naturwissenchaftliche Rundschau* **3** (1890) 37–40.

Nagy-Ilosva, L. I. de (1894) Ueber das in der Luft und in den atmosphärischen Niederschlägen vorkommende Wasserstoffhyperoxyd. *Ber.* **27**, 920–925.

Nahir, T. M. and G. A. Dawson (1987) Oxidation of sulfur dioxide by ozone in highly dispersed water droplets. *J. Atmos. Chem.* **5**, 373–383.

Narayanaswami, R. (1939) Some measurements of chloride, nitate and nitrite present in the water of the monsoon rains at Bombay. *Proceedings of the Indian Academy of Sciences* **9**, 518–525.

Nasse, O. (1869) Das Ozon. *Das Ausland* **42**, 390–394.

Nasse, O. (1870) Die sogenannten Ozonreaktionen und der Sauerstoff im thierischen Organismus. *Archiv für die gesamte Physiologie des Menschen und der Tiere* **3**, 204–214.

Nebbia, G. and G. Kauffman (2007) Prophet of solar energy: A retrospective view of Giacomo Luigi Ciamician (1857–1922), the founder of green chemistry, on the 150th anniversary of his birth. *Chemical Education* **12**, 362–369.

Neftel, A., P. Jacob and D. Klockow (1984) Measurements of hydrogen peroxide in polar ice samples. *Nature* **311**, 43–45.

Neftel, A., E. Moor, H. Oeschger and B. Stauffer (1985) Evidence from polar ice cores for the increase in atmospheric CO_2 in the past two centuries. *Nature* **315**, 45–47.
Neftel, A., P. Jacob and D. Klockow (1986) Long-term trend of hydrogen peroxide in polar ice cores. *Tellus* **38B**, 262–270.
Neftel, A., R. C. Bales and D. J. Jacob (1995) H_2O_2 and HCHO in polar snow and their relation to atmospheric chemistry. In: Ice core studies of global biogeochemical cycles (ed. R. Delmas). NATO-ASI Series I, Vol. 30, Springer, Berlin, pp. 249–264.
Negretti, H. and J. Zambra (1864) A treatise on meteorological instruments: Explanatory of their scientific principles, method of construction, and practical use. Vol. 3. Negretti & Zambra's Establ., London. 152 pp.
Nehls, H. (1939) Untersuchungen über die Menge der in den Niederschlägen enthaltenen Verunreinigungen. *Kleine Mitteilungen für die Mitglieder des Vereins für Wasser, und Lufthygiene* **15**, 122–126 and 204–208.
Neiburger, M. (1969) Artificial modification of clouds and precipitation. Technical Note No. 105, WMO, Geneve, 47 pp.
Neiburger, M. (1959) Meteorological aspects of oxidation type air pollution. In: The Rossby Memorial Volume. The Rockefeller Institute Press in assoc. with Oxford Univ., New York, pp. 158–169.
Nékám, L. A. (1890) Ueber die Untersuchung der organischen Substanz in der Luft. *Archiv für Hygiene und Bakteriologie* **11**, 396–409.
Neljubin, A. P. (1827) Chemische Untersuchung kleiner Aërolithen, die innerhalb der Hagelkörner enthalten waren, welche im Sterlitamak'schen Kreise des Orenburg'schen Gouvernement's aus der Luft niederfielen. *Archiv der gesamten Naturlehre* **10**, 378–387.
Nernst, W. (1919) Zur Anwendung des Einsteinschen Photochemischen Äquivalentgesetzes. *Z. Elektrochem.* **24**, 335–336.
Neumann, F. (1857) Ueber den Ozongehalt der Atmosphäre. *Pogg. Ann.* **178**, 614–618.
Newton, I. (1704) Opticks: or, A treatise of the reflections, refractions, inflexions and colours of light. Also two treatises of the species and magnitude of curvilinear figures. Sam Smith and Benj. Walford, printers to the Royal Society, London, 416 pp.
Nichol, S. E. (2018) Dobson spectrophotometer #17: past, present and future. *Weather and Climate (New Zealand)* **38**, 16–27.
Nichols, M. L. and C. W. Morse (1930) The reduction of nitric oxide. *J. phys. Chem.* **35**, 1239–1252.
Nicholson, A. W. (1881) Relative to atmospheric ozone and the best methods for its observation. *Annual report of the State Board of Health of Michigan 1880* **8**, 287–302.
Nicolet, M. (1945) Contribution à l'étude de la structure de l'ionosphère. *Institut royal météorologique de Belgique, Mémoires* **19**, 162 pp.
Nicolet, M. (1950) On the photodissociation of water vapour in the mesosphere. *Planetary and Space Science* **32**, 871–880.
Niebergall, C. (1858) Beobachtungen über den Ozongehalt der Luft in Arnstadt im Jahre 1855. In: Beneke, F. W. (1858) pp. 218–227.
Niki, H., P. D. Maker, C. M. Savage and L. P. Breitenbach (1977) Fourier transform IR spectroscopic observation of pernitric acid formed via HOO + NO2 → HOONO2. *Chemical Physics Letters* **45**, 564–566.
Nishimura, M. and K. Tanaka (1972) Sea water may not be a source of boron in the atmosphere. *JGR* **77**, 5239–5242.
Nobbe, F., ed. (1877) Die landwirthschaftlichen Versuchs-Stationen. Band XXII. Entwicklung und Tätigkeit der land- und forstwirthschaftlichen Versuchs-Stationen in den ersten 25 Jahren ihres Bestehens. Berlin, Wiegand, Hempel & Parey.
Nolan, J. J. and V. H. Guerrini (1936) The determination of the mass and size of the atmospheric condensation nuclei. *Trans. Faraday Soc.* **32**, 1175–1181.

Nolte, P. (1999) Ein Leben für die Chemie. 200 Jahre Christian Friedrich Schönbein. 1799–1999 (Ed. Arbeitskreis Stadtgeschichte der Volkshochschule Metzingen-Ermstal e. V., Metzingen). Leibfarth & Schwarz, Dettingen, 312 pp.

Nordenskiöld, A. E. (1874) Ueber kosmischen Staub, der mit atmosphärischem Niederschlag auf die Erdoberfläche herabfällt. *Pogg. Ann.* **151**, 154–165.

Nordenskiöld, A. E. (1892) Carl Wilhem Schele. Nachgelassene Briefe und Aufzeichnungen. P. A. Norstedt, Stockholm, 491 pp.

Norrish, R. G. W. (1929) Photochemical equilibrium in nitrogen peroxide. Part II. The dependence of quantum efficiency on wave-length. *J. Chem. Soc.* 1604–1611.

Norrish, R. G. W. (1934) Part II. Free radicals of short life: Chemical aspects. A. General and inorganic. The primary photochemical production of some free radicals. *Trans. Faraday Soc.* **30**, 103–113.

Norrish, R. G. W. and R. P. Wayne (1965) The photolysis of ozone by ultraviolet radiation. II. The photolysis of ozone mixed with certain hydrogen-containing substances. *Proc. R. Soc. Lond.* **288A**, 361–370.

Noxon, J. F. (1976) Atmospheric nitrogen fixation by lightning. *JGR* **3**, 463–465.

Obenland, E. (1953) Ozon-Untersuchungen im Hochgebirge. *Archiv für physikalische Therapie, Balneologie und Klimatologie* **5**, 91–99. *See also*: Vergleichsmnessungen des Ozongehalts in Oberstdorf und auf dem Nebelhorn. *Mitteilungen des Deutschen Wetterdienstes* **3** (1953).

Odén, S. (1968) The acidification of air and precipitation and its consequences on the natural environment. *Swedish Nat. Sci. Res. Council, Ecology Committee, Bul.* **1**, l–86.

Odén, S. (1976) The acidity problem – an outline of concepts. *Water, Air and Soil Pollution* **6**, 137–166.

O'Dowd, C. D. and M. H. Smith (1993) Physicochemical properties of aerosols over the northeast Atlantic: evidence for wind-speed-related submicron sea-salt aerosol production. *JGR* **98**(D1), 1137–1149.

O'Dowd, C. D. and G. de Leeuw (2007) Marine aerosol production: A review of the current knowledge. *Phil. Trans. A* **365**, 1753–1774.

Oddie, B. C. V. (1962) The chemical composition of precipitation at cloud levels. *Q. J. Roy. Meteor. Soc.* **80**, 535–538.

Odling, W. (1861) A manual of chemistry, descriptive and theoretical. Part 1. Longman, London, 380 pp.

OECD (1975) Photochemical oxidant air pollution: a report of the Air Management Sector Group on the problem of photochemical oxidants and their precursors in the atmosphere. Organisation for Economic Co-operation and Development, Paris, 93 pp.

Oeffinger, H. (1874) Bitte um Betheiligung an Ozonbeobachtungen. *Z. Öster. Ges. Met.* **9**, 137–138.

Ogburn, W. F. (1922) Social change with respect to culture and original nature. Viking Press, New York, 392 pp.

Ogiwara, S. and T. Okita (1952) Electron-microscope study of cloud and fog nuclei. *Tellus A* **4**, 233–240.

Ogren, J. and H. Rodhe (1986) Measurements of the chemical composition of cloudwater at a clean air site in central Scandinavia. *Tellus* **38B**, 190–196.

Ogura, Y. (1940) The chemical composition of the rainwater in Utsunomiya during 1937 and 1938. *Journal of the Meteorological Society of Japan* **18**, 107–111.

Okita, T. (1955) Interaction between aerosols and raindrops. *Journal of the Meteorological Society Japan* **33**, 220–226.

Okita, T. (1968) Concentration of sulfate and other inorganic composition in fog and cloud water and in aerosol. *Journal Meteorological Society Japan* **46**, 120–125.

Okita, T. and H. Tsuji (1958) Chloride content in precipitation at Asahikawa region in relation to air trajectories. *Journal of the Meteorological Society Japan, Ser. II* **36**, 73–75.

Okita, T., S. Morimoto, M. Izawa and S. Konno (1976) Measurement of inorganic gaseous and particulate nitrates in the atmosphere. *Atmos. Env.* **10**, 1085–1089.
Olszyna, K. J., J. F. Meagher and E. M. Bailey (1988) Gas-phase, cloud and rain-water measurements of hydrogen peroxide at a high-elevation site. *Atmos. Env.* **22**, 1699–1706.
Oltmans, S. J. and W. D. Komhyr (1986) Surface ozone distributions and variations from 1973–1994 measurements at the NOAA Geophysical Monitoring for Climatic Change baseline observatories. *JGR* **91**, 5229–5236.
Oltmans, S. J., A. S. Lefohn, D. Shadwick, J. M. Harris, H. E. Scheel, et al.(2013) Recent tropospheric ozone changes – A pattern dominated by slow or no growth. *Atmos. Env.* **67**, 331–351.
Oparin, A. I. [Опарин] (1924) Происхождение жизни [Origin of life]. Изд. Московский рабочий, Москва, 147 pp. In English: Oparin, A. I. (1938) Origin of life. Macmillan, New York, 270 pp.
Oparin, A. I. [Опарин] (1936) Возникновение жизни на Земле [Formation of life on Earth]. Издательство Академии наук СССР, Москва-Ленинград, 159 pp.; 2nd enlarged edition 1941, 267 pp.
Ortiz, V., M. A. Rubio and E. A. Lissi (2000) Hydrogen peroxide deposition and decomposition in rain and dew waters. *Atmos. Env.* **34**, 1139–1146.
Ortman, G. C. (1966) Monitoring methane in atmosphere with a flame ionization detector. *Analytical Chemistry* **98**, 644–646.
Osann, E. (1839) Physikalisch-medicinische Darstellung der bekannten Heilquellen der vorzüglichsten Länder Europa's. Erster Theil. Zweite vermehrte Auflage. F. Dümmler, Berlin, 578 pp.
Osann, G. (1848) Zur näheren Kenntnis des Ozons. *Pogg. Ann.* **75**, 386–392; and Ibid. 77 (1849) 592–596.
Osann, G. (1851a) Ueber Ozon-Reaction in der atmosphärischen Luft. *Gilb. Ann.* **82**, 158–160.
Osann, G. (1851b) Ueber den Ozon-Sauerstoff. *Pogg. Ann.* **82**, 537–544.
Osann, G. (1853) Beschreibung eines Ozonometers. *J. prakt. Chem.* **58**, 92–95.
Osann, G. (1855) Ueber die reducirende Wirkung des elektrolytisch ausgeschiedenen Wasserstoffgases. *Pogg. Ann.* **95**, 311–315.
Osann, G. (1865) Ueber Antozone. *J. prakt. Chem.* **95**, 55–58.
Osann, G. (1866) Üeber Blitzschlag. *Allgemeine medicinische Central-Zeitung* **35**, 619.
Osann, G. (1860) Ueber den Ozon-Wasserstoff und Ozon-Sauerstoff. *J. prakt. Chem.* **81**, 20–28.
Osann, G. (1864) Ueber den Ozon-Wasserstoff und Ozon-Sauerstoff. *J. prakt. Chem.* **92**, 20–31. [note that in some publications Osann is cited as H. Osann or Hofrath Osann; that is for Osann (1853, 1869, 1964).
Osman, W. A. (1958) Alessandro Volta and the inflammable-air eudiometer. *Annals of Science* **14**, 215–242.
Ostwald, Wi., ed. (1894) Chemische Abhandlung von der Luft und dem Feuer. Ostwald's Klassiker der exakten Wissenschaften. Nr. 58. Engelmann, Leipzig (1894), 112 pp. (this reprint is based on the 2nd German edition 1782; note all page information's are based on this German reprint).
Ostwald, Wo. (1909) Grundriss der Kolloidchemie. Theodor Steinkopff, Dresden, 525 pp.
Owens, J. S. (1922) Suspended impurity in the air. *Proc. R. Soc. Lond. Series A* **101**, 18–37.
Owens, J. S. (1924) The automatic measurement of atmospheric pollution. *Nature* **114**, 330–332.
Ozon (1949) Vorträge und Diskussionen anläßlich der Sondertagung "Ozon" am 17. und 18. IV. 1944 in Tharandt. Berichte des Deutschen Wetterdienstes in der US-Zone Nr. 11, Bad Kissingen, 71 pp.
Ozone (1936) Conference on atmospheric ozone held at Oxford, September 9th to 11th, 1936. Q. *J. R. Meteor. Soc.* **62**(S1). Gueron, J. and M. Pettre: Difficultiés d'ordre chemicique dans le dosage de l'ozone atmosphérique, pp. 2–6. Dobson: Measurement of atmiosphgeric ozone by photo-electric method, pp. 11–13. Chalonge, D.; Résumé of the discussion on methods of observations, pp. 13–14.

Paetzold, H. K. (1952) Erfassung der vertikalen Ozonverteilung in verschiedenen geographischen Breiten bei Mondfinsternissen. *Journal of Atmospheric and Terrestrial Physics* **2**, 183–188.
Paetzold, H. K. (1953) Die atmosphärische Ozonschicht und ihre vertikale Verteilung. *Umschau* **53**, 715–717.
Paetzold, H. K. and H. Zschörner (1955) Beobachtung eines "Ozonloches" über den Alpen. *Z. Met.* **9**, 250–251.
Paetzold, H. K. and E. Regener (1957) Ozon in der Erdatmosphäre. In: Geophysics II. Handbuch der Physik / Encyclopedia of Physics (ed. J. Bartels), Vol 10 / 48, Springer, Berlin, Heidelberg, pp. 370–426.
Pahl, S. (1996) Feuchte Deposition auf Nadelwälder in den Hochlagen der Mittelgebirge. *Berichte des Deutschen Wetterdienstes* **198**, Offenbach/M., Selbstverlag, 137 pp.
Pahl, S., P. Winkler, B. G. Arends, G. P. A. Kos, D. Schell, M. C. Facchini, S. Fuzzi, M. W. Gallagher, R. N. Colvile, T. W. Choularton, A. Berner, C. Kruisz, M. Bizjak, K. Acker and W. Wieprecht (1997) Vertical gradients of dissolved chemical constituents in evaporating clouds. *Atmos. Env.* **31**, 2577–2588.
Palmieri, L. (1863) Sull'ozono atmosferico. *Il Nuovo Cimento* **18**, 45–63.
Palmieri, L. (1865) Sull' ozono atmosferico nuove indagini. *Atti dell' Accademia delle scienze Fisiche e Matematiche*, Napoli, Vol. 2 no. 4, 16 pp.
Palmieri, L. (1866) Dell' ozono e dell' antiozono. *Rendiconto dell' Accademia delle scienze Fisiche e Matematiche*, Napoli, August, pp. 268–270.
Palmqvist, A. (1892) Undersökningar öfver atmosferens kolsyrehalt [Investigation of the atmospheric content of carbonic acid]. *Bih. t. K. Sv. Vet. Akad. Hand. [Bihang till Kungliga Svenska vetenskaps-akademiens handlingar]* **18**, Afd. II, Nr. 2, 3–39.
Paneth, F. A. (1937) The chemical composition of the atmosphere. *Q. J. R. Meteor. Soc.* **63**, 433–438.
Paneth, F. and J. Edgar (1938) Concentration and measurement of atmospheric ozone. *Nature* **142**, 112–113.
Paneth, F. and E. Glückauf (1937) Helium content of the stratosphere. *Nature* **136**, 717–718.
Paneth, F. and E. Glückauf (1941) Measurement of atmospheric ozone by a quick electrochemical method. *Nature* **147**, 614–615.
Paneth, F. and E. Glückauf (1946) The helium content of atmospheric air. *Proc. R. Soc. Lond. (A)* **185**, 89–98.
Papée, H. M. (1959) On the chemical composition of atmospheric aerosols – Part 1: some stoichiometric aspects of the problem. *Geofisica pura e applicate* **44**, 191–203.
Paping, B. J. (1796) Diss. med. de sulphure to calcis optimo contra salivationem mercurialem remedio. Groningen, 39 pp.
Paracelsus, Th. (1577) Aurora thesaurusque philosophorum. Accessit Monarchia physica per Gerardum Dorneum; in defensionem Paracelsicorum principiorum… Præterea Anatomia viva. In English: Paracelsus his Aurora, & Treasure of the Philosophers. As also The Water-Stone of The Wise Men; Describing the matter of and manner how to attain the universal Tincture. Faithfully Englished. And Published by J. H. Oxon. London, Printed for Giles Calvert and are to be sold at the Black Spred Eagle, at the West end of Pauls (1659) 229 pp.
Paracelsus, Th. (1616) AUREOLI Philippi Theophrasti Bombasts von Hohenheim Paracelsi, deß Edlen, Hochgelehrten, Fürtrefflichsten, Weitberümbtesten Philosophi und Medici OPERA: Bücher und Schrifften, so viel deren zur Hand gebracht: und vorwenig Jahren, mit und auß ihren glaubwürdigen eygener Handgeschriebenen Originalien collacioniert, vergliechen, verbessert: Und durch JOANNEM HUSERUM BRISGOIUM in zehen underschiedliche Theil, in Truck gegeben. Jetz von newem mit fleiß übersehen, auch mit etlichen bißhero unbekandten Tractaten gemehrt, und umb mehrer Bequemlichkeit willen, in zwen underschiedliche Tomos unnd Theil gebracht, deren Begriff und Ordnung, nach der Vorrede zu finden, sampt beyder Theilen fleißigen und volkommenen Registern, Band 1. Lazari Zetzners Seligen Erben, Straßburg, 1127 pp.

Parrish, D. D., K. S. Law, J. Staehelin, R. Derwent, O. R. Cooper, H. Tanimoto, A. Volz-Thomas, S. Gilge, H.-E. Scheel, M. Steinbacher and E. Chan (2012) Long-term changes in lower tropospheric baseline ozone concentrations at northern mid-latitudes. *ACP* **12**, 11485–11504.

Parrish, D. D., K. S. Law, J. Staehelin, R. Derwent, O. R. Cooper, et al. (2013) Lower tropospheric ozone at northern mid-latitudes: Changing seasonal cycle. *Geophys. Res. Lett.* **40**, 1631–1636.

Parrot, G. F. (1801) Grundzüge einer neuen Theorie der Ausdünstung und des Niederschlags des Wassers in der Atmosphäre. *Magazin für den neuesten Zustand der Naturkunde in Rücksicht auf die dazu gehörigen Hülfswissenschaften* **3**, 1–57.

Partington, J. R. (1928) The composition of water. Bell and Sons, London, 106 pp.

Partington, J. R. (1933) The scientific work of Joseph Priestley. *Nature* **131**, 348–350.

Partington, J. R. (1936) Joan Baptista van Helmont. *Annales of Science* **1**, 359–384.

Partington, J. R. (1956) The life and work of John Mayow (1641–1679): Part two. *Isis* **47**, 405–417.

Partington, J. R. (1961) A history of chemistry. Vol. 2. Macmillan, London, 795 pp.; Vol. 3 (1962) 854 pp., Vol. 4 (1964) 1007 pp. and (unfinished) Vol. 1 (1970) 370 pp.

Passerini, N. P. (1891) Sui materiali disciolti nell' aqua piovana precipitate negli annio 1888, 1889, 1890. Richerche chimiche institute presso la stazione meteologica della Scuola Agraria de Scandicci (Firenze). *Bolletino della Società Meteorologica Italiana* **2**, 11.

Passerini, N. P. (1893) Sopra la quantità di cloro contenuta nella pioggia caduta presso l'Osservatorio della Scuola Agraria de Scandicci (Firenze). *L'Agricoltura Italiana* **15**, 448–457.

Passerini, N. P. and G. D'Achiardi (1901) Sopra la pioggia melmosa (Pioggia di sangue) caduta in Firenze la sera del 10 Marzo 1901. Atti della Reale Accademia Economico-Agraria dei Georgofili di Firenze, Serie IV **24**, 137–153.

Pasteur, L. (1862) Die in der Atmosphäre vorhandenen organisierten Körperchen, Prüfung der Lehre von der Urzeugung. Ostwald's Klassiker der exakten Wiss, Nr. 39. W. Engelmann, Leipzig, 98 pp.

Pauling, L. (1932) The nature of the chemical bond. IV. The energy of single bonds and the relative electronegativity of atoms. *J. Am. Chem. Soc.* **54**, 3570–3582.

Paulot, F., D. J. Jacob, M. T. Johnson, T. G. Bell, A. R. Baker, W. C. Keene, I. D. Lima, S. C. Doney and C. A. Stock (2015) Global oceanic emission of ammonia: Constraints from seawater and atmospheric observations. *Global Biogeochemical Cycles* **29**, 1165–1178.

Pavelin, E. G., C. E. Johnson, S. Rughooputh and R. Toumi (1999) Evaluation of pre-industrial surface ozone measurements made using Schönbein's method. *Atmos. Env.* **33**, 919–929.

Pawlak, I. and J. Jarosławski (2014) Analysis of surface ozone variations based on the long-term measurement series in Kraków (1854–1878), (2005–2013) and Belsk (1995–2012). In: Achievements, history and challenges in geophysics (eds. R. Bialik, M. Majdański and M. Moskalik), GeoPlanet: Earth and Planetary Sciences, Springer, pp. 341–333.

Payen, A. (1866) Sur l'iodine de potassium. *C. r.* **61**, 466–473, 512–514.

Pease, R. T. and W. P. Munro (1934) The slow oxidation of propane. *J. Am. Chem. Soc.* **56**, 2034–2038.

Péligot, E. (1847) Sur un procede propre a determiner d'une maniere sensible rapide la quantite d'azote contenue dans les substances organique. *Journal de Pharmacie et de Chimie, ser. 3* **11**, 334–337.

Péligot, E. (1857) Ueber die Zusammenstzung der natürlichen Wässer. *J. prakt. Chem.* **71**, 393–402. *See also*: Études sur la composition des eaux *C. r.* **54**, 193–202.

Pendleton, E. M. (1874) Text-book of scientific agriculture with practical deductions. A. S. Barner & Co, New York. 419 pp.

Peng, Z., Z. Jianbo, Y. Peng, Z. Limin, H. Min and Y. Zhang (2003) Variation concentration of atmospheric peroxides in Beijing and Guangzhou. *Acta Scientiarum Naturalium Universitatis Pekinensis* **39**, 871–879.

Peng, Y. P., K. S. Chen, C. H. Lai, P. J. Lu and J. H. Kao (2006) Concentrations of H_2O_2 and HNO_3 and O_3-VOC-NO_x sensitivity in ambient air in southern Taiwan. *Atmos. Env.* **40**, 6741–6751.

Penkett, S. A. (1972) oxidation of SO_2 and other atmospheric gases by O_3 in aqueous solution. *Nature Physical Sciences* **240**, 105–106.

Penkett, S. A., B. M. R. Jones, K. A. Brice and A. E. J. Eggleton (1979) The importance of atmospheric O_3 and H_2O_2 in oxidizing SO_2 in cloud and rainwater. *Atmos. Env.* **13**, 123–137.

Perner, D. and U. Platt (1979) Detection of nitrous acid in the atmosphere by differential optical absorption. *Geophys. Res. Lett.* **6**, 917–920.

Pernter, J. M. (1881) Ozonbeobachtungen. *Z. Öster. Ges. Met.* **16**, 394–397.

Pernter, J. M. (1900) Das moderne Wetterschießen. Roth, Stuttgart, 16 pp.

Pernter, J. M. (1907) Das Ende des Wetterschießens. *Meteorologische Zeitschrift* **24**.

Pernter J. M. and W. Trabert (1901) Untersuchungen über das Wetterschiessen. *Meteorologische Zeitschrift* **17**, 385–414.

Perros, P. E. (1993) Large scale distribution of hydrogen peroxide from aircraft measurements during Tropoz II experiment. *Atmos. Env.* **27A**, 1695–1708.

Peter, R. (1849) Remarks on ozone and its supposed influence in the production of epidemic diseases, especially of epidemic cholera. *Transylvania Medical Journal* **1**, 148–157. *See also*: *The American Journal of the Medical Sciences* **37** (1850) 274–275.

Petermann, A. and J. Graftiau (1891) Recherches sur la composition de l'atmosphère. Acide carbonique. Combinaisons azotées contenues dans l'air atmosphérique et dans l'eau de pluie. Première partie: Acide carbonique contenu dans l'air atmosphérique, par A. In: *Mémoires couronnés et autres mémoires publiés par l'academie royal de Belgique*. Tom XLVII (1892–1893), 79 pp.

Petermann, A. and J. Graftiau (1893) Combinaisons azotées continue dans les eaux météoriques. *Bulletin Agronomique et des Stations de Recherches de Gembloux* **52**, 5–26.

Petrenchuk, O. P. [Петренчук] (1979a) О балансе морских солей и серы в атмосфере [The balance of sea salt and sulfur in the atmosphere]. *Труды ГГО* **418**, 43–54.

Petrenchuk, O. P. [Петренчук] (1979b) Экспериментальные исследования атмосферного аэрозоля [Experimental investigation of atmospheric aerosol]. Leningrad, Gidrometeoizdat, 264 pp.

Petrenchuk, O. P. and E. S. Selezneva [Петренчук, Селезнева] (1962) Изменение концентрации основных химических примесей в осадках в зависимости от метеорологических условий [Measurements of the fundamental chemical composition of precipitation in dependence from meteorological factors]. *Труды ГГО* **134**, 14–25.

Petrenchuk, O. P. and V. M. Drozdova (1966) On the chemical composition of cloud water. *Tellus* **18**, 280–286.

Petrenchuk, O. P. and N. V. Nesterova [Петренчук, Нестерова] (1979) Об изменении химического состава атмосферных осадков и их кислотности [On measurement of the chemical composition of precipitation and its acidity]. *Труды ГГО* **418**, 55–63.

Pettenkofer, M. (1851) Ueber den Unterschied zwischen Luftheizung und Ofenheizung in ihrer Einwirkung auf die Zusammensetzung der Luft der beheizten Räume. *Polytechnisches Journal (Dingler's J.)* **119**, 40–51.

Pettenkofer, M. (1855) Untersuchungen und Beobachtungen über die Verbreitungsart der Cholera nebst Betrachtungen über Maßregeln derselben Einhalt zu thun. J. G. Cotta, München, 374 pp.

Pettenkofer, M. (1862) Ueber eine Methode, die Kohlensäure in der atmosphärischen Luft zu bestimmen. *J. prakt. Chem.* **85**, 165–179.

Pettersson, O. (1886) Luftanalyse nach einem neuen Prinzip. *Z. anal. Chem.* **25**, S. 467.

Pettersson, O. and A. Palmqvist (1887) Ein tragbarer Apparat zur Bestimmung des Kohlensäure-Gehaltes der Luft. *Ber.* **20**, 2129.

Pfaff, C. H. (1822) Ueber das sogenannte färbende Wesen in der Ostseeluft und dem Ostseewasser, und die desoxydirende Kraft der Wasserdämpfe. *Journal für Physik und Chemie* **36**, 68–73. *See also*: Nachtrag über das sogenannte färbende Wesen der Ostseeluft. Ibid., 325–328.

Pfaff, E. R. (1862) Welchen Einfluß hat der Ozongehalt der Luft auf die Krankheiten der Menschen? Adolph Henke's Zeitschrift für die Staatsarzneikunde. *Neue Folge* **42**, No. 2, *ges. Folge* **83** 189–202.

Pfeffer, W. (1897) Pflanzenphysiologie. Ein Handbuch der Lehre vom Stoffwechsel und Kraftwechsel in der Pflanze. Erster Band: Stoffwechsel. Verlag W. Engelmann, Leipzig. 639 pp.

Pham, M., J.-F. Müller, G. P. Brasseur, C. Granier and G. Mégie (1995) A three-dimensional study of the tropospheric sulfur cycle. *JGR* **100**, 26061–26092.

Pictet, M.-A. (1822) Hagel mit metallischem Kern, angeblich gefallen in Irland im Juni 1821. *Gilb. Ann.* **72**, 456.

Pieraggi, E. (1867) Ozone à Bois de Colombes, prés Paris, an avril 1867. *Annu. soc. mét. de France* **15**, 195–196.

Pierre, I. (1851) Note relative à l'emploi du sel sur les terres. Du sel contenu dans les terres non réputées terres salées et dans les eaux de pluie. Gide et J. Baudry, Paris, 12 pp; Extrait des *Annales agronomiques* **1**, 466–478 (mai 1851).

Pierre, I. (1852) Mémoire sur l'ammoniaque de l'atmosphère. *C. r.* **34**, 878–879.

Pierre, I. (1859) Chimie agricole ou l'agriculture considéreé dans ses rapports principaux avec la chimie. Paris, Libr. Agricole de la Maison Rustique, 2nd ed., 532 pp.

Pierrou, U. (1976) The global phosphorus cycle. In: Nitrogen, phosphorus and sulphur – global cycles (eds. B. H. Svensson and R. Søderlund), Swedish Natural Science Research Council, pp. 75–88.

Pierson, W. R., W. W. Brachaczek, R. A. Gorse Jr., S. M. Japar and N. M. Norbeck (1986) On the acidity of dew. *JGR* **91**, 4983–4996.

Pierson, W. R. and W. W. Brachaczek (1990) Dew chemistry and acid deposition in Glendora, California, during the 1986 Carbonaceous Species Methods Comparison Study. *Aerosol Science and Technology* **12**, 8–27.

Pietra Santa, P. (1864) Variabilité des propriétés de l'air atmosphérique. *C. r.* **58**, 1158–1159.

Pietra Santa, P. (1878–1895) L'ozone atmosphérique. *Journal d'hygiene* **3**, 266, 282, 295; **20** (1895) 133, 145, 159.

Pincus, S. (1871) Ungewöhnliche Ozonbildung. *Pogg. Ann.* **144**, 480.

Pincus, S. and J. Rollig (1867) Untersuchungen über den Stickstoff- resp. Ammoniak- und Salpetersäuregehalt der atmosphärischen Niederschläge. *Landwirtschaftliche Versuchs-stationen* **9**, 465–476 (cited after Wulff 1920, p. 91).

Pirovarov, J. J. [Пироваров, Я. Я.] (1906) К вопросу об аэральном происхождении солей в почва [On the question of salt desosition from air into soils]. *Почвоведение* (Pedology) (1906) 1–4, p. 67 (cited after Drischel (1940) concerns sulfate) and *Ann. Geol. Min. Russie.* **56** (1908) p. 274 (cited after Clarke (1911) concerns chloride); written as *Pirovaroff.*

Pissis, A. (1878) Report upon the Desert of Atacama, its geology and mineral products. In: Nitrate and guano deposits in the Desert of Atacama, Taylor and Francis, London, pp. 1–30.

Plass, G. N. (1953) The carbon dioxide theory of climatic change. *Bulletin of the American Meteorological Society* **34**, 80.

Plass, G. N. (1956a) Carbon dioxide and the climate. *American Scientist* **44**, 302–316.

Plass, G. N. (1956b) The carbon dioxide theory of climatic change. *Tellus* **8**, 140–154.

Platt, U., D. Perner, J. Schröder, C. Kessler and A. Tönnissen (1981) The diurnal variation of NO_3. *JGR* **86**, 11965–11970.

Plesničar, B. (2005) Progress in the chemistry of dihydrogen trioxide (HOOOH). *Acta Chimica Slovenia* **52**, 1–12.

Pless, J. and V. Pierre (1856) Beiträge zur Kenntnis des Ozons and des Ozongehaltes der atmosphärischen Luft. *Sitzungsber. Wien* **22**, 211–235.

Plessow, K., K. Acker, H. Heinrichs and D. Möller (2001) Time study of trace elements and major ions during two cloud events at the Mt. Brocken. *Atmos. Env.* **35**, 367–378.

Pletcher, R. H., D. A. Anderson and J. C. Tannehill (1997) Computational fluid mechanics and heat transfer. 2nd ed. CRC Press. 816 pp.
Plieninger, Th. (1850) Beobachtungen über das atmosphärische Ozon. In: I. Angelegenheiten des Vereins. 1. Vierte Generalversammlung am 30. April 1849 zu Ulm. *Jahreshefte des Vereins für vaterländische Naturkunde in Württemberg* **5**, 168–171.
Plotnikov, J. (1910) Photochemie. Wilhelm Knapp, Halle, 182 pp.
Plotnikov, J. (1920) Lehrbuch der allgemeinen Photochemie. Berlin, 730 pp.
Poey, A. (1863a) Enumération de 214,471 observations horaires faitres à l'Observatoire physico-météorique de la Havana, pendant l'année 1862. *Annu. soc. mét. de France* **11**, 51–53.
Poey, A. (1863b) Expérience sur l'ozone ou l'oxygéne naissant exhale par les plantes er répandu dans l'air de la campagne et de la ville. *Annu. soc. mét. de France* **11**, 150–154. *See also*: *C. r.* **57**, 344–348.
Poey, A. (1865) Description d'un ozonographe et d'un actino-graph desines a enregistrer de demi heure en demi heurel'ozone atmospherique a l'action chimique de la lumiere ambiant. *C. r.* **61**, 1107–1111.
Poey, A. (1867) Remarques sur les colorations ozonoscopiques obtenues à l'aide du réactif de Jame (de Sedan) et sur l'échelle ozonométrique de M. Bérigny. *C. r.* **65**, 708–712.
Polkowska, Ż., M. Błaś, K. Klimaszewska, M. Sobik, S. Małek and J. Namieśnik (2008) Chemical characterization of dew water collected in different geographic regions of Poland. *Sensors* **8**, 4006–4032.
Polli, G. (1850) Esperienze sull'ozono dell'aria atmosferica. Annali di chimica applicati alla medicina etc. *Milano* **3**, 136–151. And: *Annali universali di medicina*, Milano **134** (1850) 155–165.
Pollock, J. A. (1909) Les ions dans l'atmosphère. *Le Radium* **6**, 129–135.
Pollock, J. A. (1915a) The nature of the large ions in the air. *Phil. Mag. Series 6* **29**, 514–526.
Pollock, J. A. (1915b) The larger Ions in the air. *Nature* **95**, 286–288.
Popper, K. R. (1962) Conjectures and Refutations. The growth of scientific knowledge. Basic Books, New York, 412 pp.
Pošepný, F. (1877) Über den Ursprung der Salze abflussloser Gebiete. *Zeitschrift der Deutschen Geologischen Gesellschaft* **29**, 636–637.
Pošepný, F. (1878) Zur Genesis der Salzablagerungen, besonders jener im nordamerikanischen Westen. *Sitzungsber. Wien* **76** (1878), 179–212.
Posnjakov, A. [Позняков] (1904) Опыт исследования химического состава осадков в зависимости от метеорологических факторов [Investigation on the chemical composition of atmospheric precipitations as dependent upon meteorological factors]. Журнал опытной агрономии [Journal of Experimental Agronomy] **5**, 740–788 (ref. in *Zentralblatt der Agrikultur-Chemie* **34**, 289–290).
Prados-Roman, C., C. A. Cuevas, R. P. Fernandez, D. E. Kinnison, J.-F. Lamarque and A. Saiz-Lopez (2015) A negative feedback between anthropogenic ozone pollution and enhanced ocean emissions of iodine. *ACP* **15**, 2215–2224.
Prather, M. J., X. Zhu, Q. Tang, J. Hsu and J. L. Neu (2011) An atmospheric chemist in search of the tropopause. *JGR* **116**, D04306, 10 pp.
Preiss, M., R. Maser, D. Schell, W. Jaeschke, K. Acker, W. Wieprecht and D. Möller (1994) Measurements of S(IV) and H_2O_2 in gas and liquid phase at Great Dun Fell. In: Proceedings of EUROTRAC Symposium 094. SPB Academic Publishing, pp. 1129–1132.
Prestel, M. A. F. (1858) Ueber den Morrrauch des Jahres 1857. *Petermann's Geographische Mittheilungen* **4**, 106–110.
Prestel, M. A. F. (1865) Die jährliche, periodische Änderung des atmosphärischen Ozons und die ozonoskopische Windrose als Ergebniss der Beobachtungen zu Emden von 1857–1864. *Novorum actorum Academiae Caesareae Leopoldino-Carolinae Germanicae Naturae Curiosorum* or: *Verhandlungen der Kaiserlichen Leopoldinisch-Carolinischen Deutschen Akademie der Naturforscher* (E. Blochmann & Sohn, Dresden), Vol. 32, 14 pp. + 2 Figs.

Prestel, M. A. F. (1866) Die jährliche Periode der Ozonreaction auf der nördlichen Hemisphäre. *Z. Öster. Ges. Met.* **1**, 289–291.
Prestel, M. A. F. (1868) Ueber den Moorrauch in seiner weiten geographischen Verteilung und die durch ihn verusachten phantasmoskopischen Erscheinungen im Luftmeere. *Z. Öster. Ges. Met.* **3**, 326–333. *See also*: Ueber die Ursache der Trübung der Luft in der ersten Hälfte des Juli. Ibid. pp. 465–471.
Prestel, M. A. F. (1872) Der Boden, das Klima und die Witterung von Ostfriesland. Th. Hahn, Emden, 438 pp.
Prestel, M. A. F. (1874a) Ueber das Unzureichende der jetzt gebräuchlichen Ozonometer. *Z. Öster. Ges. Met.* **9**, 93.
Prestel, M. A. F. (1874b) Die periodische Veränderung des Ozongehalts der Luft im Laufe des Jahres. *Z. Öster. Ges. Met.* **9**, 166–167.
Prettner, J. (1874) Ozongehalt der Luft zu Klagenfurt. *Z. Öster. Ges. Met.* **9**, 167–168.
Price, W. C. (1942) Absorption spectra and absorption coefficients of atmospheric gases. *Reports on Progress in Physics* **9**, 10–17.
Priestley, J. (1767) History and Present State of Electricity, with Original Experiments. Printed for J. Dodley, J. Johnson and B. Davenport and T. Cadell, London, 745 pp.
Priestley, J. (1772) Observations on different kinds of air. *Phil. Trans.* **62**, 147–264. Separate print by W. Bowyer and H. Nichols, London, 120 pp.
Priestley, J. (1774) Experiments and observations of different kinds of air. Vol. I. J. Johnson, London, 325 pp.
Priestley, J. (1775) Experiments and observations of different kinds of air. Vol. II. J. Johnson, London, 325 pp. *See also*: The discovery of oxygen. Part I. Experiments by Joseph Priestley. Experiments. Alembic Club Reprints No. 7, reissue edition (pp. 29–103 of Vol. II). Livingstone, Edinburgh (1947), 55 pp.
Priestley, J. (1783) Experiments to phlogiston and the seeming conversion of water into air. *Phil. Trans.* **73**, 398–434 and 426–427.
Priestley, J. (1785) Experiments and observations relating to air and water. *Phil. Trans.* **75**, 279–309.
Priestley, J. (1788) Experiments and observations relating to the principle of acidity, the composition of water and phlogiston. *Phil. Trans.* **78**, 147–157.
Priestley, J. (1789) Experiments on the phlogistication of spirit of nitre. *Phil. Trans.* **79**, 139–149.
Priestley, J. (1790) Experiments and observations on different kinds of air and other branches of natural philosophy, connected with the subject. In three volumes; being the former six volumes abridged and methodized which many additions. Th. Pearson, Birmingham, Vol. 1. 411 pp.
Priestley, J. (1799) Experiments and observations relating to the analysis of atmospherical air. *Transactions of the American Philosophical Society of the Royal Society* **4**, 1–11.
Prince, C. L. (1854) Description of a box for ozone observations. *Association Medical Journal* **2**, 195–196.
Pring, J. N. (1914) The Occurrence of ozone in the upper atmosphere. *Proc. R. Soc. Lond. A* **90**, 204–219.
Pringsheim, E. (1887) Ueber die chemische Wirkung des Lichts auf Chlorknallgas. *Wied. Ann.* **268**, 384–428.
Pringal, E. (1908) Über den wesentlichen Einfluss von Spuren nitroser Gase auf die Kondensation von Wasserdampf. *Ann. Phys.* **331**, 727–750 (Dissertation Marburg 1907).
Proeschel, F. (1860) Ozone in the air. *The Sydney Morning Herald*. Fr. 23 Nov., p. 8.
Prokofewa, I. A. [Прокофьева] (1951) Атмосферный озон [Atmospheric ozone]. Издательство Академия наук СССР, 231 pp.
Prout, W. (1834) The bridgewater treatises on the power wisdom and goodness of God as manisfested in the creation. Treatise VIII. Chemistry, meteorology and the function of digestion. 2nd ed. W. Pickering, London, 571 pp.

Prout, W. (1836) Chemie, Meteorologie und verwandte Gegenstände als Zeugnisse für die Herrlichkeit des Schöpfers (Die Wunder der Natur ihre Geheimnisse, oder die Bridgewater-Bücher, 2. Band). Paul Neff, Stuttgart, 384 pp.

Puchner, H. (1892) Untersuchungen über den Kohlensäuregehalt der Atmosphäre. *Forschungen auf dem Gebiet der Agricultur-Physik (Wollny's Forschungen)* **15**, 296–383.

Quetelet, A. (1864) Observations des phénomènes périodiques. Mémoires de l'Académie royale des sciences, des lettres et des beaux-arts de Belgique. Hayez, Bruxelles, 44 pp.

Quinn, E. L. and C. L. Jones (1936) Carbon dioxide. Amer. Chem. Monograph Ser. Reinhold Publ., New York, 294 pp.

Quitmann, E. (1935) Über die Bestimmung von Gasspuren in Luft mit dem Waschrohr nach H. Cauer. *Z. anal. Chem.* **103**, 258–261.

Quitmann, E. (1937) Bestimmung von Schwefelwasserstoffspuren in der Luft. *Z. anal. Chem.* **109**, 241.

Quitmann, E. (1938) Über den Reduktionswert der Luft. *Z. anal. Chem.* **114**, 1–8.

Quitmann, E. (1940) Über die Bestimmung von Luftbeimengungen und die Bedeutung chemischer Untersuchungsmethoden für Meteorologie und Klimatologie. *Z. angew. Chem.* **53**, 80–82.

Quitmann, E. and H. Cauer (1939) Verfahren zur chemischen Analyse der Nebelkerne der Luft. *Z. anal. Chem.* **116**, 81–91.

Radau, R. (1877) Essais at notices. Les analysis de l'air a l'observatoire de Montsouris. *Revue des Deux Mondes* **19**, 949–958.

Radford, W. H. (1938) II. An instrument for sampling and measuring liquid fog water. In: Houghton, H. G. and W. H. Radford (1938) On the measurement of drop size and liquid water content in fogs and clouds. Paper in Physical Oceanography and Meteorology, Mass. Inst. of Technol. and Woods Hole Oceanographic Institution, Vol. VI, No. 4, pp. 19–31.

Rafinesque, C. S. (1819) Thoughts on atmospheric dust. *American Journal of Science* **1**, 397–400.

Raga, G. B. and P. R. Jonas (1995) Vertical distribution of aerosol particles and CCN in clear air around the British Isles. *Atmos. Env.* **29**, 673–684.

Ragona, D. (1868) Sui coefficienti ozonometrici dell'umidita' e della temperatura. *Annuario della Società dei Naturalisti in Modena* **3**, 57–66.

Rahn, K. A., C. Brosset, B. Ottar and E. M. Patterson (1982) Black and white episodes, chemical evolution of Eurasian air masses and long-range transport of carbon to the arctic. In: Particulate carbon: atmospheric life cycle (eds. G. T. Wolff and R. L. Klimisch), Plenum Publ. Corp., pp. 327–342.

Ramann, E. (1908) Ueber den Nachweis von Rauchschäden. *Allgemeine Forst- und Jagdzeitung* **84**, 233–236.

Ramanathan, K. R. (1956) Atmospheric ozone and the general circulation of atmosphere. In: Scientific proceedings of the International Association of Meteorology, September 1954 Rome, London, pp. 3–26.

Ramazzini, B. (1691) De Costituzione Anni M.DC.LXXXX Ac de Rurali epidemia, Quae Mutitenensis Agri, et vicinarum Regionum, Colonus graviter afflixit. Dissertatio. Ubi quoque Rubiginis natura disquiritur, quae fruges, et fructus vitiando aliquam caritatem Annonae intulit. D. Antonium Maliabechium Mutinae, Typis Haeredum Cassianai Impress.

Ramazzini, B. (1700) De Morbis Artificum Diatriba. Mutinae: Typis Antonii Capponi. Translated by W. C. Wright as diseases of workers. The University of Chicago Press, 1940. (Reprinted in 1964, New York, Hafner).

Ramazzini, B. (1717) Opera omnia medica & physiologica: accessit vita autoris. Sumptibus Cramer & Perachon, Genevae, 888 pp.

Rammelsberg, C. (1873) Ueber das Verhalten des Ozons zum Wasser. *Ber.* **6**, 603–605.

Ramsay, W. (1896) The gases of the atmosphere. The history of their discovery. Macmillan, London, 264 pp. Extended 4th ed. in 1915, 306 pp.

Ramsay, W. (1907) Die Gase der Atmosphäre und die Geschichte ihrer Entdeckung. Verlag W. Knapp, Halle, 160 pp.

Rankama, K. and Th. G. Sahama (1950) Geochemistry. University of Chicago Press, 911 pp.

Raschig, F. (1904) Zur Theorie des Bleikammerprozesses. *Z. angew. Chem.* **17**, 1777–1785. *See also*: ibid., **18** (1905) 1281–1323.

Raschig, F. and W. Prahl (1929) Beitrag zur Chemie der Stickstoffoxyde. *Z. angew. Chem.* **42**, 253–257.

Rasmussen, R. A. (1972) What do the hydrocarbons from trees contribute to air pollution? *Journal Air Pollution Control Association* **22**, 537–543.

Rasmussen, R. A. and M. A. K. Khalil (1981) Increase in the concentration of atmospheric methane. *Atmos. Env.* **15**, 883–886.

Rasool, S., ed. (1973) Chemistry of the lower atmosphere. Springer, 335 pp. Also published in Russian: Химия нижней атмосферы (1976). Под ред. С. Расула. М.: Мир, 408 pp.

Rauch, H. (1920) Der Wert der zur Bestimmung des Kohlensäuregehaltes der Luft benutzten Apparate mit besonderer Berücksichtigung des Aeronom (Draeger-Werk). *Zeitschrift Für Hygiene und Infektionskrankheiten* **91**, 1–28.

Raum, J. (1889) Der gegenwärtige Stand unserer Kenntnisse über den Einfluss des Lichtes auf Bacterien und auf den thierischen Organismus. *Zeitschrift für Hygiene* **6**, 312–368.

Rayleigh, J. W. S. (1901) Spectroscopic notes concerning the gases of the atmosphere. *Philosophical Magazine* **1**, 100–105.

Rayleigh, J. W. S. and W. Ramsay (1896) Argon, a new constituent of the atmosphere. Smithsonian Institute, Washington D. C., 43 pp.

Reclus, É. (1874) The ocean, atmosphere and life. Being the second series of a descriptive history of the life on the globe. Harper & Brothers, New York, 534 pp.

Redtenbacher, J. (1842) Analyse der Meteorsteine von Ivan. Ann. 41, 308–315.

Reese, C. L. (1884) Comparative oxidation of solutions of sulphurous acid and sodium sulphite. *Chemical News* **50**, 219.

Regener, E. (1906) Über die chemische Wirkung kurzwelliger Strahlung auf gasförmige Körper. *Ann. Phys.* **325** (Vierte Folge 20), 1033–1046.

Regener, E. (1946) Über das "photochemische Klima" der Erde. *Naturwissenchaften* **33**, 163–166.

Regener, E. (1948) Das atmosphärische Ozon. In: Geophysik, Teil 2. Naturforschung und Medizin in Deutschland 1936–1946 (ed. J. Bartels), Dieterich'sche Verlagsbuchhandlung, Wiesbaden, pp. 296–307.

Regener, E. (1952) Über Schwankungen des Ozons in der Troposphäre und Stratosphäre. *Journal of Atmospheric and Terrestrial Physics* **2**, 173–182.

Regener, E. and V. H. Regener (1934) Aufnahme der ultravioletten Sonnenspektrums in der Stratosphäre und vertikale Ozonverteilung. *Physikalische Zeitschrift* **35**, 788–793.

Regener, V. H. (1938a) Messungen der Ozongehaltes der Luft in Bodennähe. *Met. Z.* **55**, 459–462.

Regener, V. H. (1938b) Neue Messungen der vertikalen Verteilung des Ozons in der Atmosphäre. *Zeitschrift für Physik* **109**, 642–670.

Regener, V. H. (1951) Vertical distribution of atmospheric ozone. *Nature* **167**, 276–277.

Regnault, H. V. (1852) Recherches sur la composition de l'air atmosphérique. *Ann. chim. phys.* **36**, 385–405. *See also* the German summary: Ludwig, H. (1853) Zusammensetzung der atmosphärischen Luft. *Arch. Pharm.* **126**, 149–156.

Reich, F. (1858) Beschreibung eines Apparates zur leichten und schnellen Bestimmung des Gehaltes einer Luft an schwefelsaurem Gase. *Berg- und Hüttenmännische Zeitung* **17**, 2–4. *See also*: Die bisherigen Versuche zu Beseitigungen des schädlichen Einflusses des Hüttenrauches bei den fiskalischen Hüttenwerken zu Freiberg. Ibid., pp. 165–168, 173–176.

Reich, A. (1917) Leitfaden für die Rauch- und Rußfrage. R. Oldenbourg, München und Berlin, 383 pp.

Reichardt, E. (1863) Untersuchung von Hagel. *Zeitschrift deutscher Landwirthe, Neue Folge* **14**, p. 368 and 1864, p. 103), cited after: *Jahresbericht über die Fortschritte auf dem Gesamtgebiete der Agricultur-Chemie* **7** (für 1864) 1866 p. 70.

Reichardt, E. (1884) Bestimmung des Kohlensäuregehaltes der Luft. *Arch. Pharm.* **222**, 414–415.

Reid, D. B. (1844) Illustrations of the theory and practice of ventilation. Longmanns, London, 451 pp.

Reimann, E. J. (1857) Das Luftmeer. Eine physikalische Darstellung für gebildete Laien. Verlag H. Scheube, Gotha, 334 pp.

Reinau, E. (1930) Über den Kohlensäuregehalt erdnaher Luftschichten im Hochgebirge. *Beiträge zur Geophysik* **25**, 178–193.

Reinsch, H. (1865) Qualitative Untersuchung der atmosphärischen Luft. *Z. anal. Chem.* **4**, 457–459; *see also*: *Neues Jahrbuch für Pharmacie und verwandte Fächer* **25** (1865) 193–202.

Reinsch, H. (1869) Ueber die quantitative Bestimmung der Nebenbestandtheile der atmosphärischen Luft. *Z. anal. Chem.* **8**, 483–485.

Reiset, J. (1879) Recherches sur la proportion de l'acide carbonique dans l'air. *C. r.* **88**, 1007–1012.

Reiset, J. (1880) Recherches sur la proportion de l'acide carbonique dans l'air. *C. r.* **90**, 1144–1148.

Reiset, J. (1882) Recherches sur la proportion de l'acide carbonique dans l'air. *Ann. chim. phys.* **26**, 145–221.

Reiss, H. (1950) The kinetics of phase transition in binary systems. *J. Phys. Chem.* **18**, 840–848.

Reiter, R. (1960) Meteorobiologie und Elektrizität der Atmosphäre. Akademische Verlagsgesellschaft, Leipzig, 424 pp.

Reiter, R. (1970) On the causal relation between nitrogen-oxygen compounds in the troposphere and atmospheric electricity. *Tellus* **22**, 122–136.

Reith, R. (2011) Umweltgeschichte der Frühen Neuzeit. Enzyklopädie deutscher Geschichte, Bd. 89. Oldenbourg, München, 196 pp.

Remsen, I. (1881) Report on organic matter in the air. *Annual Report of the National Board of Health* **1880**, 308–318.

Remsen, I. (1883) On the transformation of ozone into oxygen by heat. *American Chemical Journal* **4**, 50–53.

Remy, H. (1965) Lehrbuch der anorganischen Chemie. Vol. I. 12. Auflage. Akademische Verlagsgesellschaft, Leipzig, 1120 pp.

Renger, F. (1944) Tagesgang von Ozon und Nitrit bodennaher Luft unter gleichzeitiger Messung der UV-Strahlung [Diurnal variation of ozone and nitrite in surface-near air with simultaneous measurement of UV radiation]. Doctoral Thesis, University Breslau. Note the doctoral thesis is not available and obviously lost in the chaos of the war.

Renger, F. and O. Lucke (1953) Über die meteorologischen Bedingungen der Ozondichte in bodennaher Luft. In: Abhandlungen des meteorologischen und hydrologischen Dienstes der Deutschen Demokratischen Republik, Nr. 13 (Band II), Akademie-Verlag, Berlin, 44 pp.

Renk, F. (1886) Die Luft. In: Handbuch der Hygiene und der Gewerbekrankheiten. Erster Theil. Individuelle Hygiene. 2. Abtheilung. 2. Heft (eds. M. v. Pettenkofer and H. v. Ziemssen), F. C. W. Vogel, Leipzig, 242 pp.

Renner, E. and W. Rolle (1989) Modelling of the formation of photooxidants by a Lagrangian Grid Cell Model under characteristic conditions of central Europe. *Atmos. Env.* **23**, 1841–1847.

Reslhuber, A. (1855) Über den Ozongehalt der atmosphärischen Luft. *Sitzungsber. Wien* **14**, 336–344.

Reslhuber, A. (1856) Untersuchungen über das atmosphärische Ozon. *Sitzungsber. Wien* **21**, 351–378.

Revelle, R. and H. E. Suess (1957) Carbon dioxide exchange between atmosphere and ocean and the question of an increase of atmospheric CO_2 during the past decades. *Tellus* **9**, 18–27.

Reynolds, W. C. and W. H. Taylor (1912) The decomposition of nitric acid by light. *J. Chem. Soc., Trans.* **101**, 131–140.

Reynolds, W. C. (1923) Thunderstorms and Ozone. *Nature* **112**, 396–397.

Reynolds, W. C. (1930a) Notes on London and suburban air. *Journal of Society of the Chemical Industry* **49**, 168T–172T.

Reynolds, W. C. (1930b) Globular lightning. *Nature* **125**, 413.

Reynolds, W. C. (1938) Concentration and measurement of atmospheric ozone. *Nature* **142**, 571.

Rice, O. K. (1935) The kinetics of homogeneous gas reactions. In: Annual survey of American chemistry, Vol 9, 1934 (ed. C. J. West), Reinhold Publ. Corp., New York, pp. 35–48.

Richards, L. W. (1995) Airborne chemical measurements in nighttime stratus clouds in the Los Angeles Basin. *Atmos. Env.* **29**, 27–46.

Richardson, A. (1893) The action of light in preventing putrefactive decomposition;and in inducing the formation of hydrogen peroxide in organic liquids. *J. Chem. Soc., Transactions* **63**, 1109–1139.

Rideal, E. K. (1920) Ozone. Van Nostrand, New York, 198 pp.

Rieche, A. (1936) Die Bedeutung der organischen Peroxide für die chemische Wissenschaft und Technik. F. Enke, Stuttgart, 72 pp.

Riedel, K., R. Weller, O. Schrems and G. König-Langlo (2000) Variability of tropospheric hydroperoxides at a coastal surface site in Antarctica. *Atmos. Env.* **34**, 5225–5234.

Riesenfeld, E. H. (1927) Das Ozon, seine Bildung und Verwendung. *Die Naturwissenschaften* **28**, 777–784.

Riesenfeld, E. H. and G. M. Schwab (1922) Über Ozon. *Die Naturwissenschaften* **10**, 470–471.

Rigby, M. and E. Z. Sinha (1961) Annotated bibliography on precipitation chemistry. *Meteorological, Geoastronomical and Physical Abstracts* **12**, 1430–1495.

Risler, E. (1882) Recherches sur la quantité d'acide carbonique contenue dans l'air atmosphérique. *Archives des Sciences Physiques et Naturelles, Genève* **8**, 241–243.

Risse, O. (1929) Einige Bemerkungen zum Mechanismus chemischer Röntgenreaktionen in wässrigen Lösungen. *Strahlentherapie* **34**, 578–581. *See also*: Über die Röntgenphotolyse des Hydroperoxyds. *Z. phys. Chem.* **A 140** (1929) 133–157.

Robbins, R. C., R. D. Cadle and D. L. Eckhardt (1959) The conversion of sodium chloride to hydrogen chloride in the Atmosphere. *Journal of Meteorology* **16**, 53–56.

Robinson, E. and R. C. Robbins (1970) Gaseous sulphur pollutants from urban and natural sources. *Journal Air Pollution Control Association* **20**, 233–235.

Robert, A. J. (1855) Histoire du choléra de Neudorf 1855. *Gazette médicale de Strasbourg*, Avril (cited after Fox 1872, p. 133).

Robinet, S. (1863) Quelques faits pour servir a l'etude de l'eaux de pluie. *Comptes Rendus de l'Académie des Sciences* **57**, 493–494 and *Journal de Pharmacie et de Chimie* (3e ser.) **45**, 174–276.

Rocha, M. (1875) Del ozono: su importancia en meteorologia médica;métodos dos ozonométricos;un ozonógrafo automática. *Anales de la Asociacion Larrey* **1**, 100–105 and 132–136.

Rodhe, H. (1978) Budgets and turn-over times of atmospheric sulfur compounds. *Atmos. Env.* **12**, 671–680.

Rodhe, H. (2013) Bert Bolin (1925–2007) – a world leading climate scientist and science organizer. *Tellus B* **65**, 1, doi:10.3402/tellusb.v65i0.20583.

Rodionov, S. F., A. L. Oscherobich and E. V. Rdyltvskaka [Родионов, Ошерович, Рдултвская] (1949) О простом приборе для озонометрических исследований [On a simple instrument for ozonometric observations]. *Доклады Академий Наук СССР* **66**, 381.

Rogers, W. B. (1858) On ozone observations. *Edinburgh New Philosophical Journal, new series* **7**, 35–42.

Rogers, L. H. (1958) Report on photochemical smog. *Journal of Chemical Education* **36**, 310–313.

Römer, F. G., J. W. Viljeer, L. Van Den Beld, H. J. Slangewal, A. A. Veldkamp, N. V. Kema and H. F. R. Reijnders (1985) The chemical composition of cloud and rainwater. Results of preliminary measurements from an aircraft. *Atmos. Env.* **19**, 1847–1858.

Roscoe, H. E. (1864) Note on the amount of carbonic acid contained in the air of Manchester. *Chemical News and Journal of Industrial Sciences* **9**, 80.
Roscoe, H. E. and C. Schorlemmer (1871) Kurzes Lehrbuch der Chemie nach den neuesten Ansichten der Wissenschaft. 3. Auflage. Vieweg, Braunschweig, 426 pp.
Rose, H. (1831) A manual of analytical chemistry. Translated from the German by John Griffin. Th. Tegg, London, 454 pp.
Rosing, A. and G. Wankel (1870) Ammoniakmängden i regn sne, kilders, floders og søers vand samt Havvander. *Kongelige Norske Videnskabers Selskab Skrifter* **6**, 5–28.
Rosner, F. (1987) Moses Maimonides and air pollution. *New York State Journal of Medicine* 87(10), 574; *see also*: The life of Moses Maimonides, a prominent medieval physician (2002) *Quarterly Journal of Biology and Medicine* **19**, 125–128.
Ross, H. B., C. Johansson and C. de Serves (1992) Summertime diurnal variations of atmospheric peroxides and formaldehyde in Sweden. *J. Atmos. Chem.* **14**, 411–423.
Rossby, C. G. and H. Egnér (1955) On the chemical climate and its variation with the atmospheric circulation pattern. *Tellus* **8**, 983–987.
Roster, G. (1889) L'acido carbonico dell'aria e del suolo di Firenze. Firenze, coi tipi dei successori Le Monnier, 987 pp.
Rotch, A. L. and E. C. Pickering (1889) Observations made at the: with a statement of the local weather predictions under the direction of A. Lawrence Rotch. *Annals of the Astronomical Observatory of Harvard College* **20**, 148–267.
Roth, H. D. (1989) The beginnings of organic photochemistry. *Z. angew. Chem.* **28**, 1193–1207.
Roth, H. D. (2001) Twentieth century developments in photochemistry. Brief historical sketches. *Pure Applied Chemistry* **73**, 395–403.
Rothmund, V. (1917) Über das sogenannte Antozon. *Z. Elektrochem.* **23**, 170–173.
Rothmund, V. and A. Burgstaller (1913) Über die Geschwindigkeit der Zersetzung des Ozons in wässeriger Lösung. *Monatshefte für Chemie* **34**, 665–692.
Rothmund, V. and A. Burgstaller (1917) Die Reaktion zwischen Ozon und Wasserstoffperoxyd. *Monatshefte für Chemie* **38**, 295–303.
Rubin, M. B. (2001) The history of ozone. The Schönbein period, 1839–1868. *Bulletin for the History of Chemistry* **26**, 40–56.
Rubin, M. B. (2002) The history of ozone. II. 1868–1899. *Bulletin for the History of Chemistry* **26**, 40–56.
Rubin, M. B. (2003) The history of ozone. III. C. D. Harris and the introduction of ozone into organic chemistry. *Helvetica Chimica Acta* **86**, 930–940.
Rubin, M. B. (2007) The history of ozone. V. Formation of ozone from oxygen at high temperature. *Bulletin for the History of Chemistry* **32**, 45–56.
Rubin, M. B. (2009) The history of ozone. VII. The mythical spawn of ozone: antozone, oxozone and oxohydrogen. *Bulletin for the History of Chemistry* **34**, 39–59.
Rubio, M. A., M. J. Guerrero, V. Guillermo and E. Lissi (2006) Hydroperoxides in dew water in downtown Santiago, Chile. A comparison with gas-phase values. *Atmos. Env.* **40**, 6165–6172.
Rubner, M. (1885) Calorimetrische Untersuchungen. *Zeitschrift für Biologie* **21**, 250–334.
Rubner, M. (1906) Über trübe Wintertage nebst Untersuchungen zur sog. Rauchplage der Großstädte. *Archiv für Hygiene* **57**, 323–378.
Rubner, M. (1907) Lehrbuch der Hygiene. Systematische Darstellung der Hygiene und ihrer wichtigsten Untersuchungs-Methoden. 8. Auflage, Deuticke, Leipzig und Wien, 1029 pp.
Rubner, M., M. v. Gruber and M. Ficker (1911) Handbuch der Hygiene. 1. Band. S. Hirzel, Leipzig, 788 pp.
Rudeck, E. (1886) Ueber Ozon, mit Berücksichtigung seines Werthes für uns im Haushalt der Natur. *Chemiker Zeitung* **79**, 1225–1226. *See also*: Ozon und Ozonometer. *Jahresbericht über die Fortschritte der Pharmokognosie, Pharmacie und Toxikologie* **46** (*Neue Folge* **21**), 1887, 126–127. Furthermore: *Archiv für Pharmacie* **225** (1887) 225.

Rudeck, E. (1902) Ozon und Bestimmung ozonhaltiger Luft. *Die medicinische Woche und balneologische Centralzeitung* Nr. 35 and 86, pp. 147 and 152–153.
Russell, A. C. (1984) Whigs and professionals. *Nature* **308**, 777–778.
Russell, F. A. R. (1880) London fogs. Edward Stanford, London, 44 pp.
Russell, F. A. R. (1881) Dust and fogs. *Nature* **23**, 267–268.
Russell, F. A. R. (1888) Smoke in relation to fogs in London. *Nature* **39**, 34–36.
Russell, F. A. R. (1889) The causes and character of haze. *Nature* **41**, 60–65.
Russell, E. J. (1917) Dr. N. H. J. Miller. *Nature* **98**, 392–393.
Russell, E. J. (1919) Agricultural chemistry and vegetable physiology. *Annual Reports on the Progress of Chemistry* **16**, 171–196.
Russell, F. A. R. (1895) The atmosphere in relation to human and health. In: Annual Report of the Smithsonian Institution, Gov. Printing Office, Washington D. C., pp. 203–348.
Russell, F. A. R. (1897) Haze, fog and visibility. *Q. J. R. Meteor. Soc.* **23**, 10–24.
Russell, W. J. (1892) A sketch of the chemical history of the air. *Transactions of the Sanitary Institute* **13**, 224–242.
Russell, W. J. and W. Lapraik (1877) Experiments on the decomposition of nitric oxide by pyrogallate of potash. *J. Chem. Soc.* **32**, 35–38.
Russell, W. J. and W. Watson (1891) Town fogs and their effects. *Nature* **45**, 10–16.
Russell, E. J. and E. H. Richards (1919) The amount and composition of rain falling at Rothamsted. *Journal of Agriculture Science* **9**, 309–337.
Russell, E. J. (1942) Rothamsted and its Experimental Station. *Agricultural History* **16**, 161–183.
Rust, J. N. (1824) Recherches historiques, chemiques et médicale sur l'air marécageaux. (Historische, chemische und ärztliche Unters. über die Sumpfluft etc. Preisschrift.) Ouvrage couronné par l'académie des Sciences de Lypon. Par J. S. E. Julia, Professeur de chemie médicale, commissaire examinateur de la marine pour le service de santé etc. à paris. 1823, 155 S. 8. In: Kritisches Repertorium für die gesammte Heilkunde (ed. J. N. Rust), Dritter Band, Berlin, 138–144.
Rutherford, D. (1772) Dissertatio inauguralis de aere fixo dicto, aut mephitico: quam annuente summo numine: ex auctoritate reverendi admodum viri, Gulielmi Robertson, S. S. T. P. Academiae Edinburgenae Praefecti: nec non amplissimi senatus academici consensu, et nobilissimae facultatis medicae decreto: pro gradu doctoratus, summisque in medicina honoribus et privilegiis rite et legitime consequendis, eruditorum examini subjicit / eruditorum examini subjicit Daniel Rutherford, Britannus. Balfour & Smellie, Edinburgh. 25 pp.
Rutt, J. T. (1831) The theological and miscellaneous works of Joseph Priestley. Vol. 1. Part 1. Contains memoirs and correspondences. G. Smalefield, Hackney, 424 pp.
Ryaboshapko, A. G. (1983) Atmospheric sulfur cycle. In: The global biogeochemical sulphur cycle and influence on it of human activity (eds. M. V. Ivanov and J. R. Freeney) [in Russian; the English title is original additional printed], Nauka Publ., Moskau 1983, pp. 170–255. *See also*: Skryabin et al. (1983).
Ryaboshapko, A. G., V. V. Sukhenko and S. G. Paramanov (1994) Assesssment of wet sulphur deposition over the former USSR. *Tellus* **46B**, 205–219.
Sachs, K. (1911) Enzyklopädisches Französisch-Deutsches und Deutsch-Französisches Wörterbuch. Langenscheidt, Berlin, 856 und 1160 pp.
Sachse, K. T. (1861) Ueber das Vorkommen des atmosphärischen Ozons im Jahr 1859 nach Dresdener Beobachtungen. *Jahresberichte der Gesellschaft für Natur- und Heilkunde zu Dresden 1858–1861*, pp. 56–68 (cited after Hellmann 1883, p. 420); volume not available.
Sainte-Claire Deville, Ch. (1865a) Des perturbations périodiques de la temperature dans les mois de février, mai, août et novembre. *C. r.* **61**, 350–357. *See also*: *Annu. soc. mét. de France* **19** (1871) 95–100.

Sainte-Claire Deville, Ch. (1865b) Remarques sur une communication de M. Houzeau, relative à l'ozone atmosphérique. *C. r.* **61**, 1150–1152.
Sakugawa, H. and I. R. Kaplan (1987) Collection of atmospheric H_2O_2: comparison of cold trap method with impinger bubbling method. *Atmos. Env.* **21**, 1791–1798.
Sakugawa, H., I. R. Kaplan, W. Tsai and Y. Cohen (1990) Atmospheric hydrogen peroxide. *Environ. Sci. Technol.* **24**, 1452–1462.
Sakugawa, H. and I. R. Kaplan (1993) Comparison of H_2O_2 and O_3 content in atmospheric samples in the San Bernardino mountains, Southern California. *Atmos. Env.* **27**, 1509–1515.
Sala, A. (1613) Anatomia Vitrioli, in duos tractatus divisa. Officina Aureliae, Fabriana, Allobrogum. 75 pp.
Salisbury, R. A. (1807) An account of a Storm of salt which fell in January 1803. *Transactions of the Linnean Society* **8**, 289–290.
Saltzman, B. E. (1954) Colorimetric microdetermination of nitrogen dioxide in atmosphere. *Analytical Chemistry* **26**(12), 1949–1955.
Samiha, R., M. Ahmed, M. Shohel and A. Salam (2018) Chemical composition and source characterization of hailstones in Dhaka, Bangladesh. *Journal of Geoscience and Environment Protection* **6**, 71–82.
Sauer, F., G. Schuster, C. Schäer and G. K. Moortgat (1996) Determination of H_2O_2 and organic peroxides in cloud- and rain-water on the Kleiner Feldberg during FELDEX. *Geophys. Res. Lett.* **23**, 2605–2608.
Sauer, F., S. Limbach and G. K. Moortgat (1997) Measurements of hydrogen peroxide and individual organic peroxides in the marine troposphere. *Atmos. Env.* **31**, 1173–1184.
Sauer, F., C. Schäfer, P. Neeb, O. Horie and G. K. Moortgat (1999) Formation of hydrogen peroxide in the ozonolysis of isoprene and simple alkenes under humid conditions. *Atmos. Env.* **33**, 229–241.
Sauer, F., J. Beck, G. Schuster and G. K. Moortgat (2001) Hydrogen peroxide, organic peroxides and organic acids in a forested area during FIELDVOC94. *Chemosphere, Global Change Science* **3**, 295–307.
Saussure, H. B. de (1783) Essais sur l'hygrométrie. S. Fauché, Neuchatel, pp. 524. German translation and edition by A. J. von Oettingen: Versuch über die Hygrometrie. I. Heft (1. Versuch. Beschreibung eines neuen vergleichbaren Hygrometers. 2. Versuch. Theorie der Hygrometrie). Ostwald's Klassiker der exakten Naturwiss. Bd. 115. W. Engelmann, Leipzig. 168 pp. and: Versuch über die Hygrometrie. II. Heft (3. Versuch. Theorie der Ausdünstung. 4. Versuch. Anwendung der vorhergehenden Theorie auf einige Phänomene in der Meteorologie). Ostwald's Klassiker der exakten Naturwissenschaften. Vol. 119. W. Engelmann, Leipzig, 170 pp.
Saussure, H. B. de (1796) Voyages dans les Alpes, précédés d'un essai sur l'histoire naturelle des environs de Genève. Tome IV. Fouche-Boree, Neuchâtel, pp. 199–201.
Saussure, Th. de (1802) Observations sur le changement qu'éprouve le gaz carbonique par l'étincelle et sur la décomposition du même gaz par le gaz hydrogène. *Journal de Mines* **12**, 103–109.
Saussure, Th. de (1804) Recherches chimiques sur la végétation. Didot jeune, Paris, 327 pp. German translation by A. Wieler: Chemische Untersuchungen über die Vegetation. Erste Hälfte and zweite Hälfte. Ostwald's Klassiker der exakten Naturwissenschaften (1890), Vol. 15, W. Engelmann, Leipzig, 96 pp. and Vol. 16, 113 pp.; *see also* English translation: Hill (2013). Note, page references in the text are related to the French original edition.
Saussure, Th. de (1816) Ueber den verschiedenen Gehalt der atmosphärischen Luft an Kohlensäure, im Winter und im Sommer. *Gilb. Ann.* **54**, 217–231. *See also*: *Bibliothèque Universelle de Genève* **1**.
Saussure, Th. de (1830) Ueber die Schwankungen des Kohlensäure-Gehalts der Atmosphäre. *Neues Journal für Chemie und Physik* (*Schweiggers J.*) **61**, 17–43 and 129–164. *See also*: *Pogg. Ann.* **19**, 391–433.

Saxena, V. K., R. E. Stogner, A. H. Hendler, T. P. DeFelice, R. J. Y. Yeh and N. H. Lin (1989) Monitoring the chemical climate of the Mt. Mitchell State Park for evaluation of its impact on forest decline. *Tellus* **41B**, 92–109.

Saxena, V. K. and N.-H. Lin (1990) Cloud chemistry measurements and estimates of acidic deposition on an above cloud base coniferous forest. *Atmos. Env.* **24A**, 329–352.

Schaefer, V. J. (1946) The production of ice crystals in a cloud of supercooled water droplets. *Science* **104**, 457–459.

Schaper, C. W. L. (1867) Ueber das Ozon mit Rücksicht auf die Meteorologie und Heilkunde. *Zeitschrift des Königlich Preussischen Statistischen Bureaus* **7**, 167–171.

Scharnow, U., W. Berth and W. Keller (1965) Wetterkunde. VEB Verlag für Verkehrswesen, Berlin, 415 pp.

Scharrer, K. and W. Schropp (1940) Ueber den Stickstoffgehalt der Niederschläge. *Forschungsdienst* **3**, 469–472.

Scheel, H. E., H. Areskoug, H. Gieß, B. Gomiscek, K. Granby, et al.(1997) On the spatial distribution and seasonal variation of lower-troposphere ozone over Europe. *J. Atmos. Chem.* **28**, 11–28.

Scheele, C. W. (1777) Chemische Abhandlung von der Luft und dem Feuer. Nebst einem Vorbericht von Tobern Bergmann. Verlegt von Magn. Swederus, Buchhändler; zu finden bey S. L. Crusius, Upsala und Leipzig, 155 pp.

Scheele, C. W. (1780) Chemical observations and experiments on air and fire. With a preparatory introduction, By Tobern Bergman. Translated from the German by J. R. Forster. J. Johnson, London, 259 pp. *See also*: Discovery of oxygen. Part 2. Experiments by Carl Wilhelm Scheele (1777). Alembic Club Reprints No. 8, Edinburgh (1901), 46 pp.

Scheele, C. W. (1782) Chemische Abhandlung von der Luft und dem Feuer. Nebst einem Vorbericht von Tobern Bergmann. Zweyte verbesserte Auflage. Mit einer eigenen Abhandlung über die Luftgattungen über die Luftgattungen, wie auch mit der Herren Kirwan und Priestley Bemerkungen und Herrn Scheeles Erfahrungen über die Menge der im Dunstkreise befindlichen reinsten Luft vermehrt und mit einem Register versehen von D. Johann Gottfried Leonhardi. Siegried Lebrecht Crusius, Leipzig, 286 pp. *See also Reprint*: Ostwald (1894).

Schelenz, H. E. (1886) Nordseebad St. Peter, Gehalt der Luft an Kochsalz, Ozon und Mikroorganismen, Kochsalzgehalt des Wassers, Sandes und des Dünengrases. *Arch. Pharm.* **224**, 1015–1021.

Schell, D., H. W. Georgii, R. Maser, W. Jaeschke, B. G. Arends, G. P. A. Kos, P. Winkler, T. Schneider, A. Berner and C. Kruisz (1992) Intercomparison of fog water samplers. *Tellus* **446**, 612–631.

Schellnhuber, H.-J. (1999) Earth system analysis and the second Copernican Revolution. *Nature, Suppl.* **402**, C19–C23.

Schemenauer, R. S. (1986) Acidic deposition to forests: The chemistry of high elevation fog (CHEF) Project. *Atmosphere-Ocean* **24**, 303–328.

Schemenauer, R. S., P. H. Schuepp, S. Kermasha and P. Cereceda (1988) Measurements of the properties of high elevation fog in Quebec, Canada. In: Acid deposition at high elevation sites (eds. M. H. Unsworth and D. Fowler), Kluwer Academic Publishers, Boston, pp. 359–374.

Schemenauer, R. S. and P. Cereceda (1994) A proposed standard fog collector for use in high-elevation regions. *Journal of Applied Meteorology* **33**, 1313–1322.

Schemenauer, R. S., C. M. Banic and N. Urquizo (1995) High elevation fog and precipitation chemistry in southern Quebec, Canada. *Atmos. Env.* **29**, 2235–2252.

Scherer, J. A. (1785) Geschichte der Luftgüteprüfungslehre für Aerzte und Naturfreunde. Wappler, Wien, 228 pp.

Scherer, J. A. (1792) Versuch einer neuen Nomenclatur für Deutsche Chymisten. Chr. Fr. Woppler, Wien, 224 pp.

Schiefferdecker, W. (1855) Bericht über die vom Verein für wissenschaftliche Heilkunde in Königsberg in Preussen angestellten Beobachtungen über den Ozongehalt der atmosphärischen Luft und sein Verhältnis zu den herrschenden Krankheiten. In: *Sitzungsber. Wien* **22**, 191–237.

Schlagintweit H. und A. (1854) Neue Untersuchungen über die physicalische Geographie und die Geologie der Alpen. T. O. Weigel, Leipzig, 630 pp.

Schlesinger, W. H. and A. E. Hartley (1992) A global budget for ammonia. *Biogeochemistry* **15**, 191–211.

Schloemer, W. (1938) Spektrale Messungen der ultravioletten Himmelstrahlung (Dissertation). Veröffentlichungen des meteorologischen Instituts der Universität Berlin. Band III Heft 2, 30 pp.

Schloesing, T. (1875) Sur l'ammonique de l'atmosphere. *C. r.* **80**, 175–178 (and: *Journal de Pharmacie et de Chimie* (ser. 4) **23**, 209–212); *See also*: Sur les lois des échanges d'ammoniaque entre les mers, l'atmosphère et les continents. *C. r.* **81**, 81–84.

Schloesing, T. (1876) Sur les échanges d'ammoniaque entre les eaux naturelles at l'atmosphere. *Journal de Pharmacie et de Chimie (ser. 4)* **23**, 427–430. 1876, Ammoniak im Regenwasser. *Arch. Pharm.*, **211** (1876), 270–171.

Schloesing, T. (1880) Sur la constance de proportion d'acide carbonique dans l'air. *C. r.* **90**, 1410–1413.

Schloesing, T. and A. Müntz (1877) Sur la nitrification par les ferments organisés. *C. r.* **84**, 301–304.

Schlotfeldt (1849) Beobachtungen über das Ozon und über den Einfluss desselben auf die Veränderungen der Atmosphäre. *Arch. Pharm.* **110**, 277–282.

Schmauß, A. (1920) Kolloidchemie und Meteorologie. *Met. Z.* **37**, 1–8.

Schmauß, A. and A. Wigand (1929) Die Atmosphäre als Kolloid. Sammlung Vieweg, Tagesfragen aus den Gebieten der Naturwissenschaften und der Technik, Heft 96. Fr. Vieweg & Sohn, Braunschweig, 74 pp.

Schmid, W. (1866) Ueber die Phosphornnebel. *J. prakt. Chem.* **98**, 414–418.

Schmid, W. (1869) Wasserstoffsuperoxyd in der Atmosphäre. *J. prakt. Chem.* **107**, 60.

Schmidt, M. (1988a) Von Christian Friedrich Schönbein bis zum Ozonloch. Ein Abriss der Geschichte der Ozonforschung. Katlenburg, Lindau, 67 pp.

Schmidt, M. (1988b) Pioneers of Ozone Research. A Historical Study. Kaltenburg-Landau, MPI for Aeronomy 1988. 59 p.

Schmitt, G. (1987) Methoden und Ergebnisse der Nebelanalyse. Berichte des Institutes für Meteorologie und Geophysik, Universität Frankfurt/Main, Nr. 72.

Schmitz, G. (2001) The oxidation of iodine to iodate by hydrogen peroxide. *Physical Chemistry and Chemical Physics* **4**, 4731–4746.

Schneider-Carius, K. (1955) Wetterkunde – Wetterforschung. Geschichte ihrer Probleme und Erkenntnisse in Dokumenten aus drei Jahrtausenden. Verlag Karl Albert, Freiburg, 423 pp.

Schneider-Carius, K. (1961) Das Klima, seine Definition und Darstellung. Zwei Grundsatzfragen der Klimatologie. In: Veröffentlichungen der Geophysikalischen Instituts der Universität Leipzig, 2. Serie, Bd. 17, Heft 2, pp. 143–222.

Schofield, R. (1964) Still more on the water controversy. *Chymia* **9**, 71–76.

Scholz, J. (1931) Ein neuer Apparat zur Bestimmung der Zahl geladener und ungeladener Kerne. *Zeitschrift Instrumentenkunde* **51**, 505–522. *See also*: Vereinfachter Bau eines Kernzähler. *Met. Z.* **49** (1932) 381–388.

Schönbein, C. F. (1838) Ueber die Ursachen der Farbenveränderung, welche manche Körper unter dem Einflusse der Wärme erleiden. *Pogg. Ann.* **45**, 263–281.

Schönbein, C. F. (1840) Beobachtungen über den bei der Elektrolyse des Wassers und dem Ausströmen der gewöhnlichen Elektricität aus Spitzen sich entwickelnden Geruch. *Pogg. Ann.* **50**, 616–635. *See also*: In: *Abhandlungen der Kgl. Bayerischen Akademie der Wissenschaften*, III. Bd., Abt. 1. München, Akad. d. Wiss. (1840), Erste Ausgabe, pp. 255–278; *Furthermore*: On the odour accompanying electricity and on the probability of its dependence on the presence of a new substance. *Philosophical Magazine, Series 3*, **17**, 293–294. *And*: Recherches sur la nature de l'odeur qui se manifest dans certaines actions chimiques. *C. r.* **10**, 706–710.
Schönbein, C. F. (1843) Ueber die Natur des eigenthümlichen Geruches, welcher sich sowohl am positiven Pole einer Säule während der Wasserelectrolyse, wie auch beim Ausströmen der gewöhnlichen Electricität aus Spitzen entwickelt. In: *Abhandlungen der Kgl. Bayerischen Akademie der Wissenschaften*, III. Bd., Abt. 3. München, Akad. d. Wiss. (1843), pp. 587–605.
Schönbein, C. F. (1844) Ueber die Erzeugung des Ozons auf chemischem Wege. Basel. 159 pp.
Schönbein, C. F. (1845a) Einige Bemerkungen über die Anwesenheit des Ozons in der atmosphärischen Luft und die Rolle, welche es bei der langsamen Oxidation spielen dürfte. *Pogg. Ann.* **65**, 161–173.
Schönbein, C. F. (1845b) Beleuchtung der Meinung des Hrn. Fischer betreffend das Ozon. *Pogg. Ann.* **65**, 190–196; also *J. prakt. Chem.* **34** (1845) 492; Erwiderung auf Hrn. Fischer's Replik. *Pogg. Ann.* **66** (1845) 593–594; Einige Notizen über das Jodkalium. *J. prakt. Chem.* **34** (1845) 42–45; Weitere Notizen über das Jodkalium. Ibid., **35** (1845) 181.
Schönbein, C. F. (1846) On the relation of ozone to hyponitric acid. *Phil. Mag. Series 3* **28**, 432–437.
Schönbein, C. F. (1847a) Chemische Mitteilungen. *J. prakt. Chem.* **41**, 225–234.
Schönbein, C. (1847b) Ueber Gewitterwasser. *Bericht ueber die Verhandlungen der naturwissenschaftlichen Gesellschaft in Basel für die Jahre 1844 bis 1846* **7** (1847), 3.
Schönbein, C. (1849a) Ueber Gewitterwasser. *Bericht ueber die Verhandlungen der naturwissenschaftlichen Gesellschaft in Basel für die Jahre 1846 bis 1848* **8** (1847), 4–5.
Schönbein, C. (1849b) Ueber verschiedene chemische Zustände des Sauerstoffes. *Bericht ueber die Verhandlungen der naturwissenschaftlichen Gesellschaft in Basel für die Jahre 1846 bis 1848* **8** (1847), 6–8.
Schönbein, C. E. (1849c) Denkschrift über das Ozon. Zur Einweihung des neuen Museums. Basel. 15 pp.
Schönbein, C. F. (1850) Ueber das Ozon. *J. prakt. Chem.* **51**, 321–338. *See also*: Ueber das Ozon. *Arch. Pharm.* **117** (1851) 257–272.
Schönbein, C. F. (1851a) Ueber einige mittelbaren physiologischen Wirkungen der atmosphärischen Electricität. *Zeitschrift für rationelle Medicin NF* **2**, 384–399.
Schönbein, C. F. (1851b) Ob die Atmosphäre freie Salpetersäure als regelmässigen Bestandtheil enthalte? *Zeitschrift für rationelle Medicin NF* **2**, 400–405. *See also*: Aus einer am 29. Oct. und 12. Nov. 1851 der naturforschenden Gesellschaft in Basel gemachten Mittheilung. III: Ueber die Anwesenheit freier Salpetersäure in der Atmosphäre. *J. prakt. Chem.* **55**, 14–20.
Schönbein, C. F. (1852) Ueber den im Eisenoxid und in der Untersalpetersäure enthaltenen erregten Sauerstoff. *Berichte ueber die Verhandlungen der Naturforschenden Gesellschaft in Basel* **10**, 42–50.
Schönbein, C. F. (1854) Ueber verschiedene Zustände des Ozons. *Ann.* **89**, 257–300.
Schönbein, C. F. (1857a) Ueber die eigenthümliche Bildungsweise der salpetrichten Säure. *Verhandlungen der Naturforschenden Gesellschaft in Basel* **1**, 482–487.
Schönbein, C. F. (1857b) Ueber den Zusammenhang der katalytischen Erscheinungen bei der Allotropie. *Pogg. Ann.* **100**, 1–40.
Schönbein, C. F. (1858a) Beiträge zur nähern Kenntniss des Sauerstoffes. *Abhandlungen der Königlichen Bayerischen Akademie der Wissenschaften* **VIII**, II Abth, 381–412.
Schönbein, C. F. (1858b) Ueber den riechenden Flussspath von Wesersdorf. *J. prakt. Chem.* **74**, 325–327.

Schönbein, C. F. (1861a) Ueber die Bildung des Wasserstoffsuperoxydes während der langsamen Oxydation der Metalle in feuchtem gewöhnlichem Sauerstoff oder atmosphärischer Luft. *Pogg. Ann.* **112**, 445–451.
Schönbein, C. F. (1861b) Beiträge zur näheren Kenntnis des Sauerstoffes. Theil II: Ueber das Vorkommen des freien-activen Sauerstoffes in dem Wölsendorfer Flußspath. *J. prakt. Chem.* **83**, 95–106.
Schönbein, C. F. (1861c) Ueber die Nitrification. *J. prakt. Chem.* **84**, 193–231.
Schönbein, C. F. (1862) Die Bildung des salpetersauren Ammoniaks aus Wasser und atmosphärischer Luft. *Polytechnisches Journal (Dingler's J.)* **166**, 147–150. *See also*: Schönbein's neueste Beobachtung über die Erzeugung des salpetersauren Ammoniakes. *Neues Repertorium für Pharmacie* **11**, 212–214.
Schönbein, C. F. (1864a) Ueber das Vorkommen von Wasserstoffhyperoxyd im menschlichen Körper. *Z. anal. Chem.* **3**, 245–247.
Schönbein, C. F. (1864b) Ueber ein neues höchst empfindliches Reagens auf das Wasserstoffsuperoxyd und die salpetrige Säure. *Sitzungsber. München Jahrgang 1862* **1**, 113–115.
Schönbein, C. F. (1864c) Weitere Beiträge zur näheren Kenntnis des Sauerstoffes. Sitzungsber. München Jahrgang 1862 **2**, 261–290.
Schönbein, C. F. (1866) Mittheilungen. *J. prakt. Chem.* **98**, 257–283 and **99**, 11–21.
Schönbein, C. F. (1867a) Ueber die Anwesenheit des Ozones in der atmosphärischen Luft. *J. prakt. Chem.* **101**, 321–333. *See also*: Sur la présence de l'ozone dans l'air atmosphérique. *Annuaire de la société météorologique de France* **15**, 244–251.
Schönbein, C. F. (1867b) Mittheilungen von C. F. Schönbein [36 original works upon oxygen]. *Verhandlungen der Naturforschenden Gesellschaft in Basel* **4**, 1–820.
Schönbein, C. F. (1869) Letzte Arbeiten. III. Ueber das Vorkommen des Wasserstoffsuperoxyds in der Atmosphäre. *J. prakt. Chem.* **106**, 270–273.
Schöne, E. (1873) Ueber das Verhalten von Ozon und Wasser zueinander. *Ber.* **6**, 1224–1229.
Schöne, E. (1874) Ueber das atmosphärische Wasserstoffhyperoxyd. *Ber.* **7**, 1693–1708.
Schöne, E. (1875) Zur Nachweisung und Bestimmung des Wasserstoffhyperoxyds. *Z. anal. Chem.* **14**, 137–184.
Schöne, E. (1878) Ueber das atmosphärische Wasserstoffhyperoxyd. *Ber.* **11**, 482–492, 561–566, 874–877, 1028–1031. Berichtigung zu der vorjährigen Mittheilung über das atmosphärische Wasserstoffhyperoxyd **12**, 346.
Schöne, E. [Шене] (1884) спектр озона и нахождения озона в атмосферном воздухе [Spectrum of ozone and the presence of ozone in the atmosphere]. *Журнал Русского физико-химического общества* Journal de la Société Physico-Chimique Russe **2**, 250–252. *See also*: *Journal of the Chemical Society, Abstracts* **48** (1885) 713.
Schöne, E. (1880a) Zur Frage über das Verhalten des Wasserstoffhyperoxyds zu Jodkalium. *Ber.* **13**, 627–629.
Schöne, E. (1880b) Ueber die Beweise, welche man für die Anwesenheit des Ozons in der atmosphärischen Luft angeführt hat. *Ber.* **13**, 1503–1508.
Schöne, E. (1880c) Ueber Beobachtungen in der atmosphärischen Luft mit Thalliumpapieren. *Ber.* **13**, 1508–1514.
Schöne, E. (1893) Zur Frage über das Vorkommen des Wasserstoffhyperoxyds in der atmosphärischen Luft und den atmosphärischen Niederschlägen. *Ber.* **26**, 3011–3027.
Schöne, E. (1894a) Zur Frage über das atmosphärische Wasserstoffhyperoxyd. *Ber.* **27**, 1233–1235.
Schöne (1894b) Über das Absorptionsverhalten des Ozons (Referat). *Z. anorg. Chem.* **6**, 333. *See also*: Absorption-spectrum of ozone. *Chemical News* **69** (1894) 289 (original title in Russian unknown. In: Дневник IX Съезда русских естествоиспытателей и врачей в Москве 3–11 января 1894 [Journal of the 9th congress of natural scientists and physicians, to be held in Moscow, January 3–11, 1894, No. 10]. *See also*: Über das atmosphärische Ozon (Berichtigung). Ibid. **7** (1894) 49.

Schoof, C. L. (1860) Beiträge zur Klimatologie des Harzes. Ergebnisse der zu Clausthal angestellten meteorologischen Beobachtungen vom 1. Dec. 1854 bis 1. Dec. 1859. Pieper, Clausthal, 38 pp.

Schorlemmer, C. (1882) On the origin of the world "chemistry". In: Memoirs of the Manchester Literary and Philosophical Society, 3rd Ser., 7th Vol., Ballière, London, pp. 75–80.

Schossberger, F. (1942) Über die Umwandlung des Titandioxyds. *Zeitschrift für Kristallographie* **104**, 358–374.

Schreiber, J. (1877) Die Meteorologie im Dienste der Medizin. *Z. Öster. Ges. Met.* **12**, 159–166.

Schrimpff, E. (1983) Waldsterben infolge hoher Schadstoffkonzentrationen im Nebel? *Staub-Reinhaltung Luft* **43**, 240.

Schroder, L. J., T. C. Willoughby, R. B. See and B. A. Malo (1989) The chemical composition of precipitation, dew and frost and fog in Dencer, Colorado. In: Atmospheric Deposition (Proceedings of the Baltimore Symposium, May 1989). IAHS Publ. No. 179, pp. 83–90.

Schröder, J. (1915) Composicion quimico-agricola de las agua de lluvia recogidas en Montevideo 1909–1912. *La Revista de la Asociación Rural del Uruguay* **49**, 381–391; and: *Revista del Instituto Nacional de Agronomia* **15** (1918) 29–48.

Schroeder, J. (1873) Ueber die Mineralbestandthteile des Regenwassers. *Tharander forstliche Berichte* **23**, 68–87. *See also*: Untersuchung über die Mineralbestandtheile des Regenwassers. *Chemischer Ackersmann* **19** (1873) 81–97; and: *Jahresbericht über die Fortschritte auf dem Gesamtgebiete der Agricultur-Chemie* **16–17** für die Jahre 1873–1874 (1876) 175–179.

Schroeder, J. v. and C. Reuß (1883) Die Beschädigung der Vegetation durch Rauch und die Oberharzer Hüttenrauchschäden. Paul Parey, Berlin, 333 pp.

Schroeder, W. H. and P. Urone (1974) of nitrosyl chloride from salt particles in air. *Environ. Sci. Technol.* **8**, 445–451.

Schröer, E. (1944/49) Theorie der Entstehung, Zersetzung und Verteilung des atmosphärischen Ozons. *Berichte des Wetterdienstes US-Zone* **11**, 13–33. Note that is was a lecture held in 1944 at the Tharandt ozone symposium (*Schröer* died in 1945 by unknown reason).

Schroeter, L. C. (1963) Kinetics of air oxidation of sulfurous acid salts. *Journal of Pharmaceutical Science* **52**, 559–563.

Schrötter, A. (1860) Über das Vorkommen des Ozons im Mineralreiche. *Sitzungsber. Wien* **41**, 725–734.

Schubert, E. L. (1857) Ueber die sauren Gase, welche Schwefelsäure- und Sodafabriken verbreiten, und die Mittel, dieselben unschädlich zu machen. Nach einer belgischen Staatsschrift auszugsweise bearbeitet von Prof. Dr. E. L. Schubarth in Berlin. *Polytechnisches Journal (Dingler's J.)* **145**, 375–287.

Schultz, A. (1854) Ueber den Ozongehalt der Luft in Berlin im November 1853 und während der Cholera-Epidemie daselbst in demselben Jahr. *Medicinische Zeitung (Berlin)* **23**, 39–40.

Schultz, H. (1892) Ueber einfache Methoden zur Kohlensäurebestimmung in der Luft mit besonderer Berücksichtigung der Verfahren von Rosenthal und Wolpert. Rostock, 75 pp.

Schulz, L. (1939) Kernzahlmessungen in Braunlage im Harz. In: Burckhardt et al. (1939), pp. 113–120.

Schulze, F. (1871) Tägliche Beobachtungen über den Kohlen-Säuregehalt der Atmosphäre zu Rostock vom 18. Oktober 1868 bis 31. Juli 1871. Festschrift für die 44. Versammlung Deutscher Naturforscher und Aerzte. Leopoldsche Universitätsbuchhandlung, Rostock, 34 pp.

Schumacher, H.-J. and G. Sprengler (1928) Die Reaktion zwischen Stickstoffpentoxyd und Ozon. I. *Z. phys. Chem.* **136**, 77–92. *See also*: II. Ibid. **2B** (1929) 267–281.

Schumacher, H.-J. (1932) Photokinetik des Ozons. I. Der Zerfall in rotem Licht. *Z. phys. Chem.*, Abteilung B: Chemie der Elementarprozesse, Aufbau der Materie B **17**, 405–416.

Schumann, V. (1893) Über die Photographie der Lichtstrahlen kleinster Wellenlängen. *Sitzungsber. Wien* **102**, 415–475.

Schumb, W. C., C. N. Satterfield and R. L. Wentworth (1955) Hydrogen peroxide. Reinhold Publ. Series, New York, 759 pp.

Schurath, U. (1977) Die Chemie des photochemischen Smogs. VDI Berichte Nr. 270, VDI-Verlag GmbH, Düsseldorf, pp. 13–18.
Schürmann (1875) Das Ozon. *Gaea. Natur und Leben* 11, 598–614, 655–669, 726–738. Und: Ueber das Ozon. *Sitzungsberichte und Abhandlungen der Naturwissenschaftlichen Gesellschaft Isis in Dresden* 1875, pp. 230–266.
Schütt, H.-W., ed. (1996) Christian Friedrich Schönbein. Kleine Abhandlungen. Olms-Weidmann, Hildesheim (reprint of editions Basel 1844–1856).
Schwab, G.-M. (1957) Photochemical and kinetic studies of electronic reactions. *Advances in Catalysis* **9**, 229–237.
Schwartz, S. E. (1984) Gas- and aqueous-phase chemistry of HO_2 in liquid water clouds. *JGR* **89**, 11589–11598.
Schwartz, S. E. (2003) Cloud chemistry. In: Handbook of Weather, Climate, and Water. Eds T. D. Potter and B. R. Colman. John Wiley & Sons, New York, pp. 331–355.
Schwarzenbach, V. (1850) Ueber die Einwirkung des Ozons auf Thiere. *Verhandlungen der physikalisch-medicinischen Gesellschaft zu Würzburg* **1**, 322–3232.
Scott, A. (1893) On the Composition of Water by Volume. *Phil. Trans. A* **184**, 543–568.
Scott, R. H. (1893) Fifteen years' fogs in the British Islands, 1876–1890. *Q. J. R. Meteor. Soc.* **19**, 229–238.
Scott, R. H. (1896) Notes on some of the differences between fogs, as related to the weather systems which accompany them. Submitted to the fog committee. *Q. J. R. Meteor. Soc.* **22**, 41–65.
Scott, W. D. and P. V. Hobbs (1967) The formation of sulfate in water droplets. *Journal of Atmospheric Sciences* **24**, 54–57.
Scotti, G. (1853) L'ozono. *Gazzetta medica italiana, Lombardia* **3**, 129–134.
Scoutetten, R. J. H. (1856) L'ozone, ou recherches chimiques, météorologiques, physiologiques et médicales sur l'oxygène électrisé. Victor Masson & Metz, Paris, 287+8 pp.
Scurlock, J. M. O. and D. O. Hall (1991) The carbon cycle. *New Scientist* **132**, No. 1793 (Inside Science suppl. No. 51) S1–S4.
Sedlak, D. L., J. Hoigné, M. M. David, R. N. Colvile, E. Seyffer, K. Acker, W. Wieprecht, J. A. Lind and S. Fuzzi (1997) The cloudwater chemistry of iron and copper at Great Dun Fell, U. K. *Atmos. Env.* **31**, 2515–2526.
Séguin, A. (1790) Observations générales sur la respiration et sur la chaleur animale. *Journal de Physique* **37**, 467–472.
Seibt, G. (1991) Forstrat Constantin Gerlach – ein Pionier der Rauchschadensforschung in Sachsen. *Forst und Holz* **46**, 299–301.
Seiler, W. and J. Fishman (1981) The distribution of carbon monoxide and ozone in the free troposphere. *JGR* **86**, 7255–7265.
Seinfeld, J. H. (1986) Atmospheric chemistry and physics of air pollution. John Wiley & Sons, New York, 739 pp.
Seinfeld, J. H. and S. N. Pandis (1998) Atmospheric chemistry and physics – from air pollution to climate change. John Wiley & Sons, New York, 1326 pp.
Seitz, F. (1865) Catarrh und Influenza. Eine medizinische Studie. J. G. Cotta, München, 465 pp.
Sekihara, K. (1954) On the chemical determination of atmospheric ozone at the summit of Mt. Fuji [in Japanes with English summary]. *Journal of the Meteorological Society of Japan* Ser. II **32**, 111–116.
Selander, N. E. (1888) Luftundersökningar vid Vaxholms fästning okt. 1885 – juli 1886. Kungliga Svenska Vetenskapsakad. Stockholm, Vol. 13, Afd. II, No. 9, 38 pp.
Selezneva, E. S. [Селезнева] (1945a) Ядрах конденсации в атмосфере [Condensation nuclei in the atmosphere]. *Тр. НИУ ГУГМС* [Труды Национального исследовательского университета главного управления гидрометеороло-гической службы СССР – Transactions of the National Research University of the Hydrometeorological Service of the USSR], Ser I **7**, 22–34.

Selezneva, E. S. [Селезнева] (1945b) Структура облаков по наблюдениям в Эльбрусской экспедиции [Structure of clouds at the Elbrus expedition by observations]. *Труды НИУ ГУГМС (Национальный исследовательский университет Главного управления гидрометеорологической службы)*, сер. 1, No. 7.

Selezneva, E. S. [Селезнева] (1947a) Микроструктура облаков [Microstructure of clouds]. *Метеорология и гидрологи* [Meteorology and Hydrology], No. 2.

Selezneva, E. S. [Селезнева] (1947b) Условия образования облаков по наблюдениям летом 1946 г [Conditions for formation of Cumulus clouds based on observation in summer 1946]. *Труды ГГО* **7**, 69.

Selezneva, E. S. [Селезнева] (1967) Исследования ядер конденсации и химического состава атмосфеных осадков [Investigation of condensations nuclei and chemical composition of atmospheric precipitation]. In: Главная геофизическая обсерваторая имени А. И. Воейкова за 50 лет советской власт [Main Geophysical Observatory 50 years Soviet power], Gidrometeoizdat, Leningrad, pp. 241–250.

Selezneva, E. S. [Селезнева] (1968) О связи между минерализацией и электро-проводностью атмосферных осадков [On the relation between mineralisation and electric conductivity of atmospheric clouds]. *Труды ГГО* **234**, 208–212.

Selezneva, E. S., V. M. Drozdova and O. P. Petrenchuk [Селезнева, Дроздова, Петренчук] (1963) Химический состав атмосферных осадков на ETC [Chemicial composition of atmospheric precipitation about the European Territory of the USSR]. *Тр. Всесоюзного научно-методического совещания, секция физики свободной атмосферы* [Trudy All-Union Scientific Methodological Session, chapter Physics of the Free Atmosphere] **5**, 56–63.

Selezneva, E. S. and V. M. Drozdova [Селезнева, Дроздова] (1966) О естественном фоне загрязнения атмосферы и состав осадков на территории СССР [Natural background air pollution and composition of rainfall over the territory of the USSR]. In Современные проблемы климатологии [Topical problems in climatology], Gidrometeoizdat, Leningrad, pp. 292–299.

Selezneva, E. S. and Drozdova [Селезнева, Дроздова] (1974) О влиянии методики сбора осадков на их химический состав [On the influence of sampling method of precipitation on its chemical composition]. *Труды ГГО* **314**, 158–167.

Selezneva, E. S. and O. P. Petrenchuk [Селезнева, Петренчук] (1971) Об удалении примесей из атмосферы облаками и осадками [Removal of pollutants from atmospheric clouds and precipitation]. In Метеорологические аспекты загрязнения атмосферы [Meteorology of air pollution], Gidrometeoizdat, Leningrad, pp. 253–259.

Selezneva, E. S. and Lavrinenko (1971) The significance of measurements of pH and electrical conductivity in investigations of chemical compositions of precipitation. *Idöjaras* **5–6**, 288–293.

Sertürner, F. (1818) Bemerkungen über die Verbindungen der Säuren mit basischen und indifferenten Verbindungen. *Gilb. Ann.* **30**, 33–53. *See also translation*: Remarks on the combination of acids with bases and indifferent substances. *Annals of Philosophy* **14** (1819) 37–46.

Shalamyansky, A. M., K. I. Romashkina, K. G. Pavlova, A. A. Solomatnikova and C. T. Talash [Шаламянский, Ромашкина, Павлова, Соломаникова, Талаш] (2018) Состояние защитного озонового слоя над территорией РФ (42 года регулярных измерений). *Труды ГГО* **590**, 144–159.

Shapter, Th. (1854) Observations on the development of ozone and a deoxidizing property in the atmosphere. *Association Medical Journal* **2**, 733.

Shapter, Th. (1862) The climate of the south of Devon and its influence upon health; with short accounts of Exeter, Torquay, Babbicombe, Teignmouth. 2nd edition. Churchill, London, 282 pp.

Shaw, N. and J. S. Owens (1925) The smoke problem of great cities. Constable & Comp. London, 301 pp.

Shaw, N. (1942) Manual of meteorology. Vol. 1: meteorology in history. Reprint of first ed. 1926. Univ. Press, Cambridge, 343 pp.

Shaw, W. N. (1901) Hailstorm Artillery. *Nature* **64**, 159–161 (see also: News and Views: 100 and 50 years ago. *Nature* **411** (2001) 751).

Shaw, W. N. (1906) Is London fog inevitable? *The Journal of the Royal Society for the Promotion of Health* **27**, 49–53.

Shenstone, W. A. and W. T. Evans (1898) Observations on the influence of silent discharges on atmospheric air. *J. Chem. Soc.* **73**, 246–254.

Shi, Z., M. D. Krom, T. D. Jickels, et al.(2012) Impacts on iron solubility in the mineral dust by processes in the source region and the atmosphere: a review. *Aeolian Research* **5**, 21–42.

Shmeter, S. M. [Шметер] (1952) О содершании хрора в вода облаков в связи с их микроструктурой [On the content of chlorine in cloud water in connection with its microstructure]. *Труды ЦАО* **9**, 3–60.

Shutt, F. T. (1914) The nitrogen compounds in rain and snow. *Transactions of the Royal Society of Canada* **8**, 83–87.

Shutt, F. T. and R. L. Dorrance (1917) The nitrogen compounds in rain and snow. *Transactions of the Royal Society of Canada* **11**, 63–72.

Shutt, F. T. and B. Hedley (1925) The nitrogen compounds in rain and snow. *Transactions of the Royal Society Canada* **19**, 1–10.

Sies, H. (1986) Biochemie des oxidativen Stress. *Z. angew. Chem.* **98**, 1061–1075.

Sigg, A. and A. Neftel (1988) Seasonal variations of hydrogen peroxide in polar ice cores. *Annual Glaciology* **10**, 157–162.

Sigg, A. and A. Neftel (1991) Evidence for a 50 % increase in H_2O_2 over the past 200 years from a Greenland ice core. *Nature* **351**, 557–559.

Sigg, A., T. Staffelbach and A. Neftel (1992) Gas phase measurements of hydrogen peroxide in Greenland and their meaning for the interpretation of H_2O_2 records in ice core. *J. Atmos. Chem.* **14**, 223–232.

Sigurdsson, H. (1982) Volcanic pollution and climate: the 1783 Laki eruption. *Eos* (Transactions of the American Geophysical Union) **63**, 601–602.

Sillman, S. (1995) The use of NO_y, H_2O_2 and HNO_3 as indicators for ozone-NO_y-hydrocarbon sensitivity in urban locations. *JGR* **100**, 14175–14188.

Silvestri, O. (1861) Recherches ozonomètrique faites à Pisa. *C. r.* **53**, 247–249.

Simon, S., O. Klemm, T. El-Madany, J. Walk, K. Amelung, P.-H. Lin, S.-C. Chang, N.-H. Lin, G. Engling, S.-C. Hsu, T.-H. Wey, Y.-N. Wang and Y.-C. Lee (2016) Chemical Composition of Fog Water at Four Sites in Taiwan. *Aerosol and Air Quality Research* **16**, 618–631.

Simonin, J. B. (1854) Résumé des observations météorologiques faites à Nancy pendant l' année 1853. *Mémoires de l'Académie de Stanislas* **1853**, 437–458.

Simonin, J. B. (1869) Proportions de l'ozone atmospérique pendant les diverses saisons. *Mémoires de l'Académie de Stanislas* **1868**, 229–230.

Simpson, G. C. (1941a) On the formation of cloud and rain. *Q. J. Roy. Meteor. Soc.* **67**, 99–133.

Simpson, G. C. (1941b) Sea salt and condensation nuclei. *Q. J. Roy. Meteor. Soc.* **67**, 163–169.

Singh, H. B., L. J. Sala, H. Shigeishi and E. Scribner (1979) Atmospheric halocarbons, hydrocarbons and sulfur hexafluoride: global distributions, sources and sinks. *Science* **803**, 899–903.

Singh, O. N. and P. Fabian (2003) Atmospheric ozone: a millennium issue. EGU Special Publ. Series, Kathlenburg-Lindau.

Skeib, G. and Ch. Popp (1961) Messungen des Gesamtozongehaltes der Atmosphäre in Mirny, Antarktika. *Z. Met.* **15**, 287–291.

Skryabin, G. K., M. V. Ivanov and J. R. Freney [Скрябин, Иванов, Френей], Eds. (1983) Глобальный биогеохимический цикл серы и влияние на него деятельности человека [The global biogeochemical sulphur cycle and influence on it of human activity]. Наука, Москва, 424 pp.; English edition see Ivanov and Freney (1983).

Slanina, J., J. J. Möls, J. H. Baard, H. A. Van Der Sloot, J. G. Van Raaphorst and W. Aaman (1979) Collection and analysis of rainwater;experimental problems and the interpretation of results. *International Journal of Environmental Analytical Chemistry* **7**, 161–176.

Slemr, F., G. W. Harris, D. R. Hastle, G. I. MacKay and H. I. Schiff (1986) Measurement of gas phase hydrogen peroxide in air by tunable diode laser absorption spectrometry. *JGR* **91D**, 5371–5378.

Slocum, G. (1955) Has the amount of carbon dioxide in the atmosphere changed significantly since the beginning of the twentieth century? *Monthly Weather Review* Oct., pp. 225–231.

Smallwood, Ch. (1854) Mean results of meteorological observations, made at St. Martin, Isle Jesus, Canada East (nine miles west of Montreal) for 1853. *The Canadian Journal* **2**, 230–231.

Smallwood, Ch. (1857) On ozone and the meteorology of the vicinity of Montreal. John Lovell, Montreal, 16 pp.

Smallwood, Ch. (1858) The observatory at St. Martin, Isle Jesus, Canada East. *The Canadian Journal, new series* **3**, (No. 16) 282–292.

Smallwood, Ch. (1859) Contribution to meteorology, from observations taken at St. Martin, Isle Jesus, Canada East. *The Canadian Journal, new series* **4**, (No. 22) 262–266.

Smarygin, S. N. [Смарыгин] (2018) Профессор Э. Б. Шёне – создатель и хранитель традиций преподавания неорганической и аналитической химии (к 180-летию со дня рождения [Professor H. E. Schöne – creator and keeper of traditions of inorganic and analytical chemistry teaching (to the 180th anniversary)]. *Известия ТСХА* [Izvestiya of Timiryazev Agricultural Academy] **2018**(6), 129–143.

Smil, V. (2002) The earth's biosphere. The MIT Press, Cambridge, Massachusetts (USA), 346 pp.

Smith-D'Ans (1948) Einführung in die allgemeine und anorganische Chemie auf elementarer Grundlage. XII. Auflage. Bearbeitet von J. D'Ans. G. Braun, Karlsruhe, 837 pp.

Smith, B. E. and E. J. Friedman (1982) The chemistry of dew as influenced by dry deposition: results of Sterling, VA and Champaign, IL experiments. WP-82W00141, MITRE Corp., McLean, VA.

Smith, D. A. (1972) A history of nitrous oxide and oxygen anaesthesia. Oart 1A: The discovery of nitrous oxide and of oxygen. *British Journal of Anaesthesia* **44**, 297–304.

Smith, R. A. (1845) Some remarks on the air and water of towns. *Memoirs and Proceedings, Chemical Society* **3**, 311–315.

Smith, R. A. (1852) On the air and rain of Manchester. *Memoirs of the Literary and Philosophical Society of Manchester* (Sevens Series) **10**, 207–217.

Smith, R. A. (1859a) On the estimation of the organic matter of the air. *Chemical Gazette* **17**, 176–180.

Smith, R. A. (1859b) On the air of towns. *Q. J. Chem. Soc.* **11**, 196–235.

Smith, R. A. (1872) Air and rain. The beginnings of a chemical climatology. Longmans, Green and Co., London, 600 pp.

Smith, R. A. (1873) Chemical climatology. *Journal of the Scottish Meteorological Society* **3**, 2–11.

Smith, R. A. (1877) The examination of air. *Proc. R. Soc. Lond.* **26**, 512–517.

Smith, R. A. (1879) On the distribution of Ammonia. *Memoirs of the Literary and Philosophical Society of Manchester, 3rd ser.* **6**, 267–278.

Smith, R. A. (1882) The word "Chemia" or "Chemistry". *Memoirs of the Manchester Literary and Philosophical Society*, 3rd Ser., **7**, 101–126.

Smyth, J. (1869) On the ozonometer for the observation of ozone with an aspirator, instructions for its use and the results obtained. *Proceedings of the Meteorological Society* **4**, 375–387. *See also: J. Scott. Met. Soc.* **2** (1867–69) 258.

Smyth, R. B. (1859) Ozone. *The Australian Medical Journal* **4**, 1–4.

Søderlund, R. and B. H. Svensson (1976) The global nitrogen cycle In: Nitrogen, phosphorus and sulphur – global cycles (eds. B. H. Svensson and R. Søderlund), Swedish Natural Science Research Council, pp. 23–73.

Solomon, S., R. R. Garcia, F. S. Rowland and D. J. Wuebless (1986) On the depletion of Antartic ozone. *Nature* **321**, 755–758.

Sondén, K. (1887) Zur hygienischen Luftanalyse. *Z. anal. Chem.* **26**, 592–598.

Sondén, K. and R. Tigerstedt (1895) Untersuchungen über die Respiration und den Gesammtstoffwechsel des Menschen. *Skandinavisches Archiv für Physiologie* **6**, 1–224.

Soret, J.-L. (1863) Sur la production de l'ozone par l'électrolyse et sur la ature de ce corps. *C. r.* **54**, 390–393; and: Ueber die Darstellung des Ozons durch Electrolyse und die Natur desselben. *Ann.* **127**, 38–42.

Soret, J.-L. (1864) Ueber die volumetrischen Beziehungen des Ozons. *Ann.* **130**, 95–101; Ueber das volumetrische Verhalten des Ozons. *Pogg. Ann.* **121**, 168–283.

Soret, J.-L. (1865) Recherches sur la densité de l'ozone. *C. r.* **61**, 941–944.

Soret, J.-L. (1868) Recherches sur la densité de l'ozone. *Ann. chim. phys.* **13**, 257–282. *See also*: ibid. **7** (1866) 113–118.

Spaulding, R. S., R. W. Talbot and M. J. Charles (2002) Optimization of a mist chamber (Cofer scrubber) for sampling water-soluble organics in air. *Environ. Sci. Technol.* **36**, 1798–1808.

Spelsberg, G. (1988) Rauchplage. Zur Geschichte der Luftverschmutzung. Kölner Volksblatt Verlag, 239 pp.

Spence, C. C. (1961) The Dyrenforth rainmaking experiments: A government venture in "Pluviculture". *Arizona and the West* **3**, 205–232.

Spengler, L. (1849) Influenza und Ozon. *Zeitschrift für rationelle Medicin* **7**, 70–74.

Spengler, L. (1857) Influenza und Ozon. *Froriep's Notizen aus dem Gebiete der Natur- und Heilkunde* (Jena), Jahrgang 1857, Band 4, 379–382.

Spicer, C. W. (1974) The fate of nitrogen oxides in the atmosphere. Battelle-Columbus Rep. to Coordinating Res. Council and U. S. Environ. Protection Agency. Rep. 600-3/76-030.

Spicer, C. W. (1977) Photochemical atmospheric pollutants derived from nitrogen oxides. *Atmos. Env.* **11**, 1089–1095.

Spiegel, L. (1903) der Stickstoff und seine wichtigsten Verbindungen. Vieweg & Sohn, Braunschweig, 911 pp.

Spiegelberg, F. (1984) Reinhaltung der Luft im Wandel der Zeiten. VDI-Verlag, Düsseldorf, 150 pp.

Spurny, K. R. (1996) Nikolai Albertowich Fuchs (1895–1982): aerosol scientist and humanist memories on the occasion of his 100th birthday. *J. Aerosol. Sci.* **27**, 833–852.

Spurny, K. R. (1998) Methods of aerosol measurement before the 1960s. *Aerosol Science and Technology* **29**, 329–349.

Spring, W. and L. Roland (1885) Recherches sur la proportion de l'acide carbonique contenue dans l'air. *Mémoires Couronnés et Autres Mémoires Publiés par l'Académie Royale des Sciences, des Lettres et des Beaux-Arts de Belgique* **37**, 1–94.

Staedel, W. (1898) Dichte und Molekurgewicht des Ozons. *Ber.* **34**, 3143–3134.

Staehelin, J., R. E. Bühler and J. Hoigné (1984) Ozone decomposition in water studied by pulse radiolysis -2. OH and SO_4 as chain intermediates. *J. phys. Chem.* **88**, 5999–6004.

Staehelin, J., R. Kegel and N. R. P. Harris (1998) Trend analysis of the homogenized total ozone series of Arosa (Switzerland), 1926–1996. *JGR* **103**, 8389–8399.

Staehelin, J., J. Thudium, R. Buehler, A. Volz-Thomas and W. Grabers (1994) Trends in surface ozone concentrations at Arosa (Switzerland). *Atmos. Env.* **28**, 75–101.

Staehelin, J., C. Vogler and S. Brönnimann (2009) The long history of ozone measurements: Climatological information derived from long ozone records. In: Twenty years of ozone decline (eds. C. Zerefos, G. Contopoulos and G. Skalkeas), Springer, Dordrecht, pp. 119–131.

Stamp, J. C. (1937) The science of social adjustment. London, Macmillan, 174 pp.

Stanhill, G. (1982) The Montsouris series of carbon dioxide concentration measurements, 1877–1910. *Climate Change* **4**, 221–237.
Stark, J. (1914) Ultraviolette Emissionsbanden des ein-, zwei- und dreiatomigen Sauerstoffmoleküls. *Ann. Phys.* **348** (*Vierte Folge* **43**), 319–336.
Stark, W. (1813) On rain water. *Annals of Philosophy* **3**, 140–142.
Stauffer, D. and C. S. Kiang (1974) Heteromolecular nucleation theory for multicomponent gas mixtures. *Tellus* **26**, 295–297.
Stedman, T. L. (1920) A practical medical dictionary. W. Wood, New York, 1143 pp.
Steele, L. P., P. J. Fraser, R. A. Rasmussen, M. A. K. Khalil, A. J. Crawford, T. J. Conway and R. H. Gammon (1987) Global measurements and distribution of methane. *J. Atmos. Chem.* **5**, 125–171.
Steffen, W., P. J. Crutzen and J. R. McNeill (2007) The Anthropocene: Are humans now overwhelming the Great Forces of Nature? *Ambio* **36**, 614–621.
Stein, G. (1952) Some aspects of the radiation chemistry of organic solutes. *Discussions of the Faraday Society* **12**, 227–234.
Stein, G., ed. (1968) Radiation chemistry and aqueous systems. Proceedings of the 19th L. Farkas Memorial Symposium held at the Hebrew University of Jerusalem, Israel, 27–29 December 1967. The Weizmann Science Press of Israel, Jerusalem in cooperation with John Wiley & Sons, 306 pp.
Steinhauser, F. (1959a) Über die pH-Werte des Niederschlags, der Schneedecke und des Grundwassers in Wien. *Archiv für Meteorologie, Geophysik und Bioklimatologie. Ser. B, Klimatologie, Umweltmeteorologie* **9**, 86–100.
Steinhauser, F. (1959b) Ergebnisse von Registrierungen des Ozongehaltes der Luft in Wien. *Archiv für Meteorologie, Geophysik und Bioklimatologie. Ser. A* **11**, 368–382.
Stepanova, N. A. (1952) A selective bibliography of carbon dioxide in the atmosphere. *Meteorological Abstracts and Bibliography* **3**, 137–170.
Stephens, E. R. and F. R. Burleson (1969) Distribution of light hydrocarbons in ambient air. *Journal of the Air Pollution Control Association* **19**, 929–936.
Stephens, B. B., R. F. Keeling, M. Heimann, K. D. Six, R. Murnane and K. Caldeira (1998) Testing global ocean carbon cycle models using measurements of atmospheric O_2 and CO_2 concentration. *Global Biogeochemical Cycles* **12**, 213–230.
Stevenson, W. F. (1843) On phlogiston and the decomposition of water. *Proc. R. Soc. Lond.* **5**, 629–630.
Stewart, R. (1920) Sulfur in relation to soil fertility. *Illinois Agronomic Experimental Station Bulletin* **227**, 99–108.
Stiger, A. (1898) Über das Wetterschießen am südöstlichen Abhänge des Bachergebirges nächst Windisch-Feistritz (Steiermark). Fritz Rasch, Cilli. *See also*: Hagelschießen. *Met. Z.* **14** (1897) 21–33 and 148–151.
Stöckhardt, J. A. (1846) Die Schule der Chemie, oder erster Unterricht in der Chemie, versinnlicht durch einfache Experimente. Zum Schulgebrauch und zur Selbstbelehrung, insbesondere für angehende Apotheker, Landwirte, Gewerbetreibende etc. Vieweg und Sohn, Braunschweig, 600 pp. (in 22 Editions until 1920).
Stöckhardt, J. A. (1850) Über einige durch den Bergbau und Hüttenbetrieb für die Landescultur entstehende Benachteiligungen. *Zeitschrift Deutscher Landwirthe* **4**, 33–38 and 129–137.
Stöckhardt, J. A. (1851) The principles of chemistry. Illustrated by simple experiments. Bohn, London [also published in 1851 by Bartlett, Cambrigde, Mass.], 679 pp. (note, 91 editions published in 5 languages).
Stöckhardt, J. A. (1871) Untersuchungen über die schädliche Einwirkung des Hütten- und Steinkohlenrauches auf das Wachstum der Pflanzen, insbesondere der Fichte und Tanne. *Tharandter forstliches Jahrbuch* **21**, 218–254.

Stockwell, W. R. and J. G. Calvert (1983) The mechanism of the HO-SO_2 reaction. *Atmos. Env.* **17**, 2231–2235.

Stockwell, W. R. (1995) On the HO_2 + HO_2 reaction: Its misapplication in atmospheric chemistry models. *JGR* **100**, 11695–11698.

Stohmann, F. (1879) Eine calorimetrische Methode. *J. prakt. Chem.* **19**, 115–142.

Stoklasa, J. (1882) Studie o slouceninách dusika ve vodach meteoriických [Study on nitrogen compounds in rainwater]. *Listů Chemických* **6**, 3–6, 12–15, 29–30, 43.

Stoklasa, J. (1887) Gehalt der atmosphärischen Niederschläge an Ammoniak und Salpeter-säure im Jahre 1883–1884. *Biedermanns Central-Blatt für Agrikulturchemie und rationellen Landwirtschafts-Betrieb* **16**, 361–362.

Stoklasa, J. (1906) Ueber die Menge und den Ursprung des Ammoniaks in den Producten der Vesuveruption im April 1906. *Ber.* **39**, 3530–3537.

Stoklasa, J. (1923) Die Beschädigung der Vegetation durch Rauchgase und Fabrikexhalationen. Verlag Urban & Schwarzenberg, Berlin und Wien, 487 pp.

Stolarski, R. S. and R. J. Cicerone (1974) Stratospheric chlorine: a possible sink for ozone. *Canadian Journal of Chemistry* **52**, 1610–1615.

Stolarski, R. S. (2001) History of the study of atmospheric ozone. *Ozone: Science & Engineering* **23**, 421–428.

Stone, R. G. (1936) Fog in the United States and adjacent regions. *Geographical Review* **26**, 111–134.

Stoney, G. J. (1894) Of the "electron" or atom of electricity. *Philosophical Magazine, Series 5* **38**, 418–420.

Storer, F. H. and A. H. Pearson (1871) On the amount of carbonic acid in the air. Paper presented at the Six Hundred and Thirty-First Meeting, April 1, 1871. *Proceedings of the American Academy of Arts and sciences* **8**, 310–316. *See also*: Second Annual Report Massachusetts State Board of Health, 1871, pp. 395–404).

Stothers, R. B. (1996) The great dry fog of 1783. *Climate Change* **32**, 79–89.

Strachan, R., ed. (1866) Wells, W. C.: An essay on dew (reprint from 1814). Edited, with annotations by L. P. Castella and an appendix by R. Strachan. Longmans, London pp. 153.

Stradins, J. and K. Arons (1992) Werdegang eines Wissenschaftlers; Georg Friedrich Parrots Tätigkeit in Riga (1795–1801). *Acta medico-historica Rigensia* I, **29**, 177–199.

Strambio, G. (1856) Su l'ozono atmosferico durante l'ultima epidemia colerosa in Milano. G. Chiusi, Milano, 53 pp. *See also*: *Gazetta Medica Italiana Lombardia* (Ser. 4) **1**, 217–221, 225–228 and 427–428.

Straub, D. J. and J. L. Collet, Jr. (2004) An axial-flow cyclone for aircraft-based cloud water sampling. *Journal of Atmospheric and Oceanic Technology* **21**, 1825–1839.

Strele, R. von (1898) Wetterläuten und Wetterschiessen. Zeitschrift des Deutschen und Österreichischen Alpenvereins. Bd. 28, Verlag des deutschen und österreichischen Alpenvereins, Graz, 142 pp.

Stromeyer, F. (1808) Tabellarische Uebersicht der chemisch einfachen und zusammengesetzten Stoffe. Heinrich Dieterich, Göttingen, 139 pp.

Strumpf, F. L. (1855) Systematisches Handbuch der Arzneimittellehre. Band 2. Fünfte Klasse: Metaloidica. Sechste Ordnung: Gasa atmosphaerica. Th. Ch. Frt. Englin, Berlin, pp. 921–951.

Strutt, R. J. (1918) Ultra-violet transparency of the lower atmosphere and its relative poverty in ozone. *Proceedings of the Royal Society, London* **94A**, 260–268.

Struve, H. (1869) Ueber die Gegenwart von Wasserstoffhyperoxyd in der Luft. *Z. anal. Chem.* **8**, 315–321.

Struve, H. (1871) Studien über Ozon, Wasserstoffhyperoxyd und salpetrigsaures Ammoniak. *Z. anal. Chem.* **10**, 292–298.

Struve, H. (1872) Mitteilung II. Beitrag zur Bestimmung der Salpetersäure, der salpetrigen Säure und des Wasserstoffsuperoxydes mit einer Indigolösung. *Z. anal. Chem.* **11**, 25–28.

Stewardt, R. W., S. Hameed and I. P. Pinto (1977) Photochemistry of tropospheric ozone. *JGR* **82**, 3134–3140.

Stuhl, F. (1974) Determination of the rate constant for the reaction OH + H_2S by a pulsed photolysis-resonance fluorescence method. *Berichte der Bunsen-Gesellschaft* **78**, 230–232.

Stuhlman, O. (1932) The mechanics of effervescence. *Journal of applied Physics* **2**, 457–466.

Stumm, W. and J. J. Morgan (1981) Aquatic chemistry. An introduction emphasizing chemical equilibria in natural waters. 2nd edition, John Wiley & Sons, New York, 780 pp.; *see also*: Stumm, W. and J. Morgan (1996) Aquatic Chemistry: Chemical Equilibria and Rates in Natural Waters. 3rd ed. Wiley-Interscience, New York. 1022 pp.

Stumm, W., J. J. Morgan and L. D. Schnoor (1983) Saurer Regen, eine Folge der Störung hydrogeochemischer Kreisläufe. *Naturwissenschaften* **70**, 216–223.

Stumm, W., ed. (1987) Aquatic surface chemistry: chemical processes at the particle-water interface. Wiley, New York, 520 pp.; *see also*: Stumm, W., Ed. (1990) Aquatic chemical kinetics. Reactions rates of processes in natural waters. Wiley, New York, 545 pp.

Suess, E. (1875) Die Entstehung der Alpen. W. Braunmüller, Vienna, 168 pp.

Sugawara, K. (1948) Kōsui no Kagaku [precipitation chemistry]. *Kagaku* [Nature] **18**, 485–492.

Sugawara, K., S. Oana and T. Koyama (1949, 1953) Separation of the components of atmospheric salt and their distribution. Bulletin of the Chemical Society of Japan **22**, 47–52; and 26 (1953) 123 (cited after Miyake 1965).

Sun, M., Y. Wang, T. Wang, S. Fan, W. Wang, P. Li, J. Guo and Y. Li (2010) Cloud and the corresponding precipitation chemistry in south China: Water-soluble components and pollution transport. *JGR* **115**, doi:10.1029/2010JD014315.

Sundt, L. (1921) Las teorías sobre la formación de los depósitos de nitrato de sodio. *Caliche* **3**, 97–100.

Süpfle, K., P. Hofmann and J. May (1933) Hygienische Studien über Kohlenoxyd. *Zeitschrift für Hygiene* **115**, 623–662.

Suschnig, G. (1900a) Albert Stiger's Wetterschiessen in Steiermark. Verlag von Hans Wagner, Graz, 53 pp.

Suschnig, G. (1900b) Referat über die Erfolge und Beobachtungen beim Wetterschiessen in Österreich, erstattet dem 2. intern. Wetterschiess-Congresse in Padua am 26. November 1900. Neffen, Graz, 11 pp.

Suschnig, G. (1901) Das Wetterschiessen. Verlag von Hans Wagner, Graz, 18 pp.

Sutton, M. A., J. W. Erisman, F. Dentener and D. Möller (2008) Ammonia in the environment: from ancient times to the present. *Environmental Pollution* **156**, 583–604.

Svensson, B. H. and R. Søderlund, eds. (1976) Nitrogen, phosphorus and sulphur – global cycles. Report from a project on biogeochemical cycles initiated by the Scientific Committee on Problems of the Environment (SCOPE) of ICSU, arranged by the Swedish SCOPE Committee of the Royal Swedish Academy of Sciences, Orsundsbro, Sweden, 14–18 December 1975, SCOPE report No. 7. Statens naturvetenskapliga forskningsrad, Stockholm, 192 pp.

Svistov, P. F., E. S. Semenets and M. T. Pavlova [Свистов, Семенец, Павлова] (2018a) Химия атмосфрных осадков: 60 лет регулярных наблюений [Chemistry of atmospheric precipitation: 60 years monitoring]. *Природа* [Priroda] **8**, 51–57.

Svistov, P. F., N. A. Perschina, M. T. Pavlova, A. I. Politschek, E. S. Semenets, S. S. Talasch and A. S. Talasch [Свистов, Першина, Павлова, Полищук, Семенец, Талаш] (2018b) Атлас диаграмм минерализаии влажных выпадений из воздуха и кислотности осадков в Заполянье (1958–2016 гг) [Atlas of diagrams of mineralisation of wet deposition from the atmosphere and acidity in precipitation in polar regions]. Издательство “Лема”, Sankt Petersburg, 102 pp.

Swaab, B. (1904) Der Apparat von Haldane: Eine neue Methode zur Bestimmung des Kohlensäuregehalts der Luft. *Zeitschrift für Untersuchung der Nahrungs- und Genußmittel* **8**, 524–528.

Swietlicki, E., H.-C. Hansson, B. Martinsson, B. Mentes, D. Orsini, B. Svenningsson, A. Wiedensohler, M. Wendisch, S. Pahl, P. Winkler, R. N. Colvile, R. Gieray, J. Lüttke, J. Heintzenberg, J. N. Cape, K. J. Hargreaves, R. L. Storeton-West, K. Acker, W. Wieprecht, A. Berner, C. Kruisz, M. C. Facchini, P. Laj, S. Fuzzi, B. Jones and P. Nason. (1997) Source identification during the Great Dun Fell Cloud Experiment 1993. *Atmos. Env.* **31**, 2441–2451.

Swinnerton, J. W., V. J. Linnenbom and C. H. Cheek (1969) Distribution of methane and carbon monoxide between the atmosphere and natural waters. *Environ. Sci. Technol.* **3**, 836–838.

Szabadváry, F. (1966) History of analytical chemistry. Pergamon, Oxford. 419 pp.

Szutsek, R. (1900) Das praktische Wetterschiessen. Bearb. auf Grund seiner Erfahrungen in der Schiess-Saison 1900. Graz., 22 pp.

Takenaka, N., T. Suzue, K. Ohira and Y. Maeda (1999) Natural denitrification in drying process of dew. *Environ. Sci. Technol.* **33**, 444–1447.

Takeuchi, U. (1949) The amount of chloride contained in the precipitation at Nagano in relation to the air mass over Thatarea (in Japanese; English abstract). *Journal of the Meteorological Society Japan* **27**, 25–38.

Takeuchi, U. and Z. Nakazawa (1951) The relation between the chloride content of rainwater and the air mass of central Honshu (in Japanese, English abstract). *Journal of the Meteorological Society Japan* **29**, 97–100.

Takeuchi, M., H. Okochi and M. Igawa (2003) Deposition of coarse soil particles and ambient gaseous components dominating dew water chemistry. *JGR* **108A**, ACH 4-1–4-5.

Tamm, C. O. (1958) The atmosphere. In: Encyclopaedia of plant physiology. Vol. 4 (ed. W. Ruhland), Springer, Berlin, pp. 233–242.

Tamm, C. O. (1995) Acidification research in Sweden. National and international perspectives. *Ecological Bulletin* **44**, 11–16.

Tanaka, K. (1925) Versuche zur Prüfung der Wielandschen Atmungstheorie. *Biochemische Zeitschrift* **157**, 425–433.

Tanner, R. L., G. Y. Markovits, E. M. Ferreri and T. J. Kelly (1986) Sampling and determination of gas-phase hydrogen peroxide following removal of ozone by gas-phase reaction with nitric oxide. *Journal of Analytical Chemistry* **58**, 1857–1865.

Tanner, R. L. and D. E. Schorran (1995) Measurements of gaseous peroxides near the Grand Canyon: Implication for summertime visibility impairment from aqueous-phase secondary sulfate formation. *Atmos. Env.* **29**, 1113–1122.

Tansley, A. G. (1935) The use and abuse of vegetational terms and concepts. *Ecology* **16**, 284–307.

Tarasick, D., I. E. Galbally, O. R. Cooper, M. G. Schultz, G. Ancellet, Th. Leblanc, T. J. Wallington, J. Ziemke, X. Liu, M. Steinbacher, J. Staehelin, C. Vigouroux, J. W. Hannigan, O. García, G. Foret, P. Zanis, E. Weatherhead, I. Petropavlovskikh, H. Worden, M. Osman, J. Liu, K.-L. Chang, A. Gaudel, M. Lin, M. Granados-Muñoz, A. M. Thompson, S. J. Oltmans, J. Cuesta, G. Dufour, V. Thouret, B. Hassler, Th. Trickl and J. L. Neu (2019) Tropospheric Ozone Assessment Report: Tropospheric ozone from 1877 to 2016, observed levels, trends and uncertainties *Elementa: Science of the Anthropocene* **7**, article 39, 72 pp.

Taube, H. (1957) Photochemical reactions of ozone in solution. *Trans. Faraday Soc.* **53**, 656–665.

Taylor, E. G. R. (1942) Problems of fog and air pollution. *Smokeless Air* **12**, 96–100.

Taylor, G. I. (1917) The formation of fog and mist. *Q. J. R. Meteor Soc.* **43**, 241–268.

Taylor, H. S. (1926) Photosensitisation and the mechanism of chemical reactions. *Trans. Faraday Soc.* **21**, 560–568.

Taylor, A. B. H. (1994) An episode with May-dew. *History of Science* **32**, 163–184.

Teakle, L. J. H. (1937) The salt (sodium chloride) content of rain-water. *Journal of the Department of Agriculture West Australia* **14**, 115–123.

Teich, M. (1893) Die Methode von Petterson und Palmqvist zur Bestimmung der Kohlensäure in der Luft. *Archiv für Hygiene und Bakteriologie* **19**, 38–52.

Teichert, F. (1952) Erfahrungen mit chemischen Ozonmessmethodn. *Z. Met.* **6**, 132–138.
Teichert, F. (1953) Bericht über Ozonmessversuche mit Fluoreszin. *Z. Met.* **7**, 33–34.
Teichert, F. (1955) Vergleichende Messungen des Ozongehaltes der Luft am Erdboden und in 80 m Höhe. *Z. Met.* **9**, 21–27.
Teichert, F. (1956a) Betrachtungen zur Staubuntersuchung mit Hilfe von Staubbechern. *Z. Met.* **10**, 161–173.
Teichert, F. (1956b) Ergebnisse einer regionalen Staubmessung in der Meteorologie. *Z. Met.* **10**, 369–373.
Teichert, F. (1957a) Die natürliche Radioaktivität des Regenwassers in Dresden-Wahnsdorf. *Z. Met.* **11**, 14–19.
Teichert, F. (1957b) Die Verstaubung der Luft in Abhängigkeit von meteorologischen Faktoren. *Z. Met.* **11**, 112–118.
Teichert, F. and B. Greifenhagen (1954) Physikalisch-chemische Regenwasserunter-suchungen. *Z. Met.* **8**, 275–281.
Teichert, F. and W. Warmbt (1955) Ozonuntersuchungen am meteorologischen Observatorium Wahnsdorf. In: Abhandlungen des meteorologischen und hydrologischen Dienstes der Deutschen Demokratischen Republik, Nr. 34 (Band V), Akademie-Verlag, Berlin, 67 pp.
Teichert, F. and W. Warmbt (1956) Ozonmessungen Dresden-Wahnsdorf und Fichtelberg 1954/1955. *Z. Met.* **10**, 264–277.
Teilhard de Chardin, P. (1961) The phenomenon of Man (English translation of original 1955 French edition). Harper Torchbooks, New York, 318 pp.
Thalmann, E., R. Burkard, T. Wrzesinsky, W. Eugster and O. Klemm (2002) Ion fluxes from fog and rain to an agricultural and a forest ecosystem in Europe. *Atmos. Res.* **64**, 147–158.
Than, C. (1867) Ueber das Kohlenoxysulfid. *Liebigs Ann. Supplementband* **5**, 236–247.
Than, C. (1870) Ueber Nachweisung von Ozon bei rascher Verbrennung. *Pharmaceutische Centralhalle für Deutschland* **11**, 327–328.
Thénard, L. J. (1818) Sur la combination de l'oxigène avec l'eau, et sur le propprietes extraordinaire que possède l'eau oxigènée. *Mémoires de l'Académie des sciences de l'Institut de France* **3**, 385–488.
Thénard, L. J. (1819) New results on the combination of oxygen with water. *Annals of Philosophy* **14**, 209–210; *Ann. chim. phys.* **10**, 335.
Thénard, L. J. (1827) Traité de Chimie Élémentaire, Théorique Et Pratique, Vol. 5: Suivi d'un Essai sur la Philosophie Chimique Et d'un Précis sur l'Analyse. Crochard, Paris, 556 pp.
Thénard, P. (1872) Sur un nouveau procédé de dosage de l'ozone. *C. r.* **75**, 174–177 and 351.
Thiele, H. (1907) Einige Reaktionen im ultravioletten Lichte. *Ber.* **40**, 4914–4916.
Thiele, H. (1909) Einige Reaktionen im ultravioletten Lichte. *Z. angew. Chem.* **22**, 2472–2484.
Thiele, U. (2000) Zur Geschichte der sekundären Verwertung von Stoffen in Freiberg. *Mitteilungen des Freiberger Altertumsvereins* **86**, 15–25.
Thierry, M. de (1886) La recherche médico-légale du sang: Grand spectroscope d'absorption. *La Nature; revue des sciences et de leurs applications aux arts et à l'industrie* **14** (679), 3–6.
Thierry, M. de (1897) Dosage de l'ozone atmosphérique au Mont-Blanc. *C. r.* **124**, 460–463.
Thierry, M. de (1899) Dosage du gaz carbonique au Mont Blanc. *C. r.* **129**, 315–316.
Thiriet, A. (1886) Ville de Sedan. Musée municipal. Section des sciences naturelles: Ccatalogue general. Géologie, Botanique, Zoologie, Anthropologie. Impr. A. Lahure, Paris, 32 pp.
Thomas, H. (1997) Anorganischer Kohlenstoff im Oberflächenwasser der Ostsee. Meereswissenschaftliche Berichte No. 22, Institut für Ostseeforschung, Warnemünde, 120 pp.
Thomas, J. K. (1965) The rate of reactions of the hydroxyl radical. *Trans. Faraday Soc.* **61**, 702–707.
Thomas, N. C. (1991) The early history of spectroscopy. *Journal of Chemical Education* **68**, 631–634.
Thomasson, A., S. Geffrey, E. Fréjafon, D. Weidauer, R. Fabian, Y. Godet, M. Nominé, T. Ménard, P. Rairoux, D. Möller and J. P. Wolf (2002) Lidar mapping of ozone – episode dynamics in Paris and intercomparison with spot analysers. *Applied Physics B* **74**, 453–459.

Thompson, A. M. (1992) The oxidizing capacity of the Earth's atmosphere: probable past and future change. *Science* **256**, 1157–1165.
Thompson, A. M. (1995) Photoche-mical modelling of chemical cycles: issues related to the interpretation of ice core data. In: Ice core studies of global biogeochemical cycles (ed. R. J. Delmas), NATO ASI Series, Vol. 30, Springer, pp. 265–297.
Thompson, A. M., M. A. Owens and R. W. Stewart (1989) Sensitivity of tropospheric hydrogen peroxide to global chemical and climate change. *Geophys. Res. Lett.* **16**, 53–56.
Thompson, A. M., M. A. Huntley and R. W. Stewart (1991) Perturbations to tropospheric oxidants, 1985–2035: 2. Calculations of hydrogen peroxide in chemically coherent regions. *Atmos. Env.* **25A**, 1837–1850.
Thomson, R. D. (1855) Report on the examination of certain atmosphere during the epidemic of cholera. In: Appendix to report of the committee for scientific inquiries in relation to the cholera-epidemic of 1854, G. E. Eyre and W. Spottiswoode, London, pp. 119–133.
Thomson, Th. (1807) A system of chemistry. In five volumes. The third edition. Vol. 4, Edinburgh, 62 pp.
Thomson, Th. (1820) A system of chemistry. Vol. 3, the sixth edition, London, 622 pp.
Thomson, Th. (1830) The history of chemistry. Vol. 1. Colburn and Bentley, London, 349 pp.
Thorpe, T. E. (1867) On the amount of carbonic acid contained in sea-air. *J. Chem. Soc.* **20**, 189–199.
Thorsheim, P. (2006) Inventing pollution. Coal, smoke and culture in Britain since 1800. Ohio University Press, Athens (USA), 307 pp.
Thurneisser, L. (1572) Pison. Von Kalten, Warmen Minerischen vnd Metallischen Wassern, sampt der vergleichunge der Plantarum vnd Erdgewechsen. 10 Bücher. Johan Eichorn, Frankfurt/Oder, 420 pp.
Tian, A. (1911) Radiations decomposing water and the extreme ultraviolet spectrum of the mercury arc. *C. r.* **152**, 1483–1483.
Tian, X., D. Möller, K. Acker, W. Wieprecht, R. Auel, D. Kalaß, V. Schmidt and J. Hofmeister (1999) Experimental observation of S(IV) in cloud water. *Geophysical Research Abstracts* **1**(2), 504.
Tichy, H. (1939) Gleichzeitige Messung von Ultraviolett und bodennahem Ozon. *Der Balneologe* **6**, 125–130.
Tidwell, T. (2013) Sunlight and free radicals. *Nature Chem* **5**, 637–639.
Tiggelen, A. van (1942) L'oxydation du methane photosensibilisee par l'acetone. *Annales des Mines des Belgique* **43**, 117–144.
Tilden, W. A. (1874) On aqua regia and the nitrosyl chlorides. *Chemical News* **29**, 183.
Tissandier, G. (1874) Les poussières atmospheriques. *C. r.* **78**, 821–824. *See also*: *Ann. chim. phys.* V. Ser, Tom 3, 203–208.
Tissandier, G. (1877) Les poussières de l'air. Gauthier-Villars, Paris, 106 pp.
Tissandier, G. (1878) Appendice: L'acide carbonique de l'air. Expériences exécutées a bord du ballon "Le Zènith". In: Histoire de mes ascensions; récit de vingt-quatre voyages aériens (1868–1877) précédé de simples notions sur les ballons et la navigation aérienne, M. Dreyfous, Paris (344 pp.), pp. 329–333. *See also*: Determination of carbonic acid in the air, on board the Balloon "Zenith." *Chemical News* **31** (1875) 217.
Tissandier, G. (1879) Observations météorologiques en ballon. Gauthier-Villars, Paris, 51 pp.
Tizenhauz (1848) Ueber eine aus der Atmosphäre niedergefallene Substanz. *Pogg. Ann.* **149**, 364–366. From: Note concernant une substance tombée de l'atmosphere. *C. r.* **23** (1846) 452–454.
Tobo, Y., P. J. DeMott, T. C. J. Hill, A. J. Prenni, N. G. Swoboda-Colberg, G. D. Franc and S. M. Kreidenweis (2014) Organic matter matters for ice nuclei of agricultural soil origin. *ACP* **14**, 8521–8531.
Tohjima, Y., T. Machida, T. Watai, I. Akama, T. Amari and Y. Moriwaki (2005) Preparation of gravimetric standards for measurements of atmospheric oxygen and reevaluation of atmospheric oxygen concentration. *JGR* **110**, D11302, doi:10.1029/2004JD005595.

Tomkeiev, S. I. (1944) Geochemistry in the U. S. S. R. *Nature* **154**, 814–816.
Tomlinson, Ch. (1847) The dew-drop and the mist; or, an account of the nature, properties, dangers and use of dew and mist, in various parts of the world. Society for Promoting Christian Knowledge, London, 115 pp. (this book appeared authorless).
Tomlinson, Ch. (1860) The dew-drop and the mist: An account of the phenomena and properties of atmosphere. Society for Promoting Christian Knowledge, London, 346 pp.
Torstensson, G. (1954) Stickstoff- und Schwefelverbindungen aus der Atmosphäre und ihre Bedeutung für die Pflanzen. In: Sitzungberichte der Deutschen Akademie der Landwirtschaftswissenschaften, Berlin, Band III, Heft 18, 18 pp.
Tracy, S. M. (1895) Ammonia in rainwater. Mississippi Agricultural Experimental Station 8th Ann. Rpt., p. 103.
Traube, M. (1882) Über Aktivierung des Sauerstoffs. *Ber.* **15**, 222, 659, 2421, 2423.
Traube, M. (1887) Über die elektrolytische Entstehung des Wasserstoffhyperoxyds an der Kathode. *Sitzungsber. Berlin* pp. 1041–1050. *See also*: Ueber die elektrolytische Entstehung des Wasserstoffhyperoxyds an der Anode. *Ber.* **20**, 3345–3351.
Traube, M. (1893) Ueber die Constitution des Wasserstoffhyperoxyds und Ozons. *Ber.* **26**, 1476–1481.
Treffeisen, R., K. Grunow, D. Möller and A. Hainsch (2002) Quantification of source region influences on the ozone burden. *Atmos. Env.* **36**, 3565–3582.
Tremmel, H. G., W. Junkermann, F. Slemr and U. Platt (1993) On the distribution of hydrogen peroxide in the lower troposphere over the northeastern United States during late summer 1988. *JGR* **98D**, 1083–1099.
Troili-Petersson, G. (1897) Pettersson-Palmqvists Kohlensäureapparat modifiziert für Ventilationsuntersuchungen. *Zeitschrift Für Hygiene und Infektionskrankheiten* **26**, 57–65. *See also*: Troili-Petersson, G. (1898) Über den Kohlensäuregehalt der Atmosphäre. *Sven. Vetensk. Handl.* (Svenska vetenskapsakademiens handlingar) 2, **23**, 1–17.
Trommsdorff, J. S. (1699) Ros Mellis, Non Ros, Nec Mellis Ros, der ungleich angegebene Honig-Thau. G. H. Müllerus, Erfurti, 21 pp.
Truchot, P. (1873a) Sur la proportion d'acide carbonique existant dans l'air atmosphérique. Variation de cette proportion avec l'altitude. *C. r.* **77**, 675–678.
Truchot, P. (1873b) Sur la quantité d'ammoniaque contenue dans l'air atmosphérique à différentes altitudes. *C. r.* **77**, 1159–1161.
Tschudi, J. von (1862) Der Ozongehalt der Luft im Verhältnis zum Krankenstande eines Ortes. *Wiener Medicinische Wochenschrift* **12**, 774–775.
Tsunogai, S. (1971a) Ammonia in the oceanic atmosphere and the cycle of nitrogen compounds through the atmosphere and hydrosphere. *Geochemical Journal* **5**, 57–67.
Tsunogai, S. (1971b) Oxidation rate of sulfite in water and its bearing on the origin of sulfate in meteoritic precipitation. *Geochemical Journal* **5**, 175–185.
Tsunogai, S. and K. Ikeuchi (1968) Ammonia in the atmosphere. *Geochemical Journal* **2**, 157–166.
Turco, R. P., R. C. Whitten, O. B. Toon, J. B. Pollack and P. Hamill (1980) OCS, stratospheric aerosols and climate. *Nature* **283**, 283–285.
Turner, E. and R. Christison (1827) On the effects of the poisonous gases on vegetation. *Edinburgh Medical and Surgical Journal* **28** (1827), 356–363.
Turner, H. (1991) Phytometeorologie. In: Lufthaushalt, Luftverschmutzung und Waldschäden in der Schweiz. Bd. 4 (ed. E. Schipbach), Zürich, pp. 57–70.
Twomey, S. (1968) On the composition of cloud nuclei in the northeastern United States. *Journal de Recherches Atmospheriques* **3**, 281–285.
Twomey, S. (1971) The composition of cloud nuclei. *Journal of Atmospheric Science* **28**, 377–381.
Twomey, S. (1974) Pollution and the planetary albedo. *Atmos. Env.* **8**, 1251–1256.

Tyndall, J. (1861) On the absorption and radiation of heat by gases and vapours and on the physical connection of radiation, absorption and conduction. *Philosophical Magazine, Series 4* **22**, 169–194, 273–285.

Tyndall, J. (1863) On the absorption and radiation of heat by gaseous matter. Second memoir. *Phil. Trans.* **152**, 59–98.

Tyndall, J. (1870) On haze and dust. *Nature* **1**, 339–342.

Tuxen, C. F. A. (1890) Undersøgelser over Regens Betydning her i Landet som Kvaelstofkilde for kulturplauter. *Tidsskrift for Landøkonomi* **72**, 325–350.

Udías, A. (2019) Jesuits and the natural sciences in modern times, 1814–2014. Brill Research Perspectives in Jesuit Studies **1**(3), 104 pp.

Uffelmann, J. (1888) Luftuntersuchungen, ausgeführt im hygienischen Institut der Universität Rostock. *Archiv für Hygiene und Bakteriologie* **8**, 262–350.

Uffelmann, J. (1890) Handbuch der Hygiene. Urban & Schwarzenberg, Wien/Leipzig, 852 pp.

Ule, O. (1863) Das Ozon. *Die Natur* **12**, 1–3, 17–19, 41–43, 49–52, 57–59.

Umlauft, F. (1891) Das Luftmeer. Die Grundzüge der Meteorologie und Klimatologie. A. Hartleben's Verlag, Wien, 488 pp.

Ungeheuer, H. (1950) Erste Ergebnisse der Aranmessungen in Bad Tölz. *Berichte des Wetterdienstes US-Zone* **12**, 221–224.

Urey, H. C., L. H. Dawsey and F. O. Rice (1929) The absorption spectrum and decomposition of hydrogen peroxide by light. *J. Am. Chem. Soc.* **51**, 1371–1383.

Uri, N. (1952) Inorganic free radicals in solution. *Chemical Reviews* **50**, 375–454.

Usher, F. L. and B. S. Rao (1917) The determination of ozone and oxides of nitrogen in the atmosphere. *J. Chem. Soc., Transactions* **111**, 799–809.

Usov, N. I. [Усов] (1940) Генезис и мелиорация почв Прикаспийской низменности [Genesis and melioration of soils at the Caspian lowland]. Саратовское обл. госуд. изд-во, Саратов [State Publ. Saratov], 440 pp.

Valverde-Canossa, J., W. Wieprecht, K. Acker and G. K. Moortgat (2005) H_2O_2 and organic peroxide measurements in an orographic cloud: the FEBUKO experiment. *Atmos. Env.* **39**, 4279–4290.

Van Baalen, C. and J. E. Marler (1966) Occurrence of hydrogen peroxide in sea water. *Nature* **211**, 951.

van Marum, M. (1786) In German: Beschreibung einer ungemein großen Elektrisir-Maschine und der damit im Teylerschen Museum zu Haarlem angestelten Versuche. Aus dem Holländisch übersetzt. Vol. 1. Schwickert, Leipzig 1786, 72 pp. Dutch first edition: Beschryving eener ongemeen groote Electrizeer-Machine, geplaatst in Teyler's Museum te Haarlem, in van de proefneemingen met dezelve in 't werk gesteld. J. Enschede, Haarlem 1785, 205 pp.

Van Nuys, T. C. and B. F. Adams (1887) Estimation of carbonic acid in the air. *American Chemical Journal* **9**, 64–66.

Van Valin, C. C., J. D. Ray, J. F. Boatman and R. L. Gunter (1987) Hydrogen peroxide in air during winter over the south-central United States. *Geophys. Res. Lett.* **14**, 1146–1149.

Van Valin, C. C., M. Luria, J. D. Ray and J. F. Boatman (1990) Hydrogen peroxide and ozone over the Northeastern United States in June 1987. *JGR* **95**, 5689–5695.

Varigny, H. de (1896) Air and life. *Smithsonian Miscellaneous Collections* **39**, 3–69.

Varotsos, C. and C. Cartalis (1991) Re-evaluation of surface ozone over Athens, Greece, for the period 1901–1940. *Atmos. Res.* **26**, 303–310.

Vehkamäki, H. (2006) Classical nucleation theory in multicomponent systems. Springer Science & Business Media, 188 pp.

Veley, V. H. (1892) The conditions of the formation and decomposition of nitrous acid. *Proc. R. Soc. Lond.* **52**, 27–55.

Veltmann, C. H. (1827) Zur Kenntniss des Moordampfes. *Archiv für die gesammte Naturlehre* **10**, 266–273.

Veraart, A. W. (1931) Meer Zonneschijn in het Nevelig Noorden; meer regen in de Tropen. Seyffardt's Boek en Muziekhandel, Amsterdam, 32 pp.
Verma, M. P. (2005) Revised analytical method for the determination of carbonic species in rain, ground and geothermal waters. Proceedings World Geothermal Congress, Antalya, Turkey, 24–29 April, 7 pp.
Vernadsky, V. I. [Вернадский] (1910) Парагенезис химических элементов в земной коре [Paragenesis of chemical elements in the earth's core]. Дневник XII Съезда русских естествоиспытателей и врачей в Москве с 28 декабря 1909 г. по 6 января 1910 г. [Proceedings of the 12th Congress of natural scientists and physicians in Moscow, December 28, 1909 to January 7, 1910]. *Вестник Московского университета* **7**, 73–91.
Vernadsky, V. I. [Вернадский] (1921) Записка об изучении живого вещества с геохимической точки зрения [Note on investigating living organism from geochemical point of view]. Государственное изд., Петроград, 123 pp.
Vernadsky, W. I. (1923) Plan for the establishment of the bio-geochemical laboratory. *Proceedings and Transactions of the Liverpool Biological Society* **37**, 38–48.
Vernadsky, V. I. (1924) La Géochimie. Nouvelle Collection scientifique. Félix Alcan, Paris, 404 pp.
Vernadsky, V. I. (1926a) Биосфера. Науч. хим.-техн. изд-во, Ленинград, 146 pp.; translated into French: La Biosphére (1929), Félix Alcan, Paris, 232 pp.; German see Vernadsky (1930) and English see Vernadsky (1998).
Vernadsky, V. I. (1926b) Études biogéochimiques. I. Sur la vitesse de la transmission de la vie dans la biosphère. *Известия Академии наук СССР. Серия 7. Отделение математических и естественных наук* **20**, 727–744.
Vernadsky, V. I. (1927a) Études biogéochimiques. 2. La vitesse maximum de la transmission de la vie dans la biosphère. *Известия Академии наук СССР. Серия 7. Отделение математических и естественных наук* **21**, 241–254.
Vernadsky, V. I. [Вернадский] (1927b) О проделанной Биогеохимической лабораторией работе по анализу живоговещества. *Известия Академии наук СССР. Сер. 6* **21**, 1589–1591.
Vernadsky, V. I. [Вернадский] (1927c) Очерки геохимии [Essays on geochemistry]. Государственное изд., Москва-Ленинград, 368 pp.; in Japanes: Tokyo, 1933, 553 pp.
Vernadsky, W. (1928) Über die geochemische Energie des Lebens in der Biosphere. *Zentralblatt für Mineralogie, Geologie und Paläontologie Abt. B* **11**, 583–594.
Vernadsky, V. I. (1930) Geochemie in ausgewählten Kapiteln von W. J. Vernadsky. Autorisierte Übersetzung aus dem Russischen von Dr. E. Kordes. Akademische Verlagsgesellschaft, Leipzig, 370 pp.
Vernadsky, V. I. [Вернадский] (1931) О биогеохимическом изучении явления жизни [On biogeochemical studies on phenomena of life]. Государственное изд., Москва-Ленинград, 140 pp.
Vernadsky, V. I. [Вернадский] (1932) Биогеохимия и ее значение для изучения биосферы [Biogeochemistry and ist importance for studying the biosphere]. Доклады, представл. К Торжеств. юбил. сессии Академии наук СССР, посвящ. 15-летию Октябрьской революции. Л.: Изд-во АН СССР, 17–18.
Vernadsky, V. I. [Вернадский] (1934a) Проблемы биогеохимии [Problems of biogeochemistry]. Вып. 1. Значение биогеохимии для познаниябиосферы. Л.: Изд-во АН СССР, 1934. 47 pp.; in English: Problems of biogeochemistry, the fundamental matter-energy difference between the living and the inert natural bodies of thebiosphere. *Transactions of the Connecticut Academy of Arts and Sciences* (1944) **35**, 483–517.
Vernadsky, V. I. [Вернадский] (1934b) История минералов земной коры. Том 2. История природных вод. Часть 1. Выпуск 2 [History of minerals of the earth crust. Vol. 2. History of natural waters. Part 1. Issue 2]. Leningrad, ОНТИ, Химтеорет, 212 pp. (pp. 203–403).

Vernadsky, V. I. [Вернадский] (1935) Биосфера и стратосфера [Biosphere and stratosphere]. Труды Всесоюз. конф. по изуч. Стратосферы [Proceedings of the All-Union Conference on Investigation of the Stratosphere, March 31–April 6, 1934], Изд-во АН СССР, Москва-Ленинград, pp. 575–578.

Vernadsky, V. I. [Вернадский] (1944) Несколько слов о ноосфере [Some words on noosphere. *Успехи современной биологии* [Advances in Modern Biology] **18**, 113–120.

Vernadsky, W. (1945a) La biogeochimie. *Scientia, Ser. 5* **78**, 77–84.

Vernadsky, V. I. (1945b) The biosphere and the noosphere. *Scientific American* **33**, 1–12.

Vernadsky, V. I. (1954) [Вернадский] Избранные сочинения [Selected works]. Изд-во АН СССР, Vol. 1: Очерки геохимии. Статьи по геохимии (Essays on geochemistry), 696 pp.

Vernadsky, V. I. [В. И. Вернадский] (1991) Научная мысль как планетное явление [Scientific thought as a planetary phenomenon]. Ed. A. L. Janschin [А. Л. Яншин], Nayka, Moscow. 271 pp.

Vernadsky, V. I. (1998) The biosphere. Translated by D. B. Langmuir. Opernicus. Springer-Verlag, New York, 192 pp; a reduced version in English appeared in 1986 at Synergetic-Press, 82 pp.

Vernadsky, V. I. (2000–2001) Problems of biogeochemistry. II: The fundamental matter-energy difference between the living and the inert natural bodies of the biosphere. Translated from the Russian by Jonathan Tennenbaum and Rachel Douglas. In: 21st Century. Science & Technology, Vol. 13, Issue 4., pp. 20–39.

Vernadsky, V. I. (2005) Some words about the noosphere. Translated from the Russian by Rachel Douglas. In: 21st Century. Science & Technology, Vol. 18, Issue 1. pp. 16–21.

Vernadsky, V. I. and A. I. Oparin [Вернадский, Опарин] (1930) Круговорот веществ [Cycle of matter]. In: Большая медицинская энциклопедия (БМЭ) [Great medical encyclopedia] **14**, 770–778.

Vernon, G. V. (1868) Notes of comparison of ozone papers. *Proceedings of the Literary and Philosophical Society of Manchester* **7**, 219–221.

Verver, B. (1840) Expérience sur la composition de l'air atmospherique. *Bulletin des sciences physiques et naturelles en Néerlande*, Livraison 4, pp. 191–213. *See also* (same heading): *Revue Scientifique et Industrielle* **6** (1841) 121–123. It is notable that the heading in German in *Isis* (1843, p. 790) was extended: Versuche über die Bestandtheile der atmosphärischen Luft. Sehr zahlreiche Versuche über diesen Gegenstand, der nicht bloß Sauerstoff und Stickstoff betrifft, sondern auch Wasser, Kohlensäure und Wasserstoff [Experiments upon the composition of atmospheric air. Very numerous experiments upon this issue, not only concerns oxygen and nitrogen but also on water, carbonic acid and hydrogen].

Vet, R., R. S. Artz, S. Carou, M. Shaw, C.-U. Ro, W. Aas, A. Baker, Van C. Bowersox, F. Dentener, C. Galy-Lacaux, A. Hou, J. J. Pienaar, R. Gillett, M. C. Forti, S. Gromov, H. Hara, T. Khodzher, N. M. Mahowald, S. Nickovic, P. S. P. Rao and N. W. Reid (2014) Special Issue: A global assessment of precipitation chemistry and deposition of sulphur, nitrogen, sea salt, base cations, organic acids, acidity and pH and phosphorus. *Atmos. Env.* **93**, 3–100.

Viemeister, P. E. (1960) Lightning and the origin of nitrate found in precipitation. *Journal of Meteorology* **17**, 681–683.

Vigroux, E. (1953) Contribution à l'étude expérimentale de l'absorption de l'ozone. Masson, Paris, 54 pp.

Villars, D. S. (1927) The photolysis of potassium nitrate. *J. Am. Chem. Soc.* **49**, 326–337.

Ville, G. (1852) Recherches expérimentales sur la vegetation. *C. r.* **35**, 464–468.

Vinogradov, A. P. (1943) Biogeochemical research in the U. S. S. R. *Nature* **151**, 659–661.

Vione, D., V. Maurino, C. Minero and E. Pelizzeti (2003) The atmospheric chemistry of hydrogen peroxide: A review. *Annali di chimica* **93**, 477–488.

Visser, S. (1961) Chemical composition of rainwater in Kampala, Uganda and its relation to meteorological and topographical conditions. *JGR* **66**, 3759–3766.

Visser, S. (1964) Origin of nitrates in tropical rain. *Nature* **201**, 35–36.

Vittori, O. (1960) Preliminary note on the effects of pressure waves upon hailstones. *Nubila* **3**, 34–52.
Vityn, Ja. Ja. [Витынь, Я. Я.] (1911) Количество Cl и SO_4, поступающих в почву с атмосферными осадками [The amount of Cl and SO_4, deposits to soil with atmospheric precipitation]. *Журнал опытной агрономии* [*Journal of Experimental Agronomy*] 20–32. Note that Drischel (1949) and Eriksson (1952) wrote *Wituynj*, not conforming with modern transcription.
Vogel, A. (1820) vorläufige Nachricht über die Natur der Seeluft. *Gilb. Anal.* **66**, 93–99.
Vogel, A. (1822) Versuche und Bemerkungen über die Bestandtheile der Seeluft. *Gilb. Anal.* **72**, 277–288.
Vogel, A. (1854) Chemische Untersuchung der atmosphärischen Luft während der Cholera-Epidemie zu München 1854. Ch. Kaiser, München, 8 pp.
Vogel, A. (1872) Ueber den Ammoniakgehalt des Schneewassers. *Neues Repertorium für Pharmacie* **21**, 329–339. *See also*: *Sitzungsber. München* **10** (1872) 124–133.
Vogel, A. (1883) Zur Geschichte der Liebig'schen Mineraltheorie. Verlag von Carl Habel, Berlin, 44 pp.
Vogel, H. A. (1828) Das Pyrrhin scheint keine eigenthümliche Substanz zu seyn. *Archiv für die gesammte Naturlehre* **15**, 97–101.
Vogel, H. W. (1871) Ueber die Lichtempfindlichkeit des rothen Blutlaugensalzes. *Polytechnisches Journal (Dingler's J.)* **199**, 323–325.
Vogt, H. (1939? [no year given]) Deutsche bäderwissenschaftliche Forschung 1933–1939. 62 pp. [compilation of all departments of the *Reichsanstalt* with heads and publications].
Voland, B. (1987) Die historischen Quellen der Umweltgeochemie und geochemischen Ökologie in der DDR. *Freiberger Forschungshefte Serie D* **178**, 36–74.
Volk, N. J., J. W. Tidmore and D. T. Meadows (1945) Supplements to high- analysis fertilizers with special reference to sulfur, calcium, magnesium and limestone. *Soil Science* **60**, 427–435.
Volman, D. H. (1949) The vapor-phase photo decomposition of hydrogen peroxide. *The J. Chem. Phys.* **17**, 947–950.
Volmer, M. and A. Weber (1926) Tröpfchenbildung in Dämpfen. *Z. phys. Chem.* **119**, 277–301.
Volmer, M. and H. Flood (1934) Tröpfchenbildung in Dämpfen. *Z. phys. Chem. A* **170**, 273–285.
Volta, A. (1778) Premier mémoire sur le gaz inflammable des marais. J. H. Hertz, Strassbourg, 191 pp. (translation from *Lettere sul l'aria infiammabile*, Milan, 1777). *See also*: Briefe über die entzündbare Luft der Sümpfe: nebst drey andern Briefen von dem nämlichen Verfasser, die aus dem Mayländischen Journal genommen sind, und einer Kupferplatte (1778). Stein, Straßburg, 226 pp.
Volz, A. and D. Kley (1988) Evaluation of the Montsouris series of ozone measurements made in the nineteenth century. *Nature* **332**, 240–242.
Volz-Thomas, A., D. Mihelcic, H.-W. Pätz, M. Schultz, B. Gomišcek, A. Lindskog, J. Mowrer, P. Oyola, K. Hanson, R. Schmitt, T. Nielson, A. Eggelov, F. Stordal and M. Vosbeck (1997) Photochemical ozone production rates at different TOR sites. In: Transport and chemical transformation of pollutants in the troposphere (EUROTRAC). Vol 6: Tropospheric ozone research (Ed. Ø. Hov), Springer-Verlag, Berlin, pp. 65–94.
Volz-Thomas, A., H. Geiss, A. Hofzumahaus and K. H. Becker (2003) Introduction to special section: Photochemistry Experiment in BERLIOZ (PHOBE). *JGR* **108** (D4) special issues, which contains 9 contributions.
Volz, F. (1952) Über die Zerstörung des Ozons in der Troposphäre. *Berichte des Wetterdienstes US-Zone* **35**, 257–261.
Vonnegut, B. (1947) The nucleation of ice formation by silver iodide. *Journal of Applied Physics* **18**, 593–595.
Vosmaer, A. (1916) Ozone. Its manufacture, properties and uses. D. van Nostrand Comp, New York, 197 pp.

Vuuren, D. P. van and K. Riahi (2011) The relationship between short-term emissions and long-term concentration targets. *Climate Change* **104**, 793–801.
Wafer, L. (1699) A new voyage and description of the Isthmus of America. Reprinted from the original edition of 1699. G. P. Winship, Cleveland, 1905, 209 pp.
Wagner, H. (1929) Titanweißkatalyse? *Farben-Zeitung* **34**, 1243–1245.
Wagner, H. (1939) Die Körperfarben. Chemie in Einzeldarstellungen, Vol. XIII (ed. J. Schmidt), Wissenschaftliche Verlagsgesellschaft, Stuttgart, 688 pp.
Wagner, T., C. von Friedeburg, M. Wenig, C. Otten and U. Platt (2002) UV-visible observations of atmospheric O_4 absorptions using direct moonlight and zenith-scattered sunlight for clear-sky and cloudy sky conditions. *JGR* **107**(D20), 4424, doi:10.1029/2001JD001026.
Waite, A. E. (1887) The real history of the Rosicricians. G. Reway, London, 446 pp.
Waldbauer, O. (1935) Schwefelbestimmung in der Luft von Budapest nach Liesegang's Glockenmethode. *Gesundheits-Ingenieur* **48**, 160–161.
Walden, P. (1941) Geschichte der organischen Chemie seit 1880. Zweiter Band zu C. Graebe: Geschichte der organischen Chemie. Springer-Verlag, Berlin, 946 pp.
Waldenburg, L. (1864) Die Inhalation der zerstäubten Flüssigkeiten, sowie der Dämpfe und Gase in ihrer Wirkung auf die Krankheiten. Lehrbuch der respiratorischen Therapie. G. Reimer, Berlin, 567 pp.
Waldmann, W. (1872a) Was sind und wie wirken Sauerstoff- und Ozonsauerstoff(?)-Inhalationen? Zur Klärung dieser Frage veröffentlicht. A. Hirschwald, Berlin, 27 pp.
Waldmann, W. (1872b) Zur Ozonfrage. *Deutsche Klinik* **24**, 313–316.
Waldman, J. M., J. W. Munger, D. J. Jacob, R. C. Flagan and M. R. Hoffmann (1982) Chemical composition of acid fog. *Science* **218**, 677–680.
Waldman, J. M., J. W. Munger, D. L. Jacob and M. R. Hoffmann (1985) Chemical characterization of stratus cloudwater and its role as a vector for pollutant deposition in a.
Walker, J. (1900) Estimation of atmospheric carbon dioxide. *J. Chem. Soc., Transactions* **77**, 1110–1114.
Walker, S. J., M. J. Evans, A. V. Jackson, M. Steinbacher, C. Zellweger and J. B. McQuaid (2006) Processes controlling the concentration of hydroperoxides at Jungfraujoch Observatory, Switzerland. *ACP* **6**, 5525–5536.
Waller, A. (1847a) Microscopic observations on the so-called vesicular vapours of water, as existing in the vapours of steam and in clouds & c. *Phil. Trans.* **137**, 23–30.
Waller, A. (1847b) Über Hagel und die in den Hagelkörnern enthaltenen organischen Bestandtheile. *Fortschritte der Geographie und Naturgeschichte* **41**, 345–348. Additional observations om hail and organic bodies containing in hailstones. *Philosophical Magazine* **30** (1847) 159–170.
Waller, R. E. (1964) Experiments on the calibration of smoke filters. *Journal of the Air Pollution Control Association* **14**, 323–325.
Wallerius, J. G. (1750) Mineralogie, oder Mineralreich von Ihm eingeteilt und beschrieben. Ins Deutsche übersetzt von Johann Daniel Denso. G. Nicolai, Berlin, 600 pp. [First German edition of this important mineralogical classic first published in Swedish in 1747].
Wallerius, J. G. (1770) The natural and chemical elements of agriculture. John Bell, London, 200 pp. [First English edition of agriculturae fundamenta chemica (1761), which was published as a dissertation at Uppsala with Wallerius as praeses and Count Gustaf Adolph Gyllenborg as respondens].
Walling, C. (1957) Radicals in solution. John Wiley & Sons, New York, 631 pp.
Wallington, T. J., J. H. Seinfeld and J. R. Barker (2018) 100 years of progress in gas-phase atmospheric chemistry research. In: *Meteorological Monographs* **59**, pp. 10.1–10.52.
Walshaw, C. D. (1989) Dobson – The man and his work. *Planetary and Space Science* **37**, 1485–1507.
Wanklyn, J. A. and E. T. Chapman (1891) Water analysis. 8th ed. Paul Trübner & Comp, London, 214 pp.

Wanklyn, J. A. and W. J. Cooper (1891) Air-Analysis: A practical Treatise on the examination of air with an appendix on illumination gas. Paul Trübner & Comp, London, 82 pp.
Warburg, E. (1902) Ueber die Bildung des Ozons bei der Spitzenentladung in Sauerstoff. Ann. Phys. 314 (Vierte Folge 9), 781–792. Ueber spontane Desozonisierung. Ibid., 1286–1303.
Warburg, E. (1904) Über die Ozonisierung des Sauerstoffs durch stille elektrische Entladung. *Ann. Phys.* **318** (*Vierte Folge* **13**), 464–476.
Warburg, E. (1911) Über den Energieumsatz bei photochemischen Vorgängen in Gasen. *Sitzungsber. Berlin* pp. 746–754.
Warburg, E. (1914) Über den Energieumsatz bei photochemischen Vorgängen in Gasen. IV. Einfluß der Wellenlänge und des Druckes auf die photochemische Ozonisierung. *Sitzungsber. Berlin* pp. 872–885. Further communications by Warburg: Quantenchemische Grundlagen der Photochemie. *Z. Elektrochem.* **26** (1929) 54–59; Zum Energieumsatz bei photochemischen Vorgängen in Gasen. Ibid. **27** (1921) 133–142.
Warburg, O. (1918) Über den Energieumsatz bei photochemischen Vorgängen VIII. Die Photolyse in wäßriger Lösung und das photochemische Äquivalentgesetz. *Sitzungsber. Berlin* pp. 1228–1246.
Warburg, E. and G. Leithäuser (1906) Über die Oxydation des Stickstoffs bei der Wirkung der stillen Entladung auf die atmosphärische Luft. *Ann. Phys.* **325** (*Vierte Folge* **20**), 743–750.
Warington, R. (1887) A contribution to the study of well water. *J. Chem. Soc.* **51**, 500–501.
Warington, R. (1889) The amount of nitric acid in the rain-water at Rothamsted, with notes on the analysis of rain-water. *J. Chem. Soc.* **55**, 537–545.
Warmbt, W. (1958) Klima und Wetter im Leben des Menschen. VEB Verlag Volk und Gesundheit, Berlin, 48 pp.
Warmbt, W. (1962) Luftchemische Untersuchungen des bodennahen Ozons 1952–1961. Methoden und Ergebnisse. Habilitationsschrift, Technische Universität Dresden.
Warmbt, W. (1964) Luftchemische Untersuchungen des bodennahen Ozons 1952–1961. In: Abhandlungen des meteorologischen und hydrologischen Dienstes der Deutschen Demokratischen Republik, Nr. 72 (Band X), Akademie-Verlag, Berlin, 95 pp. (Habilitation Thesis, Technical University Dresden 1963).
Warmbt, W. (1965) Ozonmessungen über der Meeresoberfläche im Nordantlantik und im Seegebiet von Westgrönland. *Z. Met.* **18**, 151–156.
Warmbt, W. (1966) Surface ozone and artificial beta-activity in Dresden-Wahnsdorf. *Tellus* **18**, 441–450.
Warmbt, W. (1979) Ergebnisse langjähriger Messungen des bodennahen Ozons in der DDR. *Z. Met.* **29**, 24–31.
Warmbt, W. (1980) Messungen des bodennahen Ozons in der Hohen Tatra. In: Abhandlungen des Meteorologischen Dienstes der DDR, Nr. 124, Band XVI, pp. 191–195.
Warmbt, W. and J. Kolbig (1978) Messungen des bodennahen Ozons in Mirny, Antarktika. *Z. Met.* **28**, 270–273.
Warmbt, W., H. Mrose and G. Philipp (1974) Vorläufige Messergebnisse des bodennahen Ozons unter Verwendung von Chromtryoxidfiltern. *Geodätische und geophysikalische Veröffentlichungen / Nationalkomitee für Geodäsie und Geophysik bei der Akademie der Wissenschaften der DDR.* Reihe II, **18**, 67–80.
Warnatz, J., U. Maas and R. W. Dibble (1996) Combustion. Physical and chemical fundamentals, modelling and simulation, experiments, pollutant formation. Springer-Verlag, Berlin Heidelberg, 265 pp.
Warneck, P. (1974) On the role of OH and HO_2 radicals in the troposphere. *Tellus* **26**, 39–46.
Warneck, P. (1988) Chemistry of the natural atmosphere. Academic Press, San Diego, 7753 pp. *See also*: 2nd ed: (1999) Academic Press, New York, 969 pp.

Warneck, P. (2009) Zur Geschichte der Luftchemie in Deutschland. *Mitteilungen der Fachgruppe Umweltchemie und Ökotoxikologie* **9**, 5–11.

Watanabe, K., Y. Ishizaka and H. Tanaka (1995) Measurements of atmospheric peroxides concentrations near the summit of Mt. Norikura in Japan. *Journal of the Meteorological Society Japan* **73**, 1153–1160.

Watanabe, K. and H. Tanaka (1995) Measurement of gaseous hydrogen peroxide concentrations in the urban atmosphere. *Journal of the Meteorological Society Japan* **73**, 839–847.

Watanabe, K., I. Nagao and H. Tanaka (1996) Atmospheric hydrogen peroxide concentration measure at Ogawara Hahajima Island in the sub-tropical Pacific Ocean. *Journal of the Meteorological Society Japan* **74**, 393–398.

Watanabe, K., K. Kamiyama and O. Watanabe (1998) Evidence for an 11-year cycle of atmospheric H_2O_2 fluctuation recorded in an ice core at the coastal region, East Antarctica. *Journal of the Meteorological Society Japan* **76**, 447–452.

Watanabe, K., Y. Ishizaka and C. Takenaka (1999) Chemical composition of fog water near the summit of Mt. Norikura in Japan. *Journal of the Meteorological Society Japan* **77**, 997–1006.

Watanabe, K., Y. Ishizaka, Y. Minami and K. Yoshida (2001) Peroxide concentrations in fog water at mountainous sites in Japan. *Water, Air and Soil Pollution* **130**, 1559–1564.

Watanabe, K., A. Iwai, N. Takeda and Y. Takebe (2005) Measurement of peroxide concentrations in precipitation in Toyama and in the snow pit at Murododaira, near the summer of Mt. Tateyama. *Bulletin of Glaciological Research* **22**, 51–55.

Watanabe, K., Y. Takebe, N. Sode, Y. Igarashi, H. Takahashi and Y. Dokiya (2006) Fog and rain water chemistry at Mt. Fuji: a case study during the September 2002 campaign. *Atmos. Res.* **82**, 652–662.

Watanabe, K., M. Aoki, N. Eda, Y. Saito, Y. Sakai, S. Tamura, M. Ohata, M. Kawabuchi, A. Takahashi, N. Miyashita and K. Yamada (2009) Measurements of peroxide concentrations in precipitation and dew water in Toyama, Japan. *Bulletin of Glaciological Research* **25**, 1–5.

Watanabe, K., H. Honoki, S. Iwama, K. Iwatake, S. Mori, D. Nishimoto, S. Komori, Y. Saito, H. Yamada and Y. Uehara (2011) Chemical composition of fog water at Mt. Tateyama near the coast of the Japan Sea in Central Japan. *Erdkunde* **65**, 233–245.

Waterman, L. S. (1983) Comments on "The Montsouris series of carbon dioxide concentration measurements, 1877–1910 by Stanhill". *Climate Change* **5**, 413–415.

Watkins, B. A., D. D. Parrish, M. Trainer, R. B. Norton, J. E. Yee, F. C. Fehsenfeld and B. G. Heikes (1995) Factors influencing the mixing ratio of gas phase hydrogen peroxide during the summer at Niwot Ridge, Colorado. *JGR* **100**, 22,831–22,840. *See also*: Watkins, B. A., D. D. Parrish, S. Buhr, R. B. Norton, M. Trainer, J. E. Yee and F. C. Fehsenfeld. Factors influencing the mixing ratio of gas phase hydrogen peroxide during the summer at Kinterbish, Alabama. Ibid. **100**, 22,841–22,851.

Watson, H. E. (1910) The densities and molecular weights of neon and helium. *Journal of the Chemical Society, Transaction* **97**, 810–833.

Watson, H. H. (1850) Ueber die Menge Kohlensäure in der atmosphärischen Luft. *J. prakt. Chem.* **6**, 76–78.

Watson, W. W. (1924) Spectrum of water vapour. *Astrophysical Journal* **60**, 145–158.

Watt, J. (1784) Thoughts on the constituent parts of water and of dephlogisticated air;with an account of some experiments on that subject. *Phil. Trans.* **74**, 329–353.

Way, J. T. (1855) The atmosphere as a source of nitrogen to plants being an account of recent researches on this subject. *Journal of the Royal Agricultural Society of England* **16**, 249–267.

Way, J. T. (1856) On the composition of the waters of land-drainage and of rain. *Journal of the Royal Agricultural Society of England* **17**, 123–162. *See also*: On the quantity of nitric acid and ammonia in rain-water. Ibid., pp. 618–621.

Weathers, K. C., G. E. Likens, F. H. Bormann, J. S. Eaton, K. D. Kimball, J. N. Galloway, T. G. Siccama and D. Smiley (1988) Chemical concentrations in cloud water from four sites in the eastern United States. In: Acid deposition at high elevation sites (eds. M. H. Unsworth and D. Fowler), Kluwer Acad. Press, pp. 345–357.

Weaver, R. (1872) On the quality of atmospheric air in public and private buildings in Leicester. *Lancet* **100**(2555), 223.

Weber, E. (1858) Ueber das Ozon als Luftbestandtheil und seine Beziehungen zu den verschiedenen Zuständen der Atmosphäre. *Jahresbericht des Mannheimer Vereins für Naturkunde* **23**/**24**, 40–65. *See also*: Das Ozon. *Mittheilungen des badischen aerztlichen Vereins* **12** (1858) 161–168.

Weber, E. (1867) Mittelwerthe der Ozonreaktion in Mannheim aus den Jahren 1858–1866. *Jahresbericht des Mannheimer Vereins für Naturkunde* **33**, 47–61.

Weedon, T. (1909) Fertilizing value of rainwater. *Annual Report of the Department of Agriculture and Stock*, Queensland (1908/09), Brisbane, pp. 59, 60, 77, 78.

Weekley, E. (1967) An etymological dictionary of modern English. Dover Publication, 448 pp.

Weeks, M. E. (1932) The discovery of the elements. IV. Three important gases. *Journal of Chemical Education* **9**, 215–225.

Weeks, M. E. (1956) Discovery of the elements. 6th edition. *Journal of Chemical Education* (special edition). 924 pp.

Wegener, A. (1911) Thermodynamik der Atmosphäre. Leipzig, J. A. Barth, 331 pp.

Weidinger, T., G. Baranka, R. Balázs and K. Tóth (2011) Comparison of 19th century and present concentrations and depositions of ozone in Central Europe. *Acta Silvatica et Lignaria Hungarica* **7**, 23–28.

Weigert, F. (1907) Über chemische Lichtwirkungen. II. Photochemisch sensibilisierte Gasreaktionen und eine Theorie der katalytischen Lichtwirkung. *Ann. Phys.* **329** (*Vierte Folge* **24**), 243–266.

Weigert, F. (1908) Über chemische Lichtwirkungen. III. Ozonzersetzung durch Licht. *Z. Elektrochem.* **14**, 591–597.

Weihan, S., L. Wei, D. Guoan and W. E. Wilson (1989) A study on hydrogen peroxide in the atmosphere. *Advances in Atmospheric Sciences* **6**, 509–515.

Weinstein-Lloyd, J. and S. E. Schwartz (1991) Low-intensity radiolysis study of free-radical reactions in cloud water: H_2O_2 production and destruction. *Environ. Sci. Technol.* **25**, 791–800.

Weinstock, B. (1969) Carbon monoxide: residence time in the atmosphere. *Science* **166**, 224–225.

Weinstock, B. and H. Niki (1972) Carbon monoxide balance in nature. *Sciences* **176**, 290–292.

Weiss, J. (1935) Investigations on the radical HO_2 in solution. *Trans. Faraday Soc.* **31**, 668–681.

Weiss, J. (1944) Radiochemistry of aqueous solutions. *Nature* **153**, 748–750.

Weiss, J. and C. W. Humphrey (1949) reaction between hydrogen peroxide and iron salts. *Nature* **163**, 691.

Welbel, B. M. [Вельбель, Б. М.] (1903) О вопросу о содержании азота в атмосферных осадках (On the content of nitrogen in atmospheric precipitation). *Журнал опытной агрономии* [*Journal of Experimental Agronomy*] **4**, 188–195. Note Miller (1905a) cites this paper in German: Zur Frage über den Stickstoffgehalt der atmosphärischen Niederschläge. *Journ. Exper. Landw.* See also reviews in: *Proc. Roy. Soc.* 1903, **71**, 421; *J. Chem. Soc., Abstract* 1903, **84**, 743–749 under the title "nitrogen in atmospheric precipitation"; *Naturwiss. Rundsch.* 1903, **18**, 399.

Welbel [Вельбель] (1903–06) Отчет по химической лабораторий [report of the chemical laboratory]. *Годичного отчета Плотянской сельскохозяйственной опытной станции* [*Annual Report of the Ploty Agricultural Experimental Station*] **8** (1903) 69–96; **9** (1904) 95–120, 130–134; **10** (1905) 75-116; **11** (1906) 128–135. Note, the reports have been published 1893–1913.

Welbel, B. [Вельбель, Б.] (1906) Состав атмосферных осадков, выпавших в 1904 г. [Composition of atmospheric precipitation, falling in the year 1904]. *Записки Императорского общества сельского хозяйства южной России* [*Notes of the Royal Society of Agriculture of Southern Russia*] **2–3**, 78–83.
Wells, S. (1854) On the practical results of quarantine. *Association Medical Journal* **2**, 831–841.
Wells, A. E. (1915) Investigation of the sulphur dioxide content of the atmosphere in the Selby "smoke area". In Holmes et al. (1915), pp. 82–212.
Wells, W. Ch. (1814) Essay on dew and several appearances connected with it. Longmans, London (reprint 1866) 120 pp. + appendix (by R. Strachan), 32 pp.
Wells, M., K. N. Bower, T. W. Choularton, J. N. Cape, M. A. Sutton, R. L. Storeton-West, D. Fowler, A. Wiedensohler, H.-C. Hansen, B. Svenningsson, W. Wieprecht, M. Preiss, B. G. Arends, S. Pahl, A. Berner, C. Kruiz, P. Laj, M. C. Fachini and S. Fuzzi (1997) The reduced Nitrogen budget of an orographic cloud. *Atmos. Env.* **31**, 2599–2614.
Weltzien, C. (1860a) Ueber die Polarisation des Sauerstoffs, die Ozonide und Antozonide. *Ann.* **115**, 121–129.
Weltzien, C. (1860b) Ueber die Sauerstoffverbindungen des Stickstoffs. *Ann.* **115**, 213–228.
Weltzien, C. (1866) Ueber das Wasserstoffperoxyd und das Ozon. *Ann.* **138**, 129–164.
Weng, H., J. Lin, R. Martin et al.(2020) Global high-resolution emissions of soil NO_x, sea salt aerosols and biogenic volatile organic compounds. *Scientific Data* **7**, 148.
Went, F. W. (1955) Air pollution. *Scientific American* **192**, 63–72.
Went, F. W. (1960) Blue hazes in the atmosphere. *Nature* **187**, 641–643.
Wesely, M. L., D. L. Sisterson and J. D. Jastrow (1990) Observations of the chemical properties of dew on vegetation that affect the dry deposition of SO_2. *JGR* **95**, 7501–7514.
Westberg, K., N. Cohen and K. W. Wilson (1971) Carbon monoxide: its role in photochemical smog formation. *Science* **171**, 1013–1015.
Wetherill, C. M. (1865) Report of experiments upon the ventilation of the Capitol extension (pp. 139–211). In: Improvement of ventilation of the Capitol. Report No. 49, 41st Congress, 3d Series (214 pp.). In: Reports of the committees of the house of representatives, 1870–71, in one volume, Washington, U. S. Government Printing Office.
Wetzel, W. (1928) Die Salzbildungen der chilenischen Wüste. *Chemie der Erde* **3**, 375–436.
Wexler, H., W. Moreland and W. Weyant (1960) A preliminary report on ozone observations at Little America. *Monthly Weather Review* **88**, 43–54.
Whelpdale, D. M. and M.-S. Kaiser, eds. (1996) Global Acid Deposition Assessment. WMO GAW Report No. 106, World Meteorological Organization, Geneva, Switzerland, 260 pp.
Whitehead, W. L. (1920) The Chilean nitrate deposits. *Economic Geology* **15**, 187–224.
Whytlaw-Gray, R. and H. S. Patterson (1932) Smoke: A study of aerial disperse systems. London, E. Arnold & Co., 192 pp. (see pp. 57–72).
Wiedeburg, J. E. B. (1784) Ueber die Erdbeben und den allgemeinen Nebel 1783. Eunoische Buchhandlung, Jena, 86 pp.
Wiberg, N. (2007) Holleman-Wiberg. Lehrbuch der anorganischen Chemie. 102. De Gruyter, Berlin, New York, 2149 pp.
Wiegmann, A. J. F. (1824) Ueber das Vorkommen von Salzen, sauren Erden, Metallen und andern Substanzen in der Atmosphäre und den atmosphärischen Niederschlägen. *Arch. Pharm.* **7**, 199–2008.
Wiegmann, A. J. F. (1826) Ueber das Vorkommen der Phosphorsäure in meteorischen Niederschlägen. *Arch. Pharm.* **16**, 151–158.
Wiegmann, A. J. F. (1827) Ueber den Höhenrauch (Heerrauch, Hehrauch, trocknen, stinkenden Nebel). *Archiv für die gesammte Naturlehre* **10**, 491–496. 1827.
Wiegmann, A. F. A. (1829) Ueber das Pyrrhin; nachträgliche Bemerkungen zu Prof. Vogels Abh. (XV. Bd., S. 97, die. Archiv). *Archiv für die gesammte Naturlehre* **16**, 196–198.

Wieland, H. and W. Franke (1929) Über den Mechanismus der Oxydationsvorgänge. XVII. Über das Verhältnis der Oxydationsgeschwindigkeiten von molekularem Sauerstoff und Hydroperoxyd. *Liebigs Ann.* **473**, 289–300.
Wienhaus, O. (1999) Julius Adolph Stöckhardt – a pioneer of applied chemistry. *Z. anal. Chem.* **363**, 139–144.
Wienhaus, O. and H.-G. Däßler (1991) 140 Jahre Immissionforschung am Institut für Pflanzenchemie und Holzchemie in Tharandt. *Staub – Reinhaltung Luft* **51**, 461–466.
Wierzbicki, D. (1882) Ozon atmosferyczny i roczny ruch jego według dwudziesto-pięcioletnych spostrzeżeń [Atmospheric ozone and its annual motions according to 25 years recorded observations]. *Rozprawy i Sprawozdania z posiedzeń Wydziała Matematyczno-Przyroniczego Akademii Umiejętności* **9**, 88–155.
Wiesner, G. H. (1914) Nitrogen and chlorine of rain water. *Chemical News* **109**, 85–87.
Wigand, A. (1913) Zur Erkenntnis der atmosphärischen Trübung. *Met. Z.* **30**, 1–2.
Wigand, A. (1919) Die vertikale Verteilung der Kondensationskerne in der freien Atmosphäre. *Ann. Phys.* **364** (*Vierte Folge* **59**), 689–741.
Wigand, A. (1930) Das atmosphärische Aerosol. *Die Naturwissenschaften* **18**, 31–33.
Wigley, T. M. L. (1983) The pre-industrial carbon dioxide level. *Climate Change* **5**, 315–320.
Wildt, R. (1934) Ozon und Sauerstoff in den Planetenatmosphären. In: Nachrichten von der Gesellschaft der Wissenschaften zu Göttingen. Mathematisch-Physikalische Klasse, Neue Folge, Band 1, Nr. 1, Weidmannsche Buchhandlung, Berlin, 9 pp.
Wilkening, K. E. (2004) Acid rain science and politics in Japan. MIT Press, 322 pp.
Wilkins, E. T. (1954) Air pollution and the London fog of December 1952. *Journal of the Royal Sanitary Institute* **74**, 1–21.
Wilks, S. S. (1959) Carbon monoxide in green plants. *Science* **129**, 964–966.
Williams, S. F. and O. K. Beddow (1932) analysis of the precipitation of rains and snows at Mount Vernon, Iowa. *Chemical News* **145**, 49–43.
Williams, W. C. (1897) Die Menge der in der Atmosphäre vorhandenen Kohlensäure. *Ber.* **30**, 1450–1456.
Williamson, A. W. (1843) CXLVI. Some experiments on ozone. *Memoirs and Proceedings of the Chemical Society* **2**, 395–398.
Willett, H. C. (1928) Fog and haze, their causes, distribution and forecasting. *Monthly Weather Review* **56**, 435–468.
Willey, J. D., R. J. Kieber and R. D. Lancaster (1996) Coastal rainwater hydrogen peroxide: Concentration and deposition. *J. Atmos. Chem.* **25**, 149–165.
Wilson, A. T. (1959) Organic nitrogen in New Zealand snows. *Nature* **183**, 318–319. *See also*: Surface of the ocean as a source of air-born nitrogenous material and other plant nutrients. Ibid., 99–100.
Wilson, B. D. (1921) Nitrogen in the rainwater at Ithaca, New York. *Soil Science* **11**, 101–110. *See also*: Sulfur supplied to the soil in rain water. *Journal of the American Society of Agronomy* **13**, 226–229.
Wilson, B. D. (1923) The quantity of sulfur in rainwater. *Journal of the American Society for Agronomy* **15**, 453–456.
Wilson, B. D. (1926) Nitrogen and sulfur in rainwater in New York. *Journal of the American Society for Agronomy* **18**, 1108–1112.
Wilson, C. T. R. (1899) On the condensation nuclei produced in gases by the action of Röntgen rays, uranium rays, ultraviolet light and other agents. *Phil. Trans. A* **192**, 403–453.
Wilson, C. T. R. (1912) On an expansion apparatus for making visible the tracks of ionizing particles in gases and some results obtained by its use. *Proc. R. Soc. Lond. A* **87**, 277–292.
Wilson, R. C., Z. L. Fleming, P. S. Monks, G. Clain, S. Henne et al.(2012) Have primary emission reduction measures reduced ozone across Europe? An analysis of European rural background ozone trends 1996–2005. *ACP* **12**, 437–454.

Winer, A. M., M. C. Dodd, D. R. Fitz, P. R. Miller, E. R. Stephens, K. Neisess, M. Meyers, D. E. Brown and C. W. Johnson (1982) Assembling a vegetative hydrocarbon emission inventory for the California South Coast Air Basin: Direct measurement of emission rates, leaf biomass and vegetative distribution, In: For presentation at the 75th Annual Meeting of the Air Pollution Control Association, New Orleans, 20 pp.

Winkler, C. (1880) Mittheilungen über die Versuche zur Beseitigung des Hüttenrauches bei der Schneeberger Ultramarinfabrik zu Schindlers Werk bei Bockau in Sachsen. *Jahresberichte für das Berg- und Hüttenwesen im Königreich Sachsen*, Freiberg, pp. 50–70.

Winkler, C. (1883) Wirkt die in unserem Zeitalter stattfindende Massenverbrennung von Steinkohle verändernd auf die Beschaffenheit der Atmosphäre (Vortrag beim Zweiten Allgemeinen Deutschen Bergmannstage in Dresden). In: Clemens Winkler's Vorträge und Abhandlungen über Abgase und Rauchschäden. Sammlung von Abhandlungen über Abgase und Rauchschäden (ed. O. Brunck), Heft 8. Parey, Berlin, 1913, pp. 62–68.

Winkler, C. (1896) Über den Einfluss des Wasserdampfgehaltes saurer Gase auf deren Vegetationsschädlichkeit. Vortrag, gehalten auf der Hauptversammlung des Vereins Deutscher Chemiker in Halle, 31. Mai bis 3. Juni 1896.

Winkler, C. (1900) Wann endet das Zeitalter der Verbrennung? Vortrag gehalten beim Allgemeinen Bergmannstage in Teplitz am 5. September 1899 von Dr. Clemens Winkler, Professor der Chemie an der Königlich Sächsischen Bergakademie Freiberg. Verlag von Craz & Gerlach (Joh. Stettner), Freiberg, 1900.

Winkler, L. W. (1901) Die Löslichkeit der Gase in Wasser [Dritte Abhandlung]. *Ber.* **34**, 1412–1422.

Winkler, L. W. (1935) Bestimmung des Reduktionsvermögens der verunreinigten Luft. *Z. anal. Chem.* **103**, 183–186.

Winkler, P. (1977) Automatic analyser for pH and electrical conductivity of precipitation. In. papers presented at the WMO Technical Conference on Instruments and methods of Observation (TECIMO), Hamburg. WMO-No. 480, Genf, pp. 191–196.

Winkler, P. (1980) Observations on acidity in continental and in marine atmospheric aerosols and in precipitation. *JGR* **85**, 4481–4486.

Winkler, P. (1982) Zur Trendentwicklung des pH-Wertes des Niederschlags in Mitteleuropa. *Zeitschrift für Pflanzenernährung und Bodenkunde* **145**, 576–685. *See also*: Deposition of acid in precipitation. In: Deposition of atmospheric pollutants (eds. H.-W. Georgii and I. Pankrath), D. Reidel Publ., pp. 67–76.

Winkler, P. (1992) Design and calibration of a fog water collector. In: Instruments and Observing Methods, Report No. 49: Instruments and methods of observation, Vienna, 11–15 May 1992 WMO/TD-No. 462, pp. 17–21.

Wislicenus, H., O. Schwarz, H. Sertz, F. Schröder, F. Müller and F. Bender (1914) Experimentelle Rauchschäden. Versuche über die äußeren und inneren Einwirkungen von Ruß, sauren Nebeln und stark verdünnten Gasen auf die Pflanze. In: Sammlung von Abhandlungen über Abgase von Rauchschäden, Vol. 10 (ed. H. Wislicenus), Paul Parey, Berlin, 168 pp.

Wisniak, J. (2004) The nature and composition of water. *Indian Journal of Chemical Technology* **11**, 434–444.

Wisniak, J. (2010) Daniel Berthelot. Part III. Contribution to photochemistry. *Educación Química* **21**, 314–323.

Wisniak, J. (2013) John William Draper. *Educación Química* **24**, 215–223.

Wisniewski, J. (1982) The potential acidity associated with dews, frosts and fogs. *Water, Air and Soil Pollution* **17**, 361–377.

Wisniewski, J. and L. E. Keitz (1983) Acid rain deposition patterns in the continental United States. *Water, Air and Soil Pollution* **19**, 327–339.

Witt, O. N. (1876) Zur Kenntniss des Baues und der Bildung färbender Kohlenstoff-verbindungen. *Ber.* **9**, 522–527.

Witting, E. (1823) Ueber einige zufällige, fremdartige Bestandtheile der atmosphärischen Luft und ihrer Ausmittelung. *Arch. Pharm.* **4**, 215–229.

Witting, E. (1824) Ueber das Vorkommen von Salzen, Säuren, Erden, Metallen etc in der Atmosphäre und in den atmosphärischen Niederschlägen. *Archiv des Apotheker-Vereins im nördlichen Teutschland für die Pharmacie und ihrer Hülfswissenchaften* **7**, 199–208.

Witting, E. (1825) Abhandlung über die Mischungsverhältnisse der atmosphärischen Luft, hinsichtlich der fremdartigen Beimischungen wie der Säuren, Salze, auch metallischer Stoffe u. s. w., welche darin vorkommen. *Arch. Pharm.* **11**, 68–88. *See also*: Zur weiteren Kenntnis der fremdartigen Beimischungen in der Erdatmosphäre. *Archiv für die gesammte Naturlehre* **5**, 189–194.

Witting, E. (1826) Nachtrag zu meinen meteorologischen Beobachtungen namentlich der Gegenwart von Phosphor- und Salzsäure während eines Höhenrauches, und Ursprung des letztern. *Arch. Pharm.* **19**, 177–189.

Witting, E. (1827) Nachtheilige Einwirkung eines Hehrrauches (Höherauch) auf den Vegetationsprozess. *Medizinisch-Chirurgische Zeitung (Salzburg)* **1**, 191–191.

Wittwer, W. C. (1855) Ueber die Einwirkung des Lichts auf Chlorwasser. *Pogg. Ann.* **94**, 597–612.

Witz, G. (1885) Sur la présence de l'acide sulfureux dans l'atmosphère des villes. *C. r.* **100**, 1385–1388.

Wobrock, W., D. Schell, R. Maser, W. Jaeschke, H. W. Georgii, W. Wieprecht, B. G. Arends, J. J. Möls, G. P. A. Kos, S. Fuzzi, M. C. Facchini, G. Orsi, A. Berner, I. Solly, C. Kruisz, I. B. Svenningsson, A. Wiedensohler, H.-C. Hansson, J. A. Ogren, K. J. Noone, A. Hallberg, S. Pahl, T. Schneider, P. Winkler, W. Winiwarter, R. N. Colvile, T. W. Choularton, A. I. Flossmann and S. Borrmann (1994) The Kleiner Feldberg Cloud Experiment 1990. An overview. *J. Atmos. Chem.* **19**, 3–35.

Wofsy, S. C., J. C. McConnell and M. B. McElroy (1972) Atmospheric CH_4, CO, and CO_2. *JGR* **77**, 4477–4493.

Wöhler, F. (1863) Grundriss der unorganischen Chemie. 13. Auflage. Duncker und Humblot, Berlin, 287 pp.

Wohlgemuth, J., D. Pfäfflin, W. Jaeschke, F. Deutsch, P. Hoffmann and H. M. Ortner (2001) Photochemical formation of hydrogen peroxide in atmospheric droplets: the role of iron, oxalate and trace metals on the H_2O_2 production. In: Dynamics and Chemistry of Hydrometeors. Final Report of the Collaborative Research Centre 233 "Dynamik und Chemie der Meteore". Collaborative Research Centers Sonderforschungsbereiche (DFG) (ed. R. Jaenicke), Wiley VCH, Weinheim, Bonn, pp. 346–362.

Wolf, R. (1854) Ueber Beobachtungen mit dem Schönbein'schen Ozonometer. *Pogg. Ann.* **93**, 14–315.

Wolf, R. (1855) Ueber den Ozongehalt der Luft und seinen Zusammenhang mit der Mortalität. Huber & Comp., Bern, 21 pp. See also: Mittheilungen der naturforschenden Gesellschaft in Bern aus dem Jahre 1855. *C. r.* **41**, 57–77. *See also*: Des variations de l'ozone considérées en ells-mémes et relativement aux variations dans l'etat hygiene du lieu d'observation. *C. r.* **41** (1855) 419–420 and 909–910.

Wolf, R. (1856) Neue Beobachtungen und Bemerkungen über den Ozongehalt der Luft. Mittheilungen der naturforschenden Gesellschaft in Bern aus dem Jahre 1856. *Pogg. Ann.* **97**, 57–68. *See also*: Ozon-Beobachtungen im Jahre 1855. *Pogg. Ann.* **97**, 640.

Wolff, E. Th. (1847) Die chemischen Forschungen auf dem Gebiet der Agricultur und Pflanzenphysiologie. Joh. Ambrosius Barth, Leipzig, 549 pp.

Wolffenstein, O. (1870) Beitrag zur Ozonfrage. *Pogg. Ann.* **139**, 320–329.

Wolffhügel, G. (1875) Ueber den sanitären Werth des atmosphärischen Ozons. *Zeitschrift für Biologie* **11**, 408–458.

Wolkowicz, A. (1894) Ozon im Sinne des periodischen Systems. *Z. anorg. Chem.* **5**, 164–165.

Wollny, E. (1885) Beiträge zur Frage der Schwankungen im Kohlensäuregehalt der atmosphärischen Luft. *Forschungen auf dem Gebiet der Agricultur-Physik (Wollny's Forschungen)* **8**, 405–423.

Wollny, E. (1896) Untersuchungen über das Verhalten atmosphärischer Niederschläge zur Pflanze und zum Boden. Forschungen auf dem Gebiet der. *Agricultur-Physik (Wollny's Forschungen)* **19**, 267–286 (appeared in 20 Vol. 1878–1898).

Wolpert, H. (1892) Eine einfache Luftprüfungsmethode auf Kohlensäure, mit wissenschaftlicher Grundlage. Baumgärtner's Buchhandlung, Leipzig, 123 pp.

Wolpert, H. (1905) Über verbrennliche gasförmige Kohlenstoffverbindungen in der Luft. *Archiv für Hygiene* **52**, 151–178.

Woodcock, A. H. (1950) Condensation nuclei and precipitation. *Journal of Meteorology* **7**, 161–162.

Woodcock, A. H. (1952) Atmospheric salt particles and raindrops. *Journal of Meteorology* **9**, 200–212.

Woodcock, A. H. (1953) Salt nuclei in marine air as a function of altitude and wind force. *Journal of Meteorology* **10**, 362–371.

Woodcock, A. H. and M. M. Gifford (1949) Sampling atmospheric sea-salt nuclei over the Ocean. *Journal of Marine Research* **8**, 177–197.

Woodcock, A. H., C. F. Kientzler, A. B. Arons and D. C. Blanchard (1953) Giant condensation nuclei from bursting bubbles. *Nature* **172**, 1144–1145.

Wourtzel, E. (1912) Synthése du gaz chlorure de nitrosyle et le poids atomique du chlore. *C. r.* **155**, 345–346.

Wragge, C. (1884) Ozone at sea. *Nature* **29**, 336.

Wright, H. L. (1932) Observations of smoke particles and condensation nuclei at Kew Observatory. *Geophysical Memoirs* Vol. 6, No. 57. Published by the authority of the Meteorological committee. 22 pp.

Wright, H. L. (1940) Sea salt nuclei. *Q. J. R. Meteor. Soc.* **66**, 3–11; *and*: The origin of sea salt nuclei. Ibid. **66**, 11–12.

Wulf, O. R. (1928) Photochemical ozonization and its relation to the polymerisation of oxygen. *J. Am. Chem. Soc.* **50**, 2596–2604. See also: Proceedings of the National Academy of Sciences USA 14 (1928) 356–359.

Wulff, A. (1920) Bibliographia agrogeologica. Essay of a systematic bibliography of agro-geology. Wageningen, 286 pp.

Würfel, H. (1965) Die wissenschaftlichen Veröffentlichungen aus dem Forschungsinstitut für Bioklimatologie des Meteorologischen Dienstes der Deutschen Demokratischen Republik. *International Journal of Bioclimatology* **9**, 67–75.

Wurster, C. (1886a) Die Griess'sche Reaktion auf Salpetrige Säure bei Gegenwart von Wassertsoffperoxyd. *Ber.* **19**, 3206–3208.

Wurster, C. (1886b) Die Activirung des Sauerstoffs der Atmosphäre und deren Zusammenhang mit den elektrischen Erscheinungen der Luft und mit der Entstehung der Gewitter. *Ber.* **19**, 3208–3217.

Wurster, C. (1886c) Die Activirung des Sauerstoffes im Papierblatte. *Ber.* **19**, 3217–3218.

Wurtz, A. (1869) Histoire des doctrines chimiques depuis Lavoisier jusqu'à nos jours. Hachette, Paris, 285 pp.

Yaalon, D. H. and E. Ganor (1968) Chemical composition of dew and dry fallout in Jerusalem, Israel. *Nature* **217**, 1139–1140.

Yeatts, Jr., R. B. and H. Taube (1949) The kinetics of the reactions of ozone and chloride ion in aqueous solution. *J. Am. Chem. Soc* **71**, 4100–4105.

Yeung, L. Y., L. T. Murray, P. Martinerie, E. Witrant, H. Hui, A. Banerjee, A. Orsi and J. Chappellaz (2019) Isotopic constraint on the twentieth-century increase in tropospheric ozone. *Nature* **570**, 224–227.

Yoshizumi, K., K. Aoki and I. Nouchi (1984) Measurements of the concentration in rainwater and of the Henry's law constant of hydrogen peroxide. *Atmos. Env.* **18**, 395–401.
Yost, D. H. and H. Russell (1946) Systematic inorganic chemistry. Prentice Hall, New York, 452 pp.
Zabelin, I. V. (1864) Ueber die Bildung von salpetrigsaurem Ammoniak aus Wasser und Stickstoff und über den Nachweis von Ammoniak im Blut, im Harn und der Expirationsluft. *Ann.* **130**, 54–95.
Zagar, K. B. and K. B. Cholodova [Жагар, Холодова] (1961) Химический состав атмосферных осадков [Chemical composition of atmospheric precipitation]. *Гидрохимические материалы АН СССР* [Hydrometeorological papers of the Academy of Sciences of the USSR] **31**, 3–111.
Zafiriou, O. C. (1983) Natural water photochemistry. In: Chemical oceanography (eds. J. P. Riley and R. Chester), Academic Press, London, Vol. 8, pp. 339–379.
Zaizev, V. A. [Зайцев] (1948a) Методика микрофотографирования капель тумана и облаков [Microphotography of fog and cloud droplets]. *Труды ГГО* No. **9**.
Zaizev, V. A. [Зайцев] (1948b) Новый метод определения водности облаков и туманов [A new method for determination of liquid water content of clouds and fogs]. *Труды ГГО* 13.
Zaizev, V. A. [Зайцев] (1950) Водность и распределение капель в кучевых облаках [Liquid water content distribution of droplets in cumulus clouds]. *Труды ГГО* No. **19**.
Zaizev, V. A. and A. A. Ledokhovich [Зайцев, Ледохович] (1960) Приборы и методика исследования туманов и облаков с самолета [Instruments and methods for investigation of fogs and clouds by aircrafts]. Гидрометеорологическое издательство, Ленинград, 545 pp.
Zaizev, V. A. and A. A. Ledokhovich [Зайцев, Ледекевич] (1962) К вопросу о регистрации водности облаков [On measurement of liquid water content of flogs and clouds]. *Труды аркт. и анаркт. инст.* [Transactions of the Arctic and Antarctic Institute] **239**, 128.
Zalasiewicz, J., M. Williams, A. Smith, T. L. Barry, A. L. Coe, P. R. Bown, P. Brenchley, D. Cantrill, A. Gale, P. Gibbard, F. J. Gregory, M. W. Hounslow, A. C. Kerr, P. Pearson, R. Knox, J. Powell, C. Waters, J. Marshall, M. Oates, P. Rawson and P. Stone (2008) Are we now living in the Anthropocene? *GSA Today* **18**, 4–8.
Zedler, J. H. (1735) Grosses vollständiges Universal-Lexicon aller Wissenschaften und Künste. Band 13 (Hi – Hz). Leipzig, 910 pp.
Zeldovich, Y. B. [Яков Борисович Зельдович] (1942) К теории образования новой фазы. Кавитация. *Журнал Экспериментальной и Теоретической Физики* (ЖЭТФ) [Journal of Experimental and Theoretical Physics] **12**, 525–538. And: On the theory of new phase formation: Cavitation.]. *Acta Physicochimica URSS* **18** (1943) 1–22.
Zeldovich, Y. B. (1946) Окисление азота при горении и взрыва [The oxidation of nitrogen in combustion and explosions]. *Доклады Академии наук СССР* **51**, 213–216 (*see also*: Acta Physicochim. USSR **21**, 577–628). *See also*: Selected works of Yakov Borisovich Zeldovich. Chemical physics and hydrodynamics. Volume I. 26. Oxidation of nitrogen in combustion and explosions. pp. 404–410.
Zenger, W. (1858) Über eine neue Bestimmungsmethode des Ozon-Sauerstoffes. *Sitzungsber. Wien* **24**, 78–91.
Zetzsch, C. and W. Behnke (1992) Heterogeneous photochemical sources of atomic Cl in the troposphere. *Berichte der Bunsengesellschaft für physikalische Chemie* **96**, 488–493.
Zhang, Q., J. Liu, Y. He and S. Hatakeyama (2017) Measurement of hydrogen peroxide and organic hydroperoxide concentrations during autumn in Beijing, China. *Journal of Environmental Sciences* **64**, doi:10.1016/j.jes.2016.12.015.
Zhao, D. and B. Sun (1986) Air pollution and acid rain in China. *Ambio* **15**, 2–5.
Zhao, D., J. Xiong, Y. Yu and W. H. Chan (1988) Acid rain in Southwestern China. *Atmos. Env.* **22**, 349–358.
Zhavoronkina, T. K. [Жаворонкина] (1955) Содержание хлоридов и сульфатов в атмосферных осадках по 33 меридиану [Chloride and sulfate content in atmospheric precipitation along the 33rd meridian]. *Труды морского гидрофизического института* [Transactions of the Ocean Hydrophysical Institute] **5**, 99–103.

Zier, M. (1992) Über die Variabilität der Spurenstoffkonzentration im Nebelwasser im Verlaufe einzelner Nebelereignisse auf dem Kamm des Erzgebirges. *Met. Z. Neue Folge* **1**, 221–228.

Zierath, R. (1981) Inhaltsstoffe atmosphärischer Niederschläge und ihr Einfluß auf die Sicker- und Grundwasserbeschaffenheit am Beispiel ausgewählter Gebiete. Dissertation, Technical University Dresden.

Zika, R. G. and E. S. Saltzman (1982) Interaction of ozone and hydrogen peroxide in water. Implications for analysis of H_2O_2 in air. *Geophys. Res. Lett.* **9**, 231–234.

Zika, R. G., E. S. Saltzman, W. L. Chameides and D. D. Davis (1982) H_2O_2 levels in rainwater collected in South Florida and the Bahama Island. *JGR* **87**, 5015–5017.

Zimmermann, E. A. W. von (1775) Beobachtungen auf einer Harzreise nebst einem Versuche die Höhe des Brockens durch das Barometer zu bestimmen. Waisenhaus-Buchdruckerei. Braunschweig, 65 pp.

Zimmermann, W. (1824) Beiträge zur näheren Kenntnis der wässrigen Meteore. *Achiv für die gesamte Naturlehre* **1**, 257–292 and 310–312.

Zimmermann, H. (1856) Ozon-Messungen in Ottweiler. *Allgemeine medicinische Central-Zeitung* **25**, 114.

Zimmermann, W. F. A. (1865) Der Erdball und seine Naturwunder. Populäres Handbuch der physischen Geographie. 1. Band: Allgemeine irdische und kosmische Verhältnisse. Die Atmosphäre. Verlag G. Hempel, Berlin, 176+383 pp.

Zimmermann, F. (1873) Die sanitären Zustände Helgolands. mit specieller Berücksichtigung des Ozongehalts der Luft. Im Selbstverlage des Verfassers, Helgoland, 46 pp.

Zimmermann, S. (1905) Zur Kenntnis der Metallnitrosoverbindungen und des Stickoxyds. *Monatshefte für Chemie* **26**, 1277–1294.

Zimmermann, G. (1953) kleinklimatische Ozonmessungen in Bad Kissingen. *Miteilungen des Deutschen Wetterdienstes* Nr. 1, 13 pp.

Zimmermann, P. R. (1979) Determination of emission rates of hydrocarbons from indigenous species of vegetation in the Tampa/St. Petersburg Florida area. Appendix C, Tampa Bay *Area* Photochemical Oxidant Study, EPA 904/9-77, U. S. E. P. A., Research Triangle Park, NC.

Zimmermann, L. and F. Zimmermann (2002) Fog Deposition to Norway Spruce Stands at high elevation sites in the Eastern Erzgebirge. *Journal of Hydrology* **256**, 166–175.

Zittel, K. (1874) Beobachtungen über Ozon in der lybischen Wüste. *Sitzun-gsber. München* **4**, 215–230. *See also*: *Z. Öster. Ges. Met.* **9**, 301–302.

Zobrist, J. (1987) Methoden zur Bestimmung der Azidität in Niederschlagsproben. VDI Berichte 608, Düsseldorf, pp. 401–420.

Zobrist, J., L. Sigg and F. Stumm (1985) Der Nebel als Träger konzentrierter Schadstoffe. In: Wasserwirtschaft im Spannungsfeld Umweltschutz. Gewässer, Wasser, Abwasser (ed. B. Böhnke), Heft 100, pp. 371–393.

Zuo, Y. and Y. Deng (1999) Evidence for lightning induced production of hydrogen peroxide during thunderstorms. *Geochimica et Cosmochimica Acta* **63**, 3451–3455.

Zuo, Y. and J. Hoigné (1992) Formation of hydrogen peroxide and depletion of oxalic acid in atmospheric water by photolysis of iron(III)-oxalato complexes. *Environ. Sci. Technol.* **26**, 1014–1022. *See also*: Evidence for photochemical formation of H_2O_2 and oxidation of SO_2 in authentic fog water. *Science* **260** (1993) 71–73.

Subject index

acid deposition 230
acid rain 158, 230, 232
acid rain, first termination 228
acidity
– atmospheric 246, 688
– rain water 243
aerial acid 155, 354, 395
aerial particles 306, 349
aerial salts 287, 680
aerosol
– biological 304, 313
– research 256, 270, 320, 330
agricultural experiment station
– Dahme 58, 184, 405
– Flahult 198, 239
– Gembloux 192
– in Canada 220
– in Illinois 218
– in Prussia 185
– in Saxony 175
– in USA 175
– Lille 713
– Oklahoma 221
– Palermo 674
– Petrowskoje Rasumowskoje 641
– Ploty 198, 205, 280
– Rostock 185
– Rothamsted 174, 186
– Torino 185
air
– different kinds 78
– goodness 385
air analysis
– at Montsouris 556
– at Munich 377
– oxygen 379
air chemical process
– first phrase in literature 288, 311
air chemistry
– first use of the word 2
– in past and future 729
– research periods 14
air cleanser 168, 470, 474, 571
air pollution
– books on history XII
– England 291
– view of 19th century 11
– view of 20th century 13, 710
air pollution chemistry 11, 118, 298
Aitken counter 294
albuminoid 97, 446, 704
alkalinity 245
ammonia
– 20th century research 678
– absorption from air 185
– aerosol research 677
– agricultural role 173
– deposition trend 187
– dry deposition 674
– early history 670
– in air 32
– in dew water 193
– in fog water 32
– in hail 167
– in rainwater 177, 218
– photooxidation 227
– sea as source 677
– trend in cloud water 268
– trend in deposition 240
– trend in rainwater 242
ammonium *see* ammonia
ammonium acetate 13, 672
ammonium chloride 168, 670
ammonium nitrate 182, 670, 678, 690
ammonium nitrite 96, 191, 329, 473, 635, 687
ammonium sulfate 329, 347, 711
animacula 145
Antarctic measurements 13, 183, 587, 617, 666
Anthropocene 48
antozone 95, 479, 485, 634
aqua regia 78, 335, 341
aquated electron 724
aran 606
argon
– in air 376, 393
arsenite method 558, 593, 599
aspiration method 398, 446, 672
aspirator 377, 397, 514, 540, 548, 555, 595, 695
Atacama desert 342
atmizone 483
atmosphere
– air chemistry 1
– discovering periods XIII
– etymology 81

https://doi.org/10.1515/9783110732467-009

atmospheric chemistry *see* air chemistry
– definition 8
atmospheric electricity 131, 317, 328, 458, 475, 501, 536, 555, 561, 608, 698
autoxidation 102, 474, 723

balloon flights 326, 406, 416, 544, 584, 596
balneological research 5, 204
barium hydroxide 397, 415
barium peroxide 481, 482, 632, 647
base neutralizing capacity 246
Bergson tube 447
bicarbonate
– in air 672
– in rainwater 159, 247
bioaerosol 304
biogeochemical cycle 46, 50
biogeochemistry
– definition 47
– future 730
– *Vernadsky*'s view 45
biosphere
– origin of the word 49
– *Vernadsky*'s view 49
bituminous matter 307, 318, 449
black carbon 305
black episodes 246
black rain 310
bleaching 94, 102, 278, 471, 638, 656
blue cloud 448
blue fog 325
blue haze 322, 448, 484
blue mist 322
Brewer spectrophotometer 588
bromine
– and ozone 462
– ozone test paper 501
– photolysis 101
– stratospheric chemistry 591
brown clouds 319

calcium
– in dust 155, 250, 306, 318
– in hail 167
– in rainwater 208, 218, 241
carbon dioxide
– 19th century measurements 425
– climate change 432
– photochemistry 488
carbon dioxide theory 429
carbon monoxide
– discovery 432
– in air 434
– oxidation 82
– ozone precursor 630
– reaction with OH 434
carbonaceous substance *see* organic matter
carbonate
– in dust 155
– in rainwater 159, 169
carbonometer 400
carbonyl sulfide 710
catalytic decomposition of H2O2 516, 641
catalytic decomposition of O3 493, 571, 584, 611
catalytic oxidation of SO2 677, 723
CCN *see* condensation nuclei
Central Aerological Observatory *see* observatory, Dolgoprudny (CAO)
Central Meteorological Observatory (CMO) *see* observatory, Tokyo
chaos 74, 727
Chapman cycle 577
chemical climatology 19, 39, 118, 613
chemical meteorology 19, 101, 729
chemical weather 386
chemical weathering 430
chemistry
– definition 60
chloride
– and ozone 493
– cycling 201
– degassing 157, 337, 340
– deposition 237
– global deposition 333
– in cloud water 262
– in fog water 258
– in lava 708
– in rainwater 152, 166, 187, 201, 205, 208, 214, 217
– in rime 256
– in sea salt 333
– in town fog 294
chloride-to-sodium ratio *see* sodium-to-chloride ratio
chlorine *see also* chloride
– and ozone 19, 101, 103, 462, 468, 591
– cycle 56, 205
– from coal combustion 346

– in aqua regia 341
– ozone test paper 501
– photolysis 98
chlorine nitrate 340
chlorine-hydrogen reaction 100
cholera epidemic 377, 388, 440, 551, 560
chromophore, coining the word 493
chromophoric substances 493, 658
CLAW hypothesis 49, 710
climate definition 8
cloud
– definition 127, 135, 252
– Greek mythology 107, 118
– research milestones 121
cloud chamber 325
cloud chemistry
– definition 252
cloud chemistry research
– in Canada 266
– in East Germany 261
– in Japan 270
– in the Soviet Union 261
– pioneers 272
cloud dissipation 317
cloud droplets *see* droplets
cloud seeding 141
cloud water
– sampler 273
clouds of smoke 254
coal gas 437
cobalt
– in dust 307
– in rainwater 160
– in snow 305
Cofer scrubber *see* mist chamber
colloid 284
column ozone 580, 588
condensation nuclei
– blue haze 449
– chemical analysis 256
– discovery 129
– droplet formation 133, 201, 255
– dust particles 325
– photochemical formation 575
– research milestones 327
– sea salt 333
– termination 326
convertibility 147
cycling matter 53, 59, 118
cycling salts 201

denitrification 57
dephlogisticated air 79, 353, 354, 370
dephlogisticated substances 355
deposition gauge 113, 220, 642
detergent of the atmosphere 470
dew
– 19th century studies 280
– 20th century studies 280
– and ozone 531
– bleaching property 633
– drying, ammonium nitrite 691
– early chemical studies 161, 278, 316, 720
– early views 110
– ecological research 281
– hydrogen peroxide 638, 644, 664
– sampler 87, 113, 279
– studies in India 227, 280
– *Well*'s explanation 127
dew condensation cup 230, 256
dew condensation method 33, 644
dimethyl sulfide 710
dinitrogen monoxide *see* N2O
dinitrogen pentoxide *see* N2O5
diseases and air pollution 77, 157, 292, 314
diseases and ozone 498, 543, 563, 568
Dobson instrument 580, 588
Dobson unit 731
droplets
– 19th century studies 132
– air craft sampler 259
– chemistry 657, 658, 677, 704
– dew chemistry 280
– electron microscopy 334
– historic views 123
– microscopy 326
– sea water 257, 339
– size 164, 263, 271, 323
– supercooled 140
dry deposition 181, 215, 571
dry fog 163, 316, 319
dry haze 316
dust
– and rainwater 159, 219, 228
– as aerosol 284
– etymology 283
– from fires 318
– historic views 301

– in air 294, 306
– pollution 151, 250, 292
– sampling 289
– volcanic 304
dust counter 323, 328
dust deposition 217, 294
dust particles
– as CCN 129, 134, 322
– biological 313
– historic views 82
– number concentration 328
– organic 304
– size 320
dust precipitation 289

ecosystem
– acidification 263
– and atmospheric chemistry 275
– organic emissions 449
Elbrus Experiment 224
electric discharges
– Atacama desert 343
– *Böttger*'s experiments 473, 694
– *Cavendish*'s experiments 357, 686
– cloud seeding 141
– CO2 decomposition 433
– H2O2 formation 96, 641
– in air 495, 641, 694
– nitrous gases formation 517, 695, 699
– ozone from pure oxygen 555
– *Schönbein*'s experiments 460, 634
electric fog 316
electrolysis of water 459, 464, 494
electron transfer 105, 494, 649, 724
electrotherapy 498
ethene 436
expedition
– Amu-Darja 1874/75 513
– Elbrus 1934/35 599
– Greenland 1888 421
– Iceland 1864 568
– Novara 1857–1859 540
– Polaris 1871/72 565
– Pommerania 1871 and 1872 499
– Vesuvius 1819 708

Fenton chemistry 656
ferric chloride 97
field experiments 141, 268, 628, 705
fixed air 350, 354
Flaschenmethode 397
fluorspar 474
fog
– definition 135, 252
– etymology 107
– formation 133, 136, 324, 484
fog chamber 325
fog dissipation 137, 141, 258
fog droplets *see* droplets

gas
– historic terms 66
gas-to-particle conversion 132, 317, 321, 325, 610
Geissler tube 575
Griess reagent 600

hail
– early chemical studies 161, 167
– first chemical investigation 148
hail shooting 138
Hartley band 573
haze
– and ozone 547
– definition 298
– formation 321, 449
– mist and fog 109, 135, 292
– particles 230
– research pioneers 327
helium
– detection in lava 542
– in air 376
historic terms
– hydrometeors 107
– of gases 66
– phlogiston theory 355
HO2
– aqueous chemistry 652
– detection 489
– in slow combustion 637
– ozone destruction 590
– photochemistry 434, 488, 625, 648, 668
honeydew 277, 312
HOx cycle 590
Huggins bands 574
humic-like matter 165, 305
hydrated electron 658

hydrochloric acid
– acid rain 214, 228, 232
– *Cauer*'s view 334, 338
– discovery 345
– in air 318, 337
– release from sea salt 336
hydrogen
– chemistry 434, 488
– discovery 355
– early studies 372
– in air 349
hydrogen peroxide
– in dew water 639
hydrogen sulfide
– and ozone test-paper 510
– early history 709
– in air 155, 444, 520
– in rainwater 145
– volcanoes 58
hydrogen-oxygen reaction 381, 487
hydroperoxyl radical *see* HO2
hydroxyl radical
– air cleanser 470
– detection *488*
– in water photolysis 227
– photochemistry 103, 486
– research pioneers 43
hypochlorous acid 66, 338, 499
hyponitrite 686
hyponitrous acid 473, 686, 692

Immissionsklimatologie 11, 526
inflammable air 79, 349, 371, 433, 437, 681
influenza and ozone 518, 520, 550
interfacial chemistry 102, 282, 651
iodide reaction 188, 512
iodide-starch paper 518, 570
iodine
– and ozone at sea 624
– *Cauer*'s research 20
– ozone test paper 511
iodometry 434
iron
– aqueous chemistry 656
– in aerolites 310
– in dust 208, 307
– in fog water 263
– in rainwater 166
– in rime 257
– in snow 305
iron-III oxalate 658

Javanese lye 185
jet dust counter 297
Jodozone 510
Junge layer 38, 710

konimeter 39, 328
krypton
– in air 376

large ion 328
lead poisoning 157
Leyden bottle 575
lightning
– air chemistry 317
– as H2O2 source 529
– as NOx source 219, 476, 517, 630, 703
– as O3 source 322, 461, 474, 495
– blue haze formation 322
– nitrite in rainwater 223
– no O3 source 536, 573
liquid water content
– cloud water, Mt. Brocken 274
litmus paper 166, 228, 635, 696, 712
London
– air pollution 286, 291
– rainwater chemistry 166, 182, 243
London fog
– coining the word 292
– early studies 291
– LWC 136
– of 1952 718
– pioneers in 19th century 135
long-range transport
– of H2O2 663
– of smoke 301, 319
Los Angeles smog 13, 299, 448, 489

M-124 spectrometer 588, 596
M-83 ozone spectrophotometer 593
magnesium
– in aerolites 310
– in rainwater 179
magnesium chloride 157
Main Geophysical Observatory *see* observatory, St. Petersburg (MGO)
manganese
– in aerolites 310

– in dust 307
manganese dioxide 509, 598
manganese peroxide 462, 474
marine hypothesis 333
Master Chemical Mechanism 43
materia prima 119
May-dew 276
mephitic air 79
methane 437, 439
methane cycle 440
methylene 437
miasma 689
mineral matter 208, 294, 306, 318
mineral theory 55, 186
minimetric method 400
mist
– and fog 135, 263
– as town fog 292
– etymology 109
mist chamber 447
Möckern station 175, 183
modeling 40, 59, 247, 265, 625, 629, 726
Mohr's method 201
molecular electrolysis 102, 493, 653
Montsouris Observatory
– CO2 measurements 405, 412
– dew water sampling 313
– ozone measurements 557, 623
– rainwater sampling 17
moor smoke 317
moorland burning 163
Mt. Brocken
– *Aßmann*'s studies 132
– cloud chemistry 259, 268
– H2O2 measurements 665
– history of studies 271
– ozone measurements 616

N2O
– discovery 682
– from H2N2O2 in water 692
– from nitroxyl 686
– from NO in water 691
– in air 702
– photolysis 591, 703
N2O3 575, 684, 703
N2O5 340, 684, 687
neon
– in air 376
Nessler's reagent 175
nickel
– in dust 307
– in meteorites 311
– in snow 305
nitrate
– Atacama desert 342
– in aerosol 339
– in cloud water 262, 269
– in dew water 193, 280
– in rainwater 172, 179, 182, 187, 190, 211, 221, 227
– in rime 257
– trend in cloud water 268
– trend in deposition 240
– trend in rainwater 242
nitrate radical 700
nitric acid
– in thunderstorm 150
nitric oxide *see* NO
nitrification 57, 174, 248
nitrite
– in cloud water, history 261
– in rainwater 222, 227, 241, 641
– oxidation by H2O2 660
– oxidation by ozone 572
nitro-aerial particles 349
nitrogen controversy 186
nitrogen cycle 59
nitrogen dioxide *see* NO2
nitrogen fixation 57, 117, 178
nitrogen monoxide *see* NO
nitrogen pentoxide *see* N2O5
nitrosyl chloride 340, 688
nitrous acid *see also* nitrous gases and nitrite
– and ozone test paper 510, 521, 525
– *Cavendish*'s experiments 352
– detection in air 705
– early views 634, 681, 684
– in combustion processes 697
– in electric discharges 517, 699
nitrous air test 364
nitrous anhydride *see* N2O3
nitrous gases 288, 352, 380, 680
nitrous oxide *see* N2O
nitroxyl 686
nitrum 679
nitryl chloride 340, 342

NO
- detection in air 702
- early views 685, 690
- lightning as source 630
- limitation, O3 formation 627
- oxidation by ozone 699
- photochemistry 625
- reaction with water 691
- soil emission 630
NO2
- in air 701
- photochemistry 625
- reaction with water 692
non-methane hydrocarbons 444

O(1D) 488, 578, 591, 629, 668
O4 574
observatory (including astronomical, astrophysical, meteorological)
- Algiers 204, 646
- Beeston, UK 545
- Blue Hill 566, 594
- Brussels 506
- Collmberg 259, 329, 585
- Cracow 539
- Dolgoprudny (CAO) 141, 224, 258, 592, 593
- Flagstaff, Australia 567
- Greenwich 322, 544
- Haldde, Finnmark 256
- Hamburg 247, 330
- Havana 500
- Hohenpeißenberg 589, 620
- Hoher Sonnblick 32, 647
- Jungfraujoch 272, 439, 582, 598, 600
- Kew 327, 422, 604
- Kobe 222
- Krásno 541
- Kremsmünster 534
- Lindenberg 581, 614
- Lyon 551
- Mauna Loa 429
- Modena 319, 542
- Moncalieri 506, 542
- Mont Blanc 560
- Montsouris 118, 179, 188, 313, 412, 555, 556, 558
- Napoli 542
- Odessa 199
- Ógyalla 192, 538
- Palermo 305
- Pic-Du-Midi 22
- Potsdam 35, 586
- Prague 536, 539
- Radcliffe, Oxford 548
- Round Hill 39, 258, 339, 677
- Saint-Martin, Canada 561
- Sonnblick 26
- St. Peterburg (MGO) 141, 224, 225, 259, 592
- Taunus 33, 38, 241, 267, 677
- Tokyo 222
- Toronto 563
- Urbino 541
- Wahnsdorf 33, 241, 327, 614, 619
- zi-ka-wei, Shanghai 581
odor *see also* smell
- electric 317, 451, 457, 459
- nitrous 473, 517
- of air pollution 123, 456, 489
- of deposit 310
- of fluorspar 474
- of fog 254
- of fresh air 570
- of ozone 456, 516
- of swamp 279
- of thunderbolt 456
- sulphureous 322, 456, 495
olefiant gas 436, 443
organic acids
- and acidity 245
- historic views 436, 442
- in air 449
organic matter
- airborne 312
- extraterrestrial origin 311
- first detection 148, 150
- in dew 161
- in dust 311
- sampling device 446
organic nitrogen 97, 167, 170, 182, 187, 211, 279, 303, 312
organic oxides, historic views 436
Ox 626
oxozone 574
oxygen-hydrogen reaction 487
ozonate 637
ozonation
- of air 484, 576, 637

ozone
– animal toxicity 498
– dry deposition 630
– history 455
– precursors 625
– pure 465
– stratosphere 577, 581, 588
– surface-near 513, 608, 621, 625
ozone hole 13, 589, 592
ozone layer 580, 581, 591
ozone meter *see* ozonometer
ozone potassium iodide reaction 596
ozone radiosonde 585
ozone test-paper
– criticism 507, 527, 547, 570
– exposing time 535, 554
– Houzeau's 509, 525, 553
– Jame de Sedan 500, 559, 568
– Lender's 532
– Lowe's 544
– Moffat's 543, 545
– Schönbein's 495, 545, 557
ozone therapy 498, 525
ozone titration 571, 611
ozone water 498
ozone-box 500, 510, 531, 542, 543
ozonized air 351, 465, 498, 572
ozonometer
– definition 495, 507
– *Houzeaus*'s 509
– *Rudeck*'s 534
– *Schönbein*'s 95, 475, 509
– *Smyth*'s 546
ozonometry 474, 507, 515

PAN 626, 705
particle, etymology 320
particulate matter *see* dust
Passatstaub 302
pernitrous acid 660
Pettenkofer's method 400, 401, 414
Pettersson's method 401
pH
– of cloud water, Russia 261
– of cloud water, USA 266
– of fog water 262
– of nucleis 23, 230
– of rainwater, China 246
– of rainwater, East Germany 250
– of rainwater, Europe 230, 247
– of rainwater, Germany before 1945 207, 228
– of rainwater, Japan before 1945 222
– of rainwater, Russia 226, 246
– of rainwater, USA 230, 244
phlogisticated air 353, 681
phlogisticated nitrous acid 681
phlogisticated substances 355
phlogisticated water 681
phlogiston 353
phosphane *see* phosphine
phosphine 171, 460, 637
phosphorus, slow combustion 637
phosphurated hydrogen 460
photocatalysis 657
photochemical smog
– chemistry 434, 625, 649
– Los Angeles 448, 607, 662
– SO2 photolysis 725
pneumatic chemistry 349
potassium arsenite 555, 595
potassium iodide 472, 500, 516, 539, 556, 597, 604, 631
potassium permanganate 21, 95, 472, 637
potassium peroxide 637
potassium tetroxide 637
pre-industrial
– CO2 values 424
– O3 values 622
precipitation chemistry research
– in Britain 208, 214
– in China 246
– in Czech Republic 216, 242
– in East Germany 241, 250
– in Germany before 1945 206, 215, 217, 228
– in Hungary 192
– in India 198, 227
– in Russia 199, 205
– in USA 219, 240
– in USSR 225
– in West Germany 241
Pyrrhin 312

rainwater
– research milestone 145
red dust 301
red rain 161
rime 256
– chemical studies 256

– pH 228
rotating arm collector 265
Rothamsted Station
– history 174, 201
– NH3 dry deposition 676
– rainwater sampling 187, 200
Ruhmkorff induction coil 698
rusting 210, 651, 688

Sahara dust storms 302
sal ammoniac 670
sal armenicum 671
sal nitrum 679
sal petrae 679
saltpeter
– Atacama desert 342
– historic views 679
– in air 144, 357
Scheidekunst 116
Schönbein reaction 524, 631
Schumann-Runge continuum 573
sea dust 203
sea salt
– as CCN 326, 333
– chlorine degassing 31, 157, 336, 338, 689
– formation 201, 333
– in rainwater 162, 223
sea salt cycle 53
sea salt fractionation 262
sea salt hypothesis 327
sea salt sulfate 162
secondary organic aerosol 312
self-cleaning 54, 78, 652
silver chloride 97
smell *see also* odor
– hepatic 709
– nitrous 680
– of air pollution 379, 445
– of burnt gunpowder 456
– of dew 316
– of dry haze 317
– of fog 81, 253, 255
– of lignite coal 318
– of London fog 291
– of rainwater 163, 308
– of vegetation 449
– sulphureous 456
smog, coining the word 296
smog chamber 43, 625, 660, 705, 729
smoke damage research 721
smoke plague 297, 711
smoky fog 254, 292
sodium chloride
– in dust particles 330, 677
– in evaporated rainwater 212
– in rainwater 160, 202, 204, 223, 239
– in rime 256
– reaction with HNO3 340
sodium-to-chloride ratio 261, 330, 339
soil dust 165, 170, 208, 246, 305
soot
– and smoke 288, 291, 298
– deposit from air 296
– in dust 311
– in rainwater 208, 286, 310, 314
– in rime 257
– in snow 214
spiritus animalis 671
spiritus igno-aereus 89
spiritus nitri 335, 680
spiritus salis marini 335
spiritus urinae 671
spiritus vitrioli 707
Spritzwasserhypothese 333
Spurenstoffhaushalt 18
starch 500
starch-iodide paper 496, 510
stinking mist 254
stratosphere-to-troposphere O3 transport 603, 625, 627
string collector 261, 273
sulfate
– first detection in CCN 31
– in CCN 329, 726
– in cloud water 258, 262, 267
– in dust deposition 217
– in lava 708
– in rainwater 187, 205, 211, 216, 222, 242, 248
– in rime 32, 257
– stratosphere 38, 710
– wet deposition 179, 205
sulfate radical 724
sulfite
– aqueous-phase oxidation 722
– in dew water 281
sulfite radical 723
sulfur cycle 205, 710

sulfur dioxide
– absorption into rain gauge 219
– air pollution 210, 238, 248, 520, 571, 707, 712
– and dew 281
– aqueous oxidation 239
– early photochemistry 321
– forest damages 721
– in air 12, 294, 605, 665, 713, 716
– interference with O3 measurements 617
– sampling apparatus 715
sulfuric acid *see* sulfate and sulfur dioxide
superoxide anion 102, 649, 657

tetraoxygen 465, 576
tetroxide 464, 467, 597, 637, 686
thallium test-paper 501, 512, 559, 635
three-body collision 577
thunderstorm
– hydrogen peroxide 639
tipping point 48, 630
tonico nervina 97
town gas 81
town pollution 17, 157, 181, 206, 217, 286, 292, 299, 407, 440, 476, 571, 613, 675, 701, 709, 718, 724
transition metal ions 660, 723
transmutation 83, 107, 156

Umkehr method 580
UV radiation 18, 223, 434, 488, 576, 654

vesicle theory 133, 550
vesicles 133, 167, 253, 314

Waldsterben 118, 265, 616
weather modification 137, 140, 141, 224
wedge constant 587
wet deposition
– ammonia 185, 675
– iodine 179
– mineralic substances 169, 179, 712
– nitrogen, trend 240
– of carbonic acid 172
– sea salt 333
– sulfur, in USA 219
– total nitrogen 185, 188
white cloud 458
white mist 299
Wurster's reagent 478

xenon
– discovery 375
– in air 376

Zeldovich mechanism 704

Bei Fragen zur Produktsicherheit wenden Sie sich bitte an:
If you have any questions regarding product safety,
please contact:

Walter de Gruyter GmbH
Genthiner Straße 13
10785 Berlin
productsafety@degruyterbrill.com